TUESDAY AT 10:00

Applied Linear Statistical Models

Regression, Analysis of Variance, and Experimental Designs

Applied Linear Statistical Models

Regression, Analysis of Variance, and Experimental Designs

John Neter
University of Georgia

William Wasserman
Syracuse University

Michael H. Kutner
Emory University

1985 Second edition

RICHARD D. IRWIN, INC.
Homewood, Illinois 60430

ISBN 0-256-02447-2

Library of Congress Catalog Card No. 84-80445

Printed in the United States of America

2 3 4 5 6 7 8 9 0 DO 2 1 0 9 8 7 6 5

1

Some basic results in probability and statistics

This chapter contains some basic results in probability and statistics. It is intended as a reference chapter to which you may refer as you read this book. Sometimes, specific references to results in this chapter are made in the text. At other times, you may wish to refer on your own to particular results in this chapter as you feel the need.

You may prefer to scan the results on probability and statistical inference in this chapter before reading Chapter 2, or you may proceed directly to the next chapter.

1.1 SUMMATION AND PRODUCT OPERATORS

Summation operator

The summation operator Σ is defined as follows:

$$\sum_{i=1}^{n} Y_i = Y_1 + Y_2 + \cdots + Y_n \tag{1.1}$$

Some important properties of this operator are:

(1.2a) $$\sum_{i=1}^{n} k = nk \qquad \text{where } k \text{ is a constant}$$

(1.2b) $$\sum_{i=1}^{n} (Y_i + Z_i) = \sum_{i=1}^{n} Y_i + \sum_{i=1}^{n} Z_i$$

(1.2c) $$\sum_{i=1}^{n} (a + cY_i) = na + c\sum_{i=1}^{n} Y_i \qquad \text{where } a \text{ and } c \text{ are constants}$$

The double summation operator $\Sigma\Sigma$ is defined as follows:

(1.3) $$\sum_{i=1}^{n}\sum_{j=1}^{m} Y_{ij} = \sum_{i=1}^{n} (Y_{i1} + \cdots + Y_{im})$$
$$= Y_{11} + \cdots + Y_{1m} + Y_{21} + \cdots + Y_{2m} + \cdots + Y_{nm}$$

An important property of the double summation operator is:

(1.4) $$\sum_{i=1}^{n}\sum_{j=1}^{m} Y_{ij} = \sum_{j=1}^{m}\sum_{i=1}^{n} Y_{ij}$$

Product operator

The product operator Π is defined as follows:

(1.5) $$\prod_{i=1}^{n} Y_i = Y_1 \cdot Y_2 \cdot Y_3 \cdots Y_n$$

1.2 PROBABILITY

Addition theorem

Let A_i and A_j be two events defined on a sample space. Then:

(1.6) $$P(A_i \cup A_j) = P(A_i) + P(A_j) - P(A_i \cap A_j)$$

where $P(A_i \cup A_j)$ denotes the probability of either A_i or A_j or both occurring; $P(A_i)$ and $P(A_j)$ denote, respectively, the probability of A_i and the probability of A_j; and $P(A_i \cap A_j)$ denotes the probability of both A_i and A_j occurring.

Multiplication theorem

Let $P(A_i|A_j)$ denote the conditional probability of A_i occurring, given that A_j has occurred. This conditional probability is defined as follows:

(1.7) $$P(A_i|A_j) = \frac{P(A_i \cap A_j)}{P(A_j)} \qquad P(A_j) \neq 0$$

The multiplication theorem states:

(1.8) $$\begin{aligned} P(A_i \cap A_j) &= P(A_i)P(A_j|A_i) \\ &= P(A_j)P(A_i|A_j) \end{aligned}$$

Complementary events

The complementary event of A_i is denoted by $\bar{A}_i$. The following results for complementary events are useful:

(1.9) $$P(\bar{A}_i) = 1 - P(A_i)$$

(1.10) $$P(\overline{A_i \cup A_j}) = P(\bar{A}_i \cap \bar{A}_j)$$

1.3 RANDOM VARIABLES

Throughout this section, we assume that the random variable Y assumes a finite number of outcomes. (If Y is a continuous random variable, the summation process is replaced by integration.)

Expected value

Let the random variable Y assume the outcomes $Y_1, \ldots, Y_k$ with probabilities given by the probability function:

(1.11) $$f(Y_s) = P(Y = Y_s) \qquad s = 1, \ldots, k$$

The expected value of Y is defined:

(1.12) $$E(Y) = \sum_{s=1}^{k} Y_s f(Y_s)$$

An important property of the expectation operator E is:

(1.13) $$E(a + cY) = a + cE(Y) \qquad \text{where } a \text{ and } c \text{ are constants}$$

Special cases of this are:

(1.13a) $$E(a) = a$$

(1.13b) $$E(cY) = cE(Y)$$

(1.13c) $$E(a + Y) = a + E(Y)$$

Variance

The variance of the random variable Y is denoted by $\sigma^2(Y)$ and is defined as follows:

(1.14) $$\sigma^2(Y) = E\{[Y - E(Y)]^2\}$$

An equivalent expression is:

(1.14a) $$\sigma^2(Y) = E(Y^2) - [E(Y)]^2$$

The variance of a linear function of Y is frequently encountered. We denote the variance of $a + cY$ by $\sigma^2(a + cY)$ and have:

(1.15) $$\sigma^2(a + cY) = c^2\sigma^2(Y) \qquad \text{where } a \text{ and } c \text{ are constants}$$

Special cases of this result are:

(1.15a) $$\sigma^2(a + Y) = \sigma^2(Y)$$

(1.15b) $$\sigma^2(cY) = c^2\sigma^2(Y)$$

Joint, marginal, and conditional probability distributions

Let the joint probability function for the two random variables Y and Z be denoted by $g(Y, Z)$:

(1.16) $$g(Y_s, Z_t) = P(Y = Y_s \cap Z = Z_t) \qquad s = 1, \ldots, k;\ t = 1, \ldots, m$$

The marginal probability function of Y, denoted by $f(Y)$, is:

(1.17a) $$f(Y_s) = \sum_{t=1}^{m} g(Y_s, Z_t) \qquad s = 1, \ldots, k$$

and the marginal probability function of Z, denoted by $h(Z)$, is:

(1.17b) $$h(Z_t) = \sum_{s=1}^{k} g(Y_s, Z_t) \qquad t = 1, \ldots, m$$

The conditional probability function of Y, given $Z = Z_t$, is:

(1.18a) $$f(Y_s|Z_t) = \frac{g(Y_s, Z_t)}{h(Z_t)} \qquad h(Z_t) \neq 0;\ s = 1, \ldots, k$$

and the conditional probability function of Z, given $Y = Y_s$, is:

(1.18b) $$h(Z_t|Y_s) = \frac{g(Y_s, Z_t)}{f(Y_s)} \qquad f(Y_s) \neq 0;\ t = 1, \ldots, m$$

Covariance

The covariance of Y and Z is denoted by $\sigma(Y, Z)$ and is defined:

(1.19) $$\sigma(Y, Z) = E\{[Y - E(Y)][Z - E(Z)]\}$$

An equivalent expression is:

(1.19a) $$\sigma(Y, Z) = E(YZ) - [E(Y)][E(Z)]$$

The covariance of $a_1 + c_1Y$ and $a_2 + c_2Z$ is denoted by $\sigma(a_1 + c_1Y, a_2 + c_2Z)$, and we have:

(1.20) $$\sigma(a_1 + c_1Y, a_2 + c_2Z) = c_1c_2\sigma(Y, Z)$$ where a_1, a_2, c_1, c_2 are constants

Special cases of this are:

(1.20a) $$\sigma(c_1Y, c_2Z) = c_1c_2\sigma(Y, Z)$$

(1.20b) $$\sigma(a_1 + Y, a_2 + Z) = \sigma(Y, Z)$$

By definition, we have:

(1.21) $$\sigma(Y, Y) = \sigma^2(Y)$$

where $\sigma^2(Y)$ is the variance of Y.

Independent random variables

(1.22) Random variables Y and Z are independent if and only if:

$$g(Y_s, Z_t) = f(Y_s)h(Z_t) \qquad s = 1, \ldots, k;\ t = 1, \ldots, m$$

If Y and Z are independent random variables:

(1.23) $$\sigma(Y, Z) = 0 \qquad \text{when } Y, Z \text{ are independent}$$

(In the special case where Y and Z are jointly normally distributed, $\sigma(Y, Z) = 0$ implies that Y and Z are independent.)

Functions of random variables

Let $Y_1, \ldots, Y_n$ be n random variables. Consider the function Σa_iY_i where the a_i are constants. We then have:

(1.24a) $$E\left(\sum_{i=1}^{n} a_iY_i\right) = \sum_{i=1}^{n} a_iE(Y_i) \qquad \text{where the } a_i \text{ are constants}$$

(1.24b) $$\sigma^2\left(\sum_{i=1}^{n} a_iY_i\right) = \sum_{i=1}^{n}\sum_{j=1}^{n} a_ia_j\sigma(Y_i, Y_j)$$ where the a_i are constants

Specifically, we have for $n = 2$:

(1.25a) $$E(a_1Y_1 + a_2Y_2) = a_1E(Y_1) + a_2E(Y_2)$$

(1.25b) $$\sigma^2(a_1Y_1 + a_2Y_2) = a_1^2\sigma^2(Y_1) + a_2^2\sigma^2(Y_2) + 2a_1a_2\sigma(Y_1, Y_2)$$

If the random variables Y_i are independent, we have:

(1.26) $$\sigma^2\left(\sum_{i=1}^{n} a_iY_i\right) = \sum_{i=1}^{n} a_i^2\sigma^2(Y_i)$$ when the Y_i are independent

Special cases of this are:

(1.26a) $$\sigma^2(Y_1 + Y_2) = \sigma^2(Y_1) + \sigma^2(Y_2)$$ when Y_1, Y_2 are independent

(1.26b) $$\sigma^2(Y_1 - Y_2) = \sigma^2(Y_1) + \sigma^2(Y_2)$$ when Y_1, Y_2 are independent

When the Y_i are independent random variables, the covariance of two linear functions Σa_iY_i and Σc_iY_i is:

(1.27) $$\sigma\left(\sum_{i=1}^{n} a_iY_i, \sum_{i=1}^{n} c_iY_i\right) = \sum_{i=1}^{n} a_ic_i\sigma^2(Y_i)$$ when the Y_i are independent

Central limit theorem

(1.28) If $Y_1, \ldots, Y_n$ are independent random observations from a population with probability function $f(Y)$ for which $\sigma^2(Y)$ is finite, the sample mean $\bar{Y}$:

$$\bar{Y} = \frac{\sum_{i=1}^{n} Y_i}{n}$$

is approximately normally distributed when the sample size n is reasonably large, with mean $E(Y)$ and variance $\sigma^2(Y)/n$.

1.4 NORMAL PROBABILITY DISTRIBUTION AND RELATED DISTRIBUTIONS

Normal probability distribution

The density function for a normal random variable Y is:

(1.29) $$f(Y) = \frac{1}{\sqrt{2\pi}\sigma}\exp\left[-\frac{1}{2}\left(\frac{Y-\mu}{\sigma}\right)^2\right] \qquad -\infty < Y < +\infty$$

where μ and σ are the two parameters of the normal distribution and $\exp(a)$ denotes e^a.

The mean and variance of a normal random variable Y are:

(1.30a) $$E(Y) = \mu$$

(1.30b) $$\sigma^2(Y) = \sigma^2$$

Function of normal random variable. A linear function of a normal random variable Y has the following property:

(1.31) If Y is a normal random variable, the transformed variable $Y' = a + cY$ (a and c are constants) is normally distributed, with mean $a + cE(Y)$ and variance $c^2\sigma^2(Y)$.

Standard normal variable. The standard normal variable z:

(1.32) $$z = \frac{Y - \mu}{\sigma} \qquad \text{where } Y \text{ is a normal random variable}$$

is normally distributed, with mean 0 and variance 1. We denote this as follows:

(1.33) $$z \text{ is } N(0, 1)$$

(↑ Mean, ↑ Variance)

Table A–1 in the Appendix contains the cumulative probabilities A for percentiles $z(A)$ where:

(1.34) $$P\{z \leq z(A)\} = A$$

For instance, when $z(A) = 2.00$, $A = .9772$. Because the normal distribution is symmetrical about 0, when $z(A) = -2.00$, $A = 1 - .9772 = .0228$.

Function of independent normal random variables. Let $Y_1, \ldots, Y_n$ be independent normal random variables. We then have:

(1.35) When $Y_1, \ldots, Y_n$ are independent normal random variables, the linear combination $a_1Y_1 + a_2Y_2 + \cdots + a_nY_n$ is normally distributed, with mean $\Sigma a_i E(Y_i)$ and variance $\Sigma a_i^2 \sigma^2(Y_i)$.

χ^2 distribution

Let $z_1, \ldots, z_\nu$ be ν independent standard normal variables. We then define:

(1.36) $$\chi^2(\nu) = z_1^2 + z_2^2 + \cdots + z_\nu^2 \qquad \text{where the } z_i \text{ are independent}$$

The χ^2 distribution has one parameter, ν, which is called the *degrees of freedom* (*df*). The mean of the χ^2 distribution with ν degrees of freedom is:

(1.37) $$E[\chi^2(\nu)] = \nu$$

Table A–3 in the Appendix contains percentiles of various χ^2 distributions. We define $\chi^2(A; \nu)$ as follows:

$$P\{\chi^2(\nu) \leq \chi^2(A; \nu)\} = A \tag{1.38}$$

Suppose $\nu = 5$. The 90th percentile of the χ^2 distribution with 5 degrees of freedom is $\chi^2(.90; 5) = 9.24$.

t distribution

Let z and $\chi^2(\nu)$ be independent random variables (standard normal and χ^2, respectively). We then define:

$$t(\nu) = \frac{z}{\left[\frac{\chi^2(\nu)}{\nu}\right]^{1/2}} \qquad \text{where } z \text{ and } \chi^2(\nu) \text{ are independent} \tag{1.39}$$

The t distribution has one parameter, the *degrees of freedom* ν. The mean of the t distribution with ν degrees of freedom is:

$$E[t(\nu)] = 0 \tag{1.40}$$

Table A–2 in the Appendix contains percentiles of various t distributions. We define $t(A; \nu)$ as follows:

$$P\{t(\nu) \leq t(A; \nu)\} = A \tag{1.41}$$

Suppose $\nu = 10$. The 90th percentile of the t distribution with 10 degrees of freedom is $t(.90; 10) = 1.372$. Because the t distribution is symmetrical about 0, we have $t(.10; 10) = -1.372$.

F distribution

Let $\chi^2(\nu_1)$ and $\chi^2(\nu_2)$ be two independent χ^2 random variables. We then define:

$$F(\underset{\substack{\text{Numerator}\\ df}}{\nu_1}, \underset{\substack{\text{Denominator}\\ df}}{\nu_2}) = \frac{\chi^2(\nu_1)}{\nu_1} \div \frac{\chi^2(\nu_2)}{\nu_2} \qquad \text{where } \chi^2(\nu_1) \text{ and } \chi^2(\nu_2) \text{ are independent} \tag{1.42}$$

The F distribution has two parameters, the *numerator degrees of freedom* and the *denominator degrees of freedom,* here ν_1 and ν_2, respectively.

Table A–4 in the Appendix contains percentiles of various F distributions. We define $F(A; \nu_1, \nu_2)$ as follows:

$$P\{F(\nu_1, \nu_2) \leq F(A; \nu_1, \nu_2)\} = A \tag{1.43}$$

Suppose $\nu_1 = 2$, $\nu_2 = 3$. The 90th percentile of the F distribution with 2 and 3 degrees of freedom, respectively, in the numerator and denominator is $F(.90; 2, 3) = 5.46$.

Percentiles below 50 percent can be obtained by utilizing the relation:

(1.44) $$F(A; \nu_1, \nu_2) = \frac{1}{F(1 - A; \nu_2, \nu_1)}$$

Thus, $F(.10; 3, 2) = 1/F(.90; 2, 3) = 1/5.46 = .183$.

The following relation exists between the t and F random variables:

(1.45a) $$[t(\nu)]^2 = F(1, \nu)$$

and the percentiles of the t and F distributions are related as follows:

(1.45b) $$[t(.5 + A/2; \nu)]^2 = F(A; 1, \nu)$$

1.5 STATISTICAL ESTIMATION

Properties of estimators

(1.46) An estimator $\hat{\theta}$ of the parameter θ is *unbiased* if:

$$E(\hat{\theta}) = \theta$$

(1.47) An estimator $\hat{\theta}$ is a *consistent estimator* of θ if:

$$\lim_{n \to \infty} P(|\hat{\theta} - \theta| \geq \varepsilon) = 0 \qquad \text{for any } \varepsilon > 0$$

(1.48) An estimator $\hat{\theta}$ is a *sufficient estimator* of θ if the conditional joint probability function of the sample observations, given $\hat{\theta}$, does not depend on the parameter θ.

(1.49) An estimator $\hat{\theta}$ is a *minimum variance estimator* of θ if for any other estimator θ^*:

$$\sigma^2(\hat{\theta}) \leq \sigma^2(\theta^*) \qquad \text{for all } \theta^*$$

Maximum likelihood estimators

The method of maximum likelihood is a general method of finding estimators. Suppose we are sampling a population whose probability function $f(Y; \theta)$ involves one parameter, θ. Given independent observations $Y_1, \ldots, Y_n$, the joint probability function of the sample observations is:

(1.50a) $$g(Y_1, \ldots, Y_n) = \prod_{i=1}^{n} f(Y_i; \theta)$$

When this joint probability function is viewed as a function of θ, with the observations given, it is called the *likelihood function* $L(\theta)$.

(1.50b) $$L(\theta) = \prod_{i=1}^{n} f(Y_i; \theta)$$

Maximizing $L(\theta)$ with respect to θ yields the maximum likelihood estimator of θ. Under quite general conditions, maximum likelihood estimators are consistent and sufficient.

Least squares estimators

The method of least squares is another general method of finding estimators. The sample observations are assumed to be of the form (for the case of a single parameter θ):

(1.51) $$Y_i = f_i(\theta) + \varepsilon_i \qquad i = 1, \dots, n$$

where $f_i(\theta)$ is a known function of the parameter θ and the ε_i are random variables, usually assumed to have expectation $E(\varepsilon_i) = 0$.

With the method of least squares, for the given sample observations, the sum of squares:

(1.52) $$Q = \sum_{i=1}^{n} [Y_i - f_i(\theta)]^2$$

is considered as a function of θ. The least squares estimator of θ is obtained by minimizing Q with respect to θ. In many instances, least squares estimators are unbiased and consistent.

1.6 INFERENCES ABOUT POPULATION MEAN—NORMAL POPULATION

We have a random sample of n observations $Y_1, \dots, Y_n$ from a normal population with mean μ and standard deviation σ. The sample mean and sample standard deviation are:

(1.53a) $$\bar{Y} = \frac{\sum_i Y_i}{n}$$

(1.53b) $$s = \left[\frac{\sum_i (Y_i - \bar{Y})^2}{n - 1}\right]^{1/2}$$

and the estimated standard deviation of the sampling distribution of $\bar{Y}$ is:

(1.53c) $$s(\bar{Y}) = \frac{s}{\sqrt{n}}$$

We then have:

(1.54) $\dfrac{\bar{Y} - \mu}{s(\bar{Y})}$ is distributed as t with $n - 1$ degrees of freedom when the random sample is from a normal population.

Calculus is not required for reading *Applied Linear Statistical Models*. In a number of instances we use calculus to demonstrate how some important results are obtained, but these demonstrations are confined to supplementary comments or notes and can be omitted without any loss of continuity. Readers who do know calculus will find these comments and notes in natural sequence so that the benefits of the mathematical developments are obtained in their immediate context. Some basic elements of matrix algebra are needed for linear models, in general, and for multiple regression, in particular. Chapter 6 introduces these elements of matrix algebra in the context of simple regression for easy learning.

Applied Linear Statistical Models is intended for use in undergraduate or graduate courses in linear statistical models and in second courses in applied statistics. The extent to which material presented in this text is used in a particular course depends upon the amount of time available and the objectives of the course. Some possible courses include:

1. A two-quarter or two-semester course in regression, analysis of variance, and basic experimental designs might be based on the following chapters:
   Regression: 2, 3, 4, 5 (Sections 5.1–5.4), 6 (Sections 6.1–6.12), 7, 8, 10 (Sections 10.1–10.3), 12.
   Analysis of variance: 16, 17, 18, 20, 21, 25.
   Experimental designs: 26, 27, 28, 31.
2. A one-quarter or one-semester course in regression analysis might be based on the following chapters:
   2, 3, 4, 5 (Sections 5.1–5.4), 6 (Sections 6.1–6.12), 7, 8, 9, 10 (Sections 10.1–10.3), 11 (selected topics), 12, 13, 14.
3. A one-quarter or one-semester course in analysis of variance might be based on the following chapters:
   16, 17, 18, 19 (selected topics), 20, 21, 22, 23 (selected topics), 24, 25.
4. A one-quarter or one-semester course in regression and analysis of variance might be based on the following chapters:
   Regression: 2, 3, 4, 5 (Sections 5.1–5.4), 6 (Sections 6.1–6.12), 7, 8, 10 (Sections 10.1–10.3).
   Analysis of variance: 16, 17, 18, 20, 21.
5. A one-quarter or one-semester course in basic experimental designs might be based on the following chapters:
   26, 27, 28, 29, 30, 31.

As time permits, the instructor could cover additional topics in the text.

This book can also be used for self-study by persons engaged in the fields of business administration, economics, and the social, health, and biological sciences who desire to obtain competence in the application of linear statistical models.

A book such as this cannot be written without substantial assistance from others. We are indebted to the many contributors who have developed the theory and practice discussed in this book. We also would like to acknowledge appreciation to our students who helped us in a variety of ways to fashion the method of

pendent indicator variables. Another chapter takes up computer-assisted selection procedures for obtaining a "best" set of independent variables to be employed in the regression model. The use of residual analysis for examining the aptness of the model is a recurring theme throughout this book. So is the use of remedial measures that may be helpful when the model is not appropriate. In the analysis of the results of a study, we emphasize the use of estimation procedures, rather than tests, because estimation is often more meaningful in practice. Also, since practical problems seldom are concerned with a single comparison, we stress the use of multiple comparison procedures.

Theoretical ideas are presented to the degree needed for good understanding in making sound applications. Proofs are given in those instances where we feel they serve to demonstrate an important method of approach. Emphasis is placed on a thorough understanding of the models, particularly the meaning of the model parameters, since such understanding is basic to proper applications. A wide variety of case examples is presented to illustrate the use of the theoretical principles, to show the great diversity of applications of linear statistical models, and to demonstrate how analyses are carried out for different problems.

We use "Notes" and "Comments" sections in each chapter to present additional discussion and matters related to the mainstream of development. In this way, the basic ideas in a chapter are presented concisely and without distraction. Similarly, optional "Topics" chapters supplement chapters containing the main development and present a variety of additional topics that in most cases can be omitted without loss of continuity.

Applications of linear statistical models frequently require extensive computations. We take the position that a computer is available in most applied work. Further, almost every computer user has access to program packages for regression analysis and analysis of variance of different types. Hence, we explain the basic mathematical steps in fitting a linear statistical model but do not dwell on computational details. This approach permits us to avoid many complex formulas and enables us to focus on basic principles. We make extensive use in this text of computer capabilities for performing computations and illustrate a variety of computer printouts and explain how these are used for analysis.

A selection of problems is provided at the end of each chapter (excepting Chapter 1). Here the reader can reinforce his or her understanding of the methodology and use the concepts learned to analyze data. We have been careful to supply data-analysis problems that typify genuine applications. In most problems the calculations are best handled on a calculator or computer, and we urge that this avenue be used when possible.

We assume that the reader of *Applied Linear Statistical Models* has had an introductory course in statistical inference, covering the material outlined in Chapter 1. Should some gaps in the reader's background exist, he or she can read the relevant portions of an introductory text, or the instructor of the class may use supplemental materials for covering the missing segments. Chapter 1 is primarily intended as a reference chapter of basic statistical results for continuing use as the reader progresses through the book.

regression analysis, these include weighted regression, selection of independent variables, and normal probability plots. In analysis of variance, we have greatly expanded the discussion of the unbalanced case, devoting a full chapter to the two-factor unbalanced case including a consideration of missing cells. In the discussion of experimental designs, we have unified the treatment of missing observations in terms of the regression approach, and have split the material on randomized block designs into two chapters to facilitate understanding. At the same time we have made extensive revisions of other materials on the basis of classroom experience to improve the clarity of the presentation.

3. We have strengthened the integration of regression analysis, analysis of variance, and experimental designs through the general linear model, and have simplified the discussion by considering the same coding of indicator variables for all analysis of variance models.

4. We have placed greater emphasis on graphic methods for diagnosis, analysis, and presentation of results, and have introduced a number of computer-generated plots to demonstrate the usefulness of computer graphics.

5. The scope of the examples has been expanded to include applications from the health and biological sciences, in addition to applications from management, economics, and the social sciences. In all cases, an application can be readily understood by the general reader, regardless of background.

6. We have added three extensive real-world data sets that can be employed in a variety of ways.

7. Finally, we have substantially expanded the problem materials at the ends of the chapters and have grouped them into three categories, namely *Problems, Exercises,* and *Projects*. The *Problems* category includes basic problems and questions, the *Exercises* category includes conceptual and theoretical questions, and the *Projects* category includes problems utilizing large data sets and/or involving extensive calculations and analysis.

The first 15 chapters of the Second Edition of *Applied Linear Statistical Models* have also been published as a separate book under the title *Applied Linear Regression Models*.

A key feature of *Applied Linear Statistical Models* is its unified approach to the application of linear statistical models in regression, analysis of variance, and experimental designs. Instead of treating these areas in isolated fashion, we seek to show the interrelationships between them. Use of a common notation for regression on the one hand and analysis of variance and experimental designs on the other facilitates a unified view. The notion of a general linear statistical model, which arises naturally in the context of regression models, is carried over to analysis of variance and experimental design models to bring out their relation to regression models. This unified approach also has the advantage of simplified presentation.

We have included in this book not only the more conventional topics in regression, analysis of variance, and basic experimental designs, but also have taken up topics that are frequently slighted though important in practice. Thus, we devote a full chapter to indicator variables, covering both dependent and inde-

Preface

Linear statistical models for regression, analysis of variance, and experimental designs are widely used today in business administration, economics, and the social, health, and biological sciences. Successful applications of these models require a sound understanding of both the underlying theory and the practical problems that are encountered in using the models in real-life situations. While *Applied Linear Statistical Models,* Second Edition, is basically an applied book, it seeks to blend theory and applications effectively, avoiding the extremes of presenting theory in isolation and of giving elements of applications without the needed understanding of the theoretical foundations.

The second edition differs from the first in a number of important respects.

1. We have added some important new topics. In the area of regression models, we have introduced discussions of detection of multicollinearity, ridge regression, and detection of influential observations. In recent years, noteworthy new developments in the detection of multicollinearity and influential observations have taken place, and a current text in linear models needs to cover these topics adequately. We have also added a chapter on nonlinear regression to introduce the reader to nonlinear statistical models.

In the area of experimental designs, we have added two chapters on nested designs, including rules for finding expected mean squares and sums of squares for a wide variety of balanced nested and crossed designs.

We have also added some specialized topics, such as P-values of test statistics, the method of least absolute deviations, analysis of variance techniques when the treatment means have unequal importance, and single degree of freedom tests.

2. We have reorganized and expanded a number of topics. In the area of

To
Dorothy, Ron, David,
  Mona, Jordan
Cathy, Christopher, Timothy,
  Randall, Erin, Fiona
Nancy, Michelle, Allison

Contents

presentation contained herein. We are grateful to the many users of the First Edition of *Applied Linear Statistical Models* who provided us with comments and suggestions based on their teaching with this text. We are also indebted to Professors James E. Holstein, University of Missouri, and David L. Sherry, University of West Florida, who carefully reviewed *Applied Linear Statistical Models* to provide suggestions for this volume. Dr. Robert L. Vogel assisted us diligently in the checking of the manuscript, for which we are most appreciative. Michael J. Lynn prepared the computer-generated plots using a Zeta model 3600 plotter, and George Cotsonis and Shizuki Yamamoto assisted us in the checking of calculations and in other ways. Rebecca Baggett, Marcia Pittard, and Jennifer Scott ably handled the typing and other preparation of a difficult manuscript. We are most grateful to all of these persons for their help and assistance.

Finally, our families bore patiently the pressures caused by our commitment to complete this revision. We are appreciative of their understanding.

John Neter
William Wasserman
Michael H. Kutner

Interval estimation

The confidence limits for μ with a confidence coefficient of $1 - \alpha$ are obtained by means of (1.54):

$$\bar{Y} \pm t(1 - \alpha/2; n - 1)s(\bar{Y}) \tag{1.55}$$

Example 1. Obtain a 95 percent confidence interval for μ when:

$$n = 10 \qquad \bar{Y} = 20 \qquad s = 4$$

We require:

$$s(\bar{Y}) = \frac{4}{\sqrt{10}} = 1.265 \qquad t(.975; 9) = 2.262$$

so that the confidence limits are $20 \pm 2.262(1.265)$. Hence, the 95 percent confidence interval for μ is:

$$17.1 \leq \mu \leq 22.9$$

Tests

One-sided and two-sided tests concerning the population mean μ are constructed by means of (1.54), based on the test statistic:

$$t^* = \frac{\bar{Y} - \mu_0}{s(\bar{Y})} \tag{1.56}$$

Table 1.1 contains the decision rules for each of three possible cases, with the risk of making a Type I error controlled at α.

TABLE 1.1 Decision rules for tests concerning mean μ of normal population

Alternatives	*Decision Rule*
	(a)
H_0: $\mu = \mu_0$	If $\lvert t^*\rvert \leq t(1 - \alpha/2; n - 1)$, conclude H_0
H_a: $\mu \neq \mu_0$	If $\lvert t^*\rvert > t(1 - \alpha/2; n - 1)$, conclude H_a
	where: $t^* = \dfrac{\bar{Y} - \mu_0}{s(\bar{Y})}$
	(b)
H_0: $\mu \geq \mu_0$	If $t^* \geq t(\alpha; n - 1)$, conclude H_0
H_a: $\mu < \mu_0$	If $t^* < t(\alpha; n - 1)$, conclude H_a
	(c)
H_0: $\mu \leq \mu_0$	If $t^* \leq t(1 - \alpha; n - 1)$, conclude H_0
H_a: $\mu > \mu_0$	If $t^* > t(1 - \alpha; n - 1)$, conclude H_a

Example 2. Choose between the alternatives:

$$H_0: \mu \leq 20$$
$$H_a: \mu > 20$$

when α is to be controlled at .05 and:

$$n = 15 \qquad \bar{Y} = 24 \qquad s = 6$$

We require:

$$s(\bar{Y}) = \frac{6}{\sqrt{15}} = 1.549$$
$$t(.95; 14) = 1.761$$

so that the decision rule is:

$$\text{If } t^* \leq 1.761, \text{ conclude } H_0$$
$$\text{If } t^* > 1.761, \text{ conclude } H_a$$

Since $t^* = (24 - 20)/1.549 = 2.58 > 1.761$, we conclude H_a.

Example 3. Choose between the alternatives:

$$H_0: \mu = 10$$
$$H_a: \mu \neq 10$$

when α is to be controlled at .02 and:

$$n = 25 \qquad \bar{Y} = 5.7 \qquad s = 8$$

We require:

$$s(\bar{Y}) = \frac{8}{\sqrt{25}} = 1.6$$
$$t(.99; 24) = 2.492$$

so that the decision rule is:

$$\text{If } |t^*| \leq 2.492, \text{ conclude } H_0$$
$$\text{If } |t^*| > 2.492, \text{ conclude } H_a$$

where the symbol $| \, |$ stands for the absolute value. Since $|t^*| = |(5.7 - 10)/1.6| = |-2.69| = 2.69 > 2.492$, we conclude H_a.

***P*-value for sample outcome.** The *P*-value for a sample outcome is the probability that the sample outcome could have been more extreme than the observed one when $\mu = \mu_0$. Large *P*-values support H_0 while small *P*-values support H_a. A test can be carried out by comparing the *P*-value with the specified α risk. If the *P*-value equals or is greater than the specified α, H_0 is concluded. If the *P*-value is less than α, H_a is concluded.

Example 4. In Example 2, $t^* = 2.58$. The P-value for this sample outcome is the probability $P[t(14) > 2.58]$. From Table A–2, we find $t(.985; 14) = 2.415$ and $t(.990; 14) = 2.624$. Hence, the P-value is between .010 and .015. In fact, it can be shown to be .011. Thus, for $\alpha = .05$, H_a is concluded.

Example 5. In Example 3, $t^* = -2.69$. We find from Table A–2 that $P[t(24) < -2.69]$ is between .005 and .0075. In fact, it can be shown to be .0064. Because the test is two-sided and the t distribution is symmetrical, the two-sided P-value is twice the one-sided value, or $2(.0064) = .013$. Hence, for $\alpha = .02$, we conclude H_a.

Relation between tests and confidence intervals. There is a direct relation between tests and confidence intervals. For example, the two-sided confidence limits (1.55) can be used for testing:

$$H_0\colon \mu = \mu_0$$
$$H_a\colon \mu \neq \mu_0$$

If μ_0 is contained within the $1 - \alpha$ confidence interval, then the two-sided decision rule in Table 1.1a, with level of significance α, will lead to conclusion H_0, and vice versa. If μ_0 is not contained within the confidence interval, the decision rule will lead to H_a, and vice versa.

There are similar correspondences between one-sided confidence intervals and one-sided decision rules.

1.7 COMPARISONS OF TWO POPULATION MEANS—NORMAL POPULATIONS

Independent samples

There are two normal populations, with means μ_1 and μ_2, respectively, and with the same standard deviation σ. The means μ_1 and μ_2 are to be compared on the basis of independent samples for each of the two populations:

$$\text{Sample 1: } Y_1, \ldots, Y_{n_1}$$
$$\text{Sample 2: } Z_1, \ldots, Z_{n_2}$$

Estimators of the two population means are the sample means:

(1.57a)
$$\bar{Y} = \frac{\sum_i Y_i}{n_1}$$

(1.57b)
$$\bar{Z} = \frac{\sum_i Z_i}{n_2}$$

and an estimator of $\mu_1 - \mu_2$ is $\bar{Y} - \bar{Z}$.

An estimator of the common variance σ^2 is:

(1.58) $$s^2 = \frac{\sum_i (Y_i - \bar{Y})^2 + \sum_i (Z_i - \bar{Z})^2}{n_1 + n_2 - 2}$$

and an estimator of $\sigma^2(\bar{Y} - \bar{Z})$, the variance of the sampling distribution of $\bar{Y} - \bar{Z}$, is:

(1.59) $$s^2(\bar{Y} - \bar{Z}) = s^2\left[\frac{1}{n_1} + \frac{1}{n_2}\right]$$

We have:

(1.60) $\dfrac{(\bar{Y} - \bar{Z}) - (\mu_1 - \mu_2)}{s(\bar{Y} - \bar{Z})}$ is distributed as t with $n_1 + n_2 - 2$ degrees of freedom when the two independent samples come from normal populations with the same standard deviation.

Interval estimation. The confidence limits for $\mu_1 - \mu_2$ with confidence coefficient $1 - \alpha$ are obtained by means of (1.60):

(1.61) $$(\bar{Y} - \bar{Z}) \pm t(1 - \alpha/2; n_1 + n_2 - 2)s(\bar{Y} - \bar{Z})$$

Example 6. Obtain a 95 percent confidence interval for $\mu_1 - \mu_2$ when:

$$n_1 = 10 \quad \bar{Y} = 14 \quad \Sigma(Y_i - \bar{Y})^2 = 105$$
$$n_2 = 20 \quad \bar{Z} = 8 \quad \Sigma(Z_i - \bar{Z})^2 = 224$$

We require:

$$s^2 = \frac{105 + 224}{10 + 20 - 2} = 11.75$$
$$s^2(\bar{Y} - \bar{Z}) = 11.75\left(\frac{1}{10} + \frac{1}{20}\right) = 1.7625$$
$$s(\bar{Y} - \bar{Z}) = 1.328$$
$$t(.975; 28) = 2.048$$

$$3.3 = (14 - 8) - 2.048(1.328) \leq \mu_1 - \mu_2 \leq (14 - 8) + 2.048(1.328) = 8.7$$

Tests. One-sided and two-sided tests concerning $\mu_1 - \mu_2$ are constructed by means of (1.60). Table 1.2 contains the decision rules for each of three possible cases, based on the test statistic:

(1.62) $$t^* = \frac{\bar{Y} - \bar{Z}}{s(\bar{Y} - \bar{Z})}$$

with the risk of making a Type I error controlled at α.

TABLE 1.2 Decision rules for tests concerning means μ_1 and μ_2 of two normal populations ($\sigma_1 = \sigma_2 = \sigma$)

Alternatives	*Decision Rule*
	(a)
H_0: $\mu_1 = \mu_2$	If $\|t^*\| \leq t(1 - \alpha/2; n_1 + n_2 - 2)$, conclude H_0
H_a: $\mu_1 \neq \mu_2$	If $\|t^*\| > t(1 - \alpha/2; n_1 + n_2 - 2)$, conclude H_a
	where: $t^* = \dfrac{\bar{Y} - \bar{Z}}{s(\bar{Y} - \bar{Z})}$
	(b)
H_0: $\mu_1 \geq \mu_2$	If $t^* \geq t(\alpha; n_1 + n_2 - 2)$, conclude H_0
H_a: $\mu_1 < \mu_2$	If $t^* < t(\alpha; n_1 + n_2 - 2)$, conclude H_a
	(c)
H_0: $\mu_1 \leq \mu_2$	If $t^* \leq t(1 - \alpha; n_1 + n_2 - 2)$, conclude H_0
H_a: $\mu_1 > \mu_2$	If $t^* > t(1 - \alpha; n_1 + n_2 - 2)$, conclude H_a

Example 7. Choose between the alternatives:

$$H_0\text{: } \mu_1 = \mu_2$$
$$H_a\text{: } \mu_1 \neq \mu_2$$

when α is to be controlled at .10 and the data are those of Example 6. We require $t(.95; 28) = 1.701$, so that the decision rule is:

$$\text{If } |t^*| \leq 1.701\text{, conclude } H_0$$
$$\text{If } |t^*| > 1.701\text{, conclude } H_a$$

Since $|t^*| = |(14 - 8)/1.328| = |4.52| = 4.52 > 1.701$, we conclude H_a.

The one-sided P-value here is the probability $P[t(28) > 4.52]$. We see from Table A–2 that this P-value is less than .0005. In fact, it can be shown to be .00005. Hence, the two-sided P-value is .0001. For $\alpha = .10$, the appropriate conclusion therefore is H_a.

Paired observations

When the observations in the two samples are paired (e.g., attitude scores Y_i and Z_i for the ith sample employee before and after a year's experience on the job), we use the differences:

$$W_i = Y_i - Z_i \qquad i = 1, \ldots, n \tag{1.63}$$

in the fashion of a sample from a single population. Thus, when the W_i can be treated as observations from a normal population, we have:

(1.64) $\dfrac{\bar{W} - (\mu_1 - \mu_2)}{s(\bar{W})}$ is distributed as t with $n - 1$ degrees of freedom when the differences W_i can be considered to be observations from a normal population and:

$$\bar{W} = \frac{\sum_i W_i}{n}$$

$$s^2(\bar{W}) = \frac{\sum_i (W_i - \bar{W})^2}{n - 1} \div n$$

1.8 INFERENCES ABOUT POPULATION VARIANCE—NORMAL POPULATION

When sampling from a normal population, the following holds for the sample variance s^2 where s is defined in (1.53b):

(1.65) $\dfrac{(n - 1)s^2}{\sigma^2}$ is distributed as χ^2 with $n - 1$ degrees of freedom when the random sample is from a normal population.

Interval estimation

The lower confidence limit L and the upper confidence limit U in a confidence interval for the population variance σ^2 with confidence coefficient $1 - \alpha$ are obtained by means of (1.65):

$$\text{(1.66)} \qquad L = \frac{(n - 1)s^2}{\chi^2(1 - \alpha/2; n - 1)} \qquad U = \frac{(n - 1)s^2}{\chi^2(\alpha/2; n - 1)}$$

Example 8. Obtain a 98 percent confidence interval for σ^2, using the data of Example 1 ($n = 10$, $s = 4$).

We require:

$$s^2 = 16 \qquad \chi^2(.01; 9) = 2.09 \qquad \chi^2(.99; 9) = 21.67$$

$$6.6 = \frac{9(16)}{21.67} \leq \sigma^2 \leq \frac{9(16)}{2.09} = 68.9$$

Tests

One-sided and two-sided tests concerning the population variance σ^2 are constructed by means of (1.65). Table 1.3 contains the decision rule for each of three possible cases, with the risk of making a Type I error controlled at α.

TABLE 1.3 Decision rules for tests concerning variance σ^2 of normal population

Alternatives	*Decision Rule*
	(a)
H_0: $\sigma^2 = \sigma_0^2$	If $\chi^2(\alpha/2; n-1) \le \dfrac{(n-1)s^2}{\sigma_0^2} \le \chi^2(1-\alpha/2; n-1)$, conclude H_0
H_a: $\sigma^2 \neq \sigma_0^2$	Otherwise conclude H_a
	(b)
H_0: $\sigma^2 \ge \sigma_0^2$	If $\dfrac{(n-1)s^2}{\sigma_0^2} \ge \chi^2(\alpha; n-1)$, conclude H_0
H_a: $\sigma^2 < \sigma_0^2$	If $\dfrac{(n-1)s^2}{\sigma_0^2} < \chi^2(\alpha; n-1)$, conclude H_a
	(c)
H_0: $\sigma^2 \le \sigma_0^2$	If $\dfrac{(n-1)s^2}{\sigma_0^2} \le \chi^2(1-\alpha; n-1)$, conclude H_0
H_a: $\sigma^2 > \sigma_0^2$	If $\dfrac{(n-1)s^2}{\sigma_0^2} > \chi^2(1-\alpha; n-1)$, conclude H_a

1.9 COMPARISONS OF TWO POPULATION VARIANCES—NORMAL POPULATIONS

Independent samples are selected from two normal populations, with means and variances of μ_1 and σ_1^2 and μ_2 and σ_2^2, respectively. Using the notation of Section 1.7, the two sample variances are:

(1.67a)
$$s_1^2 = \frac{\sum_i (Y_i - \bar{Y})^2}{n_1 - 1}$$

(1.67b)
$$s_2^2 = \frac{\sum_i (Z_i - \bar{Z})^2}{n_2 - 1}$$

We have:

(1.68) $\dfrac{s_1^2}{\sigma_1^2} \div \dfrac{s_2^2}{\sigma_2^2}$ is distributed as $F(n_1 - 1, n_2 - 1)$ when the two independent samples come from normal populations.

Interval estimation

The lower and upper confidence limits L and U for σ_1^2/σ_2^2 with confidence coefficient $1 - \alpha$ are obtained by means of (1.68):

$$(1.69)\qquad L = \frac{s_1^2}{s_2^2}\frac{1}{F(1-\alpha/2; n_1-1, n_2-1)} \qquad U = \frac{s_1^2}{s_2^2}\frac{1}{F(\alpha/2; n_1-1, n_2-1)}$$

Example 9. Obtain a 90 percent confidence interval for σ_1^2/σ_2^2 when the data are:

$$n_1 = 16 \qquad n_2 = 21$$
$$s_1^2 = 54.2 \qquad s_2^2 = 17.8$$

We require:

$$F(.05; 15, 20) = 1/F(.95; 20, 15) = 1/2.33 = .429$$
$$F(.95; 15, 20) = 2.20$$

$$1.4 = \frac{54.2}{17.8}\frac{1}{2.20} \le \frac{\sigma_1^2}{\sigma_2^2} \le \frac{54.2}{17.8}\frac{1}{.429} = 7.1$$

Tests

One-sided and two-sided tests concerning σ_1^2/σ_2^2 are constructed by means of (1.68). Table 1.4 contains the decision rules for each of three possible cases, with the risk of making a Type I error controlled at α.

TABLE 1.4 Decision rules for tests concerning variances σ_1^2 and σ_2^2 of two normal populations

Alternatives	*Decision Rule*
	(a)
H_0: $\sigma_1^2 = \sigma_2^2$	If $F(\alpha/2; n_1-1, n_2-1) \le \frac{s_1^2}{s_2^2}$ $\le F(1-\alpha/2; n_1-1, n_2-1)$, conclude H_0
H_a: $\sigma_1^2 \neq \sigma_2^2$	Otherwise conclude H_a
	(b)
H_0: $\sigma_1^2 \ge \sigma_2^2$	If $\frac{s_1^2}{s_2^2} \ge F(\alpha; n_1-1, n_2-1)$, conclude H_0
H_a: $\sigma_1^2 < \sigma_2^2$	If $\frac{s_1^2}{s_2^2} < F(\alpha; n_1-1, n_2-1)$, conclude H_a
	(c)
H_0: $\sigma_1^2 \le \sigma_2^2$	If $\frac{s_1^2}{s_2^2} \le F(1-\alpha; n_1-1, n_2-1)$, conclude H_0
H_a: $\sigma_1^2 > \sigma_2^2$	If $\frac{s_1^2}{s_2^2} > F(1-\alpha; n_1-1, n_2-1)$, conclude H_a

Example 10. Choose between the alternatives:

$$H_0: \sigma_1^2 = \sigma_2^2$$
$$H_a: \sigma_1^2 \neq \sigma_2^2$$

when α is to be controlled at .02 and the data are those of Example 9.

We require:

$$F(.01; 15, 20) = 1/F(.99; 20, 15) = 1/3.37 = .297$$
$$F(.99; 15, 20) = 3.09$$

so that the decision rule is:

$$\text{If } .297 \leq \frac{s_1^2}{s_2^2} \leq 3.09, \text{ conclude } H_0$$
$$\text{Otherwise conclude } H_a$$

Since $s_1^2/s_2^2 = 54.2/17.8 = 3.04$, we conclude H_0.

PART I

Basic regression analysis

2

Linear regression with one independent variable

Regression analysis is a statistical tool that utilizes the relation between two or more quantitative variables so that one variable can be predicted from the other, or others. For example, if one knows the relation between advertising expenditures and sales, one can predict sales by regression analysis once the level of advertising expenditures has been set.

In Part I of this book, we take up regression analysis when a single predictor variable is used for predicting the variable of interest. In this chapter specifically, we consider the basic ideas of regression analysis and discuss the estimation of the parameters of the regression model.

2.1 RELATIONS BETWEEN VARIABLES

The concept of a relation between two variables, such as between family income and family expenditures for housing, is a familiar one. We distinguish between a *functional* relation and a *statistical* relation, and consider each of these in turn.

Functional relation between two variables

A functional relation between two variables is expressed by a mathematical formula. If X is the *independent* variable and Y the *dependent* variable, a functional relation is of the form:

$$Y = f(X)$$

Given a particular value of X, the function f indicates the corresponding value of Y.

Example. Consider the relation between dollar sales (Y) of a product sold at a fixed price and number of units sold (X). If the selling price is \$2 per unit, the relation is expressed by the equation:

$$Y = 2X$$

This functional relation is shown in Figure 2.1. Number of units sold and dollar sales during three recent periods (while the unit price remained constant at \$2) were as follows:

Period	*Number of Units Sold*	*Dollar Sales*
1	75	\$150
2	25	50
3	130	260

These observations are plotted also in Figure 2.1. Note that all fall directly on the line of functional relationship. This is characteristic of all functional relations.

FIGURE 2.1 Example of functional relation

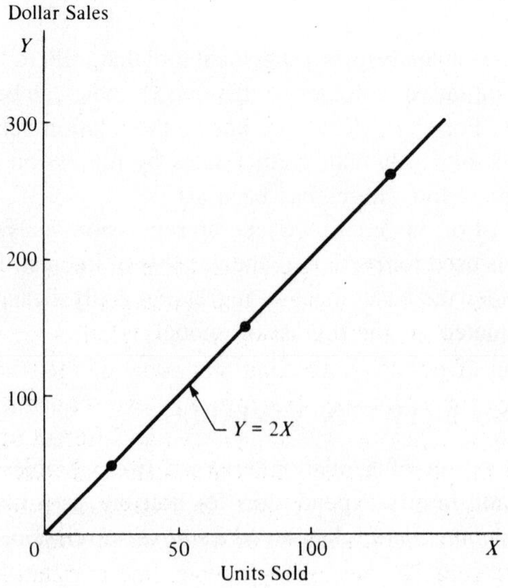

Statistical relation between two variables

A statistical relation, unlike a functional relation, is not a perfect one. In general, the observations for a statistical relation do not fall directly on the curve of relationship.

Example 1. A certain spare part is manufactured by the Westwood Company once a month in lots which vary in size as demand fluctuates. Table 2.1, page 36, contains data on lot size and number of man-hours of labor for 10 recent production runs performed under similar production conditions. These data are plotted in Figure 2.2a. Man-hours are taken as the *dependent* or *response* variable Y, and lot size as the *independent* or *predictor* variable X. The plotting is done as before. For instance, the first production run results are plotted as $X = 30$, $Y = 73$.

FIGURE 2.2 Statistical relation between lot size and number of man-hours—Westwood Company example

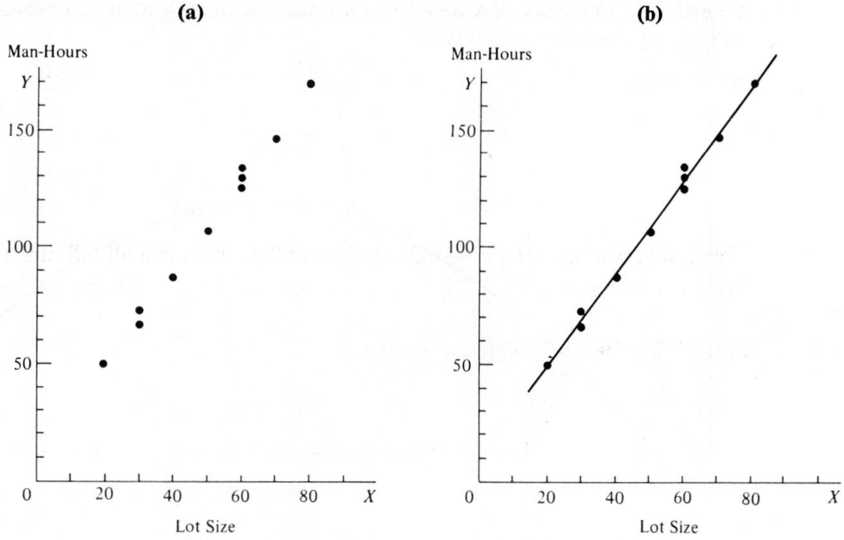

Figure 2.2a clearly suggests that there is a relation between lot size and number of man-hours, in the sense that the larger the lot size, the greater tends to be the number of man-hours. However, the relation is not a perfect one. There is a scattering of points, suggesting that some of the variation in man-hours is not accounted for by lot size. For instance, two production runs (1 and 8) consisted of 30 parts, yet they required somewhat different numbers of man-hours. Because of the scattering of points in a statistical relation, Figure 2.2a is called a *scatter diagram* or *scatter plot*. In statistical terminology, each point in the scatter diagram represents an *observation* or *trial*.

In Figure 2.2b, we have plotted a line of relationship which describes the statistical relation between man-hours and lot size. It indicates the general tendency by which man-hours vary with changes in lot size. Note that most of the points do not fall directly on the line of statistical relationship. This scattering of points around the line represents variation in man-hours which is not associated with the lot size, and which is usually considered to be of a random nature. Statistical relations can be highly useful, even though they do not have the exactitude of a functional relation.

Example 2. Figure 2.3 presents data on age and level of a steroid in plasma for 17 healthy females between 8 and 25 years old. The data strongly suggest that the statistical relationship is *curvilinear* (not linear). The curve of relationship has also been drawn in Figure 2.3. It implies that as age becomes increasingly higher, steroid level increases up to a point and then begins to decline. Note again the scattering of points around the curve of statistical relationship, typical of all statistical relations.

FIGURE 2.3 Curvilinear statistical relation between age and steroid level in healthy females aged 8 to 25

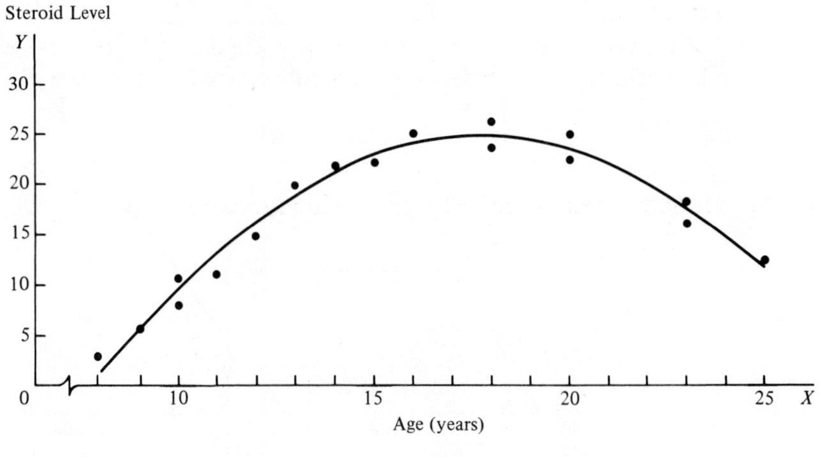

2.2 REGRESSION MODELS AND THEIR USES

Basic concepts

A regression model is a formal means of expressing the two essential ingredients of a statistical relation:

1. A tendency of the dependent variable Y to vary with the independent variable or variables in a systematic fashion.
2. A scattering of observations around the curve of statistical relationship.

These two characteristics are embodied in a regression model by postulating that:

1. In the population of observations associated with the sampled process, there is a probability distribution of Y for each level of X.
2. The means of these probability distributions vary in some systematic fashion with X.

Example. Consider again the Westwood Company lot size example. The number of man-hours Y is treated in a regression model as a random variable. For each lot size, there is postulated a probability distribution of Y. Figure 2.4 shows such a probability distribution for $X = 30$, which is the lot size for the first production run in Table 2.1. The actual number of man-hours Y (73 in our example in Table 2.1) is then viewed as a random selection from this probability distribution.

Figure 2.4 also shows probability distributions of Y for lot sizes $X = 50$ and $X = 70$. Note that the means of the probability distributions have a systematic relation to the level of X. This systematic relationship is called the *regression function of Y on X*. The graph of the regression function is called the *regression curve*. Note that in Figure 2.4 the regression function is linear. This would imply for our example that the expected (mean) number of man-hours varies linearly with lot size.

There is of course no a priori reason why man-hours need be linearly related to lot size. Figure 2.5 shows another possible regression model for our example.

FIGURE 2.4 Pictorial representation of linear regression model

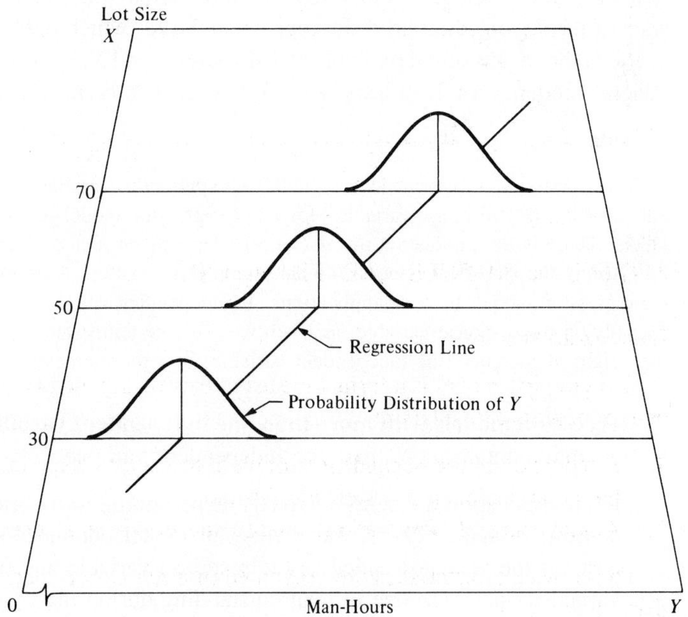

Here the regression function is curvilinear, with a shape reflecting economies of scale with larger lot sizes. Figure 2.5 differs in orientation from Figure 2.4 in that the X and Y axes are plotted conventionally in Figure 2.5. While this makes it not quite as easy to view the probability distributions, the orientation of Figure 2.5 shows the regression curve in the perspective to be utilized from here on.

FIGURE 2.5 Pictorial representation of curvilinear regression model

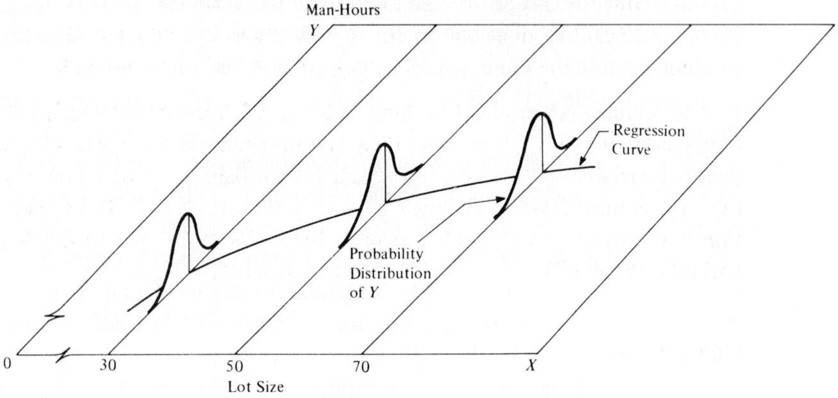

Regression models may differ in the form of the regression function as in Figures 2.4 and 2.5, in the shape of the probability distributions of the Y's, and in still other ways. Whatever the variation, the concept of a probability distribution of Y for given X is the formal counterpart to the empirical scatter in a statistical relation. Similarly, the regression curve, which describes the relation between the means of the probability distributions and X, is the counterpart to the general tendency of Y to vary with X systematically in a statistical relation.

Note

The expressions "independent variable" or "predictor variable" for X and "dependent variable" or "response variable" for Y in a regression model simply are conventional labels. There is no implication that Y causally depends on X in a given case. No matter how strong the statistical relation, no cause-and-effect pattern is necessarily implied by the regression model. In some applications, an independent variable actually is dependent causally on the response variable, as when we estimate temperature (the response) from the height of mercury (the independent variable) in a thermometer.

Regression models with more than one independent variable. Regression models may contain more than one independent variable.

1. In an application of regression analysis pertaining to 67 branch offices of a consumer finance chain, the regression model contained direct operating cost for the year just ended as the response variable and four independent variables—average size of loan outstanding during the year, average num-

ber of loans outstanding, total number of new loan applications processed, and office salary scale index.

2. In a tractor purchase study, the response variable was volume (in horsepower) of tractor purchases in each sales territory of a farm equipment firm. There were nine independent variables, including average age of tractors on farms in the territory, number of farms in the territory, and a quantity index of crop production in the territory.
3. In a medical study of short children, the response variable was the peak plasma growth hormone level. There were 14 independent variables, including age, sex, height, weight, and 10 skinfold measurements.

The features represented in Figures 2.4 and 2.5 must be extended into further dimensions when there is more than one independent variable. With two independent variables X_1 and X_2, for instance, a probability distribution of Y for each (X_1, X_2) combination is assumed by the regression model. The systematic relation between the means of these probability distributions and the independent variables X_1 and X_2 is then given by a regression surface.

Construction of regression models

Selection of independent variables. Since reality must be reduced to manageable proportions whenever we construct models, only a limited number of independent or predictor variables can—or should—be included in a regression model for any situation of interest. A central problem therefore is that of choosing, for a regression model, a set of independent variables which is ''good'' in some sense for the purposes of the analysis. A major consideration in making this choice is the extent to which a chosen variable contributes to reducing the remaining variation in Y after allowance is made for the contributions of other independent variables that have tentatively been included in the regression model. Other considerations include the importance of the variable as a causal agent in the process under analysis; the degree to which observations on the variable can be obtained more accurately, or quickly, or economically than on competing variables; and the degree to which the variable can be preset by management. In Chapter 12, we shall discuss procedures and problems in choosing the independent variables to be included in a regression model.

Functional form of regression equation. The choice of the functional form of the regression equation is related to the choice of the independent variables. Sometimes, relevant theory may indicate the appropriate functional form. Learning theory, for instance, may indicate that the regression function relating unit production costs to the number of previous times the item has been produced should have a specified shape with particular asymptotic properties.

More frequently, however, the functional form of the regression equation is not known in advance and must be decided upon once the data have been collected and analyzed. Thus, linear or quadratic regression functions are often used as satisfactory first approximations to regression functions of unknown nature. Indeed, these simple types of regression functions may be used even when theory provides the relevant functional form, notably when the known form is highly complex but can be reasonably approximated by a linear or quadratic regression function. Figure 2.6a illustrates a case where a complex regression function may be reasonably approximated by a linear regression function. Figure 2.6b provides an example where two linear regression functions may be used "piecewise" to approximate a complex regression function.

FIGURE 2.6 Uses of linear regression function to approximate complex regression functions

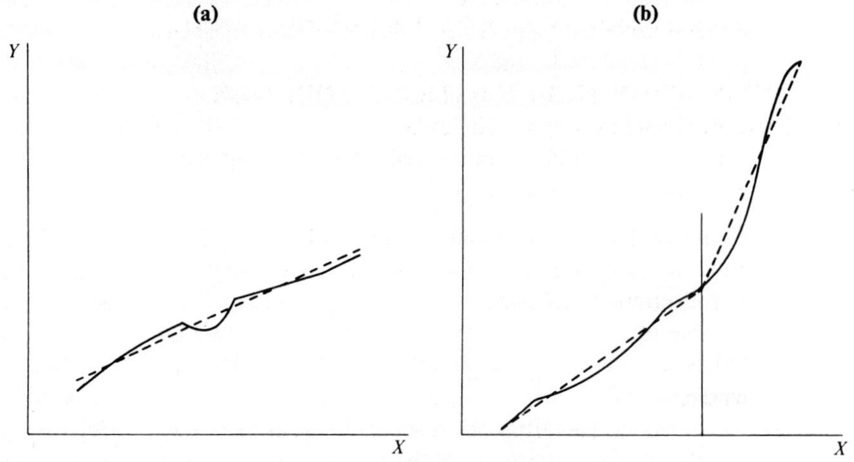

Scope of model. In formulating a regression model, we usually need to restrict the coverage of the model to some interval or region of values of the independent variable or variables. The scope is determined either by the design of the investigation or by the range of data at hand. For instance, a company studying the effect of price on sales volume investigated six price levels, ranging from \$4.95 to \$6.95. Here, the scope of the model would be limited to price levels ranging from near \$5 to near \$7. The shape of the regression function would be in serious doubt substantially outside this range because the investigation provided no evidence as to the nature of the statistical relation below \$4.95 or above \$6.95.

Uses of regression analysis

Regression analysis serves three major purposes: (1) description, (2) control, and (3) prediction, as illustrated by the three examples cited earlier. The tractor purchase study served a descriptive purpose. In the study of branch office operating costs, the purpose was administrative control; management was able by developing a usable statistical relation between costs and independent variables in the system, to set cost standards for each branch office in the company chain. In the medical study of short children, the purpose was prediction. Clinicians were able to use the statistical relation to predict growth hormone deficiencies in short children using simple measurements of the children.

The several purposes of regression analysis frequently overlap in practice. The Westwood Company lot size example provides a case in point. Knowledge of the relation between lot size and man-hours in past production runs enables management to predict the man-hour requirements for the next production run of given lot size, for purposes of cost estimation and production scheduling. After the run is completed, management can compare the actual man-hours against the predicted hours for purposes of administrative control.

2.3 REGRESSION MODEL WITH DISTRIBUTION OF ERROR TERMS UNSPECIFIED

Formal statement of model

In Part I of this book, we consider a basic regression model where there is only one independent variable and the regression function is linear. The model can be stated as follows:

$$Y_i = \beta_0 + \beta_1 X_i + \varepsilon_i \tag{2.1}$$

where:

Y_i is the value of the response variable in the ith trial

β_0 and β_1 are parameters

X_i is a known constant, namely, the value of the independent variable in the ith trial

ε_i is a random error term with mean $E(\varepsilon_i) = 0$ and variance $\sigma^2(\varepsilon_i) = \sigma^2$; ε_i and ε_j are uncorrelated so that the covariance $\sigma(\varepsilon_i, \varepsilon_j) = 0$ for all i, j; $i \neq j$

$i = 1, \ldots, n$

Model (2.1) is said to be *simple, linear in the parameters,* and *linear in the independent variable*. It is "simple" in that there is only one independent variable, "linear in the parameters" because no parameter appears as an exponent or is multiplied or divided by another parameter, and "linear in the independent variable" because this variable appears only in the first power. A model which is linear in the parameters and the independent variable is also called a *first-order model*.

Important features of model

1. The observed value of Y in the ith trial is the sum of two components: (1) the constant term $\beta_0 + \beta_1 X_i$ and (2) the random term ε_i. Hence, Y_i is a random variable.

2. Since $E(\varepsilon_i) = 0$, it follows from (1.13c) that:

$$E(Y_i) = E(\beta_0 + \beta_1 X_i + \varepsilon_i) = \beta_0 + \beta_1 X_i + E(\varepsilon_i) = \beta_0 + \beta_1 X_i$$

Note that $\beta_0 + \beta_1 X_i$ plays the role of the constant a in theorem (1.13c).

Thus, the response Y_i, when the level of X existing in the ith trial is X_i, comes from a probability distribution whose mean is:

(2.2) $$E(Y_i) = \beta_0 + \beta_1 X_i$$

We therefore know that the regression function for model (2.1) is:

(2.3) $$E(Y) = \beta_0 + \beta_1 X$$

since the regression function relates the means of the probability distributions of Y for any given X to the level of X.

3. The observed value of Y in the ith trial exceeds or falls short of the value of the regression function by the error term amount ε_i.

4. The error terms ε_i are assumed to have constant variance σ^2. It therefore follows that the variance of the response Y_i is:

(2.4) $$\sigma^2(Y_i) = \sigma^2$$

since, using theorem (1.15a), we have:

$$\sigma^2(\beta_0 + \beta_1 X_i + \varepsilon_i) = \sigma^2(\varepsilon_i) = \sigma^2$$

Thus, model (2.1) assumes that the probability distributions of Y have the same variance σ^2, regardless of the level of the independent variable X.

5. The error terms are assumed to be uncorrelated. Hence, the outcome in any one trial has no effect on the error term for any other trial—as to whether it is positive or negative, or small or large. Since the error terms ε_i and ε_j are uncorrelated, so are the responses Y_i and Y_j.

6. In summary, model (2.1) implies that the response variable observations Y_i come from probability distributions whose means are $E(Y_i) = \beta_0 + \beta_1 X_i$ and whose variances are σ^2, the same for all levels of X. Further, any two observations Y_i and Y_j are uncorrelated.

Example

Suppose that regression model (2.1) is applicable for the Westwood Company lot size application and is as follows:

$$Y_i = 9.5 + 2.1X_i + \varepsilon_i$$

Figure 2.7 contains a presentation of the regression function:

$$E(Y) = 9.5 + 2.1X$$

Suppose that in the ith trial, a lot of $X_i = 45$ units is produced and the actual number of man-hours is $Y_i = 108$. In that case, the error term value is $\varepsilon_i = +4$, for we have:

$$E(Y_i) = 9.5 + 2.1(45) = 104$$

and:

$$Y_i = 108 = 104 + 4$$

Figure 2.7 displays the probability distribution of Y when $X = 45$, and indicates from where in this distribution the observation $Y_i = 108$ came. Note again that the error term ε_i is simply the deviation of Y_i from its mean value $E(Y_i)$.

FIGURE 2.7 Illustration of linear regression model (2.1)

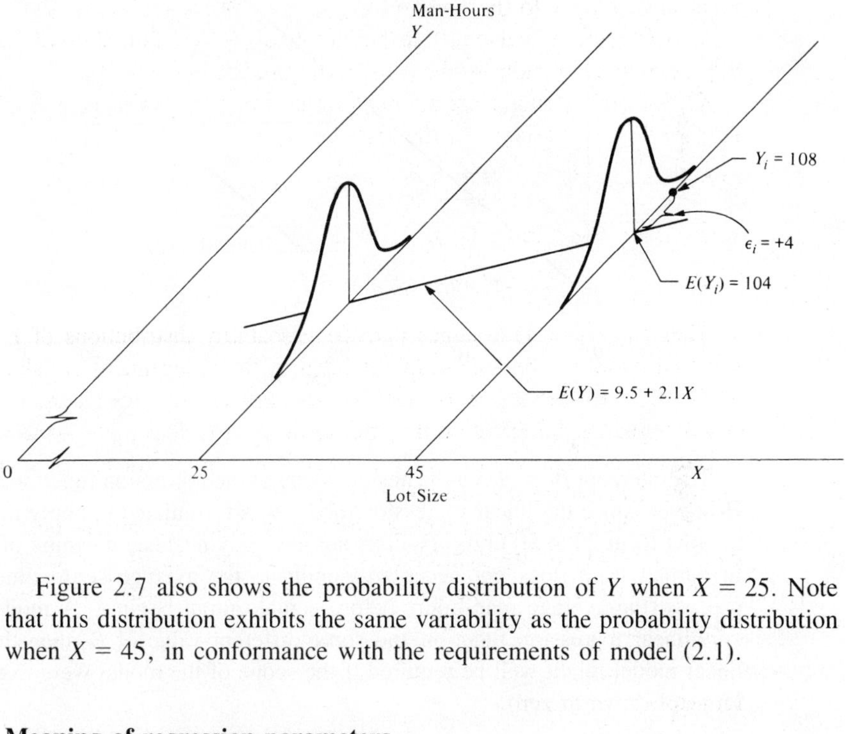

Figure 2.7 also shows the probability distribution of Y when $X = 25$. Note that this distribution exhibits the same variability as the probability distribution when $X = 45$, in conformance with the requirements of model (2.1).

Meaning of regression parameters

The parameters β_0 and β_1 in regression model (2.1) are called *regression coefficients*. β_1 is the slope of the regression line. It indicates the change in the mean of the probability distribution of Y per unit increase in X. The parameter β_0 is the Y intercept of the regression line. If the scope of the model includes $X = 0$, β_0 gives the mean of the probability distribution of Y at $X = 0$. When the scope of the model does not cover $X = 0$, β_0 does not have any particular meaning as a separate term in the regression model.

Example. Figure 2.8 shows the regression function:

$$E(Y) = 9.5 + 2.1X$$

for the previous Westwood Company lot size example. The slope $\beta_1 = 2.1$ indicates that an increase of one unit in lot size leads to an increase in the mean of the probability distribution of Y of 2.1 man-hours.

FIGURE 2.8 Meaning of linear regression parameters

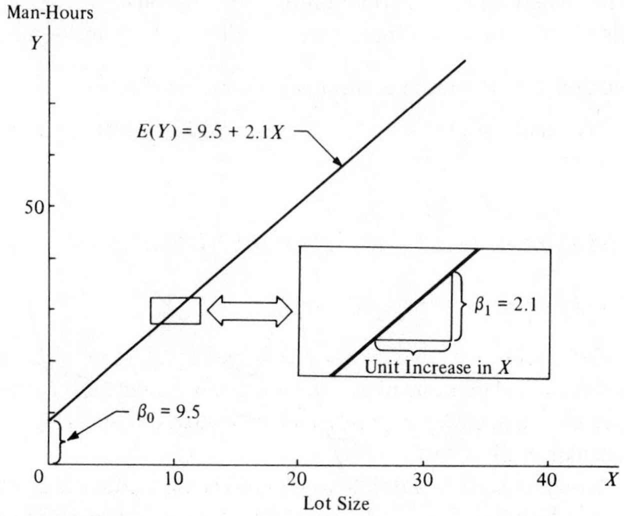

The intercept $\beta_0 = 9.5$ indicates the value of the regression function at $X = 0$. However, since the linear regression model was formulated to apply to lot sizes ranging from 20 to 80 units, β_0 does not have any intrinsic meaning of its own. In particular, it does not necessarily indicate the average setup time for the process (the average man-hours before actual output begins). A model with a curvilinear regression function and some different value of β_0 than that in the linear model might well be required if the scope of the model were to extend to lot sizes down to zero.

Alternative versions of model

Sometimes it is convenient to write model (2.1) in somewhat different, though equivalent, forms. Let X_0 be a *dummy variable* identically equal to one. Then, we can write (2.1) as follows:

$$Y_i = \beta_0 X_0 + \beta_1 X_i + \varepsilon_i \qquad \text{where } X_0 \equiv 1 \tag{2.5}$$

Another modification sometimes helpful is to use for the independent variable the deviation $X_i - \bar{X}$ rather than X_i. To leave model (2.1) unchanged, we need to write:

$$\begin{aligned} Y_i &= \beta_0 + \beta_1(X_i - \bar{X}) + \beta_1\bar{X} + \varepsilon_i \\ &= (\beta_0 + \beta_1\bar{X}) + \beta_1(X_i - \bar{X}) + \varepsilon_i \\ &= \beta_0^* + \beta_1(X_i - \bar{X}) + \varepsilon_i \end{aligned}$$

$y = \beta_0 + \beta_1 \ell + \beta_1 \tilde{X} + e$

Thus, an alternative model version is:

(2.6) $$Y_i = \beta_0^* + \beta_1(X_i - \bar{X}) + \varepsilon_i$$

where:

(2.6a) $$\beta_0^* = \beta_0 + \beta_1\bar{X}$$

We shall use models (2.1), (2.5), and (2.6) interchangeably as convenience dictates.

2.4 ESTIMATION OF REGRESSION FUNCTION

Obtaining needed sample data

Ordinarily, we do not know the values of the regression parameters β_0 and β_1 in model (2.1) and need to estimate them from sample data. Such sample data may be obtained by experimental or nonexperimental means. We shall briefly consider each in turn.

Sometimes, it is possible to conduct a controlled experiment to provide data from which the regression parameters can be estimated. Consider, for instance, an insurance company that wishes to study the relation between productivity of its clerks in processing claims and amount of training. Five clerks selected at random are trained for two weeks, five for three weeks, five for four weeks, and five for five weeks, and the productivity of the clerks is then observed. These data on length of training (X) and productivity (Y) to serve as a basis for estimating the regression parameters are *experimental data*.

Often it is not practical or feasible to conduct controlled experiments, in which case *nonexperimental data,* also called *observational data,* will need to be utilized. Such data are obtained without controlling the independent variable of interest. For example, public health officials wishing to study the relation between age of person (X) and number of days of illness last year (Y) would probably use data obtained from health records or from a survey of the population since they cannot assign ages at random to persons. Such data are nonexperimental data since the independent variable is not controlled. Similarly, the Westwood Company in our earlier lot size example needed to rely on nonexperimental data since the lot size at any given time was dictated by the demand for the product, which was not under the control of the company.

Once the data have been obtained, either by experiment or nonexperimentally, they can be assembled in a table such as Table 2.1 for the Westwood

Company example. We shall denote the (X, Y) observations for the first trial as (X_1, Y_1), for the second trial (X_2, Y_2), and in general for the ith trial (X_i, Y_i) where $i = 1, \ldots, n$. For the data in Table 2.1, $X_1 = 30$, $Y_1 = 73$, and so on, and $n = 10$.

TABLE 2.1 Data on lot size and number of man-hours—Westwood Company example

Production Run i	*Lot Size* X_i	*Man-Hours* Y_i
1	30	73
2	20	50
3	60	128
4	80	170
5	40	87
6	50	108
7	60	135
8	30	69
9	70	148
10	60	132

Method of least squares

To find "good" estimators of the regression parameters β_0 and β_1, we shall employ the method of least squares. For each sample observation (X_i, Y_i), the method of least squares considers the deviation of Y_i from its expected value:

$$Y_i - (\beta_0 + \beta_1 X_i) \tag{2.7}$$

In particular, the method of least squares requires that we consider the sum of the n squared deviations. This criterion is denoted by Q:

$$Q = \sum_{i=1}^{n} (Y_i - \beta_0 - \beta_1 X_i)^2 \tag{2.8}$$

According to the method of least squares, the estimators of β_0 and β_1 are those values b_0 and b_1, respectively, that minimize the criterion Q for the given sample observations (X_i, Y_i).

Example. Figure 2.9a contains a scatter plot of the sample data of Table 2.1 for the Westwood Company example. In Figure 2.9b is plotted a fitted regression line using the arbitrary estimates:

$$b_0 = 30$$
$$b_1 = 0$$

Also shown in Figure 2.9b are the deviations $Y_i - 30 - (0)X_i$. Note that each deviation corresponds to the vertical distance between Y_i and the fitted regression line. Clearly, the fit is poor. Hence, the deviations are large and so are the

FIGURE 2.9 Example of deviations from different fitted regression lines

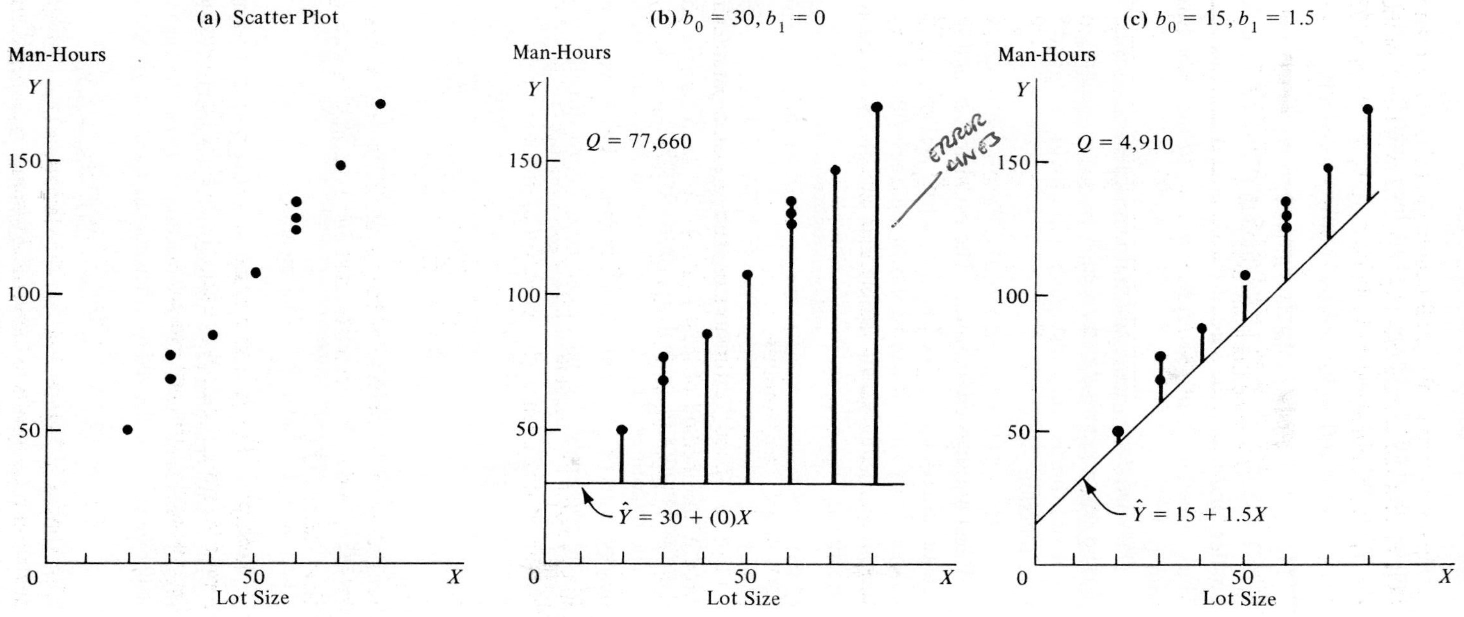

squared deviations. The sum Q of the squared deviations is (observations in ascending order):

$$Q = (50 - 30)^2 + (69 - 30)^2 + \cdots + (170 - 30)^2 = 77{,}660$$

Figure 2.9c contains the deviations $Y_i - b_0 - b_1X_i$ for the estimates $b_0 = 15$, $b_1 = 1.5$. Here, the fit is better (though still not good), the deviations are much smaller, and hence the sum of the squared deviations is reduced to $Q = 4{,}910$. Thus, a better fit of the regression line to the data corresponds to a smaller sum Q.

The objective of the method of least squares is to find estimates b_0 and b_1 for β_0 and β_1, respectively, for which Q is a minimum. In a certain sense, to be discussed shortly, these estimates will provide a "good" fit of the linear regression function.

Least squares estimators. The estimators b_0 and b_1 which satisfy the least squares criterion could be found by a trial and error procedure. However, this is not necessary since it can be shown that the values b_0 and b_1 which minimize Q for any particular set of sample data are given by the following simultaneous equations:

(2.9a) $$\Sigma Y_i = nb_0 + b_1\Sigma X_i$$

(2.9b) $$\Sigma X_iY_i = b_0\Sigma X_i + b_1\Sigma X_i^2$$

The equations (2.9a) and (2.9b) are called *normal equations;* b_0 and b_1 are called *point estimators* of β_0 and β_1, respectively.

The quantities ΣY_i, ΣX_i, and so on in (2.9) are calculated from the sample observations (X_i, Y_i). The equations then can be solved simultaneously for b_0 and b_1. Alternatively, b_0 and b_1 can be obtained directly as follows:

(2.10a) $$b_1 = \frac{\Sigma X_iY_i - \dfrac{(\Sigma X_i)(\Sigma Y_i)}{n}}{\Sigma X_i^2 - \dfrac{(\Sigma X_i)^2}{n}} = \frac{\Sigma(X_i - \bar{X})(Y_i - \bar{Y})}{\Sigma(X_i - \bar{X})^2}$$

(2.10b) $$b_0 = \frac{1}{n}(\Sigma Y_i - b_1\Sigma X_i) = \bar{Y} - b_1\bar{X}$$

where $\bar{X}$ and $\bar{Y}$ have the usual meaning.

Note

The normal equations (2.9) can be derived by calculus. For given sample observations (X_i, Y_i), the quantity Q in (2.8) is a function of β_0 and β_1. The values of β_0 and β_1 which minimize Q can be derived by differentiating (2.8) with respect to β_0 and β_1. We obtain:

$$\frac{\partial Q}{\partial \beta_0} = -2\Sigma(Y_i - \beta_0 - \beta_1X_i)$$

$$\frac{\partial Q}{\partial \beta_1} = -2\Sigma X_i(Y_i - \beta_0 - \beta_1X_i)$$

We then set these partial derivatives equal to zero, using b_0 and b_1 to denote the particular values of β_0 and β_1, respectively, which minimize Q:

$$-2\Sigma(Y_i - b_0 - b_1X_i) = 0$$
$$-2\Sigma X_i(Y_i - b_0 - b_1X_i) = 0$$

Simplifying, we obtain:

$$\sum_{i=1}^{n}(Y_i - b_0 - b_1X_i) = 0$$

$$\sum_{i=1}^{n}X_i(Y_i - b_0 - b_1X_i) = 0$$

Expanding out, we have:

$$\Sigma Y_i - nb_0 - b_1\Sigma X_i = 0$$
$$\Sigma X_iY_i - b_0\Sigma X_i - b_1\Sigma X_i^2 = 0$$

from which the normal equations (2.9) are obtained by rearranging terms.

A test of the second partial derivatives will show that a minimum is obtained with the least squares estimators b_0 and b_1.

Properties of least squares estimators. An important theorem, called the *Gauss-Markov theorem,* states:

(2.11) Under the conditions of model (2.1), the least squares estimators b_0 and b_1 in (2.10) are unbiased and have minimum variance among all unbiased linear estimators.

This theorem, which is proven in the next chapter, states first that both b_0 and b_1 are unbiased estimators. Hence:

$$E(b_0) = \beta_0$$
$$E(b_1) = \beta_1$$

so that neither estimator tends to overestimate or underestimate systematically.

Second, the theorem states that the sampling distributions of b_0 and b_1 have smaller variability than those of any other estimators belonging to a particular class of estimators. Thus, the least squares estimators are more precise than any of these other estimators. The class of estimators for which the least squares estimators are "best" consists of all unbiased estimators which are linear functions of the observations $Y_1, \ldots, Y_n$. The estimators b_0 and b_1 are such linear functions of the Y's. Consider, for instance, b_1. We have from (2.10a):

$$b_1 = \frac{\Sigma(X_i - \bar{X})(Y_i - \bar{Y})}{\Sigma(X_i - \bar{X})^2}$$

It will be shown in (3.5) that this expression is equal to:

$$b_1 = \frac{\Sigma(X_i - \bar{X})Y_i}{\Sigma(X_i - \bar{X})^2} = \Sigma k_iY_i$$

where:

$$k_i = \frac{X_i - \bar{X}}{\Sigma(X_i - \bar{X})^2}$$

Since the k_i are known constants (because the X_i are known constants), b_1 is a linear combination of the Y_i and hence is a linear estimator.

In the same fashion, it can be shown that b_0 is a linear estimator.

Among all linear estimators that are unbiased then, b_0 and b_1 have the smallest variability in repeated samples in which the X levels remain unchanged.

Example. To illustrate the calculation of the least squares estimators b_0 and b_1, we will use the Westwood Company case discussed earlier. The sample data are given in Table 2.1 and plotted in Figure 2.9a. Table 2.2 gives the basic results required to calculate b_0 and b_1. We have: $\Sigma Y_i = 1{,}100$, $\Sigma X_i = 500$, $\Sigma X_i Y_i = 61{,}800$, $\Sigma X_i^2 = 28{,}400$, $n = 10$. Using (2.10) we obtain:

$$b_1 = \frac{\Sigma X_i Y_i - \dfrac{(\Sigma X_i)(\Sigma Y_i)}{n}}{\Sigma X_i^2 - \dfrac{(\Sigma X_i)^2}{n}} = \frac{61{,}800 - \dfrac{(500)(1{,}100)}{10}}{28{,}400 - \dfrac{(500)^2}{10}} = 2.0$$

$$b_0 = \frac{1}{n}(\Sigma Y_i - b_1 \Sigma X_i) = \frac{1}{10}[1{,}100 - 2.0(500)] = 10.0$$

Thus, we estimate that the mean number of man-hours increases by 2.0 hours for each unit increase in lot size.

TABLE 2.2 Basic calculations to obtain b_0 and b_1—Westwood Company example

	Y_i	X_i	$X_i Y_i$	X_i^2	*(for later use)* Y_i^2
	73	30	2,190	900	5,329
	50	20	1,000	400	2,500
	128	60	7,680	3,600	16,384
	170	80	13,600	6,400	28,900
	87	40	3,480	1,600	7,569
	108	50	5,400	2,500	11,664
	135	60	8,100	3,600	18,225
	69	30	2,070	900	4,761
	148	70	10,360	4,900	21,904
	132	60	7,920	3,600	17,424
Total	1,100	500	61,800	28,400	134,660

Point estimation of mean response

Estimated regression function. Given sample estimators b_0 and b_1 of the parameters in the regression function (2.3):

$$E(Y) = \beta_0 + \beta_1 X$$

it is natural that we estimate the regression function as follows:

$$\hat{Y} = b_0 + b_1 X \tag{2.12}$$

where $\hat{Y}$ (read Y hat) is the value of the estimated regression function at the level X of the independent variable.

We will call a *value* of the response variable a *response* and will call $E(Y)$ the *mean response*. Thus, the mean response is the mean of the probability distribution of Y corresponding to the level X of the independent variable. $\hat{Y}$ then is a point estimator of the mean response when the level of the independent variable is X. It can be shown as an extension of the Gauss-Markov theorem (2.11) that $\hat{Y}$ is an unbiased estimator of $E(Y)$, with minimum variance in the class of unbiased linear estimators.

For the observations in the sample, we will call $\hat{Y}_i$:

$$\hat{Y}_i = b_0 + b_1 X_i \qquad i = 1, \ldots, n \tag{2.13}$$

the *fitted value* for the ith observation. Thus, the fitted value $\hat{Y}_i$ is to be viewed in distinction to the *observed value* Y_i.

Example. For the Westwood Company case, we found that the least squares estimates of the regression coefficients were:

$$b_0 = 10.0 \qquad b_1 = 2.0$$

Hence, the estimated regression function is:

$$\hat{Y} = 10.0 + 2.0X$$

If we are interested in the mean number of man-hours when the lot size is $X = 55$, our point estimate would be:

$$\hat{Y} = 10.0 + 2.0(55) = 120$$

Thus, we would estimate that the mean number of man-hours for production runs of $X = 55$ units is 120. We interpret this to mean that if many runs of size 55 are produced under the conditions of the 10 runs in the sample, the mean labor time for these many runs is about 120 hours. Of course, the labor time for any one run of size 55 is likely to fall above or below the mean response because of inherent variability in the system, as represented by the error term in the model.

Figure 2.10 contains a computer plot of the estimated regression function $\hat{Y} = 10.0 + 2.0X$, as well as the original data. Note the improved fit of the least squares regression line over the arbitrary lines in Figure 2.9. Indeed, the criterion

Q for the least squares regression line now is only $Q = 60$, as will be shown shortly, a much smaller value than the values of Q for the arbitrary fitted lines in Figure 2.9.

FIGURE 2.10 Observations and least squares regression line for Westwood Company example: $b_0 = 10.0$, $b_1 = 2.0$

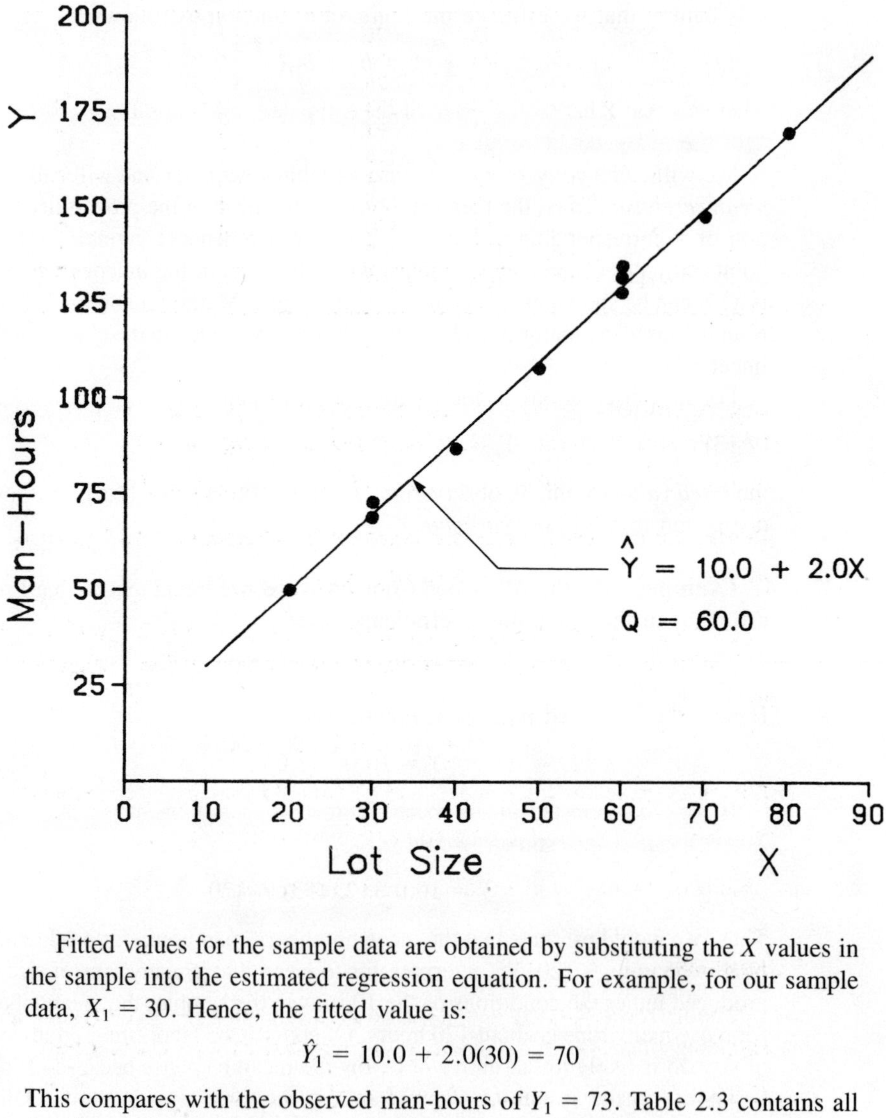

Fitted values for the sample data are obtained by substituting the X values in the sample into the estimated regression equation. For example, for our sample data, $X_1 = 30$. Hence, the fitted value is:

$$\hat{Y}_1 = 10.0 + 2.0(30) = 70$$

This compares with the observed man-hours of $Y_1 = 73$. Table 2.3 contains all the observed and fitted values for the Westwood Company data.

Alternative model (2.6). If the alternative regression model (2.6):

$$Y_i = \beta_0^* + \beta_1(X_i - \bar{X}) + \varepsilon_i$$

TABLE 2.3 Fitted values, residuals, and squared residuals—Westwood Company example

Observation Number i	*Lot Size* X_i	*Man-Hours* Y_i	*Estimated Mean Response* $\hat{Y}_i$	*Residual* $(Y_i - \hat{Y}_i) = e_i$	*Squared Residual* $(Y_i - \hat{Y}_i)^2 = e_i^2$
1	30	73	70	+3	9
2	20	50	50	0	0
3	60	128	130	−2	4
4	80	170	170	0	0
5	40	87	90	−3	9
6	50	108	110	−2	4
7	60	135	130	+5	25
8	30	69	70	−1	1
9	70	148	150	−2	4
10	60	132	130	+2	4
Total	500	1,100	1,100	0	60

is to be utilized, the least squares estimator b_1 of β_1 is the same as before. The least squares estimator of $\beta_0^* = \beta_0 + \beta_1\bar{X}$ is, using (2.10b):

$$b_0^* = b_0 + b_1\bar{X} = (\bar{Y} - b_1\bar{X}) + b_1\bar{X} = \bar{Y} \tag{2.14}$$

Hence, the estimated regression equation for alternative model (2.6) is:

$$\hat{Y} = \bar{Y} + b_1(X - \bar{X}) \tag{2.15}$$

In our Westwood Company example, $\bar{Y} = 1{,}100/10 = 110$ and $\bar{X} = 500/10 = 50$ (Table 2.2). Hence, the estimated regression equation in alternative form is:

$$\hat{Y} = 110.0 + 2.0(X - 50)$$

For our sample data, $X_1 = 30$; hence, we estimate the mean response to be:

$$\hat{Y}_1 = 110.0 + 2.0(30 - 50) = 70$$

which, of course, is identical to our earlier result.

Residuals

The ith *residual* is the difference between the observed value Y_i and the corresponding fitted value $\hat{Y}_i$. Denoting this residual by e_i, we can write:

$$e_i = Y_i - \hat{Y}_i = Y_i - b_0 - b_1X_i \tag{2.16}$$

Figure 2.11 shows the 10 residuals for the Westwood Company example. The magnitudes of the residuals are shown by the vertical lines between an observation and the fitted value on the estimated regression line. The residuals are calculated in Table 2.3 above.

FIGURE 2.11 Least squares regression line and residuals—Westwood Company example (observed values and residuals not plotted to scale)

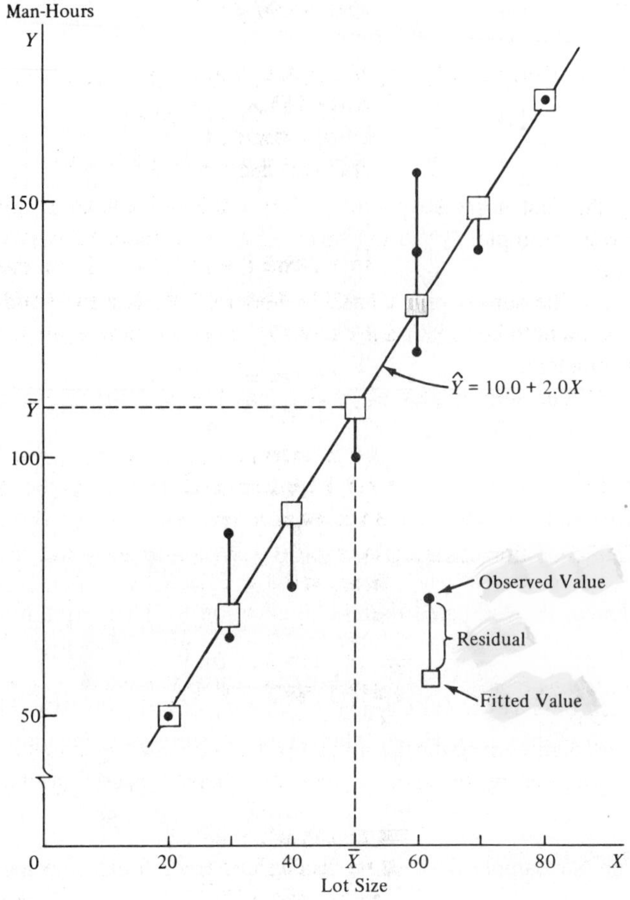

We need to distinguish between the model error term value $\varepsilon_i = Y_i - E(Y_i)$ and the residual $e_i = Y_i - \hat{Y}_i$. The former involves the vertical deviation of Y_i from the unknown population regression line, and hence is unknown. On the other hand, the residual is the observed vertical deviation of Y_i from the fitted regression line.

Residuals are highly useful for studying whether a given regression model is appropriate for the data at hand. We shall discuss this use in Chapter 4.

Properties of fitted regression line

The regression line fitted by the method of least squares has a number of properties worth noting:

1. The sum of the residuals is zero:

$$\sum_{i=1}^{n} e_i = 0 \tag{2.17}$$

This property can be proven easily. We have:

$$\begin{aligned}\Sigma e_i &= \Sigma(Y_i - b_0 - b_1X_i)\\ &= \Sigma Y_i - nb_0 - b_1\Sigma X_i = 0\end{aligned}$$

by the first normal equation (2.9a). Table 2.3 illustrates this property for our earlier example. Rounding errors may, of course, be present in any particular case.

2. The sum of the squared residuals, Σe_i^2, is a minimum. This was the requirement to be satisfied in deriving the least squares estimators of the regression parameters.

3. The sum of the observed values Y_i equals the sum of the fitted values $\hat{Y}_i$:

$$\sum_{i=1}^{n} Y_i = \sum_{i=1}^{n} \hat{Y}_i \tag{2.18}$$

This condition is implicit in the first normal equation (2.9a):

$$\begin{aligned}\Sigma Y_i &= nb_0 + b_1\Sigma X_i\\ &= \Sigma b_0 + \Sigma b_1X_i = \Sigma(b_0 + b_1X_i) = \Sigma\hat{Y}_i\end{aligned}$$

It follows from (2.18) that the mean of the $\hat{Y}_i$ is the same as the mean of the Y_i, namely, $\bar{Y}$.

4. The sum of the weighted residuals is zero when the residual in the ith trial is weighted by the level of the independent variable in the ith trial:

$$\sum_{i=1}^{n} X_ie_i = 0 \tag{2.19}$$

This follows from the second normal equation (2.9b):

$$\begin{aligned}\Sigma X_ie_i &= \Sigma X_i(Y_i - b_0 - b_1X_i)\\ &= \Sigma X_iY_i - b_0\Sigma X_i - b_1\Sigma X_i^2 = 0\end{aligned}$$

5. The sum of the weighted residuals is zero when the residual in the ith trial is weighted by the fitted value of the response variable for the ith trial:

$$\sum_{i=1}^{n} \hat{Y}_ie_i = 0 \tag{2.20}$$

This property is a consequence of (2.17) and (2.19).

6. The regression line always goes through the point $(\bar{X}, \bar{Y})$. This can be readily seen from the alternative form of the estimated regression line in (2.15). If $X = \bar{X}$, we have:

$$\hat{Y} = \bar{Y} + b_1(X - \bar{X}) = \bar{Y} + b_1(\bar{X} - \bar{X}) = \bar{Y}$$

Figure 2.11 demonstrates this property for our lot size example.

2.5 ESTIMATION OF ERROR TERMS VARIANCE σ^2

The variance σ^2 of the error terms ε_i in the regression model (2.1) needs to be estimated for a variety of purposes. Frequently, we would like to obtain an indication of the variability of the probability distributions of Y. Further, as we shall see in the next chapter, a variety of inferences concerning the regression function and the prediction of Y require an estimate of σ^2.

Point estimator of σ^2

Single population. To lay the basis for developing an estimator of σ^2 for the regression model (2.1), let us consider for a moment the simpler problem of sampling from a single population. In obtaining the sample variance s^2, we begin by considering the deviation of an observation Y_i from the estimated mean $\bar{Y}$, squaring it, and then summing all such deviations:

$$\sum_{i=1}^{n} (Y_i - \bar{Y})^2$$

Such a sum is called a *sum of squares*. The sum of squares is then divided by the degrees of freedom associated with it. This number is $n - 1$ here, because one degree of freedom is lost by using the estimate $\bar{Y}$ instead of the population mean μ. The resulting estimator is the usual sample variance:

$$s^2 = \frac{\sum_{i=1}^{n} (Y_i - \bar{Y})^2}{n - 1}$$

which is an unbiased estimator of the variance σ^2 of an infinite population. The sample variance is often called a *mean square*, because a sum of squares has been divided by the appropriate number of degrees of freedom.

Regression model. The logic of developing an estimator of σ^2 for the regression model is the same as when sampling a single population. Recall in this connection from (2.4) that the variance of each observation Y_i is σ^2, the same as that of each error term ε_i. We again need to calculate a sum of squared deviations, but must recognize that the Y_i come from different probability distributions with different means, depending upon the level X_i. Thus, the deviation of an observation Y_i must be calculated around its own estimated mean $\hat{Y}_i$. Hence, the deviations are the residuals:

$$Y_i - \hat{Y}_i = e_i$$

and the appropriate sum of squares, denoted by *SSE*, is:

$$SSE = \sum_{i=1}^{n} (Y_i - \hat{Y}_i)^2 = \sum_{i=1}^{n} (Y_i - b_0 - b_1 X_i)^2 = \sum_{i=1}^{n} e_i^2 \tag{2.21}$$

where *SSE* stands for *error sum of squares* or *residual sum of squares*.

The sum of squares *SSE* has $n - 2$ degrees of freedom associated with it. Two degrees of freedom are lost because both β_0 and β_1 had to be estimated in obtaining $\hat{Y}_i$. Hence, the appropriate mean square, denoted by *MSE*, is:

$$MSE = \frac{SSE}{n-2} = \frac{\Sigma(Y_i - \hat{Y}_i)^2}{n-2} \tag{2.22}$$

$$= \frac{\Sigma(Y_i - b_0 - b_1 X_i)^2}{n-2} = \frac{\Sigma e_i^2}{n-2}$$

where *MSE* stands for *error mean square* or *residual mean square*.

It can be shown that *MSE* is an unbiased estimator of σ^2 for the regression model (2.1):

$$E(MSE) = \sigma^2 \tag{2.23}$$

An estimator of the standard deviation σ is simply the positive square root of *MSE*.

Alternative computational formulas

There are a number of alternative computational formulas for *SSE*. Three of these are as follows:

$$SSE = \Sigma Y_i^2 - b_0 \Sigma Y_i - b_1 \Sigma X_i Y_i \tag{2.24a}$$

$$SSE = \Sigma(Y_i - \bar{Y})^2 - \frac{[\Sigma(X_i - \bar{X})(Y_i - \bar{Y})]^2}{\Sigma(X_i - \bar{X})^2} \tag{2.24b}$$

$$SSE = \left[\Sigma Y_i^2 - \frac{(\Sigma Y_i)^2}{n}\right] - \frac{\left[\Sigma X_i Y_i - \dfrac{\Sigma X_i \Sigma Y_i}{n}\right]^2}{\Sigma X_i^2 - \dfrac{(\Sigma X_i)^2}{n}} \tag{2.24c}$$

Comments

1. Formula (2.24a) is useful if b_0 and b_1 have already been calculated. Otherwise, (2.24b) and (2.24c) are more direct.

2. In (2.24a), the estimates b_0 and b_1 should be carried to a large number of digits in order to yield reliable results for *SSE*.

3. To obtain (2.24a), recall that by (2.21) we have:

$$SSE = \Sigma(Y_i - b_0 - b_1 X_i)^2$$

Thus:

$$\begin{aligned} SSE &= \Sigma Y_i^2 - 2b_0\Sigma Y_i - 2b_1\Sigma X_iY_i + nb_0^2 + 2b_0b_1\Sigma X_i + b_1^2\Sigma X_i^2 \\ &= \Sigma Y_i^2 - 2b_0\Sigma Y_i - 2b_1\Sigma X_iY_i + b_0(nb_0 + b_1\Sigma X_i) \\ &\qquad + b_1(b_0\Sigma X_i + b_1\Sigma X_i^2) \end{aligned}$$

The expressions in parentheses are equal to ΣY_i and ΣX_iY_i, respectively, by the normal equations (2.9). Substituting these terms within the parentheses yields an expression which reduces directly to (2.24a).

4. None of the three alternative formulas explicitly provides the residuals e_i. As noted earlier, the residuals are useful for studying the appropriateness of the model.

Example

Returning to our Westwood Company lot size example, we will calculate *SSE* by (2.21). The residuals were obtained earlier in Table 2.3. This table also shows the squared residuals. From these results, we obtain:

$$SSE = 60$$

Since $10 - 2 = 8$ degrees of freedom are associated with *SSE*, we find:

$$MSE = \frac{60}{8} = 7.5$$

Finally, a point estimate of σ, the standard deviation of the probability distribution of Y for any X, is $\sqrt{7.5} = 2.74$ man-hours.

Consider again the case where the lot size is $X = 55$ units. We estimated earlier that the probability distribution of Y for this lot size has a mean of 120 man-hours. Now, we have the additional information that the standard deviation of this distribution is estimated to be 2.74 man-hours.

If we wished to use, say, (2.24a) for calculating *SSE*, we would need ΣY_i^2. This sum is calculated in Table 2.2. We then obtain, using the results in Table 2.2 and the estimates $b_0 = 10.0$, $b_1 = 2.0$:

$$\begin{aligned} SSE &= \Sigma Y_i^2 - b_0\Sigma Y_i - b_1\Sigma X_iY_i \\ &= 134{,}660 - 10.0(1{,}100) - 2.0(61{,}800) = 60 \end{aligned}$$

which is, of course, the same result (except sometimes for rounding errors) as obtained earlier.

2.6 NORMAL ERROR REGRESSION MODEL

No matter what may be the functional form of the distribution of ε_i (and hence of Y_i), the least squares method provides unbiased point estimators of β_0 and β_1 which have minimum variance among all unbiased linear estimators. To set up interval estimates and make tests, however, we do need to make an assumption about the functional form of the distribution of the ε_i. The standard assumption is that the error terms are normally distributed, and we will adopt it here. A normal error term greatly simplifies the theory of regression analysis and is justifiable in many real world situations where regression analysis is applied.

Normal error model

The normal error model is as follows:

$$Y_i = \beta_0 + \beta_1 X_i + \varepsilon_i \tag{2.25}$$

where:

Y_i is the observed response in the ith trial
X_i is a known constant, the level of the independent variable in the ith trial
β_0 and β_1 are parameters
ε_i are independent $N(0, \sigma^2)$
$i = 1, \ldots, n$

Comments

1. The symbol $N(0, \sigma^2)$ stands for "normally distributed, with mean 0 and variance σ^2."

2. The normal error model (2.25) is the same as the regression model (2.1) with unspecified error distribution, except that model (2.25) assumes that the errors ε_i are normally distributed.

3. Because model (2.25) assumes that the errors are normally distributed, the assumption of uncorrelatedness of the ε_i in model (2.1) becomes one of independence in the normal error model.

4. Model (2.25) implies that the Y_i are independent normal random variables, with mean $E(Y_i) = \beta_0 + \beta_1 X_i$ and variance σ^2. Figure 2.4 (p. 27) pictures this normal error model. Each of the probability distributions of Y there is normally distributed, with constant variability, and the regression function is linear.

5. A major reason why the normality assumption for the error terms is justifiable in many situations is that the error terms frequently represent the effects of many factors omitted explicitly from the model, which do affect the response to some extent and which vary at random without reference to the independent variable X. For instance, in the lot size example, such factors as time lapse since the last production run, particular machines used, season of the year, and personnel employed, could vary more or less at random from run to run, independent of lot size. Also, there might be random measurement errors in recording Y. Insofar as these random effects have a degree of mutual independence, the composite error term ε_i representing all these factors would tend to comply with the central limit theorem and the error term distribution would approach normality as the number of factor effects becomes large.

A second reason why the normality assumption for the error terms is frequently justifiable is that the estimation and testing procedures to be discussed in the next chapter are based on the t distribution, which is not sensitive to moderate departures from normality. Thus, unless the departures from normality are serious, particularly with respect to skewness, the actual confidence coefficients and risks of errors will be close to the levels for exact normality.

Estimation of parameters by method of maximum likelihood

When the functional form of the probability distribution of the error terms is specified, estimators of the parameters β_0, β_1, and σ^2 can be obtained by the *method of maximum likelihood*. This method utilizes the joint probability distribution of the sample observations. When this joint probability distribution is viewed as a function of the parameters, given the particular sample observations, it is called the *likelihood function*. The likelihood function for the normal error model (2.25), given the sample observations $Y_1, \ldots, Y_n$, is:

$$(2.26) \quad L(\beta_0, \beta_1, \sigma^2) = \prod_{i=1}^{n} \frac{1}{(2\pi\sigma^2)^{1/2}} \exp\left[-\frac{1}{2\sigma^2}(Y_i - \beta_0 - \beta_1 X_i)^2\right]$$

$$= \frac{1}{(2\pi\sigma^2)^{n/2}} \exp\left[-\frac{1}{2\sigma^2}\sum_{i=1}^{n}(Y_i - \beta_0 - \beta_1 X_i)^2\right]$$

The values of β_0, β_1, and σ^2 which maximize this likelihood function are the maximum likelihood estimators. These are:

(2.27)

Parameter	*Maximum Likelihood Estimator*	
β_0	b_0	same as (2.10b)
β_1	b_1	same as (2.10a)
σ^2	$\hat{\sigma}^2 =$	$\dfrac{\Sigma(Y_i - \hat{Y}_i)^2}{n}$

Thus, the maximum likelihood estimators of β_0 and β_1 are the same estimators as provided by the method of least squares. The maximum likelihood estimator $\hat{\sigma}^2$ is biased, and ordinarily the unbiased estimator *MSE* as given in (2.22) is used. Note that the unbiased estimator *MSE* differs but slightly from the maximum likelihood estimator $\hat{\sigma}^2$, especially if n is not small:

$$(2.28) \quad MSE = \frac{n}{n-2}\hat{\sigma}^2$$

Comments

1. Since the maximum likelihood estimators b_0 and b_1 are the same as the least squares estimators, they have the properties of all least squares estimators:

a. They are unbiased.
b. They have minimum variance among all unbiased linear estimators.

In addition, the maximum likelihood estimators b_0 and b_1 for the normal error model (2.25) have other desirable properties:

c. They are consistent, as defined in (1.47).
d. They are sufficient, as defined in (1.48).
e. They are minimum variance unbiased; i.e., they have minimum variance in the class of all unbiased estimators (linear or otherwise).

Thus, for the normal error model, the estimators b_0 and b_1 have many desirable properties.

2. We find the values of β_0, β_1, and σ^2 which maximize the likelihood function L in (2.26) by taking partial derivatives of L with respect to β_0, β_1, and σ^2, equating each of the partials to zero, and solving the system of equations thus obtained. We can work with $\log_e L$, rather than L, because both L and $\log_e L$ are maximized for the same values of β_0, β_1, and σ^2:

$$\log_e L = -\frac{n}{2}\log_e 2\pi - \frac{n}{2}\log_e \sigma^2 - \frac{1}{2\sigma^2}\Sigma(Y_i - \beta_0 - \beta_1 X_i)^2 \tag{2.29}$$

Partial differentiation of this logarithmic likelihood is much easier; it yields:

$$\frac{\partial(\log_e L)}{\partial\beta_0} = \frac{1}{\sigma^2}\Sigma(Y_i - \beta_0 - \beta_1 X_i)$$

$$\frac{\partial(\log_e L)}{\partial\beta_1} = \frac{1}{\sigma^2}\Sigma X_i(Y_i - \beta_0 - \beta_1 X_i)$$

$$\frac{\partial(\log_e L)}{\partial\sigma^2} = -\frac{n}{2\sigma^2} + \frac{1}{2\sigma^4}\Sigma(Y_i - \beta_0 - \beta_1 X_i)^2$$

We now set these partial derivatives equal to zero, replacing β_0, β_1, and σ^2 by the estimators b_0, b_1, and $\hat{\sigma}^2$. We obtain after some simplification:

$$\Sigma(Y_i - b_0 - b_1 X_i) = 0 \tag{2.30a}$$

$$\Sigma X_i(Y_i - b_0 - b_1 X_i) = 0 \tag{2.30b}$$

$$\frac{\Sigma(Y_i - b_0 - b_1 X_i)^2}{n} = \hat{\sigma}^2 \tag{2.30c}$$

Formulas (2.30a) and (2.30b) are identical to the earlier least squares normal equations (2.9), and (2.30c) is the biased estimator of σ^2 given earlier in (2.27).

2.7 COMPUTER INPUTS AND OUTPUTS

Regression calculations used to be tedious chores, especially when the number of observations was large and when there were several independent variables. Today computers can be used quite easily to perform regression calculations, with one of many available packaged programs. Also, a number of calculators contain regression routines.

The inputting of data will vary from program to program. With some, the X and Y observations are entered as separate sets. For our Westwood Company data in Table 2.1, one form of data input would be:

X_i	Y_i
30	73
20	50
etc.	etc.

In some other cases, the inputting of the data would be in the form: X_1, Y_1, X_2, Y_2, etc. For our example, this input form would be: 30, 73, 20, 50, etc.

The computer output will also vary from one program package to another. Figure 2.12 illustrates a typical output format when the linear regression model is fitted to the Westwood Company data in Table 2.1 by the SPSS computer program (Ref. 2.1). The computed values of b_0, b_1, and $\sqrt{MSE}$ are annotated in Figure 2.12, and agree with our earlier results. In subsequent chapters, we shall explain the additional output in Figure 2.12.

FIGURE 2.12 Segment of computer output for regression run on Westwood Company data (SPSS, Ref. 2.1)

```
                    VARIABLES
1 SIZE             2 HOURS
   30.0000             73.0000
   20.0000             50.0000
   60.0000            128.0000
   80.0000            170.0000
   40.0000             87.0000
   50.0000            108.0000
   60.0000            135.0000
   30.0000             69.0000
   70.0000            148.0000
   60.0000            132.0000


DEPENDENT VARIABLE..     HOURS


VARIABLE(S) ENTERED ON STEP NUMBER  1..     S!ZE


MULTIPLE R            0.99780
R SQUARE              0.99561

STANDARD ERROR        2.73861 <- √MSE
----------------- VARIABLES IN THE EQUATION -----------------
VARIABLE                B                  STD ERROR B           F
SIZE             2.000000 <- b1              0.04697      1813.333
(CONSTANT)       10.00000 <- b0


VARIABLE              MEAN       STANDARD DEV      CASES
SIZE               50.0000            19.4365         10
HOURS             110.0000            38.9587         10


ANALYSIS OF VARIANCE        DF       SUM OF SQUARES           MEAN SQUARE
REGRESSION                  1.        13600.00000           13600.00000
RESIDUAL                    8.           60.00000               7.50000
```

PROBLEMS

2.1. Refer to the sales volume example on page 24. Suppose that the number of units sold is measured accurately but clerical errors are frequently made in determining the dollar sales. Would the relation between the number of units sold and dollar sales still be a functional one? Discuss.

2.2. The members of a health spa pay annual membership dues of \$300 plus a charge of \$2 for each visit to the spa. Let Y denote the total dollar cost for the year for a member and X the number of visits by the member during the year. Express the relation between X and Y mathematically. Is it a functional or a statistical relation?

2.3. Experience with a certain type of plastic indicates that a relation exists between the hardness (measured in Brinell units) of items molded from the plastic (Y) and the elapsed time since termination of the molding process (X). It is proposed to study this relation by means of regression analysis. A participant in the discussion objects, pointing out that the hardening of the plastic "is the result of a natural chemical process that doesn't leave anything to chance, so the relation must be mathematical and regression analysis is not appropriate." Evaluate this objection.

2.4. In Table 2.1, the lot size X is the same in production runs 1 and 8 but the man-hours Y differ. What feature of regression model (2.1) is illustrated by this?

2.5. When asked to state the simple linear regression model, a student wrote it as follows: $E(Y_i) = \beta_0 + \beta_1 X_i + \varepsilon_i$. Do you agree?

2.6. Consider the normal error regression model (2.25). Suppose that the parameter values are $\beta_0 = 200$, $\beta_1 = 5.0$, and $\sigma = 4$.

a. Plot this normal error regression model in the fashion of Figure 2.7. Show the distributions of Y for $X = 10$, 20, and 40.

b. Explain the meaning of the parameters β_0 and β_1. Assume that the scope of the model includes $X = 0$.

2.7. In a simulation exercise, regression model (2.1) applies with $\beta_0 = 100$, $\beta_1 = 20$, and $\sigma^2 = 25$. An observation on Y will be made for $X = 5$.

a. Can you state the exact probability that Y will fall between 195 and 205? Explain.

b. If the normal error regression model (2.25) is applicable, can you now state the exact probability that Y will fall between 195 and 205? If so, state it.

2.8. In Figure 2.7, suppose another observation were obtained at $X = 45$. Would $E(Y)$ for this new observation still be 104? Would the Y value for this new observation again be 108?

2.9. A student in accounting enthusiastically declared: "Regression is a very powerful tool. We can isolate fixed and variable costs by fitting a linear regression model, even when we have no data for small lots." Discuss.

2.10. An analyst in a large corporation studied the relation between current annual salary (Y) and age (X) for the 46 computer programmers presently employed in the company. She concluded that the relation is curvilinear, reaching a maximum at 47 years. Does this imply that the salary for a programmer increases until age 47 and then decreases? Explain.

2.11. The regression function relating production output by an employee after taking a training program (Y) to the production output before the training program (X) is $E(Y) = 20 + .95X$, where X ranges from 40 to 100. An observer concludes that the training program does not raise production output on the average because β_1 is not greater than 1.0. Comment.

2.12. Evaluate the following statement: "For the least squares method to be fully valid, it is required that the distribution of Y is normal."

2.13. A person states that b_0 and b_1 in the fitted regression equation (2.12) can be estimated by the method of least squares. Comment.

2.14. According to (2.17), $\Sigma e_i = 0$ when model (2.1) is fitted to a set of n observations by the method of least squares. Is it also true that $\Sigma \varepsilon_i = 0$? Comment.

2.15. Grade point average. The director of admissions of a small college administered a newly designed entrance test to 20 students selected at random from the new freshman class in a study to determine whether a student's grade point average (GPA) at the end of the freshman year (Y) can be predicted from the entrance test score (X). The results of the study follow. Assume that the first-order regression model (2.1) is appropriate.

i:	1	2	3	4	5	6	7	8	9	10
X_i:	5.5	4.8	4.7	3.9	4.5	6.2	6.0	5.2	4.7	4.3
Y_i:	3.1	2.3	3.0	1.9	2.5	3.7	3.4	2.6	2.8	1.6

i:	11	12	13	14	15	16	17	18	19	20
X_i:	4.9	5.4	5.0	6.3	4.6	4.3	5.0	5.9	4.1	4.7
Y_i:	2.0	2.9	2.3	3.2	1.8	1.4	2.0	3.8	2.2	1.5

Summary calculational results are: $\Sigma X_i = 100.0$, $\Sigma Y_i = 50.0$, $\Sigma X_i^2 = 509.12$, $\Sigma Y_i^2 = 134.84$, $\Sigma X_i Y_i = 257.66$.

a. Obtain the least squares estimates of β_0 and β_1, and state the estimated regression function.
b. Plot the estimated regression function and the data. Does the estimated regression function appear to fit the data well?
c. Obtain a point estimate of the mean freshman GPA when the entrance test score is $X = 5.0$.
d. What is the point estimate of the change in the mean response when the entrance test score increases by one point?

2.16. Calculator maintenance. The Tri-City Office Equipment Corporation sells an imported desk calculator on a franchise basis and performs preventive maintenance and repair service on this calculator. The data below have been collected from 18 recent calls on users to perform routine preventive maintenance service; for each call, X is the number of machines serviced and Y is the total number of minutes spent by the service person. Assume that the first-order regression model (2.1) is appropriate.

i:	1	2	3	4	5	6	7	8	9
X_i:	7	6	5	1	5	4	7	3	4
Y_i:	97	86	78	10	75	62	101	39	53

i:	10	11	12	13	14	15	16	17	18
X_i:	2	8	5	2	5	7	1	4	5
Y_i:	33	118	65	25	71	105	17	49	68

Summary calculational results are: $\Sigma Y_i = 1{,}152$, $\Sigma X_i = 81$, $\Sigma(Y_i - \bar{Y})^2 = 16{,}504$, $\Sigma(X_i - \bar{X})^2 = 74.5$, $\Sigma(X_i - \bar{X})(Y_i - \bar{Y}) = 1{,}098$.

a. Obtain the estimated regression function.
b. Plot the estimated regression function and the data. How well does the estimated regression function fit the data?
c. Interpret b_0 in your estimated regression function. Does b_0 provide any relevant information here? Explain.

d. Obtain a point estimate of the mean service time when $X = 5$ machines are serviced.

2.17. Airfreight breakage. A substance used in biological and medical research is shipped by airfreight to users in cartons of 1,000 ampules. The data below, involving 10 shipments, were collected on the number of times the carton was transferred from one aircraft to another over the shipment route (X) and the number of ampules found to be broken upon arrival (Y). Assume that the first-order regression model (2.1) is appropriate.

i:	1	2	3	4	5	6	7	8	9	10
X_i:	1	0	2	0	3	1	0	1	2	0
Y_i:	16	9	17	12	22	13	8	15	19	11

a. Obtain the estimated regression function. Plot the estimated regression function and the data. Does a linear regression function appear to give a good fit here?

b. Obtain a point estimate of the expected number of broken ampules when $X = 1$ transfer is made.

c. Estimate the increase in the expected number of ampules broken when there are 2 transfers as compared to 1 transfer.

d. Verify that your fitted regression line goes through the point $(\bar{X}, \bar{Y})$.

2.18. Plastic hardness. Refer to Problem 2.3. Twelve batches of the plastic were made, and from each batch one test item was molded and the hardness measured at some specific point in time. The results are shown below; X is elapsed time in hours, and Y is hardness in Brinell units. Assume that the first-order regression model (2.1) is appropriate.

i:	1	2	3	4	5	6	7	8	9	10	11	12
X_i:	32	48	72	64	48	16	40	48	48	24	80	56
Y_i:	230	262	323	298	255	199	248	279	267	214	359	305

a. Obtain the estimated regression function. Plot the estimated regression function and the data. Does a linear regression function appear to give a good fit here?

b. Obtain a point estimate of the mean hardness when $X = 48$ hours.

c. Obtain a point estimate of the change in mean hardness when X increases by one hour.

2.19. Refer to **Grade point average** Problem 2.15.

a. Obtain the residuals e_i. Do they sum to zero in accord with (2.17)?

b. Estimate σ^2 and σ. In what units is σ expressed?

2.20. Refer to **Calculator maintenance** Problem 2.16.

a. Obtain the residuals e_i and the sum of the squared residuals Σe_i^2. What is the relation between the sum of the squared residuals here and the quantity Q in (2.8)?

b. Obtain point estimates of σ^2 and σ. In what units is σ expressed?

2.21. Refer to **Airfreight breakage** Problem 2.17.

a. Obtain the residual for the first observation. What is its relation to ε_1?

b. Compute Σe_i^2 and MSE. What is estimated by MSE?

2.22. Refer to **Plastic hardness** Problem 2.18.

a. Obtain the residuals e_i. Do they sum to zero in accord with (2.17)?

b. Estimate σ^2 and σ. In what units is σ expressed?

2.23. Muscle mass. A person's muscle mass is expected to decrease with age. To explore this relationship in women, a nutritionist randomly selected four women from each 10-year age group, beginning with age 40 and ending with age 79. The results follow; X is age, and Y is a measure of muscle mass. Assume that the first-order regression model (2.1) is appropriate.

i:	1	2	3	4	5	6	7	8
X_i:	71	64	43	67	56	73	68	56
Y_i:	82	91	100	68	87	73	78	80

i:	9	10	11	12	13	14	15	16
X_i:	76	65	45	58	45	53	49	78
Y_i:	65	84	116	76	97	100	105	77

a. Obtain the estimated regression function. Plot the estimated regression function and the data. Does a linear regression function appear to give a good fit here? Does your plot support the anticipation that muscle mass decreases with age?

b. Obtain the following: (1) a point estimate of the difference in the mean muscle mass for women differing in age by one year, (2) a point estimate of the mean muscle mass for women aged $X = 60$ years, (3) the value of the residual for the eighth observation, (4) a point estimate of σ^2.

2.24. Robbery rate. A criminologist studying the relationship between population density and robbery rate in medium-sized U.S. cities collected the following data for a random sample of 16 cities; X is the population density of the city (number of people per unit area), and Y is the robbery rate last year (number of robberies per 100,000 people). Assume that the first-order regression model (2.1) is appropriate.

i:	1	2	3	4	5	6	7	8
X_i:	59	49	75	54	78	56	60	82
Y_i:	209	180	195	192	215	197	208	189

i:	9	10	11	12	13	14	15	16
X_i:	69	83	88	94	47	65	89	70
Y_i:	213	201	214	212	205	186	200	204

a. Obtain the estimated regression function. Plot the estimated regression function and the data. Does the linear regression function appear to give a good fit here? Discuss.

b. Obtain point estimates of the following: (1) the difference in the mean robbery rate in cities that differ by one unit in population density, (2) the mean robbery rate last year in cities with population density $X = 60$, (3) ε_{10}, (4) σ^2.

EXERCISES

2.25. Refer to regression model (2.1). Assume that $X = 0$ is within the scope of the model. What is the implication for the regression function if $\beta_0 = 0$ so that the model is $Y_i = \beta_1 X_i + \varepsilon_i$? How would the regression function plot on a graph?

2.26. Refer to regression model (2.1). What is the implication for the regression function if $\beta_1 = 0$ so that the model is $Y_i = \beta_0 + \varepsilon_i$? How would the regression function plot on a graph?

2.27. Refer to **Plastic hardness** Problem 2.18. Suppose one test item was molded from a single batch of plastic and the hardness of this one item was measured at 12 different points in time. Would the error term in the model for this case still reflect the same effects as for the experiment initially described? Would you expect the error terms for the different points in time to be uncorrelated? Discuss.

2.28. Derive the expression for b_1 in (2.10a) from the normal equations in (2.9).

2.29. (Calculus needed.) Refer to the model $Y_i = \beta_0 + \varepsilon_i$ in Exercise 2.26. Using the method of least squares, derive the estimator of β_0 for this model.

2.30. Prove that the least squares estimator of β_0 obtained in Exercise 2.29 is unbiased.

2.31. Prove the result in (2.20)—that the sum of the residuals weighted by the fitted values is zero.

2.32. Refer to Table 2.1 for the Westwood Company example. When asked to present a point estimate of the expected man-hours for runs of 60 pieces, a person gave the estimate 131.7 because this is the mean number of man-hours in the three runs of size 60 in the study. A critic states that this person's approach "throws away" most of the data in the study because observations on lot sizes other than 60 are ignored. Comment.

2.33. In **Airfreight breakage** Problem 2.17, the least squares estimates are $b_0 = 10.20$ and $b_1 = 4.00$, and $\Sigma e_i^2 = 17.60$. Evaluate the least squares criterion Q in (2.8) for the estimates: (1) $b_0 = 9, b_1 = 3$; (2) $b_0 = 11, b_1 = 5$. Is the criterion Q larger for these estimates than for the least squares estimates?

2.34. Two observations on Y were obtained at each of three X levels—namely, $X = 5$, $X = 10$, and $X = 15$.

a. Show that the least squares regression line fitted to the *three* points $(5, \overline{Y}_1)$, $(10, \overline{Y}_2)$, and $(15, \overline{Y}_3)$, where $\overline{Y}_1$, $\overline{Y}_2$, and $\overline{Y}_3$ denote the means of the Y observations at the three X levels, is identical to the least squares regression line fitted to the original six observations.

b. In this study, could the error term variance σ^2 be estimated without fitting a regression line? Explain.

2.35. In the Westwood Company example, the observations at $X = 20$ and $X = 80$ fall directly on the fitted regression line (Table 2.3 and Figure 2.2b). If these two observations were deleted, would the least squares regression line fitted to the remaining eight observations be changed? [*Hint:* What is the contribution of the two observations to the least squares criterion Q in (2.8)?]

2.36. (Calculus needed.) Refer to the model $Y_i = \beta_1 X_i + \varepsilon_i$, $i = 1, \ldots, n$, in Exercise 2.25.

a. Find the least squares estimator of β_1.
b. Assume that the error terms ε_i are independent $N(0, \sigma^2)$ and that σ^2 is known. State the likelihood function for the n sample observations and obtain the maximum likelihood estimator of β_1. Is it the same as the least squares estimator?
c. Show that the maximum likelihood estimator is unbiased.

2.37. Shown below are the number of galleys of type set (X) and the dollar cost of correcting typographical errors (Y) in a random sample of recent orders handled by a firm specializing in technical printing. Assume that the regression model $Y_i = \beta_1 X_i + \varepsilon_i$ is appropriate, with normally distributed independent error terms whose variance is $\sigma^2 = 16$.

i:	1	2	3	4	5	6
X_i:	7	12	4	14	25	30
Y_i:	128	213	75	250	446	540

a. State the likelihood function for the six observations, for $\sigma^2 = 16$.
b. Evaluate the likelihood function for $\beta_1 = 17$, 18, and 19. For which of these β_1 values is the likelihood function largest?
c. The maximum likelihood estimator is $b_1 = \Sigma X_i Y_i / \Sigma X_i^2$. Find the maximum likelihood estimate. Are your results in part (b) consistent with this estimate?

PROJECTS

2.38. Refer to the SMSA data set (pp. 1109–13). The number of active physicians in an SMSA (Y) is expected to be related to total population, land area, and total personal income. Assume that the first-order regression model (2.1) is appropriate in each case.

a. Regress the number of active physicians in turn on each of the three independent variables. State the estimated regression functions.
b. Plot the three estimated regression functions and data on separate graphs. Does a linear regression relation appear to provide a good fit in each of the three cases?
c. Calculate *MSE* for each of the three cases. Which independent variable leads to the smallest variability around the fitted regression line?

2.39. Refer to the SMSA data set (pp. 1109–13).

a. For each geographic region, regress total serious crimes in an SMSA (Y) against total population (X). Assume that the first-order regression model (2.1) is appropriate for each region. State the estimated regression functions.
b. Are the estimated regression functions similar for the four regions? Discuss.
c. Calculate *MSE* for each region. Is the variability around the fitted regression line approximately the same for the four regions? Discuss.

2.40. Refer to the SENIC data set (pp. 1105–8). The average length of stay in a hospital (Y) is anticipated to be related to infection risk, available facilities and services, and routine chest X-ray ratio. Assume that the first-order regression model (2.1) is appropriate in each case.

a. Regress average length of stay on each of the three independent variables. State the estimated regression functions.
b. Plot the three estimated regression functions and data on separate graphs. Does a linear relation appear to provide a good fit in each of the three cases?
c. Calculate *MSE* for each of the three cases. Which independent variable leads to the smallest variability around the fitted regression line?

2.41. Refer to the SENIC data set (pp. 1105–8).
a. For each geographic region, regress average length of stay in hospital (Y) against infection risk (X). Assume that the first-order regression model (2.1) is appropriate for each region. State the estimated regression functions.
b. Are the estimated regression functions similar for the four regions? Discuss.
c. Calculate *MSE* for each region. Is the variability around the fitted regression line approximately the same for the four regions? Discuss.

CITED REFERENCE

2.1 Nie, N. H.; C. H. Hull; J. G. Jenkins; K. Steinbrenner; and D. H. Bent. *SPSS: Statistical Package for the Social Sciences*. 2d ed. New York: McGraw-Hill, 1975.

3

Inferences in regression analysis

In this chapter, we first take up inferences concerning the regression parameters β_0 and β_1, considering both interval estimation of these parameters and tests about them. We then discuss interval estimation of the mean $E(Y)$ of the probability distribution of Y, for given X, and prediction intervals for a new observation Y, given X. Finally, we take up the analysis of variance approach to regression analysis.

Throughout this chapter, and in the remainder of Part I unless otherwise stated, we assume that the normal error model (2.25) *is applicable*. This model is:

(3.1) $$Y_i = \beta_0 + \beta_1 X_i + \varepsilon_i$$

where:

β_0 and β_1 are parameters
X_i are known constants
ε_i are independent $N(0, \sigma^2)$

3.1 INFERENCES CONCERNING β_1

Frequently, we are interested in drawing inferences about β_1, the slope of the regression line in model (3.1). For instance, a market research analyst studying the relation between sales (Y) and advertising expenditures (X) may wish to obtain an interval estimate of β_1 because it will provide information as to how many additional sales dollars, on the average, are generated by an additional dollar of advertising expenditure.

At times, tests concerning β_1 are of interest, particularly one of the form:

$$H_0: \beta_1 = 0$$
$$H_a: \beta_1 \neq 0$$

The reason for interest in testing whether or not $\beta_1 = 0$ is that $\beta_1 = 0$ indicates that there is no linear association between Y and X. Figure 3.1 illustrates the case when $\beta_1 = 0$ for the normal error model (3.1). Note that the regression line is horizontal and that the means of the probability distributions of Y are therefore all equal, namely:

$$E(Y) = \beta_0 + (0)X = \beta_0$$

Since model (3.1) assumes normality of the probability distributions of Y with constant variance, and since the means are equal when $\beta_1 = 0$, it follows that the probability distributions of Y are identical when $\beta_1 = 0$. This is shown in Figure 3.1. Thus, $\beta_1 = 0$ for model (3.1) implies not only that there is no linear association between Y and X but also that there is no relation of any type between Y and X, since the probability distributions of Y are then identical at all levels of X.

Before discussing inferences concerning β_1 further, we need to consider the sampling distribution of b_1, our point estimator of β_1.

FIGURE 3.1 Model (3.1) when $\beta_1 = 0$

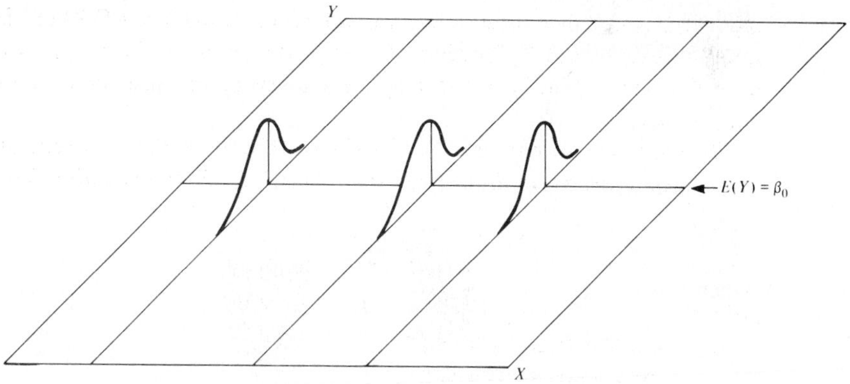

Sampling distribution of b_1

The point estimator b_1 was given in (2.10a) as follows:

$$b_1 = \frac{\Sigma(X_i - \bar{X})(Y_i - \bar{Y})}{\Sigma(X_i - \bar{X})^2} \tag{3.2}$$

The sampling distribution of b_1 refers to the different values of b_1 that would be obtained with repeated sampling when the levels of the independent variable X are held constant from sample to sample.

(3.3) For model (3.1), the sampling distribution of b_1 is normal, with mean and variance:

(3.3a) $$E(b_1) = \beta_1$$

(3.3b) $$\sigma^2(b_1) = \frac{\sigma^2}{\Sigma(X_i - \bar{X})^2}$$

To show this, we need to recognize that b_1 is a linear combination of the observations Y_i.

***b_1* as linear combination of the *Y_i*.** We will show that b_1, as defined in (3.2), can be written:

$$b_1 = \Sigma k_i Y_i$$

Here the k_i are constants; hence, b_1 is a linear combination of the Y_i. We first prove:

(3.4) $$\Sigma(X_i - \bar{X})(Y_i - \bar{Y}) = \Sigma(X_i - \bar{X})Y_i$$

This follows since:

$$\Sigma(X_i - \bar{X})(Y_i - \bar{Y}) = \Sigma(X_i - \bar{X})Y_i - \Sigma(X_i - \bar{X})\bar{Y}$$

But $\Sigma(X_i - \bar{X})\bar{Y} = \bar{Y}\Sigma(X_i - \bar{X}) = 0$ since $\Sigma(X_i - \bar{X}) = 0$. Hence, (3.4) holds.

We now express b_1, using (3.4):

$$b_1 = \frac{\Sigma(X_i - \bar{X})(Y_i - \bar{Y})}{\Sigma(X_i - \bar{X})^2} = \frac{\Sigma(X_i - \bar{X})Y_i}{\Sigma(X_i - \bar{X})^2}$$

We can rewrite this as follows:

(3.5) $$b_1 = \Sigma k_i Y_i$$

where:

(3.5a) $$k_i = \frac{X_i - \bar{X}}{\Sigma(X_i - \bar{X})^2}$$

Observe that the k_i are fixed quantities since the X_i are fixed.

Note

The quantities k_i have a number of interesting properties that will be used later:

1.

(3.6) $$\Sigma k_i = 0$$

because $\Sigma k_i = \Sigma(X_i - \bar{X})/\Sigma(X_i - \bar{X})^2 = 0/\Sigma(X_i - \bar{X})^2 = 0$.

2.

$$\Sigma k_i X_i = 1 \tag{3.7}$$

because $\Sigma k_i X_i = \Sigma(X_i - \bar{X})X_i/\Sigma(X_i - \bar{X})^2 = \Sigma(X_i - \bar{X})(X_i - \bar{X})/\Sigma(X_i - \bar{X})^2 = \Sigma(X_i - \bar{X})^2/\Sigma(X_i - \bar{X})^2 = 1$. The result $\Sigma(X_i - \bar{X})(X_i - \bar{X}) = \Sigma(X_i - \bar{X})X_i$ is obtained in the same way as (3.4).

3.

$$\Sigma k_i^2 = \frac{1}{\Sigma(X_i - \bar{X})^2} \tag{3.8}$$

because:

$$\Sigma k_i^2 = \Sigma\left[\frac{(X_i - \bar{X})}{\Sigma(X_i - \bar{X})^2}\right]^2 = \frac{1}{[\Sigma(X_i - \bar{X})^2]^2}\Sigma(X_i - \bar{X})^2$$
$$= \frac{1}{\Sigma(X_i - \bar{X})^2}$$

Normality. We return now to the sampling distribution of b_1 for the normal error model (3.1). The normality of the sampling distribution of b_1 follows at once from the fact that b_1 is a linear combination of the Y_i. The Y_i are independently, normally distributed according to model (3.1), and theorem (1.35) states that a linear combination of independent normal random variables is normally distributed.

Mean. The unbiasedness of the point estimator b_1, stated earlier in the Gauss-Markov theorem (2.11), is easy to show:

$$E(b_1) = E(\Sigma k_i Y_i) = \Sigma k_i E(Y_i) = \Sigma k_i(\beta_0 + \beta_1 X_i)$$
$$= \beta_0 \Sigma k_i + \beta_1 \Sigma k_i X_i$$

Hence, by (3.6) and (3.7) $E(b_1) = \beta_1$.

Variance. The variance of b_1 can be derived readily. We need only remember that the Y_i are independent random variables, each with variance σ^2, and that the k_i are constants. Hence, we obtain by (1.26):

$$\sigma^2(b_1) = \sigma^2(\Sigma k_i Y_i) = \Sigma k_i^2 \sigma^2(Y_i)$$
$$= \Sigma k_i^2 \sigma^2 = \sigma^2 \Sigma k_i^2$$
$$= \sigma^2 \frac{1}{\Sigma(X_i - \bar{X})^2}$$

The last step follows from (3.8).

Estimated variance. We can estimate the variance of the sampling distribution of b_1:

$$\sigma^2(b_1) = \frac{\sigma^2}{\Sigma(X_i - \bar{X})^2}$$

by replacing the parameter σ^2 with the unbiased estimator of σ^2, namely, *MSE:*

$$(3.9) \qquad s^2(b_1) = \frac{MSE}{\Sigma(X_i - \bar{X})^2} = \frac{MSE}{\Sigma X_i^2 - \frac{(\Sigma X_i)^2}{n}}$$

The point estimator $s^2(b_1)$ is an unbiased estimator of $\sigma^2(b_1)$. Taking the square root, we obtain $s(b_1)$, our point estimator of $\sigma(b_1)$.

Note

We stated in theorem (2.11) that b_1 has minimum variance among all unbiased linear estimators of the form:

$$\hat{\beta}_1 = \Sigma c_i Y_i$$

where the c_i are arbitrary constants. We shall now prove this. Since $\hat{\beta}_1$ must be unbiased, the following must hold:

$$E(\hat{\beta}_1) = E(\Sigma c_i Y_i) = \Sigma c_i E(Y_i) = \beta_1$$

Now $E(Y_i) = \beta_0 + \beta_1 X_i$ by (2.2) so that the above condition becomes:

$$E(\hat{\beta}_1) = \Sigma c_i(\beta_0 + \beta_1 X_i) = \beta_0 \Sigma c_i + \beta_1 \Sigma c_i X_i = \beta_1$$

For the unbiasedness condition to hold, the c_i must follow the restrictions:

$$\Sigma c_i = 0 \qquad \Sigma c_i X_i = 1$$

Now the variance of $\hat{\beta}_1$ is by (1.26):

$$\sigma^2(\hat{\beta}_1) = \Sigma c_i^2 \sigma^2(Y_i) = \sigma^2 \Sigma c_i^2$$

Let us define $c_i = k_i + d_i$, where the k_i are the least squares constants in (3.5) and the d_i are arbitrary constants. We can then write:

$$\sigma^2(\hat{\beta}_1) = \sigma^2 \Sigma c_i^2 = \sigma^2 \Sigma(k_i + d_i)^2 = \sigma^2[\Sigma k_i^2 + \Sigma d_i^2 + 2\Sigma k_i d_i]$$

We know that $\sigma^2 \Sigma k_i^2 = \sigma^2(b_1)$ from our proof above. Further, $\Sigma k_i d_i = 0$; this follows from the restrictions on the k_i and the c_i. Hence, we have:

$$\sigma^2(\hat{\beta}_1) = \sigma^2(b_1) + \sigma^2 \Sigma d_i^2$$

Note that the smallest value of Σd_i^2 is zero. Hence, the variance of $\hat{\beta}_1$ is at a minimum when $\Sigma d_i^2 = 0$. But this can only occur if all $d_i = 0$, which implies $c_i \equiv k_i$. Thus, the least squares estimator b_1 has minimum variance among all unbiased linear estimators.

Sampling distribution of $(b_1 - \beta_1)/s(b_1)$

Since b_1 is normally distributed, we know that the standardized statistic $(b_1 - \beta_1)/\sigma(b_1)$ is a standard normal variable. Ordinarily, of course, we need to estimate $\sigma(b_1)$ by $s(b_1)$, and hence are interested in the distribution of the standardized statistic $(b_1 - \beta_1)/s(b_1)$. An important theorem in statistics states:

$$(3.10) \qquad \frac{(b_1 - \beta_1)}{s(b_1)} \text{ is distributed as } t(n-2) \text{ for model (3.1)}$$

Intuitively, this result should not be unexpected. We know that if the observations Y_i come from the same normal population, $(\bar{Y} - \mu)/s(\bar{Y})$ follows the t distribution with $n - 1$ degrees of freedom. The estimator b_1, like $\bar{Y}$, is a linear combination of the observations Y_i. The reason for the difference in the degrees of freedom is that two parameters (β_0 and β_1) need to be estimated for the regression model; hence, two degrees of freedom are lost here.

Note

We can show that $(b_1 - \beta_1)/s(b_1)$ is distributed as t with $n - 2$ degrees of freedom by relying on the following theorem:

(3.11) For model (3.1), SSE/σ^2 is distributed as χ^2 with $n - 2$ degrees of freedom, and is independent of b_0 and b_1.

First, let us rewrite $(b_1 - \beta_1)/s(b_1)$ as follows:

$$\frac{b_1 - \beta_1}{\sigma(b_1)} \div \frac{s(b_1)}{\sigma(b_1)}$$

The numerator is a standard normal variable z. The nature of the denominator can be seen by first considering:

$$\frac{s^2(b_1)}{\sigma^2(b_1)} = \frac{\dfrac{MSE}{\Sigma(X_i - \bar{X})^2}}{\dfrac{\sigma^2}{\Sigma(X_i - \bar{X})^2}} = \frac{MSE}{\sigma^2} = \frac{\dfrac{SSE}{n-2}}{\sigma^2}$$

$$= \frac{SSE}{\sigma^2(n-2)} \sim \frac{\chi^2(n-2)}{n-2}$$

where the symbol $\sim$ stands for "is distributed as." The last step follows from (3.11). Hence, we have:

$$\frac{b_1 - \beta_1}{s(b_1)} \sim \frac{z}{\sqrt{\dfrac{\chi^2(n-2)}{n-2}}}$$

But by theorem (3.11), z and χ^2 are independent, since z is a function of b_1 and b_1 is independent of $SSE/\sigma^2 \sim \chi^2$. Hence, by definition (1.39), it follows that:

$$\frac{b_1 - \beta_1}{s(b_1)} \sim t(n-2)$$

This result places us in a position to readily make inferences concerning β_1.

Confidence interval for β_1

Since $(b_1 - \beta_1)/s(b_1)$ follows a t distribution, we can make the following probability statement:

(3.12) $P\{t(\alpha/2; n-2) \leq (b_1 - \beta_1)/s(b_1) \leq t(1 - \alpha/2; n-2)\} = 1 - \alpha$

Here, $t(\alpha/2; n - 2)$ denotes the $(\alpha/2)100$ percentile of the t distribution with

$n - 2$ degrees of freedom. Because of the symmetry of the t distribution, it follows that:

$$t(\alpha/2; n-2) = -t(1-\alpha/2; n-2) \tag{3.13}$$

Rearranging the inequalities in (3.12) and using (3.13), we obtain:

$$P\{b_1 - t(1-\alpha/2; n-2)s(b_1) \leq \beta_1 \leq b_1 + t(1-\alpha/2; n-2)s(b_1)\} = 1 - \alpha \tag{3.14}$$

Since (3.14) holds for all possible values of β_1, the $1 - \alpha$ confidence limits for β_1 are:

$$b_1 \pm t(1-\alpha/2; n-2)s(b_1) \tag{3.15}$$

Example. Let us return to the Westwood Company lot size example of Chapter 2. Management wishes an estimate of β_1 with a 95 percent confidence coefficient. We summarize in Table 3.1 the needed results obtained earlier. First, we need to obtain $s(b_1)$:

$$s^2(b_1) = \frac{MSE}{\Sigma(X_i - \bar{X})^2} = \frac{7.5}{3,400} = .002206$$

and:

$$s(b_1) = .04697$$

For a 95 percent confidence coefficient, we require $t(.975; 8)$. From Table A–2 in the Appendix, we find $t(.975; 8) = 2.306$. The 95 percent confidence interval, by (3.15), then is:

$$2.0 - 2.306(.04697) \leq \beta_1 \leq 2.0 + 2.306(.04697)$$
$$1.89 \leq \beta_1 \leq 2.11$$

Thus, with confidence coefficient .95, we estimate that the mean number of man-hours increases by somewhere between 1.89 and 2.11 for each increase of one part in the lot size.

TABLE 3.1 Results for Westwood Company example obtained in Chapter 2

$n = 10$	$\bar{X} = 50$
$b_0 = 10.0$	$b_1 = 2.0$
$\hat{Y} = 10.0 + 2.0X$	$SSE = 60$
$\sum X_i^2 = 28,400$	$MSE = 7.5$

$$\sum X_i^2 - \frac{(\sum X_i)^2}{n} = \sum (X_i - \bar{X})^2 = 3,400$$

$$\sum X_i Y_i - \frac{\sum X_i \sum Y_i}{n} = \sum (X_i - \bar{X})(Y_i - \bar{Y}) = 6,800$$

$$\sum (Y_i)^2 - \frac{(\sum Y_i)^2}{n} = \sum (Y_i - \bar{Y})^2 = 13,660$$

Note

In Chapter 2, we noted that the scope of a regression model is restricted ordinarily to some interval of values of the independent variable. This is particularly important to keep in mind when using estimates of the slope β_1. In our lot size example, a linear regression model appeared appropriate for lot sizes between 20 and 80, the range of the independent variable in the recent past. It may not be reasonable to use the estimate of the slope to infer the effect of lot size on number of man-hours far outside this range since the regression relation may not be linear there.

Tests concerning β_1

Since $(b_1 - \beta_1)/s(b_1)$ is distributed as t with $n - 2$ degrees of freedom, tests concerning β_1 can be set up in ordinary fashion using the t distribution.

Example 1: Two-sided test. Suppose a cost analyst in the Westwood Company is interested in testing whether or not there is a linear association between man-hours and lot size, using regression model (3.1). The two alternatives then are:

$$\begin{aligned} &H_0: \beta_1 = 0 \\ &H_a: \beta_1 \neq 0 \end{aligned} \tag{3.16}$$

If the analyst wishes to control the risk of a Type I error at .05, he could indeed conclude H_a at once by referring to the 95 percent confidence interval for β_1 constructed earlier, since the interval does not include 0.

An explicit test of the alternatives (3.16) is based on the test statistic:

$$t^* = \frac{b_1}{s(b_1)} \tag{3.17}$$

The decision rule with this test statistic when controlling the level of significance at α is:

$$\begin{aligned} &\text{If } |t^*| \leq t(1 - \alpha/2; n - 2), \text{ conclude } H_0 \\ &\text{If } |t^*| > t(1 - \alpha/2; n - 2), \text{ conclude } H_a \end{aligned} \tag{3.17a}$$

For the Westwood Company example, where $\alpha = .05$, $b_1 = 2.0$, and $s(b_1) = .04697$, we require $t(.975; 8) = 2.306$. Thus, the decision rule for testing alternatives (3.16) is:

$$\begin{aligned} &\text{If } |t^*| \leq 2.306, \text{ conclude } H_0 \\ &\text{If } |t^*| > 2.306, \text{ conclude } H_a \end{aligned}$$

Since $|t^*| = |2.0/.04697| = 42.58 > 2.306$, we conclude H_a, that $\beta_1 \neq 0$ or that there is a linear association between man-hours and lot size.

The P-value for the sample outcome is obtained by finding the probability $P[t(8) > t^* = 42.58]$. We see from Table A–2 that this probability is less than .0005. Indeed, it can be shown to be almost 0, to be denoted by 0+. Thus, the two-sided P-value is $2(0+) = 0+$. Since the two-sided P-value is less than the specified level of significance $\alpha = .05$, we could conclude H_a directly.

Example 2: One-sided test. If the analyst had wished to test whether or not β_1 is positive, controlling the level of significance at $\alpha = .05$, the alternatives would have been:

$$H_0: \beta_1 \leq 0$$
$$H_a: \beta_1 > 0$$

and the decision rule based on test statistic (3.17) would have been:

$$\text{If } t^* \leq t(1 - \alpha; n - 2), \text{ conclude } H_0$$
$$\text{If } t^* > t(1 - \alpha; n - 2), \text{ conclude } H_a$$

For $\alpha = .05$, we require $t(.95; 8) = 1.860$. Since $t^* = 42.58 > 1.860$, we would conclude H_a, that β_1 is positive.

This same conclusion could be reached directly from the one-sided P-value, which was noted in Example 1 to be $0+$. Since this P-value is less than .05, we would conclude H_a.

Comments

1. Many computer packages and scientific publications commonly report the P-value together with the value of the test statistic. In this way, one can conduct a test at any desired level of significance α by comparing the P-value with the specified level α. Users of computer packages need to be careful to ascertain whether one-sided or two-sided P-values are furnished.

2. Occasionally, it is desired to test whether or not β_1 equals some specified nonzero value β_{10}, which may be a historical norm, the value for a comparable process, or an engineering specification. For such a test, the appropriate test statistic is:

$$t^* = \frac{b_1 - \beta_{10}}{s(b_1)} \tag{3.18}$$

The decision rule to be used for the alternatives:

$$H_0: \beta_1 = \beta_{10}$$
$$H_a: \beta_1 \neq \beta_{10}$$

is still (3.17a), but it now is based on t^* defined in (3.18).

Note that test statistic (3.18) simplifies to test statistic (3.17) when the test involves $H_0: \beta_1 = \beta_{10} = 0$.

3.2 INFERENCES CONCERNING β_0

As noted in Chapter 2, there are only infrequent occasions when we wish to make inferences concerning β_0, the intercept of the regression line. These occur when the scope of the model includes $X = 0$.

Sampling distribution of b_0

The point estimator b_0 was given in (2.10b) as follows:

(3.19) $$b_0 = \bar{Y} - b_1\bar{X}$$

The sampling distribution of b_0 refers to the different values of b_0 that would be obtained with repeated sampling when the levels of the independent variable X are held constant from sample to sample.

(3.20) For model (3.1), the sampling distribution of b_0 is normal, with mean and variance:

(3.20a) $$E(b_0) = \beta_0$$

(3.20b) $$\sigma^2(b_0) = \sigma^2 \frac{\Sigma X_i^2}{n\Sigma(X_i - \bar{X})^2} = \sigma^2\left[\frac{1}{n} + \frac{\bar{X}^2}{\Sigma(X_i - \bar{X})^2}\right]$$

The normality of the sampling distribution of b_0 follows because b_0, like b_1, is a linear combination of the observations Y_i. The results for the mean and variance of the sampling distribution of b_0 can be obtained in similar fashion as those for b_1.

An estimator of $\sigma^2(b_0)$ is obtained by replacing σ^2 by its point estimator MSE:

(3.21) $$s^2(b_0) = MSE\frac{\Sigma X_i^2}{n\Sigma(X_i - \bar{X})^2} = MSE\left[\frac{1}{n} + \frac{\bar{X}^2}{\Sigma(X_i - \bar{X})^2}\right]$$

The square root, $s(b_0)$, is an estimator of $\sigma(b_0)$.

Sampling distribution of $(b_0 - \beta_0)/s(b_0)$

Analogous to theorem (3.10) for b_1, there is a theorem for b_0 that states:

(3.22) $$\frac{b_0 - \beta_0}{s(b_0)} \text{ is distributed as } t(n-2) \text{ for model (3.1)}$$

Hence, confidence intervals for β_0 and tests concerning β_0 can be set up in ordinary fashion, using the t distribution.

Confidence interval for β_0

The $1 - \alpha$ confidence limits for β_0 are obtained in the same manner as those for β_1 derived earlier. They are:

(3.23) $$b_0 \pm t(1 - \alpha/2; n-2)s(b_0)$$

Example. As noted earlier, the scope of the model for the Westwood Company example does not extend to lot sizes of $X = 0$. Hence, the regression parameter β_0 may not have intrinsic meaning. If, nevertheless, a 90 percent confidence interval for β_0 were desired, we would proceed by finding $t(.95; 8)$ and $s(b_0)$. From Table A–2, we find $t(.95; 8) = 1.860$. Using the earlier results summarized in Table 3.1, we obtain by (3.21):

$$s^2(b_0) = MSE\frac{\Sigma X_i^2}{n\Sigma(X_i - \bar{X})^2} = (7.5)\frac{28{,}400}{10(3{,}400)} = 6.26471$$

or:

$$s(b_0) = 2.50294$$

Hence, the 90 percent confidence interval for β_0 is:

$$10.0 - 1.860(2.50294) \leq \beta_0 \leq 10.0 + 1.860(2.50294)$$
$$5.34 \leq \beta_0 \leq 14.66$$

We caution again that this confidence interval does not necessarily provide meaningful information. For instance, it does not necessarily provide information about the "setup" costs of producing a lot of parts (costs incurred in setting up the production process, no matter what is the lot size), since we are not certain whether a linear regression model is appropriate when the scope of the model is extended to $X = 0$.

3.3 SOME CONSIDERATIONS ON MAKING INFERENCES CONCERNING β_0 AND β_1

Effect of departures from normality

If the probability distributions of Y are not exactly normal but do not depart seriously, the sampling distributions of b_0 and b_1 will be approximately normal, and the use of the t distribution will provide approximately the specified confidence coefficient or level of significance. Even if the distributions of Y are far from normal, the estimators b_0 and b_1 generally have the property of *asymptotic normality*—their distributions approach normality under very general conditions as the sample size increases. Thus, with sufficiently large samples, the confidence intervals and decision rules given earlier still apply even if the probability distributions of Y depart far from normality. For large samples, the t value is, of course, replaced by the z value for the standard normal distribution.

Interpretation of confidence coefficient and risks of errors

Since model (3.1) assumes that the X_i are known constants, the confidence coefficient and risks of errors are interpreted with respect to taking repeated samples in which the X observations are kept at the same levels as in the observed sample. For instance, we constructed a confidence interval for β_1 with a confidence coefficient of .95 in the Westwood Company example. This coefficient is interpreted to mean that if many independent samples are taken where the levels of X (the lot sizes) in the first sample are repeated in these other samples and a 95 percent confidence interval is constructed for each sample, 95 percent of the intervals will contain the true value of β_1.

Spacing of the X levels

Inspection of formulas (3.3b) and (3.20b) for the variances of b_1 and b_0, respectively, indicates that for given n and σ^2, these variances are affected by the spacing of the X levels in the observed data. For example, the more the spread in the X levels, the larger is the quantity $\Sigma(X_i - \bar{X})^2$ and the smaller is the variance of b_1. We will discuss in Section 5.9 how the X observations should be spaced in experiments where spacing can be controlled.

Power of tests

The power of tests on β_0 and β_1 can be obtained from Table A–5 in the Appendix, which contains charts of the power function of the t test. Consider, for example, the general decision problem:

(3.24)
$$H_0: \beta_1 = \beta_{10}$$
$$H_a: \beta_1 \neq \beta_{10}$$

for which the general test statistic (3.18) is employed:

(3.24a)
$$t^* = \frac{b_1 - \beta_{10}}{s(b_1)}$$

and the decision rule for level of significance α is:

(3.24b)
$$\text{If } |t^*| \leq t(1 - \alpha/2; n - 2), \text{ conclude } H_0$$
$$\text{If } |t^*| > t(1 - \alpha/2; n - 2), \text{ conclude } H_a$$

The power of the test is the probability that the decision rule will lead to conclusion H_a when H_a in fact holds. Specifically, the power is given by:

(3.25)
$$\text{Power} = P\{|t^*| > t(1 - \alpha/2; n - 2) | \delta\}$$

where δ is a measure of *noncentrality*—i.e., how far the true value of β_1 is from β_{10}:

(3.26)
$$\delta = \frac{|\beta_1 - \beta_{10}|}{\sigma(b_1)}$$

Table A–5 presents the power of the two-sided t test (in percent) for $\alpha = .01$ and $\alpha = .05$, for various degrees of freedom *df*. To illustrate the use of this table, let us return to the Westwood Company example where we tested:

$$H_0: \beta_1 = \beta_{10} = 0$$
$$H_a: \beta_1 \neq \beta_{10} = 0$$

Suppose we wish to know the power of the test when $\beta_1 = .25$. To ascertain this, we need to know σ^2, the variance of the error terms. Assume that $\sigma^2 = 10.0$ so that $\sigma^2(b_1)$ for our example would be:

$$\sigma^2(b_1) = \frac{\sigma^2}{\Sigma(X_i - \bar{X})^2} = \frac{10.0}{3{,}400} = .002941$$

or $\sigma(b_1) = .05423$. Then $\delta = |.25 - 0| \div .05423 = 4.6$. We enter the graph for $\alpha = .05$ (the level of significance used in the test) and approximate visually the curve for eight degrees of freedom. Reading the ordinate at $\delta = 4.6$, we obtain 97 percent approximately. Thus, if $\beta_1 = .25$, the probability would be about .97 that we would be led to conclude H_a ($\beta_1 \neq 0$). In other words, if $\beta_1 = .25$, we would be almost certain to conclude that there is a relation between man-hours and lot size.

The power of tests concerning β_0 can be obtained from Table A–5 in completely analogous fashion. For one-sided tests, Table A–5 should be entered so that one half the level of significance shown there is the level of significance of the one-sided test.

3.4 INTERVAL ESTIMATION OF $E(Y_h)$

In regression analysis, one of the major goals usually is to estimate the mean for one or more probability distributions of Y. Consider, for example, a study of the relation between level of piecework pay (X) and worker productivity (Y). The mean productivity at high and medium levels of piecework pay may be of particular interest for purposes of analyzing the benefits obtained from an increase in the pay. As another example, the Westwood Company may be interested in the mean response (mean number of man-hours) for lot sizes of $X = 40$ parts, $X = 55$ parts, and $X = 70$ parts for purposes of choosing appropriate lot sizes for production.

Let X_h denote the level of X for which we wish to estimate the mean response. X_h may be a value which occurred in the sample, or it may be some other value of the independent variable within the scope of the model. The mean response when $X = X_h$ is denoted by $E(Y_h)$. Formula (2.12) gives us the point estimator $\hat{Y}_h$ of $E(Y_h)$:

(3.27) $$\hat{Y}_h = b_0 + b_1 X_h$$

We consider now the sampling distribution of $\hat{Y}_h$.

Sampling distribution of $\hat{Y}_h$

The sampling distribution of $\hat{Y}_h$, like the earlier sampling distributions discussed, refers to the different values of $\hat{Y}_h$ which would be obtained if repeated samples were selected, each holding the levels of the independent variable X constant, and calculating $\hat{Y}_h$ for each sample.

(3.28) For model (3.1), the sampling distribution of $\hat{Y}_h$ is normal, with mean and variance:

(3.28a) $$E(\hat{Y}_h) = E(Y_h)$$

(3.28b) $$\sigma^2(\hat{Y}_h) = \sigma^2\left[\frac{1}{n} + \frac{(X_h - \bar{X})^2}{\Sigma(X_i - \bar{X})^2}\right]$$

Normality. The normality of the sampling distribution of $\hat{Y}_h$ follows directly from the fact that $\hat{Y}_h$ is a linear combination of the observations Y_i.

Mean. To prove that $\hat{Y}_h$ is an unbiased estimator of $E(Y_h)$, we proceed as follows:

$$E(\hat{Y}_h) = E(b_0 + b_1X_h) = E(b_0) + X_hE(b_1)$$
$$= \beta_0 + \beta_1X_h$$

by (3.3a) and (3.20a).

Variance. First, we show that b_1 and $\bar{Y}$ are uncorrelated and hence, for model (3.1), independent:

(3.29) $$\sigma(\bar{Y}, b_1) = 0$$

where $\sigma(\bar{Y}, b_1)$ denotes the covariance between $\bar{Y}$ and b_1. We begin with the definitions:

$$\bar{Y} = \Sigma\left(\frac{1}{n}\right)Y_i$$

$$b_1 = \Sigma k_iY_i$$

where k_i is as defined in (3.5a). We now use theorem (1.27), with $a_i = 1/n$ and $c_i = k_i$; remember that the Y_i are independent random variables:

$$\sigma(\bar{Y}, b_1) = \Sigma\left(\frac{1}{n}\right)k_i\sigma^2(Y_i) = \frac{\sigma^2}{n}\Sigma k_i$$

But we know from (3.6) that $\Sigma k_i = 0$. Hence, the covariance is 0.

Now we are ready to find the variance of $\hat{Y}_h$. We shall use the estimator in the alternative form (2.15):

$$\sigma^2(\hat{Y}_h) = \sigma^2(\bar{Y} + b_1[X_h - \bar{X}])$$

Since $\bar{Y}$ and b_1 are independent and X_h and $\bar{X}$ are constants, we obtain:

$$\sigma^2(\hat{Y}_h) = \sigma^2(\bar{Y}) + (X_h - \bar{X})^2\sigma^2(b_1)$$

Now $\sigma^2(b_1)$ is given in (3.3b) and:

$$\sigma^2(\bar{Y}) = \frac{\sigma^2(Y_i)}{n} = \frac{\sigma^2}{n}$$

Hence:

$$\sigma^2(\hat{Y}_h) = \frac{\sigma^2}{n} + (X_h - \bar{X})^2 \frac{\sigma^2}{\Sigma(X_i - \bar{X})^2}$$

which, upon a slight rearrangement of terms, yields (3.28b).

Note the effect of the term $(X_h - \bar{X})^2$ on $\sigma^2(\hat{Y}_h)$. The further X_h is from $\bar{X}$, the greater is the quantity $(X_h - \bar{X})^2$ and the larger is the variance of $\hat{Y}_h$. An intuitive explanation of this effect is found in Figure 3.2. Shown there are two sample regression lines, based on two samples for the same set of X values. The two regression lines are assumed to go through the same $(\bar{X}, \bar{Y})$ point to isolate the effect of interest, namely, the effect of variation in the estimated slope b_1 from sample to sample. Note that at X_1, near $\bar{X}$, the fitted values $\hat{Y}_1$ for the two sample regression lines are close to each other. At X_2, which is far from $\bar{X}$, the situation is different. Here, the fitted values $\hat{Y}_2$ differ substantially. Thus, variation in the slope b_1 from sample to sample has a much more pronounced effect on $\hat{Y}_h$ for X levels far from the mean $\bar{X}$ than for X levels near $\bar{X}$. Hence, the variation in the $\hat{Y}_h$ values from sample to sample will be greater when X_h is far from the mean than when X_h is near the mean.

FIGURE 3.2 Effect on $\hat{Y}_h$ of variation in b_1 from sample to sample in two samples with same means $\bar{Y}$ and $\bar{X}$

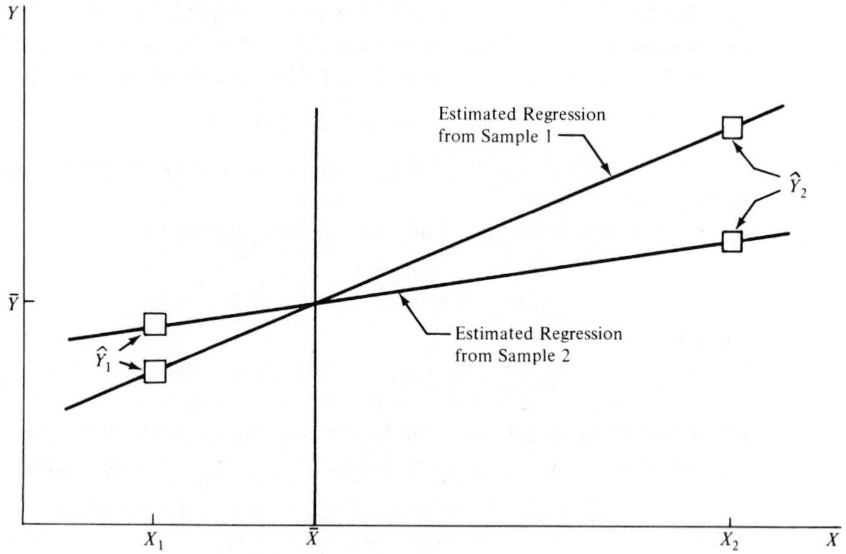

When MSE is substituted for σ^2 in (3.28b), we obtain $s^2(\hat{Y}_h)$, the estimated variance of $\hat{Y}_h$:

$$s^2(\hat{Y}_h) = MSE\left[\frac{1}{n} + \frac{(X_h - \bar{X})^2}{\Sigma(X_i - \bar{X})^2}\right] \tag{3.30}$$

The estimated standard deviation of $\hat{Y}_h$ then is $s(\hat{Y}_h)$, the square root of $s^2(\hat{Y}_h)$.

Sampling distribution of $[\hat{Y}_h - E(Y_h)]/s(\hat{Y}_h)$

Since we have encountered the t distribution in each type of inference for regression model (3.1) considered up to this point, it should not be surprising that:

$$\frac{\hat{Y}_h - E(Y_h)}{s(\hat{Y}_h)} \text{ is distributed as } t(n-2) \text{ for model (3.1)} \tag{3.31}$$

Hence, all inferences concerning $E(Y_h)$ are carried out in the usual fashion with the t distribution. We illustrate the construction of confidence intervals, since in practice these are more frequently used than tests.

Confidence interval for $E(Y_h)$

A confidence interval for $E(Y_h)$ is constructed in the standard fashion, making use of the t distribution as indicated by theorem (3.31). The $1 - \alpha$ confidence limits are:

$$\hat{Y}_h \pm t(1 - \alpha/2; n-2)s(\hat{Y}_h) \tag{3.32}$$

Example 1. Returning to the Westwood Company lot size example, let us find a 90 percent confidence interval for $E(Y_h)$ when the lot size is $X_h = 55$ parts. Using the earlier results in Table 3.1, we find the point estimate $\hat{Y}_h$:

$$\hat{Y}_{55} = 10.0 + 2.0(55) = 120$$

Next, we need to find the estimated standard deviation $s(\hat{Y}_h)$. We obtain, using (3.30):

$$s^2(\hat{Y}_{55}) = 7.5\left[\frac{1}{10} + \frac{(55-50)^2}{3{,}400}\right] = .80515$$

so that:

$$s(\hat{Y}_{55}) = .89730$$

For a 90 percent confidence coefficient, we require $t(.95; 8) = 1.860$. Hence, our confidence interval with confidence coefficient .90 is by (3.32):

$$120 - 1.860(.89730) \leq E(Y_{55}) \leq 120 + 1.860(.89730)$$
$$118.3 \leq E(Y_{55}) \leq 121.7$$

We conclude with confidence coefficient .90 that the mean number of man-hours required when lots of 55 parts are produced is somewhere between 118.3 and 121.7.

Example 2. Suppose the Westwood Company wishes to estimate $E(Y_h)$ when $X_h = 80$ parts with a 90 percent confidence interval. We require:

$$\hat{Y}_{80} = 10.0 + 2.0(80) = 170$$
$$s^2(\hat{Y}_{80}) = 7.5\left[\frac{1}{10} + \frac{(80-50)^2}{3{,}400}\right] = 2.73529$$
$$s(\hat{Y}_{80}) = 1.65387$$
$$t(.95; 8) = 1.860$$

Hence, the 90 percent confidence interval is:

$$170 - 1.860(1.65387) \leq E(Y_{80}) \leq 170 + 1.860(1.65387)$$
$$166.9 \leq E(Y_{80}) \leq 173.1$$

Note that this confidence interval is somewhat wider than that for example 1, since the X_h level here ($X_h = 80$) is substantially farther from the mean $\bar{X} = 50$ than the X_h level for example 1 ($X_h = 55$).

Comments

1. Since the X_i are known constants in model (3.1), the interpretation of confidence intervals and risks of errors in inferences on the mean response is in terms of taking repeated samples in which the X observations are at the same levels as in the sample actually taken. We noted this same point earlier in connection with inferences on β_0 and β_1.

2. We see from formula (3.28b) that for given sample results, the variance of $\hat{Y}_h$ is smallest when $X_h = \bar{X}$. Thus, in an experiment to estimate the mean response at a particular level X_h of the independent variable, the precision of the estimate will be greatest if (everything else remaining equal) the observations on X are spaced so that $\bar{X} = X_h$.

3. When the sample size is large, the t value in the confidence limits (3.32) may be replaced by the standard normal z value, since the t distribution approaches the standard normal distribution with increasing sample size.

4. The usual relationship between confidence intervals and tests applies in inferences concerning the mean response. Thus, the two-sided confidence limits (3.32) can be utilized for two-sided tests concerning the mean response at X_h. Alternatively, a regular decision rule can be set up.

5. Confidence limits (3.32) apply when a single mean response is to be estimated from the sample. We discuss in Chapter 5 how to proceed when a number of mean responses are to be estimated from the same sample.

3.5 PREDICTION OF NEW OBSERVATION

We consider now the prediction of a new observation Y corresponding to a given level X of the independent variable. In our Westwood Company illustration, for instance, the next lot to be produced consists of 55 parts and management wishes to predict the number of man-hours for this particular lot. As another example, an economist has estimated the regression relation between company sales and number of persons 16 or more years old, based on data for the past 10 years. Given a reliable demographic projection of the number of persons 16 or more years old for next year, the economist wishes to predict next year's company sales.

The new observation on Y is viewed as the result of a new trial, independent of the trials on which the regression analysis is based. We shall denote the level of X for the new trial as X_h and the new observation on Y as $Y_{h(\text{new})}$. Of course, we assume that the underlying regression model applicable for the basic sample data continues to be appropriate for the new observation.

The distinction between estimation of the mean response $E(Y_h)$, discussed in the preceding section, and prediction of a new response $Y_{h(\text{new})}$, discussed now, is basic. In the former case, we estimate the *mean* of the distribution of Y. In the present case, we predict an *individual outcome* drawn from the distribution of Y. Of course, the great majority of individual outcomes deviate from the mean response, and this must be allowed for in the procedure for predicting $Y_{h(\text{new})}$.

Prediction interval when parameters known

To illustrate the nature of a *prediction interval* for a new observation $Y_{h(\text{new})}$ in as simple a fashion as possible, we shall first assume that all regression parameters are known. Later we shall drop this assumption and make appropriate modifications.

Suppose that the Westwood Company plans to produce a lot of $X_h = 40$ parts in a few weeks and that the relevant parameters of the regression model are known to be:

$$\beta_0 = 9.5 \qquad \beta_1 = 2.1$$
$$E(Y) = 9.5 + 2.1X$$
$$\sigma^2 = 10.0$$

Thus, for $X_h = 40$ parts, we have:

$$E(Y_{40}) = 9.5 + 2.1(40) = 93.5$$

Figure 3.3 shows the probability distribution of Y for $X_h = 40$ parts. Its mean is $E(Y_{40}) = 93.5$, and its standard deviation is $\sigma = \sqrt{10.0} = 3.162$. Further, the distribution is normal in accord with model (3.1).

Suppose we were to predict that the number of man-hours for the next lot of $X_h = 40$ parts will be between:

$$E(Y_{40}) \pm 3\sigma$$
$$93.5 \pm 3(3.162)$$

so that the prediction interval would be:

$$84.0 \leq Y_{40(\text{new})} \leq 103.0$$

Since 99.7 percent of the area in a normal probability distribution falls within three standard deviations from the mean, the probability is .997 that this prediction interval will give a correct prediction for the next production run of 40 parts.

The basic idea of a prediction interval is thus to choose a range in the distribution of Y wherein most of the observations will fall, and to declare that the next

FIGURE 3.3 Prediction of $Y_{h(\text{new})}$ when parameters known

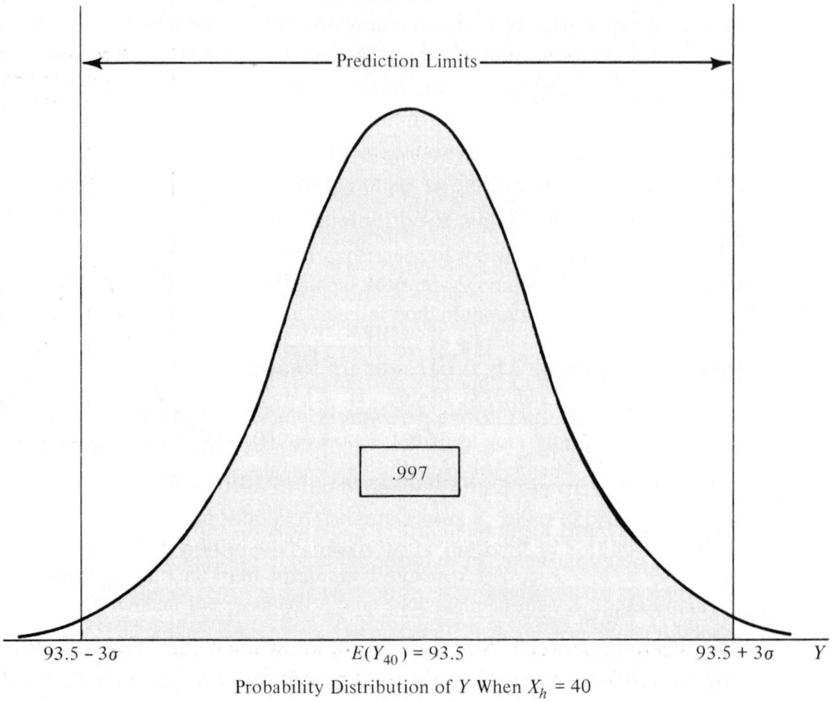

observation will fall in this range. The usefulness of the prediction interval depends, as always, on the width of the interval and the needs for precision by the user.

In general, when the regression parameters are known, the $1 - \alpha$ prediction limits for $Y_{h(\text{new})}$ are:

$$E(Y_h) \pm z(1 - \alpha/2)\sigma \tag{3.33}$$

In centering the limits around $E(Y_h)$, we obtain the narrowest interval consistent with the specified probability of a correct prediction.

Prediction interval for $Y_{h(\text{new})}$ when parameters unknown

When the regression parameters are unknown, they must be estimated. The mean of the distribution of Y is estimated by $\hat{Y}_h$, as usual, and the variance of the distribution of Y is estimated by MSE. We cannot, however, simply use the prediction limits (3.33) with the parameters replaced by the corresponding point estimators. The reason is illustrated intuitively in Figure 3.4. Shown there are two probability distributions of Y, corresponding to the upper and lower limits of a confidence interval for $E(Y_h)$. In other words, the distribution of Y could be located as far left as the one shown, as far right as the other one shown, or

FIGURE 3.4 Prediction of $Y_{h(\text{new})}$ when parameters unknown

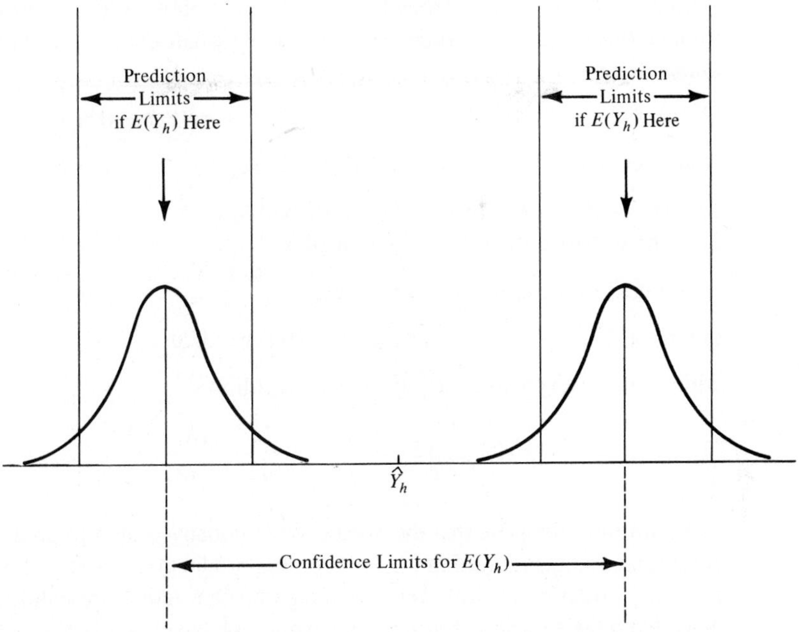

anywhere in between. Since we do not know the mean $E(Y_h)$ and only estimate it by a confidence interval, we cannot be certain of the location of the distribution of Y.

Figure 3.4 also shows the prediction limits for each of the two probability distributions of Y presented there. Since we cannot be certain of the location of the distribution of Y, prediction limits for $Y_{h(\text{new})}$ clearly must take account of two elements, as shown in Figure 3.4:

1. Variation in possible location of the distribution of Y.
2. Variation within the probability distribution of Y.

Prediction limits for a new observation Y at a given level X_h are obtained by means of the following theorem:

(3.34) $$\frac{\hat{Y}_h - Y}{s(Y_{h(\text{new})})} \text{ is distributed as } t(n-2) \text{ for model (3.1)}$$

Note that the standardized statistic (3.34) uses the point estimator $\hat{Y}_h$ in the numerator rather than the true mean $E(Y_h)$ because the true mean is unknown and cannot be used in making a prediction. The estimated standard deviation $s(Y_{h(\text{new})})$ in the denominator of the standardized statistic will be defined shortly.

From theorem (3.34), it follows in the usual fashion that the $1 - \alpha$ prediction limits for a new observation are [for instance, compare (3.34) to (3.10) and relate $\hat{Y}_h$ to b_1 and Y to β_1]:

(3.35) $$\hat{Y}_h \pm t(1 - \alpha/2; n-2)s(Y_{h(\text{new})})$$

The variance of the numerator of the standardized statistic (3.34) is readily obtained, utilizing the independence of the new observation Y and the original sample observations on which $\hat{Y}_h$ is based. We shall denote the variance of the numerator by $\sigma^2(Y_{h(\text{new})})$, and we obtain:

(3.36) $$\sigma^2(Y_{h(\text{new})}) = \sigma^2(\hat{Y}_h - Y) = \sigma^2(\hat{Y}_h) + \sigma^2$$

Note that $\sigma^2(Y_{h(\text{new})})$ has two components:

1. The variance of the sampling distribution of $\hat{Y}_h$.
2. The variance of the distribution of Y.

An unbiased estimator of $\sigma^2(Y_{h(\text{new})})$ is:

(3.37) $$s^2(Y_{h(\text{new})}) = s^2(\hat{Y}_h) + MSE$$

which can be expressed, using (3.30), as follows:

(3.37a) $$s^2(Y_{h(\text{new})}) = MSE\left[1 + \frac{1}{n} + \frac{(X_h - \bar{X})^2}{\Sigma(X_i - \bar{X})^2}\right]$$

Example. Suppose that the Westwood Company wishes to predict the number of man-hours required in the forthcoming production run of size 55 with a 90 percent prediction interval, and that the parameter values are unknown. We require $t(.95; 8) = 1.860$. From earlier work, we have:

$$\hat{Y}_{55} = 120 \qquad s^2(\hat{Y}_{55}) = .80515$$
$$MSE = 7.5$$

Using (3.37), we obtain:

$$s^2(Y_{55(\text{new})}) = .80515 + 7.5 = 8.30515$$

so that:

$$s(Y_{55(\text{new})}) = 2.88187$$

Hence, the 90 percent prediction interval for $Y_{55(\text{new})}$ is by (3.35):

$$120 - 1.860(2.88187) \leq Y_{55(\text{new})} \leq 120 + 1.860(2.88187)$$
$$114.6 \leq Y_{55(\text{new})} \leq 125.4$$

With confidence coefficient .90, we predict that the number of man-hours for the next production run of 55 parts will be somewhere between 114.6 and 125.4.

Comments

1. The 90 percent prediction interval for $Y_{55(\text{new})}$ just obtained is wider than the 90 percent confidence interval for $E(Y_{55})$ obtained on page 75. The reason is that when predicting a new observation, we encounter both the variability in $\hat{Y}_h$ from sample to sample as well as the variation within the probability distribution of Y.

2. Formula (3.37a) indicates that the prediction interval is wider the further X_h is from $\bar{X}$. The reason for this is that the estimate of the mean $\hat{Y}_h$, as noted earlier, is less precise as X_h is located further away from $\bar{X}$.

3. The confidence coefficient for the prediction limits (3.35) refers to the taking of repeated samples based on the same set of X values, and calculating prediction limits for $Y_{h(\text{new})}$ for each sample.

4. Prediction limits lend themselves to statistical control uses. In our example, suppose that the new production run of 55 parts, for which the prediction limits were 114.6 and 125.4 hours, actually required 135 hours. Management here would have an indication that a change in the production process may have occurred, and may wish to initiate a search for the assignable cause.

5. When the sample size is large, the last two terms inside the brackets in (3.37a) are small compared to 1, the first term in the brackets. Also, of course, the t distribution is then approximately normal. Hence, approximate $1 - \alpha$ prediction limits for $Y_{h(\text{new})}$ when n is large are:

(3.38) $$\hat{Y}_h \pm z(1 - \alpha/2)\sqrt{MSE}$$

6. Prediction limits (3.35) apply for a single prediction based on the sample data. Next, we discuss how to predict the mean of several new observations at a given X_h; and in Chapter 5, we take up how to make several predictions at different X_h values.

7. Prediction intervals resemble confidence intervals. However, they differ conceptually. A confidence interval represents an inference on a parameter, and is an interval which is intended to cover the value of the parameter. A prediction interval, on the other hand, is a statement about the value to be taken by a random variable.

Prediction of mean of m new observations for given X_h

Occasionally, one would like to predict the mean of m new observations on Y for a given level of the independent variable. Suppose the Westwood Company has been asked to bid on a contract that calls for $m = 3$ independent production runs of $X_h = 55$ parts during the next few months. Management would like to predict the mean man-hours per run for these three runs, and then convert this into a prediction of the total man-hours required to fill the contract.

We shall denote the mean value of Y to be predicted as $\bar{Y}_{h(\text{new})}$. It can be shown that the appropriate $1 - \alpha$ prediction limits are:

(3.39) $$\hat{Y}_h \pm t(1 - \alpha/2; n - 2)s(\bar{Y}_{h(\text{new})})$$

where:

(3.39a) $$s^2(\bar{Y}_{h(\text{new})}) = s^2(\hat{Y}_h) + \frac{MSE}{m}$$

or equivalently:

(3.39b) $$s^2(\bar{Y}_{h(\text{new})}) = MSE\left[\frac{1}{m} + \frac{1}{n} + \frac{(X_h - \bar{X})^2}{\Sigma(X_i - \bar{X})^2}\right]$$

Note from (3.39a) that the variance $s^2(\bar{Y}_{h(\text{new})})$ has two components:

1. The variance of the sampling distribution of $\hat{Y}_h$.
2. The variance of the mean of m observations from the probability distribution of Y.

Example. In the Westwood Company example, let us find the 90 percent prediction interval for the mean number of man-hours $\bar{Y}_{h(\text{new})}$ in three new production runs, each for $X_h = 55$ parts. From previous work, we have:

$$\hat{Y}_{55} = 120 \qquad s^2(\hat{Y}_{55}) = .80515$$
$$MSE = 7.5 \qquad t(.95; 8) = 1.860$$

Hence, we obtain:

$$s^2(\bar{Y}_{55(\text{new})}) = .80515 + \frac{7.5}{3} = 3.30515$$

or:

$$s(\bar{Y}_{55(\text{new})}) = 1.81801$$

The prediction interval for the mean man-hours per run then is:

$$120 - 1.860(1.81801) \leq \bar{Y}_{55(\text{new})} \leq 120 + 1.860(1.81801)$$
$$116.6 \leq \bar{Y}_{55(\text{new})} \leq 123.4$$

Note that these prediction limits are somewhat narrower than those for predicting the man-hours for a single lot of 55 parts because they involve a prediction of the mean man-hours for three lots.

We obtain the prediction interval for the total number of man-hours in the three production runs by multiplying the prediction limits for $\bar{Y}_{55(\text{new})}$ by three:

$$349.8 = 3(116.6) \leq \text{Total man-hours} \leq 3(123.4) = 370.2$$

Thus, it can be predicted with 90 percent confidence that between 350 and 370 man-hours will be needed to fill the contract for three lots of 55 parts each.

3.6 CONSIDERATIONS IN APPLYING REGRESSION ANALYSIS

We have now discussed the major uses of regression analysis—to make inferences about the regression parameters, to estimate the mean response for a given X, and to predict a new observation Y for a given X. It remains to make a few cautionary remarks about implementing applications of regression analysis.

1. Frequently, regression analysis is used to make inferences for the future. For instance, the Westwood Company may wish to estimate expected man-hours for given lot sizes for purposes of planning future production. In applications of this type, it is important to remember that the validity of the regression application depends upon whether basic causal conditions in the period ahead will be similar to those in existence during the period upon which the regression analysis is based. This caution applies whether mean responses are to be estimated, new observations predicted, or regression parameters estimated.

2. In predicting new observations on Y, the independent variable X itself often has to be predicted. For instance, we mentioned earlier the prediction of company sales for next year from a demographic projection of the number of

persons 16 years of age or older next year. A prediction of company sales under these circumstances is a conditional prediction, dependent upon the correctness of the population projection. It is easy to forget the conditional nature of this type of prediction.

3. Another caution deals with inferences pertaining to levels of the independent variable which fall outside the range of observations. Unfortunately, this situation frequently occurs in practice. A company which predicts its sales from a regression relation of company sales to disposable personal income will often find the level of disposable personal income of interest (e.g., for the year ahead) to fall beyond the range of past data. If the X level does not fall far beyond this range, one may have reasonable confidence in the application of the regression analysis. On the other hand, if the X level falls far beyond the range of past data, extreme caution should be exercised since one cannot be sure that the regression function which fits the past data is appropriate over the wider range of the independent variable.

4. A statistical test which leads to the conclusion that $\beta_1 \neq 0$ does not establish a cause-and-effect relation between the independent and dependent variables. For instance, with nonexperimental data, both the X and Y variables may be simultaneously influenced by other variables not included in the regression model. Thus, data on grade school children's vocabulary (X) and writing speed (Y) may show a clear linear association, but this could be largely the result of a child's age, amount of education, and similar factors that affect both X and Y. On the other hand, the existence of a regression relation in controlled experiments is often good evidence of a cause-and-effect relation.

5. Finally, we should note again that special problems arise when one wishes to estimate the mean response or predict a new observation for a number of different levels of the independent variable, as is frequently the case. The confidence coefficients for the limits (3.32) for estimating $E(Y)$ and for the prediction limits (3.35) for a new observation apply for a single level of X for a given sample. In Chapter 5, we discuss how to make multiple inferences from a given sample.

3.7 CASE WHEN X IS RANDOM

The normal error model (3.1), which has been used throughout this chapter and will continue to be used, assumes that the X values are known constants. As a consequence of this, the confidence coefficients and risks of errors refer to repeated sampling when the X values are kept the same from sample to sample.

Frequently, it may not be appropriate to consider the X values as known constants. For instance, consider regressing daily bathing suit sales by a department store on mean daily temperature. Surely, the department store cannot control daily temperatures, so it would not be meaningful to think of repeated sampling when the temperature levels are the same from sample to sample.

In this type of situation, it may be preferable to consider both Y and X as random variables. Does this mean then that all of our earlier results are not

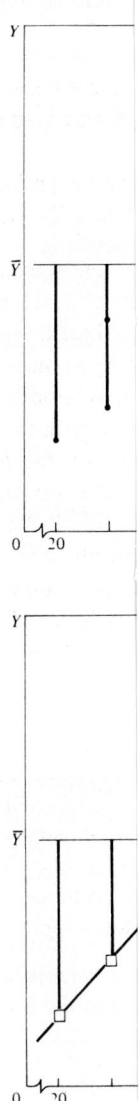

When we utilize the regression approach, the variation reflecting the uncertainty in the data is that of the Y observations around the regression line:

$$Y_i - \hat{Y}_i \tag{3.43}$$

These deviations are shown in Figure 3.5b. The measure of variation in the data with the regression model is the sum of the squared deviations (3.43), which is the familiar SSE of (2.21):

$$SSE = \Sigma(Y_i - \hat{Y}_i)^2 \tag{3.44}$$

Again, SSE denotes *error sum of squares*. If $SSE = 0$, all observations fall on the fitted regression line. The larger is SSE, the greater is the variation of the Y observations around the regression line.

For the Westwood Company example, we know from earlier work (Table 3.1) that:

$$SSTO = 13{,}660$$
$$SSE = 60$$

What accounts for the substantial difference between these two sums of squares? The difference, as we shall show shortly, is another sum of squares:

$$SSR = \Sigma(\hat{Y}_i - \bar{Y})^2 \tag{3.45}$$

where SSR stands for *regression sum of squares*. Note that SSR is a sum of squared deviations, the deviations being:

$$\hat{Y}_i - \bar{Y} \tag{3.46}$$

These deviations are shown in Figure 3.5c. Each deviation is simply the difference between the fitted value on the regression line and the mean of the fitted values $\bar{Y}$. [Recall from (2.18) that the mean of the fitted values $\hat{Y}_i$ is $\bar{Y}$.] If the regression line is horizontal so that $\hat{Y}_i - \bar{Y} \equiv 0$, $SSR = 0$. Otherwise, SSR is positive.

SSR may be considered a measure of the variability of the Y's associated with the regression line. The larger is SSR in relation to $SSTO$, the greater is the effect of the regression relation in accounting for the total variation in the Y observations.

For our lot size example, we have:

$$SSR = SSTO - SSE = 13{,}660 - 60 = 13{,}600$$

which indicates that most of the total variability in man-hours is accounted for by the relation between lot size and man-hours.

Formal development of partitioning. Consider the deviation $Y_i - \bar{Y}$, the basic quantity measuring the variation of the observations Y_i. We can decompose this deviation as follows:

(3.47)
$$\underbrace{Y_i - \bar{Y}}_{\substack{\text{Total}\\ \text{deviation}}} = \underbrace{\hat{Y}_i - \bar{Y}}_{\substack{\text{Deviation}\\ \text{of fitted}\\ \text{regression}\\ \text{value}\\ \text{around mean}}} + \underbrace{Y_i - \hat{Y}_i}_{\substack{\text{Deviation}\\ \text{around}\\ \text{regression}\\ \text{line}}}$$

Thus, the total deviation $Y_i - \bar{Y}$ can be viewed as the sum of two components:

1. The deviation of the fitted value $\hat{Y}_i$ around the mean $\bar{Y}$.
2. The deviation of Y_i around the regression line.

Figure 3.5d shows this decomposition for one of the observations.

It is a remarkable property that the sums of these squared deviations have the same relationship:

(3.48)
$$\Sigma(Y_i - \bar{Y})^2 = \Sigma(\hat{Y}_i - \bar{Y})^2 + \Sigma(Y_i - \hat{Y}_i)^2$$

or, using the notation in (3.42), (3.44), and (3.45):

(3.48a)
$$SSTO = SSR + SSE$$

To prove this basic result in the analysis of variance, we proceed as follows:

$$\begin{aligned}\Sigma(Y_i - \bar{Y})^2 &= \Sigma[(\hat{Y}_i - \bar{Y}) + (Y_i - \hat{Y}_i)]^2\\ &= \Sigma[(\hat{Y}_i - \bar{Y})^2 + (Y_i - \hat{Y}_i)^2 + 2(\hat{Y}_i - \bar{Y})(Y_i - \hat{Y}_i)]\\ &= \Sigma(\hat{Y}_i - \bar{Y})^2 + \Sigma(Y_i - \hat{Y}_i)^2 + 2\Sigma(\hat{Y}_i - \bar{Y})(Y_i - \hat{Y}_i)\end{aligned}$$

The last term on the right is zero, as can be seen by expanding it out:

$$2\Sigma(\hat{Y}_i - \bar{Y})(Y_i - \hat{Y}_i) = 2\Sigma\hat{Y}_i(Y_i - \hat{Y}_i) - 2\bar{Y}\Sigma(Y_i - \hat{Y}_i)$$

The first summation on the right is zero by (2.20), and the second is zero by (2.17). Hence, (3.48) follows.

Computational formulas. The definitional formulas for *SSTO, SSR,* and *SSE* presented above are often not convenient for hand computation. Useful computational formulas for *SSTO* and *SSR,* which are algebraically equivalent to the definitional formulas, are:

(3.49)
$$SSTO = \Sigma Y_i^2 - \frac{(\Sigma Y_i)^2}{n} = \Sigma Y_i^2 - n\bar{Y}^2$$

(3.50a)
$$\begin{aligned}SSR &= b_1\left[\Sigma X_i Y_i - \frac{\Sigma X_i \Sigma Y_i}{n}\right]\\ &= b_1[\Sigma(X_i - \bar{X})(Y_i - \bar{Y})]\end{aligned}$$

or:

(3.50b)
$$SSR = b_1^2\Sigma(X_i - \bar{X})^2$$

Computational formulas for *SSE* were given earlier in (2.24).

Using the results for the Westwood Company example summarized in Table 3.1, we obtain for *SSR* by (3.50a):

$$SSR = 2.0(6{,}800) = 13{,}600$$

This, of course, is the same result obtained previously by taking the difference $SSTO - SSE$, except sometimes for a slight difference due to rounding effects.

Breakdown of degrees of freedom

Corresponding to the partitioning of the total sum of squares *SSTO*, there is a partitioning of the associated degrees of freedom (abbreviated *df*). We have $n - 1$ degrees of freedom associated with *SSTO*. One degree of freedom is lost because the deviations $Y_i - \bar{Y}$ are not independent in that they must sum to zero. Equivalently, one degree of freedom is lost because the sample mean $\bar{Y}$ is used to estimate the population mean.

SSE, as noted earlier, has $n - 2$ degrees of freedom associated with it. Two degrees of freedom are lost because the two parameters β_0 and β_1 were estimated in obtaining the fitted values $\hat{Y}_i$.

SSR has one degree of freedom associated with it. There are two parameters in the regression equation, but the deviations $\hat{Y}_i - \bar{Y}$ are not independent because they must sum to zero; hence, one degree of freedom is lost from the possible degrees of freedom.

Note that the degrees of freedom are additive:

$$n - 1 = 1 + (n - 2)$$

For our Westwood Company example, these degrees of freedom are:

$$9 = 1 + 8$$

Mean squares

A sum of squares divided by its associated degrees of freedom is called a *mean square* (abbreviated *MS*). For instance, an ordinary sample variance is a mean square since a sum of squares, $\Sigma(Y_i - \bar{Y})^2$, is divided by its associated degrees of freedom, $n - 1$. We are interested here in the *regression mean square*, denoted by *MSR*:

(3.51) $$MSR = \frac{SSR}{1} = SSR$$

and in the *error mean square*, *MSE*, defined earlier in (2.22):

(3.52) $$MSE = \frac{SSE}{n - 2}$$

For our Westwood Company example, we have $SSR = 13{,}600$ and $SSE = 60$.

Hence:

$$MSR = \frac{13{,}600}{1} = 13{,}600$$

Also, we obtained earlier:

$$MSE = \frac{60}{8} = 7.5$$

Note

The two mean squares MSR and MSE do not add to $SSTO \div (n-1) = 13{,}660 \div 9 = 1{,}518$. Thus, mean squares are not additive.

Analysis of variance table

Basic table. The breakdowns of the total sum of squares and associated degrees of freedom are displayed in the form of an analysis of variance table (ANOVA table) in Table 3.2. Mean squares of interest also are shown. In addition, there is a column of expected mean squares which will be utilized below. The ANOVA table for our Westwood Company example is shown in Table 3.3.

TABLE 3.2 ANOVA table for simple regression

Source of Variation	*SS*	*df*	*MS*	*E(MS)*
Regression	$SSR = \sum(\hat{Y}_i - \bar{Y})^2$	1	$MSR = \frac{SSR}{1}$	$\sigma^2 + \beta_1^2 \sum(X_i - \bar{X})^2$
Error	$SSE = \sum(Y_i - \hat{Y}_i)^2$	$n-2$	$MSE = \frac{SSE}{n-2}$	σ^2
Total	$SSTO = \sum(Y_i - \bar{Y})^2$	$n-1$		

TABLE 3.3 ANOVA table for Westwood Company example

Source of Variation	*SS*	*df*	*MS*
Regression	13,600	1	13,600
Error	60	8	7.5
Total	13,660	9	

Modified table. Sometimes, an ANOVA table showing one additional element of decomposition is utilized. Recall that by (3.49):

$$SSTO = \Sigma(Y_i - \bar{Y})^2 = \Sigma Y_i^2 - n\bar{Y}^2$$

In the modified ANOVA table, the *total uncorrected sum of squares,* denoted by *SSTOU,* is defined as:

(3.53) $$SSTOU = \Sigma Y_i^2$$

and the *correction for the mean sum of squares,* denoted by *SS*(correction for mean), is defined as:

(3.54) $$SS(\text{correction for mean}) = n\bar{Y}^2$$

Table 3.4 shows this modified ANOVA table. The general format is presented in part (a) and the Westwood Company results in part (b). Both types of ANOVA tables are widely used. Ordinarily, we shall utilize the basic type of table.

TABLE 3.4 Modified ANOVA table for simple regression and results for Westwood Company example

(a) General

Source of Variation	*SS*	*df*	*MS*
Regression	$SSR = \sum(\hat{Y}_i - \bar{Y})^2$	1	$MSR = \frac{SSR}{1}$
Error	$SSE = \sum(Y_i - \hat{Y}_i)^2$	$n-2$	$MSE = \frac{SSE}{n-2}$
Total	$SSTO = \sum(Y_i - \bar{Y})^2$	$n-1$	
Correction for mean	*SS*(correction for mean) $= n\bar{Y}^2$	1	
Total, uncorrected	$SSTOU = \sum Y_i^2$	n	

(b) Westwood Company Example

Source of Variation	*SS*	*df*	*MS*
Regression	13,600	1	13,600
Error	60	8	7.5
Total	13,660	9	
Correction for mean	121,000	1	
Total, uncorrected	134,660	10	

Expected mean squares

We now find the expected value of each of the mean squares in the ANOVA table so that we can know what quantity each mean square estimates.

We stated earlier that MSE is an unbiased estimator of the error variance σ^2:

(3.55) $$E(MSE) = \sigma^2$$

This follows from theorem (3.11), which states that $SSE/\sigma^2 \sim \chi^2(n-2)$ for model (3.1). Hence, it follows from property (1.37) of the chi-square distribution that:

$$E\left[\frac{SSE}{\sigma^2}\right] = n - 2$$

or that:

$$E\left[\frac{SSE}{n-2}\right] = E(MSE) = \sigma^2$$

To find the expected value of MSR, we begin with (3.50b):

$$SSR = b_1^2 \Sigma(X_i - \bar{X})^2$$

Now by (1.14a), we have:

(3.56) $$\sigma^2(b_1) = E(b_1^2) - [E(b_1)]^2$$

We know from (3.3a) that $E(b_1) = \beta_1$, and from (3.3b) that:

$$\sigma^2(b_1) = \frac{\sigma^2}{\Sigma(X_i - \bar{X})^2}$$

Hence, substituting into (3.56), we obtain:

$$E(b_1^2) = \frac{\sigma^2}{\Sigma(X_i - \bar{X})^2} + \beta_1^2$$

It now follows that:

$$E(SSR) = E(b_1^2)\Sigma(X_i - \bar{X})^2 = \sigma^2 + \beta_1^2 \Sigma(X_i - \bar{X})^2$$

Finally, $E(MSR)$ is:

(3.57) $$E(MSR) = E\left(\frac{SSR}{1}\right) = \sigma^2 + \beta_1^2 \Sigma(X_i - \bar{X})^2$$

Table 3.2 contains the expected mean squares which we have just derived.

Comments

1. The expectation of MSE is σ^2, whether or not X and Y are linearly related, i.e., whether or not $\beta_1 = 0$.

2. The expectation of MSR is also σ^2 when $\beta_1 = 0$. On the other hand, when $\beta_1 \neq 0$, $E(MSR)$ is greater than σ^2 since the term $\beta_1^2 \Sigma(X_i - \bar{X})^2$ in (3.57) then must be positive. Thus, for testing whether or not $\beta_1 = 0$, a comparison of MSR and MSE suggests itself. If MSR and MSE are of the same order of magnitude, this would suggest that $\beta_1 = 0$. On the other hand, if MSR is substantially greater than MSE, this would suggest that $\beta_1 \neq 0$. This indeed is the basic idea underlying the analysis of variance test to be discussed next.

F test of $\beta_1 = 0$ versus $\beta_1 \neq 0$

The general analysis of variance approach provides us with a battery of highly useful tests for regression models (and other linear statistical models). For the simple regression case considered here, the analysis of variance provides us with a test for:

$$H_0: \beta_1 = 0 \qquad (3.58)$$
$$H_a: \beta_1 \neq 0$$

Test statistic. The test statistic for the analysis of variance approach is denoted by F^*. As just mentioned, it compares MSR and MSE in the following fashion:

$$F^* = \frac{MSR}{MSE} \qquad (3.59)$$

The earlier motivation, based on the expected mean squares in Table 3.2, suggests that large values of F^* support H_a and values of F^* near 1 support H_0. In other words, the appropriate test is an upper-tail one.

Distribution of F^*. In order to be able to construct a statistical decision rule and examine its properties, we need to know the sampling distribution of F^*. We begin by considering the sampling distribution of F^* when H_0 ($\beta_1 = 0$) holds. The famous *Cochran's theorem* will be most helpful in this connection. For our purposes, this theorem can be put as follows:

(3.60) If all n observations Y_i come from the same normal distribution with mean μ and variance σ^2, and $SSTO$ is decomposed into k sums of squares SS_r, each with degrees of freedom df_r, then the SS_r/σ^2 terms are independent χ^2 variables with df_r degrees of freedom if:

$$\sum_{r=1}^{k} df_r = n - 1$$

Note from Table 3.2 that we have decomposed $SSTO$ into the two sums of squares SSR and SSE, and that their degrees of freedom are additive. Hence:

If $\beta_1 = 0$ so that all Y_i have the same mean $\mu = \beta_0$ and the same variance σ^2, $\frac{SSE}{\sigma^2}$ and $\frac{SSR}{\sigma^2}$ are independent χ^2 variables.

Now consider the test statistic F^*, which we can write as follows:

$$F^* = \frac{\frac{SSR}{\sigma^2}}{1} \div \frac{\frac{SSE}{\sigma^2}}{n-2} = \frac{MSR}{MSE}$$

But by Cochran's theorem, we have when H_0 holds:

$$F^* \sim \frac{\chi^2(1)}{1} \div \frac{\chi^2(n-2)}{n-2}$$

where the χ^2 variables are independent. Thus, when H_0 holds, F^* is the ratio of two independent χ^2 variables, each divided by its degrees of freedom. But this is the definition of an F random variable in (1.42).

We have thus established that if H_0 holds, F^* follows the F distribution, specifically the $F(1, n-2)$ distribution.

When H_a holds, it can be shown that F^* follows the noncentral F distribution, a complex distribution that we need not consider further at this time.

Note

SSR and SSE are independent and $SSE/\sigma^2 \sim \chi^2$ even if $\beta_1 \neq 0$. But that both SSR/σ^2 and SSE/σ^2 are χ^2 random variables requires that $\beta_1 = 0$.

Construction of decision rule. Since the test is upper-tailed and F^* is distributed as $F(1, n-2)$ when H_0 holds, the decision rule is as follows when the risk of a Type I error is to be controlled at α:

(3.61)
$$\begin{aligned} &\text{If } F^* \leq F(1-\alpha; 1, n-2), \text{ conclude } H_0 \\ &\text{If } F^* > F(1-\alpha; 1, n-2), \text{ conclude } H_a \end{aligned}$$

where $F(1-\alpha; 1, n-2)$ is the $(1-\alpha)100$ percentile of the appropriate F distribution.

Example. Using our Westwood Company lot size example again, let us repeat the earlier test on β_1. This time we will use the F test. The alternative conclusions are:

$$\begin{aligned} &H_0\text{: } \beta_1 = 0 \\ &H_a\text{: } \beta_1 \neq 0 \end{aligned}$$

As before, let $\alpha = .05$. Since $n = 10$, we require $F(.95; 1, 8)$. We find from Table A–4 in the Appendix that $F(.95; 1, 8) = 5.32$. The decision rule is:

$$\begin{aligned} &\text{If } F^* \leq 5.32, \text{ conclude } H_0 \\ &\text{If } F^* > 5.32, \text{ conclude } H_a \end{aligned}$$

We have from Table 3.3 that $MSR = 13{,}600$ and $MSE = 7.5$. Hence, F^* is:

$$F^* = \frac{13{,}600}{7.5} = 1{,}813$$

Since $F^* = 1{,}813 > 5.32$, we conclude H_a, that $\beta_1 \neq 0$, or that there is a linear association between man-hours and lot size. This is the same result as when the t test was employed, as it must be according to our discussion below.

The P-value for the test statistic is the probability $P[F(1, 8) > F^* = 1{,}813]$. From Table A–4 we see that this P-value is less than .001 since $F(.999; 1, 8) = 25.4$.

Equivalence of F test and t test. For a given α level, the F test of $\beta_1 = 0$ versus $\beta_1 \neq 0$ is equivalent algebraically to the two-tailed t test. To see this, recall from (3.50b) that:

$$SSR = b_1^2 \Sigma(X_i - \bar{X})^2$$

Thus, we can write:

$$F^* = \frac{SSR \div 1}{SSE \div (n-2)} = \frac{b_1^2 \Sigma(X_i - \bar{X})^2}{MSE}$$

But since $s^2(b_1) = MSE/\Sigma(X_i - \bar{X})^2$, we obtain:

(3.62) $$F^* = \frac{b_1^2}{s^2(b_1)} = \left[\frac{b_1}{s(b_1)}\right]^2$$

Now, we know from earlier discussion that the t^* statistic for testing whether or not $\beta_1 = 0$ is by (3.17):

$$t^* = \frac{b_1}{s(b_1)}$$

In squaring, we obtain the expression for F^* in (3.62). Thus:

$$(t^*)^2 = \left[\frac{b_1}{s(b_1)}\right]^2 = F^*$$

In our illustrative problem, we just calculated that $F^* = 1{,}813$. From earlier work, we have: $b_1 = 2.0$, $s(b_1) = .04697$. Thus:

$$(t^*)^2 = \left[\frac{2.0}{.04697}\right]^2 = 1{,}813$$

Corresponding to the relation between t^* and F^*, we have the following relation between the required percentiles of the t and F distributions in the tests: $[t(1 - \alpha/2; n - 2)]^2 = F(1 - \alpha; 1, n - 2)$. In our tests on β_1, these percentiles were: $[t(.975; 8)]^2 = (2.306)^2 = 5.32 = F(.95; 1, 8)$. Remember that the t test is two-tailed while the F test is one-tailed.

Thus, at a given α level, we can use either the t test or the F test for testing $\beta_1 = 0$ versus $\beta_1 \neq 0$. Whenever one test leads to H_0, so will the other, and correspondingly for H_a. The t test, however, is more flexible since it can be used for one-sided alternatives involving $\beta_1 (\leq \geq) 0$ versus $\beta_1 (> <) 0$, while the F test cannot.

General linear test. The analysis of variance test of $\beta_1 = 0$ versus $\beta_1 \neq 0$ is an example of a general test of a linear statistical model. We shall briefly explain this general test approach in terms of our simple regression model. We do so at this time because of the generality of the approach and the wide use we shall make of it, and because of the simplicity of understanding the approach in terms of our present problem.

We begin with the model, which in this context is called the *full* or *unrestricted* model. For our simple regression case, the full model is:

(3.63) $$Y_i = \beta_0 + \beta_1 X_i + \varepsilon_i \qquad \text{Full model}$$

We fit this full model by the method of least squares and obtain the error sum of squares SSE. In this context, we shall call this sum of squares $SSE(F)$ to indicate that it measures the variation of the Y_i around the regression line for the full model.

Next, we consider H_0. In this instance, we have:

(3.64) $$H_0\colon \beta_1 = 0$$

The model when H_0 holds is called the *reduced* or *restricted* model. Here, it is:

(3.65) $$Y_i = \beta_0 + \varepsilon_i \qquad \text{Reduced model}$$

We fit this reduced model by the method of least squares and obtain the error sum of squares for this reduced model, denoted by $SSE(R)$. When we fit the particular reduced model (3.65), it can be shown that the least squares estimator of β_0 is $\bar{Y}$. Hence, $\hat{Y}_i \equiv \bar{Y}$ and the error sum of squares for this reduced model is:

(3.66) $$SSE(R) = \Sigma(Y_i - \bar{Y})^2 = SSTO$$

The logic now is to compare $SSE(F)$ and $SSE(R)$. It can be shown that $SSE(F)$ never is greater than $SSE(R)$:

(3.67) $$SSE(F) \leq SSE(R)$$

The reason is that the more parameters there are in the model, the better one can fit the data and the smaller are the deviations around the fitted regression line. If $SSE(F)$ is not much less than $SSE(R)$, using the full model does not account for much more of the variability of the Y_i than the reduced model, in which case the data suggest H_0 holds. To put this another way, if $SSE(F)$ is close to $SSE(R)$, the variation of the observations around the regression line for the full model is almost as great as the variation around the regression line for the reduced model, in which case the added parameters in the full model really do not help to reduce the variation in the Y_i. Thus, a small difference $SSE(R) - SSE(F)$ suggests that H_0 holds. On the other hand, a large difference suggests that H_a holds because the additional parameters in the model do help to reduce substantially the variation of the observations Y_i around the fitted regression line.

The actual test statistic used is a function of $SSE(R) - SSE(F)$, namely:

(3.68) $$F^* = \frac{SSE(R) - SSE(F)}{df_R - df_F} \div \frac{SSE(F)}{df_F}$$

which follows the F distribution if H_0 holds. The degrees of freedom df_R and df_F are those associated with the reduced and full model error sums of squares, respectively. Large values of F^* lead to H_a.

For our application, we have:

$$SSE(R) = SSTO \qquad SSE(F) = SSE$$
$$df_R = n - 1 \qquad df_F = n - 2$$

so that we obtain when substituting into (3.68):

$$F^* = \frac{SSTO - SSE}{(n-1) - (n-2)} \div \frac{SSE}{n-2} = \frac{SSR}{1} \div \frac{SSE}{n-2} = \frac{MSR}{MSE}$$

which is our old test statistic (3.59).

This general approach can be used for highly complex tests of linear statistical models, as well as for simpler tests. The basic steps again are:

1. Fit the full model and obtain the error sum of squares $SSE(F)$.
2. Fit the reduced model under H_0 and obtain the error sum of squares $SSE(R)$.
3. Use the test statistic (3.68).

3.9 DESCRIPTIVE MEASURES OF ASSOCIATION BETWEEN *X* AND *Y* IN REGRESSION MODEL

We have discussed the major uses of regression analysis—estimation of parameters and means and prediction of new observations—without mentioning the "degree of linear association" between X and Y, or similar terms. The reason is that the usefulness of estimates or predictions depends upon the width of the interval and the user's needs for precision, which vary from one application to another. Hence, no single descriptive measure of the "degree of linear association" can capture the essential information as to whether a given regression relation is useful in any particular application.

Nevertheless, there are times when the degree of linear association is of interest in its own right. We shall now briefly discuss two descriptive measures that are frequently used in practice to describe the degree of linear association between X and Y.

Coefficient of determination

We saw earlier that $SSTO$ measures the variation in the observations Y_i, or the uncertainty in predicting Y, when no account of the independent variable X is taken. Thus, $SSTO$ is a measure of the uncertainty in predicting Y when X is not considered. Similarly, SSE measures the variation in the Y_i when a regression model utilizing the independent variable X is employed. A natural measure of the effect of X in reducing the variation in Y, i.e., the uncertainty in predicting Y, is therefore:

(3.69) $$r^2 = \frac{SSTO - SSE}{SSTO} = \frac{SSR}{SSTO} = 1 - \frac{SSE}{SSTO}$$

The measure r^2 is called the *coefficient of determination*. Since $0 \leq SSE \leq SSTO$, it follows that:

$$0 \leq r^2 \leq 1 \tag{3.70}$$

We may interpret r^2 as the proportionate reduction of total variation associated with the use of the independent variable X. Thus, the larger is r^2, the more is the total variation of Y reduced by introducing the independent variable X. The limiting values of r^2 occur as follows:

1. If all observations fall on the fitted regression line, $SSE = 0$ and $r^2 = 1$. This case is shown in Figure 3.6a. Here, the independent variable X accounts for all variation in the observations Y_i.

FIGURE 3.6 Scatter plots when $r^2 = 0$ and $r^2 = 1$

(a) $r^2 = 1$ **(b)** $r^2 = 0$

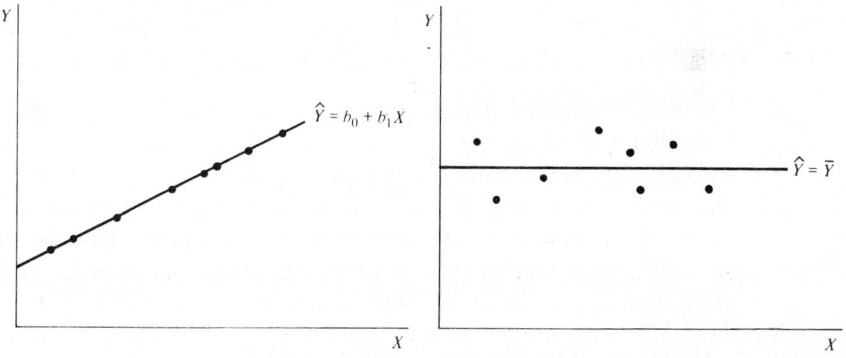

2. If the slope of the fitted regression line is $b_1 = 0$ so that $\hat{Y}_i \equiv \bar{Y}$, $SSE = SSTO$ and $r^2 = 0$. This case is shown in Figure 3.6b. Here, there is no linear association between X and Y in the sample data, and the independent variable X is of no help in reducing the variation in the observations Y_i with linear regression.

In practice, r^2 is not likely to be 0 or 1, but rather somewhere in between these limits. The closer it is to 1, the greater is said to be the degree of linear association between X and Y.

Coefficient of correlation

The square root of r^2:

$$r = \pm\sqrt{r^2} \tag{3.71}$$

is called the *coefficient of correlation*. A plus or minus sign is attached to this measure according to whether the slope of the fitted regression line is positive or negative. Thus, the range of r is:

$$-1 \leq r \leq 1 \tag{3.72}$$

Whereas r^2 indicates the proportional reduction in the variability of Y attained by the use of information about X, the square root, r, does not have such a clear-cut operational interpretation. Nevertheless, there is a tendency to use r instead of r^2 in much applied work.

It is worth noting that since for any r^2 other than 0 or 1, $r^2 < |r|$, r may give the impression of a "closer" relationship between X and Y than does the corresponding r^2. For instance, $r^2 = .10$ indicates that the total variation in Y is reduced by only 10 percent when X is introduced, yet $|r| = .32$ may give an impression of greater linear association between X and Y.

Example

For the Westwood Company example, we obtained $SSTO = 13{,}660$ and $SSE = 60$. Hence:

$$r^2 = \frac{13{,}660 - 60}{13{,}660} = .996$$

Thus, the variation in man-hours is reduced by 99.6 percent when lot size is considered.

The correlation coefficient in this example is:

$$r = +\sqrt{.996} = +.998$$

The plus sign is affixed since b_1 is positive.

Computational formula for *r*

A direct computational formula for r, which automatically furnishes the proper sign, is:

$$r = \frac{\Sigma(X_i - \bar{X})(Y_i - \bar{Y})}{[\Sigma(X_i - \bar{X})^2 \Sigma(Y_i - \bar{Y})^2]^{1/2}} \tag{3.73}$$

$$= \frac{\Sigma X_i Y_i - \dfrac{\Sigma X_i \Sigma Y_i}{n}}{\left[\left(\Sigma X_i^2 - \dfrac{(\Sigma X_i)^2}{n}\right)\left(\Sigma Y_i^2 - \dfrac{(\Sigma Y_i)^2}{n}\right)\right]^{1/2}}$$

Comments

1. The following relation between b_1 and r is worth noting:

$$b_1 = \left[\frac{\Sigma(Y_i - \overline{Y})^2}{\Sigma(X_i - \overline{X})^2}\right]^{1/2} r = \left(\frac{s_Y}{s_X}\right) r \tag{3.74}$$

where $s_Y = [\Sigma(Y_i - \overline{Y})^2/(n-1)]^{1/2}$ and $s_X = [\Sigma(X_i - \overline{X})^2/(n-1)]^{1/2}$ are the sample standard deviations for the Y and X observations, respectively. Note that $b_1 = 0$ when $r = 0$, and vice versa. Thus, $r = 0$ implies a horizontal fitted regression line, and vice versa.

2. The value taken by r^2 in a given sample tends to be affected by the spacing of the X observations. This is implied in (3.69). SSE is not affected systematically by the spacing of the X's since for model (3.1), $\sigma^2(Y_i) = \sigma^2$ at all X levels. However, the wider the spacing is of the X's in the sample when $b_1 \neq 0$, the greater will tend to be the spread of the observed Y's around $\overline{Y}$ and hence the greater will be $SSTO$. Consequently, the wider the X's are spaced, the higher will tend to be r^2.

3. The regression sum of squares SSR is often called the "explained variation" in Y. The residual sum of squares SSE is then called the "unexplained variation," and the total sum of squares the "total variation." The coefficient r^2 then is interpreted in terms of the proportion of the total variation in Y which has been "explained" by X. Unfortunately, this terminology frequently is taken literally, hence misunderstood. Remember that in a regression model there is no implication that Y necessarily depends on X in a causal or explanatory sense.

4. A value of r or r^2 relatively close to 1 sometimes is taken as an indication that sufficiently precise inferences on Y can be made from knowledge of X. As mentioned earlier, the usefulness of the regression relation depends upon the width of the confidence or prediction interval and the particular needs for precision, which vary from one application to another. Hence, no single measure is an adequate indicator of the usefulness of the regression relation.

5. Regression models do not contain any parameter to be estimated by r or r^2. These coefficients simply are descriptive measures of the degree of linear association between X and Y in the sample observations which may, or may not, be useful in any one instance. In a later chapter, we discuss correlation models which do contain a parameter for which r is an estimator.

3.10 COMPUTER OUTPUT

Figure 3.7 shows again the computer printout for the Westwood Company case presented in Figure 2.12. We referred to selected items in this printout in Chapter 2. Now, we are in a position to consider the printout as a whole.

The 10 observations on lot size and man-hours are printed at the top. This enables us to verify that the observations were entered into the computer accurately. We have annotated the output in Figure 3.7 in terms of the notation used in this book. The computer package output illustrated in Figure 3.7 does not provide $s(b_0)$, the estimated standard deviation of b_0. However, this estimate can be easily calculated from the data given in the computer output. Note in this connection that the denominator term $\Sigma(X_i - \overline{X})^2$ in (3.21) is equal to $(n-1)s_X^2$, and that s_X is given in the computer output.

FIGURE 3.7 Segment of computer output for regression run on Westwood Company data (SPSS, Ref. 3.1)

```
  X_i                                Y_i
                VARIABLES
1 SIZE        2 HOURS
  30.0000        73.0000
  20.0000        50.0000
  60.0000       128.0000
  80.0000       170.0000
  40.0000        87.0000
  50.0000       108.0000
  60.0000       135.0000
  30.0000        69.0000
  70.0000       148.0000
  60.0000       132.0000

DEPENDENT VARIABLE..     HOURS

VARIABLE(S) ENTERED ON STEP NUMBER  1..    SIZE

MULTIPLE R              0.99780 <- r
R SQUARE                0.99561 <- r^2

STANDARD ERROR          2.73861 <- √MSE
----------------- VARIABLES IN THE EQUATION -----------------
VARIABLE            B                  STD ERROR B            F
SIZE          2.000000 <- b_1    s(b_1) -> 0.04697     1813.333 <- F*
(CONSTANT)    10.00000 <- b_0

VARIABLE            MEAN          STANDARD DEV       CASES
SIZE      X̄ -> 50.0000       s_X -> 19.4365          10 \
HOURS     Ȳ -> 110.0000      s_Y -> 38.9587          10 <- n

ANALYSIS OF VARIANCE       DF      SUM OF SQUARES             MEAN SQUARE
REGRESSION                 1.   SSR -> 13600.00000     MSR -> 13600.00000
RESIDUAL <- Error          8.      SSE -> 60.00000        MSE -> 7.50000
```

Computer printouts for regression analysis programs differ substantially in format from one program to another. In addition, differences in the computed results may occur because different program packages do not control roundoff errors equally well. Before using a computer program the first time, it is a good idea to check it on a set of data for which the exact results are known.

PROBLEMS

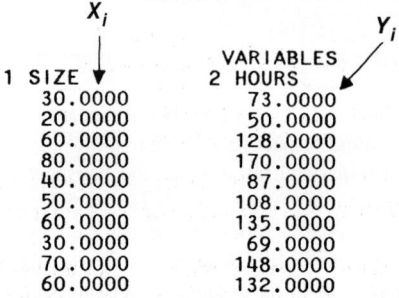

3.1. A student, working on a summer internship in the economic research office of a large corporation, studied the relation between sales of a product (Y, in million dollars) and population (X, in million persons) in the firm's 50 marketing districts. Regression model (3.1) was employed. The student first wished to test whether or not a linear association between Y and X existed. Using a time-sharing computer service available to the firm, the student accessed an interactive simple linear regression program and obtained the following information on the regression coefficients:

Parameter	*Estimated Value*	*95 Percent Confidence Limits*	
Intercept	7.43119	−1.18518	16.0476
Slope	.755048	.452886	1.05721

a. The student concluded from these results that there is a linear association between Y and X. Is the conclusion warranted? What is the implied level of significance?
b. Someone questioned the negative lower confidence limit for the intercept, pointing out that dollar sales cannot be negative even if the population in a district is zero. Discuss.

3.2. In a test of the alternatives H_0: $\beta_1 \leq 0$ versus H_a: $\beta_1 > 0$, an analyst concluded H_0. Does this conclusion imply that there is no linear association between X and Y? Explain.

3.3. A member of a student team playing an interactive marketing game received the following computer output when studying the relation between advertising expenditures (X) and sales (Y) for one of the team's products:

Estimated regression equation: $\hat{Y} = 350.7 - .18X$
Two-sided P-value for estimated slope: .91

The student stated: "The message I get here is that the more we spend on advertising this product, the fewer units we sell!" Comment.

3.4. Refer to **Grade point average** Problem 2.15. Some additional results are: $b_0 = -1.700$, $s(b_0) = .7267$, $b_1 = .8399$, $s(b_1) = .144$, $MSE = .1892$.

a. Obtain a 99 percent confidence interval for β_1. Interpret your confidence interval. Does it include zero? Why might the director of admissions be interested in whether the confidence interval includes zero?
b. Test, using the test statistic t^*, whether or not a linear association exists between student's entrance test score (X) and GPA at the end of the freshman year (Y). Use a level of significance of .01. State the alternatives, decision rule, and conclusion.
c. What is the P-value of your test in part (b)? How does it support the conclusion reached in part (b)?

3.5. Refer to **Calculator maintenance** Problem 2.16. Some additional results are: $b_0 = -2.3221$, $s(b_0) = 2.564$, $b_1 = 14.738$, $s(b_1) = .519$, $MSE = 20.086$.

a. Estimate the change in the mean service time when the number of machines serviced increases by one. Use a 90 percent confidence interval. Interpret your confidence interval.
b. Conduct a t test to determine whether or not there is a linear association between X and Y here; control the α risk at .10. State the alternatives, decision rule, and conclusion. What is the P-value of your test?
c. Are your results in parts (a) and (b) consistent? Explain.
d. The manufacturer has suggested that the mean required time should not increase by more than 14 minutes for each additional machine that is serviced on a service call. Conduct a test to decide whether this standard is being

satisfied by Tri-City. Control the risk of a Type I error at .05. State the alternatives, decision rule, and conclusion. What is the P-value of the test?

e. Does b_0 give any relevant information here about the "start-up" time on calls—i.e., about the time required before service work is begun on the machines at a customer location?

3.6. Refer to **Airfreight breakage** Problem 2.17.

a. Estimate β_1 with a 95 percent confidence interval. Interpret your interval estimate.

b. Conduct a t test to decide whether or not there is a linear association between number of times a carton is transferred (Y) and number of broken ampules (X). Use a level of significance of .05. State the alternatives, decision rule, and conclusion. What is the P-value of the test?

c. β_0 represents here the mean number of ampules broken when no transfers of the shipment are made—i.e., when $X = 0$. Obtain a 95 percent confidence interval for β_0 and interpret it.

d. A consultant has suggested, based on previous experience, that the mean number of broken ampules should not exceed 9 when no transfers are made. Conduct an appropriate test using $\alpha = .025$. State the alternatives, decision rule, and conclusion. What is the P-value of the test?

e. Obtain the power of your test in part (b) if actually $\beta_1 = 2.0$. Assume $\sigma(b_1) = .50$. Also obtain the power of your test in part (d) if actually $\beta_0 = 11$. Assume $\sigma(b_0) = .75$.

3.7. Refer to **Plastic hardness** Problem 2.18.

a. Estimate the change in the mean hardness when the elapsed time increases by one hour. Use a 99 percent confidence interval. Interpret your interval estimate.

b. The plastic manufacturer has stated that the mean hardness should increase by 2 Brinell units per hour. Conduct a two-sided test to decide whether this standard is being satisfied; use $\alpha = .01$. State the alternatives, decision rule, and conclusion. What is the P-value of the test?

c. Obtain the power of your test in part (b) if the standard actually is being exceeded by .5 Brinell units per hour. Assume $\sigma(b_1) = .16$.

3.8. Refer to Figure 3.7 for the Westwood Company example. A consultant has advised that an increase of one unit in lot size should require an increase of 1.8 in the expected number of man-hours for the given production item.

a. Conduct a test to decide whether or not the increase in the expected number of man-hours in the Westwood Company equals this standard. Use $\alpha = .05$. State the alternatives, decision rule, and conclusion.

b. Obtain the power of your test in part (a) if the consultant's standard actually is being exceeded by .1 hour. Assume $\sigma(b_1) = .05$.

c. Why is $F^* = 1813.333$, given in the printout, not relevant for the test in part (a)?

3.9. Refer to Figure 3.7. A student, noting that $s(b_1)$ is furnished in the printout, asks why $s(\hat{Y}_h)$ is not also given. Discuss.

3.10. For each of the following questions, explain whether a confidence interval for a mean response or a prediction interval for a new observation is appropriate.

a. What will be the humidity level in this greenhouse tomorrow when we set the temperature level at 31° C?

b. How much do families whose disposable income is \$23,500 spend, on the average, for meals away from home?

c. How many kilowatt-hours of electricity will be consumed next month by commercial and industrial users in the Twin Cities service area, given that the index of business activity for the area remains at its present level?

3.11. A person asks if there is a difference between the "mean response at $X = X_h$" and the "mean of m new observations at $X = X_h$." Reply.

3.12. Can $\sigma^2(Y_{h(\text{new})})$ in (3.36) be brought increasingly close to 0 as n becomes large? Is this also the case for $\sigma^2(\hat{Y}_h)$ in (3.28b)? What is the implication of this difference?

3.13. Refer to **Grade point average** Problems 2.15 and 3.4.

a. Obtain a 95 percent interval estimate of the mean freshman GPA for students whose entrance test score is 4.7. Interpret your confidence interval.

b. Mary Jones obtained a score of 4.7 on the entrance test. Predict her freshman GPA using a 95 percent prediction interval. Interpret your prediction interval.

c. Is the prediction interval in part (b) wider than the confidence interval in part (a)? Should it be?

3.14. Refer to **Calculator maintenance** Problems 2.16 and 3.5.

a. Obtain a 90 percent confidence interval for the mean service time on calls in which six machines are serviced. Interpret your confidence interval.

b. Obtain a 90 percent prediction interval for the service time on the next call in which six machines are serviced. Is your prediction interval wider than the corresponding confidence interval in part (a)? Should it be?

c. Suppose that management wishes to estimate the expected service time *per machine* on calls in which six machines are serviced. Obtain an appropriate confidence interval by converting the interval obtained in part (a). Interpret the converted confidence interval.

3.15. Refer to **Airfreight breakage** Problem 2.17.

a. Because of changes in airline routes, shipments may have to be transferred more frequently than in the past. Estimate the mean breakage for the following numbers of transfers: $X = 2, 4$. Use separate 99 percent confidence intervals. Interpret your results.

b. The next shipment will entail two transfers. Obtain a 99 percent prediction interval for the number of broken ampules for this shipment. Interpret your prediction interval.

c. In the next several days, three independent shipments will be made, each entailing two transfers. Obtain a 99 percent prediction interval for the mean number of ampules broken in the three shipments. Convert this interval into a 99 percent prediction interval for the total number of ampules broken in the three shipments.

3.16. Refer to **Plastic hardness** Problem 2.18.

a. Obtain a 98 percent confidence interval for the mean hardness of molded items with an elapsed time of 60 hours. Interpret your confidence interval.

b. Obtain a 98 percent prediction interval for the hardness of a newly molded test item with an elapsed time of 60 hours.

c. Obtain a 98 percent prediction interval for the mean hardness of 10 newly molded test items, each with an elapsed time of 60 hours.

d. Is the prediction interval in part (c) narrower than the one in part (b)? Should it be?

3.17. An analyst fitted regression model (3.1) and conducted an F test of $\beta_1 = 0$ versus $\beta_1 \neq 0$. The P-value of the test was .033, and the analyst concluded H_a: $\beta_1 \neq 0$. Was the α level used by the analyst greater than or smaller than .033? If the α level had been .01, what would have been the appropriate conclusion?

3.18. For conducting statistical tests concerning the parameter β_1, why is the t test more versatile than the F test?

3.19. When testing whether or not $\beta_1 = 0$, why is the F test a one-sided test even though H_a includes both $\beta_1 < 0$ and $\beta_1 > 0$? [*Hint:* Refer to (3.57).]

3.20. A student asks whether r^2 is a point estimator of any parameter in regression model (3.1). Respond.

3.21. A value of r^2 near 1 is sometimes interpreted to imply that the relation between Y and X is sufficiently close so that suitably precise predictions of Y can be made from knowledge of X. Is this implication a necessary consequence of the definition of r^2?

3.22. Using regression model (3.1) in an engineering safety experiment, a researcher found for the first 10 observations that r^2 was zero. Is it possible that for the complete set of 30 observations r^2 will not be zero? Could r^2 not be zero for the first 10 observations, yet equal zero for all 30 observations? Explain.

3.23. Refer to **Grade point average** Problems 2.15 and 3.4. Some additional calculational results are: $SSE = 3.406$, $SSR = 6.434$.

a. Set up the ANOVA table.
b. What is estimated by MSR in your ANOVA table? By MSE? Under what condition do MSR and MSE estimate the same quantity?
c. Conduct an F test of whether or not $\beta_1 = 0$. Control the α risk at .01. State the alternatives, decision rule, and conclusion.
d. What is the absolute magnitude of the reduction in the variation of Y when X is introduced into the regression model? What is the relative reduction? What is the name of the latter measure?
e. Obtain r and attach the appropriate sign.
f. Which measure, r^2 or r, has the more clear-cut operational interpretation? Explain.

3.24. Refer to **Calculator maintenance** Problems 2.16 and 3.5. Some additional calculational results are: $SSE = 321.396$, $SSR = 16{,}182.604$.

a. Set up the basic ANOVA table in the format of Table 3.2. Which elements of your table are additive? Also set up the ANOVA table in the format of Table 3.4a. How do the two tables differ?
b. Conduct an F test to determine whether or not there is a linear association between time spent and number of machines serviced; use $\alpha = .10$. State the alternatives, decision rule, and conclusion.
c. By how much, relatively, is the total variation in number of minutes spent on a call reduced when the number of machines serviced is introduced into the analysis? Is this a relatively small or large reduction? What is the name of this measure?
d. Calculate r and attach the appropriate sign.
e. Which measure, r or r^2, has the more clear-cut operational interpretation?

3.25. Refer to **Airfreight breakage** Problem 2.17.

a. Set up the ANOVA table. Which elements are additive?

b. Conduct an F test to decide whether or not there is a linear association between the number of times a carton is transferred and the number of broken ampules; control the α risk at .05. State the alternatives, decision rule, and conclusion.

c. Obtain the t^* statistic for the test in part (b) and demonstrate its equivalence to the F^* statistic obtained in part (b).

d. Calculate r^2 and r. What proportion of the variation in Y is accounted for by introducing X into the regression model?

3.26. Refer to **Plastic hardness** Problem 2.18.

a. Set up the ANOVA table.

b. Test by means of an F test whether or not there is a linear association between the hardness of the plastic and the elapsed time. Use $\alpha = .01$. State the alternatives, decision rule, and conclusion.

c. Plot the deviations $Y_i - \hat{Y}_i$ against X_i on a graph. Plot the deviations $\hat{Y}_i - \bar{Y}$ against X_i on another graph. From your two graphs, does SSE or SSR appear to be the larger component of $SSTO$?

d. Calculate r^2 and r.

3.27. Refer to **Muscle mass** Problem 2.23.

a. Conduct a test to decide whether or not there is a negative linear association between amount of muscle mass and age. Control the risk of Type I error at .05. State the alternatives, decision rule, and conclusion. What is the P-value of the test?

b. The two-sided P-value for b_0 is $0+$. Can it now be concluded that b_0 provides relevant information on the amount of muscle mass at birth for a female child?

c. Estimate with a 95 percent confidence interval the difference in expected muscle mass for women whose ages differ by one year. Why is it not necessary to know the specific ages to make this estimate?

3.28. Refer to **Muscle mass** Problem 2.23.

a. Obtain a 95 percent confidence interval for the mean muscle mass for women of age 60. Interpret your confidence interval.

b. Obtain a 95 percent prediction interval for the muscle mass of a woman whose age is 60. Is the prediction interval relatively precise?

3.29. Refer to **Muscle mass** Problem 2.23.

a. Plot the deviations $Y_i - \hat{Y}_i$ against X_i on one graph. Plot the deviations $\hat{Y}_i - \bar{Y}$ against X_i on another graph. From your graphs, does SSE or SSR appear to be the larger component of $SSTO$?

b. Set up the ANOVA table.

c. Test whether or not $\beta_1 = 0$ using an F test with $\alpha = .10$. State the alternatives, decision rule, and conclusion.

d. What proportion of the total variation in muscle mass remains "unexplained" when age is introduced into the analysis? Is this proportion relatively small or large?

e. Obtain r^2 and r.

3.30. Refer to **Robbery rate** Problem 2.24.

a. Test whether or not there is a linear association between robbery rate and population density using a t test with $\alpha = .01$. State the alternatives, decision rule, and conclusion. What is the P-value of the test?

b. Test whether or not $\beta_0 = 0$; control the risk of Type I error at .01. State the alternatives, decision rule, and conclusion. Why might there be interest in testing whether or not $\beta_0 = 0$?

c. Estimate β_1 with a 99 percent confidence interval. Interpret your interval estimate.

3.31. Refer to **Robbery rate** Problem 2.24.

a. Set up the ANOVA table.

b. Carry out the test in Problem 3.30a by means of the F test. Show the equivalence of the two test statistics and decision rules. Is the P-value for the F test the same as that for the t test?

c. By how much is the total variation in robbery rate reduced when population density is introduced into the analysis? Is this a relatively large or small reduction?

d. Obtain r.

3.32. Refer to **Robbery rate** Problems 2.24 and 3.30. Suppose that the test in Problem 3.30a is to be carried out by means of a general linear test.

a. State the full and reduced models.

b. Obtain: (1) $SSE(F)$, (2) $SSE(R)$, (3) df_F, (4) df_R, (5) test statistic F^* for the general linear test, (6) decision rule.

c. Are the test statistic F^* and the decision rule for the general linear test equivalent to those in Problem 3.30a?

3.33. In empirically developing a cost function from observed data on a complex chemical experiment, an analyst employed regression model (3.1). β_0 was interpreted here as the cost of setting up the experiment. The analyst hypothesized that this cost should be $7.5 thousand and wished to test the hypothesis by means of a general linear test.

a. Indicate the alternative conclusions for the test.

b. Specify the full and reduced models.

c. Without additional information, can you tell what the quantity $df_R - df_F$ in test statistic (3.68) will equal in the analyst's test? Explain.

3.34. Refer to **Grade point average** Problem 2.15.

a. Would it be more reasonable to consider the X_i as known constants or as random variables here? Explain.

b. If the X_i were considered to be random variables, would this have any effect on prediction intervals for new applicants? Explain.

3.35. Refer to **Calculator maintenance** Problems 2.16 and 3.5. How would the meaning of the confidence coefficient in Problem 3.5a change if the independent variable were considered a random variable and the conditions in (3.40) were applicable?

EXERCISES

3.36. Show that b_0 as defined in (3.19) is an unbiased estimator of β_0.

3.37. Derive the expression in (3.20b) for the variance of b_0, making use of theorem (3.29). Also explain how variance (3.20b) is a special case of variance (3.28b).

3.38. (Calculus needed.)

a. Obtain the likelihood function for the sample observations $Y_1, \ldots, Y_n$ given $X_1, \ldots, X_n$, if the conditions in (3.40) apply.

b. Obtain the maximum likelihood estimators of β_0, β_1, and σ^2. Are the estimators of β_0 and β_1 the same as those in (2.27) when the X_i are fixed?

3.39. Suppose that the normal error regression model (3.1) is applicable except that the error variance is not constant; rather the variance is larger, the larger is X. Does $\beta_1 = 0$ still imply that there is no linear association between X and Y? That there is no association between X and Y? Explain.

3.40. Derive the expression for SSR in (3.50b).

3.41. In a small-scale regression study, five observations on Y were obtained corresponding to $X = 1, 4, 10, 11$, and 14. Assume that $\sigma = .6$, $\beta_0 = 5$, and $\beta_1 = 3$.

a. What are the expected values of MSR and MSE here?

b. For purposes of determining whether or not a regression relation exists, would it have been better or worse to have made the five observations at $X = 6, 7, 8, 9$, and 10? Why? Would the same answer apply if the principal purpose were to estimate the mean response for $X = 8$? Discuss.

3.42. The simple linear regression model (3.1) is assumed to be applicable.

a. When testing H_0: $\beta_1 = 5$ versus H_a: $\beta_1 \neq 5$ by means of a general linear test, what is the reduced model? df_R?

b. When testing H_0: $\beta_0 = 2$, $\beta_1 = 5$ versus H_a: not both $\beta_0 = 2$ and $\beta_1 = 5$, what is the reduced model? df_R?

PROJECTS

3.43. Refer to the **SMSA** data set and Project 2.38. Using r^2 as the criterion, which independent variable accounts for the largest reduction in the variability in the number of active physicians?

3.44. Refer to the **SMSA** data set and Project 2.39. Obtain a separate interval estimate of β_1 for each region. Use a 90 percent confidence coefficient in each case. Do the regression lines for the different regions appear to have similar slopes?

3.45. Refer to the **SENIC** data set and Project 2.40. Using r^2 as the criterion, which independent variable accounts for the largest reduction in the variability of the average length of stay?

3.46. Refer to the **SENIC** data set and Project 2.41. Obtain a separate interval estimate of β_1 for each region. Use a 95 percent confidence coefficient in each case. Do the regression lines for the different regions appear to have similar slopes?

3.47. Five observations on Y are to be taken when $X = 4, 8, 12, 16$, and 20, respectively. The true regression function is $E(Y) = 20 + 4X$, and the ε_i are independent $N(0, 25)$.

a. Generate five normal random numbers, with mean 0 and variance 25. Consider these random numbers as the error terms for the five observations at $X = 4, 8, 12, 16$, and 20, and calculate Y_1, Y_2, Y_3, Y_4, and Y_5. Obtain the least squares estimates b_0 and b_1 when fitting a straight line to the five observations. Also calculate $\hat{Y}_h$ when $X_h = 10$.

b. Repeat part (a) 200 times, generating new random numbers each time.

c. Make a frequency distribution of the 200 estimates b_1. Calculate the mean and standard deviation of the 200 estimates b_1. Are the results consistent with theoretical expectations?

d. For each of the 200 replications, calculate a 95 percent confidence interval for $E(Y_h)$ when $X_h = 10$. What proportion of the 200 confidence intervals include $E(Y_h)$? Is this result consistent with theoretical expectations?

CITED REFERENCE

3.1 Nie, N. H.; C. H. Hull; J. G. Jenkins; K. Steinbrenner; and D. H. Bent. *SPSS: Statistical Package for the Social Sciences*. 2d ed. New York: McGraw-Hill, 1975.

4

Aptness of model and remedial measures

When a regression model, such as the simple linear regression model (3.1), is selected for an application, one can usually not be certain in advance that the model is appropriate for that application. Any one, or several, of the features of the model, such as linearity of the regression function or normality of the error terms, may not be appropriate for the particular data at hand. Hence, it is important to examine the aptness of the model for the data before further analysis based on that model is undertaken. In this chapter, we discuss some simple graphic methods for studying the aptness of a model, as well as some formal statistical tests for doing so. We conclude with a consideration of some techniques whereby the simple regression model (3.1) can be made appropriate when the data do not accord with the conditions of the model.

While the discussion in this chapter is in terms of the aptness of the simple regression model (3.1), the basic principles apply to all statistical models discussed in this book. In later chapters, additional material concerning the aptness of the model and remedial measures will be presented.

4.1 RESIDUALS

A residual e_i, as defined in (2.16), is the difference between the observed value and the fitted value:

$$e_i = Y_i - \hat{Y}_i \tag{4.1}$$

As such, it may be regarded as the observed error, in distinction to the unknown true error ε_i in the regression model:

$$\varepsilon_i = Y_i - E(Y_i) \tag{4.2}$$

For regression model (3.1), the ε_i are assumed to be independent normal random variables, with mean 0 and constant variance σ^2. If the model is appropriate for the data at hand, the observed residuals e_i should then reflect the properties assumed for the ε_i. This is the basic idea underlying *residual analysis*, a highly useful means of examining the aptness of a model.

Properties of residuals

The mean of the n residuals e_i is by (2.17):

$$\bar{e} = \frac{\Sigma e_i}{n} = 0 \tag{4.3}$$

where $\bar{e}$ denotes the mean of the residuals. Thus, since $\bar{e}$ is always 0, it provides no information as to whether the true errors ε_i have expected value $E(\varepsilon_i) = 0$.

The variance of the n residuals e_i is defined as follows for model (3.1):

$$\frac{\Sigma(e_i - \bar{e})^2}{n-2} = \frac{\Sigma e_i^2}{n-2} = \frac{SSE}{n-2} = MSE \tag{4.4}$$

If the model is appropriate, *MSE* is, as noted earlier, an unbiased estimator of the variance of the error terms σ^2.

Standardized residuals

For analytical convenience, standardized residuals are used at times in residual analysis. Since the standard deviation of the error terms ε_i is σ, which is estimated by $\sqrt{MSE}$, we shall define here the standardized residual as follows:

$$\frac{e_i - \bar{e}}{\sqrt{MSE}} = \frac{e_i}{\sqrt{MSE}} \tag{4.5}$$

We shall explain residual analysis mainly in terms of the residuals e_i, but occasionally will employ the standardized residuals.

Nonindependence of residuals

The residuals e_i are not independent random variables because they involve the fitted values $\hat{Y}_i$ which are based on the sample estimates b_0 and b_1. Thus, the residuals are associated with only $n - 2$ degrees of freedom. As a result, we know from (2.17) that the sum of the e_i must be 0 and from (2.19) that the products $X_i e_i$ must sum to 0. The same lack of independence holds for the standardized residuals.

When the sample size is large in comparison to the number of parameters in the regression model, the dependency effect among the e_i is relatively unimportant and can be ignored for most purposes.

Departures from model to be studied by residuals

We shall consider the use of residuals for examining six important types of departures from model (3.1), the simple linear regression model with normal errors:

1. The regression function is not linear.
2. The error terms do not have constant variance.
3. The error terms are not independent.
4. The model fits all but one or a few outlier observations.
5. The error terms are not normally distributed.
6. One or several important independent variables have been omitted from the model.

4.2 GRAPHIC ANALYSIS OF RESIDUALS

We take up now some informal ways in which graphs of residuals can be analyzed to provide information on whether any of the six types of departures from the simple linear regression model (3.1) just mentioned are present.

Nonlinearity of regression function

Whether or not a linear regression function is appropriate for the data being analyzed can often be studied from a scatter plot of the data, with the fitted regression function plotted on it. Figure 4.1a contains the data and the fitted regression line for a study of the relation between amount of transit information and bus ridership in eight comparable test cities, where X is the number of bus transit maps distributed free to residents of the city at the beginning of the test period and Y is the increase during the test period in average daily bus ridership during nonpeak hours. The original data and fitted values are given in Table 4.1. The graph suggests strongly that a linear regression function is not appropriate.

Figure 4.1b presents for this same example the residuals e, shown in Table 4.1, plotted against the independent variable X. The lack of fit of a linear regression function is also strongly suggested by the residual plot against X in Figure 4.1b, since the residuals depart from 0 in a systematic fashion. Note that they are negative for smaller X values, positive for medium size X values, and negative again for large X values.

In this case, both Figures 4.1a and 4.1b are effective means of examining the appropriateness of the linearity of the regression function. Figure 4.1b, the residual plot, in general has some advantages over Figure 4.1a, the scatter plot. First, the residual plot can easily be used for examining other facets of the aptness of the model. Second, there are occasions when the scaling of the scatter plot places

FIGURE 4.1 Scatter plot and residual plot for transit example illustrating nonlinear regression function

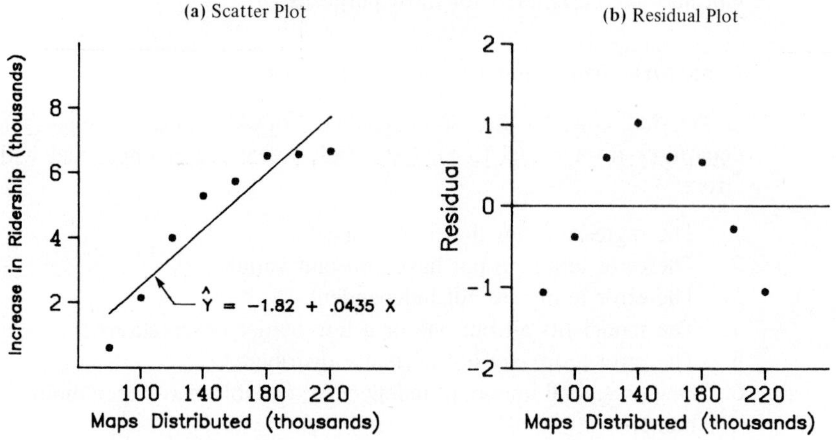

TABLE 4.1 Number of maps distributed and increase in ridership—transit example

City i	Increase in Ridership (thousands) Y_i	Maps Distributed (thousands) X_i	Fitted Value $\hat{Y}_i$	Residual $Y_i - \hat{Y}_i = e_i$
1	.60	80	1.66	−1.06
2	6.70	220	7.75	−1.05
3	5.30	140	4.27	+1.03
4	4.00	120	3.40	+.60
5	6.55	180	6.01	+.54
6	2.15	100	2.53	−.38
7	6.60	200	6.88	−.28
8	5.75	160	5.14	+.61

$$\hat{Y} = -1.82 + .0435X$$

the Y_i observations close to the fitted values $\hat{Y}_i$, for instance, when there is a steep slope. It then becomes more difficult to study the appropriateness of a linear regression function from the scatter plot. A residual plot, on the other hand, can clearly show any systematic pattern in the deviations around the regression line under these conditions.

Figure 4.2a shows a prototype situation of the residual plot against X if the linear model is appropriate. The residuals should tend to fall within a horizontal band centered around 0, displaying no systematic tendencies to be positive and negative.

Figure 4.2b shows a prototype situation of a departure from the linear regression model indicating the need for a curvilinear regression function. Here the

FIGURE 4.4 Residual plot with outlier

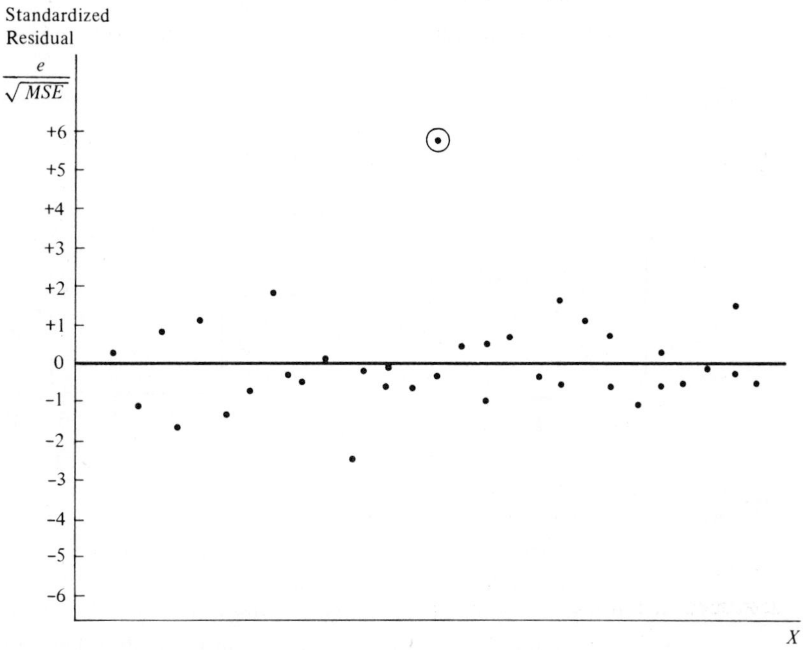

mistake or other extraneous cause. On the other hand, outliers may convey significant information, as when an outlier occurs because of an interaction with another independent variable omitted from the model. A safe rule frequently suggested is to discard an outlier only if there is direct evidence that it represents an error in recording, a miscalculation, a malfunctioning of equipment, or a similar type of circumstance.

Note

When a linear regression model is fitted to a data set with a small number of observations and an outlier is present, the fitted regression may be so distorted by the outlier that the residual plot suggests a lack of fit of the linear regression model in addition to flagging the outlier. Figure 4.5 illustrates this situation. The scatter plot in Figure 4.5a presents a case where all observations except the outlier fall around a straight-line statistical relationship. When a linear regression function is fitted to these data, the outlier causes such a shift in the fitted regression line as to lead to a systematic pattern of deviations from the fitted line for the other observations, as evidenced by the residual plot in Figure 4.5b.

Nonindependence of error terms

Whenever data are obtained in a time sequence, it is a good idea to plot the residuals against time, even though time has not been explicitly incorporated as a variable into the model. The purpose is to see if there is any correlation between the error terms over time. In an experiment to study the relation between the

FIGURE 4.5 Distorting effect on residuals caused by an outlier when remaining data follow linear regression

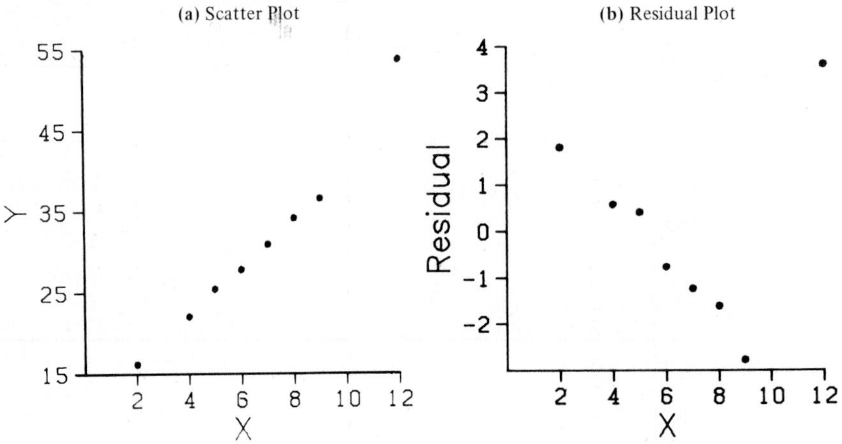

diameter of a weld (X) and the shear strength of the weld (Y), the residual plot against X as shown in Figure 4.6a appears to indicate no departures from the simple regression model, either with respect to linearity or constancy of error variance. When the residuals are plotted in the time order in which the welds were made in Figure 4.6b, however, an evident correlation between the error terms stands out. Negative residuals are associated mainly with the early trials, and positive residuals with the later trials. Apparently, some effect connected with time was present, such as learning by the welder or a gradual change in the welding equipment, so that the shear strength tended to be greater in the later welds on account of this effect.

A prototype of a time-related effect is shown in Figure 4.2d, which portrays a linear time-related effect. It is sometimes useful to view the problem of non-independence of the error terms as one in which an important variable (in this case, time) has been omitted from the model. We shall discuss this type of problem shortly.

When the error terms are independent, we would expect the residuals to fluctuate in a more or less random pattern around the base line 0, such as the scattering shown in Figure 4.7. Lack of randomness can take the form of too much alternation of points around the zero line, or too little alternation. In practice, there is little concern with the former case except in situations where the error term is subject to periodic or offsetting effects in observations at successive levels of X. Too little alternation, in contrast, frequently is a matter of concern, as in the welding example in Figure 4.6b.

Note

When the residuals are plotted against X, as in Figure 4.1b, the scatter may not appear to be random. For this plot, however, the basic problem is probably not lack of independence of the error terms but rather a poorly fitting regression function. This, indeed, is the situation portrayed in Figure 4.1a.

FIGURE 4.6 Residual plots for welding example illustrating nonindependence of error terms

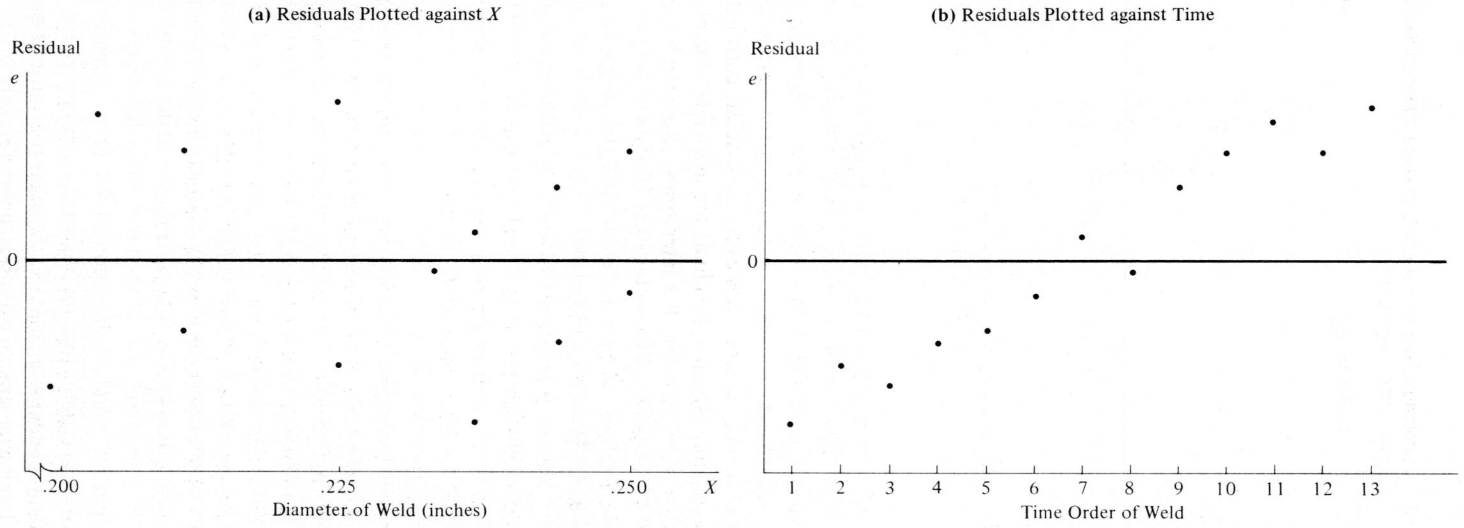

FIGURE 4.7 Residual plot against time suggesting independence of error terms

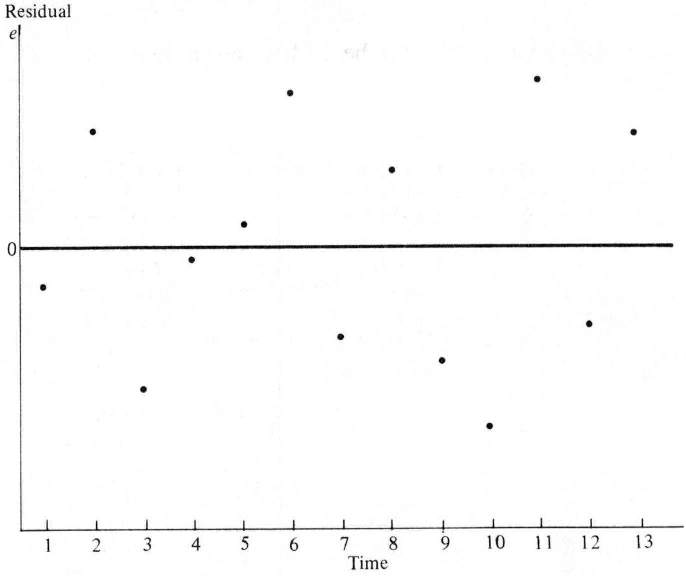

Nonnormality of error terms

As we noted earlier, small departures from normality do not create any serious problems. Major departures, on the other hand, should be of concern. The normality of the error terms can be studied informally by examining the residuals in a variety of graphic ways. One can construct a histogram of the residuals and see if gross departures from normality are shown by it. Another possibility is to determine whether, say, about 68 percent of the standardized residuals $e_i/\sqrt{MSE}$ fall between -1 and $+1$, or about 90 percent between -1.64 and $+1.64$. (If the sample size is small, the corresponding t values would be used.)

Still another possibility is to prepare a normal probability plot of the residuals. Here the residuals are plotted against their expected values when the distribution is normal. A plot which is nearly linear suggests agreement with normality whereas a plot which departs substantially from linearity suggests that the error distribution is not normal.

Table 4.2, column 1, contains the residuals, in ascending order, for a regression study of per capita library usage (Y) and size of city (X) in 10 cities. To find the expected values of the ordered residuals under normality, we utilize the facts that (1) the expected value of the error terms for regression model (3.1) is zero and (2) that the standard deviation of the error terms is estimated by $\sqrt{MSE}$. Statistical theory states that for a normal random variable with mean 0 and estimated standard deviation $\sqrt{MSE}$, the expected value of the ith smallest observation in a random sample of n is given approximately by the following expression:

(4.6) $$\sqrt{MSE}\left[z\left(\frac{i - .375}{n + .25}\right)\right]$$

where $z(A)$ as usual denotes the $(A)100$ percentile of the standard normal distribution.

TABLE 4.2 Residuals and expected values under normality for library usage example

Ascending Order i	(1) *Ordered Residual* e_i	(2) *Expected Value under Normality*
1	−1.33	−1.08
2	−.52	−.70
3	−.27	−.46
4	−.19	−.26
5	−.09	−.08
6	.09	.08
7	.18	.26
8	.44	.46
9	.66	.70
10	1.03	1.08

Squaring the residuals in Table 4.2, summing, and dividing by $n - 2$ we obtain $MSE = .4859$, so $\sqrt{MSE} = .697$. For the smallest residual, we have $i = 1$. Hence, $(i - .375)/(n + .25) = (1 - .375)/(10 + .25) = .061$, and the expected value of the smallest residual under normality is:

$$.697[z(.061)] = .697(-1.55) = -1.08$$

Similarly, the expected value of the second smallest residual under normality is obtained by finding, for $i = 2$, $(i - .375)/(n + .25) = (2 - .375)/(10 + .25) = .159$ so that:

$$.697[z(.159)] = .697(-1.00) = -.70$$

Because of the symmetry of a normal probability distribution, the expected values of the largest and second largest residuals are 1.08 and .70, respectively.

Table 4.2, column 2, contains all 10 expected values under the assumption of normality. Figure 4.8 presents a plot of the residuals against their expected values under normality. This plot is called a *normal probability plot*. Note that the points in Figure 4.8 fall reasonably close to a straight line, suggesting that the error terms are approximately normally distributed.

Many computer packages will prepare normal probability plots at the option of the user. Some of these plots utilize standardized residuals, but this does not affect the basic nature of the plot.

FIGURE 4.8 Normal probability plot of residuals—library usage example

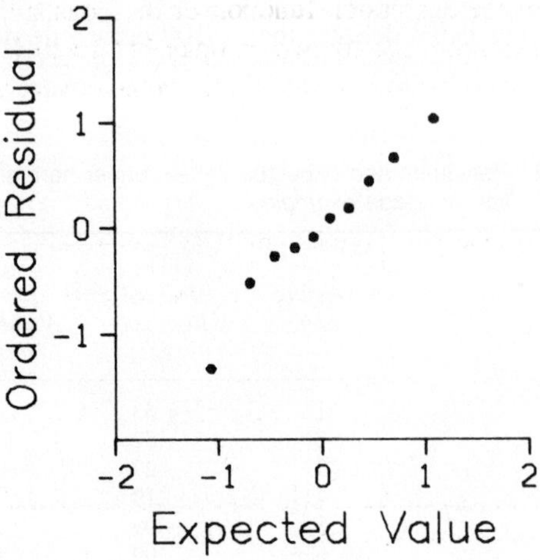

One method of assessing the linearity of the normal probability plot is to calculate the coefficient of correlation (3.73) relating the residuals e_i to their expected values under normality. A high value of the correlation coefficient, say, .90 or more, is indicative of normality. In our library usage example in Table 4.2, the coefficient of correlation is .981, supporting the conclusion of approximate normality for the error terms.

The analysis for model departures with respect to normality is, in many respects, more difficult than that for other types of departures. In the first place, random variation can be particularly mischievous when one studies the nature of a probability distribution unless the sample size is quite large. Even worse, other types of departures can and do affect the distribution of the residuals. For instance, residuals may appear to be not normally distributed because an inappropriate regression function is used or because the error variance is not constant. Hence, it is usually a good strategy to investigate these other types of departures first, before concerning oneself with the normality of the error terms.

Omission of important independent variables

Residuals should be plotted against variables omitted from the model that might have important effects on the response, data being available. The time variable cited earlier in the welding application is an example. The purpose of this additional analysis is to determine whether there are any other key independent variables that could provide important additional descriptive and predictive power to the model.

As another example, in a study to predict output by piece rate workers in an assembling operation, the relation between output (Y) and age (X) of worker was studied for a sample of employees. The plot of the residuals against X is shown in Figure 4.9a, and indicates no ground for suspecting the appropriateness of the linearity of the regression function or the constancy of the error variance.

Machines produced by two companies (A and B) are used in the assembling operation. Residual plots by type of machine were undertaken and are shown in

FIGURE 4.9 Residual plots for productivity example

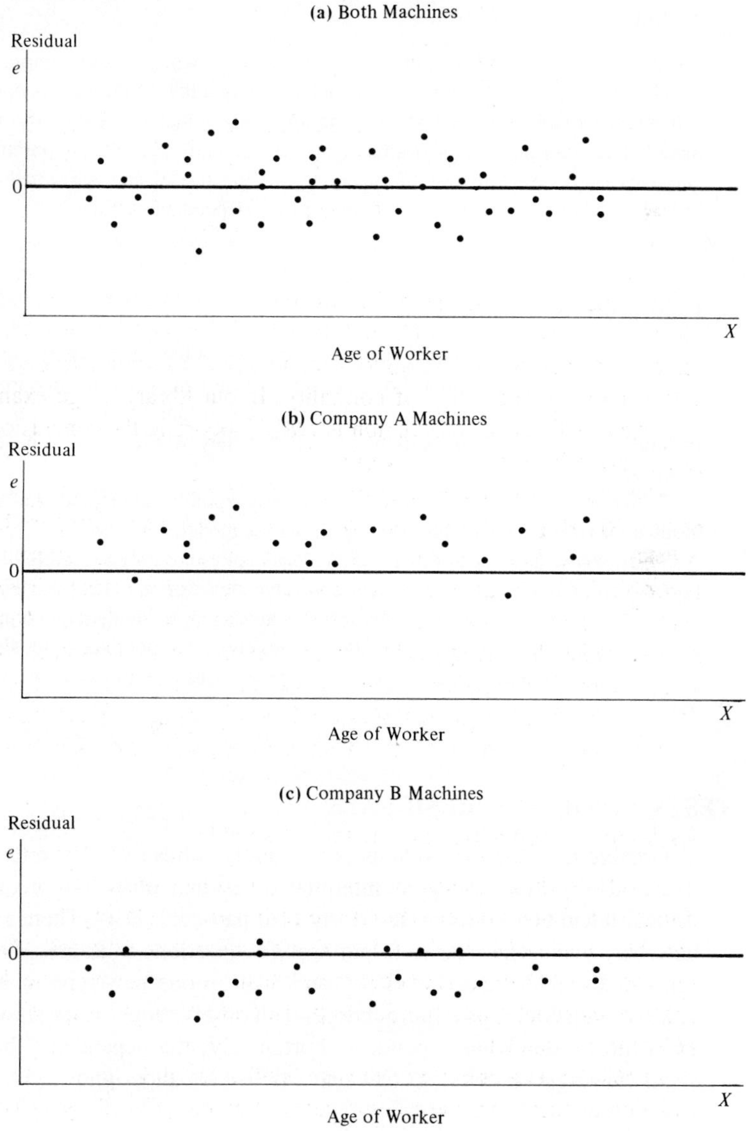

Figures 4.9b and c. Note that the residuals for machines made by Company A tend to be positive, while those for machines made by Company B tend to be negative. Thus, type of machine appears to have a definite effect on productivity, and output predictions may turn out to be far superior when this independent variable is added to the model. While this example dealt with a classification variable (type of machine), the residual analysis for an additional quantitative variable is completely analogous. One would simply plot the residuals against the additional independent variable and see whether or not the residuals tend to vary systematically with the level of the additional independent variable.

Note

We do not say that the original model is "wrong" when it can be improved materially by adding one or more independent variables. Only a few of the factors operating on any dependent variable Y in real-world situations can be included explicitly in a regression model. The chief purpose of residual analysis in identifying other important independent variables is therefore to test the adequacy of the model and see whether it could be improved materially by adding one or a few independent variables.

Comments

1. We discussed the model departures one at a time. In actuality, several types of departures may occur together. For instance, a linear regression function may be a poor fit and the variance of the error terms may not be constant. In these cases, the prototype patterns of Figure 4.2 can still be useful, but they would need to be combined into composite patterns.
2. While graphic analysis of residuals is only an informal method of analysis, in many cases it suffices for examining the aptness of a model.
3. The basic approach to residual analysis applies not only to simple linear regression but also to more complex regression and other types of statistical models.
4. Most of the routine work in residual analysis can be handled on computers. Almost all regression programs supply the fitted values and corresponding residuals, and routines are generally available whereby the various types of residual plots can be obtained optionally on printout.

4.3 TESTS INVOLVING RESIDUALS

Graphic analysis of residuals is inherently subjective. Nevertheless, subjective analysis of a variety of interrelated residual plots will frequently reveal difficulties in the model more clearly than particular tests. There are occasions, however, when one wishes to put specific questions to a test. We now review some of the relevant tests briefly, and take up one new type of test.

Most statistical tests require independent observations. As we have seen, however, the residuals are dependent. Fortunately, the dependency becomes quite small for large samples, so that one can usually then ignore it.

Tests for randomness

The runs test is frequently used to test for lack of randomness in the residuals arranged in time order. Another test, specifically designed for lack of randomness in least squares residuals, is the Durbin-Watson test. This test will be discussed in Chapter 13.

Tests for constancy of variance

When a residual plot gives the impression that the variance may be increasing or decreasing in a systematic manner related to X or $E(Y)$, a simple test is to fit separate regression functions to each half of the observations arranged by level of X, calculate error mean squares for each, and test for equality of the error variances by an F test. Another simple test is by means of rank correlation between the absolute value of the residual and the value of the independent variable.

Tests for outliers

A simple test for an outlier observation involves fitting a new regression line to the other $n - 1$ observations. The suspect observation, which was not used in fitting the new line, can now be regarded as a new observation. One can calculate the probability that in n observations, a deviation from the fitted line as great as that of the outlier will be obtained by chance. If this probability is sufficiently small, the outlier can be rejected as not having come from the same population as the other $n - 1$ observations. Otherwise, the outlier is retained.

Many other tests to aid in evaluating outliers have been developed. These are discussed in specialized references such as Reference 4.2 and in statistical journals.

Tests for normality

Goodness of fit tests can be used for examining the normality of the error terms. For instance, the chi-square test or the Kolmogorov-Smirnov test can be employed for testing the normality of the error terms by analyzing the residuals.

Note

The runs test, rank correlation, and goodness of fit tests, mentioned above, are commonly used statistical procedures which are discussed in many basic statistics texts.

4.4 *F* TEST FOR LACK OF FIT

We now take up a formal test for determining whether or not a specified regression function adequately fits the data. This lack of fit test assumes that the observations Y for given X are (1) independent and (2) normally distributed, and (3) the distributions of Y have the same variance σ^2. We illustrate this test for ascertaining whether or not a linear regression function is a good fit for the data.

Replications

The lack of fit test requires repeat observations at one or more X levels. In nonexperimental data, these may occur fortuitously, as when in a productivity study relating workers' output and age, several workers of the same age happen to be included in the study. In an experiment, one can assure by design that there are repeat observations. For instance, in an experiment on the effect of size of salesperson bonus on sales, three salespersons can be offered a particular size of bonus, for each of six bonus sizes, and their sales then observed.

Repeated trials for the same level of the independent variable, of the type described, are called *replications*. The resulting observations are called *replicates*.

Example

In an experiment involving 12 similar but scattered suburban branch offices of a commercial bank, holders of checking accounts at the offices were offered gifts for setting up savings accounts at these same offices. The initial deposit in the new savings account had to be for a specified minimum amount to qualify for the gift. The value of the gift was directly proportional to the specified minimum deposit. Various levels of minimum deposit and related gift values were used in the experiment in order to ascertain the relation between the specified minimum deposit and gift value on the one hand and number of accounts opened at the office on the other. Altogether, six levels of minimum deposit and proportional gift value were used, with two of the branch offices assigned at random to each level. One branch office had a fire during the period and was dropped from the study. Table 4.3 contains the results, where X is the amount of minimum deposit and Y is the number of new savings accounts that were opened and qualified for the gift during the test period.

TABLE 4.3 Data for bank example

Observation i	*Size of Minimum Deposit (dollars)* X_i	*Number of New Accounts* Y_i	*Observation* i	*Size of Minimum Deposit (dollars)* X_i	*Number of New Accounts* Y_i
1	125	160	7	75	42
2	100	112	8	175	124
3	200	124	9	125	150
4	75	28	10	200	104
5	150	152	11	100	136
6	175	156			

A linear regression function was fitted in the usual fashion; it is (calculations not shown):

$$\hat{Y} = 50.72251 + .48670X$$

The analysis of variance table also was obtained and is shown in Table 4.4. A scatter plot, together with the fitted regression line, is shown in Figure 4.10. The indications are strong that a linear regression function is inappropriate. To test this formally, we need to perform a decomposition of the error sum of squares *SSE* in Table 4.4 into two components called the pure error and lack of fit components.

TABLE 4.4 ANOVA table for bank example

Source of Variation	*SS*	*df*	*MS*
Regression	$SSR = 5{,}141.3$	1	$MSR = 5{,}141.3$
Error	$SSE = 14{,}741.6$	9	$MSE = 1{,}638.0$
Total	$SSTO = 19{,}882.9$	10	

FIGURE 4.10 Scatter plot and fitted regression line—bank example

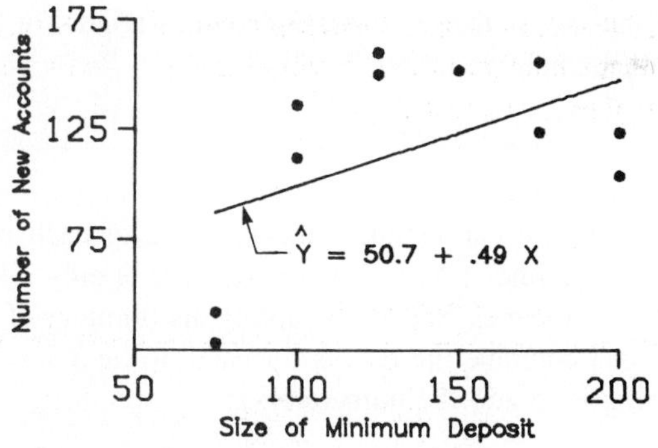

Decomposition of *SSE*

Pure error component. The basic idea for the first component of *SSE* rests on the fact that there are replications at some levels of X. Let us denote the different X levels in the study, whether or not replicated observations are present, as $X_1, \ldots, X_c$. For our example, $c = 6$ since there are six minimum deposit size levels in the study, for five of which there are two observations and for one there is a single observation. We shall let $X_1 = 75$ (the smallest minimum deposit level), $X_2 = 100, \ldots, X_6 = 200$. Further, we shall denote the number of observations for the jth level of X as n_j; for our example, $n_1 = n_2 = n_3 = n_5 = n_6 = 2$ and $n_4 = 1$. Thus, the total number of observations n is given by:

$$n = \sum_{j=1}^{c} n_j \tag{4.7}$$

If we make no assumption about the nature of the regression function but assume all other elements of model (3.1), we can still estimate the error variance σ^2 because of the repeated observations. Table 4.5 presents the same data as Table 4.3, but in a different arrangement. Table 4.5 also shows the mean of the Y observations for each minimum deposit size. We shall denote the mean of the Y observations when $X = X_j$ by $\overline{Y}_j$. Thus $\overline{Y}_1 = 35$ for the two branches with minimum deposits of $X_1 = \$75$, and so on.

Since the two Y observations for $X_1 = \$75$ come from the same probability distribution, we can estimate the variance of this distribution by calculating the usual sample variance, using the deviations around $\overline{Y}_1 = 35$:

$$\frac{(28 - 35)^2 + (42 - 35)^2}{2 - 1} = 98$$

Likewise, the two observations for $X_2 = \$100$ come from the same probability distribution, so that we can estimate the variance of this distribution by calculating the sample variance:

$$\frac{(112 - 124)^2 + (136 - 124)^2}{2 - 1} = 288$$

Similarly, we can estimate the variance of each of the other distributions except for the one at $X_4 = 150$ where there is only a single observation. Since model (3.1) assumes that all probability distributions of Y have the same variance σ^2, we can combine the results for each of the X levels. The optimum way of combining is to add the numerators:

$$\begin{aligned}(28 - 35)^2 &+ (42 - 35)^2 + (112 - 124)^2 + (136 - 124)^2 \\ &+ (160 - 155)^2 + (150 - 155)^2 + (156 - 140)^2 + (124 - 140)^2 \\ &+ (124 - 114)^2 + (104 - 114)^2 = 1{,}148\end{aligned}$$

then add the denominators:

$$1 + 1 + 1 + 1 + 1 = 5$$

and finally take the ratio:

$$\frac{1{,}148}{5} = 229.6$$

TABLE 4.5 Data for bank example, arranged by observation number and minimum deposit

	Size of Minimum Deposit (dollars)					
Observation	$X_1 = 75$	$X_2 = 100$	$X_3 = 125$	$X_4 = 150$	$X_5 = 175$	$X_6 = 200$
$i = 1$	28	112	160	152	156	124
$i = 2$	42	136	150		124	104
Mean $\bar{Y}_j$	35	124	155	152	140	114

To generalize, let us denote the ith observation for the jth level of X by Y_{ij}, where $i = 1, \ldots, n_j$; $j = 1, \ldots, c$. For our example (Table 4.5), $Y_{11} = 28$, $Y_{21} = 42$, $Y_{12} = 112$, and so on. First, we calculate the sum of squares of the deviations from the mean at any given level of X. For $X = X_j$, this sum of squares is:

$$\sum_{i=1}^{n_j} (Y_{ij} - \bar{Y}_j)^2 \tag{4.8}$$

We then add these sums of squares over all levels of X and denote this sum of the sums of squares by $SSPE$:

$$SSPE = \sum_{j=1}^{c} \sum_{i=1}^{n_j} (Y_{ij} - \bar{Y}_j)^2 \tag{4.9}$$

Here $SSPE$ stands for *pure error sum of squares*. Note that when there is only a single observation at X_j, we have $Y_{1j} = \bar{Y}_j$ so that $Y_{1j} - \bar{Y}_j = 0$. Hence, such X_j levels do not contribute to the pure error sum of squares, as was illustrated by our example.

The degrees of freedom associated with $SSPE$ are $n - c$. This is easy to see since there are as usual $n_1 - 1$ degrees of freedom associated with the sum of squares for X_1, $n_2 - 1$ degrees of freedom with the sum of squares for X_2, and so on. The sum of the degrees of freedom is:

$$\sum_{j=1}^{c} (n_j - 1) = \Sigma n_j - c = n - c \tag{4.10}$$

Again, we see that X_j levels for which $n_j = 1$ do not contribute to the degrees of freedom since $n_j - 1 = 0$ then.

The *pure error mean square MSPE* is given by:

$$MSPE = \frac{SSPE}{n - c} \tag{4.11}$$

The reason for the term "pure error" is that $MSPE$ is an unbiased estimator of the error variance σ^2 no matter what is the nature of the regression function. $MSPE$ measures the variability of the distributions of Y without relying on any assumptions about the nature of the regression relation; hence, it is a "pure" measure of the error variance.

Lack of fit component. The second component of SSE is:

(4.12) $$SSLF = SSE - SSPE$$

where $SSLF$ denotes *lack of fit sum of squares*. It can be shown that:

(4.12a) $$SSLF = \sum_{j=1}^{c} n_j(\bar{Y}_j - \hat{Y}_j)^2$$

where $\hat{Y}_j$ denotes the fitted value when $X = X_j$. Thus, $SSLF$ is a weighted sum of squares (the weights are the sample sizes n_j) of the deviations:

(4.13) $$\bar{Y}_j - \hat{Y}_j$$

Note that these deviations represent the difference between the mean $\bar{Y}_j$ and the fitted value $\hat{Y}_j$ based on the regression model. The closer the $\bar{Y}_j$ are to the $\hat{Y}_j$, the greater is the evidence that the fitted regression function is a good fit and therefore appropriate. The further the $\bar{Y}_j$ deviate from the $\hat{Y}_j$, the more the indication that the fitted regression function is inappropriate.

Figure 4.11 illustrates for the observation $X_3 = 125$, $Y_{13} = 160$, the partitioning of the error deviation $Y_{13} - \hat{Y}_3 = 48$ into the pure error deviation $Y_{13} - \bar{Y}_3 = 5$ and the lack of fit deviation $\bar{Y}_3 - \hat{Y}_3 = 43$ for testing whether or not a linear regression function is a good fit.

FIGURE 4.11 Illustration of decomposition of $Y_{ij} - \hat{Y}_j$

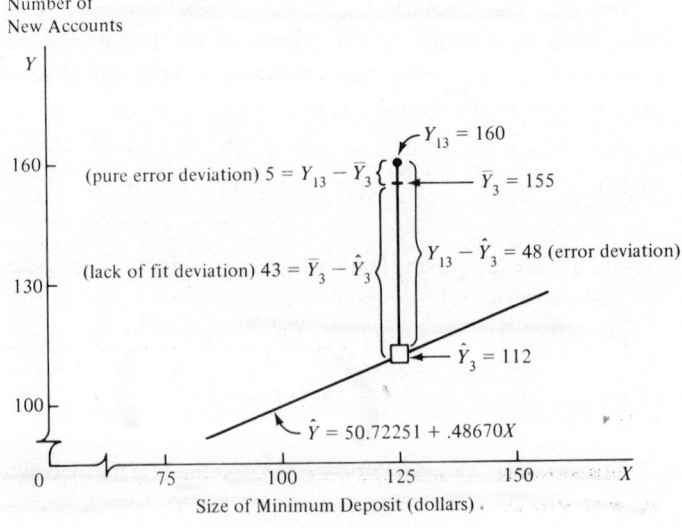

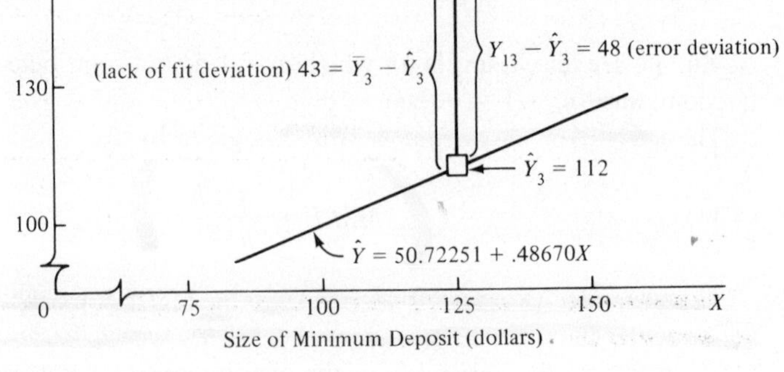

There are $c - 2$ degrees of freedom associated with $SSLF$ when testing for lack of fit of a linear regression function. The reason is that there are c levels of X and two degrees of freedom are lost because two parameters (β_0 and β_1) are estimated in obtaining the fitted values $\hat{Y}_j$. Thus, the lack of fit mean square $MSLF$ is:

$$MSLF = \frac{SSLF}{c - 2} \tag{4.14}$$

For our example, using (4.12) and the earlier results ($SSE = 14{,}741.6$ from Table 4.4, $SSPE = 1{,}148.0$ from p. 126), we obtain:

$$SSLF = 14{,}741.6 - 1{,}148.0 = 13{,}593.6$$

and:

$$MSLF = \frac{13{,}593.6}{6 - 2} = 3{,}398.4$$

Table 4.6a contains the general ANOVA table including the decomposition of SSE just explained and the mean squares of interest, and Table 4.6b contains the ANOVA decomposition for our example.

TABLE 4.6 ANOVA table for testing lack of fit of simple linear regression function

(a) General

Source of Variation	*SS*	*df*	*MS*
Regression	SSR	1	MSR
Error	SSE	$n - 2$	MSE
Lack of fit	$SSLF$	$c - 2$	$MSLF$
Pure error	$SSPE$	$n - c$	$MSPE$
Total	$SSTO$	$n - 1$	

(b) Bank Example

Source of Variation	*SS*	*df*	*MS*
Regression	$SSR = 5{,}141.3$	1	$MSR = 5{,}141.3$
Error	$SSE = 14{,}741.6$	9	$MSE = 1{,}638.0$
Lack of fit	$SSLF = 13{,}593.6$	4	$MSLF = 3{,}398.4$
Pure error	$SSPE = 1{,}148.0$	5	$MSPE = 229.6$
Total	$SSTO = 19{,}882.9$	10	

F test

Test statistic. For testing lack of fit of the regression function, the appropriate test statistic is:

$$F^* = \frac{MSLF}{MSPE} \tag{4.15}$$

We noted that $MSPE$ has expectation σ^2 no matter what is the nature of the regression function. It can be shown that for testing lack of fit of a simple linear regression function:

$$E(MSLF) = \sigma^2 + \frac{\Sigma n_j[E(Y_j) - (\beta_0 + \beta_1 X_j)]^2}{c - 2} \tag{4.16}$$

where $E(Y_j)$ denotes the true mean of the distribution of Y when $X = X_j$ and $\beta_0 + \beta_1 X_j$ is the mean response indicated by the linear regression model. If the regression function is linear, the second term in (4.16) is 0 so that $E(MSLF) = \sigma^2$ then. On the other hand, if the regression function is not linear, $E(Y_j) \neq \beta_0 + \beta_1 X_j$ so that $E(MSLF)$ will be greater than σ^2. Hence, a value of F^* near 1 accords with a linear regression function; large values of F^* indicate that the regression function is not linear.

Decision rule. Since $SSLF$ and $SSPE$ are additive, as are the degrees of freedom, we know from Cochran's theorem that F^* follows the $F(c - 2; n - c)$ distribution if the regression function is linear and all other conditions of model (3.1) hold. To decide between:

$$\begin{aligned} &H_0: E(Y) = \beta_0 + \beta_1 X \\ &H_a: E(Y) \neq \beta_0 + \beta_1 X \end{aligned} \tag{4.17}$$

we use the test statistic (4.15). The decision rule to control the risk of a Type I error at α is:

$$\begin{aligned} &\text{If } F^* \leq F(1 - \alpha; c - 2, n - c), \text{ conclude } H_0 \\ &\text{If } F^* > F(1 - \alpha; c - 2, n - c), \text{ conclude } H_a \end{aligned} \tag{4.18}$$

Example. For our example, the test statistic can be constructed easily from the results in Table 4.6b:

$$F^* = \frac{3{,}398.4}{229.6} = 14.80$$

If the level of significance is to be $\alpha = .01$, we require $F(.99; 4, 5) = 11.4$. Since $F^* = 14.80 > 11.4$, we conclude H_a, that the regression function is not linear. This, of course, accords with our visual impression from Figure 4.10. To report the P-value for the test statistic, we note that $F^* = 14.80$ lies between $F(.99; 4, 5) = 11.4$ and $F(.995; 4, 5) = 15.6$ and thus the P-value must be between .005 and .01. The exact P-value can be shown to be .006.

Comments

1. As was shown by our example, not all levels of X need have repeat observations for the F test for lack of fit to be applicable. Repeat observations at only one or some levels of X are adequate.

2. The F test for lack of fit falls into the framework of a general linear test discussed in Section 3.8. The full model is:

(4.19) For each X_j, Y is normal with mean $E(Y_j)$ and variance σ^2.

The least squares estimator of $E(Y_j)$ for the full model is $\overline{Y}_j$ so that the error sum of squares is:

$$SSE(F) = \sum_i \sum_j (Y_{ij} - \overline{Y}_j)^2 = SSPE \tag{4.20}$$

which has associated with it $n - c$ degrees of freedom.

Since H_0 states:

$$H_0\colon E(Y) = \beta_0 + \beta_1 X \tag{4.21}$$

the error sum of squares for the reduced model is:

$$SSE(R) = \sum_i \sum_j (Y_{ij} - \hat{Y}_j)^2 = SSE \tag{4.22}$$

which has associated with it $n - 2$ degrees of freedom. Substituting into (3.68) and utilizing (4.12), we obtain:

$$\frac{SSE - SSPE}{(n-2)-(n-c)} \div \frac{SSPE}{n-c} = \frac{SSLF}{c-2} \div \frac{SSPE}{n-c} = \frac{MSLF}{MSPE} \tag{4.23}$$

the same test statistic as in (4.15).

3. Suppose that prior to any analysis of the aptness of the model, we had wished to test whether or not $\beta_1 = 0$ for the data underlying Table 4.4. The test statistic (3.59) would be:

$$F^* = \frac{MSR}{MSE} = \frac{5{,}141.3}{1{,}638.0} = 3.14$$

For $\alpha = .10$, $F(.90; 1, 9) = 3.36$, and we would conclude H_0, that $\beta_1 = 0$ or that there is no *linear association* between minimum deposit size (and value of gift) and number of new accounts. A conclusion that there is no *relation* between these variables would be improper, however. Such an inference requires that model (3.1) is appropriate. Here it is not, as we have seen, because the regression function is not linear. There exists indeed a (curvilinear) relation between minimum deposit size and number of new accounts, and testing whether or not $\beta_1 = 0$ under these circumstances has entirely different implications. This illustrates the importance of always examining the aptness of a model before further inferences are drawn.

4. The F test approach just explained can be used to test the aptness of other regression functions, not just the simple linear one in (4.17). Only the degrees of freedom for $SSLF$ will need be modified. In general, $c - p$ degrees of freedom are associated with $SSLF$, where p is the number of parameters in the regression function. For the test of a simple linear regression function, $p = 2$ because there are two parameters, β_0 and β_1, in the regression function.

5. The alternative H_a in (4.17) includes all regression functions other than a linear one. For instance, it includes a quadratic regression function or a logarithmic one. If H_a is concluded, a study of residuals can be helpful in identifying an appropriate function.

6. Clearly, repeat observations are most valuable whenever we are not certain of the nature of the regression function. If at all possible, provision should be made for some replications.

7. If we conclude that the employed model in H_0 is appropriate, the usual practice is to use the error mean square *MSE* as an estimator of σ^2 in preference to the pure error mean square *MSPE*, since the former contains more degrees of freedom.

8. Observations at the same level of X are genuine repeats only if they involve independent trials with respect to the error term. Suppose that in a regression analysis of the relation between hardness (Y) and amount of carbon (X) in specimens of an alloy, the error term in the model covers, among other things, random errors in the measurement of hardness by the analyst and effects of uncontrolled production factors which vary at random from specimen to specimen and affect hardness. If the analyst takes two readings on the hardness of a specimen, this will not provide genuine replication because the effects of random variation in the production factors are fixed in any given specimen. For genuine replication, different specimens with the same carbon content (X) would have to be measured by the analyst so that *all* the effects covered in the error term could vary at random from one repeated observation to the next.

4.5 REMEDIAL MEASURES

If the simple linear regression model (3.1) is not appropriate for the data at hand, there are two basic choices:

1. Abandon model (3.1) and search for a more appropriate model.
2. Use some transformation on the data so that model (3.1) is appropriate for the transformed data.

Each approach has advantages and disadvantages. The first approach may entail a more complex model which may yield better insights, but may also lead into serious difficulties in estimating the parameters. Successful use of transformations, on the other hand, leads to relatively simple methods of estimation and may involve fewer parameters than a complex model, an advantage when the sample size is small. Yet transformations may obscure the fundamental interconnections between the variables, though at other times may illuminate them.

We shall consider the use of transformations in this chapter and the use of more complex models in later chapters. First, we provide a brief overview of remedial measures.

Nonlinearity of regression function

If the regression function is not linear, a direct approach is to modify model (3.1) with respect to the nature of the regression function. For instance, a quadratic regression function might be used:

$$E(Y) = \beta_0 + \beta_1 X + \beta_2 X^2 \tag{4.24}$$

or an exponential regression function:

$$E(Y) = \beta_0 \beta_1^X \tag{4.25}$$

In Chapter 9, we discuss models where the regression function is a polynomial; and in Chapter 14, we discuss exponential regression functions.

The transformation approach uses a transformation to linearize, at least approximately, a nonlinear regression function. For instance, the transformation:

$$Y' = \log Y \tag{4.26}$$

where Y' is the transformed variable, is often useful. We discuss the use of transformations to linearize regression functions in Section 4.6.

Nonconstancy of error variance

If the error variance is not constant but varies in a systematic fashion, a direct approach is to modify the model to allow for this and use the method of *weighted least squares* to obtain the estimators of the parameters. We discuss the use of weighted least squares for this purpose in Section 5.7.

Transformations can also be effective in stabilizing the variance. Some of these are discussed in Section 4.6. For instance, the transformation:

$$Y' = \sqrt{Y} \tag{4.27}$$

is useful in a number of applications for stabilizing the variance.

Nonindependence of error terms

If the error terms are correlated, a direct remedial measure is to work with a model which calls for correlated error terms. We discuss such a model in Chapter 13. A simple remedial transformation which is often helpful is to work with first differences, a topic also discussed in Chapter 13.

Nonnormality of error terms

Lack of normality and nonconstant error variances frequently go hand in hand. Fortunately, it is often the case that the same transformation which helps stabilize the variance, such as a logarithmic or a square root transformation, is also helpful in normalizing the error terms. It is therefore desirable that the transformation for stabilizing the error variance be utilized first, and then the residuals studied to see if serious departures from normality are still present. We discuss transformations to achieve normality in Section 4.6.

Omission of important independent variables

When residual analysis indicates that an important independent variable has been omitted from the model, the solution is to modify the model. In Chapter 7 and following chapters of Part II, we discuss multiple regression analysis in which two or more independent variables are utilized.

4.6 TRANSFORMATIONS

We now consider in more detail the use of transformations of one or both of the original variables before carrying out the regression analysis. Simple transformations of either the dependent variable Y or the independent variable X, or of both, are often sufficient to make the simple linear regression model appropriate for the transformed data. We shall illustrate the use of simple transformations by three examples.

Example 1

In columns 1 and 2 of Table 4.7 are presented data on number of days of training (X) and performance score (Y) for 10 sales trainees in a battery of simulated sales situations in an experiment. These observations are shown as a scatter plot in Figure 4.12a. Clearly the regression relation appears to be curvilinear so that the simple linear regression model (3.1) does not seem to be appropriate.

TABLE 4.7 Regression calculations with square root transformation—sales training example

	(1) *Days of Training* X	(2) *Performance Score* Y	(3) $\sqrt{Y} = Y'$	(4) XY'	(5) X^2
	.5	43	6.5574	3.2787	.25
	.5	40	6.3246	3.1623	.25
	1.0	71	8.4261	8.4261	1.00
	1.0	74	8.6023	8.6023	1.00
	1.5	107	10.3441	15.5162	2.25
	1.5	109	10.4403	15.6605	2.25
	2.0	158	12.5698	25.1396	4.00
	2.5	209	14.4568	36.1420	6.25
	3.0	270	16.4317	49.2951	9.00
	3.5	341	18.4662	64.6317	12.25
Total	17.0	1,422	112.6193	229.8545	38.50

In Figure 4.12b, the same data are plotted but the dependent variable has been transformed as follows:

$$Y' = \sqrt{Y}$$

where Y' denotes the transformed variable. Note that the scatter plot now shows a reasonably linear relation and that the variability of the scatter is reasonably constant at the different X levels. Hence, the simple linear regression model (3.1) now appears to be appropriate.

FIGURE 4.12 Scatter plots of original and transformed observations—sales training example

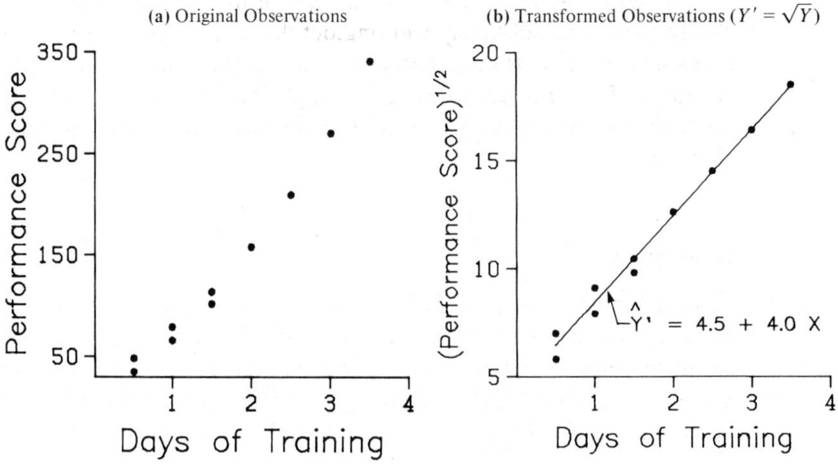

Example 2

At times, a curvilinear regression relationship is accompanied by systematic changes in the variability of the error terms and/or by error terms which follow a highly skewed distribution. In columns 1 and 2 of Table 4.8 are presented data on age (X) and plasma level of a polyamine (Y) for 14 healthy children. These data are plotted in Figure 4.13a as a scatter plot. Note the distinct curvilinear regression relationship, as well as the greater extent of scatter for younger children than for older ones.

TABLE 4.8 Regression calculations with logarithmic transformation—plasma levels example

	(1) *Age* X	(2) *Plasma Level* Y	(3) $\log_{10} Y = Y'$	(4) XY'	(5) X^2
	0 (newborn)	17.0	1.23045	0	0
	0 (newborn)	11.2	1.04922	0	0
	1	9.2	.96379	.96379	1
	1	12.6	1.10037	1.10037	1
	2	7.4	.86923	1.73846	4
	2	10.5	1.02119	2.04238	4
	3	8.3	.91908	2.75724	9
	3	5.8	.76343	2.29029	9
	4	4.6	.66276	2.65104	16
	4	6.5	.81291	3.25164	16
	5	5.3	.72428	3.62140	25
	5	3.8	.57978	2.89890	25
	6	3.2	.50515	3.03090	36
	6	4.5	.65321	3.91926	36
Total	42	109.9	11.85485	30.26567	182

FIGURE 4.13 Scatter plots of original and transformed observations—plasma levels example

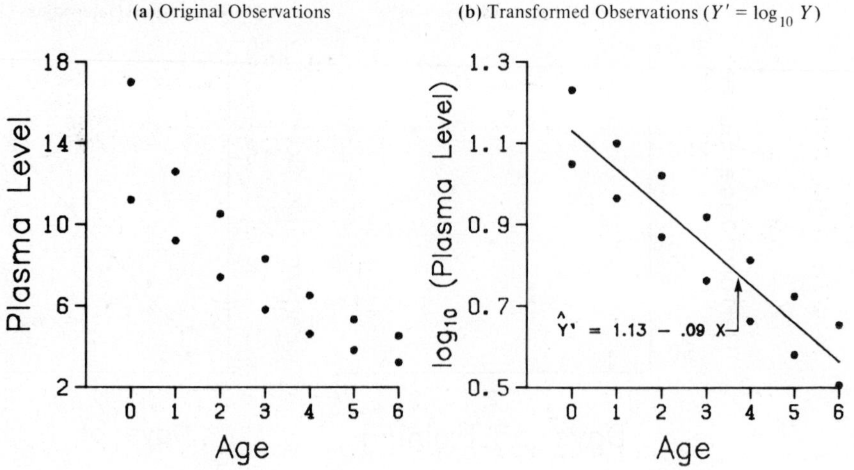

In Figure 4.13b, the same data are plotted but with the dependent variable transformed as follows:

$$Y' = \log_{10} Y$$

Note that the transformation not only has led to a reasonably linear regression relation but also that the extent of scatter at the different levels of X has become reasonably constant. Hence, use of the simple linear regression model (3.1) with the transformed data now appears to be appropriate.

Example 3

In Figure 4.14a, we present a scatter plot of data on number of years experience (X) and current hourly earnings (Y) of five employees in a shop that makes hairpieces to order. The data suggest strongly that the regression function is curvilinear. Since the Y values fall within a relatively small range, a transformation on Y is not likely to be effective and we consider a transformation on the independent variable:

$$X' = \frac{1}{X}$$

Figure 4.14b contains a scatter plot with the transformed variable X'. Note that this transformation has been successful since the points tend to fall in a linear pattern.

FIGURE 4.14 Scatter plots of original and transformed observations—hairpiece earnings example

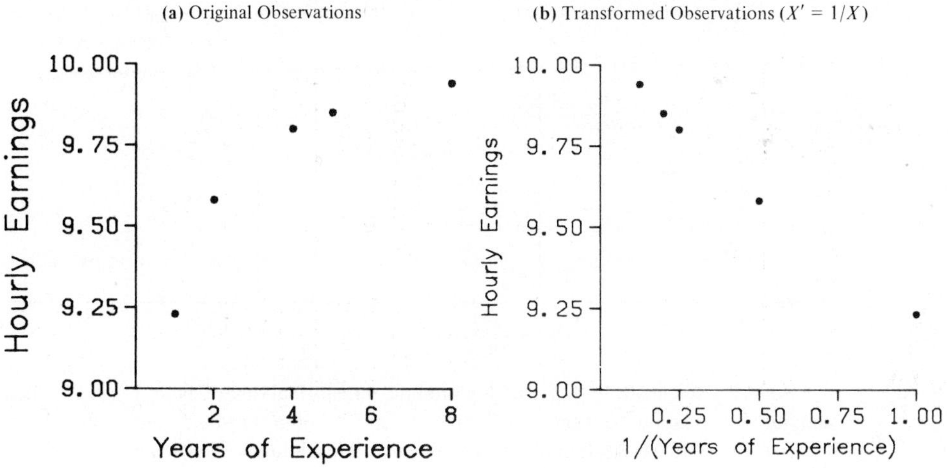

Useful transformations

For many situations, the few simple transformation types just illustrated suffice to remedy the departures from the simple linear regression model (3.1). These transformations can be applied either to the dependent variable Y, the independent variable X, or occasionally to both variables, as follows:

(4.28a) $$Y' = \sqrt{Y} \qquad X' = \sqrt{X}$$

(4.28b) $$Y' = \log_{10} Y \qquad X' = \log_{10} X$$

(4.28c) $$Y' = \frac{1}{Y} \qquad X' = \frac{1}{X}$$

Figure 4.15 is a guide for the selection of the transformation type. Note that in one case the transformation can be applied either to the Y variable or to the X variable, or to both. Use of an interactive computer package for preparing scatter plots based on the different transformations can be most helpful in deciding on an appropriate transformation.

Comments

1. At times, theoretical or a priori considerations can be utilized to help in choosing an appropriate transformation. For example, when the shape of the scatter in a study of the relation between price of a commodity (X) and quantity demanded (Y) is that in Figure 4.15c, economists may prefer a logarithmic transformation on both Y and X to linearize the relation because the slope of the regression line for the transformed variables then measures the price elasticity of demand. The slope is then commonly interpreted as showing the percent change in quantity demanded per 1 percent change in price, where it is understood that the changes are in opposite directions.

FIGURE 4.15 Potential transformations for different curvilinear patterns

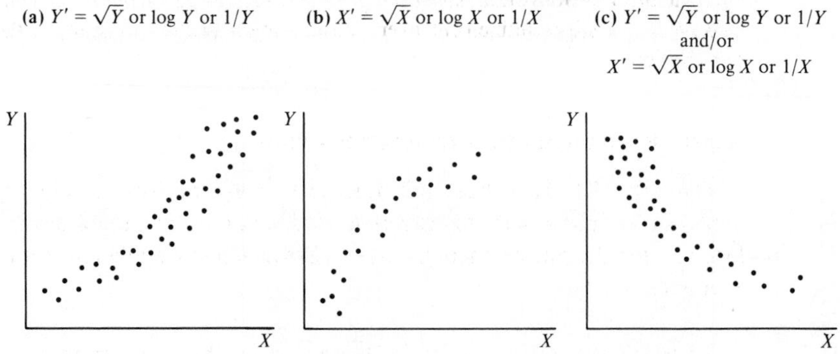

Similarly, scientists may prefer logarithmic transformations of both Y and X when studying the relation between radioactive decay (Y) of a substance and time (X) to linearize a curvilinear relation of the type illustrated in Figure 4.15c because the slope of the regression line for the transformed variables then measures the decay rate.

2. Transformations of X do not affect the variability or shape of the error distribution, while transformations of Y do. Hence, when curvilinearity is accompanied by unequal variability or skewness of the error distributions, the dependent variable needs to be transformed to remedy these additional problems. On the other hand, when curvilinearity is accompanied by constant variability, a transformation of Y may lead to significant unequal variability of the error terms and it is important to check for this possibility by examining residual plots after the transformation.

3. When either variable can be transformed (the case of Figure 4.15c), the variable for which the observations have a wider range should be considered first since a transformation is not likely to be effective when the range of the observations is narrow.

4. After a transformation has been tentatively selected, residual plots and other analyses described earlier need to be employed to ascertain that the simple linear regression model (3.1) is appropriate for the transformed data.

5. When the variance of the error terms is not constant but has a particular relation to the level of the mean response $E(Y)$ for given X, statistical theory indicates an appropriate transformation to stabilize the variance. Three important cases are [where σ_i^2 denotes the error term variance and $E(Y_i)$ denotes the mean response when $X = X_i$]:

(4.29a) If σ_i^2 is proportional to $E(Y_i)$, use $Y' = \sqrt{Y}$

(4.29b) If σ_i is proportional to $E(Y_i)$, use $Y' = \log Y$

(4.29c) If $\sqrt{\sigma_i}$ is proportional to $E(Y_i)$, use $Y' = \dfrac{1}{Y}$

At times, the observed variable Y is a proportion, such as the proportion of families with income X who are planning to purchase a new car next month. An appropriate transformation for this case is:

(4.29d) If observation is a proportion, use $Y' = 2 \arcsin \sqrt{Y}$

This transformation can be made readily on many calculator models. Also, tables to facilitate this transformation have been prepared, such as the one in Reference 4.3 which incorporates a slight refinement over transformation (4.29d) to improve the variance stabilization.

Regression analysis with transformed data

Once the data have been transformed to make the simple linear regression model (3.1) appropriate, the regression calculations are carried out in the usual fashion with the transformed data. We illustrate these calculations for two earlier examples.

Example 1. Table 4.7 (p. 134) contains for the sales training example the basic calculations required for the least squares estimators b_0 and b_1. Since $Y' = \sqrt{Y}$ now plays the role of Y in all earlier formulas, we obtain:

$$b_1 = \frac{\Sigma X_i Y_i' - \dfrac{\Sigma X_i \Sigma Y_i'}{n}}{\Sigma X_i^2 - \dfrac{(\Sigma X_i)^2}{n}} = \frac{229.8545 - \dfrac{(17)(112.6193)}{10}}{38.5 - \dfrac{(17)^2}{10}} = 4.00017$$

$$b_0 = \frac{1}{n}(\Sigma Y_i' - b_1 \Sigma X_i) = \frac{1}{10}[112.6193 - 4.00017(17)] = 4.46164$$

The fitted regression function is:

$$\hat{Y}' = 4.46164 + 4.00017X$$

where $\hat{Y}'$ is the point estimator of $E(Y')$, the mean of the probability distribution of Y' for given X.

If we wish to obtain the fitted regression equation in the original units, we simply take squares:

$$\hat{Y} = (4.46164 + 4.00017X)^2$$

For example, when $X = 3$, we have:

$$\hat{Y} = [4.46164 + 4.00017(3)]^2 = 271.0$$

Figure 4.16 presents a residual plot for the fitted regression model based on the transformed data. This plot shows no evidence of lack of fit and, in view of the few observations for large X values, no strong evidence of unequal error variances. Hence, the square root transformation of Y appears to have been effective here.

Example 2. Table 4.8 (p. 135) contains the necessary least squares calculations for our plasma levels example where the transformation $Y' = \log_{10} Y$ was found to be effective in linearizing a curvilinear relationship and in stabilizing the

FIGURE 4.16 Residual plot for transformed observations—sales training example

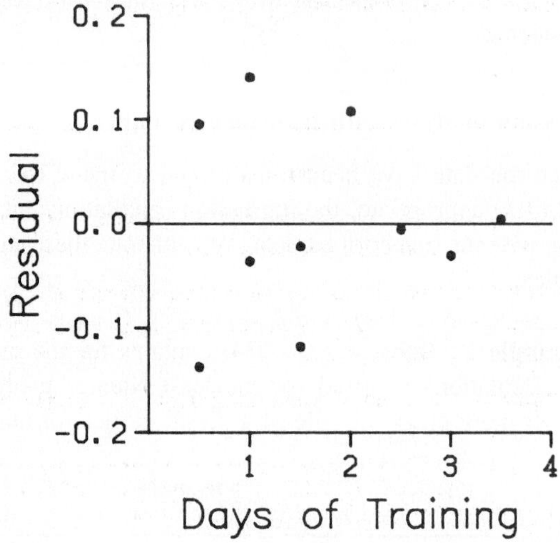

error variance. Since Y' now plays the role of Y in all earlier formulas, we obtain:

$$b_1 = \frac{\Sigma X_i Y_i' - \dfrac{\Sigma X_i \Sigma Y_i'}{n}}{\Sigma X_i^2 - \dfrac{(\Sigma X_i)^2}{n}} = \frac{30.26567 - \dfrac{(42)(11.85485)}{14}}{182 - \dfrac{(42)^2}{14}} = -.094623$$

$$b_0 = \frac{1}{n}(\Sigma Y_i' - b_1 \Sigma X_i) = \frac{1}{14}[11.85485 - (-.094623)(42)] = 1.130644$$

The fitted regression function therefore is:

$$\hat{Y}' = 1.130644 - .094623X$$

where $\hat{Y}'$ is the point estimator of $E(Y')$, the mean of the probability distribution of Y' for given X. This fitted regression function is plotted in Figure 4.13b.

If we wish to obtain the fitted regression equation in the original units, we simply take antilogs:

$$\hat{Y} = \text{antilog}_{10}\,(1.130644 - .094623X)$$

To find the fitted value $\hat{Y}$ in the original units when $X = 3$, for example, we have:

$$\hat{Y} = \text{antilog}_{10}\,[1.130644 - .094623(3)] = 7.03$$

Comments

1. We reiterate the importance of checking the model (3.1) assumptions if a transformation on Y is employed. For instance, when the transformation $Y' = \log Y$ is employed with model (3.1), it is assumed that the distribution of $\log Y$ for given X is normal with

constant variance. This needs to be checked after the transformation has been made.

2. When transformed models are employed, the estimators b_0 and b_1 obtained by least squares have the least squares properties with respect to the transformed observations, not the original ones.

PROBLEMS

4.1. Distinguish between: (1) residual and standardized residual, (2) $E(\varepsilon_i) = 0$ and $\bar{e} = 0$, (3) error term and residual.

4.2. Prepare a prototype residual plot for each of the following cases: (1) error variance decreases with X, (2) true regression function is U shaped but a linear regression function is fitted.

4.3. Refer to **Grade point average** Problem 2.15. The fitted values and residuals are:

i:	1	2	3	4	5	6	7	8	9	10
$\hat{Y}_i$:	2.92	2.33	2.25	1.58	2.08	3.51	3.34	2.67	2.25	1.91
e_i:	.18	−.03	.75	.32	.42	.19	.06	−.07	.55	−.31

i:	11	12	13	14	15	16	17	18	19	20
$\hat{Y}_i$:	2.42	2.84	2.50	3.59	2.16	1.91	2.50	3.26	1.74	2.25
e_i:	−.42	.06	−.20	−.39	−.36	−.51	−.50	.54	.46	−.75

a. Plot the residuals e_i against the fitted values $\hat{Y}_i$. What departures from regression model (3.1) can be studied from this plot? What are your findings?

b. Prepare a normal probability plot of the residuals. Also obtain the coefficient of correlation between the ordered residuals and their expected values under normality. Does the normality assumption appear to be reasonable here?

c. Information is given below for each student on two variables not included in the model, namely, intelligence test score (X_2) and high school average (X_3). Prepare additional residual plots to ascertain whether the model can be improved by including either of these variables. What do you conclude?

i:	1	2	3	4	5	6	7	8	9	10
X_2:	105	113	118	107	110	125	115	121	117	111
X_3:	2.9	2.8	3.1	2.4	3.0	2.4	3.5	3.1	3.1	2.9

i:	11	12	13	14	15	16	17	18	19	20
X_2:	123	114	120	132	122	110	119	109	116	108
X_3:	3.2	3.3	3.4	2.6	3.0	2.8	3.3	3.4	2.6	2.7

4.4. Refer to **Calculator maintenance** Problem 2.16. The fitted values and residuals are:

i:	1	2	3	4	5	6	7	8	9
$\hat{Y}_i$:	100.8	86.1	71.4	12.4	71.4	56.6	100.8	41.9	56.6
e_i:	−3.8	−.1	6.6	−2.4	3.6	5.4	.2	−2.9	−3.6

i:	10	11	12	13	14	15	16	17	18
$\hat{Y}_i$:	27.2	115.6	71.4	27.2	71.4	100.8	12.4	56.6	71.4
e_i:	5.8	2.4	−6.4	−2.2	−.4	4.2	4.6	−7.6	−3.4

a. Prepare residual plots of e_i versus $\hat{Y}_i$ and e_i versus X_i. What departures from regression model (3.1) can be studied from this plot? State your findings.
b. Prepare a normal probability plot of the residuals. Also obtain the coefficient of correlation between the ordered residuals and their expected values under normality. Does the normality assumption appear to be tenable here?
c. The observations are given in time order. Plot the residuals against time to ascertain whether or not the error terms are correlated over time. What is your conclusion?
d. Information is given below on two variables not included in the model, namely, mean operational age of machines serviced on the call (X_2, in months) and years of experience of the service person making the call (X_3). Make additional residual plots to ascertain whether the model can be improved by including either or both of these variables. What do you conclude?

i:	1	2	3	4	5	6	7	8	9
X_2:	12	21	38	16	25	32	18	14	12
X_3:	3	6	2	2	3	4	5	2	3

i:	10	11	12	13	14	15	16	17	18
X_2:	35	20	8	15	17	28	29	9	14
X_3:	6	5	3	5	6	3	5	3	6

4.5. Refer to **Airfreight breakage** Problem 2.17.
a. Obtain the residuals e_i and plot them against X_i to ascertain whether any departures from regression model (3.1) are evident. What is your conclusion?
b. Prepare a normal probability plot of the residuals. Also obtain the coefficient of correlation between the ordered residuals and their expected values under normality to ascertain whether the normality assumption is reasonable here. What do you conclude?

4.6. Refer to **Plastic hardness** Problem 2.18.
a. Obtain the residuals e_i and plot them against the fitted values $\hat{Y}_i$ to ascertain whether any departures from regression model (3.1) are evident. State your findings.
b. Prepare a normal probability plot of the residuals. Also obtain the coefficient of correlation between the ordered residuals and their expected values under normality. Does the normality assumption appear to be reasonable here?

4.7. Refer to **Muscle mass** Problem 2.23.
a. Obtain the residuals e_i and plot them against $\hat{Y}_i$ and also against X_i to ascertain whether any departures from regression model (3.1) are evident. State your conclusions.
b. Prepare a normal probability plot of the residuals. Also obtain the coefficient of correlation between the ordered residuals and their expected values under normality to ascertain whether the normality assumption is tenable here. What do you conclude?
c. The observations are given in time order. Plot the residuals against time to see whether or not the error terms are uncorrelated over time. What is your finding?

4.8. Refer to **Robbery rate** Problem 2.24.
a. Obtain the residuals and make a residual plot of e_i versus $\hat{Y}_i$. What does the plot show?

b. Prepare a normal probability plot of the residuals. Also obtain the coefficient of correlation between the ordered residuals and their expected values under normality. What do you conclude?

4.9. Electricity consumption. An economist studying the relation between household electricity consumption (Y) and number of rooms in the home (X) employed linear regression model (3.1) and obtained the following residuals:

i:	1	2	3	4	5	6	7	8	9	10
X_i:	2	3	4	5	6	7	8	9	10	11
e_i:	3.2	2.9	−1.7	−2.0	−2.3	−1.2	−.9	.8	.7	.5

Plot the residuals e_i against X_i. What problem appears to be present here? Might a transformation alleviate this problem?

4.10. Per capita earnings. A sociologist employed linear regression model (3.1) to relate per capita earnings (Y) to average number of years of schooling (X) for 12 cities. The fitted values $\hat{Y}_i$ and the standardized residuals $e_i/\sqrt{MSE}$ follow.

i:	1	2	3	4	5	6	7	8	9	10	11	12
$\hat{Y}_i$:	9.9	9.3	10.2	9.6	10.2	12.4	14.3	9.6	9.2	15.6	11.2	13.1
$e_i/\sqrt{MSE}$:	−1.12	.81	−.76	.43	.65	−.17	1.62	1.79	−.53	−3.78	.74	.32

Plot the standardized residuals against the fitted values. What does the plot suggest?

4.11. Drug concentration. A pharmacologist employed linear regression model (3.1) to study the relation between the concentration of a drug in plasma (Y) and the log-dose of the drug (X). The residuals and log-dose levels follow.

i:	1	2	3	4	5	6	7	8	9
X_i:	−1	0	1	−1	0	1	−1	0	1
e_i:	.5	2.1	−3.4	.3	−1.7	4.2	−.6	2.6	−4.0

Plot the residuals e_i against X_i. What conclusions do you draw from the plot?

4.12. A student states that she doesn't understand why the sum of squares defined in (4.9) is called a pure error sum of squares "since the formula looks like one for an ordinary sum of squares." Explain.

4.13. Refer to **Calculator maintenance** Problem 2.16. Some additional calculational results are: $SSR = 16{,}182.6$, $SSE = 321.4$.

a. In an F test for lack of fit of a linear regression function, what are the alternative conclusions?

b. Perform the test indicated in part (a). Control the risk of Type I error at .05. State the decision rule and conclusion.

c. Does your test in part (b) detect other departures from model (3.1), such as lack of constant variance or lack of normality in the error terms? Could the results of the test be affected by such departures? Discuss.

4.14. Refer to **Plastic hardness** Problem 2.18.

a. Perform an F test to determine whether or not there is lack of fit of a linear regression function. Use a level of significance of .01. State the alternatives, decision rule, and conclusion.

b. Assuming that the number of replications here was limited in advance to four, is there any advantage in conducting these all at the same level of X? Is there any disadvantage?

c. Does the test in part (a) indicate what regression function is appropriate when it leads to the conclusion that the regression function is not linear? How would you proceed?

4.15. Solution concentration. A chemist studied the concentration of a solution (Y) over time (X). Fifteen identical solutions were prepared. The 15 solutions were randomly divided into five sets of three, and the five sets were measured, respectively, after 1, 3, 5, 7, and 9 hours. The results follow.

i:	1	2	3	4	5	6	7	8	9	10	11	12	13	14	15
X_i:	9	9	9	7	7	7	5	5	5	3	3	3	1	1	1
Y_i:	.07	.09	.08	.16	.17	.21	.49	.58	.53	1.22	1.15	1.07	2.84	2.57	3.10

a. Fit a linear regression function.

b. Perform an F test to determine whether or not there is lack of fit of a linear regression function; use $\alpha = .025$. State the alternatives, decision rule, and conclusion.

c. Does the test in part (b) indicate what regression function is appropriate when it leads to the conclusion that lack of fit of a linear regression function exists? Explain.

4.16. Refer to **Solution concentration** Problem 4.15.

a. Prepare a scatter plot of the data. What transformations might you try to achieve linearity?

b. Use transformation $Y' = \log_{10} Y$ and obtain the estimated linear regression function for the transformed data.

c. Plot the estimated regression line and the transformed data. Does the regression line appear to be a good fit to the transformed data?

d. Obtain the residuals and plot them against the fitted values. Also prepare a normal probability plot. What do your plots show?

e. Express the estimated regression equation in the original units.

4.17. Sales growth. A marketing researcher studied annual sales of a product that had been introduced 10 years ago. The data were as follows, where X is the year (coded) and Y is sales in thousands of units:

i:	1	2	3	4	5	6	7	8	9	10
X_i:	0	1	2	3	4	5	6	7	8	9
Y_i:	98	135	162	178	221	232	283	300	374	395

a. Prepare a scatter plot of the data. Does a linear relation appear adequate here?

b. Use transformation $Y' = \sqrt{Y}$ and obtain the estimated linear regression function for the transformed data.

c. Plot the estimated regression line and the transformed data. Does the regression line appear to be a good fit to the transformed data?

d. Obtain the residuals and plot them against the fitted values. Also prepare a normal probability plot. What do your plots show?

e. Express the estimated regression equation in the original units.

EXERCISES

4.18. A student fitted a linear regression function for a class assignment. Some results follow.

i:	1	2	3	4	5
Y_i:	35	17	42	28	53
$\hat{Y}_i$:	42	29	32	32	40
e_i:	−7	−12	10	−4	13

The student plotted the residuals e_i against Y_i and found a positive relation. When he plotted the residuals against the fitted values $\hat{Y}_i$, he found no relation. Why is there this difference, and which is the more meaningful plot?

4.19. If the error terms in a regression model are independent $N(0, \sigma^2)$, what can be said about the error terms after transformation $X' = 1/X$ is used? Is the situation the same after transformation $Y' = 1/Y$ is used?

4.20. Using theorems (1.65), (1.36), and (1.37), show that $E(MSPE) = \sigma^2$ for the normal error regression model (3.1).

PROJECTS

4.21. Machine speed. The number of defective items produced by a machine (Y) is known to be linearly related to the speed setting of the machine (X). The data below were collected from recent quality control records.

i:	1	2	3	4	5	6	7	8	9	10	11	12
X_i:	200	400	300	400	200	300	300	400	200	400	200	300
Y_i:	28	75	37	53	22	58	40	96	46	52	30	69

a. Assuming regression model (3.1) is appropriate, obtain the estimated regression function and plot the residuals against X. What does the residual plot show?

b. Calculate the sample variance s^2 of the Y observations for each of the three machine speeds: $X = 200, 300, 400$. What is suggested by these three sample variances about whether or not the true variances at the three X levels are equal?

c. Compute $\bar{Y}/\sqrt{s}$, $\bar{Y}/s$, and $\bar{Y}/s^2$ for each of the three X levels. Suggest an appropriate transformation from (4.29) to stabilize the variance based on your computed ratios.

d. Make the transformation suggested in part (c) and obtain the estimated regression line for the transformed data. Plot the residuals against X. Does it appear from your plot that the purpose of the transformation has been attained?

4.22. Blood pressure. The following data were obtained in a study of the relation between diastolic blood pressure (Y) and age (X) for boys 5 to 13 years old.

i:	1	2	3	4	5	6	7	8
X_i:	5	8	11	7	13	12	12	6
Y_i:	63	67	74	64	75	69	90	60

a. Assuming regression model (3.1) is appropriate, obtain the estimated regression function and plot the residuals e_i against X_i. What does your residual plot show?

b. Omit observation 7 from the data and obtain the estimated regression line based on the remaining seven observations. Compare this estimated regression function to that obtained in part (a). What can you conclude about the effect of observation 7?

c. Using your fitted regression function in part (b), obtain a 99 percent prediction interval for a new Y observation at $X = 12$. Does observation Y_7 fall outside this prediction interval? What is the significance of this?

4.23. Refer to the **SMSA** data set and Project 2.38. For each of the three fitted regression models, obtain the residuals and prepare a residual plot against X and a normal probability plot. Summarize your conclusions. Is linear regression model (3.1) more apt in one case than in the others?

4.24. Refer to the **SMSA** data set and Project 2.39. For each geographic region, obtain the residuals and prepare a residual plot against X and a normal probability plot. Do the four regions appear to have similar error variances? What other conclusions do you draw from your plots?

4.25. Refer to the **SENIC** data set and Project 2.40.

a. For each of the three fitted regression models, obtain the residuals and prepare a residual plot against X and a normal probability plot. Summarize your conclusions. Is linear regression model (3.1) more apt in one case than in the others?

b. Obtain the fitted regression line for the relation between length of stay and infection risk after deleting observations 47 ($X_{47} = 6.5$, $Y_{47} = 19.56$) and 112 ($X_{112} = 5.9$, $Y_{112} = 17.94$). From this fitted regression line obtain separate 95 percent prediction intervals for new Y observations at $X = 6.5$ and $X = 5.9$, respectively. Do observations Y_{47} and Y_{112} fall outside these prediction intervals? Discuss the significance of this.

4.26. Refer to the **SENIC** data set and Project 2.41. For each geographic region, obtain the residuals and prepare a residual plot against X and a normal probability plot. Do the four regions appear to have similar error variances? What other conclusions do you draw from your plots?

CITED REFERENCES

4.1 Dixon, W. J., and M. B. Brown, eds. *BMDP-81, Biomedical Computer Programs, P-Series*. Berkeley, Calif.: University of California Press, 1981.

4.2 Barnett, Vic, and Toby Lewis. *Outliers in Statistical Data*. New York: John Wiley & Sons, 1978.

4.3 Owen, Donald B. *Handbook of Statistical Tables*. Reading, Mass.: Addison-Wesley Publishing, 1962.

5

Simultaneous inferences and other topics in regression analysis—I

In this chapter, we take up a variety of topics in simple regression analysis. Several of the topics pertain to the problem of how to make simultaneous inferences from the same set of sample observations.

5.1 JOINT ESTIMATION OF β_0 AND β_1

Need for joint estimation

A market research analyst conducted a study of the relation between level of advertising (X) and sales (Y), in which there was no advertising ($X = 0$) for some observations while for other observations the level of advertising was varied. The scatter plot suggested a linear regression in the range of the advertising expenditures levels studied. The analyst wished to draw inferences about both the intercept β_0 and the slope β_1. One means of doing this is by constructing a joint confidence region for β_0 and β_1 so that with confidence level $1 - \alpha$ both β_0 and β_1 are contained in this region.

Joint confidence region

A joint $1 - \alpha$ confidence region for β_0 and β_1 is illustrated in Figure 5.1. It can be shown that such a region is given by:

$$(5.1)\quad \frac{n(b_0 - \beta_0)^2 + 2(\Sigma X_i)(b_0 - \beta_0)(b_1 - \beta_1) + (\Sigma X_i^2)(b_1 - \beta_1)^2}{2MSE} \le F(1 - \alpha; 2, n - 2)$$

The confidence coefficient $1 - \alpha$ indicates that with repeated sampling, the confidence region (5.1) will contain both β_0 and β_1 in $(1 - \alpha)100$ percent of the cases. The confidence region consists of all points (β_0, β_1) which satisfy the inequality (5.1). The boundary of the confidence region is obtained from the equality in (5.1):

$$(5.2)\quad \frac{n(b_0 - \beta_0)^2 + 2(\Sigma X_i)(b_0 - \beta_0)(b_1 - \beta_1) + (\Sigma X_i^2)(b_1 - \beta_1)^2}{2MSE} = F(1 - \alpha; 2, n - 2)$$

The boundary is an ellipse centered around the point (b_0, b_1), as illustrated in Figure 5.1. We explain now by an example how the boundary is calculated.

Example. Let us return to the Westwood Company lot size example of the previous chapters. Suppose that we require a joint confidence region for β_0 and

FIGURE 5.1 Elliptical joint 90 percent confidence region for β_0 and β_1—Westwood Company example

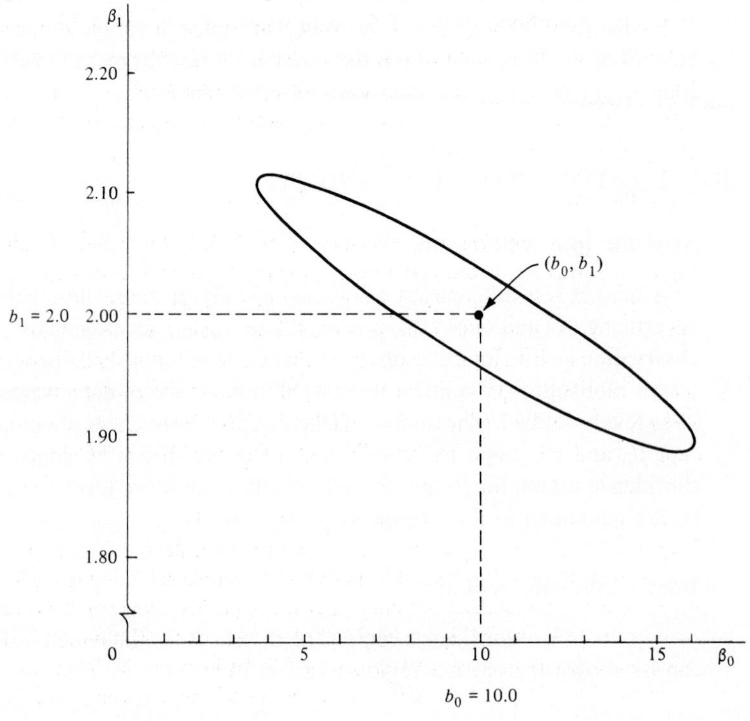

To illustrate the meaning of a family confidence coefficient further, let us return to the joint estimation of β_0 and β_1. A family confidence coefficient of, say, .95 would indicate for this situation that if repeated samples are selected and interval estimates for both β_0 and β_1 are calculated for each sample by specified procedures, 95 percent of the samples would lead to a family of estimates where *both* confidence intervals are correct. For 5 percent of the samples, either one or both of the interval estimates would be incorrect.

Clearly, a procedure which provides a family confidence coefficient is often highly desirable since it permits the analyst to weave the separate results together into an integrated set of conclusions, with an assurance that the entire set of estimates is correct. The Bonferroni method of developing joint confidence intervals with a specified family confidence coefficient is a very simple one: each statement confidence coefficient is adjusted to be higher than $1 - \alpha$ so that the family confidence coefficient is $1 - \alpha$. The method is a general one which can be applied in many cases, as we shall see, not just for the joint estimation of β_0 and β_1. Here, we explain the Bonferroni method as it applies for estimating β_0 and β_1 jointly.

Development of joint confidence intervals. We start with ordinary confidence limits for β_0 and β_1 with statement confidence coefficients $1 - \alpha$ each. These are:

$$b_0 \pm t(1 - \alpha/2; n - 2)s(b_0)$$
$$b_1 \pm t(1 - \alpha/2; n - 2)s(b_1)$$

We then ask what is the probability that both sets of limits are correct. Let A_1 denote the event that the first confidence interval does not cover β_0 and A_2 denote the event that the second confidence interval does not cover β_1. We know:

$$P(A_1) = \alpha \qquad P(A_2) = \alpha$$

Probability theorem (1.6) states:

$$P(A_1 \cup A_2) = P(A_1) + P(A_2) - P(A_1 \cap A_2)$$

and hence:

$$1 - P(A_1 \cup A_2) = 1 - P(A_1) - P(A_2) + P(A_1 \cap A_2) \tag{5.4}$$

Now by probability theorems (1.9) and (1.10), we have:

$$1 - P(A_1 \cup A_2) = P(\overline{A_1 \cup A_2}) = P(\bar{A}_1 \cap \bar{A}_2)$$

$P(\bar{A}_1 \cap \bar{A}_2)$ is the probability that both confidence intervals are correct. We thus have from (5.4):

$$P(\bar{A}_1 \cap \bar{A}_2) = 1 - P(A_1) - P(A_2) + P(A_1 \cap A_2) \tag{5.5}$$

Since $P(A_1 \cap A_2) \geq 0$, we obtain from (5.5) the Bonferroni inequality:

$$P(\bar{A}_1 \cap \bar{A}_2) \geq 1 - P(A_1) - P(A_2) \tag{5.6}$$

which for our situation is:

$$P(\bar{A}_1 \cap \bar{A}_2) \geq 1 - \alpha - \alpha = 1 - 2\alpha \tag{5.6a}$$

Thus, if β_0 and β_1 are separately estimated with, say, 95 percent confidence intervals, the Bonferroni inequality guarantees us a family confidence coefficient of at least 90 percent that both intervals based on the same sample are correct.

We can easily use the Bonferroni inequality (5.6a) to obtain a family confidence coefficient of at least $1 - \alpha$ for estimating β_0 and β_1. We do this by estimating β_0 and β_1 separately with statement confidence coefficients of $1 - \alpha/2$ each. Thus, the $1 - \alpha$ family confidence limits for β_0 and β_1, often called a *confidence set*, are by the Bonferroni procedure:

$$\begin{aligned} &b_0 \pm Bs(b_0) \\ &b_1 \pm Bs(b_1) \end{aligned} \tag{5.7}$$

where:

$$B = t(1 - \alpha/4; n - 2) \tag{5.7a}$$

Note that a statement confidence coefficient of $1 - \alpha/2$ requires the $(1 - \alpha/4)100$ percentile of the t distribution for a two-sided confidence interval.

Example. For the Westwood Company lot size application, 90 percent family confidence intervals for β_0 and β_1 require $B = t(1 - .10/4; 8) = t(.975; 8) = 2.306$. We have from before:

$$\begin{aligned} b_0 &= 10.0 \qquad s(b_0) = 2.50294 \\ b_1 &= \ \ 2.0 \qquad s(b_1) = \ .04697 \end{aligned}$$

Hence, the two pairs of confidence limits are $10.0 \pm 2.306(2.50294)$ and $2.0 \pm 2.306(.04697)$, and the joint confidence intervals are:

$$\begin{aligned} 4.2282 \leq \beta_0 \leq 15.7718 \\ 1.8917 \leq \beta_1 \leq \ \ 2.1083 \end{aligned}$$

Thus, we conclude that β_0 is between 4.23 and 15.77 *and* β_1 is between 1.89 and 2.11. The family confidence coefficient is at least .90 that the procedure leads to correct pairs of interval estimates.

Comments

1. We reiterate that the Bonferroni $1 - \alpha$ family confidence coefficient is actually a lower bound on the true (but often unknown) family confidence coefficient. To the extent that incorrect interval estimates of β_0 and β_1 tend to pair up in the family (particularly when the covariance between b_0 and b_1 is large), the families of statements will tend to be correct more than $(1 - \alpha)100$ percent of the time.

2. The Bonferroni inequality (5.6a) can easily be extended to g simultaneous confidence intervals with family confidence coefficient $1 - \alpha$:

$$P\left(\bigcap_{i=1}^{g} \bar{A}_i\right) \geq 1 - g\alpha \tag{5.8}$$

Thus, if g interval estimates are desired with a family confidence coefficient $1 - \alpha$, constructing each interval estimate with statement confidence coefficient $1 - \alpha/g$ will suffice.

3. For a given family confidence coefficient, the larger the number of confidence intervals in the family, the greater becomes the multiple B, which may make some or all of the confidence intervals too wide to be helpful. The Bonferroni technique is ordinarily most useful when the number of simultaneous estimates is not too large.

4. It is not necessary with the Bonferroni procedure that the confidence intervals have the same statement confidence coefficient. Different statement confidence coefficients can be used, depending on the importance of each estimate. For instance, in our earlier illustration β_0 might be estimated with a 92 percent confidence interval and β_1 with a 98 percent confidence interval. The family confidence coefficient by (5.6) will still be at least 90 percent.

Comparison of two approaches. Figure 5.2 contains the joint 90 percent confidence region by the Bonferroni approach for our example. Note that the region is a rectangle since the Bonferroni approach does not utilize the existing relationship between b_0 and b_1. The rectangle is centered at (b_0, b_1). Also shown in Figure 5.2 is the elliptical 90 percent joint confidence region which we obtained earlier, which is also centered at (b_0, b_1). The elliptical region is more efficient in that it covers fewer (β_0, β_1) points for the same confidence coeffi-

FIGURE 5.2 Bonferroni and elliptical joint 90 percent confidence regions for β_0 and β_1—Westwood Company example

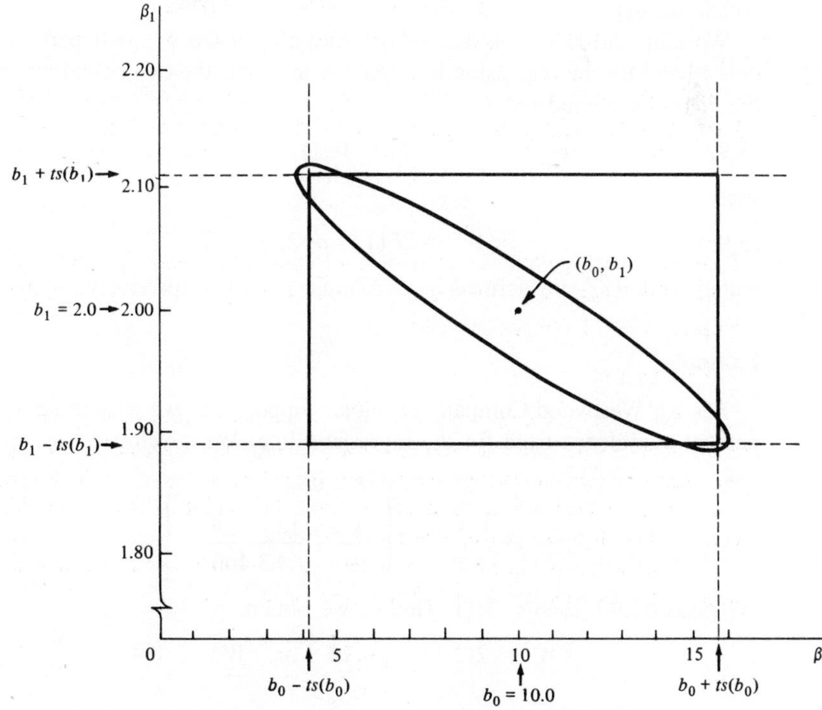

cient. Nevertheless, the Bonferroni approach can be highly useful in many cases. In multiple regression applications, in particular, the Bonferroni method comes into its own because of the ease of obtaining and interpreting the joint estimates.

5.2 CONFIDENCE BAND FOR REGRESSION LINE

At times we would like to obtain a confidence band for the regression line $E(Y) = \beta_0 + \beta_1 X$ so that we can see in what region the entire regression line lies. This differs from estimating $E(Y_h) = \beta_0 + \beta_1 X_h$ for a particular value of X_h by an interval estimate, which we took up in Section 3.4.

To obtain a confidence band for the entire regression line, we essentially need to consider the regression lines for all possible (β_0, β_1) combinations in the elliptical joint confidence region (5.1) for β_0 and β_1. For our Westwood Company example, three possible (β_0, β_1) combinations (Figure 5.1) and their corresponding regression lines are:

β_0	β_1	$E(Y) = \beta_0 + \beta_1 X$
4.00	2.11	$E(Y) = 4.00 + 2.11X$
9.00	2.00	$E(Y) = 9.00 + 2.00X$
16.00	1.89	$E(Y) = 16.00 + 1.89X$

Figure 5.3 contains a plot of these three possible regression lines. As additional lines for other possible (β_0, β_1) combinations in the confidence region in Figure 5.1 are plotted, we will fill out a confidence band for the regression line. The boundaries of the confidence band are sketched in Figure 5.3 by the broken lines, which are hyperbolas.

Working and Hotelling derived the formula for the $1 - \alpha$ hyperbolic confidence band for the regression line. At any level X_h, the two boundary values of the confidence band are:

(5.9)
$$\hat{Y}_h \pm W s(\hat{Y}_h)$$

where:

(5.9a)
$$W^2 = 2F(1 - \alpha; 2, n - 2)$$

and $\hat{Y}_h$ and $s(\hat{Y}_h)$ are defined in (3.27) and (3.30), respectively.

Example

For our Westwood Company example, suppose that we wish to set up the 90 percent confidence band for the regression line. We developed earlier:

$$\hat{Y}_h = 10.0 + 2.0X_h \qquad \text{(p. 41)}$$

$$s^2(\hat{Y}_h) = 7.5\left[\frac{1}{10} + \frac{(X_h - 50)^2}{3{,}400}\right] \qquad \text{(p. 75)}$$

We need $F(.90; 2, 8) = 3.11$. Hence, we obtain:

$$W^2 = 2(3.11) = 6.22 \quad \text{or} \quad W = 2.494$$

FIGURE 5.3 Plot of three possible regression lines for joint confidence region in Figure 5.1 (Y values not plotted to scale)

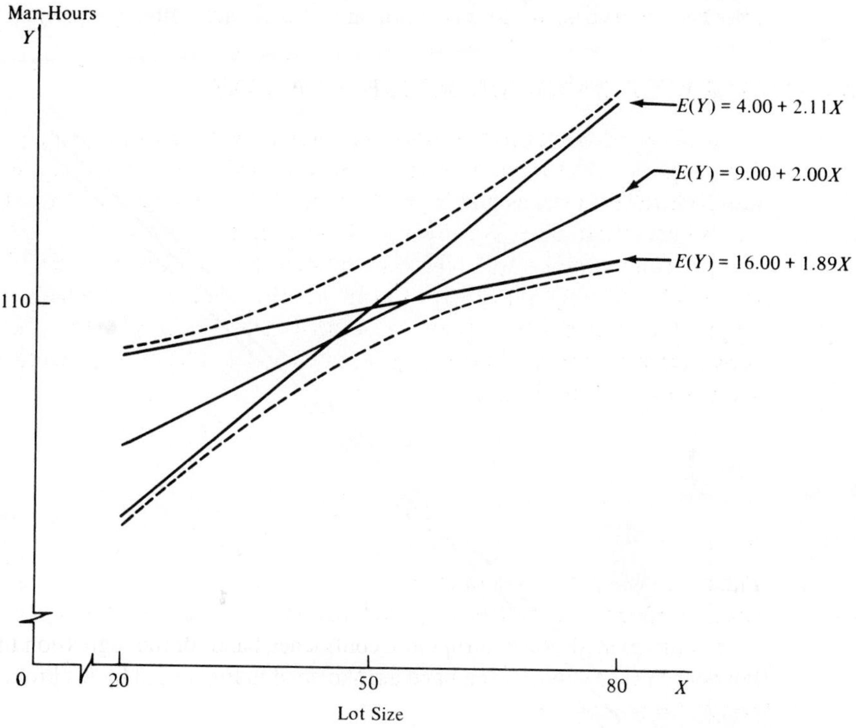

Let us find the boundary points of the confidence band at $X_h = 55$:

$$\hat{Y}_{55} = 10.0 + 2.0(55) = 120.0$$

$$s^2(\hat{Y}_{55}) = 7.5\left[\frac{1}{10} + \frac{(55 - 50)^2}{3{,}400}\right] = .80515$$

or:

$$s(\hat{Y}_{55}) = .89730$$

Hence, the 90 percent boundary points of the confidence band for the regression line at $X_h = 55$ are $120.0 \pm 2.494(.89730)$ or:

$$117.8 \leq \beta_0 + \beta_1 X_h \leq 122.2$$

In similar fashion, the boundary points at a number of other values of X_h can be developed and the boundary curves then sketched in. This has been done in Figure 5.4, which contains the 90 percent confidence band for the regression line for the Westwood Company example.

FIGURE 5.4 90 percent confidence band for regression line—Westwood Company example

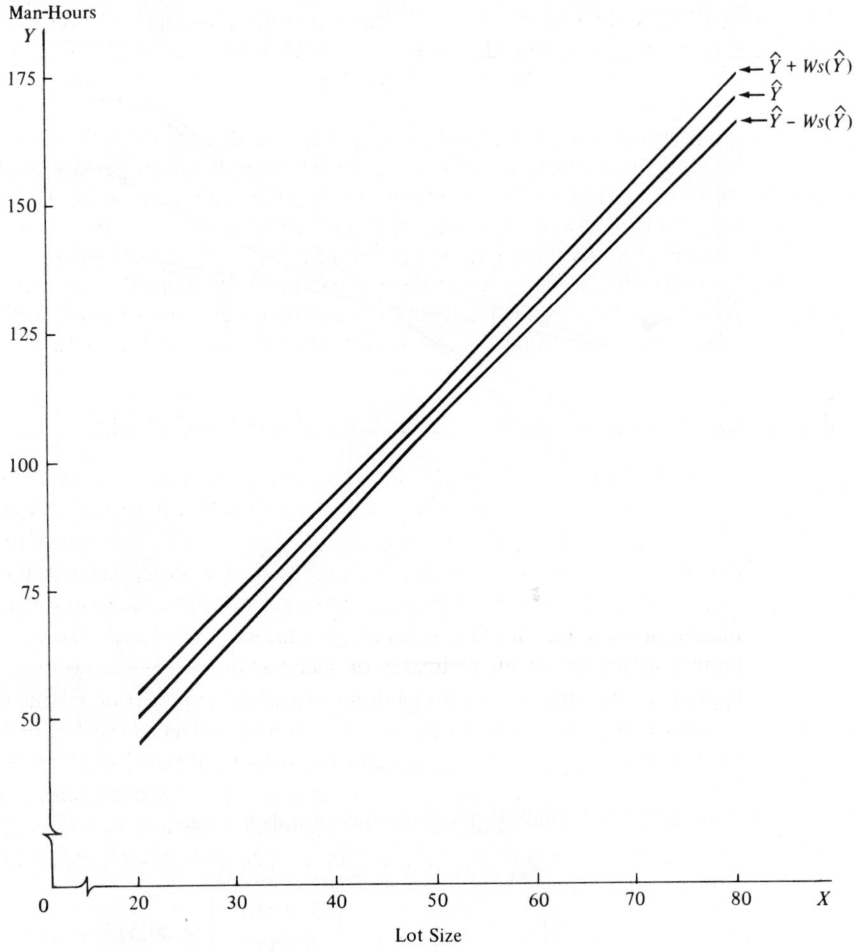

Comments

1. The boundary points (5.9) of the confidence band for the regression line are of exactly the same form as the confidence limits for a single mean response $E(Y_h)$ in (3.32) except that the t multiple has been replaced by the W multiple (named after Working).

2. If we had wished to estimate a single mean response with a 90 percent confidence coefficient, the required t value would have been $t(.95; 8) = 1.860$. Note that this t multiple, though smaller than the multiple $W = 2.494$ for the entire regression line, does not differ by any major extent. This is typically the case, so that the boundary points of the regression line band usually are not very much further apart than the confidence limits for a single mean response $E(Y_h)$ at a given X_h value. With the somewhat wider limits for the entire regression line, one is able to draw conclusions about any and all mean responses

for the entire regression line and not just about the mean response at a given X level. One use of this broader base for inferences will be explained in the next section.

3. The confidence band (5.9) applies to the entire regression line over all real-numbered values of X from $-\infty$ to $+\infty$. The confidence coefficient indicates the percent of time the estimating procedure will yield a band which covers the entire line, in a long series of samples in which the X observations are kept at the same level as in the sample actually taken.

In applications, the confidence band is ignored for that part of the regression line which is not of interest in the problem at hand. In the Westwood Company example, for instance, negative lot sizes would be ignored. The confidence coefficient for a limited segment of the band of interest is somewhat higher than $1 - \alpha$, so that $1 - \alpha$ serves then as a lower bound to the confidence coefficient.

4. Research continues on confidence banding of the regression line. An alternative procedure to the Working-Hotelling one, for instance, gives a confidence band of uniform width over a finite interval on the X axis centered around $\bar{X}$ (Ref. 5.1).

5.3 SIMULTANEOUS ESTIMATION OF MEAN RESPONSES

Often one would like to estimate the mean responses at a number of X levels from the same sample data. The Westwood Company, for instance, may wish to estimate the mean number of man-hours for lots of 30, 55, and 80 parts. We already know how to do this for any one level of X with given statement confidence coefficient. Now we shall discuss two approaches for simultaneous estimation of mean responses with a family confidence coefficient, so that there is a known assurance of all estimates of mean responses being correct. The two approaches are the Working-Hotelling approach and the Bonferroni approach.

The reason for concern with a family confidence coefficient is that separate interval estimates of $E(Y_h)$ at various levels X_h need not all be correct or all be incorrect, even though they are all based on the same sample data and fitted regression line. The combination of sampling errors in b_0 and b_1 may be such that the interval estimates of $E(Y_h)$ will be correct over some range of X levels and incorrect elsewhere.

Working-Hotelling approach

Since the Working-Hotelling confidence band for the entire regression line holds for all values of X, it certainly must hold for selected levels of the independent variable. Hence, to obtain with the Working-Hotelling approach a family of interval estimates of mean responses at different levels of X, with a $1 - \alpha$ family confidence coefficient, we simply use formula (5.9) repetitively to calculate boundary points of the confidence band at the various X levels being considered. These boundary points serve then as the family confidence limits for the interval estimates of the mean responses of interest.

Example. For the Westwood Company lot size example, suppose that we require a family of estimates of the mean number of man-hours at the following levels of lot size: 30, 55, 80. The family confidence coefficient is to be .90. We obtained earlier $\hat{Y}_h$ and $s(\hat{Y}_h)$ for $X_h = 55$, and found $W = 2.494$. In similar fashion, we can obtain the needed results for the other lot sizes. We summarize them here, without showing the calculations:

X_h	$\hat{Y}_h$	$s(\hat{Y}_h)$	$Ws(\hat{Y}_h)$
30	70.0	1.27764	3.1864
55	120.0	.89730	2.2379
80	170.0	1.65387	4.1248

Thus, the boundary points of the regression line band at $X_h = 30$, 55, and 80 are:

$$66.8 = 70.0 - 3.1864 \leq E(Y_{30}) \leq 70.0 + 3.1864 = 73.2$$
$$117.8 = 120.0 - 2.2379 \leq E(Y_{55}) \leq 120.0 + 2.2379 = 122.2$$
$$165.9 = 170.0 - 4.1248 \leq E(Y_{80}) \leq 170.0 + 4.1248 = 174.1$$

With family confidence coefficient .90, we conclude that the mean number of man-hours for lots of 30 parts is between 66.8 and 73.2, for lots of 55 parts is between 117.8 and 122.2, and for lots of 80 parts is between 165.9 and 174.1. The family confidence coefficient .90 provides assurance that the procedure leads to all correct estimates in the family of estimates.

Bonferroni approach

The Bonferroni approach, discussed earlier for simultaneous estimation of β_0 and β_1, is a completely general approach. To construct a family of confidence intervals for mean responses at different X levels, we calculate the usual confidence limits for a single mean response $E(Y_h)$, given in (3.32), and adjust the statement confidence coefficient to yield the specified family confidence coefficient.

If $E(Y_h)$ is to be estimated for g levels of X, with a family confidence coefficient of $1 - \alpha$, the Bonferroni confidence limits are:

(5.10) $$\hat{Y}_h \pm Bs(\hat{Y}_h)$$

where:

(5.10a) $$B = t(1 - \alpha/2g; n - 2)$$

g is the number of confidence intervals in the family

Example. The estimates of the mean man-hours for lot sizes of 30, 55, and 80 parts with a family confidence coefficient of .90 by the Bonferroni approach require the same data as the Working-Hotelling approach presented above. In addition, we require $B = t(1 - .10/2(3); 8) = t(.983; 8)$. By linear interpolation, we obtain $t(.983; 8) = 2.56$ (see the following Comment 4).

We thus obtain the confidence intervals, with a 90 percent family confidence coefficient:

$$66.7 = 70.0 - 2.56(1.27764) \leq E(Y_{30}) \leq 70.0 + 2.56(1.27764) = 73.3$$
$$117.7 = 120.0 - 2.56(.89730) \leq E(Y_{55}) \leq 120.0 + 2.56(.89730) = 122.3$$
$$165.8 = 170.0 - 2.56(1.65387) \leq E(Y_{80}) \leq 170.0 + 2.56(1.65387) = 174.2$$

Comments

1. In this instance the Working-Hotelling confidence limits are slightly tighter than the Bonferroni limits. In other cases where the number of statements is small, the Bonferroni limits may be tighter. For larger families, the Working-Hotelling confidence limits will always be the tighter, since W in (5.9a) stays the same for any number of statements in the family whereas B in (5.10a) becomes larger as the number of statements increases. In practice, once the family confidence coefficient has been decided upon, one can calculate the W and B multiples to determine which procedure leads to tighter confidence limits.

2. Both the Working-Hotelling and Bonferroni approaches to multiple estimation of mean responses provide lower bounds to the actual family confidence coefficient. The reason why the Working-Hotelling approach furnishes a lower bound is that the confidence coefficient $1 - \alpha$ actually applies to the entire line from $-\infty$ to $+\infty$.

3. Sometimes it is not known in advance for which levels of the independent variable to estimate the mean response. That is determined as the analysis proceeds. In such cases, it is better to use the Working-Hotelling approach.

4. To obtain an untabled percentile of the t distribution, linear interpolation in Table A–2 ordinarily will give a reasonably close approximation as long as the degrees of freedom are not minimal. In our illustration of the Bonferroni method, we required $t(.983; 8)$. From Table A–2, we know that:

$$t(.980; 8) = 2.449 \qquad t(.985; 8) = 2.634$$

Linear interpolation therefore gives:

$$t(.983; 8) = 2.449 + \left(\frac{.983 - .980}{.985 - .980}\right)(2.634 - 2.449) = 2.56$$

5.4 SIMULTANEOUS PREDICTION INTERVALS FOR NEW OBSERVATIONS

Now we consider the simultaneous prediction of g new observations on Y in g independent trials at g different levels of X. To illustrate this type of application, let us suppose the Westwood Company plans to produce the next three lots in sizes of 30, 55, and 80 parts, and wishes to predict the man-hours for each of these lots with a family confidence coefficient of .95.

Two procedures will be considered here, the Scheffé procedure and the Bonferroni procedure. Both utilize the same type of limits as for predicting a single observation, given in (3.35), and only the multiple of the estimated standard deviation is changed. The Scheffé procedure uses the F distribution, while the Bonferroni procedure uses the t distribution. The simultaneous prediction limits for g predictions with the Scheffé procedure with family confidence coefficient $1 - \alpha$ are:

$$\hat{Y}_h \pm Ss(Y_{h(\text{new})}) \tag{5.11}$$

where:

(5.11a) $$S^2 = gF(1 - \alpha; g, n - 2)$$

With the Bonferroni procedure, the $1 - \alpha$ simultaneous prediction limits are:

(5.12) $$\hat{Y}_h \pm Bs(Y_{h(\text{new})})$$

where:

(5.12a) $$B = t(1 - \alpha/2g; n - 2)$$

We can evaluate the S and B multiples to see which procedure provides tighter prediction limits. For our example, we have:

$$S^2 = 3F(.95; 3, 8) = 3(4.07) = 12.21 \quad \text{or} \quad S = 3.49$$
$$B = t(1 - .05/2(3); 8) = t(.992; 8) = 3.04$$

so that the Bonferroni method will be used here. From earlier results, we obtain (calculations not shown):

X_h	$\hat{Y}_h$	$s(Y_{h(\text{new})})$	$Bs(Y_{h(\text{new})})$
30	70.0	3.02198	9.18682
55	120.0	2.88187	8.76088
80	170.0	3.19926	9.72575

and the simultaneous prediction limits are:

$$60.8 = 70.0 - 9.18682 \le Y_{30(\text{new})} \le 70.0 + 9.18682 = 79.2$$
$$111.2 = 120.0 - 8.76088 \le Y_{55(\text{new})} \le 120.0 + 8.76088 = 128.8$$
$$160.3 = 170.0 - 9.72575 \le Y_{80(\text{new})} \le 170.0 + 9.72575 = 179.7$$

With family confidence coefficient at least .95, we can predict that the man-hours for the next three production runs all will be within the above limits.

Comments

1. Simultaneous prediction intervals for g new observations on Y at g different levels of X with a $1 - \alpha$ family confidence coefficient are wider than the corresponding single prediction intervals of (3.35). When the number of simultaneous predictions is not large, however, the difference in the width is only moderate. For instance, a single 95 percent prediction interval for our example would have utilized the t multiple $t(.975; 8) = 2.306$, which is only moderately smaller than the multiple $B = 3.04$ for three simultaneous predictions.

2. Note that both B and S become larger as g increases. This contrasts with simultaneous estimation of mean responses where B becomes larger but not W. When g is large, both the B and S multiples may become so large that the prediction intervals will be too wide to be useful. Other simultaneous estimation techniques could then be considered, as discussed in Reference 5.2.

5.5 REGRESSION THROUGH THE ORIGIN

Sometimes the regression line is known to go through the origin at (0, 0). This occurs, for instance, when X is units of output and Y is variable cost, so Y is zero by definition when X is zero. Another example is where X is the number of brands of cigarettes stocked in a supermarket in an experiment (including some supermarkets with no brands stocked) and Y is the volume of cigarette sales in the supermarket. The normal error model for these cases is the same as the general model (3.1) except $\beta_0 = 0$:

$$Y_i = \beta_1 X_i + \varepsilon_i \tag{5.13}$$

where:

β_1 is a parameter
X_i are known constants
ε_i are independent $N(0, \sigma^2)$

The regression function for model (5.13) is:

$$E(Y) = \beta_1 X \tag{5.14}$$

The least squares estimator of β_1 is obtained by minimizing:

$$Q = \Sigma(Y_i - \beta_1 X_i)^2 \tag{5.15}$$

with respect to β_1. The resulting normal equation is:

$$\Sigma X_i(Y_i - b_1 X_i) = 0 \tag{5.16}$$

leading to the point estimator:

$$b_1 = \frac{\Sigma X_i Y_i}{\Sigma X_i^2} \tag{5.17}$$

b_1 as given in (5.17) is also the maximum likelihood estimator.

An unbiased estimator of $E(Y)$ is:

$$\hat{Y} = b_1 X \tag{5.18}$$

Also, an unbiased estimator of σ^2 is:

$$MSE = \frac{\Sigma(Y_i - \hat{Y}_i)^2}{n-1} = \frac{\Sigma(Y_i - b_1 X_i)^2}{n-1} \tag{5.19}$$

The reason for the denominator $n - 1$ is that only one degree of freedom is lost in estimating the single parameter of the regression equation (5.14).

Confidence intervals for β_1, $E(Y_h)$, and a new observation $Y_{h(\text{new})}$ are shown in Table 5.1. Note that the t multiple has $n - 1$ degrees of freedom here, the

TABLE 5.1 Confidence limits for regression through origin

Estimate of—	*Estimated Variance*		*Confidence Limits*
β_1	$s^2(b_1) = \dfrac{MSE}{\Sigma X_i^2}$	(5.20)	$b_1 \pm ts(b_1)$
$E(Y_h)$	$s^2(\hat{Y}_h) = \dfrac{X_h^2 MSE}{\Sigma X_i^2}$	(5.21)	$\hat{Y}_h \pm ts(\hat{Y}_h)$
$Y_{h(\text{new})}$	$s^2(Y_{h(\text{new})}) = MSE\left[1 + \dfrac{X_h^2}{\Sigma X_i^2}\right]$	(5.22)	$\hat{Y}_h \pm ts(Y_{h(\text{new})})$
		where:	$t = t(1 - \alpha/2; n - 1)$

degrees of freedom associated with MSE. The results in Table 5.1 are derived in analogous fashion to the earlier results for our general model (3.1). Whereas for the general case, we encounter terms $(X_i - \bar{X})^2$ or $(X_h - \bar{X})^2$, here we find X_i^2 and X_h^2 because of the regression through the origin.

Example

The Charles Plumbing Supplies Company operates 12 warehouses. In an attempt to tighten procedures for planning and control, a consultant studied the relation between number of work units performed (X) and total variable labor cost (Y) in the warehouses during a test period. The data are given in Table 5.2, and the observations are shown as a scatter plot in Figure 5.5.

Model (5.13) for regression through the origin was employed since Y involves variable costs only and the other conditions of the model appeared to be satisfied as well. From Table 5.2, we have $\Sigma X_i Y_i = 894{,}714$ and $\Sigma X_i^2 = 190{,}963$. Hence:

$$b_1 = \frac{\Sigma X_i Y_i}{\Sigma X_i^2} = \frac{894{,}714}{190{,}963} = 4.68527$$

and the estimated regression function is:

$$\hat{Y} = 4.68527X$$

The fitted regression line is plotted in Figure 5.5.

To illustrate inferences for regression through the origin, suppose an interval estimate of β_1 is desired with a 95 percent confidence coefficient. We obtain (calculations not shown):

$$MSE = \frac{\Sigma(Y_i - b_1 X_i)^2}{n - 1} = \frac{2{,}457.66}{11} = 223.42$$

TABLE 5.2 Data for regression through origin—warehouse example

Warehouse i	*Work Units Performed* X_i	*Variable Labor Cost (dollars)* Y_i	$X_i Y_i$	X_i^2
1	20	114	2,280	400
2	196	921	180,516	38,416
3	115	560	64,400	13,225
4	50	245	12,250	2,500
5	122	575	70,150	14,884
6	100	475	47,500	10,000
7	33	138	4,554	1,089
8	154	727	111,958	23,716
9	80	375	30,000	6,400
10	147	670	98,490	21,609
11	182	828	150,696	33,124
12	160	762	121,920	25,600
Total	1,359	6,390	894,714	190,963

FIGURE 5.5 Scatter plot and fitted regression through origin—warehouse example

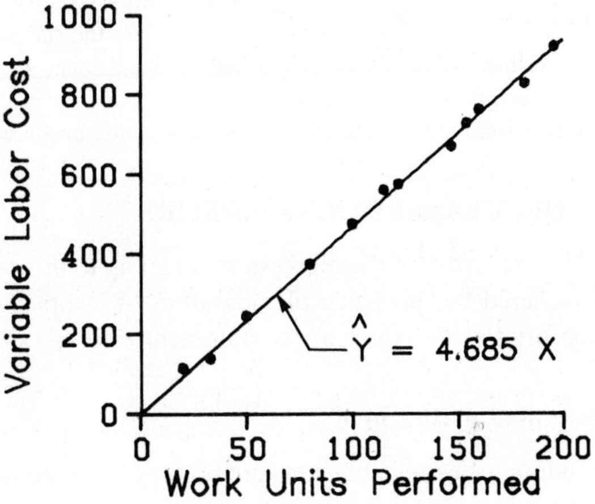

From Table 5.2, we have $\Sigma X_i^2 = 190{,}963$. Hence:

$$s^2(b_1) = \frac{MSE}{\Sigma X_i^2} = \frac{223.42}{190{,}963} = .0011700 \quad \text{or} \quad s(b_1) = .034205$$

For a 95 percent confidence coefficient, we require $t(.975; 11) = 2.201$. The confidence limits, by (5.20) in Table 5.1, are $b_1 \pm ts(b_1)$ or $4.68527 \pm 2.201(.034205)$. The 95 percent confidence interval for β_1 therefore is:

$$4.61 \leq \beta_1 \leq 4.76$$

Thus, with 95 percent confidence, it is estimated that the mean of the distribution of total variable labor costs increases by somewhere between \$4.61 and \$4.76 for each additional work unit performed.

Comments

1. In linear regression through the origin, there is no property of the form $\Sigma(Y_i - b_1X_i) = \Sigma e_i = 0$. Consequently, the residuals usually will not sum to zero here. The only property comes from the normal equation (5.16), namely, $\Sigma X_i e_i = 0$.

2. In interval estimation of $E(Y_h)$ or $Y_{h(\text{new})}$, note that the intervals (5.21) and (5.22) in Table 5.1 widen, the further X_h is from the origin. The reason is that the value of the true regression function is known precisely at the origin, so that the effect of the sampling error in the slope b_1 becomes increasingly important the farther X_h is from the origin.

3. Since only one regression parameter, β_1, must be estimated for the regression function (5.14), simultaneous estimation methods are not required to make a family of statements about several mean responses. For a given confidence coefficient $1 - \alpha$, formula (5.21) can be used repetitively with the given sample results to generate a family of statements for which the family confidence coefficient is still $1 - \alpha$.

4. Like any other model, model (5.13) should be evaluated for aptness. Even when it is known that the regression function must go through the origin, the function might not be

linear or the variance of the error terms might not be constant. Often one cannot be sure in advance that the regression function goes through the origin, and it is then safe practice to use the general model (3.1). If the regression does go through the origin, b_0 will differ from 0 only by a small sampling error, and unless the sample size is very small, use of model (3.1) has no disadvantages of any consequence. If the regression does not go through the origin, use of the general model (3.1) will avoid potentially serious difficulties resulting from forcing the regression through the origin when this is not appropriate.

5.6 EFFECT OF MEASUREMENT ERRORS

In our discussion of the regression model up to this point, we have not explicitly considered the presence of measurement errors in either X or Y. We now examine briefly the effect of measurement errors.

Measurement errors in Y

If random measurement errors are present in the dependent variable Y, no new problems are created if these errors are uncorrelated and not biased (positive and negative measurement errors tend to cancel out). Consider, for example, a study of the relation between the time required to complete a task (Y) and the complexity of the task (X). The time to complete the task may not be measured accurately because the person operating the stopwatch may not do so at the precise instants called for. As long as such measurement errors are of a random nature, uncorrelated, and not biased, these measurement errors can simply be absorbed in the model error term ε. The model error term reflects the composite effects of a large number of factors not considered in the model, one of which simply would be random errors due to inaccuracy in the process of measuring Y.

Measurement errors in X

Unfortunately, a different situation holds if the independent variable X is known only with measurement error. Frequently, to be sure, X is known without measurement error, as when the independent variable is price of a product, number of variables in an optimization problem, or wage rate for a class of employees. At other times, however, measurement errors may enter the value observed for the independent variable, for instance, when it is pressure, temperature, production line speed, or person's age.

We shall use the latter illustration in our development of the nature of the problem. Suppose we are regressing employees' piecework earnings on age. Let X_i denote the true age of the ith employee and X_i^* the age given by the employee on his or her employment record. Needless to say, the two are not always the same. We define the measurement error δ_i as follows:

(5.23) $$\delta_i = X_i^* - X_i$$

The regression model we would like to study is:

(5.24) $$Y_i = \beta_0 + \beta_1 X_i + \varepsilon_i$$

Since, however, we only observe X_i^*, model (5.24) becomes:

$$Y_i = \beta_0 + \beta_1(X_i^* - \delta_i) + \varepsilon_i \tag{5.25}$$

where we make use of (5.23) in replacing X_i. We can rewrite (5.25) as follows:

$$Y_i = \beta_0 + \beta_1 X_i^* + (\varepsilon_i - \beta_1\delta_i) \tag{5.26}$$

Model (5.26) may appear like an ordinary regression model, with independent variable X^* and error term $\varepsilon - \beta_1\delta$, but it is not. The independent variable observation X_i^* is a random variable, which, as we shall see, is correlated with the error term $\varepsilon_i - \beta_1\delta_i$. Theorem (3.40) for the case of random independent variables requires that the error term be independent of the independent variable. Hence, the standard regression results are not applicable for model (5.26).

Intuitively, we know that $\varepsilon_i - \beta_1\delta_i$ is not independent of X_i^* since (5.23) constrains $X_i^* - \delta_i$ to equal X_i. To determine the dependence formally, let us assume:

$$E(\delta_i) = 0 \tag{5.27a}$$

$$E(\varepsilon_i) = 0 \tag{5.27b}$$

$$E(\delta_i\varepsilon_i) = 0 \tag{5.27c}$$

Note that (5.27a) implies that $E(X_i^*) = E(X_i + \delta_i) = X_i$, and that (5.27c) assumes the measurement error δ_i is not correlated with the model error ε_i because by (1.19a) we have $\sigma(\delta_i, \varepsilon_i) = E(\delta_i\varepsilon_i)$ since $E(\delta_i) = E(\varepsilon_i) = 0$ by (5.27a) and (5.27b). We now wish to find the covariance:

$$\begin{aligned}
\sigma(X_i^*, \varepsilon_i - \beta_1\delta_i) &= E\{[X_i^* - E(X_i^*)][(\varepsilon_i - \beta_1\delta_i) - E(\varepsilon_i - \beta_1\delta_i)]\} \\
&= E[(X_i^* - X_i)(\varepsilon_i - \beta_1\delta_i)] \\
&= E[\delta_i(\varepsilon_i - \beta_1\delta_i)] \\
&= E(\delta_i\varepsilon_i - \beta_1\delta_i^2)
\end{aligned}$$

Now $E(\delta_i\varepsilon_i) = 0$ by (5.27c), and $E(\delta_i^2) = \sigma^2(\delta_i)$ by (1.14a) because $E(\delta_i) = 0$ by (5.27a). We therefore obtain:

$$\sigma(X_i^*, \varepsilon_i - \beta_1\delta_i) = -\beta_1\sigma^2(\delta_i) \tag{5.28}$$

This covariance is not zero if there is a linear regression relation between X and Y.

If standard least squares procedures are applied to model (5.26), the estimators b_0 and b_1 are biased and also lack the property of consistency. Great difficulties are encountered in developing unbiased estimators when there are measurement errors in X. One approach is to impose severe conditions on the problem—for example, to make fairly strong assumptions about the properties of the distributions of δ_i, the covariance of δ_i and ε_j, and so on. Another approach is to use additional variables which are known to be related to the true value of X but not with the errors of measurement δ. Such variables are called "instrumental" variables because they are used as an instrument in studying the relation between X and Y. Instrumental variables make it possible to obtain consistent estimators of the regression parameters.

Discussions of possible approaches and further references will be found in specialized works such as Reference 5.3 and in statistical journals.

Note

It may be asked what is the distinction between the case when X is a random variable, considered in Chapter 3, and the case when X is subject to random measurement errors, and why are there special problems with the latter. When X is a random variable, it is not under the control of the analyst and will vary at random from trial to trial, as when X is the number of persons entering a store in a day. If this random variable X is not subject to measurement errors, however, it can be accurately ascertained for a given trial. Thus, if there are no measurement errors in counting the number of persons entering a store in a day, the analyst has accurate information to study the relation between number of persons entering the store and sales, even though the levels of number of customers which actually occur cannot be controlled. If, on the other hand, measurement errors are present in the number of persons entering the store, a distorted picture of the relation between number of persons and sales occurs because the sales observations will frequently be matched against an incorrect number of customers. This distorting effect of measurement errors is present whether X is fixed or random.

Berkson model

There is one situation where measurement errors in X are no problem. This case was first noted by Berkson (Ref. 5.4). Frequently, the independent variable in an experiment is set at a target value. For instance, in an experiment on the effect of temperature on typist productivity, the temperature may be set at target levels of 68°F, 70°F, and so on, according to the temperature control on the thermostat. The observed temperature X_i^* is fixed here, while the actual temperature X_i is a random variable since the thermostat may not be completely accurate. Similar situations exist when water pressure is set according to a gauge, or employees of specified ages according to their employment records are selected for a study.

In all of these cases, the observation X_i^* is a fixed quantity, while the unobserved true value X_i is a random variable. The measurement error is, as before:

$$\delta_i = X_i^* - X_i \tag{5.29}$$

Here, however, there is no constraint on the relation between X_i^* and δ_i, since X_i^* is a fixed quantity. Again, we assume that $E(\delta_i) = 0$.

Model (5.26), which we obtained when replacing X_i by $X_i^* - \delta_i$, is still applicable for the Berkson case:

$$Y_i = \beta_0 + \beta_1 X_i^* + (\varepsilon_i - \beta_1 \delta_i) \tag{5.30}$$

The expected value of the error term, $E(\varepsilon_i - \beta_1\delta_i)$, is zero as before, since $E(\varepsilon_i) = 0$ and $E(\delta_i) = 0$. Further, $\varepsilon_i - \beta_1\delta_i$ is now independent of X_i^*, since X_i^*

is a constant for the Berkson case. Hence, the conditions of an ordinary regression model are met:

1. The error terms have expectation 0.
2. The independent variable is a constant, and hence the error terms are independent of it.

Thus, standard least squares procedures can be applied for the Berkson case without modification, and the estimators b_0 and b_1 will be unbiased. If we can make the standard normality and constant variance assumptions for the errors $\varepsilon_i - \beta_1\delta_i$, the usual tests and interval estimates can be utilized.

5.7 WEIGHTED LEAST SQUARES

General approach

The least squares criterion (2.8):

$$Q = \sum_{i=1}^{n} (Y_i - \beta_0 - \beta_1 X_i)^2$$

weights each observation equally. There are times, however, when some observations should receive greater weight and others smaller weight. The weighted least squares criterion for simple linear regression is:

$$Q_w = \sum_{i=1}^{n} w_i(Y_i - \beta_0 - \beta_1 X_i)^2 \tag{5.31}$$

where w_i is the given weight of the ith observation. Minimizing Q_w with respect to β_0 and β_1 leads to the normal equations:

$$\begin{aligned} \Sigma w_i Y_i &= b_0 \Sigma w_i + b_1 \Sigma w_i X_i \\ \Sigma w_i X_i Y_i &= b_0 \Sigma w_i X_i + b_1 \Sigma w_i X_i^2 \end{aligned} \tag{5.32}$$

In turn, these can be solved for the weighted least squares estimators b_0 and b_1:

$$b_1 = \frac{\Sigma w_i X_i Y_i - \dfrac{\Sigma w_i X_i \Sigma w_i Y_i}{\Sigma w_i}}{\Sigma w_i X_i^2 - \dfrac{(\Sigma w_i X_i)^2}{\Sigma w_i}} \tag{5.33a}$$

$$b_0 = \frac{\Sigma w_i Y_i - b_1 \Sigma w_i X_i}{\Sigma w_i} \tag{5.33b}$$

Note that if all weights are equal so w_i is identically equal to a constant, the normal equations (5.32) for weighted least squares reduce to the ones for unweighted least squares in (2.9) and the weighted least squares estimators (5.33) reduce to the ones for unweighted least squares in (2.10).

Weighted least squares should be used when the error term variance is not constant for all observations. The weight for an observation should then be the reciprocal of the observation's error term variance:

$$w_i = \frac{1}{\sigma_i^2} \tag{5.34}$$

where σ_i^2 is the variance of the error term for the ith observation. Thus, observations whose error terms are subject to large variation receive less weight and observations whose error terms are subject to small variation receive more weight.

Unfortunately, the error term variances σ_i^2 are usually unknown. However, when the error term variance varies with the level of the independent variable in a systematic fashion, this relation can be exploited. For instance, if the error term variance σ_i^2 is proportional to X_i^2 so that $\sigma_i^2 = kX_i^2$ where k is a proportionality factor, the weights w_i would be:

$$w_i = \frac{1}{kX_i^2} \tag{5.35}$$

Since the proportionality constant k drops out of the normal equations (5.32), we can simply use the weights:

$$w_i = \frac{1}{X_i^2} \tag{5.35a}$$

This particular weight relation of the error term variance being proportional to X^2 is frequently encountered in business, economic, and biological applications, as in studies of savings regressed on income.

We illustrate these concepts by an example.

Example

The Nielsen Construction Company studied the relation between the size of a bid in million dollars (X) and the cost to the firm of preparing the bid in thousand dollars (Y), for 12 recent bids. The data are presented in Table 5.3, columns 1 and 2, and are shown in a scatter plot in Figure 5.6. The scatter plot strongly suggests that the error variance increases with X. An analyst after conducting a preliminary residual analysis concluded that the error variance is approximately proportional to X^2. (See the following Comment 4 for a discussion of how to study the relation between the error variance and the level of X.) The analyst, therefore, decided to employ the regression model:

$$Y_i = \beta_0 + \beta_1 X_i + \varepsilon_i \qquad \text{where } \sigma_i^2 = kX_i^2 \tag{5.36}$$

Column 3 in Table 5.3 contains the weights $w_i = 1/X_i^2$, and columns 4–6 contain the needed calculations. The least squares estimators (5.33a) and (5.33b) are, using the data in Table 5.3 and noting that $w_iX_i^2 \equiv 1$ here so that $\Sigma w_iX_i^2 = n = 12$:

TABLE 5.3 Regression calculations for weighted least squares—bid preparation example

	(1) X_i	(2) Y_i	(3) $w_i = \frac{1}{X_i^2}$	(4) $w_i X_i = \frac{1}{X_i}$	(5) $w_i Y_i$	(6) $w_i X_i Y_i = \frac{Y_i}{X_i}$
	2.13	15.5	.220415	.46948	3.41643	7.27700
	1.21	11.1	.683013	.82645	7.58144	9.17355
	11.00	62.6	.008264	.09091	.51733	5.69091
	6.00	35.4	.027778	.16667	.98334	5.90000
	5.60	24.9	.031888	.17857	.79401	4.44643
	6.91	28.1	.020943	.14472	.58850	4.06657
	2.97	15.0	.113367	.33670	1.70051	5.05051
	3.35	23.2	.089107	.29851	2.06728	6.92537
	10.39	42.0	.009263	.09625	.38905	4.04235
	1.10	10.0	.826446	.90909	8.26446	9.09091
	4.36	20.0	.052605	.22936	1.05210	4.58716
	8.00	47.5	.015625	.12500	.74219	5.93750
Total	63.02	335.3	2.098714	3.87171	28.09664	72.18826

$$b_1 = \frac{72.18826 - \dfrac{(3.87171)(28.09664)}{2.098714}}{12 - \dfrac{(3.87171)^2}{2.098714}} = 4.19057$$

$$b_0 = \frac{28.09664 - 4.19057(3.87171)}{2.098714} = 5.65678$$

Hence, the fitted regression line is:

$$\hat{Y} = 5.6568 + 4.1906X$$

FIGURE 5.6 Scatter plot and fitted weighted regression line—bid preparation example

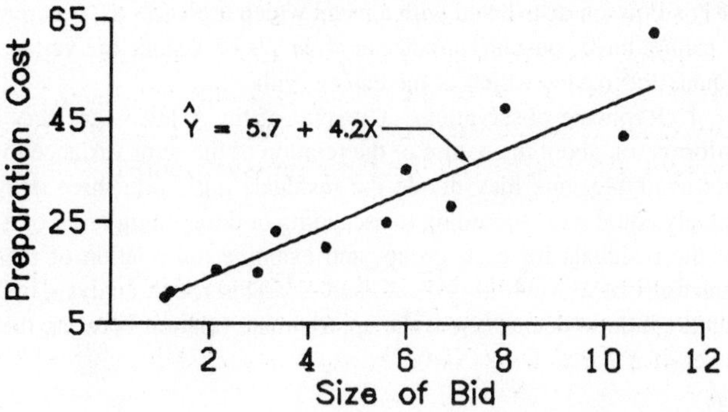

The fitted regression line is shown in Figure 5.6 and appears to be a reasonably good fit to the data.

If we had fitted a straight line by unweighted least squares to the original data in our example, we would have obtained the regression function:

$$\hat{Y} = 4.2289 + 4.5153X$$

This differs from the weighted least squares results, as will generally be the case. The reason is that the weights $w_i = 1/X_i^2$ give more emphasis to observations for smaller X (for which the distribution of Y has a smaller variance) and less emphasis to observations for larger X (for which the distribution of Y has a larger variance).

While the estimates obtained by unweighted least squares are unbiased, as are the estimates obtained by weighted least squares, the unweighted least squares estimates are subject to greater sampling variation. In our example, the estimated standard deviations of the regression coefficients for the two methods are:

Unweighted Least Squares	*Weighted Least Squares*
$s(b_0) = 3.2517$	$s(b_0) = .9652$
$s(b_1) = .5285$	$s(b_1) = .4037$

Comments

1. The condition of the error variance not being constant over all observations is called *heteroscedasticity,* in contrast to the condition of equal error variances, called *homoscedasticity*.

2. When heteroscedasticity prevails but the other conditions of model (3.1) are met, the estimators b_0 and b_1 obtained by ordinary least squares procedures are still unbiased and consistent, but they are no longer minimum variance unbiased estimators, as illustrated in the previous example.

3. Heteroscedasticity is inherent when the response in regression analysis follows a distribution in which the variance is functionally related to the mean. (Significant nonnormality in Y is encountered as well in most such cases.) Consider, in this connection, a regression analysis where X is the speed of a machine which puts a plastic coating on cable and Y is the number of blemishes in the coating per thousand feet drum of cable. If Y is Poisson distributed with a mean which increases as X increases, the distributions of Y cannot have constant variance at all levels of X since the variance of a Poisson variable equals the mean, which is increasing with X.

4. Replicate observations at several of the X levels are very desirable for obtaining information about the nature of the relation of the error variance to X. When replicates are not available, one may divide the residuals into, say, three or four groups of approximately equal size, according to ascending or descending order of X, calculate the variance of the residuals for each group, and examine the relation of these variances to various functions of X, such as $\sqrt{X}$, X, and X^2. This rough analysis frequently will be an adequate guide to decide about the approximate relation between the error variance and the level of X.

Weighted least squares by means of transformations

Many regression packages allow the user to perform a weighted least squares analysis as an option, with the user specifying the weights. When this option is not available, weighted least squares estimators can still be obtained by employing unweighted least squares on specially transformed observations.

To illustrate that weighted least squares is equivalent to unweighted least squares of specially transformed data, we consider again the weighted least squares criterion (5.31) for the case $\sigma_i^2 = kX_i^2$ where $w_i = 1/X_i^2$:

$$Q_w = \Sigma w_i(Y_i - \beta_0 - \beta_1 X_i)^2$$
$$= \Sigma \frac{1}{X_i^2}(Y_i - \beta_0 - \beta_1 X_i)^2$$

so that:

(5.37) $$Q_w = \Sigma\left(\frac{Y_i}{X_i} - \frac{\beta_0}{X_i} - \beta_1\right)^2$$

We shall use the following notation:

(5.38) $$Y_i' = \frac{Y_i}{X_i} \qquad \beta_0' = \beta_1$$
$$X_i' = \frac{1}{X_i} \qquad \beta_1' = \beta_0$$

We can then express (5.37) as follows:

(5.39) $$Q_w = \Sigma(Y_i' - \beta_0' - \beta_1' X_i')^2$$

which is the unweighted least squares criterion for the observations X_i' and Y_i' transformed according to (5.38).

The error variances for the model with the transformed variables are now constant. This can be seen by dividing the terms in the original model:

$$Y_i = \beta_0 + \beta_1 X_i + \varepsilon_i$$

by $\sqrt{w_i} = 1/X_i$:

$$\frac{Y_i}{X_i} = \frac{\beta_0}{X_i} + \beta_1 + \frac{\varepsilon_i}{X_i}$$

so that we obtain, using the notation in (5.38):

(5.40) $$Y_i' = \beta_0' + \beta_1' X_i' + \varepsilon_i'$$

where:

(5.40a) $$\varepsilon_i' = \frac{\varepsilon_i}{X_i}$$

Note that (5.40) is the simple linear regression model using the transformed variables. The variances of the error terms ε_i' in model (5.40) are now constant, as can be seen using (1.15b):

$$\sigma^2(\varepsilon_i') = \sigma^2\left(\frac{\varepsilon_i}{X_i}\right) = \frac{1}{X_i^2}\sigma^2(\varepsilon_i) = \frac{1}{X_i^2}(kX_i^2) = k \tag{5.41}$$

Hence, using unweighted least squares for the dependent variable $Y' = Y/X$ and the independent variable $X' = 1/X$ will yield the same results as a weighted least squares analysis with weights $w_i = 1/X_i^2$. Once the unweighted least squares estimators b_0' and b_1' are obtained for the transformed variables, they are related to the weighted least squares estimators for the original variables using (5.38), so that $b_0' = b_1$ and $b_1' = b_0$.

Appropriate transformations of the observations for other variance relationships, e.g., $\sigma_i^2 = kX_i$, can be found in the same fashion as explained here for $\sigma_i^2 = kX_i^2$.

5.8 INVERSE PREDICTIONS

At times, a regression model of Y on X is used to make a prediction of the value of X which gave rise to a new observation Y. This is known as an *inverse prediction*. We illustrate inverse predictions by two examples:

1. A trade association analyst has regressed the selling price of a product (Y) on its cost (X) for the 15 member firms of the association. The selling price $Y_{h(\text{new})}$ for another firm not belonging to the trade association is known, and it is desired to estimate the cost $X_{h(\text{new})}$ for this firm.

2. A regression analysis of the decrease in cholesterol level (Y) against dosage of a new drug (X) has been conducted, based on observations for 50 patients. A physician is treating a new patient for whom the cholesterol level should decrease by $Y_{h(\text{new})}$. It is desired to estimate the appropriate dosage level $X_{h(\text{new})}$ to be administered to bring about the desired cholesterol level decrease $Y_{h(\text{new})}$.

In inverse predictions, model (3.1) is assumed as before:

$$Y_i = \beta_0 + \beta_1 X_i + \varepsilon_i \tag{5.42}$$

The estimated regression function based on n observations is obtained as usual:

$$\hat{Y} = b_0 + b_1 X \tag{5.43}$$

A new observation $Y_{h(\text{new})}$ becomes available, and it is desired to estimate the level $X_{h(\text{new})}$ which gave rise to this new observation. A natural point estimator is obtained by solving (5.43) for X, given $Y_{h(\text{new})}$:

$$\hat{X}_{h(\text{new})} = \frac{Y_{h(\text{new})} - b_0}{b_1} \qquad b_1 \neq 0 \tag{5.44}$$

where $\hat{X}_{h(\text{new})}$ denotes the point estimator of the new level $X_{h(\text{new})}$. Figure 5.7

contains a representation of this point estimator for an example to be discussed shortly. $\hat{X}_{h(\text{new})}$ is, indeed, the maximum likelihood estimator of $X_{h(\text{new})}$ for regression model (3.1).

It can be shown that approximate $1 - \alpha$ confidence limits for $X_{h(\text{new})}$ are:

$$\hat{X}_{h(\text{new})} \pm t(1 - \alpha/2; n - 2)s(\hat{X}_{h(\text{new})}) \tag{5.45}$$

where:

$$s^2(\hat{X}_{h(\text{new})}) = \frac{MSE}{b_1^2}\left[1 + \frac{1}{n} + \frac{(\hat{X}_{h(\text{new})} - \bar{X})^2}{\Sigma(X_i - \bar{X})^2}\right] \tag{5.45a}$$

Example

A medical researcher in an experiment employed a new method for measuring low concentration of galactose (sugar) in the blood (Y) on 12 samples containing known concentrations (X). Altogether, four concentration levels were used in the experiment. Linear regression model (3.1) was fitted with the following results:

$$n = 12 \qquad b_0 = -.100 \qquad b_1 = 1.017 \qquad MSE = .0272$$
$$s(b_1) = .0142 \qquad \bar{X} = 5.500 \qquad \bar{Y} = 5.492 \qquad \Sigma(X_i - \bar{X})^2 = 135$$
$$\hat{Y} = -.100 + 1.017X$$

The data and the estimated regression line are plotted in Figure 5.7.

FIGURE 5.7 Scatter plot and fitted regression line—calibration example

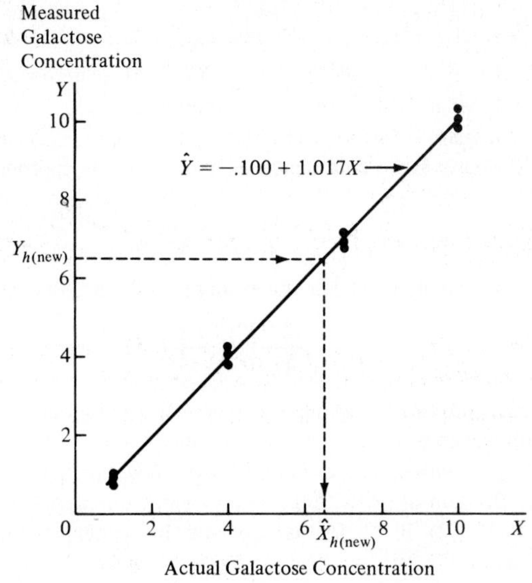

The researcher first wished to make sure that there is a linear association between the two variables. A test of H_0: $\beta_1 = 0$ versus H_a: $\beta_1 \neq 0$ utilizing the test statistic $b_1/s(b_1) = 1.017/.0142 = 71.6$ was conducted for $\alpha = .05$. Since $t(.975; 10) = 2.228$, $|t^*| = 71.6 > 2.228$ and it was concluded that $\beta_1 \neq 0$, or that a linear association exists between the measured concentration and the actual concentration.

The researcher now wishes to use the regression relation for a new patient for whom the new measurement procedure yielded $Y_{h(\text{new})} = 6.52$. It is desired to estimate the actual concentration $X_{h(\text{new})}$ for this patient by means of a 95 percent confidence interval.

Using (5.44) and (5.45a), we obtain:

$$\hat{X}_{h(\text{new})} = \frac{6.52 - (-.100)}{1.017} = 6.509$$

$$s^2(\hat{X}_{h(\text{new})}) = \frac{.0272}{(1.017)^2}\left[1 + \frac{1}{12} + \frac{(6.509 - 5.500)^2}{135}\right] = .0287$$

so that $s(\hat{X}_{h(\text{new})}) = \sqrt{.0287} = .1694$. We require $t(.975; 10) = 2.228$, and using (5.45), we obtain the confidence limits $6.509 \pm 2.228(.1694)$. Hence, the 95 percent confidence interval is:

$$6.13 \leq X_{h(\text{new})} \leq 6.89$$

Thus, it can be concluded with 95 percent confidence that the actual galactose concentration is between 6.13 and 6.89. This is approximately a ± 6 percent error, which was considered reasonable by the researcher.

Comments

1. The inverse prediction problem is also known as a *calibration problem* since it is applicable when inexpensive, quick, and approximate measurements (Y) are related to precise, often expensive, and time-consuming measurements (X) based on n observations. The resulting regression model is then used to estimate for a new approximate measurement $Y_{h(\text{new})}$ what is the precise measurement $X_{h(\text{new})}$. We illustrated this use in our previous example.

2. The approximate confidence interval (5.45) is appropriate if the quantity:

$$\frac{[t(1 - \alpha/2; n - 2)]^2 MSE}{b_1^2 \Sigma(X_i - \bar{X})^2} \tag{5.46}$$

is small, say less than .1. For our example, this quantity is:

$$\frac{(2.228)^2(.0272)}{(1.017)^2(135)} = .00097$$

so that the approximate confidence interval is appropriate.

3. Simultaneous prediction intervals based on g different new observed measurements $Y_{h(\text{new})}$, with a $1 - \alpha$ family confidence coefficient, are easily obtained using either the Bonferroni or the Scheffé procedures discussed in Section 5.4. The value of $t(1 - \alpha/2; n - 2)$ in (5.45) is replaced by either $B = t(1 - \alpha/2g; n - 2)$ or $S = [gF(1 - \alpha; g, n - 2)]^{1/2}$.

5.9 CHOICE OF *X* LEVELS

When regression data are obtained by experiment, the levels of X at which observations on Y are to be taken are under the control of the experimenter. Among other things, the experimenter will have to consider:

1. How many levels of X should be investigated?
2. What shall the two extreme levels be?
3. How shall the other levels of X, if any, be spaced?
4. How many observations should be taken at each level of X?

There is no single answer to these questions, since different purposes of the regression analysis lead to different answers. The possible objectives in regression analysis are varied, as we have noted earlier. The main objective may be to estimate the slope of the regression line, or in some cases to estimate the intercept. In many cases, the main objective is to predict one or more new observations or to estimate one or more mean responses. When the regression function is curvilinear, the main objective may be to locate the maximum or minimum mean response. At still other times, the main purpose is to determine the nature of the regression function.

To illustrate how the purpose affects the design, consider the variances of b_0, b_1, $\hat{Y}_h$, and $Y_{h(\text{new})}$ which were developed earlier:

$$\sigma^2(b_0) = \sigma^2 \frac{\Sigma X_i^2}{n\Sigma(X_i - \bar{X})^2} = \sigma^2\left[\frac{1}{n} + \frac{\bar{X}^2}{\Sigma(X_i - \bar{X})^2}\right] \tag{5.47}$$

$$\sigma^2(b_1) = \frac{\sigma^2}{\Sigma(X_i - \bar{X})^2} \tag{5.48}$$

$$\sigma^2(\hat{Y}_h) = \sigma^2\left[\frac{1}{n} + \frac{(X_h - \bar{X})^2}{\Sigma(X_i - \bar{X})^2}\right] \tag{5.49}$$

$$\sigma^2(Y_{h(\text{new})}) = \sigma^2\left[1 + \frac{1}{n} + \frac{(X_h - \bar{X})^2}{\Sigma(X_i - \bar{X})^2}\right] \tag{5.50}$$

The variance of the slope is minimized if $\Sigma(X_i - \bar{X})^2$ is maximized. This is accomplished by using two levels of X, at the two extremes for the scope of the model, and placing half of the observations at each of the two levels. Of course, if one were not sure of the linearity of the regression function, one would be hesitant to use only two levels since they would provide no information about possible departures from linearity.

If the main purpose is to estimate β_0, the number and placement of levels does not matter as long as $\bar{X} = 0$. On the other hand, to estimate the mean response or predict a new observation at X_h, it is best to use X levels so that $\bar{X} = X_h$. If a number of mean responses are to be estimated or a number of new observations are to be predicted, it would be best to spread out the X levels such that $\bar{X}$ is in the center of the X_h levels of interest.

Although the number and spacing of X levels depends very much on the major purpose of the regression analysis, some general advice can be given, at least to be used as a point of departure. D. R. Cox, a well-known statistician, suggests as follows:

> Use two levels when the object is primarily to examine whether or not . . . (the independent variable) . . . has an effect and in which direction that effect is. Use three levels whenever a description of the response curve by its slope and curvature is likely to be adequate; this should cover most cases. Use four levels if further examination of the shape of the response curve is important. Use more than four levels when it is required to estimate the detailed shape of the response curve, or when the curve is expected to rise to an asymptotic value, or in general to show features not adequately described by slope and curvature. Except in these last cases it is generally satisfactory to use equally spaced levels with equal numbers of observations per level [Ref. 5.5].

PROBLEMS

5.1. When joint confidence intervals for β_0 and β_1 are developed by the Bonferroni method with a family confidence coefficient of 90 percent, does this imply that 10 percent of the time the confidence intervals for β_0 will be incorrect? That 5 percent of the time the confidence interval for β_0 will be incorrect and 5 percent of the time that for β_1 will be incorrect? Discuss.

5.2. Refer to Problem 3.1. Suppose the student combines the two confidence intervals into a confidence set. What can you say about the family confidence coefficient for this set?

5.3. Refer to **Calculator maintenance** Problems 2.16 and 3.5.

a. Will b_0 and b_1 tend to err in the same direction or in opposite directions here? What does this imply about the tilt of the elliptical joint confidence region for β_0 and β_1 here?

b. Obtain a few boundary points of the joint confidence region for β_0 and β_1 and plot the boundary. Use a 95 percent confidence coefficient. Interpret your joint confidence region.

c. A consultant has suggested that β_0 should be zero and β_1 should equal 14.0. Does your joint confidence region support this view?

d. Obtain Bonferroni joint confidence intervals for β_0 and β_1 using a 95 percent family confidence coefficient. Do these intervals support the view of the consultant in part (c)?

5.4. Refer to **Airfreight breakage** Problem 2.17.

a. Will b_0 and b_1 tend to err in the same direction or in opposite directions here? What does this imply about the tilt of the elliptical joint confidence region for β_0 and β_1 here?

b. Obtain a few boundary points of the joint confidence region for β_0 and β_1 and plot the boundary. Use a 99 percent confidence coefficient. Interpret your joint confidence region.

c. Obtain Bonferroni joint confidence intervals for β_0 and β_1 using a 99 percent family confidence coefficient. Do the Bonferroni joint confidence intervals provide substantially less precise information than the joint confidence region in part (b)?

5.5. Refer to **Plastic hardness** Problem 2.18.

a. Obtain Bonferroni joint confidence intervals for β_0 and β_1 using a 90 percent family confidence coefficient. Interpret your confidence intervals.

b. Are b_0 and b_1 positively or negatively correlated here? Is this reflected in your joint confidence intervals in part (a)?

c. What is the meaning of the family confidence coefficient in part (a)?

5.6. Refer to **Muscle mass** Problem 2.23.

a. Obtain Bonferroni joint confidence intervals for β_0 and β_1 using a 99 percent family confidence coefficient. Interpret your confidence intervals.

b. Will b_0 and b_1 tend to err in the same direction or in opposite directions here? Explain.

c. A researcher has suggested that β_0 should equal approximately 160 and that β_1 should be between -1.9 and -1.5. Do the joint confidence intervals in part (a) support this expectation?

5.7. Refer to **Airfreight breakage** Problem 2.17. Obtain selected boundary points for the 90 percent confidence band for the true regression line. Plot the confidence band, together with the estimated regression line. How precisely do we estimate the location of the true regression line here?

5.8. Refer to **Plastic hardness** Problem 2.18. Obtain selected boundary points for the 95 percent confidence band for the true regression line. Plot the confidence band, together with the estimated regression line. How precisely do we estimate the location of the true regression line here?

5.9. Refer to **Calculator maintenance** Problems 2.16 and 3.5.

a. Estimate the expected number of minutes spent when there are 3, 5, and 7 machines to be serviced, respectively. Use interval estimates with a 90 percent family confidence coefficient based on the Working-Hotelling approach.

b. Two service calls for preventive maintenance are scheduled in which the numbers of machines to be serviced are 4 and 7, respectively. A family of prediction intervals for the times to be spent on these calls is desired with a 90 percent family confidence coefficient. Which procedure, Scheffé or Bonferroni, will provide tighter prediction limits here?

c. Obtain the family of prediction limits required in part (b) using the more efficient procedure.

5.10. Refer to **Airfreight breakage** Problem 2.17.

a. It is desired to obtain interval estimates of the mean number of broken ampules when there are 0, 1, and 2 transfers for the shipment using a 95 percent family confidence coefficient. Obtain the desired confidence intervals using the Working-Hotelling approach.

b. Are the confidence intervals obtained in part (a) more efficient than Bonferroni intervals here? Explain.

c. The next three shipments will make 0, 1, and 2 transfers, respectively. Obtain prediction limits for the number of broken ampules for each of these three shipments using the Scheffé procedure and a 95 percent family confidence coefficient.

d. Would the Bonferroni procedure have been more efficient in developing the prediction intervals in part (c)? Explain.

5.11. Refer to **Plastic hardness** Problem 2.18.

a. Management wishes to obtain interval estimates of the mean hardness when the elapsed time is 40, 50, and 60 hours, respectively. Calculate the desired confidence intervals using the Bonferroni procedure and a 90 percent family confidence coefficient. What is the meaning of the family confidence coefficient here?

b. Is the Bonferroni procedure employed in part (a) the most efficient one that could be employed here? Explain.

c. The next two test items will be measured after 30 and 40 hours of elapsed time, respectively. Predict the hardness for each of these two items using the most efficient procedure and a 90 percent family confidence coefficient.

5.12. Refer to **Muscle mass** Problem 2.23.

a. The nutritionist is particularly interested in the mean muscle mass for women aged 45, 55, and 65. Obtain joint confidence intervals for the means of interest using the Working-Hotelling procedure and a 95 percent family confidence coefficient.

b. Is the Working-Hotelling approach the most efficient one to be employed in part (a)? Explain.

c. Three additional women aged 48, 59, and 74 have contacted the nutritionist. Predict the muscle mass for each of these three women using the Bonferroni approach and a 95 percent family confidence coefficient.

d. Subsequently, the nutritionist wishes to predict the muscle mass for a fourth woman aged 64, with a family confidence coefficient of 95 percent for the four predictions. Will the three prediction intervals in part (c) have to be recalculated? Would this also be true if the Scheffé method had been used in constructing the prediction intervals?

5.13. A behavioral scientist stated recently: "I am never sure whether the regression line goes through the origin. Hence, I will not use such a model." Comment.

5.14. **Typographical errors.** Shown below are the number of galleys of type set (X) and the dollar cost of correcting typographical errors (Y) in a random sample of recent printing orders handled by a firm specializing in technical printing. Since Y involves variable costs only, an analyst wished to determine whether the regression through the origin model (5.13) is apt for studying the relation between the two variables.

i:	1	2	3	4	5	6	7	8	9	10	11	12
X_i:	7	12	10	10	14	25	30	25	18	10	4	6
Y_i:	128	213	191	178	250	446	540	457	324	177	75	107

a. Fit model (5.13) and state the estimated regression function.

b. Plot the estimated regression function and the data. Does a linear regression function through the origin appear to provide a good fit here? Comment.

c. In estimating costs of handling prospective orders, management has used a standard of \$17.50 per galley for the cost of correcting typographical errors. Test whether or not this standard should be revised; use $\alpha = .02$. State the alternatives, decision rule, and conclusion.

d. Obtain a prediction interval for the correction cost on a forthcoming job involving 10 galleys. Use a confidence coefficient of 98 percent.

5.15. Refer to **Typographical errors** Problem 5.14. Conduct a formal test for lack of fit of linear regression through the origin; use $\alpha = .01$. State the alternatives, decision rule, and conclusion.

5.16. Refer to **Grade point average** Problem 2.15. Assume that linear regression through the origin model (5.13) is appropriate.

a. Fit model (5.13) and state the estimated regression function.

b. Estimate β_1 with a 95 percent confidence interval. Interpret your interval estimate.

c. Estimate the mean freshman GPA for students whose entrance test score is 5.7. Use a 95 percent confidence interval.

5.17. Refer to **Grade point average** Problem 5.16.

a. Plot the fitted regression line and the data. Does the linear regression function through the origin appear to be a good fit here?

b. Conduct a formal test for lack of fit of linear regression through the origin; use $\alpha = .005$. State the alternatives, decision rule, and conclusion. What is the P-value of the test?

5.18. Refer to **Calculator maintenance** Problem 2.16. Assume that linear regression through the origin model (5.13) is appropriate.

a. Obtain the estimated regression function.

b. Estimate β_1 with a 90 percent confidence interval. Interpret your interval estimate.

c. Predict the service time on a new call in which six machines are to be serviced. Use a 90 percent prediction interval.

5.19. Refer to **Calculator maintenance** Problem 5.18.

a. Plot the fitted regression line and the data. Does the linear regression function through the origin appear to be a good fit here?

b. Conduct a formal test for lack of fit of linear regression through the origin; use $\alpha = .01$. State the alternatives, decision rule, and conclusion. What is the P-value of the test?

5.20. Refer to **Plastic hardness** Problem 2.18. Suppose that errors arise in X because the laboratory technician is instructed to measure the hardness of the ith specimen (Y_i) at a prerecorded elapsed time (X_i), but the timing is imperfect so the true elapsed time varies at random from the prerecorded elapsed time. Will ordinary least squares estimates be biased here? Discuss.

5.21. **Computer-assisted learning.** Data from a study of computer-assisted learning by 12 students, showing the total number of responses in completing a lesson (X) and the cost of computer time (Y, in cents), follow.

i:	1	2	3	4	5	6	7	8	9	10	11	12
X_i:	16	14	22	10	14	17	10	13	19	12	18	11
Y_i:	77	70	85	50	62	70	52	63	88	57	81	54

a. Fit a linear regression function by ordinary least squares, obtain the residuals, and prepare a plot of the residuals against X. What does the residual plot suggest?

b. Assume that $\sigma_i^2 = kX_i^2$, and use weighted least squares to fit the linear regression function. Are the regression coefficients similar to those obtained in part (a) with ordinary least squares? How do the standard deviations of the regression coefficients compare?

c. What transformation of the variables, if used with ordinary least squares, would have yielded the same regression coefficients as weighted least squares here?

5.22. Refer to **Machine speed** Problem 4.21.

a. Verify that the relation $\sigma_i = kX_i$ holds approximately here.

b. Use weighted least squares to fit the linear regression function.

c. Obtain an interval estimate of β_1 with a 95 percent confidence coefficient.

d. What transformation of the variables when used with ordinary least squares would have yielded the same regression coefficients as weighted least squares here?

5.23. Refer to **Grade point average** Problems 2.15 and 3.4. A new student earned a grade point average of 3.4 in the freshman year.

a. Obtain a 90 percent confidence interval for the student's entrance test score. Interpret your confidence interval.

b. Is criterion (5.46) met as to the appropriateness of the approximate confidence interval?

5.24. Refer to **Plastic hardness** Problem 2.18. The measurement of a new test item showed 298 Brinell units of hardness.

a. Obtain a 99 percent confidence interval for the elapsed time before the hardness was measured. Interpret your confidence interval.

b. Is criterion (5.46) met as to the appropriateness of the approximate confidence interval?

EXERCISES

5.25. What simplification takes place in the boundary of the joint confidence region (5.2) for β_0 and β_1 if the independent variable is so coded that $\bar{X} = 0$? What happens to the shape of the boundary?

5.26. If the independent variable is so coded that $\bar{X} = 0$ and the normal error model (3.1) applies, are b_0 and b_1 independent? Are the confidence intervals for β_0 and β_1 then independent?

5.27. Derive an extension of the Bonferroni inequality (5.6a) for the case of three statements, each with statement confidence coefficient $1 - \alpha$.

5.28. Show that for the fitted least squares regression line through the origin (5.18), $\Sigma X_i e_i = 0$.

5.29. Show that $\hat{Y}$ as defined in (5.18) for linear regression through the origin is an unbiased estimator of $E(Y)$.

5.30. Derive the formula for $s^2(\hat{Y}_h)$ given in Table 5.1 for linear regression through the origin.

5.31. (Calculus needed.) Derive the weighted least squares normal equations for fitting a linear regression function when the variance of the error terms is proportional to

X—i.e., $\sigma^2(\varepsilon_i) = kX_i$. Check your results by making appropriate substitutions in (5.32).

5.32. Express the weighted least squares estimator b_1 in (5.33a) in terms of the deviations $X_i - \bar{X}_w$ and $Y_i - \bar{Y}_w$, where $\bar{X}_w$ and $\bar{Y}_w$ are weighted means.

PROJECTS

5.33. Refer to the **SMSA** data set and Project 2.38. Consider the regression relation of number of active physicians to total population.

a. Obtain a joint confidence region for β_0 and β_1 and plot it; use a 95 percent confidence coefficient.

b. Plot the 95 percent confidence band for the regression line. Does it provide a fairly precise indication of the location of the regression line?

5.34. Refer to the **SENIC** data set and Project 2.40. Consider the regression relation of length of stay to infection risk.

a. Obtain a joint confidence region for β_0 and β_1 and plot it; use a 90 percent confidence coefficient.

b. Plot the 90 percent confidence band for the regression line. Does it provide a fairly precise indication of the location of the regression line?

5.35. Five observations on Y are to be taken when $X = 10, 20, 30, 40,$ and 50, respectively. The true regression function is $E(Y) = 20 + 10X$. The error terms are independent and normally distributed, with $E(\varepsilon_i) = 0$ and $\sigma^2(\varepsilon_i) = .8X_i$.

a. Generate a random Y observation for each X level, and calculate both the ordinary and weighted least squares estimates of the regression coefficient β_1 in linear regression model (3.1).

b. Repeat part (a) 200 times, generating new random numbers each time.

c. Calculate the mean and variance of the 200 ordinary least squares estimates of β_1, and do the same for the 200 weighted least squares estimates.

d. Do both the ordinary least squares and weighted least squares estimators appear to be unbiased? Explain. Which estimator appears to be more precise here? Comment.

CITED REFERENCES

5.1 Gafarian, A. V. "Confidence Bands in Straight Line Regression." *Journal of the American Statistical Association* 59 (1964), pp. 182–213.

5.2 Miller, Rupert G., Jr. *Simultaneous Statistical Inference*. 2d ed. New York: Springer-Verlag, 1981, pp. 114–16.

5.3 Johnston, J. *Econometric Methods*. 2d ed. New York: McGraw-Hill, 1971.

5.4 Berkson, J. "Are There Two Regressions?" *Journal of the American Statistical Association* 45 (1950), pp. 164–80.

5.5 Cox, D. R. *Planning of Experiments*. New York: John Wiley & Sons, 1958, pp. 141–42.

PART II

General regression and correlation analysis

6

Matrix approach to simple regression analysis

Matrix algebra is widely used for mathematical and statistical analysis. The matrix approach is practically a necessity in multiple regression analysis, since it permits extensive systems of equations and large arrays of data to be denoted compactly and operated upon efficiently.

In this chapter, we first take up a brief introduction to matrix algebra. (A fuller treatment of matrix algebra may be found in specialized texts such as Reference 6.1.) Then we apply matrix methods to the simple linear regression model discussed in Part I. While matrix algebra is not really required for simple regression with one independent variable, the application of matrix methods to this case will provide a transition to multiple regression which will be taken up in succeeding chapters.

Readers who are familiar with matrix algebra may wish to scan the introductory parts of this chapter and focus upon the parts dealing with the use of matrix methods in regression analysis.

6.1 MATRICES

Definition of matrix

A matrix is a rectangular array of elements arranged in rows and columns. An example of a matrix is:

$$
\begin{array}{c} \\ \text{Row 1} \\ \text{Row 2} \\ \text{Row 3} \end{array}
\begin{array}{c}
\begin{array}{cc} \text{Column} & \text{Column} \\ 1 & 2 \end{array} \\
\begin{bmatrix} 6{,}000 & 23 \\ 13{,}000 & 47 \\ 11{,}000 & 35 \end{bmatrix}
\end{array}
$$

The *elements* of this particular matrix are numbers representing income (column 1) and age (column 2) of three persons. The elements are arranged by row (person) and column (characteristic of person). Thus, the element in the first row and first column (6,000) represents the income of the first person. The element in the first row and second column (23) represents the age of the first person. The *dimension* of the matrix is 3×2, i.e., 3 rows by 2 columns. If we wanted to present income and age for 1,000 persons in a matrix with the same format as the one earlier, we would require a $1{,}000 \times 2$ matrix.

Other examples of matrices are:

$$
\begin{bmatrix} 1 & 0 \\ 5 & 10 \end{bmatrix} \qquad \begin{bmatrix} 4 & 7 & 12 & 16 \\ 3 & 15 & 9 & 8 \end{bmatrix}
$$

These two matrices have dimensions of 2×2 and 2×4, respectively. Note that we always specify the number of rows first and then the number of columns in giving the dimension of a matrix.

As in ordinary algebra, we may use symbols to identify the elements of a matrix:

$$
\begin{array}{c} \\ i = 1 \\ i = 2 \end{array}
\begin{array}{c}
\begin{array}{ccc} j = 1 & j = 2 & j = 3 \end{array} \\
\begin{bmatrix} a_{11} & a_{12} & a_{13} \\ a_{21} & a_{22} & a_{23} \end{bmatrix}
\end{array}
$$

Note that the first subscript identifies the row number and the second the column number. We shall use the general notation a_{ij} for the element in the ith row and the jth column. In our above example, $i = 1, 2$ and $j = 1, 2, 3$.

A matrix may be denoted by a symbol such as **A**, **X**, or **Z**. The symbol is in **boldface** to identify that it refers to a matrix. Thus, we might define for the above matrix:

$$
\mathbf{A} = \begin{bmatrix} a_{11} & a_{12} & a_{13} \\ a_{21} & a_{22} & a_{23} \end{bmatrix}
$$

Reference to the matrix **A** then implies reference to the 2×3 array just given.

Another notation for the matrix **A** just given is:

$$
\mathbf{A} = [a_{ij}] \qquad i = 1, 2; j = 1, 2, 3
$$

which avoids the need for writing out all elements of the matrix by stating only the general element. This notation can only be used, of course, when the elements of a matrix are symbols.

To summarize, a matrix with r rows and c columns will be represented either in full:

(6.1)
$$\mathbf{A} = \begin{bmatrix} a_{11} & a_{12} & \cdots & a_{1j} & \cdots & a_{1c} \\ a_{21} & a_{22} & \cdots & a_{2j} & \cdots & a_{2c} \\ \cdot & \cdot & & \cdot & & \cdot \\ \cdot & \cdot & & \cdot & & \cdot \\ \cdot & \cdot & & \cdot & & \cdot \\ a_{i1} & a_{i2} & \cdots & a_{ij} & \cdots & a_{ic} \\ \cdot & \cdot & & \cdot & & \cdot \\ \cdot & \cdot & & \cdot & & \cdot \\ \cdot & \cdot & & \cdot & & \cdot \\ a_{r1} & a_{r2} & \cdots & a_{rj} & \cdots & a_{rc} \end{bmatrix}$$

or in the abbreviated form:

(6.2)
$$\mathbf{A} = [a_{ij}] \qquad i = 1, \ldots, r; j = 1, \ldots, c$$

or simply by a boldface symbol, such as **A**.

Comments

1. Do not think of a matrix as a number. It is a set of elements arranged in an array. Only when the matrix has dimension 1×1 is there a single number in the matrix, in which case one *can* think of it interchangeably as either a matrix or a number.
2. The following is *not* a matrix:

$$\begin{bmatrix} & 14 & \\ & 8 & \\ 10 & & 15 \\ 9 & & 16 \end{bmatrix}$$

since the numbers are not arranged in columns and rows.

Square matrix

A matrix is said to be square if the number of rows equals the number of columns. Two examples are:

$$\begin{bmatrix} 4 & 7 \\ 3 & 9 \end{bmatrix} \qquad \begin{bmatrix} a_{11} & a_{12} & a_{13} \\ a_{21} & a_{22} & a_{23} \\ a_{31} & a_{32} & a_{33} \end{bmatrix}$$

Vector

A matrix containing only one column is called a *column vector* or simply a *vector*. Two examples are:

$$\mathbf{A} = \begin{bmatrix} 4 \\ 7 \\ 10 \end{bmatrix} \qquad \mathbf{C} = \begin{bmatrix} c_1 \\ c_2 \\ c_3 \\ c_4 \\ c_5 \end{bmatrix}$$

The vector **A** is a 3×1 matrix, and the vector **C** is a 5×1 matrix.

A matrix containing only one row is called a *row vector*. Two examples are:

$$\mathbf{B}' = [15 \quad 25 \quad 50] \qquad \mathbf{F}' = [f_1 \quad f_2]$$

We use the prime symbol for row vectors for reasons to be seen shortly. Note that the row vector $\mathbf{B}'$ is a 1×3 matrix and the row vector $\mathbf{F}'$ is a 1×2 matrix.

A single subscript suffices to identify the elements of a vector.

Transpose

The transpose of a matrix $\mathbf{A}$ is another matrix, denoted by $\mathbf{A}'$, that is obtained by interchanging corresponding columns and rows of the matrix $\mathbf{A}$.

For example, if:

$$\underset{3\times 2}{\mathbf{A}} = \begin{bmatrix} 2 & 5 \\ 7 & 10 \\ 3 & 4 \end{bmatrix}$$

then the transpose $\mathbf{A}'$ is:

$$\underset{2\times 3}{\mathbf{A}'} = \begin{bmatrix} 2 & 7 & 3 \\ 5 & 10 & 4 \end{bmatrix}$$

Note that the first column of $\mathbf{A}$ is the first row of $\mathbf{A}'$, and similarly the second column of $\mathbf{A}$ is the second row of $\mathbf{A}'$. Correspondingly, the first row of $\mathbf{A}$ has become the first column of $\mathbf{A}'$, and so on. Note that the dimension of $\mathbf{A}$, indicated under the symbol $\mathbf{A}$, becomes reversed for the dimension of $\mathbf{A}'$.

As another example, consider:

$$\underset{3\times 1}{\mathbf{C}} = \begin{bmatrix} 4 \\ 7 \\ 10 \end{bmatrix} \qquad \underset{1\times 3}{\mathbf{C}'} = [4 \quad 7 \quad 10]$$

Thus, the transpose of a column vector is a row vector, and vice versa. This is the reason why we used the symbol $\mathbf{B}'$ earlier to identify a row vector, since it may be thought of as the transpose of a column vector $\mathbf{B}$.

In general, we have:

(6.3)

$$\underset{r\times c}{\mathbf{A}} = \begin{bmatrix} a_{11} & \cdots & a_{1c} \\ \cdot & & \cdot \\ \cdot & & \cdot \\ \cdot & & \cdot \\ a_{r1} & \cdots & a_{rc} \end{bmatrix} = \underset{\substack{\text{Row index} \quad \text{Column index}}}{[a_{ij}]} \qquad i = 1, \ldots, r; j = 1, \ldots, c$$

$$\underset{c\times r}{\mathbf{A}'} = \begin{bmatrix} a_{11} & \cdots & a_{r1} \\ \cdot & & \cdot \\ \cdot & & \cdot \\ \cdot & & \cdot \\ a_{1c} & \cdots & a_{rc} \end{bmatrix} = \underset{\substack{\text{Row index} \quad \text{Column index}}}{[a_{ji}]} \qquad j = 1, \ldots, c; i = 1, \ldots, r$$

Thus, the element in the ith row and jth column in **A** is found in the jth row and ith column in $\mathbf{A}'$.

Equality of matrices

Two matrices **A** and **B** are said to be equal if they have the same dimension and if all corresponding elements are equal. Conversely, if two matrices are equal, their corresponding elements are equal. For example, if:

$$\mathbf{A} = \begin{bmatrix} a_1 \\ a_2 \\ a_3 \end{bmatrix} \qquad \mathbf{B} = \begin{bmatrix} 4 \\ 7 \\ 3 \end{bmatrix}$$

then $\mathbf{A} = \mathbf{B}$ implies:

$$a_1 = 4$$
$$a_2 = 7$$
$$a_3 = 3$$

Similarly, if:

$$\mathbf{A} = \begin{bmatrix} a_{11} & a_{12} \\ a_{21} & a_{22} \\ a_{31} & a_{32} \end{bmatrix} \qquad \mathbf{B} = \begin{bmatrix} 17 & 2 \\ 14 & 5 \\ 13 & 9 \end{bmatrix}$$

then $\mathbf{A} = \mathbf{B}$ implies:

$$a_{11} = 17 \qquad a_{12} = 2$$
$$a_{21} = 14 \qquad a_{22} = 5$$
$$a_{31} = 13 \qquad a_{32} = 9$$

Regression examples

In regression analysis, one basic matrix is the vector **Y**, consisting of the n observations on the dependent variable:

(6.4)
$$\underset{n\times 1}{\mathbf{Y}} = \begin{bmatrix} Y_1 \\ Y_2 \\ \cdot \\ \cdot \\ \cdot \\ Y_n \end{bmatrix}$$

Note that the transpose $\mathbf{Y}'$ is the row vector:

(6.5)
$$\underset{1\times n}{\mathbf{Y}'} = [Y_1 \quad Y_2 \quad \cdots \quad Y_n]$$

Another basic matrix in regression analysis is the **X** matrix, which is defined as follows for simple regression analysis:

(6.6)
$$\underset{n\times 2}{\mathbf{X}} = \begin{bmatrix} 1 & X_1 \\ 1 & X_2 \\ \cdot & \cdot \\ \cdot & \cdot \\ \cdot & \cdot \\ 1 & X_n \end{bmatrix}$$

The matrix **X** consists of a column of 1's and a column containing the n values of the independent variable X. Note that the transpose of **X** is:

(6.7)
$$\underset{2\times n}{\mathbf{X}'} = \begin{bmatrix} 1 & 1 & \cdots & 1 \\ X_1 & X_2 & \cdots & X_n \end{bmatrix}$$

For the Westwood Company lot size example, the **Y** and **X** matrices are (Table 2.1):

$$\mathbf{Y} = \begin{bmatrix} 73 \\ 50 \\ \cdot \\ \cdot \\ \cdot \\ 132 \end{bmatrix} \qquad \mathbf{X} = \begin{bmatrix} 1 & 30 \\ 1 & 20 \\ \cdot & \cdot \\ \cdot & \cdot \\ \cdot & \cdot \\ 1 & 60 \end{bmatrix}$$

6.2 MATRIX ADDITION AND SUBTRACTION

Adding or subtracting two matrices requires that they have the same dimension. The sum, or difference, of two matrices is another matrix whose elements each consists of the sum, or difference, of the corresponding elements of the two matrices. Suppose:

$$\underset{3\times 2}{\mathbf{A}} = \begin{bmatrix} 1 & 4 \\ 2 & 5 \\ 3 & 6 \end{bmatrix} \qquad \underset{3\times 2}{\mathbf{B}} = \begin{bmatrix} 1 & 2 \\ 2 & 3 \\ 3 & 4 \end{bmatrix}$$

then:

$$\underset{3\times 2}{\mathbf{A} + \mathbf{B}} = \begin{bmatrix} 1+1 & 4+2 \\ 2+2 & 5+3 \\ 3+3 & 6+4 \end{bmatrix} = \begin{bmatrix} 2 & 6 \\ 4 & 8 \\ 6 & 10 \end{bmatrix}$$

Similarly:

$$\underset{3\times 2}{\mathbf{A} - \mathbf{B}} = \begin{bmatrix} 1-1 & 4-2 \\ 2-2 & 5-3 \\ 3-3 & 6-4 \end{bmatrix} = \begin{bmatrix} 0 & 2 \\ 0 & 2 \\ 0 & 2 \end{bmatrix}$$

In general, if:

$$\underset{r\times c}{\mathbf{A}} = [a_{ij}] \qquad \underset{r\times c}{\mathbf{B}} = [b_{ij}] \qquad i = 1, \ldots, r; j = 1, \ldots, c$$

then:

$$\underset{r\times c}{\mathbf{A}+\mathbf{B}} = [a_{ij} + b_{ij}] \quad \text{and} \quad \underset{r\times c}{\mathbf{A}-\mathbf{B}} = [a_{ij} - b_{ij}] \tag{6.8}$$

Formula (6.8) generalizes in an obvious way to addition and subtraction of more than two matrices. Note also that $\mathbf{A} + \mathbf{B} = \mathbf{B} + \mathbf{A}$, as in ordinary algebra.

Regression example

The regression model:

$$Y_i = E(Y_i) + \varepsilon_i \qquad i = 1, \ldots, n$$

can be written compactly in matrix notation. First, let us define the vector of mean responses:

$$\underset{n\times 1}{\mathbf{E(Y)}} = \begin{bmatrix} E(Y_1) \\ E(Y_2) \\ \cdot \\ \cdot \\ \cdot \\ E(Y_n) \end{bmatrix} \tag{6.9}$$

and the vector of the error terms:

$$\underset{n\times 1}{\boldsymbol{\varepsilon}} = \begin{bmatrix} \varepsilon_1 \\ \varepsilon_2 \\ \cdot \\ \cdot \\ \cdot \\ \varepsilon_n \end{bmatrix} \tag{6.10}$$

Recalling the definition of the $\mathbf{Y}$ observation vector (6.4), we can write the regression model as follows:

$$\mathbf{Y} = \mathbf{E(Y)} + \boldsymbol{\varepsilon}$$

because:

$$\begin{bmatrix} Y_1 \\ Y_2 \\ \cdot \\ \cdot \\ \cdot \\ Y_n \end{bmatrix} = \begin{bmatrix} E(Y_1) \\ E(Y_2) \\ \cdot \\ \cdot \\ \cdot \\ E(Y_n) \end{bmatrix} + \begin{bmatrix} \varepsilon_1 \\ \varepsilon_2 \\ \cdot \\ \cdot \\ \cdot \\ \varepsilon_n \end{bmatrix} = \begin{bmatrix} E(Y_1) + \varepsilon_1 \\ E(Y_2) + \varepsilon_2 \\ \cdot \\ \cdot \\ \cdot \\ E(Y_n) + \varepsilon_n \end{bmatrix}$$

Thus, the observations vector $\mathbf{Y}$ equals the sum of two vectors, a vector containing the expected values and another containing the error terms.

6.3 MATRIX MULTIPLICATION

Multiplication of a matrix by a scalar

A scalar is an ordinary number or a symbol representing a number. We frequently encounter multiplication of a matrix by a scalar. In this, every element of the matrix is multiplied by the scalar. For example, suppose the matrix **A** is given by:

$$\mathbf{A} = \begin{bmatrix} 2 & 7 \\ 9 & 3 \end{bmatrix}$$

Then 4**A**, where 4 is the scalar, equals:

$$4\mathbf{A} = 4\begin{bmatrix} 2 & 7 \\ 9 & 3 \end{bmatrix} = \begin{bmatrix} 8 & 28 \\ 36 & 12 \end{bmatrix}$$

Similarly, $\lambda\mathbf{A}$ equals:

$$\lambda\mathbf{A} = \lambda\begin{bmatrix} 2 & 7 \\ 9 & 3 \end{bmatrix} = \begin{bmatrix} 2\lambda & 7\lambda \\ 9\lambda & 3\lambda \end{bmatrix}$$

where λ denotes the scalar.

If every element of a matrix has a common factor, this factor can be taken outside the matrix and treated as a scalar. For example:

$$\begin{bmatrix} 9 & 27 \\ 15 & 18 \end{bmatrix} = 3\begin{bmatrix} 3 & 9 \\ 5 & 6 \end{bmatrix}$$

Similarly:

$$\begin{bmatrix} \frac{5}{\lambda} & \frac{2}{\lambda} \\ \frac{3}{\lambda} & \frac{8}{\lambda} \end{bmatrix} = \frac{1}{\lambda}\begin{bmatrix} 5 & 2 \\ 3 & 8 \end{bmatrix}$$

In general, if $\mathbf{A} = [a_{ij}]$ and λ is a scalar, we have:

(6.11) $$\lambda\mathbf{A} = \mathbf{A}\lambda = [\lambda a_{ij}]$$

Multiplication of a matrix by a matrix

Multiplication of a matrix by a matrix may appear somewhat complicated at first, but a little practice will make it into a routine operation.

Consider the two matrices:

$$\underset{2\times2}{\mathbf{A}} = \begin{bmatrix} 2 & 5 \\ 4 & 1 \end{bmatrix} \qquad \underset{2\times2}{\mathbf{B}} = \begin{bmatrix} 4 & 6 \\ 5 & 8 \end{bmatrix}$$

The product **AB** will be a 2×2 matrix whose elements are obtained by finding the cross products of rows of **A** with columns of **B** and summing the cross products. For instance, to find the element in the first row and first column of the product **AB**, we work with the first row of **A** and the first column of **B**, as follows:

$$\begin{matrix} & \mathbf{A} & \mathbf{B} & & \mathbf{AB} \\ \begin{matrix}\text{Row 1}\\[1em] \text{Row 2}\end{matrix} & \begin{bmatrix} \boxed{2 \quad 5} \\ 4 \quad\ 1 \end{bmatrix} & \begin{bmatrix} \boxed{\begin{matrix}4\\5\end{matrix}} & \begin{matrix}6\\8\end{matrix} \end{bmatrix} & \text{Row 1} & \begin{bmatrix} 33 & \ \\ \ & \ \end{bmatrix} \\ & & \begin{matrix}\text{Column} & \text{Column}\\ 1 & 2\end{matrix} & & \begin{matrix}\text{Column}\\ 1\end{matrix} \end{matrix}$$

We take the cross products and sum:

$$2(4) + 5(5) = 33$$

The number 33 is the element in the first row and first column of the matrix **AB**.

To find the element in the first row and second column of **AB**, we work with the first row of **A** and the second column of **B**:

$$\begin{matrix} & \mathbf{A} & \mathbf{B} & & \mathbf{AB} \\ \begin{matrix}\text{Row 1}\\[1em] \text{Row 2}\end{matrix} & \begin{bmatrix} \boxed{2 \quad 5} \\ 4 \quad\ 1 \end{bmatrix} & \begin{bmatrix} \begin{matrix}4\\5\end{matrix} & \boxed{\begin{matrix}6\\8\end{matrix}} \end{bmatrix} & \text{Row 1} & \begin{bmatrix} 33 & 52 \\ \ & \ \end{bmatrix} \\ & & \begin{matrix}\text{Column} & \text{Column}\\ 1 & 2\end{matrix} & & \begin{matrix}\text{Column} & \text{Column}\\ 1 & 2\end{matrix} \end{matrix}$$

The sum of the cross products is:

$$2(6) + 5(8) = 52$$

Continuing this process, we find the product **AB** to be:

$$\underset{2\times 2}{\mathbf{AB}} = \begin{bmatrix} 2 & 5 \\ 4 & 1 \end{bmatrix} \begin{bmatrix} 4 & 6 \\ 5 & 8 \end{bmatrix} = \begin{bmatrix} 33 & 52 \\ 21 & 32 \end{bmatrix}$$

Let us consider another example:

$$\underset{2\times 3}{\mathbf{A}} = \begin{bmatrix} 1 & 3 & 4 \\ 0 & 5 & 8 \end{bmatrix} \qquad \underset{3\times 1}{\mathbf{B}} = \begin{bmatrix} 3 \\ 5 \\ 2 \end{bmatrix}$$

$$\underset{2\times 1}{\mathbf{AB}} = \begin{bmatrix} 1 & 3 & 4 \\ 0 & 5 & 8 \end{bmatrix} \begin{bmatrix} 3 \\ 5 \\ 2 \end{bmatrix} = \begin{bmatrix} 26 \\ 41 \end{bmatrix}$$

When obtaining the product **AB**, we say that **A** is *postmultiplied* by **B** or **B** is *premultiplied* by **A**. The reason for this precise terminology is that multiplication rules for ordinary algebra do not apply to matrix algebra. In ordinary algebra, $xy = yx$. In matrix algebra, $\mathbf{AB} \neq \mathbf{BA}$ usually. In fact, even though the product **AB** may be defined, the product **BA** may not be defined at all.

In general, the product **AB** is only defined when the number of columns in **A** equals the number of rows in **B** so that there will be corresponding terms in the cross products. Thus, in our previous two examples, we had:

Equal (columns of **A**, rows of **B**); Dimension of product (rows of **A**, columns of **B**):

$$\underset{2\times 2}{\mathbf{A}}\ \underset{2\times 2}{\mathbf{B}} = \underset{2\times 2}{\mathbf{AB}} \qquad \underset{2\times 3}{\mathbf{A}}\ \underset{3\times 1}{\mathbf{B}} = \underset{2\times 1}{\mathbf{AB}}$$

Note that the dimension of the product **AB** is given by the number of rows in **A** and the number of columns in **B**. Note also that in the second case the product **BA** would not be defined since the number of columns in **B** is not equal to the number of rows in **A**:

Unequal

$$\underset{3\times 1}{\mathbf{B}}\ \underset{2\times 3}{\mathbf{A}}$$

Here is another example of matrix multiplication:

$$\mathbf{AB} = \begin{bmatrix} a_{11} & a_{12} & a_{13} \\ a_{21} & a_{22} & a_{23} \end{bmatrix} \begin{bmatrix} b_{11} & b_{12} \\ b_{21} & b_{22} \\ b_{31} & b_{32} \end{bmatrix}$$

$$= \begin{bmatrix} a_{11}b_{11} + a_{12}b_{21} + a_{13}b_{31} & a_{11}b_{12} + a_{12}b_{22} + a_{13}b_{32} \\ a_{21}b_{11} + a_{22}b_{21} + a_{23}b_{31} & a_{21}b_{12} + a_{22}b_{22} + a_{23}b_{32} \end{bmatrix}$$

In general, if **A** has dimension $r \times c$ and **B** has dimension $c \times s$, the product **AB** is a matrix of dimension $r \times s$ whose element in the ith row and jth column is:

$$\sum_{k=1}^{c} a_{ik}b_{kj}$$

so that:

$$\underset{r\times s}{\mathbf{AB}} = \left[\sum_{k=1}^{c} a_{ik}b_{kj}\right] \qquad i = 1, \ldots, r; j = 1, \ldots, s \tag{6.12}$$

Thus, in the foregoing example, the element in the first row and second column of the product **AB** is:

$$\sum_{k=1}^{3} a_{1k}b_{k2} = a_{11}b_{12} + a_{12}b_{22} + a_{13}b_{32}$$

as indeed we found by taking the cross products of the elements in the first row of **A** and second column of **B** and summing.

Additional examples

1.

$$\begin{bmatrix} 4 & 2 \\ 5 & 8 \end{bmatrix} \begin{bmatrix} a_1 \\ a_2 \end{bmatrix} = \begin{bmatrix} 4a_1 + 2a_2 \\ 5a_1 + 8a_2 \end{bmatrix}$$

2.

$$[2 \quad 3 \quad 5] \begin{bmatrix} 2 \\ 3 \\ 5 \end{bmatrix} = [2^2 + 3^2 + 5^2] = [38]$$

Here, the product is a 1×1 matrix, which is equivalent to a scalar. Thus, the matrix product here equals the number 38.

3.

$$\begin{bmatrix} 1 & X_1 \\ 1 & X_2 \\ 1 & X_3 \end{bmatrix} \begin{bmatrix} \beta_0 \\ \beta_1 \end{bmatrix} = \begin{bmatrix} \beta_0 + \beta_1 X_1 \\ \beta_0 + \beta_1 X_2 \\ \beta_0 + \beta_1 X_3 \end{bmatrix}$$

Regression examples. Let us define the vector $\boldsymbol{\beta}$ of the regression coefficients as follows:

$$\underset{2\times 1}{\boldsymbol{\beta}} = \begin{bmatrix} \beta_0 \\ \beta_1 \end{bmatrix} \tag{6.13}$$

Then the product $\mathbf{X}\boldsymbol{\beta}$, where $\mathbf{X}$ is defined in (6.6), is an $n \times 1$ matrix:

$$\underset{n\times 1}{\mathbf{X}\boldsymbol{\beta}} = \begin{bmatrix} 1 & X_1 \\ 1 & X_2 \\ \cdot & \cdot \\ \cdot & \cdot \\ \cdot & \cdot \\ 1 & X_n \end{bmatrix} \begin{bmatrix} \beta_0 \\ \beta_1 \end{bmatrix} = \begin{bmatrix} \beta_0 + \beta_1 X_1 \\ \beta_0 + \beta_1 X_2 \\ \cdot \\ \cdot \\ \cdot \\ \beta_0 + \beta_1 X_n \end{bmatrix} \tag{6.14}$$

Since $\beta_0 + \beta_1 X_i = E(Y_i)$, we see that $\mathbf{X}\boldsymbol{\beta}$ is the vector of expected values $E(Y_i)$ for the simple linear regression model, i.e., $\mathbf{E(Y)} = \mathbf{X}\boldsymbol{\beta}$, where $\mathbf{E(Y)}$ is defined in (6.9).

Another product frequently needed is $\mathbf{Y'Y}$, where $\mathbf{Y}$ is the vector of observations on the dependent variable as defined in (6.4):

$$\underset{1\times 1}{\mathbf{Y'Y}} = [Y_1 \quad Y_2 \quad \cdots \quad Y_n] \begin{bmatrix} Y_1 \\ Y_2 \\ \cdot \\ \cdot \\ \cdot \\ Y_n \end{bmatrix} = [Y_1^2 + Y_2^2 + \cdots + Y_n^2] = [\Sigma Y_i^2] \tag{6.15}$$

Note that $\mathbf{Y'Y}$ is a 1×1 matrix, or a scalar. We thus have a compact way of writing a sum of squares: $\mathbf{Y'Y} = \Sigma Y_i^2$.

We also will need $\mathbf{X'X}$, which is a 2 × 2 matrix:

$$\underset{2\times 2}{\mathbf{X'X}} = \begin{bmatrix} 1 & 1 & \cdots & 1 \\ X_1 & X_2 & \cdots & X_n \end{bmatrix} \begin{bmatrix} 1 & X_1 \\ 1 & X_2 \\ \cdot & \cdot \\ \cdot & \cdot \\ \cdot & \cdot \\ 1 & X_n \end{bmatrix} = \begin{bmatrix} n & \Sigma X_i \\ \Sigma X_i & \Sigma X_i^2 \end{bmatrix} \tag{6.16}$$

and $\mathbf{X'Y}$, which is a 2 × 1 matrix:

$$\underset{2\times 1}{\mathbf{X'Y}} = \begin{bmatrix} 1 & 1 & \cdots & 1 \\ X_1 & X_2 & \cdots & X_n \end{bmatrix} \begin{bmatrix} Y_1 \\ Y_2 \\ \cdot \\ \cdot \\ \cdot \\ Y_n \end{bmatrix} = \begin{bmatrix} \Sigma Y_i \\ \Sigma X_i Y_i \end{bmatrix} \tag{6.17}$$

6.4 SPECIAL TYPES OF MATRICES

Certain special types of matrices arise regularly in regression analysis. We shall consider the most important of these.

Symmetric matrix

If $\mathbf{A} = \mathbf{A}'$, $\mathbf{A}$ is said to be symmetric. Thus, $\mathbf{A}$ below is symmetric:

$$\mathbf{A} = \begin{bmatrix} 1 & 4 & 6 \\ 4 & 2 & 5 \\ 6 & 5 & 3 \end{bmatrix} \qquad \mathbf{A}' = \begin{bmatrix} 1 & 4 & 6 \\ 4 & 2 & 5 \\ 6 & 5 & 3 \end{bmatrix}$$

Clearly, a symmetric matrix necessarily is square. Symmetric matrices arise typically in regression analysis when we premultiply a matrix, say, $\mathbf{X}$, by its transpose, $\mathbf{X}'$. The resulting matrix, $\mathbf{X'X}$, is symmetric, as can readily be seen from (6.16).

Diagonal matrix

A diagonal matrix is a square matrix whose off-diagonal elements are all zeros, such as:

$$\mathbf{A} = \begin{bmatrix} a_1 & 0 & 0 \\ 0 & a_2 & 0 \\ 0 & 0 & a_3 \end{bmatrix} \qquad \mathbf{B} = \begin{bmatrix} 4 & 0 & 0 & 0 \\ 0 & 1 & 0 & 0 \\ 0 & 0 & 10 & 0 \\ 0 & 0 & 0 & 5 \end{bmatrix}$$

We will often not show all zeros for a diagonal matrix, presenting it in the form:

$$\mathbf{A} = \begin{bmatrix} a_1 & & 0 \\ & a_2 & \\ 0 & & a_3 \end{bmatrix} \qquad \mathbf{B} = \begin{bmatrix} 4 & & & \\ & 1 & & 0 \\ & & 10 & \\ 0 & & & 5 \end{bmatrix}$$

Two important types of diagonal matrices are the identity matrix and the scalar matrix.

Identity matrix. The identity matrix or unit matrix is denoted by **I**. It is a diagonal matrix whose elements on the main diagonal are all 1's. Premultiplying or postmultiplying any $r \times r$ matrix **A** by the $r \times r$ identity matrix **I** leaves **A** unchanged. For example:

$$\mathbf{IA} = \begin{bmatrix} 1 & 0 & 0 \\ 0 & 1 & 0 \\ 0 & 0 & 1 \end{bmatrix} \begin{bmatrix} a_{11} & a_{12} & a_{13} \\ a_{21} & a_{22} & a_{23} \\ a_{31} & a_{32} & a_{33} \end{bmatrix} = \begin{bmatrix} a_{11} & a_{12} & a_{13} \\ a_{21} & a_{22} & a_{23} \\ a_{31} & a_{32} & a_{33} \end{bmatrix}$$

Similarly, we have:

$$\mathbf{AI} = \begin{bmatrix} a_{11} & a_{12} & a_{13} \\ a_{21} & a_{22} & a_{23} \\ a_{31} & a_{32} & a_{33} \end{bmatrix} \begin{bmatrix} 1 & 0 & 0 \\ 0 & 1 & 0 \\ 0 & 0 & 1 \end{bmatrix} = \begin{bmatrix} a_{11} & a_{12} & a_{13} \\ a_{21} & a_{22} & a_{23} \\ a_{31} & a_{32} & a_{33} \end{bmatrix}$$

Note that the identity matrix **I** therefore corresponds to the number 1 in ordinary algebra, since we have there that $1 \cdot x = x \cdot 1 = x$.

In general, we have for any $r \times r$ matrix **A**:

(6.18)
$$\mathbf{AI} = \mathbf{IA} = \mathbf{A}$$

Thus, the identity matrix can be inserted or dropped from a matrix expression whenever it is convenient to do so.

Scalar matrix. A scalar matrix is a diagonal matrix whose main-diagonal elements are the same. Two examples of scalar matrices are:

$$\begin{bmatrix} 2 & 0 \\ 0 & 2 \end{bmatrix} \qquad \begin{bmatrix} \lambda & 0 & 0 \\ 0 & \lambda & 0 \\ 0 & 0 & \lambda \end{bmatrix}$$

A scalar matrix can be expressed $\lambda\mathbf{I}$, where λ is the scalar. For instance:

$$\begin{bmatrix} 2 & 0 \\ 0 & 2 \end{bmatrix} = 2\begin{bmatrix} 1 & 0 \\ 0 & 1 \end{bmatrix} = 2\mathbf{I}$$

$$\begin{bmatrix} \lambda & 0 & 0 \\ 0 & \lambda & 0 \\ 0 & 0 & \lambda \end{bmatrix} = \lambda\begin{bmatrix} 1 & 0 & 0 \\ 0 & 1 & 0 \\ 0 & 0 & 1 \end{bmatrix} = \lambda\mathbf{I}$$

Multiplying an $r \times r$ matrix $\mathbf{A}$ by the $r \times r$ scalar matrix $\lambda\mathbf{I}$ is equivalent to multiplying $\mathbf{A}$ by the scalar λ.

Vector and matrix with all elements 1

A column vector with all elements 1 will be denoted by $\mathbf{1}$:

(6.19)
$$\underset{r\times 1}{\mathbf{1}} = \begin{bmatrix} 1 \\ 1 \\ \cdot \\ \cdot \\ \cdot \\ 1 \end{bmatrix}$$

and a square matrix with all elements 1 will be denoted by $\mathbf{J}$:

(6.20)
$$\underset{r\times r}{\mathbf{J}} = \begin{bmatrix} 1 & \cdots & 1 \\ 1 & & 1 \\ \cdot & & \cdot \\ \cdot & & \cdot \\ \cdot & & \cdot \\ 1 & \cdots & 1 \end{bmatrix}$$

For instance, we have:

$$\underset{3\times 1}{\mathbf{1}} = \begin{bmatrix} 1 \\ 1 \\ 1 \end{bmatrix} \qquad \underset{3\times 3}{\mathbf{J}} = \begin{bmatrix} 1 & 1 & 1 \\ 1 & 1 & 1 \\ 1 & 1 & 1 \end{bmatrix}$$

Note that for an $n \times 1$ vector $\mathbf{1}$ we obtain:

$$\underset{1\times 1}{\mathbf{1}'\mathbf{1}} = [1 \quad \cdots \quad 1] \begin{bmatrix} 1 \\ \cdot \\ \cdot \\ \cdot \\ 1 \end{bmatrix} = [n] = n$$

and:

$$\underset{n\times n}{\mathbf{1}\mathbf{1}'} = \begin{bmatrix} 1 \\ \cdot \\ \cdot \\ \cdot \\ 1 \end{bmatrix} [1 \quad \cdots \quad 1] = \begin{bmatrix} 1 & \cdots & 1 \\ \cdot & & \cdot \\ \cdot & & \cdot \\ \cdot & & \cdot \\ 1 & \cdots & 1 \end{bmatrix} = \underset{n\times n}{\mathbf{J}}$$

Zero vector

A zero vector is a column vector containing only zeros. It will be denoted by $\mathbf{0}$:

(6.21)
$$\underset{r\times 1}{\mathbf{0}} = \begin{bmatrix} 0 \\ 0 \\ \cdot \\ \cdot \\ \cdot \\ 0 \end{bmatrix}$$

For example, we have:

$$\underset{3\times 1}{\mathbf{0}} = \begin{bmatrix} 0 \\ 0 \\ 0 \end{bmatrix}$$

6.5 LINEAR DEPENDENCE AND RANK OF MATRIX

Linear dependence

Consider the following matrix:

$$\mathbf{A} = \begin{bmatrix} 1 & 2 & 5 & 1 \\ 2 & 2 & 10 & 6 \\ 3 & 4 & 15 & 1 \end{bmatrix}$$

Let us think now of the columns of this matrix as vectors. Thus, we view **A** as being made up of four column vectors. It happens here that the columns are interrelated in a special manner. Note that the third column vector is a multiple of the first column vector:

$$\begin{bmatrix} 5 \\ 10 \\ 15 \end{bmatrix} = 5\begin{bmatrix} 1 \\ 2 \\ 3 \end{bmatrix}$$

We say that the columns of **A** are linearly dependent. They contain redundant information, so to speak, since one column can be obtained as a linear combination of the others.

We define a set of column vectors to be linearly dependent if one vector can be expressed as a linear combination of the others. If no vector in the set can be so expressed, we define the set of vectors to be linearly independent. A more general, though equivalent, definition for the c column vectors $\mathbf{C}_1, \ldots, \mathbf{C}_c$ in an $r \times c$ matrix is:

(6.22) When c scalars $\lambda_1, \ldots, \lambda_c$, not all zero, can be found such that:

$$\lambda_1\mathbf{C}_1 + \lambda_2\mathbf{C}_2 + \cdots + \lambda_c\mathbf{C}_c = \mathbf{0}$$

where $\mathbf{0}$ denotes the zero column vector, the c column vectors are linearly dependent. If the only set of scalars for which the equality holds is $\lambda_1 = 0, \ldots, \lambda_c = 0$, the set of c column vectors is linearly independent.

To illustrate for our example, $\lambda_1 = 5$, $\lambda_2 = 0$, $\lambda_3 = -1$, $\lambda_4 = 0$ leads to:

$$5\begin{bmatrix}1\\2\\3\end{bmatrix} + 0\begin{bmatrix}2\\2\\4\end{bmatrix} - 1\begin{bmatrix}5\\10\\15\end{bmatrix} + 0\begin{bmatrix}1\\6\\1\end{bmatrix} = \begin{bmatrix}0\\0\\0\end{bmatrix}$$

Hence, the column vectors are linearly dependent. Note that some of the $\lambda_j = 0$ here. It is only required for linear dependence that not all λ_j are zero.

Rank of a matrix

The rank of a matrix is defined to be the maximum number of linearly independent columns in the matrix. We know that the rank of **A** in our earlier example cannot be 4, since the four columns are linearly dependent. We can, however, find 3 columns (1, 2, and 4) which are linearly independent. There are no scalars λ_1, λ_2, λ_4 such that $\lambda_1\mathbf{C}_1 + \lambda_2\mathbf{C}_2 + \lambda_4\mathbf{C}_4 = \mathbf{0}$ other than $\lambda_1 = \lambda_2 = \lambda_4 = 0$. Thus, the rank of **A** in our example is 3.

The rank of a matrix is unique and can equivalently be defined as the maximum number of linearly independent rows. It follows that the rank of an $r \times c$ matrix cannot exceed $\min(r, c)$, the minimum of the two values r and c.

6.6 INVERSE OF A MATRIX

In ordinary algebra, the inverse of a number is its reciprocal. Thus, the inverse of 6 is $\frac{1}{6}$. A number multiplied by its inverse always equals 1:

$$6 \cdot \tfrac{1}{6} = 1$$

$$x \cdot \frac{1}{x} = x \cdot x^{-1} = x^{-1} \cdot x = 1$$

In matrix algebra, the inverse of a matrix **A** is another matrix, denoted by $\mathbf{A}^{-1}$, such that:

$$\mathbf{A}^{-1}\mathbf{A} = \mathbf{A}\mathbf{A}^{-1} = \mathbf{I} \tag{6.23}$$

where **I** is the identity matrix. Thus, again, the identity matrix **I** plays the same role as the number 1 in ordinary algebra. An inverse of a matrix is defined only for square matrices. Even so, many square matrices do not have an inverse. If a square matrix does have an inverse, the inverse is unique.

Examples

1. The inverse of the matrix:

$$\mathbf{A} = \begin{bmatrix}2 & 4\\3 & 1\end{bmatrix}$$

is:

$$\mathbf{A}^{-1} = \begin{bmatrix} -.1 & .4 \\ .3 & -.2 \end{bmatrix}$$

since:

$$\mathbf{A}^{-1}\mathbf{A} = \begin{bmatrix} -.1 & .4 \\ .3 & -.2 \end{bmatrix} \begin{bmatrix} 2 & 4 \\ 3 & 1 \end{bmatrix} = \begin{bmatrix} 1 & 0 \\ 0 & 1 \end{bmatrix}$$

or:

$$\mathbf{A}\mathbf{A}^{-1} = \begin{bmatrix} 2 & 4 \\ 3 & 1 \end{bmatrix} \begin{bmatrix} -.1 & .4 \\ .3 & -.2 \end{bmatrix} = \begin{bmatrix} 1 & 0 \\ 0 & 1 \end{bmatrix}$$

2. The inverse of the matrix:

$$\mathbf{A} = \begin{bmatrix} 3 & 0 & 0 \\ 0 & 4 & 0 \\ 0 & 0 & 2 \end{bmatrix}$$

is:

$$\mathbf{A}^{-1} = \begin{bmatrix} \frac{1}{3} & 0 & 0 \\ 0 & \frac{1}{4} & 0 \\ 0 & 0 & \frac{1}{2} \end{bmatrix}$$

since:

$$\mathbf{A}^{-1}\mathbf{A} = \begin{bmatrix} \frac{1}{3} & 0 & 0 \\ 0 & \frac{1}{4} & 0 \\ 0 & 0 & \frac{1}{2} \end{bmatrix} \begin{bmatrix} 3 & 0 & 0 \\ 0 & 4 & 0 \\ 0 & 0 & 2 \end{bmatrix} = \begin{bmatrix} 1 & 0 & 0 \\ 0 & 1 & 0 \\ 0 & 0 & 1 \end{bmatrix}$$

Note that the inverse of a diagonal matrix is a diagonal matrix consisting simply of the reciprocals of the elements on the diagonal.

Finding the inverse

Up to this point, the inverse of a matrix **A** has been given, and we have only checked to make sure it is the inverse by seeing whether or not $\mathbf{A}^{-1}\mathbf{A} = \mathbf{I}$. But how does one find the inverse, and when does it exist?

An inverse of a square $r \times r$ matrix exists if the rank of the matrix is r. Such a matrix is said to be *nonsingular*. An $r \times r$ matrix with rank less than r is said to be *singular*, and does not have an inverse.

Finding the inverse of a matrix can often require a tremendous amount of computing. We shall take the approach in this book that the inverse of a 2×2 matrix and a 3×3 matrix can be calculated by hand. For any larger matrix, one ordinarily uses a computer or a programmable calculator to find the inverse, unless the matrix is of a special form such as a diagonal matrix. It can be shown that the inverses for 2×2 and 3×3 matrices are as follows:

1. If:

$$\mathbf{A} = \begin{bmatrix} a & b \\ c & d \end{bmatrix}$$

then:

$$\mathbf{A}^{-1} = \begin{bmatrix} a & b \\ c & d \end{bmatrix}^{-1} = \begin{bmatrix} \frac{d}{D} & \frac{-b}{D} \\ \frac{-c}{D} & \frac{a}{D} \end{bmatrix} \tag{6.24}$$

where:

$$D = ad - bc$$

D is called the *determinant* of the matrix **A**. If **A** were singular, its determinant would equal zero and no inverse of **A** would exist.

2. If:

$$\mathbf{B} = \begin{bmatrix} a & b & c \\ d & e & f \\ g & h & k \end{bmatrix}$$

then:

$$\mathbf{B}^{-1} = \begin{bmatrix} a & b & c \\ d & e & f \\ g & h & k \end{bmatrix}^{-1} = \begin{bmatrix} A & B & C \\ D & E & F \\ G & H & K \end{bmatrix} \tag{6.25}$$

where:

$$A = (ek - fh)/Z \qquad B = -(bk - ch)/Z \qquad C = (bf - ce)/Z$$
$$D = -(dk - fg)/Z \qquad E = (ak - cg)/Z \qquad F = -(af - cd)/Z$$
$$G = (dh - eg)/Z \qquad H = -(ah - bg)/Z \qquad K = (ae - bd)/Z$$

and:

$$Z = a(ek - fh) - b(dk - fg) + c(dh - eg)$$

Z is called the determinant of the matrix **B**.

Let us use (6.24) to find the inverse of:

$$\mathbf{A} = \begin{bmatrix} 2 & 4 \\ 3 & 1 \end{bmatrix}$$

We have:

$$a = 2 \qquad b = 4$$
$$c = 3 \qquad d = 1$$
$$D = ad - bc = 2(1) - 4(3) = -10$$

Hence:

$$\mathbf{A}^{-1} = \begin{bmatrix} \frac{1}{-10} & \frac{-4}{-10} \\ \frac{-3}{-10} & \frac{2}{-10} \end{bmatrix} = \begin{bmatrix} -.1 & .4 \\ .3 & -.2 \end{bmatrix}$$

as was given in an earlier example.

When an inverse $\mathbf{A}^{-1}$ has been obtained, either by hand calculations or from a computer run, it is usually wise to compute $\mathbf{A}^{-1}\mathbf{A}$ to check whether the product equals the identity matrix, allowing for minor rounding departures from 0 and 1.

Regression example

The principal inverse matrix encountered in regression analysis is the inverse of the matrix $\mathbf{X}'\mathbf{X}$ in (6.16):

$$\mathbf{X}'\mathbf{X} = \begin{bmatrix} n & \Sigma X_i \\ \Sigma X_i & \Sigma X_i^2 \end{bmatrix}$$

Using rule (6.24), we have:

$$a = n \qquad b = \Sigma X_i$$
$$c = \Sigma X_i \qquad d = \Sigma X_i^2$$

so that:

$$D = n\Sigma X_i^2 - (\Sigma X_i)(\Sigma X_i) = n\left(\Sigma X_i^2 - \frac{(\Sigma X_i)^2}{n}\right) = n\Sigma(X_i - \bar{X})^2$$

Hence:

$$(\mathbf{X}'\mathbf{X})^{-1} = \begin{bmatrix} \frac{\Sigma X_i^2}{n\Sigma(X_i - \bar{X})^2} & \frac{-\Sigma X_i}{n\Sigma(X_i - \bar{X})^2} \\ \frac{-\Sigma X_i}{n\Sigma(X_i - \bar{X})^2} & \frac{n}{n\Sigma(X_i - \bar{X})^2} \end{bmatrix} \tag{6.26}$$

Since $\Sigma X_i = n\bar{X}$, we can simplify (6.26):

$$(\mathbf{X}'\mathbf{X})^{-1} = \begin{bmatrix} \frac{\Sigma X_i^2}{n\Sigma(X_i - \bar{X})^2} & \frac{-\bar{X}}{\Sigma(X_i - \bar{X})^2} \\ \frac{-\bar{X}}{\Sigma(X_i - \bar{X})^2} & \frac{1}{\Sigma(X_i - \bar{X})^2} \end{bmatrix} \tag{6.27}$$

Uses of inverse matrix

In ordinary algebra, we solve an equation of the type:

$$5y = 20$$

by multiplying both sides of the equation by the inverse of 5, namely:

$$\tfrac{1}{5}(5y) = \tfrac{1}{5}(20)$$

We obtain:

$$y = \tfrac{1}{5}(20) = 4$$

In matrix algebra, if we have an equation:

$$\mathbf{AY} = \mathbf{C}$$

we correspondingly premultiply both sides by $\mathbf{A}^{-1}$, assuming $\mathbf{A}$ has an inverse, and obtain:

$$\mathbf{A}^{-1}\mathbf{AY} = \mathbf{A}^{-1}\mathbf{C}$$

Since $\mathbf{A}^{-1}\mathbf{AY} = \mathbf{IY} = \mathbf{Y}$, we obtain:

$$\mathbf{Y} = \mathbf{A}^{-1}\mathbf{C}$$

To illustrate this use, suppose we have two simultaneous equations:

$$2y_1 + 4y_2 = 20$$
$$3y_1 + y_2 = 10$$

which can be written as follows in matrix notation:

$$\begin{bmatrix} 2 & 4 \\ 3 & 1 \end{bmatrix} \begin{bmatrix} y_1 \\ y_2 \end{bmatrix} = \begin{bmatrix} 20 \\ 10 \end{bmatrix}$$

The solution of these equations then is:

$$\begin{bmatrix} y_1 \\ y_2 \end{bmatrix} = \begin{bmatrix} 2 & 4 \\ 3 & 1 \end{bmatrix}^{-1} \begin{bmatrix} 20 \\ 10 \end{bmatrix}$$

Earlier we found the required inverse, so we obtain:

$$\begin{bmatrix} y_1 \\ y_2 \end{bmatrix} = \begin{bmatrix} -.1 & .4 \\ .3 & -.2 \end{bmatrix} \begin{bmatrix} 20 \\ 10 \end{bmatrix} = \begin{bmatrix} 2 \\ 4 \end{bmatrix}$$

Hence, $y_1 = 2$ and $y_2 = 4$ satisfy these two equations.

6.7 SOME BASIC THEOREMS FOR MATRICES

We list here, without proof, some basic theorems for matrices which we will utilize in later work.

(6.28) $$\mathbf{A} + \mathbf{B} = \mathbf{B} + \mathbf{A}$$

(6.29) $$(\mathbf{A} + \mathbf{B}) + \mathbf{C} = \mathbf{A} + (\mathbf{B} + \mathbf{C})$$

(6.30) $$(\mathbf{AB})\mathbf{C} = \mathbf{A}(\mathbf{BC})$$

(6.31) $$\mathbf{C}(\mathbf{A}+\mathbf{B}) = \mathbf{CA}+\mathbf{CB}$$

(6.32) $$\lambda(\mathbf{A}+\mathbf{B}) = \lambda\mathbf{A}+\lambda\mathbf{B}$$

(6.33) $$(\mathbf{A}')' = \mathbf{A}$$

(6.34) $$(\mathbf{A}+\mathbf{B})' = \mathbf{A}'+\mathbf{B}'$$

(6.35) $$(\mathbf{AB})' = \mathbf{B}'\mathbf{A}'$$

(6.36) $$(\mathbf{ABC})' = \mathbf{C}'\mathbf{B}'\mathbf{A}'$$

(6.37) $$(\mathbf{AB})^{-1} = \mathbf{B}^{-1}\mathbf{A}^{-1}$$

(6.38) $$(\mathbf{ABC})^{-1} = \mathbf{C}^{-1}\mathbf{B}^{-1}\mathbf{A}^{-1}$$

(6.39) $$(\mathbf{A}^{-1})^{-1} = \mathbf{A}$$

(6.40) $$(\mathbf{A}')^{-1} = (\mathbf{A}^{-1})'$$

6.8 RANDOM VECTORS AND MATRICES

A random vector or a random matrix contains elements which are random variables. Thus, the observation vector $\mathbf{Y}$ in (6.4) is a random vector since the Y_i elements are random variables.

Expectation of random vector or matrix

Suppose we have $n = 3$ observations and are concerned with the observation vector:

$$\mathbf{Y} = \begin{bmatrix} Y_1 \\ Y_2 \\ Y_3 \end{bmatrix}$$

The expected value of $\mathbf{Y}$ is a vector, denoted by $\mathbf{E}(\mathbf{Y})$, which is defined as follows:

$$\mathbf{E}(\mathbf{Y}) = \begin{bmatrix} E(Y_1) \\ E(Y_2) \\ E(Y_3) \end{bmatrix}$$

Thus, the expected value of a random vector is a vector whose elements are the expected values of the random variables which are the elements of the random vector. Similarly, the expectation of a random matrix is a matrix whose elements are the expected values of the corresponding random variables in the original matrix. We encountered a vector of expected values earlier in (6.9).

In general, for a random vector $\mathbf{Y}$ the expectation is:

(6.41) $$\underset{n\times 1}{\mathbf{E}(\mathbf{Y})} = [E(Y_i)] \qquad i = 1,\ldots,n$$

and for a random matrix $\mathbf{Y}$ with dimension $n \times p$, the expectation is:

(6.42) $$\underset{n\times p}{\mathbf{E}(\mathbf{Y})} = [E(Y_{ij})] \qquad i = 1,\ldots,n; j = 1,\ldots,p$$

Regression example. Suppose the number of observations in a regression application is $n = 3$. The three error terms $\varepsilon_1, \varepsilon_2, \varepsilon_3$ each have expectation zero. For the error vector:

$$\boldsymbol{\varepsilon} = \begin{bmatrix} \varepsilon_1 \\ \varepsilon_2 \\ \varepsilon_3 \end{bmatrix}$$

we have:

$$\mathbf{E}(\boldsymbol{\varepsilon}) = \mathbf{0}$$

since:

$$\mathbf{E}(\boldsymbol{\varepsilon}) = \begin{bmatrix} E(\varepsilon_1) \\ E(\varepsilon_2) \\ E(\varepsilon_3) \end{bmatrix} = \begin{bmatrix} 0 \\ 0 \\ 0 \end{bmatrix} = \mathbf{0}$$

Variance-covariance matrix of a random vector

Consider again the random vector $\mathbf{Y}$ consisting of three observations Y_1, Y_2, Y_3. Each random variable has a variance, $\sigma^2(Y_i)$, and any two random variables have a covariance, $\sigma(Y_i, Y_j)$. We can assemble these in a matrix called the *variance-covariance matrix of* $\mathbf{Y}$, denoted by $\boldsymbol{\sigma}^2(\mathbf{Y})$:

(6.43)
$$\boldsymbol{\sigma}^2(\mathbf{Y}) = \begin{bmatrix} \sigma^2(Y_1) & \sigma(Y_1, Y_2) & \sigma(Y_1, Y_3) \\ \sigma(Y_2, Y_1) & \sigma^2(Y_2) & \sigma(Y_2, Y_3) \\ \sigma(Y_3, Y_1) & \sigma(Y_3, Y_2) & \sigma^2(Y_3) \end{bmatrix}$$

Note that the variances are on the main diagonal and the covariance $\sigma(Y_i, Y_j)$ is found in the ith row and jth column of the matrix. Thus, $\sigma(Y_2, Y_1)$ is found in the second row, first column, and $\sigma(Y_1, Y_2)$ is found in the first row, second column. Remember, of course, that $\sigma(Y_2, Y_1) = \sigma(Y_1, Y_2)$. Since in general $\sigma(Y_i, Y_j) = \sigma(Y_j, Y_i)$ for $i \neq j$, $\boldsymbol{\sigma}^2(\mathbf{Y})$ is a symmetric matrix.

It follows readily that:

(6.44)
$$\boldsymbol{\sigma}^2(\mathbf{Y}) = \mathbf{E}\{[\mathbf{Y} - \mathbf{E}(\mathbf{Y})]\ [\mathbf{Y} - \mathbf{E}(\mathbf{Y})]'\}$$

For our illustration, we have:

$$\boldsymbol{\sigma}^2(\mathbf{Y}) = \mathbf{E}\begin{bmatrix} Y_1 - E(Y_1) \\ Y_2 - E(Y_2) \\ Y_3 - E(Y_3) \end{bmatrix} [Y_1 - E(Y_1) \quad Y_2 - E(Y_2) \quad Y_3 - E(Y_3)]$$

Multiplying the two matrices and then taking expectations, we obtain:

Location in Product	*Term*	*Expected Value*
Row 1, column 1	$[Y_1 - E(Y_1)]^2$	$\sigma^2(Y_1)$
Row 1, column 2	$[Y_1 - E(Y_1)][Y_2 - E(Y_2)]$	$\sigma(Y_1, Y_2)$
Row 1, column 3	$[Y_1 - E(Y_1)][Y_3 - E(Y_3)]$	$\sigma(Y_1, Y_3)$
Row 2, column 1	$[Y_2 - E(Y_2)][Y_1 - E(Y_1)]$	$\sigma(Y_2, Y_1)$
etc.	etc.	etc.

This, of course, leads to the variance-covariance matrix in (6.43). Remember the definitions of variance and covariance in (1.14) and (1.19), respectively, when taking expectations.

To generalize, the variance-covariance matrix for an $n \times 1$ random vector $\mathbf{Y}$ is:

$$\underset{n\times n}{\boldsymbol{\sigma}^2(\mathbf{Y})} = \begin{bmatrix} \sigma^2(Y_1) & \sigma(Y_1, Y_2) & \cdots & \sigma(Y_1, Y_n) \\ \sigma(Y_2, Y_1) & \sigma^2(Y_2) & \cdots & \sigma(Y_2, Y_n) \\ \cdot & \cdot & & \cdot \\ \cdot & \cdot & & \cdot \\ \cdot & \cdot & & \cdot \\ \sigma(Y_n, Y_1) & \sigma(Y_n, Y_2) & \cdots & \sigma^2(Y_n) \end{bmatrix} \tag{6.45}$$

Note again that $\boldsymbol{\sigma}^2(\mathbf{Y})$ is a symmetric matrix.

Regression example. Let us return to the example based on $n = 3$ observations. Suppose that the three error terms have constant variance, $\sigma^2(\varepsilon_i) = \sigma^2$, and are uncorrelated so that $\sigma(\varepsilon_i, \varepsilon_j) = 0$ for $i \neq j$. We can then write the variance-covariance matrix for the random vector $\boldsymbol{\varepsilon}$ of the previous example as follows:

$$\boldsymbol{\sigma}^2(\boldsymbol{\varepsilon}) = \sigma^2\mathbf{I}$$

since:

$$\sigma^2\mathbf{I} = \sigma^2\begin{bmatrix} 1 & 0 & 0 \\ 0 & 1 & 0 \\ 0 & 0 & 1 \end{bmatrix} = \begin{bmatrix} \sigma^2 & 0 & 0 \\ 0 & \sigma^2 & 0 \\ 0 & 0 & \sigma^2 \end{bmatrix}$$

Note that all variances are σ^2 and all covariances are zero.

Some basic theorems

Frequently, we shall encounter a random vector $\mathbf{W}$ which is obtained by premultiplying the random vector $\mathbf{Y}$ by a constant matrix $\mathbf{A}$ (a matrix whose elements are fixed):

$$\mathbf{W} = \mathbf{AY} \tag{6.46}$$

Some basic theorems for this case are:

$$\mathbf{E}(\mathbf{A}) = \mathbf{A} \tag{6.47}$$

$$\mathbf{E}(\mathbf{W}) = \mathbf{E}(\mathbf{AY}) = \mathbf{AE}(\mathbf{Y}) \tag{6.48}$$

$$\boldsymbol{\sigma}^2(\mathbf{W}) = \boldsymbol{\sigma}^2(\mathbf{AY}) = \mathbf{A}[\boldsymbol{\sigma}^2(\mathbf{Y})]\mathbf{A}' \tag{6.49}$$

where $\boldsymbol{\sigma}^2(\mathbf{Y})$ is the variance-covariance matrix of $\mathbf{Y}$.

Example. As a simple illustration of the use of these theorems, consider:

$$\underset{2\times 1}{\begin{bmatrix} W_1 \\ W_2 \end{bmatrix}} = \underset{2\times 2}{\begin{bmatrix} 1 & -1 \\ 1 & 1 \end{bmatrix}} \underset{2\times 1}{\begin{bmatrix} Y_1 \\ Y_2 \end{bmatrix}} = \begin{bmatrix} Y_1 - Y_2 \\ Y_1 + Y_2 \end{bmatrix}$$

$$\mathbf{W} \qquad \mathbf{A} \qquad \mathbf{Y}$$

We then have by (6.48):

$$\mathbf{E(W)} = \begin{bmatrix} 1 & -1 \\ 1 & 1 \end{bmatrix} \begin{bmatrix} E(Y_1) \\ E(Y_2) \end{bmatrix} = \begin{bmatrix} E(Y_1) - E(Y_2) \\ E(Y_1) + E(Y_2) \end{bmatrix}$$

and by (6.49):

$$\boldsymbol{\sigma}^2(\mathbf{W}) = \begin{bmatrix} 1 & -1 \\ 1 & 1 \end{bmatrix} \begin{bmatrix} \sigma^2(Y_1) & \sigma(Y_1, Y_2) \\ \sigma(Y_2, Y_1) & \sigma^2(Y_2) \end{bmatrix} \begin{bmatrix} 1 & 1 \\ -1 & 1 \end{bmatrix}$$

$$= \begin{bmatrix} \sigma^2(Y_1) + \sigma^2(Y_2) - 2\sigma(Y_1, Y_2) & \sigma^2(Y_1) - \sigma^2(Y_2) \\ \sigma^2(Y_1) - \sigma^2(Y_2) & \sigma^2(Y_1) + \sigma^2(Y_2) + 2\sigma(Y_1, Y_2) \end{bmatrix}$$

Thus:

$$\begin{aligned} \sigma^2(W_1) &= \sigma^2(Y_1 - Y_2) = \sigma^2(Y_1) + \sigma^2(Y_2) - 2\sigma(Y_1, Y_2) \\ \sigma^2(W_2) &= \sigma^2(Y_1 + Y_2) = \sigma^2(Y_1) + \sigma^2(Y_2) + 2\sigma(Y_1, Y_2) \\ \sigma(W_1, W_2) &= \sigma(Y_1 - Y_2, Y_1 + Y_2) = \sigma^2(Y_1) - \sigma^2(Y_2) \end{aligned}$$

6.9 SIMPLE LINEAR REGRESSION MODEL IN MATRIX TERMS

We are now ready to develop simple linear regression in matrix terms. Remember again that we will not present any new results, but shall only state in matrix terms the results obtained earlier. We shall begin with the regression model (3.1):

(6.50) $$Y_i = \beta_0 + \beta_1 X_i + \varepsilon_i \qquad i = 1, \ldots, n$$

This implies:

(6.51) $$\begin{aligned} Y_1 &= \beta_0 + \beta_1 X_1 + \varepsilon_1 \\ Y_2 &= \beta_0 + \beta_1 X_2 + \varepsilon_2 \\ &\vdots \\ Y_n &= \beta_0 + \beta_1 X_n + \varepsilon_n \end{aligned}$$

We defined earlier the observation vector **Y** in (6.4), the **X** matrix in (6.6), the $\boldsymbol{\varepsilon}$ vector in (6.10), and the $\boldsymbol{\beta}$ vector in (6.13). Let us repeat these definitions:

(6.52) $$\mathbf{Y} = \begin{bmatrix} Y_1 \\ Y_2 \\ \vdots \\ Y_n \end{bmatrix} \qquad \mathbf{X} = \begin{bmatrix} 1 & X_1 \\ 1 & X_2 \\ \vdots & \vdots \\ 1 & X_n \end{bmatrix} \qquad \boldsymbol{\beta} = \begin{bmatrix} \beta_0 \\ \beta_1 \end{bmatrix} \qquad \boldsymbol{\varepsilon} = \begin{bmatrix} \varepsilon_1 \\ \varepsilon_2 \\ \vdots \\ \varepsilon_n \end{bmatrix}$$

Now we can write (6.51) in matrix terms compactly as follows:

(6.53) $$\underset{n\times 1}{\mathbf{Y}} = \underset{n\times 2}{\mathbf{X}}\ \underset{2\times 1}{\boldsymbol{\beta}} + \underset{n\times 1}{\boldsymbol{\varepsilon}}$$

since:

$$\begin{bmatrix} Y_1 \\ Y_2 \\ \cdot \\ \cdot \\ \cdot \\ Y_n \end{bmatrix} = \begin{bmatrix} 1 & X_1 \\ 1 & X_2 \\ \cdot & \cdot \\ \cdot & \cdot \\ \cdot & \cdot \\ 1 & X_n \end{bmatrix} \begin{bmatrix} \beta_0 \\ \beta_1 \end{bmatrix} + \begin{bmatrix} \varepsilon_1 \\ \varepsilon_2 \\ \cdot \\ \cdot \\ \cdot \\ \varepsilon_n \end{bmatrix} = \begin{bmatrix} \beta_0 + \beta_1 X_1 + \varepsilon_1 \\ \beta_0 + \beta_1 X_2 + \varepsilon_2 \\ \cdot \\ \cdot \\ \cdot \\ \beta_0 + \beta_1 X_n + \varepsilon_n \end{bmatrix}$$

The column of 1's in the **X** matrix may be viewed as consisting of the dummy variable $X_0 \equiv 1$ in the alternative regression model (2.5):

$$Y_i = \beta_0 X_0 + \beta_1 X_i + \varepsilon_i \qquad \text{where } X_0 \equiv 1$$

Thus, the **X** matrix may be considered to contain a column vector of the dummy variable X_0 and another column vector consisting of the independent variable observations X_i.

With respect to the error terms, model (3.1) assumes that $E(\varepsilon_i) = 0$, $\sigma^2(\varepsilon_i) = \sigma^2$, and that the ε_i are independent normal random variables. The condition $E(\varepsilon_i) = 0$ in matrix terms is:

$$\mathbf{E}(\boldsymbol{\varepsilon}) = \mathbf{0} \tag{6.54}$$

since (6.54) states:

$$\mathbf{E} \begin{bmatrix} \varepsilon_1 \\ \varepsilon_2 \\ \cdot \\ \cdot \\ \cdot \\ \varepsilon_n \end{bmatrix} = \begin{bmatrix} E(\varepsilon_1) \\ E(\varepsilon_2) \\ \cdot \\ \cdot \\ \cdot \\ E(\varepsilon_n) \end{bmatrix} = \begin{bmatrix} 0 \\ 0 \\ \cdot \\ \cdot \\ \cdot \\ 0 \end{bmatrix}$$

The condition that the error terms have constant variance σ^2 and that all covariances $\sigma(\varepsilon_i, \varepsilon_j)$ for $i \neq j$ are zero (since the ε_i are independent) is expressed in matrix terms through the variance-covariance matrix:

$$\underset{n \times n}{\boldsymbol{\sigma}^2(\boldsymbol{\varepsilon})} = \underset{n \times n}{\sigma^2 \mathbf{I}} \tag{6.55}$$

since (6.55) states:

$$\boldsymbol{\sigma}^2(\boldsymbol{\varepsilon}) = \sigma^2 \begin{bmatrix} 1 & 0 & 0 & \cdots & 0 \\ 0 & 1 & 0 & \cdots & 0 \\ \cdot & \cdot & \cdot & & \cdot \\ \cdot & \cdot & \cdot & & \cdot \\ \cdot & \cdot & \cdot & & \cdot \\ 0 & 0 & 0 & \cdots & 1 \end{bmatrix} = \begin{bmatrix} \sigma^2 & 0 & 0 & \cdots & 0 \\ 0 & \sigma^2 & 0 & \cdots & 0 \\ \cdot & \cdot & \cdot & & \cdot \\ \cdot & \cdot & \cdot & & \cdot \\ \cdot & \cdot & \cdot & & \cdot \\ 0 & 0 & 0 & \cdots & \sigma^2 \end{bmatrix}$$

Thus, the normal error model (3.1) in matrix terms is:

$$\mathbf{Y} = \mathbf{X}\boldsymbol{\beta} + \boldsymbol{\varepsilon} \tag{6.56}$$

where:

$\boldsymbol{\varepsilon}$ is a vector of independent normal random variables with $\mathbf{E}(\boldsymbol{\varepsilon}) = \mathbf{0}$ and $\boldsymbol{\sigma}^2(\boldsymbol{\varepsilon}) = \sigma^2\mathbf{I}$

6.10 LEAST SQUARES ESTIMATION OF REGRESSION PARAMETERS

Normal equations

The normal equations (2.9):

$$nb_0 + b_1\Sigma X_i = \Sigma Y_i$$

(6.57)
$$b_0\Sigma X_i + b_1\Sigma X_i^2 = \Sigma X_i Y_i$$

in matrix terms are:

(6.58)
$$\mathbf{X'Xb} = \mathbf{X'Y}$$

where $\mathbf{b}$ is the vector of the least squares regression coefficients:

(6.58a)
$$\underset{2\times 1}{\mathbf{b}} = \begin{bmatrix} b_0 \\ b_1 \end{bmatrix}$$

To see this, recall that we obtained $\mathbf{X'X}$ in (6.16) and $\mathbf{X'Y}$ in (6.17). Equation (6.58) thus states:

$$\begin{bmatrix} n & \Sigma X_i \\ \Sigma X_i & \Sigma X_i^2 \end{bmatrix} \begin{bmatrix} b_0 \\ b_1 \end{bmatrix} = \begin{bmatrix} \Sigma Y_i \\ \Sigma X_i Y_i \end{bmatrix}$$

or:

$$\begin{bmatrix} nb_0 + b_1\Sigma X_i \\ b_0\Sigma X_i + b_1\Sigma X_i^2 \end{bmatrix} = \begin{bmatrix} \Sigma Y_i \\ \Sigma X_i Y_i \end{bmatrix}$$

These are precisely the normal equations in (6.57).

Estimated regression coefficients

To obtain the estimated regression coefficients from the normal equations:

$$\mathbf{X'Xb} = \mathbf{X'Y}$$

by matrix methods, we premultiply both sides by the inverse of $\mathbf{X'X}$ (we assume this exists):

$$(\mathbf{X'X})^{-1}\mathbf{X'Xb} = (\mathbf{X'X})^{-1}\mathbf{X'Y}$$

so that we find, since $(\mathbf{X'X})^{-1}\mathbf{X'X} = \mathbf{I}$ and $\mathbf{Ib} = \mathbf{b}$:

(6.59)
$$\mathbf{b} = (\mathbf{X'X})^{-1}\mathbf{X'Y}$$

The estimators b_0 and b_1 in $\mathbf{b}$ are the same as those given earlier in (2.10a) and (2.10b). We shall demonstrate this by an example.

Example. Let us find the estimated regression coefficients for the Westwood Company lot size example by matrix methods. From earlier work, we have (Table 2.2):

$$n = 10 \qquad \Sigma Y_i = 1{,}100 \qquad \Sigma X_i = 500 \qquad \Sigma X_i^2 = 28{,}400$$
$$\Sigma X_i Y_i = 61{,}800$$

Let us now use (6.26) to evaluate $(\mathbf{X}'\mathbf{X})^{-1}$. We have:

$$n\Sigma(X_i - \bar{X})^2 = n\left[\Sigma X_i^2 - \frac{(\Sigma X_i)^2}{n}\right] = 10\left[28{,}400 - \frac{(500)^2}{10}\right] = 34{,}000$$

Therefore:

$$(\mathbf{X}'\mathbf{X})^{-1} = \begin{bmatrix} \dfrac{\Sigma X_i^2}{n\Sigma(X_i - \bar{X})^2} & \dfrac{-\Sigma X_i}{n\Sigma(X_i - \bar{X})^2} \\ \dfrac{-\Sigma X_i}{n\Sigma(X_i - \bar{X})^2} & \dfrac{n}{n\Sigma(X_i - \bar{X})^2} \end{bmatrix} = \begin{bmatrix} \dfrac{28{,}400}{34{,}000} & \dfrac{-500}{34{,}000} \\ \dfrac{-500}{34{,}000} & \dfrac{10}{34{,}000} \end{bmatrix}$$

$$= \begin{bmatrix} .83529412 & -.01470588 \\ -.01470588 & .00029412 \end{bmatrix}$$

We also wish to make use of (6.17) to evaluate $\mathbf{X}'\mathbf{Y}$:

$$\mathbf{X}'\mathbf{Y} = \begin{bmatrix} \Sigma Y_i \\ \Sigma X_i Y_i \end{bmatrix} = \begin{bmatrix} 1{,}100 \\ 61{,}800 \end{bmatrix}$$

Hence, by (6.59):

$$\mathbf{b} = \begin{bmatrix} b_0 \\ b_1 \end{bmatrix} = (\mathbf{X}'\mathbf{X})^{-1}\mathbf{X}'\mathbf{Y} = \begin{bmatrix} .83529412 & -.01470588 \\ -.01470588 & .00029412 \end{bmatrix} \begin{bmatrix} 1{,}100 \\ 61{,}800 \end{bmatrix}$$

$$= \begin{bmatrix} 10.0 \\ 2.0 \end{bmatrix}$$

or $b_0 = 10.0$ and $b_1 = 2.0$. This agrees with the results in Chapter 2. Any difference would have been due to rounding errors.

To reduce the effect of rounding errors when obtaining the vector **b** by hand calculations, it is often desirable to move the constant in the denominator of the elements of $(\mathbf{X}'\mathbf{X})^{-1}$ outside the matrix, and do the division as the last step. For our example, this would lead to:

$$(\mathbf{X}'\mathbf{X})^{-1} = \frac{1}{n\Sigma(X_i - \bar{X})^2}\begin{bmatrix} \Sigma X_i^2 & -\Sigma X_i \\ -\Sigma X_i & n \end{bmatrix}$$

$$= \frac{1}{34{,}000}\begin{bmatrix} 28{,}400 & -500 \\ -500 & 10 \end{bmatrix}$$

$$\mathbf{b} = \frac{1}{34{,}000}\begin{bmatrix} 28{,}400 & -500 \\ -500 & 10 \end{bmatrix}\begin{bmatrix} 1{,}100 \\ 61{,}800 \end{bmatrix}$$

$$= \frac{1}{34{,}000}\begin{bmatrix} 340{,}000 \\ 68{,}000 \end{bmatrix} = \begin{bmatrix} 10.0 \\ 2.0 \end{bmatrix}$$

In this instance, the two methods of calculation lead to identical results. Often, however, postponing division by $n\Sigma(X_i - \bar{X})^2$ until the end yields more accurate results.

Comments

1. To derive the normal equations by the method of least squares, we minimize the quantity:

$$Q = \Sigma[Y_i - (\beta_0 + \beta_1 X_i)]^2$$

In matrix notation:

$$Q = (\mathbf{Y} - \mathbf{X}\boldsymbol{\beta})'(\mathbf{Y} - \mathbf{X}\boldsymbol{\beta}) \tag{6.60}$$

Expanding out, we obtain:

$$Q = \mathbf{Y}'\mathbf{Y} - \boldsymbol{\beta}'\mathbf{X}'\mathbf{Y} - \mathbf{Y}'\mathbf{X}\boldsymbol{\beta} + \boldsymbol{\beta}'\mathbf{X}'\mathbf{X}\boldsymbol{\beta}$$

since $(\mathbf{X}\boldsymbol{\beta})' = \boldsymbol{\beta}'\mathbf{X}'$ by (6.35). Note now that $\mathbf{Y}'\mathbf{X}\boldsymbol{\beta}$ is 1×1, hence is equal to its transpose, which according to (6.36) is $\boldsymbol{\beta}'\mathbf{X}'\mathbf{Y}$. Thus, we find:

$$Q = \mathbf{Y}'\mathbf{Y} - 2\boldsymbol{\beta}'\mathbf{X}'\mathbf{Y} + \boldsymbol{\beta}'\mathbf{X}'\mathbf{X}\boldsymbol{\beta} \tag{6.61}$$

To find the value of $\boldsymbol{\beta}$ which minimizes Q, we differentiate with respect to β_0 and β_1. Let:

$$\frac{\boldsymbol{\partial}}{\boldsymbol{\partial\beta}}(Q) = \begin{bmatrix} \dfrac{\partial Q}{\partial \beta_0} \\ \dfrac{\partial Q}{\partial \beta_1} \end{bmatrix} \tag{6.62}$$

Then it follows that:

$$\frac{\boldsymbol{\partial}}{\boldsymbol{\partial\beta}}(Q) = -2\mathbf{X}'\mathbf{Y} + 2\mathbf{X}'\mathbf{X}\boldsymbol{\beta} \tag{6.63}$$

Equating to zero and substituting $\mathbf{b}$ for $\boldsymbol{\beta}$ gives the matrix form of the least squares normal equations:

$$\mathbf{X}'\mathbf{X}\mathbf{b} = \mathbf{X}'\mathbf{Y}$$

2. A comparison of the normal equations and $\mathbf{X}'\mathbf{X}$ shows that whenever the columns of $\mathbf{X}'\mathbf{X}$ are linearly dependent, the normal equations will be linearly dependent also. No unique solutions can be obtained for b_0 and b_1 in that case. Fortunately, in most regression applications, the columns of $\mathbf{X}'\mathbf{X}$ are linearly independent, leading to unique solutions for b_0 and b_1.

6.11 ANALYSIS OF VARIANCE RESULTS

Fitted values and residuals

Let the vector of the fitted values $\hat{Y}_i$ be denoted by $\hat{\mathbf{Y}}$:

(6.64)
$$\underset{n\times 1}{\hat{\mathbf{Y}}} = \begin{bmatrix} \hat{Y}_1 \\ \hat{Y}_2 \\ \cdot \\ \cdot \\ \cdot \\ \hat{Y}_n \end{bmatrix}$$

and the vector of the residuals $e_i = Y_i - \hat{Y}_i$ be denoted by $\mathbf{e}$:

(6.65)
$$\underset{n\times 1}{\mathbf{e}} = \begin{bmatrix} e_1 \\ e_2 \\ \cdot \\ \cdot \\ \cdot \\ e_n \end{bmatrix}$$

In matrix notation, we then have:

(6.66)
$$\underset{n\times 1}{\hat{\mathbf{Y}}} = \underset{n\times 2}{\mathbf{X}} \; \underset{2\times 1}{\mathbf{b}}$$

because:

$$\begin{bmatrix} \hat{Y}_1 \\ \hat{Y}_2 \\ \cdot \\ \cdot \\ \cdot \\ \hat{Y}_n \end{bmatrix} = \begin{bmatrix} 1 & X_1 \\ 1 & X_2 \\ \cdot & \cdot \\ \cdot & \cdot \\ \cdot & \cdot \\ 1 & X_n \end{bmatrix} \begin{bmatrix} b_0 \\ b_1 \end{bmatrix} = \begin{bmatrix} b_0 + b_1X_1 \\ b_0 + b_1X_2 \\ \cdot \\ \cdot \\ \cdot \\ b_0 + b_1X_n \end{bmatrix}$$

Similarly:

(6.67)
$$\underset{n\times 1}{\mathbf{e}} = \underset{n\times 1}{\mathbf{Y}} - \underset{n\times 1}{\hat{\mathbf{Y}}} = \underset{n\times 1}{\mathbf{Y}} - \underset{n\times 1}{\mathbf{Xb}}$$

Sums of squares

To see how the sums of squares are expressed in matrix notation, we begin with *SSTO*. We know from (3.49) that:

(6.68)
$$SSTO = \Sigma Y_i^2 - n\bar{Y}^2 = \Sigma Y_i^2 - \frac{(\Sigma Y_i)^2}{n}$$

We also know from (6.15) that:

$$\mathbf{Y'Y} = \Sigma Y_i^2$$

The subtraction term $n\bar{Y}^2 = (\Sigma Y_i)^2/n$ in matrix form uses $\mathbf{1}$, the vector of 1's defined in (6.19), as follows:

(6.69)
$$\frac{(\Sigma Y_i)^2}{n} = \left(\frac{1}{n}\right)\mathbf{Y'11'Y}$$

For instance, if $n = 2$, we have:

$$\left(\frac{1}{n}\right)[Y_1 \quad Y_2]\begin{bmatrix}1\\1\end{bmatrix}[1 \quad 1]\begin{bmatrix}Y_1\\Y_2\end{bmatrix} = \left(\frac{1}{n}\right)(\Sigma Y_i)(\Sigma Y_i) = \frac{(\Sigma Y_i)^2}{n}$$

Hence, it follows that:

(6.70a) $$SSTO = \mathbf{Y}'\mathbf{Y} - \left(\frac{1}{n}\right)\mathbf{Y}'\mathbf{1}\mathbf{1}'\mathbf{Y}$$

Just as ΣY_i^2 is represented by $\mathbf{Y}'\mathbf{Y}$ in matrix terms, so $SSE = \Sigma e_i^2 = \Sigma(Y_i - \hat{Y}_i)^2$ can be represented as follows:

(6.70b) $$SSE = \mathbf{e}'\mathbf{e} = (\mathbf{Y} - \mathbf{X}\mathbf{b})'(\mathbf{Y} - \mathbf{X}\mathbf{b})$$

which can be shown to equal:

(6.70c) $$SSE = \mathbf{Y}'\mathbf{Y} - \mathbf{b}'\mathbf{X}'\mathbf{Y}$$

Finally, it can be shown that:

(6.70d) $$SSR = \mathbf{b}'\mathbf{X}'\mathbf{Y} - \left(\frac{1}{n}\right)\mathbf{Y}'\mathbf{1}\mathbf{1}'\mathbf{Y}$$

Example. Let us find SSE for the Westwood Company lot size example by matrix methods, using (6.70c). We know from earlier results:

$$\mathbf{Y}'\mathbf{Y} = \Sigma Y_i^2 = 134{,}660$$

We also know from earlier:

$$\mathbf{b} = \begin{bmatrix}10.0\\2.0\end{bmatrix} \qquad \mathbf{X}'\mathbf{Y} = \begin{bmatrix}1{,}100\\61{,}800\end{bmatrix}$$

Hence:

$$\mathbf{b}'\mathbf{X}'\mathbf{Y} = [10.0 \quad 2.0]\begin{bmatrix}1{,}100\\61{,}800\end{bmatrix} = 134{,}600$$

and:

$$SSE = \mathbf{Y}'\mathbf{Y} - \mathbf{b}'\mathbf{X}'\mathbf{Y} = 134{,}660 - 134{,}600 = 60$$

which is the same result as that obtained in Chapter 2. Any difference would have been due to rounding errors.

Similarly, we can find SSR using (6.70d):

$$SSR = \mathbf{b}'\mathbf{X}'\mathbf{Y} - \left(\frac{1}{n}\right)\mathbf{Y}'\mathbf{1}\mathbf{1}'\mathbf{Y}$$
$$= 134{,}600 - 10(110)^2 = 13{,}600$$

since the subtraction term in SSR equals $n\bar{Y}^2$, and $\bar{Y} = 110$ for the Westwood Company example.

Note

To illustrate the derivation of the sums of squares expressions in matrix notation, consider *SSE*:

$$SSE = \mathbf{e}'\mathbf{e} = (\mathbf{Y} - \mathbf{Xb})'(\mathbf{Y} - \mathbf{Xb}) = \mathbf{Y}'\mathbf{Y} - 2\mathbf{b}'\mathbf{X}'\mathbf{Y} + \mathbf{b}'\mathbf{X}'\mathbf{Xb}$$

In substituting for the right-most $\mathbf{b}$ we obtain by (6.59):

$$\begin{aligned} SSE &= \mathbf{Y}'\mathbf{Y} - 2\mathbf{b}'\mathbf{X}'\mathbf{Y} + \mathbf{b}'\mathbf{X}'\mathbf{X}(\mathbf{X}'\mathbf{X})^{-1}\mathbf{X}'\mathbf{Y} \\ &= \mathbf{Y}'\mathbf{Y} - 2\mathbf{b}'\mathbf{X}'\mathbf{Y} + \mathbf{b}'\mathbf{I}\mathbf{X}'\mathbf{Y} \end{aligned}$$

In dropping $\mathbf{I}$ and subtracting, we obtain the result in (6.70c):

$$SSE = \mathbf{Y}'\mathbf{Y} - \mathbf{b}'\mathbf{X}'\mathbf{Y}$$

Sums of squares as quadratic forms

The ANOVA sums of squares can be shown to be *quadratic forms*. An example of a quadratic form of the observations Y_i when $n = 2$ is:

(6.71) $$5Y_1^2 + 6Y_1Y_2 + 4Y_2^2$$

Note that this expression is a second-degree polynomial containing terms involving the squares of the observations and the cross product. We can express (6.71) in matrix terms as follows:

(6.71a) $$[Y_1 \quad Y_2]\begin{bmatrix} 5 & 3 \\ 3 & 4 \end{bmatrix}\begin{bmatrix} Y_1 \\ Y_2 \end{bmatrix} = \mathbf{Y}'\mathbf{AY}$$

where $\mathbf{A}$ is a symmetric matrix of coefficients.

In general, a quadratic form is defined as:

(6.72) $$\underset{1\times 1}{\mathbf{Y}'\mathbf{AY}} = \sum_{i=1}^{n}\sum_{j=1}^{n} a_{ij}Y_iY_j \qquad \text{where } a_{ij} = a_{ji}$$

$\mathbf{A}$ is a symmetric $n \times n$ matrix and is called the *matrix of the quadratic form*.

The ANOVA sums of squares *SSTO*, *SSR*, and *SSE* are all quadratic forms. To see this, we need to express the matrix forms for these sums of squares in (6.70) still more compactly. We do this by noting that:

(6.73) $$\underset{n\times 1}{\mathbf{1}} \; \underset{1\times n}{\mathbf{1}'} = \underset{n\times n}{\mathbf{J}}$$

where $\mathbf{J}$ is the $n \times n$ matrix all of whose elements are 1's, as defined in (6.20). Also, the transpose of $\mathbf{b}$ in (6.59) can be obtained using (6.36) and (6.33):

(6.74) $$\mathbf{b}' = [(\mathbf{X}'\mathbf{X})^{-1}\mathbf{X}'\mathbf{Y}]' = \mathbf{Y}'\mathbf{X}(\mathbf{X}'\mathbf{X})^{-1}$$

by noting that $(\mathbf{X}'\mathbf{X})^{-1}$ is a symmetric matrix so that it equals its transpose. Hence:

(6.75a) $$SSTO = \mathbf{Y}'\left[\mathbf{I} - \left(\frac{1}{n}\right)\mathbf{J}\right]\mathbf{Y}$$

(6.75b) $$SSR = \mathbf{Y}'\left[\mathbf{X}(\mathbf{X}'\mathbf{X})^{-1}\mathbf{X}' - \left(\frac{1}{n}\right)\mathbf{J}\right]\mathbf{Y}$$

(6.75c) $$SSE = \mathbf{Y}'[\mathbf{I} - \mathbf{X}(\mathbf{X}'\mathbf{X})^{-1}\mathbf{X}']\mathbf{Y}$$

Each of these sums of squares can now be seen to be of the form $\mathbf{Y}'\mathbf{A}\mathbf{Y}$. It can be shown that the three $\mathbf{A}$ matrices:

(6.76a) $$\mathbf{I} - \left(\frac{1}{n}\right)\mathbf{J}$$

(6.76b) $$\mathbf{X}(\mathbf{X}'\mathbf{X})^{-1}\mathbf{X}' - \left(\frac{1}{n}\right)\mathbf{J}$$

(6.76c) $$\mathbf{I} - \mathbf{X}(\mathbf{X}'\mathbf{X})^{-1}\mathbf{X}'$$

are symmetric. Hence, *SSTO*, *SSR*, and *SSE* are quadratic forms, with the matrices of the quadratic forms given in (6.76). Quadratic forms play an important role in statistics because all sums of squares in the analysis of variance for linear statistical models can be expressed as quadratic forms.

6.12 INFERENCES IN REGRESSION ANALYSIS

As we saw in earlier chapters, all interval estimates are of the following form: point estimator plus and minus a certain number of estimated standard deviations of the point estimator. Similarly, all tests require the point estimator and the estimated standard deviation of the point estimator or, in the case of analysis of variance tests, various sums of squares. Matrix algebra is of principal help in inference making when obtaining the estimated standard deviations and sums of squares. We have already given the matrix equivalents of the sums of squares for the analysis of variance. Hence, we focus here chiefly on the matrix expressions for the estimated standard deviations of point estimators of interest.

Regression coefficients

The variance-covariance matrix of $\mathbf{b}$:

(6.77) $$\boldsymbol{\sigma}^2(\mathbf{b}) = \begin{bmatrix} \sigma^2(b_0) & \sigma(b_0, b_1) \\ \sigma(b_1, b_0) & \sigma^2(b_1) \end{bmatrix}$$

is:

(6.78) $$\underset{2\times 2}{\boldsymbol{\sigma}^2(\mathbf{b})} = \sigma^2(\mathbf{X}'\mathbf{X})^{-1}$$

or, using (6.27):

(6.78a) $$\boldsymbol{\sigma}^2(\mathbf{b}) = \begin{bmatrix} \dfrac{\sigma^2\Sigma X_i^2}{n\Sigma(X_i - \bar{X})^2} & \dfrac{-\bar{X}\sigma^2}{\Sigma(X_i - \bar{X})^2} \\ \dfrac{-\bar{X}\sigma^2}{\Sigma(X_i - \bar{X})^2} & \dfrac{\sigma^2}{\Sigma(X_i - \bar{X})^2} \end{bmatrix}$$

When MSE is substituted for σ^2 in (6.78a) we have:

$$\text{(6.79)} \qquad \underset{2\times 2}{\mathbf{s}^2(\mathbf{b})} = MSE(\mathbf{X}'\mathbf{X})^{-1} = \begin{bmatrix} \dfrac{MSE\Sigma X_i^2}{n\Sigma(X_i - \bar{X})^2} & \dfrac{-\bar{X}MSE}{\Sigma(X_i - \bar{X})^2} \\ \dfrac{-\bar{X}MSE}{\Sigma(X_i - \bar{X})^2} & \dfrac{MSE}{\Sigma(X_i - \bar{X})^2} \end{bmatrix}$$

where $\mathbf{s}^2(\mathbf{b})$ is the estimated variance-covariance matrix of $\mathbf{b}$. In (6.78a), you will recognize the variances of b_0 (3.20b) and b_1 (3.3b) and the covariance of b_0 and b_1 (5.3). Likewise, the estimated variances in (6.79) are familiar from earlier chapters.

Joint confidence region for β_0 and β_1

The boundary for the joint confidence region for β_0 and β_1, given in (5.2), is expressed in matrix terms as follows:

$$\text{(6.80)} \qquad \frac{(\mathbf{b} - \boldsymbol{\beta})'\mathbf{X}'\mathbf{X}(\mathbf{b} - \boldsymbol{\beta})}{2MSE} = F(1 - \alpha; 2, n - 2)$$

Mean response

To estimate the mean response at X_h, let us define the vector:

$$\text{(6.81)} \qquad \underset{2\times 1}{\mathbf{X}_h} = \begin{bmatrix} 1 \\ X_h \end{bmatrix} \quad \text{or} \quad \mathbf{X}_h' = [1 \quad X_h]$$

The fitted value in matrix notation then is:

$$\text{(6.82)} \qquad \hat{Y}_h = \underset{1\times 1}{\mathbf{X}_h'\mathbf{b}}$$

since:

$$\mathbf{X}_h'\mathbf{b} = [1 \quad X_h] \begin{bmatrix} b_0 \\ b_1 \end{bmatrix} = [b_0 + b_1 X_h] = [\hat{Y}_h] = \hat{Y}_h$$

Note that $\mathbf{X}_h'\mathbf{b}$ is a 1×1 matrix; hence, we can write the final result as a scalar.

The variance of $\hat{Y}_h$, given earlier in (3.28b), is in matrix notation:

$$\text{(6.83)} \qquad \sigma^2(\hat{Y}_h) = \sigma^2\mathbf{X}_h'(\mathbf{X}'\mathbf{X})^{-1}\mathbf{X}_h = \mathbf{X}_h'\boldsymbol{\sigma}^2(\mathbf{b})\mathbf{X}_h$$

where $\boldsymbol{\sigma}^2(\mathbf{b})$ is the variance-covariance matrix of the regression coefficients in (6.78). Note, therefore, that $\sigma^2(\hat{Y}_h)$ is a function of the variances $\sigma^2(b_0)$ and $\sigma^2(b_1)$ and of the covariance $\sigma(b_0, b_1)$.

The estimated variance of $\hat{Y}_h$, given earlier in (3.30), is in matrix notation:

$$\text{(6.84)} \qquad s^2(\hat{Y}_h) = MSE(\mathbf{X}_h'(\mathbf{X}'\mathbf{X})^{-1}\mathbf{X}_h) = \mathbf{X}_h'\mathbf{s}^2(\mathbf{b})\mathbf{X}_h$$

where $\mathbf{s}^2(\mathbf{b})$ is the estimated variance-covariance matrix of the regression coefficients in (6.79).

Prediction of new observation

The estimated variance $s^2(Y_{h(\text{new})})$, given earlier in (3.37), is in matrix notation:

$$s^2(Y_{h(\text{new})}) = MSE + s^2(\hat{Y}_h) = MSE + \mathbf{X}_h' \mathbf{s}^2(\mathbf{b})\mathbf{X}_h \tag{6.85}$$
$$= MSE(1 + \mathbf{X}_h'(\mathbf{X}'\mathbf{X})^{-1}\mathbf{X}_h)$$

Examples

1. We wish to find $s^2(b_0)$ and $s^2(b_1)$ for the Westwood Company lot size example by matrix methods. We found earlier that $MSE = 7.5$ and:

$$(\mathbf{X}'\mathbf{X})^{-1} = \begin{bmatrix} .83529412 & -.01470588 \\ -.01470588 & .00029412 \end{bmatrix}$$

Hence, by (6.79):

$$\mathbf{s}^2(\mathbf{b}) = MSE(\mathbf{X}'\mathbf{X})^{-1} = 7.5\begin{bmatrix} .83529412 & -.01470588 \\ -.01470588 & .00029412 \end{bmatrix}$$
$$= \begin{bmatrix} 6.264706 & -.1102941 \\ -.1102941 & .0022059 \end{bmatrix}$$

Thus, $s^2(b_0) = 6.26471$ and $s^2(b_1) = .002206$. These are the same as the results obtained in Chapter 3.

Note how simple it is to find the estimated variances of the regression coefficients as soon as $(\mathbf{X}'\mathbf{X})^{-1}$ has been obtained. This inverse is needed in the first place to find the regression coefficients, so that practically no extra work is required to obtain their estimated variances.

2. We wish to find $s^2(\hat{Y}_h)$ for the Westwood Company example when $X_h = 55$. We define:

$$\mathbf{X}_h' = [1 \quad 55]$$

and obtain by (6.84):

$$s^2(\hat{Y}_{55}) = \mathbf{X}_h'\mathbf{s}^2(\mathbf{b})\mathbf{X}_h$$
$$= [1 \quad 55]\begin{bmatrix} 6.264706 & -.1102941 \\ -.1102941 & .0022059 \end{bmatrix}\begin{bmatrix} 1 \\ 55 \end{bmatrix} = .80520$$

This is the same result as that obtained in Chapter 3, except for a minor difference due to rounding.

Comments

1. To illustrate a derivation in matrix terms, let us find the variance-covariance matrix of $\mathbf{b}$. Recall that:

$$\mathbf{b} = (\mathbf{X}'\mathbf{X})^{-1}\mathbf{X}'\mathbf{Y} = \mathbf{A}\mathbf{Y}$$

where $\mathbf{A}$ is a constant matrix:

$$\mathbf{A} = (\mathbf{X}'\mathbf{X})^{-1}\mathbf{X}'$$

Hence by (6.49), we have:

$$\boldsymbol{\sigma}^2(\mathbf{b}) = \mathbf{A}[\boldsymbol{\sigma}^2(\mathbf{Y})]\mathbf{A}'$$

Now $\boldsymbol{\sigma}^2(\mathbf{Y}) = \sigma^2\mathbf{I}$. Further, it follows from (6.74) that:

$$\mathbf{A}' = \mathbf{X}(\mathbf{X}'\mathbf{X})^{-1}$$

We find therefore:

$$\begin{aligned}\boldsymbol{\sigma}^2(\mathbf{b}) &= (\mathbf{X}'\mathbf{X})^{-1}\mathbf{X}'\sigma^2\mathbf{I}\mathbf{X}(\mathbf{X}'\mathbf{X})^{-1}\\ &= \sigma^2(\mathbf{X}'\mathbf{X})^{-1}\mathbf{X}'\mathbf{X}(\mathbf{X}'\mathbf{X})^{-1}\\ &= \sigma^2(\mathbf{X}'\mathbf{X})^{-1}\mathbf{I}\\ &= \sigma^2(\mathbf{X}'\mathbf{X})^{-1}\end{aligned}$$

2. Since $\hat{Y}_h = \mathbf{X}_h'\mathbf{b}$, it follows at once from (6.49) that:

$$\sigma^2(\hat{Y}_h) = \mathbf{X}_h'[\boldsymbol{\sigma}^2(\mathbf{b})]\mathbf{X}_h$$

Hence:

$$\sigma^2(\hat{Y}_h) = [1 \quad X_h] \begin{bmatrix} \sigma^2(b_0) & \sigma(b_0, b_1) \\ \sigma(b_1, b_0) & \sigma^2(b_1) \end{bmatrix} \begin{bmatrix} 1 \\ X_h \end{bmatrix}$$

or:

$$\sigma^2(\hat{Y}_h) = \sigma^2(b_0) + 2X_h\sigma(b_0, b_1) + X_h^2\sigma^2(b_1) \tag{6.86}$$

Using the results from (6.78a), we obtain:

$$\sigma^2(\hat{Y}_h) = \frac{\sigma^2\Sigma X_i^2}{n\Sigma(X_i - \bar{X})^2} + \frac{2X_h(-\bar{X})\sigma^2}{\Sigma(X_i - \bar{X})^2} + \frac{X_h^2\sigma^2}{\Sigma(X_i - \bar{X})^2}$$

which reduces to the familiar expression:

$$\sigma^2(\hat{Y}_h) = \sigma^2\left[\frac{1}{n} + \frac{(X_h - \bar{X})^2}{\Sigma(X_i - \bar{X})^2}\right] \tag{6.87}$$

Thus, we see explicitly that the variance expression in (6.87) contains contributions from $\sigma^2(b_0)$, $\sigma^2(b_1)$, and $\sigma(b_0, b_1)$, which it must according to theorem (1.25b) since $\hat{Y}_h$ is a linear combination of b_0 and b_1:

$$\hat{Y}_h = b_0 + b_1X_h$$

3. We do not show the results in matrix terms for other types of inferences, such as simultaneous prediction of several new observations on Y at different X_h levels, since these are based on results we have developed.

6.13 WEIGHTED LEAST SQUARES

The regression results for weighted least squares can be stated compactly with matrix algebra. Let the matrix $\mathbf{W}$ be a diagonal matrix containing the weights w_i.

(6.88)
$$\underset{n\times n}{\mathbf{W}} = \begin{bmatrix} w_1 & & & & 0 \\ & w_2 & & & \\ & & \cdot & & \\ & & & \cdot & \\ 0 & & & & w_n \end{bmatrix}$$

The weighted least squares normal equations (5.32) can then be expressed as follows:

(6.89)
$$\mathbf{X'WXb} = \mathbf{X'WY}$$

and the weighted least squares estimators are:

(6.90)
$$\underset{2\times 1}{\mathbf{b}} = (\mathbf{X'WX})^{-1}\mathbf{X'WY}$$

Note that if $\mathbf{W} = \mathbf{I}$, as it would for unweighted least squares, (6.90) reduces to the unweighted estimators (6.59).

Other results for weighted least squares bear a similar relation to the earlier results for unweighted least squares. For instance, when the error term variances σ_i^2 are not equal, the weights w_i are chosen to be inversely proportional to σ_i^2, so that $\sigma_i^2 = \sigma^2/w_i$. The variance-covariance matrix of the weighted least squares estimators, then, is:

(6.91)
$$\underset{2\times 2}{\boldsymbol{\sigma}^2(\mathbf{b})} = \sigma^2(\mathbf{X'WX})^{-1}$$

and the estimated variance-covariance matrix is:

(6.92)
$$\underset{2\times 2}{\mathbf{s}^2(\mathbf{b})} = MSE_w(\mathbf{X'WX})^{-1}$$

where MSE_w is based on the weighted squared deviations:

(6.92a)
$$MSE_w = \frac{\Sigma w_i(Y_i - \hat{Y}_i)^2}{n-2}$$

6.14 RESIDUALS

For later analysis of residuals, it will be useful to recognize that each residual e_i can be expressed as a linear combination of the observations Y_i. It can be shown that the vector of the residuals $\mathbf{e}$, defined in (6.65), equals:

(6.93)
$$\underset{n\times 1}{\mathbf{e}} = (\underset{n\times n}{\mathbf{I}} - \underset{n\times n}{\mathbf{H}})\underset{n\times 1}{\mathbf{Y}}$$

where:

(6.93a)
$$\underset{n\times n}{\mathbf{H}} = \mathbf{X}(\mathbf{X'X})^{-1}\mathbf{X'}$$

Note from (6.76c) that the matrix $\mathbf{I} - \mathbf{H}$ is the matrix of the quadratic form (6.75c) for $SSE = \Sigma e_i^2$.

The square $n \times n$ matrix $\mathbf{H}$ is called the *hat matrix* and plays an important role in regression analysis, as we shall see in Chapter 11 when we consider whether or not regression results are unduly influenced by one or a few observations. The matrix $\mathbf{I} - \mathbf{H}$ is symmetric and has the special property (called idempotency):

$$(\mathbf{I} - \mathbf{H})(\mathbf{I} - \mathbf{H}) = \mathbf{I} - \mathbf{H} \tag{6.94}$$

In general, a matrix $\mathbf{M}$ is said to be idempotent if $\mathbf{MM} = \mathbf{M}$.

It can be shown that the variance-covariance matrix of the vector of residuals $\mathbf{e}$ also involves the matrix $\mathbf{I} - \mathbf{H}$:

$$\boldsymbol{\sigma}^2(\mathbf{e}) = \sigma^2(\mathbf{I} - \mathbf{H}) \tag{6.95}$$

and is estimated by:

$$\mathbf{s}^2(\mathbf{e}) = MSE(\mathbf{I} - \mathbf{H}) \tag{6.96}$$

Note

The variance-covariance matrix of $\mathbf{e}$ can be derived by means of (6.49). Since $\mathbf{e} = (\mathbf{I} - \mathbf{H})\mathbf{Y}$, we obtain:

$$\boldsymbol{\sigma}^2(\mathbf{e}) = (\mathbf{I} - \mathbf{H})\boldsymbol{\sigma}^2(\mathbf{Y})(\mathbf{I} - \mathbf{H})'$$

Now $\boldsymbol{\sigma}^2(\mathbf{Y}) = \boldsymbol{\sigma}^2(\boldsymbol{\varepsilon}) = \sigma^2\mathbf{I}$ for the normal error model according to (6.55). Also, $(\mathbf{I} - \mathbf{H})' = \mathbf{I} - \mathbf{H}$ because of the symmetry of the matrix. Hence:

$$\begin{aligned}\boldsymbol{\sigma}^2(\mathbf{e}) &= \sigma^2(\mathbf{I} - \mathbf{H})\mathbf{I}(\mathbf{I} - \mathbf{H}) \\ &= \sigma^2(\mathbf{I} - \mathbf{H})(\mathbf{I} - \mathbf{H})\end{aligned}$$

In view of property (6.94), we obtain formula (6.95):

$$\boldsymbol{\sigma}^2(\mathbf{e}) = \sigma^2(\mathbf{I} - \mathbf{H})$$

PROBLEMS

6.1. For the matrices below, obtain: (1) $\mathbf{A} + \mathbf{B}$, (2) $\mathbf{A} - \mathbf{B}$, (3) $\mathbf{AC}$, (4) $\mathbf{AB}'$, (5) $\mathbf{B}'\mathbf{A}$.

$$\mathbf{A} = \begin{bmatrix} 1 & 4 \\ 2 & 6 \\ 3 & 8 \end{bmatrix} \qquad \mathbf{B} = \begin{bmatrix} 1 & 3 \\ 1 & 4 \\ 2 & 5 \end{bmatrix} \qquad \mathbf{C} = \begin{bmatrix} 3 & 8 & 1 \\ 5 & 4 & 0 \end{bmatrix}$$

State the dimension of each resulting matrix.

6.2. For the matrices below, obtain: (1) $\mathbf{A} + \mathbf{C}$, (2) $\mathbf{A} - \mathbf{C}$, (3) $\mathbf{B}'\mathbf{A}$, (4) $\mathbf{AC}'$, (5) $\mathbf{C}'\mathbf{A}$.

$$\mathbf{A} = \begin{bmatrix} 2 & 1 \\ 3 & 5 \\ 5 & 7 \\ 4 & 8 \end{bmatrix} \qquad \mathbf{B} = \begin{bmatrix} 6 \\ 9 \\ 3 \\ 1 \end{bmatrix} \qquad \mathbf{C} = \begin{bmatrix} 3 & 8 \\ 8 & 6 \\ 5 & 1 \\ 2 & 4 \end{bmatrix}$$

State the dimension of each resulting matrix.

6.3. Show how the following expressions are written in terms of matrices: (1) $Y_i - \hat{Y}_i = e_i$, (2) $\Sigma X_i e_i = 0$. Assume $i = 1, \ldots, 4$.

6.4. Flavor deterioration. The results shown below were obtained in a small-scale experiment to study the relation between °F of storage temperature (X) and number of weeks before flavor deterioration of a food product begins to occur (Y).

i:	1	2	3	4	5
X_i:	+8	+4	0	−4	−8
Y_i:	7.8	9.0	10.2	11.0	11.7

Assume that the first-order regression model (3.1) is applicable. Using matrix methods, find: (1) $\mathbf{Y}'\mathbf{Y}$, (2) $\mathbf{X}'\mathbf{X}$, (3) $\mathbf{X}'\mathbf{Y}$.

6.5. Consumer finance. The data below show for a consumer finance company operating in six cities, the number of competing loan companies operating in the city (X) and the number per thousand of the company's loans made in that city that are currently delinquent (Y):

i:	1	2	3	4	5	6
X_i:	4	1	2	3	3	4
Y_i:	16	5	10	15	13	22

Assume that the first-order regression model (3.1) is applicable. Using matrix methods, find: (1) $\mathbf{Y}'\mathbf{Y}$, (2) $\mathbf{X}'\mathbf{X}$, (3) $\mathbf{X}'\mathbf{Y}$.

6.6. Refer to **Airfreight breakage** Problem 2.17. Using matrix methods, find: (1) $\mathbf{Y}'\mathbf{Y}$, (2) $\mathbf{X}'\mathbf{X}$, (3) $\mathbf{X}'\mathbf{Y}$.

6.7. Refer to **Plastic hardness** Problem 2.18. Using matrix methods, find: (1) $\mathbf{Y}'\mathbf{Y}$, (2) $\mathbf{X}'\mathbf{X}$, (3) $\mathbf{X}'\mathbf{Y}$.

6.8. Let $\mathbf{B}$ be defined as follows:

$$\mathbf{B} = \begin{bmatrix} 1 & 5 & 0 \\ 1 & 0 & 5 \\ 1 & 0 & 5 \end{bmatrix}$$

a. Are the column vectors of $\mathbf{B}$ linearly dependent?
b. What is the rank of $\mathbf{B}$?
c. What must be the determinant of $\mathbf{B}$?

6.9. Let $\mathbf{A}$ be defined as follows:

$$\mathbf{A} = \begin{bmatrix} 0 & 1 & 8 \\ 0 & 3 & 1 \\ 0 & 5 & 5 \end{bmatrix}$$

a. Are the column vectors of $\mathbf{A}$ linearly dependent?
b. Restate definition (6.22) in terms of row vectors. Are the row vectors of $\mathbf{A}$ linearly dependent?
c. What is the rank of $\mathbf{A}$?
d. Calculate the determinant of $\mathbf{A}$.

6.10. Find the inverse of each of the following matrices:

$$\mathbf{A} = \begin{bmatrix} 2 & 4 \\ 3 & 1 \end{bmatrix} \qquad \mathbf{B} = \begin{bmatrix} 4 & 3 & 2 \\ 6 & 5 & 10 \\ 10 & 1 & 6 \end{bmatrix}$$

Check in each case that the resulting matrix is indeed the inverse.

6.11. Find the inverse of the following matrix:

$$\mathbf{A} = \begin{bmatrix} 5 & 1 & 3 \\ 4 & 0 & 5 \\ 1 & 9 & 6 \end{bmatrix}$$

Check that the resulting matrix is indeed the inverse.

6.12. Refer to **Flavor deterioration** Problem 6.4. Find $(\mathbf{X}'\mathbf{X})^{-1}$.

6.13. Refer to **Consumer finance** Problem 6.5. Find $(\mathbf{X}'\mathbf{X})^{-1}$.

6.14. Consider the simultaneous equations:

$$4y_1 + 7y_2 = 25$$
$$2y_1 + 3y_2 = 12$$

a. Write these equations in matrix notation.
b. Using matrix methods, find the solutions for y_1 and y_2.

6.15. Consider the simultaneous equations:

$$5y_1 + 2y_2 = 8$$
$$23y_1 + 7y_2 = 28$$

a. Write these equations in matrix notation.
b. Using matrix methods, find the solutions for y_1 and y_2.

6.16. Consider the estimated linear regression function in the form of (2.15). Write expressions in this form for the fitted values $\hat{Y}_i$ in matrix terms for $i = 1, \ldots, 5$.

6.17. Consider the following functions of the random variables Y_1, Y_2, and Y_3:

$$W_1 = Y_1 + Y_2 + Y_3$$
$$W_2 = Y_1 - Y_2$$
$$W_3 = Y_1 - Y_2 - Y_3$$

a. State the above in matrix notation.
b. Find the expectation of the random vector $\mathbf{W}$.
c. Find the variance-covariance matrix of $\mathbf{W}$.

6.18. Consider the following functions of the random variables Y_1, Y_2, Y_3, and Y_4:

$$W_1 = \frac{1}{4}(Y_1 + Y_2 + Y_3 + Y_4)$$
$$W_2 = \frac{1}{2}(Y_1 + Y_2) - \frac{1}{2}(Y_3 + Y_4)$$

a. State the above in matrix notation.
b. Find the expectation of the random vector $\mathbf{W}$.
c. Find the variance-covariance matrix of $\mathbf{W}$.

6.19. Find the matrix $\mathbf{A}$ of the quadratic form:

$$3Y_1^2 + 10Y_1Y_2 + 17Y_2^2$$

6.20. Find the matrix $\mathbf{A}$ of the quadratic form:

$$7Y_1^2 - 8Y_1Y_2 + 8Y_2^2$$

6.21. For the matrix:

$$\mathbf{A} = \begin{bmatrix} 5 & 2 \\ 2 & 1 \end{bmatrix}$$

find the quadratic form of the observations Y_1 and Y_2.

6.22. For the matrix:

$$\mathbf{A} = \begin{bmatrix} 1 & 0 & 4 \\ 0 & 3 & 0 \\ 4 & 0 & 9 \end{bmatrix}$$

find the quadratic form of the observations Y_1, Y_2, and Y_3.

6.23. Refer to **Flavor deterioration** Problems 6.4 and 6.12.

a. Using matrix methods, obtain the following: (1) vector of estimated regression coefficients, (2) vector of residuals, (3) SSR, (4) SSE, (5) estimated variance-covariance matrix of $\mathbf{b}$, (6) point estimate of $E(Y_h)$ when $X_h = -6$, (7) estimated variance of $\hat{Y}_h$ when $X_h = -6$.

b. What simplifications arose from the spacing of the X levels in the experiment?

c. Using matrix methods, obtain the numerator of the left term in (6.80).

6.24. Refer to **Consumer finance** Problems 6.5 and 6.13.

a. Using matrix methods, obtain the following: (1) vector of estimated regression coefficients, (2) vector of residuals, (3) SSR, (4) SSE, (5) estimated variance-covariance matrix of $\mathbf{b}$, (6) point estimate of $E(Y_h)$ when $X_h = 4$, (7) estimated variance of $Y_{h(\text{new})}$ when $X_h = 4$.

b. From your estimated variance-covariance matrix in part (a5), obtain the following: (1) $s(b_0, b_1)$; (2) $s^2(b_0)$; (3) $s(b_1)$.

6.25. Refer to **Airfreight breakage** Problems 2.17 and 6.6.

a. Using matrix methods, obtain the following: (1) $(\mathbf{X}'\mathbf{X})^{-1}$, (2) $\mathbf{b}$, (3) $\mathbf{e}$, (4) SSR, (5) SSE, (6) $\mathbf{s}^2(\mathbf{b})$, (7) $\hat{Y}_h$ when $X_h = 2$, (8) $s^2(\hat{Y}_h)$ when $X_h = 2$.

b. From part (a6), obtain the following: (1) $s^2(b_1)$; (2) $s(b_0, b_1)$; (3) $s(b_0)$.

6.26. Refer to **Plastic hardness** Problems 2.18 and 6.7.

a. Using matrix methods, obtain the following: (1) $(\mathbf{X}'\mathbf{X})^{-1}$, (2) $\mathbf{b}$, (3) $\hat{\mathbf{Y}}$, (4) SSR, (5) SSE, (6) $\mathbf{s}^2(\mathbf{b})$, (7) $s^2(Y_{h(\text{new})})$ when $X_h = 30$.

b. From part (a6), obtain the following: (1) $s^2(b_0)$; (2) $s(b_0, b_1)$; (3) $s(b_1)$.

EXERCISES

6.27. Refer to regression model (5.13). Set up the expectation vector for $\boldsymbol{\varepsilon}$. Assume that $i = 1, \ldots, 4$.

6.28. Consider model (5.13) for regression through the origin and the estimator b_1 given in (5.17). Obtain (5.17) by utilizing (6.59) with $\mathbf{X}$ suitably defined.

6.29. Consider the least squares estimator $\mathbf{b}$ given in (6.59). Using matrix methods, show that $\mathbf{b}$ is an unbiased estimator.

6.30. Show that $\hat{Y}_h$ in (6.82) can be expressed in matrix terms as $\mathbf{b}'\mathbf{X}_h$.

6.31. Refer to regression model (5.36). Set up the variance-covariance matrix for the error terms when $i = 1, \ldots, 4$. Assume $\sigma(\varepsilon_i, \varepsilon_j) = 0$ for $i \neq j$.

6.32. Derive the variance-covariance matrix $\boldsymbol{\sigma}^2(\mathbf{b})$ in (6.91) for the weighted least squares estimators when the variance-covariance matrix of the observation Y_i is $\sigma^2\mathbf{W}^{-1}$, where $\mathbf{W}$ is given in (6.88).

6.33. a. Obtain an expression for $\hat{\mathbf{Y}}$ in terms of the $\mathbf{H}$ matrix defined in (6.93a). [*Hint:* Use (6.93).]

b. Obtain an expression for the variance-covariance matrix of the fitted values $\hat{Y}_i$, $i = 1, \ldots, n$, in terms of the hat matrix.

CITED REFERENCE

6.1 Graybill, Franklin A. *Introduction to Matrices with Applications in Statistics*. Belmont, Calif.: Wadsworth, 1969.

7

Multiple regression—I

Multiple regression analysis is one of the most widely used of all statistical tools. In this chapter, we first discuss a variety of multiple regression models. Then we present the basic statistical results for multiple regression in matrix form. Since the matrix expressions for multiple regression are the same as for simple regression, we state the results without much discussion. We then give an example, illustrating a variety of inferences in multiple regression analysis. Finally, we take up some additional facets of multiple regression analysis.

7.1 MULTIPLE REGRESSION MODELS

Need for several independent variables

When we first introduced regression analysis in Chapter 2, we spoke of regression models containing a number of independent variables. We mentioned a regression model where the dependent variable was direct operating cost for a branch office of a consumer finance chain, and four independent variables were considered, including average number of loans outstanding at the branch and total number of new loan applications processed by the branch. We also mentioned a tractor purchase study where the response variable was volume of tractor purchases in a sales territory, and the nine independent variables included number of farms in the territory and quantity of crop production in the territory. In addition, we mentioned a study of short children where the response variable was the peak plasma growth hormone level, and the 14 independent variables included sex, age, and various body measurements. In all these examples, one independent variable in the model would have provided an inadequate description since a number of key independent variables affect the response variable in important and distinctive ways. Furthermore, in situations of this type, one will frequently find that predictions of the response variable based on a model containing only a single independent variable are too imprecise to be useful. A more complex model, containing additional independent variables, typically is more helpful in providing sufficiently precise predictions of the response variable.

In each of the examples mentioned, the analysis is based on observational data because some or all of the independent variables are not susceptible to direct control. Multiple regression analysis is also highly useful in experimental situations where the experimenter can control the independent variables. An experimenter typically will wish to investigate a number of independent variables simultaneously because almost always more than one key independent variable influences the response. For example, in a study on productivity of work crews, the experimenter may wish to control both the size of the crew and the level of bonus pay. Similarly, in a study on responsiveness to a drug, the experimenter may wish to control both the dose of the drug and the body surface area of the subject.

First-order model with two independent variables

When there are two independent variables X_1 and X_2, the model:

$$Y_i = \beta_0 + \beta_1 X_{i1} + \beta_2 X_{i2} + \varepsilon_i \tag{7.1}$$

is called a first-order model with two independent variables. A first-order model, it will be recalled from Chapter 2, is linear in the parameters and linear in the independent variables. Y_i denotes as usual the response in the ith trial, and X_{i1} and X_{i2} are the values of the two independent variables in the ith trial. The parameters of the model are β_0, β_1, and β_2, and the error term is ε_i.

Assuming that $E(\varepsilon_i) = 0$, the regression function for model (7.1) is:

$$E(Y) = \beta_0 + \beta_1 X_1 + \beta_2 X_2 \tag{7.2}$$

Analogous to simple linear regression, where the regression function $E(Y) = \beta_0 + \beta_1 X$ is a line, the regression function (7.2) is a plane. Figure 7.1 contains a representation of a portion of the response plane:

$$E(Y) = 20.0 + .95X_1 - .50X_2 \tag{7.3}$$

Note that a point on the response plane (7.3) corresponds to the mean response $E(Y)$ at the given combination of levels of X_1 and X_2.

Figure 7.1 also shows a series of observations Y_i corresponding to given levels of the two independent variables (X_{i1}, X_{i2}). Note that each vertical rule in Figure 7.1 represents the difference between Y_i and the mean $E(Y_i)$ of the probability distribution for (X_{i1}, X_{i2}) on the response plane. Hence, the vertical distance from Y_i to the response plane represents the error term $\varepsilon_i = Y_i - E(Y_i)$.

Frequently the regression function in multiple regression is called a *regression*

FIGURE 7.1 Example of response surface—a response plane with observations scattered about it

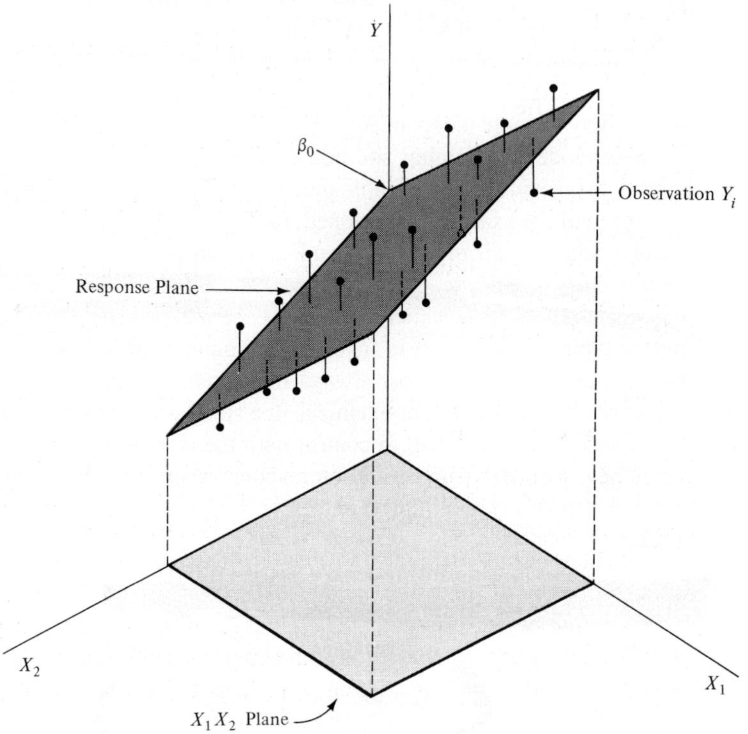

surface or a *response surface*. In Figure 7.1, the response surface is just a simple plane, but in other cases the response surface may be complex in nature.

Meaning of regression coefficients. Let us now consider the meaning of the regression parameters in the multiple regression function (7.2). The parameter β_0 is the Y intercept of the regression plane. If the scope of the model includes $X_1 = 0, X_2 = 0$, β_0 gives the mean response at $X_1 = 0, X_2 = 0$. Otherwise, β_0 does not have any particular meaning as a separate term in the regression model.

The parameter β_1 indicates the change in the mean response per unit increase in X_1 when X_2 is held constant. Likewise, β_2 indicates the change in the mean response per unit increase in X_2 when X_1 is held constant. To see this for our example, suppose X_2 is held at the level $X_2 = 20$. The regression function (7.3) now is:

$$
\begin{aligned}
(7.4) \qquad E(Y) &= 20.0 + .95X_1 - .50(20) = (20.0 - 10.0) + .95X_1 \\
&= 10.0 + .95X_1
\end{aligned}
$$

Note that for $X_2 = 20$, the response function is a straight line with slope .95. The same is true for any other value of X_2; only the intercept of the response function differs. Hence, $\beta_1 = .95$ indicates that the mean response increases by .95 with a unit increase in X_1 when X_2 is constant, no matter what the level of X_2. More loosely speaking, we state that β_1 indicates the change in $E(Y)$ with a unit increase in X_1 when X_2 is held constant.

Similarly, $\beta_2 = -.50$ in model (7.3) indicates that the mean response decreases by .50 with a unit increase in X_2 when X_1 is held constant.

When the effect of X_1 on the mean response does not depend on the level of X_2, and correspondingly the effect of X_2 does not depend on the level of X_1, the two independent variables are said to have *additive effects* or *not to interact*. Thus, the first-order model (7.1) is designed for independent variables whose effects on the mean response are additive or do not interact.

The parameters β_1 and β_2 are frequently called *partial regression coefficients* because they reflect the partial effect of one independent variable when the other independent variable is included in the model and is held constant.

Example. Suppose that the response surface in (7.3) pertains to urban full-service stations of a major oil company and shows the effect of variety and adequacy of services (X_1) and average time taken to reach car (X_2) on the ratio of actual gallonage of gasoline sold to potential gallonage (Y), where X_1 is expressed as an index with 100 = average, X_2 is in seconds, and Y is stated as a percent. Increasing the index of adequacy of services by one point while holding average time to reach car constant leads to an increase of .95 percent point in the expected ratio of actual to potential gallonage. If the index of adequacy of services is held constant and the average time to reach car is increased by one second, the expected ratio of actual to potential gallonage decreases by .50 percent point.

Comments

1. A regression model for which the response surface is a plane can be used either in its own right when it is appropriate, or as an approximation to a more complex response surface. Many complex response surfaces can be approximated well by a plane for limited ranges of X_1 and X_2.

2. We can readily establish the meaning of β_1 and β_2 by calculus, taking partial derivatives of the response surface (7.2) with respect to X_1 and X_2 in turn:

$$\frac{\partial E(Y)}{\partial X_1} = \beta_1 \qquad \frac{\partial E(Y)}{\partial X_2} = \beta_2$$

The partial derivatives measure the rate of change in $E(Y)$ with respect to one independent variable when the other is held constant.

First-order model with more than two independent variables

We consider now the case where there are $p - 1$ independent variables $X_1, \ldots, X_{p-1}$. The model:

$$Y_i = \beta_0 + \beta_1 X_{i1} + \beta_2 X_{i2} + \cdots + \beta_{p-1} X_{i,p-1} + \varepsilon_i \tag{7.5}$$

is called a first-order model with $p - 1$ independent variables. It can also be written:

(7.5a) $$Y_i = \beta_0 + \sum_{k=1}^{p-1} \beta_k X_{ik} + \varepsilon_i$$

or, if we let $X_{i0} \equiv 1$, it can be written as:

(7.5b) $$Y_i = \sum_{k=0}^{p-1} \beta_k X_{ik} + \varepsilon_i \qquad \text{where } X_{i0} \equiv 1$$

Assuming that $E(\varepsilon_i) = 0$, the response function for model (7.5) is:

(7.6) $$E(Y) = \beta_0 + \beta_1 X_{i1} + \beta_2 X_{i2} + \cdots + \beta_{p-1} X_{i,p-1}$$

This response function is a *hyperplane*, which is a plane in more than two dimensions. It is no longer possible to picture this response surface, as we were able to do in Figure 7.1 for the case of two independent variables. Nevertheless, the meaning of the parameters is analogous to the two independent variables case. The parameter β_k indicates the change in the mean response $E(Y)$ with a unit increase in the independent variable X_k, when all other independent variables X_1, X_2, etc., included in the model are held constant. Note again that the effect of any independent variable on the mean response is the same for model (7.5), no matter what are the levels at which the other independent variables are held. Hence, the first-order model (7.5) is designed for independent variables whose effects on the mean response are additive and therefore do not interact.

Note

If $p - 1 = 1$, model (7.5) reduces to:

$$Y_i = \beta_0 + \beta_1 X_{i1} + \varepsilon_i$$

which is the simple linear regression model considered in earlier chapters.

General linear regression model

In general, the variables $X_1, \ldots, X_{p-1}$ in a regression model do not have to represent different independent variables, as we shall shortly see. We therefore define the general linear regression model, with normal error terms, simply in terms of X variables:

(7.7) $$Y_i = \beta_0 + \beta_1 X_{i1} + \beta_2 X_{i2} + \cdots + \beta_{p-1} X_{i,p-1} + \varepsilon_i$$

where:

$\beta_0, \beta_1, \ldots, \beta_{p-1}$ are parameters
$X_{i1}, \ldots, X_{i,p-1}$ are known constants
ε_i are independent $N(0, \sigma^2)$
$i = 1, \ldots, n$

If we let $X_{i0} \equiv 1$, model (7.7) can be written as follows:

(7.7a) $$Y_i = \beta_0 X_{i0} + \beta_1 X_{i1} + \beta_2 X_{i2} + \cdots + \beta_{p-1} X_{i,p-1} + \varepsilon_i$$

where $X_{i0} \equiv 1$

or:

(7.7b)
$$Y_i = \sum_{k=0}^{p-1} \beta_k X_{ik} + \varepsilon_i \qquad \text{where } X_{i0} \equiv 1$$

The response function for model (7.7) is, since $E(\varepsilon_i) = 0$:

(7.8)
$$E(Y) = \beta_0 + \beta_1 X_1 + \beta_2 X_2 + \cdots + \beta_{p-1} X_{p-1}$$

Thus, the general linear regression model implies that the observations Y_i are independent normal variables, with mean $E(Y_i)$ as given by (7.8) and with constant variance σ^2.

This general linear model encompasses a vast variety of situations. We shall consider a few of these now:

$p - 1$ independent variables. When $X_1, \ldots, X_{p-1}$ represent $p - 1$ different independent variables, the general linear model (7.7) is, as we have seen, a first-order model in which there are no interacting effects between the independent variables.

Polynomial regression. Consider the curvilinear regression model with one independent variable:

(7.9)
$$Y_i = \beta_0 + \beta_1 X_i + \beta_2 X_i^2 + \varepsilon_i$$

If we let $X_{i1} = X_i$ and $X_{i2} = X_i^2$, we can write (7.9) as follows:

$$Y_i = \beta_0 + \beta_1 X_{i1} + \beta_2 X_{i2} + \varepsilon_i$$

so that model (7.9) is a particular case of the general linear regression model. While (7.9) illustrates a curvilinear model where the response function is quadratic, models with higher degree polynomial response functions are also particular cases of the general linear regression model.

Transformed variables. Consider the model:

(7.10)
$$\log Y_i = \beta_0 + \beta_1 X_{i1} + \beta_2 X_{i2} + \beta_3 X_{i3} + \varepsilon_i$$

Here, the response surface is a highly complex one, yet model (7.10) can be treated as a general linear regression model. If we let $Y_i' = \log Y_i$, we can write model (7.10) as follows:

$$Y_i' = \beta_0 + \beta_1 X_{i1} + \beta_2 X_{i2} + \beta_3 X_{i3} + \varepsilon_i$$

which is in the form of the general linear regression model. The dependent variable just happens to be measured as the logarithm of Y.

Many models can be transformed into general linear regression models. Thus, the model:

(7.11)
$$\left(Y_i = \frac{1}{\beta_0 + \beta_1 X_{i1} + \beta_2 X_{i2} + \varepsilon_i} \right)$$

can be transformed to a general linear regression model by letting $Y_i' = 1/Y_i$. We then have:

$$Y_i' = \beta_0 + \beta_1 X_{i1} + \beta_2 X_{i2} + \varepsilon_i$$

Interaction effects. Consider the model in two independent variables X_1 and X_2:

$$Y_i = \beta_0 + \beta_1 X_{i1} + \beta_2 X_{i2} + \beta_3 X_{i1} X_{i2} + \varepsilon_i \tag{7.12}$$

The meaning of β_1 and β_2 here is not the same as that given earlier because of the cross-product term $\beta_3 X_{i1} X_{i2}$. It can be shown that the change in the mean response with a unit increase in X_1 when X_2 is held constant is:

$$\beta_1 + \beta_3 X_2 \tag{7.13}$$

Similarly, the change in the mean response with a unit change in X_2 when X_1 is held constant is:

$$\beta_2 + \beta_3 X_1 \tag{7.14}$$

Hence, in model (7.12) both the effect of X_1 for given level of X_2 and the effect of X_2 for given level of X_1 depend on the level of the other independent variable.

In Figure 7.2, we illustrate the effect of the cross-product term in model (7.12). In Figure 7.2a, we consider a response function without a cross-product term:

$$E(Y) = 10 + 2X_1 + 5X_2$$

and show there the response function $E(Y)$ when $X_2 = 1$ and when $X_2 = 3$. Note that the mean response increases by the amount $\beta_1 = 2$ with a unit increase of X_1, whether $X_2 = 1$ or $X_2 = 3$.

In Figure 7.2b, we consider the same response function but with the cross-product term $.5X_1X_2$ added:

$$E(Y) = 10 + 2X_1 + 5X_2 + .5X_1X_2$$

and show the response function $E(Y)$ when $X_2 = 1$ and when $X_2 = 3$. Note that the slope of the response function when plotted against X_1 now differs for $X_2 = 1$ and $X_2 = 3$. The slope of the response function when $X_2 = 1$ is by (7.13):

$$\beta_1 + \beta_3 X_2 = 2 + .5(1) = 2.5$$

and when $X_2 = 3$, the slope is:

$$\beta_1 + \beta_3 X_2 = 2 + .5(3) = 3.5$$

Hence, β_1 in model (7.12) containing a cross-product term no longer indicates the change in the mean response for a unit increase in X_1 for any given X_2 level. That effect in this model depends on the level of X_2. Model (7.12) with the cross-product term is therefore designed for independent variables whose effects on the dependent variable *interact*. The cross-product term $\beta_3 X_{i1} X_{i2}$ is called an *interaction term*. While the mean response in model (7.12) when X_2 is constant is

FIGURE 7.2 Effect of cross-product term in response function with two independent variables

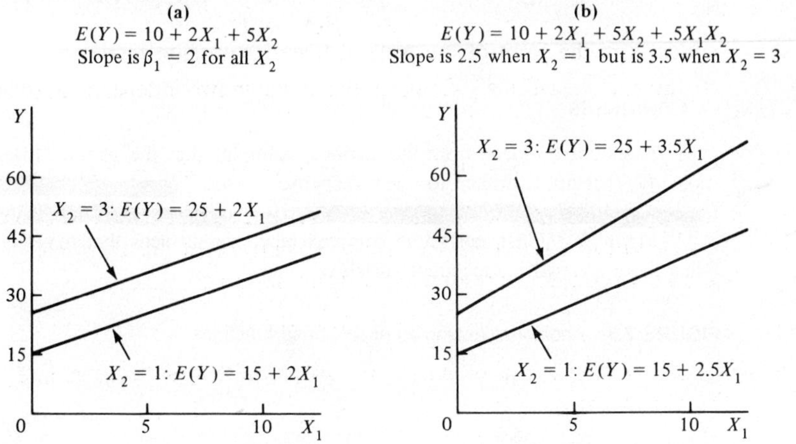

still a linear function of X_1, now both the intercept and the slope of the response function change as the level at which X_2 is held constant is varied. The same holds when the mean response is regarded as a function of X_2, with X_1 constant.

Despite these complexities of model (7.12), it can still be regarded as a general linear regression model. Let $X_{i3} = X_{i1}X_{i2}$. We can then write (7.12) as follows:

$$Y_i = \beta_0 + \beta_1 X_{i1} + \beta_2 X_{i2} + \beta_3 X_{i3} + \varepsilon_i$$

which is in the form of the general linear regression model.

Note

To derive (7.13) and (7.14), we differentiate:

$$E(Y) = \beta_0 + \beta_1 X_1 + \beta_2 X_2 + \beta_3 X_1 X_2$$

with respect to X_1 and X_2, respectively:

$$\frac{\partial E(Y)}{\partial X_1} = \beta_1 + \beta_3 X_2 \qquad \frac{\partial E(Y)}{\partial X_2} = \beta_2 + \beta_3 X_1$$

Combination of cases. A regression model may combine a number of the elements we have just noted and still can be treated as a general linear regression model. Consider a model with two independent variables, each in quadratic form, with an interaction term:

$$(7.15) \qquad Y_i = \beta_0 + \beta_1 X_{i1} + \beta_2 X_{i1}^2 + \beta_3 X_{i2} + \beta_4 X_{i2}^2 + \beta_5 X_{i1} X_{i2} + \varepsilon_i$$

Let us define:

$$Z_{i1} = X_{i1} \qquad Z_{i2} = X_{i1}^2 \qquad Z_{i3} = X_{i2} \qquad Z_{i4} = X_{i2}^2 \qquad Z_{i5} = X_{i1}X_{i2}$$

We can then write model (7.15) as follows:

$$Y_i = \beta_0 + \beta_1 Z_{i1} + \beta_2 Z_{i2} + \beta_3 Z_{i3} + \beta_4 Z_{i4} + \beta_5 Z_{i5} + \varepsilon_i$$

which is in the form of the general linear regression model.

Comments

1. It should be clear from the various examples that the general linear regression model (7.7) is not restricted to linear response surfaces. The term *linear model* refers to the fact that (7.7) is linear in the parameters, not to the shape of the response surface.
2. Figure 7.3 illustrates some complex response surfaces that may be encountered when there are two independent variables.

FIGURE 7.3 Additional examples of response functions

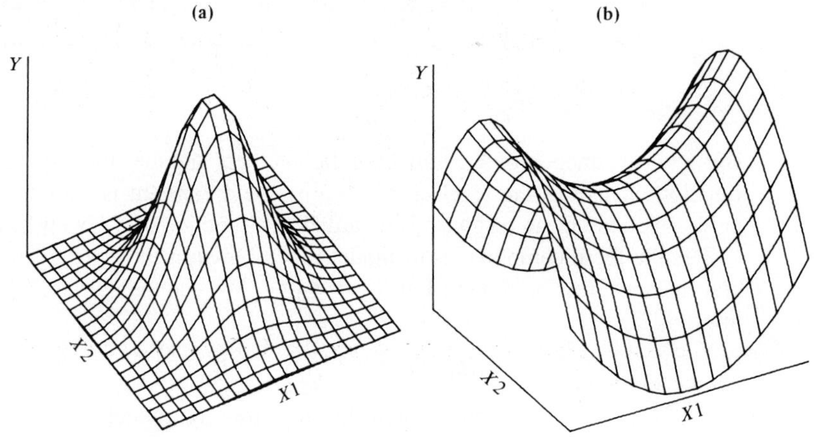

Interactions and nature of response surface

We introduced the concept of interacting independent variables earlier, and now shall illustrate further how the response surface differs when two independent variables do not interact and when they do interact.

Figure 7.4a contains a representation of a response surface in which the two independent variables (mean season temperature, amount of rainfall) do not interact on the dependent variable (corn yield). The absence of interactions can be seen by considering the corn yield curves for given mean season temperatures as a function of rainfall. These curves all have the same shape and differ only by a constant. Thus, each ordinate of the corn yield curve when the mean temperature is 70° is a constant number of units higher than the corresponding ordinate for the corn yield curve when the mean temperature is 78°.

Equivalently, one can note the absence of interactions by considering the corn yield curves for given amounts of rainfall as a function of temperature. Again, these curves are the same in shape and differ only by a constant.

FIGURE 7.4 Response surfaces for additive and interacting independent variables

(a) Independent Variables Do Not Interact

Yield of corn as function of season rainfall and mean temperature

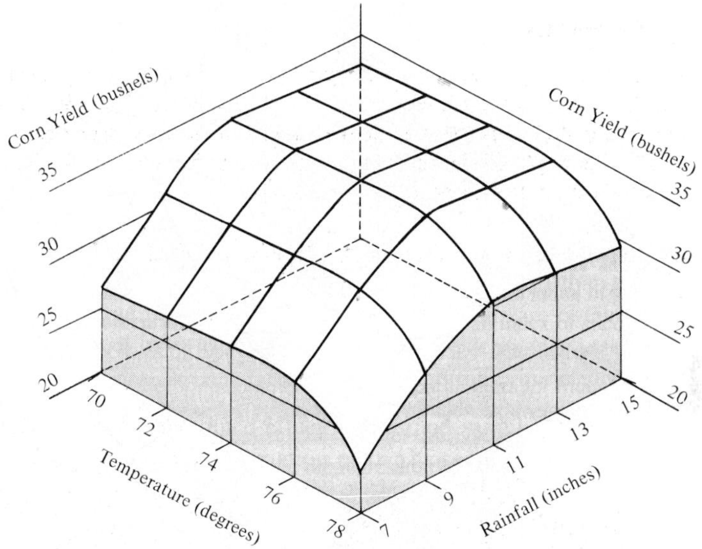

Absence of interactions therefore implies that the mean response $E(Y)$ can be expressed in the form:

(7.16) $$E(Y) = f_1(X_1) + f_2(X_2)$$

where f_1 and f_2 can be any functions, not necessarily simple ones.

Figure 7.4b illustrates a case where the two independent variables (age, percent of normal weight) interact on the dependent variable (mortality ratio). Here, the shape of the mortality ratio curve as a function of percent of normal weight varies for different ages. For men 22 years old, both underweight and overweight persons have higher mortality rates than normal (normal = 100) for that age. On the other hand, for men 52 years old, the mortality rate is above normal for that age for overweight persons but not for underweight persons. Similarly, the mortality ratio curves as a function of age vary in shape for different weights.

We can illustrate the difference in the shape of the response function when the two independent variables do and do not interact in yet another way, namely, by representing the response surface by means of a contour diagram. Such a diagram shows, for a number of different response levels, the various combinations of the two independent variables which yield the same level of response. Figure 7.5a shows a contour diagram for the response surface portrayed in Figure 7.1:

FIGURE 7.4 ***(concluded)***

(**b**) Independent Variables Interact

Mortality ratio for men as function of age and percent of normal weight

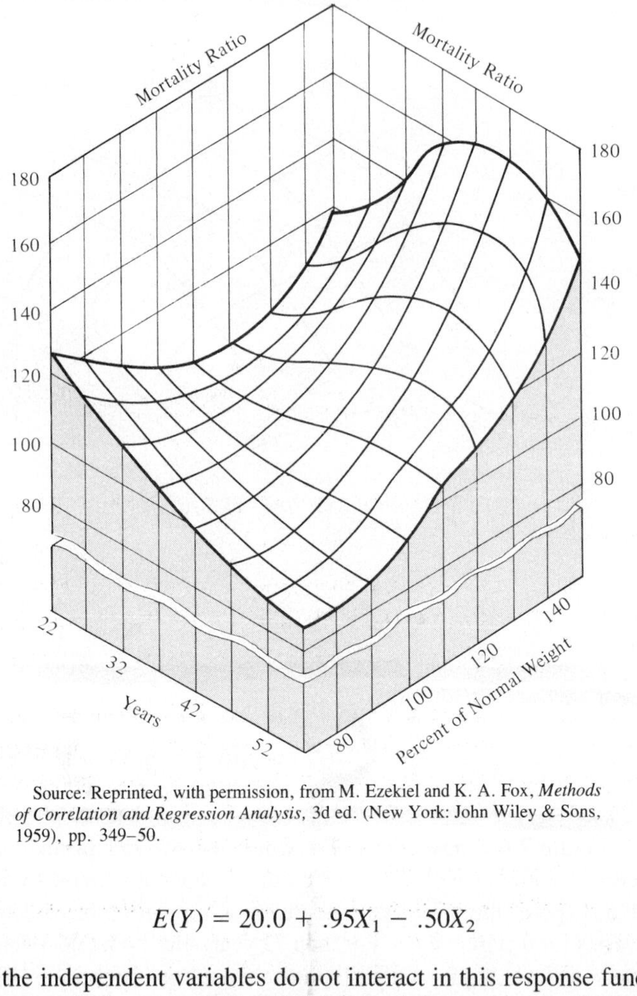

Source: Reprinted, with permission, from M. Ezekiel and K. A. Fox, *Methods of Correlation and Regression Analysis,* 3d ed. (New York: John Wiley & Sons, 1959), pp. 349–50.

$$E(Y) = 20.0 + .95X_1 - .50X_2$$

Note that the independent variables do not interact in this response function and that the contour lines are parallel. Figure 7.5b shows a contour diagram for the response function:

$$E(Y) = 5X_1 + 7X_2 + 3X_1X_2$$

where the two independent variables interact and the contour curves are not parallel.

In general, additive or noninteracting independent variables lead to parallel contour curves while interacting independent variables lead to nonparallel contour curves.

FIGURE 7.5   Response contour diagrams

(a) $E(Y) = 20.0 + .95X_1 - .50X_2$
Noninteracting Independent Variables

(b) $E(Y) = 5X_1 + 7X_2 + 3X_1X_2$
Interacting Independent Variables

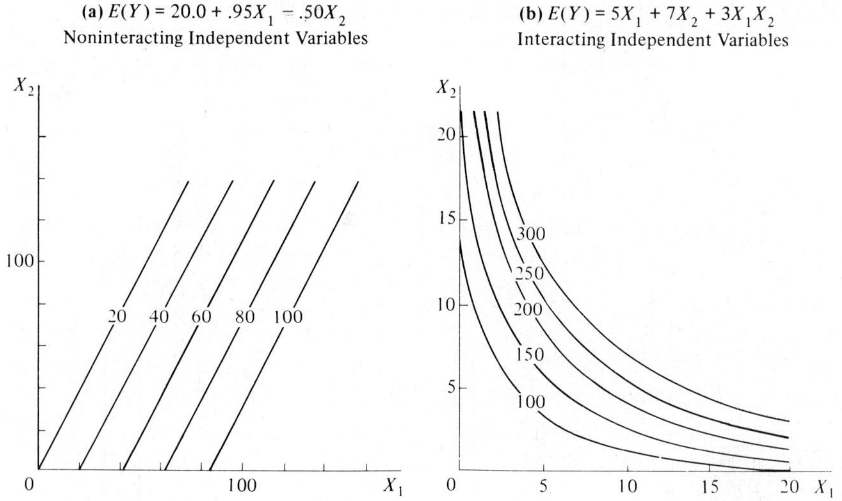

7.2   GENERAL LINEAR REGRESSION MODEL IN MATRIX TERMS

We shall now present the principal results for the general linear regression model (7.7) in matrix terms. This model, as we have noted, encompasses a wide variety of particular cases. The results to be presented are applicable to all of these.

It is a remarkable property of matrix algebra that the results for the general linear regression model (7.7) appear exactly the same in matrix notation as those for the simple linear regression model (6.56). Only the degrees of freedom and other constants related to the number of independent variables and the dimensions of some matrices will be different. Hence, we shall be able to present the results very concisely.

The matrix notation, to be sure, may hide enormous computational complexities. The inverse of a 10×10 matrix **A** requires tremendous amounts of computation, yet is simply represented as $\mathbf{A}^{-1}$. Our reason for emphasizing matrix algebra is that it indicates the essential conceptual steps in the solution. The actual computations will in all but the very simplest cases be done by programmable calculator or computer. Hence, it does not matter for us whether $(\mathbf{X'X})^{-1}$ represents finding the inverse of a 2×2 or a 10×10 matrix. The important point is to know what the inverse of the matrix represents.

To express the general linear regression model (7.7):

$$Y_i = \beta_0 + \beta_1 X_{i1} + \beta_2 X_{i2} + \cdots + \beta_{p-1}X_{i,p-1} + \varepsilon_i$$

in matrix terms, we need to define the following matrices:

(7.17)

(7.17a)
$$\underset{n\times 1}{\mathbf{Y}} = \begin{bmatrix} Y_1 \\ Y_2 \\ \cdot \\ \cdot \\ \cdot \\ Y_n \end{bmatrix}$$

(7.17b)
$$\underset{n\times p}{\mathbf{X}} = \begin{bmatrix} 1 & X_{11} & X_{12} & \cdots & X_{1,p-1} \\ 1 & X_{21} & X_{22} & \cdots & X_{2,p-1} \\ \cdot & \cdot & \cdot & & \cdot \\ \cdot & \cdot & \cdot & & \cdot \\ \cdot & \cdot & \cdot & & \cdot \\ 1 & X_{n1} & X_{n2} & \cdots & X_{n,p-1} \end{bmatrix}$$

(7.17c)
$$\underset{p\times 1}{\boldsymbol{\beta}} = \begin{bmatrix} \beta_0 \\ \beta_1 \\ \cdot \\ \cdot \\ \cdot \\ \beta_{p-1} \end{bmatrix}$$

(7.17d)
$$\underset{n\times 1}{\boldsymbol{\varepsilon}} = \begin{bmatrix} \varepsilon_1 \\ \varepsilon_2 \\ \cdot \\ \cdot \\ \cdot \\ \varepsilon_n \end{bmatrix}$$

Note that the $\mathbf{Y}$ and $\boldsymbol{\varepsilon}$ vectors are the same as for simple regression. The $\boldsymbol{\beta}$ vector contains additional regression parameters, and the $\mathbf{X}$ matrix contains a column of 1's as well as a column of the n values for each of the $p - 1$ X variables in the regression model. The row subscript for each element X_{ik} in the $\mathbf{X}$ matrix identifies the trial, and the column subscript identifies the X variable.

In matrix terms, the general linear regression model (7.7) is:

(7.18)
$$\underset{n\times 1}{\mathbf{Y}} = \underset{n\times p}{\mathbf{X}}\ \underset{p\times 1}{\boldsymbol{\beta}} + \underset{n\times 1}{\boldsymbol{\varepsilon}}$$

where:

- $\mathbf{Y}$ is a vector of observations
- $\boldsymbol{\beta}$ is a vector of parameters
- $\mathbf{X}$ is a matrix of constants
- $\boldsymbol{\varepsilon}$ is a vector of independent normal random variables with expectation $\mathbf{E}(\boldsymbol{\varepsilon}) = \mathbf{0}$ and variance-covariance matrix $\boldsymbol{\sigma}^2(\boldsymbol{\varepsilon}) = \sigma^2\mathbf{I}$

Consequently, the random vector $\mathbf{Y}$ has expectation:

(7.18a)
$$\mathbf{E}(\mathbf{Y}) = \mathbf{X}\boldsymbol{\beta}$$

and the variance-covariance matrix of $\mathbf{Y}$ is:

(7.18b)
$$\boldsymbol{\sigma}^2(\mathbf{Y}) = \sigma^2\mathbf{I}$$

7.3 LEAST SQUARES ESTIMATORS

Let us denote the vector of estimated regression coefficients $b_0, b_1, \ldots, b_{p-1}$ as $\mathbf{b}$:

$$\underset{p\times 1}{\mathbf{b}} = \begin{bmatrix} b_0 \\ b_1 \\ b_2 \\ \cdot \\ \cdot \\ \cdot \\ b_{p-1} \end{bmatrix} \tag{7.19}$$

The least squares normal equations for the general linear regression model (7.18) are:

$$\underset{p\times p}{(\mathbf{X}'\mathbf{X})}\ \underset{p\times 1}{\mathbf{b}} = \underset{p\times n}{\mathbf{X}'}\ \underset{n\times 1}{\mathbf{Y}} \tag{7.20}$$

and the least squares estimators are:

$$\underset{p\times 1}{\mathbf{b}} = \underset{p\times p}{(\mathbf{X}'\mathbf{X})^{-1}}\underset{p\times 1}{\mathbf{X}'\mathbf{Y}} \tag{7.21}$$

For model (7.18), these least squares estimators are also maximum likelihood estimators and have all the properties mentioned in Chapter 2: they are unbiased, minimum variance unbiased, consistent, and sufficient.

7.4 ANALYSIS OF VARIANCE RESULTS

Let the vector of the fitted values $\hat{Y}_i$ be denoted by $\hat{\mathbf{Y}}$ and the vector of the residual terms $e_i = Y_i - \hat{Y}_i$ be denoted by $\mathbf{e}$:

(7.22)

$$\underset{n\times 1}{\hat{\mathbf{Y}}} = \begin{bmatrix} \hat{Y}_1 \\ \hat{Y}_2 \\ \cdot \\ \cdot \\ \cdot \\ \hat{Y}_n \end{bmatrix} \tag{7.22a}$$

$$\underset{n\times 1}{\mathbf{e}} = \begin{bmatrix} e_1 \\ e_2 \\ \cdot \\ \cdot \\ \cdot \\ e_n \end{bmatrix} \tag{7.22b}$$

The fitted values are represented by:

$$\hat{\mathbf{Y}} = \mathbf{X}\mathbf{b} \tag{7.23}$$

and the residual terms by:

$$\mathbf{e} = \mathbf{Y} - \hat{\mathbf{Y}} = \mathbf{Y} - \mathbf{X}\mathbf{b} \tag{7.24}$$

Sums of squares and mean squares

The sums of squares for the analysis of variance are:

$$SSTO = \mathbf{Y}'\mathbf{Y} - \left(\frac{1}{n}\right)\mathbf{Y}'\mathbf{1}\mathbf{1}'\mathbf{Y} \tag{7.25}$$

$$SSR = \mathbf{b}'\mathbf{X}'\mathbf{Y} - \left(\frac{1}{n}\right)\mathbf{Y}'\mathbf{1}\mathbf{1}'\mathbf{Y} \tag{7.26}$$

$$SSE = \mathbf{e}'\mathbf{e} = (\mathbf{Y} - \mathbf{X}\mathbf{b})'(\mathbf{Y} - \mathbf{X}\mathbf{b}) = \mathbf{Y}'\mathbf{Y} - \mathbf{b}'\mathbf{X}'\mathbf{Y} \tag{7.27}$$

where **1** is an $n \times 1$ vector of 1's as defined in (6.19).

SSTO, as usual, has $n - 1$ degrees of freedom associated with it. *SSE* has $n - p$ degrees of freedom associated with it since p parameters need to be estimated in the regression function for model (7.18). Finally, *SSR* has $p - 1$ degrees of freedom associated with it, representing the number of X variables $X_1, \ldots, X_{p-1}$.

Table 7.1 shows these analysis of variance results, as well as the mean squares *MSR* and *MSE*:

(7.28)
$$MSR = \frac{SSR}{p-1}$$

(7.29)
$$MSE = \frac{SSE}{n-p}$$

The expectation of *MSE* is σ^2, as for simple regression. The expectation of *MSR* is σ^2 plus a quantity that is nonnegative. For instance, when $p - 1 = 2$, we have:

$$E(MSR) = \sigma^2 + [\beta_1^2 \Sigma(X_{i1} - \overline{X}_1)^2 + \beta_2^2 \Sigma(X_{i2} - \overline{X}_2)^2 + 2\beta_1\beta_2 \Sigma(X_{i1} - \overline{X}_1)(X_{i2} - \overline{X}_2)]/2$$

Note that if both β_1 and β_2 equal zero, $E(MSR) = \sigma^2$. Otherwise $E(MSR) > \sigma^2$.

TABLE 7.1 ANOVA table for general linear regression model (7.18)

Source of Variation	*SS*	*df*	*MS*
Regression	$SSR = \mathbf{b'X'Y} - \left(\frac{1}{n}\right)\mathbf{Y'11'Y}$	$p - 1$	$MSR = \frac{SSR}{p-1}$
Error	$SSE = \mathbf{Y'Y} - \mathbf{b'X'Y}$	$n - p$	$MSE = \frac{SSE}{n-p}$
Total	$SSTO = \mathbf{Y'Y} - \left(\frac{1}{n}\right)\mathbf{Y'11'Y}$	$n - 1$	

F test for regression relation

To test whether there is a regression relation between the dependent variable Y and the set of X variables $X_1, \ldots, X_{p-1}$, i.e., to choose between the alternatives:

(7.30a)
$$\begin{aligned} &H_0: \beta_1 = \beta_2 = \cdots = \beta_{p-1} = 0 \\ &H_a: \text{not all } \beta_k \; (k = 1, \ldots, p-1) \text{ equal } 0 \end{aligned}$$

we use the test statistic:

(7.30b)
$$F^* = \frac{MSR}{MSE}$$

The decision rule to control the Type I error at α is:

(7.30c)
$$\begin{aligned} &\text{If } F^* \leq F(1-\alpha; p-1, n-p), \text{ conclude } H_0 \\ &\text{If } F^* > F(1-\alpha; p-1, n-p), \text{ conclude } H_a \end{aligned}$$

The existence of a regression relation by itself does not of course assure that useful predictions can be made by using it.

Note that when $p - 1 = 1$, this test reduces to the F test in (3.61) for testing in simple linear regression whether or not $\beta_1 = 0$.

Coefficient of multiple determination

The coefficient of multiple determination, denoted by R^2, is defined as follows:

(7.31)
$$R^2 = \frac{SSR}{SSTO} = 1 - \frac{SSE}{SSTO}$$

It measures the proportionate reduction of total variation in Y associated with the use of the set of X variables $X_1, \ldots, X_{p-1}$. The coefficient of multiple determination R^2 reduces to the coefficient of determination r^2 in (3.69) for simple linear regression when $p - 1 = 1$, i.e., when one independent variable is in model (7.18). Just as for r^2, we have:

(7.32)
$$0 \leq R^2 \leq 1$$

R^2 assumes the value 0 when all $b_k = 0$ $(k = 1, \ldots, p-1)$. R^2 takes on the value 1 when all observations fall directly on the fitted response surface, i.e., when $Y_i = \hat{Y}_i$ for all i.

Comments

1. To distinguish between the coefficients of determination for simple and multiple regression, we shall from now on call r^2 the *coefficient of simple determination.*

2. It can be shown that the coefficient of multiple determination R^2 can be viewed as a coefficient of simple determination r^2 between the responses Y_i and the fitted values $\hat{Y}_i$.

3. A large R^2 does not necessarily imply that the fitted model is a useful one. For instance, observations may have been taken at only a few levels of the independent variables. Despite a high R^2 in this case, the fitted model may not be useful because most predictions would require extrapolations outside the region of observations. Again, even though R^2 is large, MSE may still be too large for inferences to be useful in a case where high precision is required.

4. Adding more independent variables to the model can only increase R^2 and never reduce it, because SSE can never become larger with more independent variables and $SSTO$ is always the same for a given set of responses. Since R^2 often can be made large by including a large number of independent variables, it is sometimes suggested that a modified measure be used which recognizes the number of independent variables in the model. This *adjusted coefficient of multiple determination*, denoted by R_a^2, is defined:

(7.33)
$$R_a^2 = 1 - \left(\frac{n-1}{n-p}\right)\frac{SSE}{SSTO}$$

This adjusted coefficient of multiple determination may actually become smaller when another independent variable is introduced into the model, because the decrease in *SSE* may be more than offset by the loss of a degree of freedom in the denominator $n - p$.

Coefficient of multiple correlation

The coefficient of multiple correlation R is the positive square root of R^2:

(7.34) $$R = \sqrt{R^2}$$

It equals in absolute value the correlation coefficient r in (3.71) for simple correlation when $p - 1 = 1$, i.e., when there is one independent variable in model (7.18).

Note

From now on, we shall call r the *coefficient of simple correlation* to distinguish it from the coefficient of multiple correlation.

7.5 INFERENCES ABOUT REGRESSION PARAMETERS

The least squares estimators in $\mathbf{b}$ are unbiased:

(7.35) $$\mathbf{E}(\mathbf{b}) = \boldsymbol{\beta}$$

The variance-covariance matrix $\boldsymbol{\sigma}^2(\mathbf{b})$:

(7.36) $$\boldsymbol{\sigma}^2(\mathbf{b}) = \begin{bmatrix} \sigma^2(b_0) & \sigma(b_0, b_1) & \cdots & \sigma(b_0, b_{p-1}) \\ \sigma(b_1, b_0) & \sigma^2(b_1) & \cdots & \sigma(b_1, b_{p-1}) \\ \vdots & \vdots & & \vdots \\ \sigma(b_{p-1}, b_0) & \sigma(b_{p-1}, b_1) & \cdots & \sigma^2(b_{p-1}) \end{bmatrix}$$

is given by:

(7.37) $$\underset{p\times p}{\boldsymbol{\sigma}^2(\mathbf{b})} = \sigma^2(\mathbf{X}'\mathbf{X})^{-1}$$

The estimated variance-covariance matrix $\mathbf{s}^2(\mathbf{b})$:

(7.38) $$\mathbf{s}^2(\mathbf{b}) = \begin{bmatrix} s^2(b_0) & s(b_0, b_1) & \cdots & s(b_0, b_{p-1}) \\ s(b_1, b_0) & s^2(b_1) & \cdots & s(b_1, b_{p-1}) \\ \vdots & \vdots & & \vdots \\ s(b_{p-1}, b_0) & s(b_{p-1}, b_1) & \cdots & s^2(b_{p-1}) \end{bmatrix}$$

is given by:

(7.39) $$\underset{p\times p}{\mathbf{s}^2(\mathbf{b})} = MSE(\mathbf{X}'\mathbf{X})^{-1}$$

From $\mathbf{s}^2(\mathbf{b})$, one can obtain $s^2(b_0)$, $s^2(b_1)$ or whatever other variance is needed, or any needed covariances.

Interval estimation of β_k

For the normal error model (7.18), we have:

(7.40) $$\frac{b_k - \beta_k}{s(b_k)} \sim t(n-p) \qquad k = 0, 1, \ldots, p-1$$

Hence, the confidence limits for β_k with $1 - \alpha$ confidence coefficient are:

(7.41) $$b_k \pm t(1 - \alpha/2; n-p)s(b_k)$$

Tests for β_k

Tests for β_k are set up in the usual fashion. To test:

(7.42a) $$\begin{aligned} H_0&: \beta_k = 0 \\ H_a&: \beta_k \neq 0 \end{aligned}$$

we may use the test statistic:

(7.42b) $$t^* = \frac{b_k}{s(b_k)}$$

and the decision rule:

(7.42c) $$\begin{aligned} &\text{If } |t^*| \leq t(1 - \alpha/2; n-p), \text{ conclude } H_0 \\ &\text{Otherwise conclude } H_a \end{aligned}$$

The power of the t test can be obtained as explained in Chapter 3, with the degrees of freedom modified to $n - p$.

As with simple regression, the test whether or not $\beta_k = 0$ in multiple regression models can also be conducted by means of an F test. We discuss this test in Chapter 8.

Joint inferences

The boundary of the joint confidence region for all p of the β_k regression parameters ($k = 0, 1, \ldots, p-1$) with confidence coefficient $1 - \alpha$ is:

(7.43) $$\frac{(\mathbf{b} - \boldsymbol{\beta})'\mathbf{X}'\mathbf{X}(\mathbf{b} - \boldsymbol{\beta})}{pMSE} = F(1 - \alpha; p, n-p)$$

The region defined by this boundary is generally difficult to obtain and interpret.

The Bonferroni joint confidence intervals, on the other hand, are easy to obtain and interpret. If g parameters are to be estimated jointly (where $g \leq p$), the confidence limits with family confidence coefficient $1 - \alpha$ are:

(7.44) $$b_k \pm Bs(b_k)$$

where:

(7.44a) $$B = t(1 - \alpha/2g; n-p)$$

In Section 8.4, we discuss tests concerning a subset of the regression parameters.

7.6 INFERENCES ABOUT MEAN RESPONSE

Interval estimation of $E(Y_h)$

For given values of $X_1, \ldots, X_{p-1}$, denoted by $X_{h1}, \ldots, X_{h,p-1}$, the mean response is denoted by $E(Y_h)$. We define the vector $\mathbf{X}_h$:

$$\mathbf{X}_h = \begin{bmatrix} 1 \\ X_{h1} \\ X_{h2} \\ \cdot \\ \cdot \\ \cdot \\ X_{h,p-1} \end{bmatrix} \tag{7.45}$$

so that the mean response to be estimated is:

$$E(Y_h) = \mathbf{X}_h'\boldsymbol{\beta} \tag{7.46}$$

The estimated mean response corresponding to $\mathbf{X}_h$, denoted by $\hat{Y}_h$, is:

$$\hat{Y}_h = \mathbf{X}_h'\mathbf{b} \tag{7.47}$$

This estimator is unbiased:

$$E(\hat{Y}_h) = \mathbf{X}_h'\boldsymbol{\beta} = E(Y_h) \tag{7.48}$$

and its variance is:

$$\sigma^2(\hat{Y}_h) = \sigma^2\mathbf{X}_h'(\mathbf{X}'\mathbf{X})^{-1}\mathbf{X}_h = \mathbf{X}_h'\boldsymbol{\sigma}^2(\mathbf{b})\mathbf{X}_h \tag{7.49}$$

Note that the variance $\sigma^2(\hat{Y}_h)$ is a function of the variances $\sigma^2(b_k)$ of the regression coefficients and of the covariances $\sigma(b_k, b_{k'})$ between pairs of regression coefficients, just as in simple linear regression. The estimated variance $s^2(\hat{Y}_h)$ is given by:

$$s^2(\hat{Y}_h) = MSE(\mathbf{X}_h'(\mathbf{X}'\mathbf{X})^{-1}\mathbf{X}_h) = \mathbf{X}_h'\mathbf{s}^2(\mathbf{b})\mathbf{X}_h \tag{7.50}$$

The $1 - \alpha$ confidence limits for $E(Y_h)$ are:

$$\hat{Y}_h \pm t(1 - \alpha/2; n - p)s(\hat{Y}_h) \tag{7.51}$$

Confidence region for regression surface

The $1 - \alpha$ confidence region for the entire regression surface is an extension of the Working-Hotelling confidence band for the regression line when there is one independent variable. Boundary points of the confidence region at $\mathbf{X}_h$ are:

$$\hat{Y}_h \pm Ws(\hat{Y}_h) \tag{7.52}$$

where:

$$W^2 = pF(1 - \alpha; p, n - p) \tag{7.52a}$$

The confidence coefficient is $1 - \alpha$ that the region contains the entire regression surface over all combinations of real-numbered values of the X variables.

Simultaneous confidence intervals for several mean responses

When it is desired to estimate a number of mean responses $E(Y_h)$ corresponding to different $\mathbf{X}_h$ vectors, one can employ two basic approaches:

1. Use the Working-Hotelling type confidence region bounds from (7.52) for the several $\mathbf{X}_h$ vectors of interest. Since these bounds cover the mean responses for all possible $\mathbf{X}_h$ vectors with confidence coefficient $1 - \alpha$, they will cover the mean responses for selected $\mathbf{X}_h$ vectors with confidence coefficient greater than $1 - \alpha$.

2. Use Bonferroni simultaneous confidence intervals. When g statements are to be made with family confidence coefficient $1 - \alpha$, the Bonferroni confidence limits are:

(7.53)
$$\hat{Y}_h \pm Bs(\hat{Y}_h)$$

where:

(7.53a)
$$B = t(1 - \alpha/2g; n - p)$$

For any particular application, one should compare W and B to see which procedure will lead to narrower confidence intervals. If the $\mathbf{X}_h$ levels are not specified in advance but are determined as the analysis proceeds, it is better to use the Working-Hotelling type limits (7.52).

F test for lack of fit

To test whether the response function:

(7.54)
$$E(Y) = \beta_0 + \beta_1 X_1 + \cdots + \beta_{p-1} X_{p-1}$$

is an appropriate response surface for the data at hand requires repeat observations, as for simple regression analysis. Repeat observations in multiple regression are replicate observations on Y corresponding to levels of each of the X variables which are constant from trial to trial. Thus, with two independent variables repeat observations require that X_1 and X_2 each remain at given levels from trial to trial.

The procedures described in Chapter 4 for the F test for lack of fit are applicable to multiple regression. Once the ANOVA table, shown in Table 7.1, has been obtained, SSE is decomposed into pure error and lack of fit components. The pure error sum of squares $SSPE$ is obtained by first calculating for each replicate group the sum of squared deviations of the Y observations around the group mean, where a replicate group has the same values for the X variables. Suppose there are c replicate groups with distinct sets of levels for the X variables, and let the mean of the Y observations for the jth group be denoted by $\bar{Y}_j$. Then the sum of squares for the jth group is given by (4.8), and the pure error sum of squares is the sum of these sums of squares, as given by (4.9). The lack of fit sum of squares $SSLF$ equals the difference $SSE - SSPE$, as indicated by (4.12).

The number of degrees of freedom associated with $SSPE$ is $n - c$, and the number of degrees of freedom associated with $SSLF$ is $(n - p) - (n - c) = c - p$.

The F test is conducted as described in Chapter 4, but with the degrees of freedom modified to those just stated.

7.7 PREDICTIONS OF NEW OBSERVATIONS

Prediction of new observation $Y_{h(\text{new})}$

The prediction limits with $1 - \alpha$ confidence coefficient for a new observation $Y_{h(\text{new})}$ corresponding to $\mathbf{X}_h$, the specified values of the X variables, are:

$$\hat{Y}_h \pm t(1 - \alpha/2; n - p)s(Y_{h(\text{new})}) \tag{7.55}$$

where:

$$\begin{aligned} s^2(Y_{h(\text{new})}) &= MSE + s^2(\hat{Y}_h) = MSE + \mathbf{X}_h'\mathbf{s}^2(\mathbf{b})\mathbf{X}_h \\ &= MSE(1 + \mathbf{X}_h'(\mathbf{X}'\mathbf{X})^{-1}\mathbf{X}_h) \end{aligned} \tag{7.55a}$$

Prediction of mean of m new observations at $\mathbf{X}_h$

When m new observations are to be selected at $\mathbf{X}_h$ and their mean $\bar{Y}_{h(\text{new})}$ is to be predicted, the $1 - \alpha$ prediction limits are:

$$\hat{Y}_h \pm t(1 - \alpha/2; n - p)s(\bar{Y}_{h(\text{new})}) \tag{7.56}$$

where:

$$\begin{aligned} s^2(\bar{Y}_{h(\text{new})}) &= \frac{MSE}{m} + s^2(\hat{Y}_h) = \frac{MSE}{m} + \mathbf{X}_h'\mathbf{s}^2(\mathbf{b})\mathbf{X}_h \\ &= MSE\left(\frac{1}{m} + \mathbf{X}_h'(\mathbf{X}'\mathbf{X})^{-1}\mathbf{X}_h\right) \end{aligned} \tag{7.56a}$$

Predictions of g new observations

Simultaneous prediction limits for g new observations at g different levels of $\mathbf{X}_h$ with family confidence coefficient $1 - \alpha$ are given by:

$$\hat{Y}_h \pm Ss(Y_{h(\text{new})}) \tag{7.57}$$

where:

$$S^2 = gF(1 - \alpha; g, n - p) \tag{7.57a}$$

and $s^2(Y_{h(\text{new})})$ is given by (7.55a).

Alternatively, the Bonferroni simultaneous prediction limits can be used. For g predictions with a $1 - \alpha$ family confidence coefficient, they are:

$$\hat{Y}_h \pm Bs(Y_{h(\text{new})}) \tag{7.58}$$

where:

(7.58a) $$B = t(1 - \alpha/2g; n - p)$$

A comparison of S and B in advance of any particular use will indicate which procedure will lead to narrower prediction intervals.

7.8 AN EXAMPLE—MULTIPLE REGRESSION WITH TWO INDEPENDENT VARIABLES

In this section, we shall develop a multiple regression application with two independent variables. We shall illustrate a number of different types of inferences which might be made for this application but will not take up every possible type of inference.

Setting

The Zarthan Company sells a special skin cream through fashion stores exclusively. It operates in 15 marketing districts and is interested in predicting district sales. Table 7.2 contains data on sales by district, as well as district data on target population and per capita discretionary income. Sales are to be treated as the dependent variable Y, and target population and per capita discretionary income as independent variables X_1 and X_2, respectively, in an exploration of the feasibility of predicting district sales from target population and per capita discretionary income. The first-order model:

(7.59) $$Y_i = \beta_0 + \beta_1 X_{i1} + \beta_2 X_{i2} + \varepsilon_i$$

with normal error terms is expected to be appropriate.

TABLE 7.2 Basic data—Zarthan Company example

District i	*Sales (gross of jars; 1 gross = 12 dozen)* Y_i	*Target Population (thousands of persons)* X_{i1}	*Per Capita Discretionary Income (dollars)* X_{i2}
1	162	274	2,450
2	120	180	3,254
3	223	375	3,802
4	131	205	2,838
5	67	86	2,347
6	169	265	3,782
7	81	98	3,008
8	192	330	2,450
9	116	195	2,137
10	55	53	2,560
11	252	430	4,020
12	232	372	4,427
13	144	236	2,660
14	103	157	2,088
15	212	370	2,605

Basic calculations

The **Y** and **X** matrices for the Zarthan Company illustration are shown in Table 7.3. We shall require:

1.

$$\mathbf{X}'\mathbf{X} = \begin{bmatrix} 1 & 1 & \cdots & 1 \\ 274 & 180 & \cdots & 370 \\ 2{,}450 & 3{,}254 & \cdots & 2{,}605 \end{bmatrix} \begin{bmatrix} 1 & 274 & 2{,}450 \\ 1 & 180 & 3{,}254 \\ \cdot & \cdot & \cdot \\ \cdot & \cdot & \cdot \\ \cdot & \cdot & \cdot \\ 1 & 370 & 2{,}605 \end{bmatrix}$$

which yields:

$$\mathbf{X}'\mathbf{X} = \begin{bmatrix} 15 & 3{,}626 & 44{,}428 \\ 3{,}626 & 1{,}067{,}614 & 11{,}419{,}181 \\ 44{,}428 & 11{,}419{,}181 & 139{,}063{,}428 \end{bmatrix} \tag{7.60}$$

2.

$$\mathbf{X}'\mathbf{Y} = \begin{bmatrix} 1 & 1 & \cdots & 1 \\ 274 & 180 & \cdots & 370 \\ 2{,}450 & 3{,}254 & \cdots & 2{,}605 \end{bmatrix} \begin{bmatrix} 162 \\ 120 \\ \cdot \\ \cdot \\ \cdot \\ 212 \end{bmatrix}$$

which yields:

$$\mathbf{X}'\mathbf{Y} = \begin{bmatrix} 2{,}259 \\ 647{,}107 \\ 7{,}096{,}619 \end{bmatrix} \tag{7.61}$$

TABLE 7.3 **Y** and **X** matrices—Zarthan Company example

$$\mathbf{Y} = \begin{bmatrix} 162 \\ 120 \\ 223 \\ 131 \\ 67 \\ 169 \\ 81 \\ 192 \\ 116 \\ 55 \\ 252 \\ 232 \\ 144 \\ 103 \\ 212 \end{bmatrix} \qquad \mathbf{X} = \begin{bmatrix} 1 & 274 & 2{,}450 \\ 1 & 180 & 3{,}254 \\ 1 & 375 & 3{,}802 \\ 1 & 205 & 2{,}838 \\ 1 & 86 & 2{,}347 \\ 1 & 265 & 3{,}782 \\ 1 & 98 & 3{,}008 \\ 1 & 330 & 2{,}450 \\ 1 & 195 & 2{,}137 \\ 1 & 53 & 2{,}560 \\ 1 & 430 & 4{,}020 \\ 1 & 372 & 4{,}427 \\ 1 & 236 & 2{,}660 \\ 1 & 157 & 2{,}088 \\ 1 & 370 & 2{,}605 \end{bmatrix}$$

3.

$$(\mathbf{X}'\mathbf{X})^{-1} = \begin{bmatrix} 15 & 3{,}626 & 44{,}428 \\ 3{,}626 & 1{,}067{,}614 & 11{,}419{,}181 \\ 44{,}428 & 11{,}419{,}181 & 139{,}063{,}428 \end{bmatrix}^{-1}$$

Using (6.25), we define:

$$\begin{array}{lll} a = 15 & b = 3{,}626 & c = 44{,}428 \\ d = 3{,}626 & e = 1{,}067{,}614 & f = 11{,}419{,}181 \\ g = 44{,}428 & h = 11{,}419{,}181 & k = 139{,}063{,}428 \end{array}$$

so that:

$$\begin{aligned} Z &= 14{,}497{,}044{,}060{,}000 \\ A &= 1.246348416 \\ B &= .0002129664176 \end{aligned}$$

and so on. We obtain:

(7.62) $$(\mathbf{X}'\mathbf{X})^{-1} =$$

$$\begin{bmatrix} 1.2463484 & 2.1296642\text{E}-4 & -4.1567125\text{E}-4 \\ 2.1296642\text{E}-4 & 7.7329030\text{E}-6 & -7.0302518\text{E}-7 \\ -4.1567125\text{E}-4 & -7.0302518\text{E}-7 & 1.9771851\text{E}-7 \end{bmatrix}$$

Note that some of the results in the $(\mathbf{X}'\mathbf{X})^{-1}$ matrix are given in the E format, where, say, E − 4 stands for $10^{-4} = 1/10^4$. Thus, 2.1296642E − 4 stands for .00021296642.

Algebraic equivalents. Note that $\mathbf{X}'\mathbf{X}$ for the first-order model (7.59) with two independent variables is:

$$\mathbf{X}'\mathbf{X} = \begin{bmatrix} 1 & 1 & \cdots & 1 \\ X_{11} & X_{21} & \cdots & X_{n1} \\ X_{12} & X_{22} & \cdots & X_{n2} \end{bmatrix} \begin{bmatrix} 1 & X_{11} & X_{12} \\ 1 & X_{21} & X_{22} \\ \cdot & \cdot & \cdot \\ \cdot & \cdot & \cdot \\ \cdot & \cdot & \cdot \\ 1 & X_{n1} & X_{n2} \end{bmatrix}$$

or:

(7.63) $$\mathbf{X}'\mathbf{X} = \begin{bmatrix} n & \Sigma X_{i1} & \Sigma X_{i2} \\ \Sigma X_{i1} & \Sigma X_{i1}^2 & \Sigma X_{i1}X_{i2} \\ \Sigma X_{i2} & \Sigma X_{i2}X_{i1} & \Sigma X_{i2}^2 \end{bmatrix}$$

Thus, for our example:

$$\begin{aligned} n &= 15 \\ \Sigma X_{i1} &= 274 + 180 + \cdots = 3{,}626 \\ \Sigma X_{i1}X_{i2} &= 274(2{,}450) + 180(3{,}254) + \cdots = 11{,}419{,}181 \\ &\text{etc.} \end{aligned}$$

These elements are found in (7.60).

Also note that $\mathbf{X'Y}$ for the first-order model with two independent variables is:

$$\text{(7.64)} \qquad \mathbf{X'Y} = \begin{bmatrix} 1 & 1 & \cdots & 1 \\ X_{11} & X_{21} & \cdots & X_{n1} \\ X_{12} & X_{22} & \cdots & X_{n2} \end{bmatrix} \begin{bmatrix} Y_1 \\ Y_2 \\ \vdots \\ Y_n \end{bmatrix} = \begin{bmatrix} \Sigma Y_i \\ \Sigma X_{i1} Y_i \\ \Sigma X_{i2} Y_i \end{bmatrix}$$

For our example, we have:

$$\Sigma Y_i = 162 + 120 + \cdots = 2{,}259$$
$$\Sigma X_{i1} Y_i = 274(162) + 180(120) + \cdots = 647{,}107$$
$$\Sigma X_{i2} Y_i = 2{,}450(162) + 3{,}254(120) + \cdots = 7{,}096{,}619$$

These are the elements found in (7.61).

Estimated regression function

The least squares estimates $\mathbf{b}$ are readily obtained by (7.21), given our basic calculations in (7.61) and (7.62):

$$\mathbf{b} = (\mathbf{X'X})^{-1}\mathbf{X'Y}$$

$$= \begin{bmatrix} 1.2463484 & 2.1296642\text{E}-4 & -4.1567125\text{E}-4 \\ 2.1296642\text{E}-4 & 7.7329030\text{E}-6 & -7.0302518\text{E}-7 \\ -4.1567125\text{E}-4 & -7.0302518\text{E}-7 & 1.9771851\text{E}-7 \end{bmatrix} \times \begin{bmatrix} 2{,}259 \\ 647{,}107 \\ 7{,}096{,}619 \end{bmatrix}$$

$$= \begin{bmatrix} 3.4526127900 \\ .4960049761 \\ .009199080867 \end{bmatrix}$$

Thus:

$$\begin{bmatrix} b_0 \\ b_1 \\ b_2 \end{bmatrix} = \begin{bmatrix} 3.4526127900 \\ .4960049761 \\ .009199080867 \end{bmatrix}$$

and the estimated regression function is:

$$\hat{Y} = 3.453 + .496X_1 + .00920X_2$$

This estimated regression function indicates that mean sales are expected to increase by .496 gross when the target population increases by one thousand, holding per capita discretionary income constant, and that mean sales are expected to increase by .0092 gross when per capita discretionary income increases by one dollar, holding population constant.

Algebraic version of normal equations. The normal equations in algebraic form for the case of two independent variables can be obtained readily from (7.63) and (7.64). We have:

$$(\mathbf{X}'\mathbf{X})\mathbf{b} = \mathbf{X}'\mathbf{Y}$$

$$\begin{bmatrix} n & \Sigma X_{i1} & \Sigma X_{i2} \\ \Sigma X_{i1} & \Sigma X_{i1}^2 & \Sigma X_{i1}X_{i2} \\ \Sigma X_{i2} & \Sigma X_{i2}X_{i1} & \Sigma X_{i2}^2 \end{bmatrix} \begin{bmatrix} b_0 \\ b_1 \\ b_2 \end{bmatrix} = \begin{bmatrix} \Sigma Y_i \\ \Sigma X_{i1}Y_i \\ \Sigma X_{i2}Y_i \end{bmatrix}$$

from which we obtain the normal equations:

$$\begin{aligned} \Sigma Y_i &= nb_0 + b_1\Sigma X_{i1} + b_2\Sigma X_{i2} \\ \Sigma X_{i1}Y_i &= b_0\Sigma X_{i1} + b_1\Sigma X_{i1}^2 + b_2\Sigma X_{i1}X_{i2} \\ \Sigma X_{i2}Y_i &= b_0\Sigma X_{i2} + b_1\Sigma X_{i1}X_{i2} + b_2\Sigma X_{i2}^2 \end{aligned} \tag{7.65}$$

Analysis of aptness of model

To examine the aptness of regression model (7.59) with independent variables X_1 and X_2 for the data at hand, we require the fitted values $\hat{Y}_i$ and the residuals $e_i = Y_i - \hat{Y}_i$. We obtain by (7.23):

$$\hat{\mathbf{Y}} = \mathbf{Xb}$$

$$\begin{bmatrix} \hat{Y}_1 \\ \hat{Y}_2 \\ \cdot \\ \cdot \\ \cdot \\ \hat{Y}_{15} \end{bmatrix} = \begin{bmatrix} 1 & 274 & 2{,}450 \\ 1 & 180 & 3{,}254 \\ \cdot & \cdot & \cdot \\ \cdot & \cdot & \cdot \\ \cdot & \cdot & \cdot \\ 1 & 370 & 2{,}605 \end{bmatrix} \begin{bmatrix} 3.4526127900 \\ .4960049761 \\ .009199080867 \end{bmatrix} = \begin{bmatrix} 161.896 \\ 122.667 \\ \cdot \\ \cdot \\ \cdot \\ 210.938 \end{bmatrix}$$

Further, by (7.24) we find:

$$\mathbf{e} = \mathbf{Y} - \hat{\mathbf{Y}}$$

$$\begin{bmatrix} e_1 \\ e_2 \\ \cdot \\ \cdot \\ \cdot \\ e_{15} \end{bmatrix} = \begin{bmatrix} 162 \\ 120 \\ \cdot \\ \cdot \\ \cdot \\ 212 \end{bmatrix} - \begin{bmatrix} 161.896 \\ 122.667 \\ \cdot \\ \cdot \\ \cdot \\ 210.938 \end{bmatrix} = \begin{bmatrix} .104 \\ -2.667 \\ \cdot \\ \cdot \\ \cdot \\ 1.062 \end{bmatrix}$$

Figures 7.6, 7.7, and 7.8 contain plots of the residuals e_i against the fitted values $\hat{Y}_i$, against X_{i1}, and against X_{i2}, respectively. These plots were generated by the BMDP computer package. There are no suggestions in any of these plots that systematic deviations from the fitted response plane are present, nor that the error variance varies either with the level of $\hat{Y}$ or with the levels of X_1 or X_2. We do not show a normal probability plot, but it does not indicate any major departure from normality. Hence, model (7.59) appears to be apt for this application.

FIGURE 7.6 Residual plot against $\hat{Y}$—Zarthan Company example

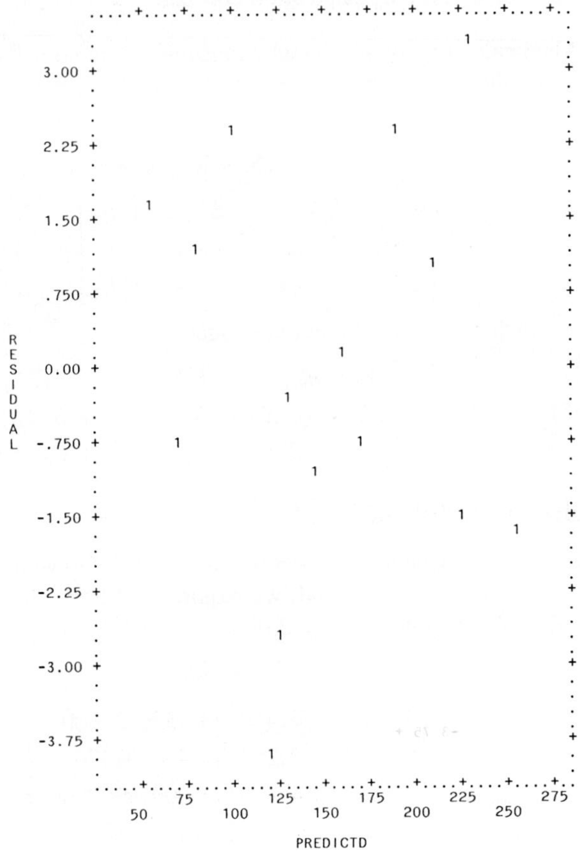

Analysis of variance

To test whether sales are related to population and per capita discretionary income, we construct the ANOVA table in Table 7.4. (p. 255). The basic quantities needed are:

$$\mathbf{Y'Y} = [162 \quad 120 \quad \cdots \quad 212]\begin{bmatrix}162\\120\\ \cdot \\ \cdot \\ \cdot \\212\end{bmatrix}$$

$$= (162)^2 + (120)^2 + \cdots + (212)^2$$
$$= 394{,}107.000$$

FIGURE 7.7 Residual plot against X_1—Zarthan Company example

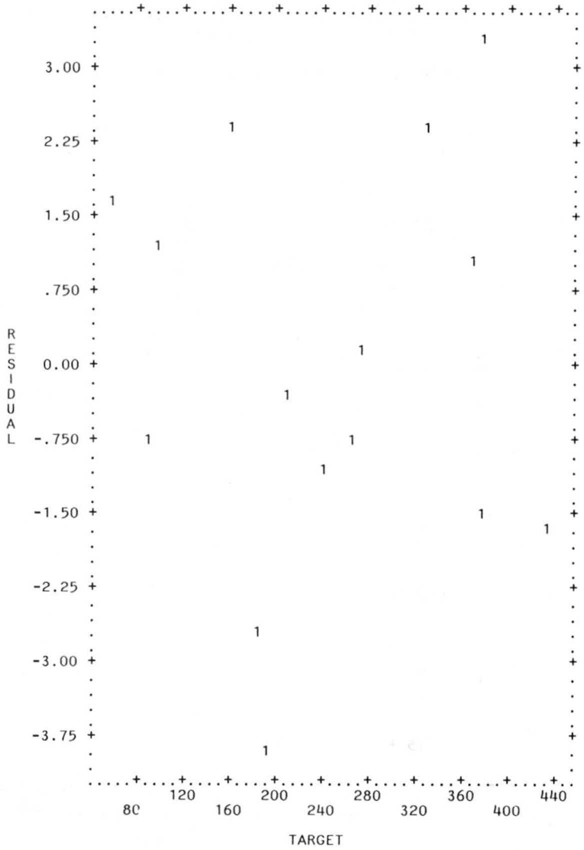

$$\left(\frac{1}{n}\right)\mathbf{Y'11'Y} = \frac{1}{15}[162 \quad 120 \quad \cdots \quad 212]\begin{bmatrix}1\\1\\ \cdot \\ \cdot \\ \cdot \\1\end{bmatrix}[1 \quad 1 \quad \cdots \quad 1]\begin{bmatrix}162\\120\\ \cdot \\ \cdot \\ \cdot \\212\end{bmatrix}$$

$$= \frac{1}{n}(\Sigma Y_i)(\Sigma Y_i) = \frac{(2{,}259)^2}{15} = 340{,}205.400$$

Thus:

$$SSTO = \mathbf{Y'Y} - \left(\frac{1}{n}\right)\mathbf{Y'11'Y} = 394{,}107.000 - 340{,}205.400 = 53{,}901.600$$

FIGURE 7.8 Residual plot against X_2—Zarthan Company example

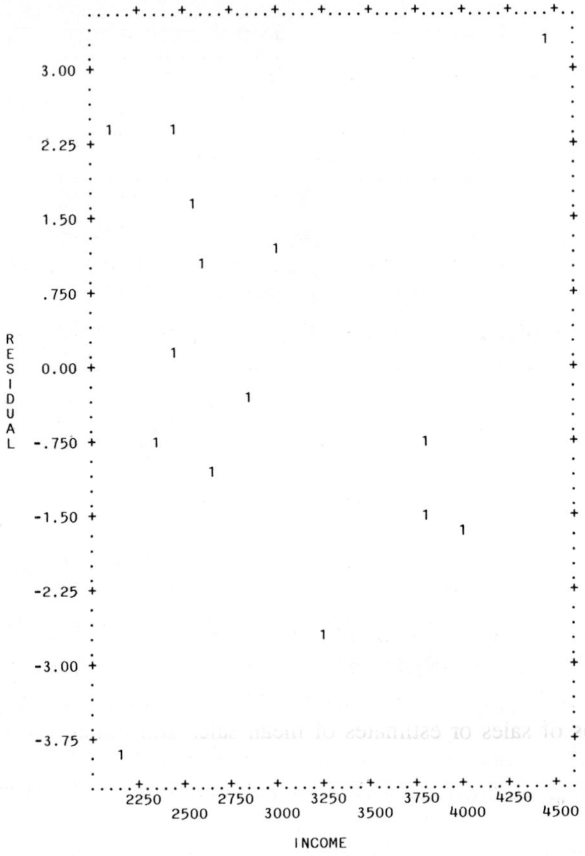

and using our result in (7.61):

$$\begin{aligned} SSE &= \mathbf{Y'Y} - \mathbf{b'X'Y} \\ &= 394{,}107.000 - [3.4526127900 \quad .4960049761 \quad .009199080867] \times \begin{bmatrix} 2{,}259 \\ 647{,}107 \\ 7{,}096{,}619 \end{bmatrix} \\ &= 394{,}107.000 - 394{,}050.116 = 56.884 \end{aligned}$$

Finally, we obtain by subtraction:

$$SSR = SSTO - SSE = 53{,}901.600 - 56.884 = 53{,}844.716$$

The degrees of freedom and mean squares are entered in Table 7.4. Note that three regression parameters had to be estimated; hence, 15 − 3 = 12 degrees of

TABLE 7.4 ANOVA table—Zarthan Company example

Source of Variation	*SS*	*df*	*MS*
Regression	$SSR = 53{,}844.716$	2	$MSR = 26{,}922.358$
Error	$SSE = 56.884$	12	$MSE = 4.740$
Total	$SSTO = 53{,}901.600$	14	

freedom are associated with *SSE*. Also, the number of degrees of freedom associated with *SSR* are two—the number of *X* variables in the model.

Test of regression relation. To test whether sales are related to population and per capita discretionary income:

$$H_0: \beta_1 = 0 \text{ and } \beta_2 = 0$$
$$H_a: \text{not both } \beta_1 \text{ and } \beta_2 \text{ equal } 0$$

we use test statistic (7.30b):

$$F^* = \frac{MSR}{MSE} = \frac{26{,}922.358}{4.740} = 5{,}680$$

Assuming α is to be .05, we require $F(.95; 2, 12) = 3.89$. Since $F^* = 5{,}680 > 3.89$, we conclude H_a, that sales are related to population and per capita discretionary income. Whether this relation is useful for making predictions of sales or estimates of mean sales still remains to be seen.

The *P*-value for this test is less than .001 since we note from Table A–4 that $F(.999; 2, 12) = 13.0$. In fact, it can be shown that the *P*-value is 0+.

Coefficient of multiple determination. For our example, we have by (7.31):

$$R^2 = \frac{SSR}{SSTO} = \frac{53{,}844.716}{53{,}901.600} = .9989$$

Thus, when the two independent variables population and per capita discretionary income are considered, the variation in sales is reduced by 99.9 percent.

Algebraic expression for *SSE*. The error sum of squares for the case of two independent variables in algebraic terms is:

$$SSE = \mathbf{Y}'\mathbf{Y} - \mathbf{b}'\mathbf{X}'\mathbf{Y} = \Sigma Y_i^2 - [b_0 \quad b_1 \quad b_2]\begin{bmatrix} \Sigma Y_i \\ \Sigma X_{i1} Y_i \\ \Sigma X_{i2} Y_i \end{bmatrix}$$

or:

(7.66)
$$SSE = \Sigma Y_i^2 - b_0 \Sigma Y_i - b_1 \Sigma X_{i1} Y_i - b_2 \Sigma X_{i2} Y_i$$

Note how this expression is a straightforward extension of (2.24a) for the case of one independent variable.

Estimation of regression parameters

The Zarthan Company is not interested in the parameter β_0 since it falls far outside the scope of the model. It is desired to estimate β_1 and β_2 jointly with a family confidence coefficient .90. We shall use the simultaneous Bonferroni confidence limits in (7.44), since these are easy to develop and interpret.

First, we need the estimated variance-covariance matrix $\mathbf{s}^2(\mathbf{b})$:

$$\mathbf{s}^2(\mathbf{b}) = MSE(\mathbf{X}'\mathbf{X})^{-1}$$

MSE is given in Table 7.4, and $(\mathbf{X}'\mathbf{X})^{-1}$ was obtained in (7.62). Hence:

(7.67)

$$\mathbf{s}^2(\mathbf{b}) = 4.7403$$

$$\times \begin{bmatrix} 1.2463484 & 2.1296642\text{E} - 4 & -4.1567125\text{E} - 4 \\ 2.1296642\text{E} - 4 & 7.7329030\text{E} - 6 & -7.0302518\text{E} - 7 \\ -4.1567125\text{E} - 4 & -7.0302518\text{E} - 7 & 1.9771851\text{E} - 7 \end{bmatrix}$$

$$= \begin{bmatrix} 5.9081 & 1.0095\text{E} - 3 & -1.9704\text{E} - 3 \\ 1.0095\text{E} - 3 & 3.6656\text{E} - 5 & -3.3326\text{E} - 6 \\ -1.9704\text{E} - 3 & -3.3326\text{E} - 6 & 9.3725\text{E} - 7 \end{bmatrix}$$

The two elements we require are:

$$s^2(b_1) = .000036656 \quad \text{or} \quad s(b_1) = .006054$$
$$s^2(b_2) = .00000093725 \quad \text{or} \quad s(b_2) = .0009681$$

Next, we require for $g = 2$ simultaneous estimates:

$$B = t(1 - .10/2(2); 12) = t(.975; 12) = 2.179$$

Now we are ready to obtain the two simultaneous confidence intervals:

$$.4960 - 2.179(.006054) \leq \beta_1 \leq .4960 + 2.179(.006054)$$

or:

$$.483 \leq \beta_1 \leq .509$$

$$.009199 - 2.179(.0009681) \leq \beta_2 \leq .009199 + 2.179(.0009681)$$

or:

$$.0071 \leq \beta_2 \leq .0113$$

With family confidence coefficient .90, we conclude that β_1 falls between .483 and .509 and that β_2 falls between .0071 and .0113.

Note that the simultaneous confidence intervals suggest that both β_1 and β_2

are positive, which is in accord with theoretical expectations that sales should increase with either higher target population or higher per capita discretionary income, the other variable being held constant.

Estimation of mean response

Suppose the Zarthan Company would like to estimate expected (mean) sales in a district with target population $X_{h1} = 220$ thousand persons and per capita discretionary income $X_{h2} = 2{,}500$ dollars. We define:

$$\mathbf{X}_h = \begin{bmatrix} 1 \\ 220 \\ 2{,}500 \end{bmatrix}$$

The point estimate of mean sales is by (7.47):

$$\hat{Y}_h = \mathbf{X}_h'\mathbf{b} = [1 \quad 220 \quad 2{,}500] \begin{bmatrix} 3.4526 \\ .4960 \\ .009199 \end{bmatrix} = 135.57$$

The estimated variance by (7.50) and using the results in (7.67) is:

$$\begin{aligned} s^2(\hat{Y}_h) &= \mathbf{X}_h'\mathbf{s}^2(\mathbf{b})\mathbf{X}_h \\ &= [1 \quad 220 \quad 2{,}500] \\ &\quad \times \begin{bmatrix} 5.9081 & 1.0095E-3 & -1.9704E-3 \\ 1.0095E-3 & 3.6656E-5 & -3.3326E-6 \\ -1.9704E-3 & -3.3326E-6 & 9.3725E-7 \end{bmatrix} \begin{bmatrix} 1 \\ 220 \\ 2{,}500 \end{bmatrix} \\ &= .46638 \end{aligned}$$

or:

$$s(\hat{Y}_h) = .68292$$

Assume that the confidence coefficient for the interval estimate of $E(Y_h)$ is to be .95. We then need $t(.975; 12) = 2.179$, and obtain by (7.51):

$$135.57 - 2.179(.68292) \le E(Y_h) \le 135.57 + 2.179(.68292)$$

or:

$$134.1 \le E(Y_h) \le 137.1$$

Thus, with confidence coefficient .95, we estimate that mean sales in a district with target population of 220 thousand and per capita discretionary income of $2,500 are somewhere between 134.1 and 137.1 gross.

Algebraic version of estimated variance $s^2(\hat{Y}_h)$. Since by (7.50):

$$s^2(\hat{Y}_h) = \mathbf{X}_h'\mathbf{s}^2(\mathbf{b})\mathbf{X}_h$$

it follows for the case of two independent variables:

$$(7.68)\quad s^2(\hat{Y}_h) = s^2(b_0) + X_{h1}^2 s^2(b_1) + X_{h2}^2 s^2(b_2) + 2X_{h1}s(b_0, b_1) + 2X_{h2}s(b_0, b_2) + 2X_{h1}X_{h2}s(b_1, b_2)$$

When we substitute in (7.68), utilizing the estimated variances and covariances from (7.67), we obtain the same result as before, namely, $s^2(\hat{Y}_h) = .46638$.

Prediction limits for new observations

Suppose the Zarthan Company would like to predict sales in two districts. The two districts have the following characteristics:

	District A	*District B*
X_{h1}	220	375
X_{h2}	2,500	3,500

To determine which simultaneous prediction intervals are best here, we shall find S as given in (7.57a) and B as given in (7.58a) for $g = 2$, assuming the family confidence coefficient is to be .90:

$$S^2 = 2F(.90; 2, 12) = 2(2.81) = 5.62$$

or:

$$S = 2.37$$

and

$$B = t(1 - .10/2(2); 12) = t(.975; 12) = 2.179$$

Hence, the Bonferroni limits are more efficient here.

For district A, we shall use the results we found when estimating mean sales, since the levels of the independent variables are the same as before. We have from earlier:

$$\hat{Y}_A = 135.57 \qquad s^2(\hat{Y}_A) = .46638 \qquad MSE = 4.7403$$

Hence, by (7.55a):

$$s^2(Y_{A(new)}) = MSE + s^2(\hat{Y}_A) = 4.7403 + .46638 = 5.20668$$

or:

$$s(Y_{A(new)}) = 2.28182$$

In similar fashion, we obtain:

$$\hat{Y}_B = 221.65 \qquad s(Y_{B(new)}) = 2.34536$$

We found before that $B = 2.179$. Hence, the simultaneous Bonferroni prediction intervals with family confidence coefficient .90 are by (7.58):

$$135.57 - 2.179(2.28182) \leq Y_{A(new)} \leq 135.57 + 2.179(2.28182)$$

or:

$$130.6 \leq Y_{A(new)} \leq 140.5$$

$$221.65 - 2.179(2.34536) \leq Y_{B(new)} \leq 221.65 + 2.179(2.34536)$$

or:

$$216.5 \leq Y_{B(new)} \leq 226.8$$

With family confidence coefficient .90, we predict that sales in the two districts will be within the indicated limits. The Zarthan Company considers these prediction limits sufficiently precise, and hence useful.

Computer printout

Figure 7.9 contains an illustrative computer printout for the Zarthan Company example, obtained by using the GLM (general linear model) program of the SAS (Statistical Analysis System) computer package (Ref. 7.1). Regression analysis printouts differ in format from one computer program to another, as may be seen by comparing the output in Figure 7.9 with other outputs presented in earlier chapters. However, the basic information presented in the different outputs is essentially the same for the major statistical regression packages.

We have annotated the output in Figure 7.9 to tie in with the notation of this book. The first two blocks of information contain intermediate regression analysis results in matrix form, specifically the $\mathbf{X'X}$ and $(\mathbf{X'X})^{-1}$ matrices. The label "intercept" in these matrices refers to $X_{i0} \equiv 1$ in the alternative regression model (7.7b).

The next block presents information about the estimated regression coefficients b_k. Shown, in turn, are the estimates b_k, the test statistics $t_k^* = b_k/s(b_k)$ for testing whether or not $\beta_k = 0$, the two-sided P-values for the test statistics, and the estimated standard deviations $s(b_k)$.

The fourth block contains ANOVA information: the ANOVA table, the F^* value for the test of whether or not a regression relation exists, the P-value for this test, $\sqrt{MSE}$, and R^2.

The final block shows the observed values Y_i, the fitted values $\hat{Y}_i$, and the residuals e_i.

Because of rounding, some results in Figure 7.9 do not coincide precisely with the corresponding results given earlier. In this connection, it should be noted that different computer regression packages may lead to somewhat different results because final results are rounded to different extents, and even more importantly because rounding errors are not handled equally well by all packages. Particularly when there are a number of independent variables, some of which are highly correlated, rounding errors can be a serious source of difficulty. It is a wise policy to investigate a computer regression package before using it, for instance, by comparing its output for a test problem against results known to be accurate.

FIGURE 7.9 Computer printout for Zarthan Company example (SAS, Ref. 7.1)

THE X'X MATRIX

	INTERCEPT	TARGTP	INCOME
INTERCEPT ← X_0	15.00	3626.00	44428.00
TARGTP ← X_1	3626.00	1067614.00	11419181.00
INCOME ← X_2	44428.00	11419181.00	139063428.00

X'X INVERSE MATRIX

	INTERCEPT	TARGTP	INCOME
INTERCEPT	1.24634842	0.00021297	-0.00041567
TARGTP	0.00021297	0.00000773	-0.00000070
INCOME	-0.00041567	-0.00000070	0.00000020

PARAMETER	ESTIMATE	T FOR H0: PARAMETER=0	PR > |T|	STD ERROR OF ESTIMATE
INTERCEPT	3.45261279	1.42	0.1809	2.43065049
TARGTP	0.49600498	81.92	0.0001	0.00605444
INCOME	0.00919908	9.50	0.0001	0.00096811
	↑ b_k	↑ $t_k^* = b_k/s(b_k)$	↑ Two-sided P-value	↑ $s(b_k)$

SOURCE	DF	SUM OF SQUARES	MEAN SQUARE	F VALUE
MODEL	2	SSR → 53844.71643444	MSR → 26922.35821722	F^* → 5679.47
ERROR	12	SSE → 56.88356556	MSE → 4.74029713	
CORRECTED TOTAL	14	53901.60000000 ↑ $SSTO$		

One sided P-value

PR > F	R-SQUARE
→ 0.0001	0.998945 ← R^2

STD DEV

2.17722234 ← $\sqrt{MSE}$

OBSERVATION	Y_i OBSERVED VALUE	$\hat{Y}_i$ PREDICTED VALUE	e_i RESIDUAL
1	162.00000000	161.89572437	0.10427563
2	120.00000000	122.66731763	-2.66731763
3	223.00000000	224.42938429	-1.42938429
4	131.00000000	131.24062439	-0.24062439
5	67.00000000	67.69928353	-0.69928353
6	169.00000000	169.68485530	-0.68485530
7	81.00000000	79.73193570	1.26806430
8	192.00000000	189.67200303	2.32799697
9	116.00000000	119.83201895	-3.83201895
10	55.00000000	53.29052354	1.70947646
11	252.00000000	253.71505760	-1.71505760
12	232.00000000	228.69079490	3.30920510
13	144.00000000	144.97934226	-0.97934226
14	103.00000000	100.53307489	2.46692511
15	212.00000000	210.93805961	1.06194039

Caution about hidden extrapolations

Before concluding this illustration of multiple regression analysis, we should caution again about making estimates or predictions outside the scope of the model. The danger, of course, is that the model may not be appropriate when extended outside the region of the observations. In multiple regression, it is particularly easy to lose track of this region since the levels of $X_1, \ldots, X_{p-1}$ *jointly* define the region. Thus, one cannot merely look at the ranges of each independent variable. Consider Figure 7.10, where the shaded region is the region of observations for a multiple regression application with two independent variables. The circled dot is within the ranges of the independent variables X_1 and X_2 individually, yet is well outside the joint region of observations.

FIGURE 7.10 Region of observations on X_1 and X_2 jointly, compared with ranges of X_1 and X_2 individually

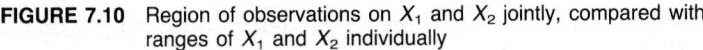

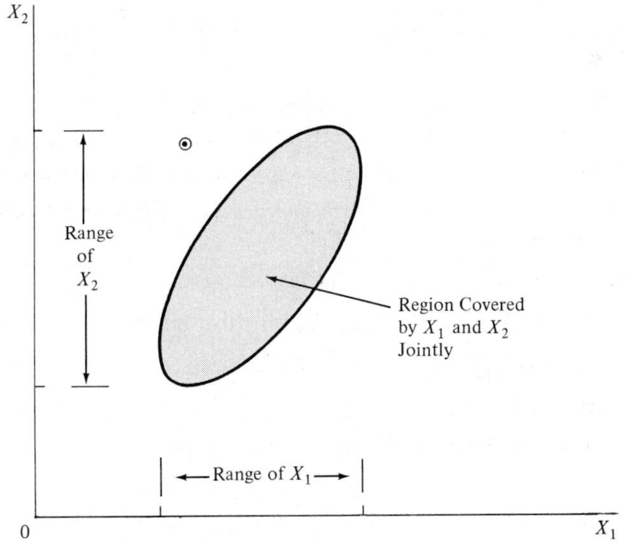

7.9 STANDARDIZED REGRESSION COEFFICIENTS

Standardized regression coefficients have been proposed to facilitate comparisons between regression coefficients. Ordinarily, it is difficult to compare regression coefficients because of differences in the units involved. We cite two examples.

1. When considering the fitted response function:

$$\hat{Y} = 200 + 20{,}000X_1 + .2X_2$$

one may be tempted to conclude that X_1 is the only important independent variable, and that X_2 has little effect on the dependent variable Y. A little reflection should make one wary of this conclusion. The reason is that we do not know the units involved. Suppose the units are:

Y in dollars
X_1 in thousand dollars
X_2 in cents

In that event, the effect on the mean response of a \$1,000 increase in X_1 when X_2 is constant would be exactly the same as the effect of a \$1,000 increase in X_2 when X_1 is constant, despite the difference in the regression coefficients.

2. In our Zarthan Company example, we cannot make any comparison between b_1 and b_2 because b_1 is in units of one gross of jars per thousand persons while b_2 is in units of one gross of jars per dollar of per capita discretionary income.

Standardized regression coefficients, also sometimes called *beta coefficients,* are defined as follows:

$$B_k = b_k\left[\frac{s_k}{s_Y}\right] = b_k\left[\frac{\dfrac{\sum_i (X_{ik}-\bar{X}_k)^2}{n-1}}{\dfrac{\sum_i (Y_i-\bar{Y})^2}{n-1}}\right]^{1/2} = b_k\left[\frac{\sum_i (X_{ik}-\bar{X}_k)^2}{\sum_i (Y_i-\bar{Y})^2}\right]^{1/2} \tag{7.69}$$

where s_k and s_Y are the standard deviations of the X_k and Y observations, respectively. The effect of the term in brackets is to make B_k dimensionless. The coefficient B_k reflects the change in the mean response (in units of standard deviations of Y) per unit change in the independent variable X_k (in units of standard deviations of X_k) when all other independent variables are held constant.

For our Zarthan Company example, we obtain the following standardized regression coefficients, using the final expression in (7.69):

$$B_1 = .496\left[\frac{191{,}089}{53{,}902}\right]^{1/2} = .934$$

$$B_2 = .00920\left[\frac{7{,}473{,}616}{53{,}902}\right]^{1/2} = .108$$

Sometimes, these standardized regression coefficients are interpreted as showing that target population (X_1) has a greater impact than per capita discretionary income (X_2) on sales because B_1 is much larger than B_2. However, as we will see in the next chapter, one must be cautious about interpreting regression coefficients, whether standardized or not. The reason is that when the independent variables are correlated among themselves, the regression coefficients are affected by the other independent variables in the model. We shall discuss this problem in Chapter 8.

Not only does the presence of correlations among the independent variables affect the magnitude of the standardized regression coefficients but the spacings of the observations on the independent variables also affect the standardized regression coefficients. Sometimes the spacings of the observations on the independent variables may be quite arbitrary.

Hence, it is ordinarily not wise to interpret a standardized regression coefficient as reflecting the importance of the independent variable.

7.10 WEIGHTED LEAST SQUARES

Unequal weighting of the observations can be carried out in multiple regression as in simple regression. Let the weights w_i be contained in the diagonal matrix **W** of weights:

(7.70)
$$\underset{n\times n}{\mathbf{W}} = \begin{bmatrix} w_1 & & & 0 \\ & w_2 & & \\ & & \ddots & \\ 0 & & & w_n \end{bmatrix}$$

The weighted least squares estimators of the regression coefficients then are:

(7.71)
$$\underset{p\times 1}{\mathbf{b}} = (\mathbf{X}'\mathbf{W}\mathbf{X})^{-1}\mathbf{X}'\mathbf{W}\mathbf{Y}$$

When the error term variances σ_i^2 are not equal, the weights w_i are chosen to be inversely proportional to σ_i^2, so that $\sigma_i^2 = \sigma^2/w_i$. The estimated variance-covariance matrix of the regression coefficients then is:

(7.72)
$$\underset{p\times p}{\mathbf{s}^2(\mathbf{b})} = MSE_w(\mathbf{X}'\mathbf{W}\mathbf{X})^{-1}$$

where MSE_w is based on the weighted squared deviations:

(7.72a)
$$MSE_w = \frac{\Sigma w_i(Y_i - \hat{Y}_i)^2}{n - p}$$

As for simple regression, the appropriate weights w_i may often be found from basic relationships. For instance, it may be found that the error term variances σ_i^2 are proportional to X_{ik}^2, the level of the kth independent variable squared. Hence, $\sigma_i^2 = \sigma^2 X_{ik}^2$ then and the weights $w_i = 1/X_{ik}^2$ would be used.

PROBLEMS

7.1. Refer to Figure 7.4a. By how much approximately does mean yield increase when rainfall increases from 9 to 11 inches and temperature is held constant? Could you have answered this question if rainfall and temperature interact in their effects on crop yield?

7.2. Consider the response function $E(Y) = 25 + 3X_1 + 4X_2 + 1.5X_1X_2$.

a. Plot the response function against X_1 when $X_2 = 3$ and when $X_2 = 6$. How is the interaction effect of X_1 and X_2 on Y apparent from this graph?

b. Sketch a set of contour curves for the response surface. How is the interaction effect of X_1 and X_2 on Y apparent from this graph?

7.3. Consider the response function $E(Y) = 14 + 7X_1 - 5X_2$.

a. Plot the response function against X_2 when $X_1 = 1$ and when $X_1 = 4$. How does the graph indicate that the effects of X_1 and X_2 on Y are additive?

b. Sketch a set of contour curves for the response surface. How does the graph indicate that the effects of X_1 and X_2 on Y are additive?

7.4. Set up the $\mathbf{X}$ matrix and $\boldsymbol{\beta}$ vector for each of the following models (assume $i = 1, \ldots, 4$):

a. $Y_i = \beta_0 + \beta_1 X_{i1} + \beta_2 X_{i1} X_{i2} + \varepsilon_i$

b. $\log Y_i = \beta_0 + \beta_1 X_{i1} + \beta_2 X_{i2} + \varepsilon_i$

7.5. Set up the $\mathbf{X}$ matrix and $\boldsymbol{\beta}$ vector for each of the following models (assume $i = 1, \ldots, 5$):

a. $Y_i = \beta_1 X_{i1} + \beta_2 X_{i2} + \beta_3 X_{i1}^2 + \varepsilon_i$

b. $\sqrt{Y_i} = \beta_0 + \beta_1 X_{i1} + \beta_2 \log_{10} X_{i2} + \varepsilon_i$

7.6. A student stated: "Adding independent variables to a regression model can never reduce R^2, so we should include all available independent variables in the model." Comment.

7.7. Why is it not meaningful to attach a sign to the coefficient of multiple correlation R, although we do so for the coefficient of simple correlation r?

7.8. **Brand preference.** In a small-scale study of the relation between degree of brand liking (Y) and moisture content (X_1) and sweetness (X_2) of the product, the following results were obtained (data are coded):

i:	1	2	3	4	5	6	7	8	9	10	11	12	13	14	15	16
X_{i1}:	4	4	4	4	6	6	6	6	8	8	8	8	10	10	10	10
X_{i2}:	2	4	2	4	2	4	2	4	2	4	2	4	2	4	2	4
Y_i:	64	73	61	76	72	80	71	83	83	89	86	93	88	95	94	100

Assume that regression model (7.1) with independent normal error terms is appropriate.

a. Find the estimated regression coefficients. State the estimated regression function. How is b_1 interpreted here?

b. Test whether there is a regression relation using a level of significance of .01. State the alternatives, decision rule, and conclusion. What does your test imply about β_1 and β_2?

c. What is the P-value of the test in part (b)?

d. Estimate β_1 and β_2 jointly by the Bonferroni procedure using a 99 percent family confidence coefficient. Interpret your results.

7.9. Refer to **Brand preference** Problem 7.8.

a. Calculate the coefficient of multiple determination R^2. How is it interpreted here?

b. Calculate the coefficient of simple determination r^2 between Y_i and $\hat{Y}_i$. Does it equal R^2?

7.10. Refer to **Brand preference** Problem 7.8.

a. Obtain an interval estimate of $E(Y_h)$ when $X_{h1} = 5$ and $X_{h2} = 4$. Use a 99 percent confidence coefficient. Interpret your interval estimate.

b. Obtain a prediction interval for a new observation $Y_{h(\text{new})}$ when $X_{h1} = 5$ and $X_{h2} = 4$. Use a 99 percent confidence coefficient.

7.11. Refer to **Brand preference** Problem 7.8.

a. Obtain the residuals.

b. Plot the residuals against $\hat{Y}$, X_1, and X_2 on separate graphs. Also prepare a normal probability plot. Analyze the plots and summarize your findings.

c. Conduct a formal test for lack of fit of the first-order regression function; use $\alpha = .01$. State the alternatives, decision rule, and conclusion.

7.12. Chemical shipment. The observations to follow, taken on 20 incoming shipments of chemicals in drums arriving at a warehouse, show number of drums in shipment (X_1), total weight of shipment (X_2, in hundred pounds), and number of minutes required to handle shipment (Y).

i:	1	2	3	4	5	6	7	8	9	10
X_{i1}:	7	18	5	14	11	5	23	9	16	5
X_{i2}:	5.11	16.72	3.20	7.03	10.98	4.04	22.07	7.03	10.62	4.76
Y_i:	58	152	41	93	101	38	203	78	117	44

i:	11	12	13	14	15	16	17	18	19	20
X_{i1}:	17	12	6	12	8	15	17	21	6	11
X_{i2}:	11.02	9.51	3.79	6.45	4.60	13.86	13.03	15.21	3.64	9.57
Y_i:	121	112	50	82	48	127	140	155	39	90

Assume that regression model (7.1) with independent normal error terms is appropriate.

a. Obtain the estimated regression function. How is b_1 here interpreted? How is b_2 here interpreted?

b. Test whether there is a regression relation, using a level of significance of .05. State the alternatives, decision rule, and conclusion. What does your test result imply about β_1 and β_2? What is the P-value of the test?

c. Estimate β_1 and β_2 jointly by the Bonferroni procedure using a 95 percent family confidence coefficient. Interpret your results.

d. Calculate the coefficient of multiple determination R^2. How is this measure interpreted here?

7.13. Refer to **Chemical shipment** Problem 7.12.

a. Management desires simultaneous interval estimates of the mean handling times for five typical shipments specified to be as follows:

	1	2	3	4	5
X_1:	5	6	10	14	20
X_2:	3.20	4.80	7.00	10.00	18.00

Obtain the family of estimates using a 95 percent family confidence coefficient. Employ the Working-Hotelling type bounds or the Bonferroni procedure, whichever is more efficient.

b. For the observations in Problem 7.12, would you consider a shipment of 20 drums with a weight of 5 hundred pounds to be within the scope of the model? What about a shipment of 20 drums with a weight of 19 hundred pounds? Support your answer by preparing a relevant plot.

7.14. Refer to **Chemical shipment** Problem 7.12. Four separate shipments with the following characteristics will arrive in the next day or two:

	1	2	3	4
X_1:	9	12	15	18
X_2:	7.20	9.00	12.50	16.50

Management desires predictions of the handling times for these shipments so that the actual handling times can be compared with the predicted times to determine whether any are "out of line." Develop the needed predictions using the most efficient approach and a family confidence coefficient of 95 percent.

7.15. Refer to **Chemical shipment** Problem 7.12.

a. Obtain the residuals and plot them against $\hat{Y}$, X_1, and X_2 on separate graphs. Also prepare a normal probability plot. Analyze the plots and summarize your findings.

b. Can you conduct a formal test for lack of fit here?

7.16. Refer to **Chemical shipment** Problem 7.12. Three new shipments are to be received, each with $X_{h1} = 7$ and $X_{h2} = 6$.

a. Obtain a 95 percent prediction interval for the mean handling time for these shipments.

b. Convert the interval obtained in part (a) into a 95 percent prediction interval for the total handling time for the three shipments.

7.17. **Patient satisfaction.** A hospital administrator wished to study the relation between patient satisfaction (Y) and patient's age (X_1, in years), severity of illness (X_2, an index), and anxiety level (X_3, an index). She randomly selected 23 patients and collected the data presented below, where larger values of Y, X_2, and X_3 are, respectively, associated with more satisfaction, increased severity of illness, and more anxiety.

i:	1	2	3	4	5	6	7	8	9	10	11	12
X_{i1}:	50	36	40	41	28	49	42	45	52	29	29	43
X_{i2}:	51	46	48	44	43	54	50	48	62	50	48	53
X_{i3}:	2.3	2.3	2.2	1.8	1.8	2.9	2.2	2.4	2.9	2.1	2.4	2.4
Y_i:	48	57	66	70	89	36	46	54	26	77	89	67

i:	13	14	15	16	17	18	19	20	21	22	23
X_{i1}:	38	34	53	36	33	29	33	55	29	44	43
X_{i2}:	55	51	54	49	56	46	49	51	52	58	50
X_{i3}:	2.2	2.3	2.2	2.0	2.5	1.9	2.1	2.4	2.3	2.9	2.3
Y_i:	47	51	57	66	79	88	60	49	77	52	60

Assume that regression model (7.5) for three independent variables with independent normal error terms is appropriate.

a. Obtain the estimated regression function.

b. Test whether there is a regression relation; use a .10 level of significance. State the alternatives, decision rule, and conclusion. What does your test imply about β_1, β_2, and β_3? What is the P-value of the test?

c. Obtain joint interval estimates of β_1, β_2, and β_3 using a 90 percent family confidence coefficient. Interpret your results.

d. Calculate the coefficient of multiple correlation. What does it indicate here?

7.18. Refer to **Patient satisfaction** Problem 7.17.

a. Obtain an interval estimate of the mean satisfaction when $X_{h1} = 35$, $X_{h2} = 45$, and $X_{h3} = 2.2$. Use a 90 percent confidence coefficient. Interpret your confidence interval.

b. Obtain a prediction interval for a new patient's satisfaction when $X_{h1} = 35$, $X_{h2} = 45$, and $X_{h3} = 2.2$. Use a 90 percent confidence coefficient. Interpret your prediction interval.

7.19. Refer to **Patient satisfaction** Problem 7.17.

a. Obtain the residuals and plot them against $\hat{Y}$ and each of the independent variables on separate graphs. Also prepare a normal probability plot. Analyze your plots and summarize your findings.

b. Can you conduct a formal test for lack of fit here?

7.20. **Mathematicians salaries.** A researcher in a scientific foundation wished to evaluate the relation between intermediate and senior level annual salaries of research mathematicians (Y, in thousand dollars) and an index of publication quality (X_1), number of years of experience (X_2), and an index of success in obtaining grant support (X_3). The data for a sample of 24 intermediate and senior level research mathematicians follow.

i:	1	2	3	4	5	6	7	8	9	10	11	12
X_{i1}:	3.5	5.3	5.1	5.8	4.2	6.0	6.8	5.5	3.1	7.2	4.5	4.9
X_{i2}:	9	20	18	33	31	13	25	30	5	47	25	11
X_{i3}:	6.1	6.4	7.4	6.7	7.5	5.9	6.0	4.0	5.8	8.3	5.0	6.4
Y_i:	33.2	40.3	38.7	46.8	41.4	37.5	39.0	40.7	30.1	52.9	38.2	31.8

i:	13	14	15	16	17	18	19	20	21	22	23	24
X_{i1}:	8.0	6.5	6.6	3.7	6.2	7.0	4.0	4.5	5.9	5.6	4.8	3.9
X_{i2}:	23	35	39	21	7	40	35	23	33	27	34	15
X_{i3}:	7.6	7.0	5.0	4.4	5.5	7.0	6.0	3.5	4.9	4.3	8.0	5.0
Y_i:	43.3	44.1	42.8	33.6	34.2	48.0	38.0	35.9	40.4	36.8	45.2	35.1

Assume that regression model (7.5) for three independent variables with independent normal error terms is appropriate.

a. Obtain the estimated regression function.

b. Test whether there is a regression relation; use $\alpha = .05$. State the alternatives, decision rule, and conclusion. What does your test imply about β_1, β_2, and β_3? What is the P-value of the test?

c. Estimate β_1, β_2, and β_3 jointly by the Bonferroni procedure using a 95 percent family confidence coefficient. Interpret your results.

d. Calculate R^2 and interpret this measure.

7.21. Refer to **Mathematicians salaries** Problem 7.20. The researcher wishes to obtain simultaneous interval estimates of the mean salary levels for four typical research mathematicians specified as follows:

	1	2	3	4
X_1	5.0	6.0	4.0	7.0
X_2	20	30	10	50
X_3	5.0	6.0	4.0	7.0

Obtain the family of estimates using a 95 percent family confidence coefficient. Employ the most efficient procedure.

7.22. Refer to **Mathematicians salaries** Problem 7.20. Three research mathematicians with the following characteristics did not provide any salary information in the study.

	1	2	3
X_1	5.4	6.2	6.4
X_2	17	12	21
X_3	6.0	5.8	6.1

Develop separate prediction intervals for the annual salaries of these mathematicians using a 95 percent statement confidence coefficient in each case. Can the salaries of these three mathematicians be predicted fairly precisely?

7.23. Refer to **Mathematicians salaries** Problem 7.20.

a. Obtain the residuals and plot them against $\hat{Y}$ and each of the independent variables on separate graphs. Also prepare a normal probability plot. Analyze your plots and summarize your findings.

b. Can you conduct a formal test for lack of fit here?

7.24. Refer to **Chemical shipment** Problem 7.12.

a. Obtain the standardized regression coefficients.

b. Calculate the coefficient of determination between the two independent variables. Is it meaningful here to consider the standardized regression coefficients to reflect the effect of one independent variable when the other is held constant?

7.25. Refer to **Patient satisfaction** Problem 7.17.

a. Obtain the standardized regression coefficients.

b. Calculate the coefficients of determination between all pairs of independent variables. Do these indicate that it is meaningful here to consider the standardized regression coefficients as indicating the effect of one independent variable when the others are held constant?

EXERCISES

7.26. For each of the following models, indicate whether it is a general linear regression model. If it is not, state whether it can be expressed in the form of (7.7) by a suitable transformation:

a. $Y_i = \beta_0 + \beta_1 X_{i1} + \beta_2 \log_{10} X_{i2} + \beta_3 X_{i1}^2 + \varepsilon_i$

b. $Y_i = \varepsilon_i \exp(\beta_0 + \beta_1 X_{i1} + \beta_2 X_{i2}^2)$

c. $Y_i = \beta_0 + \log_{10}(\beta_1 X_{i1}) + \beta_2 X_{i2} + \varepsilon_i$

d. $Y_i = \beta_0 \exp(\beta_1 X_{i1}) + \varepsilon_i$

e. $Y_i = [1 + \exp(\beta_0 + \beta_1 X_{i1} + \varepsilon_i)]^{-1}$

7.27. (Calculus needed.) Consider the multiple regression model:

$$Y_i = \beta_1 X_{i1} + \beta_2 X_{i2} + \varepsilon_i \qquad i = 1, \ldots, n$$

where the ε_i are uncorrelated, with $E(\varepsilon_i) = 0$ and $\sigma^2(\varepsilon_i) = \sigma^2$.

a. Derive the least squares estimators of β_1 and β_2.

b. Assuming that the ε_i are independent normal random variables, state the likelihood function and obtain the maximum likelihood estimators of β_1 and β_2. Are these the same as the least squares estimators?

7.28. (Calculus needed.) Consider the multiple regression model:

$$Y_i = \beta_0 + \beta_1 X_{i1} + \beta_2 X_{i1}^2 + \beta_3 X_{i2} + \varepsilon_i \qquad i = 1, \ldots, n$$

where the ε_i are independent $N(0, \sigma^2)$. Derive the least squares normal equations. Will these yield the same estimators of the regression coefficients as the maximum likelihood estimators?

7.29. An analyst wanted to fit the regression model $Y_i = \beta_0 + \beta_1 X_{i1} + \beta_2 X_{i2} + \beta_3 X_{i3} + \varepsilon_i$, $i = 1, \ldots, n$, by the method of least squares when it is known that $\beta_2 = 4$. How can the analyst obtain the desired fit using a multiple regression computer program?

7.30. For regression model (7.1), show that the coefficient of simple determination r^2 between Y_i and $\hat{Y}_i$ equals the coefficient of multiple determination R^2.

7.31. In a small-scale regression study, the following data were obtained:

i:	1	2	3	4	5	6
X_{i1}:	7	4	16	3	21	8
X_{i2}:	33	41	7	49	5	31
Y_i:	42	33	75	28	91	55

Assume that regression model (7.1) with independent normal error terms is appropriate. Using matrix methods, obtain (a) **b**; (b) **e**; (c) *SSE*; (d) *SSR*; (e) $\mathbf{s}^2(\mathbf{b})$; (f) $\hat{Y}_h$ when $X_{h1} = 10$, $X_{h2} = 30$; (g) $s^2(\hat{Y}_h)$ when $X_{h1} = 10$, $X_{h2} = 30$.

PROJECTS

7.32. Refer to the **SMSA** data set. You have been asked to evaluate two alternative models for predicting the number of active physicians (Y) in an SMSA. Proposed model I includes as independent variables total population (X_1), land area (X_2), and total personal income (X_3). Proposed model II includes as independent variables population density (X_1, total population divided by land area), percent of population in central cities (X_2), and total personal income (X_3).

a. For each of the two proposed models, fit the first-order regression model (7.5) with three independent variables.

b. Calculate R^2 for each model. Is one model clearly preferable in terms of this measure?

c. For each model, obtain the residuals and plot them against $\hat{Y}$ and against each of the three independent variables. Also prepare a normal probability plot for each of the two fitted models. Analyze your plots and state your findings. Is one model clearly preferable in terms of aptness?

7.33. Refer to the **SMSA** data set.

a. For each geographic region, regress the number of serious crimes in an SMSA (Y) against population density (X_1, total population divided by land area), total personal income (X_2), and percent high school graduates (X_3).

Use the first-order regression model (7.5) with three independent variables. State the estimated regression functions.

b. Are the estimated regression functions similar for the four regions? Discuss.

c. Calculate MSE and R^2 for each region. Are these measures similar for the four regions? Discuss.

7.34. Refer to the **SENIC** data set. Two models have been proposed for predicting the average length of patient stay in a hospital (Y). Model I utilizes as independent variables age (X_1), infection risk (X_2), and available facilities and services (X_3). Model II uses as independent variables number of beds (X_1), infection risk (X_2), and available facilities and services (X_3).

a. For each of the two proposed models, fit the first-order regression model (7.5) with three independent variables.

b. Calculate R^2 for each model. Is one model clearly preferable in terms of this measure?

c. For each model, obtain the residuals and plot them against $\hat{Y}$ and against each of the three independent variables. Also prepare a normal probability plot for each of the two fitted models. Analyze your plots and state your findings. Is one model clearly preferable in terms of aptness?

7.35. Refer to the **SENIC** data set.

a. For each geographic region, regress infection risk (Y) against the independent variables age (X_1), routine culturing ratio (X_2), average daily census (X_3), and available facilities and services (X_4). Use the first-order regression model (7.5) with four independent variables. State the estimated regression functions.

b. Are the estimated regression functions similar for the four regions? Discuss.

c. Calculate MSE and R^2 for each region. Are these measures similar for the four regions? Discuss.

CITED REFERENCE

7.1 *SAS User's Guide*. 1979 ed. Raleigh, N.C.: SAS Institute, 1979.

8

Multiple regression—II

In this chapter, we continue our discussion of multiple regression by first considering multicollinearity and its effects in multiple regression models. Then we take up several other topics in multiple regression, including additional tests of hypotheses concerning the regression coefficients.

8.1 MULTICOLLINEARITY AND ITS EFFECTS

In multiple regression analysis, one is often concerned with the nature and significance of the relations between the independent variables and the dependent variable. Questions that are frequently asked include:

1. What is the relative importance of the effects of the different independent variables?
2. What is the magnitude of the effect of a given independent variable on the dependent variable?
3. Can any independent variable be dropped from the model because it has little or no effect on the dependent variable?
4. Should any independent variables not yet included in the model be considered for possible inclusion?

If the independent variables included in the model are (1) uncorrelated among themselves and (2) uncorrelated with any other independent variables that are related to the dependent variable but omitted from the model, relatively simple answers can be given to these questions. Unfortunately, in many nonexperimental situations in business, economics, and the social and biological sciences, the independent variables tend to be correlated among themselves and with other variables that are related to the dependent variable but are not included in the model.

When the independent variables are correlated among themselves, *intercorrelation* or *multicollinearity* among them is said to exist. (Sometimes the latter term is reserved for those instances when the correlation among independent variables is very high or even perfect.) We shall explore now a variety of interrelated problems that are created by multicollinearity among the independent variables. First, however, we examine the situation when the independent variables are not correlated.

Example of uncorrelated independent variables

Table 8.1 contains data for a small-scale experiment on the effect of work crew size (X_1) and level of bonus pay (X_2) on crew productivity score (Y). It is easy to show that X_1 and X_2 are uncorrelated here, i.e., $r_{12}^2 = 0$, where r_{12}^2 denotes the coefficient of simple determination between X_1 and X_2. Table 8.2a contains the fitted regression function and analysis of variance table when both X_1 and X_2 are included in the model. Table 8.2b contains the same information when only X_1 is included in the model, and Table 8.2c contains this information when only X_2 is in the model. In Table 8.2a, we use the notation $SSR(X_1, X_2)$ and $SSE(X_1, X_2)$ to indicate explicitly the two independent variables in the model. Similarly, in Table 8.2b, we use the notation $SSR(X_1)$ and $SSE(X_1)$ to show that only independent variable X_1 is in the model, so that the regression here is a simple regression. We do likewise in Table 8.2c.

TABLE 8.1 Work crew productivity example with uncorrelated independent variables

Trial i	*Crew Size* X_{i1}	*Bonus Pay* X_{i2}	*Crew Productivity Score* Y_i
1	4	$2	42
2	4	2	39
3	4	3	48
4	4	3	51
5	6	2	49
6	6	2	53
7	6	3	61
8	6	3	60

An important feature to note in Table 8.2 is that the regression coefficients for X_1 and X_2 are the same, whether only the given independent variable is included

TABLE 8.2 ANOVA tables for work crew productivity example with uncorrelated independent variables

(a) Regression of Y on X_1 and X_2

$\hat{Y} = .375 + 5.375X_1 + 9.250X_2$

Source of Variation	SS	df	MS
Regression	$SSR(X_1, X_2) = 402.250$	2	$MSR(X_1, X_2) = 201.125$
Error	$SSE(X_1, X_2) = 17.625$	5	$MSE(X_1, X_2) = 3.525$
Total	$SSTO = 419.875$	7	

(b) Regression of Y on X_1

$\hat{Y} = 23.500 + 5.375X_1$

Source of Variation	SS	df	MS
Regression	$SSR(X_1) = 231.125$	1	$MSR(X_1) = 231.125$
Error	$SSE(X_1) = 188.750$	6	$MSE(X_1) = 31.458$
Total	$SSTO = 419.875$	7	

(c) Regression of Y on X_2

$\hat{Y} = 27.250 + 9.250X_2$

Source of Variation	SS	df	MS
Regression	$SSR(X_2) = 171.125$	1	$MSR(X_2) = 171.125$
Error	$SSE(X_2) = 248.750$	6	$MSE(X_2) = 41.458$
Total	$SSTO = 419.875$	7	

in the model or both independent variables are included. This is a result of the two independent variables being uncorrelated.

Thus, if the independent variables are uncorrelated, the effects ascribed to them by a first-order regression model are the same no matter which other independent variables are included in the model. This is a strong argument for controlled experiments whenever possible, since experimental control permits making the independent variables uncorrelated.

Another important feature of Table 8.2 is related to the error sums of squares. Note from Table 8.2a that the error sum of squares when both X_1 and X_2 are included in the model is $SSE(X_1, X_2) = 17.625$. When only X_1 is included in the model, however, the error sum of squares is $SSE(X_1) = 188.750$ according to Table 8.2b. Since the variation in Y when X_1 alone is considered is 188.750 but is only 17.625 when both X_1 and X_2 are considered, we may ascribe the difference:

$$SSE(X_1) - SSE(X_1, X_2) = 188.750 - 17.625 = 171.125$$

to the effect of X_2. We shall denote this difference by $SSR(X_2|X_1)$:

$$SSR(X_2|X_1) = SSE(X_1) - SSE(X_1, X_2) \tag{8.1}$$

When we fit a regression function containing only X_2, we also obtain a measure of the reduction in the variation of Y associated with X_2, namely, $SSR(X_2)$. Table 8.2c indicates that $SSR(X_2) = 171.125$, which is the same as $SSR(X_2|X_1) = 171.125$. The reason for this is that X_1 and X_2 are uncorrelated.

The story is the same for independent variable X_1. Let:

$$SSR(X_1|X_2) = SSE(X_2) - SSE(X_1, X_2) \tag{8.2}$$

For our example, we have:

$$SSR(X_1|X_2) = 248.750 - 17.625 = 231.125$$

This sum of squares is the same as $SSR(X_1) = 231.125$ from Table 8.2b, obtained when Y is regressed only on X_1.

In general, when two independent variables are uncorrelated, the marginal contribution of an independent variable in reducing the error sum of squares when the other independent variable is in the model is exactly the same as when this independent variable is in the model alone.

Note

To show that the regression coefficient of X_1 is unchanged when X_2 is added to the regression model in the case where X_1 and X_2 are uncorrelated, consider the algebraic expression for b_1 in the multiple regression model with two independent variables:

$$b_1 = \frac{\dfrac{\Sigma(X_{i1} - \bar{X}_1)(Y_i - \bar{Y})}{\Sigma(X_{i1} - \bar{X}_1)^2} - \left[\dfrac{\Sigma(Y_i - \bar{Y})^2}{\Sigma(X_{i1} - \bar{X}_1)^2}\right]^{1/2} r_{Y2}r_{12}}{1 - r_{12}^2} \tag{8.3}$$

where r_{Y2} denotes the coefficient of simple correlation between Y and X_2, and r_{12}, as before, denotes the coefficient of simple correlation between X_1 and X_2.

If X_1 and X_2 are uncorrelated, $r_{12} = 0$, and (8.3) reduces to:

$$b_1 = \frac{\Sigma(X_{i1} - \bar{X}_1)(Y_i - \bar{Y})}{\Sigma(X_{i1} - \bar{X}_1)^2} \tag{8.3a}$$

But (8.3a) is the estimator of the slope for the simple regression of Y on X_1, per (2.10a).

Hence, if X_1 and X_2 are uncorrelated, adding X_2 to the regression model does not change the regression coefficient for X_1; correspondingly, adding X_1 to the regression model does not change the regression coefficient for X_2.

Example of correlated independent variables

As mentioned earlier, nonexperimental data in many disciplines frequently consist of correlated independent variables. For example, in a regression of family food expenditures on the independent variables family income, family savings, and age of head of the household, the independent variables will be correlated among themselves. Further, the independent variables will also be correlated with other socioeconomic variables not included in the model that do affect family food expenditures, such as family size. We shall now consider the effects of multicollinearity, i.e., correlated independent variables, upon the regression coefficients and upon the regression sums of squares.

Effect of multicollinearity on regression coefficients. Table 8.3 contains data for a study of the relation of body fat (Y) to triceps skinfold thickness (X_1) and thigh circumference (X_2), based on a sample of 20 healthy females 25–34 years old. The triceps skinfold thicknesses and thigh circumferences for these subjects are highly correlated, as the scatter plot in Figure 8.1 suggests. Indeed, the coefficient of simple correlation between these two independent variables for the 20 subjects is $r_{12} = +.92$.

TABLE 8.3 Body fat example with correlated independent variables

Subject i	*Triceps Skinfold Thickness* X_{i1}	*Thigh Circumference* X_{i2}	*Body Fat* Y_i
1	19.5	43.1	11.9
2	24.7	49.8	22.8
3	30.7	51.9	18.7
4	29.8	54.3	20.1
5	19.1	42.2	12.9
6	25.6	53.9	21.7
7	31.4	58.5	27.1
8	27.9	52.1	25.4
9	22.1	49.9	21.3
10	25.5	53.5	19.3
11	31.1	56.6	25.4
12	30.4	56.7	27.2
13	18.7	46.5	11.7
14	19.7	44.2	17.8
15	14.6	42.7	12.8
16	29.5	54.4	23.9
17	27.7	55.3	22.6
18	30.2	58.6	25.4
19	22.7	48.2	14.8
20	25.2	51.0	21.1

Table 8.4a contains the results of a computer run for regressing Y on X_1 and X_2, and presents the fitted regression function and the analysis of variance table. Again, we use the notation $SSR(X_1, X_2)$ and $SSE(X_1, X_2)$ to indicate explicitly that both independent variables are in the fitted model. Table 8.4b contains the fitted regression and the analysis of variance table for the regression of Y on X_1 only, and Table 8.4c contains the results for the regression of Y on X_2 only.

Note first that the regression coefficient for X_1, triceps skinfold thickness, is not the same in Tables 8.4a and b. Thus, the effect ascribed to X_1 by the fitted response function varies here, depending upon whether only X_1 or both X_1 and X_2

FIGURE 8.1 Scatter plot of thigh circumference against triceps skinfold thickness—body fat example

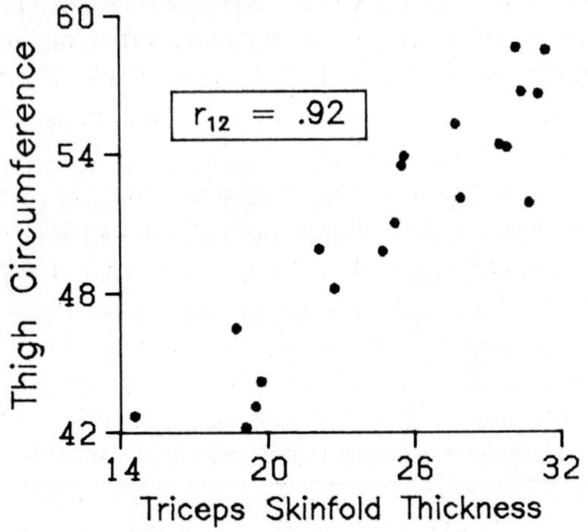

TABLE 8.4 ANOVA table for body fat example with correlated independent variables

(**a**) Regression of Y on X_1 and X_2

$$\hat{Y} = -19.174 + .2224X_1 + .6594X_2$$

Source of Variation	*SS*	*df*	*MS*
Regression	$SSR(X_1, X_2) = 385.44$	2	$MSR(X_1, X_2) = 192.72$
Error	$SSE(X_1, X_2) = 109.95$	17	$MSE(X_1, X_2) = 6.47$
Total	$SSTO = 495.39$	19	

(**b**) Regression of Y on X_1

$$\hat{Y} = -1.496 + .8572X_1$$

Source of Variation	*SS*	*df*	*MS*
Regression	$SSR(X_1) = 352.27$	1	$MSR(X_1) = 352.27$
Error	$SSE(X_1) = 143.12$	18	$MSE(X_1) = 7.95$
Total	$SSTO = 495.39$	19	

(**c**) Regression of Y on X_2

$$\hat{Y} = -23.634 + .8566X_2$$

Source of Variation	*SS*	*df*	*MS*
Regression	$SSR(X_2) = 381.97$	1	$MSR(X_2) = 381.97$
Error	$SSE(X_2) = 113.42$	18	$MSE(X_2) = 6.30$
Total	$SSTO = 495.39$	19	

are being considered in the model. The reason for the different regression coefficients is that the independent variables are correlated here, as we saw earlier from Figure 8.1. If we were to consider as a third independent variable X_3, midarm circumference, the regression coefficient for X_1 in the fitted response function with three independent variables would be different again from the two coefficients in Table 8.4 since midarm circumference is moderately correlated with triceps skinfold thickness.

Next, we turn to the regression coefficient for X_2, thigh circumference. We see again from Table 8.4 that the regression coefficient for X_2 when X_1 is also in the model is different from the regression coefficient when X_2 is the only independent variable in the model.

The important conclusion we must draw is: When independent variables are correlated, the regression coefficient of any independent variable depends on which other independent variables are included in the model and which ones are left out. Thus, a regression coefficient does not reflect any inherent effect of the particular independent variable on the dependent variable but only a marginal or partial effect, given whatever other correlated independent variables are included in the model.

Effect of multicollinearity on regression sums of squares. Note from Table 8.4a that the error sum of squares when both X_1 and X_2 are included in the model is $SSE(X_1, X_2) = 109.95$. When only X_2 is included in the model, the error sum of squares is $SSE(X_2) = 113.42$ as seen from Table 8.4c. Since the variation in Y when X_2 alone is considered is 113.42 but is 109.95 when both X_1 and X_2 are considered, we ascribe the difference to the effect of X_1. Using (8.2), we obtain:

$$SSR(X_1|X_2) = SSE(X_2) - SSE(X_1, X_2) = 113.42 - 109.95 = 3.47$$

When we fit a regression function containing only X_1, we also obtain a measure of the reduction in variation of Y associated with X_1, namely, $SSR(X_1)$. For our example, Table 8.4b indicates that $SSR(X_1) = 352.27$ which is not the same as $SSR(X_1|X_2) = 3.47$. The reason for the large difference is the high positive correlation between X_1 and X_2.

The story is the same for the other independent variable. Using (8.1), we obtain:

$$SSR(X_2|X_1) = SSE(X_1) - SSE(X_1, X_2) = 143.12 - 109.95 = 33.17$$

This sum of squares is not the same as $SSR(X_2) = 381.97$ from Table 8.4c, obtained when Y is regressed only on X_2.

The important conclusion is: When independent variables are correlated, there is no unique sum of squares which can be ascribed to an independent variable as reflecting its effect in reducing the total variation in Y. The reduction in the total variation ascribed to an independent variable must be viewed in the context of the other independent variables included in the model, whenever the independent variables are correlated.

Let us consider the meaning of $SSR(X_1)$ and $SSR(X_1|X_2)$ further. $SSR(X_1)$ measures the reduction in the variation of Y when X_1 is introduced into the regression model and no other independent variable is present. $SSR(X_1|X_2)$ measures the further reduction in the variation of Y when X_2 is already in the regression model and X_1 is introduced as a second independent variable. Hence, the notation $SSR(X_1|X_2)$ is used to show that the sum of squares measures a reduction in variation of Y associated with X_1, given that X_2 is already included in the model.

The reason why $SSR(X_1|X_2) = 3.47$ is less than $SSR(X_1) = 352.27$ in our example should now be apparent. Since X_1 is highly positively correlated with X_2, much of the power of X_1 to reduce the variation in Y is already accounted for by $SSR(X_2)$ when X_2 alone is included in the model. Hence, the marginal effect of X_1 in reducing the variation in Y, given that X_2 is in the model, is less than the effect if X_1 were introduced into the model without X_2 being present. For the same reason, $SSR(X_2|X_1)$ is less than $SSR(X_2)$.

The terms $SSR(X_2|X_1)$ and $SSR(X_1|X_2)$ are called *extra sums of squares,* since they indicate the additional or extra reduction in the error sum of squares achieved by introducing an additional independent variable.

Simultaneous tests on regression coefficients

Just as multicollinearity among the independent variables leads to regression coefficients that vary depending on which correlated independent variables are included in the model, so does multicollinearity also cause difficulties in statistical tests of the regression coefficients. A not infrequent abuse in the analysis of multiple regression models is to examine the t^* statistic in (7.42b):

$$t^* = \frac{b_k}{s(b_k)}$$

for each regression coefficient in turn to decide whether or not $\beta_k = 0$ for $k = 1, \ldots, p - 1$. Even if a simultaneous inference procedure is used, and often it is not, problems still exist.

Let us consider the first-order regression model with two independent variables:

(8.4) $$Y_i = \beta_0 + \beta_1 X_{i1} + \beta_2 X_{i2} + \varepsilon_i \qquad \text{Full model}$$

If the test on β_1 indicates it is zero, the regression model (8.4) would be:

$$Y_i = \beta_0 + \beta_2 X_{i2} + \varepsilon_i$$

If the test on β_2 indicates it is zero, the regression model (8.4) would be:

$$Y_i = \beta_0 + \beta_1 X_{i1} + \varepsilon_i$$

However, if the separate tests indicate that $\beta_1 = 0$ and $\beta_2 = 0$, that does not necessarily jointly imply that:

$$Y_i = \beta_0 + \varepsilon_i$$

since neither of the tests considers this alternative.

For an example, consider the data in Table 8.5 for 10 ski resorts in New England during a period of normal snow conditions. The computer output for the regression of visitor days (Y) on miles of intermediate trails (X_1) and lift capacity (X_2) is summarized in Table 8.6a. The proper test for the existence of a regression relation:

$$H_0\colon \beta_1 = 0 \text{ and } \beta_2 = 0$$
$$H_a\colon \text{not both } \beta_1 \text{ and } \beta_2 \text{ equal } 0$$

is the F test of (7.30). The test statistic for our example is (Table 8.6a):

$$F^* = \frac{MSR}{MSE} = \frac{811{,}865{,}088}{2{,}757{,}701} = 294$$

→ REJECT H_0 SOME REGRESSION RELATIONSHIP.

Controlling the level of significance at .05, we require $F(.95; 2, 7) = 4.74$. Since $F^* = 294 > 4.74$, we conclude H_a, that there is a regression relation between Y and the independent variables X_1 and X_2. Hence, at least one of the two regression coefficients does not equal zero. The P-value for the test is less than .001 because $F(.999; 2, 7) = 21.7$.

Let us now examine the t^* statistics at a 5 percent family level of significance by the Bonferroni technique. We require $t(.9875; 7) = 2.84$. Since both t^* statistics have absolute values that do not exceed 2.84 (Table 8.6a), we would conclude $\beta_1 = 0$ and $\beta_2 = 0$, contrary to the earlier conclusion that not both coefficients equal zero.

To understand this apparently paradoxical result, let us investigate the test of, say:

$$H_0\colon \beta_2 = 0$$
$$H_a\colon \beta_2 \neq 0$$

by the general linear test approach. We first obtain the error sum of squares

TABLE 8.5 Ski resort example with highly correlated independent variables

Ski Resort i	*Miles of Intermediate Trails* X_{i1}	*Lift Capacity (skiers per hour)* X_{i2}	*Total Visitor Days during Sample Period* Y_i
1	10.5	2,200	19,929
2	2.5	1,000	5,839
3	13.1	3,250	23,696
4	4.0	1,475	9,881
5	14.7	3,800	30,011
6	3.6	1,200	7,241
7	7.1	1,900	11,634
8	22.5	5,575	45,684
9	17.0	4,200	36,476
10	6.4	1,850	12,068

TABLE 8.6 Selected computer outputs for ski resort example with highly correlated independent variables

(a) Regression of Y on X_1 and X_2

$\hat{Y} = -1{,}806.82 + 1{,}131.03X_1 + 4.00X_2$

Source of Variation	SS	df	MS
Regression	$SSR(X_1, X_2) = 1{,}623{,}730{,}176$	2	$MSR(X_1, X_2) = 811{,}865{,}088$
Error	$SSE(X_1, X_2) = 19{,}303{,}908$	7	$MSE(X_1, X_2) = 2{,}757{,}701$
Total	$SSTO = 1{,}643{,}034{,}084$	9	

Variable	Estimated Regression Coefficient	Estimated Standard Deviation	t^*
X_1	$b_1 = 1{,}131.03$	$s(b_1) = 615.76$	1.837
X_2	$b_2 = 4.002$	$s(b_2) = 2.71$	1.477

(b) Regression of Y on X_1

$\hat{Y} = -363.98 + 2{,}032.53X_1$

Source of Variation	SS	df	MS
Regression	$SSR(X_1) = 1{,}617{,}707{,}400$	1	$MSR(X_1) = 1{,}617{,}707{,}400$
Error	$SSE(X_1) = 25{,}326{,}684$	8	$MSE(X_1) = 3{,}165{,}836$
Total	$SSTO = 1{,}643{,}034{,}084$	9	

$SSE(F)$ for the full model in (8.4), associated with $n - 3$ degrees of freedom. Next, we formulate the reduced model under H_0:

$$Y_i = \beta_0 + \beta_1 X_{i1} + \varepsilon_i \qquad \text{Reduced model} \tag{8.5}$$

and obtain the error sum of squares $SSE(R)$ for the reduced model. Associated with $SSE(R)$ are $n - 2$ degrees of freedom. The test statistic (3.68) then leads to:

$$F^* = \frac{SSE(R) - SSE(F)}{(n-2) - (n-3)} \div \frac{SSE(F)}{n-3}$$

In our notation recognizing explicitly the independent variables in the model, we have from the two parts of Table 8.6:

$$SSE(F) = SSE(X_1, X_2) = 19{,}303{,}908$$
$$SSE(R) = SSE(X_1) = 25{,}326{,}684$$

so that by (8.1):

$$SSE(R) - SSE(F) = SSE(X_1) - SSE(X_1, X_2) = SSR(X_2|X_1)$$

Hence:

(8.6)
$$F^* = \frac{SSR(X_2|X_1)}{1} \div \frac{SSE(X_1, X_2)}{n-3}$$

For our example in Table 8.6, we have:

$$SSR(X_2|X_1) = 25{,}326{,}684 - 19{,}303{,}908 = 6{,}022{,}776$$

and hence:

$$F^* = \frac{6{,}022{,}776}{1} \div \frac{19{,}303{,}908}{7} = 2.18$$

Recall from Table 8.6a that the t^* statistic for testing $\beta_2 = 0$ is $t^* = 1.477$. Since:

$$(t^*)^2 = (1.477)^2 = 2.18 = F^*$$

we see that the two test statistics are equivalent. We already knew this holds for simple regression, and now we can see it also holds for multiple regression.

The F^* test statistic (8.6) to test whether or not $\beta_2 = 0$ is called a *partial F test* statistic to distinguish it from the F^* statistic in (7.30b) for testing whether *all* $\beta_k = 0$, i.e., whether or not there is a regression relation between Y and the set of independent variables. The latter test is called the *overall F test.*

The test statistic (8.6) for the partial F test indicates clearly that a test on whether or not $\beta_2 = 0$ is a *marginal* test, given that X_1 is in the model. Similarly, a test on whether or not $\beta_1 = 0$ is a *marginal* test, given that X_2 is in the model. It is apparent now why the simultaneous tests with the t^* statistics for the two different regression coefficients both led to the conclusion that the regression coefficient equals zero. The independent variables X_1 and X_2 in Table 8.5 are highly positively correlated; indeed, the coefficient of simple correlation is $r_{12} = +.99$. Hence, if X_1 is already in the model, adding X_2 achieves little more reduction in the variation of Y, and $SSR(X_2|X_1)$ is small. When $SSR(X_2|X_1)$ is small, the test for β_2 leads to a small partial F test statistic F^* and therefore to the conclusion that $\beta_2 = 0$.

Similarly, the explanation why the t test led to the conclusion that $\beta_1 = 0$ is that X_1 is highly positively correlated with X_2, and X_2 is assumed to be in the model when the t test for β_1 is employed.

Thus, despite the fact that there is a clear relation between the dependent variable Y and the set of independent variables X_1 and X_2, the separate t tests indicated the respective regression coefficients equal zero because each test considers only the marginal contribution of the independent variable, given that the other is included in the model.

Note

We have just seen that it is possible that a set of independent variables is related to the dependent variable, yet all of the individual tests on the regression coefficients will lead to the conclusion that they equal zero because of the multicollinearity among the independendent variables. This apparently paradoxical result is also possible under special circumstances when there is no multicollinearity among the independent variables. The special circumstances are not likely to be found in practice, however.

A final comment

The discussion in this section has indicated that the choice of the particular set of independent variables which are to be included in the model is highly important and that in the presence of multicollinearity, the interpretation of regression coefficients and regression sums of squares must be undertaken with caution. The regression coefficients are affected not only by the other intercorrelated variables in the model but also by intercorrelated variables omitted from the model. For instance, an analyst was perplexed to find in a regression of territory company sales on territory population size, per capita income, and some other independent variables that the confidence interval for the regression coefficient for population size indicated it is negative. The analyst should have considered some of the omitted independent variables in search of an explanation. A consultant noted that the analyst did not include the major competitor's market penetration in the model. Since the competitor was most active and effective in territories with large populations, and thereby kept company sales down in these territories, the result of the omission of this independent variable from the model was a negative coefficient for the population size variable.

In view of the importance of the problems caused by multicollinearity among the independent variables, we take up this subject in more detail in Chapter 11. We consider there how to identify the presence of multicollinearity and take up several measures which may help to overcome some of the problems caused by multicollinearity.

8.2 DECOMPOSITION OF *SSR* INTO EXTRA SUMS OF SQUARES

Extra sums of squares

We defined the extra sum of squares $SSR(X_1|X_2)$ in (8.2):

(8.7a) $$SSR(X_1|X_2) = SSE(X_2) - SSE(X_1, X_2)$$

Likewise, we defined in (8.1):

(8.7b) $$SSR(X_2|X_1) = SSE(X_1) - SSE(X_1, X_2)$$

These extra sums of squares reflect the reduction in the error sum of squares by adding an independent variable to the model, given that another independent variable is already in the model.

Any reduction in the error sum of squares, of course, is equal to the same increase in the regression sum of squares since always:

$$SSTO = SSR + SSE$$

Hence, an extra sum of squares can also be thought of as the increase in the regression sum of squares achieved by introducing the new variable. We can therefore state, equivalently:

(8.8a) $$SSR(X_1|X_2) = SSR(X_1, X_2) - SSR(X_2)$$

(8.8b) $$SSR(X_2|X_1) = SSR(X_1, X_2) - SSR(X_1)$$

We show in Figure 8.2 a schematic representation of the extra sums of squares for our body fat example. The total bar on the left represents $SSTO$. The unshaded component of this bar is $SSR(X_2)$, and the combined shaded area represents $SSE(X_2)$. The latter area in turn is the combination of the extra sum of squares $SSR(X_1|X_2)$ and the error sum of squares $SSE(X_1, X_2)$ when both X_1 and X_2 are included in the model. Similarly, the bar on the right in Figure 8.2 shows the decomposition containing $SSR(X_2|X_1)$. Note in both cases how the extra sum of squares can be viewed either as a reduction in the error sum of squares or as an increase in the regression sum of squares when the second variable is added to the regression model.

FIGURE 8.2 Schematic representation of extra sums of squares—body fat example

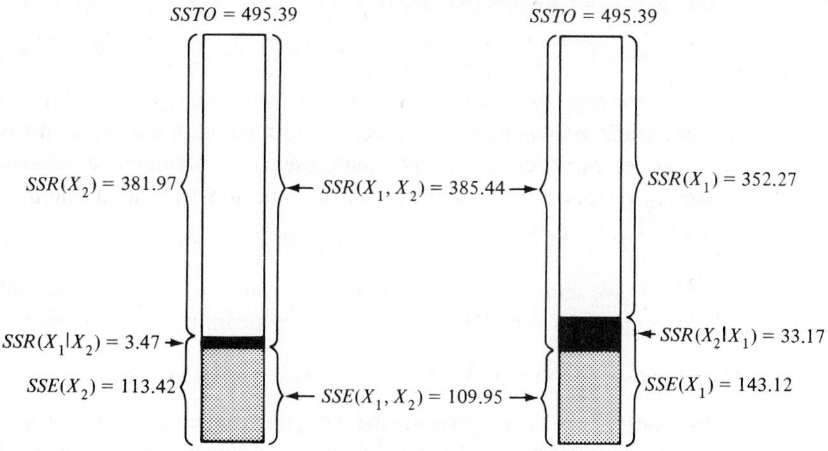

Extensions for three or more X variables are straightforward. For instance, we define:

(8.9) $$SSR(X_3|X_1, X_2) = SSE(X_1, X_2) - SSE(X_1, X_2, X_3)$$

$SSR(X_3|X_1, X_2)$ measures the reduction in the remaining variation of Y which is achieved by introducing X_3 into the regression model when X_1 and X_2 are already in the model.

Each extra sum of squares involving the addition of one independent variable into the regression model has associated with it one degree of freedom.

Decomposition of *SSR*

We can now obtain a variety of decompositions for the regression sum of squares *SSR*. Let us consider the case of three *X* variables. We begin with the identity (3.48a) for variable X_1:

$$SSTO = SSR(X_1) + SSE(X_1) \tag{8.10}$$

where the notation now shows explicitly that X_1 is the *X* variable in the model. Replacing $SSE(X_1)$ by its equivalent in (8.7b), we obtain:

$$SSTO = SSR(X_1) + SSR(X_2|X_1) + SSE(X_1, X_2) \tag{8.10a}$$

Replacing $SSE(X_1, X_2)$ by its equivalent in (8.9), we obtain:

$$SSTO = SSR(X_1) + SSR(X_2|X_1) + SSR(X_3|X_1, X_2) + SSE(X_1, X_2, X_3) \tag{8.10b}$$

Since we have the same identity for multiple regression with three independent variables as in (8.10) for a single independent variable, namely:

$$SSTO = SSR(X_1, X_2, X_3) + SSE(X_1, X_2, X_3) \tag{8.11}$$

equation (8.10b) therefore reduces to:

$$SSR(X_1, X_2, X_3) = SSR(X_1) + SSR(X_2|X_1) + SSR(X_3|X_1, X_2) \tag{8.12}$$

Thus, the regression sum of squares has been decomposed into marginal components, each associated with one degree of freedom. Of course, the order of the independent variables is arbitrary, and other orders can be developed. For instance:

$$SSR(X_1, X_2, X_3) = SSR(X_3) + SSR(X_1|X_3) + SSR(X_2|X_1, X_3) \tag{8.13}$$

Indeed, we can define extra sums of squares for two or more independent variables at a time and obtain still other decompositions. For instance, we define:

$$SSR(X_2, X_3|X_1) = SSE(X_1) - SSE(X_1, X_2, X_3) \tag{8.14}$$

Thus, $SSR(X_2, X_3|X_1)$ represents the reduction in the variation of *Y* gained when X_2 and X_3 are added to the model already containing X_1. There are two degrees of freedom associated with $SSR(X_2, X_3|X_1)$, as may be seen from the following relation which is based directly on the definitions of the extra sums of squares:

$$SSR(X_2, X_3|X_1) = SSR(X_2|X_1) + SSR(X_3|X_1, X_2) \tag{8.14a}$$

Thus, extra sums of squares for two or more *X* variables can be obtained from extra sums of squares where one *X* variable is added at a time.

With $SSR(X_2, X_3|X_1)$ we can then make use of the decomposition:

$$SSR(X_1, X_2, X_3) = SSR(X_1) + SSR(X_2, X_3|X_1) \tag{8.15}$$

It is obvious that the number of possible decompositions becomes vast as the number of independent variables increases. Table 8.7 contains the ANOVA table for one possible decomposition for the case of three independent variables, and Table 8.8 contains two possible decompositions for our earlier body fat example.

Uses of extra sums of squares for tests concerning regression coefficients

One of the major uses of extra sums of squares is for conducting tests concerning regression coefficients without having to fit both the full and reduced models

TABLE 8.7 Example of ANOVA table with decomposition of *SSR* for three independent variables

Source of Variation	*SS*	*df*	*MS*
Regression	$SSR(X_1, X_2, X_3)$	3	$MSR(X_1, X_2, X_3)$
X_1	$SSR(X_1)$	1	$MSR(X_1)$
$X_2 \mid X_1$	$SSR(X_2 \mid X_1)$	1	$MSR(X_2 \mid X_1)$
$X_3 \mid X_1, X_2$	$SSR(X_3 \mid X_1, X_2)$	1	$MSR(X_3 \mid X_1, X_2)$
Error	$SSE(X_1, X_2, X_3)$	$n-4$	$MSE(X_1, X_2, X_3)$
Total	$SSTO$	$n-1$	

TABLE 8.8 ANOVA tables with different decompositions of *SSR*—body fat example

(a)

Source of Variation	*SS*	*df*	*MS*
Regression	$SSR(X_1, X_2) = 385.44$	2	192.72
X_1	$SSR(X_1) = 352.27$	1	352.27
$X_2 \mid X_1$	$SSR(X_2 \mid X_1) = 33.17$	1	33.17
Error	$SSE(X_1, X_2) = 109.95$	17	6.47
Total	$SSTO = 495.39$	19	

(b)

Source of Variation	*SS*	*df*	*MS*
Regression	$SSR(X_1, X_2) = 385.44$	2	192.72
X_2	$SSR(X_2) = 381.97$	1	381.97
$X_1 \mid X_2$	$SSR(X_1 \mid X_2) = 3.47$	1	3.47
Error	$SSE(X_1, X_2) = 109.95$	17	6.47
Total	$SSTO = 495.39$	19	

separately. Many computer packages provide extra sums of squares in any desired order when fitting a regression model. Often, the decomposition provided is that corresponding to the order in which the X variables are entered in the regression fit.

In our ski resort example, for instance, the independent variables were entered in the order X_1 and X_2 and the computer output yielded the following results:

Source of Variation	*SS*	*df*	*MS*
Regression	1,623,730,176	2	811,865,088
X_1	1,617,707,400	1	1,617,707,400
$X_2 \mid X_1$	6,022,776	1	6,022,776
Error	19,303,908	7	2,757,701

Hence, to test whether or not $\beta_2 = 0$, we need not actually fit the reduced model since the partial F test statistic (8.6) can be calculated immediately from the above results:

$$F^* = \frac{SSR(X_2|X_1)}{1} \div \frac{SSE(X_1, X_2)}{n-3}$$

$$= \frac{6,022,776}{1} \div \frac{19,303,908}{7} = 2.18$$

As we shall see later, judicious choices in obtaining extra sums of squares will permit a variety of tests on the regression coefficients without requiring extra computer runs for fitting reduced models.

8.3 COEFFICIENTS OF PARTIAL DETERMINATION

A coefficient of multiple determination R^2, it will be recalled, measures the proportionate reduction in the variation of Y achieved by the introduction of the entire set of X variables considered in the model. A *coefficient of partial determination,* in contrast, measures the marginal contribution of one X variable, when all others are already included in the model.

Two independent variables

Let us consider a first-order multiple regression model with two independent variables, as given in (7.1):

$$Y_i = \beta_0 + \beta_1 X_{i1} + \beta_2 X_{i2} + \varepsilon_i$$

$SSE(X_2)$ measures the variation in Y when X_2 is included in the model. $SSE(X_1, X_2)$ measures the variation in Y when both X_1 and X_2 are included in the model. Hence, the relative marginal reduction in the variation in Y associated with X_1 when X_2 is already in the model is:

$$\frac{SSE(X_2) - SSE(X_1, X_2)}{SSE(X_2)}$$

This measure is the coefficient of partial determination between Y and X_1, given that X_2 is in the model. We denote this measure by $r^2_{Y1.2}$:

$$r^2_{Y1.2} = \frac{SSE(X_2) - SSE(X_1, X_2)}{SSE(X_2)} = 1 - \frac{SSE(X_1, X_2)}{SSE(X_2)} \tag{8.16}$$

Using (8.7a), we can express the coefficient of partial determination in terms of an extra sum of squares:

$$r^2_{Y1.2} = \frac{SSR(X_1|X_2)}{SSE(X_2)} \tag{8.16a}$$

$r^2_{Y1.2}$ thus measures the proportionate reduction in the variation of Y remaining after X_2 is included in the model which is gained by also including X_1 in the model.

The coefficient of partial determination between Y and X_2, given that X_1 is in the model, is defined:

$$r^2_{Y2.1} = \frac{SSR(X_2|X_1)}{SSE(X_1)} \tag{8.17}$$

For our earlier body fat example, these two coefficients of partial determination are (Tables 8.4 and 8.8):

$$r^2_{Y1.2} = \frac{3.47}{113.42} = .031$$

$$r^2_{Y2.1} = \frac{33.17}{143.12} = .232$$

Thus, when X_1 here is added to the model containing X_2, the error sum of squares $SSE(X_2)$ is reduced by 3.1 percent. Correspondingly, when X_2 is added to the model containing X_1, the error sum of squares $SSE(X_1)$ is reduced by 23.2 percent.

General case

The generalization of coefficients of partial determination to three or more independent variables in the model is immediate. For instance:

$$r^2_{Y1.23} = \frac{SSR(X_1|X_2, X_3)}{SSE(X_2, X_3)} \tag{8.18a}$$

$$r^2_{Y2.13} = \frac{SSR(X_2|X_1, X_3)}{SSE(X_1, X_3)} \tag{8.18b}$$

$$r^2_{Y3.12} = \frac{SSR(X_3|X_1, X_2)}{SSE(X_1, X_2)} \tag{8.18c}$$

Note that in the subscripts to r^2, the entries to the left of the dot show in turn the variable taken as the response and the variable being added. The entries to the right of the dot show the X variables already in the model.

Comments

1. The coefficients of partial determination can take on values between 0 and 1, as the definitions readily indicate.

2. A coefficient of partial determination can be interpreted as a coefficient of simple determination. Consider a multiple regression model with two independent variables. Suppose we regress Y on X_2 and obtain the residuals:

$$Y_i - \hat{Y}_i(X_2)$$

where $\hat{Y}_i(X_2)$ denotes the fitted values of Y when X_2 is in the model.

Suppose we further regress X_1 on X_2 and obtain the residuals:

$$X_{i1} - \hat{X}_{i1}(X_2)$$

where $\hat{X}_{i1}(X_2)$ denotes the fitted values of X_1 in the regression of X_1 on X_2. The coefficient of simple determination r^2 between these two sets of residuals equals the coefficient of partial determination $r^2_{Y1.2}$. Thus, this coefficient measures the relation between Y and X_1 when both of these have been adjusted for their linear relationships to X_2.

Coefficients of partial correlation

The square root of the coefficient of partial determination is called the *coefficient of partial correlation.* This coefficient is frequently used in practice, although it does not have as clear a meaning as the coefficient of partial determination.

For our body fat example, we have:

$$r_{Y1.2} = \sqrt{.031} = .176$$

$$r_{Y2.1} = \sqrt{.232} = .482$$

Usually the partial correlation coefficient is given the same sign as that of the corresponding regression coefficient in the fitted regression function. For our body fat example, we had (Table 8.4) $b_1 = +.2224$ and $b_2 = +.6594$. Since these are both positive, each of the partial correlation coefficients above is taken as positive.

Partial correlation coefficients are frequently used in computer routines for finding the best independent variable to be selected next for inclusion in the regression model. We shall discuss this use in Chapter 12.

Note

The coefficients of partial determination can be expressed in terms of simple or other partial correlation coefficients. For example:

(8.19) $$r_{Y2.1}^2 = \frac{(r_{Y2} - r_{12}r_{Y1})^2}{(1 - r_{12}^2)(1 - r_{Y1}^2)}$$

(8.20) $$r_{Y2.13}^2 = \frac{(r_{Y2.3} - r_{12.3}r_{Y1.3})^2}{(1 - r_{12.3}^2)(1 - r_{Y1.3}^2)}$$

Extensions are straightforward.

8.4 TESTING HYPOTHESES CONCERNING REGRESSION COEFFICIENTS IN MULTIPLE REGRESSION

We have already discussed how to conduct two types of tests concerning the regression coefficients in a multiple regression model. For completeness, we summarize these tests here and then take up some additional types of tests.

Test whether all $\beta_k = 0$

This is the *overall F test* (7.30) of whether or not there is a regression relation between the dependent variable Y and the set of independent variables. The alternatives are:

(8.21) $$\begin{aligned} &H_0\text{: } \beta_1 = \beta_2 = \cdots = \beta_k = 0 \\ &H_a\text{: not all } \beta_k \ (k = 1, \ldots, p-1) \text{ equal } 0 \end{aligned}$$

and the test statistic is:

(8.22) $$\begin{aligned} F^* &= \frac{SSR(X_1, \ldots, X_{p-1})}{p-1} \div \frac{SSE(X_1, \ldots, X_{p-1})}{n-p} \\ &= \frac{MSR}{MSE} \end{aligned}$$

If H_0 holds, $F^* \sim F(p-1, n-p)$. Large values of F^* lead to conclusion H_a.

Test whether a single $\beta_k = 0$

This is the *partial F test* of whether a particular regression coefficient β_k equals zero. The alternatives are:

(8.23) $$\begin{aligned} &H_0\text{: } \beta_k = 0 \\ &H_a\text{: } \beta_k \neq 0 \end{aligned}$$

and the test statistic is:

(8.24) $$\begin{aligned} F^* &= \frac{SSR(X_k | X_1, \ldots, X_{k-1}, X_{k+1}, \ldots, X_{p-1})}{1} \div \frac{SSE(X_1, \ldots, X_{p-1})}{n-p} \\ &= \frac{MSR(X_k | X_1, \ldots, X_{k-1}, X_{k+1}, \ldots, X_{p-1})}{MSE} \end{aligned}$$

If H_0 holds, $F^* \sim F(1, n-p)$. Large values of F^* lead to conclusion H_a. Computer packages which provide extra sums of squares permit use of this test without having to fit the reduced model.

An equivalent test statistic is, as we have seen, (7.42b):

$$t^* = \frac{b_k}{s(b_k)} \tag{8.25}$$

If H_0 holds, $t^* \sim t(n-p)$. Large values of $|t^*|$ lead to conclusion H_a.

Since the two tests are equivalent, the choice is usually made in terms of available information provided by the computer package output.

Test whether some regression coefficients equal zero

Sometimes we wish to determine whether or not some regression coefficients equal zero. The approach is that of a general linear test, and no new problems arise. We first illustrate the approach by an example and then state the general test statistic.

Example. We consider the multiple regression application in Table 8.9. The data pertain to 14 different computer simulations conducted in designing the layout for a chemical warehouse. The dependent variable is CPU time (the operating time required by the computer's central processing unit to run the simulation), the first independent variable is number of trials in the simulation, and the second independent variable is number of statements in the computer program. Since it is hypothesized that the effect of number of statements (X_2) on CPU time (Y) may be curvilinear, the model being considered is:

$$Y_i = \beta_0 + \beta_1 X_{i1} + \beta_2 X_{i2} + \beta_3 X_{i2}^2 + \varepsilon_i \qquad \text{Full model} \tag{8.26}$$

TABLE 8.9 Data for computer simulation example

Simulation i	*Number of Trials* X_{i1}	*Number of Statements in Program* X_{i2}	X_{i2}^2	*CPU Time (seconds)* Y_i
1	550	458	209,764	445.37
2	600	152	23,104	408.88
3	200	635	403,225	264.61
4	50	128	16,384	73.90
5	350	100	10,000	246.07
6	300	550	302,500	312.00
7	200	577	332,929	250.36
8	175	234	54,756	179.95
9	200	500	250,000	243.76
10	200	491	241,081	243.63
11	175	580	336,400	240.72
12	700	135	18,225	462.60
13	800	162	26,244	531.45
14	650	176	30,976	445.48

It is desired that we test whether the number of statements variable can be dropped from the model. Hence, we wish to choose between the alternatives:

$$H_0: \beta_2 = \beta_3 = 0$$
$$H_a: \text{not both } \beta_2 \text{ and } \beta_3 \text{ equal zero}$$

We first fit the full model and obtain (results are shown in Table 8.10a):

$$SSE(F) = SSE(X_1, X_2, X_2^2) = 44.1$$

The model under H_0 is:

(8.27) $$Y_i = \beta_0 + \beta_1 X_{i1} + \varepsilon_i \qquad \text{Reduced model}$$

When we fit the reduced model, we obtain (Table 8.10b):

$$SSE(R) = SSE(X_1) = 17{,}240.3$$

The general test statistic (3.68):

$$F^* = \frac{SSE(R) - SSE(F)}{(n-2) - (n-4)} \div \frac{SSE(F)}{n-4}$$

can be simplified here because:

$$SSE(R) - SSE(F) = SSE(X_1) - SSE(X_1, X_2, X_2^2) = SSR(X_2, X_2^2|X_1)$$

Hence, we can write:

(8.28) $$F^* = \frac{SSR(X_2, X_2^2|X_1)}{2} \div \frac{SSE(F)}{n-4}$$

TABLE 8.10 Computer results for computer simulation example

(a) Regression of Y on X_1, X_2, and X_2^2

$\hat{Y} = 7.028 + .595X_1 + .323X_2 - .000173X_2^2$

Source of Variation	*SS*	*df*	*MS*
Regression	$SSR(X_1, X_2, X_2^2) =$ 214,691.2	3	71,563.7
Error	$SSE(X_1, X_2, X_2^2) =$ 44.1	10	4.41
Total	$SSTO =$ 214,735.3	13	

(b) Regression of Y on X_1

$\hat{Y} = 122.71 + .511X_1$

Source of Variation	*SS*	*df*	*MS*
Regression	$SSR(X_1) =$ 197,495.1	1	197,495.1
Error	$SSE(X_1) =$ 17,240.3	12	1,436.7
Total	$SSTO =$ 214,735.4	13	

For our example, we have (Table 8.10):

$$SSR(X_2, X_2^2|X_1) = 17{,}240.3 - 44.1 = 17{,}196.2$$

$$F^* = \frac{17{,}196.2}{2} \div \frac{44.1}{10} = 1{,}950$$

Suppose the α risk is to be .01. We require $F(.99; 2, 10) = 7.56$. Since $F^* = 1{,}950 > 7.56$, we conclude H_a, that the number of statements variable should not be dropped from the model.

Actually, we did not need to fit the reduced model separately to conduct the test; we used this approach only to explain the logic of the test. The computer run for fitting the full model provided the following decomposition of *SSR*:

Component	*SS*
X_1	197,495
$X_2\|X_1$	17,053
$X_2^2\|X_1, X_2$	144

Since, by (8.14a):

$$SSR(X_2, X_2^2|X_1) = SSR(X_2|X_1) + SSR(X_2^2|X_1, X_2)$$

we can obtain directly:

$$SSR(X_2, X_2^2|X_1) = 17{,}053 + 144 = 17{,}197$$

The difference in the final digit is due to rounding effects.

In our example, the regression model contained both X_2 and X_2^2 terms. Computational difficulties can at times arise in such cases when obtaining the least squares regression coefficients. Chapter 9 discusses polynomial regression models and how to avoid these computational difficulties.

Test statistic. For the general multiple regression model:

(8.29) $$Y_i = \beta_0 + \beta_1 X_{i1} + \cdots + \beta_{p-1} X_{i,p-1} + \varepsilon_i \qquad \text{Full model}$$

we wish to test:

(8.30)
$$H_0: \beta_q = \beta_{q+1} = \cdots = \beta_{p-1} = 0$$
$$H_a: \text{not all of the } \beta\text{'s in } H_0 \text{ equal zero}$$

where for convenience, we arrange the model so that the last $p - q$ coefficients are the ones to be tested. We first fit the full model and thereby obtain $SSE(X_1, \ldots, X_{p-1})$. Then we fit the reduced model:

(8.31) $$Y_i = \beta_0 + \beta_1 X_{i1} + \cdots + \beta_{q-1} X_{i,q-1} + \varepsilon_i \qquad \text{Reduced model}$$

and obtain $SSE(X_1, \ldots, X_{q-1})$. Finally, we set up the general linear test statistic (3.68), which here is:

(8.32) $$F^* = \frac{SSE(X_1, \ldots, X_{q-1}) - SSE(X_1, \ldots, X_{p-1})}{(n-q) - (n-p)} \div \frac{SSE(X_1, \ldots, X_{p-1})}{n-p}$$

or equivalently:

$$F^* = \frac{SSR(X_q, \ldots, X_{p-1}|X_1, \ldots, X_{q-1})}{p - q} \div MSE(F) \tag{8.32a}$$

Note that test statistic (8.32a) actually encompasses the two earlier cases. If $q = 1$, the test is whether all regression coefficients equal zero. If $q = p - 1$, the test is whether a single regression coefficient equals zero. Also note that test statistic (8.32a) can be calculated without having to fit the reduced model if the computer program provides the needed extra sums of squares:

$$\begin{aligned} SSR(X_q, \ldots, X_{p-1}|X_1, \ldots, X_{q-1}) &= SSR(X_q|X_1, \ldots, X_{q-1}) \\ &\quad + \cdots + SSR(X_{p-1}|X_1, \ldots, X_{p-2}) \end{aligned} \tag{8.33}$$

Other tests

Other types of tests are occasionally required. For instance, for the full model containing three X variables:

$$Y_i = \beta_0 + \beta_1 X_{i1} + \beta_2 X_{i2} + \beta_3 X_{i3} + \varepsilon_i \qquad \text{Full model} \tag{8.34}$$

we might wish to test:

$$\begin{aligned} H_0&: \beta_1 = \beta_2 \\ H_a&: \beta_1 \neq \beta_2 \end{aligned} \tag{8.35}$$

The procedure would be to fit the full model (8.34), and then the reduced model:

$$Y_i = \beta_0 + \beta_c(X_{i1} + X_{i2}) + \beta_3 X_{i3} + \varepsilon_i \qquad \text{Reduced model} \tag{8.36}$$

where β_c denotes the common coefficient for β_1 and β_2 and $X_{i1} + X_{i2}$ is the corresponding new independent variable. We then use the general F^* test statistic (3.68) with 1 and $n - 4$ degrees of freedom. Since this test does not involve alternatives where one or more regression coefficients equal zero, extra sums of squares are not applicable and the reduced model must be fitted for the test.

8.5 MATRIX FORMULATION OF GENERAL LINEAR TEST

The general linear test is based on the test statistic (3.68):

$$F^* = \frac{SSE(R) - SSE(F)}{df_R - df_F} \div \frac{SSE(F)}{df_F} \tag{8.37}$$

which, when H_0 holds, follows the F distribution with $df_R - df_F$ degrees of freedom for the numerator and df_F degrees of freedom for the denominator. We now explain how to represent this test statistic in matrix form.

Full model

The full regression model with $p - 1$ predictor variables is given by (7.18):

$$\mathbf{Y} = \mathbf{X}\boldsymbol{\beta} + \boldsymbol{\varepsilon} \tag{8.38}$$

The least squares estimators for the full model will now be denoted by $\mathbf{b}_F$ and are, as before, given by (7.21):

$$\mathbf{b}_F = (\mathbf{X}'\mathbf{X})^{-1}\mathbf{X}'\mathbf{Y} \tag{8.39}$$

Also, the error sum of squares is given by (7.27):

$$SSE(F) = (\mathbf{Y} - \mathbf{X}\mathbf{b}_F)'(\mathbf{Y} - \mathbf{X}\mathbf{b}_F) = \mathbf{Y}'\mathbf{Y} - \mathbf{b}_F'\mathbf{X}'\mathbf{Y} \tag{8.40}$$

which has associated with it $df_F = n - p$ degrees of freedom.

Statement of hypothesis H_0

A linear test hypothesis H_0 is represented in matrix form as follows:

$$H_0\text{: } \underset{s\times p}{\mathbf{C}} \ \underset{p\times 1}{\boldsymbol{\beta}} = \underset{s\times 1}{\mathbf{h}} \tag{8.41}$$

where $\mathbf{C}$ is a specified $s \times p$ matrix of rank s and $\mathbf{h}$ is a specified $s \times 1$ vector.

Example 1. The regression model contains two X variables, and we wish to test H_0: $\beta_1 = 2$. Then:

$$\underset{1\times 3}{\mathbf{C}} = [0 \quad 1 \quad 0] \qquad \underset{1\times 1}{\mathbf{h}} = [2]$$

and we have:

$$H_0\text{: } \mathbf{C}\boldsymbol{\beta} = [0 \quad 1 \quad 0]\begin{bmatrix}\beta_0\\ \beta_1\\ \beta_2\end{bmatrix} = [2]$$

or H_0: $\beta_1 = 2$.

Example 2. The regression model contains two X variables, and we wish to test H_0: $\beta_1 = \beta_2 = 0$. Then:

$$\underset{2\times 3}{\mathbf{C}} = \begin{bmatrix}0 & 1 & 0\\ 0 & 0 & 1\end{bmatrix} \qquad \underset{2\times 1}{\mathbf{h}} = \begin{bmatrix}0\\ 0\end{bmatrix}$$

and we have:

$$H_0\text{: } \mathbf{C}\boldsymbol{\beta} = \begin{bmatrix}0 & 1 & 0\\ 0 & 0 & 1\end{bmatrix}\begin{bmatrix}\beta_0\\ \beta_1\\ \beta_2\end{bmatrix} = \begin{bmatrix}0\\ 0\end{bmatrix}$$

or H_0: $\beta_1 = 0$, $\beta_2 = 0$.

Example 3. The regression model contains three X variables, and we wish to test H_0: $\beta_1 = \beta_2$. Then:

$$\underset{1\times 4}{\mathbf{C}} = [0 \quad 1 \quad -1 \quad 0] \qquad \underset{1\times 1}{\mathbf{h}} = [0]$$

and we have:

$$H_0\text{: } \mathbf{C}\boldsymbol{\beta} = [0 \quad 1 \quad -1 \quad 0] \begin{bmatrix} \beta_0 \\ \beta_1 \\ \beta_2 \\ \beta_3 \end{bmatrix} = [0]$$

or H_0: $\beta_1 - \beta_2 = 0$.

Reduced model

The reduced model is:

$$\mathbf{Y} = \mathbf{X}\boldsymbol{\beta} + \boldsymbol{\varepsilon} \qquad \text{where } \mathbf{C}\boldsymbol{\beta} = \mathbf{h} \tag{8.42}$$

It can be shown that the least squares estimators under the reduced model, to be denoted by $\mathbf{b}_R$, are:

$$\mathbf{b}_R = \mathbf{b}_F - (\mathbf{X}'\mathbf{X})^{-1}\mathbf{C}'(\mathbf{C}(\mathbf{X}'\mathbf{X})^{-1}\mathbf{C}')^{-1}(\mathbf{C}\mathbf{b}_F - \mathbf{h}) \tag{8.43}$$

and the error sum of squares is:

$$SSE(R) = (\mathbf{Y} - \mathbf{X}\mathbf{b}_R)'(\mathbf{Y} - \mathbf{X}\mathbf{b}_R) \tag{8.44}$$

which has associated with it $df_R = n - (p - s)$ degrees of freedom.

Test statistic

It can be shown that the difference $SSE(R) - SSE(F)$ can be expressed as follows:

$$SSE(R) - SSE(F) = (\mathbf{C}\mathbf{b}_F - \mathbf{h})'(\mathbf{C}(\mathbf{X}'\mathbf{X})^{-1}\mathbf{C}')^{-1}(\mathbf{C}\mathbf{b}_F - \mathbf{h}) \tag{8.45}$$

which has associated with it $df_R - df_F = (n - p + s) - (n - p) = s$ degrees of freedom.

Hence, the test statistic is:

$$F^* = \frac{SSE(R) - SSE(F)}{s} \div \frac{SSE(F)}{n - p} \tag{8.46}$$

where $SSE(R) - SSE(F)$ is given by (8.45) and $SSE(F)$ is given by (8.40).

To confirm that the degrees of freedom associated with $SSE(R) - SSE(F)$ are s, consider the earlier three examples.

1. In Example 1, $s = 1$. This is consistent with the numerator degrees of freedom in test statistic (8.24).

2. In Example 2, $s = 2$. This is consistent with the numerator degrees of freedom in test statistic (8.22).
3. In Example 3, $s = 1$. This is consistent with the numerator degrees of freedom in the test statistic for the example on page 293.

Note

The least squares estimators $\mathbf{b}_R$ under the reduced model, given in (8.43), can be derived by minimizing the least squares criterion $Q = (\mathbf{Y} - \mathbf{X}\boldsymbol{\beta})'(\mathbf{Y} - \mathbf{X}\boldsymbol{\beta})$ subject to the constraint $\mathbf{C}\boldsymbol{\beta} - \mathbf{h} = \mathbf{0}$, using Lagrangian multipliers.

PROBLEMS

8.1. A speaker stated in a workshop on applied regression analysis: "In business and the social sciences, some degree of multicollinearity in survey data is practically inevitable." Does this statement apply equally to experimental data?

8.2. Refer to the Zarthan Company example on page 247. The company's sales manager has suggested that the predictive ability of the model could be greatly improved if promotional expenditures were added to the model, since these expenditures are known to have a substantial impact on sales. The company allocates its total promotional budget proportionately to the target population in the districts. Thus, a district containing 4.7 percent of the total target population receives 4.7 percent of the total promotional budget. Evaluate the sales manager's suggestion.

8.3. Refer to **Brand preference** Problem 7.8.

a. Fit the first-order simple linear regression model (3.1) for relating brand liking (Y) to moisture content (X_1). State the fitted regression function.
b. Compare the estimated regression coefficient for moisture content obtained in part (a) with the corresponding coefficient obtained in Problem 7.8a. What do you find?
c. Does $SSR(X_1)$ equal $SSR(X_1 \mid X_2)$ here? If not, is the difference substantial?
d. Calculate the coefficient of simple correlation between X_1 and X_2. What bearing does this have on your findings in parts (b) and (c)?

8.4. Refer to **Chemical shipment** Problem 7.12.

a. Fit the first-order simple linear regression model (3.1) for relating number of minutes required to handle shipment (Y) to total weight of shipment (X_2). State the fitted regression function.
b. Compare the estimated regression coefficient for total weight of shipment obtained in part (a) with the corresponding coefficient obtained in Problem 7.12a. What do you find?
c. Does $SSR(X_2)$ equal $SSR(X_2 \mid X_1)$ here? If not, is the difference substantial?
d. Calculate the coefficient of simple correlation between X_1 and X_2. What bearing does this have on your findings in parts (b) and (c)?

8.5. Refer to **Patient satisfaction** Problem 7.17.

a. Fit the first-order linear regression model (7.1) for relating patient satisfaction (Y) to patient's age (X_1) and severity of illness (X_2). State the fitted regression function.

b. Compare the estimated regression coefficients for patient's age and severity of illness obtained in part (a) with the corresponding coefficients obtained in Problem 7.17a. What do you find?

c. Does $SSR(X_1)$ equal $SSR(X_1 \mid X_3)$ here? Does $SSR(X_2)$ equal $SSR(X_2 \mid X_3)$?

d. Calculate the coefficients of simple correlation between pairs of X_1, X_2, and X_3. What bearing do these have on your findings in parts (b) and (c)?

8.6. Refer to **Chemical shipment** Problem 7.12. Does $SSR(X_1) + SSR(X_2 \mid X_1)$ equal $SSR(X_2) + SSR(X_1 \mid X_2)$ here? Must this always be the case?

8.7. Refer to **Brand preference** Problem 7.8. Test whether X_2 can be dropped from the regression model given that X_1 is retained. Use the F^* test statistic and level of significance .01. State the alternatives, decision rule, and conclusion.

8.8. Refer to **Chemical shipment** Problem 7.12. Test whether X_1 can be dropped from the regression model given that X_2 is retained. Use the F^* test statistic and $\alpha =$ .05. State the alternatives, decision rule, and conclusion.

8.9. Refer to **Patient satisfaction** Problem 7.17. Test whether X_3 can be dropped from the regression model given that X_1 and X_2 are retained. Use the F^* test statistic and level of significance .025. State the alternatives, decision rule, and conclusion.

8.10. Refer to the work crew productivity example on page 272.

a. Calculate r_{Y1}^2, r_{Y2}^2, r_{12}^2, $r_{Y1.2}^2$, $r_{Y2.1}^2$, and R^2. Explain what each coefficient measures and interpret your results.

b. Are any of the results obtained in part (a) special because the two independent variables are uncorrelated?

c. Obtain the standardized regression coefficients. How do you interpret these coefficients? Do they have a special meaning here because the independent variables are uncorrelated? (*Hint:* Obtain r_{Y1}, r_{Y2}.)

8.11. Refer to **Brand preference** Problem 7.8. Calculate r_{Y1}^2, r_{Y2}^2, r_{12}^2, $r_{Y1.2}^2$, $r_{Y2.1}^2$, and R^2. Explain what each coefficient measures and interpret your results.

8.12. Refer to **Chemical shipment** Problem 7.12. Calculate r_{Y1}^2, r_{Y2}^2, r_{12}^2, $r_{Y1.2}^2$, $r_{Y2.1}^2$, and R^2. Explain what each coefficient measures and interpret your results.

8.13. Refer to **Patient satisfaction** Problem 7.17.

a. Calculate r_{Y1}^2, $r_{Y1.2}^2$, and $r_{Y1.23}^2$. How is the degree of linear association between Y and X_1 affected as other independent variables have already been included in the model?

b. Make a similar analysis as in part (a) for the degree of linear association between Y and X_2. Are your findings similar to those in part (a) for Y and X_1?

8.14. Refer to the computer simulation example on page 290. An observer states that both the number of trials variable (X_1) and the second-order term for the number of statements variable (X_2^2) can be dropped from model (8.26). Conduct the appropriate test using a level of significance of .01. State the alternatives, decision rule, and conclusion.

8.15. Refer to **Patient satisfaction** Problem 7.17. Test whether both X_2 and X_3 can be dropped from the regression model given that X_1 is retained. Use $\alpha = .025$. State the alternatives, decision rule, and conclusion.

8.16. Refer to **Mathematicians salaries** Problem 7.20. Test whether both X_1 and X_3 can be dropped from the regression model given that X_2 is retained; use $\alpha = .01$. State the alternatives, decision rule, and conclusion.

EXERCISES

8.17. a. Define each of the following extra sums of squares: (1) $SSR(X_5 \mid X_1)$; (2) $SSR(X_3, X_4 \mid X_1)$; (3) $SSR(X_4 \mid X_1, X_2, X_3)$.

b. For a multiple regression model with five X variables, what is the relevant extra sum of squares for testing whether or not $\beta_5 = 0$? Whether or not $\beta_2 = \beta_4 = 0$?

8.18. Show that:

a. $SSR(X_1, X_2, X_3, X_4) = SSR(X_1) + SSR(X_2, X_3 \mid X_1) + SSR(X_4 \mid X_1, X_2, X_3)$

b. $SSR(X_1, X_2, X_3, X_4) = SSR(X_2, X_3) + SSR(X_1 \mid X_2, X_3) + SSR(X_4 \mid X_1, X_2, X_3)$

8.19. Refer to **Brand preference** Problem 7.8.

a. Regress Y on X_2 using the simple linear regression model (3.1) and obtain the residuals.

b. Regress X_1 on X_2 using the simple linear regression model (3.1) and obtain the residuals.

c. Calculate the coefficient of simple correlation between the two sets of residuals and show that it equals $r_{Y1.2}$.

8.20. An undergraduate working for a campus apparel shop serving student customers studied the relation between monthly allowance received by customer (X_1), number of years customer is in college (X_2), and dollar sales to customer to date (Y). The predictive model considered was:

$$Y_i = \beta_0 + \beta_1 X_{i1} + \beta_2 X_{i2} + \beta_3 X_{i1}^2 + \varepsilon_i$$

State the reduced models for testing whether or not: (1) $\beta_1 = \beta_3 = 0$, (2) $\beta_0 = 0$, (3) $\beta_3 = 5$, (4) $\beta_0 = 10$, (5) $\beta_1 = \beta_2$.

8.21. Refer to Exercise 8.20. For each of the cases, state the hypothesis H_0 using the matrix formulation (8.41).

8.22. The following regression model is being considered in a water resources study:

$$Y_i = \beta_0 + \beta_1 X_{i1} + \beta_2 X_{i2} + \beta_3 X_{i1} X_{i2} + \beta_4 \sqrt{X_{i3}} + \varepsilon_i$$

State the reduced models for testing whether or not: (1) $\beta_3 = \beta_4 = 0$, (2) $\beta_3 = 0$, (3) $\beta_1 = \beta_2 = 5$, (4) $\beta_4 = 7$.

8.23. Refer to Exercise 8.22. For each of the cases, state the hypothesis H_0 using the matrix formulation (8.41).

8.24. (Calculus needed.) Derive the least squares estimator under the reduced model (8.43), where $\mathbf{C\beta} = \mathbf{h}$. [*Hint:* The Lagrangian function is:

$$L = (\mathbf{Y} - \mathbf{X\beta})'(\mathbf{Y} - \mathbf{X\beta}) + \boldsymbol{\lambda}'(\mathbf{C\beta} - \mathbf{h}), \text{ where } \boldsymbol{\lambda}' = (\lambda_1, \ldots, \lambda_s).]$$

8.25. Derive (8.45). [*Hint:* Show that $SSE(R) - SSE(F) = (\mathbf{b}_F - \mathbf{b}_R)'\mathbf{X}'\mathbf{X}(\mathbf{b}_F - \mathbf{b}_R)$ and obtain an expression for $\mathbf{b}_F - \mathbf{b}_R$ from (8.43).]

PROJECTS

8.26. Refer to the **SMSA** data set. For predicting the number of active physicians (Y) in an SMSA, it has been decided to include total population (X_1) and total personal income (X_2) as independent variables. The question now is whether an additional independent variable would be helpful in the model, and if so, which variable would be most helpful. Assume that a first-order multiple regression model is appropriate.

a. For each of the following variables, calculate the coefficient of partial determination given that X_1 and X_2 are included in the model: land area (X_3), percent of population 65 or older (X_4), number of hospital beds (X_5), and total serious crimes (X_6).

b. On the basis of the results in part (a), which of the four additional independent variables is best? Is the extra sum of squares associated with this variable larger than those for the other three variables?

c. Using the F^* test statistic, test whether or not the variable determined to be best in part (b) is helpful in the regression model when X_1 and X_2 are included in the model; use $\alpha = .01$. State the alternatives, decision rule, and conclusion. Would the F^* test statistics for the other three potential independent variables be as large as the one here? Discuss.

8.27. Refer to the **SENIC** data set. For predicting the average length of stay of patients in a hospital (Y), it has been decided to include age (X_1) and infection risk (X_2) as independent variables. The question now is whether an additional independent variable would be helpful in the model, and if so, which variable would be most helpful. Assume that a first-order multiple regression model is appropriate.

a. For each of the following variables, calculate the coefficient of partial determination given that X_1 and X_2 are included in the model: routine culturing ratio (X_3), average daily census (X_4), number of nurses (X_5), and available facilities and services (X_6).

b. On the basis of the results in part (a), which of the four additional independent variables is best? Is the extra sum of squares associated with this variable larger than those for the other three variables?

c. Using the F^* test statistic, test whether or not the variable determined to be best in part (b) is helpful in the regression model when X_1 and X_2 are included in the model; use $\alpha = .05$. State the alternatives, decision rule, and conclusion. Would the F^* test statistics for the other three potential independent variables be as large as the one here? Discuss.

8.28. Refer to Exercise 7.31. It is desired to test whether or not $\beta_1 = \beta_2$. Using matrix methods, obtain $SSE(R) - SSE(F)$ according to (8.45).

9

Polynomial regression

In this chapter, we consider one important type of curvilinear response model, namely, the polynomial regression model. This is the most frequently used curvilinear response model in practice, because of its ease in handling as a special case of the general linear regression model (7.18). First, we discuss some commonly used polynomial regression models. Then we present two cases to illustrate some of the major problems encountered with polynomial regression.

9.1 POLYNOMIAL REGRESSION MODELS

Polynomial regression models can contain one, two, or more than two independent variables. Further, the independent variable can be present in various powers. We illustrate now some major possibilities.

One independent variable—second order

The model:

$$Y_i = \beta_0 + \beta_1 x_i + \beta_2 x_i^2 + \varepsilon_i \qquad (9.1)$$

where:

$$x_i = X_i - \overline{X}$$

is called a *second-order model with one independent variable* because the single independent variable appears to the first and second powers. Note that the independent variable is expressed as a deviation around its mean $\bar{X}$, and that the ith observation deviation is denoted by x_i. The reason for using deviations around the mean in polynomial regression models is that X, X^2, and higher-power terms often will be highly correlated. This can cause serious computational difficulties when the $\mathbf{X'X}$ matrix is inverted for estimating the regression coefficients. Expressing the independent variable as a deviation from its mean reduces the multicollinearity substantially and tends to avoid computational difficulties.

The regression coefficients in polynomial regression are frequently written in a slightly different fashion, to reflect the pattern of the exponents:

(9.1a) $$Y_i = \beta_0 + \beta_1 x_i + \beta_{11} x_i^2 + \varepsilon_i$$

We shall employ this latter notation in this chapter.

The response function for model (9.1a) is:

(9.2) $$E(Y) = \beta_0 + \beta_1 x + \beta_{11} x^2$$

which is a parabola and is frequently called a *quadratic* response function. Figure 9.1 contains two examples of second-order polynomial response functions.

The regression coefficient β_0 represents the mean response of Y when $x = 0$, i.e., when $X = \bar{X}$. The regression coefficient β_1 is often called the *linear effect coefficient* while β_{11} is called the *quadratic effect coefficient*.

FIGURE 9.1 Examples of second-order polynomial response functions

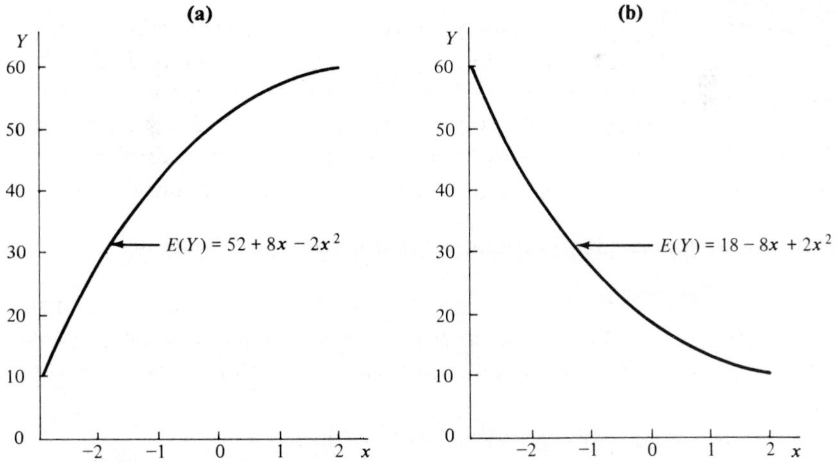

Uses of second-order model. The second-order polynomial response function (9.2) has two basic types of uses:

1. When the true response function is indeed a second-degree polynomial, containing additive linear and quadratic effect components.

2. When the true response function is unknown (or complex) but a second-order polynomial is a good approximation to the true function.

The second type of use is the more common one, but it entails a special danger, that of extrapolation. Consider again Figure 9.1a. This response function may fit the data at hand very well. If, however, information about $E(Y)$ is sought for a larger value of x, extrapolation of this response function leads to the result shown in Figure 9.2, namely, a turning down of the response function, which may not be in accord with reality. Polynomial regressions of all types, especially those of higher order, share this danger of extrapolation. They may provide good fits for the data at hand, but may turn in unexpected directions when extrapolated beyond the range of the data.

FIGURE 9.2 Extrapolation of second-order polynomial response function in Figure 9.1a

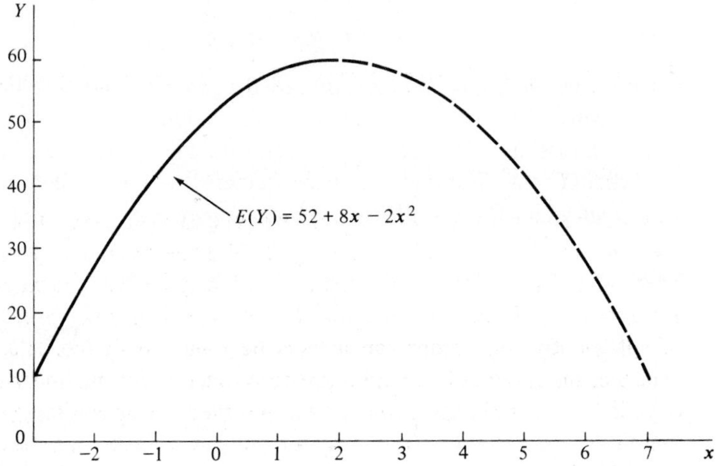

One independent variable—third order

The model:

$$Y_i = \beta_0 + \beta_1 x_i + \beta_{11} x_i^2 + \beta_{111} x_i^3 + \varepsilon_i \tag{9.3}$$

where:

$$x_i = X_i - \bar{X}$$

is a *third-order model with one independent variable*. The response function for model (9.3) is:

$$E(Y) = \beta_0 + \beta_1 x + \beta_{11} x^2 + \beta_{111} x^3 \tag{9.4}$$

Figure 9.3 contains two examples of third-order polynomial response functions.

FIGURE 9.3 Examples of third-order polynomial response functions

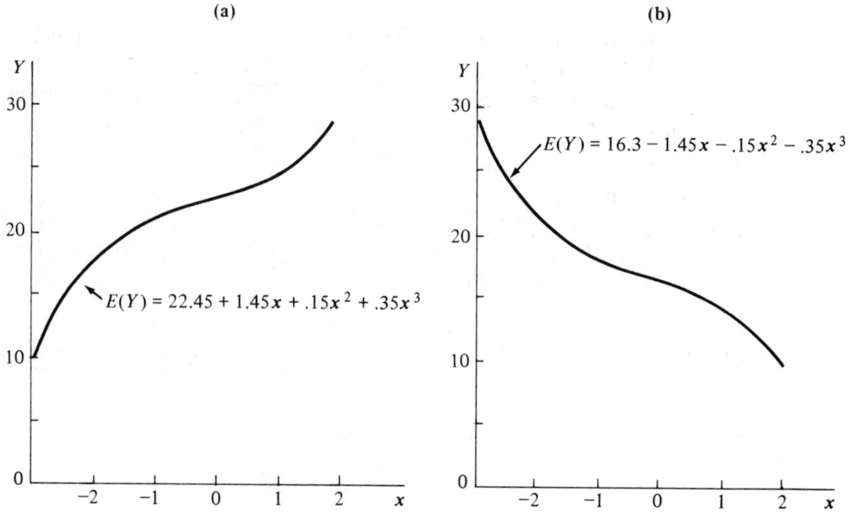

One independent variable—higher orders

Polynomial models with the independent variable present in higher powers than the third are not often employed. The interpretation of the coefficients becomes difficult for such models, and they may be highly erratic for even small extrapolations. It must be recognized in this connection that a polynomial model of sufficiently high order can always be found to fit the data perfectly. For instance, the fitted polynomial regression function for one independent variable of order $n - 1$ will pass through all n observed Y values. One needs to be wary therefore of using high order polynomials for the sole purpose of obtaining a good fit. Such regression functions may not show clearly the basic elements of the regression relation between X and Y and may lead to erratic extrapolations.

Two independent variables—second order

The model:

$$(9.5) \qquad Y_i = \beta_0 + \beta_1 x_{i1} + \beta_2 x_{i2} + \beta_{11} x_{i1}^2 + \beta_{22} x_{i2}^2 + \beta_{12} x_{i1} x_{i2} + \varepsilon_i$$

where:

$$x_{i1} = X_{i1} - \bar{X}_1$$
$$x_{i2} = X_{i2} - \bar{X}_2$$

is a *second-order model with two independent variables*. The response surface is:

$$(9.6) \qquad E(Y) = \beta_0 + \beta_1 x_1 + \beta_2 x_2 + \beta_{11} x_1^2 + \beta_{22} x_2^2 + \beta_{12} x_1 x_2$$

which is the equation of a conic section. Note that model (9.5) contains separate linear and quadratic components for each of the two independent variables and a cross-product term. The latter represents the interaction effects between x_1 and x_2, as we noted in Chapter 7. The coefficient β_{12} is often called the *interaction effect coefficient*.

The second-order response surface for two independent variables in (9.6) represents the two basic types of surfaces illustrated in Figure 7.3. Stationary and rising ridges constitute limiting cases of these two basic types of response surfaces.

Usually, it is easiest to portray the second-order response surface (9.6) in terms of contour lines. Figure 9.4 contains a representation of the response function in terms of contour curves:

$$E(Y) = 1{,}740 - 4x_1^2 - 3x_2^2 - 3x_1x_2 \tag{9.7}$$

Note that this response surface has a maximum at $x_1 = 0$ and $x_2 = 0$.

FIGURE 9.4 Example of a quadratic response surface: $E(Y) = 1{,}740 - 4x_1^2 - 3x_2^2 - 3x_1x_2$

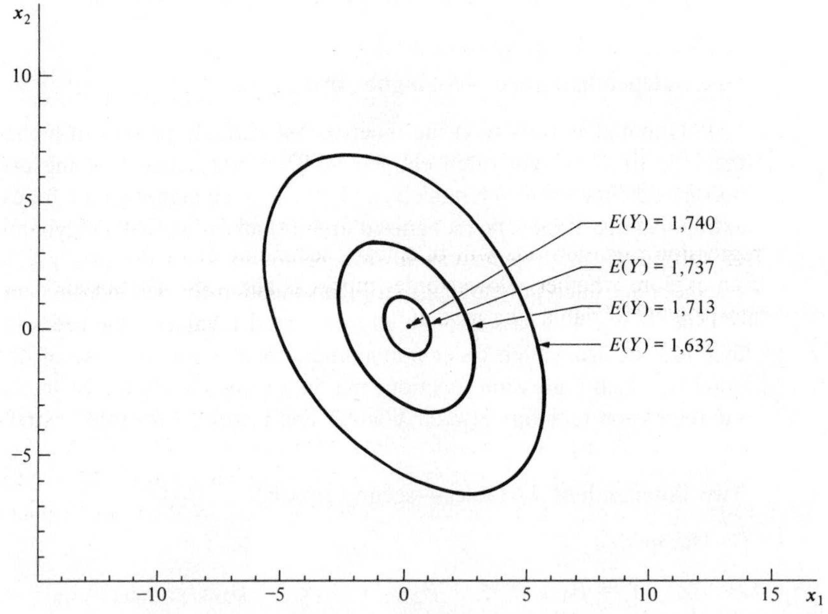

Polynomial models in two (or more) independent variables are well adapted to situations where the response function is unknown and a suitable model is to be developed empirically.

Note

The cross-product term $\beta_{12}x_1x_2$ in (9.6) is considered to be a second-order term, the same as $\beta_{11}x_1^2$ or $\beta_{22}x_2^2$. The reason can be seen readily by writing the latter terms as $\beta_{11}x_1x_1$ and $\beta_{22}x_2x_2$, respectively.

Three independent variables—second order

The *second-order model with three independent variables* is:

$$Y_i = \beta_0 + \beta_1 x_{i1} + \beta_2 x_{i2} + \beta_3 x_{i3} + \beta_{11} x_{i1}^2 + \beta_{22} x_{i2}^2 + \beta_{33} x_{i3}^2 + \beta_{12} x_{i1} x_{i2} + \beta_{13} x_{i1} x_{i3} + \beta_{23} x_{i2} x_{i3} + \varepsilon_i \quad (9.8)$$

where:

$$x_{i1} = X_{i1} - \bar{X}_1$$
$$x_{i2} = X_{i2} - \bar{X}_2$$
$$x_{i3} = X_{i3} - \bar{X}_3$$

The response surface for this model is:

$$E(Y) = \beta_0 + \beta_1 x_1 + \beta_2 x_2 + \beta_3 x_3 + \beta_{11} x_1^2 + \beta_{22} x_2^2 + \beta_{33} x_3^2 + \beta_{12} x_1 x_2 + \beta_{13} x_1 x_3 + \beta_{23} x_2 x_3 \quad (9.9)$$

The coefficients β_{12}, β_{13}, and β_{23} are interaction effects coefficients for interactions between pairs of independent variables.

Use of polynomial regression models

Fitting of polynomial regression models presents no new problems since, as we have seen in Chapter 7, they are special cases of the general linear regression model (7.18). Hence, all earlier results on fitting apply, as do the earlier results on making inferences.

When using a polynomial regression model as an approximation to the true regression function, one will often fit a second-order or third-order model and then explore whether a lower-order model is adequate. For instance, with one independent variable, the model:

$$Y_i = \beta_0 + \beta_1 x_i + \beta_{11} x_i^2 + \beta_{111} x_i^3 + \varepsilon_i$$

may be fitted with the hope that the cubic term and perhaps even the quadratic term can be dropped. Thus, one would wish to test whether or not $\beta_{111} = 0$, or whether or not both $\beta_{11} = 0$ and $\beta_{111} = 0$. Similar tests would often be conducted with polynomial regression models for two or more independent variables.

9.2 EXAMPLE 1—ONE INDEPENDENT VARIABLE

We illustrate now some of the major types of analyses usually conducted with polynomial regression models with one independent variable.

Setting

A staff analyst for a cafeteria chain wishes to investigate the relation between the number of self-service coffee dispensers in a cafeteria line and sales of coffee. Fourteen cafeterias that are similar in such respects as volume of business, type of clientele, and location are chosen for the experiment. The number of self-service dispensers that are placed in the test cafeterias varies from zero (coffee is dispensed here by a line attendant) to six and is assigned randomly to each cafeteria.

Table 9.1 contains the results of the experimental study. Sales are measured in hundreds of gallons of coffee sold.

TABLE 9.1 Data for cafeteria coffee sales example

Cafeteria i	*Number of Dispensers* X_i	*Coffee Sales (hundred gallons)* Y_i
1	0	508.1
2	0	498.4
3	1	568.2
4	1	577.3
5	2	651.7
6	2	657.0
7	3	713.4
8	3	697.5
9	4	755.3
10	4	758.9
11	5	787.6
12	5	792.1
13	6	841.4
14	6	831.8

Fitting of model

The analyst believes that the relation between sales and number of self-service dispensers is quadratic in the range of observations; sales should increase as the number of dispensers is greater, but if the space is cluttered with dispensers, this increase becomes retarded. Hence, she would like to fit the quadratic model:

$$Y_i = \beta_0 + \beta_1 x_i + \beta_{11} x_i^2 + \varepsilon_i \tag{9.10}$$

where:

$$x_i = X_i - \bar{X}$$

She further anticipates that the error terms ε_i will be fairly normally distributed with constant variance.

The **Y** and **X** matrices for this application are given in Table 9.2. Note that the **X** matrix contains a column of 1's, a column of the independent variable observations x (expressed as deviations around their mean $\bar{X} = 3$), and a column of the x^2 values. From this point on, the calculations are routine. We could do the matrix calculations manually, as illustrated in Chapter 7, or use a computer multiple regression program. Since no new problems are encountered, we simply present the basic computer output in Table 9.3, including needed extra sums of squares and the $\mathbf{s}^2(\mathbf{b})$ matrix.

TABLE 9.2 Data matrices for cafeteria coffee sales example

$$
\mathbf{Y} = \begin{bmatrix} 508.1 \\ 498.4 \\ 568.2 \\ 577.3 \\ 651.7 \\ 657.0 \\ 713.4 \\ 697.5 \\ 755.3 \\ 758.9 \\ 787.6 \\ 792.1 \\ 841.4 \\ 831.8 \end{bmatrix}
\qquad
\mathbf{X} = \begin{bmatrix}
 & x & x^2 \\
1 & -3 & 9 \\
1 & -3 & 9 \\
1 & -2 & 4 \\
1 & -2 & 4 \\
1 & -1 & 1 \\
1 & -1 & 1 \\
1 & 0 & 0 \\
1 & 0 & 0 \\
1 & 1 & 1 \\
1 & 1 & 1 \\
1 & 2 & 4 \\
1 & 2 & 4 \\
1 & 3 & 9 \\
1 & 3 & 9
\end{bmatrix}
$$

TABLE 9.3 Regression results for cafeteria coffee sales example

(a) Regression Coefficients

Regression Coefficient	Estimated Regression Coefficient	Estimated Standard Deviation	t^*
β_0	705.474	3.208	219.91
β_1	54.893	1.050	52.28
β_{11}	−4.249	.606	−7.01

(b) Analysis of Variance

Source of Variation	SS	df	MS
Regression	171,773	2	85,887
x	168,741	1	168,741
$x^2\|x$	3,033	1	3,033
Error	679	11	61.7
Total	172,453	13	

(c) $s^2(\mathbf{b})$ Matrix

$$
\begin{bmatrix}
10.2912 & 0 & -1.4702 \\
0 & 1.1026 & 0 \\
-1.4702 & 0 & .3675
\end{bmatrix}
$$

The fitted regression function is:

(9.11) $$\hat{Y} = 705.47 + 54.89x - 4.25x^2$$

This response function is plotted in Figure 9.5, together with the original data. We show the horizontal scale expressed at the bottom in the deviation units x and at the top in the original units X.

FIGURE 9.5 Fitted second-order polynomial regression—cafeteria coffee sales example

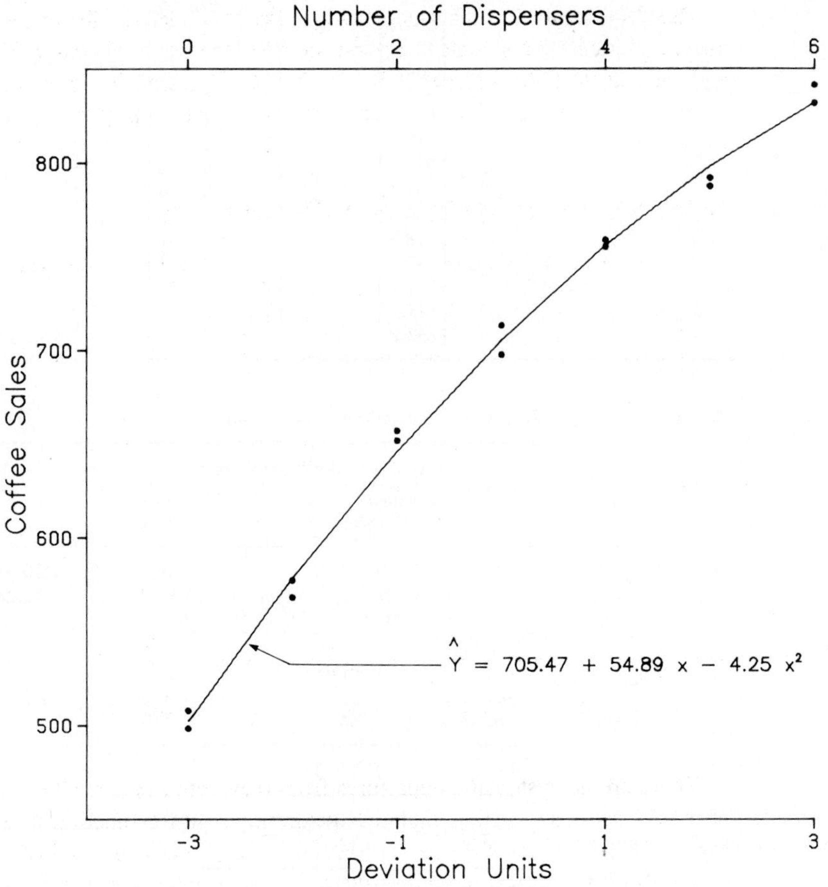

Algebraic version of normal equations. The algebraic version of the least squares normal equations:

$$\mathbf{X'Xb} = \mathbf{X'Y}$$

for the second-order polynomial model (9.10) can be readily obtained from (7.65) by replacing X_{i1} by x_i and X_{i2} by x_i^2. Since $\Sigma x_i = 0$, this yields the normal equations:

$$\begin{aligned} \Sigma Y_i &= nb_0 + b_{11}\Sigma x_i^2 \\ \Sigma x_i Y_i &= b_1 \Sigma x_i^2 + b_{11}\Sigma x_i^3 \\ \Sigma x_i^2 Y_i &= b_0 \Sigma x_i^2 + b_1 \Sigma x_i^3 + b_{11}\Sigma x_i^4 \end{aligned} \tag{9.12}$$

Analysis of aptness of model

Residual analysis. To study the aptness of model (9.10) for her data, the analyst plotted the residuals e_i against the fitted values, as shown in Figure 9.6a, and also against the independent variable x_i expressed in deviation units, as shown in Figure 9.6b. We do not present the calculations of the e_i, as these are routine.

FIGURE 9.6 Residual plots for cafeteria coffee sales example

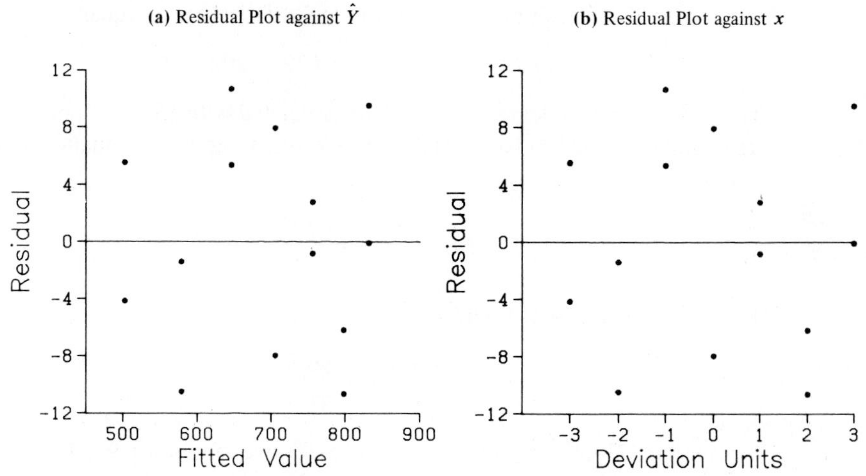

There are no systematic departures from 0 evident in the residuals as either $\hat{Y}$ or x increases, suggesting that the quadratic response function is a good fit. Figure 9.5 makes this point also. Further, there is no tendency in Figures 9.6a and 9.6b for the spread in the residuals to vary systematically, so it appears that the constant error variance assumption is reasonable. A normal probability plot, not shown here, did not provide any strong evidence that the distribution of the error terms is far from normal.

Based on this study of the aptness of the model, the analyst was willing to conclude that the normal error model (9.10) with constant error variance is appropriate here.

Test for quadratic response function. Since there are two repeat observations for each level of x, the analyst could have made a formal test of the aptness of the model, the alternatives being:

(9.13)
$$H_0\colon E(Y) = \beta_0 + \beta_1 x + \beta_{11} x^2$$
$$H_a\colon E(Y) \neq \beta_0 + \beta_1 x + \beta_{11} x^2$$

The basic ANOVA results were presented earlier in Table 9.3b. The pure error sum of squares is obtained as follows from the data in Table 9.2:

$$SSPE = (508.1 - 503.25)^2 + (498.4 - 503.25)^2 + (568.2 - 572.75)^2 + \cdots + (831.8 - 836.6)^2 = 292$$

Note that $\bar{Y}_1 = 503.25$ for $x = -3$, $\bar{Y}_2 = 572.75$ for $x = -2$, and so on. There are $14 - 7 = 7$ degrees of freedom associated with $SSPE$. Hence, we have:

$$MSPE = \frac{SSPE}{7} = \frac{292}{7} = 41.7$$

Now we are in a position to obtain the lack of fit sum of squares by (4.12):

$$SSLF = SSE - SSPE = 679 - 292 = 387$$

There are $7 - 3 = 4$ degrees of freedom associated with $SSLF$. (Remember that three parameters had to be estimated for the fitted regression equation.) Hence, we have:

$$MSLF = \frac{SSLF}{4} = \frac{387}{4} = 96.8$$

Thus, test statistic (4.15) here is:

$$F^* = \frac{MSLF}{MSPE} = \frac{96.8}{41.7} = 2.32$$

Assuming the level of significance is to be .05, we require $F(.95; 4, 7) = 4.12$. Since $F^* = 2.32 \leq 4.12$, we conclude H_0, that the quadratic response function is appropriate.

Test whether β_{11} equals zero

***t* test.** The analyst next studied whether the quadratic term could be dropped from the model. She therefore wished to test:

$$\begin{aligned} &H_0: \beta_{11} = 0 \\ &H_a: \beta_{11} \neq 0 \end{aligned} \tag{9.14}$$

H_0 implies that there is no quadratic effect in the response function.

Table 9.3a indicates that:

$$t^* = \frac{b_{11}}{s(b_{11})} = \frac{-4.249}{.606} = -7.012$$

For a level of significance of .05, we require $t(.975; 11) = 2.201$. The decision rule is:

$$\begin{aligned} &\text{If } |t^*| \leq 2.201, \text{ conclude } H_0 \\ &\text{If } |t^*| > 2.201, \text{ conclude } H_a \end{aligned}$$

Since $|t^*| = 7.012 > 2.201$, we conclude H_a, that a quadratic effect does exist, so that the quadratic term should be retained in the model.

Partial *F* test. The analyst could also have used the partial F test to choose the appropriate conclusion in (9.14). Indeed, she had specified the order of entering the variables x and x^2 into the computer regression fit so that she would obtain the extra sums of squares $SSR(x)$ and $SSR(x^2|x)$ in the output. Utilizing the partial F test statistic (8.24) and the results in Table 9.3b, we obtain:

$$F^* = \frac{MSR(x^2|x)}{MSE} = \frac{3{,}033}{61.7} = 49.2$$

For a 5 percent level of significance, we need $F(.95; 1, 11) = 4.84$. Since $F^* = 49.2 > 4.84$, we are led to conclude H_a, as by the t test.

Note

One may observe here the relation discussed in the previous chapter between the t and partial F tests as to whether a regression coefficient equals zero. We have for the two test statistics:

$$(t^*)^2 = (-7.012)^2 = 49.2 = F^*$$

Estimation of regression coefficients

The analyst next wished to obtain confidence bounds on the two regression coefficients β_1 and β_{11} with family confidence coefficient .90. The Bonferroni method is to be used, in view of its simplicity and the ease of interpreting the results.

Here $g = 2$ statements are desired; hence, by (7.44a), we have:

$$B = t(1 - .10/2(2); 11) = t(.975; 11) = 2.201$$

From Table 9.3a, we find:

$$b_1 = 54.893 \qquad s(b_1) = 1.050$$
$$b_{11} = -4.249 \qquad s(b_{11}) = .606$$

The Bonferroni confidence intervals therefore are by (7.44):

$$54.893 - 2.201(1.050) \le \beta_1 \le 54.893 + 2.201(1.050)$$

or:

$$52.58 \le \beta_1 \le 57.20$$

$$-4.249 - 2.201(.606) \le \beta_{11} \le -4.249 + 2.201(.606)$$

or:

$$-5.58 \le \beta_{11} \le -2.92$$

The analyst was satisfied with the precision of these two statements, feeling that the intervals are narrow enough to give her reliable simultaneous information about the comparative magnitudes of the linear and quadratic effects.

Coefficient of multiple determination

For a descriptive measure of the degree of relation between coffee sales and number of dispensing machines, the analyst calculated the coefficient of multiple determination using the data in Table 9.3b:

$$R^2 = \frac{SSR}{SSTO} = \frac{171{,}773}{172{,}453} = .996$$

This measure shows that the variation in coffee sales is reduced by 99.6 percent when the quadratic relation to the number of dispensing machines is utilized.

Note that the coefficient of multiple determination R^2 is the relevant measure here, not the coefficient of simple determination r^2, since model (9.10) is a multiple regression model even though it contains only one independent variable. Sometimes in curvilinear regression, the coefficient of multiple correlation R is called the *correlation index*.

Estimation of mean response

The analyst was particularly interested in the mean response for $X_h = 3$ dispensing machines. She wished to estimate this mean response with a 98 percent confidence coefficient. The proper interval estimate is given by (7.51). For our example, where $x_h = X_h - \bar{X} = 3 - 3 = 0$, we have:

$$\mathbf{X}_h = \begin{bmatrix} 1 \\ x_h \\ x_h^2 \end{bmatrix} = \begin{bmatrix} 1 \\ 0 \\ 0 \end{bmatrix}$$

The estimated mean response $\hat{Y}_h$ corresponding to $\mathbf{X}_h$ is by (7.47):

$$\hat{Y}_h = \mathbf{X}_h'\mathbf{b} = [1 \quad 0 \quad 0] \begin{bmatrix} 705.474 \\ 54.893 \\ -4.249 \end{bmatrix} = 705.474$$

Next, using the results in Table 9.3c for $\mathbf{s}^2(\mathbf{b})$, we obtain when substituting into (7.50):

$$\begin{aligned} s^2(\hat{Y}_h) &= \mathbf{X}_h'\mathbf{s}^2(\mathbf{b})\mathbf{X}_h \\ &= [1 \quad 0 \quad 0] \begin{bmatrix} 10.2912 & 0 & -1.4702 \\ 0 & 1.1026 & 0 \\ -1.4702 & 0 & .3675 \end{bmatrix} \begin{bmatrix} 1 \\ 0 \\ 0 \end{bmatrix} \\ &= 10.2912 \end{aligned}$$

or:

$$s(\hat{Y}_h) = \sqrt{10.2912} = 3.208$$

We require $t(.99; 11) = 2.718$. Hence, we obtain:

$$705.474 - 2.718(3.208) \le E(Y_h) \le 705.474 + 2.718(3.208)$$

or:

$$696.8 \le E(Y_h) \le 714.2$$

With confidence coefficient .98, the analyst can conclude that the mean of the distribution of coffee sales when three dispensing machines are used is somewhere between 696.8 and 714.2 hundred gallons.

Regression function in terms of X

The analyst wishes for reporting purposes to express the fitted regression function in terms of X rather than in terms of deviations $x = X - \bar{X}$. The following formulas provide the appropriate coefficients for model (9.10); the primes on the coefficients denote the new coefficients in terms of X:

(9.15a) $$b_0' = b_0 - b_1\bar{X} + b_{11}\bar{X}^2$$

(9.15b) $$b_1' = b_1 - 2b_{11}\bar{X}$$

(9.15c) $$b_{11}' = b_{11}$$

For our example, where $\bar{X} = 3$, we obtain:

$$b_0' = 705.474 - 54.893(3) + (-4.249)(3)^2 = 502.554$$
$$b_1' = 54.893 - 2(-4.249)(3) = 80.387$$
$$b_{11}' = -4.249$$

so that the fitted regression function in terms of X is:

$$\hat{Y} = 502.554 + 80.387X - 4.249X^2$$

The fitted values and residuals for the regression function in terms of X are exactly the same as for the regression function in terms of the deviations x. The reason for utilizing model (9.10), which is expressed in terms of deviations x, is to avoid potential calculational difficulties due to multicollinearity between X and X^2, inherent in polynomial regression.

Note

The estimated standard deviations of the regression coefficients in Table 9.3a do not apply to the regression coefficients in terms of X obtained by (9.15). If the estimated standard deviations for the regression coefficients in terms of X are desired, they may be obtained from $\mathbf{s}^2(\mathbf{b})$ in Table 9.3c by using theorem (6.49), where the transformation matrix $\mathbf{A}$ is developed from (9.15).

9.3 EXAMPLE 2—TWO INDEPENDENT VARIABLES

We shall discuss now another example of polynomial regression, this one involving two independent variables. Rather than carrying this example through all of the various analytical stages as we did the first example, we shall focus here on the analysis of interaction effects and quadratic effects.

Setting

For a sample of 18 managers in the 35–44 age group, Table 9.4 shows the average annual income during the past two years (X_1), risk aversion score (X_2), and amount of life insurance carried (Y). Risk aversion was measured by a standard questionnaire administered to each manager; the higher the score the greater the degree of risk aversion.

TABLE 9.4 Data for life insurance example

Manager i	*Average Annual Income (thousand dollars)* X_{i1}	*Risk Aversion Score* X_{i2}	*Amount of Life Insurance Carried (thousand dollars)* Y_i
1	66.290	7	196
2	40.964	5	63
3	72.996	10	252
4	45.010	6	84
5	57.204	4	126
6	26.852	5	14
7	38.122	4	49
8	35.840	6	49
9	75.796	9	266
10	37.408	5	49
11	54.376	2	105
12	46.186	7	98
13	46.130	4	77
14	30.366	3	14
15	39.060	5	56
16	79.380	1	245
17	52.766	8	133
18	55.916	6	133
	$\bar{X}_1 = 50.037$	$\bar{X}_2 = 5.389$	

A researcher was studying the relation of average annual income and risk aversion to amount of life insurance carried by managers in the given age group. He expected that a quadratic relation would hold between income and amount of life insurance carried. However, he would not have been surprised if aversion to risk showed only linear effects and no quadratic effects on amount of life insurance carried, and he was quite uncertain whether or not the two variables interact in their effects on amount of life insurance carried. Hence, he fitted the second-order polynomial regression model:

$$Y_i = \beta_0 + \beta_1 x_{i1} + \beta_2 x_{i2} + \beta_{11} x_{i1}^2 + \beta_{22} x_{i2}^2 + \beta_{12} x_{i1} x_{i2} + \varepsilon_i \tag{9.16}$$

where:

$$x_{i1} = X_{i1} - \bar{X}_1$$
$$x_{i2} = X_{i2} - \bar{X}_2$$

with the intention of first testing for the presence of interaction effects and then for quadratic effects of aversion to risk.

Development of model

Table 9.5a contains the basic results for the fit of model (9.16). Since the researcher wished to test first for the interaction effects ($\beta_{12}x_1x_2$), he entered the variables for the regression fit so as to obtain the extra sum of squares $SSR(x_1x_2|x_1, x_2, x_1^2, x_2^2)$ for the partial F test. The ANOVA table and decomposition of SSR into extra sums of squares is shown in Table 9.5b.

TABLE 9.5 Regression results for model (9.16)—life insurance example

(a) Regression Coefficients

Regression Coefficient	*Estimated Regression Coefficient*	*Estimated Standard Deviation*	t^*
β_0	102.768	.662	155.15
β_1	4.4930	.0475	94.54
β_2	6.028	.301	20.00
β_{11}	.03579	.00219	16.34
β_{22}	.166	.120	1.38
β_{12}	−.0196	.0140	−1.40

(b) Analysis of Variance

Source of Variation	*SS*	*df*	*MS*
Regression	108,006	5	21,601
x_1	104,474	1	104,474
$x_2\|x_1$	2,284	1	2,284
$x_1^2\|x_1, x_2$	1,238	1	1,238
$x_2^2\|x_1, x_2, x_1^2$	3	1	3
$x_1x_2\|x_1, x_2, x_1^2, x_2^2$	6	1	6
Error	36	12	3.00
Total	108,042	17	

The test for the presence of interaction effects involves the alternatives:

$$H_0: \beta_{12} = 0 \qquad (9.17)$$
$$H_a: \beta_{12} \neq 0$$

Using the partial F test statistic (8.24), the researcher obtained:

$$F^* = \frac{MSR(x_1x_2|x_1, x_2, x_1^2, x_2^2)}{MSE} = \frac{6}{3} = 2.00$$

For level of significance $\alpha = .05$, we require $F(.95; 1, 12) = 4.75$. Since $F^* = 2.00 \leq 4.75$, we conclude H_0, that no interaction effects exist. This result was welcome to the researcher, as it simplifies the interpretation of the effects of the two independent variables.

Note that the researcher could also have tested whether or not $\beta_{12} = 0$ by using $|t^*| = |-1.40| = 1.40$ (Table 9.5a).

At this point, the researcher tentatively decided to adopt the no-interaction model:

(9.18) $$Y_i = \beta_0 + \beta_1 x_{i1} + \beta_2 x_{i2} + \beta_{11} x_{i1}^2 + \beta_{22} x_{i2}^2 + \varepsilon_i$$

but he still wished to examine whether a quadratic effect for risk aversion exists. This test can be conducted by a partial F test without fitting a new model. We utilize the definition of $SSR(x_1x_2|x_1, x_2, x_1^2, x_2^2)$:

$$SSR(x_1x_2|x_1, x_2, x_1^2, x_2^2) = SSE(x_1, x_2, x_1^2, x_2^2) - SSE(x_1, x_2, x_1^2, x_2^2, x_1x_2)$$

Hence:

$$\begin{aligned} SSE(x_1, x_2, x_1^2, x_2^2) &= SSR(x_1x_2|x_1, x_2, x_1^2, x_2^2) + SSE(x_1, x_2, x_1^2, x_2^2, x_1x_2) \\ &= 6 + 36 = 42 \end{aligned}$$

When testing model (9.18) for:

(9.19) $$\begin{aligned} H_0&: \beta_{22} = 0 \\ H_a&: \beta_{22} \neq 0 \end{aligned}$$

the partial F test statistic (8.24) is:

$$\begin{aligned} F^* &= \frac{SSR(x_2^2|x_1, x_2, x_1^2)}{1} \div \frac{SSE(x_1, x_2, x_1^2, x_2^2)}{18 - 5} \\ &= \frac{3}{1} \div \frac{42}{13} = .93 \end{aligned}$$

For a 5 percent level of significance, we require $F(.95; 1, 13) = 4.67$. Since $F^* = .93 \leq 4.67$, we conclude H_0, that there is no quadratic effect for aversion to risk.

Hence, the researcher decided to adopt the revised model:

(9.20) $$Y_i = \beta_0 + \beta_1 x_{i1} + \beta_2 x_{i2} + \beta_{11} x_{i1}^2 + \varepsilon_i$$

where:

$$\begin{aligned} x_{i1} &= X_{i1} - \bar{X}_1 \\ x_{i2} &= X_{i2} - \bar{X}_2 \end{aligned}$$

and fitted this model to the data. He obtained the estimated response function:

$$\hat{Y} = 103.136 + 4.551x_1 + 5.685x_2 + .0371x_1^2$$

which in the original units is:

$$\hat{Y} = -74.583 + .8383X_1 + 5.685X_2 + .0371X_1^2$$

Figure 9.7 contains a three-dimensional computer-generated plot of this fitted response surface. The researcher then used this fitted response function for further investigation of the effects of average annual income and aversion to risk on amount of life insurance carried in the population under study.

FIGURE 9.7 Three-dimensional computer-generated plot of response function $\hat{Y} = -74.583 + .8383X_1 + 5.685X_2 + .0371X_1^2$—life insurance example

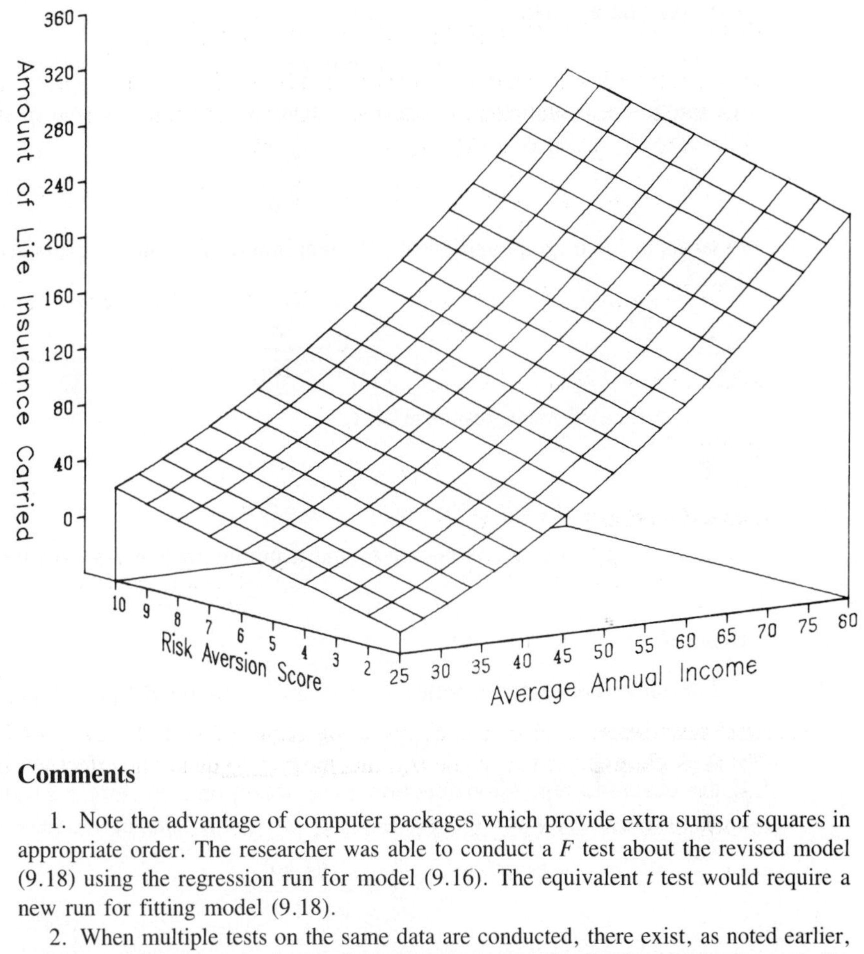

Comments

1. Note the advantage of computer packages which provide extra sums of squares in appropriate order. The researcher was able to conduct a F test about the revised model (9.18) using the regression run for model (9.16). The equivalent t test would require a new run for fitting model (9.18).

2. When multiple tests on the same data are conducted, there exist, as noted earlier, problems with respect to the level of significance for the family of inferences. In the example just cited, the researcher was willing to conduct two tests on the same data, one for interaction effects and one for quadratic effects of aversion to risk. The reason was that he knew by the Bonferroni inequality (5.6a) that the family level of significance for the two tests (each conducted at the 5 percent level of significance) could not exceed 10 percent.

In Chapter 12, we shall discuss more fully the empirical determination of an appropriate regression model.

9.4 ESTIMATING THE MAXIMUM OR MINIMUM OF A QUADRATIC REGRESSION FUNCTION

Sometimes in quadratic regression, we wish to estimate the maximum (or minimum) mean response of the regression function, and/or the level of X at which the maximum (or minimum) occurs. Figure 9.2 illustrates a quadratic response function with a maximum mean response.

Given the estimated quadratic response function:

$$\hat{Y} = b_0 + b_1 x + b_{11} x^2 \tag{9.21}$$

the maximum (minimum) occurs at the level x_m:

$$x_m = -\frac{b_1}{2b_{11}} \tag{9.22}$$

In terms of the original variable X, the maximum (minimum) occurs at the level X_m:

$$X_m = \bar{X} - \frac{b_1}{2b_{11}} \tag{9.22a}$$

The estimated mean response at X_m is:

$$\hat{Y}_m = b_0 - \frac{b_1^2}{4b_{11}} \tag{9.23}$$

$\hat{Y}_m$ is a maximum if b_{11} is negative, and a minimum if b_{11} is positive.

Example

For our earlier cafeteria coffee sales example, the fitted regression curve was:

$$\hat{Y} = 705.47 + 54.89x - 4.25x^2$$

If the quadratic regression function were appropriate for larger x values than those in the study, we could estimate that maximum mean coffee sales occur at:

$$X_m = 3 - \frac{54.89}{2(-4.25)} = 9$$

and the estimated mean response there is:

$$\hat{Y}_m = 705.47 + 54.89(9) - 4.25(9)^2 = 855$$

Comments

1. To derive (9.22), we differentiate $\hat{Y}$ in (9.21) with respect to x, and set this derivative equal to 0:

$$\frac{d\hat{Y}}{dx} = \frac{d}{dx}(b_0 + b_1 x + b_{11} x^2) = b_1 + 2b_{11}x = 0$$

and obtain:

$$x_m = -\frac{b_1}{2b_{11}}$$

Substituting this value into the fitted response function (9.21), we find:

$$\hat{Y}_m = b_0 + b_1\left(\frac{-b_1}{2b_{11}}\right) + b_{11}\left(\frac{-b_1}{2b_{11}}\right)^2$$
$$= b_0 - \frac{b_1^2}{4b_{11}}$$

2. For large samples, the approximate estimated variance of X_m is:

$$s^2(X_m) = \frac{b_1^2}{4b_{11}^2}\left[\frac{s^2(b_1)}{b_1^2} + \frac{s^2(b_{11})}{b_{11}^2} - \frac{2s(b_1, b_{11})}{b_1 b_{11}}\right] \tag{9.24}$$

This approximate estimated variance can be used to construct a confidence interval for the true X level at which the maximum (minimum) occurs. Approximate confidence intervals for $E(Y_m)$ can also be obtained. These are discussed in Reference 9.1.

9.5 SOME FURTHER COMMENTS ON POLYNOMIAL REGRESSION

1. The use of polynomial models in X is not without drawbacks. Such models can be more expensive in degrees of freedom than alternative nonlinear models or linear models with transformed variables. Another potential drawback is that multicollinearity is unavoidable. Indeed, if the levels of X are restricted to a narrow range, the degree of multicollinearity in the columns of the **X** matrix can be quite high, especially for higher-degree polynomials. It is for this reason that all polynomial regression models in this chapter are formulated in terms of deviations $x_i = X_i - \bar{X}$. To illustrate how helpful the use of deviation variables can be, in our life insurance example the coefficient of correlation between X_1 and X_1^2 is .991, but it is only .445 between x_1 and x_1^2. As noted earlier, when multicollinearity is high, serious calculational difficulties in inverting the $\mathbf{X'X}$ matrix can arise.

2. An alternative to using variables expressed in deviations from the mean in polynomial regression is to use *orthogonal polynomials*. Orthogonal polynomials are uncorrelated. Some computer packages use orthogonal polynomials in their polynomial regression routines and present the final fitted results in terms of both the orthogonal polynomials and the original polynomials. Orthogonal polynomials are discussed in specialized texts such as Reference 9.2.

3. Sometimes a quadratic response function is fitted for the purpose of establishing the linearity of the response function when repeat observations are not available for directly testing the linearity of the response function. Fitting the quadratic model:

$$Y_i = \beta_0 + \beta_1 x_i + \beta_{11} x_i^2 + \varepsilon_i \tag{9.25}$$

and testing whether $\beta_{11} = 0$ does not, however, necessarily establish that a linear response function is appropriate. Figure 9.8 provides an example. If sample data were obtained for the response function in Figure 9.8, model (9.25) fitted, and a test on β_{11} made, it likely would lead to the conclusion that $\beta_{11} = 0$. Yet a linear

FIGURE 9.8 Example of curvilinear response function

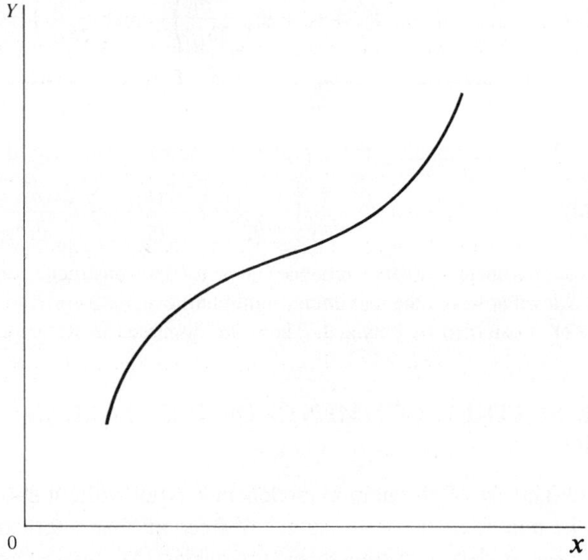

response function clearly might not be appropriate. Examination of residuals would disclose this lack of fit, and should always accompany formal testing of polynomial regression coefficients.

4. When a polynomial regression model with one independent variable is employed, one ordinarily fits a polynomial to the highest power expected to be appropriate, and the decomposition of *SSR* into extra sum of squares components proceeds as follows:

$$SSR(x)$$
$$SSR(x^2|x)$$
$$SSR(x^3|x, x^2)$$
$$\text{etc.}$$

The reason for this approach is that generally one is most interested in whether higher-order terms can be dropped from the model. Thus, when a cubic model is fitted because it is expected that a third-order model will be sufficient and one wishes to test whether $\beta_{111} = 0$, the appropriate extra sum of squares is $SSR(x^3|x, x^2)$. If, instead, one wishes to test whether a linear term is adequate so that $\beta_{11} = \beta_{111} = 0$, the appropriate extra sum of squares is $SSR(x^2, x^3|x) = SSR(x^2|x) + SSR(x^3|x, x^2)$.

Ordinarily, one would not fit a third-order model and test first whether a lower-order coefficient is zero, say, whether $\beta_{11} = 0$. The reason is that it is usually desired to employ as simple a regression model as possible, which in the case of polynomial regression means a lower-order model.

PROBLEMS

9.1. A speaker stated: "In developing third-order or higher-order polynomial regression models in social science and managerial applications, inferences on the β's usually take the form of direct tests. There is relatively little interest in estimating the β's to assess effects of the individual polynomial terms." Why might this be so?

9.2. Plot several contour curves for the quadratic response surface $E(Y) = 140 + 4x_1^2 - 2x_2^2 + 5x_1x_2$.

9.3. A junior investment analyst used a polynomial regression model of relatively high order in a research seminar on municipal bonds. She obtained an R^2 of .991 in the regression of net interest yield of bond (Y) on industrial diversity index of municipality (X) for seven bond issues. A classmate, unimpressed, said: "You overfitted. Your curve follows the random effects in the data."

a. Comment on the criticism.

b. Might R_a^2 defined in (7.33) be more appropriate than R^2 as a descriptive measure here?

9.4. A student in a class demonstration of how to fit a second-order polynomial model in one independent variable entered the X variables in the form X, X^2. He was disturbed when the computer program would not include X^2 in the regression model and regressed Y on X only. The output contained the message: X-SQUARE IS REDUNDANT VARIABLE. X′X IS NEAR-SINGULAR WHEN X-SQUARE IS INCLUDED. Explain the situation. What should the student have done?

9.5. Refer to the life insurance example on page 314. A student observed that the interaction term $\beta_{12}x_1x_2$ and the quadratic effect term $\beta_{22}x_2^2$ were each dropped from model (9.16) at an .05 level of significance and suggested that this same result could have been obtained from a glance at Table 9.5a, since each relevant $|t^*|$ statistic does not exceed $t(.975; 12) = 2.179$. Do you agree with the student's suggestion? Explain.

9.6. **Mileage study.** The effectiveness of a new experimental overdrive gear in reducing gasoline consumption was studied in 12 trials with a light truck equipped with this gear. In the data that follow, X_i denotes the constant speed (in miles per hour) on the test track in the ith trial and Y_i denotes miles per gallon obtained.

i:	1	2	3	4	5	6	7	8	9	10	11	12
X_i:	35	35	40	40	45	45	50	50	55	55	60	60
Y_i:	22	20	28	31	37	38	41	39	34	37	27	30

The second-order regression model (9.1a) with independent normal error terms is expected to be appropriate.

a. Fit regression model (9.1a). Plot the fitted regression function and the data. Does the quadratic regression function appear to be a good fit here? Find R^2.

b. Test whether or not there is a regression relation. Control the risk of a Type I error at .05. State the alternatives, decision rule, and conclusion.

c. Estimate the mean miles per gallon for test runs at a speed of 48 miles per hour; use a 95 percent confidence interval.

d. Predict the miles per gallon in the next test run at 48 miles per hour; use a 95 percent prediction interval.

e. Test whether the quadratic term can be dropped from the regression model; use $\alpha = .05$. State the alternatives, decision rule, and conclusion.

f. Express the fitted regression function obtained in part (a) in terms of the original variable X.

9.7. Refer to **Mileage study** Problem 9.6.

a. Obtain the residuals and plot them against $\hat{Y}$. Also prepare a normal probability plot. Interpret your plots.

b. Test formally for lack of fit of the quadratic regression function; use $\alpha = .05$. State the alternatives, decision rule, and conclusion. What assumptions did you implicitly make in this test?

c. Fit the third-order model (9.3) and test whether or not $\beta_{111} = 0$; use $\alpha = .05$. State the alternatives, decision rule, and conclusion. Is your conclusion consistent with your finding in part (b)?

9.8. **Piecework operation.** An operations analyst in a multinational electronics firm studied factors affecting production in a piecework operation where earnings are based on the number of pieces produced. Two employees each were selected from various age groups and data on their productivity last year were obtained (X is age of employee, in years; Y is employee's productivity, coded):

i:	1	2	3	4	5	6	7	8	9
X_i:	20	20	25	25	30	30	35	35	40
Y_i:	97	93	99	105	109	106	109	111	100

i:	10	11	12	13	14	15	16	17	18
X_i:	40	45	45	50	50	55	55	60	60
Y_i:	105	97	101	105	103	105	109	112	110

The analyst recognized that the relation between age and productivity is complex, in part because earnings targets (which he could not measure) shift in complex ways with age. However, he believed that for purposes of estimating mean responses, the response function can be approximated suitably by a polynomial of third order and that the error terms are independent and approximately normally distributed.

a. Fit regression model (9.3). Plot the fitted regression function and the data. Does the cubic regression function appear to be a good fit here? Find R^2.

b. Test whether or not there is a regression relation; use a level of significance of .01. State the alternatives, decision rule, and conclusion. What is the P-value of the test?

c. Obtain joint interval estimates for the mean productivity of employees aged 53, 58, and 62, respectively. Use the most efficient simultaneous estimation procedure and a 99 percent family confidence coefficient.

d. Predict the productivity of an employee aged 53 using a 99 percent prediction interval.

e. Express the fitted regression function obtained in part (a) in terms of the original variable X.

9.9. Refer to **Piecework operation** Problem 9.8.

a. Test whether both the quadratic and cubic terms can be dropped from the regression model; use $\alpha = .01$. State the alternatives, decision rule, and conclusion.

b. Test whether the cubic term alone can be dropped from the regression model; use $\alpha = .01$. State the alternatives, decision rule, and conclusion.

9.10. Refer to **Piecework operation** Problem 9.8.

a. Obtain the residuals and plot them against the fitted values. Also prepare a normal probability plot. What do your plots show?

b. Test formally for lack of fit. Control the risk of a Type I error at .01. State the alternatives, decision rule, and conclusion. What assumptions did you implicitly make in this test?

9.11. **Sales forecasting.** The Wheaton Company introduced a new product in 1975. Annual sales of this product (Y, in thousand units) follow; the time period (X) is coded, with $X = 1$ for 1975.

i:	1	2	3	4	5	6	7	8	9
X_i:	1	2	3	4	5	6	7	8	9
Y_i:	3.49	3.78	4.05	4.41	4.73	5.12	5.56	5.99	6.44

Assume that the second-order polynomial regression model (9.1a) with independent normal error terms is appropriate.

a. Fit regression model (9.1a). Plot the fitted regression function and the data. Does the quadratic regression function appear to be a good fit here? What is R^2? Do you believe that the quadratic regression function is appropriate for making projections to 1995? Discuss.

b. Obtain simultaneous Bonferroni confidence intervals for β_1 and β_{11} with a 90 percent family confidence coefficient.

c. Predict sales of the product for 1985 using a 90 percent prediction interval.

d. Express the fitted regression function obtained in part (a) in the original X units.

9.12. Refer to **Sales forecasting** Problem 9.11.

a. Test whether the quadratic term can be dropped from the regression model. Control the risk of a Type I error at .10. State the alternatives, decision rule, and conclusion. What is the P-value of the test?

b. Obtain the residuals. Plot the residuals against the fitted values. Also plot them against time. What do these plots show?

9.13. **Crop yield.** An agronomist studied the effects of moisture (X_1, in inches) and temperature (X_2, in °C) on the yield of a new hybrid tomato (Y). The experimental data follow.

i:	1	2	3	4	5	6	7	8	9	10	11	12	13
X_{i1}:	6	6	6	6	6	8	8	8	8	8	10	10	10
X_{i2}:	20	21	22	23	24	20	21	22	23	24	20	21	22
Y_i:	49.2	48.1	48.0	49.6	47.0	51.5	51.7	50.4	51.2	48.4	51.1	51.5	50.3

i:	14	15	16	17	18	19	20	21	22	23	24	25
X_{i1}:	10	10	12	12	12	12	12	14	14	14	14	14
X_{i2}:	23	24	20	21	22	23	24	20	21	22	23	24
Y_i:	48.9	48.7	48.6	47.0	48.0	46.4	46.2	43.2	42.6	42.1	43.9	40.5

The agronomist expects that the second-order polynomial regression model (9.5) with independent normal error terms is appropriate here.

a. Fit regression model (9.5). Plot the Y observations against the fitted values. Does the response function provide a good fit?

b. Calculate R^2. What information does this measure provide?
c. Test whether or not there is a regression relation; use $\alpha = .05$. State the alternatives, decision rule, and conclusion. What is the P-value of the test?
d. Estimate the mean yield when $X_1 = 7$ and $X_2 = 22$; use a 95 percent confidence interval.
e. Express the fitted response function obtained in part (a) in the original X variables.

9.14. Refer to **Crop yield** Problem 9.13.
a. Test whether or not the interaction term can be dropped from the model. Control the α risk at .005. State the alternatives, decision rule, and conclusion.
b. Assuming that the interaction term has been dropped from the model, test whether or not the quadratic effect term for temperature can be dropped from the model; control the α risk at .005. State the alternatives, decision rule, and conclusion. What is the combined α risk for both the test here and the one in part (a)?
c. Fit the second-order polynomial model omitting the interaction term and the quadratic effect term for temperature. Obtain the residuals and plot them against the fitted values. What does your plot show?

9.15. Computerized game. Students comprising firm A in a computerized marketing game have approached you for assistance in analyzing the relation between promotional expenditures (X) and demand for their firm's product (Y) in the firm's home territory. They believe that the following characteristics hold in this relation: (1) demand in the home territory is affected primarily by promotional expenditures, (2) the relation is either quadratic or linear within the range of X levels of interest to the firm. The team has provided the observations shown below for the 14 periods covered in the game to date (X in thousand dollars, Y in thousand units), and has stated that these observations span the X levels of interest.

i:	1	2	3	4	5	6	7
X_i:	17	15	25	10	18	15	20
Y_i:	56.15	54.50	55.27	52.54	56.23	55.97	55.55

i:	8	9	10	11	12	13	14
X_i:	25	17	13	20	23	25	16
Y_i:	54.32	55.14	54.28	55.78	55.65	54.96	55.06

Assume that the second-order model (9.1a) with independent normal error terms applies.
a. Fit this model and test whether a regression relation exists. Use a level of significance of .01. State the alternatives, decision rule, and conclusion.
b. Test whether the quadratic term can be dropped from the model. Use a level of significance of .01. State the alternatives, decision rule, and conclusion.
c. Obtain the residuals and plot them against $\hat{Y}$. Also obtain a normal probability plot. What do your plots show?
d. Conduct a formal test for lack of fit using a level of significance of .01. State the alternatives, decision rule, and conclusion. Does your conclusion imply that the model cannot be improved further? Discuss.

9.16. Refer to **Computerized game** Problem 9.15. Someone who is familiar with this computerized marketing game enters the discussion. She states that in the system of equations on which the game is based, a quadratic relation does hold between promotional expenditures and mean demand in the firm's home territory. She believes that another significant variable related to expected demand in the home territory is the ratio of the firm's selling price to the average competitive selling price; however she does not recall whether this price ratio has both linear and quadratic effects. She also does not recall whether price ratio and promotional expenditures interact in affecting demand. The firm's price ratios for the 14 periods are as follows:

i:	1	2	3	4	5	6	7
Ratio:	.931	.976	1.045	.939	1.010	1.059	1.000

i:	8	9	10	11	12	13	14
Ratio:	.950	.995	1.011	1.008	.947	1.000	1.017

a. Fit the second-order polynomial regression model (9.5) with promotional expenditures (X_1) and price ratio (X_2) as independent variables. How much has R^2 increased by adding the price ratio as an independent variable?

b. Test whether the price ratio variable should be retained in the model. Control the risk of Type I error at .05. State the alternatives, decision rule, and conclusion.

c. Assuming that the price ratio variable is to be retained in the model, test whether the interaction term is needed in the model; use $\alpha = .01$. State the alternatives, decision rule, and conclusion. What is the P-value of the test?

d. The team has decided to adopt regression model (9.5) without interaction effects. Fit this model and obtain the residuals. Plot the residuals against $\hat{Y}$ and against the time order of the observations. Also, prepare a normal probability plot. Analyze these plots and state your findings.

9.17. Refer to **Mileage study** Problem 9.6.

a. At what speed is the estimated quadratic response function a maximum? What is the estimated mean mileage at this speed?

b. Does the maximum of the response function occur within the scope of the model?

9.18. Refer to **Sales forecasting** Problem 9.11.

a. In what year does the minimum of the estimated quadratic response function occur? What is the estimated mean sales for this year?

b. Does the minimum of the response function occur within the scope of the model?

EXERCISES

9.19. Consider the second-order regression model with one independent variable (9.1a) and the following two sets of X values:

Set 1:	1.0	1.5	1.1	1.3	1.9	.8	1.2	1.4
Set 2:	12	1	123	17	415	71	283	38

For each set, calculate the coefficient of correlation between X and X^2, then between x and x^2. Also calculate the coefficients of correlation between X and X^3 and between x and x^3. What generalizations are suggested by your results?

9.20. (Calculus needed.) Refer to the second-order response function (9.2). Explain precisely the meaning of the linear effect coefficient β_1 and the quadratic effect coefficient β_{11}.

9.21. a. Derive the expressions for b_0', b_1', and b_{11}' in (9.15).

b. Using theorem (6.49), obtain the variance-covariance matrix for the regression coefficients pertaining to the original X variable in terms of the variance-covariance matrix for the regression coefficients pertaining to the transformed x variable.

9.22. How are the normal equations (9.12) simplified if the X values are equally spaced, such as the time series representation $X_1 = 1, X_2 = 2, \ldots, X_n = n$?

PROJECTS

9.23. Refer to the **SMSA** data set. It is desired to fit the second-order regression model (9.1a) for relating number of active physicians (Y) against total population (X).

a. Fit the second-order regression model. Plot the residuals against the fitted values. How well does the second-order model appear to fit the data?

b. Obtain R^2 for the second-order regression model. Also obtain the coefficient of simple determination r^2 for the first-order regression model. Has the addition of the quadratic term in the regression model substantially increased the coefficient of determination?

c. Test whether the quadratic term can be dropped from the regression model; use $\alpha = .05$. State the alternatives, decision rule, and conclusion.

d. Omit observation 1 (New York City) from the data set. Fit the second-order regression model (9.1a) based on the remaining 140 SMSA's. Repeat the test in part (c). Has the omission of the outlying observation affected your conclusion about whether the quadratic term can be dropped from the model?

9.24. Refer to the **SMSA** data set. A regression model relating serious crime rate (Y, total serious crimes divided by total population) to population density (X_1, total population divided by land area) and percent of population in central cities (X_3) is to be constructed.

a. Fit the second-order regression model (9.5). Plot the residuals against the fitted values. How well does the second-order model appear to fit the data? What is R^2?

b. Test whether or not all quadratic and interaction terms can be dropped from the model; use $\alpha = .01$. State the alternatives, decision rule, and conclusion.

c. Instead of using the independent variable population density, total population (X_1) and land area (X_2) are to be employed as separate independent variables, in addition to percent of population in central cities (X_3). The regression model should contain linear and quadratic terms for total population, and linear terms only for land area and percent of population in central cities. (No interaction terms are to be included in this model.) Fit this regression model and obtain R^2. Is this coefficient of multiple determination substantially different from the one for the model in part (a)?

9.25. Refer to the **SENIC** data set. The second-order regression model (9.1a) is to be fitted for relating number of nurses (Y) to available facilities and services (X).

a. Fit the second-order regression model. Plot the residuals against the fitted values. How well does the second-order model appear to fit the data?

b. Obtain R^2 for the second-order regression model. Also obtain the coefficient of simple determination r^2 for the first-order regression model. Has the addition of the quadratic term in the regression model substantially increased the coefficient of determination?

c. Test whether the quadratic term can be dropped from the regression model; use $\alpha = .10$. State the alternatives, decision rule, and conclusion.

9.26. Refer to **Sales forecasting** Problem 9.11. Instead of using a polynomial regression model here, it has been suggested that a transformation of variables might yield an equally good fit and would be more desirable since forecasting requires extrapolation.

a. Fit a linear regression model for relating $Y' = \sqrt{Y}$ against X. Plot the fitted regression function and the transformed data. How effective does the use of the transformed variable appear to be here?

b. Obtain the fitted values and transform them to the original variable Y. Then calculate the residuals in the original variable. Plot these residuals against X and analyze your plot.

c. Square the residuals in the original variable obtained in part (b), sum, and obtain MSE. Compare this with MSE for the quadratic model in Problem 9.11a. How does the variability around the fitted regression function, as measured by MSE, compare for the two approaches?

d. Repeat parts (a) through (c) using the transformation $Y' = \log_{10} Y$. Is either the square-root or the logarithmic transformation clearly preferable here?

CITED REFERENCES

9.1 Williams, E. J. *Regression Analysis*. New York: John Wiley & Sons, 1959.

9.2 Draper, N. R., and H. Smith. *Applied Regression Analysis*. 2d ed. New York: John Wiley & Sons, 1981.

10

Indicator variables

Throughout the previous chapters on regression analysis, we have utilized quantitative variables in the regression models considered. Quantitative variables take on values on a well-defined scale; examples are income, age, temperature, and amounts of liquid assets.

Many variables of interest in business, economics, and the social and biological sciences, however, are not quantitative but are qualitative. Examples of qualitative variables are sex (male, female), purchase status (purchase, no purchase), and disability status (not disabled, partly disabled, fully disabled).

Qualitative variables can be used in a multiple regression model just as quantitative variables can, as we shall explain in this chapter. First, we take up the case where some or all of the independent variables are qualitative. Then we turn to the case where the dependent variable is qualitative.

10.1 ONE INDEPENDENT QUALITATIVE VARIABLE

An economist wished to relate the speed with which a particular insurance innovation is adopted (Y) to the size of the insurance firm (X_1) and the type of firm. The dependent variable is measured by the number of months elapsed between the time the first firm adopted the innovation and the time the given firm adopted the innovation. The first independent variable, size of firm, is quantitative, and is measured by the amount of total assets of the firm. The second independent variable, type of firm, is qualitative and is composed of two classes—stock companies and mutual companies. In order that such a qualitative variable can be used in a regression model, quantitative indicators for the classes of the qualitative variable must be found.

Indicator variables

There are many ways of quantitatively identifying the classes of a qualitative variable. We shall use indicator variables that take on the values 0 and 1. These indicator variables are easy to use and are widely employed, but they are by no means the only way to quantify a qualitative variable.

For our example, where the qualitative variable has two classes, we might define two indicator variables X_2 and X_3 as follows:

$$(10.1) \qquad \begin{aligned} X_2 &= \begin{matrix} 1 \text{ if stock company} \\ 0 \text{ otherwise} \end{matrix} \\ X_3 &= \begin{matrix} 1 \text{ if mutual company} \\ 0 \text{ otherwise} \end{matrix} \end{aligned}$$

Assuming that a first-order model is to be employed, it would be:

$$(10.2) \qquad Y_i = \beta_0 X_{i0} + \beta_1 X_{i1} + \beta_2 X_{i2} + \beta_3 X_{i3} + \varepsilon_i \qquad \text{where } X_{i0} \equiv 1$$

This intuitive approach of setting up an indicator variable for each class of the qualitative variable unfortunately leads to computational difficulties. To see why, suppose we have $n = 4$ observations, the first two being stock firms for which $X_2 = 1$ and $X_3 = 0$, and the second two being mutual firms for which $X_2 = 0$ and $X_3 = 1$. The **X** matrix would then be:

$$\begin{matrix} & X_0 & X_1 & X_2 & X_3 \end{matrix}$$
$$\mathbf{X} = \begin{bmatrix} 1 & X_{11} & 1 & 0 \\ 1 & X_{21} & 1 & 0 \\ 1 & X_{31} & 0 & 1 \\ 1 & X_{41} & 0 & 1 \end{bmatrix}$$

Note that the X_0 column is equal to the sum of the X_2 and X_3 columns, so that the columns are linearly dependent according to definition (6.22). This has a serious effect on the $\mathbf{X'X}$ matrix:

$$\mathbf{X'X} = \begin{bmatrix} 1 & 1 & 1 & 1 \\ X_{11} & X_{21} & X_{31} & X_{41} \\ 1 & 1 & 0 & 0 \\ 0 & 0 & 1 & 1 \end{bmatrix} \begin{bmatrix} 1 & X_{11} & 1 & 0 \\ 1 & X_{21} & 1 & 0 \\ 1 & X_{31} & 0 & 1 \\ 1 & X_{41} & 0 & 1 \end{bmatrix}$$

$$
= \begin{bmatrix}
4 & \sum_{i=1}^{4} X_{i1} & 2 & 2 \\
\sum_{i=1}^{4} X_{i1} & \sum_{i=1}^{4} X_{i1}^2 & \sum_{i=1}^{2} X_{i1} & \sum_{i=3}^{4} X_{i1} \\
2 & \sum_{i=1}^{2} X_{i1} & 2 & 0 \\
2 & \sum_{i=3}^{4} X_{i1} & 0 & 2
\end{bmatrix}
$$

It is quickly apparent that the first column of the $\mathbf{X}'\mathbf{X}$ matrix equals the sum of the last two columns, so that the columns are linearly dependent. Hence, the $\mathbf{X}'\mathbf{X}$ matrix does not have an inverse, and no unique estimators of the regression coefficients can be found.

A simple way out of this difficulty is to drop one of the indicator variables. In our example, for instance, we might drop X_3. While dropping one indicator variable is not the only way out of the difficulty, it leads to simple interpretations of the parameters. In general, therefore, we shall follow the principle:

(10.3) A qualitative variable with c classes will be represented by $c - 1$ indicator variables, each taking on the values 0 and 1.

Note

Indicator variables are frequently also called *dummy variables* or *binary variables*. The latter term has reference to the binary number system containing only 0 and 1.

Interpretation of regression parameters

Returning to our example, suppose that we drop the indicator variable X_3 from model (10.2) so that the model becomes:

(10.4) $$Y_i = \beta_0 + \beta_1 X_{i1} + \beta_2 X_{i2} + \varepsilon_i$$

where:

$$X_{i1} = \text{Size of firm}$$

$$X_{i2} = \begin{matrix} 1 \text{ if stock company} \\ 0 \text{ otherwise} \end{matrix}$$

The response function for this model is:

(10.5) $$E(Y) = \beta_0 + \beta_1 X_1 + \beta_2 X_2$$

To understand the meaning of the parameters of this model, consider first the case of a mutual firm. For such a firm, $X_2 = 0$ and we have:

(10.5a) $$E(Y) = \beta_0 + \beta_1 X_1 + \beta_2(0) = \beta_0 + \beta_1 X_1 \qquad \text{Mutual firms}$$

Thus, the response function for mutual firms is a straight line, with Y intercept β_0 and slope β_1. This response function is shown in Figure 10.1.

For a stock firm, $X_2 = 1$ and the response function (10.5) is:

$$(10.5b) \qquad E(Y) = \beta_0 + \beta_1 X_1 + \beta_2(1) = (\beta_0 + \beta_2) + \beta_1 X_1 \qquad \text{Stock firms}$$

This also is a straight line, with the same slope β_1 but with Y intercept $\beta_0 + \beta_2$. This response function is also shown in Figure 10.1.

The meaning of the parameters in the response function (10.5) is now clear. With reference to our earlier example, the mean time elapsed before the innovation is adopted, $E(Y)$, is a linear function of size of firm (X_1), with the same slope β_1 for both types of firms. β_2 indicates how much higher (lower) the response function for stock firms is than the one for mutual firms. Thus, β_2 measures the differential effect of type of firm. In general, β_2 shows how much higher (lower) the mean response line is for the class coded 1 than the line for the class coded 0.

FIGURE 10.1 Illustration of meaning of regression parameters for model (10.4) with indicator variable X_2—insurance innovation example

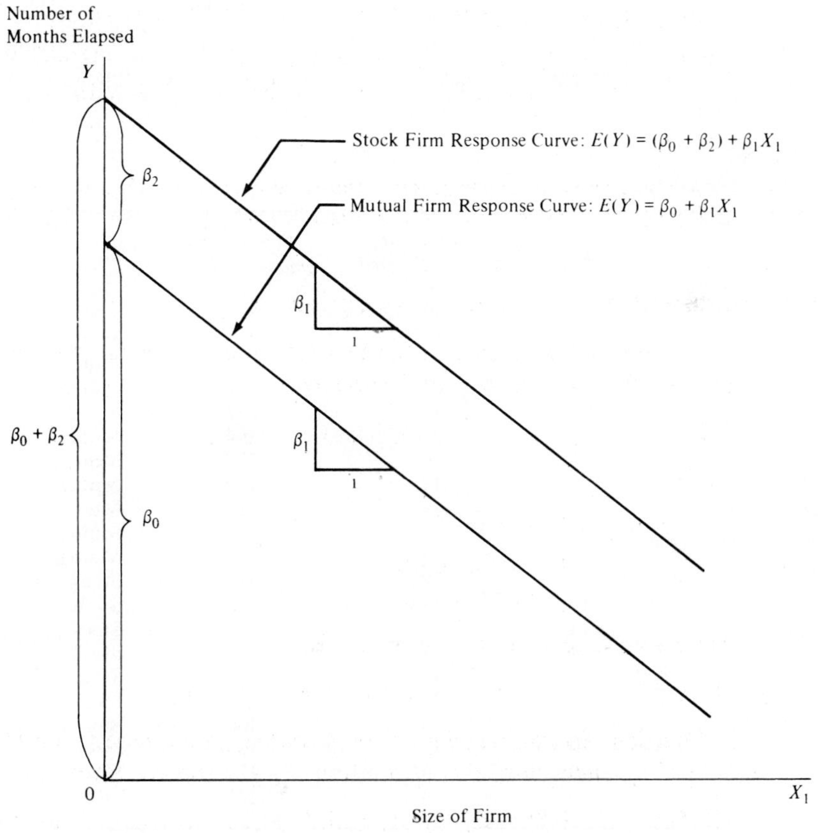

Example

With reference to our earlier illustration, the economist studied 10 mutual firms and 10 stock firms. The data are shown in Table 10.1. The **Y** and **X** data matrices are shown in Table 10.2. Note that $X_2 = 1$ for each stock firm and $X_2 = 0$ for each mutual firm.

Given the **Y** and **X** matrices in Table 10.2, fitting the regression model (10.4) is straightforward. Table 10.3 presents the key results from a computer run. The fitted response function is:

$$\hat{Y} = 33.87407 - .10174X_1 + 8.05547X_2$$

Figure 10.2 (p. 334) contains the fitted response function for each type of firm, together with the actual observations.

The economist was most interested in the effect of type of firm (X_2) on the elapsed time for the innovation to be adopted. He therefore desired to obtain a 95 percent confidence interval for β_2. We require $t(.975; 17) = 2.110$ and obtain from the data in Table 10.3:

$$4.97675 = 8.05547 - 2.110(1.45911) \leq \beta_2 \leq 8.05547 + 2.110(1.45911) = 11.13419$$

Thus, with 95 percent confidence, we conclude that stock companies tend to adopt the particular innovation studied somewhere between 5 and 11 months later, on the average, than mutual companies.

TABLE 10.1 Data for insurance innovation study

Firm i	*Number of Months Elapsed* Y_i	*Size of Firm (million dollars)* X_{i1}	*Type of Firm*
1	17	151	Mutual
2	26	92	Mutual
3	21	175	Mutual
4	30	31	Mutual
5	22	104	Mutual
6	0	277	Mutual
7	12	210	Mutual
8	19	120	Mutual
9	4	290	Mutual
10	16	238	Mutual
11	28	164	Stock
12	15	272	Stock
13	11	295	Stock
14	38	68	Stock
15	31	85	Stock
16	21	224	Stock
17	20	166	Stock
18	13	305	Stock
19	30	124	Stock
20	14	246	Stock

TABLE 10.2 Data matrices for insurance innovation study

$$\mathbf{Y} = \begin{bmatrix} 17 \\ 26 \\ 21 \\ 30 \\ 22 \\ 0 \\ 12 \\ 19 \\ 4 \\ 16 \\ 28 \\ 15 \\ 11 \\ 38 \\ 31 \\ 21 \\ 20 \\ 13 \\ 30 \\ 14 \end{bmatrix} \qquad \mathbf{X} = \begin{bmatrix} 1 & 151 & 0 \\ 1 & 92 & 0 \\ 1 & 175 & 0 \\ 1 & 31 & 0 \\ 1 & 104 & 0 \\ 1 & 277 & 0 \\ 1 & 210 & 0 \\ 1 & 120 & 0 \\ 1 & 290 & 0 \\ 1 & 238 & 0 \\ 1 & 164 & 1 \\ 1 & 272 & 1 \\ 1 & 295 & 1 \\ 1 & 68 & 1 \\ 1 & 85 & 1 \\ 1 & 224 & 1 \\ 1 & 166 & 1 \\ 1 & 305 & 1 \\ 1 & 124 & 1 \\ 1 & 246 & 1 \end{bmatrix}$$

(Columns of **X**: X_0, X_1, X_2)

TABLE 10.3 Regression results for model (10.4) fit—insurance innovation example

(a) Regression Coefficients

Regression Coefficient	*Estimated Regression Coefficient*	*Estimated Standard Deviation*	t^*
β_0	33.87407	1.81386	18.68
β_1	−.10174	.00889	−11.44
β_2	8.05547	1.45911	5.52

(b) Analysis of Variance

Source of Variation	*SS*	*df*	*MS*
Regression	1,504.41	2	752.20
Error	176.39	17	10.38
Total	1,680.80	19	

A formal test of:

$$H_0: \beta_2 = 0$$
$$H_a: \beta_2 \neq 0$$

with level of significance .05 would lead to H_a, that type of firm has an effect, since the confidence interval for β_2 does not include zero.

FIGURE 10.2 Fitted regression functions for model (10.4)—insurance innovation example

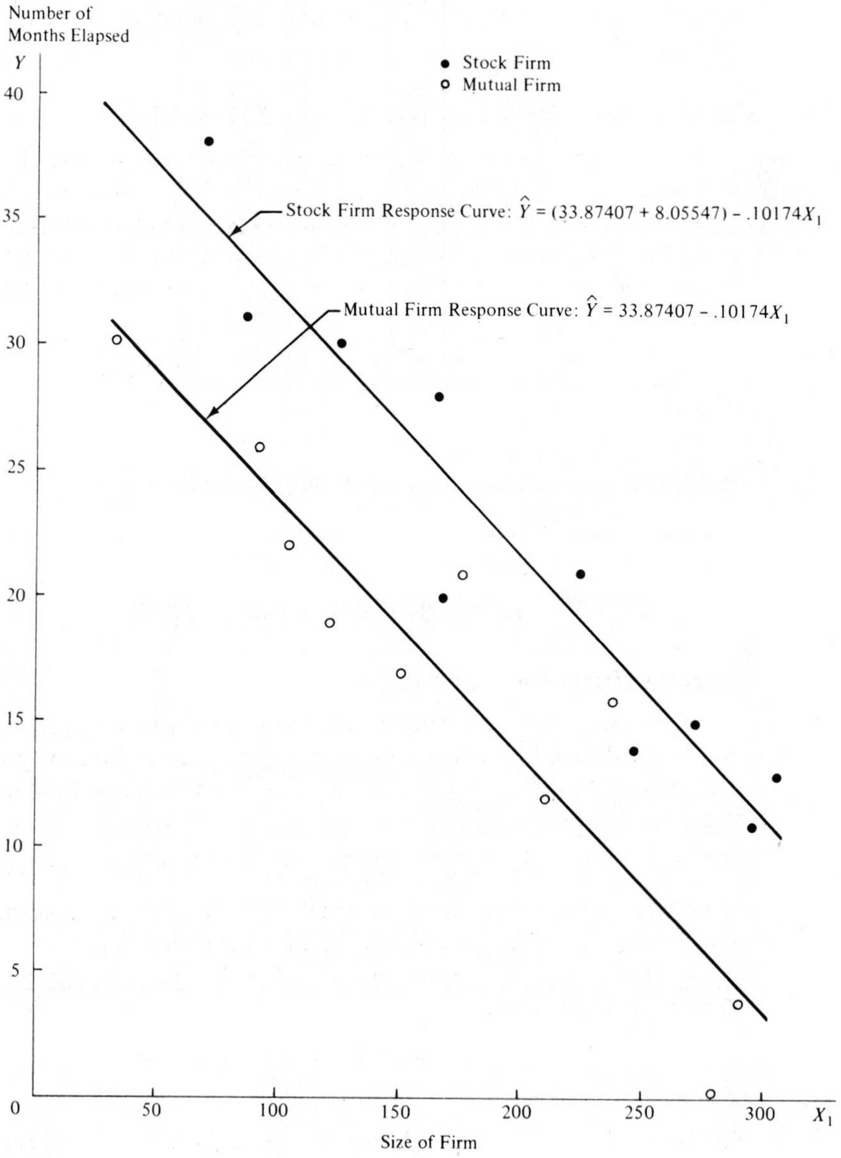

The economist also carried out other analyses, some of which will be described shortly.

Note

The reader may wonder why we did not simply fit separate regressions for stock firms and mutual firms in our example, and instead adopted the approach of fitting one regression with an indicator variable. There are two reasons for this. Since the model assumed equal slopes and the same constant error term variance for each type of firm, the common slope β_1 can best be estimated by pooling the two types of firms. Also, other inferences, such as ones pertaining to β_0 and β_2, can be made more precisely by working with one regression model containing an indicator variable since more degrees of freedom will then be associated with *MSE*.

10.2 MODEL CONTAINING INTERACTION EFFECTS

In our earlier illustration, the economist actually did not begin his analysis with model (10.4) since he expected interaction effects between size and type of firm. Even though one of the independent variables in the regression model is qualitative, interaction effects are introduced into the model in the usual manner, by including cross-product terms. A first-order model with an interaction term for our example is:

(10.6) $$Y_i = \beta_0 + \beta_1 X_{i1} + \beta_2 X_{i2} + \beta_3 X_{i1} X_{i2} + \varepsilon_i$$

where:

$$X_{i1} = \text{Size of firm}$$

$$X_{i2} = \begin{matrix} 1 \text{ if stock company} \\ 0 \text{ otherwise} \end{matrix}$$

The response function for this model is:

(10.7) $$E(Y) = \beta_0 + \beta_1 X_1 + \beta_2 X_2 + \beta_3 X_1 X_2$$

Meaning of regression parameters

The meaning of the regression parameters in the response function (10.7) can best be understood by examining the nature of this function for each type of firm. For a mutual firm, $X_2 = 0$ and hence $X_1 X_2 = 0$. The response function for mutual firms therefore is:

(10.7a) $$E(Y) = \beta_0 + \beta_1 X_1 + \beta_2(0) + \beta_3(0) = \beta_0 + \beta_1 X_1 \quad \text{Mutual firms}$$

This response function is shown in Figure 10.3. Note that the Y intercept is β_0 and the slope is β_1 for the response function for mutual firms.

For stock firms, $X_2 = 1$ and hence $X_1 X_2 = X_1$. The response function for stock firms therefore is:

$$E(Y) = \beta_0 + \beta_1 X_1 + \beta_2(1) + \beta_3 X_1$$

or:

(10.7b) $$E(Y) = (\beta_0 + \beta_2) + (\beta_1 + \beta_3) X_1 \qquad \text{Stock firms}$$

FIGURE 10.3 Illustration of meaning of regression parameters for model (10.6) with indicator variable X_2 and interaction term—insurance innovation example

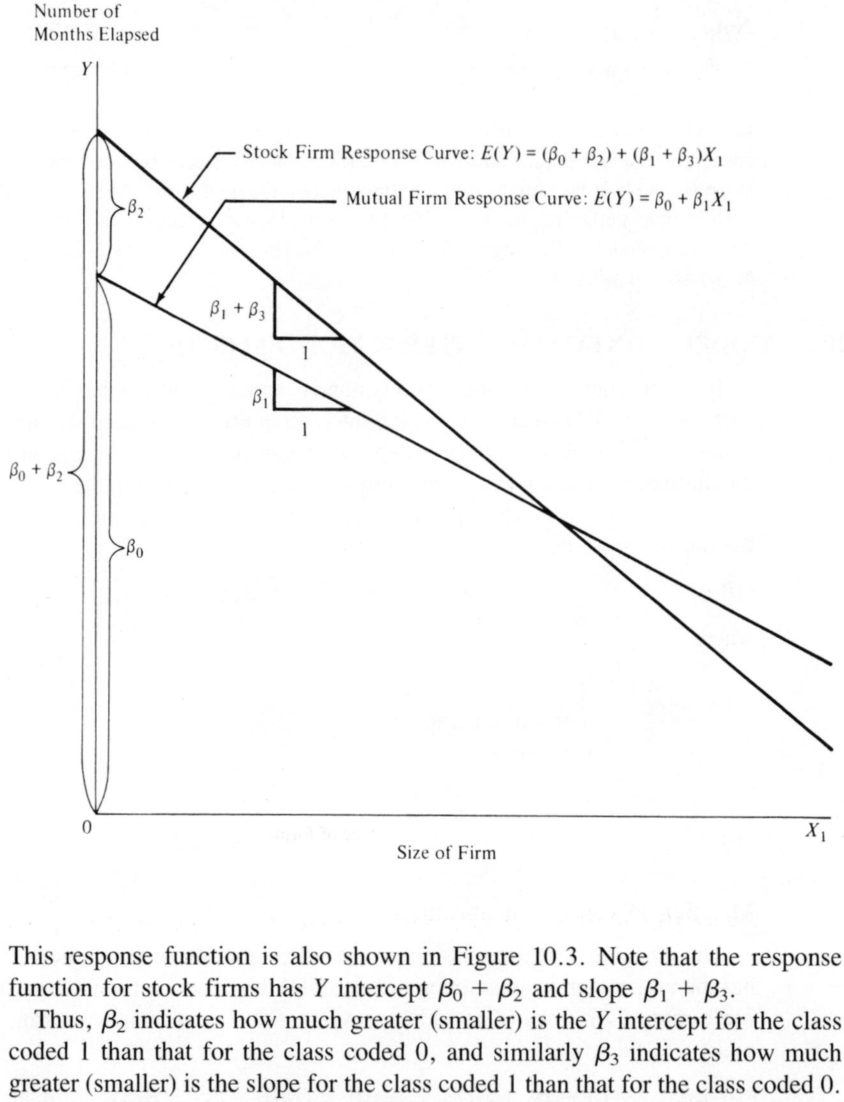

This response function is also shown in Figure 10.3. Note that the response function for stock firms has Y intercept $\beta_0 + \beta_2$ and slope $\beta_1 + \beta_3$.

Thus, β_2 indicates how much greater (smaller) is the Y intercept for the class coded 1 than that for the class coded 0, and similarly β_3 indicates how much greater (smaller) is the slope for the class coded 1 than that for the class coded 0. Because both the intercept and the slope differ for the two classes in model (10.6), it is no longer true that β_2 indicates how much higher (lower) one response line is than the other. Figure 10.3 makes it clear that the effect of type of firm with model (10.6) depends on the size of the firm. For smaller firms, according to Figure 10.3, mutual firms tend to innovate more quickly, but for larger firms stock firms tend to innovate more quickly. Thus, when interaction effects are present, the effect of the qualitative variable can only be studied by comparing the regression functions for each class of the qualitative variable.

FIGURE 10.4 Another illustration of model (10.6) with indicator variable X_2 and interaction term—insurance innovation example

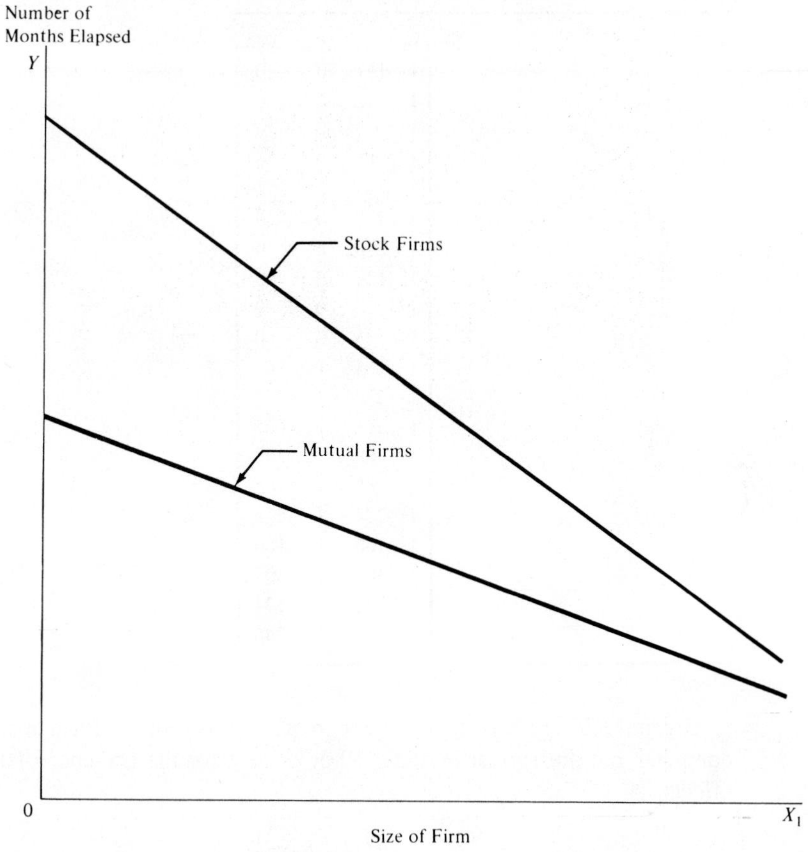

Figure 10.4 illustrates another possible interaction situation. Here, mutual firms tend to introduce the innovation more quickly than stock firms for all sizes of firms in the scope of the model, but the differential effect is much smaller for large firms than for small ones.

Example

Since the economist anticipated that interaction effects between size and type of firm may be present, he actually first wished to fit model (10.6):

$$Y_i = \beta_0 + \beta_1 X_{i1} + \beta_2 X_{i2} + \beta_3 X_{i1} X_{i2} + \varepsilon_i$$

Table 10.4 shows the **X** matrix for this model. The **Y** matrix is the same as in Table 10.2. Note that the X_1X_2 column in the **X** matrix in Table 10.4 contains 0 for mutual companies and X_{i1} for stock companies.

TABLE 10.4 **X** matrix for fitting model (10.6) with interaction term—insurance innovation example

$$
\mathbf{X} = \begin{array}{c} \\ \end{array}
\begin{bmatrix}
X_0 & X_1 & X_2 & X_1X_2 \\
1 & 151 & 0 & 0 \\
1 & 92 & 0 & 0 \\
1 & 175 & 0 & 0 \\
1 & 31 & 0 & 0 \\
1 & 104 & 0 & 0 \\
1 & 277 & 0 & 0 \\
1 & 210 & 0 & 0 \\
1 & 120 & 0 & 0 \\
1 & 290 & 0 & 0 \\
1 & 238 & 0 & 0 \\
1 & 164 & 1 & 164 \\
1 & 272 & 1 & 272 \\
1 & 295 & 1 & 295 \\
1 & 68 & 1 & 68 \\
1 & 85 & 1 & 85 \\
1 & 224 & 1 & 224 \\
1 & 166 & 1 & 166 \\
1 & 305 & 1 & 305 \\
1 & 124 & 1 & 124 \\
1 & 246 & 1 & 246
\end{bmatrix}
$$

Given the **Y** and **X** matrices, the regression fit is routine. Basic results from a computer run are shown in Table 10.5. To test for the presence of interaction effects:

$$
\begin{aligned}
H_0&: \beta_3 = 0 \\
H_a&: \beta_3 \neq 0
\end{aligned}
$$

the economist used the t^* statistic from Table 10.5a:

$$
t^* = \frac{b_3}{s(b_3)} = \frac{-.0004171}{.01833} = -.02
$$

For level of significance .05, we require $t(.975; 16) = 2.120$. Since $|t^*| = .02 \leq 2.120$, we conclude H_0, i.e., $\beta_3 = 0$, so that no interaction effects are present. The two-sided P-value for the test is very high, namely, .98. It was because of this result that the economist adopted model (10.4) with no interaction term, which we discussed earlier.

Note

Fitting model (10.6) yields the same response functions as fitting separate regressions for stock firms and mutual firms. An advantage of using model (10.6) with an indicator

TABLE 10.5 Regression results for fit of model (10.6) with interaction term—insurance innovation example

(a) Regression Coefficients

Regression Coefficient	*Estimated Regression Coefficient*	*Estimated Standard Deviation*	t^*
β_0	33.83837	2.44065	13.86
β_1	−.10153	.01305	−7.78
β_2	8.13125	3.65405	2.23
β_3	−.0004171	.01833	−.02

(b) Analysis of Variance

Source of Variation	*SS*	*df*	*MS*
Regression	1,504.42	3	501.47
Error	176.38	16	11.02
Total	1,680.80	19	

variable is that one regression run on the computer will yield both fitted regressions.

Another advantage is that tests for comparing the regression functions for the different classes of the qualitative variable can be clearly seen to be tests of regression coefficients in a general linear model. For instance, Figure 10.3 makes it clear for our example that the test whether the two regression functions have the same slope involves:

$$H_0: \beta_3 = 0$$
$$H_a: \beta_3 \neq 0$$

Similarly, the test whether the two regression functions in our example are identical would involve:

$$H_0: \beta_2 = \beta_3 = 0$$
$$H_a: \text{not both } \beta_2 = 0 \text{ and } \beta_3 = 0$$

10.3 MORE COMPLEX MODELS

We now briefly consider more complex models involving qualitative independent variables.

Qualitative variable with more than two classes

If a qualitative independent variable has more than two classes, we require additional indicator variables in the regression model. Consider the regression of tool wear (Y) on tool speed (X_1), where we wish to include also tool model (M1, M2, M3, M4) as an independent variable. Since the qualitative variable (tool model) has four classes, we require three indicator variables. Let us define them as follows:

$$X_2 = \begin{cases} 1 \text{ if tool model M1} \\ 0 \text{ otherwise} \end{cases}$$

(10.8)
$$X_3 = \begin{cases} 1 \text{ if tool model M2} \\ 0 \text{ otherwise} \end{cases}$$

$$X_4 = \begin{cases} 1 \text{ if tool model M3} \\ 0 \text{ otherwise} \end{cases}$$

First-order model. A first-order model is:

(10.9) $$Y_i = \beta_0 + \beta_1 X_{i1} + \beta_2 X_{i2} + \beta_3 X_{i3} + \beta_4 X_{i4} + \varepsilon_i$$

For this model, the data input for the **X** matrix would be as follows:

Tool Model	X_0	X_1	X_2	X_3	X_4
M1	1	X_{i1}	1	0	0
M2	1	X_{i1}	0	1	0
M3	1	X_{i1}	0	0	1
M4	1	X_{i1}	0	0	0

The response function for model (10.9) is:

(10.10) $$E(Y) = \beta_0 + \beta_1 X_1 + \beta_2 X_2 + \beta_3 X_3 + \beta_4 X_4$$

To see the meaning of the regression parameters, consider first the response function for tool models M4 for which $X_2 = 0$, $X_3 = 0$, and $X_4 = 0$:

(10.10a) $$E(Y) = \beta_0 + \beta_1 X_1 \qquad \text{Tool models M4}$$

For tool models M1, $X_2 = 1$, $X_3 = 0$, and $X_4 = 0$, and the response function is:

(10.10b) $$E(Y) = \beta_0 + \beta_1 X_1 + \beta_2 = (\beta_0 + \beta_2) + \beta_1 X_1 \qquad \text{Tool models M1}$$

Similarly, the response functions for tool models M2 and M3 are:

(10.10c) $$E(Y) = (\beta_0 + \beta_3) + \beta_1 X_1 \qquad \text{Tool models M2}$$

(10.10d) $$E(Y) = (\beta_0 + \beta_4) + \beta_1 X_1 \qquad \text{Tool models M3}$$

Thus, the response function (10.10) implies that the regression of tool wear on tool speed is linear, with the same slope for all types of tool models. The coefficients β_2, β_3, and β_4 indicate, respectively, how much higher (lower) the response functions for tool models M1, M2, and M3 are than the one for tool models M4. Thus, β_2, β_3, and β_4 measure the differential effects of the qualitative variable classes on the height of the response function, always compared with the class for which $X_2 = X_3 = X_4 = 0$. Figure 10.5 illustrates a possible arrangement of the response functions.

When using model (10.9), one may wish to estimate differential effects other than against tool models M4. For instance, $\beta_4 - \beta_3$ measures how much higher (lower) the response function for tool models M3 is than the response function for tool models M2, as may be seen by comparing (10.10c) and (10.10d). The point estimator of this quantity is, of course, $b_4 - b_3$, and the estimated variance

FIGURE 10.5 Illustration of model (10.9)—tool wear example

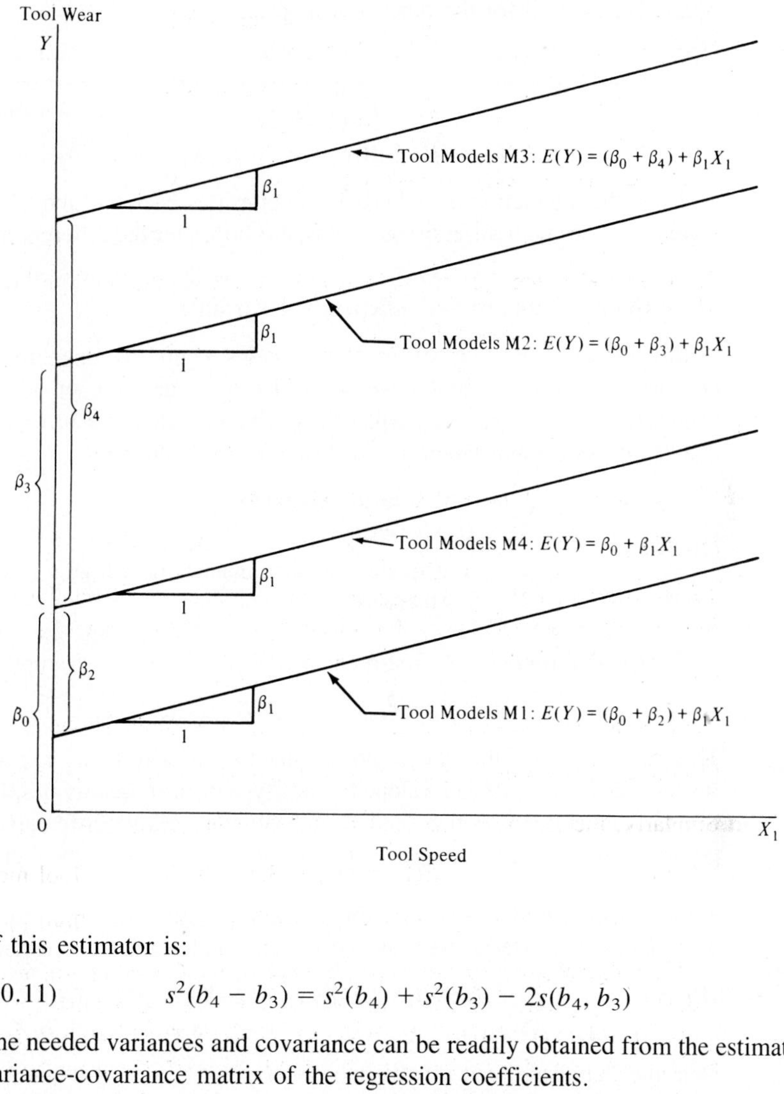

of this estimator is:

$$s^2(b_4 - b_3) = s^2(b_4) + s^2(b_3) - 2s(b_4, b_3) \tag{10.11}$$

The needed variances and covariance can be readily obtained from the estimated variance-covariance matrix of the regression coefficients.

First-order model with interactions added. If interaction effects between tool speed and tool model are present in our previous illustration, model (10.9) would be modified as follows:

$$Y_i = \beta_0 + \beta_1 X_{i1} + \beta_2 X_{i2} + \beta_3 X_{i3} + \beta_4 X_{i4} + \beta_5 X_{i1} X_{i2} + \beta_6 X_{i1} X_{i3} + \beta_7 X_{i1} X_{i4} + \varepsilon_i \tag{10.12}$$

The response function for tool models M4, for which $X_2 = 0$, $X_3 = 0$, and $X_4 = 0$, is as follows:

(10.13a) $$E(Y) = \beta_0 + \beta_1 X_1 \qquad \text{Tool models M4}$$

Similarly, we find for the other tool models:

(10.13b) $$E(Y) = (\beta_0 + \beta_2) + (\beta_1 + \beta_5)X_1 \qquad \text{Tool models M1}$$

(10.13c) $$E(Y) = (\beta_0 + \beta_3) + (\beta_1 + \beta_6)X_1 \qquad \text{Tool models M2}$$

(10.13d) $$E(Y) = (\beta_0 + \beta_4) + (\beta_1 + \beta_7)X_1 \qquad \text{Tool models M3}$$

Thus, the interaction model (10.12) implies that each tool model has its own regression line, with different intercepts and slopes for the different tool models.

More than one qualitative independent variable

Models can readily be constructed for cases where two or more of the independent variables are qualitative. Consider the regression of advertising expenditures (Y) on sales (X_1), type of firm (incorporated, not incorporated), and quality of sales management (high, low). We may define:

(10.14)
$$X_2 = \begin{cases} 1 & \text{if firm incorporated} \\ 0 & \text{otherwise} \end{cases}$$
$$X_3 = \begin{cases} 1 & \text{if quality of sales management high} \\ 0 & \text{otherwise} \end{cases}$$

First-order model. A first-order model for the above example is:

(10.15) $$Y_i = \beta_0 + \beta_1 X_{i1} + \beta_2 X_{i2} + \beta_3 X_{i3} + \varepsilon_i$$

This model implies that the response function of advertising expenditures on sales is linear, with the same slope for all "type of firm–quality of sales management" combinations, and β_2 and β_3 indicate the additive differential effects of type of firm and quality of sales management on the height of the regression line.

First-order model with certain interactions added. A first-order model with interaction effects between pairs of the independent variables added is:

(10.16) $$Y_i = \beta_0 + \beta_1 X_{i1} + \beta_2 X_{i2} + \beta_3 X_{i3} + \beta_4 X_{i1} X_{i2} + \beta_5 X_{i1} X_{i3} + \beta_6 X_{i2} X_{i3} + \varepsilon_i$$

Note the implications of this model:

Type of Firm	*Quality of Sales Management*	*Response Function*
Incorp.	High	$E(Y) = (\beta_0 + \beta_2 + \beta_3 + \beta_6) + (\beta_1 + \beta_4 + \beta_5)X_1$
Not incorp.	High	$E(Y) = (\beta_0 + \beta_3) + (\beta_1 + \beta_5)X_1$
Incorp.	Low	$E(Y) = (\beta_0 + \beta_2) + (\beta_1 + \beta_4)X_1$
Not incorp.	Low	$E(Y) = \beta_0 + \beta_1 X_1$

Not only are all response functions different for the various "type of firm–quality of sales management" combinations, but the differential effects of one qualitative variable on the intercept depend on the particular class of the other qualitative variable. For instance, when we move from "not incorporated–low quality" to "incorporated–low quality," the intercept changes by β_2. But if we move from "not incorporated–high quality" to "incorporated–high quality," the intercept changes by $\beta_2 + \beta_6$.

Qualitative independent variables only

Regression models containing only qualitative independent variables can also be constructed. With reference to our previous example, we could regress advertising expenditures only on type of firm and quality of sales management. The first-order model then would be:

(10.17) $$Y_i = \beta_0 + \beta_2 X_{i2} + \beta_3 X_{i3} + \varepsilon_i$$

where X_{i2} and X_{i3} are defined in (10.14).

Comments

1. Models in which all independent variables are qualitative are called *analysis of variance models*.

2. Models containing some quantitative and some qualitative independent variables, where the chief independent variables of interest are qualitative and the quantitative independent variables are introduced primarily to reduce the variance of the error terms, are called *covariance models*.

10.4 OTHER USES OF INDEPENDENT INDICATOR VARIABLES

Comparison of two or more regression functions

Frequently we encounter regressions for two or more populations and wish to examine their similarities and differences. We present two examples.

1. An economist is studying the relation between amount of savings and level of income for middle-income families from urban and rural areas, based on independent samples from the two populations. Each of the two relations can be modeled by linear regression. She wishes to compare whether, at given income levels, urban and rural families tend to save the same amount—i.e., whether the two regression lines are the same. If they are not, she wishes to explore whether at least the amounts of savings out of an additional dollar of income are the same for the two groups—i.e., whether the slopes of the two regression lines are the same.

2. A company has two instruments constructed to identical specifications to measure pressure in an industrial process. A study has been made for each instrument of the relation between its gauge readings and actual pressures as determined by an almost exact but slow and costly method. If the two regression lines are the same, a single calibration schedule can be developed for the two instruments; otherwise, two different calibration schedules would be required.

When it is reasonable to assume that the error term variances in the regression models for the different populations are equal, use of indicator variables permits us to test the equality of the different regression functions. If the error variances are not equal, transformations may equalize them at least approximately.

We have already seen that regression models with indicator variables that contain interaction terms permit testing of the equality of regression functions for the different classes of a qualitative variable. This methodology can be used directly for testing the equality of regression functions for different populations. One simply considers the different populations studied as classes of an independent variable, defines indicator variables for the different populations, and develops a regression model containing appropriate interaction terms. Since no new principles arise in the testing of the equality of regression functions for different populations, we shall utilize the two earlier examples to illustrate the approach.

Example 1—Savings study. In the savings study example, the economist was willing to assume that the regression relation between savings (Y) and income (X_1) for middle-income families is linear for both rural and urban families, and that the error term variances for the two populations are the same. Since she wished to test whether the two regression lines are the same, she fitted model (10.6) which permits both the slopes and the intercepts to be different in the two regressions:

$$Y_i = \beta_0 + \beta_1 X_{i1} + \beta_2 X_{i2} + \beta_3 X_{i1} X_{i2} + \varepsilon_i \tag{10.18}$$

where:

$$X_{i1} = \text{Family income}$$

$$X_{i2} = \begin{matrix} 1 \text{ if urban family} \\ 0 \text{ otherwise} \end{matrix}$$

Identity of the regression functions is tested by considering the alternatives:

$$\begin{aligned} &H_0\colon \beta_2 = \beta_3 = 0 \\ &H_a\colon \text{Not both } \beta_2 = 0 \text{ and } \beta_3 = 0 \end{aligned} \tag{10.19}$$

The appropriate test statistic is given by (8.32a):

$$F^* = \frac{SSR(X_2, X_1X_2|X_1)}{2} \div \frac{SSE(X_1, X_2, X_1X_2)}{n - 4} \tag{10.19a}$$

where n represents the combined sample size for both populations.

If the economist only wished to test whether the slopes of the regression lines are the same, the alternatives would be:

$$\begin{aligned} &H_0\colon \beta_3 = 0 \\ &H_a\colon \beta_3 \neq 0 \end{aligned} \tag{10.20}$$

and the appropriate test statistic is either the t^* statistic (8.25) or the partial F test statistic (8.24):

(10.20a) $$F^* = \frac{SSR(X_1X_2|X_1, X_2)}{1} \div \frac{SSE(X_1, X_2, X_1X_2)}{n-4}$$

If the economist wishes to estimate the difference in the slopes of the regression lines for urban and rural families, she would construct a confidence interval for β_3 in the usual fashion.

Example 2—Instrument calibration study. The engineer making the calibration study believed that the regression functions relating gauge reading (Y) and actual pressure (X_1) for both instruments are second-order polynomials:

$$E(Y) = \beta_0 + \beta_1X_1 + \beta_2X_1^2$$

but that they might differ for the two instruments. Hence, he employed the model (using a deviation variable for X_1 to reduce multicollinearity problems—see Chapter 9):

(10.21) $$Y_i = \beta_0 + \beta_1x_{i1} + \beta_2x_{i1}^2 + \beta_3X_{i2} + \beta_4x_{i1}X_{i2} + \beta_5x_{i1}^2X_{i2} + \varepsilon_i$$

where:

$$x_{i1} = X_{i1} - \overline{X}_1 = \text{deviation of actual pressure}$$

$$X_{i2} = \begin{cases} 1 & \text{if instrument B} \\ 0 & \text{otherwise} \end{cases}$$

Note that for instrument A, where $X_2 = 0$, the response function is:

(10.22a) $$E(Y) = \beta_0 + \beta_1x_1 + \beta_2x_1^2$$

and for instrument B, where $X_2 = 1$, the response function is:

(10.22b) $$E(Y) = (\beta_0 + \beta_3) + (\beta_1 + \beta_4)x_1 + (\beta_2 + \beta_5)x_1^2$$

Hence, the test for equality of the two response functions involves the alternatives:

(10.23) $$\begin{aligned} &H_0: \beta_3 = \beta_4 = \beta_5 = 0 \\ &H_a: \text{Not all three } \beta_k = 0 \; (k = 3, 4, 5) \end{aligned}$$

and the appropriate test statistic is (8.32a):

(10.23a) $$F^* = \frac{SSR(X_2, x_1X_2, x_1^2X_2|x_1, x_1^2)}{3} \div \frac{SSE(x_1, x_1^2, X_2, x_1X_2, x_1^2X_2)}{n-6}$$

where n represents the combined sample size for both populations.

Comments

1. The approach just described is completely general. If three or more populations are involved, additional indicator variables would simply be added to the model.

2. The use of indicator variables for testing whether two or more regression functions are the same is equivalent to the general linear test approach where fitting the full model involves fitting separate regressions to the data from each population and fitting the reduced model involves fitting one regression to the combined data.

Piecewise linear regression

Sometimes the regression of Y on X follows a particular linear relation in some range of X, but follows a different linear relation elsewhere. For instance, unit cost (Y) regressed on lot size may follow a certain linear regression up to $X_p =$ 500, at which point the slope changes because of some operating efficiencies only possible with lot sizes of more than 500. Figure 10.6 illustrates this situation.

FIGURE 10.6 Illustration of piecewise linear regression

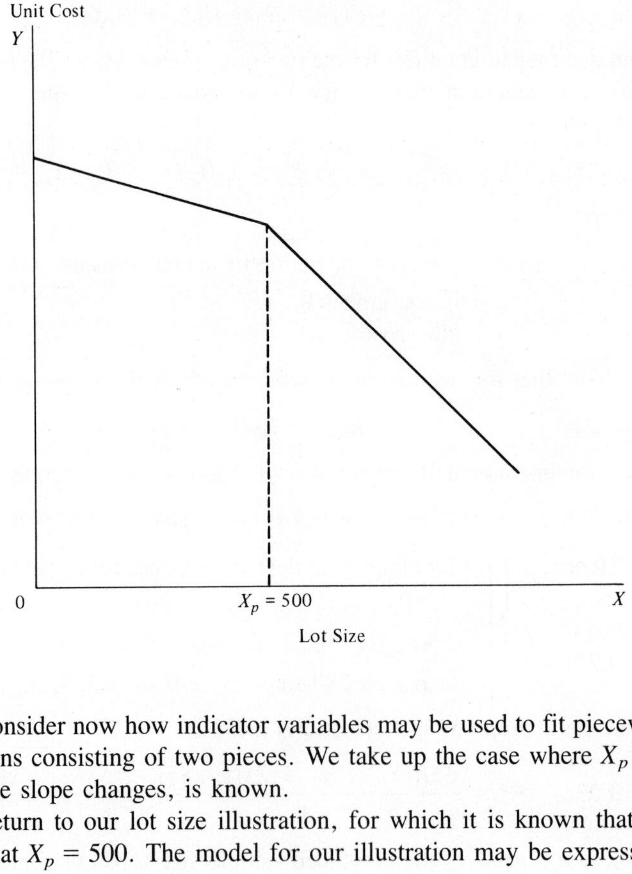

We consider now how indicator variables may be used to fit piecewise linear regressions consisting of two pieces. We take up the case where X_p, the point where the slope changes, is known.

We return to our lot size illustration, for which it is known that the slope changes at $X_p = 500$. The model for our illustration may be expressed as follows:

$$Y_i = \beta_0 + \beta_1 X_{i1} + \beta_2 (X_{i1} - 500) X_{i2} + \varepsilon_i \tag{10.24}$$

where:

$$X_{i1} = \text{Lot size}$$

$$X_{i2} = \begin{matrix} 1 \text{ if } X_{i1} > 500 \\ 0 \text{ otherwise} \end{matrix}$$

To check that model (10.24) does provide a two-piecewise linear regression, consider the response function:

(10.25) $$E(Y) = \beta_0 + \beta_1 X_1 + \beta_2(X_1 - 500)X_2$$

When $X_1 \leq 500$, $X_2 = 0$ so that (10.25) becomes:

(10.25a) $$E(Y) = \beta_0 + \beta_1 X_1 \qquad X_1 \leq 500$$

On the other hand, when $X_1 > 500$, $X_2 = 1$ and we obtain:

(10.25b) $$E(Y) = (\beta_0 - 500\beta_2) + (\beta_1 + \beta_2)X_1 \qquad X_1 > 500$$

Thus, β_1 and $\beta_1 + \beta_2$ are the slopes of the two regression lines, and β_0 and $(\beta_0 - 500\beta_2)$ are the two Y intercepts. These parameters are shown in Figure 10.7.

Example. Table 10.6a contains eight observations on unit costs for given lot sizes. It is known that the response function slope changes at $X_p = 500$ so that

FIGURE 10.7 Illustration of parameters of piecewise linear regression model (10.24)

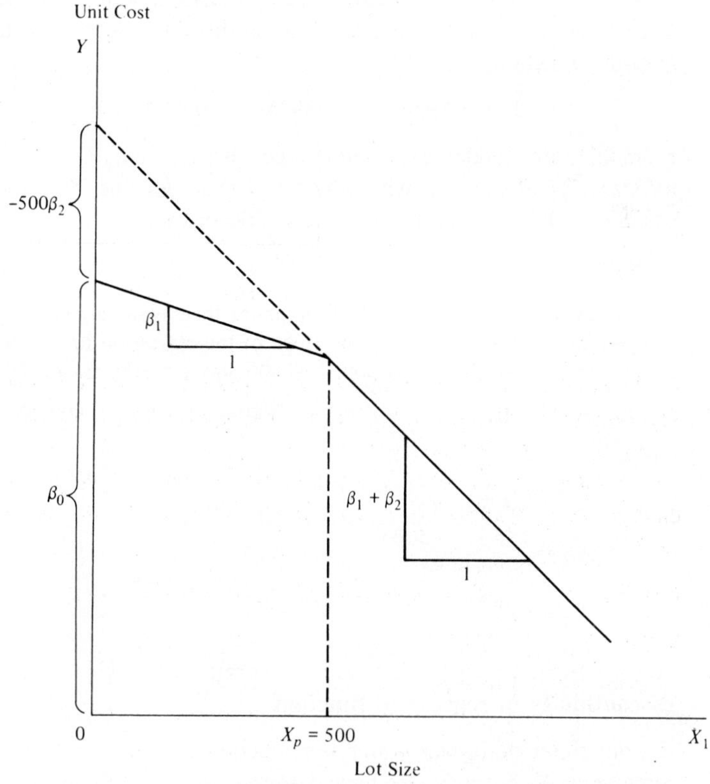

TABLE 10.6 Data and **X** matrix for piecewise linear regression—lot size example

(a)

Lot i	Unit Cost (dollars) Y_i	Lot Size X_i
1	2.57	650
2	4.40	340
3	4.52	400
4	1.39	800
5	4.75	300
6	3.55	570
7	2.49	720
8	3.77	480

(b)

$$\mathbf{X} = \begin{bmatrix} 1 & 650 & 150 \\ 1 & 340 & 0 \\ 1 & 400 & 0 \\ 1 & 800 & 300 \\ 1 & 300 & 0 \\ 1 & 570 & 70 \\ 1 & 720 & 220 \\ 1 & 480 & 0 \end{bmatrix}$$

(columns: X_0, X_1, $(X_1 - 500)X_2$)

model (10.24) is to be employed. Table 10.6b contains the **X** matrix for our example. The left column of **X** is a column of 1's as usual. The next column contains X_{i1}. The final column on the right contains $(X_{i1} - 500)X_{i2}$, which consists of 0's for all lot sizes up to 500, and of $X_{i1} - 500$ for all lot sizes above 500. The fitting of regression model (10.24) at this point becomes routine. The fitted response function is:

$$\hat{Y} = 5.89545 - .00395X_1 - .00389(X_1 - 500)X_2$$

From this fitted model, expected unit cost is estimated to decline by .00395 for a lot size increase of one when the lot size is less than 500 and by .00395 + .00389 = .00784 when the lot size is 500 or more.

Note

The extension of model (10.24) to more than two-piecewise regression lines is straightforward. For instance, if the slope of the regression line in our earlier lot size illustration actually changes at both $X = 500$ and $X = 800$, the model would be:

$$Y_i = \beta_0 + \beta_1 X_{i1} + \beta_2(X_{i1} - 500)X_{i2} + \beta_3(X_{i1} - 800)X_{i3} + \varepsilon_i \qquad (10.26)$$

where:

X_{i1} = Lot size

$X_{i2} = \begin{array}{l} 1 \text{ if } X_{i1} > 500 \\ 0 \text{ otherwise} \end{array}$

$X_{i3} = \begin{array}{l} 1 \text{ if } X_{i1} > 800 \\ 0 \text{ otherwise} \end{array}$

Discontinuity in regression function

Sometimes the linear regression function may not only change its slope at some value X_p but may also have a jump point there. Figure 10.8 illustrates this case. Another indicator variable must now be introduced to take care of the jump. Suppose time required to solve a task successfully (Y) is to be regressed on complexity of task (X), when complexity of task is measured on a quantitative scale from 0 to 100. It is known that the slope of the response line changes at $X_p = 40$, and it is believed that the regression relation may be discontinuous there. We therefore set up the model:

$$Y_i = \beta_0 + \beta_1 X_{i1} + \beta_2(X_{i1} - 40)X_{i2} + \beta_3 X_{i3} + \varepsilon_i \tag{10.27}$$

where:

$$X_{i1} = \text{Complexity of task}$$

$$X_{i2} = \begin{matrix} 1 \text{ if } X_{i1} > 40 \\ 0 \text{ otherwise} \end{matrix}$$

$$X_{i3} = \begin{matrix} 1 \text{ if } X_{i1} > 40 \\ 0 \text{ otherwise} \end{matrix}$$

The response function for model (10.27) is:

$$E(Y) = \beta_0 + \beta_1 X_1 + \beta_2(X_1 - 40)X_2 + \beta_3 X_3 \tag{10.28}$$

When $X_1 \leq 40$, then $X_2 = 0$ and $X_3 = 0$, so (10.28) becomes:

$$E(Y) = \beta_0 + \beta_1 X_1 \qquad X_1 \leq 40 \tag{10.28a}$$

FIGURE 10.8 Illustration of model (10.27) for discontinuous piecewise linear regression

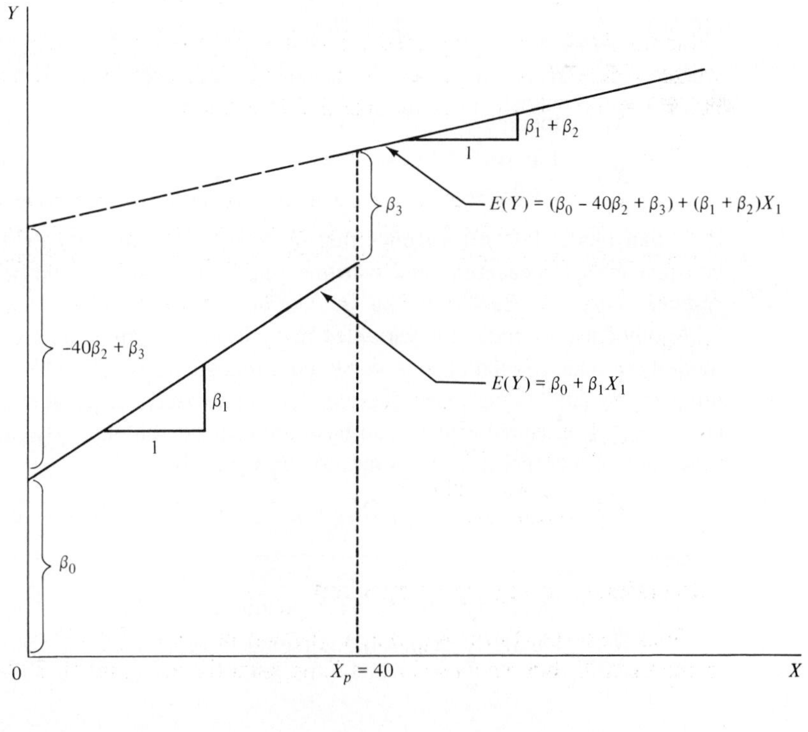

Similarly, when $X_1 > 40$, then $X_2 = 1$ and $X_3 = 1$, so (10.28) becomes:

$$(10.28b) \qquad E(Y) = (\beta_0 - 40\beta_2 + \beta_3) + (\beta_1 + \beta_2)X_1 \qquad X_1 > 40$$

These two response functions are shown in Figure 10.8, together with the parameters involved. Note that β_3 represents the difference in the mean responses for the two regression lines at $X_p = 40$ and β_2 represents the difference in the two slopes.

The estimation of the regression coefficients for model (10.27) presents no new problems. One may test whether or not $\beta_3 = 0$ in the usual manner. If it is concluded that $\beta_3 = 0$, the regression function is continuous at X_p so that the earlier piecewise linear regression model applies.

Time series applications

Economists and business analysts frequently use time series data in regression analysis. For instance, savings (Y) may be regressed on income (X), where both the savings and income data pertain to a number of years. The model employed might be:

$$(10.29) \qquad Y_t = \beta_0 + \beta_1 X_t + \varepsilon_t \qquad t = 1, \ldots, n$$

where Y_t and X_t are savings and income, respectively, for time period t. Suppose that the period covered includes both peacetime and wartime years, and that this factor should be recognized since it is anticipated that savings in wartime years tend to be higher. The following model might then be appropriate:

$$(10.30) \qquad Y_t = \beta_0 + \beta_1 X_{t1} + \beta_2 X_{t2} + \varepsilon_t$$

where:

$$X_{t1} = \text{Income}$$

$$X_{t2} = \begin{matrix} 1 \text{ if period } t \text{ peacetime} \\ 0 \text{ otherwise} \end{matrix}$$

Note that model (10.30) assumes that the marginal propensity to save (β_1) is constant in both peacetime and wartime years, and that only the height of the response curve is affected by this qualitative variable.

Another use of indicator variables in time series applications occurs when monthly or quarterly data are used. Suppose that quarterly sales (Y) are regressed on quarterly advertising expenditures (X_1) and quarterly disposable personal income (X_2). If seasonal effects also have an influence on quarterly sales, a first-order model incorporating seasonal effects would be:

$$(10.31) \qquad Y_t = \beta_0 + \beta_1 X_{t1} + \beta_2 X_{t2} + \beta_3 X_{t3} + \beta_4 X_{t4} + \beta_5 X_{t5} + \varepsilon_t$$

where:

$$X_{t1} = \text{Quarterly advertising expenditures}$$

$$X_{t2} = \text{Quarterly disposable personal income}$$

$$X_{t3} = \begin{array}{l} 1 \text{ if first quarter} \\ 0 \text{ otherwise} \end{array}$$

$$X_{t4} = \begin{array}{l} 1 \text{ if second quarter} \\ 0 \text{ otherwise} \end{array}$$

$$X_{t5} = \begin{array}{l} 1 \text{ if third quarter} \\ 0 \text{ otherwise} \end{array}$$

10.5 SOME CONSIDERATIONS IN USING INDEPENDENT INDICATOR VARIABLES

Indicator variables versus allocated codes

An alternative to the use of indicator variables for a qualitative independent variable is to employ *allocated codes*. Consider, for instance, the independent variable "frequency of product use" which has three classes: frequent user, occasional user, nonuser. With the allocated codes approach, a single independent variable is employed and values are assigned to the classes; for instance:

Class	X_1
Frequent user	3
Occasional user	2
Nonuser	1

The allocated codes are, of course, arbitrary and could be other sets of numbers. The model with allocated codes for our example, assuming no other independent variables, would be:

$$Y_i = \beta_0 + \beta_1 X_{i1} + \varepsilon_i \tag{10.32}$$

The basic difficulty with allocated codes is that they define a metric for the classes of the qualitative variable which may not be reasonable. To see this concretely, consider the mean responses with model (10.32) for the three classes of the qualitative variable:

Class	$E(Y)$
Frequent user	$E(Y) = \beta_0 + \beta_1(3) = \beta_0 + 3\beta_1$
Occasional user	$E(Y) = \beta_0 + \beta_1(2) = \beta_0 + 2\beta_1$
Nonuser	$E(Y) = \beta_0 + \beta_1(1) = \beta_0 + \beta_1$

Note the key implication:

$$E(Y|\text{frequent user}) - E(Y|\text{occasional user}) = E(Y|\text{occasional user}) - E(Y|\text{nonuser}) = \beta_1$$

Thus, the coding 1, 2, 3 implies equal distances between the three user classes, which may not be in accord with reality. Other allocated codes may, of course, imply different spacings of the classes of the qualitative variable, but these would ordinarily still be arbitrary.

Indicator variables, in contrast, make no assumptions about the spacing of the classes and rely on the data to show the differential effects that occur. If, for the same example, two indicator variables, say, X_1 and X_2, are employed to represent the qualitative variable, as follows:

Class	X_1	X_2
Frequent user	1	0
Occasional user	0	1
Nonuser	0	0

the model would be:

$$Y_i = \beta_0 + \beta_1 X_{i1} + \beta_2 X_{i2} + \varepsilon_i \tag{10.33}$$

Here β_1 measures the differential effect:

$$E(Y|\text{frequent user}) - E(Y|\text{nonuser})$$

and β_2 measures:

$$E(Y|\text{occasional user}) - E(Y|\text{nonuser})$$

Thus, β_1 and β_2 measure the differential effects of user status relative to the class of nonusers without any arbitrary restrictions to be satisfied by these differential effects.

Indicator variables versus quantitative variables

If an independent variable is quantitative, such as age, one can nevertheless use indicator variables instead. For instance, the quantitative variable age may be transformed by grouping ages into classes such as under 21, 21–34, 35–49, etc. Indicator variables may then be used for the classes of this new independent variable. At first sight, this may seem to be a questionable approach because information about the actual ages is thrown away. Furthermore, additional parameters are placed into the model, which leads to a reduction of the degrees of freedom associated with *MSE*.

Nevertheless, there are occasions when replacement of a quantitative variable by indicator variables may be appropriate. Consider, for example, a large-scale survey in which the relation between liquid assets (Y) and age (X) of head of household is to be studied. Two thousand households will be included in the study, so that the loss of 10 or 20 degrees of freedom is immaterial. The analyst is very much in doubt about the shape of the regression function, which could be highly complex, and hence may prefer the indicator variable approach in order to obtain information about the shape without making any assumptions about the functional form of the regression function.

Another alternative, also utilizing indicator variables, is available to the analyst in doubt about the functional form of a possibly complex regression function. He or she could use the quantitative variable age, but employ piecewise linear regression with a number of pieces. Again, this approach loses degrees of freedom for estimating *MSE*, but this is of no concern in large-scale studies. The benefit would be that information about the shape of the regression function is obtained without making strong assumptions about its functional form.

Other codings for indicator variables

As stated earlier, many different codings of indicator variables are possible. We mention here two alternative codings to our 0, 1 coding with $c - 1$ indicator variables for a qualitative variable with c classes.

For the insurance innovation example, where Y is time to adopt an innovation, X_1 is size of insurance firm, and the second independent variable is type of company (stock, mutual), we could use the coding:

(10.34)
$$X_2 = \begin{matrix} 1 \text{ if stock company} \\ -1 \text{ if mutual company} \end{matrix}$$

In that case, the first-order linear regression model would be:

(10.35)
$$Y_i = \beta_0 + \beta_1 X_{i1} + \beta_2 X_{i2} + \varepsilon_i$$

which has response function:

(10.36)
$$E(Y) = \beta_0 + \beta_1 X_1 + \beta_2 X_2$$

This response function is as follows for the two types of companies:

(10.36a)
$$E(Y) = (\beta_0 + \beta_2) + \beta_1 X_1 \qquad \text{Stock firms}$$

(10.36b)
$$E(Y) = (\beta_0 - \beta_2) + \beta_1 X_1 \qquad \text{Mutual firms}$$

Thus, β_0 here may be viewed as an "average" intercept of the regression line, from which the stock company and mutual company intercepts differ by β_2 in opposite directions. A test whether the regression lines are the same for both types of companies involves H_0: $\beta_2 = 0$, H_a: $\beta_2 \neq 0$.

A second alternative coding scheme is to use a 0, 1 indicator variable for each of the c classes of the qualitative variable and drop the intercept term in the regression model. For our insurance innovation example, we would have:

(10.37)
$$Y_i = \beta_1 X_{i1} + \beta_2 X_{i2} + \beta_3 X_{i3} + \varepsilon_i$$

where:

$$X_{i1} = \text{Size of firm}$$

$$X_{i2} = \begin{matrix} 1 \text{ if stock company} \\ 0 \text{ otherwise} \end{matrix}$$

$$X_{i3} = \begin{matrix} 1 \text{ if mutual company} \\ 0 \text{ otherwise} \end{matrix}$$

Here, the two response functions would be:

(10.38a)
$$E(Y) = \beta_2 + \beta_1 X_1 \qquad \text{Stock firms}$$

(10.38b)
$$E(Y) = \beta_3 + \beta_1 X_1 \qquad \text{Mutual firms}$$

A test of whether or not the two regression lines are the same would involve the alternatives H_0: $\beta_2 = \beta_3$, H_a: $\beta_2 \neq \beta_3$. This type of test is discussed in Section 8.4.

10.6 DEPENDENT INDICATOR VARIABLE

In a variety of applications, the dependent variable of interest has only two possible outcomes, and therefore can be represented by an indicator variable taking on values 0 and 1.

1. In an analysis of whether or not business firms have an industrial relations department, according to size of firm, the dependent variable was defined to have the two possible outcomes: firm has industrial relations department, firm does not have industrial relations department. These outcomes may be coded 1 and 0, respectively (or vice versa).
2. In a study of labor force participation of wives, as a function of age of wife, number of children, and husband's income, the dependent variable Y was defined to have the two possible outcomes: wife in labor force, wife not in labor force. Again, these outcomes may be coded 1 and 0, respectively.
3. In a study of liability insurance possession, according to age of head of household, amount of liquid assets, and type of occupation of head of household, the dependent variable Y was defined to have the two possible outcomes: household has liability insurance policy, household does not have liability insurance policy. These outcomes again can be coded 1 and 0, respectively.

These examples show the wide range of applications in which the dependent variable is dichotomous, and hence may be represented by an indicator variable. A dependent dichotomous variable, taking on the values 0 and 1, is sometimes said to involve *quantal responses* or *binary responses*.

Meaning of response function when dependent variable is binary

Consider the simple linear regression model:

$$Y_i = \beta_0 + \beta_1 X_i + \varepsilon_i \qquad Y_i = 0, 1 \tag{10.39}$$

where the responses Y_i are binary 0, 1 observations. The expected response $E(Y_i)$ has a special meaning in this case. Since $E(\varepsilon_i) = 0$ as usual, we have:

$$E(Y_i) = \beta_0 + \beta_1 X_i \tag{10.40}$$

Consider now Y_i as an ordinary Bernoulli random variable for which we can state the probability distribution as follows:

Y_i	*Probability*
1	$P(Y_i = 1) = \pi_i$
0	$P(Y_i = 0) = 1 - \pi_i$

Thus, π_i is the probability that $Y_i = 1$ and $1 - \pi_i$ is the probability that $Y_i = 0$. By the ordinary definition of expected value of a random variable in (1.12), we obtain:

$$E(Y_i) = 1(\pi_i) + 0(1 - \pi_i) = \pi_i \tag{10.41}$$

Equating (10.40) and (10.41), we thus find:

$$E(Y_i) = \beta_0 + \beta_1 X_i = \pi_i \tag{10.42}$$

The mean response $E(Y_i) = \beta_0 + \beta_1 X_i$ as given by the response function is therefore simply the probability that $Y_i = 1$ when the level of the independent variable is X_i. This interpretation of the mean response applies whether the response function is a simple linear one, as here, or a complex multiple regression one. The mean response, when the dependent variable is a 0,1 indicator variable, always represents the probability that $Y = 1$ for the given levels of the independent variables. Figure 10.9 illustrates a simple linear response function for a dependent indicator variable. Here, the indicator variable Y refers to whether or

FIGURE 10.9 Illustration of response function when dependent variable is binary—industrial relations department example

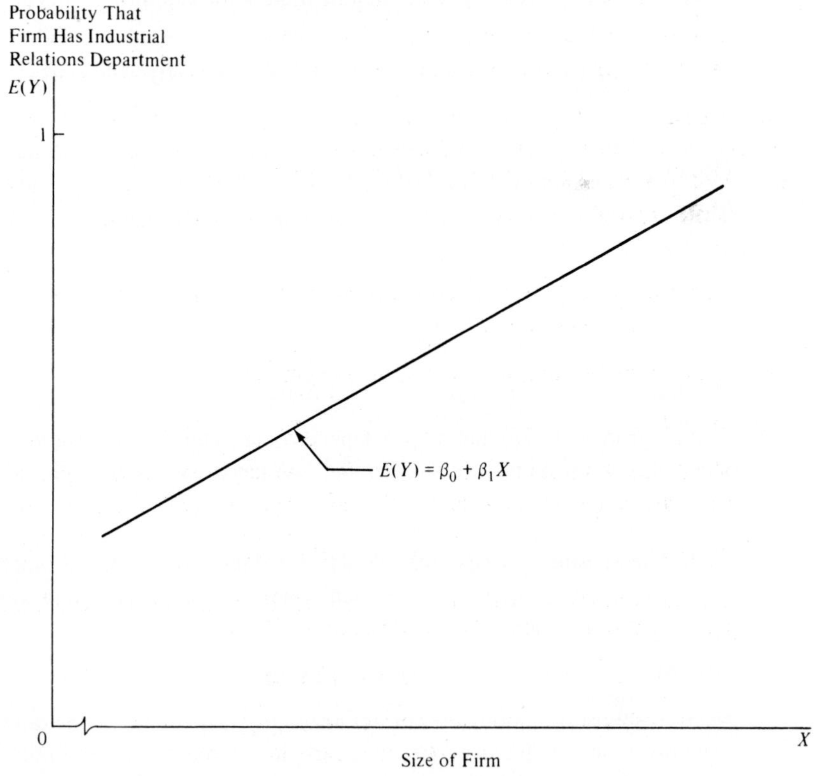

not a firm has an industrial relations department, and the independent variable X is size of firm. The response function in Figure 10.9 shows the probability that firms of given size have an industrial relations department.

Special problems when dependent variable is binary

Special problems arise, unfortunately, when the dependent variable is an indicator variable. We shall consider three of these now, using a simple linear regression model as an illustration.

1. Nonnormal error terms. For a binary 0,1 dependent variable, the error terms $\varepsilon_i = Y_i - (\beta_0 + \beta_1 X_i)$ can take on only two values:

(10.43a) $$\text{When } Y_i = 1\text{: } \varepsilon_i = 1 - \beta_0 - \beta_1 X_i$$

(10.43b) $$\text{When } Y_i = 0\text{: } \varepsilon_i = -\beta_0 - \beta_1 X_i$$

Clearly, the normal error regression model (3.1), assuming that the ε_i are normally distributed, is not appropriate.

2. Nonconstant error variance. Another problem with the error terms ε_i is that they do not have equal variances when the dependent variable is an indicator variable. To see this, we shall obtain $\sigma^2(Y_i)$ for the simple linear regression model (10.39):

$$\sigma^2(Y_i) = E\{[Y_i - E(Y_i)]^2\} = (1 - \pi_i)^2\pi_i + (0 - \pi_i)^2(1 - \pi_i)$$

or:

(10.44) $$\sigma^2(Y_i) = \pi_i(1 - \pi_i) = [E(Y_i)][1 - E(Y_i)]$$

The variance of ε_i is, of course, the same as that of Y_i because $\varepsilon_i = Y_i - \pi_i$ and π_i is a constant:

(10.45) $$\sigma^2(\varepsilon_i) = \pi_i(1 - \pi_i) = [E(Y_i)][1 - E(Y_i)]$$

or:

(10.45a) $$\sigma^2(\varepsilon_i) = (\beta_0 + \beta_1 X_i)(1 - \beta_0 - \beta_1 X_i)$$

Note from (10.45a) that $\sigma^2(\varepsilon_i)$ depends on X_i. Hence, the error term variances will differ at different levels of X, and ordinary least squares will no longer be optimal.

3. Constraints on response function. Since the response function represents probabilities when the dependent variable is a 0,1 indicator variable, the mean responses should be constrained as follows:

(10.46) $$0 \leq E(Y) = \pi \leq 1$$

Many response functions do not automatically possess this constraint. A linear response function, for instance, may fall outside the constraint limits within the range of the independent variable in the scope of the model.

These three problems create difficulties, but solutions can be found. Problem 2 concerning unequal error variances, for instance, can be handled by using weighted least squares. Problem 3 about the constraints on the response function can be handled by making sure that the mean responses for the fitted model do not fall below 0 or above 1 for levels of X within the scope of the model, or else by using a model which automatically meets the constraints.

Finally, even though the error terms are not normal when the dependent variable is binary, the method of least squares still provides unbiased estimators that are asymptotically normal under quite general conditions. Hence, when the sample size is large, inferences concerning the regression coefficients and mean responses can be made in the same fashion as when the error terms are assumed to be normally distributed.

10.7 LINEAR REGRESSION WITH DEPENDENT INDICATOR VARIABLE

We consider now the fitting of linear response functions when the dependent variable is binary. Then we shall explain the fitting of curvilinear response functions.

Illustration

A systems analyst studied the effect of computer programming experience on ability to complete a complex programming task, including debugging, within a specified time. Twenty-five persons were selected for the study. They had varying amounts of programming experience (measured in months of experience), as shown in Table 10.7, column 1. All persons were given the same programming task, and the results of their success in the task are shown in Table 10.7, column 2. The results are coded in binary fashion: if the task was completed successfully in the allotted time, $Y = 1$, and if the task was not completed successfully, $Y = 0$. Figure 10.10 contains a scatter plot of the data. This plot is not too informative because of the nature of the dependent variable, other than to indicate that ability to complete the task successfully appears to increase with amount of experience. At this point, it was decided to fit the linear regression model (10.39).

Ordinary least squares fit

One approach to fitting model (10.39) is to fit it by ordinary least squares despite the unequal error variances. The estimated regression coefficients will still be unbiased, but they will no longer have the minimum variance property among the class of unbiased linear estimators. Thus, the use of ordinary least squares may lead to inefficient estimates, i.e., estimates with larger variances than could be obtained with weighted procedures.

TABLE 10.7 Data for programming task example

Person i	(1) Months of Experience X_i	(2) Task Success Y_i	(3) $\hat{Y}_i$	(4) $\hat{w}_i$
1	14	0	.34920	4.4003
2	29	0	.82212	6.8382
3	6	0	.09697	11.4196
4	25	1	.69601	4.7263
5	18	1	.47531	4.0098
6	4	0	.03392	30.5196
7	18	0	.47531	4.0098
8	12	0	.28614	4.8956
9	22	1	.60142	4.1717
10	6	0	.09697	11.4196
11	30	1	.85365	8.0044
12	11	0	.25461	5.2691
13	30	1	.85365	8.0044
14	5	0	.06544	16.3502
15	20	1	.53837	4.0237
16	13	0	.31767	4.6135
17	9	0	.19156	6.4573
18	32	1	.91671	13.0967
19	24	0	.66448	4.4854
20	13	1	.31767	4.6135
21	19	0	.50684	4.0007
22	4	0	.03392	30.5196
23	28	1	.79059	6.0403
24	22	1	.60142	4.1717
25	8	1	.16003	7.4394

The ordinary least squares fit to the data in Table 10.7 leads to the following results:

Coefficient	Estimated Coefficient	Estimated Standard Deviation
β_0	$b_0 = -.092197$	$s(b_0) = .183272$
β_1	$b_1 = .031528$	$s(b_1) = .009606$

The estimated response function therefore is:

$$\hat{Y} = -.092197 + .031528X \tag{10.47}$$

This response function is shown in Figure 10.10. It may be used in the ordinary manner. For instance, the estimated mean response for persons with $X_h = 14$ months experience is:

$$\hat{Y}_h = -.092197 + .031528(14) = .34920$$

Thus, we estimate that the probability is .349 that a person with 14 months experience will successfully complete the programming task.

FIGURE 10.10 Scatter plot and fitted regression line—programming task example

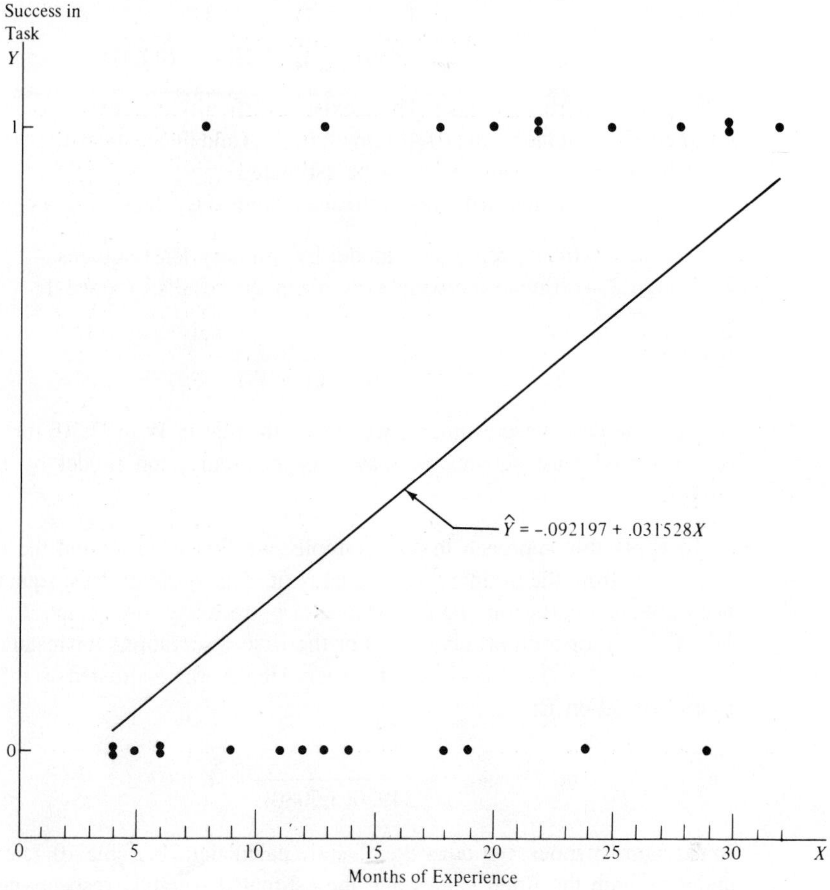

As noted earlier from the scatter plot, this probability increases with increasing experience. The coefficient b_1 indicates that the estimated probability increases by .0315 for each additional month of experience. Clearly, the model cannot be extrapolated much beyond the range of the data here or else the estimated probabilities would be negative or exceed 1, meaningless results. For $X_h = 40$ months experience, for instance, $\hat{Y}_h = 1.17$.

Weighted least squares fit

It was pointed out in Chapter 5 that weighted least squares provides efficient estimates when the error variances are unequal. Since we know from (10.45) that:

$$\sigma^2(\varepsilon_i) = \pi_i(1 - \pi_i) = [E(Y_i)][1 - E(Y_i)]$$

it would appear that using weighted least squares with weights:

$$w_i = \frac{1}{\pi_i(1 - \pi_i)} = \frac{1}{[E(Y_i)][1 - E(Y_i)]} \tag{10.48}$$

is the proper approach to take. There exists a difficulty, however, in carrying this out, namely, that the w_i in (10.48) involve $E(Y_i)$ and hence the parameters β_0 and β_1, which are unknown and must be estimated.

A way out of this difficulty is to use a two-stage least squares procedure:

1. Stage 1—fit the regression model by ordinary least squares.
2. Stage 2—estimate the weights w_i from the results of stage 1:

$$\hat{w}_i = \frac{1}{\hat{Y}_i(1 - \hat{Y}_i)} \tag{10.49}$$

   and then use these estimated weights in the matrix **W** in (7.70) for obtaining weighted least squares estimates for the regression model by means of (7.71).

To apply this approach in our example, we first need to find the estimated weights $\hat{w}_i$ from the ordinary least squares fit. This ordinary least squares fit has been obtained earlier in (10.47). Hence, we are ready to calculate $\hat{Y}_i$ and then $1/\hat{Y}_i(1 - \hat{Y}_i)$ for each observation. For the first observation, for instance, $X_1 = 14$ so that $\hat{Y}_1 = .34920$, as found earlier. Hence, the estimated weight for the first observation is:

$$\hat{w}_1 = \frac{1}{.34920(.65080)} = 4.4003$$

In the same manner, the other weights are calculated. In Table 10.7, columns 3 and 4 contain the fitted values and the estimated weights, respectively.

To obtain the weighted least squares estimates for our example, we utilized a computer run with the weights as given in Table 10.7. This led to the following results:

Regression Coefficient	*Estimated Regression Coefficient*	*Estimated Standard Deviation*
β_0	$b_0 = -.117113$	$s(b_0) = .111841$
β_1	$b_1 = .032672$	$s(b_1) = .006644$

The fitted response function by the two-stage least squares approach therefore is:

$$\hat{Y} = -.117113 + .032672X \tag{10.50}$$

Note that this estimated response function does not differ markedly from the one obtained by unweighted least squares in (10.47), although the estimated standard deviations of the regression coefficients are now somewhat smaller.

Comments

1. If the mean responses range between about .2 and .8 for the scope of the model, there is little to be gained from weighted least squares since the error variances will differ but little. Only if the mean responses range below .2 and/or above .8 will the error variances be sufficiently unequal to make weighted least squares worthwhile.

2. In our example, the fitted response function does not fall below 0 or above 1 within the range of the data (4 months to 32 months experience). If it did, a curvilinear response function would have to be employed. We discuss one important curvilinear model below.

3. The weighted least squares approach could employ additional stages, with the weights refined at each stage. Usually, however, the gain from additional iterations is not large. For our example, for instance, another iteration led to the following results:

$$b_0 = -.125665 \qquad s(b_0) = .084555$$
$$b_1 = .033075 \qquad s(b_1) = .005885$$

Note that there is relatively little decrease from the previous stage in the estimated standard deviations and only small changes in the regression coefficients.

4. If there are repeat observations at the different X levels, the procedure of fitting a linear response function can be simplified. Let $X_1, \ldots, X_c$ denote the X levels, p_j the proportion of 1's at X_j $(j = 1, \ldots, c)$, and n_j the number of observations at X_j. Fitting the sample proportions p_j with weights:

$$w_j = n_j \tag{10.51}$$

leads to the identical estimated response function as ordinary least squares applied to the individual Y observations. If there are hundreds of Y observations located at a limited number of X_j levels, much computational effort can be saved by fitting the sample proportions p_j.

If weighted least squares estimates are to be obtained, only one stage is required when sample proportions p_j are available. The weights would be as follows:

$$\hat{w}_j = \frac{n_j}{p_j(1 - p_j)} \tag{10.52}$$

where $\hat{w}_j$ denotes the estimated weight.

We discuss fitting a regression model to grouped data more fully below.

5. While we have illustrated only simple linear regression, the extension to multiple regression is straightforward.

10.8 LOGISTIC RESPONSE FUNCTION

Both theoretical and empirical considerations suggest that when the dependent variable is an indicator variable, the shape of the response function will frequently be curvilinear. Figure 10.11 contains a curvilinear response function which has been found appropriate in many instances involving a binary dependent variable. Note that this response function is shaped like a tilted S, and that it has asymptotes at 0 and 1. The latter feature assures that the constraints on $E(Y)$ in (10.46) are automatically met. The response function plotted in Figure 10.11 is called the *logistic* function and is given by:

FIGURE 10.11 Example of logistic response function

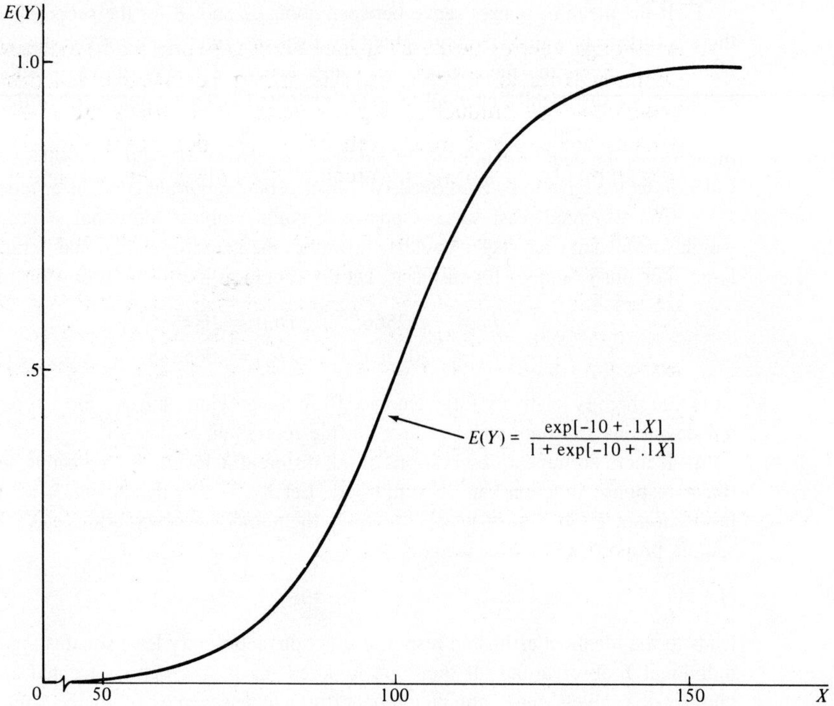

(10.53)
$$E(Y) = \frac{\exp(\beta_0 + \beta_1 X)}{1 + \exp(\beta_0 + \beta_1 X)}$$

An interesting property of the logistic function is that it can easily be linearized. Let us denote $E(Y)$ by π, since the mean response is a probability when the dependent variable is a 0,1 indicator variable. Then if we make the transformation:

(10.54)
$$\pi' = \log_e\left(\frac{\pi}{1 - \pi}\right)$$

we obtain from (10.53):

(10.55)
$$\pi' = \beta_0 + \beta_1 X$$

The transformation (10.54) is called the *logistic* or *logit* transformation of the probability π.

Fitting of logistic function

The fitting of the transformed logistic response function (10.55) is relatively simple when there are repeat observations at each X level. We now explain the fitting procedure for this case, which arises frequently in practice. We cite two examples:

1. A pricing experiment involves showing a new product to a consumer, providing information about it, and then asking the consumer whether he or she would buy the product at a given price. Five prices are studied, and n persons are exposed to a given price. The dependent variable is binary (would purchase, would not purchase); the independent variable is price and has five classes.
2. Four hundred heads of households are asked to indicate on a 10-point scale their intent to buy a new car within the next 12 months. A year later, each household is interviewed to determine whether a new car was purchased. The dependent variable is binary (did purchase, did not purchase). The independent variable is the measure of intent to buy and has 10 classes.

We shall denote the X levels at which observations are obtained by $X_1, \ldots, X_c$. The number of observations at level X_j will be denoted by n_j ($j = 1, \ldots, c$). It can be shown that we need consider only the total number of 1's at each X level, and not the individual Y values. Let R_j be the number of 1's at the level X_j. Hence, the proportion of 1's at the level X_j, denoted by p_j, is:

$$p_j = \frac{R_j}{n_j} \tag{10.56}$$

Table 10.8 on page 365 contains the X_j, n_j, R_j, and p_j values for an example we shall discuss shortly. X_j refers to the price reduction offered by a coupon, n_j is the number of households which received a coupon with price reduction X_j, R_j is the number of these households that redeemed the coupon, and p_j is the proportion of households receiving a coupon with price reduction X_j that did redeem the coupon.

We fit the transformed logistic response function (10.55):

$$\pi' = \beta_0 + \beta_1 X$$

by making the logistic transformation (10.54) on the sample proportions:

$$p_j' = \log_e\left(\frac{p_j}{1 - p_j}\right) \tag{10.57}$$

and using p_j' as the dependent variable. If transformation (10.57) is to be carried out manually, pocket calculators with $\log_e$ function keys or tables of natural logarithms can be used.

The logistic transformation, while linearizing the response function, does not eliminate the unequal variances of the error terms. Hence, weighted least squares should be used. It can be shown that when n_j is reasonably large, the approximate variance of p_j' is:

$$\sigma^2(p_j') = \frac{1}{n_j \pi_j (1 - \pi_j)} \tag{10.58}$$

which is estimated by:

(10.59) $$s^2(p_j') = \frac{1}{n_j p_j(1 - p_j)}$$

Hence, the estimated weights to be used in the weighted least squares computations are:

(10.60) $$\hat{w}_j = n_j p_j(1 - p_j)$$

Note that the use of the estimated weights (10.60) requires the sample sizes n_j to be reasonably large.

The fitting of a linear regression model with independent variable X and dependent variable p_j', using weighted least squares, is straightforward. Once the fitted response function has been obtained:

(10.61) $$\hat{\pi}' = b_0 + b_1 X$$

it can be transformed back into the original units, if desired:

(10.62) $$\hat{\pi} = \frac{\exp(b_0 + b_1 X)}{1 + \exp(b_0 + b_1 X)}$$

Example

In a study of the effectiveness of coupons offering a price reduction on a given product, 1,000 homes were selected and a coupon and advertising material for the product were mailed to each. The coupons offered different price reductions (5, 10, 15, 20, and 30 cents), and 200 homes were assigned at random to each of the price reduction categories. The independent variable X in this study is the amount of price reduction, and the dependent variable Y is a binary variable indicating whether or not the coupon was redeemed within a six-month period.

It was expected that the logistic response function would be an appropriate description of the relation between price reduction and probability that the coupon is utilized. Since there were repeat observations at each X_j, and since the number of repeat observations at each X_j was large ($n_j = 200$, for all j), the procedure described earlier could be used for fitting the logistic response function.

Table 10.8 contains the basic data for this experimental study in columns 1 through 4. The transformed proportions p_j' are shown in column 5. For instance, for $X_1 = 5$, we have:

$$p_1' = \log_e\left(\frac{.160}{1 - .160}\right) = -1.65823$$

Finally, the approximate weights $\hat{w}_j$ are shown in column 6 of Table 10.8. For instance, for $X_1 = 5$, we have:

$$\hat{w}_1 = n_1 p_1(1 - p_1) = 200(.160)(.840) = 26.880$$

TABLE 10.8 Data for coupon effectiveness example

(1) Price Reduction X_j	(2) Number of Households n_j	(3) Number of Coupons Redeemed R_j	(4) Proportion of Coupons Redeemed p_j	(5) Transformed Proportion p'_j	(6) Weight $\hat{w}_j$
5	200	32	.160	−1.6582	26.880
10	200	51	.255	−1.0721	37.995
15	200	70	.350	−.6190	45.500
20	200	103	.515	.0600	49.955
30	200	148	.740	1.0460	38.480

Prior to the fitting of the logistic model, the transformed proportions p'_j were plotted against X_j. This plot is shown in Figure 10.12. It appears from there that a linear response function would fit the transformed proportions well. Hence, it was decided to proceed with fitting the transformed logistic model (10.55).

The fitting of the transformed logistic model by weighted least squares is straightforward. A computer run using weighted least squares, with the weights those in Table 10.8, column 6, led to the following fitted response function:

$$\hat{\pi}' = -2.18506 + .108700X \tag{10.63}$$

This response function is plotted in Figure 10.12, and it appears to fit the data well.

The fitted response function (10.63) is used in the ordinary manner. To estimate the probability of a coupon redemption if the price reduction is, say, 25 cents, we first substitute in (10.63):

$$\hat{\pi}' = -2.18506 + .108700(25) = .53244$$

and then transform to the original variable:

$$\text{antilog}_e(.53244) = 1.70308 = \frac{\hat{\pi}}{1 - \hat{\pi}}$$

so that:

$$\hat{\pi} = .630$$

when $X = 25$ cents. Hence, we estimate that about 63 percent of 25-cent coupons will be redeemed within six months.

The interpretation of the slope $b_1 = .1087$ in the fitted logistic response function (10.63) is not simple, unlike the straightforward interpretation of the slope in a linear regression model. The reason is that the effect of increasing X by a unit varies for the logistic model according to the location of the starting point on the X scale. One interpretation of b_1 is found in the property of the logistic function that the "odds" $\hat{\pi}/(1 - \hat{\pi})$ are multiplied by $\exp(b_1)$ for any unit increase in X.

FIGURE 10.12 Plot of transformed data and fitted logistic response function—coupon effectiveness example

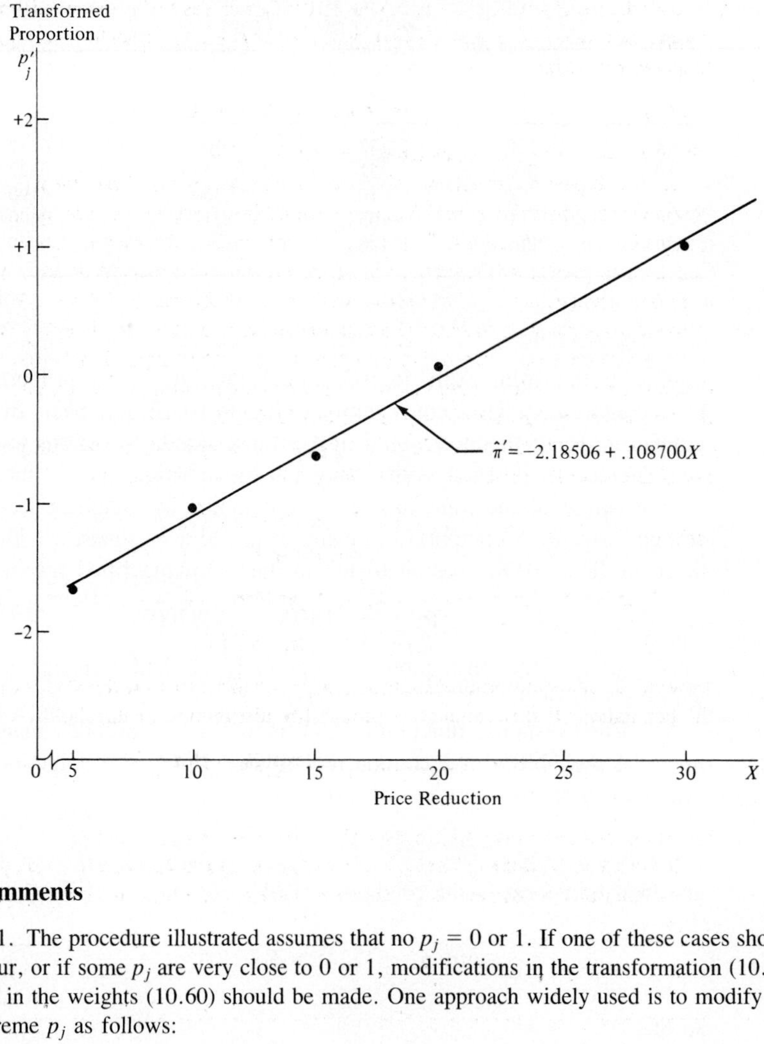

Comments

1. The procedure illustrated assumes that no $p_j = 0$ or 1. If one of these cases should occur, or if some p_j are very close to 0 or 1, modifications in the transformation (10.57) and in the weights (10.60) should be made. One approach widely used is to modify the extreme p_j as follows:

(10.64a) $$p_j = \frac{1}{2n_j} \quad \text{if } R_j = 0$$

(10.64b) $$p_j = 1 - \frac{1}{2n_j} \quad \text{if } R_j = n_j$$

Reference 10.1 discusses several appropriate modifications for the case when p_j equals 0 or 1.

2. A curvilinear response function of almost the same shape as the logistic function (10.53) is obtained by transforming the p_j by means of the cumulative normal distribution. This transformation is called a *probit* transformation. Reference 10.2 describes this trans-

formation. The probit transformation leads to more complex calculations than the logistic transformation, but it does have the desirable property that inferences can be made more readily with it than with the logistic transformation.

3. The transformed logistic model (10.55) can easily be extended into a multiple regression model. For, say, two independent variables, the transformed logistic response function would be:

$$\pi' = \beta_0 + \beta_1 X_1 + \beta_2 X_2 \tag{10.65}$$

where π' is the logistic transformation (10.54) of π.

4. The logistic model sometimes is motivated by threshold theory. Consider the breaking strength of concrete blocks, measured in pounds per square inch. Each block is assumed to have a threshold T_i, such that it will break if pressure equal to or greater than T_i is applied and it will not break if smaller pressure is applied. A block can be tested for only one pressure level, since some weakening of the block occurs with the first test. Thus, it is not possible to find the actual threshold of each block, but only whether or not the threshold level is above or below a particular pressure applied to the block. Of course, different pressures can be applied to different blocks in order to obtain information about the breaking strength thresholds of the blocks in the population.

Let $Y_i = 1$ if the block breaks when pressure X_i is applied, and $Y_i = 0$ if the block does not break. The earlier statement of threshold theory then implies:

$$\begin{aligned} Y_i &= 1 \text{ whenever } T_i \le X_i \\ Y_i &= 0 \text{ whenever } T_i > X_i \end{aligned} \tag{10.66}$$

It follows that for given pressure X_i applied to a block selected at random:

$$\pi_i = P(Y_i = 1 | X_i) = P(T_i \le X_i) \tag{10.67}$$

Now $P(T \le X)$ is the cumulative probability distribution of the thresholds of all blocks in the population. If this cumulative probability distribution of thresholds is logistic:

$$P(T \le X) = \frac{\exp(\beta_0 + \beta_1 X)}{1 + \exp(\beta_0 + \beta_1 X)}$$

we arrive at the logistic response model (10.53).

Incidentally, if the probability distribution of thresholds were normal, the probit response model in comment 2 above would be the relevant one according to threshold theory.

PROBLEMS

10.1. A student who used a regression model that included indicator variables was upset when receiving only the following output on the multiple regression printout: XTRANSPOSE X SINGULAR. What is a likely source of the difficulty?

10.2. Refer to regression model (10.4). Portray graphically the response curves for this model if $\beta_0 = 25.3$, $\beta_1 = .20$, and $\beta_2 = -12.1$.

10.3. In a regression study of factors affecting learning time for a certain task (measured in minutes), sex of learner was included as an independent variable (X_2) that was coded $X_2 = 1$ if male and 0 if female. It was found that $b_2 = 22.3$ and

$s(b_2) = 3.8$. An observer questioned whether the coding scheme for sex is fair because it results in a positive coefficient, leading to longer learning times for males than females. Comment.

10.4. Refer to **Calculator maintenance** Problem 2.16. The users of the desk calculators are either training institutions that use a student model, or business firms that employ a commercial model. An analyst at Tri-City wishes to fit a regression model including both number of calculators serviced (X_1) and type of calculator (X_2) as independent variables and estimate the effect of calculator model (S—student, C—commercial) on number of minutes spent on the service call. Records show that the models serviced in the 18 calls were:

i:	1	2	3	4	5	6	7	8	9
Model:	C	S	C	C	C	S	S	C	C

i:	10	11	12	13	14	15	16	17	18
Model:	C	C	S	S	C	S	C	C	C

Assume that regression model (10.4) is appropriate, and let $X_2 = 1$ if student model and 0 if commercial model.

a. Explain the meaning of all regression coefficients in the model.
b. Fit the regression model and state the estimated regression function.
c. Estimate the effect of calculator model on mean service time with a 95 percent confidence interval. Interpret your interval estimate.
d. Why would the analyst wish to include the number of calculators variable in the model when interest is in estimating the effect of type of calculator model on service time?
e. Obtain the residuals and plot them against X_1X_2. Is there any indication that an interaction term in the model would be helpful?

10.5. Refer to **Grade point average** Problem 2.15. An assistant to the director of admissions conjectures that the predictive power of the model could be improved by adding information on whether the student had chosen a major field of concentration at the time the application was submitted. Assume that regression model (10.4) is appropriate, where X_1 is entrance test score and $X_2 = 1$ if student had indicated a major field of concentration at the time of application and 0 if the major field was undecided. Students 3, 4, 5, 6, 9, and 17 had indicated a major field at the time of application.

a. Explain how each regression coefficient in model (10.4) is interpreted here.
b. Fit the regression model and state the estimated regression function.
c. Test whether the X_2 variable can be dropped from the regression model; use $\alpha = .01$. State the alternatives, decision rule, and conclusion.
d. Obtain the residuals for model (10.4) and plot them against X_1X_2. Is there any evidence in your plot that it would be helpful to include an interaction term in the model?

10.6. Refer to regression models (10.4) and (10.6). Would the conclusion that $\beta_2 = 0$ have the same implication for each of these models? Explain.

10.7. Refer to regression model (10.6). Portray graphically the response curves for this model if $\beta_0 = 25$, $\beta_1 = .30$, $\beta_2 = -12.5$, and $\beta_3 = .05$. Describe the nature of the interaction effect.

10.8. Refer to **Calculator maintenance** Problems 2.16 and 10.4. Fit regression model (10.6) and test whether the interaction term can be dropped from the model; control the α risk at .10. State the alternatives, decision rule, and conclusion. If the interaction term cannot be dropped from the model, describe the nature of the interaction effect.

10.9. Refer to **Grade point average** Problems 2.15 and 10.5.

a. Fit regression model (10.6) and state the estimated regression function.

b. Test whether the interaction term can be dropped from the model; use $\alpha =$ .05. State the alternatives, decision rule, and conclusion. If the interaction term cannot be dropped from the model, describe the nature of the interaction effect.

10.10. In a regression analysis of on-the-job head injuries to warehouse laborers caused by falling objects, Y is a measure of severity of the injury, X_1 is an index reflecting both the weight of the object and the distance it fell, and X_2 and X_3 are indicator variables for nature of head protection worn at the time of the accident, coded as follows:

Type of Protection	X_2	X_3
Hard hat	1	0
Bump cap	0	1
None	0	0

The response function to be used in the study is $E(Y) = \beta_0 + \beta_1 X_1 + \beta_2 X_2 + \beta_3 X_3$.

a. Develop the response function for each type-of-protection category.

b. For each of the following questions, specify the alternatives H_0 and H_a for the appropriate test: (1) With X_1 fixed, does wearing a bump cap reduce the expected severity of injury as compared with wearing no protection? (2) With X_1 fixed, is the expected severity of injury the same when wearing a hard hat as when wearing a bump cap?

10.11. Refer to the tool wear regression model (10.12). Suppose the indicator variables had been defined as follows: $X_2 = 1$ if tool model M2 and 0 otherwise, $X_3 = 1$ if tool model M3 and 0 otherwise, $X_4 = 1$ if tool model M4 and 0 otherwise. Indicate the meaning of each of the following: (1) β_3, (2) $\beta_4 - \beta_3$, (3) β_1, (4) β_7.

10.12. Refer to the advertising expenditures regression model (10.16).

a. How is β_4 interpreted here? What is the meaning of $\beta_3 - \beta_2$ here?

b. State the alternatives for a test of whether the response functions are the same in incorporated and not-incorporated firms with high-quality sales management.

10.13. A marketing research trainee in the national office of a chain of shoe stores used the following response function to study seasonal (winter, spring, summer, fall) effects on sales of a certain line of shoes: $E(Y) = \beta_0 + \beta_1 X_1 + \beta_2 X_2 + \beta_3 X_3$. The X's are indicator variables defined as follows: $X_1 = 1$ if winter and 0 otherwise, $X_2 = 1$ if spring and 0 otherwise, $X_3 = 1$ if fall and 0 otherwise. After fitting the model, she tested the regression coefficients β_j $(j = 0, \ldots, 3)$ and came to the following set of conclusions at an .05 family level of significance: $\beta_0 \neq 0$, $\beta_1 = 0$, $\beta_2 \neq 0$, $\beta_3 \neq 0$. In her report she then wrote: "Results of regression analysis show that climatic and other seasonal factors have no influence in determining sales of this shoe line in the winter. Seasonal influences do exist in the other seasons.'' Do you agree with this interpretation of the test results?

10.14. Assessed valuations. A tax consultant studied the current relation between selling price and assessed valuation of one-family residential dwellings in a large tax district. He obtained data for a random sample of nine recent ''arm's-length'' sales transactions of one-family dwellings located on corner lots and also obtained data for a random sample of 14 recent sales of one-family dwellings not located on corner lots. In the data that follow, both selling price (Y) and assessed valuation (X_1) are expressed in thousand dollars. Assume that the error term variances in the two populations are equal and that regression model (10.6) is appropriate.

Corner Lots

i:	1	2	3	4	5	6	7	8	9
X_{i1}:	17.5	12.5	20.0	16.0	15.0	14.7	17.5	12.3	11.5
Y_i:	56.2	42.5	68.6	54.8	50.0	47.5	56.9	34.0	39.0

Noncorner Lots

i:	1	2	3	4	5	6	7	8	9	10	11	12	13	14
X_{i1}:	10.0	13.8	15.0	19.5	17.0	12.5	14.5	12.8	12.0	16.0	10.0	17.0	10.8	15.0
Y_i:	31.2	36.9	41.0	51.8	48.0	33.3	38.0	35.9	32.0	44.3	29.0	46.1	30.0	42.0

a. Plot the sample data for the two populations on one graph, using different symbols for the two samples. Does the regression relation appear to be the same for the two populations?

b. Test for identity of the regression functions for dwellings on corner lots and dwellings in other locations; control the risk of Type I error at .10. State the alternatives, decision rule, and conclusion.

c. Plot the estimated regression functions for the two populations and describe the nature of the differences between them.

d. Estimate the difference in the slopes of the two regression functions using a 90 percent confidence interval.

e. Prepare a residual plot for each sample. Does the assumption of equal error term variances appear to be reasonable here?

10.15. Tire testing. A testing laboratory with equipment that simulates highway driving studied for two makes (A, B) of a certain type of truck tire the relation between operating cost per mile (Y) and cruising speed (X_1). The observations are shown below (all data are coded). An engineer now wishes to decide whether or not the regression of operating cost on cruising speed is the same for the two makes. Assume that the error term variances for the two makes of tires are the same and that model (10.6) is appropriate.

Make A

i:	1	2	3	4	5	6	7	8	9	10
X_{i1}:	10	20	20	30	40	40	50	60	60	70
Y_i:	9.8	12.5	14.2	14.9	19.0	16.5	20.9	22.4	24.1	25.8

Make B

i:	1	2	3	4	5	6	7	8	9	10
X_{i1}:	10	20	20	30	40	40	50	60	60	70
Y_i:	15.0	14.5	16.1	16.5	16.4	19.1	20.9	22.3	19.8	21.4

a. Plot the sample data for the two populations on one graph, using different symbols for the two samples. Does the relation between speed and cost appear to be the same for the two makes of tires?

b. Test whether or not the regression functions are the same for the two makes of tires. Control the risk of Type I error at .05. State the alternatives, decision rule, and conclusion.

c. Suppose the question of interest simply had been whether the two regression lines have equal slopes. Answer this question by setting up a 95 percent confidence interval for the difference between the two slopes. What do you find?

d. Prepare a residual plot for each make of tire. Does the assumption of equal error term variances appear to be reasonable here?

10.16. Refer to **Muscle mass** Problem 2.23. The nutritionist conjectures that the regression of muscle mass on age follows a two-piece linear relation, with the slope changing at age 60 without discontinuity.

a. State the regression model that applies if the nutritionist's conjecture is correct. What are the respective response functions when age is 60 or less and when age is over 60?

b. Fit the regression model specified in part (a) and state the estimated regression function.

c. Test whether a two-piece linear regression function is needed; use $\alpha = .01$. State the alternatives, decision rule, and conclusion. What is the P-value of the test?

10.17. **Shipment handling.** Global Electronics periodically imports shipments of a certain large part used as a component in several of its products. The size of the shipment varies depending upon production schedules. For handling and distribution to assembling plants, shipments of size 250 thousand parts or less are sent to warehouse A; larger shipments are sent to warehouse B since this warehouse has specialized equipment that provides greater economies of scale for large shipments. The data below were collected on the 10 most recent shipments; X is the size of the shipment (in thousand parts), and Y is the direct cost of handling the shipment in the warehouse (in thousand dollars).

i:	1	2	3	4	5	6	7	8	9	10
X_i:	225	350	150	200	175	180	325	290	400	125
Y_i:	11.95	14.13	8.93	10.98	10.03	10.13	13.75	13.30	15.00	7.97

A two-piecewise linear regression model with a possible discontinuity at $X =$ 250 is to be fitted.

a. Specify the regression model to be used.

b. Fit this regression model. Plot the fitted response function and the data. Is there any indication that greater economies of scale are obtained in handling relatively large shipments than relatively small ones?

c. Test whether or not both the two separate slopes and the discontinuity can be dropped from the model. Control the level of significance at .025. State the alternatives, decision rule, and conclusion.

d. For relatively small shipments, what is the point estimate of the increase in expected handling cost for each increase of one thousand in the size of the shipment? What is the corresponding estimate for relatively large shipments?

10.18. In time series analysis, the X variable representing time usually is defined to take on values 1, 2, etc., for the successive time periods. Does this represent an allocated code when the time periods are actually 1979, 1980, etc.?

10.19. An analyst wishes to include number of older siblings in family as an independent variable in a regression analysis of factors affecting maturation in eighth graders. The number of older siblings in the sample observations ranges from zero to four. Discuss whether this variable should be placed in the model as an ordinary quantitative variable or by means of four 0,1 indicator variables.

10.20. Refer to regression model (10.2) for the insurance innovation study. Suppose X_0 were dropped to eliminate the linear dependence in the $\mathbf{X}$ matrix so that the model is $Y_i = \beta_1 X_{i1} + \beta_2 X_{i2} + \beta_3 X_{i3} + \varepsilon_i$. What is the meaning here of the regression coefficients β_1, β_2, and β_3?

10.21. Refer to Figure 10.10. A student stated: "The least squares line cannot be correct as it stands because it is obvious that when Y can take on only values of 0 and 1 the least squares line has to be horizontal." Comment. What would be the implication if the least squares line were horizontal?

10.22. **Performance ability.** A psychologist made a small-scale study to examine the nature of the relation, if any, between an employee's emotional stability (X) and the employee's ability to perform in a task group (Y). Emotional stability was measured by a written test and ability to perform in a task group ($Y = 1$ if able, $Y = 0$ if unable) was evaluated by the supervisor. The results were:

i:	1	2	3	4	5	6	7	8
X_i:	474	619	584	638	399	481	624	582
Y_i:	0	1	0	1	0	1	1	1

a. Fit a linear regression function by ordinary least squares.

b. Now use the two-stage weighted least squares procedure to fit the linear regression function. Was use of weighted least squares helpful here in terms of the precision of the regression coefficients?

c. Is there any evidence available from this small-scale study whether or not a linear response function is appropriate for the range of X values encountered?

10.23. **Annual dues.** The board of directors of a professional association conducted a random sample survey of 30 members to assess the effects of several possible amounts of dues increase. The sample results follow. X denotes the dollar increase in annual dues posited in the survey interview and $Y = 1$ if the interviewee indicated that the membership will not be renewed at that amount of dues increase and 0 if the membership will be renewed.

i:	1	2	3	4	5	6	7	8	9	10	11	12	13	14	15
X_i:	30	30	30	31	32	33	34	35	35	35	36	37	38	39	40
Y_i:	0	1	0	0	0	0	1	0	0	1	1	0	0	1	0

i:	16	17	18	19	20	21	22	23	24	25	26	27	28	29	30
X_i:	40	40	41	42	43	44	45	45	45	46	47	48	49	50	50
Y_i:	1	1	0	1	1	1	0	1	1	0	1	1	0	1	1

a. Fit a linear regression function by ordinary least squares. State the estimated regression function. Does the estimated regression function fall below 0 or above 1 within the scope of the fitted model?

b. Construct a 95 percent confidence interval for the slope of the response function. Interpret your interval estimate.

c. Estimate the mean response for a $40 dues increase. Use a 95 percent confidence interval. Interpret your interval estimate.

d. Fit a linear regression function by the two-stage weighted least squares procedure. Was weighted least squares helpful here in terms of the precision of the regression coefficients? Explain.

10.24. Bottle return. A carefully controlled experiment was conducted to study the effect of the size of the deposit level on the likelihood that a returnable one-liter soft-drink bottle will be returned. A bottle return was scored 1, and no return was scored 0. The data to follow show the number of bottles that were returned (R_j) out of 500 sold (n_j) at each of six deposit levels (X_j, in cents):

j:	1	2	3	4	5	6
Deposit level X_j:	2	5	10	20	25	30
Number sold n_j:	500	500	500	500	500	500
Number returned R_j:	72	103	170	296	406	449

An analyst believes that the logistic response function (10.53) is appropriate for studying the relation between size of deposit and the probability a bottle will be returned.

a. Apply transformation (10.57) and plot the transformed proportions against X. Does the plot support the use of the transformed model (10.55)?

b. Fit the logistic response function using the transformed proportions and weighted least squares, and state the fitted response function.

c. What is the estimated probability that a bottle will be returned when the deposit is 10 cents? When the deposit is 30 cents?

d. Estimate the amount of deposit for which 75 percent of the bottles will be returned.

10.25. Toxicity experiment. In an experiment testing the effect of a toxic substance, 1,500 experimental insects were divided at random into six groups of 250 each. The insects in each group were exposed to a fixed dose of the toxic substance. A day later, each insect was observed. Death from exposure was scored 1, and survival was scored 0. The results are shown below; X_j denotes the dose level (on a logarithmic scale) received by the insects in group j and R_j denotes the number of insects that died out of the 250 (n_j) in the group.

j:	1	2	3	4	5	6
X_j:	1	2	3	4	5	6
R_j:	28	53	93	126	172	197
n_j:	250	250	250	250	250	250

a. Fit the logistic response function (10.53) using transformation (10.57) and weighted least squares. State the fitted response function in the original units.
b. Plot the fitted response function in the original units and the original data. Does the fit appear to be a good one?
c. What is the estimated median dose—i.e., the dose for which 50 percent of the experimental insects would be expected to die?

EXERCISES

10.26. Refer to the instrument calibration study Example 2 in Section 10.4. Suppose that three instruments (A, B, C) had been developed to identical specifications, that the regression functions relating gauge reading (Y) to actual pressure (X_1) are second-order polynomials for each instrument, that the error term variances are the same, and that the polynomial coefficients may differ from one instrument to the next. Let X_3 denote a second indicator variable, where $X_3 = 1$ if instrument C and 0 otherwise.
a. Expand model (10.21) to cover this situation.
b. State the alternatives, define the test statistic, and give the decision rule for each of the following tests when the level of significance is .01: (1) test whether the second-order regression functions for the three instruments are identical, (2) test whether all three regression functions have the same intercept, (3) test whether both the linear and quadratic effects are the same in all three regression functions.

10.27. Refer to **Muscle mass** Problem 10.16. Specify the model for the case where the slope changes at age 40 and again at age 60 with no discontinuities.

10.28. In a regression study, three types of banks were involved, namely, commercial, mutual savings, and savings and loan. Consider the following system of indicator variables for type of bank:

Type of Bank	X_2	X_3
Commercial	1	0
Mutual savings	0	1
Savings and loan	−1	−1

a. Develop a first-order linear regression model for relating last year's profit or loss (Y) to size of bank (X_1) and type of bank (X_2, X_3).
b. State the response functions for the three types of banks.
c. Interpret each of the following quantities: (1) β_2, (2) β_3, (3) $-\beta_2 - \beta_3$.

10.29. Refer to regression model (10.17) and exclude variable X_3.
a. Obtain the $\mathbf{X'X}$ matrix for this special case of a single qualitative independent variable, for $i = 1, \ldots, n$ and n_1 firms that are not incorporated.
b. Using (7.21), find $\mathbf{b}$.
c. Using (7.26) and (7.27), find SSR and SSE.

PROJECTS

10.30. Refer to the **SMSA** data set. The number of active physicians (Y) is to be regressed against total population (X_1), total personal income (X_2), and geographic region.

a. Fit a first-order regression model. Let $X_3 = 1$ if NE and 0 otherwise, $X_4 = 1$ if NC and 0 otherwise, and $X_5 = 1$ if S and 0 otherwise.

b. Examine whether the effect for the Northeastern region on number of active physicians differs from the effect for the North Central region by constructing an appropriate 90 percent confidence interval. Interpret your interval estimate.

c. Test whether any geographic effects are present; use $\alpha = .10$. State the alternatives, decision rule, and conclusion. What is the P-value of the test?

10.31. Refer to the **SENIC** data set. Infection risk (Y) is to be regressed against length of stay (X_1), age (X_2), routine chest X-ray ratio (X_3), and medical school affiliation (X_4).

a. Fit a first-order regression model. Let $X_4 = 1$ if hospital has medical school affiliation and 0 if not.

b. Estimate the effect of medical school affiliation on infection risk using a 98 percent confidence interval. Interpret your interval estimate.

c. It has been suggested that the effect of medical school affiliation on infection risk may interact with the effects of age and routine chest X-ray ratio. Add appropriate interaction terms to the model, fit the revised regression model, and test whether the interaction terms are helpful; use $\alpha = .10$. State the alternatives, decision rule, and conclusion.

10.32. Refer to the **SENIC** data set. Length of stay (Y) is to be regressed on age (X_1), routine culturing ratio (X_2), average daily census (X_3), available facilities and services (X_4), and region (X_5, X_6, X_7).

a. Fit a first-order regression model. Let $X_5 = 1$ if NE and 0 otherwise, $X_6 = 1$ if NC and 0 otherwise, and $X_7 = 1$ if S and 0 otherwise.

b. Test whether the routine culturing ratio can be dropped from the model; use a level of significance of .05. State the alternatives, decision rule, and conclusion.

c. Examine whether the effect on length of stay for hospitals located in the Western region differs from that for hospitals located in the other three regions by constructing an appropriate confidence interval for each pairwise comparison. Use the Bonferroni procedure with a 95 percent family confidence coefficient. Summarize your findings.

10.33. Refer to the **SENIC** data set. A public health official has requested a study of whether medical school affiliation (Y) is related to available facilities and services (X). Let $Y = 1$ if the hospital has a medical school affiliation and 0 if not.

a. Fit a linear response function, using ordinary least squares.

b. Plot the data and the fitted regression line. Is the regression line consistent with probability characteristics (i.e., is it between 0 and 1) within the scope of the observations?

CITED REFERENCES

10.1 Cox, D. R. *The Analysis of Binary Data*. London: Methuen & Co. Ltd., 1970.

10.2 Finney, D. J. *Probit Analysis*. 3d ed. Cambridge: Cambridge University Press, 1971.

11

Multicollinearity, influential observations, and other topics in regression analysis—II

In this chapter, we take up selected topics in regression analysis dealing with reparameterization to improve computational accuracy, multicollinearity, and detection of influential observations.

11.1 REPARAMETERIZATION TO IMPROVE COMPUTATIONAL ACCURACY

Roundoff errors in least squares calculations

Least squares results can be sensitive to rounding of data in intermediate stages of calculations. When the number of independent variables is small—say, three or less—roundoff effects can be controlled by carrying a sufficient number of digits in intermediate calculations. Indeed, most computer regression programs use double precision arithmetic (e.g., use of 16 digits instead of 8 digits) in all computations to control roundoff effects. Still, with a large number of independent variables, serious roundoff effects can arise despite the use of many digits in intermediate calculations.

Roundoff errors tend to enter into least squares calculations primarily when the inverse of $\mathbf{X'X}$ is taken. Of course, any errors in $(\mathbf{X'X})^{-1}$ may be magnified when calculating $\mathbf{b}$ or making other subsequent calculations. The danger of serious roundoff errors in $(\mathbf{X}'\mathbf{X})^{-1}$ is particularly great when (1) $\mathbf{X}'\mathbf{X}$ has a determinant which is close to zero and/or (2) the elements of $\mathbf{X}'\mathbf{X}$ differ substantially in order of magnitude. The first condition arises when some or all of the independent variables are highly intercorrelated. Remedial measures for this situation will be discussed in Section 11.4. One such measure is to discard one or several of the intercorrelated independent variables in an effort to shift the determinant away from near zero so that roundoff errors will not have as severe an effect.

The second condition arises when the variables have substantially different magnitudes so that the entries in the $\mathbf{X}'\mathbf{X}$ matrix cover a wide range, say, from 15 to 49,000,000. A solution for this condition is to transform the variables and thereby reparameterize the regression model. We now address this second problem. The transformation which we take up is called the *correlation transformation*. It makes all entries in the $\mathbf{X}'\mathbf{X}$ matrix for the transformed variables fall between -1 and $+1$ inclusive, so that the calculation of the inverse matrix becomes much less subject to roundoff errors due to dissimilar orders of magnitudes than with the original variables. Many computer regression packages automatically use this transformation to obtain the basic regression results and then retransform to the original variables. In any case, users of computer programs are well advised to check that the regression package to be employed makes appropriate provisions to prevent roundoff errors from getting out of hand.

Correlation transformation

We shall illustrate the correlation transformation for the case of two independent variables. The basic regression model which we shall assume is the usual first-order one:

$$Y_i = \beta_0 + \beta_1 X_{i1} + \beta_2 X_{i2} + \varepsilon_i \tag{11.1}$$

The first step in order to make the entries in the $\mathbf{X}'\mathbf{X}$ matrix of similar magnitude is to express all observations on the independent variables as deviations from their respective means. To illustrate this for $n = 2$, if:

$$X_{11} = 3,\ X_{21} = 5 \quad \text{and} \quad X_{12} = 1{,}006,\ X_{22} = 1{,}010$$

the deviations would be:

$$X_{11} - \bar{X}_1 = -1,\ X_{21} - \bar{X}_1 = 1 \quad \text{and} \quad X_{12} - \bar{X}_2 = -2,\ X_{22} - \bar{X}_2 = 2$$

Note that the deviations are of similar magnitude whereas the original observations were not.

To use the deviations $X_{i1} - \bar{X}_1$ and $X_{i2} - \bar{X}_2$, model (11.1) must be modified by adding and subtracting the same terms:

$$Y_i = (\beta_0 + \beta_1\bar{X}_1 + \beta_2\bar{X}_2) + \beta_1(X_{i1} - \bar{X}_1) + \beta_2(X_{i2} - \bar{X}_2) + \varepsilon_i$$

or:

$$Y_i = \beta_0' + \beta_1(X_{i1} - \bar{X}_1) + \beta_2(X_{i2} - \bar{X}_2) + \varepsilon_i \tag{11.2}$$

where:

(11.2a) $$\beta_0' = \beta_0 + \beta_1\bar{X}_1 + \beta_2\bar{X}_2$$

It can be shown that the least squares estimator of β_0' is always $\bar{Y}$. Hence, we can rewrite (11.2) as follows:

(11.3) $$Y_i - \bar{Y} = \beta_1(X_{i1} - \bar{X}_1) + \beta_2(X_{i2} - \bar{X}_2) + \varepsilon_i$$

While it might appear that by eliminating one parameter from the model we have been able to increase the degrees of freedom available for *MSE* by one, this is not so since the dependent variable observations $Y_i - \bar{Y}$ are now subject to the restriction $\Sigma(Y_i - \bar{Y}) = 0$.

The second step in developing the correlation transformation is to express each deviation variable in units of its standard deviation:

(11.4) $$\frac{Y_i - \bar{Y}}{s_Y} \qquad \frac{X_{i1} - \bar{X}_1}{s_1} \qquad \frac{X_{i2} - \bar{X}_2}{s_2}$$

where s_Y, s_1, and s_2 are the respective standard deviations of Y, X_1, and X_2 defined as follows:

(11.5a) $$s_Y = \sqrt{\frac{\Sigma(Y_i - \bar{Y})^2}{n-1}}$$

(11.5b) $$s_1 = \sqrt{\frac{\Sigma(X_{i1} - \bar{X}_1)^2}{n-1}}$$

(11.5c) $$s_2 = \sqrt{\frac{\Sigma(X_{i2} - \bar{X}_2)^2}{n-1}}$$

The final step in obtaining the correlation transformation is to use the following function of the standardized variables in (11.4):

(11.6a) $$Y_i' = \frac{1}{\sqrt{n-1}}\left(\frac{Y_i - \bar{Y}}{s_Y}\right)$$

(11.6b) $$X_{i1}' = \frac{1}{\sqrt{n-1}}\left(\frac{X_{i1} - \bar{X}_1}{s_1}\right)$$

(11.6c) $$X_{i2}' = \frac{1}{\sqrt{n-1}}\left(\frac{X_{i2} - \bar{X}_2}{s_2}\right)$$

Reparameterized model

The regression model with the transformed variables Y', X_1', and X_2' as defined by the correlation transformation in (11.6) is a simple extension of model (11.3):

(11.7) $$Y_i' = \beta_1'X_{i1}' + \beta_2'X_{i2}' + \varepsilon_i'$$

It is easy to show that the new parameters β_1' and β_2' in (11.7) and the original parameters β_0, β_1, and β_2 in (11.1) are related as follows:

(11.8a) $$\beta_1 = \left(\frac{s_Y}{s_1}\right)\beta_1'$$

(11.8b) $$\beta_2 = \left(\frac{s_Y}{s_2}\right)\beta_2'$$

(11.8c) $$\beta_0 = \bar{Y} - \beta_1\bar{X}_1 - \beta_2\bar{X}_2$$

Thus, the new regression coefficients β_1' and β_2' and the original regression coefficients β_1 and β_2 are related by simple scaling factors involving ratios of standard deviations.

X′X matrix for transformed variables

The **X** matrix for the transformed variables in model (11.7) is:

$$\mathbf{X} = \begin{bmatrix} X_{11}' & X_{12}' \\ X_{21}' & X_{22}' \\ \cdot & \cdot \\ \cdot & \cdot \\ \cdot & \cdot \\ X_{n1}' & X_{n2}' \end{bmatrix}$$

Remember that model (11.7) does not contain an intercept term; hence, there is no column of 1's in the **X** matrix. The **X′X** matrix then is:

(11.9) $$\mathbf{X}'\mathbf{X} = \begin{bmatrix} X_{11}' & X_{21}' & \cdots & X_{n1}' \\ X_{12}' & X_{22}' & \cdots & X_{n2}' \end{bmatrix} \begin{bmatrix} X_{11}' & X_{12}' \\ X_{21}' & X_{22}' \\ \cdot & \cdot \\ \cdot & \cdot \\ \cdot & \cdot \\ X_{n1}' & X_{n2}' \end{bmatrix}$$

$$= \begin{bmatrix} \Sigma(X_{i1}')^2 & \Sigma X_{i1}'X_{i2}' \\ \Sigma X_{i2}'X_{i1}' & \Sigma(X_{i2}')^2 \end{bmatrix}$$

Let us now consider the elements of this **X′X** matrix. First, we have:

$$\Sigma(X_{i1}')^2 = \Sigma\left(\frac{X_{i1} - \bar{X}_1}{\sqrt{n-1}\, s_1}\right)^2 = \frac{\Sigma(X_{i1} - \bar{X}_1)^2}{n-1} \div s_1^2 = 1$$

Similarly:

$$\Sigma(X_{i2}')^2 = 1$$

Finally:

$$\Sigma X_{i1}'X_{i2}' = \Sigma\left(\frac{X_{i1} - \bar{X}_1}{\sqrt{n-1}\, s_1}\right)\left(\frac{X_{i2} - \bar{X}_2}{\sqrt{n-1}\, s_2}\right)$$

$$= \frac{1}{n-1} \frac{\Sigma(X_{i1} - \bar{X}_1)(X_{i2} - \bar{X}_2)}{s_1 s_2}$$

$$= \frac{\Sigma(X_{i1} - \bar{X}_1)(X_{i2} - \bar{X}_2)}{[\Sigma(X_{i1} - \bar{X}_1)^2]^{1/2}[\Sigma(X_{i2} - \bar{X}_2)^2]^{1/2}}$$

But this equals r_{12}, the coefficient of correlation between X_1 and X_2 by (3.73). Since $\Sigma X'_{i1}X'_{i2} = \Sigma X'_{i2}X'_{i1}$, we find that the $\mathbf{X'X}$ matrix for the transformed variables, denoted by $\mathbf{r}_{XX}$, is:

(11.10)
$$\mathbf{r}_{XX} = \mathbf{X'X} = \begin{bmatrix} 1 & r_{12} \\ r_{12} & 1 \end{bmatrix}$$

By (3.72), r_{12} must fall between -1 and $+1$ inclusive. Hence, it follows that the elements of the $\mathbf{r}_{XX}$ matrix must have values between -1 and $+1$ inclusive. While our example dealt with the case of two independent variables, the same result follows for any number of independent variables transformed according to the principle of (11.6).

The matrix:

$$\mathbf{r}_{XX} = \begin{bmatrix} 1 & r_{12} \\ r_{12} & 1 \end{bmatrix}$$

is called the *correlation matrix of the independent variables*. For $p - 1$ independent variables, $\mathbf{r}_{XX}$ is a $(p - 1) \times (p - 1)$ matrix containing 1's on the main diagonal and the correlation coefficients r_{ij} off the main diagonal.

Regression calculations

The least squares estimators b'_1 and b'_2 for the transformed model (11.7) are obtained in the usual fashion. The inverse of the $\mathbf{r}_{XX}$ matrix is:

(11.11)
$$\mathbf{r}_{XX}^{-1} = \frac{1}{1 - r_{12}^2} \begin{bmatrix} 1 & -r_{12} \\ -r_{12} & 1 \end{bmatrix}$$

It is easy to show that the $\mathbf{X'Y}$ matrix for the transformed variables is:

(11.12)
$$\mathbf{X'Y} = \begin{bmatrix} r_{Y1} \\ r_{Y2} \end{bmatrix}$$

where r_{Y1} and r_{Y2} are the coefficients of correlation between Y and X_1 and between Y and X_2, respectively. Hence, the estimated regression coefficients for the reparameterized model (11.7) are by (7.21):

(11.13)
$$\mathbf{b} = \frac{1}{1 - r_{12}^2} \begin{bmatrix} 1 & -r_{12} \\ -r_{12} & 1 \end{bmatrix} \begin{bmatrix} r_{Y1} \\ r_{Y2} \end{bmatrix} = \frac{1}{1 - r_{12}^2} \begin{bmatrix} r_{Y1} - r_{12}r_{Y2} \\ r_{Y2} - r_{12}r_{Y1} \end{bmatrix}$$

The return to the estimated regression coefficients for the original model is accomplished by employing the relations in (11.8):

(11.14a)
$$b_1 = \left(\frac{s_Y}{s_1}\right) b'_1$$

(11.14b) $$b_2 = \left(\frac{s_Y}{s_2}\right)b'_2$$

(11.14c) $$b_0 = \bar{Y} - b_1\bar{X}_1 - b_2\bar{X}_2$$

Comments

1. Some computer packages present both the regression coefficients b_k for the original model as well as the coefficients b'_k for the transformed model. The latter are sometimes labeled *beta coefficients* in printouts.

2. The regression coefficients b'_k for the reparameterized model are the same as the standardized regression coefficients B_k discussed in Section 7.9, as a comparison of (7.69) with (11.14) makes clear. Thus, use of the transformed variables in (11.6) automatically leads to the standardized regression coefficients.

3. Some computer printouts show the magnitude of the determinant of the correlation matrix of the independent variables. A near-zero value for this determinant implies both a high degree of linear association among the independent variables and a high potential for roundoff errors. For the case of two independent variables, this determinant is seen to be $1 - r_{12}^2$, which approaches 0 as r_{12}^2 approaches 1.

4. When the correlation matrix of the independent variables is augmented by a row and column for Y, it is called the *correlation matrix*. A correlation matrix shows the coefficients of correlation for all pairs of dependent and independent variables. This information is useful in a variety of tasks—for instance, in selecting the final independent variables to be included in the model. Many computer programs display the correlation matrix in the printout.

For the case of two independent variables, the correlation matrix is as follows:

$$\begin{bmatrix} 1 & r_{Y1} & r_{Y2} \\ r_{Y1} & 1 & r_{12} \\ r_{Y2} & r_{12} & 1 \end{bmatrix}$$

Since the correlation matrix is symmetric, the lower (or upper) triangular block of elements is frequently omitted in computer printouts.

5. It is possible to use the correlation transformation with a computer package that does not permit regression through the origin, because the intercept coefficient b'_0 will always be zero for data so transformed. The other regression coefficients will also be correct, as will be all sums of squares. The degrees of freedom and mean squares will not all be correct, however, and will need to be corrected for the regression through the origin.

11.2 PROBLEMS OF MULTICOLLINEARITY

When we discussed multiple regression in Chapter 8, we noted some key problems that typically arise when the independent variables which are being considered for the model are highly correlated among themselves:

1. Adding or deleting an independent variable changes the regression coefficients.
2. The extra sum of squares associated with an independent variable varies,

depending upon which independent variables already are included in the model.

3. The estimated regression coefficients individually may not be statistically significant even though a definite statistical relation exists between the dependent variable and the set of independent variables.

These problems can also arise without substantial multicollinearity being present, but only under unusual circumstances not likely to be found in practice.

We shall now expand on the topic of multicollinearity because high intercorrelations among independent variables are frequently found in nonexperimental data in management and the social and biological sciences. For example, such pairs of independent variables as family income and liquid assets and store sales and number of employees would tend to be correlated highly.

Nature of problem

To see the essential nature of the problem of multicollinearity, we shall employ a simple example. The data in Table 11.1 refer to four sample observations on a dependent variable and two independent variables. Mr. A was asked to fit the multiple regression model:

$$E(Y) = \beta_0 + \beta_1 X_1 + \beta_2 X_2 \tag{11.15}$$

He returned in a short time with the fitted model:

$$\hat{Y} = -87 + X_1 + 18X_2 \tag{11.16}$$

He was proud of this model because it fit the data perfectly. The fitted values are shown in Table 11.1.

It so happened that Ms. B also was asked to fit model (11.15) to the same data, and she arrived at the fitted model:

$$\hat{Y} = -7 + 9X_1 + 2X_2 \tag{11.17}$$

Again, this model fits perfectly, as shown in Table 11.1.

TABLE 11.1 Example of perfectly correlated independent variables

Observation i	X_{i1}	X_{i2}	Y_i	Fitted Values: Model (11.16)	Fitted Values: Model (11.17)
1	2	6	23	23	23
2	8	9	83	83	83
3	6	8	63	63	63
4	10	10	103	103	103

Model (11.16): $\hat{Y} = -87 + X_1 + 18X_2$

Model (11.17): $\hat{Y} = -7 + 9X_1 + 2X_2$

Indeed, it can be shown that infinitely many models will fit the data in Table 11.1 perfectly. The reason is that the independent variables X_1 and X_2 are perfectly related, according to the relation:

$$X_2 = 5 + .5X_1 \tag{11.18}$$

Note carefully that fitted models (11.16) and (11.17) are entirely different response surfaces. The regression coefficients are different, and the fitted values will differ when X_1 and X_2 do not follow relation (11.18). For example, the fitted value for model (11.16) when $X_1 = 5$ and $X_2 = 5$ is:

$$\hat{Y} = -87 + 5 + 18(5) = 8$$

while the fitted value for model (11.17) is:

$$\hat{Y} = -7 + 9(5) + 2(5) = 48$$

Thus, when X_1 and X_2 are perfectly related and, as in our example, the data do not contain any random error component, many different response functions will lead to the same perfectly fitted values for the observations and to the same fitted values for any other (X_1, X_2) combinations following the relation between X_1 and X_2. Yet these response functions are not the same and will lead to different fitted values for (X_1, X_2) combinations that do not follow the relation between X_1 and X_2.

Two key implications of this example are:

1. The perfect relation between X_1 and X_2 did not inhibit our ability to obtain a good fit to the data.
2. Since many different models provide the same good fit, one cannot interpret any one set of regression coefficients as reflecting the effects of the different independent variables. Thus, in fitted model (11.16), $b_1 = 1$ and $b_2 = 18$ do not imply that X_2 is the key independent variable and X_1 plays little role, because model (11.17) provides an equally good fit and its regression coefficients have opposite comparative magnitudes.

Effects of multicollinearity

In actual practice, we seldom find independent variables that are perfectly related or data that do not contain some random error component. Nevertheless, the implications just noted for our idealized example still have relevance.

1. The fact that some or all independent variables are correlated among themselves does not, in general, inhibit our ability to obtain a good fit nor does it tend to affect inferences about mean responses or predictions of new observations, provided these inferences are made within the region of observations. (Figure 7.10 on p. 261 provides an illustration of the concept of the region of observations for the case of two independent variables.)

2. The counterpart in real life to the many different regression functions providing equally good fits to the data in our idealized example is that the estimated regression coefficients tend to have large sampling variability when the inde-

pendent variables are highly correlated. Thus, the estimated regression coefficients tend to vary widely from one sample to the next when the independent variables are highly correlated. As a result, only imprecise information may be available about the individual true regression coefficients. Indeed, each of the estimated regression coefficients individually may be statistically not significant even though a definite statistical relation exists between the dependent variable and the set of independent variables.

3. The common interpretation of regression coefficients as measuring the change in the expected value of the dependent variable when the corresponding independent variable is increased by one unit while all other independent variables are held constant is not fully applicable when multicollinearity exists. While it may be conceptually possible to vary one independent variable and hold the others constant, it may not be possible in practice to do so for independent variables that are highly correlated. For example, in a regression model for predicting crop yield from amount of rainfall and hours of sunshine, the relation between the two independent variables makes it unrealistic to consider varying one while holding the other constant. Therefore, the simple interpretation of the regression coefficients as measuring marginal effects is often unwarranted with highly correlated independent variables.

Example. To illustrate these basic points, consider the data in Table 11.2 for the body fat example which was discussed earlier in Chapter 8. We shall now

TABLE 11.2 Body fat example with three independent variables, two of which are highly correlated

Subject i	*Triceps Skinfold Thickness* X_{i1}	*Thigh Circumference* X_{i2}	*Midarm Circumference* X_{i3}	*Body Fat* Y_i
1	19.5	43.1	29.1	11.9
2	24.7	49.8	28.2	22.8
3	30.7	51.9	37.0	18.7
4	29.8	54.3	31.1	20.1
5	19.1	42.2	30.9	12.9
6	25.6	53.9	23.7	21.7
7	31.4	58.5	27.6	27.1
8	27.9	52.1	30.6	25.4
9	22.1	49.9	23.2	21.3
10	25.5	53.5	24.8	19.3
11	31.1	56.6	30.0	25.4
12	30.4	56.7	28.3	27.2
13	18.7	46.5	23.0	11.7
14	19.7	44.2	28.6	17.8
15	14.6	42.7	21.3	12.8
16	29.5	54.4	30.1	23.9
17	27.7	55.3	25.7	22.6
18	30.2	58.6	24.6	25.4
19	22.7	48.2	27.1	14.8
20	25.2	51.0	27.5	21.1

consider also a third independent variable—X_3, the midarm circumference—in addition to triceps skinfold thickness (X_1) and thigh circumference (X_2), the two independent variables previously considered.

Suppose that we regress body fat (Y) on triceps skinfold thickness (X_1) only. The results of a least squares fit of the response function:

$$E(Y) = \beta_0 + \beta_1 X_1 \tag{11.19}$$

are shown in Table 11.3a. The coefficient of determination r_{Y1}^2 (the notation shows that the relation between Y and X_1 is being considered) is:

$$r_{Y1}^2 = \frac{SSR(X_1)}{SSTO} = \frac{352.27}{495.39} = .711$$

which indicates that the variability of Y is reduced by 71 percent by considering independent variable X_1. Also note that $s(b_1)$ is relatively small:

$$\frac{s(b_1)}{b_1} = \frac{.1288}{.8572} = .1503$$

Let us now add independent variable X_2 to the model. This variable is highly correlated with X_1. The coefficient of determination between the two independent variables X_1 and X_2, denoted by r_{12}^2, is $r_{12}^2 = .853$. The results of fitting the response function:

$$E(Y) = \beta_0 + \beta_1 X_1 + \beta_2 X_2 \tag{11.20}$$

are shown in Table 11.3b. Note the following:

1. The fit of model (11.20) has not been made worse in the sense of a higher error sum of squares SSE, despite the introduction of a highly correlated independent variable. Indeed, we noted earlier that SSE can never increase as the result of introducing another independent variable. (MSE can increase if the reduction in SSE is not adequate to compensate for the loss of one degree of freedom.) For our example, the coefficient of multiple determination is:

$$R^2 = \frac{SSR(X_1, X_2)}{SSTO} = \frac{385.44}{495.39} = .778$$

indicating that the variability of Y is reduced by 78 percent when both X_1 and X_2 are considered. Further, $MSE = 6.47$ now, as compared with $MSE = 7.95$ when only X_1 is included in the model.

2. The estimate of the regression coefficient β_1 for model (11.20) has larger sampling variability than before; the sampling variation of b_1 now is $s(b_1) = .3034$ as compared to .1288 when only X_1 is included in the model. Also, the relative sampling variation of b_2 is quite large:

$$\frac{s(b_2)}{b_2} = \frac{.2912}{.6594} = .4416$$

TABLE 11.3 Regression results for body fat example

(a) Regression of Y on X_1

$$\hat{Y} = -1.496 + .8572X_1$$

Source of Variation	*SS*	*df*	*MS*
Regression	352.27	1	352.27
Error	143.12	18	7.95
Total	495.39	19	

Variable	*Estimated Regression Coefficient*	*Estimated Standard Deviation*	t^*
X_1	$b_1 = .8572$	$s(b_1) = .1288$	6.66

(b) Regression of Y on X_1 and X_2

$$\hat{Y} = -19.174 + .2224X_1 + .6594X_2$$

Source of Variation	*SS*	*df*	*MS*
Regression	385.44	2	192.72
Error	109.95	17	6.47
Total	495.39	19	

Variable	*Estimated Regression Coefficient*	*Estimated Standard Deviation*	t^*
X_1	.2224	.3034	.73
X_2	.6594	.2912	2.26

(c) Regression of Y on X_1, X_2, and X_3

$$\hat{Y} = 117.08 + 4.334X_1 - 2.857X_2 - 2.186X_3$$

Source of Variation	*SS*	*df*	*MS*
Regression	396.98	3	132.33
Error	98.41	16	6.15
Total	495.39	19	

Variable	*Estimated Regression Coefficient*	*Estimated Standard Deviation*	t^*
X_1	4.334	3.016	1.44
X_2	−2.857	2.582	−1.11
X_3	−2.186	1.596	−1.37

Indeed, separate tests of β_1 and β_2, each at the level of significance .01, would lead to the conclusion that $\beta_1 = 0$ and $\beta_2 = 0$, whereas a test of the entire regression relation, based on:

$$F^* = \frac{MSR(X_1, X_2)}{MSE(X_1, X_2)} = \frac{192.72}{6.47} = 29.787$$

would lead to the conclusion, for level of significance .01, that a regression relation does exist.

Let us finally add independent variable X_3 to the model. This variable is not highly correlated with either of the other two independent variables, the coefficients of determination between X_3 and the other two independent variables being $r_{13}^2 = .210$ and $r_{23}^2 = .007$, respectively. The results of fitting the response function:

$$E(Y) = \beta_0 + \beta_1 X_1 + \beta_2 X_2 + \beta_3 X_3 \tag{11.21}$$

are shown in Table 11.3c. Note the following:

1. The fit of the model has been improved somewhat further, the coefficient of multiple determination being:

$$R^2 = \frac{SSR(X_1, X_2, X_3)}{SSTO} = \frac{396.98}{495.39} = .801$$

as compared to $R^2 = .778$ for model (11.20). Further, $MSE(X_1, X_2, X_3) = 6.15$ now, as compared with $MSE(X_1, X_2) = 6.47$ for the model with X_1 and X_2 only.

2. The estimate of the regression coefficient β_2 has actually changed signs (the estimate changed from .6594 to -2.857). In addition, the sampling variations of b_1 and b_2 in model (11.21) both increased dramatically; $s(b_1) = 3.016$ now whereas it was .3034 for model (11.20), and $s(b_2) = 2.582$ now whereas before it was .2912. Again the high degree of multicollinearity among the independent variables X_1 and X_2 is responsible for the inflated variability of the estimates of the regression coefficients. Separate tests of β_1, β_2, and β_3, each at the level of significance .01, would lead to the conclusion that $\beta_1 = 0$, $\beta_2 = 0$, and $\beta_3 = 0$, whereas a test for the entire regression relation, based on:

$$F^* = \frac{MSR(X_1, X_2, X_3)}{MSE(X_1, X_2, X_3)} = \frac{132.33}{6.15} = 21.517$$

would lead to the conclusion, at the level of significance .01, that a regression relation does exist.

3. To the extent that a change in thigh circumference (X_2) is almost always accompanied by a corresponding change in triceps skinfold thickness (X_1), the usefulness of the measures β_1 and β_2 is diminished because they reflect the effect of a change in one variable with no change in the other.

Comments

1. It was noted in Section 11.1 that a near-zero determinant of $\mathbf{X'X}$ is a potential source of serious roundoff errors in least squares results. Severe multicollinearity has the

effect of making this determinant come close to zero. Thus, under severe multicollinearity, the regression coefficients may be subject to large roundoff errors as well as large sampling variances. Hence, it is particularly advisable to employ the correlation transformation (11.6) when multicollinearity is present.

2. Just as high intercorrelations between the independent variables tend to make the estimated regression coefficients imprecise (i.e., erratic from sample to sample), so do the coefficients of partial correlation between the dependent variable and each of the independent variables tend to become erratic from sample to sample when the independent variables are highly correlated.

3. The effect of intercorrelations between the independent variables on the standard deviations of the estimated regression coefficients can be seen readily when the variables in the model are transformed by means of the correlation transformation (11.6). Consider the model with two independent variables:

$$Y_i = \beta_0 + \beta_1 X_{i1} + \beta_2 X_{i2} + \varepsilon_i \tag{11.22}$$

This model in the transformed variables is given by (11.7) and is:

$$Y_i' = \beta_1' X_{i1}' + \beta_2' X_{i2}' + \varepsilon_i' \tag{11.23}$$

The $(\mathbf{X}'\mathbf{X})^{-1}$ matrix for this transformed model is given by (11.11):

$$\mathbf{r}_{XX}^{-1} = (\mathbf{X}'\mathbf{X})^{-1} = \frac{1}{1 - r_{12}^2}\begin{bmatrix} 1 & -r_{12} \\ -r_{12} & 1 \end{bmatrix} \tag{11.24}$$

where r_{12} is the coefficient of correlation between X_1 and X_2. Hence, the variance-covariance matrix of the estimated regression coefficients is by (7.37):

$$\boldsymbol{\sigma}^2(\mathbf{b}) = (\sigma')^2 \mathbf{r}_{XX}^{-1} = (\sigma')^2 \frac{1}{1 - r_{12}^2}\begin{bmatrix} 1 & -r_{12} \\ -r_{12} & 1 \end{bmatrix} \tag{11.25}$$

where $(\sigma')^2$ is the error term variance for the transformed model (11.23).

Thus, the estimated regression coefficients b_1' and b_2' have the same variance:

$$\sigma^2(b_1') = \sigma^2(b_2') = \frac{(\sigma')^2}{1 - r_{12}^2} \tag{11.26}$$

which becomes larger as the correlation between X_1 and X_2 increases. Indeed, as X_1 and X_2 approach perfect correlation (i.e., as r_{12}^2 approaches 1), the variances of b_1' and b_2' become larger without limit.

4. We have noted that high multicollinearity is usually not a problem when the purpose of the regression analysis is to make inferences on the response function or predictions of new observations, provided these inferences are made within the range of observations. In our body fat example, for instance, the estimated mean body fat when the only independent variable included in the model is triceps skinfold thickness (X_1), together with its estimated standard deviation, are as follows for $X_{h1} = 25.0$ (calculations not shown):

$$\hat{Y}_h = 19.93 \qquad s(\hat{Y}_h) = .632$$

When the highly correlated independent variable thigh circumference (X_2) is also included in the model, the estimated mean body fat, together with its estimated standard deviation, are as follows for $X_{h1} = 25.0$ and $X_{h2} = 50.0$:

$$\hat{Y}_h = 19.36 \qquad s(\hat{Y}_h) = .624$$

Thus, the precision of the estimated mean response is equally good as before, despite the addition of the second independent variable which is highly correlated with the first one. This stability in the precision of the estimated mean response occurred despite the fact that the estimated standard deviation of b_1 became substantially larger when X_2 was added to the model (Table 11.3). The essential reason for the stability is that the covariance between b_1 and b_2 is negative, and plays a strong counteracting influence to the increase in $s^2(b_1)$ in determining the value of $s^2(\hat{Y}_h)$ as given in (7.68).

When all three independent variables are included in the model, the estimated mean body fat, together with its estimated standard deviation, are as follows for $X_{h1} = 25.0$, $X_{h2} = 50.0$, and $X_{h3} = 29.0$:

$$\hat{Y}_h = 19.19 \qquad s(\hat{Y}_h) = .621$$

Thus, the addition of the third independent variable, which is not highly correlated with the first two independent variables, does not materially affect the precision of the estimated mean response either.

11.3 VARIANCE INFLATION FACTORS AND OTHER METHODS OF DETECTING MULTICOLLINEARITY

A variety of informal and formal methods have been developed for detecting the presence of serious multicollinearity.

Informal methods

Indications of the presence of serious multicollinearity are given by the following diagnostics:

1. Large changes in the estimated regression coefficients when a variable is added or deleted, or when an observation is altered or deleted.
2. Nonsignificant results in individual tests on the regression coefficients for important independent variables.
3. Estimated regression coefficients with an algebraic sign that is the opposite of that expected from theoretical considerations or prior experience.
4. Large coefficients of correlation between pairs of independent variables in the correlation matrix $\mathbf{r}_{XX}$.
5. Wide confidence intervals for the regression coefficients representing important independent variables.

Example. In the body fat example, the independent variables triceps skinfold thickness and thigh circumference are highly correlated with each other. Also, we noted large changes in the estimated regression coefficients and their estimated standard errors when a variable was added, nonsignificant results in individual tests on anticipated important variables, and an estimated negative coefficient when a positive coefficient was expected. Therefore, serious multicollinearity among the independent variables is suspected.

Note

The informal methods just described have important limitations. They do not provide quantitative measurements of the impact of multicollinearity nor may they identify the nature of the multicollinearity. For instance, if independent variables X_1, X_2, and X_3 have low pairwise correlations, the correlation matrix $\mathbf{r}_{XX}$ will provide no indication of the presence of multicollinearity even though the three variables may be closely related as a group. Thus, examination of simple correlation coefficients will not necessarily disclose the existence of relations among groups of independent variables.

Another limitation of the informal diagnostic methods is that sometimes the observed behavior may occur without multicollinearity being present.

Variance inflation factors

One formal method of detecting the presence of multicollinearity is by means of variance inflation factors. These factors measure how much the variances of the estimated regression coefficients are inflated as compared to when the independent variables are not linearly related.

To understand the significance of variance inflation factors, we begin with the precision of least squares estimated regression coefficients, which is measured by their variances. We know from (7.37) that the variance-covariance matrix of the estimated regression coefficients is:

(11.27) $$\boldsymbol{\sigma}^2(\mathbf{b}) = \sigma^2(\mathbf{X}'\mathbf{X})^{-1}$$

To reduce roundoff errors in calculating $(\mathbf{X}'\mathbf{X})^{-1}$, we noted in Section 11.1 that it is desirable to first transform the variables by means of the correlation transformation (11.6). In the transformed model, the estimated regression coefficients b_k' are, as we have seen, the standardized coefficients defined in (7.69). The variance-covariance matrix of the estimated standardized regression coefficients is according to (11.25):

(11.28) $$\boldsymbol{\sigma}^2(\mathbf{b}) = (\sigma')^2\mathbf{r}_{XX}^{-1}$$

where $\mathbf{r}_{XX}$ is the matrix of the pairwise simple correlation coefficients among the independent variables, as illustrated in (11.10) for $p - 1 = 2$ independent variables, and $(\sigma')^2$ is the error term variance for the transformed model.

Note from (11.28) that the variance of b_k' $(k = 1, \ldots, p - 1)$ is equal to the product of the error term variance $(\sigma')^2$ and the kth diagonal element of the matrix $\mathbf{r}_{XX}^{-1}$. This second factor is called the *variance inflation factor (VIF)*. It can be shown that the variance inflation factor for b_k', denoted by $(VIF)_k$, is:

(11.29) $$(VIF)_k = (1 - R_k^2)^{-1} \qquad k = 1, 2, \ldots, p - 1$$

where R_k^2 is the coefficient of multiple determination when X_k is regressed on the $p - 2$ other X variables in the model. Hence, we have:

(11.30) $$\sigma^2(b_k') = (\sigma')^2(VIF)_k = \frac{(\sigma')^2}{1 - R_k^2}$$

We presented in (11.26) the special results for $\sigma^2(b_k')$ when $p - 1 = 2$, for which $R_k^2 = r_{12}^2$, the coefficient of simple determination between X_1 and X_2.

The variance inflation factor $(VIF)_k$ is equal to 1 when $R_k^2 = 0$, i.e., when X_k is not linearly related to the other X variables. When $R_k^2 \neq 0$, then $(VIF)_k$ is greater than 1, indicating an inflated variance for b'_k. This is evident from (11.30) since the denominator becomes smaller as R_k^2 becomes larger, leading to a larger variance. When X_k has a perfect linear association with the other X variables in the model so that $R_k^2 = 1$, then $(VIF)_k$ and $\sigma^2(b'_k)$ are unbounded.

The largest $(VIF)_k$ among all X variables is often used as an indicator of the severity of multicollinearity. A maximum $(VIF)_k$ in excess of 10 is often taken as an indication that multicollinearity may be unduly influencing the least squares estimates.

The mean of the $(VIF)_k$'s also provides information about the severity of the multicollinearity in terms of how far the estimated standardized regression coefficients b'_k are from the true values β'_k. It can be shown that the expected value of the sum of these squared errors $(b'_k - \beta'_k)^2$ is given by:

$$E\left[\sum_{k=1}^{p-1} (b'_k - \beta'_k)^2\right] = (\sigma')^2 \sum_{k=1}^{p-1} (VIF)_k \tag{11.31}$$

Thus, large $(VIF)_k$ values result, on the average, in larger differences between the estimated and true standardized regression coefficients.

When no X variable is linearly related to the others in the model, $R_k^2 \equiv 0$; hence, $(VIF)_k \equiv 1$ and:

$$E\left[\sum_{k=1}^{p-1} (b'_k - \beta'_k)^2\right] = (\sigma')^2(p-1) \qquad \text{when } (VIF)_k \equiv 1 \tag{11.31a}$$

A ratio of the results in (11.31) and (11.31a) provides useful information about the effect of multicollinearity on the sum of the squared errors:

$$\frac{(\sigma')^2 \Sigma (VIF)_k}{(\sigma')^2 (p-1)} = \frac{\Sigma (VIF)_k}{p-1}$$

Note that this ratio is simply the mean of the $(VIF)_k$ factors, to be denoted by $(\overline{VIF})$:

$$(\overline{VIF}) = \frac{\sum_{i=1}^{p-1} (VIF)_k}{p-1} \tag{11.32}$$

Mean VIF values considerably larger than 1 are indicative of serious multicollinearity problems.

Example. Table 11.4 contains the estimated standardized regression coefficients and the $(VIF)_k$ values for our body fat example. The maximum $(VIF)_k$ is 708.84 and $(\overline{VIF}) = 459.26$. Thus, the expected sum of the squared errors in the least squares regression coefficients is nearly 460 times as large as it would be if the X variables were uncorrelated. In addition, all three $(VIF)_k$ factors greatly exceed 10, which again indicates that serious multicollinearity problems exist.

TABLE 11.4 Variance inflation factors for body fat example

Variable	b'_k	$(VIF)_k$
X_1	4.2637	708.84
X_2	−1.5614	104.61
X_3	−2.9287	564.34
Maximum $(VIF)_k = 708.84$		$\overline{(VIF)} = 459.26$

It is interesting to note that $(VIF)_3 = 564$ despite the fact that both r_{13}^2 and r_{23}^2 are small. Here is an instance where X_3 is strongly related to X_1 and X_2 together ($R_3^2 = .998$), even though the pairwise coefficients of simple determination are small. Examination of the correlation matrix $\mathbf{r}_{XX}$ would not have disclosed this multicollinearity.

Comments

1. A number of computer regression programs use the reciprocal of the variance inflation factor to detect instances where an X variable should not be allowed into the fitted regression model because of excessively high interdependence between this variable and the other X variables in the model. Tolerance limits for $1/(VIF)_k = 1 - R_k^2$ frequently used are .01, .001, or .0001, below which the variable is not entered into the model.

2. A limitation of variance inflation factors for detecting multicollinearities is that they cannot distinguish between several simultaneous multicollinearities.

3. A number of other methods for detecting multicollinearity have been proposed. These are more complex than variance inflation factors and are discussed in specialized texts such as Reference 11.1.

11.4 RIDGE REGRESSION AND OTHER REMEDIAL MEASURES FOR MULTICOLLINEARITY

A variety of remedial measures have been proposed for the difficulties caused by multicollinearity. Some of these leave intact the method of least squares for estimating the regression coefficients while others introduce modifications in the method of estimation.

Remedial measures with ordinary least squares

We consider first remedial measures that may be employed with ordinary least squares.

1. As we have seen, the presence of serious multicollinearity often does not affect the usefulness of the fitted model for making inferences about mean responses or making predictions, provided that the values of the independent variables for which inferences are to be made follow the same multicollinearity pattern as the data on which the regression model is based. Hence, one remedial

measure is to restrict the use of the fitted regression model to inferences for values of the independent variables which follow the same pattern of multicollinearity.

2. In polynomial regression models, as we noted in Chapter 9, expressing the independent variable(s) in the form of deviations from the mean serves to reduce substantially the multicollinearity among the first-order, second-order, and higher-order terms for any given independent variable.

3. One or several independent variables may be dropped from the model in order to lessen the multicollinearity and thereby reduce the standard errors of the estimated regression coefficients of the independent variables remaining in the model. This remedial measure has two important limitations. First, no direct information is obtained about the dropped independent variables. Second, the magnitudes of the regression coefficients for the independent variables remaining in the model are affected by the correlated independent variables not included in the model.

4. Sometimes it is possible to add some observations which break the pattern of multicollinearity. Often, however, this option is not available. In business and economics, for instance, many independent variables cannot be controlled, so that new observations will tend to show the same intercorrelation patterns as the earlier observations.

5. In some economic studies, it is possible to estimate the regression coefficients for different independent variables from different sets of data to avoid the problems of multicollinearity. Demand studies, for instance, may use both cross-section and time series data to this end. Suppose the independent variables in a demand study are price and income, and the relation to be estimated is:

$$Y_i = \beta_0 + \beta_1 X_{i1} + \beta_2 X_{i2} + \varepsilon_i \tag{11.33}$$

where Y is demand, X_1 is income, and X_2 is price. The income coefficient β_1 may then be estimated from cross-section data. The demand variable Y is thereupon adjusted:

$$Y_i' = Y_i - b_1 X_{i1} \tag{11.34}$$

Finally, the price coefficient β_2 is estimated by regressing the adjusted demand variable Y' on X_2.

Ridge regression

Biased estimation. Ridge regression is one of several methods that have been proposed to remedy multicollinearity problems by modifying the method of least squares to allow biased estimators of the regression coefficients. When an estimator has only a small bias and is substantially more precise than an unbiased estimator, it may well be the preferred estimator since it will have a larger probability of being close to the true parameter value. Figure 11.1 illustrates this situation. Estimator b is unbiased but imprecise, while estimator b^R is much

FIGURE 11.1 Biased estimator with small variance may be preferable to unbiased estimator with large variance

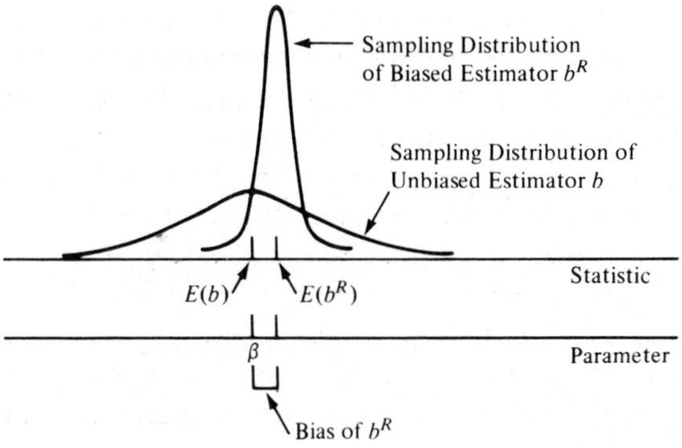

more precise but has a small bias. The probability that b^R falls near the true value β is much greater than that for the unbiased estimator b.

A measure of the combined effect of bias and sampling variation is the expected value of the squared deviation of the biased estimator b^R from the true parameter β. This measure is called the *mean squared error*, and it can be shown to equal:

$$E(b^R - \beta)^2 = \sigma^2(b^R) + [E(b^R) - \beta]^2 \tag{11.35}$$

Thus, the mean squared error equals the variance of the estimator plus the squared bias. Note that if the estimator is unbiased, the mean squared error is identical to the variance of the estimator.

Ridge estimators. For ordinary least squares, the normal equations are given by (7.20):

$$(\mathbf{X}'\mathbf{X})\mathbf{b} = \mathbf{X}'\mathbf{Y} \tag{11.36}$$

When all variables are transformed by the correlation transformation (11.6), the transformed regression model is given by:

$$Y_i' = \beta_1' X_{i1}' + \beta_2' X_{i2}' + \cdots + \beta_{p-1}' X_{i,p-1}' + \varepsilon_i' \tag{11.37}$$

and the least squares normal equations become:

$$\mathbf{r}_{XX}\mathbf{b} = \mathbf{r}_{YX} \tag{11.38}$$

where $\mathbf{r}_{XX}$ is a $(p - 1) \times (p - 1)$ matrix containing the pairwise coefficients of simple correlation between the independent variables:

$$\underset{(p-1)\times(p-1)}{\mathbf{r}_{XX}} = \begin{bmatrix} 1 & r_{12} & \cdots & r_{1,p-1} \\ r_{21} & 1 & \cdots & r_{2,p-1} \\ \vdots & \vdots & & \vdots \\ r_{p-1,1} & r_{p-1,2} & \cdots & 1 \end{bmatrix} \tag{11.39a}$$

and $\mathbf{r}_{YX}$ is a $(p-1) \times 1$ vector containing the coefficients of simple correlation between the dependent variable and each of the independent variables:

$$\underset{(p-1)\times 1}{\mathbf{r}_{YX}} = \begin{bmatrix} r_{Y1} \\ r_{Y2} \\ \cdot \\ \cdot \\ \cdot \\ r_{Y,p-1} \end{bmatrix} \tag{11.39b}$$

Formulas (11.7), (11.10), and (11.12) are illustrative of the case $p - 1 = 2$ independent variables.

The ridge standardized regression estimators are obtained by introducing into the least squares normal equations (11.38) a biasing constant $c \geq 0$, in the following form:

$$(\mathbf{r}_{XX} + c\mathbf{I})\mathbf{b}^R = \mathbf{r}_{YX} \tag{11.40}$$

where $\mathbf{b}^R$ is the vector of the standardized ridge regression coefficients b_k^R:

$$\mathbf{b}^R = \begin{bmatrix} b_1^R \\ b_2^R \\ \cdot \\ \cdot \\ \cdot \\ b_{p-1}^R \end{bmatrix} \tag{11.41}$$

and $\mathbf{I}$ is the $(p-1) \times (p-1)$ identity matrix. Solution of the normal equations (11.40) yields the ridge standardized regression coefficients:

$$\mathbf{b}^R = (\mathbf{r}_{XX} + c\mathbf{I})^{-1}\mathbf{r}_{YX} \tag{11.42}$$

The constant c reflects the amount of bias in the estimators. When $c = 0$, (11.42) reduces to the ordinary least squares regression coefficients in standardized form, as given in (7.69). When $c > 0$, the ridge regression coefficients are biased but tend to be more stable (i.e., less variable) than ordinary least squares estimators.

Choice of biasing constant c. It can be shown that the bias component of the total mean squared error of the ridge regression estimator $\mathbf{b}^R$ increases as c gets larger (with all b_k^R tending toward zero), while at the same time the variance component becomes smaller. It can further be shown that there always exists some value c for which the ridge regression estimator $\mathbf{b}^R$ has a smaller total mean squared error than the ordinary least squares estimator $\mathbf{b}$. The difficulty is that the optimum value of c varies from one application to another and is unknown.

A commonly used method of determining the biasing constant c is based on the *ridge trace* and the variance inflation factors $(VIF)_k$ in (11.29). The ridge trace is a simultaneous plot of the values of the $p - 1$ estimated ridge standardized regression coefficients for different values of c, usually between 0 and 1.

Extensive experience has indicated that an estimated regression coefficient b_k^R may fluctuate widely as c is changed slightly from 0, and may even change signs. Gradually, however, these wide fluctuations cease and the magnitude of the regression coefficient tends to change only slowly as c is increased further. At the same time, the value of $(VIF)_k$ tends to fall rapidly as c is changed from 0, and gradually $(VIF)_k$ also tends to change only moderately as c is increased further. One therefore examines the ridge trace and the variance inflation factors and chooses the smallest value of c where it is deemed that the regression coefficients first become stable in the ridge trace and the variance inflation factors have become sufficiently small. The choice is thus a judgmental one.

Example. We noted previously the severe multicollinearity in the data for our body fat example. Indeed, in the model with three independent variables (Table 11.3c, p. 387), the estimated regression coefficient b_2 is negative even though it was expected that amount of body fat is positively related to thigh circumference. Ridge regression calculations were made for the body fat example data in Table 11.2 (calculations not shown). The ridge standardized regression coefficients for selected values of c are presented in Table 11.5, and the

TABLE 11.5 Ridge estimated standardized regression coefficients for different biasing constants c— body fat example

c	b_1^R	b_2^R	b_3^R
.000	4.264	−2.929	−1.561
.001	2.035	−.9408	−.7087
.002	1.441	−.4113	−.4813
.003	1.165	−.1661	−.3758
.004	1.006	−.0248	−.3149
.005	.9028	.0670	−.2751
.006	.8300	.1314	−.2472
.007	.7760	.1791	−.2264
.008	.7343	.2158	−.2103
.009	.7012	.2448	−.1975
.010	.6742	.2684	−.1870
.020	.5463	.3774	−.1369
.030	.5004	.4134	−.1181
.040	.4760	.4302	−.1076
.050	.4605	.4392	−.1005
.060	.4494	.4443	−.0952
.070	.4409	.4472	−.0909
.080	.4341	.4486	−.0873
.090	.4283	.4491	−.0841
.100	.4234	.4490	−.0812
.200	.3914	.4347	−.0613
.300	.3703	.4154	−.0479
.400	.3529	.3966	−.0376
.500	.3377	.3791	−.0295
.600	.3240	.3629	−.0229
.700	.3116	.3481	−.0174
.800	.3002	.3344	−.0129
.900	.2896	.3218	−.0091
1.000	.2798	.3101	−.0059

TABLE 11.6 *VIF* values for regression coefficients and R^2 for different biasing constants c — body fat example

c	$(VIF)_1$	$(VIF)_2$	$(VIF)_3$	R^2
.000	708.84	564.34	104.61	.8014
.001	125.73	100.27	19.28	.7888
.002	50.56	40.45	8.28	.7852
.003	27.18	21.84	4.86	.7832
.004	16.98	13.73	3.36	.7819
.005	11.64	9.48	2.58	.7809
.006	8.50	6.98	2.19	.7801
.007	6.50	5.38	1.82	.7794
.008	5.15	4.30	1.62	.7787
.009	4.19	3.54	1.48	.7781
.010	3.49	2.98	1.38	.7775
.020	1.10	1.08	1.01	.7726
.030	.63	.70	.92	.7682
.040	.45	.56	.88	.7639
.050	.37	.49	.85	.7597
.060	.32	.45	.83	.7556
.070	.30	.42	.81	.7515
.080	.28	.40	.79	.7475
.090	.26	.39	.78	.7436
.100	.25	.37	.76	.7397
.200	.21	.31	.63	.7031
.300	.18	.27	.54	.6702
.400	.17	.24	.46	.6405
.500	.15	.21	.40	.6134
.600	.14	.19	.35	.5887
.700	.13	.18	.31	.5659
.800	.12	.16	.28	.5449
.900	.11	.15	.25	.5254
1.000	.11	.14	.23	.5073

variance inflation factors are given in Table 11.6. The coefficients of multiple determination R^2 are also shown in the latter table. Figure 11.2 presents the ridge trace of the estimated standardized regression coefficients. To facilitate the analysis, the horizontal c scale in Figure 11.2 is logarithmic.

Note the instability in Figure 11.2 of the regression coefficients for very small values of c. The estimated regression coefficient b_2^R, in fact, changes signs. Also note the rapid decrease in the variance inflation factors in Table 11.6. It was decided to employ $c = .02$ here because for this value of the biasing constant the ridge regression coefficients have *VIF* values near 1 and the estimated regression coefficients appear to have become reasonably stable. The resulting model for $c = .02$ is:

$$\hat{Y}' = .5463X_1' + .3774X_2' - .1369X_3'$$

Transforming back to the original variables by (11.8), as extended to three independent variables, we obtain:

$$\hat{Y} = -7.3978 + .5553X_1 + .3681X_2 - .1917X_3$$

FIGURE 11.2 Ridge trace of estimated standardized regression coefficients—body fat example

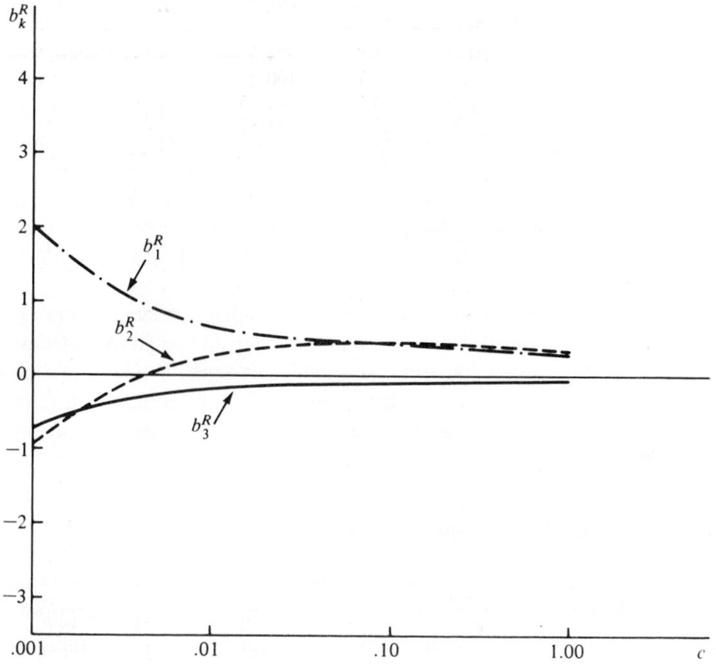

where $\bar{Y} = 20.195$, $\bar{X}_1 = 25.305$, $\bar{X}_2 = 51.170$, $\bar{X}_3 = 27.620$, $s_Y = 5.106$, $s_1 = 5.023$, $s_2 = 5.235$, and $s_3 = 3.647$.

The improper sign on the estimate for β_2 has now been eliminated, and the estimated regression coefficients are more in line with prior expectations. The sum of the squared residuals for the transformed variables, which increases with c, has only increased from .1986 at $c = 0$ to .2274 at $c = .02$ while R^2 decreased from .8014 to .7726. These changes are relatively modest. The estimated mean body fat when $X_{h1} = 25.0$, $X_{h2} = 50.0$, and $X_{h3} = 29.0$ is 19.33 for the ridge regression at $c = .02$ compared to 19.19 utilizing the ordinary least squares solution. Thus, the ridge solution at $c = .02$ appears to be quite satisfactory here and a reasonable alternative to the ordinary least squares solution.

Comments

1. The normal equations (11.40) for the ridge estimators are as follows:

$$
\begin{aligned}
(1+c)b_1^R + r_{12}b_2^R + \cdots + r_{1,p-1}b_{p-1}^R &= r_{Y1} \\
r_{21}b_1^R + (1+c)b_2^R + \cdots + r_{2,p-1}b_{p-1}^R &= r_{Y2} \\
\vdots \qquad\qquad &\quad \vdots \\
r_{p-1,1}b_1^R + r_{p-1,2}b_2^R + \cdots + (1+c)b_{p-1}^R &= r_{Y,p-1}
\end{aligned}
\tag{11.43}
$$

where r_{ij} is the coefficient of simple correlation between the ith and jth independent variables and r_{Yj} is the coefficient of simple correlation between the dependent variable Y and the jth independent variable.

2. Ridge regression estimates tend to be stable in the sense that they are usually little affected by small changes in the data on which the fitted regression is based. In contrast, ordinary least squares estimates may be highly unstable under these conditions when the independent variables are highly multicollinear. Also, the ridge estimated regression function at times will provide good estimates of mean responses or predictions of new observations for levels of the independent variables outside the region of the observations on which the regression function is based. In contrast, the estimated regression function based on ordinary least squares may perform quite poorly in such instances. Of course, any estimation or prediction well outside the region of the observations should always be made with great caution.

3. A major limitation of ridge regression is that ordinary inference procedures are not applicable and exact distributional properties are not known. Another limitation is that the choice of the biasing constant c is a judgmental one. While formal methods have been developed for making this choice, these methods have their own limitations.

4. The ridge regression procedures have been generalized to allow for differing biasing constants for the different estimated regression coefficients.

Other remedial measures

Still other approaches to remedying the problems of multicollinearity have been developed. These include regression with principal components, where the independent variables are linear combinations of the original independent variables, and Bayesian regression, where prior information about the regression coefficients is incorporated into the estimation procedure. More information about these approaches, as well as about ridge regression and generalized ridge regression, may be obtained from specialized works such as Reference 11.1.

11.5 IDENTIFICATION OF OUTLYING OBSERVATIONS

Frequently in regression analysis applications, the data set contains some observations which are outlying or extreme, i.e., observations which are well separated from the remainder of the data. These outlying observations may involve large residuals and often have dramatic effects on the fitted least squares regression function. It is therefore important to study the outlying observations carefully and decide whether they should be retained or eliminated, and if retained, whether their influence should be reduced in the fitting process and/or the regression model revised.

An observation may be outlying or extreme with respect to its Y value, its X value(s), or both. Figure 11.3 illustrates this for the case of regression with a single independent variable. In the scatter plot in Figure 11.3, observation 1 is outlying with respect to its Y value. Note that this point falls far outside the scatter, although its X value is near the middle of the range of the observations on the independent variable. Observations 2 and 3 are outlying with respect to their X values since they have much larger X values than those for the other observations; observation 3 is also outlying with respect to its Y value.

FIGURE 11.3 Scatter plot for regression with one independent variable illustrating outlying observations

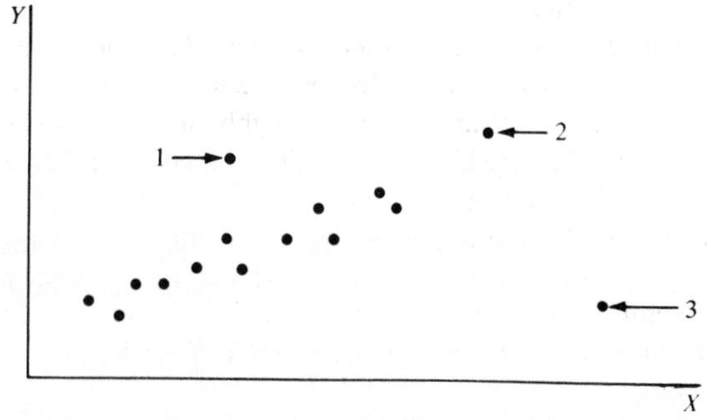

Not all outlying observations have a strong influence on the fitted regression function. Observation 1 in Figure 11.3 may not be too influential because there are a number of other observations that have similar X values, which will keep the fitted regression function from being displaced too far by the outlying observation. Likewise, observation 2 may not be too influential because its Y value is consistent with the regression relation displayed by the nonextreme observations. Observation 3, on the other hand, is likely to be very influential in affecting the fit of the regression function because it is outlying with regard to its X value, and its Y value is not consistent with the regression relation for the other observations.

In the case of regression with one or two independent variables, it is relatively simple to identify outlying observations with respect to their X or Y values and to study whether or not they are influential in affecting the fitted regression function. When more than two independent variables are included in the regression model, however, the identification of outlying observations by simple graphic means, such as scatter and residual plots, becomes difficult and more powerful methods are required. We now discuss some methods for identifying observations that are outlying with respect to their X or Y values.

Use of hat matrix for identifying outlying X observations

We encountered the hat matrix $\mathbf{H}$ in Chapter 6 where we noted in (6.93) that the least squares residuals can be expressed as a linear combination of the observations Y_i by means of the hat matrix:

(11.44) $$\mathbf{e} = (\mathbf{I} - \mathbf{H})\mathbf{Y}$$

The hat matrix $\mathbf{H}$ is given by (6.93a):

(11.45) $$\underset{n\times n}{\mathbf{H}} = \mathbf{X}(\mathbf{X}'\mathbf{X})^{-1}\mathbf{X}'$$

Similarly, the fitted values $\hat{Y}_i$ can be expressed as linear combinations of the observations Y_i through the hat matrix:

(11.46) $$\hat{\mathbf{Y}} = \mathbf{H}\mathbf{Y}$$

Further, we noted in (6.95) that the variances and covariances of the residuals involve the hat matrix:

(11.47) $$\boldsymbol{\sigma}^2(\mathbf{e}) = \sigma^2(\mathbf{I} - \mathbf{H})$$

so that the variance of residual e_i, denoted by $\sigma^2(e_i)$, is:

(11.48) $$\sigma^2(e_i) = \sigma^2(1 - h_{ii})$$

where h_{ii} is the ith element on the main diagonal of the hat matrix.

The diagonal element h_{ii} of the hat matrix can be obtained directly from:

(11.49) $$h_{ii} = \mathbf{X}_i'(\mathbf{X}'\mathbf{X})^{-1}\mathbf{X}_i$$

where $\mathbf{X}_i$ corresponds to the $\mathbf{X}_h$ vector in (7.45) except that $\mathbf{X}_i$ pertains to the ith sample observation:

(11.49a) $$\underset{p\times 1}{\mathbf{X}_i} = \begin{bmatrix} 1 \\ X_{i,1} \\ \cdot \\ \cdot \\ \cdot \\ X_{i,p-1} \end{bmatrix}$$

Note that $\mathbf{X}_i'$ is simply the ith row of the $\mathbf{X}$ matrix, pertaining to the ith sample observation.

The diagonal elements h_{ii} have some useful properties:

(11.50) $$0 \le h_{ii} \le 1 \qquad \sum_{i=1}^{n} h_{ii} = p$$

where p is the number of regression parameters in the regression function including the intercept term.

The diagonal element h_{ii} in the hat matrix is called the *leverage* (in terms of the X values) of the ith observation. It indicates whether or not the X values for the ith observation are outlying, because it can be shown that h_{ii} is a measure of the distance between the X values for the ith observation and the means of the X values for all n observations. Thus, a large leverage value h_{ii} indicates that the ith observation is distant from the center of the X observations. Figure 11.4 illustrates this for the case of two independent variables. Observation 1 is distant from the center $(\bar{X}_1, \bar{X}_2)$ and has a large leverage value $h_{11} = .812$ while observation 2 is near the center and has a small leverage value $h_{22} = .253$.

If the ith observation is an outlying X observation—i.e., one with a large leverage value h_{ii}—it exercises substantial leverage in determining the fitted value $\hat{Y}_i$. This is so for the following reasons:

1. The fitted value $\hat{Y}_i$ is a linear combination of the observed Y values, as shown in (11.46), and h_{ii} is the weight of observation Y_i in determining this fitted value. Thus, the larger is h_{ii}, the more important is Y_i in determining

FIGURE 11.4 Illustration of observations with X values near and far from center

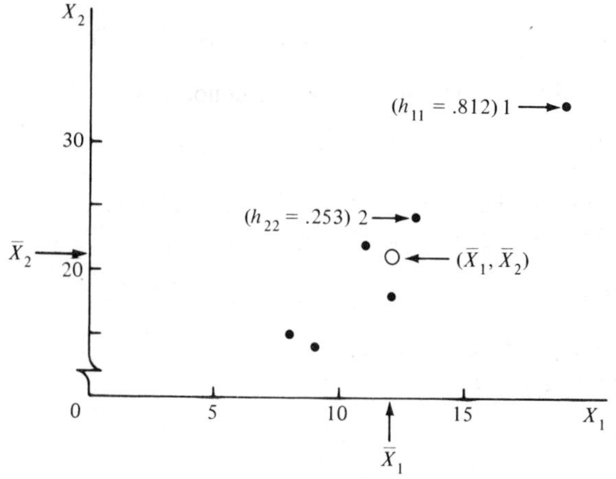

$\hat{Y}_i$. Remember that h_{ii} is a function only of the X values, so h_{ii} measures the role of the X values in determining how important Y_i is in affecting the fitted value $\hat{Y}_i$.

2. The larger is h_{ii}, the smaller is the variance of the residual e_i, as may be seen from (11.48). Hence, the larger is h_{ii}, the closer the fitted value $\hat{Y}_i$ will tend to be to the observed value Y_i. In the extreme case where $h_{ii} = 1$, $\sigma^2(e_i) = 0$ so that the fitted value $\hat{Y}_i$ is then forced to equal the observed value Y_i. Since observations with high leverage tend to have smaller residuals, it may not be possible to detect them by an examination of the residuals alone.

A leverage value h_{ii} is usually considered to be large if it is more than twice as large as the mean leverage value, denoted by $\bar{h}$, which according to (11.50) is:

$$\bar{h} = \frac{\sum_{i=1}^{n} h_{ii}}{n} = \frac{p}{n} \tag{11.51}$$

Hence, leverage values greater than $2p/n$ are considered by this rule to indicate outlying observations with regard to the X values. Additional evidence of an extreme observation is the existence of a gap between the leverage values for most of the observations and the unusually large leverage value(s).

Example. We continue with the body fat example, using only the two independent variables triceps skinfold thickness (X_1) and thigh circumference (X_2) so that the results using the hat matrix can be compared to simple graphic plots. The data for this example were presented earlier in Table 8.3 (p. 275). Figure 11.5 contains a scatter plot of X_2 against X_1, where the observations are shown by their observation number. We note from Figure 11.5 that observations 15 and 3 appear to be outlying ones with respect to the pattern of the X values. Observa-

FIGURE 11.5 Scatter plot of thigh circumference against triceps skinfold thickness—body fat example

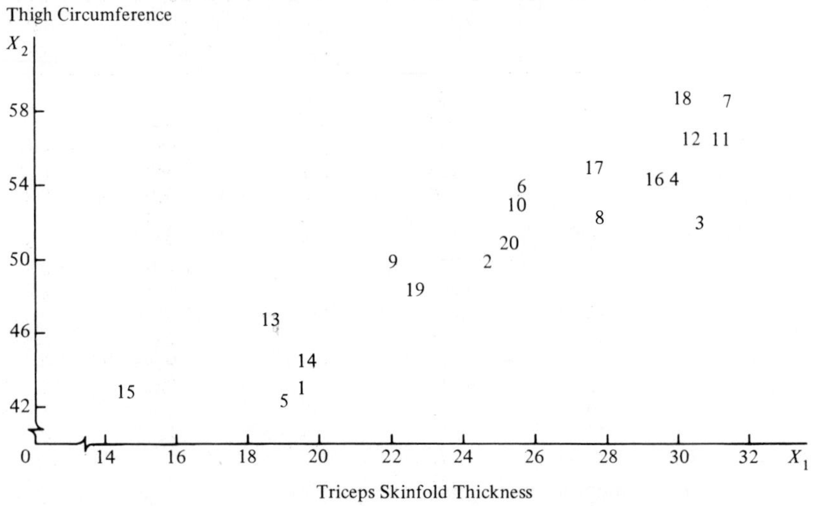

tion 15 is outlying for X_1 and at the low end of the range for X_2, while observation 3 is outlying in terms of the pattern of multicollinearity, though it is not outlying for either of the independent variables separately. Observations 1 and 5 also appear to be somewhat extreme.

Calculation of the hat matrix (11.45) confirms these impressions. Table 11.7, column 2, contains the leverage values h_{ii} for the body fat example. Note that the two largest leverage values are $h_{33} = .372$ and $h_{15,15} = .333$. Both exceed the criterion of twice the mean leverage value, $2p/n = 2(3)/20 = .30$, and are separated by a substantial gap from the next largest leverage values $h_{55} = .248$ and $h_{11} = .201$. Having identified observations 3 and 15 as outlying observations in terms of their X values, we shall need to ascertain how influential these observations are in the fitting of the regression function. We consider this question after taking up the identification of outlying Y observations.

Use of studentized deleted residuals for identifying outlying Y observations

The detection of outlying or extreme Y observations based on an examination of the residuals has been considered in earlier chapters. We utilized there either the residuals e_i:

$$e_i = Y_i - \hat{Y}_i \tag{11.52}$$

or the standardized residuals:

$$\frac{e_i}{\sqrt{MSE}} \tag{11.53}$$

TABLE 11.7 Residuals, diagonal elements of the hat matrix, studentized deleted residuals, and Cook's distances—body fat example

i	(1) e_i	(2) h_{ii}	(3) d_i^*	(4) D_i
1	−1.683	.201	−.730	.046
2	3.643	.059	1.534	.046
3	−3.176	.372	−1.656	.490
4	−3.158	.111	−1.348	.072
5	.000	.248	.000	.000
6	−.361	.129	−.148	.001
7	.716	.156	.298	.006
8	4.015	.096	1.760	.098
9	2.655	.115	1.117	.053
10	−2.475	.110	−1.034	.044
11	.336	.120	.137	.001
12	2.226	.109	.923	.035
13	−3.947	.178	−1.825	.211
14	3.447	.148	1.524	.125
15	.571	.333	.267	.013
16	.642	.095	.258	.002
17	−.851	.106	.344	.005
18	−.783	.197	.335	.010
19	−2.857	.067	−1.176	.032
20	1.040	.050	.409	.003

We introduce now two refinements to make the analysis of residuals more effective for identifying outlying Y observations.

When the residuals e_i have substantially different variances $\sigma^2(e_i)$, as given in (11.48), it is better to consider the magnitude of e_i relative to $\sigma(e_i)$ instead of $\sqrt{MSE}$ to give recognition to differences in their sampling errors. Since by (11.48) we have:

$$\sigma^2(e_i) = \sigma^2(1 - h_{ii})$$

an unbiased estimator of this variance is:

$$s^2(e_i) = MSE(1 - h_{ii}) \tag{11.54}$$

The ratio of e_i to $s(e_i)$ is called the *studentized residual* and will be denoted by e_i^*:

$$e_i^* = \frac{e_i}{s(e_i)} \tag{11.55}$$

Note that the residuals e_i will have substantially different sampling variations if the leverage values h_{ii} differ markedly, but the studentized residuals have constant variance (when the model is appropriate).

The second refinement is to measure the ith residual $e_i = Y_i - \hat{Y}_i$ when the fitted regression is based on the observations excluding the ith one. In this way, the fitted value $\hat{Y}_i$ cannot be influenced by the ith observation to be close to Y_i because this observation is not part of the data set on which the fitted value is based. Thus, the ith observation is deleted, the regression function is fitted to the remaining $n - 1$ observations, and the point estimate of the expected value when the X levels are those of the ith observation, to be denoted by $\hat{Y}_{(i)}$, will be compared with the actual Y_i observed value. The notation $\hat{Y}_{(i)}$ reminds us that the ith observation was omitted when fitting the regression function. The residual:

$$d_i = Y_i - \hat{Y}_{(i)} \tag{11.56}$$

is called a *deleted residual* and is denoted by d_i.

Note that a deleted residual corresponds to the prediction error in the numerator of (3.34) when predicting a new observation from the fitted regression function based on earlier observations, except that in (3.34) the difference considered is $\hat{Y}_{(i)} - Y_i$ and the notation differs from the present one. Hence, we know from (7.55a) that the estimated variance of d_i is:

$$s^2(d_i) = MSE_{(i)}(1 + \mathbf{X}_i'(\mathbf{X}_{(i)}'\mathbf{X}_{(i)})^{-1}\mathbf{X}_i) \tag{11.57}$$

where $\mathbf{X}_i$ is the X observations vector (11.49a) for the ith observation, $MSE_{(i)}$ is the mean square error when the ith observation is omitted in fitting the regression function, and $\mathbf{X}_{(i)}$ is the $\mathbf{X}$ matrix with the ith observation deleted. Also, it follows from (7.55) that:

$$\frac{d_i}{s(d_i)} \sim t(n - p - 1) \tag{11.58}$$

Remember that $n - 1$ observations are used here in predicting the ith observation; hence, the degrees of freedom are $(n - 1) - p = n - p - 1$.

Combining the two refinements, we shall use for diagnosis of outlying or extreme Y observations the deleted residual d_i in (11.56) and studentize it by dividing it by its estimated standard deviation given by (11.57). The *studentized deleted residual*, to be denoted by d_i^*, therefore is:

$$d_i^* = \frac{d_i}{s(d_i)} \tag{11.59}$$

We know from (11.58) that each studentized deleted residual d_i^* follows the t distribution with $n - p - 1$ degrees of freedom. The d_i^*, however, are not independent.

Fortunately, the studentized deleted residuals d_i^* in (11.59) can be calculated without having to fit regression functions with the ith observation omitted. It can be shown that an algebraically equivalent expression for d_i^* is:

$$d_i^* = e_i\left[\frac{n - p - 1}{SSE(1 - h_{ii}) - e_i^2}\right]^{1/2} \tag{11.59a}$$

Thus, the studentized deleted residual d_i^* can be calculated from the residual e_i, the error sum of squares SSE, and the leverage value h_{ii}, all for the fitted regression based on the n observations.

To identify outlying Y observations, we examine the studentized deleted residuals for large absolute values and use the appropriate t distribution to ascertain how far in the tails such outlying values fall.

Example. We illustrate the calculation of studentized deleted residuals for the first observation in the body fat example. The X values for this observation are $X_{11} = 19.5$ and $X_{12} = 43.1$. Using the fitted regression function from Table 8.4a, we obtain:

$$\hat{Y}_1 = -19.174 + .2224(19.5) + .6594(43.1) = 13.583$$

Since $Y_1 = 11.9$, the residual for this observation is $e_1 = 11.9 - 13.583 = -1.683$. We also know from Table 8.4a that $SSE = 109.95$ and from Table 11.7 that $h_{11} = .201$. Hence, by (11.59a), we find:

$$d_1^* = -1.683\left[\frac{20 - 3 - 1}{109.95(1 - .201) - (-1.683)^2}\right]^{1/2} = -.730$$

The studentized deleted residuals for all 20 observations are shown in column 3 of Table 11.7.

Note that observations 3, 8, and 13 have the largest absolute studentized deleted residuals. If we consider tail areas of .05 on each side to be extreme, we will need to compare the absolute values of the studentized deleted residuals with $t(.95; 16) = 1.746$. Based on this comparison, we should consider observations 8 and 13 extreme enough to warrant studying whether or not they are influential observations. Incidentally, consideration of the residuals e_i (shown in Table 11.7, column 1) here would also have identified observations 8 and 13 as the most outlying ones.

11.6 IDENTIFICATION OF INFLUENTIAL OBSERVATIONS AND REMEDIAL MEASURES

After identifying outlying observations with respect to their X values and/or their Y values, the next step is to ascertain whether or not they are influential in affecting the fit of the regression function, possibly leading to serious distortion effects.

One measure of the influence of the ith observation on the fit of the regression function is the difference between the vector $\mathbf{b}$ of the estimated regression coefficients based on all n observations and the vector $\mathbf{b}_{(i)}$ based on the $n - 1$ observations with the ith observation deleted:

$$\mathbf{b} - \mathbf{b}_{(i)} \tag{11.60}$$

Another possible measure of the influence of the ith observation is the difference between the fitted value $\hat{Y}_i$ based on the regression with all n observations and the fitted value $\hat{Y}_{(i)}$ obtained when the ith observation is deleted:

$$\hat{Y}_i - \hat{Y}_{(i)} \tag{11.61}$$

Cook's distance measure

An overall measure of the impact of the ith observation on the estimated regression coefficients is *Cook's distance measure* D_i. Recall from (7.43) that the boundary of the confidence region for all p regression coefficients β_k ($k = 0, 1, \ldots, p-1$) is given by:

$$\frac{(\mathbf{b}-\boldsymbol{\beta})'\mathbf{X}'\mathbf{X}(\mathbf{b}-\boldsymbol{\beta})}{pMSE} = F(1-\alpha; p, n-p) \tag{11.62}$$

Cook's distance measure D_i uses the same structure for measuring the combined impact of the differences in the estimated regression coefficients when the ith observation is deleted:

$$D_i = \frac{(\mathbf{b}-\mathbf{b}_{(i)})'\mathbf{X}'\mathbf{X}(\mathbf{b}-\mathbf{b}_{(i)})}{pMSE} \tag{11.63}$$

While D_i does not follow the F distribution, it has been found useful to relate the value D_i to the corresponding F distribution according to (11.62) and ascertain the percentile value. If the percentile value is less than about 10 or 20 percent, the ith observation has little apparent influence on the fitted regression function. If, on the other hand, the percentile value is near 50 percent or more, the distance between the vectors $\mathbf{b}$ and $\mathbf{b}_{(i)}$ should be considered large, implying that the ith observation has a substantial influence on the fit of the regression function.

Fortunately, Cook's distance measure D_i can be calculated without fitting new regression functions where the ith observation is deleted. An algebraically equivalent expression is:

$$D_i = \frac{e_i^2}{pMSE}\left[\frac{h_{ii}}{(1-h_{ii})^2}\right] \tag{11.63a}$$

Note from (11.63a) that D_i depends on two factors: (1) the size of the residual e_i and (2) the leverage value h_{ii}. The larger is either e_i or h_{ii}, the larger is D_i. Thus, the ith observation can be influential: (1) by having a large residual e_i and only a moderate leverage value h_{ii}, or (2) by having a large leverage value h_{ii} with only a moderately sized residual e_i, or (3) by having both a large residual e_i and a large leverage value h_{ii}.

Example. In the body fat example, we had identified observations 3 and 15 as outlying X observations and observations 8 and 13 as outlying Y observations. We now calculate Cook's distance measure for each of these observations. To illustrate the calculations, we shall consider observation 3. From Table 11.7, we have $e_3 = -3.176$ and $h_{33} = .372$; and from Table 8.4a, we have $MSE = 6.47$. Since $p = 3$, we obtain using (11.63a):

$$D_3 = \frac{(-3.176)^2}{3(6.47)}\left[\frac{.372}{(1-.372)^2}\right] = .490$$

The distance measures for all of the observations are presented in Table 11.7, column 4. We note from column 4 that observation 3 clearly is the most influential observation, with the next largest distance measure $D_{13} = .211$ being substantially smaller.

To assess the magnitude of $D_3 = .490$, we refer to the corresponding F distribution, namely, $F(p, n-p) = F(3, 17)$. It can be shown that .490 is about the 31st percentile of this distribution. Hence, it appears that observation 3 does influence the regression fit, but the extent of the influence may not be large enough to call for consideration of remedial measures.

Additional insights about the extent of the influence of observation 3 may be obtained by comparing the fitted value $\hat{Y}_3$ when all observations are utilized with the fitted value $\hat{Y}_{(3)}$ when observation 3 is deleted. It can be shown that $\hat{Y}_3 = 21.877$ and $\hat{Y}_{(3)} = 23.756$, so that omission of observation 3 increases the fitted value by 8.6 percent. This change indicates that observation 3, although it influences the fitted regression, may not play such a strong role as to require consideration of remedial measures.

Comments

1. Cook's distance measure D_i may be viewed as reflecting in the aggregate the differences between the fitted values for each observation when all n observations are used in the data base and the fitted values when the ith observation is deleted, since it can be shown that an equivalent expression for D_i is:

$$D_i = \frac{(\hat{\mathbf{Y}} - \hat{\mathbf{Y}}_{(i)})'(\hat{\mathbf{Y}} - \hat{\mathbf{Y}}_{(i)})}{pMSE} \tag{11.64}$$

Here, $\hat{\mathbf{Y}}$ as usual is the vector of the fitted values when all n observations are used in the data base for the regression fit and $\hat{\mathbf{Y}}_{(i)}$ is the vector of the fitted values when the ith observation is deleted from the data base.

2. Analysis of outlying and influential observations is a necessary component of good regression analysis. However, it is neither automatic nor foolproof and requires good judgment by the analyst. The methods which have been described often work well but at other times will be ineffective. For example, if two influential outlying observations are nearly coincident, an analysis which deletes one observation at a time and estimates the change in fit will result in virtually no change for these two outlying observations. The reason is that the retained outlying observation will mask the effect of the deleted outlying observation.

Remedial measures

After using the hat matrix, studentized deleted residuals, and Cook's distances to identify outlying influential observations that have a substantial impact on the least squares regression fit, one must decide what to do about such observations. Clearly, an outlying influential observation should not be automatically discarded, because it may be entirely correct and simply represents an unlikely event. Discarding of such an outlying observation could lead to the undesirable consequence of increased variances of some of the estimated regression coefficients.

If, on the other hand, the circumstances surrounding the data provide an explanation of the unusual observation which indicates an exceptional situation not to be covered by the model, the discarding of the observation may be appropriate. Thus, when an outlying influential observation can definitely be shown to be the result of a gross measurement error, it would be appropriate to discard that observation.

When the outlying influential observation is accurate, it may not represent an unlikely event but rather a failure of the model. The failure may be either the omission of an important independent variable or the choice of an incorrect functional form, such as omission of a curvature effect for an independent variable included in the model. Often, identification of outlying influential observations leads to valuable insights for strengthening the model.

When an outlying influential observation is accurate but no explanation can be found for it, a less severe alternative than discarding the observation is to dampen its influence. We shall now discuss one method of doing this.

Method of least absolute deviations. This method is one of a variety of robust methods that have the property of being insensitive to both outlying data values and inadequacies of the model employed. The method of least absolute deviations estimates the regression coefficients by minimizing the sum of the absolute deviations of the observations from their means. The criterion to be minimized is:

$$\sum_{i=1}^{n} |Y_i - (\beta_0 + \beta_1 X_{i1} + \cdots + \beta_{p-1} X_{i,p-1})| \tag{11.65}$$

Since absolute deviations rather than squared ones are involved here, the method of least absolute deviations places less emphasis on outlying observations than does the method of least squares.

The estimated regression coefficients according to the method of least absolute deviations can be obtained by linear programming techniques. Details about the computational aspects may be found in specialized texts, such as Reference 11.2.

Example. In the body fat example, observation 3 was identified as having a substantial influence on the fitted regression function. If observation 3 were deleted, the fitted regression function would be:

$$\hat{Y} = -12.428 + .5641X_1 + .3635X_2$$

An alternative to deleting observation 3 would be to reduce its influence. Using the method of least absolute deviations, the fitted regression function is:

$$\hat{Y} = -17.027 + .4173X_1 + .5203X_2$$

Since the fitted regression based on all observations is (Table 8.4a):

$$\hat{Y} = -19.174 + .2224X_1 + .6594X_2$$

it is seen that the method of least absolute deviations leads to more modest changes than dropping observation 3 entirely. An analysis of the residuals shows that the method of least absolute deviations resulted in reductions of those residuals that are largest absolutely with the method of least squares.

Comments

1. The residuals for the method of least absolute deviations ordinarily will not sum to zero.

2. The solution for the estimated regression coefficients with the method of least absolute deviations may not be unique.

3. The method of least absolute deviations is also called *minimum absolute deviations, minimum sum of absolute deviations,* and *minimum L_1-norm.*

4. Numerous other robust procedures besides the method of least absolute deviations have been proposed. Reference 11.3 discusses a number of these procedures.

PROBLEMS

11.1. Refer to the example of perfectly correlated independent variables in Table 11.1.
- a. Develop another model, like models (11.16) and (11.17), that fits the data perfectly.
- b. What is the intersection of the infinitely many response surfaces that fit the data perfectly?

11.2. The progress report of a research analyst to the supervisor stated: "All the estimated regression coefficients in our model with three independent variables to predict sales are statistically significant. Our new preliminary model with seven independent variables, which includes the three variables of our smaller model, is less satisfactory because only two of the seven regression coefficients are statistically significant. Yet in some initial trials the expanded model is giving more precise sales predictions than the smaller model. The reasons for this anomaly are now being investigated." Comment.

11.3. Two authors wrote as follows: "Our research utilized a multiple regression model. Two of the independent variables important in our theory turned out to be highly correlated in our observations. This made it difficult to assess the individual effects of each of these variables separately. We retained both variables in our model, however, because the high coefficient of multiple determination made this difficulty unimportant." Comment.

11.4. A student asked: "Why is it necessary to perform diagnostic checks of the fit when R^2 is large?" Comment.

11.5. **Cosmetics sales.** An assistant in the district sales office of a national cosmetics firm obtained data, shown below, on advertising expenditures and sales last year in the district's 14 territories. X_1 denotes expenditures for point-of-sale displays in beauty salons and department stores (in thousand dollars) while X_2 and X_3 represent the corresponding expenditures for local media advertising and prorated share of national media advertising, respectively. Y denotes sales (in thousand cases). The assistant was instructed to estimate the increase in expected sales when X_1 is increased by one thousand dollars and X_2 and X_3 are held constant, and was told to use an ordinary multiple regression model with linear terms for the independent variables and with independent normal error terms.

i:	1	2	3	4	5	6	7
X_{i1}:	4.2	6.5	3.0	2.1	2.9	7.2	4.8
X_{i2}:	4.0	6.5	3.5	2.0	3.0	7.0	5.0
X_{i3}:	3.0	5.0	4.0	3.0	4.0	3.0	4.5
Y_i:	8.26	14.70	9.73	5.62	7.84	12.18	8.56

i:	8	9	10	11	12	13	14
X_{i1}:	4.3	2.6	3.1	6.2	5.5	2.2	3.0
X_{i2}:	4.0	2.5	3.0	6.0	5.5	2.0	2.8
X_{i3}:	5.0	5.0	4.0	4.5	5.0	4.0	3.0
Y_i:	10.77	7.56	8.90	12.51	10.46	7.15	6.74

a. State the regression model to be employed and fit it to the data.

b. Test whether there is a regression relation between sales and the three independent variables. Use a level of significance of .05. State the alternatives, decision rule, and conclusion.

c. Test for each of the regression coefficients β_k ($k = 1, 2, 3$) individually whether or not $\beta_k = 0$. Use a level of significance of .05 each time. Do the conclusions of these tests correspond to that obtained in part (b)?

d. Obtain the correlation matrix.

e. What do the results in parts (b), (c), and (d) suggest about the suitability of the data for the research objective?

11.6. Refer to **Cosmetics sales** Problem 11.5.

a. Verify that the variance inflation factor for variable X_1 is $(VIF)_1 = 66.29$. The other variance inflation factors are $(VIF)_2 = 66.99$ and $(VIF)_3 = 1.09$. What do these suggest about the effects of multicollinearity here?

b. The assistant eventually decided to drop variables X_2 and X_3 from the model "to clear up the picture." Fit the assistant's revised model. Is the assistant now in a better position to achieve the research objective?

c. Why would an experiment here be more effective in providing suitable data to meet the research objective? How would you design such an experiment? What model would you employ?

11.7. Refer to **Patient satisfaction** Problem 7.17.

a. Obtain the correlation matrix. What does it show about pairwise linear associations among the independent variables?

b. The variance inflation factors are $(VIF)_1 = 1.35$, $(VIF)_2 = 2.76$, and $(VIF)_3 = 2.87$. What do these results suggest about the effects of multicollinearity here? Are these results more revealing than those in part (a)?

11.8. Refer to **Brand preference** Problem 7.8.

a. Obtain the correlation matrix. What does it show about the linear association between the two independent variables?

b. Find the two variance inflation factors. Why are they both equal to 1?

11.9. Refer to **Mathematicians salaries** Problem 7.20.

a. Obtain the correlation matrix. What does it show about the pairwise linear associations among the independent variables?

b. Obtain the variance inflation factors. Do they indicate that a serious multicollinearity problem exists here?

11.10. Refer to **Cosmetics sales** Problem 11.5. Given below are the estimated ridge standardized regression coefficients, the variance inflation factors, and R^2 for selected biasing constants c.

c:	.000	.005	.01	.02	.03	.04	.05	.06
b_1^R:	.273	.327	.349	.368	.376	.380	.382	.383
b_2^R:	.549	.494	.470	.447	.435	.427	.422	.417
b_3^R:	.260	.260	.260	.259	.257	.256	.254	.253
$(VIF)_1$:	66.29	24.11	12.45	5.20	2.92	1.91	1.38	1.07
$(VIF)_2$:	66.99	24.36	12.57	5.25	2.94	1.92	1.39	1.07
$(VIF)_3$:	1.09	1.06	1.04	1.01	.99	.97	.95	.93
R^2:	.840	.838	.836	.832	.828	.824	.821	.816

a. Make a ridge trace plot for the given c values. Do the ridge regression coefficients exhibit substantial changes near $c = 0$?
b. Suggest a reasonable value for the biasing constant c based on the ridge trace, the (VIF)'s, and R^2.
c. Transform the estimated standardized regression coefficients selected in part (b) back to the original variables and obtain the fitted values for the 14 observations. How similar are these fitted values to those obtained with the ordinary least squares fit in Problem 11.5a?

11.11. Refer to **Chemical shipment** Problem 7.12. Given below are the estimated ridge standardized regression coefficients, variance inflation factors, and R^2 for selected biasing constants c.

c:	.000	.005	.01	.05	.07	.09	.10	.20
b_1^R:	.451	.453	.455	.460	.460	.459	.458	.444
b_2^R:	.561	.556	.552	.526	.517	.508	.504	.473
$(VIF)_1 = (VIF)_2$:	7.03	6.20	5.51	2.65	2.03	1.61	1.46	.71
R^2:	.987	.984	.982	.962	.952	.943	.940	.894

a. Make a ridge trace plot for the given c values. Do the regression coefficients exhibit substantial changes near $c = 0$?
b. Why are the $(VIF)_1$ values the same as the $(VIF)_2$ values here?
c. Suggest a reasonable value for the biasing constant c based on the ridge trace in part (a), the (VIF)'s, and R^2.
d. Transform the estimated standardized regression coefficients selected in part (c) back to the original variables and obtain the fitted values for the 20 observations. How similar are these fitted values to those obtained with the ordinary least squares fit in Problem 7.12a?

11.12. Refer to **Brand preference** Problem 7.8. The diagonal elements of the hat matrix are: $h_{55} = h_{66} = h_{77} = h_{88} = h_{99} = h_{10,10} = h_{11,11} = h_{12,12} = .137$ and $h_{11} = h_{22} = h_{33} = h_{44} = h_{13,13} = h_{14,14} = h_{15,15} = h_{16,16} = .237$.
a. Explain the reason for the pattern in the diagonal elements of the hat matrix.
b. According to the rule of thumb stated in the chapter, are any of the observations outlying with regard to their X values?
c. Obtain the studentized deleted residuals and identify any outlying Y observations.
d. Calculate Cook's distance D_i for each observation. Are any observations influential according to this measure?

11.13. Refer to **Chemical shipment** Problem 7.12. The diagonal elements of the hat matrix are:

i:	1	2	3	4	5	6	7	8	9	10
h_{ii}:	.091	.194	.131	.268	.149	.141	.429	.067	.135	.165

i:	11	12	13	14	15	16	17	18	19	20
h_{ii}:	.179	.051	.110	.156	.095	.128	.097	.230	.112	.073

a. Identify any outlying X observations using the rule of thumb presented in the chapter.
b. Obtain the studentized deleted residuals and identify any outlying Y observations.
c. Calculate Cook's distance D_i for each observation. Are any observations influential according to this measure? How much is the fitted value for observation $i = 7$ changed when all observations are included and when observation 7 is omitted from the fit? [*Hint:* $\hat{Y}_{(i)} = Y_i - e_i/(1 - h_{ii})$.]

11.14. Refer to **Patient satisfaction** Problem 7.17. The diagonal elements of the hat matrix are:

i:	1	2	3	4	5	6	7	8	9	10	11	12
h_{ii}:	.134	.193	.070	.235	.204	.319	.060	.174	.339	.104	.209	.143

i:	13	14	15	16	17	18	19	20	21	22	23
h_{ii}:	.078	.231	.158	.238	.059	.057	.254	.137	.313	.072	.230

a. Identify any outlying X observations.
b. Obtain the studentized deleted residuals and identify any outlying Y observations.
c. Calculate Cook's distance D_i for each observation. Are any observations influential according to this measure?

11.15. Refer to **Mathematicians salaries** Problem 7.20. The diagonal elements of the hat matrix are:

i:	1	2	3	4	5	6	7	8	9	10	11	12
h_{ii}:	.184	.059	.132	.071	.214	.146	.115	.179	.241	.288	.083	.128

i:	13	14	15	16	17	18	19	20	21	22	23	24
h_{ii}:	.320	.098	.186	.151	.267	.146	.198	.206	.118	.136	.225	.110

a. Identify any outlying X observations.
b. Obtain the studentized deleted residuals and identify any outlying Y observations.
c. Calculate Cook's distance D_i for each observation. Are any observations influential according to this measure?

EXERCISES

11.16. Refer to the work crew productivity example data in Table 8.1.
a. For the variables transformed according to (11.6), obtain: (1) $\mathbf{X'X}$, (2) $\mathbf{X'Y}$, (3) $\mathbf{b}$, (4) $\mathbf{s}^2(\mathbf{b})$.

b. Show that the regression coefficients obtained in part (a3) are the same as the standardized regression coefficients according to (7.69).

11.17. Show that the least squares estimator of β_0' in (11.2a) is $\bar{Y}$.

11.18. Derive the relations between the β_k and β_k' ($k = 1, 2$) in (11.8a) and (11.8b).

11.19. Derive the expression for $\mathbf{X}'\mathbf{Y}$ in (11.12) for the transformed model (11.7).

11.20. Derive the mean squared error in (11.35).

11.21. Refer to the least absolute deviations estimates for the body fat example on page 410—namely, $b_0 = -17.027$, $b_1 = .4173$, and $b_2 = .5203$.

a. Find the sum of the absolute deviations of the sample observations from the fitted values based on the least absolute deviations estimates.

b. For the least squares estimated regression coefficients $b_0 = -19.174$, $b_1 = .2224$, and $b_2 = .6594$, find the sum of the absolute deviations. Is this sum larger than the sum obtained in part (a)?

PROJECTS

11.22. Refer to **Patient satisfaction** Problem 7.17.

a. Obtain the estimated ridge standardized regression coefficients, variance inflation factors, and R^2 for the following biasing constants: $c = .000, .005, .01, .02, .03, .04, .05$.

b. Make a ridge trace plot for the given c values. Do the ridge regression coefficients exhibit substantial changes near $c = 0$?

c. Suggest a reasonable value for the biasing constant c based on the ridge trace, the (VIF)'s, and R^2.

d. Transform the estimated standardized regression coefficients selected in part (c) back to the original variables and obtain the fitted values for the 23 observations. How similar are these fitted values to those obtained with the ordinary least squares fit in Problem 7.17a?

11.23. Refer to **Mathematicians salaries** Problem 7.20.

a. Obtain the estimated ridge standardized regression coefficients, variance inflation factors, and R^2 for the following biasing constants: $c = .000, .005, .01, .02, .03, .04, .05$.

b. Make a ridge trace plot for the given c values. Do the ridge regression coefficients exhibit substantial changes near $c = 0$?

c. Suggest a reasonable value for the biasing constant c based on the ridge trace, the (VIF)'s, and R^2.

d. Transform the estimated standardized regression coefficients selected in part (c) back to the original variables and obtain the fitted values for the 24 observations. How similar are these fitted values to those obtained with the ordinary least squares fit in Problem 7.20a?

11.24. Refer to the **SENIC** data set.

a. Regress the logarithm of length of stay (Y') on infection risk (X_1), number of beds (X_2), and average daily census (X_3).

b. Obtain the residuals and identify outliers.

c. Obtain the correlation matrix and the variance inflation factors. What do these suggest about the effects of multicollinearity?
d. Obtain the estimated ridge regression coefficients, variance inflation factors, and R^2 for the values of the biasing constant c given in Table 11.6.
e. Make a ridge trace plot and determine a reasonable value for the biasing constant c based on this plot, the (VIF)'s, and R^2.

11.25. Refer to the **SMSA** data set.
a. Regress number of active physicians (Y) on number of hospital beds (X_1), total personal income (X_2), and total serious crimes (X_3).
b. Obtain the residuals and identify outliers.
c. Obtain the correlation matrix and the variance inflation factors. What do these suggest about the effects of multicollinearity?
d. Obtain the estimated ridge regression coefficients, variance inflation factors, and R^2 for the values of the biasing constant c in Table 11.6.
e. Make a ridge trace plot and determine a reasonable value for the biasing constant c based on this plot, the (VIF)'s, and R^2.

CITED REFERENCES

11.1 Belsley, David A.; Edwin Kuh; and Roy E. Welsch. *Regression Diagnostics: Identifying Influential Data and Sources of Collinearity*. New York: John Wiley & Sons, 1980.

11.2 Kennedy, William J., Jr., and James E. Gentle. *Statistical Computing*. New York: Marcel Dekker, 1980.

11.3 Hogg, Robert V. "Statistical Robustness: One View of Its Use in Applications Today." *The American Statistician* 33 (1979), pp. 108–15.

12

Selection of independent variables

One of the most difficult problems in regression analysis often is the selection of the set of independent variables to be employed in the model. In this chapter, we take up several computer-assisted search methods for helping to identify one or a number of possible sets of independent variables to be included in the regression model.

12.1 NATURE OF PROBLEM

As we have seen in previous chapters, regression analysis has three major uses: (1) description, (2) control, and (3) prediction. For each of these uses, the investigator must specify the set of independent variables to be employed for describing, controlling, and/or predicting the dependent variable.

In some fields, theory can aid in selecting the independent variables to be employed and in specifying the functional form of the regression relation. Often in these fields, controlled experiments can be undertaken to furnish data on the basis of which the regression parameters can be estimated and the theoretical form of the regression function tested.

In many other subject matter fields, however, including the social and behavioral sciences and management, serviceable theoretical models are relatively rare. To complicate matters further, the available theoretical models may involve independent variables that are not directly measurable, such as a family's future earnings over the next 10 years. Under these conditions, investigators are often forced to prospect for independent variables that could conceivably be related to the dependent variable under study. Obviously, such a set of independent variables will be large. For example, a company's sales of portable dishwashers in a district may be affected by population size, per capita income, percent of population in urban areas, percent of population under 50 years old, percent of families with children at home, etc. etc.!

After such a lengthy list has been compiled, some of the independent variables can be screened out. An independent variable (1) may not be fundamental to the problem, (2) may be subject to large measurement errors, and/or (3) may effectively duplicate another independent variable in the list. Other independent variables that cannot be measured may either be deleted or replaced by proxy variables that are highly correlated with them.

Typically, the number of independent variables that remain after this initial screening is still large. Further, many of these variables will be highly intercorrelated. Hence, the investigator usually will wish to reduce the number of independent variables to be used in the final model. There are several reasons for this. A regression model with a large number of independent variables is expensive to maintain. Further, regression models with a limited number of independent variables are easier to analyze and understand. Finally, the presence of many highly intercorrelated independent variables may add little to the predictive power of the model while substantially increasing the sampling variation of the regression coefficients, detracting from the model's descriptive abilities, and increasing the problem of roundoff errors (as we have seen in Chapter 11).

The investigator must be careful, however, not to eliminate key explanatory variables because that could seriously damage the explanatory power of the model and lead to biased estimates of regression coefficients, mean responses, and predictions of new observations.

The problem then is how to shorten the list of independent variables so as to obtain, in some sense, a "good" selection of independent variables. This subset of independent variables needs to be small enough so that maintenance costs are manageable and analysis is facilitated, yet it must be large enough so that adequate description, control, or prediction is possible.

Since the purposes of regression analysis vary, no one subset of independent variables is usually "best" for all uses. For instance, descriptive uses of a regression model typically will emphasize precise estimation of the regression coefficients, while predictive uses will focus on the prediction errors. Often, different subsets of the potential independent variables will best serve these varying purposes. Even for a given purpose, it is often found that several subsets are about equally "good" according to a given criterion, and the choice of the subset of independent variables to be employed in the regression model needs to be made on the basis of additional considerations. The entire selection process is, and should be, pragmatic, with large doses of subjective judgment. The selection procedures to be discussed in this chapter are aids to the investigator's judgment and should not be used in a purely mechanical fashion. A mechanical approach, for instance, might omit an important independent variable just because it occurred in the sample within a narrow range of values and therefore turned out to be statistically nonsignificant.

We shall discuss in this chapter two approaches to the selection of independent variables. The first approach considers all possible regression models that can be developed from the pool of potential independent variables and identifies subsets of the independent variables which are "good" according to a criterion specified by the investigator. The second approach employs automatic search procedures to arrive at a single subset of the independent variables.

The choice of the independent variables to be employed in the regression model does not, of course, fully determine the model to be utilized. Other determinations must also be made, such as whether an independent variable appears in linear form, in a transformed fashion, or with a quadratic term added, and whether interaction terms should be included in the model. Our discussion of the choice of independent variables in this chapter assumes that the investigator has already considered the functional form of the regression relation (whether given variables are to appear in linear form, quadratic form, etc.), whether the independent variables or the dependent variable are first transformed (e.g., by a logarithmic transformation), and whether any interaction terms are to be included. At this point, a selection procedure is employed to reduce the number of X variables, which include not only the potential independent variables in first-order form but also quadratic and other curvature terms and interaction terms.

Note

All too often, unwary investigators will screen the set of independent variables by fitting the regression model containing the entire set of potential X variables and then simply dropping those for which the t^* statistic (8.25):

$$t_k^* = \frac{b_k}{s(b_k)} \tag{12.1}$$

has a small absolute value. As we know from Chapter 11, this procedure can lead to the dropping of important intercorrelated independent variables. Clearly, a good search procedure must be able to handle important intercorrelated independent variables in such a way that not all of them will be dropped.

12.2 EXAMPLE

In order to illustrate the selection procedures to be discussed in the following sections, we shall use a relatively simple example which has four potential independent variables. By limiting the number of potential independent variables, we shall be able to explain the selection procedures without overwhelming the reader with masses of computer printouts.

A hospital surgical unit was interested in predicting survival in patients under-

TABLE 12.1 Potential independent variables and dependent variable—surgical unit example

Case Number	*Blood Clotting Score* X_1	*Prognostic Index* X_2	*Enzyme Function Test* X_3	*Liver Function Test* X_4	*Survival Time* Y	$Y' = \log_{10} Y$
1	6.7	62	81	2.59	200	2.3010
2	5.1	59	66	1.70	101	2.0043
3	7.4	57	83	2.16	204	2.3096
4	6.5	73	41	2.01	101	2.0043
5	7.8	65	115	4.30	509	2.7067
6	5.8	38	72	1.42	80	1.9031
7	5.7	46	63	1.91	80	1.9031
8	3.7	68	81	2.57	127	2.1038
9	6.0	67	93	2.50	202	2.3054
10	3.7	76	94	2.40	203	2.3075
11	6.3	84	83	4.13	329	2.5172
12	6.7	51	43	1.86	65	1.8129
13	5.8	96	114	3.95	830	2.9191
14	5.8	83	88	3.95	330	2.5185
15	7.7	62	67	3.40	168	2.2253
16	7.4	74	68	2.40	217	2.3365
17	6.0	85	28	2.98	87	1.9395
18	3.7	51	41	1.55	34	1.5315
19	7.3	68	74	3.56	215	2.3324
20	5.6	57	87	3.02	172	2.2355
21	5.2	52	76	2.85	109	2.0374
22	3.4	83	53	1.12	136	2.1335
23	6.7	26	68	2.10	70	1.8451
24	5.8	67	86	3.40	220	2.3424
25	6.3	59	100	2.95	276	2.4409
26	5.8	61	73	3.50	144	2.1584
27	5.2	52	86	2.45	181	2.2577
28	11.2	76	90	5.59	574	2.7589
29	5.2	54	56	2.71	72	1.8573
30	5.8	76	59	2.58	178	2.2504
31	3.2	64	65	0.74	71	1.8513
32	8.7	45	23	2.52	58	1.7634
33	5.0	59	73	3.50	116	2.0645
34	5.8	72	93	3.30	295	2.4698
35	5.4	58	70	2.64	115	2.0607
36	5.3	51	99	2.60	184	2.2648
37	2.6	74	86	2.05	118	2.0719
38	4.3	8	119	2.85	120	2.0792
39	4.8	61	76	2.45	151	2.1790
40	5.4	52	88	1.81	148	2.1703
41	5.2	49	72	1.84	95	1.9777
42	3.6	28	99	1.30	75	1.8751
43	8.8	86	88	6.40	483	2.6840
44	6.5	56	77	2.85	153	2.1847
45	3.4	77	93	1.48	191	2.2810
46	6.5	40	84	3.00	123	2.0899
47	4.5	73	106	3.05	311	2.4928
48	4.8	86	101	4.10	398	2.5999
49	5.1	67	77	2.86	158	2.1987
50	3.9	82	103	4.55	310	2.4914
51	6.6	77	46	1.95	124	2.0934
52	6.4	85	40	1.21	125	2.0969
53	6.4	59	85	2.33	198	2.2967
54	8.8	78	72	3.20	313	2.4955

going a particular type of liver operation. A random selection of 54 patients was available for analysis. From each patient record, the following information was extracted from the preoperational evaluation:

X_1 blood clotting score
X_2 prognostic index, which includes the age of patient
X_3 enzyme function test score
X_4 liver function test score

These constitute the potential independent variables for a predictive regression model. The dependent variable was survival time, which was ascertained in a follow-up study. The data on the potential independent variables and the dependent variable are presented in Table 12.1. Since the survival time distribution is substantially skewed to the right, the logarithm of the survival time $Y' = \log_{10} Y$ was taken as the dependent variable.

The surgical unit wished to obtain a subset of independent variables for predicting Y'. The task was assigned to an analyst, who first obtained the correlation matrix for all the variables from a computer run. This matrix provides valuable basic information on the nature of the problems to be encountered. Table 12.2 contains the correlation matrix based on a computer run, omitting the duplicate terms below the main diagonal.

Table 12.2 indicates that all of the independent variables are linearly associated with Y', X_4 showing the highest degree of association and X_1 the lowest. The correlation matrix further shows intercorrelations among the potential independent variables. In particular, the individual pairwise correlations between X_4 and X_1, X_2, and X_3 are moderately high. The task now is to determine whether all four independent variables are required for the prediction model.

TABLE 12.2 Correlation matrix for surgical unit example

	Y'	X_1	X_2	X_3	X_4
Y'	1.000	.346	.593	.665	.726
X_1		1.000	.090	−.150	.502
X_2			1.000	−.024	.369
X_3				1.000	.416
X_4					1.000

12.3 ALL POSSIBLE REGRESSION MODELS

The *all-possible-regressions selection procedure* calls for an examination of all possible regression models involving the potential X variables and identifying "good" subsets according to some criterion. When this selection procedure is used with our surgical unit example, for instance, 16 different regression models are to be considered, as shown in Table 12.3. First, there is the regression model with no X variables, i.e., the model $Y_i = \beta_0 + \varepsilon_i$. Then there are the regression models with one X variable (X_1, X_2, X_3, X_4), with two X variables (X_1 and X_2, X_1 and X_3, X_1 and X_4, X_2 and X_3, X_2 and X_4, X_3 and X_4), and so on.

Different criteria for comparing the various regression models may be used with the all-possible-regressions selection procedure. We shall discuss three—R_p^2, MSE_p, and C_p. Before doing so, we need to develop some notation. Let us denote the number of potential X variables in the pool by $P - 1$. We assume throughout this chapter that all regression models contain an intercept term β_0. Hence, the regression function containing all potential X variables contains P parameters, and the function with no X variables contains one parameter (β_0).

The number of X variables in a subset will be denoted by $p - 1$, as always, so that there are p parameters in the regression function for this subset of X variables. Thus, we have:

$$1 \leq p \leq P \tag{12.2}$$

The all-possible-regressions approach assumes that the number of observations n exceeds the maximum number of potential parameters:

$$n > P \tag{12.3}$$

and, indeed, it is highly desirable that n be substantially larger than P so that sound results can be obtained.

R_p^2 criterion

The R_p^2 criterion calls for an examination of the coefficient of multiple determination R^2, defined in (7.31), in order to select one or several subsets of X variables. We show the number of parameters in the regression model as a subscript of R^2. Thus, R_p^2 indicates that there are p parameters, or $p - 1$ predictor variables, in the regression equation on which R_p^2 is based.

Since R_p^2 is a ratio of sums of squares:

$$R_p^2 = \frac{SSR_p}{SSTO} = 1 - \frac{SSE_p}{SSTO} \tag{12.4}$$

and the denominator is constant for all possible regressions, R_p^2 varies inversely with the error sums of squares SSE_p. But we know that SSE_p can never increase as additional independent variables are included in the model. Thus, R_p^2 will be a maximum when all $P - 1$ potential X variables are included in the regression model. The reason for using the R_p^2 criterion with the all-possible-regressions approach therefore cannot be to maximize R_p^2. Rather, the intent is to find the point where adding more X variables is not worthwhile because it leads to a very small increase in R_p^2. Often, this point is reached when only a limited number of X variables is included in the regression model. Clearly, the determination of where diminishing returns set in is a judgmental one.

Example. Table 12.3, column 4, contains the R_p^2 values for all possible regression models for our surgical unit example. The data were obtained from a series of computer runs. For instance, when X_4 is the only X variable in the

TABLE 12.3 R_p^2, MSE_p, and C_p values for all possible regression models—surgical unit example

	(1)	(2)	(3)	(4)	(5)	(6)
X Variables in Model	p	*df*	SSE_p	R_p^2	MSE_p	C_p
None	1	53	3.9728	0	.0750	1,721.6
X_1	2	52	3.4960	.120	.0672	1,510.7
X_2	2	52	2.5762	.352	.0495	1,100.1
X_3	2	52	2.2154	.442	.0426	939.0
X_4	2	52	1.8777	.527	.0361	788.3
X_1, X_2	3	51	2.2324	.438	.0438	948.6
X_1, X_3	3	51	1.4073	.646	.0276	580.3
X_1, X_4	3	51	1.8759	.528	.0368	789.5
X_2, X_3	3	51	.7431	.813	.0146	283.7
X_2, X_4	3	51	1.3922	.650	.0273	573.5
X_3, X_4	3	51	1.2455	.687	.0244	508.0
X_1, X_2, X_3	4	50	.1099	.972	.00220	3.1
X_1, X_2, X_4	4	50	1.3905	.650	.0278	574.8
X_1, X_3, X_4	4	50	1.1157	.719	.0223	452.1
X_2, X_3, X_4	4	50	.4653	.883	.00931	161.7
X_1, X_2, X_3, X_4	5	49	.1098	.972	.00224	5.0

regression model, we obtain:

$$R_2^2 = 1 - \frac{SSE(X_4)}{SSTO} = 1 - \frac{1.8777}{3.9728} = .527$$

Note that $SSTO = SSE_1 = 3.9728$.

The R_p^2 values are plotted in Figure 12.1. The maximum R_p^2 value for the possible subsets of $p - 1$ predictor variables, denoted by $\max(R_p^2)$, appears at the top of the graph for each p. These points are connected by dashed lines to show the impact of adding additional X variables. Figure 12.1 makes it clear that little increase in $\max(R_p^2)$ takes place after three X variables are included in the model. Hence, the use of the subset (X_1, X_2, X_3) in the regression model appears to be reasonable according to the R_p^2 criterion.

Note that variable X_4, which singly correlates most highly with the dependent variable, is not in the $\max(R_p^2)$ models for $p = 3$ and $p = 4$, indicating that X_2 and X_3 contain much of the information presented by X_4. If it were desired that X_4 be retained in the model and that the subset model be limited to three X variables, the subset (X_2, X_3, X_4) should then be considered as next best according to the R_p^2 criterion for $p = 4$. The coefficient of multiple determination associated with this subset, $R_4^2 = .883$, would be somewhat smaller than $R_4^2 = .972$ for the subset (X_1, X_2, X_3).

MSE_p or R_a^2 criterion

Since R_p^2 does not take account of the number of parameters in the model, and since $\max(R_p^2)$ can never decrease as p increases, the use of the adjusted coeffi-

FIGURE 12.1 R_p^2 plot for surgical unit example

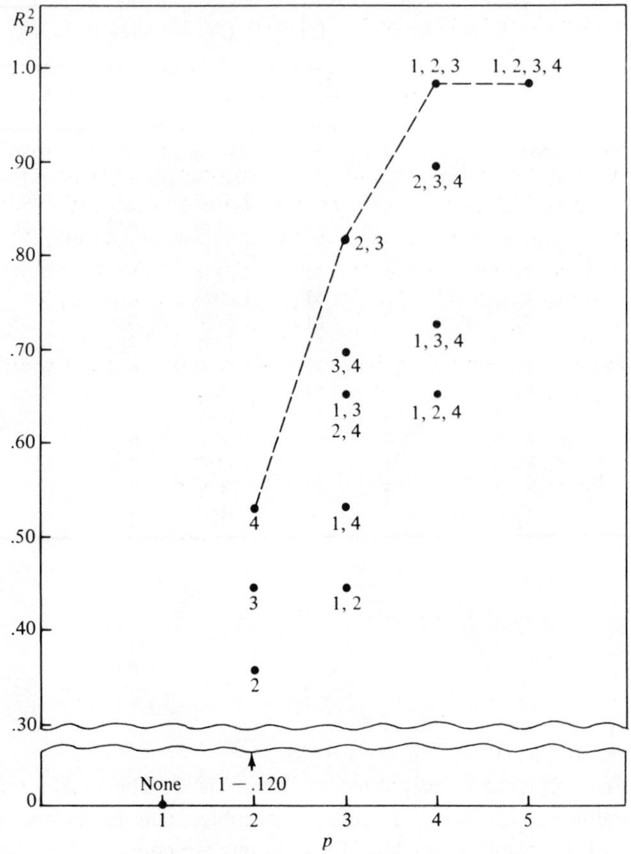

cient of multiple determination R_a^2 in (7.33):

$$R_a^2 = 1 - \left(\frac{n-1}{n-p}\right)\frac{SSE}{SSTO} = 1 - \frac{MSE}{\dfrac{SSTO}{n-1}} \tag{12.5}$$

has been suggested as a criterion which takes the number of parameters in the model into account through the degrees of freedom. It can be seen from (12.5) that R_a^2 increases if and only if MSE decreases since $SSTO/(n - 1)$ is fixed for the given Y observations. Hence, R_a^2 and MSE are equivalent criteria. We shall consider here the criterion MSE_p. Min(MSE_p) can, indeed, increase as p increases when the reduction in SSE_p becomes so small that it is not sufficient to offset the loss of an additional degree of freedom. Users of the MSE_p criterion either seek to find the subset of X variables that minimizes MSE_p, or one or several subsets for which MSE_p is so close to the minimum that adding more variables is not worthwhile.

Example. The MSE_p values for all possible regression models for our surgical unit example are shown in Table 12.3, column 5. For instance, if the regression model contains only X_4, we have:

$$MSE_2 = \frac{SSE(X_4)}{n-2} = \frac{1.8777}{52} = .0361$$

Figure 12.2 contains the MSE_p plot for our example. We have connected the $\min(MSE_p)$ values for each p by dashed lines. The story which Figure 12.2 tells is very similar to that told by Figure 12.1. The subset (X_1, X_2, X_3) appears to be best. Indeed, the mean square error achieved with this subset is practically the same as that with (X_1, X_2, X_3, X_4), which uses all potential X variables.

If X_4 were to be included in the model with $p = 4$, the subset (X_2, X_3, X_4) would be best, involving $MSE_4 = .009$ which is somewhat higher than $MSE_4 = .002$ for subset (X_1, X_2, X_3).

FIGURE 12.2 *MSE_p plot for surgical unit example*

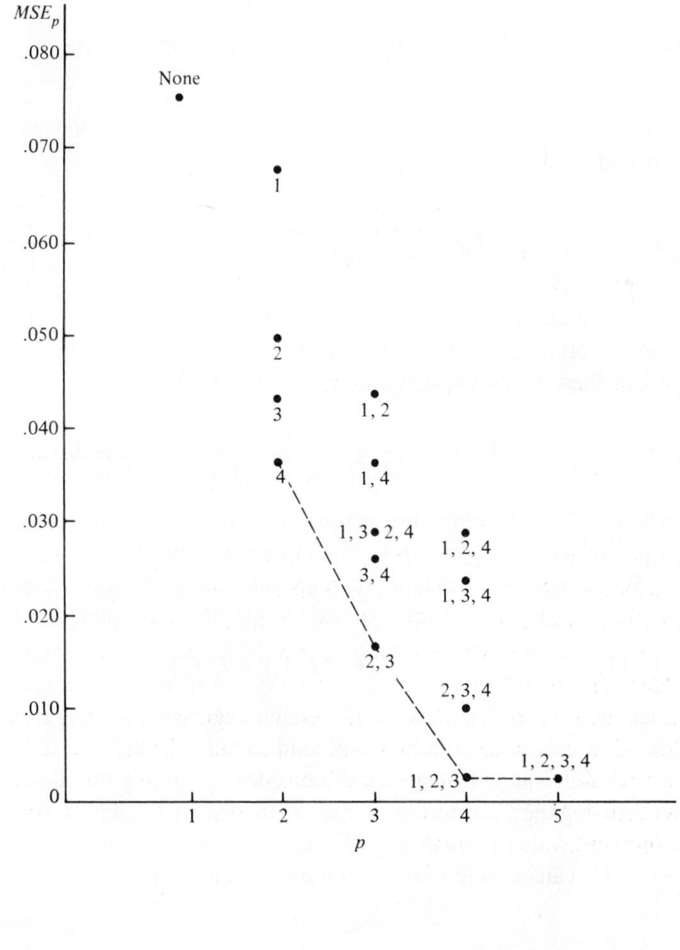

C_p criterion

This criterion is concerned with the *total mean squared error* of the n fitted values for each of the various subset regression models. The mean squared error concept involves a bias component and a random error component. The mean squared error for an estimated regression coefficient was defined in (11.35). Here, the mean squared error pertains to the fitted values $\hat{Y}_i$ for the regression model employed. The bias component for the ith fitted value $\hat{Y}_i$ is:

$$E(\hat{Y}_i) - E(Y_i) \tag{12.6}$$

where $E(\hat{Y}_i)$ is the expectation of the ith fitted value for the given regression model and $E(Y_i)$ is the true mean response. The random error component for $\hat{Y}_i$ is simply $\sigma^2(\hat{Y}_i)$, its variance. The mean squared error for $\hat{Y}_i$ is then the sum of the squared bias and the variance:

$$[E(\hat{Y}_i) - E(Y_i)]^2 + \sigma^2(\hat{Y}_i) \tag{12.7}$$

The total mean squared error for all n fitted values $\hat{Y}_i$ is the sum of the n individual mean squared errors:

$$\sum_{i=1}^{n} [E(\hat{Y}_i) - E(Y_i)]^2 + \sum_{i=1}^{n} \sigma^2(\hat{Y}_i) \tag{12.8}$$

The criterion measure, denoted by Γ_p, is simply the total mean squared error divided by σ^2, the true error variance:

$$\Gamma_p = \frac{1}{\sigma^2}\left\{\sum_{i=1}^{n} [E(\hat{Y}_i) - E(Y_i)]^2 + \sum_{i=1}^{n} \sigma^2(\hat{Y}_i)\right\} \tag{12.9}$$

The model which includes all $P - 1$ potential X variables is assumed to have been carefully chosen so that $MSE(X_1, \ldots, X_{P-1})$ is an unbiased estimator of σ^2. It can then be shown that an estimator of Γ_p is C_p:

$$C_p = \frac{SSE_p}{MSE(X_1, \ldots, X_{P-1})} - (n - 2p) \tag{12.10}$$

where SSE_p is the error sum of squares for the fitted subset regression model with p parameters (i.e., with $p - 1$ predictor variables).

When there is no bias in the regression model with $p - 1$ predictor variables so that $E(\hat{Y}_i) \equiv E(Y_i)$, the expected value of C_p is approximately p:

$$E[C_p \mid E(\hat{Y}_i) \equiv E(Y_i)] \simeq p \tag{12.11}$$

Thus, when the C_p values for all possible regression models are plotted against p, those models with little bias will tend to fall near the line $C_p = p$. Models with substantial bias will tend to fall considerably above this line.

In using the C_p criterion, one seeks to identify subsets of X variables for which (1) the C_p value is small and (2) the C_p value is near p. Sets of X variables with small C_p values have a small total mean squared error, and when the C_p value is also near p, the bias of the regression model is small. It may sometimes occur that the regression model based on the subset of X variables with the smallest C_p value involves substantial bias. In that case, one may at times prefer a regression model based on a somewhat larger subset of X variables for which the C_p value is slightly larger but which does not involve a substantial bias component. Reference 12.1 contains extended discussions of applications of the C_p criterion.

Example. Table 12.3, column 6, contains the C_p values for all possible regression models for our surgical unit example. For instance, when X_4 is the only X variable in the regression model, the C_p value is:

$$C_2 = \frac{SSE(X_4)}{MSE(X_1, X_2, X_3, X_4)} - [n - 2(2)]$$

$$= \frac{1.8777}{.00224} - (54 - 4) = 788.3$$

The C_p values for all possible regression models are plotted in Figure 12.3. We

FIGURE 12.3 C_p plot for surgical unit example

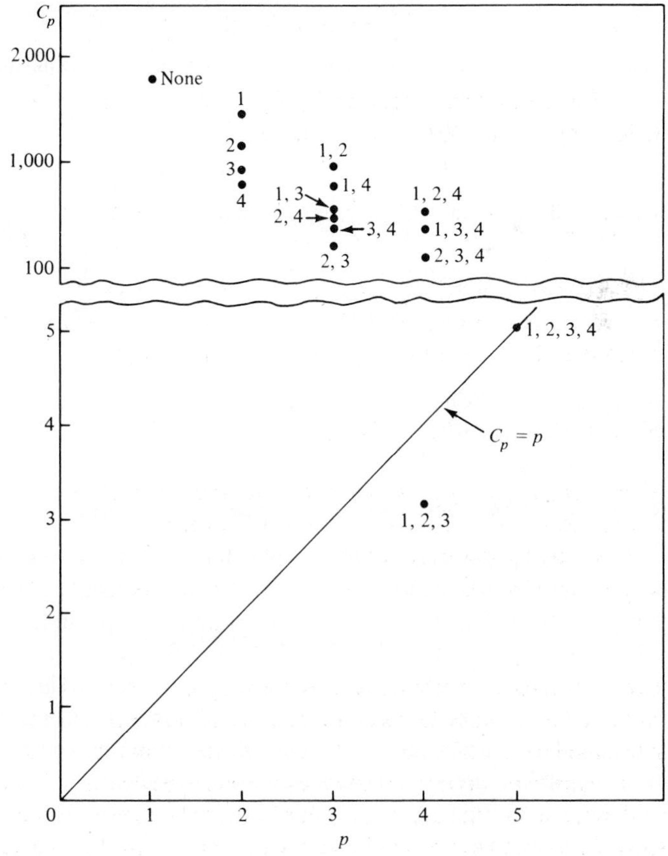

find again that subset (X_1, X_2, X_3) is suggested. This subset has the smallest C_p value, with no indication of any bias in the regression model. The fact that the C_p measure for this model, $C_4 = 3.1$, is below $p = 4$ is the result of random variation in the C_p measure.

Note that use of all potential X variables (X_1, X_2, X_3, X_4) would involve a larger total mean squared error. Also, use of subset (X_2, X_3, X_4) with C_p value $C_4 = 161.7$ would be poor because of the substantial bias with that model.

Comments

1. Effective use of the C_p criterion requires careful development of the pool of $P - 1$ potential X variables, with the independent variables expressed in appropriate form (linear, quadratic, transformed) and useless variables excluded so that $MSE(X_1, \ldots, X_{P-1})$ provides an unbiased estimate of the error variance σ^2.

2. The C_p criterion places major emphasis on the fit of the subset model for the n sample observations. At times, a modification of the C_p criterion which emphasizes new observations to be predicted might be preferable.

3. To see why C_p as defined in (12.10) is an estimator of Γ_p, we need to utilize two results that we shall simply state. First, it can be shown that:

$$\sum_{i=1}^{n} \sigma^2(\hat{Y}_i) = p\sigma^2 \tag{12.12}$$

Thus, the total random error of the n fitted values $\hat{Y}_i$ increases as the number of variables in the regression model increases.

Further, it can be shown that:

$$E(SSE_p) = \Sigma[E(\hat{Y}_i) - E(Y_i)]^2 + (n - p)\sigma^2 \tag{12.13}$$

Hence, Γ_p in (12.9) can be expressed as follows:

$$\begin{aligned} \Gamma_p &= \frac{1}{\sigma^2}[E(SSE_p) - (n - p)\sigma^2 + p\sigma^2] \\ &= \frac{E(SSE_p)}{\sigma^2} - (n - 2p) \end{aligned} \tag{12.14}$$

Replacing $E(SSE_p)$ by the estimator SSE_p and using $MSE(X_1, \ldots, X_{P-1})$ as an estimator of σ^2 yields C_p in (12.10).

Identification of "best" subsets by use of algorithm

A major disadvantage of the all-possible-regressions selection procedure is the amount of computation required. Since each potential independent variable either can be included or excluded, there are $2^{(P-1)}$ possible regression models when there are $P - 1$ potential X variables. When $P - 1 = 10$, for instance, there are 1,024 possible regression models. With the availability of large computers today, running all possible regression models for as many as 10 potential X variables is not too time consuming. Beyond that, however, the development of all possible regression models becomes inefficient.

Even when all possible regression models can easily be calculated, as when there are eight variables in the pool of potential X variables, it may be difficult for the investigator to evaluate carefully all of the regression models fitted. In the case of 8 potential X variables, for instance, there would be 256 models to consider, a major task for an investigator.

Thus, whenever the pool of potential X variables is not very small, it is highly desirable that the investigator be able to concentrate on a limited number of regression models which are the "best" ones according to a specified criterion. This limited number might consist of the "best" 5 or 10 subsets according to the criterion employed, so that the investigator can then carefully study these regression models for choosing the final model to be employed.

Time-saving algorithms have been developed in which the "best" subsets according to a specified criterion are identified without requiring the fitting of most of the possible subset regression models. Thus, if, say, the C_p criterion is to be employed and the five "best" subsets according to this criterion are to be identified, these algorithms search for the five subsets of X variables with the smallest C_p values using much less computational effort than when all possible subsets are evaluated. These algorithms are called *"best" subsets algorithms*. Not only do these algorithms provide the best subsets according to the specified criterion, but they often will also provide a number of "good" subsets for each possible number of X variables in the model to give the investigator helpful information in making the final selection of the subset of X variables to be employed in the regression model.

Example. In our surgical unit example, use of one of the "best" subsets algorithms will provide a portion of the information in Table 12.3. Suppose that the C_p criterion is to be employed and that the "best" three subsets are to be identified. The algorithm will then identify subsets (X_1, X_2, X_3), (X_1, X_2, X_3, X_4), and (X_2, X_3, X_4) as the three subsets with the smallest C_p values. In addition, information about three "good" subsets for each level of p may be provided.

Some final comments

The all-possible-regressions selection approach or a "best" subsets algorithm leads to the identification of a small number of subsets which are "good" according to a specified criterion. While in our surgical unit example, each of the three criteria pointed to the same "best" subset, this is not always the case. It therefore may be desirable at times to consider more than one criterion in evaluating possible subsets of X variables.

Once the investigator has identified a few subsets as "good" ones, a final choice of the model variables must be made. This choice is aided by residual analyses and examinations of influential observations for each of the competing models. Information gained by these analyses, together with knowledge by the invesigator about the phenomenon under study, will be helpful in choosing the final regression model to be employed.

12.4 STEPWISE REGRESSION

In those occasional cases when the pool of potential X variables contains 40 to 60 or even more variables, use of a "best" subsets algorithm may not be feasible. An automatic search procedure that develops sequentially the subset of X variables to be included in the regression model may be helpful in those cases. The *stepwise regression procedure* is probably the most widely used of the automatic search methods. It was developed to economize on computational efforts, as compared with the all-possible-regressions approach, while arriving at a reasonably "good" subset of independent variables. Essentially, this search method develops a sequence of regression models, at each step adding or deleting an X variable. The criterion for adding or deleting an X variable can be stated equivalently in terms of error sum of squares reduction, coefficient of partial correlation, or F^* statistic.

Search algorithm

We shall describe the stepwise regression search algorithm in terms of the F^* statistic for the partial F test.

1. The stepwise regression routine first fits a simple regression model for each of the $P - 1$ potential X variables. For each simple regression model, the F^* statistic (3.59) for testing whether or not the slope is zero is obtained:

$$F_k^* = \frac{MSR(X_k)}{MSE(X_k)} \tag{12.15}$$

Recall that $MSR(X_k) = SSR(X_k)$ measures the reduction in the total variation of Y associated with the use of the variable X_k. The X variable with the largest F^* value is the candidate for first addition. If this F^* value exceeds a predetermined level, the X variable is added. Otherwise, the program terminates with no X variable considered sufficiently helpful to enter the regression model.

2. Assume X_7 is the variable entered at step 1. The stepwise regression routine now fits all regression models with two X variables, where X_7 is one of the pair. For each such regression model, the partial F test statistic (8.24):

$$F_k^* = \frac{MSR(X_k \mid X_7)}{MSE(X_7, X_k)} = \left[\frac{b_k}{s(b_k)}\right]^2 \tag{12.16}$$

is obtained. This is the statistic for testing whether or not $\beta_k = 0$ when X_7 and X_k are the variables in the model. The X variable with the largest F^* value is the candidate for addition at the second stage. If this F^* value exceeds a predetermined level, the second X variable is added. Otherwise the program terminates.

3. Suppose X_3 is added at the second stage. Now the stepwise regression routine examines whether any of the other X variables already in the model should be dropped. For our illustration, there is at this stage only one other X variable in the model, X_7, so that only one partial F test statistic is obtained:

$$F_7^* = \frac{MSR(X_7|X_3)}{MSE(X_3, X_7)} \qquad (12.17)$$

At later stages, there would be a number of these F^* statistics, for each of the variables in the model besides the one last added. The variable for which this F^* value is smallest is the candidate for deletion. If this F^* value falls below a predetermined limit, the variable is dropped from the model; otherwise, it is retained.

4. Suppose X_7 is retained so that both X_3 and X_7 are now in the model. The stepwise regression routine now examines which X variable is the next candidate for addition, then examines whether any of the variables already in the model should now be dropped, and so on until no further X variables can either be added or deleted, at which point the search terminates.

It should be noted that the stepwise regression algorithm allows an X variable, brought into the model at an earlier stage, to be dropped subsequently if it is no longer helpful in conjunction with variables added at later stages.

Example

Figure 12.4 shows the computer printout obtained when a particular stepwise regression routine (BMDP2R, Ref. 12.2) was applied to our surgical unit example. The minimum acceptable F limit for adding a variable and the maximum acceptable F limit for removing a variable were specified to be 4.0 and 3.9, respectively, as shown at the top left of Figure 12.4. Since the degrees of freedom associated with MSE vary, depending on the number of X variables in the model, and since repeated tests on the same data are undertaken, fixed F limits for adding or deleting a variable have no precise probabilistic meaning. Note, however, that $F(.95; 1, 60) = 4.00$, so that the specified F limits of 4.0 and 3.9 would correspond roughly to a level of significance of .05 for any single test based on approximately 50 degrees of freedom.

The minimum acceptable tolerance of .01 shown in the upper left of Figure 12.4 is a specification to guard against the entry of a variable that is highly correlated with the other X variables already in the model. As explained in Section 11.3, the tolerance is defined as $1 - R_k^2$, where R_k^2 is the coefficient of multiple determination when X_k is regressed on the other X variables in the regression model. The tolerance specification of .01 in Figure 12.4 provides that no variable is to be added to the model which has a coefficient of multiple determination with the other X variables already in the model which exceeds $1 - .01 = .99$ or which would cause the R_k^2 for any variable in the model to exceed .99.

We shall now follow through the steps.

1. At step 0, no X variable is in the model so that the model to be fitted is $Y_i = \beta_0 + \varepsilon_i$. The residual or error sum of squares shown in the ANOVA table in Figure 12.4 for step 0 is therefore $\Sigma(Y_i - \bar{Y})^2 = SSTO = 3.9728$. For each potential X variable, the F^* statistic (12.15) is calculated. In Figure 12.4, these F_k^*

FIGURE 12.4 Stepwise regression for surgical unit example (BMDP2R, Ref. 12.2)

```
REGRESSION TITLE. . . . . . . . . . . . . . . . .SURGICAL UNIT DATA
STEPPING ALGORITHM. . . . . . . . . . . . . . . .F
MAXIMUM NUMBER OF STEPS . . . . . . . . . . . .      10
DEPENDENT VARIABLE. . . . . . . . . . . . . . .       5 Y
MINIMUM ACCEPTABLE F TO ENTER . . . . . . . . .   4.000,    4.000
MAXIMUM ACCEPTABLE F TO REMOVE. . . . . . . . .   3.900,    3.900
MINIMUM ACCEPTABLE TOLERANCE. . . . . . . . . . 0.01000
SUBSCRIPTS OF THE INDEPENDENT VARIABLES . . . .   1    2    3    4


STEP NO.    0

STD. ERROR OF EST.      0.2738

ANALYSIS OF VARIANCE
                  SUM OF SQUARES     DF    MEAN SQUARE
        RESIDUAL  3.9727688          53  0.7495785E-01  <-- SSTO/(n-1)
  SSTO --^
                  VARIABLES IN EQUATION                                   .            VARIABLES NOT IN EQUATION
                         STD. ERROR  STD REG                 F TO         .               PARTIAL                 F TO
   VARIABLE      COEFFICIENT  OF COEFF   COEFF    TOLERANCE   REMOVE   LEVEL.  VARIABLE    CORR.   TOLERANCE     ENTER  LEVEL
(Y-INTERCEPT         2.206 )                                               .
                       ^                                                   . X1        1  0.34640   1.00000       7.09    1
                       Y-bar                                               . X2        2  0.59289   1.00000      28.19    1
                                                                           . X3        3  0.66513   1.00000      41.26    1
                                                                           . X4        4  0.72621   1.00000      58.02    1
                                                                                              ^                     ^
                                                                                             r_Yk                  F*_k

STEP NO.    1
VARIABLE ENTERED    4 X4

MULTIPLE R            0.7262  <-- R
MULTIPLE R-SQUARE     0.5274  <-- R^2
ADJUSTED R-SQUARE     0.5183  <-- R^2_a
STD. ERROR OF EST.    0.1900  <-- sqrt(MSE)

ANALYSIS OF VARIANCE
SSR(X4) -->       SUM OF SQUARES     DF    MEAN SQUARE            F RATIO
      REGRESSION  2.0951490           1    2.095149  <-- MSR(X4)  58.02  <-- F*
        RESIDUAL  1.8776188          52  0.3610805E-01  <-- MSE(X4)
SSE(X4) -->
                  VARIABLES IN EQUATION                                   .            VARIABLES NOT IN EQUATION
                         STD. ERROR  STD REG                 F TO         .               PARTIAL                 F TO
   VARIABLE      COEFFICIENT  OF COEFF   COEFF    TOLERANCE   REMOVE   LEVEL.  VARIABLE    CORR.   TOLERANCE     ENTER  LEVEL
(Y-INTERCEPT         1.696 )                                               .
X4          4        0.186      0.024     0.726    1.00000     58.02      1 . X1        1 -0.03105   0.74758       0.05    1
                                                                           . X2        2  0.50849   0.86382      17.79    1
                                                                           . X3        3  0.58031   0.82659      25.90    1
                       ^          ^         ^                    ^                            ^          ^          ^
                      b_k       s(b_k)     B_k                  F*_k                       r_Yk.4    1-r^2_4k     F*_k

STEP NO.    2
VARIABLE ENTERED    3 X3

MULTIPLE R            0.8286
MULTIPLE R-SQUARE     0.6865

ADJUSTED R-SQUARE     0.6742
STD. ERROR OF EST.    0.1563

ANALYSIS OF VARIANCE
                  SUM OF SQUARES     DF    MEAN SQUARE            F RATIO
      REGRESSION  2.7274637           2    1.363731                55.85
        RESIDUAL  1.2453051          51  0.2441774E-01

                  VARIABLES IN EQUATION                                   .            VARIABLES NOT IN EQUATION
                         STD. ERROR  STD REG                 F TO         .               PARTIAL                 F TO
   VARIABLE      COEFFICIENT  OF COEFF   COEFF    TOLERANCE   REMOVE   LEVEL.  VARIABLE    CORR.   TOLERANCE     ENTER  LEVEL
(Y-INTERCEPT         1.389 )                                               .
X3          3        0.006      0.001     0.439    0.82659     25.90      1 . X1        1  0.32276   0.59179       5.81    1
X4          4        0.139      0.022     0.543    0.82659     39.72      1 . X2        2  0.79149   0.82580      83.85    1
                                                                                                         ^
                                                                                                      1-R^2_k
```

values are shown under the heading "Variables not in equation" and are called "F to enter" values. We see that F_4^* is the largest one:

$$F_4^* = \frac{MSR(X_4)}{MSE(X_4)} = \frac{2.09515}{.03611} = 58.02$$

Since this value exceeds the minimum acceptable F-to-enter value 4.0, X_4 is added to the model.

The column headed "Level" refers to a package option which permits the user to give different priorities to the various potential X variables. Note that in the present example all X variables have the same priority.

2. At this stage, step 1 has been completed. The current regression model contains X_4, and the estimated regression coefficients, the analysis of variance table, and selected other information about the current model are provided.

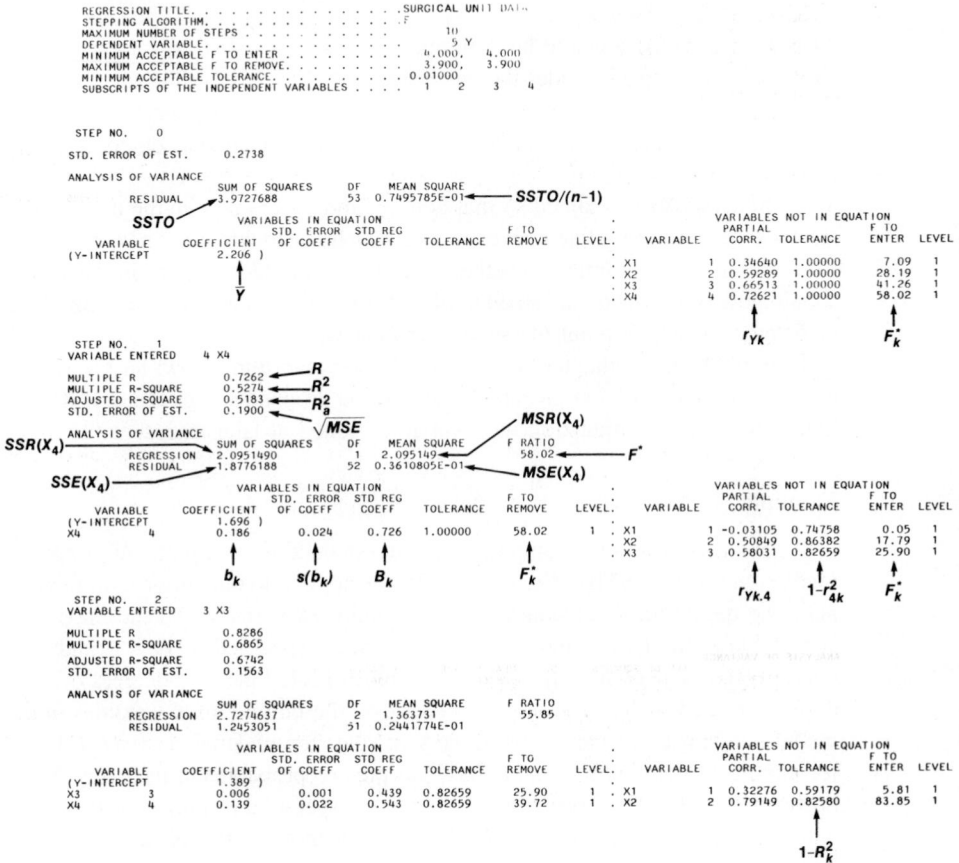

FIGURE 12.4 ***(concluded)***

```
 STEP NO.    3
VARIABLE ENTERED    2 X2

MULTIPLE R             0.9396
MULTIPLE R-SQUARE      0.8829
ADJUSTED R-SQUARE      0.8759
STD. ERROR OF EST.     0.0965

ANALYSIS OF VARIANCE
                  SUM OF SQUARES      DF    MEAN SQUARE       F RATIO
       REGRESSION  3.5075836           3   1.169194           125.67
       RESIDUAL   0.46518236          50  0.9303644E-02

                    VARIABLES IN EQUATION                                     .            VARIABLES NOT IN EQUATION
                             STD. ERROR  STD REG                   F TO        .               PARTIAL                F TO
   VARIABLE       COEFFICIENT  OF COEFF   COEFF      TOLERANCE     REMOVE    LEVEL.    VARIABLE      CORR.  TOLERANCE    ENTER  LEVEL
(Y-INTERCEPT           0.942 )                                                 .
X2           2         0.008      0.001    0.488    0.82580        83.85      1  . X1          1  0.87411  0.55586    158.69   1
X3           3         0.007      0.001    0.543    0.79021        99.63      1  .
X4           4         0.082      0.015    0.320    0.68298        29.86      1  .


 STEP NO.    4
VARIABLE ENTERED    1 X1

MULTIPLE R             0.9861
MULTIPLE R-SQUARE      0.9724
ADJUSTED R-SQUARE      0.9701
STD. ERROR OF EST.     0.0473

ANALYSIS OF VARIANCE
                  SUM OF SQUARES      DF    MEAN SQUARE       F RATIO
       REGRESSION  3.8630199           4  0.9657550           431.19
       RESIDUAL   0.10974741          49  0.2239743E-02

                    VARIABLES IN EQUATION                                     .            VARIABLES NOT IN EQUATION
                             STD. ERROR  STD REG                   F TO        .               PARTIAL                F TO
   VARIABLE       COEFFICIENT  OF COEFF   COEFF      TOLERANCE     REMOVE    LEVEL.    VARIABLE      CORR.  TOLERANCE    ENTER  LEVEL
(Y-INTERCEPT           0.489 )                                                 .
X1           1         0.069      0.005    0.401    0.55586       158.69      1  .
X2           2         0.009      0.000    0.571    0.77567       449.08      1  .
X3           3         0.009      0.000    0.736    0.59594       571.83      1  .
X4           4         0.002      0.010    0.008    0.39134         0.04      1  .


 STEP NO.    5
VARIABLE REMOVED    4 X4

MULTIPLE R             0.9861
MULTIPLE R-SQUARE      0.9724
ADJUSTED R-SQUARE      0.9707
STD. ERROR OF EST.     0.0469

ANALYSIS OF VARIANCE
                  SUM OF SQUARES      DF    MEAN SQUARE       F RATIO
       REGRESSION  3.8629322           3   1.287643           586.17
       RESIDUAL   0 10983557          50  0.2196711E-02

                    VARIABLES IN EQUATION                                     .            VARIABLES NOT IN EQUATION
                             STD. ERROR  STD REG                   F TO        .               PARTIAL                F TO
   VARIABLE       COEFFICIENT  OF COEFF   COEFF      TOLERANCE     REMOVE    LEVEL.    VARIABLE      CORR.  TOLERANCE    ENTER  LEVEL
(Y-INTERCEPT           0.484 )                                                 .
X1           1         0.069      0.004    0.405    0.97011       288.23      1  . X4          4  0.02833  0.39134      0.04   1
X2           2         0.009      0.000    0.574    0.99178       590.58      1  .
X3           3         0.010      0.000    0.739    0.97751       966.28      1  .

* * * * * F-LEVELS(   4.000,    3.900) OR TOLERANCE INSUFFICIENT FOR FURTHER STEPPING
```

Next, all regression models containing X_4 and another independent variable are fitted, and the F^* statistics calculated. They are now:

$$F_k^* = \frac{MSR(X_k \mid X_4)}{MSE(X_4, X_k)}$$

These statistics are shown in step 1 under the heading "Variables not in equation." X_3 has the highest F^* value, which exceeds 4.0, so that X_3 now enters the model.

3. Step 2 in Figure 12.4 summarizes the situation at this point. X_3 and X_4 are now in the model, and information about this model is provided. Next, a test whether X_4 should be dropped is undertaken. The F^* statistic is shown under the heading "Variables in equation" and is called "F to remove":

$$F_4^* = \frac{MSR(X_4 \mid X_3)}{MSE(X_3, X_4)} = 39.72$$

Since this F^* value exceeds the maximum acceptable F-to-remove value 3.9, X_4 is not dropped.

4. Next, all regression models containing X_3, X_4, and one of the remaining potential X variables are fitted. The appropriate F^* statistics now are:

$$F_k^* = \frac{MSR(X_k \mid X_3, X_4)}{MSE(X_3, X_4, X_k)}$$

These statistics are shown in step 2 under the heading "Variables not in equation." X_2 has the highest F^* value, which exceeds 4.0, so that X_2 now enters the model.

5. Step 3 in Figure 12.4 summarizes the situation at this point. X_2, X_3, and X_4 are now in the model. Next, a test is undertaken whether X_3 or X_4 should be dropped. The F^* statistics to remove a variable are shown under the heading "Variables in equation" in step 3. F_4^* is smallest:

$$F_4^* = \frac{MSR(X_4 \mid X_2, X_3)}{MSE(X_2, X_3, X_4)} = 29.86$$

Since its value exceeds 3.9, X_4 is not dropped from the model.

6. At this point, only X_1 remains in the potential pool. Its F^* value to enter exceeds 4.0 (see "Variables not in equation" under step 3), so X_1 is entered into the model.

7. Step 4 in Figure 12.4 summarizes the addition of variable X_1 into the model containing variables X_2, X_3, and X_4. Next, a test is undertaken to determine whether either X_2, X_3, or X_4 should be dropped. The F^* statistics are shown under the heading "Variables in equation" in step 4. Note that:

$$F_4^* = \frac{MSR(X_4 \mid X_1, X_2, X_3)}{MSE(X_1, X_2, X_3, X_4)} = .04$$

is smallest, and that its value is less than 3.9; hence, X_4 is deleted.

8. Step 5 summarizes the dropping of X_4 from the model. Since the only potential variable remaining is X_4, which has just been dropped, it cannot enter the regression now. The algorithm therefore next considers the "F to remove" values in step 5 which indicate that F_1^* is smallest. However, since its value exceeds 3.9, X_1 is not dropped from the model and the search process is terminated.

Thus, the stepwise search algorithm identifies (X_1, X_2, X_3) as the "best" subset of X variables, a result that happens to be consistent with our previous analyses based on the all-possible-regressions approach.

Comments

1. In the surgical unit example, the tolerance requirement was always met; hence, no variable was excluded from the model as a result of too high a correlation with the other X variables in the model.

2. Variations of the rules for entering and removing variables illustrated in the example are possible. For instance, different F-to-enter and F-to-remove values can be employed in accordance with the degrees of freedom associated with MSE in the F^* statistic. However, this refinement often is not utilized, and fixed values are employed instead since the repeated testing in the search procedure does not permit precise probabilistic interpretations.

3. The minimum acceptable F-to-enter value should never be smaller than the maximum acceptable F-to-remove value; otherwise cycling is possible where a variable is continually entered and removed.

4. The order in which variables enter the regression model does not reflect their importance. In our surgical unit example, for instance, X_4 was the first variable to enter the model yet it was eventually dropped.

5. The stepwise regression routine we employed prints out the partial correlation coefficients at each stage. These could be used equivalently to the F^* values for screening the X variables, and indeed some routines actually use the partial correlation coefficients for screening.

6. The F limits for adding and deleting a variable need not be selected in terms of approximate significance levels, but may be determined descriptively in terms of error reduction. For instance, an F limit of 2.0 for adding a variable may be specified with the thought that the marginal error reduction associated with the added variable should be at least twice as great as the remaining error mean square once that variable has been added.

7. A limitation of the stepwise regression search approach is that it presumes there is a single "best" subset of X variables and seeks to identify it. As noted earlier, there is often no unique "best" subset. Hence, some statisticians suggest that all possible regression models with a similar number of X variables as in the stepwise regression solution be fitted subsequently to study whether some other subsets of X variables might be better.

Another limitation of the stepwise regression routine is that it sometimes arrives at an unreasonable "best" subset when the X variables are very highly correlated.

Other automatic search procedures

There are a number of other automatic search procedures which have been proposed to find a "best" subset of independent variables. We mention two of these. Neither of the two methods, however, has gained the acceptance of the stepwise search procedure.

Forward selection. This search procedure is a simplified version of stepwise regression, omitting the test whether a variable once entered into the model should be dropped.

Backward elimination. This search procedure is the opposite of forward selection. It begins with the model containing all potential X variables and identifies the one with the smallest F^* value. For instance, the F^* value for X_1 is:

$$F_1^* = \frac{MSR(X_1 \mid X_2, \ldots, X_{P-1})}{MSE(X_1, \ldots, X_{P-1})} \tag{12.18}$$

If the minimum F_k^* value is less than a predetermined limit, that independent variable is dropped. The model with the remaining $P - 2$ predictor variables is then fitted, and the next candidate for dropping is identified. This process continues until no further independent variables can be dropped.

The backward elimination procedure requires more computations than the forward selection method since it starts with the biggest possible model. However, it does have the advantage of showing users the implications of models with many variables.

12.5 SELECTION OF VARIABLES WITH RIDGE REGRESSION

In Section 11.4, we discussed the use of ridge regression for helping to overcome problems related to multicollinearities among the X variables. The ridge trace mentioned there (Figure 11.2, p. 399) can also be used to identify variables which might be dropped from the regression model. It has been suggested that variables be dropped whose ridge trace is unstable, with the coefficient tending toward the value of zero. Also, variables should be dropped whose ridge trace is stable but at a very small value. Finally, variables with unstable ridge traces that do not tend toward zero should be considered as candidates for dropping.

12.6 IMPLEMENTATION OF SELECTION PROCEDURES

Options and refinements

Our discussion of the major selection procedures for identifying "good" sets of X variables has focused on the main conceptual issues and not on options, variations, and refinements available with particular computer packages. It is essential that the specific features of the package to be employed are fully understood so that intelligent use of the package can be made. In some packages, there is an option for regression models through the origin. Some packages permit variables to be brought into the model and tested in pairs or other groupings instead of singly, to save computing time or for other reasons. Some packages, once a "best" regression model is identified, will fit all the possible regression models in the same number of variables and will develop information for each model so that a final choice can be made by the user. Some stepwise programs have options for forcing variables into the regression model; such variables are not removed even if their F^* values become too low.

The diversity of these options and special features serves to emphasize a point made earlier: there is no unique way of searching for "good" subsets of X variables, and subjective elements must play an important role in the search process.

Completion of model building process

The screening of variables by a computerized selection process is only one step in the building of a regression model. Once the set of X variables has been identified, the resulting model needs to be studied for its aptness by the methods of Chapters 4 and 11. If repeat observations are available, a formal test for lack of fit can be made. In any case, a variety of residual plots and analyses can be employed to identify the nature of lack of fit, outliers, and influential observations. When the original set of $P - 1$ potential X variables excludes cross-product terms and powers of the independent variables to keep the selection problem within reasonable bounds, residual plots against such "missing" variables, or augmenting the model of "best" independent variables by adding cross-product and/or power terms, can be useful in identifying ways in which the model fit can be improved further.

Cautions in use of final model

The model-building process, as we have just noted, requires repeated analyses on the same set of data in order to arrive at a model which fits the data well. A consequence is that the model may be subject to *prediction bias*, i.e., the indicated predictive ability of the model for the data on which the model is based may be greater than the model's predictive ability for new data. The prediction bias arises because the choice of the final model is so uniquely related to the observations at hand. The prediction bias may be particularly large when the effects of independent variables are small.

It is good statistical practice to measure the prediction bias by observing the predictive power of the model on a new set of data. If necessary, some of the original data can be kept aside for this calibration of predictive power and the model derived only from the remaining data.

Often, a predictive model is desired for values of the independent variables which cover only a portion of the entire observation space. In that case, it is good practice to test the stability of the regression model by fitting it to that portion of the observations which fall in the space of future interest and comparing the regression results with those for the model based on all observations. Similarly, if the data are time series, it is often desirable to study the stability of the regression model over time by fitting the model also to the most recent data alone, and comparing results.

In this connection, it is worthwhile repeating an earlier caution. When the independent variables are highly intercorrelated, use of the model for prediction for values of the independent variables that do not follow the past pattern of multicollinearity becomes highly suspect.

PROBLEMS

12.1. A speaker stated: "In a well-designed experiment involving quantitative independent variables, a procedure for screening the independent variables after the observations are obtained is not necessary." Discuss.

12.2. An educational researcher wishes to predict the grade point average in graduate work for applicants to the Graduate School. List a dozen variables that might be useful independent variables here.

12.3. Agency revenues. An economic consultant was retained by a large employment agency in a metropolitan area to develop a regression model for predicting monthly agency revenues (Y). She decided that three economic indicators for the area were potentially useful as independent variables, namely, average weekly overtime hours of production workers in manufacturing (X_1), number of job vacancies in manufacturing (X_2), and index of help wanted advertising in newspapers (X_3). Monthly observations on agency revenues and the three independent variables (all seasonally adjusted) were obtained for the past 25 months. The ANOVA table for the model $Y_i = \beta_0 + \beta_1 X_{i1} + \beta_2 X_{i2} + \beta_3 X_{i3} + \varepsilon_i$ is as follows:

Source of Variation	*SS*	*df*	*MS*
Regression	5,409.89	3	1,803.30
Error	16.35	21	.78
Total	5,426.24	24	

a. Test to determine whether a regression relation exists. Use a level of significance of .05. State the alternatives, decision rule, and conclusion.
b. If a regression relation had not existed, what would this imply about screening the independent variables?

12.4. Refer to **Agency revenues** Problem 12.3. The consultant decided to screen the independent variables to determine the best set for predicting agency revenues. The regression sums of squares for all possible regression models were found to be as follows:

Independent Variables in Model	*SSR*	*Independent Variables in Model*	*SSR*
X_1	2,970.64	X_1, X_2	5,123.80
X_2	3,654.85	X_1, X_3	5,409.59
X_3	3,584.54	X_2, X_3	3,741.30
		X_1, X_2, X_3	5,409.89

a. Indicate which subset of independent variables is best for predicting Y according to each of the following criteria: (1) R_p^2, (2) MSE_p, (3) C_p. Support your recommendations with appropriate graphs.
b. Did the three criteria in part (a) identify the same best subset? Does this always happen?
c. Would stepwise regression have any advantages here as a screening procedure over all possible regressions?
d. An observer states: "There are only three variables, so why screen? You might as well use all three." Discuss.

12.5. Refer to **Patient satisfaction** Problem 7.17. The ANOVA table for the model $Y_i = \beta_0 + \beta_1 X_{i1} + \beta_2 X_{i2} + \beta_3 X_{i3} + \varepsilon_i$ is as follows:

Source of Variation	*SS*	*df*	*MS*
Regression	4,133.62	3	1,377.87
Error	2,011.59	19	105.87
Total	6,145.21	22	

The hospital administrator decided to screen the independent variables to determine the best subset for predicting patient satisfaction. The regression sums of squares for all possible regression models are as follows:

Independent Variables in Model	*SSR*	*Independent Variables in Model*	*SSR*
X_1	3,678.44	X_1, X_2	4,081.21
X_2	2,120.61	X_1, X_3	4,063.98
X_3	2,229.32	X_2, X_3	2,426.92
		X_1, X_2, X_3	4,133.62

a. Indicate which subset of independent variables you would recommend as best for predicting Y according to each of the following criteria: (1) R_p^2, (2) MSE_p, (3) C_p. Support your recommendations with appropriate graphs.
b. Did the three criteria in part (a) identify the same best subset? Does this always happen?
c. Would stepwise regression have any advantages here as a screening procedure over all possible regressions?

12.6. Roofing shingles. Data on sales last year (Y, in thousand squares) in 20 sales districts are given below for a maker of asphalt roofing shingles. Shown also are promotional expenditures (X_1, in thousand dollars), number of active accounts (X_2), number of competing brands (X_3), and district potential (X_4, coded) for each of the districts.

District i	X_{i1}	X_{i2}	X_{i3}	X_{i4}	Y_i
1	5.5	31	10	8	79.3
2	2.5	55	8	6	200.1
3	8.0	67	12	9	163.2
4	3.0	50	7	16	200.1
5	3.0	38	8	15	146.0
6	2.9	71	12	17	177.7
7	8.0	30	12	8	30.9
8	9.0	56	5	10	291.9
9	4.0	42	8	4	160.0
10	6.5	73	5	16	339.4
11	5.5	60	11	7	159.6
12	5.0	44	12	12	86.3
13	6.0	50	6	6	237.5
14	5.0	39	10	4	107.2
15	3.5	55	10	4	155.0
16	8.0	70	6	14	291.4
17	6.0	40	11	6	100.2
18	4.0	50	11	8	135.8
19	7.5	62	9	13	223.3
20	7.0	59	9	11	195.0

It is believed that a regression model containing only first-order terms and no interaction terms will be appropriate.

a. Find the three best subsets according to the C_p criterion. Is there relatively little bias in the subset model with the smallest C_p value?
b. For the subset model with the smallest C_p value, obtain the residuals and plot them against $\hat{Y}$ and each of the independent variables in the subset model on separate graphs. Also prepare a normal probability plot. On the

basis of your plots, do you suggest any modifications in the model to be employed?

12.7. Job proficiency. A personnel officer in a governmental agency administered four newly developed aptitude tests to each of 25 applicants for entry-level clerical positions in the agency. For purposes of the study, all 25 applicants were accepted for positions irrespective of their test scores. After a probationary period, each applicant was rated for proficiency on the job. The scores on the four tests (X_1, X_2, X_3, X_4) and the job proficiency score (Y) for the 25 employees were as follows:

	Test Score				*Job Proficiency Score*
Subject	X_1	X_2	X_3	X_4	Y
1	86	110	100	87	88
2	62	97	99	100	80
3	110	107	103	103	96
4	101	117	93	95	76
5	100	101	95	88	80
6	78	85	95	84	73
7	120	77	80	74	58
8	105	122	116	102	116
9	112	119	106	105	104
10	120	89	105	97	99
11	87	81	90	88	64
12	133	120	113	108	126
13	140	121	96	89	94
14	84	113	98	78	71
15	106	102	109	109	111
16	109	129	102	108	109
17	104	83	100	102	100
18	150	118	107	110	127
19	98	125	108	95	99
20	120	94	95	90	82
21	74	121	91	85	67
22	96	114	114	103	109
23	104	73	93	80	78
24	94	121	115	104	115
25	91	129	97	83	83

It is expected that a regression model containing only first-order terms and no interaction terms will be appropriate.

a. Find the three best subsets according to the C_p criterion. Is there relatively little bias in the subset model with the smallest C_p value?

b. For the subset model with the smallest C_p value, obtain the residuals and plot them against $\hat{Y}$ and each of the independent variables in the subset model on separate graphs. Also prepare a normal probability plot. On the basis of your plots, do you suggest any modifications in the model to be employed?

12.8. Two researchers investigated factors affecting summer attendance at privately operated beaches on Lake Ontario, and collected information on attendance and 11 explanatory variables for 42 beaches. Two summers were studied, of relatively hot and relatively cool weather, respectively. A "best" subsets algorithm now is to be used to screen the potential independent variables.

a. Should the screening be done for both summers combined or should it be done separately for each summer? Explain the problems involved and how you might handle them.

b. Will the "best" subsets screening procedure select those independent variables that are most important in a causal sense for determining beach attendance?

12.9. In stepwise regression, what advantage is there in using a relatively large F limit for adding variables? What advantage is there in using a smaller F limit for adding variables?

12.10. In stepwise regression, why should the F limit for deleting variables never exceed the F limit for adding variables?

12.11. Draw a flowchart of each of the following selection methods: (1) stepwise regression, (2) forward selection, (3) backward elimination.

12.12. Refer to **Agency revenues** Problems 12.3 and 12.4. The consultant was interested to learn how the stepwise selection procedure and some of its variations would perform in this application.

a. Determine the subset of variables that is selected as best by the stepwise regression procedure using F limits of 4.2 and 4.1 to add or delete a variable, respectively. Show your steps.

b. To what level of significance in any individual test is the F limit of 4.2 for adding a variable approximately equivalent here?

c. Determine the subset of variables that is selected as best by the forward selection procedure using an F limit of 4.2 to add a variable. Show your steps.

d. Determine the subset of variables that is selected as best by the backward elimination procedure using an F limit of 4.1 to delete a variable. Show your steps.

e. Compare the results of the three selection procedures. How consistent are these results? How do the results compare with those for all possible regressions in Problem 12.4?

12.13. Refer to **Agency revenues** Problem 12.12a. Suppose the consultant "forced" X_2 into the best subset for administrative reasons by arbitrarily entering it first and not removing it even if its F^* value becomes too low. Which subset of variables (including X_2) is now selected as best by the stepwise regression procedure if F limits of 4.2 and 4.1 are used to add or delete a variable, respectively? Did the forced inclusion of X_2 affect the selection of the other variables included in the best subset? Will this always happen?

12.14. Refer to **Patient satisfaction** Problems 7.17 and 12.5. The hospital administrator was interested to learn how the stepwise selection procedure and some of its variations would perform here.

a. Determine the subset of variables that is selected as best by the stepwise regression procedure using F limits of 3.0 and 2.9 to add or delete a variable, respectively. Show your steps.

b. To what level of significance in any individual test is the F limit of 3.0 for adding a variable approximately equivalent here?

c. Determine the subset of variables that is selected as best by the forward selection procedure using an F limit of 3.0 to add a variable. Show your steps.

d. Determine the subset of variables that is selected as best by the backward elimination procedure using an F limit of 2.9 to delete a variable. Show your steps.

e. Compare the results of the three selection procedures. How consistent are these results? How do the results compare with those for all possible regressions in Problem 12.5?

12.15. Refer to **Roofing shingles** Problem 12.6.

a. Using stepwise regression, find the best subset of independent variables to predict sales. Use F limits for adding or deleting a variable of 4.0 and 3.9, respectively.

b. How does the best subset according to stepwise regression compare with the best subset according to the C_p criterion obtained in Problem 12.6a?

12.16. Refer to **Job proficiency** Problem 12.7.

a. Using stepwise regression, find the best subset of independent variables to predict job proficiency. Use F limits of 3.5 and 3.4 for adding or deleting a variable, respectively.

b. How does the best subset according to stepwise regression compare with the best subset according to the C_p criterion obtained in Problem 12.7a?

12.17. An engineer has stated: ''Screening of variables should always be done using the objective stepwise regression procedure.'' Discuss.

EXERCISES

12.18. The true quadratic regression function is $E(Y) = 15 + 20X + 3X^2$. The fitted linear regression function is $\hat{Y} = 13 + 40X$, for which $E(b_0) = 10$ and $E(b_1) = 45$. What are the bias and sampling error components of the mean squared error for $X_i = 10$? For $X_i = 20$?

12.19. Prove (12.12). [*Hint:* Use Exercise 6.33 and (11.50).]

12.20. Refer to (12.16). Show that the same variable X_k that maximizes the test statistic F_k^* also maximizes the coefficient of partial determination $r_{Yk.7}^2$.

PROJECTS

12.21. Refer to the **SENIC** data set. Length of stay (Y) is to be predicted, and the pool of potential independent variables includes all other variables in the data set except medical school affiliation and region. It is believed that a model with $\log_{10} Y$ as the dependent variable and the independent variables in first-order terms with no interaction terms will be appropriate.

a. Using the C_p criterion, obtain the three best subsets. Which of these subset models appears to have the smallest bias? Which of the three models would you recommend as best?

b. Divide the data set into two halves by considering the first 56 observations as one half and the remaining 57 observations as the other half. Fit the regression model recommended as best in part (a) using the first 56 observations. Then obtain the deviations of the remaining 57 observations from their respective "predicted" values, i.e., obtain $Y_i - \hat{Y}_i$. How well does the model perform on the hold-out (validation) sample? Calculate $\Sigma(Y_i - \hat{Y}_i)^2/n$ for the last 57 observations and compare it with *MSE* for the first 56 observations. Is there any evidence of a large prediction bias?

c. For the recommended subset model in part (a), obtain the residuals and plot them against the fitted values and each of the independent variables in the subset model. Also prepare a normal probability plot. Do these plots suggest any modifications in the model?

d. Would stepwise regression with F limits of 3.5 and 3.4 for adding or deleting a variable, respectively, lead to the same best model as the all-possible-regressions approach?

12.22. Refer to the **SMSA** data set. A public safety official wishes to predict the rate of serious crimes in an SMSA (Y, total number of serious crimes per 100,000 population). The pool of potential independent variables includes all other variables in the data set except region.

a. Using the C_p criterion, obtain the three best subsets. Which of these subset models appears to have the smallest bias? Which of the three models would you recommend as best?

b. Divide the data set into two halves by considering the odd-numbered observations as one half and the even-numbered observations as the other half. Fit the regression model recommended as best in part (a) using the odd-numbered observations. Then obtain the deviations of the even-numbered observations from their respective "predicted" values, i.e., obtain $Y_i - \hat{Y}_i$. How well does the model perform on the hold-out (validation) sample? Calculate $\Sigma(Y_i - \hat{Y}_i)^2/n$ for the even-numbered observations and compare it with *MSE* for the odd-numbered observations. Is there any evidence of a large prediction bias?

c. For the recommended subset model in part (a), obtain the residuals and plot them against the fitted values and each of the independent variables in the subset model. Also prepare a normal probability plot. Do these plots suggest any modifications in the model?

d. Would stepwise regression with F-to-enter and F-to-remove values of 4.0 and 3.9, respectively, lead to the same best model as the all-possible-regressions approach?

CITED REFERENCES

12.1 Daniel, Cuthbert, and Fred S. Wood. *Fitting Equations to Data*. 2d ed. New York: Wiley-Interscience, 1980.

12.2 Dixon, W. J., and M. B. Brown, eds. *BMDP-81, Biomedical Computer Programs, P-Series*. Berkeley, Calif.: University of California Press, 1981.

13

Autocorrelation in time series data

The basic regression models considered so far have assumed that the random error terms ε_i are either uncorrelated random variables or independent normal random variables. In business and economics, many regression applications involve time series data. For such data, the assumption of uncorrelated or independent error terms is often not appropriate; rather, the error terms are frequently correlated positively over time. Error terms correlated over time are said to be *autocorrelated* or *serially correlated*.

A major cause of positively autocorrelated error terms in business and economic regression applications involving time series data is the omission of one or several key variables from the model. When time-ordered effects of such "missing" key variables are positively correlated, the error terms in the regression model will tend to be positively autocorrelated since the error terms include effects of missing variables. Suppose, for example, that annual sales of a product are regressed against average yearly price over a period of 30 years. If population size has an important effect on sales, its omission from the model may lead to the error terms being positively autocorrelated because the effect of population size on sales likely is positively correlated over time.

Another common cause of positively autocorrelated error terms in economic data is systematic coverage errors in the dependent variable time series, which errors often tend to be positively correlated over time.

13.1 PROBLEMS OF AUTOCORRELATION

If the error terms in the regression model are positively autocorrelated, the use of ordinary least squares procedures has a number of important consequences. We summarize these first, and then discuss them in more detail:

1. The ordinary least squares regression coefficients are still unbiased, but they no longer have the minimum variance property and may be quite inefficient.
2. *MSE* may seriously underestimate the variance of the error terms.
3. $s(b_k)$ calculated according to ordinary least squares procedures may seriously underestimate the true standard deviation of the estimated regression coefficient with that procedure.
4. The confidence intervals and tests using the t and F distributions, discussed earlier, are no longer strictly applicable.

To illustrate these problems intuitively, we shall consider the simple linear regression model with time series data:

$$Y_t = \beta_0 + \beta_1 X_t + \varepsilon_t$$

Here, Y_t and X_t are observations for period t. Let us assume that the error terms ε_t are positively autocorrelated as follows:

$$\varepsilon_t = \varepsilon_{t-1} + u_t$$

The u_t, called *disturbances*, are independent normal random variables. Thus, any error term ε_t is the sum of the previous error term ε_{t-1} and a new disturbance term u_t. We shall assume here that the u_t have mean 0 and variance 1.

In Table 13.1, column 1, we show 10 random observations on the normal variable u_t with mean 0 and variance 1, obtained from a standard normal random numbers generator. Suppose now that $\varepsilon_0 = 3.0$; we obtain then:

$$\varepsilon_1 = \varepsilon_0 + u_1 = 3.0 + .5 = 3.5$$
$$\varepsilon_2 = \varepsilon_1 + u_2 = 3.5 - .7 = 2.8$$
$$\text{etc.}$$

TABLE 13.1 Example of positively autocorrelated error terms

t	(1) u_t	(2) $\varepsilon_{t-1} + u_t = \varepsilon_t$
0	—	3.0
1	+.5	3.0 + .5 = 3.5
2	−.7	3.5 − .7 = 2.8
3	+.3	2.8 + .3 = 3.1
4	0	3.1 + 0 = 3.1
5	−2.3	3.1 − 2.3 = .8
6	−1.9	.8 − 1.9 = −1.1
7	+.2	−1.1 + .2 = − .9
8	−.3	− .9 − .3 = −1.2
9	+.2	−1.2 + .2 = −1.0
10	−.1	−1.0 − .1 = −1.1

The error terms ε_t are shown in Table 13.1, column 2, and they are plotted in Figure 13.1. Note the systematic pattern in these error terms. Their positive relation over time is shown by the fact that adjacent error terms tend to be of the same magnitude.

Suppose that X_t in the regression model represents time, such that $X_1 = 1$, $X_2 = 2$, etc. Further, suppose we know that $\beta_0 = 2$ and $\beta_1 = .5$. Figure 13.2a contains the true regression line and the observed Y values based on the error terms in Figure 13.1. Figure 13.2b contains the estimated regression line, fitted by ordinary least squares methods, and the observed Y values. Notice that the fitted regression line differs sharply from the true regression line because the initial ε_0 value was large and the succeeding positively autocorrelated error terms tended to be large for some time. This persistency pattern in the positively autocorrelated error terms leads to a fitted regression line far from the true one. Had the initial ε_0 value been small, say, $\varepsilon_0 = -.2$, and the disturbances different, a sharply different fitted regression line might have been obtained because of the persistency pattern, as shown in Figure 13.2c. This variation from sample to sample in the fitted regression lines because of the positively autocorrelated error terms may be so substantial as to lead to large variances of the estimated regression coefficients when ordinary least squares methods are used.

Another key problem with applying ordinary least squares methods when the error terms are positively autocorrelated, as mentioned before, is that MSE may

FIGURE 13.1 Example of positively autocorrela.ed error terms

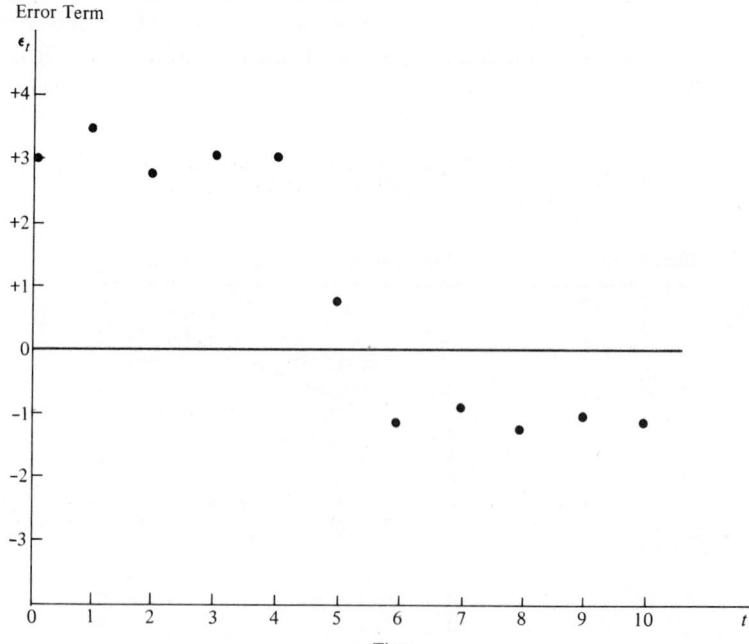

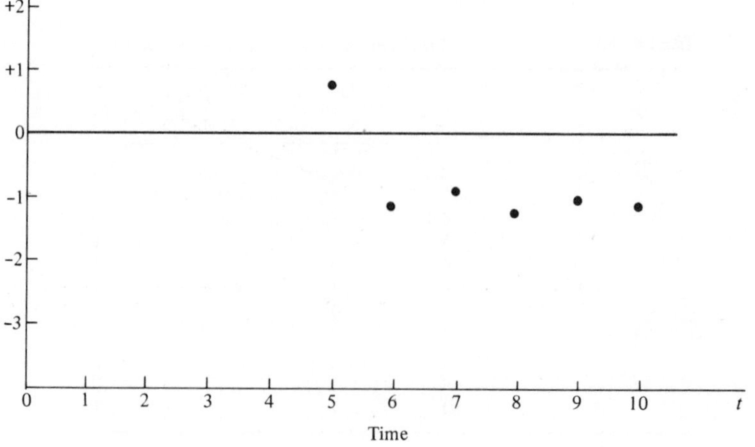

FIGURE 13.2 Regression with positively autocorrelated error terms

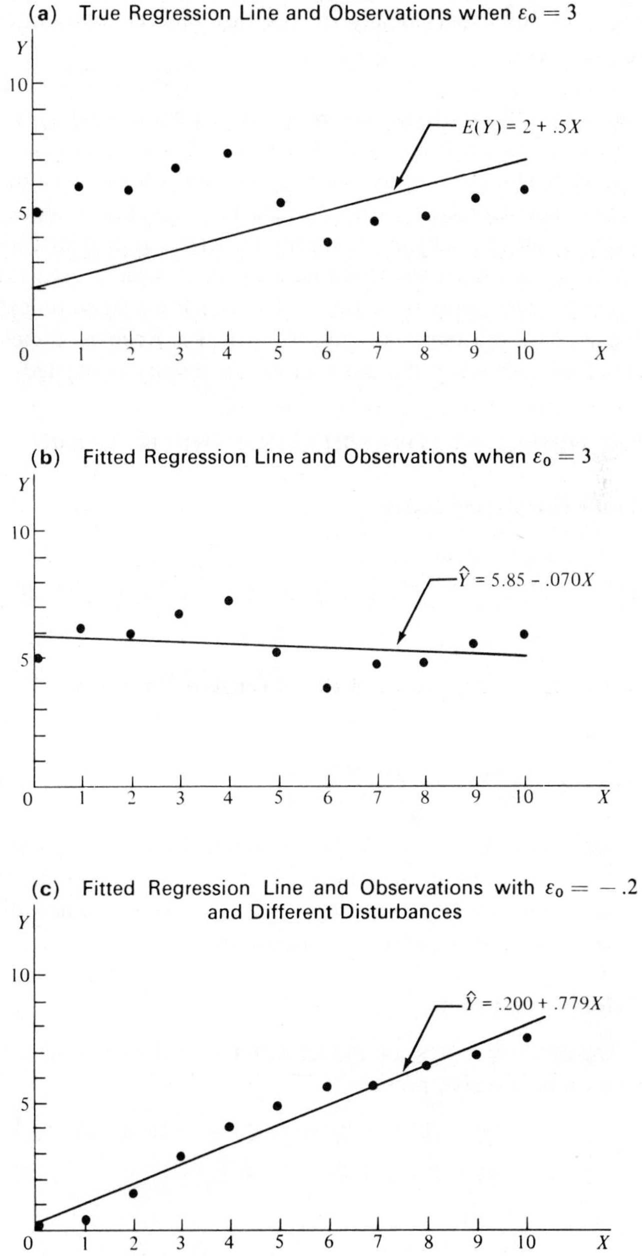

seriously underestimate the variance of the ε_t. Figure 13.2 makes this clear. Note that the variability of the Y values around the fitted regression line in Figure 13.2b is substantially smaller than the variability of the Y's around the true regression line in Figure 13.2a. This is one of the factors leading to an indication of greater precision of the regression coefficients than is actually the case when ordinary least squares methods are used in the presence of positively autocorrelated errors.

In view of the seriousness of the problems created by autocorrelated errors, it is important that their presence be detected. A plot of residuals against time is an effective, though subjective, means of detecting autocorrelated errors. Formal statistical tests have also been developed. A widely used test is based on the first-order autoregressive error model, which we take up next. This model is a simple one, yet experience suggests that it is frequently applicable in business and economics when the error terms are serially correlated.

13.2 FIRST-ORDER AUTOREGRESSIVE ERROR MODEL

Simple linear regression

The simple linear regression model for one independent variable with the random error terms following a first-order autoregressive process is:

$$\begin{aligned} Y_t &= \beta_0 + \beta_1 X_t + \varepsilon_t \\ \varepsilon_t &= \rho\varepsilon_{t-1} + u_t \end{aligned} \tag{13.1}$$

where:

ρ is a parameter such that $|\rho| < 1$
u_t are independent $N(0, \sigma^2)$

Note that (13.1) is identical to the simple linear regression model (3.1) except for the structure of the error terms. Each error term in model (13.1) consists of a fraction of the previous error term (when $\rho > 0$) plus a new disturbance term u_t. The parameter ρ is called the *autocorrelation parameter*.

Multiple regression

The multiple regression model with the random error terms following a first-order autoregressive process is:

$$\begin{aligned} Y_t &= \beta_0 + \beta_1 X_{t1} + \beta_2 X_{t2} + \cdots + \beta_{p-1} X_{t,p-1} + \varepsilon_t \\ \varepsilon_t &= \rho\varepsilon_{t-1} + u_t \end{aligned} \tag{13.2}$$

where:

$|\rho| < 1$
u_t are independent $N(0, \sigma^2)$

Thus, we see again that the multiple regression model (13.2) is identical to the earlier multiple regression model (7.7) except for the structure of the error terms.

Properties of error terms

It is instructive to expand the definition of the first-order autoregressive error term ε_t:

$$\varepsilon_t = \rho\varepsilon_{t-1} + u_t$$

Since $\varepsilon_{t-1} = \rho\varepsilon_{t-2} + u_{t-1}$, we obtain:

$$\varepsilon_t = \rho(\rho\varepsilon_{t-2} + u_{t-1}) + u_t = \rho^2\varepsilon_{t-2} + \rho u_{t-1} + u_t$$

Replacing now ε_{t-2} by $\rho\varepsilon_{t-3} + u_{t-2}$, we obtain:

$$\varepsilon_t = \rho^3\varepsilon_{t-3} + \rho^2 u_{t-2} + \rho u_{t-1} + u_t$$

Continuing in this fashion, we find:

$$\varepsilon_t = \sum_{s=0}^{\infty} \rho^s u_{t-s} \tag{13.3}$$

Thus, the error term ε_t in period t is a linear combination of the current and preceding disturbance terms. When $0 < \rho < 1$, (13.3) indicates that the further the period is in the past, the smaller is the weight of that disturbance term in determining ε_t.

Mean. Since $E(u_t) = 0$ according to models (13.1) and (13.2) for all t, it follows from (13.3) that:

$$E(\varepsilon_t) = 0 \tag{13.4}$$

Thus, ε_t has expectation zero, just as for regression models with uncorrelated error terms.

Variance. Since according to models (13.1) and (13.2) the u_t are independent with variance σ^2, it follows from (13.3) that:

$$\sigma^2(\varepsilon_t) = \sum_{s=0}^{\infty} \rho^{2s}\sigma^2(u_{t-s}) = \sigma^2 \sum_{s=0}^{\infty} \rho^{2s}$$

Now for $|\rho| < 1$, it is known that:

$$\sum_{s=0}^{\infty} \rho^{2s} = \frac{1}{1-\rho^2}$$

Hence, we have:

$$\sigma^2(\varepsilon_t) = \frac{\sigma^2}{1-\rho^2} \tag{13.5}$$

We thus see that the error terms have constant variance, just as for regression models with uncorrelated error terms.

Covariance. To find the covariance of ε_t and ε_{t-1}, we need to recognize that:

$$\sigma^2(\varepsilon_t) = E(\varepsilon_t^2)$$
$$\sigma(\varepsilon_t, \varepsilon_{t-1}) = E(\varepsilon_t\varepsilon_{t-1})$$

These results follow from theorems (1.14a) and (1.19a), respectively, since $E(\varepsilon_t) = 0$ by (13.4).

By (13.3), we have:

$$E(\varepsilon_t\varepsilon_{t-1}) = E[(u_t + \rho u_{t-1} + \rho^2 u_{t-2} + \cdots)(u_{t-1} + \rho u_{t-2} + \rho^2 u_{t-3} + \cdots)]$$

which can be rewritten:

$$\begin{aligned} E(\varepsilon_t\varepsilon_{t-1}) &= E\{[u_t + \rho(u_{t-1} + \rho u_{t-2} + \cdots)][u_{t-1} + \rho u_{t-2} + \rho^2 u_{t-3} + \cdots]\} \\ &= E[u_t(u_{t-1} + \rho u_{t-2} + \rho^2 u_{t-3} + \cdots)] \\ &\qquad + E[\rho(u_{t-1} + \rho u_{t-2} + \rho^2 u_{t-3} + \cdots)^2] \end{aligned}$$

Since $E(u_t u_{t-s}) = 0$ for all $s \neq 0$ by the assumed independence of the u_t, the first term drops out and we obtain:

$$E(\varepsilon_t\varepsilon_{t-1}) = \rho E(\varepsilon_{t-1}^2) = \rho\sigma^2(\varepsilon_{t-1})$$

Hence, by (13.5), which holds for all t, we have:

(13.6)
$$\sigma(\varepsilon_t, \varepsilon_{t-1}) = \rho\left(\frac{\sigma^2}{1-\rho^2}\right)$$

In general, it can be shown that:

(13.7)
$$\sigma(\varepsilon_t, \varepsilon_{t-s}) = \rho^s\left(\frac{\sigma^2}{1-\rho^2}\right) \qquad s \neq 0$$

Thus, the error terms for models (13.1) and (13.2) are autocorrelated unless the autocorrelation parameter ρ equals zero.

Note

It follows directly from (13.5) and (13.6) that the autocorrelation parameter ρ is the coefficient of correlation between ε_t and ε_{t-1}, as defined in (15.3).

13.3 DURBIN-WATSON TEST FOR AUTOCORRELATION

The Durbin-Watson test assumes the first-order autoregressive error models (13.1) or (13.2), with the values of the independent variable(s) fixed. The test consists of determining whether or not the autocorrelation parameter ρ in (13.1) or (13.2) is zero. Note that if $\rho = 0$, $\varepsilon_t = u_t$. Hence, the error terms ε_t are then independent since the u_t are independent.

Because correlated error terms in business and economic applications tend to show positive serial correlation, the usual test alternatives considered are:

$$H_0: \rho = 0 \qquad (13.8)$$
$$H_a: \rho > 0$$

The test statistic D is obtained by first fitting the ordinary least squares regression function and calculating the residuals:

$$e_t = Y_t - \hat{Y}_t \qquad (13.9)$$

and then calculating the statistic:

$$D = \frac{\sum_{t=2}^{n} (e_t - e_{t-1})^2}{\sum_{t=1}^{n} e_t^2} \qquad (13.10)$$

where n is the number of observations.

An exact test procedure is not available, but Durbin and Watson have obtained lower and upper bounds d_L and d_U such that a value of D outside these bounds leads to a definite decision. The decision rule for testing between the alternatives in (13.8) is:

(13.11)

If $D > d_U$, conclude H_0

If $D < d_L$, conclude H_a

If $d_L \leq D \leq d_U$, the test is inconclusive

Small values of D lead to the conclusion that $\rho > 0$ because the adjacent error terms ε_t and ε_{t-1} tend to be of the same magnitude when they are positively autocorrelated. Hence, the differences in the residuals, $e_t - e_{t-1}$, would tend to be small when $\rho > 0$, leading to a small numerator in D and hence to a small test statistic D.

Table A–6 contains the bounds d_L and d_U for various sample sizes (n), for two levels of significance (.05 and .01), and for various numbers of X variables ($p - 1$) in the regression model.

Example

The Blaisdell Company wished to predict its sales by using industry sales as a predictor variable. (Accurate predictions of industry sales are available from the industry's trade association.) In Table 13.2, columns 1 and 2 contain seasonally adjusted quarterly data on company sales and industry sales, respectively, for the period 1977–81. A scatter plot (not shown) suggested that a linear regression model is appropriate. The market research analyst was, however, concerned whether or not the error terms were positively autocorrelated. He therefore used the Durbin-Watson test with the alternatives:

$$H_0: \rho = 0$$
$$H_a: \rho > 0$$

TABLE 13.2 Durbin-Watson test calculations for Blaisdell Company example (company and industry sales data are seasonally adjusted)

Year and Quarter	t	(1) Company Sales ($ millions) Y_t	(2) Industry Sales ($ millions) X_t	(3) Residual e_t	(4) $e_t - e_{t-1}$	(5) $(e_t - e_{t-1})^2$	(6) e_t^2
1977: 1	1	20.96	127.3	−.026052	—	—	.0006787
2	2	21.40	130.0	−.062015	−.035963	.0012933	.0038459
3	3	21.96	132.7	.022021	.084036	.0070620	.0004849
4	4	21.52	129.4	.163754	.141733	.0200882	.0268154
1978: 1	5	22.39	135.0	.046570	−.117184	.0137321	.0021688
2	6	22.76	137.1	.046377	−.000193	.0000000	.0021508
3	7	23.48	141.2	.043617	−.002760	.0000076	.0019024
4	8	23.66	142.8	−.058435	−.102052	.0104146	.0034146
1979: 1	9	24.10	145.5	−.094399	−.035964	.0012934	.0089112
2	10	24.01	145.3	−.149142	−.054743	.0029968	.0222433
3	11	24.54	148.3	−.147991	.001151	.0000013	.0219013
4	12	24.30	146.4	−.053054	.094937	.0090130	.0028147
1980: 1	13	25.00	150.2	−.022928	.030126	.0009076	.0005257
2	14	25.64	153.1	.105852	.128780	.0165843	.0112046
3	15	26.36	157.3	.085464	−.020388	.0004157	.0073041
4	16	26.98	160.7	.106102	.020638	.0004259	.0112576
1981: 1	17	27.52	164.2	.029112	−.076990	.0059275	.0008475
2	18	27.78	165.6	.042316	.013204	.0001743	.0017906
3	19	28.24	168.7	−.044160	−.086476	.0074781	.0019501
4	20	28.78	171.7	−.033009	.011151	.0001243	.0010896
Total						.0979400	.1333018

He fitted an ordinary least squares regression line to the data in Table 13.2. The results are shown in Table 13.3a. He then calculated the residuals e_t, which are shown in column 3 of Table 13.2 and which are plotted against time in Figure 13.3. Note how the residuals consistently are above or below the fitted values for extended periods. Positive autocorrelation in the error terms is suggested by such a pattern when an appropriate regression function has been employed.

FIGURE 13.3 Residuals plotted against time—Blaisdell Company example

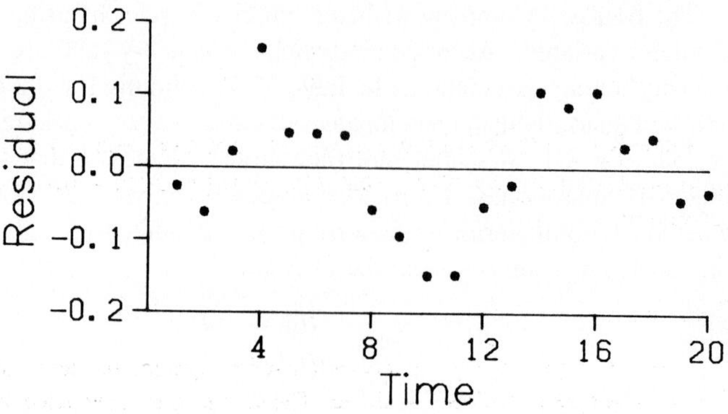

TABLE 13.3 Regression results for Blaisdell Company example

(a) Original Variables Y_t and X_t

Regression Coefficient	*Estimated Regression Coefficient*	*Estimated Standard Deviation*
β_0	$b_0 = -1.45475$	$s(b_0) = .21415$
β_1	$b_1 = .17628$	$s(b_1) = .00144$

$$\hat{Y} = -1.45475 + .17628X$$

(b) Transformed Variables $Y_t' = Y_t - rY_{t-1}$ and $X_t' = X_t - rX_{t-1}$

Regression Coefficient	*Estimated Regression Coefficient*	*Estimated Standard Deviation*
β_0'	$b_0' = -.39396$	$s(b_0') = .16704$
$\beta_1' = \beta_1$	$b_1' = .17376$	$s(b_1') = .00296$

$$b_0 = \frac{b_0'}{1-r} = \frac{-.39396}{1-.631166} = -1.06811$$

$$s(b_0) = \frac{s(b_0')}{1-r} = \frac{.16704}{1-.631166} = .45288$$

(c) First Differences $Y_t' = Y_t - Y_{t-1}$ and $X_t' = X_t - X_{t-1}$

Regression Coefficient	*Estimated Regression Coefficient*	*Estimated Standard Deviation*
$\beta_1' = \beta_1$	$b_1 = .16849$	$s(b_1) = .005096$

Columns 4, 5, and 6 of Table 13.2 contain the necessary calculations for the test statistic D. The analyst then obtained:

$$D = \frac{\sum_{t=2}^{20}(e_t - e_{t-1})^2}{\sum_{t=1}^{20} e_t^2} = \frac{.09794}{.13330} = .735$$

Using a level of significance of .01, he found in Table A–6 for $n = 20$ and $p - 1 = 1$:

$$d_L = .95$$
$$d_U = 1.15$$

Since $D = .735$ falls below $d_L = .95$, decision rule (13.11) indicates that the appropriate conclusion is H_a, namely, that the error terms are positively autocorrelated.

Comments

1. If a test for negative autocorrelation is required, the test statistic to be used is $4 - D$, where D is defined as above. The test is then conducted in the same manner

described for testing for positive autocorrelation. That is, if the quantity $4 - D$ falls below d_L, we conclude $\rho < 0$, that negative autocorrelation exists, and so on.

2. A two-sided test for H_0: $\rho = 0$ versus H_a: $\rho \neq 0$ can be made by employing both one-sided tests separately. The Type I risk with the two-sided test is 2α, where α is the Type I risk with each one-sided test.

3. When the Durbin-Watson test employing the bounds d_L and d_U gives indeterminate results, in principle more observations are required. Of course, with time series data it may be impossible to obtain more observations, or additional observations may lie in the future and be obtainable only with great delay. Durbin and Watson (Ref. 13.1) do give an approximate test which may be used when the bounds test is indeterminate, but the degrees of freedom should be larger than about 40 before this approximate test will give more than a rough indication of whether autocorrelation exists.

A reasonable procedure is to treat indeterminate results as suggesting the presence of autocorrelated errors and employ one of the remedial actions to be discussed next. If such an action does not lead to substantially different regression results, the assumption of uncorrelated error terms would appear to be satisfactory. When the remedial action does lead to substantially different regression results (such as larger estimated standard errors for the regression coefficients or the elimination of autocorrelated errors), the results obtained by means of the remedial action are probably the more useful ones.

4. The Durbin-Watson test is not robust against misspecifications of the model. For example, the Durbin-Watson test may not disclose the presence of autocorrelated errors that follow a second-order autoregressive pattern, where an error term in period t is directly related to the error term in period $t - 2$.

5. While the Durbin-Watson test is widely used, other tests for autocorrelation are available. One such alternative test, due to Theil and Nagar, is found in Reference 13.2.

13.4 REMEDIAL MEASURES FOR AUTOCORRELATION

The two principal remedial measures when autocorrelated error terms are present are to add one or more independent variables to the model or to use transformed variables.

Addition of independent variables

As noted earlier, one major cause of autocorrelated error terms is the omission from the model of one or more key independent variables that have time-ordered effects on the dependent variable. When autocorrelated error terms are found to be present, the first remedial action should always be to search for missing key independent variables. The missing variable, population size, was mentioned previously for a regression of annual sales of a product on average yearly price of the product during a 30-year period. Sometimes, use of a simple linear trend variable or use of indicator variables for seasonal effects can be helpful in eliminating or reducing autocorrelation in the error terms.

Use of transformed variables

Only when use of additional independent variables is not helpful in eliminating the problem of autocorrelated errors should a remedial action based on transformed variables be employed. We now explain two methods of transforming the variables. One is based on an iterative approach and the other uses first differences. We shall explain these methods for simple linear regression, but the same principles apply for multiple regression.

Iterative approach. The iterative approach for simple linear regression is motivated by an interesting property of model (13.1). Consider the transformed dependent variable:

$$Y_t' = Y_t - \rho Y_{t-1}$$

Substituting in this expression for Y_t and Y_{t-1} according to model (13.1), we obtain:

$$\begin{aligned} Y_t' &= (\beta_0 + \beta_1 X_t + \varepsilon_t) - \rho(\beta_0 + \beta_1 X_{t-1} + \varepsilon_{t-1}) \\ &= \beta_0(1 - \rho) + \beta_1(X_t - \rho X_{t-1}) + (\varepsilon_t - \rho\varepsilon_{t-1}) \end{aligned}$$

But by (13.1), $\varepsilon_t - \rho\varepsilon_{t-1} = u_t$. Hence:

$$Y_t' = \beta_0(1 - \rho) + \beta_1(X_t - \rho X_{t-1}) + u_t$$

where the u_t are independent error terms. Thus, when we use the transformed variables:

(13.12a) $$Y_t' = Y_t - \rho Y_{t-1}$$

(13.12b) $$X_t' = X_t - \rho X_{t-1}$$

the transformed regression model becomes:

(13.13) $$Y_t' = \beta_0' + \beta_1' X_t' + u_t$$

where:

$$\begin{aligned} \beta_0' &= \beta_0(1 - \rho) \\ \beta_1' &= \beta_1 \end{aligned}$$

Note that the transformed regression model (13.13) has independent error terms. This means that ordinary least squares methods have their usual optimum properties with this model.

The parameters in the original model (13.1) are related to the parameters in the transformed model (13.13) as follows:

(13.14a) $$\beta_0 = \frac{\beta_0'}{1 - \rho}$$

(13.14b) $$\beta_1 = \beta_1'$$

The transformed model (13.13) cannot be used directly because the autocorrelation parameter ρ needed to obtain the transformed variables in (13.12) is unknown. We can, however, estimate ρ. Note that the autoregressive error process assumed in model (13.1) can be viewed as a regression through the origin:

$$\varepsilon_t = \rho\varepsilon_{t-1} + u_t$$

where ε_t is the dependent variable, ε_{t-1} the independent variable, u_t the error term, and ρ the slope of the line through the origin. Since the ε_t and ε_{t-1} are unknown we use the residuals e_t and e_{t-1}, obtained by ordinary least squares methods, as the dependent and independent variables, respectively, and estimate ρ by fitting a straight line through the origin. From our previous discussion of regression through the origin, we know by (5.17) that the estimate of the slope ρ, denoted by r, is:

(13.15) $$r = \frac{\sum_{t=2}^{n} e_{t-1} e_t}{\sum_{t=2}^{n} e_{t-1}^2}$$

We now obtain the transformed variables:

(13.16a) $$Y_t' = Y_t - rY_{t-1}$$

(13.16b) $$X_t' = X_t - rX_{t-1}$$

and use ordinary least squares with these transformed variables.

The Durbin-Watson test is then employed to test whether the error terms for the transformed model are uncorrelated. If the test indicates that they are uncorrelated, the procedure terminates. Otherwise, the parameter ρ is reestimated from the new residuals for the regression model with the original variables, using the regression coefficients derived from the fit of the regression model with the transformed variables. A new set of transformed variables is then obtained with the new r. This process may be continued for several iterations until the Durbin-Watson test suggests that the error terms in the transformed model are uncorrelated.

Example. For our Blaisdell Company example, the necessary calculations for estimating the autocorrelation parameter ρ, based on the residuals obtained with ordinary least squares applied to the original variables, appear in Table 13.4. Column 1 repeats the residuals from Table 13.2. Column 2 contains the residuals e_{t-1}, and columns 3 and 4 contain the necessary calculations. Hence, we estimate:

$$r = \frac{.0834478}{.1322122} = .631166$$

We now obtain the transformed variables Y_t' and X_t' in (13.16a) and (13.16b):

$$Y_t' = Y_t - .631166Y_{t-1}$$
$$X_t' = X_t - .631166X_{t-1}$$

These are shown in Table 13.5. Ordinary least squares fitting of linear regression is now used with these transformed variables. The results are shown in Table 13.3b.

TABLE 13.4 Calculations for estimating ρ for Blaisdell Company example

t	(1) e_t	(2) e_{t-1}	(3) $e_t e_{t-1}$	(4) e_{t-1}^2
1	−.026052	—	—	—
2	−.062015	−.026052	.0016156	.0006787
3	.022021	−.062015	−.0013656	.0038459
4	.163754	.022021	.0036060	.0004849
5	.046570	.163754	.0076260	.0268154
6	.046377	.046570	.0021598	.0021688
7	.043617	.046377	.0020228	.0021508
8	−.058435	.043617	−.0025488	.0019024
9	−.094399	−.058435	.0055162	.0034146
10	−.149142	−.094399	.0140789	.0089112
11	−.147991	−.149142	.0220718	.0222433
12	−.053054	−.147991	.0078515	.0219013
13	−.022928	−.053054	.0012164	.0028147
14	.105852	−.022928	−.0024270	.0005257
15	.085464	.105852	.0090465	.0112046
16	.106102	.085464	.0090679	.0073041
17	.029112	.106102	.0030889	.0112576
18	.042316	.029112	.0012319	.0008475
19	−.044160	.042316	−.0018687	.0017906
20	−.033009	−.044160	.0014577	.0019501
Total			.0834478	.1322122

TABLE 13.5 Transformed variables for first iteration—Blaisdell Company example

t	(1) Y_t	(2) X_t	(3) $Y'_t = Y_t - .631166Y_{t-1}$	(4) $X'_t = X_t - .631166X_{t-1}$
1	20.96	127.3	—	—
2	21.40	130.0	8.1708	49.653
3	21.96	132.7	8.4530	50.648
4	21.52	129.4	7.6596	45.644
5	22.39	135.0	8.8073	53.327
6	22.76	137.1	8.6282	51.893
7	23.48	141.2	9.1147	54.667
8	23.66	142.8	8.8402	53.679
9	24.10	145.5	9.1666	55.369
10	24.01	145.3	8.7989	53.465
11	24.54	148.3	9.3857	56.592
12	24.30	146.4	8.8112	52.798
13	25.00	150.2	9.6627	57.797
14	25.64	153.1	9.8608	58.299
15	26.36	157.3	10.1769	60.668
16	26.98	160.7	10.3425	61.418
17	27.52	164.2	10.4911	62.772
18	27.78	165.6	10.4103	61.963
19	28.24	168.7	10.7062	64.179
20	28.78	171.7	10.9559	65.222

Based on the fitted regression for the transformed variables in Table 13.3b, residuals were obtained and the Durbin-Watson statistic calculated. The result was (calculations not shown) $D = 1.65$. From Table A–6, we find for $\alpha = .01$, $p - 1 = 1$, and $n = 19$:

$$d_L = .93 \qquad d_U = 1.13$$

Since $D = 1.65 > d_U = 1.13$, we conclude that the autocorrelation coefficient for the error terms in the model with the transformed variables is zero.

Having successfully handled the problem of autocorrelated error terms, we now transform the estimated regression coefficients and standard deviations back to the model with the original variables (Table 13.3b):

$$\begin{aligned} b_0 &= -1.06811 \qquad & s(b_0) &= .45288 \\ b_1 &= .17376 \qquad & s(b_1) &= .00296 \end{aligned}$$

Note that the estimated regression coefficients $b_0 = -1.06811$ and $b_1 = .17376$ obtained with the iterative method are close to those obtained with ordinary least squares (Table 13.3a), but that the estimated standard errors $s(b_0) = .45288$ and $s(b_1) = .00296$ with the iterative method are larger than their ordinary-least-squares counterparts. The larger standard errors are to be expected, since we noted earlier that positive autocorrelation may lead to estimated standard deviations $s(b_k)$ calculated according to ordinary least squares that seriously underestimate the true standard deviations $\sigma(b_k)$.

Comments

1. The iterative approach does not always work properly. A major reason is that when the error terms are positively autocorrelated, the estimate r in (13.15) tends to underestimate the autocorrelation parameter ρ. When this bias is serious, it can significantly reduce the effectiveness of the iterative approach.

2. There exists an approximate relation between the Durbin-Watson test statistic D in (13.10) and the estimated autocorrelation parameter r in (13.15):

(13.17) $$D \simeq 2(1 - r)$$

This relation indicates that the Durbin-Watson statistic ranges approximately between 0 and 4 since r takes on values between -1 and 1, and that D is approximately 2 when $r = 0$. Note that for the Blaisdell Company example, $D = .735$, $r = .631$, and $2(1 - r) = .738$.

First differences approach. Some economists and statisticians have suggested that instead of iterative estimation of ρ, which is not always successful, the autocorrelation parameter be assumed to equal 1. If $\rho = 1$, $\beta_0' = \beta_0(1 - \rho) = 0$, and the transformed model (13.13) becomes:

(13.18) $$Y_t' = \beta_1' X_t' + u_t$$

where:

(13.18a) $$\begin{aligned} Y_t' &= Y_t - Y_{t-1} \\ X_t' &= X_t - X_{t-1} \end{aligned}$$

Thus, the regression coefficient $\beta_1' = \beta_1$ can be directly estimated by ordinary least squares methods for regression through the origin. Note that the transformed variables (13.18a) are ordinary first differences. It has been found that this first differences approach is effective in a variety of applications in reducing the autocorrelations of the error terms, and of course it is much simpler than the iterative approach.

Example. Table 13.6 contains the transformed variables Y_t' and X_t', based on the first differences transformations in (13.18a) for our Blaisdell Company example. Application of ordinary least squares for estimating a linear regression through the origin led to the results shown in Table 13.3c. Note that the estimated regression coefficient $b_1 = .16849$ is similar to that obtained with ordinary least squares applied to the original variables ($b_1 = .17628$), but it has a larger standard error, again as expected.

TABLE 13.6 First differences for Blaisdell Company data

t	(1) Y_t	(2) X_t	(3) $Y_t' = Y_t - Y_{t-1}$	(4) $X_t' = X_t - X_{t-1}$
1	20.96	127.3	—	—
2	21.40	130.0	.44	2.7
3	21.96	132.7	.56	2.7
4	21.52	129.4	−.44	−3.3
5	22.39	135.0	.87	5.6
6	22.76	137.1	.37	2.1
7	23.48	141.2	.72	4.1
8	23.66	142.8	.18	1.6
9	24.10	145.5	.44	2.7
10	24.01	145.3	−.09	−.2
11	24.54	148.3	.53	3.0
12	24.30	146.4	−.24	−1.9
13	25.00	150.2	.70	3.8
14	25.64	153.1	.64	2.9
15	26.36	157.3	.72	4.2
16	26.98	160.7	.62	3.4
17	27.52	164.2	.54	3.5
18	27.78	165.6	.26	1.4
19	28.24	168.7	.46	3.1
20	28.78	171.7	.54	3.0

Note

Sometimes the first differences approach can overcorrect, leading to negative autocorrelations in the error terms. Hence, it may be appropriate to use a two-sided Durbin-Watson test when testing for autocorrelation with first differences data. One complication arises here. The first differences model (13.18) has no intercept term, but the Durbin-Watson test requires a fitted regression with an intercept term. A valid test for autocorrelation in a no-intercept model can be carried out by fitting for this purpose a regression function with an intercept term. Of course, the fitted no-intercept model is still the model of basic interest. In our Blaisdell Company example, the Durbin-Watson statistic for the fitted first differences regression model with an intercept term is $D = 1.75$ (calculation not shown). This indicates uncorrelated error terms for either a one-sided test (with $\alpha = .01$) or a two-sided test (with $\alpha = .02$).

Comments

1. The autoregressive error structure can also be used to advantage in situations where predictions of the dependent variable are to be made. Johnston (Ref. 13.3) discusses this problem.

2. The first-order autoregressive error process in models (13.1) and (13.2) is the simplest kind. A second-order process would be:

$$\varepsilon_t = \rho_1 \varepsilon_{t-1} + \rho_2 \varepsilon_{t-2} + u_t \tag{13.19}$$

Still higher-order processes could be postulated. Specialized approaches have been developed for complex autoregressive error processes. These are discussed in treatments of time series procedures and forecasting, such as in Reference 13.4.

PROBLEMS

13.1. Refer to Table 13.1.

a. Plot ε_t against ε_{t-1} for $t = 1, \ldots, 10$ on a graph. How is the positive first-order autocorrelation in the error terms shown by the plot?

b. If you plotted u_t against ε_{t-1} for $t = 1, \ldots, 10$, what pattern would you expect?

13.2. Refer to **Plastic hardness** Problem 2.18. If the same test item were measured at 12 different points in time, would the error terms in the regression model likely be autocorrelated? Discuss.

13.3. A student stated that the first-order autoregressive error models (13.1) and (13.2) are too simple for business time series data because the error term in period t in such data is also influenced by random effects that occurred more than one period in the past. Comment.

13.4. A student writing a term paper used ordinary least squares in fitting a simple linear regression model to some time series data containing positively autocorrelated errors, and found that the 90 percent confidence interval for β_1 was too wide to be useful. She then decided to employ model (13.1) to improve the precision of the estimate. Comment.

13.5. For each of the following tests concerning the autocorrelation parameter ρ in model (13.2) with three independent variables, state the appropriate decision rule based on the Durbin-Watson statistic for a sample of size 38: (1) H_0: $\rho = 0$, H_a: $\rho \neq 0$, $\alpha = .02$; (2) H_0: $\rho = 0$, H_a: $\rho < 0$, $\alpha = .05$; (3) H_0: $\rho = 0$, H_a: $\rho > 0$, $\alpha = .01$.

13.6. Refer to **Calculator maintenance** Problem 2.16. The observations are listed in time order. Assume that regression model (13.1) is appropriate. Test whether or not positive autocorrelation is present; use $\alpha = .01$. State the alternatives, decision rule, and conclusion.

13.7. Refer to **Chemical shipment** Problem 7.12. The observations are listed in time order. Assume that regression model (13.2) is appropriate. Test whether or not positive autocorrelation is present; use $\alpha = .05$. State the alternatives, decision rule, and conclusion.

13.8. Refer to **Crop yield** Problem 9.13. The observations are listed in time order. Assume that regression model (13.2) with first- and second-order terms for the two independent variables and no interaction term is appropriate. Test whether or not positive autocorrelation is present; use $\alpha = .01$. State the alternatives, decision rule, and conclusion.

13.9. **Microcomputer components.** A staff analyst for a manufacturer of microcomputer components has compiled monthly data for the past 16 months on the value of industry production of processing units that use these components (X, in million dollars) and the value of the firm's components used (Y, in thousand dollars). The analyst believes that a simple linear regression relation is appropriate but anticipates positive autocorrelation. The data follow.

t:	1	2	3	4	5	6	7	8
X_t:	2.052	2.026	2.002	1.949	1.942	1.887	1.986	2.053
Y_t:	102.9	101.5	100.8	98.0	97.3	93.5	97.5	102.2

t:	9	10	11	12	13	14	15	16
X_t:	2.102	2.113	2.058	2.060	2.035	2.080	2.102	2.150
Y_t:	105.0	107.2	105.1	103.9	103.0	104.8	105.0	107.2

a. Fit a simple linear regression model by ordinary least squares and obtain the residuals. Also obtain $s(b_0)$ and $s(b_1)$.

b. Plot the residuals against time and explain whether you find any evidence of positive autocorrelation.

c. Conduct a formal test for positive autocorrelation using a significance level of .05. State the alternatives, decision rule, and conclusion. Is the residual analysis in part (b) in accord with the test result?

13.10. Refer to **Microcomputer components** Problem 13.9. The analyst has decided to employ regression model (13.1) and use the iterative approach to fit the model.

a. Obtain a point estimate of the autocorrelation parameter. How well does the approximate relationship (13.17) hold here between this point estimate and the Durbin-Watson test statistic?

b. Use one iteration to obtain the estimates b_0' and b_1' of the regression coefficients β_0' and β_1' in transformed model (13.13) and state the estimated regression function. Also obtain $s(b_0')$ and $s(b_1')$.

c. Test whether any positive autocorrelation remains after the first iteration using a significance level of .05. State the alternatives, decision rule, and conclusion.

d. Restate the estimated regression function obtained in part (b) in terms of the original variables. Also obtain $s(b_0)$ and $s(b_1)$. Compare the estimated regression coefficients obtained with the iterative method and their standard errors with those obtained with ordinary least squares in Problem 13.9a.

e. Based on the results in parts (c) and (d), does the iterative method appear to have been effective here?

13.11. Refer to **Microcomputer components** Problems 13.9 and 13.10. The staff analyst wishes to try also the first differences approach.

a. Estimate the regression coefficient β_1 by the first differences approach, and obtain the estimated standard deviation of this estimate.

b. Compare the results obtained in part (a) with those in Problem 13.10d. Summarize your findings.

c. Test whether or not the error terms with the first differences approach are autocorrelated using a two-sided test and a level of significance of .10. State the alternatives, decision rule, and conclusion. Why is a two-sided test meaningful here?

13.12. Advertising agency. The managing partner of an advertising agency has become concerned about possible inefficiencies in the handling of client accounts. Monthly data on amount of billings (Y, in thousands of constant dollars) and on number of hours of staff time (X, in thousand hours) for the 20 most recent months follow. A simple linear regression model is believed to be appropriate, but positively autocorrelated error terms may be present.

t:	1	2	3	4	5	6	7	8	9	10
X_t:	2.521	2.171	2.234	2.524	2.305	2.523	3.020	3.014	3.532	3.461
Y_t:	220.4	203.9	207.2	221.9	211.3	222.7	247.6	247.6	272.9	269.1

t:	11	12	13	14	15	16	17	18	19	20
X_t:	3.737	3.801	3.576	3.586	3.447	2.723	3.019	3.117	3.623	3.618
Y_t:	283.9	287.0	275.4	275.1	269.1	232.8	248.1	252.4	278.6	278.5

a. Fit a simple linear regression model by ordinary least squares and obtain the residuals. Also obtain $s(b_0)$ and $s(b_1)$.

b. Plot the residuals against time and explain whether you find any evidence of positive autocorrelation.

c. Conduct a formal test for positive autocorrelation using a significance level of .01. State the alternatives, decision rule, and conclusion. Is the residual analysis in part (b) in accord with the test result?

13.13. Refer to **Advertising agency** Problem 13.12. Assume that regression model (13.1) is applicable.

a. Obtain a point estimate of the autocorrelation parameter. How well does the approximate relationship (13.17) hold here between the point estimate and the Durbin-Watson test statistic?

b. Use one iteration to obtain the estimates b_0' and b_1' of the regression coefficients β_0' and β_1' in transformed model (13.13) and state the estimated regression function. Also obtain $s(b_0')$ and $s(b_1')$.

c. Test whether any positive autocorrelation remains after the first iteration using a significance level of .01. State the alternatives, decision rule, and conclusion.

d. Restate the estimated regression function obtained in part (b) in terms of the original variables. Also obtain $s(b_0)$ and $s(b_1)$. Compare the estimated regression coefficients obtained with the iterative method and their standard errors with those obtained with ordinary least squares in Problem 13.12a.

e. Based on the results in parts (c) and (d), does the iterative method appear to have been effective here?

13.14. Refer to **Advertising agency** Problems 13.12 and 13.13.

a. Estimate the regression coefficient β_1 using the first differences approach by means of a 95 percent confidence interval. Interpret your interval estimate.

b. How does the estimated standard deviation of b_1 for the first differences estimate obtained in part (a) compare with that for the iterative method estimate in Problem 13.13d?

13.15. **McGill Company sales.** The data below show seasonally adjusted quarterly sales for the McGill Company (Y, in million dollars) and for the entire industry (X, in million dollars), for the most recent 20 quarters.

t:	1	2	3	4	5	6	7
X_t:	127.3	130.0	132.7	129.4	135.0	137.1	141.1
Y_t:	20.96	21.40	21.96	21.52	22.39	22.76	23.48

t:	8	9	10	11	12	13	14
X_t:	142.8	145.5	145.3	148.3	146.4	150.2	153.1
Y_t:	23.66	24.10	24.01	24.54	24.28	25.00	25.64

t:	15	16	17	18	19	20
X_t:	157.3	160.7	164.2	165.6	168.7	172.0
Y_t:	26.46	26.98	27.52	27.78	28.24	28.78

The first-order autoregressive error model (13.1) is to be employed.

a. Would you expect the autocorrelation parameter ρ to be positive, negative, or zero here?

b. Fit the linear regression model by ordinary least squares, obtain the residuals, and plot them against time. What do you find?

c. Conduct a formal test for positive autocorrelation using $\alpha = .05$. State the alternatives, decision rule, and conclusion.

13.16. Refer to **McGill Company sales** Problem 13.15.

a. Use one iteration with the iterative method to estimate the parameters β_0 and β_1 in regression model (13.1). Also obtain the estimated standard deviations of these estimates.

b. Test whether any positive autocorrelation remains after the first iteration; use $\alpha = .05$. State the alternatives, decision rule, and conclusion. Does the iterative approach appear to have been effective here?

c. Estimate β_1 with a 90 percent confidence interval. Interpret your interval estimate.

13.17. Refer to **McGill Company sales** Problems 13.15 and 13.16.

a. Estimate the regression coefficient β_1 in model (13.1) by the first differences approach using a 90 percent confidence interval.

b. Compare your result in part (a) with that in Problem 13.16c. State your findings.

c. Test whether or not the error terms with the first differences approach are positively autocorrelated using $\alpha = .05$. State the alternatives, decision rule, and conclusion.

13.18. A student applying the first differences transformation (13.18a) found that several X'_t values equaled zero but that the corresponding Y'_t values were nonzero. Does this signify that the first differences transformation is not apt for the data?

EXERCISES

13.19. Derive (13.7) for $s = 2$.

13.20. Refer to the first-order autoregressive error model (13.1). Suppose Y_t is company's percent share of the market, X_t is company's selling price as a percent of average competitive selling price, $\beta_0 = 100$, $\beta_1 = -.35$, $\rho = .6$, $\sigma^2 = 1$, and $\varepsilon_0 = 2.403$. Let X_t and u_t be as follows for $t = 1, \ldots, 10$:

t:	1	2	3	4	5	6	7	8	9	10
X_t:	100	115	120	90	85	75	70	95	105	110
u_t:	.764	.509	−.242	−1.808	−.485	.501	−.539	.434	−.299	.030

a. Plot the true regression line. Generate the observations Y_t ($t = 1, \ldots, 10$), and plot these on the same graph. Fit a least squares regression line to the observations and plot it also on the same graph. How does your fitted regression line relate to the true line?

b. Repeat the steps in part (a) but this time let $\rho = 0$. In which of the two cases does the fitted regression line come closer to the true line? Is this the expected outcome?

c. Generate the observations Y_t for $\rho = -.7$. For each of the cases $\rho = .6$, $\rho = 0$, and $\rho = -.7$, obtain the successive error term differences $\varepsilon_t - \varepsilon_{t-1}$ ($t = 1, \ldots, 10$).

d. For which of the three cases in part (c) is $\Sigma(\varepsilon_t - \varepsilon_{t-1})^2$ smallest? For which is it largest? What generalization does this suggest?

13.21. Suppose the autoregressive error process for the model $Y_t = \beta_0 + \beta_1 X_t + \varepsilon_t$ is that given by (13.19).

a. What would be the transformed variables Y_t' and X_t' to be used with the iterative method?

b. How would you estimate the parameters ρ_1 and ρ_2 for use with the iterative method?

PROJECTS

13.22. The true regression model is $Y_t = 10 + 24X_t + \varepsilon_t$, where $\varepsilon_t = .8\varepsilon_{t-1} + u_t$ and u_t are independent $N(0, 25)$.

a. Generate 11 independent random numbers from $N(0, 25)$. Use the first random number as ε_0, obtain the 10 error terms $\varepsilon_1, \ldots, \varepsilon_{10}$, and then calculate the 10 observations $Y_1, \ldots, Y_{10}$ corresponding to $X_1 = 1$, $X_2 = 2, \ldots,$ $X_{10} = 10$. Fit a linear regression function by ordinary least squares and calculate *MSE*.

b. Repeat part (a) 100 times, using new random numbers each time.

c. Calculate the mean of the 100 estimates b_1. Does it appear that b_1 is an unbiased estimator of β_1 despite the presence of positive autocorrelation?

d. Calculate the mean of the 100 estimates *MSE*. Does it appear that *MSE* is a biased estimator of σ^2? If so, does the magnitude of the bias appear to be small or large?

CITED REFERENCES

13.1 Durbin, J., and G. S. Watson. "Testing for Serial Correlation in Least Squares Regression. II." *Biometrika* 38 (1951), pp. 159–78.

13.2 Theil, H., and A. L. Nagar. "Testing the Independence of Regression Disturbances." *Journal of the American Statistical Association* 56 (1961), pp. 793–806.

13.3 Johnston, J. *Econometric Methods*. 2d ed. New York: McGraw-Hill, 1971.

13.4 Box, G. E. P., and G. M. Jenkins. *Time Series Analysis, Forecasting and Control*. Rev. ed. San Francisco: Holden-Day, 1976.

14

Nonlinear regression

The linear regression models considered up to this point are satisfactory for most regression applications. There are occasions, however, when a nonlinear regression model is most suitable. We shall consider in this chapter nonlinear regression models, how to obtain least squares estimates of the regression parameters in such models, and how to make inferences about these regression parameters. The analysis of nonlinear regression models is numerically tedious and is therefore heavily computer-oriented.

14.1 LINEAR, INTRINSICALLY LINEAR, AND NONLINEAR REGRESSION MODELS

Linear regression models

In previous chapters, we considered linear regression models, i.e., models which are linear in the parameters. Such models can be represented by:

$$Y_i = \beta_0 + \beta_1 X_{i1} + \beta_2 X_{i2} + \cdots + \beta_{p-1} X_{i,p-1} + \varepsilon_i \tag{14.1}$$

Linear regression models, as we have seen, include not only first-order models in $p-1$ independent variables but also more complex models. For instance, a polynomial regression model in one or more independent variables is linear in the parameters, such as the following model in two independent variables with linear, quadratic, and interaction terms:

$$(14.2) \qquad Y_i = \beta_0 + \beta_1 X_{i1} + \beta_2 X_{i1}^2 + \beta_3 X_{i2} + \beta_4 X_{i2}^2 + \beta_5 X_{i1} X_{i2} + \varepsilon_i$$

Also, models with transformed variables that are linear in the parameters belong to the class of linear regression models, such as the following model:

$$(14.3) \qquad \log_{10} Y_i = \beta_0 + \beta_1 \sqrt{X_{i1}} + \beta_2 \exp(X_{i2}) + \varepsilon_i$$

Intrinsically linear regression models

In addition to the multitude of models that are linear in the parameters, there are other models that, though not linear in the parameters, can be transformed so that the parameters appear in linear fashion. For example, the exponential model:

$$(14.4) \qquad Y_i = \gamma_0[\exp(\gamma_1 X_i)]\varepsilon_i$$

is nonlinear in the parameters γ_0 and γ_1. However, this model can be transformed into the linear form (14.1) by using the logarithmic transformation:

$$(14.5) \qquad \log_e Y_i = \log_e \gamma_0 + \gamma_1 X_i + \log_e \varepsilon_i$$

Letting:

$$\begin{aligned} \log_e Y_i &= Y_i' \\ \log_e \gamma_0 &= \beta_0 \\ \gamma_1 &= \beta_1 \\ \log_e \varepsilon_i &= \varepsilon_i' \end{aligned}$$

we can write model (14.5) in the usual form for a linear model:

$$(14.5a) \qquad Y_i' = \beta_0 + \beta_1 X_i + \varepsilon_i'$$

We say that model (14.4) is an *intrinsically linear model* because it can be expressed in the linear form (14.1) by a suitable transformation. Another intrinsically linear regression model is:

$$(14.6) \qquad Y_i = [\exp(\gamma)]X_i + \varepsilon_i$$

If we let $\exp(\gamma) = \beta_1$, we then have a model with the regression through the origin:

$$(14.6a) \qquad Y_i = \beta_1 X_i + \varepsilon_i$$

Note

When an intrinsically linear model has been transformed into the linear model form, such as model (14.4) transformed into model (14.5a), it is important to study the linearized model for aptness. For instance, if the error terms ε_i in (14.4) are normally distributed, the transformed error terms ε_i' in (14.5a) will not be normally distributed.

Nonlinear regression models

Nonlinear regression models are not linear in the parameters and cannot be made so by transformation. For example, *exponential model* (14.4) with an additive error term:

$$Y_i = \gamma_0 \exp(\gamma_1 X_i) + \varepsilon_i \tag{14.7}$$

is intrinsically nonlinear because no transformation exists which transforms this model into the linear form (14.1). A more general nonlinear exponential model in one independent variable with an additive error term is:

$$Y_i = \gamma_0 + \gamma_1 \exp(\gamma_2 X_i) + \varepsilon_i \tag{14.8}$$

This model is commonly used in growth studies where the rate of growth at a given time X is proportional to the amount of growth remaining as time increases, with γ_0 representing the maximum growth value. Model (14.8) is often employed to relate the concentration of a substance (Y) to elapsed time (X). Figure 14.1a shows the response function for exponential model (14.8), for $\gamma_0 = 100$, $\gamma_1 = -50$, and $\gamma_2 = -2$.

Another nonlinear regression model is the *general logistic model* with additive error term:

$$Y_i = \frac{\gamma_0}{1 + \gamma_1 \exp(\gamma_2 X_i)} + \varepsilon_i \tag{14.9}$$

This model has been used in population studies to relate number of species (Y) to time (X). Recall that logistic response function (10.53) was used in Chapter 10 for situations where the dependent variable is a 0, 1 indicator variable. Logistic response function (10.53) is a special type of logistic function. Figure 14.1b

FIGURE 14.1 Graphs of nonlinear regression response functions

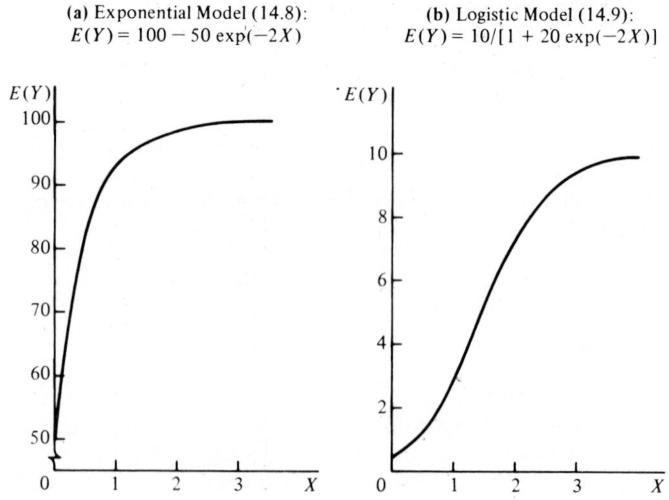

shows the response function for the general logistic model, for $\gamma_0 = 10$, $\gamma_1 = 20$, and $\gamma_2 = -2$.

Note

When nonlinear growth models are used for time series data, it is important to ascertain whether or not the error terms are uncorrelated, just as when linear regression models are applied to time series data.

14.2 EXAMPLE

In order to illustrate the analysis of nonlinear regression models, we shall use a relatively simple two-parameter example. In so doing, we shall be able to explain the concepts and procedures without overwhelming the reader with details.

A hospital administrator wished to develop a regression model for predicting the degree of long-term recovery after discharge from the hospital for severely injured patients. The predictor variable to be utilized is number of days of hospitalization (X), and the dependent variable is a prognosis index for long-term recovery (Y), with large values of the index reflecting a good prognosis. Data for 15 patients were studied; these are presented in Table 14.1.

Initially, the intrinsically linear exponential model (14.4) was fitted to the data, employing the logarithmic transformation model in (14.5). A residual analysis for this model suggested, however, that the error variance increases with X. It was therefore decided to use the two-parameter exponential model (14.7) with an additive error term, which is an intrinsically nonlinear model:

$$Y_i = \gamma_0 \exp(\gamma_1 X_i) + \varepsilon_i \tag{14.10}$$

TABLE 14.1 Data for severely injured patients example

Patient i	*Days Hospitalized* X_i	*Prognosis Index* Y_i
1	2	54
2	5	50
3	7	45
4	10	37
5	14	35
6	19	25
7	26	20
8	31	16
9	34	18
10	38	13
11	45	8
12	52	11
13	53	8
14	60	4
15	65	6

It is now desired to estimate the regression parameters γ_0 and γ_1 for this model and to study the fit of the model.

14.3 LEAST SQUARES ESTIMATION IN NONLINEAR REGRESSION

We noted in Chapter 2 that the method of least squares requires the minimization of the criterion Q which for simple linear regression is:

$$Q = \sum_{i=1}^{n}(Y_i - \beta_0 - \beta_1 X_i)^2$$

Those estimates of β_0 and β_1 which minimize Q for the given sample observations (X_i, Y_i) are the least squares estimates and are denoted by b_0 and b_1. We also noted that one can search for the least squares estimates by trying various values of b_0 and b_1 and evaluating Q each time until the minimum value of Q is found. Alternatively, one can find analytically the least squares estimates by differentiating Q with respect to β_0 and β_1, setting the derivatives equal to 0, and solving the normal equations.

The same two basic approaches apply in nonlinear regression. We shall first consider the use of the normal equations and then the use of direct search procedures.

Normal equations

In linear regression, we represent an observation Y_i as the sum of the mean response and an error term:

$$Y_i = E(Y_i) + \varepsilon_i \tag{14.11}$$

where:

$$E(Y_i) = \beta_0 + \beta_1 X_{i1} + \cdots + \beta_{p-1} X_{i,p-1}$$

Since nonlinear regression models take many different forms, as we saw earlier, we shall now simply indicate that $E(Y_i)$ is a function of p regression parameters $\gamma_0, \gamma_1, \ldots, \gamma_{p-1}$ and the ith observations on the X variables:

$$E(Y_i) = f(\mathbf{X}_i, \boldsymbol{\gamma}) \tag{14.12}$$

where:

$$\underset{q\times 1}{\mathbf{X}_i} = \begin{bmatrix} X_{i1} \\ X_{i2} \\ \cdot \\ \cdot \\ \cdot \\ X_{iq} \end{bmatrix} \qquad \underset{p\times 1}{\boldsymbol{\gamma}} = \begin{bmatrix} \gamma_0 \\ \gamma_1 \\ \cdot \\ \cdot \\ \cdot \\ \gamma_{p-1} \end{bmatrix} \tag{14.12a}$$

We denote the number of X variables by q, since the number of X variables in nonlinear regression is not directly related to the number of parameters, unlike linear regression. Thus, for model (14.10), we have:

$$f(\mathbf{X}_i, \boldsymbol{\gamma}) = \gamma_0 \exp(\gamma_1 X_i)$$

where the two ($p = 2$) parameters are γ_0 and γ_1 and $\mathbf{X}_i$ consists of one ($q = 1$) value X_i for the ith observation.

With this notation, we express a nonlinear regression model as follows:

$$Y_i = f(\mathbf{X}_i, \boldsymbol{\gamma}) + \varepsilon_i \tag{14.13}$$

The least squares criterion Q can then be written:

$$Q = \sum_{i=1}^{n} [Y_i - f(\mathbf{X}_i, \boldsymbol{\gamma})]^2 \tag{14.14}$$

The partial derivative of Q with respect to γ_k is:

$$\frac{\partial Q}{\partial \gamma_k} = \sum_{i=1}^{n} -2[Y_i - f(\mathbf{X}_i, \boldsymbol{\gamma})]\left[\frac{\partial f(\mathbf{X}_i, \boldsymbol{\gamma})}{\partial \gamma_k}\right] \tag{14.15}$$

When the p partial derivatives are each set equal to 0 and the parameters γ_k are replaced by the least squares estimates g_k, we obtain after some simplification the p normal equations:

$$\sum_{i=1}^{n} Y_i \left[\frac{\partial f(\mathbf{X}_i, \boldsymbol{\gamma})}{\partial \gamma_k}\right]_{\boldsymbol{\gamma}=\mathbf{g}} - \sum_{i=1}^{n} f(\mathbf{X}_i, \mathbf{g})\left[\frac{\partial f(\mathbf{X}_i, \boldsymbol{\gamma})}{\partial \gamma_k}\right]_{\boldsymbol{\gamma}=\mathbf{g}} = 0 \qquad k = 0, 1, \ldots, p-1 \tag{14.16}$$

where $\mathbf{g}$ is the vector of the least squares estimates g_k:

$$\mathbf{g} = \begin{bmatrix} g_0 \\ g_1 \\ \cdot \\ \cdot \\ \cdot \\ g_{p-1} \end{bmatrix} \tag{14.16a}$$

and the terms in brackets in (14.16) are partial derivatives with the parameters γ_k replaced by the least squares estimates g_k.

The normal equations (14.16) for nonlinear regression models are nonlinear in the parameter estimates g_k and are usually difficult to solve, even in the simplest of cases. Hence, numerical procedures are ordinarily required to obtain a solution iteratively. To make things still more difficult, multiple solutions may be present.

Example. In our severely injured patients example, we are employing the response function:

$$f(\mathbf{X}_i, \boldsymbol{\gamma}) = \gamma_0 \exp(\gamma_1 X_i) \tag{14.17}$$

Hence, the partial derivatives of $f(\mathbf{X}_i, \boldsymbol{\gamma})$ are:

(14.18a) $$\frac{\partial f(\mathbf{X}_i, \boldsymbol{\gamma})}{\partial \gamma_0} = \exp(\gamma_1 X_i)$$

(14.18b) $$\frac{\partial f(\mathbf{X}_i, \boldsymbol{\gamma})}{\partial \gamma_1} = \gamma_0 X_i \exp(\gamma_1 X_i)$$

Replacing γ_0 and γ_1 by the respective least squares estimates g_0 and g_1 in (14.17), (14.18a), and (14.18b), the normal equations (14.16) therefore are:

$$\begin{aligned} \Sigma Y_i \exp(g_1 X_i) \quad &- \Sigma g_0 \exp(g_1 X_i) \exp(g_1 X_i) &= 0 \\ \Sigma Y_i g_0 X_i \exp(g_1 X_i) &- \Sigma g_0 \exp(g_1 X_i) g_0 X_i \exp(g_1 X_i) &= 0 \end{aligned}$$

Upon simplification, the normal equations become:

$$\begin{aligned} \Sigma Y_i \exp(g_1 X_i) \quad &- g_0 \Sigma \exp(2 g_1 X_i) &= 0 \\ \Sigma Y_i X_i \exp(g_1 X_i) &- g_0 \Sigma X_i \exp(2 g_1 X_i) &= 0 \end{aligned}$$

These normal equations are not linear in g_0 and g_1, and no closed-form solution exists. Thus, numerical methods will be required to find the least squares estimates iteratively.

Note

When the error terms in a nonlinear regression model are independent $N(0, \sigma^2)$, the least squares estimates are the same as the maximum likelihood estimates.

Gauss-Newton method

In many nonlinear regression problems, it is more practical to find the least squares estimates by direct search procedures rather than by first obtaining the normal equations and then using numerical methods to solve these equations iteratively. The major statistical computer packages employ one or more direct search procedures for solving nonlinear regression problems.

The *Gauss-Newton method*, also called the *linearization method*, uses a Taylor series expansion to approximate the nonlinear regression model with linear terms and then employs ordinary least squares to estimate the parameters. Iteration of these steps generally leads to a solution to the nonlinear regression problem.

The Gauss-Newton method begins with initial or starting values for the regression parameters $\gamma_0, \gamma_1, \ldots, \gamma_{p-1}$. We shall denote these by $g_0^{(0)}, g_1^{(0)}, \ldots, g_{p-1}^{(0)}$, where the superscript in parentheses denotes the iteration number. The starting values $g_k^{(0)}$ may be obtained from previous or related studies, theoretical expectations, or a preliminary search for parameter values that lead to a comparatively low criterion value Q in (14.14). We shall later discuss in more detail the choice of the starting values.

Once the starting values for the parameters have been obtained, we approximate the mean responses $f(\mathbf{X}_i, \boldsymbol{\gamma})$ for the n observations by the linear terms in the Taylor series expansion around the starting values $g_k^{(0)}$. We obtain for the ith observation:

$$(14.19)\quad f(\mathbf{X}_i, \boldsymbol{\gamma}) \simeq f(\mathbf{X}_i, \mathbf{g}^{(0)}) + \sum_{k=0}^{p-1} \left[\frac{\partial f(\mathbf{X}_i, \boldsymbol{\gamma})}{\partial \gamma_k} \right]_{\boldsymbol{\gamma}=\mathbf{g}^{(0)}} (\gamma_k - g_k^{(0)})$$

where:

$$(14.19a)\quad \mathbf{g}^{(0)} = \begin{bmatrix} g_0^{(0)} \\ g_1^{(0)} \\ \cdot \\ \cdot \\ \cdot \\ g_{p-1}^{(0)} \end{bmatrix}$$

is the vector of the parameter starting values. The terms in brackets in (14.19) are the same partial derivatives of the regression function we encountered earlier in the normal equations (14.16), but here they are evaluated at $\gamma_k = g_k^{(0)}$ for $k = 0, 1, \ldots, p-1$.

Let us now simplify the notation as follows:

$$(14.20a)\quad f_i^{(0)} = f(\mathbf{X}_i, \mathbf{g}^{(0)})$$

$$(14.20b)\quad \beta_k^{(0)} = \gamma_k - g_k^{(0)}$$

$$(14.20c)\quad D_{ik}^{(0)} = \left[\frac{\partial f(\mathbf{X}_i, \boldsymbol{\gamma})}{\partial \gamma_k} \right]_{\boldsymbol{\gamma}=\mathbf{g}^{(0)}}$$

The Taylor approximation (14.19) for the ith observation mean response then becomes in this notation:

$$f(\mathbf{X}_i, \boldsymbol{\gamma}) \simeq f_i^{(0)} + \sum_{k=0}^{p-1} D_{ik}^{(0)} \beta_k^{(0)}$$

and an approximation to the nonlinear regression model (14.13):

$$Y_i = f(\mathbf{X}_i, \boldsymbol{\gamma}) + \varepsilon_i$$

is:

$$(14.21)\quad Y_i \simeq f_i^{(0)} + \sum_{k=0}^{p-1} D_{ik}^{(0)} \beta_k^{(0)} + \varepsilon_i$$

When we shift the $f_i^{(0)}$ term to the left and denote the difference $Y_i - f_i^{(0)}$ by $Y_i^{(0)}$, we obtain a linear regression model approximation:

$$(14.22)\quad Y_i^{(0)} \simeq \sum_{k=0}^{p-1} D_{ik}^{(0)} \beta_k^{(0)} + \varepsilon_i \qquad i = 1, \ldots, n$$

where:

(14.22a) $$Y_i^{(0)} = Y_i - f_i^{(0)}$$

We shall represent this approximation in matrix form as follows:

(14.23) $$\mathbf{Y}^{(0)} \simeq \mathbf{D}^{(0)}\boldsymbol{\beta}^{(0)} + \boldsymbol{\varepsilon}$$

where:

(14.23a) $$\underset{n\times 1}{\mathbf{Y}^{(0)}} = \begin{bmatrix} Y_1 - f_1^{(0)} \\ \cdot \\ \cdot \\ \cdot \\ Y_n - f_n^{(0)} \end{bmatrix}$$

(14.23b) $$\underset{n\times p}{\mathbf{D}^{(0)}} = \begin{bmatrix} D_{10}^{(0)} & \cdots & D_{1,p-1}^{(0)} \\ \cdot & & \cdot \\ \cdot & & \cdot \\ \cdot & & \cdot \\ D_{n0}^{(0)} & \cdots & D_{n,p-1}^{(0)} \end{bmatrix}$$

(14.23c) $$\underset{p\times 1}{\boldsymbol{\beta}^{(0)}} = \begin{bmatrix} \beta_0^{(0)} \\ \cdot \\ \cdot \\ \cdot \\ \beta_{p-1}^{(0)} \end{bmatrix}$$

Note that the approximate model (14.23) is precisely in the form of the general linear regression model (7.18), with the **D** matrix of partial derivatives now playing the role of the **X** matrix. We can therefore estimate the parameters $\boldsymbol{\beta}^{(0)}$ by means of the normal equations for the ordinary linear regression model and obtain according to (7.21):

(14.24) $$\mathbf{b}^{(0)} = (\mathbf{D}^{(0)\prime}\mathbf{D}^{(0)})^{-1}\mathbf{D}^{(0)\prime}\mathbf{Y}^{(0)}$$

where $\mathbf{b}^{(0)}$ is the vector of the least squares estimated regression coefficients. We use these least squares estimates to obtain revised estimated regression coefficients $g_k^{(1)}$ by means of (14.20b):

$$g_k^{(1)} = g_k^{(0)} + b_k^{(0)}$$

where $g_k^{(1)}$ denotes the revised estimate of γ_k at the end of the first iteration. In matrix form, we represent the revision process as follows:

(14.25) $$\mathbf{g}^{(1)} = \mathbf{g}^{(0)} + \mathbf{b}^{(0)}$$

At this point, we can examine whether the revised regression coefficients represent adjustments in the proper direction. The least squares criterion measure Q in (14.14) for the starting regression coefficients $\mathbf{g}^{(0)}$, to be denoted by $SSE^{(0)}$, is:

(14.26) $$SSE^{(0)} = \sum_{i=1}^{n}[Y_i - f(\mathbf{X}_i, \mathbf{g}^{(0)})]^2 = \sum_{i=1}^{n}(Y_i - f_i^{(0)})^2$$

At the end of the first iteration, the estimated regression coefficients are $\mathbf{g}^{(1)}$, and the least squares criterion measure, now denoted by $SSE^{(1)}$, is:

$$SSE^{(1)} = \sum_{i=1}^{n} [Y_i - f(\mathbf{X}_i, \mathbf{g}^{(1)})]^2 = \sum_{i=1}^{n} (Y_i - f_i^{(1)})^2 \tag{14.27}$$

If the Gauss-Newton method is working effectively in the first iteration, $SSE^{(1)}$ should be smaller than $SSE^{(0)}$ since the revised estimated regression coefficients $\mathbf{g}^{(1)}$ should be better estimates.

Note that the nonlinear regression functions $f(\mathbf{X}_i, \mathbf{g}^{(0)})$ and $f(\mathbf{X}_i, \mathbf{g}^{(1)})$ are used in calculating $SSE^{(0)}$ and $SSE^{(1)}$, and not the linear approximations from the Taylor series expansion.

The revised regression coefficients $\mathbf{g}^{(1)}$ are not, of course, the least squares estimates for the nonlinear regression problem because the fitted model (14.23) is only an approximation of the nonlinear model. The Gauss-Newton method therefore repeats the procedure just described, with $\mathbf{g}^{(1)}$ now as the starting values. This produces a new set of revised estimates, denoted by $\mathbf{g}^{(2)}$, and a new least squares criterion measure $SSE^{(2)}$. The iterative process is continued until the difference between successive coefficient estimates $\mathbf{g}^{(s+1)} - \mathbf{g}^{(s)}$ and/or the difference between successive least squares criterion measures $SSE^{(s+1)} - SSE^{(s)}$ become negligible. We shall denote the final estimates of the regression coefficients simply by $\mathbf{g}$ and the final least squares criterion measure, which is the error sum of squares, by SSE.

The Gauss-Newton method works effectively in many nonlinear regression applications. In some instances, however, the method may require numerous iterations before converging, and in a few cases it may not converge at all.

Example. In our severely injured patients example, initial values of the parameters γ_0 and γ_1 were taken to be the estimates of these parameters when the logarithmic transformation model (14.5) of the intrinsically linear exponential model (14.4) was fitted. These initial values are $g_0^{(0)} = 56.6646$ and $g_1^{(0)} = -.03797$ (calculations not shown). The least squares criterion measure at this stage requires evaluation of the nonlinear regression function (14.17) for each observation, utilizing the starting parameter values $g_0^{(0)}$ and $g_1^{(0)}$. For instance, for the first observation, for which $X_1 = 2$, we obtain:

$$\begin{aligned} f(\mathbf{X}_1, \mathbf{g}^{(0)}) = f_1^{(0)} &= g_0^{(0)} \exp(g_1^{(0)} X_1) \\ &= (56.6646)\exp[-.03797(2)] \\ &= 52.5208 \end{aligned}$$

Since $Y_1 = 54$, the deviation from the mean response is:

$$Y_1^{(0)} = Y_1 - f_1^{(0)} = 54 - 52.5208 = 1.4792$$

We see that the deviation $Y_1^{(0)}$ is actually the residual for observation 1 at the initial fitting stage since $f_1^{(0)}$ is the estimated mean response when the initial estimates $\mathbf{g}^{(0)}$ of the parameters are employed. The stage 0 residuals for this and the other sample observations are presented in Table 14.2 and constitute the $\mathbf{Y}^{(0)}$ vector.

The least squares criterion measure at this initial stage then is simply the sum of the squared stage 0 residuals:

$$SSE^{(0)} = \Sigma(Y_i - f_i^{(0)})^2 = \Sigma(Y_i^{(0)})^2$$
$$= (1.4792)^2 + \cdots + (1.1977)^2 = 56.0869$$

To revise the initial values for the parameters, we require the $\mathbf{D}^{(0)}$ matrix and the $\mathbf{Y}^{(0)}$ vector. The latter was already obtained in the process of calculating the least squares criterion measure at the start. To obtain the $\mathbf{D}^{(0)}$ matrix, we need the partial derivatives of the regression function (14.17) evaluated at $\boldsymbol{\gamma} = \mathbf{g}^{(0)}$. The partial derivatives are given in (14.18). Table 14.2 shows the $\mathbf{D}^{(0)}$ matrix entries in symbolic form and also the numerical values. To illustrate the calculations for observation $i = 1$, we know from Table 14.1 that $X_1 = 2$. Hence, evaluating the partial derivatives at $\mathbf{g}^{(0)}$, we find:

$$\left[\frac{\partial f(\mathbf{X}_1, \boldsymbol{\gamma})}{\partial \gamma_0}\right]_{\boldsymbol{\gamma}=\mathbf{g}^{(0)}} = \exp(g_1^{(0)} X_1)$$
$$= \exp[-.03797(2)] = .92687$$

$$\left[\frac{\partial f(\mathbf{X}_1, \boldsymbol{\gamma})}{\partial \gamma_1}\right]_{\boldsymbol{\gamma}=\mathbf{g}^{(0)}} = g_0^{(0)} X_1 \exp(g_1^{(0)} X_1)$$
$$= 56.6646(2)\exp[-.03797(2)] = 105.0416$$

TABLE 14.2 $\mathbf{Y}^{(0)}$ and $\mathbf{D}^{(0)}$ matrices for severely injured patients example

$$\underset{15\times 1}{\mathbf{Y}^{(0)}} = \begin{bmatrix} Y_1 - f_1^{(0)} \\ \cdot \\ \cdot \\ \cdot \\ Y_{15} - f_{15}^{(0)} \end{bmatrix} = \begin{bmatrix} Y_1 - g_0^{(0)}\exp(g_1^{(0)}X_1) \\ \cdot \\ \cdot \\ \cdot \\ Y_{15} - g_0^{(0)}\exp(g_1^{(0)}X_{15}) \end{bmatrix} = \begin{bmatrix} 1.4792 \\ 3.1337 \\ 1.5609 \\ -1.7624 \\ 1.6996 \\ -2.5422 \\ -1.1139 \\ -1.4629 \\ 2.4172 \\ -.3871 \\ -2.2625 \\ 3.1327 \\ .4259 \\ -1.8063 \\ 1.1977 \end{bmatrix}$$

$$\underset{15\times 2}{\mathbf{D}^{(0)}} = \begin{bmatrix} \exp(g_1^{(0)}X_1) & g_0^{(0)}X_1\exp(g_1^{(0)}X_1) \\ \cdot & \cdot \\ \cdot & \cdot \\ \cdot & \cdot \\ \exp(g_1^{(0)}X_{15}) & g_0^{(0)}X_{15}\exp(g_1^{(0)}X_{15}) \end{bmatrix} = \begin{bmatrix} .92687 & 105.0416 \\ .82708 & 234.3317 \\ .76660 & 304.0736 \\ .68407 & 387.6236 \\ .58768 & 466.2057 \\ .48606 & 523.3020 \\ .37261 & 548.9603 \\ .30818 & 541.3505 \\ .27500 & 529.8162 \\ .23625 & 508.7088 \\ .18111 & 461.8140 \\ .13884 & 409.0975 \\ .13367 & 401.4294 \\ .10247 & 348.3801 \\ .08475 & 312.1510 \end{bmatrix}$$

We are now ready to obtain the least squares estimates $\mathbf{b}^{(0)}$ by solving (14.24):

$$\mathbf{b}^{(0)} = (\mathbf{D}^{(0)\prime}\mathbf{D}^{(0)})^{-1}\mathbf{D}^{(0)\prime}\mathbf{Y}^{(0)}$$

We find (calculations not shown):

$$\mathbf{D}^{(0)\prime}\mathbf{D}^{(0)} = \begin{bmatrix} 3.63328 & 2{,}211.285 \\ 2{,}211.285 & 2{,}694{,}343.69 \end{bmatrix}$$

and:

$$\mathbf{D}^{(0)\prime}\mathbf{Y}^{(0)} = \begin{bmatrix} 3.4229 \\ -24.1911 \end{bmatrix}$$

so that:

$$\mathbf{b}^{(0)} = \begin{bmatrix} 3.63328 & 2{,}211.285 \\ 2{,}211.285 & 2{,}694{,}343.69 \end{bmatrix}^{-1} \begin{bmatrix} 3.4229 \\ -24.1911 \end{bmatrix}$$

$$= \begin{bmatrix} 5.499\text{E}-1 & -4.513\text{E}-4 \\ -4.513\text{E}-4 & 7.416\text{E}-7 \end{bmatrix} \begin{bmatrix} 3.4229 \\ -24.1911 \end{bmatrix} = \begin{bmatrix} 1.8932 \\ -1.563\text{E}-3 \end{bmatrix}$$

By (14.25), we obtain the revised least squares estimates $\mathbf{g}^{(1)}$:

$$\mathbf{g}^{(1)} = \mathbf{g}^{(0)} + \mathbf{b}^{(0)} = \begin{bmatrix} 56.6646 \\ -.03797 \end{bmatrix} + \begin{bmatrix} 1.8932 \\ -.001563 \end{bmatrix}$$

$$= \begin{bmatrix} 58.5578 \\ -.03953 \end{bmatrix}$$

Hence, $g_0^{(1)} = 58.5578$ and $g_1^{(1)} = -.03953$ are the revised parameter estimates at the end of the first iteration. Note that the estimated regression coefficients have been revised moderately from the initial values, as can be seen from Table 14.3a, which presents the estimated regression coefficients as well as the least squares criterion measures for the first three iterations. Note also that the least squares criterion measure has been reduced in the first iteration.

While iteration 1 led to moderate revisions in the estimated regression coefficients and a substantially better fit according to the least squares criterion, Table 14.3a indicates that iteration 2 resulted only in minor revisions of the estimated regression coefficients and little improvement in the fit. Iteration 3 led to no change in either the estimates of the coefficients or the least squares criterion measure.

Hence, the search procedure was terminated after three iterations. The final regression coefficient estimates therefore are $g_0 = 58.6065$ and $g_1 = -.03959$, and the fitted regression model is:

$$\hat{Y} = (58.6065)\exp(-.03959X) \tag{14.28}$$

The error sum of squares for this fitted model is $SSE = 49.4593$. Figure 14.2 presents a scatter plot of the data and the estimated regression function.

TABLE 14.3 Gauss-Newton method iterations to obtain nonlinear least squares estimates—severely injured patients example

(**a**) Estimates of Parameters and Least Squares Criterion Measure

Iteration	g_0	g_1	*SSE*
0	56.6646	−.03797	56.0869
1	58.5578	−.03953	49.4638
2	58.6055	−.03959	49.4593
3	58.6065	−.03959	49.4593

(**b**) Final Least Squares Estimates

k	g_k	$s(g_k)$
0	58.6065	1.472
1	−.03959	.00171

$$MSE = \frac{49.4593}{13} = 3.80456$$

(**c**) Estimated Approximate Variance-Covariance Matrix of Estimated Regression Coefficients

$$\mathbf{s}^2(\mathbf{g}) = MSE(\mathbf{D}'\mathbf{D})^{-1} = 3.80456\begin{bmatrix} 5.696E-1 & -4.682E-4 \\ -4.682E-4 & 7.697E-7 \end{bmatrix}$$

$$= \begin{bmatrix} 2.1672 & -1.781E-3 \\ -1.781E-3 & 2.928E-6 \end{bmatrix}$$

FIGURE 14.2 Plot of data and fitted nonlinear regression function—severely injured patients example

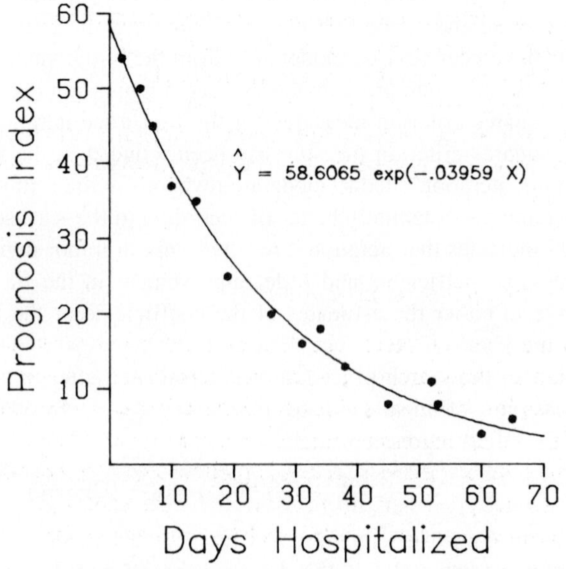

A plot of the residuals against the fitted values $\hat{Y}$ (not shown) did not suggest any serious departures from the model assumptions, and thus exponential model (14.17) and the estimated regression function (14.28) were accepted for the prognosis analysis for severely injured patients.

Comments

1. The choice of initial starting values is very important with the Gauss-Newton method because a poor choice may result in slow convergence, convergence to a local minimum, or even divergence. Good starting values will generally result in faster convergence, and if multiple minima exist, will lead to a solution that is the global minimum rather than a local minimum.

2. A variety of methods are available for obtaining starting values for the regression parameters. Often, experience can be utilized to provide good starting values for the regression parameters. Another possibility is to select p representative observations, set the regression function $f(\mathbf{X}_i, \boldsymbol{\gamma})$ equal to Y_i for each of the p observations (thereby ignoring the random error), solve the p equations for the p parameters, and use the solutions as the starting values, provided they lead to reasonably good fits of the observed data. Still another possibility is to do a grid search in the parameter space by selecting in a grid fashion various trial choices of $\mathbf{g}$, evaluate the least squares criterion Q for each of these choices, and use as the starting values that $\mathbf{g}$ vector for which Q is smallest.

3. When using the Gauss-Newton or some other direct search procedure, it is often desirable to try other sets of starting values after a solution has been obtained to make sure that the same solution will be found.

4. Some computer packages for nonlinear regression require that the user specify the starting values for the regression parameters. Others do a grid search to obtain starting values.

5. Some nonlinear regression computer programs using the Gauss-Newton method require the user to input the partial derivatives of the regression function, while others numerically calculate estimated partial derivatives from the regression function. In addition, most nonlinear computer programs have a library of commonly used regression functions which need only be specified by the user.

6. The Gauss-Newton method may produce iterations which oscillate widely or result in increases in the error sum of squares. Sometimes, these are only temporary but occasionally serious convergence problems exist. Various modifications of the Gauss-Newton method have been suggested to improve its performance, such as the Hartley modification (Ref. 14.1).

Other direct search procedures

A number of other direct search procedures besides the Gauss-Newton method are frequently used. One is the *method of steepest descent*. This method searches for the minimum least squares criterion measure Q by iteratively determining the direction in which the regression coefficients $\mathbf{g}$ should be changed. The method of steepest descent is particularly effective when the starting values $\mathbf{g}^{(0)}$ are not good and are far from the final values $\mathbf{g}$.

The *Marquardt algorithm* seeks to utilize the best features of the Gauss-

Newton method and the method of steepest descent, and occupies a middle ground between these two methods.

Additional information about direct search procedures can be found in specialized sources, such as References 14.2 and 14.3.

14.4 INFERENCES ABOUT NONLINEAR REGRESSION PARAMETERS

Estimated variances and covariances

Inferences about nonlinear regression parameters require an estimate of the error term variance σ^2. This estimate is the same as for linear regression:

$$(14.29) \qquad MSE = \frac{SSE}{n-p} = \frac{\Sigma(Y_i - \hat{Y}_i)^2}{n-p} = \frac{\Sigma[Y_i - f(\mathbf{X}_i, \mathbf{g})]^2}{n-p}$$

where $\mathbf{g}$ is the vector of the final parameter estimates. For nonlinear regression, MSE is not an unbiased estimator of σ^2, but the bias is small when the sample size is large.

When the error terms are independent and normally distributed and the sample size is reasonably large, the following theorem is helpful:

(14.30) When the error terms ε_i are independent $N(0, \sigma^2)$ and the sample size n is reasonably large, the sampling distribution of $\mathbf{g}$ is approximately normal with:

$$\mathbf{E}(\mathbf{g}) \simeq \boldsymbol{\gamma}$$

Thus, when the sample size is large the least squares estimators $\mathbf{g}$ for nonlinear regression are approximately normally distributed and unbiased. An estimate of the approximate variance-covariance matrix of the regression coefficients is:

$$(14.31) \qquad \mathbf{s}^2(\mathbf{g}) = MSE(\mathbf{D}'\mathbf{D})^{-1}$$

where $\mathbf{D}$ is the matrix of partial derivatives evaluated at the final least squares estimates $\mathbf{g}$, just as $\mathbf{D}^{(0)}$ in (14.23a) is the matrix of partial derivatives evaluated at $\mathbf{g}^{(0)}$.

Note that the estimated approximate variance-covariance matrix $\mathbf{s}^2(\mathbf{g})$ is of exactly the same form as the one for linear regression in (7.39), with $\mathbf{D}$ again playing the role of the $\mathbf{X}$ matrix.

Example. For our severely injured patients example, we know from Table 14.3a that the final error sum of squares is $SSE = 49.4593$. Since $p = 2$ parameters are present in regression model (14.17), we have:

$$MSE = \frac{SSE}{n-p} = \frac{49.4593}{15-2} = 3.80456$$

Table 14.3b presents this mean square, and Table 14.3c contains the estimated variance-covariance matrix of the regression coefficients. The matrix $(\mathbf{D}'\mathbf{D})^{-1}$ is based on the final regression coefficient estimates **g** and is shown without computational details.

We see from Table 14.3c that $s^2(g_0) = 2.1672$ and $s^2(g_1) = .000002928$. The estimated standard deviations of the regression coefficients are given in Table 14.3b.

Interval estimation of a single γ_k

When the error terms in the nonlinear regression model (14.13) are independent and normally distributed, the following approximate result holds when the sample size is large:

$$\text{(14.32)} \qquad \frac{g_k - \gamma_k}{s(g_k)} \sim t(n-p) \qquad k = 0, 1, \ldots, p-1$$

Hence, approximate $1 - \alpha$ confidence limits for any single γ_k are the usual ones:

$$\text{(14.33)} \qquad g_k \pm t(1 - \alpha/2; n-p)s(g_k)$$

Example. For our severely injured patients example, it is desired to estimate γ_1 with a 95 percent confidence interval. We require $t(.975; 13) = 2.160$, and find from Table 14.3b that $g_1 = -.03959$ and $s(g_1) = .00171$. Hence:

$$-.03959 - 2.160(.00171) \le \gamma_1 \le -.03959 + 2.160(.00171)$$
$$-.0433 \le \gamma_1 \le -.0359$$

Thus, we can conclude with 95 percent confidence that γ_1 is between $-.0433$ and $-.0359$.

Simultaneous interval estimation of several γ_k

Approximate joint confidence regions for the regression parameters in nonlinear regression can be developed, but they are difficult to interpret except when $p - 1 = 2$. Bonferroni joint confidence intervals, on the other hand, are easy to obtain and interpret, as in linear regression. If m parameters are to be estimated with family confidence coefficient $1 - \alpha$, the joint Bonferroni confidence limits are:

$$\text{(14.34)} \qquad g_k \pm Bs(g_k)$$

where:

$$\text{(14.34a)} \qquad B = t(1 - \alpha/2m; n-p)$$

Example. In our severely injured patients example, it is desired to obtain simultaneous interval estimates for γ_0 and γ_1 with a 90 percent family confidence coefficient. With the Bonferroni procedure we therefore require separate confidence intervals for the two parameters, each with a 95 percent statement confidence coefficient. We have already obtained a confidence interval for γ_1 with a 95 percent statement confidence coefficient. A 95 percent statement confidence interval for γ_0 is:

$$58.6065 - 2.160(1.472) \leq \gamma_0 \leq 58.6065 + 2.160(1.472)$$
$$55.43 \leq \gamma_0 \leq 61.79$$

Hence, the joint confidence intervals with family confidence coefficient of 90 percent are:

$$55.4 \leq \gamma_0 \leq 61.8$$
$$-.0433 \leq \gamma_1 \leq -.0359$$

Test concerning a single γ_k

A test concerning a single γ_k is set up in the usual fashion. To test:

(14.35a)
$$H_0: \gamma_k = \gamma_{k0}$$
$$H_a: \gamma_k \neq \gamma_{k0}$$

where γ_{k0} is the specified value of γ_k, we may use the usual t^* test statistic when n is reasonably large:

(14.35b)
$$t^* = \frac{g_k - \gamma_{k0}}{s(g_k)}$$

and the decision rule:

(14.35c)
If $|t^*| \leq t(1 - \alpha/2; n - p)$, conclude H_0
Otherwise conclude H_a

Example. In our severely injured patients example, we wish to test:

$$H_0: \gamma_0 = 50$$
$$H_a: \gamma_0 \neq 50$$

The test statistic (14.35b) here is:

$$t^* = \frac{58.6065 - 50}{1.472} = 5.85$$

For $\alpha = .01$, we require $t(.995; 13) = 3.012$. Since $|t^*| = 5.85 > 3.012$, we conclude H_a, that $\gamma_0 \neq 50$.

Test concerning several γ_k

When a test is desired concerning several γ_k simultaneously, we use the same approach as for the general linear test, first fitting the full model and obtaining $SSE(F)$, then fitting the reduced model and obtaining $SSE(R)$, and finally calculating the same test statistic as for linear regression:

$$F^* = \frac{SSE(R) - SSE(F)}{df_R - df_F} \div MSE(F) \tag{14.36}$$

For large n, this test statistic is distributed approximately as $F(df_R - df_F, df_F)$ when H_0 holds.

14.5 LEARNING CURVE EXAMPLE

We shall now present a second example to provide an additional illustration of the nonlinear regression concepts developed in this chapter. An electronic products manufacturer undertook the production of a new product in two locations (location A: coded $X_1 = 1$, location B: coded $X_1 = 0$). Location B has more modern facilities and hence was expected to be more efficient than location A, even after the initial learning period. An industrial engineer calculated the expected unit production cost for a modern facility after learning has occurred. Weekly unit production costs for each location were then expressed as a fraction of this expected cost. The reciprocal of this fraction is a measure of relative efficiency, and this measure was utilized as the efficiency measure in this study.

It is well known that efficiency increases over time when a new product is produced, and that the improvements eventually slow down and the process stabilizes. Hence, it was decided to employ an exponential model with an upper asymptote for expressing the relation between relative efficiency (Y) and time (X_2), and to incorporate a constant effect for the difference in the two production locations. The model decided on was:

$$Y_i = \gamma_0 + \gamma_1 X_{i1} + \gamma_3 \exp(\gamma_2 X_{i2}) + \varepsilon_i \tag{14.37}$$

Here, γ_0 is the upper asymptote for location B as X_2 gets large, and $\gamma_0 + \gamma_1$ is the upper asymptote for location A. The parameters γ_2 and γ_3 reflect the speed of learning, which was expected to be the same in the two locations.

While weekly data on relative production efficiency for each location were available, we shall only use observations for selected weeks during the first 90 weeks of production to simplify the presentation. The data on location, week, and relative efficiency are presented in Table 14.4. Note that learning was relatively rapid in both locations, and that the relative efficiency in location B toward the end of the 90-week period even exceeded 1.0, i.e., the actual unit costs then were lower than the industrial engineer's expected unit cost.

Model (14.37) is nonlinear in the parameters γ_2 and γ_3. Hence, a direct search estimation procedure was to be employed, for which starting values for the parameters are needed. These were developed partly from past experience, partly from analysis of the data. Previous studies indicated that γ_3 should be in the neighborhood of $-.5$, so $g_3^{(0)} = -.5$ was used. Since the difference in the relative efficiencies between locations A and B for a given week tended to average $-.0459$ during the 90-week period, a starting value $g_1^{(0)} = -.0459$ was specified. The largest observed relative efficiency for location B was 1.028, so

TABLE 14.4 Data for learning curve example

Observation i	*Location* X_{i1}	*Week* X_{i2}	*Relative Efficiency* Y_i
1	1	1	.483
2	1	2	.539
3	1	3	.618
4	1	5	.707
5	1	7	.762
6	1	10	.815
7	1	15	.881
8	1	20	.919
9	1	30	.964
10	1	40	.959
11	1	50	.968
12	1	60	.971
13	1	70	.960
14	1	80	.967
15	1	90	.975
16	0	1	.517
17	0	2	.598
18	0	3	.635
19	0	5	.750
20	0	7	.811
21	0	10	.848
22	0	15	.943
23	0	20	.971
24	0	30	1.012
25	0	40	1.015
26	0	50	1.007
27	0	60	1.022
28	0	70	1.028
29	0	80	1.017
30	0	90	1.023

that a starting value $g_0^{(0)} = 1.025$ was felt to be reasonable. Only a starting value for γ_2 remains to be found. This was chosen by selecting a typical relative efficiency observation in the middle of the time period, $Y_{24} = 1.012$, equating it to the response function with $X_{24,1} = 0$, $X_{24,2} = 30$, and the previous starting values for the other regression coefficients (thus ignoring the error term):

$$1.012 = 1.025 - (.5)\exp(30\gamma_2)$$

and solving for γ_2. Thereby the starting value $g_2^{(0)} = -.122$ was obtained. Tests for several other representative observations yielded similar starting values, and $g_2^{(0)} = -.122$ was therefore considered to be a reasonable initial value.

With the four starting values $g_0^{(0)} = 1.025$, $g_1^{(0)} = -.0459$, $g_2^{(0)} = -.122$, and $g_3^{(0)} = -.5$, a computer package direct search program was utilized to obtain the least squares estimates. The resulting least squares regression function was:

$$\hat{Y} = 1.0156 - .04727X_1 - (.5524)\exp(-.1348X_2) \tag{14.38}$$

and the error sum of squares was $SSE = .00329$, with $30 - 4 = 26$ degrees of

FIGURE 14.3 Plot of data and fitted nonlinear regression functions—learning curve example

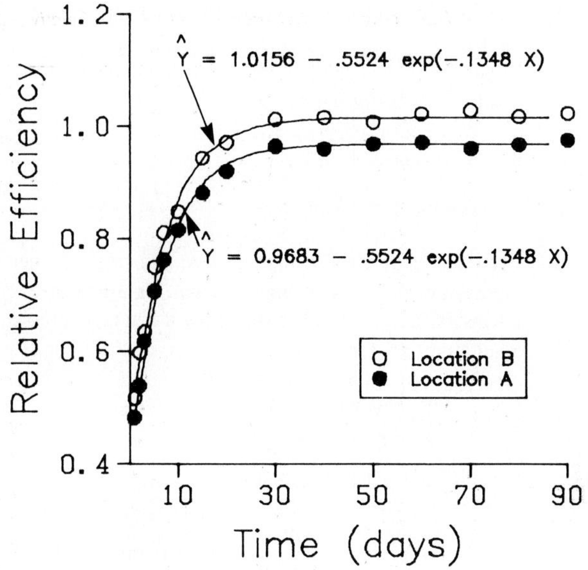

freedom. Figure 14.3 presents the scatter plot and the fitted regression functions for the two locations. Residual plots did not indicate any noticeable departures from the assumed model.

Special interest existed in the parameter γ_1, reflecting the effect of location. A 95 percent confidence interval is to be constructed. We require $t(.975; 26) = 2.056$. The computer printout contained the estimated variance-covariance matrix $\mathbf{s}^2(\mathbf{g})$ from which it was found that $s(g_1) = \sqrt{.000016885} = .00411$. Hence, the 95 percent confidence interval for γ_1 is:

$$-.04727 - 2.056(.00411) \leq \gamma_1 \leq -.04727 + 2.056(.00411)$$
$$-.0557 \leq \gamma_1 \leq -.0388$$

Since γ_1 is seen to be negative, this confidence interval confirms that location A with its less modern facilities tends to be less efficient.

Note

When growth or learning curve models are fitted to data constituting repeated observations on the same unit, such as efficiency data for the same production unit at different points in time, the error terms may be correlated. Hence, in these situations it is important to ascertain whether or not a model assuming uncorrelated error terms is reasonable. In the learning curve example, a plot of the residuals against time order did not suggest any serious correlations among the error terms.

PROBLEMS

14.1. For each of the following models, indicate whether it is a linear regression model, an intrinsically linear model, or a nonlinear model. In the case of an intrinsically linear model, state how it can be expressed in the form of (14.1) by a suitable transformation:

a. $Y_i = \exp(\gamma_0 + \gamma_1 X_i + \varepsilon_i)$

b. $Y_i = \exp(\gamma_0 + \gamma_1 X_i) + \varepsilon_i$

c. $Y_i = \gamma_0 + \dfrac{\gamma_1}{\gamma_0} X_i + \varepsilon_i$

14.2. For each of the following models, indicate whether it is a linear regression model, an intrinsically linear model, or a nonlinear model. In the case of an intrinsically linear model, state how it can be expressed in the form of (14.1) by a suitable transformation:

a. $\log_e Y_i = \gamma_0 + \gamma_1 \log_e X_i + \varepsilon_i$

b. $Y_i = \gamma_0 X_{i1}^{\gamma_1} X_{i2}^{\gamma_2} \varepsilon_i$

c. $Y_i = \gamma_0 - \gamma_1 \gamma_2^{X_i} + \varepsilon_i$

14.3. a. Plot the logistic response function:

$$E(Y) = \frac{300}{1 + (30)\exp(-1.5X)} \qquad X \geq 0$$

b. What is the asymptote of this response function? For what value of X does the response function reach 90 percent of its asymptote?

14.4. a. Plot the exponential response function:

$$E(Y) = 49 - (30)\exp(-1.1X) \qquad X \geq 0$$

b. What is the asymptote of this response function? For what value of X does the response function reach 95 percent of its asymptote?

14.5. **Home computers.** A computer manufacturer hired a market research firm to investigate the relationship between the likelihood a family will purchase a home computer and the price of the home computer. The data below are based on a survey of 1,000 heads of households who were asked if they are likely to purchase a home computer at a given price. Ten prices (X, in hundred dollars) were studied, and 100 heads randomly selected were assigned to a given price. The proportion likely to purchase at a given price is denoted by Y.

i:	1	2	3	4	5	6	7	8	9	10
X_i:	1	2.5	5	10	20	30	40	50	75	100
Y_i:	.95	.85	.58	.46	.31	.28	.19	.11	.06	.03

The following exponential model with independent normal error terms was deemed to be appropriate:

$$Y_i = \gamma_0 + \gamma_2 \exp(-\gamma_1 X_i) + \varepsilon_i$$

a. To obtain initial estimates of γ_0, γ_1, and γ_2, note that Y approaches a lower asymptote γ_0 as X increases without bound. Hence, let $g_0^{(0)} = 0$ and observe that when we ignore the error term a logarithmic transformation then yields $Y_i' = \beta_0 + \beta_1 X_i$, where $Y_i' = \log_e Y_i$, $\beta_0 = \log_e \gamma_2$, and $\beta_1 = -\gamma_1$. Therefore, fit a linear regression function based on the transformed data and use as initial estimates $g_0^{(0)} = 0$, $g_1^{(0)} = -b_1$, and $g_2^{(0)} = \exp(b_0)$.

b. Using the starting values obtained in part (a), find the least squares estimates of the parameters γ_0, γ_1, and γ_2.

c. Assume that the number of observations is reasonably large. Obtain approximate joint confidence intervals for the parameters γ_0, γ_1, and γ_2 using the Bonferroni procedure and a 90 percent family confidence coefficient.

14.6. Refer to **Home computers** Problem 14.5.

a. Plot the estimated nonlinear regression function and the data. Does the fit appear to be adequate?

b. Obtain the residuals and plot them against the fitted values and against X on separate graphs. Also obtain a normal probability plot. Does the model appear to be adequate?

14.7. **Enzyme kinetics.** In an enzyme kinetics study the velocity of a reaction (Y) is expected to be related to the concentration (X) as follows:

$$Y_i = \frac{\gamma_0 X_i}{\gamma_1 + X_i} + \varepsilon_i$$

Eleven concentrations have been studied and the results follow:

i:	1	2	3	4	5	6	7	8	9	10	11
X_i:	1	2	3	4	5	7.5	10	15	20	30	40
Y_i:	2.1	4.9	6.5	7.0	8.4	10.2	12.5	14.6	16.1	19.7	23.2

a. To obtain starting values for γ_0 and γ_1, observe that when the error term is ignored we have $Y_i' = \beta_0 + \beta_1 X_i'$, where $Y_i' = 1/Y_i$, $\beta_0 = 1/\gamma_0$, $\beta_1 = \gamma_1/\gamma_0$, and $X_i' = 1/X_i$. Therefore fit a linear regression function to the transformed data to obtain initial estimates $g_0^{(0)} = 1/b_0$ and $g_1^{(0)} = b_1/b_0$.

b. Using the starting values obtained in part (a), find the least squares estimates of the parameters γ_0 and γ_1.

c. Assume that the number of observations is reasonably large. (1) Obtain an approximate 95 percent confidence interval for γ_0. (2) Test whether or not $\gamma_1 = 20$; use $\alpha = .05$. State the alternatives, decision rule, and conclusion.

14.8. Refer to **Enzyme kinetics** Problem 14.7.

a. Plot the estimated nonlinear regression function and the data. Does the fit appear to be adequate?

b. Obtain the residuals and plot them against the fitted values and against X on separate graphs. Also obtain a normal probability plot. What do your plots show?

14.9. **Drug responsiveness.** A pharmacologist modeled the responsiveness to a drug using the following nonlinear regression model:

$$Y_i = \gamma_0 - \frac{\gamma_0}{1 + \left(\dfrac{X_i}{\gamma_2}\right)^{\gamma_1}} + \varepsilon_i$$

X denotes the dose level, in coded form, and Y the responsiveness expressed as a percent of the maximum possible responsiveness. In the model, γ_0 is the expected response at saturation, γ_2 is the concentration that produces a half maximal response, and γ_1 is related to the slope. The data for nine dose levels follow.

i:	1	2	3	4	5	6	7	8	9
X_i:	1	2	3	4	5	6	7	8	9
Y_i:	.5	2.3	3.4	24.0	54.7	82.1	94.8	96.2	96.4

a. Obtain least squares estimates of the parameters γ_0, γ_1, and γ_2 using starting values $g_0^{(0)} = 100$, $g_1^{(0)} = 5$, and $g_2^{(0)} = 4.8$.

b. Assume that the number of observations is reasonably large. Obtain approximate joint confidence intervals for the parameters γ_0, γ_1, and γ_2 using the Bonferroni procedure with a 91 percent family confidence coefficient. Interpret your results.

14.10. Refer to **Drug responsiveness** Problem 14.9.

a. Plot the estimated nonlinear regression function and the data. Does the fit appear to be adequate?

b. Obtain the residuals and plot them against the fitted values and against X on separate graphs. Also obtain a normal probability plot. What do your plots show about the adequacy of the regression model?

14.11. **Process yield.** The yield (Y) of a chemical process depends on the temperature (X_1) and pressure (X_2). The following nonlinear regression model is expected to be applicable:

$$Y_i = \gamma_0 X_{i1}^{\gamma_1} X_{i2}^{\gamma_2} + \varepsilon_i$$

Prior to beginning full-scale production, 18 tests were undertaken to study the process yield for various temperature and pressure combinations. The results follow.

i:	1	2	3	4	5	6	7	8	9
X_{i1}:	1	10	100	1	10	100	1	10	100
X_{i2}:	1	1	1	10	10	10	100	100	100
Y_i:	12	32	103	20	61	198	38	133	406

i:	10	11	12	13	14	15	16	17	18
X_{i1}:	1	10	100	1	10	100	1	10	100
X_{i2}:	1	1	1	10	10	10	100	100	100
Y_i:	8	38	98	14	56	205	43	128	398

a. To obtain starting values for γ_0, γ_1, and γ_2, note that when we ignore the random error term, a logarithmic transformation yields $Y_i' = \beta_0 + \beta_1 X_{i1}' + \beta_2 X_{i2}'$, where $Y_i' = \log_{10} Y_i$, $\beta_0 = \log_{10} \gamma_0$, $\beta_1 = \gamma_1$, $X_{i1}' = \log_{10} X_{i1}$, $\beta_2 = \gamma_2$, and $X_{i2}' = \log_{10} X_{i2}$. Fit an ordinary first-order multiple regression model to the transformed data and use as starting values $g_0^{(0)} = \text{antilog}_{10}\, b_0$, $g_1^{(0)} = b_1$, and $g_2^{(0)} = b_2$.

b. Using the starting values obtained in part (a), find the least squares estimates of the parameters γ_0, γ_1, and γ_2.

14.12. Refer to **Process yield** Problem 14.11. Assume that the number of observations is reasonably large so that large-sample theory is applicable.

a. Test the hypotheses H_0: $\gamma_1 = \gamma_2$ against H_a: $\gamma_1 \neq \gamma_2$ using the .05 level of significance. State the alternatives, decision rule, and conclusion.

b. Obtain approximate joint confidence intervals for the parameters γ_1 and γ_2 using the Bonferroni procedure and a 95 percent family confidence coefficient.

c. What do you conclude about the parameters γ_1 and γ_2 based on the results in parts (a) and (b)?

14.13. Refer to **Process yield** Problem 14.11.

a. Plot the estimated nonlinear regression function and the data. Does the fit appear to be adequate?

b. Obtain the residuals and plot them against $\hat{Y}$, X_1, and X_2 on separate graphs. Also obtain a normal probability plot. What do your plots show about the adequacy of the model?

14.14. Refer to **Process yield** Problem 14.11. Conduct a formal approximate test for lack of fit of the nonlinear regression function. Use $\alpha = .05$ and assume that the number of observations is reasonably large. State the alternatives, decision rule, and conclusion.

EXERCISES

14.15. (Calculus needed.) Refer to **Home computers** Problem 14.5. Obtain the least squares normal equations and show that they are nonlinear in the estimated regression coefficients g_0, g_1, and g_2.

14.16. (Calculus needed.) Refer to **Enzyme kinetics** Problem 14.7. Obtain the least squares normal equations and show that they are nonlinear in the estimated regression coefficients g_0 and g_1.

14.17. (Calculus needed.) Refer to **Process yield** Problem 14.11. Obtain the least squares normal equations and show that they are nonlinear in the estimated regression coefficients g_0, g_1, and g_2.

14.18. Refer to **Drug responsiveness** Problem 14.9.

a. Assuming that $E(\varepsilon_i) = 0$, show that:

$$E(Y) = \gamma_0\left(\frac{A}{1 + A}\right)$$

where:

$$A = \exp[\gamma_1(\log_e X - \log_e \gamma_2)] = \exp(\beta_0 + \beta_1 X')$$

and $\beta_0 = -\gamma_1 \log_e \gamma_2$, $\beta_1 = \gamma_1$, and $X' = \log_e X$.

b. Assuming γ_0 is known, show that:

$$\frac{E(Y')}{1 - E(Y')} = \exp(\beta_0 + \beta_1 X')$$

where $Y' = Y/\gamma_0$.

c. What transformation do these results suggest for obtaining a simple linear regression function in the transformed variables?

d. How can starting values for finding the least squares estimates of the nonlinear regression parameters be obtained from the estimates of the linear regression coefficients?

PROJECTS

14.19. Refer to **Enzyme kinetics** Problem 14.7. Starting values for finding the least squares estimates of the parameters of the nonlinear regression model are to be obtained by a grid search. The following bounds for the two parameters have been specified:

$$5 \leq \gamma_0 \leq 65$$
$$5 \leq \gamma_1 \leq 65$$

Obtain 49 grid points by using all possible combinations of the boundary values and five other equally spaced points for each parameter range. Evaluate the least squares criterion (14.14) for each grid point and identify the point providing the best fit. Does this point give reasonable starting values here?

14.20. Refer to **Process yield** Problem 14.11. Starting values for finding the least squares estimates of the nonlinear regression model coefficients are to be obtained by a grid search. The following bounds for the parameters have been postulated:

$$1 \leq \gamma_0 \leq 21$$
$$.2 \leq \gamma_1 \leq .8$$
$$.1 \leq \gamma_2 \leq .7$$

Obtain 27 gridpoints by using all possible combinations of the boundary values and the midpoint for each of the parameter ranges. Evaluate the least squares criterion (14.14) for each gridpoint and identify the point providing the best fit. Does this point give reasonable starting values here?

CITED REFERENCES

14.1 Hartley, H. O. "The Modified Gauss-Newton Method for the Fitting of Non-linear Regression Functions by Least Squares." *Technometrics* 3 (1961), pp. 269–80.

14.2 Gallant, A. R. "Nonlinear Regression." *The American Statistician* 29 (1975), pp. 73–81.

14.3 Kennedy, W. J., Jr., and J. E. Gentle. *Statistical Computing*. New York: Marcel Dekker, 1980.

15

Normal correlation models

The purpose of this chapter is to indicate the relation between regression models and their uses, discussed in Chapters 2–14, and normal correlation models. We first take up bivariate normal correlation models, and then consider multivariate normal models.

15.1 DISTINCTION BETWEEN REGRESSION AND CORRELATION MODELS

As we know, the basic regression models taken up in this book assume that the independent variables $X_1, \ldots, X_{p-1}$ are fixed constants, and primary interest exists in making inferences about the dependent variable Y on the basis of the independent variables.

We saw in Chapter 3, for the case of a single independent variable, that the regression analysis for a normal error regression model is applicable even when X is a random variable, provided that the conditional distributions of Y follow certain specifications and the marginal distribution of X does not involve the regression model parameters β_0, β_1, and σ^2. Thus, in the case where X is a random variable, only the conditional distributions of Y were specified, and a restriction was placed on the marginal distribution of X. We did not, however, seek to completely specify the joint distribution of X and Y. While the discussion in Chapter 3 dealt with only a single independent variable, all of the points apply to multiple regression models containing a number of independent variables which are random.

Correlation models, like regression models with random independent variables, consist of variables all of which are random. Correlation models differ from regression models by specifying the joint distribution of the variables completely. Furthermore, the variables in a correlation model play a symmetrical role, with no one variable automatically designated as the dependent variable. Correlation models are employed to study the nature of the relations between the variables, and also may be used for making inferences about any one of the variables on the basis of the others.

Thus, an analyst may use a correlation model for the two variables "height of person" and "weight of person" in a study of a sample of persons, each variable being taken as random. He or she might wish to study the relation between the two variables or might be interested in making inferences about weight of a person on the basis of the person's height, in making inferences about height on the basis of weight, or in both.

Other examples where a correlation model may be appropriate are:

1. To study the relations between service station sales of gasoline, auxiliary products, and repair services.
2. To study the relation between company net income determined by generally accepted accounting principles and net income according to tax regulations.
3. To study the relations between a person's blood pressure, body temperature, and weight.

The correlation model most widely employed is the normal correlation model. We discuss it now for the case of two variables.

15.2 BIVARIATE NORMAL DISTRIBUTION

The normal correlation model for the case of two variables is based on the *bivariate normal distribution*. Let us denote the two variables as Y_1 and Y_2. (We do not use the notation X and Y in this chapter because both variables play a symmetrical role in correlation analysis.) We say that Y_1 and Y_2 are *jointly normally distributed* if their joint probability distribution is the bivariate normal distribution.

Density function

The density function for the bivariate normal distribution is as follows:

$$(15.1) \quad f(Y_1, Y_2) = \frac{1}{2\pi\sigma_1\sigma_2\sqrt{1-\rho_{12}^2}} \exp\left\{-\frac{1}{2(1-\rho_{12}^2)}\left[\left(\frac{Y_1-\mu_1}{\sigma_1}\right)^2 - 2\rho_{12}\left(\frac{Y_1-\mu_1}{\sigma_1}\right)\left(\frac{Y_2-\mu_2}{\sigma_2}\right) + \left(\frac{Y_2-\mu_2}{\sigma_2}\right)^2\right]\right\}$$

Note that this density function involves five parameters: $\mu_1, \mu_2, \sigma_1, \sigma_2, \rho_{12}$. We shall explain the meaning of these parameters shortly. First, let us consider a graphic representation of the bivariate normal distribution.

Graphic representation

Figure 15.1 contains a graphic representation of a bivariate normal distribution. It is a surface in three-dimensional space. For every pair of (Y_1, Y_2) values, there is a density $f(Y_1, Y_2)$ represented by the height of the surface at that point. The surface is continuous, and probability corresponds to volume under the surface.

FIGURE 15.1 Example of bivariate normal distribution

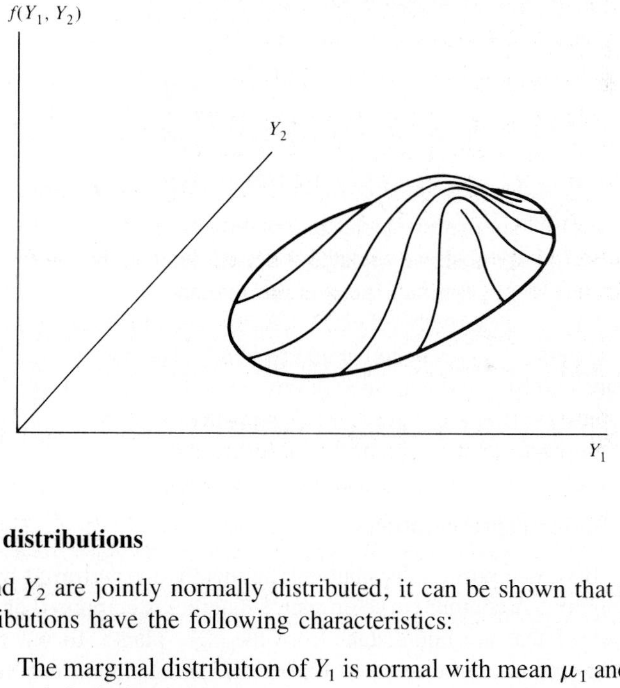

Marginal distributions

If Y_1 and Y_2 are jointly normally distributed, it can be shown that their marginal distributions have the following characteristics:

(15.2a) The marginal distribution of Y_1 is normal with mean μ_1 and standard deviation σ_1:

$$f_1(Y_1) = \frac{1}{\sqrt{2\pi}\,\sigma_1}\exp\left[-\frac{1}{2}\left(\frac{Y_1 - \mu_1}{\sigma_1}\right)^2\right]$$

(15.2b) The marginal distribution of Y_2 is normal with mean μ_2 and standard deviation σ_2:

$$f_2(Y_2) = \frac{1}{\sqrt{2\pi}\,\sigma_2}\exp\left[-\frac{1}{2}\left(\frac{Y_2 - \mu_2}{\sigma_2}\right)^2\right]$$

Thus, when Y_1 and Y_2 are jointly normally distributed, each of the two variables by itself is normally distributed. It is not generally true, however, that if Y_1 and Y_2 are each normally distributed, they must be jointly normally distributed in accord with (15.1).

Meaning of parameters

The five parameters of the bivariate normal density function (15.1) have the following meaning:

1. μ_1 and σ_1 are, respectively, the mean and standard deviation of the marginal distribution of Y_1.
2. μ_2 and σ_2 are, respectively, the mean and standard deviation of the marginal distribution of Y_2.
3. ρ_{12} is the *coefficient of correlation* between the random variables Y_1 and Y_2. It is defined as follows:

$$\rho_{12} = \frac{\sigma_{12}}{\sigma_1 \sigma_2} \tag{15.3}$$

where σ_{12} is the covariance between Y_1 and Y_2, as defined in (1.19):

$$\sigma_{12} = E[(Y_1 - \mu_1)(Y_2 - \mu_2)] \tag{15.4}$$

If Y_1 and Y_2 are independent, $\sigma_{12} = 0$ according to (1.23) so that $\rho_{12} = 0$ then. If Y_1 and Y_2 are positively related—i.e., Y_1 tends to be large when Y_2 is large, and small when Y_2 is small—σ_{12} is positive and so is ρ_{12}. On the other hand, if Y_1 and Y_2 are negatively related—i.e., Y_1 tends to be large when Y_2 is small, and vice versa—σ_{12} is negative and so is ρ_{12}. The coefficient of correlation ρ_{12} is a pure number, and can take on any value between -1 and $+1$ inclusive. It assumes $+1$ if Y_1 and Y_2 are perfectly positively related in a linear fashion, and -1 if the perfect linear relation is a negative one.

Contour representation

Bivariate normal distributions frequently are portrayed in terms of a contour diagram. A contour curve on such a diagram is composed of all the points on the surface that are equidistant from the Y_1Y_2 plane. To put this another way, a contour curve is composed of all (Y_1, Y_2) outcomes which have constant density $f(Y_1, Y_2)$. Thus we can picture a contour as the cross section obtained by slicing a bivariate normal surface horizontally at a fixed distance above the Y_1Y_2 plane, as in Figure 15.2.

Figure 15.3 presents a contour diagram for the bivariate normal surface of Figure 15.1. It is a property of the bivariate normal distribution that all contour curves are ellipses except when $\rho_{12} = 0$ and $\sigma_1 = \sigma_2$. Note that the ellipses have a common center at (μ_1, μ_2), and have common major and minor axes. Also note that the higher the horizontal cross section of the surface is above the Y_1Y_2 plane, the smaller is the corresponding contour ellipse.

FIGURE 15.2 Contour ellipse for bivariate normal surface

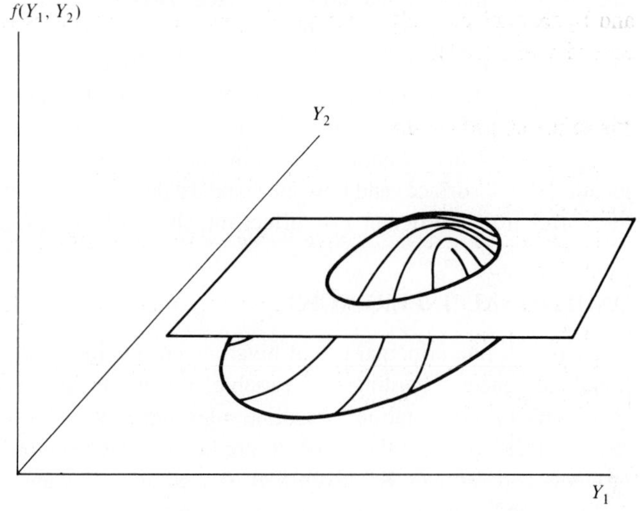

FIGURE 15.3 Contour diagram for bivariate normal surface in Figure 15.1

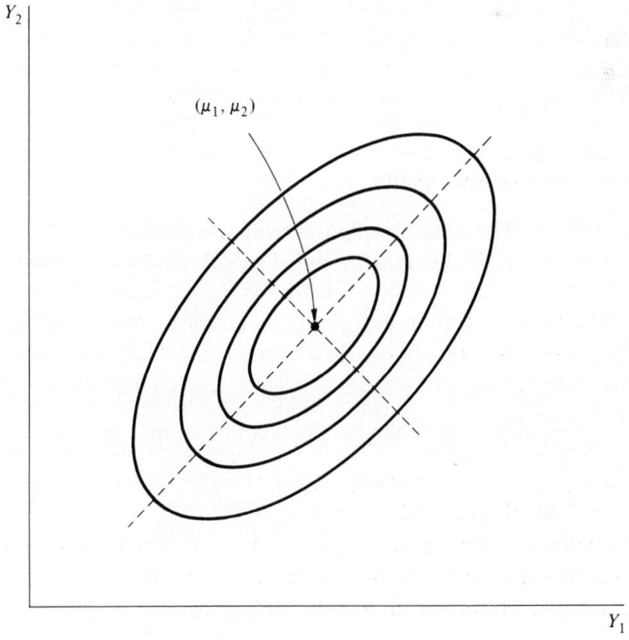

Figure 15.4 illustrates the effects of different parameter values on the location and shape of the bivariate normal surface. Note that when Y_1 and Y_2 are positively related so that $\rho_{12} > 0$, the principal axis has positive slope, implying that the surface tends to run along a line with positive slope. When Y_1 and Y_2 are negatively related so that $\rho_{12} < 0$, the principal axis has a negative slope, implying that the surface tends to run along a line with negative slope.

Figure 15.4 also demonstrates how the mean values μ_1 and μ_2 affect the location of the surface, and how the standard deviations σ_1 and σ_2, together with the correlation coefficient ρ_{12}, affect the shape of the surface.

15.3 CONDITIONAL INFERENCES

As noted, one principal use of bivariate correlation models is to make conditional inferences regarding one variable, given the other variable. Suppose Y_1 represents a service station's gasoline sales and Y_2 its sales of auxiliary products and services. We may then wish to predict a service station's sales of auxiliary products and services Y_2, given that its gasoline sales are $Y_1 = \$5{,}500$.

Such conditional inferences require the use of conditional probability distributions, which we discuss next.

Conditional probability distributions of Y_1

The conditional density function of Y_1 for any given value of Y_2 is denoted by $f(Y_1 | Y_2)$ and defined as follows:

(15.5)
$$f(Y_1 | Y_2) = \frac{f(Y_1, Y_2)}{f_2(Y_2)}$$

where $f(Y_1, Y_2)$ is the joint density function of Y_1 and Y_2, and $f_2(Y_2)$ is the marginal density function of Y_2. When Y_1 and Y_2 are jointly normally distributed according to (15.1) so that the marginal density function $f_2(Y_2)$ is given by (15.2b), it can be shown that:

(15.6) The conditional probability distribution of Y_1 for any given value of Y_2 is normal with mean $\alpha_{1.2} + \beta_{12}Y_2$ and standard deviation $\sigma_{1.2}$:

$$f(Y_1 | Y_2) = \frac{1}{\sqrt{2\pi}\,\sigma_{1.2}} \exp\left[-\frac{1}{2}\left(\frac{Y_1 - \alpha_{1.2} - \beta_{12}Y_2}{\sigma_{1.2}}\right)^2\right]$$

The parameters $\alpha_{1.2}$, β_{12}, and $\sigma_{1.2}$ of the conditional probability distributions of Y_1 are functions of the parameters of the joint probability distribution (15.1), as follows:

(15.7a)
$$\alpha_{1.2} = \mu_1 - \mu_2\rho_{12}\frac{\sigma_1}{\sigma_2}$$

(15.7b)
$$\beta_{12} = \rho_{12}\frac{\sigma_1}{\sigma_2}$$

FIGURE 15.4 Effects of parameter values on location and shape of bivariate normal distribution

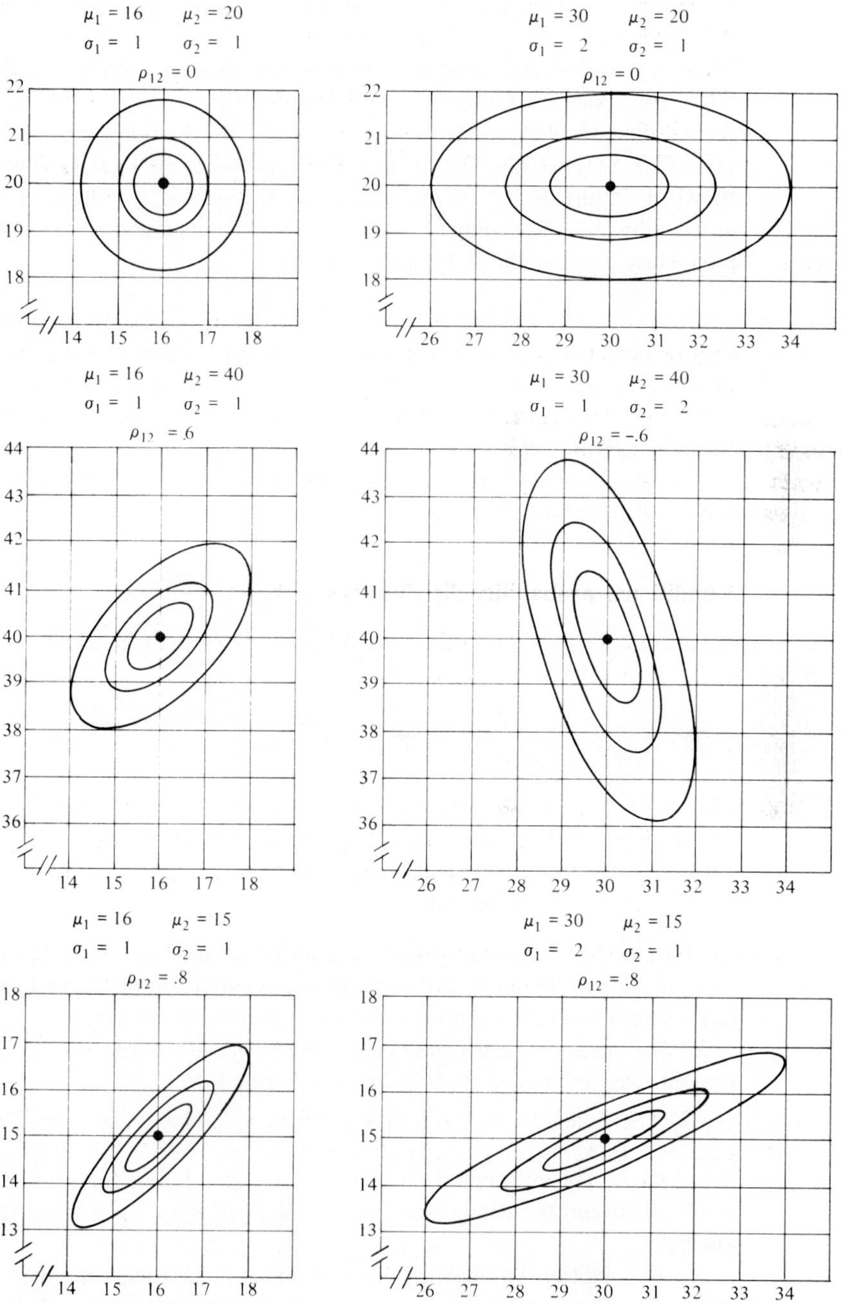

Source: Adapted, with permission, from Helen M. Walker and Joseph Lev, *Statistical Inference* (New York: Holt, Rinehart & Winston, 1953), p. 250.

(15.7c) $$\sigma_{1.2}^2 = \sigma_1^2(1 - \rho_{12}^2)$$

Important characteristics of conditional distributions. Three important characteristics of the conditional probability distributions of Y_1 are normality, linear regression, and constant variance. We take up each of these in turn.

1. The conditional probability distribution of Y_1 for any given value of Y_2 is normal. Imagine that we slice a bivariate normal distribution vertically at a given value of Y_2, say, at Y_{h2}. That is, we slice it parallel to the Y_1 axis. This slicing is shown in Figure 15.5. The exposed cross section has the shape of a normal distribution, and after being scaled so that its area is 1, it portrays the conditional probability distribution of Y_1, given that $Y_2 = Y_{h2}$.

FIGURE 15.5 Cross section of bivariate normal distribution at Y_{h2}

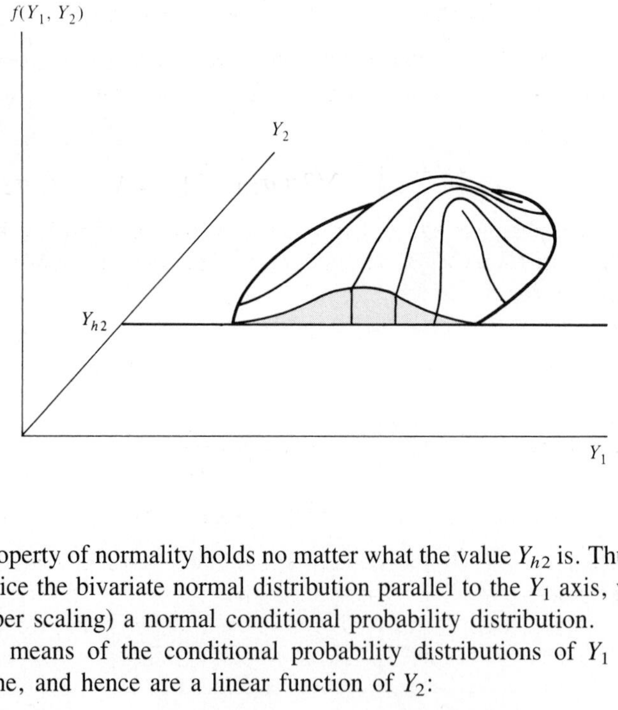

This property of normality holds no matter what the value Y_{h2} is. Thus, whenever we slice the bivariate normal distribution parallel to the Y_1 axis, we obtain (after proper scaling) a normal conditional probability distribution.

2. The means of the conditional probability distributions of Y_1 fall on a straight line, and hence are a linear function of Y_2:

(15.8) $$E(Y_1 | Y_2) = \alpha_{1.2} + \beta_{12}Y_2$$

Here $\alpha_{1.2}$ is the intercept parameter and β_{12} the slope parameter. Thus, the relation between the conditional means and Y_2 is given by a linear regression function.

3. All conditional probability distributions of Y_1 have the same standard deviation $\sigma_{1.2}$. Thus, no matter where we slice the bivariate normal distribution parallel to the Y_1 axis, the resulting conditional probability distribution (after

scaling to have an area of 1) has the same standard deviation. Hence, constant variances characterize the conditional probability distributions of Y_1.

Equivalence to normal error regression model. Suppose that we select a random sample of observations (Y_1, Y_2) from a bivariate normal population and wish to make conditional inferences about Y_1, given Y_2. The preceding discussion makes it clear that the normal error regression model (2.25) is entirely applicable because:

1. The Y_1 observations are independent.
2. The Y_1 observations when Y_2 is considered given or fixed are normally distributed with mean $E(Y_1 | Y_2) = \alpha_{1.2} + \beta_{12}Y_2$ and constant variance $\sigma_{1.2}^2$.

Conditional probability distributions of Y_2

The random variables Y_1 and Y_2 play symmetrical roles in the bivariate normal probability distribution (15.1). Hence, it follows:

(15.9) The conditional probability distribution of Y_2 for any given value of Y_1 is normal with mean $\alpha_{2.1} + \beta_{21}Y_1$ and standard deviation $\sigma_{2.1}$:

$$f(Y_2 | Y_1) = \frac{1}{\sqrt{2\pi}\,\sigma_{2.1}}\exp\left[-\frac{1}{2}\left(\frac{Y_2 - \alpha_{2.1} - \beta_{21}Y_1}{\sigma_{2.1}}\right)^2\right]$$

The parameters $\alpha_{2.1}$, β_{21}, and $\sigma_{2.1}$ of the conditional probability distributions of Y_2 are functions of the parameters of the joint probability distribution (15.1), as follows:

(15.10a) $$\alpha_{2.1} = \mu_2 - \mu_1\rho_{12}\frac{\sigma_2}{\sigma_1}$$

(15.10b) $$\beta_{21} = \rho_{12}\frac{\sigma_2}{\sigma_1}$$

(15.10c) $$\sigma_{2.1}^2 = \sigma_2^2(1 - \rho_{12}^2)$$

The parameter $\alpha_{2.1}$ is the intercept of the line of regression of Y_2 on Y_1, and the parameter β_{21} is the slope of this line.

Again, we find that the conditional correlation model of Y_2 for given Y_1 is the equivalent of the normal error regression model (2.25).

Comments

1. The notation for the parameters of the conditional correlation models departs somewhat from our previous notation for regression models. The symbol α is now used to denote the regression intercept. The subscript 1.2 to α indicates that Y_1 is regressed on Y_2. Similarly, the subscript 2.1 to α indicates that Y_2 is regressed on Y_1. The symbol β_{12} indicates that it is the slope in the regression of Y_1 on Y_2, while β_{21} is the slope in the regression of Y_2 on Y_1. Finally, $\sigma_{2.1}$ is the standard deviation of the conditional probability

distributions of Y_2 for any given Y_1, while $\sigma_{1.2}$ is the standard deviation of the conditional probability distributions of Y_1 for any given Y_2. This notation can be extended straightforwardly for multivariate correlation models.

2. Two distinct regressions are involved in a bivariate normal model, that of Y_1 on Y_2 when Y_2 is fixed and that of Y_2 on Y_1 when Y_1 is fixed. In general, the two regression lines are not the same. For instance, the two slopes β_{12} and β_{21} are the same only if $\sigma_1 = \sigma_2$, as can be seen from (15.7b) and (15.10b).

3. Figure 15.6 illustrates the relation of the two regression lines to the contour ellipses. Note that both regression lines go through the point (μ_1, μ_2). If $\rho_{12} = 0$, the two regression lines intersect at right angles. The larger absolutely is ρ_{12}, the more the two regression lines come together.

FIGURE 15.6 Illustration of relation between lines of regression and contour ellipses

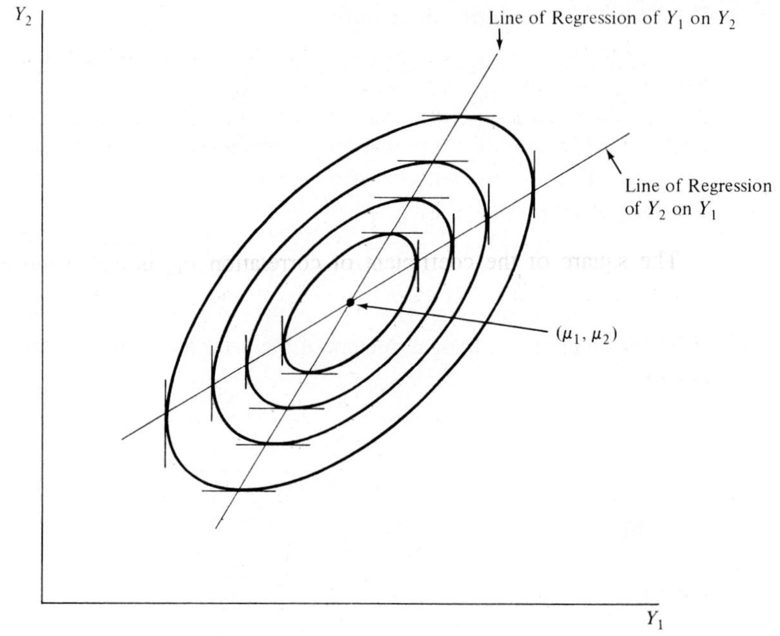

Use of regression analysis

In view of the equivalence of each of the conditional bivariate normal correlation models (15.6) and (15.9) with the normal error regression model (2.25), all conditional inferences with these correlation models can be made by means of the usual regression methods. Thus, if a researcher has data which can be appropriately described as having been generated from a bivariate normal distribution and wishes to make inferences about Y_2, given a particular value of Y_1, the ordinary regression techniques would be applicable. Thus, the regression equation of Y_2 on Y_1 would be estimated by means of (2.12), the slope of the regression equation would be estimated by means of the interval estimate (3.15), a new observation Y_2, given the value of Y_1, would be predicted by means of (3.35), and so on. Computer regression packages can be used in the usual manner. To avoid notational problems, it may be helpful to relabel the variables according to regression usage: $Y = Y_2, X = Y_1$. Of course, if conditional inferences on Y_1 for given values of Y_2 are desired, the notation correspondences would be: $Y = Y_1$, $X = Y_2$.

Note

When obtaining interval estimates for the conditional correlation models, the confidence coefficient refers to repeated samples where pairs of observations (Y_1, Y_2) are obtained from the bivariate normal population. We noted a similar point for regression models where the independent variables are random.

15.4 INFERENCES ON ρ_{12}

A principal use of correlation models is to study the relationships between the variables. In a bivariate normal model, the parameter ρ_{12} and its square, ρ_{12}^2, provide information about the degree of relationship between the two variables Y_1 and Y_2. Of the two measures, ρ_{12}^2 is the more meaningful one.

Coefficient of determination

The square of the coefficient of correlation ρ_{12} is called the *coefficient of determination*. We noted earlier in (15.7c) and (15.10c) that:

(15.11a) $$\sigma_{1.2}^2 = \sigma_1^2(1 - \rho_{12}^2)$$

(15.11b) $$\sigma_{2.1}^2 = \sigma_2^2(1 - \rho_{12}^2)$$

We can rewrite these expressions as follows:

(15.12a) $$\rho_{12}^2 = \frac{\sigma_1^2 - \sigma_{1.2}^2}{\sigma_1^2}$$

(15.12b) $$\rho_{12}^2 = \frac{\sigma_2^2 - \sigma_{2.1}^2}{\sigma_2^2}$$

The meaning of ρ_{12}^2 is now clear. Consider first (15.12a). ρ_{12}^2 measures how much smaller relatively is the variability in any conditional distribution of Y_1, for a given level of Y_2, than is the variability in the marginal distribution of Y_1. Thus, ρ_{12}^2 measures the relative reduction in the variability of Y_1 associated with the use of the variable Y_2. Correspondingly, (15.12b) shows that ρ_{12}^2 also measures the relative reduction in the variability of Y_2 associated with the use of the variable Y_1.

It can be shown that:

(15.13) $$0 \leq \rho_{12}^2 \leq 1$$

$\rho_{12}^2 = 0$ if Y_1 and Y_2 are independent, so that the variances of each variable in the conditional probability distributions are then no smaller than the variance in the marginal distribution. $\rho_{12}^2 = 1$ if there is no variability in the conditional probability distributions for each variable, so that perfect predictions of either variable can be made from the other.

Note

The interpretation of ρ_{12}^2 as measuring the relative reduction in the conditional variances as compared with the marginal variance is valid for the case of a bivariate normal population, but not for many other bivariate populations. Of course, the interpretation implies nothing in a causal sense.

Point estimators of ρ_{12} and ρ_{12}^2

The maximum likelihood estimator of ρ_{12}, denoted by r_{12}, is given by:

$$r_{12} = \frac{\Sigma(Y_{i1} - \bar{Y}_1)(Y_{i2} - \bar{Y}_2)}{[\Sigma(Y_{i1} - \bar{Y}_1)^2]^{1/2}[\Sigma(Y_{i2} - \bar{Y}_2)^2]^{1/2}} \tag{15.14}$$

This estimator is biased (unless $\rho_{12} = 0$ or 1), but the bias is small if n is large.

The coefficient of determination ρ_{12}^2 is estimated by the square of the sample coefficient of correlation, r_{12}^2.

Note

The maximum likelihood estimator (15.14) of the population correlation coefficient is the same as the descriptive coefficient of correlation in (3.73) for the regression model. With a correlation model, r_{12} is an estimator of a parameter. With a regression model, in contrast, r_{12} is only a descriptive measure which reflects the proportion of the total sum of squares that is partitioned into the regression sum of squares. Another difference is that in the standard regression case where the X_i are fixed, the magnitude of r_{12} can be arbitrarily affected by the spacing pattern chosen for the X_i. For the bivariate normal model, on the other hand, the magnitude of r_{12} cannot be so affected since neither variable is under the control of the investigator.

Test whether $\rho_{12} = 0$

When the population is a bivariate normal one, it is frequently desired to test between:

$$\begin{aligned} H_0&: \rho_{12} = 0 \\ H_a&: \rho_{12} \neq 0 \end{aligned} \tag{15.15}$$

The reason for interest in this test is that in the case where Y_1 and Y_2 are jointly normally distributed, $\rho_{12} = 0$ implies that Y_1 and Y_2 are independent.

We can use regression procedures for the test since (15.7b) implies that the following alternatives are equivalent to those in (15.15):

$$\begin{aligned} H_0&: \beta_{12} = 0 \\ H_a&: \beta_{12} \neq 0 \end{aligned} \tag{15.15a}$$

and (15.10b) implies that the following alternatives are also equivalent to the ones in (15.15):

$$\text{(15.15b)} \qquad \begin{aligned} &H_0\colon \beta_{21} = 0 \\ &H_a\colon \beta_{21} \neq 0 \end{aligned}$$

It can be shown that the statistics for testing either (15.15a) or (15.15b) can be expressed directly in terms of r_{12}:

$$\text{(15.16)} \qquad t^* = \frac{r_{12}\sqrt{n-2}}{\sqrt{1-r_{12}^2}}$$

If H_0 holds, t^* follows the $t(n-2)$ distribution. The appropriate decision rule to control the Type I error at α is:

$$\text{(15.17)} \qquad \begin{aligned} &\text{If } |t^*| \leq t(1-\alpha/2; n-2), \text{ conclude } H_0 \\ &\text{If } |t^*| > t(1-\alpha/2; n-2), \text{ conclude } H_a \end{aligned}$$

Test statistic (15.16) is identical to the regression t^* test statistic (3.17).

Interval estimation of ρ_{12}

Because the sampling distribution of r_{12} is complicated when $\rho_{12} \neq 0$, interval estimation of ρ_{12} is usually done by means of a transformation.

***z'* transformation.** This transformation, due to R. A. Fisher, is as follows:

$$\text{(15.18)} \qquad z' = \frac{1}{2}\log_e\left(\frac{1+r_{12}}{1-r_{12}}\right)$$

When n is large (25 or more is a useful rule of thumb), the distribution of z' is approximately normal with mean and variance:

$$\text{(15.19)} \qquad E(z') = \zeta = \frac{1}{2}\log_e\left(\frac{1+\rho_{12}}{1-\rho_{12}}\right)$$

$$\text{(15.20)} \qquad \sigma^2(z') = \frac{1}{n-3}$$

Note that the transformation from r_{12} to z' in (15.18) is the same as the relation in (15.19) between ρ_{12} and $E(z') = \zeta$. Also note that the variance of z' is a known constant, depending only on the sample size n.

Table A–7 gives paired values for the left and right sides of (15.18) and (15.19), thus eliminating the need for calculations. For instance, if r_{12} or ρ_{12} equals .25, Table A–7 indicates that z' or ζ equals .2554, and vice versa. The values on the two sides of the transformation always have the same sign. Thus, if r_{12} or ρ_{12} is negative, a minus sign is attached to the value in Table A–7. For instance, if $r_{12} = -.25$, $z' = -.2554$.

Interval estimate. Since z' is approximately normally distributed for large n, it follows that the standardized statistic:

$$\frac{z' - \zeta}{\sigma(z')} \tag{15.21}$$

is approximately a standard normal variable when n is large. Therefore, the $1 - \alpha$ confidence limits for ζ are:

$$z' \pm z(1 - \alpha/2)\sigma(z') \tag{15.22}$$

where $z(1 - \alpha/2)$ is the $(1 - \alpha/2)100$ percentile of the standard normal distribution. These percentiles are given in Table A–1. The $1 - \alpha$ confidence limits for ρ_{12} are then obtained by transforming the limits on ζ by means of (15.19).

Comments

1. As usual, a confidence interval for ρ_{12} can be employed to test whether or not ρ_{12} has a specified value—say, .5—by noting whether or not the specified value falls within the confidence limits.

2. Confidence limits for ρ_{12}^2 can be obtained by squaring the respective confidence limits for ρ_{12}.

Example

An economist investigated food purchasing patterns by households in a midwestern city. He selected 200 households with family incomes between \$7,500 and \$17,500, and ascertained from each household, among other things, the proportions of the food budget expended for beef and poultry, respectively. He expected these to be negatively related, and wished to estimate the coefficient of correlation with a 95 percent confidence interval. The economist had some supporting evidence which suggested that the joint distribution of the two variables does not depart markedly from a bivariate normal one.

The point estimate of ρ_{12} was $r_{12} = -.61$ (data and calculations not shown). To obtain a 95 percent confidence interval estimate, we require:

$$z' = -.7089 \text{ when } r_{12} = -.61 \qquad \text{(from Table A–7)}$$

$$\sigma(z') = \frac{1}{\sqrt{200 - 3}} = .07125$$

$$z(.975) = 1.96$$

Hence, the confidence interval for ζ, by (15.22), is:

$$-.849 = -.7089 - 1.96(.07125) \leq \zeta \leq -.7089 + 1.96(.07125) = -.569$$

Using Table A–7 to transform back to ρ_{12}, we obtain:

$$-.69 \leq \rho_{12} \leq -.51$$

This confidence interval was sufficiently precise to be useful to the economist, confirming the negative relation and indicating that the degree of linear association is moderately high.

Caution

Correlation models, like regression models, do not express any causal relations. Earlier cautions about drawing conclusions as to causality from regression findings apply equally to correlation studies. Correlation findings can be useful in analyzing causal relationships, but they do not by themselves establish causal patterns.

15.5 MULTIVARIATE NORMAL DISTRIBUTION

The normal correlation model for the case of p variables $Y_1, \ldots, Y_p$ is based on the *multivariate normal distribution*. This distribution is an extension of the bivariate normal distribution, and has corresponding properties. In particular, if $Y_1, \ldots, Y_p$ are jointly normally distributed (i.e., they follow the multivariate normal distribution), the marginal probability distribution of each variable Y_k is normal, with mean μ_k and standard deviation σ_k.

Conditional inferences

One major use of multivariate correlation models is to make conditional inferences on one variable when the other variables have given values. It can be shown in the case of the multivariate normal distribution that:

(15.23) The conditional probability distribution of Y_k for any given set of values for the other variables is normal with mean given by a linear regression function and constant variance.

Suppose Y_1, Y_2, Y_3, and Y_4 are jointly normally distributed. The conditional probability distribution of, say, Y_3, when Y_1, Y_2, and Y_4 are fixed at any specified levels, has the characteristics:

1. It is normal.
2. It has a mean given by:

$$E(Y_3 | Y_1, Y_2, Y_4) = \alpha_{3.124} + \beta_{31.24}Y_1 + \beta_{32.14}Y_2 + \beta_{34.12}Y_4$$

3. It has constant variance $\sigma^2_{3.124}$.

The notation is a straightforward extension of that for the bivariate correlation case. Thus, $\beta_{31.24}$ denotes the regression coefficient of Y_1 when Y_3 is regressed on Y_1, Y_2, and Y_4.

It is clear from the above that the conditional multivariate correlation models are equivalent to the normal error multiple regression model (7.7). Hence, in the case where the variables are jointly normally distributed, all inferences on one variable conditional on the other variables being fixed are carried out by the usual multiple regression techniques. For instance, an interval estimate of a conditional mean would be obtained by (7.51), or a prediction of a new observation on the variable of interest, given the values of the other variables, would be obtained by (7.55). To facilitate use of the regression formulas, it may be helpful to relabel the variable of interest Y and the other variables X's. Computer multiple regression packages can be used in ordinary fashion.

Coefficients of multiple correlation and determination

A major use of multivariate correlation models is to study the relationships between the variables. One set of measures useful to this end consists of the *coefficients of multiple determination* and the *coefficients of multiple correlation*.

Meaning of coefficients. A coefficient of multiple determination is associated with each variable. Suppose that Y_1, Y_2, Y_3, and Y_4 are included in the correlation model. The coefficient of multiple determination associated with, say, Y_1 is denoted by $\rho_{1.234}^2$, and defined as follows:

$$\rho_{1.234}^2 = \frac{\sigma_1^2 - \sigma_{1.234}^2}{\sigma_1^2} \tag{15.24}$$

where $\sigma_{1.234}^2$ is the variance of the conditional distributions of Y_1 when the other variables are fixed. Thus, $\rho_{1.234}^2$ measures how much smaller, relatively, is the variability in the conditional distributions of Y_1, when the other variables are fixed at given values, than is the variability in the marginal distribution of Y_1. The other coefficients of multiple determination are defined and interpreted in similar fashion.

It can be shown that coefficients of multiple determination take on values between 0 and 1 inclusive. Thus:

$$0 \leq \rho_{1.234}^2 \leq 1 \tag{15.25}$$

Let us consider the significance of the limiting values. If $\rho_{1.234}^2 = 0$, no reduction in the variability of Y_1 takes place by considering the other variables. Equivalently, $\rho_{1.234}^2 = 0$ in the case of the normal correlation model implies that Y_1 is independent of Y_2, Y_3, and Y_4 so that $\rho_{12} = \rho_{13} = \rho_{14} = 0$.

At the other extreme, if $\rho_{1.234}^2 = 1$, the conditional distributions of Y_1, given Y_2, Y_3, and Y_4, have no variability so that perfect predictions can be made of Y_1 from knowledge of the other variables.

The positive square root of a coefficient of multiple determination is the coefficient of multiple correlation. Thus, for Y_1, we have in our example:

$$\rho_{1.234} = \sqrt{\rho_{1.234}^2} \tag{15.26}$$

$\rho_{1.234}$ can be viewed as a simple correlation coefficient, namely, between Y_1 and $\alpha_{1.234} + \beta_{12.34}Y_2 + \beta_{13.24}Y_3 + \beta_{14.23}Y_4$.

Estimation of coefficients. Coefficients of multiple determination and correlation are estimated by (7.31) and (7.34), respectively, the descriptive coefficients in the regression case. Thus, to estimate $\rho_{1.234}^2$, we need the total sum of squares for Y_1, denoted by $SSTO(Y_1)$. Then we require the regression sum of squares when Y_1 is regressed on Y_2, Y_3, and Y_4, denoted by $SSR(Y_2, Y_3, Y_4)$.

The estimator, denoted by $R^2_{1.234}$, is:

(15.27)
$$R^2_{1.234} = \frac{SSR(Y_2, Y_3, Y_4)}{SSTO(Y_1)}$$

The positive square root of $R^2_{1.234}$, denoted by $R_{1.234}$, is the estimated coefficient of multiple correlation for Y_1.

Testing of coefficients. To test, say:

(15.28)
$$\begin{aligned} H_0&: \rho_{1.234} = 0 \\ H_a&: \rho_{1.234} \neq 0 \end{aligned}$$

we can utilize the equivalent test on the regression coefficients:

(15.28a)
$$\begin{aligned} &H_0: \beta_{12.34} = \beta_{13.24} = \beta_{14.23} = 0 \\ &H_a: \text{not all regression coefficients are zero} \end{aligned}$$

It turns out that the F^* statistic for this test, given in (7.30b), can be expressed directly in terms of $R^2_{1.234}$, as follows:

(15.29)
$$F^* = \frac{R^2_{1.234}}{1 - R^2_{1.234}} \quad \frac{n-4}{3}$$

In general, the factor on the right is $(n - q - 1)/q$ when there are q predictor variables.

If H_0 holds, F^* follows the $F(q, n - q - 1)$ distribution so that the decision rule to control the Type I error risk at α is set up in the usual fashion:

(15.30)
$$\begin{aligned} &\text{If } F^* \leq F(1 - \alpha; q, n - q - 1), \text{ conclude } H_0 \\ &\text{If } F^* > F(1 - \alpha; q, n - q - 1), \text{ conclude } H_a \end{aligned}$$

Coefficients of partial correlation and determination

Meaning of coefficients. Suppose again that four variables Y_1, Y_2, Y_3, and Y_4 are included in a multivariate normal correlation model. Consider now the correlation betweeen Y_1 and Y_2 in the conditional joint distribution when each of the variables Y_3 and Y_4 is fixed at a given level. When all variables are jointly normally distributed, this correlation does not depend on the levels where Y_3 and Y_4 are fixed and is given by:

(15.31)
$$\rho_{12.34} = \frac{\sigma_{12.34}}{\sigma_{1.34}\sigma_{2.34}}$$

$\rho_{12.34}$ is called the *coefficient of partial correlation* between Y_1 and Y_2 when Y_3 and Y_4 are fixed.

The square of $\rho_{12.34}$ is called the *coefficient of partial determination* and is denoted by $\rho^2_{12.34}$. The following relation holds:

(15.32)
$$\rho^2_{12.34} = \frac{\sigma^2_{1.34} - \sigma^2_{1.234}}{\sigma^2_{1.34}}$$

Thus, $\rho^2_{12.34}$ measures how much smaller, relatively, is the variability in the conditional distributions of Y_1, given Y_2, Y_3, and Y_4, than it is in the conditional distributions of Y_1, given Y_3 and Y_4 only.

The other partial correlation coefficients are defined and interpreted in similar fashion. For instance, $\rho_{12.3}$ measures the correlation between Y_1 and Y_2 when only Y_3 is fixed.

A coefficient of partial correlation $\rho_{12.3}$ is called a *first-order* coefficient, a coefficient $\rho_{12.34}$ a *second-order* coefficient, and so on. All partial correlation coefficients measure the correlation between two variables; the order of the coefficient simply indicates how many other variables are fixed in the conditional bivariate distribution.

Point estimation of coefficients. Point estimators of the coefficients of partial determination and correlation are the descriptive regression measures encountered earlier; see, for instance, (8.16). Thus, to estimate $\rho^2_{12.3}$, we regress Y_1 on Y_3 and obtain the error sum of squares $SSE(Y_3)$. Next, we regress Y_1 on Y_2 and Y_3 and obtain the error sum of squares $SSE(Y_2, Y_3)$. The estimator of $\rho^2_{12.3}$ then is:

(15.33)
$$r^2_{12.3} = \frac{SSE(Y_3) - SSE(Y_2, Y_3)}{SSE(Y_3)} = \frac{SSR(Y_2|Y_3)}{SSE(Y_3)}$$

Tests concerning coefficients. Researchers often wish to test whether two variables are correlated in the conditional probability distributions when other variables are fixed. Thus, when four variables Y_1, Y_2, Y_3, and Y_4 are being considered and the correlation between Y_1 and Y_2, when Y_3 and Y_4 are fixed, is of interest, the following test may be desired to see if Y_1 and Y_2 are correlated in the conditional joint distributions:

(15.34)
$$\begin{aligned} H_0&: \rho_{12.34} = 0 \\ H_a&: \rho_{12.34} \neq 0 \end{aligned}$$

Such tests concerning partial correlation coefficients can be carried out via the test statistic (15.16) for the bivariate case, with r_{12} replaced by the estimated partial correlation coefficient and n replaced by $n - q$, where q is the number of variables that are held fixed.

Interval estimation of coefficients. Interval estimates of partial correlation coefficients are obtained via the z' transformation (15.18) in identical fashion to that for simple correlation coefficients. Only the standard deviation $\sigma(z')$ need be modified. It is, for q variables being held fixed:

(15.35)
$$\sigma(z') = \frac{1}{\sqrt{n - q - 3}}$$

Example

An operations analyst, wishing to study the relations between three types of test scores made by applicants for entry-level clerical positions in a large insurance company, drew a sample of 250 such applicants from recent records and ascertained their scores. The variables were:

Y_1 verbal aptitude test score
Y_2 reading aptitude test score
Y_3 personal interview score

The multivariate normal model was considered to be applicable for the study.

The analyst's initial interest was in the partial correlation coefficient $\rho_{23.1}$. Since her computer program did not have an option for calculating partial correlation coefficients by a single command (many programs have such an option), the analyst ran two separate regressions, Y_2 on Y_1 and Y_2 on Y_1 and Y_3. She obtained in turn $SSE(Y_1) = 6{,}340$ and $SSE(Y_1, Y_3) = 6{,}086$. The point estimate of $\rho_{23.1}^2$ then was calculated corresponding to (15.33):

$$r_{23.1}^2 = \frac{SSE(Y_1) - SSE(Y_1, Y_3)}{SSE(Y_1)} = \frac{6{,}340 - 6{,}086}{6{,}340} = .040$$

Hence $r_{23.1} = .20$. (The sign of $r_{23.1}$ here is positive because the regression coefficient for Y_3 when Y_2 is regressed on Y_1 and Y_3 is positive.) Desiring a confidence interval with a 95 percent confidence coefficient, the analyst required:

$$z' = .2027 \qquad \text{when } r_{23.1} = .20$$

$$\sigma(z') = \frac{1}{\sqrt{250 - 1 - 3}} = \frac{1}{\sqrt{246}} = .06376$$

$$z(.975) = 1.96$$

The confidence interval for ζ by (15.22) is:

$$.0777 = .2027 - 1.96(.06376) \leq \zeta \leq .2027 + 1.96(.06376) = .3277$$

In transforming from ζ to ρ, the analyst obtained:

$$.08 \leq \rho_{23.1} \leq .32$$

Thus, the coefficient of partial correlation between reading aptitude score and interview score, with verbal aptitude score fixed, was at best relatively low and could, indeed, be close to zero.

Next the analyst ascertained from a printout of the regression of Y_2 on Y_3 that $r_{23} = .83$. For a 95 percent confidence interval, she then obtained:

$$.79 \leq \rho_{23} \leq .86$$

Thus, ρ_{23} turned out to be substantially larger than $\rho_{23.1}$.

The comparative magnitudes of $\rho_{23.1}$ and ρ_{23} suggested to the analyst (and further investigation verified) that in the company's interviewing procedure the interview scores (Y_3) depend in good part on verbal skills. Also, the reading aptitude scores (Y_2) obtained with the test used by the company tend to be heavily influenced by verbal aptitude. Thus, when verbal aptitude is not considered, the degree of relation between Y_2 and Y_3 is relatively high since applicants with relatively good (poor) verbal aptitude tend to score well (poorly) in both the reading aptitude test and the personal interview. However, among applicants at any given level of verbal aptitude score the degree of relationship between reading score and interview score tends to be low.

PROBLEMS

15.1. A management trainee in a production department wished to study the relation between weight of rough casting and machining time to produce the finished block. He selected castings so that the weights would be spaced equally apart in the sample and observed the corresponding machining times. Would you recommend that a regression or a correlation model be used? Explain.

15.2. A social scientist stated: "The conditions for the bivariate and multivariate normal distributions are so rarely met in my experience that I feel much safer using a regression model." Comment.

15.3. A student was investigating from a large sample whether variables Y_1 and Y_2 follow a bivariate normal distribution. She obtained the residuals when regressing Y_1 on Y_2, and also obtained the residuals when regressing Y_2 on Y_1. She then prepared a normal probability plot for each set of residuals. Do these two normal probability plots provide sufficient information for finding whether the two variables follow a bivariate normal distribution? Explain.

15.4. Refer to Figures 15.1 and 15.3. Where in the Y_1Y_2 plane is the height of the bivariate normal surface greatest? How can this point be ascertained from inspection of the density function (15.1)?

15.5. Plot a contour diagram for the bivariate normal distribution with parameters $\mu_1 = 50$, $\mu_2 = 100$, $\sigma_1 = 3$, $\sigma_2 = 4$, and $\rho_{12} = .80$.

15.6. Refer to Problem 15.5.

a. State the characteristics of the marginal distribution of Y_1.

b. State the characteristics of the conditional distribution of Y_2 when $Y_1 = 55$.

c. State the characteristics of the conditional distribution of Y_1 when $Y_2 = 95$.

15.7. Refer to Problem 15.5.

a. Give the two regression lines for this model.

b. Why are there two regression lines for a bivariate normal distribution and not just one? What is the meaning of each?

c. Must β_{12} and β_{21} have the same sign?

15.8. Refer to Figure 15.6. For any specified value Y_{h2}, can you identify where the conditional density $f(Y_1|Y_{h2})$ is maximized? How can this result be ascertained also from inspection of the conditional density function (15.6)?

15.9. a. Plot a contour diagram for the bivariate normal distribution with parameters $\mu_1 = 14$, $\mu_2 = 350$, $\sigma_1 = 2$, $\sigma_2 = 25$, and $\rho_{12} = .90$.

b. Give the two regression lines for this model. Why are there two regression lines and not just one?

15.10. Refer to Figure 15.6. If the two regression lines were not labeled, could you determine by inspection which line pertains to the regression of Y_2 on Y_1 and which to the regression of Y_1 on Y_2? Explain.

15.11. Explain whether any of the following would be affected if the bivariate normal model (15.1) were employed instead of the normal error regression model (3.1) with fixed levels of the independent variable: (1) point estimates of the regression coefficients, (2) confidence limits for the regression coefficients, (3) interpretation of the confidence coefficient.

15.12. Refer to **Plastic hardness** Problem 2.18. A student was analyzing these data and received the following standard query from the interactive regression and correlation computer package: CALCULATE CONFIDENCE INTERVAL FOR POPULATION CORRELATION COEFFICIENT RHO? ANSWER Y OR N. Would a "yes" response provide meaningful information here? Explain.

15.13. **Property assessments.** The observations that follow show assessed value for property tax purposes (Y_1, in thousand dollars) and sales price (Y_2, in thousand dollars) for a sample of 15 parcels of land for industrial development sold recently in "arm's-length" transactions in a tax district. Assume that bivariate normal model (15.1) is appropriate here.

i:	1	2	3	4	5	6	7	8
Y_{i1}:	13.9	16.0	10.3	11.8	16.7	12.5	10.0	11.4
Y_{i2}:	28.6	34.7	21.0	25.5	36.8	24.0	19.1	22.5

i:	9	10	11	12	13	14	15
Y_{i1}:	13.9	12.2	15.4	14.8	14.9	12.9	15.8
Y_{i2}:	28.3	25.0	31.1	29.6	35.1	30.0	36.2

a. Plot the observations in a scatter diagram. Does the bivariate normal model appear to be appropriate here? Discuss.

b. Calculate r_{12}. What parameter is estimated by r_{12}? What is the interpretation of this parameter?

c. Test whether or not Y_1 and Y_2 are statistically independent in the population, using test statistic (15.16) and level of significance .01. State the alternatives, decision rule, and conclusion.

d. To test $\rho_{12} = .6$ versus $\rho_{12} \neq .6$, would it be appropriate to use test statistic (15.16)?

15.14. **Contract profitability.** A cost analyst for a drilling and blasting contractor examined 84 contracts handled in the last two years and found that the coefficient of correlation between value of contract (Y_1) and profit contribution generated by the contract (Y_2) is $r_{12} = .61$. Assume that the bivariate normal model (15.1) applies.

a. Test whether or not Y_1 and Y_2 are statistically independent in the population; use $\alpha = .05$. State the alternatives, decision rule, and conclusion.

b. Estimate ρ_{12} with a 95 percent confidence interval.

c. Convert the confidence interval in part (b) to a 95 percent confidence interval for ρ_{12}^2. Interpret this interval estimate.

15.15. Bid preparation. A building construction consultant studied the relationship between cost of bid preparation (Y_1) and amount of bid (Y_2) for the consulting firm's clients. In a sample of 103 bids prepared by clients, $r_{12} = .87$. Assume that the bivariate normal model (15.1) applies.

a. Test whether or not $\rho_{12} = 0$; control the risk of Type I error at .10. State the alternatives, decision rule, and conclusion. What would be the implication if $\rho_{12} = 0$?

b. Obtain a 90 percent confidence interval for ρ_{12}. Interpret this interval estimate.

c. Convert the confidence interval in part (b) to a 90 percent confidence interval for ρ_{12}^2.

15.16. Water flow. An engineer, desiring to estimate the coefficient of correlation ρ_{12} between rate of water flow at point A in a stream (Y_1) and concurrent rate of flow at point B (Y_2), obtained $r_{12} = .83$ in a sample of 147 observations. Assume that the bivariate normal model (15.1) is appropriate.

a. Obtain a 99 percent confidence interval for ρ_{12}.

b. Convert the confidence interval in part (a) to a 99 percent confidence interval for ρ_{12}^2.

15.17. Pharmaceutical study. A marketing research analyst in a pharmaceutical firm wishes to study the relation between the following quantified variables in the target population:

Y_1 socioeconomic status of homemaker
Y_2 magazine media exposure to homemaker
Y_3 awareness of homemaker to firm's new product

She believes it is reasonable to assume that the variables follow a multivariate normal distribution.

a. Explain what is measured by each of the following: (1) $\sigma_{3.12}^2$, (2) $\sigma_{31.2}$, (3) $\sigma_{1.3}^2$.

b. Explain the meaning of each of the following: (1) $\rho_{2.13}^2$, (2) $\rho_{2.3}^2$, (3) $\rho_{12.3}^2$.

15.18. Household survey. In a random sample of 250 households, the coefficient of mulitiple correlation between purchases of soft drinks (Y_1) and purchases of snack foods (Y_2) and purchases of dairy products (Y_3) was $R_{1.23} = .892$. The coefficient of partial correlation $r_{13.2}$ was found to be $-.413$.

a. Test whether or not $\rho_{1.23} = 0$; use $\alpha = .05$. State the alternatives, decision rule, and conclusion.

b. Test whether or not $\rho_{13.2} = 0$; use $\alpha = .05$. State the alternatives, decision rule, and conclusion.

c. Estimate $\rho_{13.2}$ with a 95 percent confidence interval. Convert this confidence interval into a 95 percent confidence interval for $\rho_{13.2}^2$. Interpret this latter interval estimate.

15.19. Microcomputer assembly. A consultant studied the relation between visual acuity (Y_1), manual dexterity (Y_2), and tactile reflex (Y_3) in persons who had proven to be proficient as assemblers of microcomputer components. Observations on these variables were obtained for a random sample of 30 persons. Sums of squares for regressions run by the consultant follow. Assume that the multivariate normal correlation model applies.

	Regression				
Sum of Squares	Y_1 *on* Y_2	Y_1 *on* Y_3	Y_1 *on* Y_2, Y_3	Y_2 *on* Y_1	Y_2 *on* Y_3
Regression	7,436	66	7,439	6,900	42
Error	392	7,762	389	363	7,221
Total	7,828	7,828	7,828	7,263	7,263

	Regression			
Sum of Squares	Y_2 *on* Y_1, Y_3	Y_3 *on* Y_1	Y_3 *on* Y_2	Y_3 *on* Y_1, Y_2
Regression	6,901	46	32	64
Error	362	5,327	5,341	5,309
Total	7,263	5,373	5,373	5,373

a. What does $\rho_{1.23}^2$ denote here? Estimate this parameter by a point estimate. What information is provided by this estimate?

b. Test whether or not $\rho_{1.23} = 0$ using a level of significance of .01. State the alternatives, decision rule, and conclusion.

c. Test whether or not $\rho_{12.3} = 0$; use $\alpha = .01$. State the alternatives, decision rule, and conclusion.

d. Estimate $\rho_{12.3}$ with a 99 percent confidence interval. Interpret your interval estimate.

15.20. Refer to **Microcomputer assembly** Problem 15.19. Obtain r_{12}, r_{13}, and r_{23}. Also obtain $r_{12.3}$, $r_{13.2}$, and $r_{23.1}$. Summarize the information provided by these coefficients.

EXERCISES

15.21. The random variables Y_1 and Y_2 follow the bivariate normal distribution (15.1). Show that if $\rho_{12} = 0$, Y_1 and Y_2 are independent random variables.

15.22. (Calculus needed.)

a. Obtain the maximum likelihood estimators of the parameters of the bivariate normal distribution (15.1).

b. Using the results in part (a), obtain the maximum likelihood estimators of the parameters of the conditional probability distribution of Y_1 for any value of Y_2, as given in (15.7).

c. Show that the maximum likelihood estimators of $\alpha_{1.2}$ and β_{12} obtained in part (b) are the same as the least squares estimators (2.10) for the regression coefficients in the simple linear regression model.

15.23. Show that test statistics (3.17) and (15.16) are equivalent.

15.24. Show that test statistic (7.30b) for $p = 4$ is equivalent to test statistic (15.29).

15.25. Show that the ratio *SSR/SSTO* is the same whether Y_1 is regressed on Y_2 or Y_2 is regressed on Y_1. [*Hint*: Use (3.50) and (2.10a).]

PROJECT

15.26. Refer to **Pharmaceutical study** Problem 15.17. The following observations were obtained in a pilot study involving a random sample of 42 homemakers:

Home-maker i	Y_{i1}	Y_{i2}	Y_{i3}	*Home-maker* i	Y_{i1}	Y_{i2}	Y_{i3}
1	76	81	53	22	57	51	32
2	92	89	78	23	49	52	28
3	85	84	66	24	71	72	54
4	63	70	46	25	70	66	55
5	50	48	38	26	91	86	67
6	89	93	68	27	95	91	78
7	35	29	20	28	81	79	60
8	67	68	41	29	40	38	15
9	94	98	77	30	41	43	18
10	83	79	61	31	88	91	72
11	24	28	12	32	63	63	40
12	31	37	14	33	59	54	34
13	42	36	17	34	24	26	11
14	71	71	53	35	96	91	79
15	87	89	71	36	74	80	52
16	28	23	3	37	30	38	14
17	36	36	12	38	81	85	65
18	61	58	35	39	96	92	78
19	74	70	47	40	38	40	16
20	66	60	40	41	52	46	35
21	81	86	66	42	76	68	42

a. The analyst first wished to test whether or not $\rho_{3.12} = 0$. Perform the test using a level of significance of .01. State the alternatives, decision rule, and conclusion. What is the implication for the analyst if $\rho_{3.12} = 0$?

b. Estimate each coefficient of simple correlation and each coefficient of partial correlation with a separate 99 percent confidence interval. What is the family confidence coefficient for the entire set of estimates?

c. Analyze the results obtained in part (b) and summarize your findings.

PART III

Basic analysis of variance

16

Single-factor analysis of variance

Analysis of variance (ANOVA) models are versatile statistical tools for studying the relation between a dependent variable and one or more independent variables. They do not require making assumptions about the nature of the statistical relation, nor do they require that the independent variables are quantitative.

In the present chapter, we consider first the relation between analysis of variance and regression. We then take up the basic elements of single-factor analysis of variance, which is appropriate when one independent variable is being studied. In the remainder of Part III, we continue our discussion of single-factor analysis of variance. In Part IV, we shall take up multifactor analysis of variance, where two or more independent variables are under investigation.

16.1 RELATION BETWEEN REGRESSION AND ANALYSIS OF VARIANCE

Regression analysis, as we have seen, is concerned with the statistical relation between one or more independent variables and a dependent variable. Both the independent and dependent variables in ordinary regression models are quantitative. (We exclude from this discussion the use of indicator variables in regression models, described in Chapter 10.) The regression function describes the nature of the statistical relation between the mean response and the level(s) of the independent variable(s).

We encountered the analysis of variance in our consideration of regression. It was used there for a variety of tests concerning the regression coefficients, the fit of the regression model, and the like. The analysis of variance is actually much more general than its use with regression models indicated. Analysis of variance models are a basic type of statistical model. They are concerned, like regression models, with the statistical relation between one or more independent variables and a dependent variable. Like regression, the analysis of variance is appropriate for both observational data and data based on formal experiments. Further, like ordinary regression, the dependent variable for analysis of variance is a quantitative variable. Analysis of variance models differ from ordinary regression models in two key respects:

1. The independent variables in analysis of variance models may be qualitative (sex, geographic location, plant shift, etc.).
2. If the independent variables are quantitative, no assumption is made in analysis of variance models about the nature of the statistical relation between them and the dependent variable. Thus, the need to specify the nature of the regression function encountered in ordinary regression analysis does not arise in analysis of variance models.

Illustrations

Figure 16.1 illustrates the essential differences between regression and analysis of variance models for the case where the independent variable is quantitative. Shown in Figure 16.1a is the regression model for a pricing study involving three different price levels, $X = \$50, \$60, \$70$. Note that the XY plane has been rotated from its usual position so that the Y axis faces the viewer. For each level of the independent variable, there is a probability distribution of sales volumes. The means of these probability distributions fall on the regression curve, which describes the statistical relation between price level and mean sales volume.

The analysis of variance model for the same study is illustrated in Figure 16.1b. The three price levels are treated as separate populations, each leading to a probability distribution of sales volumes. The quantitative differences in the three price levels and their statistical relation to expected sales volume are not considered by the analysis of variance model.

Figure 16.2 illustrates the analysis of variance model for a study of the effects of four different types of incentive pay systems on employee productivity. Here, each type of incentive pay system corresponds to a different population, and there is associated with each a probability distribution of employee productivities. Since type of incentive pay system is a qualitative variable, Figure 16.2 does not contain a corresponding regression model representation.

FIGURE 16.1 Relation between regression and analysis of variance models

(a) Regression Model

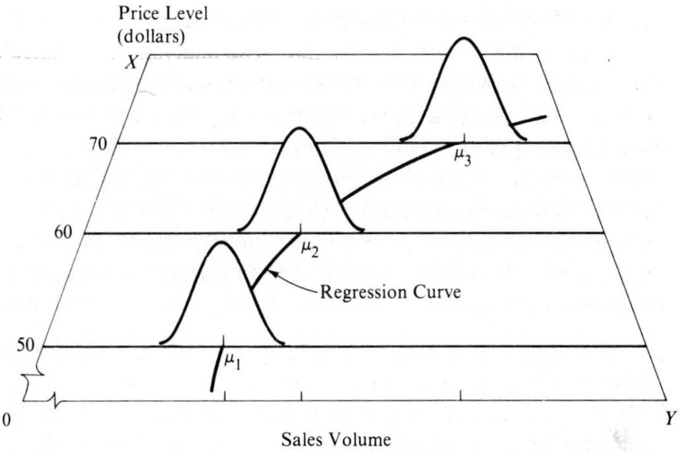

(b) Analysis of Variance Model

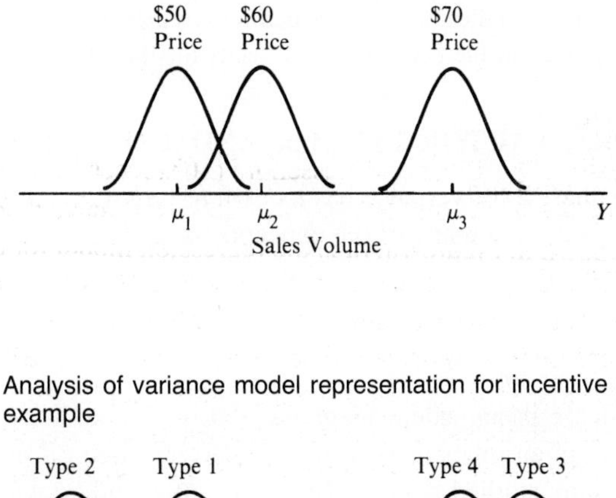

FIGURE 16.2 Analysis of variance model representation for incentive pay example

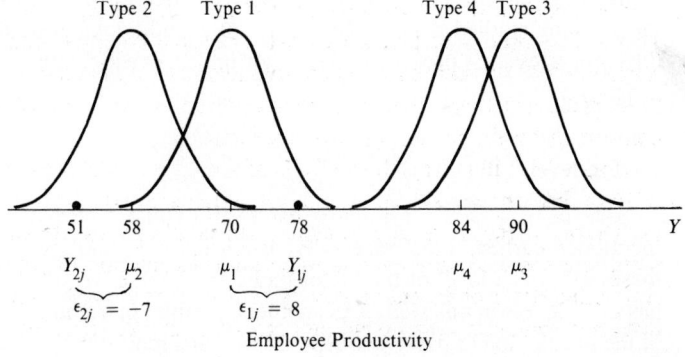

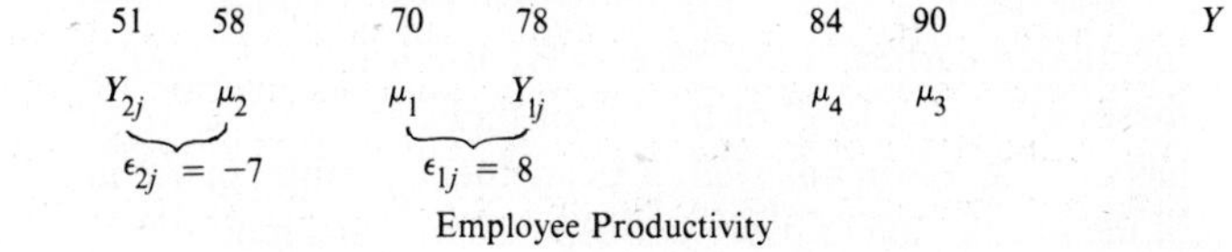

Choice between two types of models

When the independent variables are qualitative, there is no fundamental choice available between regression and analysis of variance models. The situation differs, however, when the independent variables are quantitative. Here, the choice involves on one side an analysis not requiring a specification of the nature of the statistical relation (analysis of variance) and on the other an analysis requiring this specification (ordinary regression analysis). If there is substantial doubt about the nature of the statistical relation, a strategy sometimes followed is to first employ an analysis of variance model to study the effects of the independent variables on the dependent variable without restrictive assumptions on the nature of the statistical relation. The next step, then, is to turn to regression analysis to exploit the quantitative character of the independent variables.

Note

If you have studied Chapter 10, you will know that regression analysis by means of indicator variables can handle qualitative independent variables, and it can also handle by the same means quantitative independent variables without making assumptions about the nature of the statistical relationship. When indicator variables are so used with regression models, the same results will be obtained as with analysis of variance models. The reason why analysis of variance exists as a distinct statistical methodology is that the structure of the independent indicator variables permits computational simplifications that are explicitly recognized in the statistical procedures for the analysis of variance.

16.2 FACTORS, FACTOR LEVELS, AND TREATMENTS

The analysis of variance has its own terminology. We shall introduce some of it now. Note that many of the concepts have already been encountered in regression analysis.

Factor

A *factor* is an independent variable to be studied in an investigation. For instance, in an investigation of the effect of price on sales of a luxury item, the factor being studied is price. Similarly, in a study comparing the appeal of four different television programs, the factor under investigation is type of television program.

Factor level

A *level* of a factor is a particular form of that factor. In the pricing study mentioned earlier, three prices were used, namely, $50, $60, and $70. Each of these prices is a level of the factor under study, and we say that the price factor has three levels in this study. As another example, in a study of the effect of color of the questionnaire paper on response rate in a mail survey, color of paper is the factor under study and each different color used is a level of that factor.

Single-factor and multifactor studies

Investigations differ as to the number of factors studied. Some are *single-factor* studies, where only one factor is of concern. For instance, the study of the appeal of four different television programs mentioned earlier is an example of a single-factor study. In *multifactor* investigations, two or more factors are investigated simultaneously. An example of a multifactor investigation is a study of the effects of temperature and concentration of the solvent upon the yield of a certain chemical process. Here, two factors—temperature and concentration—are studied simultaneously to obtain information about their effects. Multifactor investigations will be discussed in Part IV; now we shall focus on single-factor studies.

Experimental and classification factors

A factor may be categorized as to whether it is an experimental factor or a classification factor. In any investigation based on observational data, the factors under study are classification factors. A *classification* factor pertains to the characteristic of the units under study and is not under the control of the investigator. For instance, in a study of the effects of education and experience of salespeople on their sales volumes where the study was based on a sample of salespeople presently employed by the company, education and experience are classification factors. The reason is that these refer to the characteristics of the salespeople in the study and were not manipulated experimentally. On the other hand, an *experimental* factor is one where the level of the factor is assigned at random to the experimental unit. An example is the factor "temperature" in an experiment on the effect of five cooking temperatures on the fluffiness of omelets prepared from a mix. The experimental units here are different packages of the mix, and the choice of which particular temperature is to be used with a given package of mix is made at random.

We have pointed out before that experimental data provide firmer foundations for conclusions than do observational data. If experimental factors were only to appear in experimental studies and classification factors only in observational studies, there would be no need to distinguish between these two types of factors. However, classification factors can be found in experimental studies, and therefore it is important to recognize them as such, since the inferences pertaining to these factors will not be as clear as those for experimental factors.

Example. An appliance manufacturer operates three training centers in the United States for training mechanics to service the company's products. At each center, two different training programs were studied, with the trainees assigned at random to one of the two training programs. One may view this as a two-factor study, the factors being "training program" and "training center." If the same training program is superior to the other in all three centers, the evidence is quite clear as to the effects of the training programs since at each center the trainees were assigned at random to the two programs.

On the other hand, differences between the three training centers cannot be interpreted as clearly because "training center" is a classification factor. One center may excel for any number of reasons, such as because its staff is doing a better training job, it has better facilities, or because trainees assigned to it come from a geographic region in which better education is provided. External evidence would be required as to whether or not the education of trainees at the three centers is the same, whether or not the facilities are equal, and the like, before a clear understanding of the reasons for differences between training centers could be obtained.

Qualitative and quantitative factors

Finally, we need to distinguish between qualitative and quantitative factors. A *qualitative* factor is one where the levels differ by some qualitative attribute. Examples are type of advertisement or brand of rust inhibitor. On the other hand, a *quantitative* factor is one where each level is described by a numerical quantity on a scale. Examples are temperature in degrees centigrade, age in years, or price in dollars.

Treatments

In single-factor studies, a treatment corresponds to a factor level. Thus, in a study of five advertisements, each advertisement is a treatment. In multifactor studies, a treatment corresponds to a combination of factor levels. Thus, in a study of the effects on sales volume of price ($.25, $.29) and package color (red, blue), each combination, such as "red package color—$.25 price," is a treatment. This particular study contains four treatments since there are four "price-package color" combinations.

16.3 USES OF ANALYSIS OF VARIANCE MODELS

Analysis of variance models basically are used to analyze the effects of the independent variable or variables under study on the dependent variable. The purpose of the study affects the resulting use of the analysis. We illustrate these points with three examples:

Example 1. A hospital was using a standard type of therapy to treat a medical problem. Recently two new types of therapy were proposed. The analysis of variance model was then utilized to determine whether either of the two new types of therapy is superior to the present type and, if so, which one is best.

Example 2. Four machines in a plant were studied with respect to the diameters of ball bearings they were turning out. The purpose of the analysis of variance model study was to determine whether substantial differences between the machines existed. If so, the machines would need to be calibrated.

Example 3. In a sample study of employees of an organization, the employees' absentee rates were studied according to length of service, sex, marital status, and type of job of the employee. The analysis of variance model for this multifactor investigation was utilized to determine whether the effects of these independent variables interact in important ways and which one of the independent variables shows the greatest statistical relation to the dependent variable.

These three examples demonstrate that analysis of variance models for single-factor studies are utilized to compare different factor level effects, to ascertain the "best" factor level, and the like. In multifactor studies, analysis of variance models are employed to determine whether the different factors interact, which factors are the key ones, which factor combinations are "best," and so on.

16.4 MODEL I—FIXED EFFECTS

Distinction between models I and II

We shall consider two single-factor analysis of variance models. Model I, to be taken up here, applies to such cases as a comparison of five different advertisements or a comparison of four different rust inhibitors where the conclusions will pertain to just those factor levels included in the study. Model II, to be discussed in Chapter 19, applies to a different type of situation, namely, where the conclusions will extend to a population of factor levels of which the levels in the study are a sample. Consider, for instance, a company that owns several hundred retail stores throughout the country. Seven of these stores are selected at random, and a sample of employees from each store is then chosen and asked in a confidential interview for an evaluation of the management of the store. The seven stores in the study constitute the seven levels of the factor under study, namely, retail stores. In this case, however, management is not just interested in the seven stores included in the study but wishes to generalize the study results to all of the retail stores it owns. Another example when model II is applicable is when 3 machines out of 75 in a plant are selected at random and their output studied for a period of 10 days. The three machines constitute the three factor levels in this study, but interest is not just in the three machines in the study but in all machines in the plant.

Thus, the essential difference between situations where model I and model II are applicable is that model I is the relevant one when the factor levels are chosen because of intrinsic interest in them (e.g., five different advertisements) and they are not considered as a sample from a larger population. Model II is appropriate when the factor levels constitute a sample from a larger population (e.g., 3 machines out of 75) and interest is in this larger population.

Basic ideas

The basic elements of analysis of variance model I for a single-factor study are quite simple. Corresponding to each factor level, there is a probability distribution of responses. For example, in a study of the effects of four types of incentive pay on employee productivity, there is a probability distribution of employee productivities for each type of incentive pay. The analysis of variance model I assumes that:

1. Each of the probability distributions is normal.
2. Each probability distribution has the same variance (standard deviation).
3. The observations for each factor level are random observations from the corresponding probability distribution and are independent of the observations for any other factor level.

Figure 16.2 illustrates these conditions. Note the normality of the probability distributions and the constant variability. The probability distributions differ only with respect to their means. Differences in the means hence reflect the essential factor level effects, and it is for this reason that the analysis of variance focuses on the mean responses for the different factor levels. The analysis of the sample data from each of the factor level probability distributions therefore usually proceeds in two steps:

1. Determine whether or not the factor level means are the same.
2. If the factor level means are not the same, examine how they differ and what the implications of the differences are.

In this chapter, we consider step 1, the testing procedure for determining whether or not the factor level means are equal. In the next chapter, we take up the analysis of the factor level effects when the means are not equal.

Formulation of model I

There are r levels of the factor under study (e.g., four types of incentive pay), and we shall denote any one of these by the index i ($i = 1, \ldots, r$). The number of observations for the ith factor level is denoted by n_i, and the total number of observations in the study is denoted by n_T, where:

(16.1)
$$n_T = \sum_{i=1}^{r} n_i$$

This notation differs from that used earlier for regression models where the subscript i identifies the observation.

For analysis of variance models we shall always use the last subscript to represent the observation for a given factor level or treatment. Here, the index j will be used to identify the given observation for a particular factor level. Thus, the jth observation for the ith factor level will be represented by Y_{ij}. For instance, Y_{ij} is the productivity of the jth employee in the ith plant, or the sales volume of the jth store featuring the ith type of shelf display. Since the number of observations for the ith factor level is denoted by n_i, we have $j = 1, \ldots, n_i$.

Model I can now be stated as follows:

$$Y_{ij} = \mu_i + \varepsilon_{ij} \tag{16.2}$$

where:

- Y_{ij} is the value of the response variable in the jth trial for the ith factor level or treatment
- μ_i are parameters
- ε_{ij} are independent $N(0, \sigma^2)$
- $i = 1, \ldots, r;\ j = 1, \ldots, n_i$

Important features of model

1. The observed value of Y in the jth trial for the ith factor level or treatment is the sum of two components: (a) a constant term μ_i and (b) a random error term ε_{ij}.

2. Since $E(\varepsilon_{ij}) = 0$, it follows that:

$$E(Y_{ij}) = \mu_i \tag{16.3}$$

Thus, all observations for the ith factor level have the same expectation μ_i.

3. Since μ_i is a constant, it follows from (1.15a) that:

$$\sigma^2(Y_{ij}) = \sigma^2(\varepsilon_{ij}) = \sigma^2 \tag{16.4}$$

Thus, all observations have the same variance, regardless of factor level.

4. Since each ε_{ij} is normally distributed, so is each Y_{ij}. This follows from (1.31) because Y_{ij} is a linear function of ε_{ij}.

5. The error terms are assumed to be independent. Hence, the error term outcome on any one trial has no effect on the error term for the outcome of any other trial for the same factor level or for a different factor level. Since the ε_{ij} are independent, so are the observations Y_{ij}.

6. In view of these features, model (16.2) can be restated as follows:

$$Y_{ij} \text{ are independent } N(\mu_i, \sigma^2) \tag{16.5}$$

Note

The analysis of variance model (16.2) is a linear model because it can be expressed in matrix terms in the form (7.18), i.e., as $\mathbf{Y} = \mathbf{X}\boldsymbol{\beta} + \boldsymbol{\varepsilon}$. We illustrate this for a study involving $r = 3$ treatments for each of which two observations are made. Thus, $n_1 = n_2 = n_3 = 2$. $\mathbf{Y}$, $\mathbf{X}$, $\boldsymbol{\beta}$, and $\boldsymbol{\varepsilon}$ are defined as follows here:

$$\mathbf{Y} = \begin{bmatrix} Y_{11} \\ Y_{12} \\ Y_{21} \\ Y_{22} \\ Y_{31} \\ Y_{32} \end{bmatrix} \quad \mathbf{X} = \begin{bmatrix} 1 & 0 & 0 \\ 1 & 0 & 0 \\ 0 & 1 & 0 \\ 0 & 1 & 0 \\ 0 & 0 & 1 \\ 0 & 0 & 1 \end{bmatrix} \quad \boldsymbol{\beta} = \begin{bmatrix} \mu_1 \\ \mu_2 \\ \mu_3 \end{bmatrix} \quad \boldsymbol{\varepsilon} = \begin{bmatrix} \varepsilon_{11} \\ \varepsilon_{12} \\ \varepsilon_{21} \\ \varepsilon_{22} \\ \varepsilon_{31} \\ \varepsilon_{32} \end{bmatrix} \tag{16.6}$$

Note the simple structure of the $\mathbf{X}$ matrix and that the $\boldsymbol{\beta}$ vector consists of the means μ_i.

To see that these matrices yield the analysis of variance model (16.2), recall from (7.18a) that the vector of expected values $E(Y_{ij})$ is given by $\mathbf{E(Y)} = \mathbf{X\beta}$. We thus obtain:

$$\text{(16.7)} \qquad \mathbf{E(Y)} = \begin{bmatrix} E(Y_{11}) \\ E(Y_{12}) \\ E(Y_{21}) \\ E(Y_{22}) \\ E(Y_{31}) \\ E(Y_{32}) \end{bmatrix} = \mathbf{X\beta} = \begin{bmatrix} 1 & 0 & 0 \\ 1 & 0 & 0 \\ 0 & 1 & 0 \\ 0 & 1 & 0 \\ 0 & 0 & 1 \\ 0 & 0 & 1 \end{bmatrix} \begin{bmatrix} \mu_1 \\ \mu_2 \\ \mu_3 \end{bmatrix} = \begin{bmatrix} \mu_1 \\ \mu_1 \\ \mu_2 \\ \mu_2 \\ \mu_3 \\ \mu_3 \end{bmatrix}$$

which indicates properly that $E(Y_{ij}) = \mu_i$. Hence, model (16.2)—$Y_{ij} = \mu_i + \varepsilon_{ij}$—in matrix form is given by $\mathbf{Y} = \mathbf{X\beta} + \boldsymbol{\varepsilon}$:

$$\text{(16.8)} \qquad \mathbf{Y} = \begin{bmatrix} Y_{11} \\ Y_{12} \\ Y_{21} \\ Y_{22} \\ Y_{31} \\ Y_{32} \end{bmatrix} = \mathbf{X\beta} + \boldsymbol{\varepsilon} = \begin{bmatrix} \mu_1 \\ \mu_1 \\ \mu_2 \\ \mu_2 \\ \mu_3 \\ \mu_3 \end{bmatrix} + \begin{bmatrix} \varepsilon_{11} \\ \varepsilon_{12} \\ \varepsilon_{21} \\ \varepsilon_{22} \\ \varepsilon_{31} \\ \varepsilon_{32} \end{bmatrix}$$

Example

Suppose that analysis of variance model (16.2) is applicable to our earlier incentive pay study illustration and that the parameters are as follows:

$$\mu_1 = 70 \qquad \mu_2 = 58 \qquad \mu_3 = 90 \qquad \mu_4 = 84 \qquad \sigma = 4$$

Figure 16.2 contains a representation of this model. Note that employee productivities for incentive pay type 1 according to this model are normally distributed with mean $\mu_1 = 70$ and standard deviation $\sigma = 4$.

Suppose that in the jth trial of incentive pay type 1, the observed productivity is $Y_{1j} = 78$. In that case, the error term value is $\varepsilon_{1j} = 8$, for we have:

$$\varepsilon_{1j} = Y_{1j} - \mu_1 = 78 - 70 = 8$$

Figure 16.2 shows this observation Y_{1j}. Note that the deviation of Y_{1j} from the mean μ_1 represents the error term ε_{1j}. This figure also shows the observation $Y_{2j} = 51$, for which the error term value is $\varepsilon_{2j} = -7$.

Interpretation of factor level means

Observational data. In an observational study, the factor level means μ_i correspond to the means for the different factor level populations. For instance, in a study of the productivity of employees in each of three shifts operated in a plant, the populations consist of the employee productivities for each of the three shifts. The population mean μ_1 is the mean productivity for employees in shift 1; μ_2 and μ_3 are interpreted similarly. The variance σ^2 refers to the variability of employee productivities within a shift.

Experimental data. In an experimental study, the factor level mean μ_i stands for the mean response that would be obtained if the ith treatment were applied to all units in the population of experimental units about which inferences are to be drawn. Similarly, the variance σ^2 refers to the variability of responses if any given experimental treatment were applied to the entire population of experimental units. For instance, in a study of three different training programs in which 90 employees participate, a third of these employees is assigned at random to each of the three programs. The mean μ_1 here denotes the mean response—for instance, the mean gain in productivity—if training program 1 were given to each employee in the population of experimental units; the means μ_2 and μ_3 are interpreted correspondingly. The variance σ^2 denotes the variability in productivity gains if any one training program were given to each employee in the population of experimental units.

Comments

1. Model I, like any other model, is not likely to be met exactly by any real-world situation. However, it will be met approximately in many cases. As we shall note later, the statistical procedures based on model I are quite robust so that even if the actual conditions of the situation under study differ substantially from those of model I, the statistical analysis may still be appropriate to a high degree of approximation.

2. At times, all treatments under study are given to each of the units in the study. For instance, a person may be asked to use toothpaste A for a week and then give it a rating, which is followed by a week's use of toothpastes B and C each. In this type of case, model I is usually not appropriate since the several responses by the same person for the different treatments under study are likely to be correlated. For instance, if a person prefers tooth powder to toothpaste, all of the ratings for the different toothpastes are likely to be low. In Chapter 28, we take up models for this situation.

16.5 FITTING OF ANOVA MODEL

The parameters of the analysis of variance model (16.2) are ordinarily unknown and must be estimated from sample data. As with regression, we use the method of least squares to fit the ANOVA model and provide estimators of the model parameters. Before turning to these estimators, we shall describe an example to be used throughout the remainder of this chapter and shall develop needed notation.

Example

The Kenton Food Company wished to test four different package designs for a new breakfast cereal. Ten stores, with approximately equal sales volumes, were selected as the experimental units. Each store was randomly assigned one of the package designs, with two of the package designs assigned to three stores each and the other two designs to two stores each. Other relevant conditions besides package design, such as price, amount and location of shelf space, and special promotional efforts, were kept the same for all stores in the experiment. Sales, in number of cases, were observed for the study period, and the results are recorded in Table 16.1a.

Note

In this type of situation it would be more reasonable to assign an equal number of stores to each package design if there is equal interest in each of the four designs. We are utilizing unequal sample sizes to demonstrate the analytical procedure in full generality. Also, the small sample sizes in this example are not indicative of sample sizes needed for this type of study. Here, we simply wish to keep computations to a minimum.

Notation

Table 16.1b shows the symbolic notation for the Kenton Food Company data in Table 16.1a. Y_{ij}, as explained earlier, represents the observation on the jth sample unit for the ith factor level; here, Y_{ij} is the number of cases sold by the jth store assigned to the ith package design. For example, Y_{11} represents the sales of the first store assigned package design 1. For our example, $Y_{11} = 12$ cases. Similarly, sales of the second store assigned package design 3 are $Y_{32} = 17$ cases.

TABLE 16.1 Number of cases sold by stores for each of four package designs—Kenton Food Company example

(a) Sample Data

Package Design	*Store* 1	2	3	*Total*	*Mean*	*Number of Stores*
1	12	18		30	15	2
2	14	12	13	39	13	3
3	19	17	21	57	19	3
4	24	30		54	27	2
All designs				180	18	10

(b) Symbolic Notation

Factor Level i	*Sample Unit* (j) 1	2	3	*Total*	*Mean*	*Number of Sample Units*
1	Y_{11}	Y_{12}		$Y_{1.}$	$\bar{Y}_{1.}$	n_1
2	Y_{21}	Y_{22}	Y_{23}	$Y_{2.}$	$\bar{Y}_{2.}$	n_2
3	Y_{31}	Y_{32}	Y_{33}	$Y_{3.}$	$\bar{Y}_{3.}$	n_3
4	Y_{41}	Y_{42}		$Y_{4.}$	$\bar{Y}_{4.}$	n_4
All factor levels				$Y_{..}$	$\bar{Y}_{..}$	n_T

The total of the observations for the ith factor level is denoted by $Y_{i.}$:

$$Y_{i.} = \sum_{j=1}^{n_i} Y_{ij} \tag{16.9}$$

Thus, the dot in $Y_{i.}$ indicates an aggregation over the j index; in our example, the aggregation is over all stores assigned to the ith package design. For instance, the total sales for all stores assigned package design 1 are, according to Table 16.1, $Y_{1.} = 30$ cases. Similarly, total sales for all stores assigned package design 4 are $Y_{4.} = 54$ cases.

The sample mean for the ith factor level is denoted by $\bar{Y}_{i.}$:

$$\bar{Y}_{i.} = \frac{\sum_{j=1}^{n_i} Y_{ij}}{n_i} = \frac{Y_{i.}}{n_i} \tag{16.10}$$

In our illustration, the mean number of cases sold by stores assigned package design 1 is $\bar{Y}_{1.} = 15$. Thus, the dot in the subscript indicates that the averaging was done over j (stores).

The total of all observations in the study is denoted by $Y_{..}$:

$$Y_{..} = \sum_{i=1}^{r} \sum_{j=1}^{n_i} Y_{ij} \tag{16.11}$$

where the two dots indicate aggregation over both the j and i indexes (in our example, over all stores for any one package design and then over all package designs).

Finally, the overall mean for all observations is denoted by $\bar{Y}_{..}$:

$$\bar{Y}_{..} = \frac{\sum_i \sum_j Y_{ij}}{n_T} = \frac{Y_{..}}{n_T} \tag{16.12}$$

The dots here indicate that the averaging was done over both i and j. For our example, we have from Table 16.1:

$$\bar{Y}_{..} = \frac{180}{10} = 18$$

Least squares estimators

According to the least squares criterion, the sum of the squared deviations of the observations around their expected values must be minimized with respect to the parameters. For model (16.2), we have by (16.3) that:

$$E(Y_{ij}) = \mu_i$$

Hence, the quantity to be minimized is:

$$Q = \sum_i \sum_j (Y_{ij} - \mu_i)^2 \tag{16.13}$$

Now (16.13) can be written as follows:

$$Q = \sum_j (Y_{1j} - \mu_1)^2 + \sum_j (Y_{2j} - \mu_2)^2 + \cdots + \sum_j (Y_{rj} - \mu_r)^2 \tag{16.13a}$$

Note that each of the parameters appears in only one of the component sums in (16.13a). Hence, Q can be minimized by minimizing each of the component sums separately. It is well known that the sample mean minimizes a sum of squared deviations. Hence, the least squares estimator of μ_i, denoted by $\hat{\mu}_i$, is:

$$\hat{\mu}_i = \bar{Y}_{i.} \tag{16.14}$$

Thus, the *fitted value* for observation Y_{ij}, denoted as for regression models by $\hat{Y}_{ij}$, is simply the corresponding factor level sample mean here:

$$\hat{Y}_{ij} = \bar{Y}_{i.} \tag{16.15}$$

Example. For our Kenton Food Company illustration, the least squares estimates of the model parameters are as follows according to Table 16.1:

Parameter	*Least Squares Estimate*
μ_1	$\hat{\mu}_1 = \bar{Y}_{1.} = 15$
μ_2	$\hat{\mu}_2 = \bar{Y}_{2.} = 13$
μ_3	$\hat{\mu}_3 = \bar{Y}_{3.} = 19$
μ_4	$\hat{\mu}_4 = \bar{Y}_{4.} = 27$

Thus, the mean sales per store with package design 1 are estimated to be 15 cases for the population of stores under study, and the fitted values for the observations for package design 1 are $\hat{Y}_{11} = \hat{Y}_{12} = \bar{Y}_{1.} = 15$. Similarly, the mean sales for package design 2 are estimated to be 13 cases per store, and the fitted values for the observations for this package design are $\hat{Y}_{2j} = \bar{Y}_{2.} = 13$.

A simple graphic presentation of the sample results in Table 16.1 is a portrayal of the factor level sample means $\bar{Y}_{i.}$ on a scale, as follows:

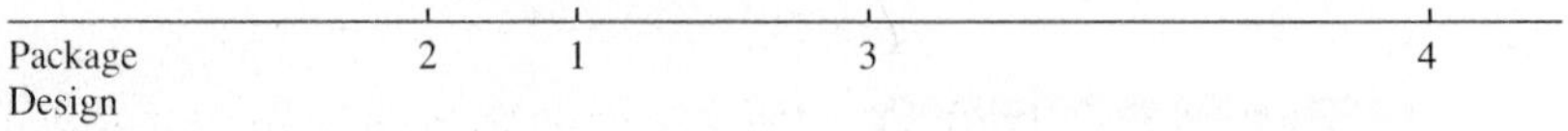

We note that the sample mean for package design 4 is substantially larger than the sample means for the other three designs. Before undertaking any analysis of the effects of the four package designs, we shall consider first whether or not the differences in the four sample means are simply the result of random variation.

Comments

1. The least squares estimators given earlier are also maximum likelihood estimators for the normal error model (16.2). Hence, they have all of the desirable properties mentioned in Chapter 2 for the regression estimators. For example, they are minimum variance unbiased estimators.

2. To derive the least squares estimator of μ_i, we need to minimize, with respect to μ_i, the ith component sum of squares in (16.13a):

$$Q_i = \sum_j (Y_{ij} - \mu_i)^2 \tag{16.16}$$

Differentiating with respect to μ_i, we obtain:

$$\frac{dQ_i}{d\mu_i} = \sum_j -2(Y_{ij} - \mu_i)$$

When we set this derivative equal to zero and replace the parameter μ_i by the least squares estimator $\hat{\mu}_i$, we obtain the result in (16.14):

$$-2\sum_{j=1}^{ni} (Y_{ij} - \hat{\mu}_i) = 0$$

$$\sum_j Y_{ij} = n_i\hat{\mu}_i$$

$$\hat{\mu}_i = \bar{Y}_{i.}$$

Residuals

Residuals are highly useful for examining the aptness of the analysis of variance model in a given application. The residual e_{ij} is again defined, as for regression models, as the difference between the observed and fitted values:

$$e_{ij} = Y_{ij} - \hat{Y}_{ij} = Y_{ij} - \bar{Y}_{i.} \tag{16.17}$$

Thus, a residual here represents the deviation of an observation from the respective factor level sample mean.

Example. Table 16.2 contains the residuals for the Kenton Food Company illustration. For instance, from Table 16.1, we find:

$$e_{11} = Y_{11} - \bar{Y}_{1.} = 12 - 15 = -3$$
$$e_{21} = Y_{21} - \bar{Y}_{2.} = 14 - 13 = +1$$

Note from Table 16.2 that the residuals sum to zero for each factor level. This illustrates an important property of the residuals for analysis of variance model (16.2)—they sum to zero for each factor level i:

$$\sum_j e_{ij} = 0 \qquad i = 1, \ldots, r \tag{16.18}$$

TABLE 16.2 Residuals for Kenton Food Company example

Package Design i	*Store* (j) 1	2	3	*Total*
1	−3	+3		0
2	+1	−1	0	0
3	0	−2	+2	0
4	−3	+3		0
All designs				0

The use of residuals for examining the aptness of an analysis of variance model will be discussed in Chapter 18.

16.6 ANALYSIS OF VARIANCE

Just as the analysis of variance for a regression model partitions the total sum of squares into the regression sum of squares and the error sum of squares, so a corresponding partitioning exists for the analysis of variance model (16.2).

Partitioning of *SSTO*

The total variability of the Y_{ij} observations, not using any information about factor levels, is measured in terms of the deviation of each observation Y_{ij} around the overall mean $\bar{Y}_{..}$:

(16.19) $$Y_{ij} - \bar{Y}_{..}$$

The conventional measure of the total variability is the sum of the squares of these deviations, denoted as for regression by *SSTO* for *total sum of squares*:

(16.20) $$SSTO = \sum_i \sum_j (Y_{ij} - \bar{Y}_{..})^2$$

When we utilize information about the factor levels, the deviations reflecting the uncertainty remaining in the data are those of each observation Y_{ij} around its respective factor level mean $\bar{Y}_{i.}$:

(16.21) $$Y_{ij} - \bar{Y}_{i.}$$

The difference between the deviations (16.19) and (16.21) reflects the difference between the factor level mean and the overall mean:

(16.22) $$(Y_{ij} - \bar{Y}_{..}) - (Y_{ij} - \bar{Y}_{i.}) = \bar{Y}_{i.} - \bar{Y}_{..}$$

Note how we can now decompose the total deviation $Y_{ij} - \bar{Y}_{..}$ into two components:

$$Y_{ij} - \bar{Y}_{..} = \bar{Y}_{i.} - \bar{Y}_{..} + Y_{ij} - \bar{Y}_{i.} \tag{16.23}$$

Total deviation	Deviation of factor level mean around overall mean	Deviation around factor level mean

Thus, the total deviation $Y_{ij} - \bar{Y}_{..}$ can be viewed as the sum of two components:

1. The deviation of the factor level mean around the overall mean.
2. The deviation of Y_{ij} around its factor level mean. This deviation is simply the residual e_{ij} according to (16.17).

Figure 16.3 illustrates this decomposition for the Kenton Food Company data.

When we square (16.23) and then sum, the cross products on the right drop out and we obtain:

$$\sum_i \sum_j (Y_{ij} - \bar{Y}_{..})^2 = \sum_i n_i(\bar{Y}_{i.} - \bar{Y}_{..})^2 + \sum_i \sum_j (Y_{ij} - \bar{Y}_{i.})^2 \tag{16.24}$$

The term on the left is *SSTO*, as defined in (16.20). The first term on the right will be denoted by *SSTR*, standing for *treatment sum of squares*:

$$SSTR = \sum_i n_i(\bar{Y}_{i.} - \bar{Y}_{..})^2 \tag{16.24a}$$

The second term on the right in (16.24) will be denoted by *SSE*, standing for *error sum of squares*:

$$SSE = \sum_i \sum_j (Y_{ij} - \bar{Y}_{i.})^2 = \sum_i \sum_j e_{ij}^2 \tag{16.24b}$$

Thus, (16.24) can be written equivalently:

$$SSTO = SSTR + SSE \tag{16.25}$$

The correspondence to the regression decomposition in (3.48a) is readily apparent.

The total sum of squares for the analysis of variance model is therefore made up of these two components:

1. *SSE*: A measure of the random variation of the observations around the respective factor level sample means. The less variation among the observations for each factor level, the smaller is *SSE*. If $SSE = 0$, all observations for a factor level are the same, and this holds for all factor levels. The more the observations for a factor level differ among themselves, the larger will be *SSE*.

2. *SSTR*: A measure of the extent of differences between factor level sample means, based on the deviations of the factor level sample means $\bar{Y}_{i.}$ around the overall mean $\bar{Y}_{..}$. If all factor level sample means $\bar{Y}_{i.}$ are the same, then $SSTR = 0$. The more the factor level sample means differ, the larger will be *SSTR*.

FIGURE 16.3 Partitioning of total deviations $Y_{ij} - \bar{Y}_{..}$—Kenton Food Company data

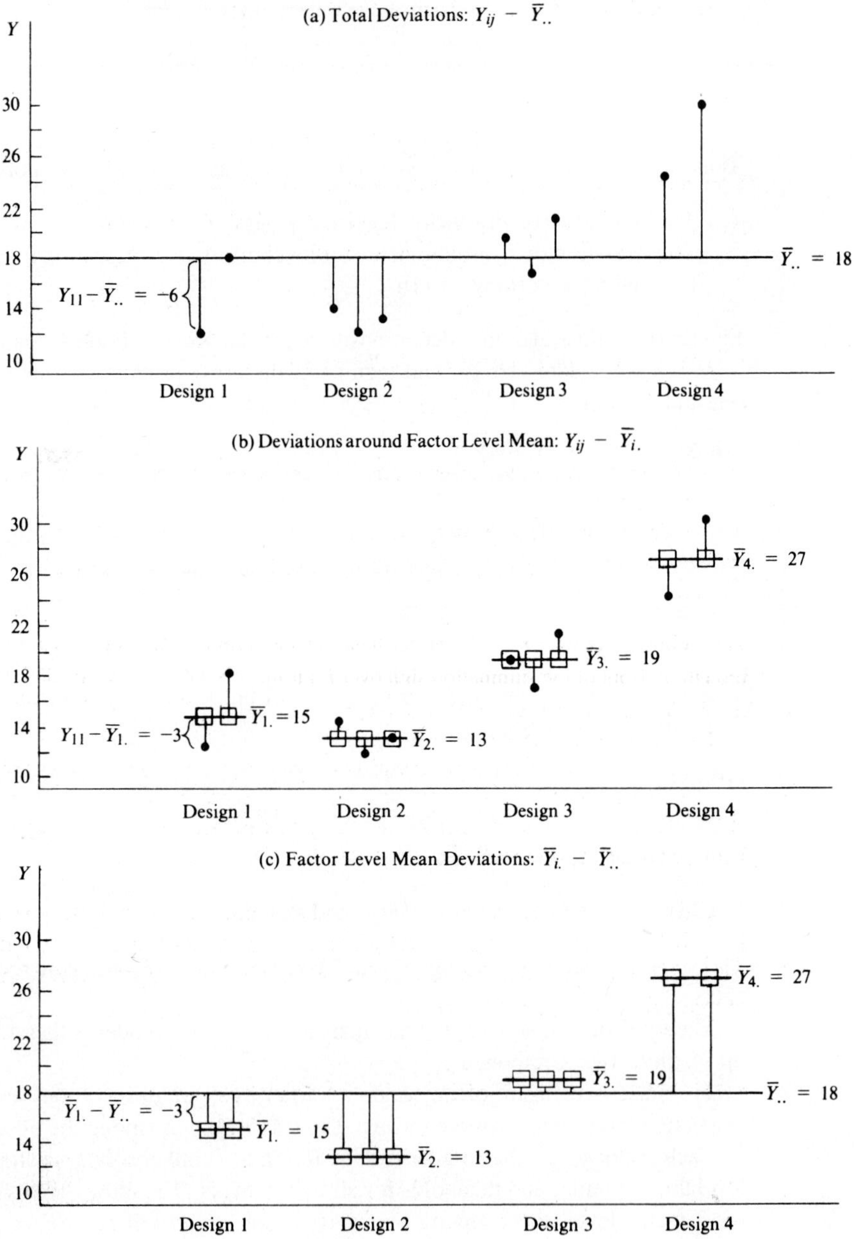

Comments

1. To prove (16.24), we begin by considering (16.23):

$$Y_{ij} - \bar{Y}_{..} = (\bar{Y}_{i.} - \bar{Y}_{..}) + (Y_{ij} - \bar{Y}_{i.})$$

Squaring both sides we obtain:

$$(Y_{ij} - \bar{Y}_{..})^2 = (\bar{Y}_{i.} - \bar{Y}_{..})^2 + (Y_{ij} - \bar{Y}_{i.})^2 + 2(\bar{Y}_{i.} - \bar{Y}_{..})(Y_{ij} - \bar{Y}_{i.})$$

If we sum over all sample observations in the study (i.e., over both i and j), we obtain:

$$\sum_i \sum_j (Y_{ij} - \bar{Y}_{..})^2 = \sum_i \sum_j (\bar{Y}_{i.} - \bar{Y}_{..})^2 + \sum_i \sum_j (Y_{ij} - \bar{Y}_{i.})^2 + \sum_i \sum_j 2(\bar{Y}_{i.} - \bar{Y}_{..})(Y_{ij} - \bar{Y}_{i.}) \tag{16.26}$$

The first term on the right in (16.26) equals:

$$\sum_i \sum_j (\bar{Y}_{i.} - \bar{Y}_{..})^2 = \sum_i n_i(\bar{Y}_{i.} - \bar{Y}_{..})^2 \tag{16.27}$$

since $(\bar{Y}_{i.} - \bar{Y}_{..})^2$ is constant when summed over j; hence, n_i such terms are picked up for the summation over j.

The third term on the right in (16.26) is zero:

$$\sum_i \sum_j 2(\bar{Y}_{i.} - \bar{Y}_{..})(Y_{ij} - \bar{Y}_{i.}) = 2\sum_i (\bar{Y}_{i.} - \bar{Y}_{..}) \sum_j (Y_{ij} - \bar{Y}_{i.}) = 0 \tag{16.28}$$

This follows because $\bar{Y}_{i.} - \bar{Y}_{..}$ is constant for the summation over j; hence, it can be brought in front of the summation sign over j. Further, $\sum_j (Y_{ij} - \bar{Y}_{i.}) = 0$, since the sum of the deviations around the arithmetic mean is zero.

Thus, (16.26) reduces to (16.24).

2. The squared factor level mean deviations $(\bar{Y}_{i.} - \bar{Y}_{..})^2$ in $SSTR$ in (16.24a) are weighted by the number of observations n_i at that factor level. The reason is that for each of the observations at factor level i, the deviation component $\bar{Y}_{i.} - \bar{Y}_{..}$ is the same, as Figure 16.3c makes clear.

Computational formulas. For hand computing, the definitional formulas for *SSTO*, *SSTR*, and *SSE* given earlier are usually not too convenient. Useful computational formulas for hand computing, which are algebraically identical to the definitional formulas, are:

$$SSTO = \sum_i \sum_j Y_{ij}^2 - \frac{Y_{..}^2}{n_T} \tag{16.29a}$$

$$SSTR = \sum_i \frac{Y_{i.}^2}{n_i} - \frac{Y_{..}^2}{n_T} \tag{16.29b}$$

$$SSE = \sum_i \sum_j Y_{ij}^2 - \sum_i \frac{Y_{i.}^2}{n_i} \tag{16.29c}$$

Example. The analysis of variance breakdown of the total sum of squares for our Kenton Food Company data in Table 16.1a is obtained as follows, using the computational formulas in (16.29):

$$SSTO = (12)^2 + (18)^2 + (14)^2 + \cdots + (30)^2 - \frac{(180)^2}{10}$$
$$= 3{,}544 - 3{,}240$$
$$= 304$$

$$SSTR = \frac{(30)^2}{2} + \frac{(39)^2}{3} + \frac{(57)^2}{3} + \frac{(54)^2}{2} - \frac{(180)^2}{10}$$
$$= 3{,}498 - 3{,}240$$
$$= 258$$

$$SSE = 3{,}544 - 3{,}498$$
$$= 46$$

Thus, the decomposition of *SSTO* for our example is:

$$304 = 258 + 46$$
$$SSTO = SSTR + SSE$$

Note that much of the total variation in the observations is associated with variation between the factor level sample means.

Breakdown of degrees of freedom

Corresponding to the decomposition of the total sum of squares, we can also obtain a breakdown of the associated degrees of freedom.

SSTO has $n_T - 1$ degrees of freedom associated with it. There are altogether n_T deviations $Y_{ij} - \bar{Y}_{..}$, but one degree of freedom is lost because the deviations are not independent in that they must sum to zero, i.e., $\Sigma\Sigma(Y_{ij} - \bar{Y}_{..}) = 0$.

SSTR has $r - 1$ degrees of freedom associated with it. There are r factor level mean deviations $\bar{Y}_{i.} - \bar{Y}_{..}$, but one degree of freedom is lost because the deviations are not independent in that the weighted sum must equal zero, i.e., $\Sigma n_i(\bar{Y}_{i.} - \bar{Y}_{..}) = 0$.

SSE has $n_T - r$ degrees of freedom associated with it. This can be readily seen by considering the component of *SSE* for the ith factor level:

$$\sum_{j=1}^{n_i} (Y_{ij} - \bar{Y}_{i.})^2 \tag{16.30}$$

The expression in (16.30) is the equivalent of a total sum of squares considering only the ith factor level. Hence, there are $n_i - 1$ degrees of freedom associated with this sum of squares. Since *SSE* is a sum of squares consisting of component sums of squares such as the one in (16.30), the degrees of freedom associated with *SSE* are the sum of the component degrees of freedom:

(16.31) $$(n_1 - 1) + (n_2 - 1) + \cdots + (n_r - 1) = n_T - r$$

Example. For the Kenton Food Company illustration, for which $n_T = 10$ and $r = 4$, the degrees of freedom associated with the three sums of squares are as follows:

SS	*df*
SSTO	10 − 1 = 9
SSTR	4 − 1 = 3
SSE	10 − 4 = 6

Note that degrees of freedom, like sums of squares, are additive:

$$9 = 3 + 6$$

Mean squares

The mean squares, as usual, are obtained by dividing each sum of squares by its associated degrees of freedom. We therefore have:

(16.32a) $$MSTR = \frac{SSTR}{r - 1}$$

(16.32b) $$MSE = \frac{SSE}{n_T - r}$$

Here, *MSTR* stands for *treatment mean square* and *MSE*, as before, stands for *error mean square*.

Example. For our Kenton Food Company example, we obtain from earlier results:

$$MSTR = \frac{258}{3} = 86$$

$$MSE = \frac{46}{6} = 7.67$$

Note that the two mean squares do not add to $SSTO/(n_T - 1) = 304/9 = 33.8$. Thus, the mean squares here, as in regression, are not additive.

Analysis of variance table

The breakdowns of the total sum of squares and degrees of freedom, together with the resulting mean squares, are presented in an ANOVA table such as Table 16.3. The ANOVA table for our Kenton Food Company example is presented in Table 16.4.

TABLE 16.3 ANOVA table for single-factor study

Source of Variation	*SS*	*df*	*MS*	*E(MS)*
Between treatments	$SSTR = \Sigma n_i(\bar{Y}_{i.} - \bar{Y}_{..})^2$	$r - 1$	$MSTR = \dfrac{SSTR}{r-1}$	$\sigma^2 + \dfrac{1}{r-1}\Sigma n_i(\mu_i - \mu_.)^2$
Error (within treatments)	$SSE = \Sigma\Sigma(Y_{ij} - \bar{Y}_{i.})^2$	$n_T - r$	$MSE = \dfrac{SSE}{n_T - r}$	σ^2
Total	$SSTO = \Sigma\Sigma(Y_{ij} - \bar{Y}_{..})^2$	$n_T - 1$		

TABLE 16.4 ANOVA table for Kenton Food Company example

Source of Variation	*SS*	*df*	*MS*
Between designs	258	3	86
Error	46	6	7.67
Total	304	9	

Expected mean squares

The expected values of *MSE* and *MSTR* can be shown to be as follows:

(16.33a) $$E(MSE) = \sigma^2$$

(16.33b) $$E(MSTR) = \sigma^2 + \frac{\Sigma n_i(\mu_i - \mu_.)^2}{r - 1}$$

where:

$$\mu_. = \frac{\Sigma n_i \mu_i}{n_T}$$

These expected values are shown in the *E(MS)* column of Table 16.3.

Two important features of the expected mean squares deserve attention:

1. *MSE* is an unbiased estimator of the variance of the error terms ε_{ij}, whether or not the factor level means μ_i are equal. This is intuitively reasonable since the variability of the observations within each factor level is not affected by the magnitudes of the factor level means.

2. If all factor level means μ_i are equal and hence equal to the weighted mean $\mu_.$, then $E(MSTR) = \sigma^2$ since the second term on the right in (16.33b) becomes zero. Hence, *MSTR* and *MSE* both estimate the error variance σ^2 if all factor level means μ_i are equal. If, however, the factor level means are not equal, *MSTR* tends on the average to be larger than *MSE*, since the second term in (16.33b) will then be positive. This is intuitively reasonable, as illustrated in Figure 16.4 for four treatments. The situation portrayed there assumes that all

FIGURE 16.4 Sampling distributions of $\bar{Y}_{i.}$ for four treatments

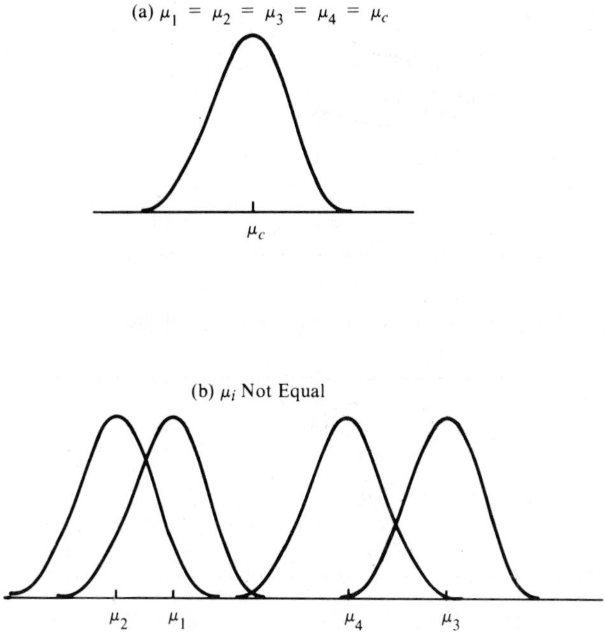

sample sizes are equal, i.e., $n_i \equiv n$. If all μ_i are equal, then all $\bar{Y}_{i.}$ follow the same sampling distribution, with common mean μ_c and variance σ^2/n; this is portrayed in Figure 16.4a. If the μ_i are not equal, on the other hand, the $\bar{Y}_{i.}$ follow different sampling distributions, each with the same variability σ^2/n but centered on different means μ_i. One such possibility is shown in Figure 16.4b. Hence, the $\bar{Y}_{i.}$ will tend to differ more from each other if the μ_i differ than if the μ_i are equal, and consequently *MSTR* will tend to be larger when the factor level means are not the same than when they are equal. This property of *MSTR* is utilized in constructing the statistical test discussed in the next section to determine whether or not the factor level means μ_i are the same. If *MSTR* and *MSE* are of the same order of magnitude, this is taken to suggest that the factor level means μ_i are equal. If *MSTR* is substantially larger than *MSE*, this is taken to suggest that the μ_i are not equal.

Comments

1. To find the expected value of *MSE*, we first note that *MSE* can be expressed as follows:

(16.34)
$$MSE = \frac{1}{n_T - r} \sum_i \sum_j (Y_{ij} - \bar{Y}_{i.})^2$$

$$= \frac{1}{n_T - r} \sum_i \left[(n_i - 1) \frac{\sum_j (Y_{ij} - \bar{Y}_{i.})^2}{n_i - 1} \right]$$

Now let us denote the ordinary sample variance of the observations for the ith factor level by s_i^2:

$$s_i^2 = \frac{\sum_j (Y_{ij} - \bar{Y}_{i.})^2}{n_i - 1} \tag{16.35}$$

Hence, (16.34) can be expressed as follows:

$$MSE = \frac{1}{n_T - r} \sum_i (n_i - 1)s_i^2 \tag{16.36}$$

Since it is well known that the sample variance (16.35) is an unbiased estimator of the population variance, which in our case is σ^2 for all factor levels, we obtain:

$$E(MSE) = \frac{1}{n_T - r} \sum_i (n_i - 1)E(s_i^2)$$

$$= \frac{1}{n_T - r} \sum_i (n_i - 1)\sigma^2$$

or:

$$E(MSE) = \sigma^2$$

2. We shall derive the expected value of $MSTR$ for the simplified case when all sample sizes n_i are the same, namely, $n_i \equiv n$. The general result in (16.33b) becomes for this special case:

$$E(MSTR) = \sigma^2 + \frac{n\Sigma(\mu_i - \mu_.)^2}{r - 1} \qquad \text{when } n_i \equiv n \tag{16.37}$$

Further, when all factor level sample sizes are n, $MSTR$ as defined in (16.24a) and (16.32a) becomes:

$$MSTR = \frac{n\Sigma(\bar{Y}_{i.} - \bar{Y}_{..})^2}{r - 1} \qquad \text{when } n_i \equiv n \tag{16.38}$$

To derive (16.37), consider the model formulation for Y_{ij} in (16.2):

$$Y_{ij} = \mu_i + \varepsilon_{ij}$$

Averaging the Y_{ij} for the ith factor level, we obtain:

$$\bar{Y}_{i.} = \mu_i + \bar{\varepsilon}_{i.} \tag{16.39}$$

where $\bar{\varepsilon}_{i.}$ is the average of the ε_{ij} for the ith factor level:

$$\bar{\varepsilon}_{i.} = \frac{\sum_j \varepsilon_{ij}}{n} \tag{16.40}$$

Averaging the Y_{ij} over all factor levels, we obtain:

$$\bar{Y}_{..} = \mu_. + \bar{\varepsilon}_{..} \tag{16.41}$$

where $\mu_.$, which is defined in (16.33b), becomes for $n_i \equiv n$:

(16.42)
$$\mu_{.} = \frac{n\Sigma\mu_i}{nr} = \frac{\Sigma\mu_i}{r} \qquad \text{when } n_i \equiv n$$

and $\bar{\varepsilon}_{..}$ is the average of all ε_{ij}:

(16.43)
$$\bar{\varepsilon}_{..} = \frac{\Sigma\Sigma\varepsilon_{ij}}{nr}$$

Since the sample sizes are equal, we also have:

(16.44)
$$\bar{Y}_{..} = \frac{\Sigma\bar{Y}_{i.}}{r} \qquad \bar{\varepsilon}_{..} = \frac{\Sigma\bar{\varepsilon}_{i.}}{r}$$

Using (16.39) and (16.41), we obtain:

(16.45)
$$\bar{Y}_{i.} - \bar{Y}_{..} = (\mu_i + \bar{\varepsilon}_{i.}) - (\mu_{.} + \bar{\varepsilon}_{..}) = (\mu_i - \mu_{.}) + (\bar{\varepsilon}_{i.} - \bar{\varepsilon}_{..})$$

If we square $\bar{Y}_{i.} - \bar{Y}_{..}$ and sum over the factor levels, we obtain:

(16.46)
$$\Sigma(\bar{Y}_{i.} - \bar{Y}_{..})^2 = \Sigma(\mu_i - \mu_{.})^2 + \Sigma(\bar{\varepsilon}_{i.} - \bar{\varepsilon}_{..})^2 + 2\Sigma(\mu_i - \mu_{.})(\bar{\varepsilon}_{i.} - \bar{\varepsilon}_{..})$$

We now wish to find $E[\Sigma(\bar{Y}_{i.} - \bar{Y}_{..})^2]$, and therefore need to find the expected value of each term on the right in (16.46):

a. Since $\Sigma(\mu_i - \mu_{.})^2$ is a constant, its expectation is:

(16.47)
$$E[\Sigma(\mu_i - \mu_{.})^2] = \Sigma(\mu_i - \mu_{.})^2$$

b. Before finding the expectation of the second term on the right, consider first the expression:

$$\frac{\Sigma(\bar{\varepsilon}_{i.} - \bar{\varepsilon}_{..})^2}{r - 1}$$

This is an ordinary sample variance, since $\bar{\varepsilon}_{..}$ is the sample mean of the r terms $\bar{\varepsilon}_{i.}$ per (16.44). We further know that the sample variance is an unbiased estimator of the variance of the variable, in this case the variable being $\bar{\varepsilon}_{i.}$. But $\bar{\varepsilon}_{i.}$ is just the mean of n independent error terms ε_{ij} by (16.40). Hence:

$$\sigma^2(\bar{\varepsilon}_{i.}) = \frac{\sigma^2(\varepsilon_{ij})}{n} = \frac{\sigma^2}{n}$$

Therefore:

$$E\left[\frac{\Sigma(\bar{\varepsilon}_{i.} - \bar{\varepsilon}_{..})^2}{r - 1}\right] = \frac{\sigma^2}{n}$$

so that:

(16.48)
$$E[\Sigma(\bar{\varepsilon}_{i.} - \bar{\varepsilon}_{..})^2] = \frac{(r - 1)\sigma^2}{n}$$

c. Since both $\bar{\varepsilon}_{i.}$ and $\bar{\varepsilon}_{..}$ are means of ε_{ij} terms, all of which have expectation 0, it follows that:

$$E(\bar{\varepsilon}_{i.}) = 0 \qquad E(\bar{\varepsilon}_{..}) = 0$$

Hence:

$$E[2\Sigma(\mu_i - \mu_{.})(\bar{\varepsilon}_{i.} - \bar{\varepsilon}_{..})] = 2\Sigma(\mu_i - \mu_{.})E(\bar{\varepsilon}_{i.} - \bar{\varepsilon}_{..}) = 0 \tag{16.49}$$

We have thus shown, by (16.47), (16.48), and (16.49), that:

$$E[\Sigma(\bar{Y}_{i.} - \bar{Y}_{..})^2] = \Sigma(\mu_i - \mu_{.})^2 + \frac{(r-1)\sigma^2}{n}$$

But then (16.37) follows at once:

$$E(MSTR) = E\left[\frac{n\Sigma(\bar{Y}_{i.} - \bar{Y}_{..})^2}{r-1}\right] = \frac{n}{r-1}\left[\Sigma(\mu_i - \mu_{.})^2 + \frac{(r-1)\sigma^2}{n}\right]$$

$$= \sigma^2 + \frac{n\Sigma(\mu_i - \mu_{.})^2}{r-1}$$

16.7 *F* TEST FOR EQUALITY OF FACTOR LEVEL MEANS

It is customary to begin the analysis of a single-factor study by determining whether or not the factor level means μ_i are equal. If, for instance, the four package designs in our Kenton Food Company example lead to the same mean sales volumes, there is no need for further analysis, such as to determine which design is best or how two particular designs compare in stimulating sales.

Thus, the alternative conclusions we wish to consider are:

$$\begin{aligned} &H_0: \mu_1 = \mu_2 = \cdots = \mu_r \\ &H_a: \text{not all } \mu_i \text{ are equal} \end{aligned} \tag{16.50}$$

Test statistic

The test statistic to be used for choosing between the alternatives in (16.50) is:

$$F^* = \frac{MSTR}{MSE} \tag{16.51}$$

Note that *MSTR* here plays the role corresponding to *MSR* for a regression model.

Large values of F^* support H_a, since *MSTR* will tend to exceed *MSE* when H_a holds, as we saw from (16.33). Values of F^* near 1 support H_0, since both *MSTR* and *MSE* have the same expected value when H_0 holds. Hence, the appropriate test is an upper-tail one.

Distribution of F^*

When all treatment means μ_i are equal, each observation Y_{ij} has the same expected value. In view of the additivity of sums of squares and degrees of freedom, Cochran's theorem (3.60) then implies:

$$\text{When } H_0 \text{ holds, } \frac{SSE}{\sigma^2} \text{ and } \frac{SSTR}{\sigma^2} \text{ are independent } \chi^2 \text{ variables}$$

It follows in the same fashion as for regression:

$$\text{When } H_0 \text{ holds, } F^* \text{ is distributed as } F(r-1, n_T - r)$$

If H_a holds, that is, if the μ_i are not all equal, F^* does *not* follow the F distribution. Rather, it follows a complex distribution called the *noncentral F distribution*. We shall make use of the noncentral F distribution when we discuss the power of the F test.

Note

SSTR and *SSE* are independent even if all μ_i are not equal. Intuitively, this can be seen because *SSE* reflects the variability within the factor level samples, and this within-sample variability is not affected by the magnitudes of the factor level means. *SSTR*, on the other hand, is solely based on the factor level means $\bar{Y}_{i.}$.

Construction of decision rule

Usually, the risk of making a Type I error is controlled in constructing the decision rule. This provides protection against making further, more detailed, analyses of the factor effects when in fact there are no differences in the factor level means. The Type II error can also be controlled, as we shall see later, through sample size determination.

Since we know that F^* is distributed as $F(r-1, n_T - r)$ if H_0 holds and that large values of F^* lead to conclusion H_a, the appropriate decision rule to control the level of significance at α is:

$$\begin{aligned} &\text{If } F^* \leq F(1-\alpha; r-1, n_T - r), \text{ conclude } H_0 \\ &\text{If } F^* > F(1-\alpha; r-1, n_T - r), \text{ conclude } H_a \end{aligned} \tag{16.52}$$

where $F(1-\alpha; r-1, n_T - r)$ is the $(1-\alpha)100$ percentile of the appropriate F distribution.

Example

For our Kenton Food Company illustration, we wish to test whether or not mean sales are the same for the four package designs:

$$\begin{aligned} &H_0\colon \mu_1 = \mu_2 = \mu_3 = \mu_4 \\ &H_a\colon \text{not all } \mu_i \text{ are equal} \end{aligned}$$

Suppose that management wishes to control the risk of making a Type I error at $\alpha = .05$. We therefore require $F(.95; 3, 6)$, where the degrees of freedom are those shown in Table 16.4. From Table A–4 in the Appendix, we find $F(.95; 3, 6) = 4.76$. Hence, the decision rule is:

$$\begin{aligned} &\text{If } F^* \leq 4.76, \text{ conclude } H_0 \\ &\text{If } F^* > 4.76, \text{ conclude } H_a \end{aligned}$$

Using the data from Table 16.4, the test statistic is:

$$F^* = \frac{MSTR}{MSE} = \frac{86}{7.67} = 11.2$$

Since $F^* = 11.2 > 4.76$, we conclude H_a, that the factor level means μ_i are not equal, or that the four different package designs do not lead to the same mean sales volume. Thus, we conclude that there is a relation between package design and sales volume.

The P-value for the test statistic is the probability $P[F(3, 6) > F^* = 11.2]$. We see from Table A–4 that the P-value is between .005 and .01.

The conclusion of a relation between package design and sales volume did not surprise the sales manager of the Kenton Food Company. He conducted the study in the first place because he expected the four package designs to have different effects on sales volume and was interested in finding out the nature of these differences. In the next chapter, we discuss the second stage of the analysis, namely, how to study the nature of the factor level effects when differences exist.

Comments

1. If there are only two factor levels so that $r = 2$, it can easily be shown that the test employing F^* in (16.51) is the equivalent of the two-population, two-sided t test in Table 1.2a. The F test here has $(1, n_T - 2)$ degrees of freedom, and the t test has $n_1 + n_2 - 2$ or $n_T - 2$ degrees of freedom; thus both tests lead to equivalent critical regions. For comparing two population means, the t test generally is to be preferred since it can be used to conduct both two-sided and one-sided tests (Table 1.2); the F test can be used only for two-sided tests.

2. Since the F test for testing the alternatives (16.50) is a test of a linear statistical model, it can be obtained by the general method explained in Chapter 3:

a. The full model is model (16.2):

(16.53) $$Y_{ij} = \mu_i + \varepsilon_{ij} \qquad \text{Full model}$$

Fitting the full model by the method of least squares leads to the fitted values $\hat{Y}_{ij} = \bar{Y}_{i.}$, per (16.15), and the resulting error sum of squares:

$$SSE(F) = \Sigma\Sigma(Y_{ij} - \hat{Y}_{ij})^2 = \Sigma\Sigma(Y_{ij} - \bar{Y}_{i.})^2$$

b. The reduced model under H_0 is:

(16.54) $$Y_{ij} = \mu_c + \varepsilon_{ij} \qquad \text{Reduced model}$$

where μ_c is the common mean for all factor levels. Fitting the reduced model leads to the least squares estimator $\hat{\mu}_c = \bar{Y}_{..}$ so that all fitted values are $\hat{Y}_{ij} = \bar{Y}_{..}$ and the resulting error sum of squares is:

$$SSE(R) = \Sigma\Sigma(Y_{ij} - \hat{Y}_{ij})^2 = \Sigma\Sigma(Y_{ij} - \bar{Y}_{..})^2$$

c. The general test statistic (3.68) then becomes:

$$F^* = \frac{SSE(R) - SSE(F)}{(n_T - 1) - (n_T - r)} \div \frac{SSE(F)}{n_T - r}$$

Note that $SSE(R)$ has $n_T - 1$ degrees of freedom associated with it because one parameter (μ_c) had to be estimated. Also, $SSE(F)$ has $n_T - r$ degrees of freedom because r parameters ($\mu_1, \ldots, \mu_r$) had to be estimated.

Since we have, according to (16.20) and (16.24b), respectively:

$$SSE(R) = SSTO$$

$$SSE(F) = SSE$$

and since by (16.25) $SSTO - SSE = SSTR$, we obtain the test statistic (16.51):

$$F^* = \frac{SSTR}{r-1} \div \frac{SSE}{n_T - r} = \frac{MSTR}{MSE}$$

16.8 COMPUTER INPUT AND OUTPUT FOR ANOVA PACKAGES

For analysis of variance as for regression analysis, packaged computer programs are readily available for handling calculations. We will assume, as in our regression analysis discussions, that computers or programmable calculators are used in handling analysis of variance calculations for all but the simplest data sets.

For regression analysis, a single ''multiple regression'' packaged program frequently suffices for a variety of applications such as simple linear regression, polynomial regression, and multiple regression. Packaged analysis of variance programs, on the other hand, are often keyed to specific analyses. Thus, a library may contain one program for single-factor analysis of variance and other programs for multifactor analyses, which we will discuss in later chapters. Formats for input of observations and output of results often differ substantially from one library to another, and may even differ somewhat among the different analysis of variance programs in a given library.

One format for input of the Kenton Food Company data in Table 16.1 requires two columns:

Treatment	*Observation*
1	12
1	18
2	14
2	12
2	13
⋮	⋮
4	30

Here the treatment identifications are entered in one column and the observations are entered in another. Other formats will also be encountered when using different ANOVA packages.

Figure 16.5 illustrates a typical output format when the single-factor analysis of variance model is used. The output illustrated in Figure 16.5 is for the Kenton Food Company data in Table 16.1 and was produced by the BMDP computer program (Ref. 16.1). We have added some annotations to facilitate understand-

FIGURE 16.5 Segment of computer output for single-factor analysis of variance—Kenton Food Company example (BMDP, Ref. 16.1)

```
C A S E           1          2
NO. LABEL       DESIGN      SOLD   ← Y_ij

    1             1          12
    2             1          18
    3             2          14
    4             2          12
    5             2          13
    6             3          19
    7             3          17
    8             3          21
    9             4          24
   10             4          30


NUMBER OF CASES PER GROUP

DESIGN1      2.
DESIGN2      3.
DESIGN3      3.  ← n_i
DESIGN4      2.
TOTAL       10.  ← n_T


ESTIMATES OF MEANS

                DESIGN1     DESIGN2     DESIGN3     DESIGN4     TOTAL

SOLD       2    15.0000     13.0000     19.0000     27.0000     18.0000 ← Ȳ..
                                  ↑
                                  Ȳ_i.

ANALYSIS OF VARIANCE TABLE                    SSTR            MSTR

SOURCE OF VARIANCE       D.F.   SUM OF SQ.    MEAN SQ.     F-VALUE   PROB(TAIL)
EQUALITY OF CELL MEANS     3    258.0000      86.0000      11.2174     0.0071
   ERROR                   6     46.0000       7.6667         ↑          ↑
Between treatments              SSE           MSE             F*     One-sided
                                                                      P-value
```

ing of the output. Note that the data format is listed at the top of Figure 16.5, followed by the number of observations for each package design, estimates of the factor level means, and the analysis of variance table.

The results in Figure 16.5 coincide with those obtained by hand calculations in Tables 16.1 and 16.4. This is because roundoff errors are no problem in calculations involving the simple data set of our Kenton example. With more complex data sets, roundoff effects can arise, so that somewhat differing results could be obtained then.

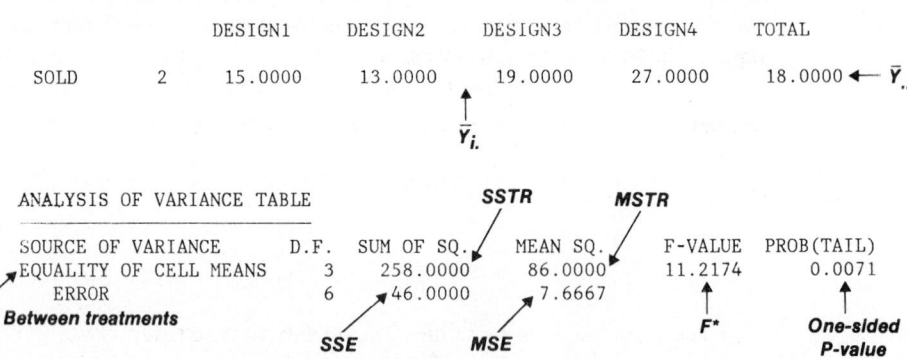

We noted, in discussing roundoff errors in regression analysis, that least squares results tend to be highly sensitive to roundoff effects and that different regression packages do not control such effects equally well. Similarly, different analysis of variance packages differ in quality, and it is a good idea to check a library program before using it for the first time. One procedure is to test the program on a complex data set for which accurate results are already known.

16.9 POWER OF *F* TEST

The power of the F test is of importance for assessing the discriminating ability of the decision rule employed, and also for determining needed sample sizes. By the power of the F test, we refer to the probability that the decision rule will lead to conclusion H_a when in fact H_a holds. Specifically, the power is given by the following expression:

$$\text{Power} = P[F^* > F(1-\alpha; r-1, n_T - r) \mid \phi] \tag{16.55}$$

where ϕ is the *noncentrality parameter*, that is, a measure of how unequal the μ_i are:

$$\phi = \frac{1}{\sigma}\sqrt{\frac{\Sigma n_i(\mu_i - \mu_.)^2}{r}} \tag{16.56}$$

where:

$$\mu_. = \frac{\Sigma n_i \mu_i}{n_T}$$

When all factor level samples are of equal size n, the parameter ϕ becomes:

$$\phi = \frac{1}{\sigma}\sqrt{\frac{n}{r}\Sigma(\mu_i - \mu_.)^2} \qquad \text{when } n_i \equiv n \tag{16.56a}$$

where:

$$\mu_. = \frac{\Sigma \mu_i}{r}$$

To determine power probabilities, we need to utilize the noncentral F distribution since this is the sampling distribution of F^* when H_a holds. The resulting calculations are quite complex. Fortunately, charts have been prepared that make the determination of power probabilities relatively simple. Table A–8 contains the Pearson-Hartley charts of the power of the F test. The proper curve to utilize depends on the number of factor levels, the sample sizes, and the level of significance employed in the decision rule. The Pearson-Hartley charts are used as follows:

1. Each page refers to a different ν_1, the number of degrees of freedom for the numerator of F^*. For our situation, $\nu_1 = r - 1$, or the number of factor

levels minus one. Table A–8 contains charts for $\nu_1 = 2, 3, 4, 5$, and 6 as shown in the upper left-hand corner of each chart.

2. Two levels of significance, denoted by α, are used in the charts, namely, $\alpha = .05$ and $\alpha = .01$. There are two X scales, depending on which level of significance is employed. Further, the left set of curves on each chart refers to $\alpha = .05$ and the right set to $\alpha = .01$.
3. There are separate curves for different values of ν_2, the degrees of freedom for the denominator of F^*. In our case, $\nu_2 = n_T - r$. The curves are indexed according to the value of ν_2 at the top of the chart. Since only selected values of ν_2 are used in the charts, one needs to interpolate for intermediate values of ν_2.
4. The X scale represents ϕ, the noncentrality parameter defined in (16.56).
5. Finally, the Y scale gives the power $1 - \beta$, where β is the risk of making a Type II error.

Examples

1. Consider the case where $\nu_1 = 2$, $\nu_2 = 10$, $\phi = 3$, and $\alpha = .05$. We then find from Table A–8 (p. 1089) that the power is $1 - \beta = .983$ approximately.

2. Suppose that for our Kenton Food Company example, the analyst wishes to know the power of the decision rule on page 543 when there are substantial differences between the factor level means. Specifically, she wishes to consider the case when $\mu_1 = 12.5$, $\mu_2 = 13$, $\mu_3 = 18$, and $\mu_4 = 21$. The weighted mean in (16.56) therefore is:

$$\mu_. = \frac{2(12.5) + 3(13) + 3(18) + 2(21)}{10} = 16$$

Thus, the specified value of ϕ is:

$$\phi = \frac{1}{\sigma}\left[\frac{2(-3.5)^2 + 3(-3)^2 + 3(2)^2 + 2(5)^2}{4}\right]^{1/2}$$

$$= \frac{1}{\sigma}(5.33)$$

Note that we still need to know σ, the variability of the error terms ε_{ij} in the model. Suppose that from past experience it is known that $\sigma = 2.5$ cases approximately. Then we have:

$$\phi = \frac{1}{2.5}(5.33) = 2.13$$

Further, we have for this example:

$$\nu_1 = r - 1 = 3$$
$$\nu_2 = n_T - r = 6$$
$$\alpha = .05$$

Hence, from the table on page 1090 we find that the power is $1 - \beta = .72$ approximately. In other words, there are about 72 chances in 100 that the decision rule will lead to the detection of differences in the mean sales volumes for the four package designs when the differences are the ones specified earlier.

Comments

1. If the decision rule in our package design example had been set up with a level of significance $\alpha = .01$, the power would have been $1 - \beta = .36$ approximately. Thus, the smaller the specified α risk, the smaller is the power for any given ϕ, and hence the larger the risk of a Type II error.

2. Any given value of ϕ encompasses many different combinations of factor level means μ_i. Thus, in our package design example, $\mu_1 = 12.5$, $\mu_2 = 13$, $\mu_3 = 18$, $\mu_4 = 21$ and $\mu_1 = 21$, $\mu_2 = 18$, $\mu_3 = 13$, $\mu_4 = 12.5$ lead to the same value of $\phi = 2.13$ and hence to the same power.

3. The effect of misspecifying σ in determining the power unfortunately can be great. In our Kenton Food Company example, if $\sigma = 3.0$ instead of 2.5, ϕ would be 1.78 and the power (for $\alpha = .05$) would be .56 instead of .72.

4. The larger ϕ, that is, the larger the differences between the factor level means, the higher the power and hence the smaller the probability of making a Type II error for a given risk α of making a Type I error.

5. Since many single-factor studies are undertaken because of the expectation that the factor level means differ and it is desired to investigate these differences, the α risk used in constructing the decision rule for determining whether or not the factor level means are equal is often set relatively high (e.g., .05 or .10 instead of .01) so as to increase the power of the test.

6. The Pearson-Hartley chart for $\nu_1 = 1$ is not reproduced in Table A–8 since this case corresponds to the comparison of two population means. As was noted earlier, the F test is the equivalent of the two-sided t test for this case, and the power charts for the two-sided t test presented in Table A–5 can then be used, with noncentrality parameter:

(16.57)
$$\delta = \frac{|\mu_1 - \mu_2|}{\sigma\sqrt{\dfrac{1}{n_1} + \dfrac{1}{n_2}}}$$

and degrees of freedom $n_1 + n_2 - 2$.

16.10 ALTERNATIVE FORMULATION OF MODEL I

Development of alternative model

At times, an alternative formulation of the parameters of the single-factor ANOVA model I in (16.2) is used. With this alternative formulation, the treatment means μ_i are expressed in an equivalent fashion by means of the identity:

(16.58)
$$\mu_i \equiv \mu_. + (\mu_i - \mu_.)$$

where $\mu_.$ is a constant that can be defined to fit the purposes of the study. We shall denote the difference $\mu_i - \mu_.$ by τ_i:

$$\tau_i = \mu_i - \mu_. \tag{16.59}$$

so that (16.58) can be expressed in equivalent fashion as:

$$\mu_i \equiv \mu_. + \tau_i \tag{16.60}$$

The difference $\tau_i = \mu_i - \mu_.$ is called the *effect of the ith factor level.*

The ANOVA model I in (16.2) can now be stated as follows:

$$Y_{ij} = \mu_. + \tau_i + \varepsilon_{ij} \tag{16.61}$$

where:

$\mu_.$ is a constant component common to all observations
τ_i is the effect of the ith factor level (a constant for each factor level)
ε_{ij} are independent $N(0, \sigma^2)$
$i = 1, \ldots, r;\ j = 1, \ldots, n_i$

Model (16.61) is a linear model, like the equivalent model (16.2). We shall demonstrate this in the next section.

Definition of $\mu_.$

The splitting up of the factor level mean μ_i into two components, an overall constant $\mu_.$ and a specific factor level effect τ_i, depends on the definition of $\mu_.$ and can be done in many ways. We shall now explain two basic ways to define $\mu_.$.

Unweighted mean. Often, a definition of $\mu_.$ as the unweighted average of all factor level means μ_i is found to be useful:

$$\mu_. = \frac{\sum_{i=1}^{r} \mu_i}{r} \tag{16.62}$$

This definition implies:

$$\sum_{i=1}^{r} \tau_i = 0 \tag{16.63}$$

because by (16.59) we have:

$$\Sigma\tau_i = \Sigma(\mu_i - \mu_.) = \Sigma\mu_i - r\mu_.$$

and:

$$\Sigma\mu_i = r\mu_.$$

by (16.62). Thus, the definition of the overall constant $\mu_.$ in (16.62) implies a restriction on the τ_i, in this case that their sum must be zero.

For our earlier incentive pay example in Figure 16.2, we have $\mu_1 = 70$, $\mu_2 = 58$, $\mu_3 = 90$, and $\mu_4 = 84$. When $\mu_.$ is defined according to (16.62), we obtain:

$$\mu_. = \frac{70 + 58 + 90 + 84}{4} = 75.5$$

Hence:

$$\begin{aligned} \tau_1 &= 70 - 75.5 = -5.5 \\ \tau_2 &= 58 - 75.5 = -17.5 \\ \tau_3 &= 90 - 75.5 = 14.5 \\ \tau_4 &= 84 - 75.5 = 8.5 \end{aligned}$$

The first factor level effect $\tau_1 = -5.5$, for instance, indicates that the mean employee productivity for incentive pay type 1 is 5.5 units less than the average productivity for all four types of incentive pay.

Weighted mean. The constant $\mu_.$ can also be defined as some weighted average of the factor level means μ_i:

(16.64) $$\mu_. = \sum_{i=1}^{r} w_i \mu_i$$

where the w_i are weights defined so that $\Sigma w_i = 1$. The restriction on the τ_i then is:

(16.65) $$\sum_{i=1}^{r} w_i \tau_i = 0$$

This follows in the same fashion as (16.63).

The choice of weights w_i should depend on the meaningfulness of the resulting measures of the factor level effects. We present now two examples where different weightings are appropriate—weighting according to a known measure of importance and weighting according to sample size.

Example 1. A car rental firm wanted to estimate the average fuel consumption (in miles per gallon) for its large fleet of cars which consists of 50 percent compacts, 30 percent sedans, and 20 percent station wagons. Here, a meaningful measure of $\mu_.$ might be in terms of the overall mean fuel consumption:

(16.66) $$\mu_. = .5\mu_1 + .3\mu_2 + .2\mu_3$$

where μ_1, μ_2, and μ_3 are the mean fuel consumptions for the three types of cars in the fleet.

An estimate of $\mu_.$ here is:

(16.67) $$\hat{\mu}_. = .5\bar{Y}_{1.} + .3\bar{Y}_{2.} + .2\bar{Y}_{3.}$$

Example 2. If the car rental firm in Example 1 used sample sizes for the three types of cars in its fleet that are approximately in the same ratios as the numbers of cars of each type in the fleet, use of the proportions n_1/n_T, n_2/n_T, and n_3/n_T, respectively, as weights might be meaningful. The resulting definition of the overall fuel consumption constant $\mu_.$ would then be:

$$\mu_. = \frac{n_1}{n_T}\mu_1 + \frac{n_2}{n_T}\mu_2 + \frac{n_3}{n_T}\mu_3 \tag{16.68}$$

This quantity would be estimated by $\bar{Y}_{..}$:

$$\hat{\mu}_. = \frac{n_1}{n_T}\bar{Y}_{1.} + \frac{n_2}{n_T}\bar{Y}_{2.} + \frac{n_3}{n_T}\bar{Y}_{3.} = \bar{Y}_{..} \tag{16.69}$$

When all sample sizes are equal, $\mu_.$ as defined in (16.68) reduces to the unweighted mean (16.62).

Test for equality of factor level means

Since the alternative model (16.61) is equivalent to the basic analysis of variance model (16.2), the test for equality of factor level means uses the same test statistic F^* in (16.51) as when model (16.2) is employed. The only difference is in the statement of the alternatives. For model (16.2), the alternatives are as specified in (16.50):

$$H_0: \mu_1 = \mu_2 = \cdots = \mu_r$$
$$H_a: \text{not all } \mu_i \text{ are equal}$$

For model (16.61), these same alternatives pertaining to the equality of factor level means are:

$$\begin{aligned} &H_0: \tau_1 = \tau_2 = \cdots = \tau_r = 0 \\ &H_a: \text{not all } \tau_i \text{ equal } 0 \end{aligned} \tag{16.70}$$

The equivalence of the two forms can be readily established. The equality of the factor level means $\mu_1 = \mu_2 = \cdots = \mu_r$ implies that all τ_i are equal. This follows from (16.60) since the constant term $\mu_.$ is common to all factor level effects τ_i. Further, the equality of the factor level means implies that all $\tau_i = 0$, whether the restriction on the τ_i is of the form in (16.63) or (16.65). In either case, the restriction can be satisfied in only one way given the equality of the τ_i, namely, that $\tau_i \equiv 0$. Thus, it is equivalent to state that all factor level means μ_i are equal or that all factor level effects τ_i equal zero.

16.11 REGRESSION APPROACH TO SINGLE-FACTOR ANALYSIS OF VARIANCE

We noted earlier that the single-factor analysis of variance model (16.2) is a linear model, as is also the equivalent alternative model (16.61). Thus, we can obtain the test statistic F^* for testing the equality of the factor level means μ_i by means of the matrix formulation in Section 8.5. More simply, we can obtain the test statistic F^* without matrix manipulations by use of a multiple regression program. We shall now explain the regression approach to single-factor analysis of variance. For this purpose, we shall utilize the alternative version (16.61) of the single-factor ANOVA model:

$$Y_{ij} = \mu_{.} + \tau_i + \varepsilon_{ij}$$

and assume that equal weightings of the factor level means are appropriate for defining the overall constant $\mu_{.}$.

To state model (16.61) as a linear model, the parameters $\mu_{.}, \tau_1, \ldots, \tau_r$ need to be represented in the model. However, constraint (16.63) for the case of equal weightings:

$$\sum_{i=1}^{r} \tau_i = 0$$

implies that:

$$\tau_r = -\tau_1 - \tau_2 - \cdots - \tau_{r-1} \tag{16.71}$$

Thus, we shall need only the parameters $\mu_{.}, \tau_1, \ldots, \tau_{r-1}$ for the linear model since τ_r is a function of $\tau_1, \ldots, \tau_{r-1}$.

To illustrate how a linear model is developed with this approach, consider a single-factor study with $r = 3$ factor levels when $n_1 = n_2 = n_3 = 2$. The **Y**, **X**, $\boldsymbol{\beta}$, and $\boldsymbol{\varepsilon}$ matrices for this case are as follows:

$$\mathbf{Y} = \begin{bmatrix} Y_{11} \\ Y_{12} \\ Y_{21} \\ Y_{22} \\ Y_{31} \\ Y_{32} \end{bmatrix} \quad \mathbf{X} = \begin{bmatrix} 1 & 1 & 0 \\ 1 & 1 & 0 \\ 1 & 0 & 1 \\ 1 & 0 & 1 \\ 1 & -1 & -1 \\ 1 & -1 & -1 \end{bmatrix} \quad \boldsymbol{\beta} = \begin{bmatrix} \mu_{.} \\ \tau_1 \\ \tau_2 \end{bmatrix} \quad \boldsymbol{\varepsilon} = \begin{bmatrix} \varepsilon_{11} \\ \varepsilon_{12} \\ \varepsilon_{21} \\ \varepsilon_{22} \\ \varepsilon_{31} \\ \varepsilon_{32} \end{bmatrix} \tag{16.72}$$

Note that the vector of expected values, $\mathbf{E}(\mathbf{Y}) = \mathbf{X}\boldsymbol{\beta}$, yields the following:

$$\mathbf{E}(\mathbf{Y}) = \begin{bmatrix} E(Y_{11}) \\ E(Y_{12}) \\ E(Y_{21}) \\ E(Y_{22}) \\ E(Y_{31}) \\ E(Y_{32}) \end{bmatrix} = \mathbf{X}\boldsymbol{\beta} = \begin{bmatrix} 1 & 1 & 0 \\ 1 & 1 & 0 \\ 1 & 0 & 1 \\ 1 & 0 & 1 \\ 1 & -1 & -1 \\ 1 & -1 & -1 \end{bmatrix} \begin{bmatrix} \mu_{.} \\ \tau_1 \\ \tau_2 \end{bmatrix} = \begin{bmatrix} \mu_{.} + \tau_1 \\ \mu_{.} + \tau_1 \\ \mu_{.} + \tau_2 \\ \mu_{.} + \tau_2 \\ \mu_{.} - \tau_1 - \tau_2 \\ \mu_{.} - \tau_1 - \tau_2 \end{bmatrix} \tag{16.73}$$

Since $\tau_3 = -\tau_1 - \tau_2$ according to (16.71), we see that $E(Y_{31}) = E(Y_{32}) = \mu_{.} + \tau_3$. Thus, the above **X** matrix and $\boldsymbol{\beta}$ vector representation provides the appropriate expected values:

$$E(Y_{ij}) = \mu_{.} + \tau_i$$

in all cases.

The illustration in (16.72) indicates how we need to define in general the multiple regression model so that it is the equivalent of the single-factor ANOVA model (16.61). We shall let X_{ij1} denote the value of independent variable X_1 for the jth observation from the ith factor level, X_{ij2} the value of independent variable X_2 for this same observation, and so on, using altogether $r - 1$ independent variables in the model. The multiple regression model then is as follows:

(16.74) $$Y_{ij} = \mu_{.} + \tau_1 X_{ij1} + \tau_2 X_{ij2} + \cdots + \tau_{r-1} X_{ij,r-1} + \varepsilon_{ij} \qquad \text{Full model}$$

where:

$$X_{ij1} = \begin{array}{l} 1 \text{ if observation from factor level } 1 \\ -1 \text{ if observation from factor level } r \\ 0 \text{ otherwise} \end{array}$$

$$\vdots \qquad\qquad \vdots$$

$$X_{ij,r-1} = \begin{array}{l} 1 \text{ if observation from factor level } r-1 \\ -1 \text{ if observation from factor level } r \\ 0 \text{ otherwise} \end{array}$$

Note how the ANOVA model parameters play the role of regression function parameters in (16.74); the intercept term is $\mu_{.}$, and the regression coefficients are $\tau_1, \tau_2, \ldots, \tau_{r-1}$. The independent variables all take on the values 1, −1, or 0. (This coding was discussed for a simpler case in Section 10.5.)

To test the equality of the treatment means μ_i by means of the regression approach, we state the alternatives in the equivalent formulation (16.70), noting that τ_r must equal zero when $\tau_1 = \tau_2 = \cdots = \tau_{r-1} = 0$ according to (16.71):

(16.75) $$\begin{array}{l} H_0\colon \tau_1 = \tau_2 = \cdots = \tau_{r-1} = 0 \\ H_a\colon \text{not all } \tau_i \text{ equal zero} \end{array}$$

Note that H_0 states that all regression coefficients in regression model (16.74) are zero. Thus, we employ the usual test statistic (7.30b) for testing whether or not there is a regression relation:

(16.76) $$F^* = \frac{MSR}{MSE}$$

Example

To test the equality of mean sales for the four cereal package designs in the Kenton Food Company example by means of the regression approach, we shall employ the regression model:

(16.77) $$Y_{ij} = \mu_{.} + \tau_1 X_{ij1} + \tau_2 X_{ij2} + \tau_3 X_{ij3} + \varepsilon_{ij}$$

where:

$$X_{ij1} = \begin{array}{l} 1 \text{ if observation from factor level 1} \\ -1 \text{ if observation from factor level 4} \\ 0 \text{ otherwise} \end{array}$$

$$X_{ij2} = \begin{array}{l} 1 \text{ if observation from factor level 2} \\ -1 \text{ if observation from factor level 4} \\ 0 \text{ otherwise} \end{array}$$

$$X_{ij3} = \begin{array}{l} 1 \text{ if observation from factor level 3} \\ -1 \text{ if observation from factor level 4} \\ 0 \text{ otherwise} \end{array}$$

The observations vector $\mathbf{Y}$ and the $\mathbf{X}$ matrix for the data in Table 16.1a are shown in Table 16.5a. For observation Y_{11}, for instance, note that $X_1 = 1$, $X_2 = 0$, and $X_3 = 0$; hence, we obtain from (16.77):

$$E(Y_{11}) = \mu_{.} + \tau_1$$

Similarly, for observation Y_{42} we have $X_1 = -1$, $X_2 = -1$, and $X_3 = -1$; hence:

TABLE 16.5 Regression approach to the analysis of variance—
Kenton Food Company example

(a) Data Matrices for Regression Model (16.77)

$$\mathbf{Y} = \begin{bmatrix} 12 \\ 18 \\ 14 \\ 12 \\ 13 \\ 19 \\ 17 \\ 21 \\ 24 \\ 30 \end{bmatrix} \qquad \mathbf{X} = \begin{bmatrix} & X_1 & X_2 & X_3 \\ 1 & 1 & 0 & 0 \\ 1 & 1 & 0 & 0 \\ 1 & 0 & 1 & 0 \\ 1 & 0 & 1 & 0 \\ 1 & 0 & 1 & 0 \\ 1 & 0 & 0 & 1 \\ 1 & 0 & 0 & 1 \\ 1 & 0 & 0 & 1 \\ 1 & -1 & -1 & -1 \\ 1 & -1 & -1 & -1 \end{bmatrix}$$

(b) Fitted Regression Function

$$\hat{Y} = 18.5 - 3.5X_1 - 5.5X_2 + .5X_3$$

(c) Regression Analysis of Variance Table

Source of Variation	SS	df	MS
Regression	$SSR = 258$	3	$MSR = 86$
Error	$SSE = 46$	6	$MSE = 7.67$
Total	$SSTO = 304$	9	

$$E(Y_{42}) = \mu_. - \tau_1 - \tau_2 - \tau_3 = \mu_. + \tau_4$$

since $\tau_4 = -\tau_1 - \tau_2 - \tau_3$.

Note that we employ the following codings in the independent variables for observations from each of the four factor levels:

	Coding		
Factor Level	X_1	X_2	X_3
1	1	0	0
2	0	1	0
3	0	0	1
4	−1	−1	−1

A computer run of a multiple regression package for the data in Table 16.5a yielded the fitted regression function and analysis of variance table presented in Tables 16.5b and 16.5c. Test statistic (16.76) therefore is:

$$F^* = \frac{MSR}{MSE} = \frac{86}{7.67} = 11.2$$

This is the same test statistic obtained earlier based on the analysis of variance calculations. Indeed, the analysis of variance table in Table 16.5c obtained with the regression approach is the same as the one in Table 16.4 obtained with the analysis of variance approach except that the treatment sum of squares and mean square in Table 16.4 are called the regression sum of squares and mean square in Table 16.5c.

From this point on, the test procedure based on the regression approach parallels the analysis of variance test procedure explained earlier.

Note

In the fitted regression function in Table 16.5b, the intercept term 18.5 is the unweighted average of the factor level sample means $\bar{Y}_{i.}$ because $\mu_.$ was defined as the unweighted average of the factor level means μ_i.

Comments

1. The regression approach has not generally been utilized for ordinary analysis of variance problems. The reason is that the **X** matrix for analysis of variance problems usually is of a very simple structure, as we have seen in Table 16.5a for the Kenton Food Company example. This simple structure permits computational simplifications that are explicitly recognized in the statistical procedures for analysis of variance. We take up the regression approach to analysis of variance here, and in later chapters, for two principal reasons. First, we see that analysis of variance models are encompassed by the general linear statistical model (7.18) considered in Chapter 7. Second, the regression approach is very useful for analyzing some multifactor studies when the structure of the **X** matrix is not simple.

2. When the analysis of variance test is to be conducted by means of the regression approach based on ANOVA model (16.2):

$$Y_{ij} = \mu_i + \varepsilon_{ij}$$

the $\boldsymbol{\beta}$ vector is defined to contain all r treatment means μ_i:

$$\boldsymbol{\beta} = \begin{bmatrix} \mu_1 \\ \cdot \\ \cdot \\ \cdot \\ \mu_r \end{bmatrix} \tag{16.78}$$

and r independent variables $X_1, X_2, \ldots, X_r$ are utilized, each defined as a 0,1 variable as illustrated in Chapter 10:

$$\begin{aligned} X_1 &= \begin{matrix} 1 \text{ if observation from factor level 1} \\ 0 \text{ otherwise} \end{matrix} \\ &\vdots \\ X_r &= \begin{matrix} 1 \text{ if observation from factor level } r \\ 0 \text{ otherwise} \end{matrix} \end{aligned} \tag{16.79}$$

The regression model therefore is:

$$Y_{ij} = \mu_1 X_{ij1} + \mu_2 X_{ij2} + \cdots + \mu_r X_{ijr} + \varepsilon_{ij} \qquad \text{Full model} \tag{16.80}$$

with the μ_i playing the role of regression coefficients.

The $\mathbf{X}$ matrix with this approach contains only 0 and 1 entries. For example, for $r = 3$ factor levels with $n_1 = n_2 = n_3 = 2$ observations, the $\mathbf{X}$ matrix (observations in order Y_{11}, Y_{12}, Y_{21}, etc.) and $\boldsymbol{\beta}$ vector would be as follows:

$$\mathbf{X} = \begin{bmatrix} 1 & 0 & 0 \\ 1 & 0 & 0 \\ 0 & 1 & 0 \\ 0 & 1 & 0 \\ 0 & 0 & 1 \\ 0 & 0 & 1 \end{bmatrix} \qquad \boldsymbol{\beta} = \begin{bmatrix} \mu_1 \\ \mu_2 \\ \mu_3 \end{bmatrix}$$

Note that regression model (16.80) has no intercept term. When a computer regression package is to be employed for this case, it is important that a fit with no intercept term be specified.

The test of whether or not the factor level means are equal, i.e., $\mu_1 = \mu_2 = \cdots = \mu_r$, asks only whether or not the regression coefficients in (16.80) are equal, not whether or not they equal zero. Hence, we need to fit the full model and then the reduced model to conduct this test. The reduced model when H_0: $\mu_1 = \cdots = \mu_r$ holds is:

$$Y_{ij} = \mu_c + \varepsilon_{ij} \qquad \text{Reduced model} \tag{16.81}$$

where μ_c is the common value of all μ_i under H_0. The $\mathbf{X}$ matrix here consists simply of a column of 1's. The $\mathbf{X}$ matrix and $\boldsymbol{\beta}$ vector for the reduced model in our example would be:

$$\mathbf{X} = \begin{bmatrix} 1 \\ 1 \\ 1 \\ 1 \\ 1 \\ 1 \end{bmatrix} \qquad \boldsymbol{\beta} = [\mu_c]$$

After the full and reduced models are fitted and the error sums of squares are obtained for each fit, the usual general linear test statistic (3.68) is then calculated.

PROBLEMS

16.1. Refer to Figure 16.1a. Could you determine the mean sales level when the price level is 68 dollars if you knew the true regression function? Could you make this determination from Figure 16.1b if you only knew the values of the parameters μ_1, μ_2, and μ_3 of the ANOVA model? What distinction between regression models and ANOVA models is demonstrated by your answers?

16.2. A market researcher, having collected data on breakfast cereal expenditures by families with 1, 2, 3, 4, and 5 children living at home, plans to use an ordinary regression model to estimate the mean expenditures at each of these five family size levels. However, she is undecided between fitting a linear or a quadratic regression model, and the data do not give clear evidence in favor of one model or the other. A colleague suggests: "For your purposes you might simply use an ANOVA model." Is this a useful suggestion? Explain.

16.3. Refer to the **SENIC** data set. An analyst has set up four age groupings for variable 3 (age) and wishes to apply ANOVA model (16.2) to determine whether the mean infection risk (variable 4) is the same in the four age groups.

a. What is the dependent variable here?
b. Identify the factor studied. What are the factor levels?
c. Is the factor quantitative or qualitative?
d. Is the factor an experimental or a classification factor?

16.4. Thirty trainees are randomly divided into three groups of 10 and each group is given instruction in the use of a different word-processing system. At the end of the training period, each trainee is given the same "benchmark" word-processing project to complete and the time required for completion is recorded. ANOVA model (16.2) will be used to test whether or not the mean time is the same for the three systems.

a. What is the dependent variable here?
b. Identify the factor studied and the factor levels.
c. Is the factor an experimental or a classification factor? Would your answer differ if each trainee had been allowed to select the word-processing system of his or her choice?

16.5. In a study of intentions to get flu-vaccine shots in an area threatened by an epidemic, 90 persons were classified into three groups of 30 according to the degree of risk of getting flu. Each group was together when the persons were asked about the likelihood of getting the shots, on a probability scale ranging from 0 to 1.0. Unavoidably, most persons overheard the answers of nearby respondents. An analyst wishes to test whether the mean intent scores are the same for the three risk groups. Consider each assumption for ANOVA model (16.2) and explain whether this assumption is likely to hold in the present situation.

16.6. A student asks: ''Why is the F test for equality of factor level means not a two-tail test since any differences among the factor level means can occur in either direction?'' Explain, utilizing the expressions for the expected mean squares in (16.33).

16.7. A company, studying the relation between job satisfaction and length of service of employees, classified employees into three length-of-service groups (less than 5 years, 5–10 years, more than 10 years). Suppose $\mu_1 = 65$, $\mu_2 = 80$, $\mu_3 = 95$, and $\sigma = 3$, and that ANOVA model (16.2) is applicable.

a. Draw a representation of this model in the format of Figure 16.2.

b. Find $E(MSTR)$ and $E(MSE)$ if 25 employees from each group are selected at random for intensive interviewing about job satisfaction. Is $E(MSTR)$ substantially larger than $E(MSE)$ here? What is the implication of this?

16.8. In a study of length of hospital stay (in number of days) of persons in four income groups, the parameters are as follows: $\mu_1 = 5.1$, $\mu_2 = 6.3$, $\mu_3 = 7.9$, $\mu_4 = 9.5$, $\sigma = 2.8$. Assume that ANOVA model (16.2) is appropriate.

a. Draw a representation of this model in the format of Figure 16.2.

b. Suppose 100 persons from each income group are randomly selected for the study. Find $E(MSTR)$ and $E(MSE)$. Is $E(MSTR)$ substantially larger than $E(MSE)$ here? What is the implication of this?

c. If $\mu_2 = 5.6$ and $\mu_3 = 9.0$, everything else remaining the same, what would $E(MSTR)$ be? Why is $E(MSTR)$ larger here than in part (b) even though the range of the factor level means is the same?

16.9. **Productivity improvement.** An economist compiled data on productivity improvements last year for a sample of firms producing electronic computing equipment. The firms were classified according to the level of their average expenditures for research and development in the past three years (low, moderate, high). The results of the study follow (productivity improvement is measured on a scale from 0 to 100). Assume that ANOVA model (16.2) is appropriate.

	j											
i	1	2	3	4	5	6	7	8	9	10	11	12
1 Low	7.6	8.2	6.8	5.8	6.9	6.6	6.3	7.7	6.0			
2 Moderate	6.7	8.1	9.4	8.6	7.8	7.7	8.9	7.9	8.3	8.7	7.1	8.4
3 High	8.5	9.7	10.1	7.8	9.6	9.5						

Summary calculational results are: $SSTR = 20.125$, $SSE = 15.362$.

a. Obtain the fitted values.

b. Obtain the residuals. Do they sum to zero in accord with (16.18)?

c. Obtain the analysis of variance table.

d. Test whether or not the mean productivity improvement differs according to the level of research and development expenditures. Control the α risk at .05. State the alternatives, decision rule, and conclusion.

e. What is the P-value of the test in part (d)? How does it support the conclusion reached in part (d)?

f. What appears to be the nature of the relationship between research and development expenditures and productivity improvement?

16.10. Questionnaire color. In an experiment to investigate the effect of color of paper (blue, green, orange) on response rates for questionnaires distributed by the "windshield method" in supermarket parking lots, 15 representative supermarket parking lots were chosen in a metropolitan area and each color was assigned at random to five of the lots. The response rates (in percent) follow. Assume that ANOVA model (16.2) is appropriate.

		j				
i		1	2	3	4	5
1	Blue	28	26	31	27	35
2	Green	34	29	25	31	29
3	Orange	31	25	27	29	28

Summary calculational results are: $SSTR = 7.60$, $SSE = 116.40$.

a. Obtain the fitted values.

b. Obtain the residuals.

c. Obtain the analysis of variance table.

d. Conduct a test to determine whether or not the mean response rates for the three colors differ. Use a level of significance of $\alpha = .10$. State the alternatives, decision rule, and conclusion. What is the P-value of the test?

e. When informed of the findings, an executive said: "See? I was right all along. We might as well print the questionnaires on plain white paper, which is cheaper." Does this conclusion follow from the findings of the study? Discuss.

16.11. Rehabilitation therapy. A rehabilitation center researcher was interested in examining the relationship between physical fitness prior to surgery of persons undergoing corrective knee surgery and time required in physical therapy until successful rehabilitation. Patient records in the rehabilitation center were examined, and 24 male subjects ranging in age from 18 to 30 years who had undergone similar corrective knee surgery during the past year were selected for the study. The number of days required for successful completion of physical therapy and the prior physical fitness status (below average, average, above average) for each patient follow.

		j									
i		1	2	3	4	5	6	7	8	9	10
1	Below average	29	42	38	40	43	40	30	42		
2	Average	30	35	39	28	31	31	29	35	29	33
3	Above average	26	32	21	20	23	22				

Assume that ANOVA model (16.2) is appropriate.

a. Obtain the fitted values.
b. Obtain the residuals. Do they sum to zero in accord with (16.18)?
c. Obtain the analysis of variance table.
d. Test whether or not the mean number of days required for successful rehabilitation is the same for the three fitness groups. Control the α risk at .01. State the alternatives, decision rule, and conclusion.
e. Obtain the P-value for the test in part (d). Explain how the same conclusion reached in part (d) can be obtained by knowing the P-value.
f. What appears to be the nature of the relationship between physical fitness status and duration of required physical therapy?

16.12. Cash offers. A consumer organization studied the effect of age of automobile owner on size of cash offer for a used car by utilizing 12 persons in each of three age groups (young, middle, elderly) who acted as the owner of a used car. A medium price, three-year-old car was selected for the experiment, and the "owners" solicited cash offers for this car from 36 dealers selected at random from the dealers in the region. Randomization was used in assigning the dealers to the "owners." The offers (in hundred dollars) follow. Assume that ANOVA model (16.2) is applicable.

		j											
i		1	2	3	4	5	6	7	8	9	10	11	12
1	Young	23	25	21	22	21	22	20	23	19	22	19	21
2	Middle	28	27	27	29	26	29	27	30	28	27	26	29
3	Elderly	23	20	25	21	22	23	21	20	19	20	22	21

a. Obtain the fitted values.
b. Obtain the residuals.
c. Obtain the analysis of variance table.
d. Conduct the F test for equality of factor level means; use $\alpha = .01$. State the alternatives, decision rule, and conclusion. What is the P-value of the test?
e. What appears to be the nature of the relationship between age of owner and mean cash offer?

16.13. Filling machines. A company uses six filling machines of the same make and model to place detergent into cartons that show a label weight of 32 ounces. The production manager has complained that the six machines do not place the same amount of fill into the cartons. A consultant requested that 20 filled cartons be selected randomly from each of the six machines and the content of each carton carefully weighed. The observations (stated for convenience as deviations from 32.00 ounces) follow. Assume that ANOVA model (16.2) is applicable.

	j									
i	1	2	3	4	5	6	7	8	9	10
1	−.14	.20	.07	.18	.38	.10	−.04	−.27	.27	−.21
2	.46	.11	.12	.47	.24	.06	−.12	.33	.06	−.03
3	.21	.78	.32	.45	.22	.35	.54	.24	.47	.62
4	.49	.58	.52	.29	.27	.55	.40	.14	.48	.34
5	−.19	.27	.06	.11	.23	.15	.01	.22	.29	.14
6	.05	−.05	.28	.47	.12	.27	.08	.17	.43	−.07

i	j = 11	12	13	14	15	16	17	18	19	20
1	.39	−.07	−.02	.28	.09	.13	.26	.07	−.01	−.19
2	.05	.53	.42	.29	.36	.04	.17	.02	.11	.12
3	.47	.55	.59	.71	.45	.48	.44	.50	.20	.61
4	.01	.33	.18	.13	.48	.54	.51	.42	.45	.20
5	.20	.30	−.11	.27	−.20	.24	.20	.14	.35	−.18
6	.20	.01	.10	.16	−.06	.13	.43	.35	−.09	.05

a. Obtain the fitted values.
b. Obtain the residuals. Do they sum to zero in accord with (16.18)?
c. Obtain the analysis of variance table.
d. Test whether or not the mean fill differs among the six machines; control the α risk at .05. State the alternatives, decision rule, and conclusion. Does your conclusion support the production manager's complaint?
e. What is the P-value of the test in part (d)? Is this value consistent with your conclusion in part (d)? Explain.
f. Does the variation between the mean fills for the six machines appear to be large relative to the variability in fills between cartons for any given machine? Explain.

16.14. **Premium distribution.** A soft-drink manufacturer uses five agents (1, 2, 3, 4, 5) to handle premium distributions for its various products. The marketing director desired to study the timeliness with which the premiums are distributed. Twenty transactions for each agent were selected at random, and the time lapse (in days) for handling each transaction was determined. The results follow. Assume that ANOVA model (16.2) is appropriate.

i	j = 1	2	3	4	5	6	7	8	9	10	11	12	13	14	15	16	17	18	19	20
1	24	24	29	20	21	25	28	27	23	21	24	26	23	24	28	23	23	27	26	25
2	18	20	20	24	22	29	23	24	28	19	24	25	21	20	24	22	19	26	22	21
3	10	11	8	12	12	10	14	9	8	11	16	12	18	14	13	11	14	9	11	12
4	15	13	18	16	12	19	10	18	11	17	15	12	13	13	14	17	16	17	14	16
5	33	22	28	35	29	28	30	31	29	28	33	30	32	33	29	35	32	26	30	29

a. Obtain the fitted values.
b. Obtain the residuals. Do they sum to zero in accord with (16.18)?
c. Obtain the analysis of variance table.
d. Test whether or not the mean time lapse differs for the five agents; use $\alpha = .10$. State the alternatives, decision rule, and conclusion.
e. What is the P-value of the test in part (d)? Explain how the same conclusion as in part (d) can be reached by knowing the P-value.
f. Based on the treatment sample means, does there appear to be much variation in the mean time lapse for the five agents? Is this variation necessarily the result of differences in the efficiency of operations of the five agents? Discuss.

16.15. Refer to Example 1 on page 548. Find the power of the test if $\alpha = .01$, everything else remaining unchanged. How does this power compare with that in Example 1?

16.16. Refer to Example 2 on page 548. The analyst is also interested in the power of the test when $\mu_1 = \mu_2 = 13$ and $\mu_3 = \mu_4 = 18$. Assume that $\sigma = 2.5$.

a. Obtain the power of the test if $\alpha = .05$.

b. What would be the power of the test if $\alpha = .01$?

16.17. Refer to **Productivity improvement** Problem 16.9. Obtain the power of the test in Problem 16.9d if $\mu_1 = 6.5$, $\mu_2 = 9.5$, and $\mu_3 = 11.5$. Assume that $\sigma = .9$.

16.18. Refer to **Rehabilitation therapy** Problem 16.11. Obtain the power of the test in Problem 16.11d if $\mu_1 = 42$, $\mu_2 = 35$, and $\mu_3 = 24$. Assume that $\sigma = 4.5$.

16.19. Refer to **Cash offers** Problem 16.12. Obtain the power of the test in Problem 16.12d if the mean cash offers are $\mu_1 = 22$, $\mu_2 = 28$, and $\mu_3 = 22$. Assume that $\sigma = 1.6$.

16.20. Refer to Problem 16.7. What are the values of τ_1, τ_2, and τ_3 if the ANOVA model is expressed in the alternative formulation (16.61) and $\mu_.$ is defined by (16.62)?

16.21. Refer to **Premium distribution** Problem 16.14. Suppose that 25 percent of all premium distributions are handled by agent 1, 20 percent by agent 2, 20 percent by agent 3, 20 percent by agent 4, and 15 percent by agent 5.

a. Obtain a point estimate of $\mu_.$ when the ANOVA model is expressed in the alternative formulation (16.61) and $\mu_.$ is defined by (16.64), with the weights being the proportions of premium distribution handled by each agent.

b. State the alternatives for the test of equality of factor level means in terms of model (16.61) for the present case. Would this statement be affected if $\mu_.$ were defined according to (16.62)? Explain.

16.22. Refer to **Productivity improvement** Problem 16.9. Regression model (16.74) is to be employed for testing the equality of the factor level means.

a. Set up the **Y**, **X**, and **β** matrices.

b. Obtain **Xβ**. Develop equivalent expressions of the elements of this vector in terms of μ_i.

c. Obtain the fitted regression function. What is estimated by the intercept term?

d. Obtain the regression analysis of variance table.

e. Conduct the test for equality of factor level means; use $\alpha = .05$. State the alternatives, decision rule, and conclusion.

16.23. Refer to **Questionnaire color** Problem 16.10. Regression model (16.74) is to be employed for testing the equality of the factor level means.

a. Set up the **Y**, **X**, and **β** matrices.

b. Obtain **Xβ**. Develop equivalent expressions of the elements of this vector in terms of μ_i.

c. Obtain the fitted regression function. What is estimated by the intercept term?

d. Obtain the regression analysis of variance table.

e. Conduct the test for equality of factor level means; use $\alpha = .10$. State the alternatives, decision rule, and conclusion.

16.24. Refer to **Cash offers** Problem 16.12.

a. Fit regression model (16.74) to the data. What is estimated by the intercept term?

b. Obtain the regression analysis of variance table and test whether or not the factor level means are equal; use $\alpha = .01$. State the alternatives, decision rule, and conclusion.

16.25. Refer to **Rehabilitation therapy** Problem 16.11.

a. Fit the full regression model (16.80) to the data. Would a fitted regression model containing an intercept term be proper here?

b. Fit the reduced model (16.81) to the data.

c. Use test statistic (3.68) for testing the equality of the factor level means; employ a level of significance of $\alpha = .01$.

EXERCISES

16.26. (Calculus needed.) State the likelihood function for ANOVA model (16.2) when $r = 3$ and $n_i \equiv 2$. Find the maximum likelihood estimators. Are they the same as the least squares estimators (16.14)?

16.27. Show that the result in (16.29b) is the algebraic equivalent of (16.24a).

16.28. Show that when test statistic t^* in Table 1.2a is squared, it is equivalent to the F^* test statistic (16.51) for $r = 2$.

16.29. Derive the restriction in (16.65) when the constant $\mu_.$ is defined according to (16.64).

16.30. a. Obtain the least squares estimators of the regression coefficients in the full regression model (16.80). What is $SSE(F)$ here?

b. Obtain the least squares estimator of μ_c in the reduced regression model (16.81). What is $SSE(R)$ here?

16.31. Consider the illustration in (16.6)–(16.8) demonstrating that ANOVA model (16.2) is a linear model. To test the equality of the three treatment means, we could therefore employ the general matrix approach and use (8.45) for constructing the test statistic. Show that if:

$$\mathbf{C} = \begin{bmatrix} 1 & -1 & 0 \\ 1 & 0 & -1 \end{bmatrix} \qquad \mathbf{h} = \begin{bmatrix} 0 \\ 0 \end{bmatrix}$$

formula (8.45) is equivalent to $SSTR$.

PROJECTS

16.32. Refer to the **SENIC** data set. Test whether or not the mean infection risk (variable 4) is the same in the four geographic regions (variable 9); use a level of significance of $\alpha = .05$. Assume that ANOVA model (16.2) is applicable. State the alternatives, decision rule, and conclusion.

16.33. Refer to the **SENIC** data set. The effect of average age of patient (variable 3) on mean infection risk (variable 4) is to be studied. For purposes of this ANOVA study, average age is to be classified into four categories: Under 50.0, 50.0–

54.9, 55.0–59.9, 60.0 and over. Assume that ANOVA model (16.2) is applicable. Test whether or not the mean infection risk differs for the four age groups. Control the α risk at .10. State the alternatives, decision rule, and conclusion.

16.34. Refer to the **SMSA** data set. The effect of geographic region (variable 12) on the crime rate (variable 11 ÷ variable 3) is to be studied. Assume that ANOVA model (16.2) is applicable. Test whether or not the mean crime rates for the four geographic regions differ; use $\alpha = .05$. State the alternatives, decision rule, and conclusion.

16.35. Consider a test involving H_0: $\mu_1 = \mu_2 = \mu_3$. Five observations are to be taken for each factor level, and a level of significance of $\alpha = .05$ is to be employed in the test.

a. Generate five random normal observations when $\mu_1 = 100$ and $\sigma = 12$ to represent the observations for treatment 1. Repeat this for the other two treatments when $\mu_2 = \mu_3 = 100$ and $\sigma = 12$. Finally, calculate the F^* test statistic (16.51).

b. Repeat part (a) 100 times.

c. Calculate the mean of the 100 F^* statistics.

d. What proportion of the F^* statistics lead to conclusion H_0? Is this consistent with theoretical expectations?

e. Repeat parts (a) and (b) when $\mu_1 = 80$, $\mu_2 = 60$, $\mu_3 = 160$, and $\sigma = 12$. Calculate the mean of the 100 F^* statistics. How does this mean compare with the mean obtained in part (c) when $\mu_1 = \mu_2 = \mu_3 = 100$? Is this result consistent with use of decision rule (16.52)?

f. What proportion of the 100 test statistics obtained in part (e) lead to conclusion H_a? Does it appear that the test has satisfactory power when $\mu_1 = 80$, $\mu_2 = 60$, and $\mu_3 = 160$?

CITED REFERENCE

16.1 Dixon, W. J., and M. B. Brown, eds. *BMDP-81, Biomedical Computer Programs, P-Series*. Berkeley, Calif.: University of California Press, 1981.

17

Analysis of factor effects

The F test for determining whether or not the factor level means μ_i differ, discussed in the previous chapter, is a preliminary test to establish whether detailed analysis of the factor level effects is warranted. If the F test leads to the conclusion that the factor level means μ_i are equal, the implication is that there is no relation between the factor and the dependent variable. On the other hand, if the F test leads to the conclusion that the factor level means μ_i differ, the implication is that there is a relation between the factor and the dependent variable. In that case, a thorough analysis of the nature of the factor level effects is usually undertaken. This is done in two principal ways:

1. A direct analysis of the factor level effects of interest using estimation techniques.
2. Statistical tests in regard to the factor level effects of interest.

We shall illustrate each of these two avenues of approach in turn, but will concentrate on the estimation approach in view of its greater usefulness. Throughout this chapter, we assume the fixed effects analysis of variance model (16.2):

$$Y_{ij} = \mu_i + \varepsilon_{ij} \tag{17.1}$$

where:

μ_i are parameters
ε_{ij} are independent $N(0, \sigma^2)$

17.1 ESTIMATION OF FACTOR LEVEL EFFECTS

If the F test indicates that the factor level means μ_i differ, one generally proceeds directly to estimate the factor level effects of interest. Estimates of factor level effects usually employed include:

1. Estimation of a factor level mean μ_i.
2. Estimation of the difference between two factor level means.
3. Estimation of a contrast among factor level means.
4. Estimation of a linear combination of factor level means.

We shall discuss each of these four types of estimation problems in turn.

Estimation of factor level mean

An unbiased point estimator of the factor level mean μ_i was obtained in (16.14):

(17.2) $$\hat{\mu}_i = \bar{Y}_{i.}$$

This estimator has mean and variance:

(17.3a) $$E(\bar{Y}_{i.}) = \mu_i$$

(17.3b) $$\sigma^2(\bar{Y}_{i.}) = \frac{\sigma^2}{n_i}$$

The latter result follows because (16.39) indicates that $\bar{Y}_{i.} = \mu_i + \bar{\varepsilon}_{i.}$, the sum of a constant plus a mean of n_i independent ε_{ij} terms, each of which has variance σ^2. Further, $\bar{Y}_{i.}$ is normally distributed because the error terms ε_{ij} are normally distributed.

The estimated variance of $\bar{Y}_{i.}$ is denoted by $s^2(\bar{Y}_{i.})$ and is obtained as usual by replacing σ^2 in (17.3b) by the unbiased point estimator MSE:

(17.4) $$s^2(\bar{Y}_{i.}) = \frac{MSE}{n_i}$$

The estimated standard deviation $s(\bar{Y}_{i.})$ is the positive square root of (17.4).

It can be shown that:

(17.5) $$\frac{\bar{Y}_{i.} - \mu_i}{s(\bar{Y}_{i.})} \text{ is distributed as } t(n_T - r) \text{ for model (17.1)}$$

where the degrees of freedom are those associated with MSE. The result (17.5) follows from the definition of t in (1.39) since: (1) $\bar{Y}_{i.}$ is normally distributed and (2) MSE/σ^2 is distributed independently of $\bar{Y}_{i.}$ as $\chi^2(n_T - r)/(n_T - r)$ according to the following theorem:

(17.6) For model (17.1), SSE/σ^2 is distributed as χ^2 with $n_T - r$ degrees of freedom, and is independent of $\bar{Y}_{1.}, \ldots, \bar{Y}_{r.}$.

It follows directly from (17.5) that the confidence limits for μ_i with confidence coefficient $1 - \alpha$ are:

$$\bar{Y}_{i.} \pm t(1 - \alpha/2; n_T - r)s(\bar{Y}_{i.}) \tag{17.7}$$

Example. In the Kenton Food Company illustration, the sales manager wished to estimate mean sales for package design 1 with a 95 percent confidence coefficient.

Using the results from Table 17.1, which summarizes the calculations from the previous chapter, we have:

$$\bar{Y}_{1.} = 15 \qquad n_1 = 2 \qquad MSE = 7.67$$

TABLE 17.1 Results for Kenton Food Company example obtained in Chapter 16

	Package Design (i) 1	2	3	4	Total
n_i	2	3	3	2	10
$Y_{i.}$	30	39	57	54	180
$\bar{Y}_{i.}$	15	13	19	27	18

Source of Variation	SS	df	MS
Between designs	258	3	86
Error	46	6	7.67
Total	304	9	

Package Design	Characteristics
1	3 colors, with cartoons
2	3 colors, without cartoons
3	5 colors, with cartoons
4	5 colors, without cartoons

We require $t(.975; 6)$. From Table A–2 in the Appendix, we obtain $t(.975; 6) = 2.447$. Finally, we need $s(\bar{Y}_{1.})$. We have:

$$s^2(\bar{Y}_{1.}) = \frac{MSE}{n_1} = \frac{7.67}{2} = 3.835$$

so that:

$$s(\bar{Y}_{1.}) = \sqrt{3.835} = 1.958$$

Hence, we obtain the confidence interval:

$$15 - 2.447(1.958) \le \mu_1 \le 15 + 2.447(1.958)$$
$$10.2 \le \mu_1 \le 19.8$$

Thus, we estimate with confidence coefficient .95 that the mean sales per store for package design 1 are between 10.2 and 19.8 cases.

Estimation of difference between two factor level means

Frequently two treatments or factor levels are to be compared by estimating the difference D between the two factor level means, say, μ_i and $\mu_{i'}$:

$$D = \mu_i - \mu_{i'} \tag{17.8}$$

Such a difference between two factor level means will be called a *pairwise comparison*. A point estimator of (17.8), denoted by $\hat{D}$, is:

$$\hat{D} = \bar{Y}_{i.} - \bar{Y}_{i'.} \tag{17.9}$$

This point estimator is unbiased:

$$E(\hat{D}) = \mu_i - \mu_{i'} \tag{17.10}$$

Since $\bar{Y}_{i.}$ and $\bar{Y}_{i'.}$ are independent, the variance of $\hat{D}$ follows from (1.26b):

$$\sigma^2(\hat{D}) = \sigma^2(\bar{Y}_{i.}) + \sigma^2(\bar{Y}_{i'.}) = \sigma^2\left(\frac{1}{n_i} + \frac{1}{n_{i'}}\right) \tag{17.11}$$

The estimated variance of $\hat{D}$, denoted by $s^2(\hat{D})$, is given by:

$$s^2(\hat{D}) = MSE\left(\frac{1}{n_i} + \frac{1}{n_{i'}}\right) \tag{17.12}$$

Finally, $\hat{D}$ is normally distributed by (1.35) because $\hat{D}$ is a linear combination of independent normal variables.

It follows from these characteristics, theorem (17.6), and the definition of t in (1.39) that:

$$\frac{\hat{D} - D}{s(\hat{D})} \text{ is distributed as } t(n_T - r) \text{ for model (17.1)} \tag{17.13}$$

Hence, the $1 - \alpha$ confidence limits for D are:

$$\hat{D} \pm t(1 - \alpha/2; n_T - r)s(\hat{D}) \tag{17.14}$$

Example. For our Kenton Food Company example of the previous chapter, package designs 1 and 2 used 3-color printing and designs 3 and 4 used 5-color printing, as shown in Table 17.1. We wish to estimate the difference in mean sales for 5-color designs 3 and 4 using a 95 percent confidence interval. That is, we wish to estimate $D = \mu_3 - \mu_4$. From Table 17.1, we have:

$$\bar{Y}_{3.} = 19 \qquad n_3 = 3 \qquad MSE = 7.67$$
$$\bar{Y}_{4.} = 27 \qquad n_4 = 2$$

Hence:

$$\hat{D} = \bar{Y}_{3.} - \bar{Y}_{4.} = 19 - 27 = -8$$

The estimated variance of $\hat{D}$ is:

$$s^2(\hat{D}) = MSE\left(\frac{1}{n_3} + \frac{1}{n_4}\right) = 7.67\left(\frac{1}{3} + \frac{1}{2}\right) = 6.392$$

so that the estimated standard deviation of $\hat{D}$ is:

$$s(\hat{D}) = \sqrt{6.392} = 2.528$$

We require $t(.975; 6) = 2.447$. The desired confidence interval therefore is:

$$-8 - 2.447(2.528) \leq \mu_3 - \mu_4 \leq -8 + 2.447(2.528)$$
$$-14.2 \leq \mu_3 - \mu_4 \leq -1.8$$

Thus, we estimate with confidence coefficient .95 that the mean sales for package design 3 fall short of those for package design 4 by somewhere between 1.8 and 14.2 cases per store.

Estimation of contrast

A *contrast* is a comparison involving two or more factor level means and includes the previous case of a pairwise difference between two factor level means in (17.8). A contrast will be denoted by L, and is defined as a linear combination of the factor level means μ_i where the coefficients c_i sum to zero:

$$L = \sum_{i=1}^{r} c_i \mu_i \tag{17.15}$$

where:

$$\sum_{i=1}^{r} c_i = 0$$

Illustrations of contrasts. We illustrate some contrasts by reference to our Kenton Food Company example. Recall that package designs 1 and 2 used 3-color printing and designs 3 and 4 used 5-color printing. In addition, as shown in Table 17.1, package designs 1 and 3 utilized cartoons while no cartoons were utilized in designs 2 and 4.

1. Comparison of the mean sales for the two 3-color designs:

   $$L = \mu_1 - \mu_2$$

   Here, $c_1 = 1$, $c_2 = -1$, $c_3 = 0$, $c_4 = 0$, and $\Sigma c_i = 0$.
2. Comparison of the mean sales for the 3-color and 5-color designs:

$$L = \frac{\mu_1 + \mu_2}{2} - \frac{\mu_3 + \mu_4}{2}$$

Here, $c_1 = 1/2$, $c_2 = 1/2$, $c_3 = -1/2$, $c_4 = -1/2$, and $\Sigma c_i = 0$.

3. Comparison of the mean sales for designs with and without cartoons:

$$L = \frac{\mu_1 + \mu_3}{2} - \frac{\mu_2 + \mu_4}{2}$$

Here, $c_1 = 1/2$, $c_2 = -1/2$, $c_3 = 1/2$, $c_4 = -1/2$, and $\Sigma c_i = 0$.

Note that the first contrast is simply a pairwise comparison. In the second and third contrasts, we compare averages of several factor level means. The averages used here are unweighted averages of the means μ_i; these are ordinarily the averages of interest. In special cases one might be interested in weighted averages of the μ_i to describe the mean response for a group of several factor levels. For example, if both 3-color and 5-color designs were to be employed, with 3-color printing used three times as often as 5-color printing, the comparison of the effect of cartoons versus no cartoons might be based on the contrast:

$$L = \frac{3\mu_1 + \mu_3}{4} - \frac{3\mu_2 + \mu_4}{4}$$

Here, $c_1 = 3/4$, $c_2 = -3/4$, $c_3 = 1/4$, $c_4 = -1/4$, and $\Sigma c_i = 0$.

Confidence interval for L. An unbiased estimator of L is:

(17.16) $$\hat{L} = \sum_{i=1}^{r} c_i \bar{Y}_{i.}$$

Since the $\bar{Y}_{i.}$ are independent, the variance of $\hat{L}$ according to (1.26) is:

(17.17) $$\sigma^2(\hat{L}) = \sum_{i=1}^{r} c_i^2 \sigma^2(\bar{Y}_{i.}) = \sigma^2 \sum_{i=1}^{r} \frac{c_i^2}{n_i}$$

An unbiased estimator of this variance is:

(17.18) $$s^2(\hat{L}) = MSE \sum_{i=1}^{r} \frac{c_i^2}{n_i}$$

$\hat{L}$ is normally distributed by (1.35) because it is a linear combination of independent normal random variables. It can be shown by theorem (17.6), the characteristics of $\hat{L}$ just mentioned, and the definition of t that:

(17.19) $$\frac{\hat{L} - L}{s(\hat{L})} \text{ is distributed as } t(n_T - r) \text{ for model (17.1)}$$

Consequently, the $1 - \alpha$ confidence limits for L are:

$$\hat{L} \pm t(1 - \alpha/2; n_T - r)s(\hat{L}) \tag{17.20}$$

Example. In the Kenton Food Company illustration, the mean sales for the 3-color designs are to be compared to the mean sales for the 5-color designs. Let us estimate this effect. We wish to estimate:

$$L = \frac{\mu_1 + \mu_2}{2} - \frac{\mu_3 + \mu_4}{2}$$

The point estimate is (see data in Table 17.1):

$$\hat{L} = \frac{\bar{Y}_{1.} + \bar{Y}_{2.}}{2} - \frac{\bar{Y}_{3.} + \bar{Y}_{4.}}{2} = \frac{15 + 13}{2} - \frac{19 + 27}{2} = -9$$

Since $c_1 = 1/2$, $c_2 = 1/2$, $c_3 = -1/2$, and $c_4 = -1/2$, we obtain:

$$\Sigma\frac{c_i^2}{n_i} = \frac{(1/2)^2}{2} + \frac{(1/2)^2}{3} + \frac{(-1/2)^2}{3} + \frac{(-1/2)^2}{2} = \frac{5}{12} = .4167$$

and:

$$s^2(\hat{L}) = MSE\Sigma\frac{c_i^2}{n_i} = 7.67(.4167) = 3.196$$

or:

$$s(\hat{L}) = 1.79$$

For a confidence coefficient of 95 percent, we require $t(.975; 6) = 2.447$, and the confidence interval for L is:

$$-9 - 2.447(1.79) \leq L \leq -9 + 2.447(1.79)$$
$$-13.4 \leq L \leq -4.6$$

Therefore, we conclude with confidence coefficient .95 that mean sales for the 3-color designs fall below those for the 5-color designs by somewhere between 4.6 and 13.4 cases per store.

Note

Theorem (17.19) enables us to test any hypothesis concerning a contrast L by means of a t test. Such tests are called single degree of freedom tests. These are discussed in Section 17.5.

Estimation of linear combination

Occasionally, we are interested in a linear combination of the factor level means that is not a contrast. For example, suppose that the Kenton Food Company will use all four package designs, one in each of its four major marketing regions, and that these marketing regions account for 35, 28, 12, and 25 percent of sales, respectively. In that case, there might be interest in the overall mean sales per store for all regions:

$$L = .35\mu_1 + .28\mu_2 + .12\mu_3 + .25\mu_4$$

Note that this linear combination is of the form $L = \Sigma c_i \mu_i$ but that the coefficients c_i do not sum to zero, as they must for a contrast.

We define a *linear combination of the factor level means* μ_i as:

$$L = \sum_{i=1}^{r} c_i \mu_i \tag{17.21}$$

with no restrictions on the coefficients c_i.

Note that for our example of a linear combination, the coefficients are $c_1 = .35$, $c_2 = .28$, $c_3 = .12$, and $c_4 = .25$, which here sum to 1.0.

Confidence limits for a linear combination L are obtained in exactly the same way as those for a contrast by means of (17.20), using the point estimator (17.16) and the estimated variance (17.18).

Need for multiple comparison procedures

The procedures for estimating the factor level effects discussed up to this point have two important limitations:

1. The confidence coefficient $1 - \alpha$ applies only to a particular estimate, not to a series of estimates.
2. The confidence coefficient $1 - \alpha$ is appropriate only if the estimate was not suggested by the data.

The first limitation is familiar from regression analysis. It is particularly serious for analysis of variance models because frequently many different comparisons are of interest here, and one needs to piece the different findings together. Consider the very simple case where three different advertisements are being compared for their effectiveness in stimulating sales. The following estimates of their comparative effectiveness have been obtained, each with a 95 percent statement confidence coefficient:

$$\begin{aligned} 59 \le \mu_2 - \mu_1 &\le 62 \\ -2 \le \mu_3 - \mu_1 &\le \ \ 3 \\ 58 \le \mu_2 - \mu_3 &\le 64 \end{aligned}$$

It would be natural here to piece the different comparisons together and conclude that advertisement 2 leads to highest mean sales, while advertisements 1 and 3 are substantially less effective and do not differ much among themselves. One would therefore like a family confidence coefficient for this family of statements, to provide known assurance that all statements in the family are correct.

The second limitation of the procedures for estimating factor level effects, namely, that the estimate must not be suggested by the data, is an important one in exploratory investigations where many new questions may be suggested once the data are being analyzed. Sometimes, the process of studying effects suggested by the data is called *data snooping*. Analysts often have the tendency to investigate comparisons where the effect appears to be large from the sample data. Now, effects may appear large because in fact they are, or because a random occurrence made them appear large even though they are not. Consequently, investigating only comparisons for which the effect appears to be large implies a smaller confidence coefficient than the specified one if in fact the effect is small or nonexistent. Thus, it can be shown that if six factor levels are being studied and the analyst will always compare the smallest and largest factor level means by using the confidence limits (17.14) with a 95 percent confidence coefficient, the interval estimate will suggest a real effect 40 percent of the time when indeed there is no difference between any of the factor level means (Ref. 17.1). With a greater number of factor levels, the likelihood of an erroneous indication of a real effect would be even greater.

One solution to this problem of making comparisons that are suggested by initial analysis of the data is to use a multiple comparison procedure where the family of statements includes all the possible statements one anticipates might be made after the data are examined. For instance, in an investigation where five factor level means are being studied, it is decided in advance that principal interest is in three pairwise comparisons. However, it is also agreed that other pairwise comparisons that will appear interesting should be studied as well. In this case, one can use the family of *all* pairwise comparisons as the basis for obtaining an appropriate family confidence coefficient for the comparisons suggested by the data.

In the next three sections, we shall discuss three multiple comparison procedures for analysis of variance models that permit the family confidence coefficient to be controlled. Two of these procedures allow data snooping to be undertaken naturally without affecting the confidence coefficient. Of the three methods, the Scheffé and Bonferroni methods have been encountered before. The third, the Tukey method, is new and will be discussed first.

17.2 TUKEY METHOD OF MULTIPLE COMPARISONS

The Tukey method of multiple comparisons that we will consider here applies when:

> The family of interest is the set of all pairwise comparisons of factor level means; in other words, the family consists of estimates of all pairs $\mu_i - \mu_{i'}$.

The Tukey method is exact when all sample sizes are equal, and it is a conservative method when the sample sizes are unequal with family confidence coefficient of at least $1 - \alpha$.

Studentized range distribution

The Tukey method utilizes the *studentized range distribution*. Suppose that we have r independent observations $Y_1, \ldots, Y_r$ from a normal distribution with mean μ and variance σ^2. Let w be the range for this set of observations; thus:

$$w = \max(Y_i) - \min(Y_i) \tag{17.22}$$

Suppose further that we have an estimate s^2 of the variance σ^2 which is based on ν degrees of freedom and is independent of the Y_i. Then, the ratio w/s is called the *studentized range*. It is denoted by:

$$q(r, \nu) = \frac{w}{s} \tag{17.23}$$

where the arguments in parentheses remind us that the distribution of q depends on r and ν. The distribution of q has been tabulated, and selected percentiles are presented in Table A–9.

This table is simple to use. Suppose that $r = 5$ and $\nu = 10$. The 95th percentile is then $q(.95; 5, 10) = 4.65$, which means:

$$P\left[\frac{w}{s} = q(5, 10) \leq 4.65\right] = .95$$

Thus, with five normal Y observations, the probability is .95 that their range is not more than 4.65 times as great as an independent sample standard deviation based on 10 degrees of freedom.

Multiple comparison confidence intervals

The Tukey multiple comparison confidence limits for all pairwise comparisons $\mu_i - \mu_{i'}$ with family confidence coefficient of at least $1 - \alpha$ are as follows:

$$\hat{D} \pm Ts(\hat{D}) \tag{17.24}$$

where:

$$\hat{D} = \bar{Y}_{i.} - \bar{Y}_{i'.} \tag{17.24a}$$

$$s^2(\hat{D}) = s^2(\bar{Y}_{i.}) + s^2(\bar{Y}_{i'.}) = MSE\left(\frac{1}{n_i} + \frac{1}{n_{i'}}\right) \tag{17.24b}$$

$$T = \frac{1}{\sqrt{2}}q(1 - \alpha; r, n_T - r) \tag{17.24c}$$

Note that the point estimator $\hat{D}$ in (17.24a) and the estimated variance in (17.24b) are the same as those in (17.9) and (17.12) for a single pairwise comparison. Thus, the only difference between the confidence limits (17.24) for simultaneous comparisons and those in (17.14) for a single comparison is the multiple of the estimated standard deviation.

The family confidence coefficient $1 - \alpha$ pertaining to the multiple pairwise comparisons refers to the proportion of correct families of pairwise comparisons when repeated sets of samples are selected and all pairwise confidence intervals are calculated each time. A family of pairwise comparisons is considered to be correct if every pairwise comparison in the family is correct. Thus, when the family confidence coefficient is $1 - \alpha$, all pairwise comparisons in the family will be correct in $(1 - \alpha)100$ percent of the families.

Example 1—Equal sample sizes

In a study of the effectiveness of different rust inhibitors, four brands (A, B, C, D) were tested, each on a different set of five units. The basic results were as follows:

i	*Inhibitor*	n_i	$\bar{Y}_{i.}$
1	A	5	43
2	B	5	89
3	C	5	67
4	D	5	40

$MSE = 4.50$

The higher is the mean, the more effective is the rust inhibitor.

The analysis of variance is shown in Table 17.2. Using a level of significance of .05 for testing whether or not the four rust inhibitors differ in effectiveness, we require $F(.95; 3, 16) = 3.24$. Using the mean squares from Table 17.2, the test statistic is:

$$F^* = \frac{MSTR}{MSE} = \frac{2{,}631.25}{4.50} = 584.72$$

TABLE 17.2 ANOVA table for rust inhibitor example

Source of Variation	*SS*	*df*	*MS*
Between rust inhibitors	7,893.75	3	2,631.25
Error	72.00	16	4.50
Total	7,965.75	19	

Since $F^* = 584.72 > 3.24$, we conclude that the four rust inhibitors differ in effectiveness.

To examine the nature of the differences, it was desired to estimate all pairwise comparisons by means of the Tukey procedure, using a family confidence coefficient of 95 percent. Since $r = 4$ and $n_T - r = 16$, the required percentile of the studentized range distribution is $q(.95; 4, 16)$. From Table A–9, we find $q(.95; 4, 16) = 4.05$. Hence, by (17.24c), we obtain:

$$T = \frac{1}{\sqrt{2}}(4.05) = 2.86$$

Further, we need $s(\hat{D})$. Using (17.24b), we find for any pairwise comparison since equal sample sizes were employed:

$$s^2(\hat{D}) = MSE\left(\frac{1}{n_i} + \frac{1}{n_{i'}}\right) = 4.50\left(\frac{1}{5} + \frac{1}{5}\right) = 1.80$$

so that:

$$s(\hat{D}) = 1.34$$

Hence, we obtain for each pairwise comparison:

$$Ts(\hat{D}) = 2.86(1.34) = 3.8$$

The pairwise confidence intervals with 95 percent family confidence coefficient therefore are:

$$42.2 = (89 - 43) - 3.8 \le \mu_2 - \mu_1 \le (89 - 43) + 3.8 = 49.8$$
$$20.2 = (67 - 43) - 3.8 \le \mu_3 - \mu_1 \le (67 - 43) + 3.8 = 27.8$$
$$-.8 = (43 - 40) - 3.8 \le \mu_1 - \mu_4 \le (43 - 40) + 3.8 = 6.8$$
$$18.2 = (89 - 67) - 3.8 \le \mu_2 - \mu_3 \le (89 - 67) + 3.8 = 25.8$$
$$45.2 = (89 - 40) - 3.8 \le \mu_2 - \mu_4 \le (89 - 40) + 3.8 = 52.8$$
$$23.2 = (67 - 40) - 3.8 \le \mu_3 - \mu_4 \le (67 - 40) + 3.8 = 30.8$$

95% SURE THAT ALL CORRECT AT THE SAME TIME

To study the composite information, we shall first order the rust inhibitors from poorest to best on a scale according to the mean response $\bar{Y}_{i.}$:

Brand	D A	C	B

The pairwise comparisons indicate that all but one of the differences (D and A) are statistically significant (confidence interval does not cover zero). We shall show this as follows:

Brand	<u>D A</u>	C	B

The line between D and A indicates that there is no clear evidence whether D or A is the better rust inhibitor. The absence of a line signifies that a difference in performance has been found. Thus, the multiple comparison procedure permits us to infer that with a 95 percent family confidence coefficient for the chain of conclusions, B is the best inhibitor (better by somewhere between 18 and 26 units than the second best), C is second best, and A and D follow substantially behind with little or no difference between them.

Example 2—Unequal sample sizes

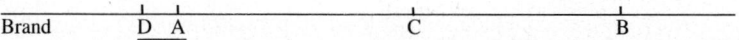

In our Kenton Food Company example, the sales manager was interested in the comparative performance of the four package designs. The analyst developed all pairwise comparisons by means of the Tukey procedure with a family confidence coefficient of at least 90 percent. Since the sample sizes are not equal here, the estimated standard deviation $s(\hat{D})$ must be recalculated for each pairwise comparison. To compare designs 1 and 2, for instance, we obtain:

$$\hat{D} = \bar{Y}_{1.} - \bar{Y}_{2.} = 15 - 13 = 2$$

$$s^2(\hat{D}) = MSE\left(\frac{1}{n_1} + \frac{1}{n_2}\right) = 7.67\left(\frac{1}{2} + \frac{1}{3}\right) = 6.39$$

$$s(\hat{D}) = 2.53$$

For a 90 percent family confidence coefficient, we require $q(.90; 4, 6) = 4.07$ so that we obtain:

$$T = \frac{1}{\sqrt{2}}(4.07) = 2.88$$

Hence, the confidence limits are $2 \pm 2.88(2.53)$ and the confidence interval for $\mu_1 - \mu_2$ is:

$$-5.3 \le \mu_1 - \mu_2 \le 9.3$$

In the same way, we obtain the other five confidence intervals:

$$-3.3 = (19 - 15) - 2.88(2.53) \le \mu_3 - \mu_1 \le (19 - 15) + 2.88(2.53) = 11.3$$
$$4.0 = (27 - 15) - 2.88(2.77) \le \mu_4 - \mu_1 \le (27 - 15) + 2.88(2.77) = 20.0$$
$$-.5 = (19 - 13) - 2.88(2.26) \le \mu_3 - \mu_2 \le (19 - 13) + 2.88(2.26) = 12.5$$
$$6.7 = (27 - 13) - 2.88(2.53) \le \mu_4 - \mu_2 \le (27 - 13) + 2.88(2.53) = 21.3$$
$$.7 = (27 - 19) - 2.88(2.53) \le \mu_4 - \mu_3 \le (27 - 19) + 2.88(2.53) = 15.3$$

We summarize the comparative performance graphically, indicating no significant differences by a rule:

| Package Design | 2 | 1 | 3 | | 4 |

Thus, we can conclude with at least 90 percent family confidence coefficient that design 4 is clearly the most effective design. However, the small-scale study does not permit any ordering among the other three designs because each of those pairwise confidence intervals encompasses the possibility of either design having larger mean sales per store.

Comments

1. The Tukey method, as noted earlier, is exact if all factor level sample sizes are the same. Recent research has shown that the Tukey method is a conservative method when the sample sizes are unequal. When used with unequal sample sizes, the method is sometimes called the *Tukey-Kramer method*.

2. If not all pairwise comparisons are of interest, the confidence coefficient for the family of comparisons being considered will be greater than the specification $1 - \alpha$ used

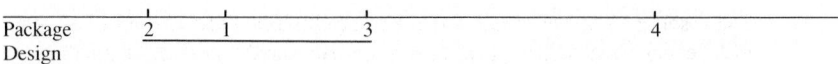

in setting up the Tukey intervals. Thus, the confidence coefficient $1 - \alpha$ with the Tukey method serves as a guaranteed minimum level when not all pairwise comparisons are of interest.

3. The Tukey method can be used for data snooping as long as the effects to be studied on the basis of preliminary data analysis are pairwise comparisons.

4. The Tukey method can be modified to handle general contrasts between factor level means. We do not discuss this modification since the Scheffé method (to be discussed next) is to be preferred for this situation.

5. To derive the Tukey simultaneous confidence intervals for the case when all sample sizes are equal, i.e., $n_i \equiv n$ so that $n_T = rn$, consider the deviations:

$$(\bar{Y}_{1.} - \mu_1), \ldots, (\bar{Y}_{r.} - \mu_r) \tag{17.25}$$

and assume that model (17.1) applies. The deviations in (17.25) are then independent variables (because the error terms are independent), they are normally distributed (because the error terms are independent normal variables), they have the same expectation zero (because μ_i is subtracted from $\bar{Y}_{i.}$), and they have the same variance σ^2/n. Further, MSE/n is an estimator of σ^2/n that is independent of the deviations $(\bar{Y}_{i.} - \mu_i)$ per theorem (17.6). Thus, it follows from the definition of the studentized range q in (17.23) that:

$$\frac{\max(\bar{Y}_{i.} - \mu_i) - \min(\bar{Y}_{i.} - \mu_i)}{\sqrt{\dfrac{MSE}{n}}} \sim q(r, n_T - r) \tag{17.26}$$

where $n_T - r$ is the number of degrees of freedom associated with SSE, $\max(\bar{Y}_{i.} - \mu_i)$ is the largest deviation, and $\min(\bar{Y}_{i.} - \mu_i)$ is the smallest deviation.

In view of (17.26), we can write the following probability statement:

$$P\left[\frac{\max(\bar{Y}_{i.} - \mu_i) - \min(\bar{Y}_{i.} - \mu_i)}{\sqrt{\dfrac{MSE}{n}}} \le q(1 - \alpha; r, n_T - r)\right] = 1 - \alpha \tag{17.27}$$

Note now that the following inequality holds for *all* pairs of factor levels i and i':

$$|(\bar{Y}_{i.} - \mu_i) - (\bar{Y}_{i'.} - \mu_{i'})| \le \max(\bar{Y}_{i.} - \mu_i) - \min(\bar{Y}_{i.} - \mu_i) \tag{17.28}$$

The absolute value at the left is needed since the factor levels i and i' are not ordered so that we may be subtracting the larger deviation from the smaller. To put this another way, we are merely concerned here with the difference between the two factor level deviations regardless of direction.

Since the inequality (17.28) holds for all pairs of factor levels i and i', it follows from (17.27) that the probability:

$$P\left[\left|\frac{(\bar{Y}_{i.} - \mu_i) - (\bar{Y}_{i'.} - \mu_{i'})}{\sqrt{\dfrac{MSE}{n}}}\right| \le q(1 - \alpha; r, n_T - r)\right] = 1 - \alpha \tag{17.29}$$

holds for all $r(r - 1)/2$ pairwise comparisons among the r factor levels. By rearranging the inequality in (17.29), using the definitions of $s^2(\hat{D})$ in (17.24b) and of T in (17.24c), and noting that for the equal sample size case $s^2(\hat{D})$ becomes:

$$s^2(\hat{D}) = MSE\left(\frac{1}{n} + \frac{1}{n}\right) = \frac{2MSE}{n}$$

we obtain the Tukey multiple comparison confidence limits in (17.24).

17.3 SCHEFFÉ METHOD OF MULTIPLE COMPARISONS

The Scheffé method of multiple comparisons was encountered previously for regression models. It is also applicable for analysis of variance models. It applies for analysis of variance models when:

> The family of interest is the set of estimates of all possible contrasts among the factor level means:

(17.30) $$L = \Sigma c_i \mu_i \qquad \text{where } \Sigma c_i = 0$$

Thus, infinitely many statements belong to this family. The Scheffé method is exact whether the factor level sample sizes are equal or unequal.

We noted earlier that an unbiased estimator of L is:

(17.31) $$\hat{L} = \Sigma c_i \overline{Y}_{i.}$$

for which the estimated variance is:

(17.32) $$s^2(\hat{L}) = MSE\Sigma\frac{c_i^2}{n_i}$$

Scheffé showed that the probability is $1 - \alpha$ that *all* confidence limits of the type:

(17.33) $$\hat{L} \pm Ss(\hat{L})$$

are correct simultaneously, where $\hat{L}$ and $s(\hat{L})$ are given by (17.31) and (17.32), respectively, and S is given by:

(17.33a) $$S^2 = (r - 1)F(1 - \alpha; r - 1, n_T - r)$$

Thus, if we were to calculate the confidence intervals for all conceivable contrasts by use of (17.33), then in $(1 - \alpha)100$ percent of repetitions of the experiment, the entire set of confidence intervals in the family would be correct.

Note that the simultaneous confidence limits in (17.33) differ from those for a single confidence limit in (17.20) only with respect to the multiple of the estimated standard deviation.

Example

Suppose that in our Kenton Food Company example dealing with the four cereal package designs, we wish to estimate the following contrasts with family confidence coefficient of 90 percent:

Comparison of 3-color and 5-color designs:

$$L_1 = \frac{\mu_1 + \mu_2}{2} - \frac{\mu_3 + \mu_4}{2}$$

Comparison of designs with and without cartoons:

$$L_2 = \frac{\mu_1 + \mu_3}{2} - \frac{\mu_2 + \mu_4}{2}$$

Comparison of the two 3-color designs:

$$L_3 = \mu_1 - \mu_2$$

Comparison of the two 5-color designs:

$$L_4 = \mu_3 - \mu_4$$

Consider the estimation of L_1. Earlier, we found:

$$\hat{L}_1 = -9$$
$$s(\hat{L}_1) = 1.79$$

Since $r = 4$ and $n_T - r = 6$ (Table 17.1), we have:

$$S^2 = (r-1)F(1-\alpha; r-1, n_T - r) = 3F(.90; 3, 6) = 3(3.29) = 9.87$$

or:

$$S = 3.14$$

Hence, the confidence interval for L_1 by the Scheffé multiple comparison method is:

$$-9 - 3.14(1.79) \leq L_1 \leq -9 + 3.14(1.79)$$
$$-14.6 \leq L_1 \leq -3.4$$

In similar fashion, we obtain the other desired confidence intervals, and the entire set is:

$$-14.6 \leq L_1 \leq -3.4$$
$$-8.6 \leq L_2 \leq 2.6$$
$$-5.9 \leq L_3 \leq 9.9$$
$$-15.9 \leq L_4 \leq -.1$$

This set of confidence intervals has a family confidence coefficient of 90 percent, so that any chain of conclusions derived from the intervals has associated with it this confidence coefficient. The principal conclusions drawn by the sales manager from the above set of estimates were as follows: 5-color designs lead to higher mean sales than 3-color designs, the increase being somewhere between 3 and 15 cases per store. No overall effect of cartoons in the package design is indicated, although for 5-color designs the use of a cartoon leads to mean sales that are lower than those when no cartoon is used.

Comments

1. If in our Kenton Food Company example we wished to estimate a single contrast with statement confidence coefficient .90, the required t value would be $t(.95; 6) = 1.943$. This t value is smaller than the Scheffé multiple $S = 3.14$, so that the single confidence interval would be somewhat narrower. The increased width of the interval with the Scheffé method is the price paid for a known confidence coefficient for a family

of statements and a chain of conclusions drawn from them, and for the possibility of making comparisons not specified in advance of the data analysis.

2. Since applications of the Scheffé method never involve all conceivable contrasts, the confidence coefficient for the finite family of statements actually considered will be greater than $1 - \alpha$. Thus, when we state the confidence coefficient is $1 - \alpha$ with the Scheffé method, we really mean it is guaranteed to be at least $1 - \alpha$. For this reason, it has been suggested that the confidence coefficient $1 - \alpha$ used with the Scheffé method be below the level ordinarily used, since $1 - \alpha$ is a lower bound and the actual confidence coefficient will be greater. Confidence coefficients of 90 percent and 95 percent with the Scheffé method are frequently mentioned.

3. The Scheffé method can be used for a wide variety of data snooping since the family of statements contains all possible contrasts.

Comparison of Scheffé method with Tukey method

1. If only pairwise comparisons are to be made, the Tukey method gives narrower confidence limits and is therefore the preferred method.
2. In the case of general contrasts, the Scheffé method tends to give narrower confidence limits and is therefore the preferred method.
3. The Scheffé method has the property that if the test based on F^* indicates that the factor level means μ_i are not equal, the corresponding Scheffé multiple comparison procedure will find at least one contrast (out of all possible contrasts) that differs significantly from zero (the confidence interval does not cover zero). It may be, though, that this contrast is not one of those estimated by the analyst.

17.4 BONFERRONI MULTIPLE COMPARISON METHOD

The Bonferroni method of multiple comparisons was encountered earlier for regression models. It is also applicable for analysis of variance models when:

> The family of interest is the particular set of pairwise comparisons, contrasts, or linear combinations specified by the user.

The Bonferroni method is applicable whether the factor level sample sizes are equal or unequal and whether pairwise comparisons, contrasts, linear combinations, or a mixture of these are to be estimated.

We shall denote the number of statements in the family by g and treat them all as linear combinations since pairwise comparisons and contrasts are special cases of linear combinations. The Bonferroni inequality (5.8) then implies that the confidence coefficient is at least $1 - \alpha$ that the following confidence limits for the g linear combinations L_i are all correct:

$$\hat{L}_i \pm Bs(\hat{L}_i) \qquad i = 1, \ldots, g \tag{17.34}$$

where:

$$B = t(1 - \alpha/2g; n_T - r)$$

Example

Suppose the sales manager of the Kenton Food Company is interested in estimating the following two contrasts with family confidence coefficient .975:

Comparison of 3-color and 5-color designs:

$$L_1 = \frac{\mu_1 + \mu_2}{2} - \frac{\mu_3 + \mu_4}{2}$$

Comparison of designs with and without cartoons:

$$L_2 = \frac{\mu_1 + \mu_3}{2} - \frac{\mu_2 + \mu_4}{2}$$

Earlier we found:

$$\hat{L}_1 = -9 \qquad s(\hat{L}_1) = 1.79$$
$$\hat{L}_2 = -3 \qquad s(\hat{L}_2) = 1.79$$

For a 97.5 percent family confidence coefficient with the Bonferroni method, we require:

$$B = t[1 - .025/2(2); 6] = t(.99375; 6) = 3.57$$

We can now complete the confidence intervals for the two contrasts. For L_1, we have:

$$-9 - 3.57(1.79) \leq L_1 \leq -9 + 3.57(1.79)$$
$$-15.4 \leq L_1 \leq -2.6$$

Similarly, we obtain the other confidence interval:

$$-9.4 \leq L_2 \leq 3.4$$

These confidence intervals have a guaranteed family confidence coefficient of 97.5 percent, which means that in at least 97.5 percent of repetitions of the experiment, both intervals will be correct.

Again, we would conclude from this family of estimates that mean sales for 5-color designs are higher than those for 3-color designs (by somewhere between 3 and 15 cases per store), and that no overall effect of cartoons in the package design is indicated.

The Scheffé multiple for a 97.5 percent family confidence coefficient in this case would have been:

$$S^2 = 3F(.975; 3, 6) = 3(6.60) = 19.8$$

or:

$$S = 4.45$$

as compared to the Bonferroni multiple $B = 3.57$. Thus, the Scheffé method here would have led to wider confidence intervals than the Bonferroni method.

Note

It is not necessary that all comparisons be estimated with statement confidence coefficients $1 - (\alpha/g)$ for the Bonferroni family confidence coefficient to be $1 - \alpha$. Different statement confidence coefficients $1 - \alpha_i$ may be used, depending upon the importance of each statement, provided that $\alpha_1 + \alpha_2 + \cdots + \alpha_g = \alpha$.

Comparison of Bonferroni method with Scheffé and Tukey methods

1. If all pairwise comparisons are of interest, the Tukey method is superior to the Bonferroni method in the sense of leading to narrower confidence intervals. If not all pairwise comparisons are to be considered, however, the Bonferroni method may be the better at times.
2. The Bonferroni method will be better than the Scheffé method when the number of contrasts to be estimated is about the same as the number of factor levels, or less. Indeed, the number of statements to be made must exceed the number of factor levels by a considerable amount before the Scheffé method becomes better.
3. In any given problem, one may compute the Bonferroni multiple as well as the Scheffé multiple and, when appropriate, the Tukey multiple, and select the one that is smallest. This choice is proper since it does not depend on the observed data.
4. The Bonferroni multiple comparison method does not lend itself to data snooping unless one can specify in advance the family of statements one may be interested in, and provided this family is not large. On the other hand, the Tukey and Scheffé methods involve families of statements that lend themselves naturally to data snooping.
5. There are still other methods of making multiple comparisons. Many of these are designed for special cases, such as comparing experimental treatments with a control treatment. A good reference book on multiple comparisons is that by Miller (Ref. 17.2).

17.5 SINGLE DEGREE OF FREEDOM TESTS

When the overall F test statistic (16.51) leads to the conclusion that the factor level means μ_i in a single-factor study are not all equal, the investigation of the nature of the factor level effects sometimes is conducted by means of tests dealing with specific questions rather than by estimation of pairwise comparisons, contrasts, or linear combinations. For example, the sales manager of the Kenton Food Company wished to know whether or not mean sales per store for the 3-color designs are the same as those for the 5-color designs. Since all factor level means μ_i are considered to be of the same importance, this question involves the alternatives:

$$H_0\colon \frac{\mu_1 + \mu_2}{2} = \frac{\mu_3 + \mu_4}{2}$$

$$H_a: \frac{\mu_1 + \mu_2}{2} \neq \frac{\mu_3 + \mu_4}{2}$$

These alternatives can be stated equivalently as:

$$H_0: \frac{\mu_1 + \mu_2}{2} - \frac{\mu_3 + \mu_4}{2} = 0$$

$$H_a: \frac{\mu_1 + \mu_2}{2} - \frac{\mu_3 + \mu_4}{2} \neq 0$$

A test that involves a linear combination of the factor level means μ_i is called a *single degree of freedom test*. In general, the two-sided alternatives for a single degree of freedom test are stated as follows:

(17.35)
$$\begin{aligned} &H_0: \Sigma c_i \mu_i = c \\ &H_a: \Sigma c_i \mu_i \neq c \end{aligned}$$

where the c_i and c are appropriate constants. In the earlier example, for instance, we have $c_1 = 1/2$, $c_2 = 1/2$, $c_3 = -1/2$, $c_4 = -1/2$, and $c = 0$.

To test the alternatives (17.35), we utilize theorem (17.19) yielding the t^* test statistic:

(17.36)
$$t^* = \frac{\Sigma c_i \overline{Y}_{i.} - c}{\sqrt{MSE \Sigma \frac{c_i^2}{n_i}}}$$

which follows the t distribution with $n_T - r$ degrees of freedom when H_0 holds. An equivalent test statistic is $(t^*)^2$, denoted by F^*:

(17.37)
$$F^* = (t^*)^2 = \frac{(\Sigma c_i \overline{Y}_{i.} - c)^2}{MSE \Sigma \frac{c_i^2}{n_i}}$$

which follows the F distribution with 1 and $n_T - r$ degrees of freedom when H_0 holds. Remember from (1.45a) that $[t(n_T - r)]^2 = F(1, n_T - r)$.

Example

To test in the Kenton Food Company example whether or not the mean sales per store for the 3-color designs and the 5-color designs are equal, the analyst used test statistic (17.37) with $\alpha = .05$. The alternatives are:

$$H_0: \frac{\mu_1 + \mu_2}{2} - \frac{\mu_3 + \mu_4}{2} = 0$$

$$H_a: \frac{\mu_1 + \mu_2}{2} - \frac{\mu_3 + \mu_4}{2} \neq 0$$

so that $c_1 = c_2 = 1/2$, $c_3 = c_4 = -1/2$, and $c = 0$. Using the sample results in Table 17.1, we obtain the test statistic:

$$F^* = \frac{\left(\dfrac{\overline{Y}_{1.} + \overline{Y}_{2.}}{2} - \dfrac{\overline{Y}_{3.} + \overline{Y}_{4.}}{2} - 0\right)^2}{MSE\left[\dfrac{(1/2)^2}{n_1} + \dfrac{(1/2)^2}{n_2} + \dfrac{(-1/2)^2}{n_3} + \dfrac{(-1/2)^2}{n_4}\right]}$$

$$= \frac{\left(\dfrac{15 + 13}{2} - \dfrac{19 + 27}{2}\right)^2}{7.67\left(\dfrac{.25}{2} + \dfrac{.25}{3} + \dfrac{.25}{3} + \dfrac{.25}{2}\right)} = 25.3$$

For $\alpha = .05$, we require $F(.95; 1, 6) = 5.99$. Hence, the decision rule is:

If $F^* \leq 5.99$, conclude H_0
If $F^* > 5.99$, conclude H_a

Since $F^* = 25.3 > 5.99$, we conclude H_a, that mean sales for the 3-color designs differ from the mean sales for the 5-color designs.

If the analyst had wished to utilize test statistic (17.36), the value of the test statistic would have been:

$$t^* = \frac{\dfrac{\overline{Y}_{1.} + \overline{Y}_{2.}}{2} - \dfrac{\overline{Y}_{3.} + \overline{Y}_{4.}}{2} - 0}{\sqrt{MSE\left[\dfrac{(1/2)^2}{n_1} + \dfrac{(1/2)^2}{n_2} + \dfrac{(-1/2)^2}{n_3} + \dfrac{(-1/2)^2}{n_4}\right]}}$$

$$= \frac{\dfrac{15 + 13}{2} - \dfrac{19 + 27}{2}}{\sqrt{7.67\left(\dfrac{.25}{2} + \dfrac{.25}{3} + \dfrac{.25}{3} + \dfrac{.25}{2}\right)}} = -5.03$$

For $\alpha = .05$, we require $t(.975; 6) = 2.447$ and the decision rule is:

If $|t^*| \leq 2.447$, conclude H_0
If $|t^*| > 2.447$, conclude H_a

Since $|t^*| = 5.03 > 2.447$, we conclude H_a, as before with the F^* test statistic.

Comments

1. Any single degree of freedom test can also be conducted by setting up a confidence interval for the appropriate pairwise comparison, contrast, or linear combination and noting whether or not the confidence interval includes the value c specified in the alternatives. For instance, we obtained earlier the 95 percent confidence interval for the difference in mean sales between the 3-color and 5-color designs:

$$-13.4 \leq \frac{\mu_1 + \mu_2}{2} - \frac{\mu_3 + \mu_4}{2} \leq -4.6$$

Since this 95 percent confidence interval does not include $c = 0$, a test of the difference between the two group means for $\alpha = .05$ leads to the conclusion that the two group means are not equal.

2. When interest exists in the P-value of a single degree of freedom test and tables are to be used for ascertaining the P-value, it is better to use t^* as the test statistic instead of F^* because t tables generally contain more percentiles than do F tables.

3. The test statistic t^* can also be used when one-sided tests are to be conducted, but the test statistic F^* is only appropriate for two-sided tests.

4. A number of single-factor analysis of variance computer packages permit the user to specify a contrast of interest and then will furnish either the t^* or the F^* test statistic.

Multiple single degree of freedom tests

When the analysis of factor effects is carried out by single degree of freedom tests, usually several single degree of freedom tests are conducted to answer related questions. For instance, the sales manager of the Kenton Food Company might wish to know not only whether the number of colors has an effect on mean sales but also whether use of cartoons has an effect. Whenever several single degree of freedom tests are conducted, the level of significance and power, insofar as the family of tests is concerned, is affected. For example, if three different single degree of freedom F or t tests are conducted, each with $\alpha = .05$, then the probability that each of the tests will lead to conclusion H_0 when indeed H_0 is correct in each case, assuming independence of the tests, is $(.95)^3 = .857$. Thus, the level of significance that at least one of the three tests leads to conclusion H_a when H_0 holds in each case would be $1 - .857 = .143$ and not .05. We see then that the level of significance and power for a *family* of tests is not the same as that for an *individual* test. Actually, the F^* or t^* statistics are dependent since they all are based on the same sample data and use the same *MSE* value. It often therefore becomes difficult to determine the actual level of significance and power for a family of tests.

Control of the family level of significance when using a number of single degree of freedom tests can be achieved by using either the Bonferroni method, the Scheffé method (when contrasts are involved), or the Tukey method (when pairwise comparisons are involved). One simply constructs the appropriate confidence intervals and draws the proper test conclusion from each of the confidence intervals. Not only does this approach permit simultaneous single degree of freedom tests with control over the family level of significance, but it also provides information about the magnitude of any effects that are present.

Example. In the Kenton Food Company example, the analyst wished to test whether or not any two factor level means μ_i and $\mu_{i'}$ differ. Thus, the analyst was interested in conducting tests involving the alternatives:

Test 1:	H_0: $\mu_1 = \mu_2$	Test 2:	H_0: $\mu_1 = \mu_3$	Test 3:	H_0: $\mu_1 = \mu_4$
	H_a: $\mu_1 \neq \mu_2$		H_a: $\mu_1 \neq \mu_3$		H_a: $\mu_1 \neq \mu_4$
Test 4:	H_0: $\mu_2 = \mu_3$	Test 5:	H_0: $\mu_2 = \mu_4$	Test 6:	H_0: $\mu_3 = \mu_4$
	H_a: $\mu_2 \neq \mu_3$		H_a: $\mu_2 \neq \mu_4$		H_a: $\mu_3 \neq \mu_4$

The family level of significance is to be controlled at $\alpha = .10$.

The analyst utilized the Tukey multiple comparison procedure for obtaining all pairwise comparisons and used the decision rule:

If the confidence interval includes 0, conclude H_0
Otherwise conclude H_a

We obtained these confidence intervals earlier and repeat the results here, together with the appropriate conclusions:

Test 1: $-5.3 \leq \mu_1 - \mu_2 \leq 9.3$ Conclude H_0

Test 2: $-3.3 \leq \mu_3 - \mu_1 \leq 11.3$ Conclude H_0

Test 3: $4.0 \leq \mu_4 - \mu_1 \leq 20.0$ Conclude H_a

Test 4: $-.5 \leq \mu_3 - \mu_2 \leq 12.5$ Conclude H_0

Test 5: $6.7 \leq \mu_4 - \mu_2 \leq 21.3$ Conclude H_a

Test 6: $.7 \leq \mu_4 - \mu_3 \leq 15.3$ Conclude H_a

Thus, all pairwise differences between μ_4 and the other factor level means are statistically significant (i.e., the differences are not zero), while all other pairwise differences are not statistically significant.

Often, the results of these pairwise tests are summarized by setting up groups of factor levels whose means do not differ according to the single degree of freedom tests. Here, there are two such groups:

Group 1		*Group 2*	
Package design 1	$\bar{Y}_{1.} = 15$	Package design 4	$\bar{Y}_{4.} = 27$
Package design 2	$\bar{Y}_{2.} = 13$		
Package design 3	$\bar{Y}_{3.} = 19$		

Comments

1. The Bonferroni procedure can be used for multiple single degree of freedom tests not only by means of confidence intervals but also with either test statistic t^* or F^*. If, say, four single degree of freedom tests are to be conducted with family level of significance $\alpha = .10$, each test with t^* or F^* is simply conducted with individual significance level $.10/4 = .025$.

2. When the Tukey multiple comparison method is used for testing pairwise differences, the tests are sometimes called *honestly significant differences tests*.

17.6 ANALYSIS OF FACTOR EFFECTS WHEN FACTOR QUANTITATIVE

When the factor under investigation is quantitative, the analysis of factor effects can be carried beyond the point of multiple comparisons to include a study of the nature of the response function. Consider an experimental study undertaken to investigate the effect of the price of a product on sales. Five different price levels are investigated (28 cents, 29 cents, 30 cents, 31 cents, and 32 cents), and the experimental unit is a store. After a preliminary test whether or not mean sales differ for the price levels studied, the analyst might use multiple comparisons to examine whether "odd pricing" at 29 cents actually leads to higher sales than "even pricing" at 28 cents, as well as other questions of interest. In addition, the analyst may wish to study whether mean sales are a specified function of price, in the range of prices studied in the experiment. Further, once the relation has been established, the analyst may wish to use it for estimating sales volumes at various price levels not studied.

The methods of regression analysis discussed in Parts I and II are, of course, appropriate for the analysis of the response function. It should only be noted that the single-factor studies discussed in this chapter almost always involve replications at the different factor levels, so that the lack of fit of a specified response function can be tested. For this purpose, it should be remembered that the analysis of variance error sum of squares in (16.24b) is identical to the pure error sum of squares of Chapter 4 in (4.9). We illustrate this relation in the following example.

Example

In a study to reduce raw material costs in a glass working firm, an operations analyst collected the experimental data in Table 17.3 on the number of acceptable units produced from equal amounts of raw material by 28 entry-level piecework employees who had received special training as part of the experiment. Four training levels were used (6, 8, 10, and 12 hours), with 7 of the employees being assigned at random to each level. The higher the number of acceptable pieces, the more efficient is the employee in utilizing the raw material.

TABLE 17.3 Data for piecework trainees example

Treatment (hours of training)		*j*						
i		1	2	3	4	5	6	7
1	6 hours	40	39	39	36	42	43	41
2	8 hours	53	48	49	50	51	50	48
3	10 hours	53	58	56	59	53	59	58
4	12 hours	63	62	59	61	62	62	61

Preliminary analysis. The analyst first tested whether or not the mean number of acceptable pieces is the same for the four training levels. Model (17.1) was employed:

$$Y_{ij} = \mu_i + \varepsilon_{ij} \tag{17.38}$$

The alternative conclusions and appropriate test statistic were:

$$H_0\colon \mu_1 = \mu_2 = \mu_3 = \mu_4$$
$$H_a\colon \text{not all } \mu_i \text{ are equal}$$
$$F^* = \frac{MSTR}{MSE}$$

A computer program for single-factor ANOVA gave the output shown in Figure 17.1. Note from the mean squares column of the ANOVA table that $MSTR = 602.8926$ and $MSE = 4.2619$.

FIGURE 17.1 Segment of computer output for piecework trainees example (SPSS, Ref. 17.3)

	GROUP	COUNT (n_i)	MEAN ($\bar{Y}_{i.}$)	STANDARD DEVIATION
Treatment →	GRP01	7	40.0000	2.3094
	GRP02	7	49.8571	1.7728
	GRP03	7	56.5714	2.6367
	GRP04	7	61.4286	1.2724
	TOTAL	28	51.9643	8.4129

ANALYSIS OF VARIANCE

SOURCE	D.F.	SUM OF SQUARES	MEAN SQUARES
BETWEEN GROUPS	3	**SSTR** → 1808.6778	602.8926 ← **MSTR**
WITHIN GROUPS	24	**SSE** → 102.2856	4.2619 ← **MSE**
TOTAL	27	**SSTO** → 1910.9634	

F RATIO	F PROB.
141.461	0.0000
↑ F^*	↑ **One-sided P-value**

MULTIPLE RANGE TEST

TUKEY-HSD PROCEDURE
RANGES FOR THE 0.050 LEVEL -

3.90 ← **q(.95; 4, 24)**

HOMOGENEOUS SUBSETS

SUBSET 1		SUBSET 3	
GROUP	GRP01	GROUP	GRP03
MEAN	40.0000	MEAN	56.5714

SUBSET 2		SUBSET 4	
GROUP	GRP02	GROUP	GRP04
MEAN	49.8571	MEAN	61.4286

Residual analysis (to be discussed in Chapter 18) showed model (17.1) to be apt. Therefore, the analyst proceeded with the test, using $\alpha = .05$ since analysis of power probabilities showed that this significance level would give an acceptable balance between risks of Type I and Type II errors. The decision rule therefore was:

$$\text{If } F^* \leq F(.95; 3, 24) = 3.01, \text{ conclude } H_0$$
$$\text{If } F^* > 3.01, \text{ conclude } H_a$$

From the printout in Figure 17.1, we have:

$$F^* = \frac{MSTR}{MSE} = 141.5$$

Since $F^* = 141.5 > 3.01$, the analyst concluded H_a, that training level effects differed and that further analysis of them is warranted. The P-value for the test statistic is 0+, as shown in Figure 17.1.

Investigation of treatment effects. The analyst's interest next centered on multiple comparisons of all pairs of treatment means. A Tukey multiple comparison option in the analyst's computer package was used. It gave the output shown in the lower portion of Figure 17.1. This output presents the results of single degree of freedom tests conducted by means of the Tukey multiple comparison procedure for all pairwise comparisons. (The confidence intervals for the pairwise comparisons are not shown in the output.) All factor levels for which the test concludes that the pairwise means are equal are placed in the same subset. This form of summary of single degree of freedom tests was illustrated earlier for the Kenton Food Company example. When a subset contains only one factor level, as is the case for all groups in the output of Figure 17.1, the implication is that all single degree of freedom tests involving this factor level and each of the other factor levels led to conclusion H_a, that the two factor level means being compared are not equal.

Two points are to be noted in particular from the results in Figure 17.1: (1) All pairwise factor level differences are statistically significant. (2) There is some indication that differences between the means for adjoining factor levels diminish as the number of hours of training increases; that is, diminishing returns appear to be obtained as the length of training is increased.

Estimation of response function. These findings were in accord with the analyst's expectations. She had surmised that the treatment means μ_i would most likely follow a quadratic response function with respect to training level. She now wished to investigate this point further by fitting a quadratic regression model. The model to be fitted and tested was:

$$Y_{ij} = \beta_0 + \beta_1 x_i + \beta_{11} x_i^2 + \varepsilon_{ij} \tag{17.39}$$

where Y_{ij} and ε_{ij} are defined as earlier, the β's are regression parameters, and x_i denotes the number of hours of training in the ith training level (X_i) expressed as a deviation around the mean of all training levels ($\bar{X} = 9$), i.e., $x_i = X_i - 9$.

TABLE 17.4 **X** and **Y** matrices for piecework trainees example

$$\mathbf{Y}=\begin{bmatrix}40\\39\\39\\36\\42\\43\\41\\53\\48\\49\\50\\51\\50\\48\\53\\58\\56\\59\\53\\59\\58\\63\\62\\59\\61\\62\\62\\61\end{bmatrix}\qquad \mathbf{X}=\begin{bmatrix} & x & x^2\\ 1&-3&9\\1&-3&9\\1&-3&9\\1&-3&9\\1&-3&9\\1&-3&9\\1&-3&9\\1&-1&1\\1&-1&1\\1&-1&1\\1&-1&1\\1&-1&1\\1&-1&1\\1&-1&1\\1&1&1\\1&1&1\\1&1&1\\1&1&1\\1&1&1\\1&1&1\\1&1&1\\1&3&9\\1&3&9\\1&3&9\\1&3&9\\1&3&9\\1&3&9\\1&3&9\end{bmatrix}$$

The **X** and **Y** matrices for the regression analysis are given in Table 17.4. A computer run of a multiple regression package yielded the estimated regression function:

$$\hat{Y} = 53.52679 + 3.55000x - .31250x^2 \tag{17.40}$$

and the analysis of variance for the regression model (17.39) shown in Table 17.5a. For completeness, we repeat in Table 17.5b the analysis of variance for the ANOVA model (17.38).

Since the data contained replicates, the analyst could test the regression model (17.39) for lack of fit. She utilized the fact that the ANOVA error sum of squares in (16.24b) is identical to the regression pure error sum of squares in (4.9). Both measure variation around the mean at any given level of X (i.e., around the treatment sample mean $\bar{Y}_{i.}$). Hence, the lack of fit sum of squares can be readily obtained from previous results:

$$SSLF = \underset{\text{(Table 17.5a)}}{SSE} - \underset{\text{(Table 17.5b)}}{SSPE} = 102.864 - 102.286 = .578 \tag{17.41}$$

Since there are $c = r = 4$ levels of X here and $p = 3$ parameters in the regression model, we divide $SSLF$ by $c - p = 4 - 3 = 1$ degree of freedom and obtain

TABLE 17.5 Analyses of variance for piecework trainees example

(a) Regression Model (17.39)

Source of Variation	SS	df	MS
Regression	1,808.100	2	904.05
Error	102.864	25	4.11
Total	1,910.964	27	

(b) Analysis of Variance Model (17.38)

Source of Variation	SS	df	MS
Treatments	1,808.678	3	602.89
Error	102.286	24	4.26
Total	1,910.964	27	

(c) ANOVA for Lack of Fit Test

Source of Variation	SS	df	MS
Regression	1,808.100	2	904.05
Error	102.864	25	4.11
Lack of fit	.578	1	.58
Pure error	102.286	24	4.26
Total	1,910.964	27	

$MSLF = .578/1 = .578$. Table 17.5c contains the analysis of variance for the regression model with the error sum of squares and degrees of freedom broken down into lack of fit and pure error components.

The alternative conclusions for the test of lack of fit are:

$$H_0: E(Y) = \beta_0 + \beta_1 x + \beta_{11} x^2$$
$$H_a: E(Y) \neq \beta_0 + \beta_1 x + \beta_{11} x^2$$

and the appropriate test statistic is:

$$F^* = \frac{MSLF}{MSPE}$$

For $\alpha = .05$, the decision rule is:

$$\text{If } F^* \leq F(.95; 1, 24) = 4.26, \text{ conclude } H_0$$
$$\text{If } F^* > 4.26, \text{ conclude } H_a$$

We calculate the test statistic from Table 17.5c:

$$F^* = \frac{.58}{4.26} = .136$$

Since $F^* = .136 \leq 4.26$, the analyst concluded that the quadratic response function is a good fit. Consequently, she used the fitted regression function in (17.40) in further evaluation of the relation between mean number of acceptable pieces produced and level of training, after expressing it in the original independent variable X (number of hours of training):

$$\hat{Y} = -3.73571 + 9.17500X - .31250X^2$$

PROBLEMS

17.1. Refer to Figure 17.1 for the piecework trainees example. To estimate the treatment mean μ_1 by means of an interval estimate, would it be valid to use for the estimated variance of $\overline{Y}_{1.}$:

$$s^2(\overline{Y}_{1.}) = \frac{s_1^2}{n_1} = \frac{(2.3094)^2}{7}$$

instead of (17.4):

$$s^2(\overline{Y}_{1.}) = \frac{MSE}{n_1} = \frac{4.2619}{7}$$

What is the advantage of using (17.4)? What is a disadvantage?

17.2. Refer to **Premium distribution** Problem 16.14. A student, asked to give a class demonstration of the use of a confidence interval for comparing two treatment means, proposed to construct a 99 percent confidence interval for the pairwise comparison $D = \mu_5 - \mu_3$. He selected this particular comparison because the sample means $\overline{Y}_{5.}$ and $\overline{Y}_{3.}$ are the largest and smallest, respectively. The student stated: "This confidence interval is particularly useful. If it does not straddle zero, it indicates, with significance level $\alpha = .01$, that the factor level means are not equal."

a. Explain why the student's assertion is not correct.

b. How should the confidence interval be constructed so that the assertion can be made with significance level $\alpha = .01$?

17.3. A trainee examined a set of experimental data to find comparisons that "look promising" and calculated a family of Bonferroni confidence intervals for these comparisons with a 90 percent family confidence coefficient. Upon being informed that the Bonferroni procedure is not applicable in this case because the comparisons had been suggested by the data, the trainee stated: "This makes no difference. I would use the same formulas for the point estimates and the estimated standard errors even if the comparisons were not suggested by the data." Respond.

17.4. Consider the following linear combinations of interest in a single-factor study involving four factor levels:

$$\text{(i)} \quad \mu_1 + 3\mu_2 - 4\mu_3$$

$$\text{(ii)} \quad .3\mu_1 + .5\mu_2 + .1\mu_3 + .1\mu_4$$

$$\text{(iii)} \quad \frac{\mu_1 + \mu_2 + \mu_3}{3} - \mu_4$$

a. Which of the linear combinations are contrasts? State the coefficients for each of the contrasts.

b. Give an unbiased estimator for each of the linear combinations. Also give the estimated variance of each estimator assuming that $n_i \equiv n$.

17.5. A single-factor ANOVA study consists of $r = 6$ treatments with sample sizes $n_i \equiv 10$.

a. Assuming that pairwise comparisons of the treatment means are to be made with a 90 percent family confidence coefficient, find the T, S, and B multiples for the following numbers of pairwise comparisons in the family: $g = 2, 5, 15$. What generalization is suggested by your results?

b. Assuming that contrasts of the treatment means are to be estimated with a 90 percent family confidence coefficient, find the S and B multiples for the following numbers of contrasts in the family: $g = 2, 5, 15$. What generalization is suggested by your results?

17.6. Consider a single-factor study with $r = 5$ treatments and sample sizes $n_i \equiv 5$.

a. Find the T, S, and B multiples if $g = 2$, 5, and 10 pairwise comparisons are to be made with a 95 percent family confidence coefficient. What generalization is suggested by your results?

b. What would be the T, S, and B multiples for sample sizes $n_i \equiv 20$? Does the generalization obtained in part (a) still hold?

17.7. In making multiple comparisons, why is it appropriate to use the multiple comparison procedure that leads to the tightest confidence intervals for the sample data obtained? Discuss.

17.8. For a single-factor study with $r = 2$ treatments and sample sizes $n_i \equiv 10$, find the T, S, and B multiples for $g = 1$ pairwise comparison with a 99 percent family confidence coefficient. What generalization is suggested by your results?

17.9. Refer to **Productivity improvement** Problem 16.9. Some additional calculational results are:

i	*Research and Development Expenditures Level*	n_i	$\bar{Y}_{i.}$	
1	Low	9	6.878	
2	Moderate	12	8.133	$MSE = .6401$
3	High	6	9.200	

a. Estimate the mean productivity improvement for firms with high research and development expenditures levels; use a 95 percent confidence interval.

b. Obtain a 95 percent confidence interval for $D = \mu_2 - \mu_1$. Interpret your interval estimate.

c. Obtain confidence intervals for all pairwise comparisons of the treatment means; use the Tukey procedure and a 90 percent family confidence coefficient. State your findings and prepare a graphic summary.

d. Is the Tukey procedure employed in part (c) the most efficient one that could be used here? Explain.

17.10. Refer to **Questionnaire color** Problem 16.10. Some additional calculational results are:

i	*Color*	n_i	$\bar{Y}_{i.}$	
1	Blue	5	29.4	
2	Green	5	29.6	$MSE = 9.70$
3	Orange	5	28.0	

a. Estimate the mean response rate for blue questionnaires; use a 90 percent confidence interval.

b. Obtain a 90 percent confidence interval for $D = \mu_3 - \mu_2$. Interpret your interval estimate. In light of the result for the ANOVA test in Problem 16.10d, is it surprising that the confidence interval straddles zero? Explain.

17.11. Refer to **Rehabilitation therapy** Problem 16.11.

a. Estimate with a 99 percent confidence interval the mean number of days required in therapy for persons of average physical fitness.

b. Obtain confidence intervals for $D_1 = \mu_2 - \mu_3$ and $D_2 = \mu_1 - \mu_2$; use the Bonferroni procedure with a 95 percent family confidence coefficient. Interpret your results.

c. Would the Tukey procedure have been more efficient to use in part (b)? Explain.

d. If the researcher also wished to estimate $D_3 = \mu_1 - \mu_3$, still with a 95 percent family confidence coefficient, would the B multiple in part (b) need to be modified? Would this also be the case if the Tukey procedure had been employed?

17.12. Refer to **Cash offers** Problem 16.12.

a. Estimate the mean cash offer for young owners; use a 99 percent confidence interval.

b. Construct a 99 percent confidence interval for $D = \mu_3 - \mu_1$. Interpret your interval estimate.

c. Obtain confidence intervals for all pairwise comparisons between the treatment means; use the Tukey procedure and a 90 percent family confidence coefficient. Interpret your results and provide a graphic summary.

d. Would the Bonferroni procedure have been more efficient to use in part (c) than the Tukey procedure? Explain.

17.13. Refer to **Filling machines** Problem 16.13.

a. Construct a 95 percent confidence interval for the mean fill for machine 1.

b. Obtain a 95 percent confidence interval for $D = \mu_2 - \mu_1$. Interpret your interval estimate.

c. The consultant is particularly interested in the mean fills for machines 1, 4, and 5. Construct confidence intervals for all pairwise comparisons among these three treatment means; use the Bonferroni method with a 90 percent family confidence coefficient. Interpret your results and provide a graphic summary.

d. Would the Tukey procedure have been more efficient to use in part (c) than the Bonferroni procedure? Explain.

17.14. Refer to **Premium distribution** Problem 16.14.

a. Construct a 90 percent confidence interval for the mean time lapse for agent 1.

b. Obtain a 90 percent confidence interval for $D = \mu_2 - \mu_1$. Interpret your interval estimate.

c. The marketing director wishes to compare the mean time lapses for agents 1, 3, and 5. Obtain confidence intervals for all pairwise comparisons among these three treatment means; use the Bonferroni procedure with a 90 percent family confidence coefficient. Interpret your results and present a graphic summary.

d. Would the Tukey procedure have been more efficient to use in part (c) than the Bonferroni procedure? Explain.

17.15. Refer to **Productivity improvement** Problems 16.9 and 17.9.

a. Estimate the difference in mean productivity improvement between firms with low or moderate research and development expenditures and firms with high expenditures; use a 95 percent confidence interval. Employ an unweighted mean for the low and moderate expenditures groups. Interpret your interval estimate.

b. The sample sizes for the three factor levels are proportional to the population sizes. The economist wishes to estimate the mean productivity gain last year for all firms in the population. Estimate this overall mean productivity improvement with a 95 percent confidence interval.

c. Using the Scheffé procedure, obtain confidence intervals for the following comparisons with 90 percent family confidence coefficient:

$$D_1 = \mu_3 - \mu_2 \qquad D_3 = \mu_2 - \mu_1$$

$$D_2 = \mu_3 - \mu_1 \qquad L_1 = \frac{\mu_1 + \mu_2}{2} - \mu_3$$

Analyze your results and describe your findings.

d. Would the Bonferroni procedure have been more efficient to use in part (c) than the Scheffé procedure? Explain.

17.16. Refer to **Rehabilitation therapy** Problem 16.11.

a. Estimate the contrast $L = (\mu_1 - \mu_2) - (\mu_2 - \mu_3)$ with a 99 percent confidence interval. Interpret your interval estimate.

b. Estimate the following comparisons using the Bonferroni procedure with a 95 percent family confidence coefficient:

$$D_1 = \mu_1 - \mu_2 \qquad D_3 = \mu_2 - \mu_3$$

$$D_2 = \mu_1 - \mu_3 \qquad L_1 = D_1 - D_3$$

Analyze your results and describe your findings.

c. Would the Scheffé procedure have been more efficient to use in part (b) than the Bonferroni procedure? Explain.

17.17. Refer to **Cash offers** Problem 16.12.

a. Estimate the contrast $L = (\mu_3 - \mu_2) - (\mu_2 - \mu_1)$ with a 99 percent confidence interval. Interpret your interval estimate.

b. Estimate the following comparisons with a 90 percent family confidence coefficient; employ the most efficient multiple comparison procedure:

$$D_1 = \mu_2 - \mu_1 \qquad D_3 = \mu_3 - \mu_1$$

$$D_2 = \mu_3 - \mu_2 \qquad L_1 = D_2 - D_1$$

Interpret your results.

17.18. Refer to **Filling machines** Problem 16.13. Machines 1 and 2 were bought new five years ago, machines 3 and 4 were bought in a reconditioned state five years ago, and machines 5 and 6 were bought new last year.

a. Estimate the contrast:

$$L = \frac{\mu_1 + \mu_2}{2} - \frac{\mu_3 + \mu_4}{2}$$

with a 95 percent confidence interval. Interpret your interval estimate.

b. Estimate the following comparisons with a 90 percent family confidence coefficient; use the most efficient multiple comparison procedure:

$$D_1 = \mu_1 - \mu_2 \qquad L_2 = \frac{\mu_1 + \mu_2}{2} - \frac{\mu_5 + \mu_6}{2}$$

$$D_2 = \mu_3 - \mu_4 \qquad L_3 = \frac{\mu_1 + \mu_2 + \mu_5 + \mu_6}{4} - \frac{\mu_3 + \mu_4}{2}$$

$$D_3 = \mu_5 - \mu_6 \qquad L_4 = \frac{\mu_1 + \mu_2 + \mu_3 + \mu_4}{4} - \frac{\mu_5 + \mu_6}{2}$$

$$L_1 = \frac{\mu_1 + \mu_2}{2} - \frac{\mu_3 + \mu_4}{2}$$

Interpret your results. What can the consultant learn from these results about the differences between the six filling machines?

17.19. Refer to **Premium distribution** Problem 16.14. Agents 1 and 2 distribute merchandise only, agents 3 and 4 distribute cash-value coupons only, and agent 5 distributes both merchandise and coupons.

a. Estimate the contrast:

$$L = \frac{\mu_1 + \mu_2}{2} - \frac{\mu_3 + \mu_4}{2}$$

with a 90 percent confidence interval. Interpret your interval estimate.

b. Estimate the following comparisons with 90 percent family confidence coefficient; use the Scheffé procedure:

$$D_1 = \mu_1 - \mu_2 \qquad L_2 = \frac{\mu_3 + \mu_4}{2} - \mu_5$$

$$D_2 = \mu_3 - \mu_4 \qquad L_3 = \frac{\mu_1 + \mu_2}{2} - \frac{\mu_3 + \mu_4}{2}$$

$$L_1 = \frac{\mu_1 + \mu_2}{2} - \mu_5$$

Interpret your results.

c. Would the Bonferroni procedure have been more efficient to use in part (b) than the Scheffé procedure? Explain.

d. Of all premium distributions, 25 percent are handled by agent 1, 20 percent by agent 2, 20 percent by agent 3, 20 percent by agent 4, and 15 percent by agent 5. Estimate the overall mean time lapse for premium distributions with a 90 percent confidence interval.

17.20 Refer to **Productivity improvement** Problems 16.9 and 17.9.

a. Use a single degree of freedom test to determine whether or not $(\mu_1 + \mu_2)/2 = \mu_3$; control the α risk at .05 and employ test statistic (17.36). State the alternatives, decision rule, and conclusion.

b. Test for all pairs of factor level means whether or not they differ; use single degree of freedom tests based on the Tukey procedure with $\alpha = .05$. Set up groups of factor levels whose means do not differ.

17.21. Refer to **Cash offers** Problem 16.12.

a. Use a single degree of freedom test to determine whether or not $\mu_2 - \mu_1 = \mu_3 - \mu_2$; control the α risk at .01 and employ test statistic (17.37). State the alternatives, decision rule, and conclusion.

b. Test for all pairs of factor level means whether or not they differ; use single degree of freedom tests based on the Tukey procedure with $\alpha = .01$. Set up groups of factor levels whose means do not differ.

17.22. Refer to **Premium distribution** Problem 16.14.

a. Use a single degree of freedom test to determine whether or not $(\mu_1 + \mu_2)/2 = (\mu_3 + \mu_4)/2$; control the α risk at .10 and employ test statistic (17.36). State the alternatives, decision rule, and conclusion.

b. Test for all pairs of factor level means whether or not they differ; use single degree of freedom tests based on the Tukey procedure with $\alpha = .10$. Set up groups of factor levels whose means do not differ.

17.23. Refer to **Solution concentration** Problem 4.15. Suppose the chemist initially wished to employ ANOVA model (16.2) to determine whether or not the concentration of the solution is affected by the amount of time that has elapsed since preparation.

a. State the analysis of variance model.

b. Obtain the analysis of variance table.

c. Test whether or not the factor level means are equal; use a level of significance of $\alpha = .025$. State the alternatives, decision rule, and conclusion.

d. Make all pairwise comparisons of factor level means between adjacent lengths of time; use the Bonferroni procedure with a 95 percent family confidence coefficient. Do your results suggest that the regression relation is not linear?

17.24. Refer to **Rehabilitation therapy** Problem 16.11. A biometrician has developed a scale for physical fitness status, as follows:

Physical Fitness Status	*Scale Value*
Below average	83
Average	100
Above average	121

a. Using this physical fitness status scale, fit the first-order regression model (2.1) for regressing number of days required for therapy (Y) on physical fitness status (X).

b. Obtain the residuals and plot them against X. Does a linear regression model appear to fit the data?

c. Perform an F test to determine whether or not there is lack of fit of a linear regression function; use $\alpha = .05$. State the alternatives, decision rule, and conclusion.

d. Could you test for lack of fit of a quadratic regression function here? Explain.

17.25. Refer to **Filling machines** Problem 16.13. A maintenance engineer has suggested that the differences in mean fills for the six machines are largely related to the length of time since a machine last received major servicing. Service records indicate these lengths of time to be as follows (in months):

Filling Machine	*Number of Months*	*Filling Machine*	*Number of Months*
1	.4	4	5.3
2	3.7	5	1.4
3	6.1	6	2.1

a. Fit the second-order polynomial regression model (9.1) for regressing amount of fill (Y) on number of months since major servicing (X).

b. Obtain the residuals and plot them against X. Does a quadratic regression function appear to fit the data?

c. Perform an F test to determine whether or not there is lack of fit of a quadratic regression function; use $\alpha = .01$. State the alternatives, decision rule, and conclusion.

d. Test whether or not the quadratic term in the response function can be dropped from the model; use $\alpha = .01$. State the alternatives, decision rule, and conclusion.

EXERCISES

17.26. Show that when $r = 2$ and $n_i \equiv n$, q defined in (17.26) is equivalent to $\sqrt{2}|t^*|$, where t^* is defined in (1.60).

17.27. Starting with (17.29), complete the derivation of (17.24).

17.28. Show that when $r = 2$, S^2 defined in (17.33a) is equivalent to $[t(1 - \alpha/2; n_T - r)]^2$.

17.29. Set up the test alternatives (17.35) in the matrix format $\mathbf{C\beta} = \mathbf{h}$ of (8.41) and show that test statistic (8.46) reduces to F^* as defined in (17.37).

PROJECTS

17.30. Refer to the **SENIC** data set and Project 16.32. Obtain confidence intervals for all pairwise comparisons between the four regions; use the Tukey procedure and a 90 percent family confidence coefficient. Interpret your results and state your findings.

17.31. Refer to the **SMSA** data set and Project 16.34. Obtain confidence intervals for all pairwise comparisons between the four regions; use the Tukey procedure and a 90 percent family confidence coefficient. Interpret your results and state your findings.

17.32. Refer to Project 16.35d.

a. For each replication, construct confidence intervals for all pairwise comparisons among the three treatment means; use the Tukey procedure with a 95 percent family confidence coefficient. Then determine whether all confidence intervals for the replication are correct, given that $\mu_1 = 80$, $\mu_2 = 60$, and $\mu_3 = 160$.

b. For what proportion of the 100 replications are all confidence intervals correct? Is this proportion close to theoretical expectations? Discuss.

CITED REFERENCES

17.1 Cochran, William G., and Gertrude M. Cox. *Experimental Designs*. 2d ed. New York: John Wiley & Sons, 1957, p. 74.

17.2 Miller, Rupert G., Jr. *Simultaneous Statistical Inference*. 2d ed. New York: Springer-Verlag, 1981.

17.3 Nie, N. H.; C. H. Hull; J. G. Jenkins; K. Steinbrenner; and D. H. Bent. *SPSS: Statistical Package for the Social Sciences*. 2d ed. New York: McGraw-Hill, 1975.

18

Implementation of ANOVA model

Implementation of an analysis of variance model requires determination of appropriate sample sizes and, once the data are at hand, an examination of the aptness of the model. In this chapter, we first consider the planning of sample sizes. Then we take up a number of topics related to examining the aptness of the analysis of variance model.

18.1 PLANNING OF SAMPLE SIZES WITH POWER APPROACH

For analysis of variance problems, as for other statistical problems, it is important to plan the sample sizes so that needed protection against both Type I and Type II errors can be obtained, or so that the estimates of interest have sufficient precision to be useful. This planning is necessary to ensure that the sample sizes are large enough to detect important differences with high probability. At the same time, the sample sizes should not be so large that the cost of the study becomes excessive and that unimportant differences become statistically significant with high probability.

We shall generally assume in this discussion that all factor levels are to have equal sample sizes, reflecting that they are about equally important. Indeed, if major interest lies in pairwise comparisons of all factor level means, it can be shown that equal sample sizes maximize the precision of the various comparisons. Another reason for equal sample sizes is that certain departures from the assumed ANOVA model are not troublesome if all factor levels have the same sample size, as will be noted in Section 18.7. There will be times, however, when unequal sample sizes are appropriate. For instance, when four experimental treatments are each to be compared to a control, it may be reasonable to make the sample size for the control larger. We shall comment later on the planning of sample sizes for such a case.

Planning of sample sizes can be approached in terms of controlling the risks of making Type I and Type II errors, in terms of controlling the widths of desired confidence intervals, or in terms of a combination of these two. In this section, we consider planning of sample sizes with the power approach, which permits controlling the risks of making Type I and Type II errors.

One procedure for implementing the power approach is to use the power charts for the F test presented in Table A–8. A trial and error process is required, however, with these charts. Fortunately, tables are available that furnish the appropriate sample sizes directly. These are presented in Table A–10. They are applicable when all factor levels are to have equal sample sizes, that is, when $n_i \equiv n$.

Use of Table A–10

The planning of sample sizes using Table A–10 is done in terms of the testing framework for deciding whether or not the factor level means differ:

(18.1)
$$\begin{aligned} &H_0\colon \mu_1 = \mu_2 = \cdots = \mu_r \\ &H_a\colon \text{not all } \mu_i \text{ are equal} \end{aligned}$$

When analysis of variance model (16.2) is appropriate and H_0 does not hold, test statistic F^* in (16.51) follows the noncentral F distribution. This distribution has noncentrality parameter (16.56a) when $n_i \equiv n$:

(18.2)
$$\phi = \frac{1}{\sigma}\sqrt{\frac{n}{r}\Sigma(\mu_i - \mu_.)^2}$$

where:

$$\mu_. = \frac{\Sigma\mu_i}{r}$$

The larger is ϕ, the greater is the power.

Instead of requiring a direct specification of the levels of μ_i for which it is important to control the risk of making a Type II error, Table A–10 only requires a specification of the minimum range of factor level means for which it is important to detect differences between the μ_i with high probability. This minimum range is denoted by Δ:

$$\Delta = \max(\mu_i) - \min(\mu_i) \tag{18.3}$$

The following specifications need be made in using Table A–10:

1. The level α at which the risk of making a Type I error is to be controlled.
2. The magnitude of the minimum range Δ of the μ_i which it is important to detect with high probability. The magnitude of σ, the standard deviation of the probability distributions of Y, must also be specified since entry into Table A–10 is in terms of Δ in units of σ:

$$\frac{\Delta}{\sigma} \tag{18.4}$$

3. The level β at which the risk of making a Type II error is to be controlled for the specification given in 2. Entry into Table A–10 is in terms of the power $1 - \beta$.

When using Table A–10, four α levels are available at which the risk of making a Type I error can be controlled ($\alpha = .2, .1, .05, .01$). The Type II error risk can be controlled at one of four β levels ($\beta = .3, .2, .1, .05$) through the specification on the power $1 - \beta$. Table A–10 provides necessary sample sizes for studies consisting of $r = 2, \ldots, 10$ factor levels or treatments.

Example. A company owning a large fleet of trucks wishes to determine whether or not four different brands of snow tires have the same mean tread life (in thousands of miles). It is important to conclude that the four brands of snow tires have different mean tread lives when the difference between the means of the best and worst brands is 3 (thousand miles) or more. Thus, the minimum range specification is $\Delta = 3$. Suppose it is known from past experience that the standard deviation of the tread lives of these tires is 2 (thousand miles), approximately. Management would like to control the risks of making incorrect decisions at the following levels:

$$\alpha = .05$$
$$\beta = .10 \qquad \text{or} \qquad \text{Power} = 1 - \beta = .90$$

Entering Table A–10 for $\Delta/\sigma = 3/2 = 1.5$, $\alpha = .05$, $1 - \beta = .90$, and $r = 4$, we find $n = 14$. Hence, 14 snow tires of each brand need to be tested in order to control the risks of making incorrect decisions at the desired levels.

Specification of Δ/σ directly

Table A–10 can also be used when the minimum range is specified directly in units of the standard deviation σ. Let the specification of Δ in this case be $k\sigma$ so that we have by (18.4):

$$\frac{\Delta}{\sigma} = \frac{k\sigma}{\sigma} = k$$

Hence, Table A–10 is entered directly for the specified value k with this approach.

Example. Suppose it is specified for our snow tires example that it is important to detect differences between the mean tread lives if the range of the mean tread lives is $k = 2$ standard deviations or more. Suppose also that the other specifications are:

$$\alpha = .10$$

$$\beta = .05 \qquad \text{or} \qquad \text{Power} = 1 - \beta = .95$$

From Table A–10, we find for $k = 2$ and $r = 4$ that $n = 9$ tires will need to be tested for each brand in order that the specified risk protection will be achieved.

Note

While specifying Δ/σ directly does not require an advance planning value of the standard deviation σ, this is not of as much advantage as it might seem because a meaningful specification of Δ in units of σ will frequently require knowledge of the magnitude of the standard deviation.

Comments

1. The exact specification of Δ/σ has great effect on the sample sizes n when Δ/σ is small, but it has much less effect when Δ/σ is large. For instance, when $r = 3$, $\alpha = .05$, and $\beta = .10$, we have from Table A–10:

Δ/σ	n
1.0	27
1.5	13
2.0	8
2.5	6

Thus, unless Δ/σ is quite small, one need not be too concerned about some imprecision in specifying Δ/σ.

2. Reducing either the specified α or β risks or both increases the required sample sizes. For instance, when $r = 4$, $\alpha = .10$, and $\Delta/\sigma = 1.25$, we have:

β	$1 - \beta$	n
.20	.80	13
.10	.90	16
.05	.95	20

3. An error in the advance planning value of σ can cause a substantial miscalculation of needed sample sizes, although the needed sample sizes will still be ''in the same ball park.'' For instance, when $r = 5$, $\alpha = .05$, $\beta = .10$, and $\Delta = 3$, we have:

σ	Δ/σ	n
1	3.0	5
2	1.5	15
3	1.0	32

In view of the usual approximate nature of the advance planning value of σ, it is generally desirable to investigate the needed sample sizes for a range of likely values of σ before deciding on the sample sizes to be employed.

4. Table A–10 is based on the noncentrality parameter ϕ in (18.2) even though no specification is made of the individual factor level means μ_i for which it is important to conclude that the factor level means differ. Consider again the snow tires example where $r = 4$ brands are to be tested and a minimum range of $\Delta = 3$ (thousand miles) of the four mean tread lives μ_i is to be detected with high probability. The following are possible sets of values of the μ_i, each of which has range $\Delta = 3$:

Case	μ_1	μ_2	μ_3	μ_4	$\Sigma(\mu_i - \mu_.)^2$
1	24	27	25	26	5.00
2	25	25	26	23	4.75
3	25	25	25	28	6.75
4	25	25	26.5	23.5	4.50

The term $\Sigma(\mu_i - \mu_.)^2$ of the noncentrality parameter ϕ in (18.2) differs for each of these four possibilities and hence the power differs, even though the range is the same in all cases. Note that the term $\Sigma(\mu_i - \mu_.)^2$ is smallest for case 4, where two factor level means are at $\mu_.$ and the other two are equally spaced around $\mu_.$. Indeed, it can be shown that for a given range Δ, the term $\Sigma(\mu_i - \mu_.)^2$ is minimized when all but two factor level means are at $\mu_.$ and the two remaining factor level means are equally spaced around $\mu_.$. Thus, we have:

$$(18.5) \qquad \min \sum_{i=1}^{r} (\mu_i - \mu_.)^2 = \left(\frac{\Delta}{2}\right)^2 + \left(-\frac{\Delta}{2}\right)^2 + 0 + \cdots + 0 = \frac{\Delta^2}{2}$$

Since the power of the test varies directly with $\Sigma(\mu_i - \mu_.)^2$, use of (18.5) in calculating Table A–10 assures that the power is at least $1 - \beta$ for any combination of μ_i values with range Δ.

18.2 PLANNING OF SAMPLE SIZES WITH ESTIMATION APPROACH

The estimation approach to planning sample sizes may be used either in conjunction with the control of Type I and Type II errors or by itself. The essence of the approach is to specify the major comparisons of interest and to determine the expected widths of the confidence intervals for various sample sizes, given an advance planning value for the standard deviation. The approach is iterative, starting with an initial judgment of needed sample sizes. This initial judgment may be based on the needed sample sizes to control the risks of Type I and Type II errors when these have been previously obtained. If the anticipated widths of the confidence intervals based on the initial sample sizes are satisfactory, the iteration process is terminated. If one or more widths are too great, larger sample sizes should be tried next. If the widths are unnecessarily tight, smaller sample sizes should be tried next. This process is continued until those sample sizes are found that yield satisfactory anticipated widths.

Example

We are to plan sample sizes for our earlier snow tires example by means of the estimation approach, given that the sample sizes for each tire brand are to be equal, that is, $n_i \equiv n$. Management has indicated it wishes three types of estimates:

1. A comparison of the mean tread lives for each pair of brands:

$$\mu_i - \mu_{i'}$$

2. A comparison of the mean tread lives for the two high-priced brands (1 and 4) and the two low-priced brands (2 and 3):

$$\frac{\mu_1 + \mu_4}{2} - \frac{\mu_2 + \mu_3}{2}$$

3. A comparison of the mean tread lives for the national brands (1, 2, and 4) and the local brand (3):

$$\frac{\mu_1 + \mu_2 + \mu_4}{3} - \mu_3$$

Management further indicated that it wishes a family confidence coefficient of .95 for the entire set of comparisons.

We first need a planning value for the standard deviation of the tread lives of tires. Suppose that from past experience we judge the standard deviation to be approximately $\sigma = 2$ (thousand miles). Next, we require an initial judgment of needed sample size and shall consider $n = 10$ as a starting point.

We know from (17.17) that the variance of an estimated contrast $\hat{L}$ when $n_i \equiv n$ is:

$$\sigma^2(\hat{L}) = \frac{\sigma^2}{n}\Sigma c_i^2 \qquad \text{when } n_i \equiv n$$

Hence, given $\sigma = 2$ and $n = 10$, our anticipations of the standard deviations of the required estimators are:

Contrast	*Anticipated Variance*	*Anticipated Standard Deviation*
Pairwise comparisons	$\frac{(2)^2}{10}[(1)^2 + (-1)^2] = .80$	.89
High- and low-priced brands	$\frac{(2)^2}{10}\left[\left(\frac{1}{2}\right)^2 + \left(\frac{1}{2}\right)^2 + \left(-\frac{1}{2}\right)^2 + \left(-\frac{1}{2}\right)^2\right] = .40$	.63
National and local brands	$\frac{(2)^2}{10}\left[\left(\frac{1}{3}\right)^2 + \left(\frac{1}{3}\right)^2 + \left(\frac{1}{3}\right)^2 + (-1)^2\right] = .53$	.73

We shall employ the Scheffé multiple comparison procedure and therefore require the Scheffé multiple S in (17.33a) for $r = 4$, $n_T = nr = 10(4) = 40$, and $1 - \alpha = .95$:

$$S^2 = (r - 1)F(1 - \alpha; r - 1, n_T - r) = 3F(.95; 3, 36) = 3(2.87) = 8.61$$

or:

$$S = 2.93$$

Hence, the anticipated widths of the confidence intervals are:

Contrast	*Anticipated Width of Confidence Interval* $= \pm S\sigma(\hat{L})$
Pairwise comparisons	$\pm 2.93(.89) = \pm 2.61$ (thousand miles)
High- and low-priced brands	$\pm 2.93(.63) = \pm 1.85$ (thousand miles)
National and local brands	$\pm 2.93(.73) = \pm 2.14$ (thousand miles)

Management was satisfied with these anticipated widths. However, it was decided to increase the sample sizes from 10 to 15 in case the actual standard deviation of the tread lives of tires is somewhat greater than the anticipated value $\sigma = 2$ (thousand miles).

Comments

1. Since one cannot usually be certain that the planning value for the standard deviation is correct, it is advisable to study a range of values for the standard deviation before making a decision on sample size.

2. If the sample sizes for the factor levels are to be unequal, one can still use the iterative procedure with the estimation approach just described. For instance, suppose that four new flavors of a fruit syrup are each to be compared with a control, the present flavor. One may therefore wish to make the control sample size larger in order to increase the precision of these key comparisons. Suppose the control sample size is to be twice as large as the other factor level sample sizes. Then we can represent the control sample size as $2n$ and the other sample sizes as n. We then proceed as before with an initial specification for n, but now utilize formula (17.17) for the variance of an estimated contrast in its general form since the sample sizes are unequal.

18.3 PLANNING OF SAMPLE SIZES TO FIND "BEST" TREATMENT

There are occasions when the chief purpose of the study is to ascertain the factor level or treatment with the highest or lowest mean. In our snow tires example, for instance, it may be desired to determine which of the four brands has the longest mean tread life.

Bechhofer has developed a table, shown in Table A–11, that enables us to determine the necessary sample size so that with probability $1 - \alpha$ the highest (lowest) factor level sample mean is from the factor level with the highest (lowest) population mean. We need to specify the probability $1 - \alpha$, the standard deviation σ, and the smallest difference λ between the highest (lowest) and second highest (second lowest) factor level means that it is important to recognize. Table A–11 assumes that equal sample sizes are to be used for all factor levels, and that the analysis of variance model (16.2) is appropriate.

Example

Suppose that in our snow tires example, the chief objective is to identify the brand with the longest mean tread life. There are $r = 4$ brands. We anticipate, as before, that $\sigma = 2$ (thousand miles). Further, we are informed that a difference $\lambda = 1$ (thousand miles) between the highest and second highest brand means is important to recognize, and that the probability is to be $1 - \alpha = .90$ or greater that we identify correctly the brand with the highest mean tread life when $\lambda \geq 1$.

The entry in Table A–11 is $\lambda\sqrt{n}/\sigma$. For $r = 4$ and probability $1 - \alpha = .90$, we find from Table A–11 that $\lambda\sqrt{n}/\sigma = 2.4516$. Hence, since the λ specification is $\lambda = 1$, we obtain:

$$\frac{(1)\sqrt{n}}{2} = 2.4516$$

$$\sqrt{n} = 4.9032 \qquad \text{or} \qquad n = 25$$

Thus, when the mean tread life for the best brand exceeds that of the second best by at least 1 (thousand miles) and when $\sigma = 2$ (thousand miles), sample sizes of 25 tires for each brand provide an assurance of at least .90 that the brand with the highest sample mean is the brand with the highest population mean.

Note

If the planning value for the standard deviation is not accurate, the probability of identifying the population with the highest (lowest) mean correctly is, of course, affected. This is no different from the other approaches, where a misjudgment of the standard deviation affects the risks of making a Type II error or the widths of the confidence intervals actually obtained.

18.4 RESIDUAL ANALYSIS

When discussing regression analysis, we emphasized the importance of examining the aptness of the regression model and pointed out the effectiveness of residual plots for spotting major departures from the assumed model. Examination of aptness is no less important with an analysis of variance model, and residual plots again are most helpful for spotting major departures from the assumed model.

Since residual plots for ANOVA models are used in a similar manner as those for regression analysis, we shall give only a brief discussion of them here. Before doing so, we should note again the proper sequence of using any statistical model:

1. First, one should examine whether the proposed model is appropriate for the set of data at hand.
2. If the proposed model is not appropriate, corrective measures such as transformations of the data may have to be taken, or the model may need to be modified.
3. Only after this review of the aptness of the model has been made and any necessary corrective measures completed and their effectiveness ascertained, should the analysis of the data be undertaken.

It is not necessary, nor is it usually possible, that the ANOVA model fits perfectly. As will be noted later, the ANOVA model is reasonably robust against certain types of departures from the model, such as the error terms not being exactly normally distributed. The major purpose of the examination of the aptness of the model is therefore to detect serious departures from the conditions assumed by the model.

Residuals

The residuals e_{ij} for the ANOVA model (16.2) were defined in (16.17):

(18.6) $$e_{ij} = Y_{ij} - \hat{Y}_{ij} = Y_{ij} - \bar{Y}_{i.}$$

As in regression, standardized residuals:

(18.7) $$\frac{e_{ij}}{\sqrt{MSE}}$$

are sometimes helpful. An alternative standardized residual is useful at times, as will be seen shortly:

(18.8) $$\frac{e_{ij}}{s_i}$$

Here, s_i is the sample standard deviation of the observations from the ith factor level, as defined in (16.35).

Residual plots

Residual plots for analysis of variance models take the form of residual sequence plots or residual frequency plots. *Residual sequence plots* should be used whenever a logical sequence of the residuals exists, such as time order or geographic order of the observations. Figure 18.1 illustrates residual sequence plots for the Kenton Food Company residuals of Table 16.2, where the observation order represents the geographic sequence of the stores assigned to a treatment. It is helpful to plot the residuals for each treatment separately, either side by side as illustrated in Figure 18.1 or in a vertical arrangement.

FIGURE 18.1 Residual sequence plots for Kenton Food Company example

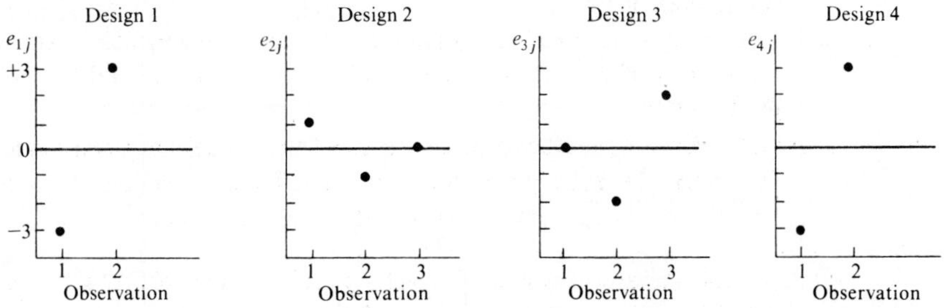

FIGURE 18.2 Residual frequency plots for Kenton Food Company example

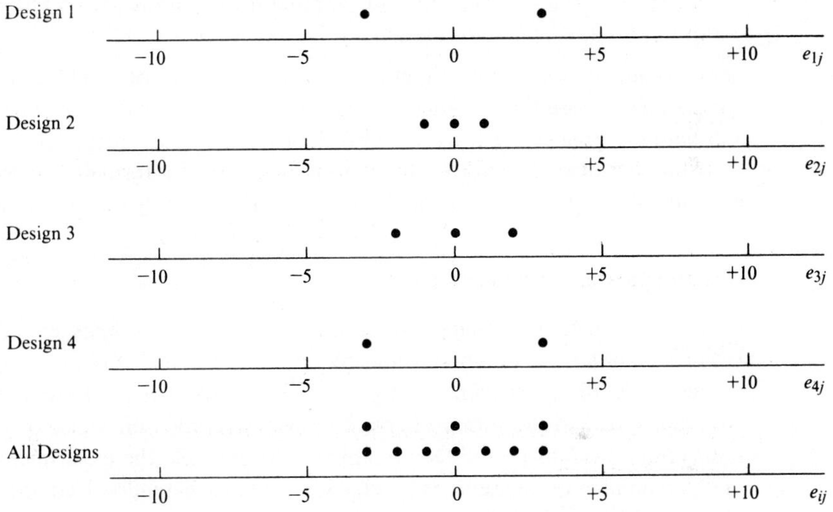

Residual frequency plots are useful when no logical order of the observations exists. Figure 18.2 illustrates residual frequency plots for the Kenton Food Company data. Here, the frequency plots are arranged vertically.

We consider now briefly how residual plots reveal various possible departures from analysis of variance model (16.2). None of these possible departures is strongly suggested in Figures 18.1 or 18.2 for the Kenton Food Company data.

Nonconstancy of error variance

ANOVA model (16.2) requires that the error terms ε_{ij} have constant variance for all factor levels. When this condition is met, the residual plots should show about the same extent of scatter of the residuals around 0 for each factor level. Figure 18.3 illustrates residual sequence plots that suggest the error variances are

FIGURE 18.3 Residual sequence plots suggesting unequal factor level variances

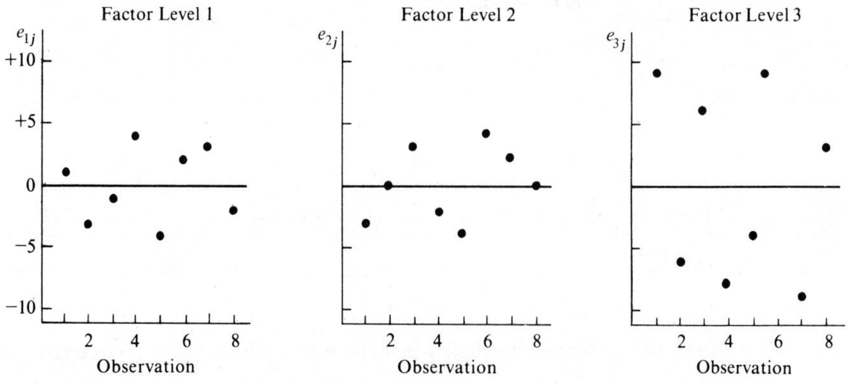

not constant. It would appear that the error terms for factor level 3 have a larger variance than those for the other two factor levels, because the extent of scatter is substantially greater for factor level 3 than for the other factor levels.

A number of statistical tests have been developed for formally examining the equality of r variances; two of these tests will be discussed in Section 18.6.

Nonindependence of error terms

Whenever data are obtained in a time sequence, the residuals should be plotted in time order to examine if they are serially correlated. Figure 18.4 contains the residuals for an experiment on group interactions. Three different treatments were applied, and the group interactions were recorded on video tapes. Seven replications were made for each treatment. Afterwards, the experimenter measured the number of interactions by viewing the tapes in randomized order. Figure 18.4 strongly suggests that the experimenter discerned a larger number of interactions as he gained experience in viewing the tapes, as a result of which the residuals in Figure 18.4 appear to be serially correlated. In this instance, an inclusion in the model of a linear term for the time effect might be sufficient to assure independence of the error terms in the revised model.

Time-related effects may also lead to increases or decreases in the error variance over time. For instance, an experimenter may make more precise measurements over time. Figure 18.5 portrays residual sequence plots where the error variance decreases over time.

Outliers

Residual plots easily show an outlier observation, that is, an observation that differs from the fitted value $\overline{Y}_{i.}$ by far more than do other observations. As noted in Chapter 4, it would appear to be wise practice to reject outlier observations only if they can be identified as being due to such specific causes as instrumentation malfunctioning, observer measurement blunder, or recording error.

FIGURE 18.4 Residual sequence plots for group interaction study illustrating time-related effect

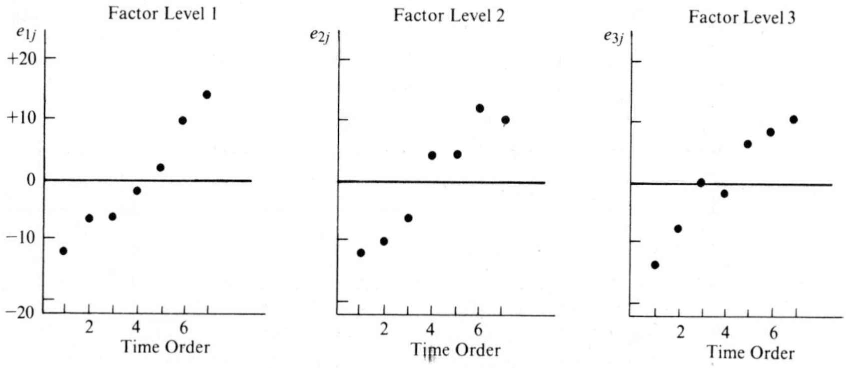

FIGURE 18.5 Residual sequence plots illustrating decreasing error variance over time

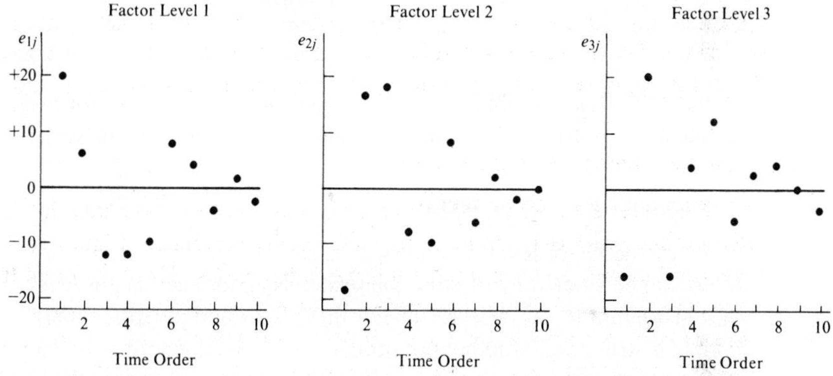

Omission of important independent variables

Residual analysis may also be used to study whether or not the single-factor ANOVA model is too coarse a model. In a learning experiment involving three motivational treatments, the residuals shown in Figure 18.6 were obtained. The residual sequence plots in Figure 18.6 show no unusual patterns. The experimenter wondered, however, whether the treatment effects differ according to the sex of the subject. The residuals for male subjects are circled in Figure 18.6; those for females are not. The results in Figure 18.6 suggest strongly that for each of the motivational treatments studied, the treatment effects do differ according to sex. Hence, a multifactor model recognizing both motivational treatment and sex of subject as independent variables might be more useful.

It should be noted that residual analysis here does not invalidate the original single-factor model. Rather, the residual analysis points out that the original model overlooks differences in treatment effects that may be important to recognize. Since there are usually many independent variables that have some effect

FIGURE 18.6 Residual sequence plots illustrating omission of important independent variable

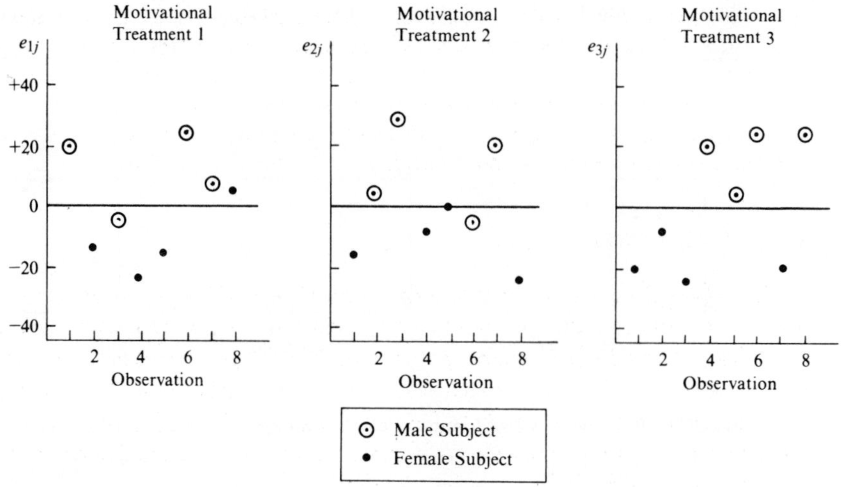

on the dependent variable, the analyst needs to identify those for residual analysis most likely to have an important effect on the dependent variable.

Nonnormality of error terms

When the factor level sample sizes n_i are not small, nonnormality of the error terms can be studied by plotting the residuals e_{ij} for each factor level in the form of a histogram to see whether or not these distributions differ markedly from a normal distribution. Another possibility is to prepare a normal probability plot of the residuals and see whether or not the points fall approximately on a straight line. The linearity can also be assessed by calculating the coefficient of correlation relating the residuals to their expected values under normality (Section 4.2). Chi-square goodness of fit tests or related tests may also be employed. When the factor level sample sizes are not large, one can combine the residuals e_{ij} for all treatments into one group, provided that the evidence suggests that there are no major differences in the error term variances for the treatments studied.

Alternatively, one can work with the standardized residuals (18.7). When normality exists and the number of residuals is not small, one would expect roughly 95 percent of the standardized residuals to fall between -2 and $+2$, and so on. When the number of residuals is small, one should use the appropriate t distribution for the analysis.

It should be noted for all of these analyses that one reason why nonnormality may be indicated is failure of the error terms ε_{ij} to have equal variances. When the factor level sample sizes are not large and unequal variances of the error terms for the different factor levels are indicated, the standardized residuals (18.8) should be used before combining all the residuals for a study of normality.

Note

As we pointed out for regression models, the residuals e_{ij} are not independent random variables. For the ANOVA model (16.2), they are subject to restrictions stated earlier in (16.18). Consequently, statistical tests that require independent observations are not exactly appropriate for residuals. If, however, the number of residuals for each factor level is large, the effect of the correlations will be small. It has been noted that graphic plots of residuals are less subject to the effects of correlation than are statistical tests because graphic plots contain the individual residuals and not simply functions of them.

18.5 TRANSFORMATIONS

If residual or other analysis indicates that the ANOVA model is not appropriate for the data at hand, a choice of possible corrective measures is available. One is to modify the model. This approach may have the disadvantage at times of leading to fairly complex analysis. Another approach is to use transformations on the data. A third approach, useful when serious lack of normality is the basic difficulty, is to employ nonparametric tests such as the median test or the Kruskal-Wallis test based on ranks (discussed in Chapter 19).

The use of transformations is the principal topic of this section. Since transformations were discussed in Chapter 4 in connection with regression analysis, our discussion here will be brief.

Transformations to stabilize variances

There are several types of situations in which the variances of the error terms are not constant, each of which requires a different type of transformation to stabilize the variance.

Variance proportional to μ_i. When the variance of the error terms for each factor level (denoted by σ_i^2) varies proportionally to the factor level mean μ_i, the sample statistics $s_i^2/\bar{Y}_{i.}$ will tend to be constant. Here s_i^2 is the sample variance for the ith factor level observations, as defined in (16.35). This type of situation is often found when the observed variable Y is a count, such as the number of attempts by a subject before the correct solution is found. For this case, a square root transformation is helpful for stabilizing the variances:

(18.9) If σ_i^2 proportional to μ_i:

$$Y' = \sqrt{Y} \qquad \text{or} \qquad Y' = \sqrt{Y} + \sqrt{Y+1}$$

Standard deviation proportional to μ_i. When the standard deviation of the error terms for each factor level is proportional to the mean, $s_i/\bar{Y}_{i.}$ tends to be constant for the different factor levels. A helpful transformation in this case for stabilizing the variances is the logarithmic transformation:

(18.10) If σ_i proportional to μ_i: $\quad Y' = \log Y$

Standard deviation proportional to μ_i^2. When the error term standard deviation is proportional to the square of the factor level mean, $s_i/\bar{Y}_{i.}^2$ tends to be constant. An appropriate transformation here is the reciprocal transformation:

$$\text{(18.11)} \qquad \text{If } \sigma_i \text{ proportional to } \mu_i^2\text{:} \qquad Y' = \frac{1}{Y}$$

Dependent variable is a proportion. At times, the observed variable Y_{ij} is a proportion p_{ij}. For instance, the treatments may be different training procedures, the unit of observation is a company training class, and the observed variable Y_{ij} is the proportion of employees in the jth class for the ith training procedure who benefited substantially by the training. Note that n_i here refers to the number of classes receiving the ith training procedure, not to the number of students.

It is well known that in the binomial case the variance of the sample proportion depends on the true proportion. When the number of cases on which each sample proportion is based is the same, this variance is:

$$\text{(18.12)} \qquad \sigma^2(p_{ij}) = \frac{\pi_i(1-\pi_i)}{m}$$

Here π_i is the population proportion for the ith treatment and m is the common number of cases on which each sample proportion is based. Since $\sigma^2(p_{ij})$ depends on the treatment proportion π_i, the variances of the error terms will not be stable if the treatment proportions π_i differ. An appropriate transformation for this case is the arc sine transformation:

$$\text{(18.13)} \qquad \text{If observation is a proportion: } Y' = 2 \arcsin \sqrt{Y}$$

When the proportions p_{ij} are based on different numbers of cases (for instance, in our earlier illustration there may be different numbers of employees in each training class), transformation (18.13) should be employed together with a weighted least squares analysis as described in Section 5.7.

Transformations to correct lack of normality

When the error terms are normally distributed but with unequal variances, a transformation of the observations to stabilize the variances will destroy the normality. Fortunately, in practice, lack of normality and unequal variances tend to go hand in hand. Further, the transformation that helps to correct the lack of constancy of error variances usually also is effective in making the distribution of the error terms more normal. It is wise policy, however, as mentioned in Chapter 4 to check the residuals after a transformation has been applied to make sure that the transformation has been effective in both stabilizing the variances and making the distribution of the error terms reasonably normal.

Comments

1. When a transformation of the observations is required, one can work completely with the transformed data for testing the equality of factor level means. On the other hand, it is often desirable when making estimates of factor effects to change a confidence interval based on the transformed variable back to an interval in the original variable for easier understanding of the significance of the results. For instance, in a pricing experiment in which the dependent variable sales is transformed using logarithms, most persons would have difficulty interpreting a confidence interval statement about $\log \mu_2 - \log \mu_1$. Since the difference of the two logs is equivalent to the log of the ratio of the means, $\log(\mu_2/\mu_1)$, the antilogs of the confidence limits for the difference between the means of the transformed variable provide information about the ratio of the mean sales for the two treatments, a quantity that is readily understood.

2. It is generally advisable to plot s_i against $\bar{Y}_{i.}$ so that one can study the nature of the relationship between them, for instance, whether it is linear or quadratic. The choice of an appropriate transformation is thereby facilitated.

3. The transformations suggested earlier are obtained by the following argument. Suppose that Y is a random variable with mean μ and variance σ_Y^2 that is a function of μ, say, $\sigma_Y^2 = g(\mu)$. For a transformation of Y, say, $Y' = h(Y)$, it can be shown using a Taylor series expansion that:

$$\sigma_{Y'}^2 \cong [h'(\mu)]^2 g(\mu) \tag{18.14}$$

where $h'(\mu)$ is the first derivative of $h(Y)$ at μ. We seek that transformation h for which $\sigma_{Y'}^2$ is a constant, for convenience, say, 1. Hence, we wish:

$$[h'(\mu)]^2 g(\mu) = 1$$

or:

$$h'(\mu) = \frac{1}{\sqrt{g(\mu)}} \tag{18.15}$$

Formula (18.15) is a differential equation whose solution is (neglecting the arbitrary constant):

$$h(\mu) = \int \frac{d\mu}{\sqrt{g(\mu)}} \tag{18.16}$$

To illustrate the use of (18.16), suppose σ_Y^2 is proportional to μ, say, $\sigma_Y^2 = k\mu = g(\mu)$. We obtain from (18.16):

$$h(\mu) = \int \frac{d\mu}{\sqrt{k\mu}} = \frac{2}{\sqrt{k}}\sqrt{\mu}$$

Thus, a square root transformation $Y' = \sqrt{Y}$ will yield a constant variance of Y', and $Y' = (2/\sqrt{k})\sqrt{Y}$ will yield a constant variance equal to 1.

Formula (18.14) makes it clear that the transformations obtained by this procedure only approximately stabilize the variances. It is therefore important to inspect the residuals for the transformed variable to see how effectively the variances have actually been stabilized.

18.6 TESTS FOR EQUALITY OF VARIANCES

Several formal tests are available for studying whether or not r populations have equal variances, as required by the ANOVA model. We shall consider two of these, the Bartlett test and the Hartley test. Both assume that each of the r populations is normal. Also, it is assumed by both tests that independent random samples are obtained from each population. The Bartlett test is an all-purpose test and can be used for equal and unequal sample sizes. The Hartley test is applicable only if the sample sizes are equal, and it is designed to be sensitive against substantial differences between the largest and the smallest population variances.

Bartlett test

The basic idea underlying the Bartlett test is simple. Let $s_1^2, \ldots, s_r^2$ denote the sample variances from r normal populations, and let df_i denote the degrees of freedom associated with the sample variance s_i^2. Then the weighted arithmetic average of the sample variances using the associated degrees of freedom df_i as weights is the mean square error:

$$MSE = \frac{1}{df_T} \sum_{i=1}^{r} df_i s_i^2 \tag{18.17}$$

where:

$$df_T = \sum_{i=1}^{r} df_i$$

Similarly, the weighted geometric average of the s_i^2, denoted by $GMSE$, is:

$$GMSE = [(s_1^2)^{df_1}(s_2^2)^{df_2} \cdots (s_r^2)^{df_r}]^{1/df_T} \tag{18.18}$$

It can be shown that for any given set of s_i^2 values, the following relation holds between these two averages:

$$GMSE \leq MSE \tag{18.19}$$

The two averages are equal if all s_i^2 are equal; the greater the variation between the s_i^2, the further apart the two averages will be. Hence, if the ratio $MSE/GMSE$ is close to 1, we have evidence that the population variances are equal. If the ratio $MSE/GMSE$ is large, it indicates that the population variances are unequal. The same conclusions follow if we consider $\log(MSE/GMSE) = \log MSE - \log GMSE$.

Bartlett has shown that a function of $\log MSE - \log GMSE$ follows, for large sample sizes, approximately the χ^2 distribution with $r - 1$ degrees of freedom when the population variances are equal. This test statistic is:

$$B = \frac{df_T}{C}(\log_e MSE - \log_e GMSE) \tag{18.20}$$

where:

$$C = 1 + \frac{1}{3(r-1)}\left[\left(\sum_{i=1}^{r} \frac{1}{df_i}\right) - \frac{1}{df_T}\right]$$

The term C is always greater than 1.

Test statistic (18.20) reduces to:

(18.20a) $$B = \frac{1}{C}\left[(df_T)\log_e MSE - \sum_{i=1}^{r}(df_i)\log_e s_i^2\right]$$

For deciding between:

(18.21) $$\begin{aligned} &H_0: \sigma_1^2 = \sigma_2^2 = \cdots = \sigma_r^2 \\ &H_a: \text{not all } \sigma_i^2 \text{ are equal} \end{aligned}$$

we calculate test statistic B. Since B is approximately distributed as χ^2 with $r - 1$ degrees of freedom when H_0 holds, and since we saw that large values of B lead to conclusion H_a, the appropriate decision rule for controlling the risk of a Type I error at α is:

(18.22) $$\begin{aligned} &\text{If } B \leq \chi^2(1-\alpha; r-1), \text{ conclude } H_0 \\ &\text{If } B > \chi^2(1-\alpha; r-1), \text{ conclude } H_a \end{aligned}$$

where $\chi^2(1-\alpha; r-1)$ is the $(1-\alpha)100$ percentile of the χ^2 distribution with $r - 1$ degrees of freedom. Percentiles of the χ^2 distribution are given in Table A–3. The chi-square approximation may be considered to be appropriate when all degrees of freedom df_i are four or more.

When the Bartlett test is used for the single-factor analysis of variance model (16.2), we have:

$$df_i = n_i - 1 \qquad df_T = \sum_{i=1}^{r}(n_i - 1) = n_T - r$$

Example. Table 18.1 contains data from a study on the time required to complete a certain production operation for each of the three plant shifts. Twenty cycles of the operation were performed by shift one, 17 by shift two, and 21 by

TABLE 18.1 Bartlett test for equality of three population variances

Population i	s_i^2	$df_i = n_i - 1$	$(df_i)s_i^2$	$\log_e s_i^2$	$(df_i)\log_e s_i^2$
1	415	19	7,885	6.02828	114.53732
2	698	16	11,168	6.54822	104.77152
3	384	20	7,680	5.95064	119.01280
Total		$df_T = 55$	26,733		338.32164

$MSE = 26{,}733 \div 55 = 486.05$

$\log_e MSE = 6.18631$

shift three. It has been ascertained that the three populations are close to normal ones. We now wish to determine by the Bartlett test whether the shift variances σ_i^2 are the same for all three shifts.

The needed calculations for the Bartlett test are shown in Table 18.1. Substituting into (18.20) for C and (18.20a) for B, we obtain:

$$C = 1 + \frac{1}{3(3-1)}\left[\left(\frac{1}{19} + \frac{1}{16} + \frac{1}{20}\right) - \frac{1}{55}\right] = 1.02449$$

and:

$$B = \frac{1}{1.02449}[55(6.18631) - 338.32164] = \frac{1.92541}{1.02449} = 1.8794$$

Assume that we are to control the risk of making a Type I error at $\alpha = .05$. We therefore require $\chi^2(.95; 3-1)$. From Table A–3, we find $\chi^2(.95; 2) = 5.99$. Hence, the decision rule is:

$$\text{If } B \leq 5.99, \text{ conclude } H_0$$
$$\text{If } B > 5.99, \text{ conclude } H_a$$

Since $B = 1.88 \leq 5.99$, we conclude H_0, that the three population variances are equal.

Note

In the above example, we could have avoided calculating the denominator C. Even before dividing by C, we can see that the numerator of the test statistic, 1.92541, falls below the action limit 5.99. Since $C > 1$ always, the effect of dividing by C is to make the test statistic B still smaller. Thus, one may compute the numerator of B first, and only calculate the denominator C if it can affect the outcome.

Box transformation. As noted earlier, the chi-square approximation to the distribution of the Bartlett test statistic (18.20) when the population variances are equal should not be used when any of the df_i are less than four. An approximation that can be used when some of the degrees of freedom are small and that also is appropriate for larger numbers of degrees of freedom utilizes the F distribution and is based on the modified Bartlett test statistic B' developed by Box:

(18.23)
$$B' = \frac{f_2 BC}{f_1(A - BC)}$$

where:

$$f_1 = r - 1$$
$$f_2 = \frac{r+1}{(C-1)^2}$$
$$A = \frac{f_2}{2 - C + \dfrac{2}{f_2}}$$

For testing the alternatives (18.21), the appropriate decision rule for controlling the risk of a Type I error at α is:

$$\text{If } B' \leq F(1-\alpha; f_1, f_2), \text{ conclude } H_0$$
$$\text{If } B' > F(1-\alpha; f_1, f_2), \text{ conclude } H_a \tag{18.24}$$

where $F(1-\alpha; f_1, f_2)$ is the $(1-\alpha)100$ percentile of the F distribution with f_1 and f_2 degrees of freedom. Percentiles of the F distribution are given in Table A–4. The value of f_2 will usually not be an integer so that it will then be necessary to interpolate in the F table. Reciprocal interpolation should be employed, as illustrated in the following example.

Example. We shall employ again the example in Table 18.1. We found earlier:

$$C = 1.02449 \qquad B = 1.8794$$

We now require:

$$f_1 = 3 - 1 = 2$$

$$f_2 = \frac{3+1}{(1.02449 - 1.0)^2} = 6{,}669.3$$

$$A = \frac{6{,}669.3}{2 - 1.02449 + \dfrac{2}{6{,}669.3}} = 6{,}834.63$$

$$B' = \frac{6{,}669.3(1.8794)(1.02449)}{2[6{,}834.63 - 1.8794(1.02449)]} = .94$$

For controlling the level of significance at $\alpha = .05$, we require $F(.95; 2, 6{,}669.3)$. From Table A–4, we find:

$$F(.95; 2, 120) = 3.07 \qquad F(.95; 2, \infty) = 3.00$$

Reciprocal interpolation is similar to linear interpolation except the reciprocals are used for determining the fraction of the difference between 3.00 and 3.07, as follows:

$$F(.95; 2, 6{,}669.3) = 3.07 + \frac{\dfrac{1}{6{,}669.3} - \dfrac{1}{120}}{\dfrac{1}{\infty} - \dfrac{1}{120}}(3.00 - 3.07)$$

$$= 3.001$$

Hence, the decision rule is:

$$\text{If } B' \leq 3.001, \text{ conclude } H_0$$
$$\text{If } B' > 3.001, \text{ conclude } H_a$$

Since $B' = .94 \leq 3.001$, we conclude H_0, that the three population variances are equal. This is the same conclusion as we obtained with the Bartlett test statistic B and the chi-square approximation.

Comments

1. The Bartlett test is quite sensitive to departures from normality. That is, if the populations in fact are not normal, the actual level of significance may differ substantially from the specified one. Hence, if the populations depart substantially from normality, it is not recommended that the Bartlett test be employed for testing equality of variances. Instead, a robust nonparametric test should be employed. Reference 18.1 cites several such tests.

2. As will be noted in Section 18.7, the F test for equality of factor level means is not much affected by unequal variances when the factor level sample sizes are approximately equal, as long as the differences in the variances are not unusually large. Hence, if the populations are reasonably normal so that the Bartlett test can be employed and the sample sizes do not differ greatly, a fairly low α level may be justified in testing the equality of variances for determining the aptness of the ANOVA model, since only large differences between variances need be detected.

Hartley test

If each of the r sample variances s_i^2 has the same number of degrees of freedom so that $df_i \equiv df$, a simple test for deciding between:

$$\begin{aligned} &H_0\colon \sigma_1^2 = \sigma_2^2 = \cdots = \sigma_r^2 \\ &H_a\colon \text{not all } \sigma_i^2 \text{ are equal} \end{aligned} \tag{18.25}$$

is due to Hartley. It is based solely on the largest sample variance, denoted by $\max(s_i^2)$, and the smallest sample variance, denoted by $\min(s_i^2)$. The test statistic is:

$$H = \frac{\max(s_i^2)}{\min(s_i^2)} \tag{18.26}$$

Clearly, values of H near 1 support H_0, and large values of H support H_a. The distribution of H when H_0 holds has been tabulated, and selected percentiles are presented in Table A–12. The distribution of H depends on the number of populations r and the common number of degrees of freedom df. As we noted earlier, the Hartley test, like the Bartlett test, assumes normal populations.

The appropriate decision rule for controlling the risk of making a Type I error at α is:

$$\begin{aligned} &\text{If } H \leq H(1-\alpha; r, df), \text{ conclude } H_0 \\ &\text{If } H > H(1-\alpha; r, df), \text{ conclude } H_a \end{aligned} \tag{18.27}$$

where $H(1-\alpha; r, df)$ is the $(1-\alpha)100$ percentile of the distribution of H when H_0 holds, for r populations and df degrees of freedom for each sample variance.

When the Hartley test is used for the single-factor ANOVA model (16.2) with equal sample sizes $n_i \equiv n$, we have $df = n - 1$.

Example. In a study of the appeal of four different television commercials, 10 observations were made for each commercial. The sample variances were as follows:

$$s_1^2 = 193 \qquad s_2^2 = 146 \qquad s_3^2 = 215 \qquad s_4^2 = 128$$

Before proceeding with the analysis of variance, it was determined that the populations are close to normal and it is now desired to use the Hartley test to ascertain whether or not the four treatment variances are equal:

$$H_0\text{: } \sigma_1^2 = \sigma_2^2 = \sigma_3^2 = \sigma_4^2$$
$$H_a\text{: not all } \sigma_i^2 \text{ are equal}$$

The level of significance is to be controlled at $\alpha = .05$. For $r = 4$ and $df = n - 1 = 9$, we require from Table A–12 $H(.95; 4, 9) = 6.31$. Hence, the appropriate decision rule is:

$$\text{If } H \leq 6.31, \text{ conclude } H_0$$
$$\text{If } H > 6.31, \text{ conclude } H_a$$

We have $\max(s_i^2) = 215$ and $\min(s_i^2) = 128$. Hence:

$$H = \frac{215}{128} = 1.68$$

Since $H = 1.68 \leq 6.31$, we conclude H_0, that all four treatment variances are equal.

Comments

1. The Hartley test strictly requires equal sample sizes. If the sample sizes are unequal but do not differ greatly, the Hartley test may still be used as an approximate test. For this purpose, the average number of degrees of freedom would be used for entering Table A–12.

2. The Hartley test, like the Bartlett test, is quite sensitive to departures from the assumption of normal populations and should not be used when substantial departures from normality exist.

3. For the same reasons noted for the Bartlett test, a low α level may be justified when the Hartley test is used for determing the aptness of the ANOVA model with respect to equal treatment variances.

18.7 EFFECTS OF DEPARTURES FROM MODEL

In preceding sections, we considered how residual analysis and other statistical techniques can be helpful in assessing the aptness of the ANOVA model for the data at hand. We also discussed the use of transformations, chiefly for stabilizing variances but also for obtaining as a by-product error distributions more nearly normal. The question now arises what are the effects of any remaining departures from the model on the inferences made. A thorough review of the many studies investigating these effects has been made by Scheffé (Ref. 18.2). Here, we shall summarize the findings.

Nonnormality

For the fixed effects model, lack of normality is not an important matter, provided the departure from normality is not of extreme form. It may be noted in this connection that *kurtosis* of the error distribution (either more or less peaked than a normal distribution) is more important than skewness of the distribution in terms of the effects on inferences.

The point estimators of factor level means and contrasts are unbiased whether or not the populations are normal. The F test for the equality of factor level means is but little affected by lack of normality, either in terms of the level of significance or power of the test. Hence, the F test is a *robust* test against departures from normality. For instance, the specified level of significance might be .05 whereas the actual level for a nonnormal error distribution might be .04 or .065. Typically, the achieved level of significance in the presence of nonnormality is slightly higher than the specified one, and the achieved power of the test is slightly less than the calculated one. Single interval estimates of factor level means and contrasts and the Scheffé multiple comparison method also are not much affected by lack of normality provided the sample sizes are not extremely small.

For the random effects model (to be discussed in the next chapter), lack of normality has more serious implications. The estimators of the variance components are still unbiased, but the actual confidence coefficient for interval estimates may be substantially different from the specified one.

Unequal error variances

When the error variances are unequal, the F test for the equality of means with the fixed effects model is only slightly affected if all factor level sample sizes are equal or do not differ greatly. Specifically, unequal error variances then raise the actual level of significance only slightly higher than the specified level. Similarly, the Scheffé multiple comparison procedure based on the F distribution is not affected to any substantial extent by unequal variances when the sample sizes are equal or are approximately the same. Thus, the F test and related analyses are robust against unequal variances when the sample sizes are approximately equal. Single comparisons between factor level means, on the other hand, can be substantially affected by unequal variances, so that the actual and specified confidence coefficients may differ markedly in these cases.

The use of equal sample sizes for all factor levels not only tends to minimize the effects of unequal variances on inferences with the F distribution but also simplifies calculational procedures. Thus, here at least, simplicity and robustness go hand in hand.

For the random effects model, unequal error variances can have pronounced effects on inferences about the variance components, even with equal sample sizes.

Nonindependence of error terms

Lack of independence of the error terms can have serious effects on inferences in the analysis of variance, for both fixed and random effects models. Since this defect is often difficult to correct, it is important to prevent it in the first place whenever feasible. The use of randomization in those stages of a study that are likely to lead to correlated error terms can be a most important insurance policy. In the case of observational data, however, randomization may not be possible. Here, in the presence of correlated error terms, one may be able to modify the model. For instance, in the earlier discussion based on Figure 18.4, it was noted that inclusion in the model of a linear term for the learning effect of the analyst might remove the correlation of the error terms.

Modification of the model because of correlated error terms may also be necessary in experimental studies. In one case, the experimenter asked each of 10 subjects to give ratings to four new flavors of a fruit syrup and to the standard flavor, on a scale from 0 to 100. She applied the single-factor analysis of variance model but found high degrees of correlation in the residuals for each subject. She thereupon modified her model to a randomized complete block design model (Chapter 28), which is often appropriate when the same subject is given each of the different treatments and differences between subjects are expected.

PROBLEMS

18.1. A market researcher stated in a seminar: "The power approach to determining sample sizes for analysis of variance problems is not meaningful; only the estimation approach should be used. We never conduct a study where all treatment means are expected to be equal, so we are always interested in a variety of estimates." Discuss.

18.2. Why do you think that the approach to planning sample sizes to find the best treatment by means of Table A–11 does not consider the risk of an incorrect identification when the best two treatment means are the same or practically the same?

18.3. Consider a single-factor study where $r = 5$, $\alpha = .01$, $\beta = .05$, and $\sigma = 10$, and equal treatment sample sizes are desired by means of the approach in Table A–10.

a. What are the required sample sizes if $\Delta = 10, 15, 20, 30$? What generalization is suggested by your results?

b. What are the required sample sizes for the same values of Δ as in part (a) if $\alpha = .05$, all other specifications remaining the same? How do these sample sizes compare with those in part (a)?

18.4. Consider a single-factor study where $r = 6$, $\alpha = .05$, $\beta = .10$, and $\Delta = 50$, and equal treatment sample sizes are desired by means of the approach in Table A–10.

a. What are the required sample sizes if $\sigma = 50, 25, 20$? What generalization is suggested by your results?

b. What are the required sample sizes for the same values of σ as in part (a) if $r = 4$, all other specifications remaining the same? How do these sample sizes compare with those in part (a)?

18.5. Consider a single-factor study where $r = 5$, $1 - \alpha = .95$, and $\sigma = 20$, and equal sample sizes are desired by means of the approach in Table A–11.

a. What are the required sample sizes if $\lambda = 20, 10, 5$? What generalization is suggested by your results?

b. What are the required sample sizes for the same values of λ as in part (a) if $\sigma = 30$, all other specifications remaining the same? How do these sample sizes compare with those in part (a)?

18.6. Refer to **Questionnaire color** Problem 16.10. Suppose that the sample sizes have not yet been determined but it has been decided to sample the same number of supermarket parking lots for each questionnaire color. A reasonable planning value for the error term standard deviation is $\sigma = 3.0$.

a. What would be the required sample sizes if: (1) differences in the response rates are to be detected with probability .90 or more when the range of the treatment means is 4.5, and (2) the α risk is to be controlled at .05?

b. If the sample sizes determined in part (a) were employed, what would be the minimum power of the test for treatment mean differences (using $\alpha = .05$) when the range of the treatment means is 6.0?

c. Suppose pairwise comparisons are of primary importance. What would be the required sample sizes if the precision of all pairwise comparisons is to be ± 3.0, using the Tukey procedure with a 95 percent family confidence coefficient?

d. Suppose the chief objective is to identify the color with the highest mean response rate. The probability should be at least .99 that the best color is recognized correctly when the difference between the response rates for the best and second best colors is 1.5 percent points or more. What are the required sample sizes?

18.7. Refer to **Rehabilitation therapy** Problem 16.11. Suppose that the sample sizes have not yet been determined but is has been decided to use the same number of patients for each physical fitness group. Assume that a reasonable planning value for the error term standard deviation is $\sigma = 4.5$ days.

a. What would be the required sample sizes if: (1) differences in the mean times for the three physical fitness categories are to be detected with probability .80 or more when the range of the treatment means is 5.63 days, and (2) the α risk is to be controlled at .01?

b. If the sample sizes determined in part (a) were employed, what would be the power of the test for treatment mean differences when $\mu_1 = 37$, $\mu_2 = 32$, and $\mu_3 = 28$?

c. Suppose primary interest is in estimating the two pairwise comparisons:

$$D_1 = \mu_1 - \mu_2 \qquad D_2 = \mu_3 - \mu_2$$

What would be the required sample sizes if the precision of each comparison is to be ± 3.0 days, using the most efficient multiple comparison procedure with a 95 percent family confidence coefficient?

d. Suppose the chief objective is to identify the physical fitness group with the smallest mean required time for therapy. The probability should be at least .90 that the correct group is identified when the mean required time for the second best group differs by 2.0 days or more. What are the required sample sizes?

18.8. Refer to **Filling machines** Problem 16.13. Suppose that the sample sizes have not yet been determined but it has been decided to sample the same number of cartons for each filling machine. Assume that a reasonable planning value for the error term standard deviation is $\sigma = .15$ ounce.

a. What would be the required sample sizes if: (1) differences in the mean amount of fill for the six filling machines are to be detected with probability .70 or more when the range of the treatment means is .15 ounce, and (2) the α risk is to be controlled at .05?

b. For the sample sizes determined in part (a), what would be the power of the test if $\mu_1 = .09$, $\mu_2 = .18$, $\mu_3 = .30$, $\mu_4 = .20$, $\mu_5 = .10$, and $\mu_6 = .20$?

c. Suppose primary interest is in estimating the following comparisons:

$$D_1 = \mu_1 - \mu_2 \qquad L_1 = \frac{\mu_1 + \mu_2}{2} - \frac{\mu_3 + \mu_4}{2}$$

$$D_2 = \mu_3 - \mu_4 \qquad L_2 = \frac{\mu_1 + \mu_2 + \mu_3 + \mu_4}{4} - \frac{\mu_5 + \mu_6}{2}$$

What would be the required sample sizes if the precision of these comparisons is not to exceed $\pm.08$ ounce, using the best multiple comparison procedure with a 95 percent family confidence coefficient?

d. Suppose the chief objective is to identify the filling machine with the smallest mean fill. The probability should be at least .95 that the filling machine with the smallest mean fill is recognized correctly when the filling machine with the next smallest amount of mean fill differs by .10 ounce or more. What are the required sample sizes?

18.9. Refer to **Premium distribution** Problem 16.14. Suppose that the sample sizes have not yet been determined but it has been decided to sample the same number of premium distributions for each agent. Assume that a reasonable planning value for the error term standard deviation is $\sigma = 3.0$ days.

a. What would be the required sample sizes if: (1) differences in the mean time lapse for the five agents are to be detected with probability .95 or more when the range of the treatment means is 3.75 days, and (2) the α risk is to be controlled at .10?

b. Suppose primary interest is in estimating the following comparisons:

$$D_1 = \mu_1 - \mu_2 \qquad L_1 = \frac{\mu_1 + \mu_2}{2} - \mu_5$$

$$D_2 = \mu_3 - \mu_4 \qquad L_2 = \frac{\mu_1 + \mu_2}{2} - \frac{\mu_3 + \mu_4}{2}$$

What would be the required sample sizes if the precision of the estimated comparisons is not to exceed ±1.0 day, using the most efficient multiple comparison procedure with a 90 percent family confidence coefficient?

c. Suppose the chief objective is to identify the best agent, i.e., the one with the smallest mean time lapse. The probability should be at least .90 that the best agent is recognized correctly when the mean time lapse for the second best agent differs by 1.0 day or more. What are the required sample sizes?

18.10. Refer to **Rehabilitation therapy** Problem 16.11. Suppose that primary interest is in comparing the below-average and above-average physical fitness groups, respectively, with the average physical fitness group. Thus, two comparisons are of interest:

$$D_1 = \mu_1 - \mu_2 \qquad D_2 = \mu_3 - \mu_2$$

Assume that a reasonable planning value for the error term standard deviation is 4.5 days.

a. It has been decided to use equal sample sizes (n) for the below-average and above-average groups. If twice this sample size ($2n$) were to be used for the average physical fitness group, what would be the required sample sizes if the precision of each pairwise comparison is to be ± 2.5 days, using the Bonferroni procedure and a 90 percent family confidence coefficient?

b. Repeat the calculations in part (a) if the sample size for the average physical fitness group is to be: (1) n and (2) $3n$, all other specifications remaining the same.

c. Compare your results in parts (a) and (b). Which design leads to the smallest total sample size here?

18.11. Refer to Figures 18.4 and 18.5. What feature of the residual sequence plots enables you to diagnose that in one case the error variance changes over time whereas in the other case the effect is of a different nature? Could you make a diagnosis about time effects from a residual frequency plot?

18.12. A student proposed in class that deviations of the observations Y_{ij} around the overall mean $\bar{Y}_{..}$ be plotted to assist in evaluating the aptness of ANOVA model (16.2). Would these deviations be helpful in studying the independence of the error terms? The constancy of the variance of the error terms? The normality of the error terms? Discuss.

18.13. A consultant discussing ANOVA applications stated: "Sometimes I find that treatment effects in an experiment do not show up through differences in the treatment means. Hence, it is important to compare the residual plots for the treatments." Later a member of the audience said: "I don't think I understood the reference to residual plots." Explain.

18.14. Refer to **Productivity improvement** Problem 16.9. The residuals are as follows:

		j											
i		1	2	3	4	5	6	7	8	9	10	11	12
1	Low	.72	1.32	−.08	−1.08	.02	−.28	−.58	.82	−.88			
2	Moderate	−1.43	−.03	1.27	.47	−.33	−.43	.77	−.23	.17	.57	−1.03	.27
3	High	−.70	.50	.90	−1.40	.40	.30						

a. Prepare residual frequency plots. What departures from ANOVA model (16.2) can be studied from these plots? What are your findings?

b. Prepare a normal probability plot of the residuals. Also obtain the coefficient of correlation between the ordered residuals and their expected values under normality. Does the normality assumption appear to be reasonable here?

c. The economist wishes to investigate whether location of the firm's home office is related to productivity improvement. The home office locations are as follows (U: U.S.; E: Europe):

	j											
i	1	2	3	4	5	6	7	8	9	10	11	12
1	U	E	E	E	E	U	U	U	U			
2	E	E	E	E	U	U	U	U	U	E	E	E
3	E	U	E	U	U	E						

Make residual frequency plots in which the location of the home office is identified. Does it appear that ANOVA model (16.2) could be improved by adding location of home office as a second factor? Explain.

18.15. Refer to **Questionnaire color** Problem 16.10. The residuals are as follows:

	j				
i	1	2	3	4	5
1	−1.4	−3.4	1.6	−2.4	5.6
2	4.4	−.6	−4.6	1.4	−.6
3	3.0	−3.0	−1.0	1.0	0.0

a. Prepare residual frequency plots. What departures from ANOVA model (16.2) can be studied from these plots? What are your findings?

b. Prepare a normal probability plot of the residuals. Also obtain the coefficient of correlation between the ordered residuals and their expected values under normality. Does the normality assumption appear to be reasonable here?

c. The observations within each factor level are in geographic sequence. Prepare residual sequence plots. What can be studied from these plots? What are your findings?

d. Calculate the standardized residuals (18.7). Obtain the centered intervals within which approximately 50 percent and 90 percent of the standardized residuals should fall if the error terms are normally distributed with constant variance. What are the actual percents of residuals within these intervals? Are these results consistent with those in part (b)?

18.16. Refer to **Rehabilitation therapy** Problem 16.11.

a. Obtain the residuals and prepare residual frequency plots. What departures from ANOVA model (16.2) can be studied from these plots? What are your findings?

b. Prepare a normal probability plot of the residuals. Also obtain the coefficient of correlation between the ordered residuals and their expected values under normality. Does the normality assumption appear to be reasonable here?

c. The observations within each factor level are in time order. Prepare residual sequence plots and analyze them. What are your findings?

d. Calculate the standardized residuals (18.7). Obtain the centered intervals within which approximately 50 percent and 90 percent of the standardized residuals should fall if the error terms are normally distributed with constant variance. What are the actual percents of the residuals within these intervals? Are these results consistent with those in part (b)?

18.17. Refer to **Cash offers** Problem 16.12.

a. Obtain the residuals and prepare residual frequency plots. What departures from ANOVA model (16.2) can be studied from these plots? What are your findings?

b. Prepare a normal probability plot of the residuals. Also obtain the coefficient of correlation between the ordered residuals and their expected values under normality. Does the normality assumption appear to be reasonable here?

c. The observations within each factor level are in time order. Prepare residual sequence plots and analyze them. What are your findings?

d. An executive in the consumer organization has been told that used-car dealers in the region tend to make lower cash offers during weekends (Friday evening through Sunday) than at other times. The times when offers were obtained are as follows (W: weekend; O: other time):

	j											
i	1	2	3	4	5	6	7	8	9	10	11	12
1	O	O	W	O	W	O	W	O	W	O	W	W
2	O	W	W	O	W	O	W	O	O	W	W	O
3	O	W	O	W	O	O	O	W	W	W	O	W

Make residual sequence plots in which the time of the offer is identified. Does it appear that ANOVA model (16.2) could be improved by adding time of offer as a second factor? Explain.

18.18. Refer to **Filling machines** Problem 16.13.

a. Obtain the residuals and prepare residual frequency plots. What departures from ANOVA model (16.2) can be studied from these plots? What are your findings?

b. Prepare a normal probability plot of the residuals. Also obtain the coefficient of correlation between the ordered residuals and their expected values under normality. Does the normality assumption appear to be reasonable here?

c. The observations within each factor level are in time order. Prepare residual sequence plots and analyze them. What are your findings?

18.19. Refer to **Premium distribution** Problem 16.14.

a. Obtain the residuals and prepare residual frequency plots. What departures from ANOVA model (16.2) can be studied from these plots? What are your findings?

b. Prepare a normal probability plot of the residuals. Also obtain the coefficient of correlation between the ordered residuals and their expected values under normality. Does the normality assumption appear to be reasonable here?

c. The observations within each factor level are in time order. Prepare residual sequence plots and analyze them. What are your findings?

18.20. **Computerized game.** Four teams competed in 20 trials of a computerized business game. Each trial involved a new game, the objective for each team being to maximize profits in the given trial. A researcher fitted ANOVA model (16.2) to determine whether or not the mean profits for the four teams are the same and obtained the following residuals:

i	j: 1	2	3	4	5	6	7	8	9	10
1	.10	.28	.10	.47	.83	.65	.28	−.08	.10	−.26
2	−1.44	−1.44	−1.12	−1.28	−.95	−.62	−.29	−.46	−.29	.03
3	−.93	−.70	−.81	−.59	−.25	−.36	−.14	.00	−.14	.09
4	−.15	.11	.25	−.02	−.15	−.29	−.02	.11	.38	.25

i	j: 11	12	13	14	15	16	17	18	19	20
1	−.45	−.63	−1.00	−.63	−.45	−.08	.10	.10	.28	.28
2	.20	.20	.36	.69	.85	1.02	.85	1.02	1.18	1.51
3	.20	.43	.31	.09	.20	.43	.54	.54	.43	.65
4	−.02	−.42	−.29	−.42	−.15	−.15	.25	.11	.25	.38

The residuals are given in time order. Construct appropriate residual plots to study whether the error terms are independent from trial to trial for each team. What are your findings?

18.21. **Helicopter service.** An operations analyst in a sheriff's department studied how frequently their emergency helicopter was used during a recent 20-day period, by time of day (shift 1: 2 A.M.–8 A.M.; shift 2: 8 A.M.–2 P.M.; shift 3: 2 P.M.–8 P.M.; shift 4: 8 P.M.–2 A.M.). The data are as follows (in time order):

i	j: 1	2	3	4	5	6	7	8	9	10	11	12	13	14	15	16	17	18	19	20
1	4	3	5	4	6	3	2	5	7	1	2	5	4	7	4	5	0	4	1	6
2	0	2	0	3	2	1	0	3	1	0	0	1	1	0	1	3	1	2	2	0
3	2	1	0	3	4	1	3	4	2	0	1	3	2	4	0	1	3	0	2	4
4	5	2	4	4	6	5	3	5	7	3	1	0	2	3	3	4	1	5	2	3

Since the data involved are counts, the analyst was concerned about the normality and equal variances assumptions of ANOVA model (16.2).

a. Obtain suitable residual plots to study whether or not the error term variances are equal for the four shifts. What are your findings?

b. For each shift, calculate $\bar{Y}_{i.}$ and s_i. Examine whether relation (18.9), (18.10), or (18.11) is most appropriate here. What do you conclude?

c. Obtain the standardized residuals (18.8) and construct a normal probability plot. Also obtain the coefficient of correlation between the ordered standardized residuals and their expected values under normality. Does the normality assumption appear to be reasonable here?

d. The analyst decided to apply the square root transformation (18.9). Obtain the transformed data $Y' = \sqrt{Y}$, and then calculate the residuals.

e. Prepare suitable plots of the residuals obtained in part (d) to study the equality of the error term variances for the four shifts. Also obtain a normal probability plot and the coefficient of correlation between the ordered residuals and their expected values under normality. What are your findings?

18.22. **Winding speeds.** In an experiment to study the effect of the speed of winding thread (1: slow; 2: normal; 3: fast; 4: maximum) onto 75-yard spools, 16 runs of 10,000 spools each were made at each of the four winding speeds. The dependent variable is the number of thread breaks during the production run. The results (in time order) are as follows:

	j															
i	1	2	3	4	5	6	7	8	9	10	11	12	13	14	15	16
1	4	3	2	3	4	4	3	6	5	4	2	4	4	2	3	4
2	7	6	4	6	7	2	9	5	5	9	3	8	6	4	7	6
3	12	6	14	12	10	9	12	17	7	6	12	11	6	13	10	14
4	17	15	7	20	13	11	16	25	11	24	18	21	16	19	9	23

Since the data involved are counts, the researcher was concerned about the normality and equal variances assumptions of ANOVA model (16.2).

a. Obtain suitable residual plots to study whether or not the error term variances are equal for the four winding speeds. What are your findings?
b. For each winding speed, calculate $\overline{Y}_{i.}$ and s_i. Examine whether relation (18.9), (18.10), or (18.11) is most appropriate here. What do you conclude?
c. Obtain the standardized residuals (18.8) and construct a normal probability plot. Also obtain the coefficient of correlation between the ordered standardized residuals and their expected values under normality. Does the normality assumption appear to be reasonable here?
d. The researcher decided to apply the logarithmic transformation (18.10). Obtain the transformed data $Y' = \log_{10} Y$, and then calculate the residuals.
e. Prepare suitable plots of the residuals obtained in part (d) to study the equality of the error term variances for the four winding speeds. Also obtain a normal probability plot and the coefficient of correlation between the ordered residuals and their expected values under normality. What are your findings?

18.23. Use reciprocal interpolation to find the following percentiles:

a. $F(.95; 3, 360)$.
b. $F(.99; 200, 4)$.
c. $F(.90; 400, 500)$.

18.24. Refer to **Productivity improvement** Problem 16.9. Some additional calculational results are:

i:	1	2	3
s_i:	.8136	.7572	.8672

Assume that the error terms are approximately normally distributed.

a. Examine by means of the Bartlett test whether or not the treatment error variances are equal; use $\alpha = .05$. State the alternatives, decision rule, and conclusion. What is the P-value of the test?
b. Would you reach the same conclusion as in part (a) with the modified Bartlett test?
c. Would the tests in parts (a) and (b) be appropriate if the distributions of the error terms were far from normal?

18.25. Refer to **Rehabilitation therapy** Problem 16.11. Assume that the distributions of the error terms are approximately normal.

a. Examine by means of the Bartlett test whether or not the treatment error variances are equal; use $\alpha = .10$. State the alternatives, decision rule, and conclusion. What is the P-value of the test?
b. Would you reach the same conclusion as in part (a) with the modified Bartlett test?

c. Would the tests in parts (a) and (b) be appropriate if the distributions of the error terms were far from normal?

18.26. Refer to **Cash offers** Problem 16.12. Assume that the error terms are approximately normally distributed.

a. Examine by means of the Bartlett test whether or not the treatment error variances are equal; use $\alpha = .01$. State the alternatives, decision rule, and conclusion. What is the P-value of the test?

b. Would you reach the same conclusion as in part (a) with the Hartley test?

18.27. Refer to **Filling machines** Problem 16.13. Assume that the error terms are approximately normally distributed.

a. Examine by means of the Bartlett test whether or not the treatment error variances are equal; use $\alpha = .01$. State the alternatives, decision rule, and conclusion. What is the P-value of the test?

b. Would you reach the same conclusion as in part (a) with the Hartley test statistic?

18.28. Refer to **Helicopter service** Problem 18.21. Assume that the error terms are approximately normally distributed.

a. For the untransformed data, test by means of the Bartlett test whether or not the treatment error variances are equal; use $\alpha = .10$. What is the P-value of the test? Are your results consistent with the diagnosis in Problem 18.21a?

b. Repeat the Bartlett test for the transformed data in Problem 18.21d. What are your findings now?

18.29. Refer to **Winding speeds** Problem 18.22. Assume that the error terms are approximately normally distributed.

a. For the untransformed data, test by means of the Hartley test whether or not the treatment error variances are equal; use $\alpha = .05$. What is the P-value of the test? Are your results consistent with the diagnosis in Problem 18.22a?

b. Repeat the Hartley test for the transformed data in Problem 18.22d. What are your findings now?

EXERCISES

18.30. Refer to Figure 18.4. Modify ANOVA model (16.2) to include a linear trend term for the time effect. Is this modified model still an ANOVA model? A linear model?

18.31. (Calculus needed.) Given $\mu_1 = 0$, $\mu_3 = 1$, and $0 \leq \mu_2 \leq 1$, show that $\Sigma(\mu_i - \mu_.)^2$ is minimized when $\mu_2 = .5$, where $\mu_. = (\mu_1 + \mu_2 + \mu_3)/3$.

18.32. (Calculus needed.) Use (18.16) to find the appropriate transformation when: (1) $\sigma_i = k\mu_i$, (2) $\sigma_i = k\mu_i^2$.

18.33. (Calculus needed.) Refer to **Rehabilitation therapy** Problem 16.11. The sample sizes for the below-average, average, and above-average physical fitness groups are to be n, kn, and n, respectively. Assuming that model (16.2) is appropriate, find the optimal value of k to minimize the variances of $\hat{D}_1 = \bar{Y}_{1.} - \bar{Y}_{2.}$ and $\hat{D}_2 = \bar{Y}_{3.} - \bar{Y}_{2.}$ for a given total sample size n_T.

PROJECTS

18.34. Refer to the **SENIC** data set and Project 16.32.

a. Obtain the residuals and prepare residual frequency plots. Are any serious departures from ANOVA model (16.2) suggested by your plots?

b. Obtain a normal probability plot of the residuals and calculate the coefficient of correlation between the ordered residuals and their expected values under normality. Is the normality assumption reasonable here?

c. Assume that the error terms are approximately normally distributed. Examine by means of the Bartlett test whether or not the geographic region error variances are equal; use $\alpha = .05$. State the alternatives, decision rule, and conclusion.

18.35. Refer to the **SENIC** data set. A test of whether or not mean length of stay (variable 2) is the same in the four geographic regions (variable 9) is desired but concern exists about the normality and equal variances assumptions of ANOVA model (16.2).

a. Obtain the residuals and plot them against the fitted values to study whether or not the error term variances are equal for the four geographic regions. What are your findings?

b. For each geographic region, calculate $\overline{Y}_{i.}$ and s_i. Examine whether relation (18.9), (18.10), or (18.11) is the most appropriate one here. What do you conclude?

c. Use the reciprocal transformation (18.11) to obtain transformed data $Y' = 1/Y$.

d. Obtain the residuals when ANOVA model (16.2) is fitted to the transformed data. Plot these residuals against the fitted values to study the equality of the error term variances for the four regions. Also obtain a normal probability plot of the residuals and the coefficient of correlation between the ordered residuals and their expected values under normality. What are your findings?

e. Assume that the transformed error terms are approximately normally distributed. Examine by means of the Bartlett test whether or not the geographic region variances are equal; use $\alpha = .01$. State the alternatives, decision rule, and conclusion. What is the P-value of the test?

f. Assume that ANOVA model (16.2) is appropriate for the transformed data Y'. Test whether or not the mean length of stay is the same in the four geographic regions. Control the α risk at .01. State the alternatives, decision rule, and conclusion.

18.36. Refer to the **SMSA** data set and Project 16.34.

a. Obtain the residuals and prepare residual frequency plots. Are any serious departures from ANOVA model (16.2) suggested by your plots?

b. Obtain a normal probability plot of the residuals and calculate the coefficient of correlation between the ordered residuals and their expected values under normality. Is the normality assumption reasonable here?

c. Assume that the error terms are approximately normally distributed. Examine by means of the Bartlett test whether or not the geographic region error variances are equal; use $\alpha = .01$. State the alternatives, decision rule, and conclusion.

CITED REFERENCES

18.1 Glaser, R. E. "Bartlett's Test of Homogeneity of Variances." In *Encyclopedia of Statistical Sciences*, vol. 1, ed. S. Kotz and N. L. Johnson. New York: John Wiley & Sons, 1982, pp. 189–91.

18.2 Scheffé, Henry. *The Analysis of Variance*. New York: John Wiley & Sons, 1959.

19

Nonparametric tests, random effects model, and other topics in ANOVA—I

In this chapter, we discuss some alternative tests to the F test for deciding whether or not the treatment means are equal. We also present model II for single-factor analysis of variance, which is appropriate when the treatment effects are random.

19.1 STUDENTIZED RANGE TEST

There are two chief reasons for considering alternatives to the F test in (16.51) for studying whether or not r treatment means are equal: (1) greater simplicity of the alternative test and (2) less restrictive model assumptions required by the alternative test. In this section, we discuss a simple competitor to the F test that is based on the same model assumptions. In the next two sections, we consider two nonparametric tests that require less restrictive model assumptions.

Studentized range test statistic

The studentized range test, which utilizes the studentized range distribution discussed in Section 17.2, has the merit of simplicity. It requires that all factor level sample sizes are equal, that is, $n_i \equiv n$. Analysis of variance model (16.2) is assumed, as before, to be appropriate.

For choosing between the two alternatives:

$$\begin{aligned} &H_0\text{: } \mu_1 = \mu_2 = \cdots = \mu_r \\ &H_a\text{: not all } \mu_i \text{ are equal} \end{aligned} \tag{19.1}$$

the studentized range test uses the test statistic:

$$q^* = \frac{\max(\bar{Y}_{i.}) - \min(\bar{Y}_{i.})}{\sqrt{\dfrac{MSE}{n}}} \tag{19.2}$$

where $\max(\bar{Y}_{i.})$ is the largest factor level sample mean $\bar{Y}_{i.}$ and $\min(\bar{Y}_{i.})$ is the smallest, and n is the common sample size.

When H_0 holds, q^* is distributed as $q(r, n_T - r)$ as defined in (17.23). As is intuitively clear, large values of q^* lead to conclusion H_a, that the treatment means are unequal, and smaller values of q^* lead to conclusion H_0, that the treatment means are equal. Thus, the decision rule for controlling the risk of a Type I error at α is:

$$\begin{aligned} &\text{If } q^* \leq q(1 - \alpha; r, n_T - r), \text{ conclude } H_0 \\ &\text{If } q^* > q(1 - \alpha; r, n_T - r), \text{ conclude } H_a \end{aligned} \tag{19.3}$$

where $q(1 - \alpha; r, n_T - r)$ is the $(1 - \alpha)100$ percentile of the appropriate q distribution.

Example

We consider again the rust inhibitor example of page 576. The sample results were:

$$\begin{aligned} \bar{Y}_{1.} &= 43 & MSE &= 4.5 \\ \bar{Y}_{2.} &= 89 & n &= 5 \\ \bar{Y}_{3.} &= 67 & n_T &= 20 \\ \bar{Y}_{4.} &= 40 \end{aligned}$$

We wish to test:

$$\begin{aligned} &H_0\text{: } \mu_1 = \mu_2 = \mu_3 = \mu_4 \\ &H_a\text{: not all } \mu_i \text{ are equal} \end{aligned}$$

by means of the studentized range test. The test statistic for our example is:

$$q^* = \frac{89 - 40}{\sqrt{\dfrac{4.5}{5}}} = \frac{49}{\sqrt{.9}} = 52$$

Suppose a level of significance of $\alpha = .05$ is specified. We then require $q(.95; 4, 20 - 4)$. From Table A–9 we find $q(.95; 4, 16) = 4.05$. Hence, the appropriate decision rule is:

$$\text{If } q^* \leq 4.05, \text{ conclude } H_0$$
$$\text{If } q^* > 4.05, \text{ conclude } H_a$$

Since $q^* = 52 > 4.05$, we conclude H_a, that the four rust inhibitors have different means.

Comments

1. The studentized range test has the merit of being somewhat simpler to perform than the F test.

2. If one is interested in testing H_0 primarily against an alternative H_a based only on the maximum difference between any two treatment means, then the studentized range test is more efficient than the F test.

19.2 KRUSKAL-WALLIS RANK TEST

We have noted that the F test for the analysis of variance is robust against departures from normality, provided the departures are not extreme. In the occasional instances where the F test would not be appropriate because of major departures from normality, a nonparametric test may be employed. We shall now discuss two such tests, the Kruskal-Wallis test in this section and the median test in the next section.

Kruskal-Wallis test statistic

The Kruskal-Wallis test is based on the ranks of the observations. To test whether the treatment means are equal with this test, the only assumption required about the population distributions is that they are continuous and of the same shape. Thus, the population distributions must have the same variability, skewness, etc., but may differ as to the location of the mean. It is also assumed that the samples from the different populations are independent random ones.

First, all n_T observations are ranked from 1 to n_T. Let $\bar{R}_{i.}$ be the mean of the ranks for the ith factor level and $\bar{R}_{..}$ the overall mean rank. The test statistic is then simply:

$$X^2_{KW} = \frac{SSTR}{\dfrac{SSTO}{n_T - 1}} = \frac{\sum_{i=1}^{r} n_i(\bar{R}_{i.} - \bar{R}_{..})^2}{\dfrac{n_T(n_T + 1)}{12}} \tag{19.4}$$

The numerator is the usual treatment sum of squares, but with the data expressed in ranks, and the denominator is the variance of the ranks $1, 2, 3, \ldots, n_T$. The test statistic X^2_{KW} can be expressed equivalently as follows:

$$X^2_{KW} = \left(\frac{12}{n_T(n_T + 1)}\sum_{i=1}^{r} n_i \bar{R}^2_{i.}\right) - 3(n_T + 1) \tag{19.4a}$$

If the n_i are reasonably large (5 or more is the usual advice), X^2_{KW} is approximately a χ^2 random variable with $r - 1$ degrees of freedom when H_0 (all μ_i are equal) holds. Large values of X^2_{KW}, as expected, lead to H_a (not all μ_i are equal). Thus, in choosing between:

$$\begin{aligned} &H_0\colon \mu_1 = \mu_2 = \cdots = \mu_r \\ &H_a\colon \text{not all } \mu_i \text{ are equal} \end{aligned} \tag{19.5}$$

the appropriate decision rule for controlling the risk of making a Type I error at α is:

$$\begin{aligned} &\text{If } X^2_{KW} \leq \chi^2(1 - \alpha; r - 1), \text{ conclude } H_0 \\ &\text{If } X^2_{KW} > \chi^2(1 - \alpha; r - 1), \text{ conclude } H_a \end{aligned} \tag{19.6}$$

Example

Servo-Data, Inc., operates three computers at different locations. The computers are identical as to make and model, but are subject to different degrees of voltage fluctuation in the power lines serving the respective installations. Management wishes to test whether or not the mean lengths of operating time between failures are the same for the three computers. The alternative conclusions are:

$$\begin{aligned} &H_0\colon \mu_1 = \mu_2 = \mu_3 \\ &H_a\colon \text{not all } \mu_i \text{ are equal} \end{aligned}$$

Table 19.1 contains the lengths of time between failures for the three computers, for five failure intervals each. Even though the sample sizes are small,

TABLE 19.1 Times between failure for three computers (in hours)—Servo-Data example

Computer		*Observation (j)*					
i		1	2	3	4	5	*Mean*
A							
	Time	105	3	90	217	22	
	Rank	11	2	10	14	4	8.2
B							
	Time	56	43	1	37	14	
	Rank	8	7	1	5	3	4.8
C							
	Time	183	144	219	86	39	
	Rank	13	12	15	9	6	11.0

they suggest highly skewed distributions. In the same table, the data are ranked from 1 to 15 and the mean ranks are shown.

Using the computational formula (19.4a), we obtain for the test statistic:

$$X_{KW}^2 = \left\{\frac{12}{15(16)} 5[(8.2)^2 + (4.8)^2 + (11.0)^2]\right\} - 3(16) = 4.8$$

A level of significance of $\alpha = .10$ has been specified. Since $r = 3$, we require $\chi^2(.90; 2)$. From Table A–3 we find $\chi^2(.90; 2) = 4.61$. The decision rule for choosing between H_0 and H_a therefore is:

$$\begin{aligned}&\text{If } X_{KW}^2 \leq 4.61, \text{ conclude } H_0\\ &\text{If } X_{KW}^2 > 4.61, \text{ conclude } H_a\end{aligned}$$

Since $X_{KW}^2 = 4.8 > 4.61$, we conclude H_a, that the mean times between failures differ for the three computers.

Comments

1. The Kruskal-Wallis test, like the F test, does not require equal sample sizes.
2. If the n_i are small so that the χ^2 approximation is not appropriate, special tables should be used for conducting the Kruskal-Wallis test; see, for instance, the tables by Owen in Reference 19.1.
3. In case of ties among some observations, each of the tied observations is given the mean of the ranks involved. Thus, if two observations are tied for what would otherwise have been the third and fourth ranked positions, each would be given the mean value 3.5. If a large number of ties exist, the test statistic in (19.4) needs to be modified.
4. The Kruskal-Wallis test can also be used to choose among the alternatives:

$$\begin{aligned}&H_0\text{: all populations are identical}\\ &H_a\text{: all populations are not identical}\end{aligned} \tag{19.7}$$

This statement of the decision problem avoids the earlier assumption that all populations are identical except for the location of the means. If conclusion H_a in (19.7) is reached, however, one cannot identify the reason for the difference. For example, the means might differ, or the variances, or the nature of the skewness, or some combination of these.

Multiple pairwise testing procedure

If the Kruskal-Wallis test leads to the conclusion that the factor level or treatment means μ_i are not equal, it is frequently desired to obtain information about the comparative magnitudes of these means. A large-sample testing analogue of the Bonferroni pairwise comparison procedure discussed in Section 17.4 based on the ranks of the observations may be employed for this purpose, provided that the sample sizes are not too small. Testing limits for all $g = r(r - 1)/2$ pairwise tests using the mean ranks $\bar{R}_{i.}$ are set up as follows for family level of significance α:

$$(\bar{R}_{i.} - \bar{R}_{i'.}) \pm B\left[\frac{n_T(n_T+1)}{12}\left(\frac{1}{n_i} + \frac{1}{n_{i'}}\right)\right]^{1/2} \tag{19.8}$$

where:

$$B = z(1 - \alpha/2g)$$

$$g = \frac{r(r-1)}{2}$$

If the testing limits include zero, we conclude that the corresponding treatment means μ_i and $\mu_{i'}$ do not differ. If the testing limits do not include zero, we conclude that the two corresponding treatment means differ. Based on all pairwise tests, we then set up groups of treatment means whose members do not differ according to the simultaneous testing procedure. In this way, we obtain information about the comparative magnitudes of the treatment means μ_i.

Example. For the Servo-Data study results in Table 19.1, we wish to ascertain, if possible, which computer has the longest mean time between failures. We shall base the analysis on ranks in view of the serious departures from normality of the distributions of times between failures. For a family significance level of $\alpha = .10$ and $g = r(r-1)/2 = 3(2)/2 = 3$ pairwise tests, we require $B = z(.9833) = 2.13$. Since all treatment sample sizes are equal, we need calculate the right term in (19.8) only once:

$$B\left[\frac{n_T(n_T+1)}{12}\left(\frac{1}{n_i} + \frac{1}{n_{i'}}\right)\right]^{1/2} = 2.13\left[\frac{15(16)}{12}\left(\frac{1}{5} + \frac{1}{5}\right)\right]^{1/2} = 6.02$$

Hence, the testing limits for the three pairwise tests are:

A and B:	$(8.2 - 4.8) \pm 6.02$	or	-2.6 and 9.4
C and B:	$(11.0 - 4.8) \pm 6.02$	or	$.2$ and 12.2
C and A:	$(11.0 - 8.2) \pm 6.02$	or	-3.2 and 8.8

The only test showing a significant difference is between computers B and C. Hence, we obtain two groupings:

Group 1	*Group 2*
Computer A	Computer A
Computer B	Computer C

These groupings indicate that the mean times between failures differ for computers B and C, but not for any other pairs of computers. Thus, the only conclusion about comparative magnitudes of the means possible from this analysis is that the mean time between failures for computer C is larger than that for computer B.

19.3 MEDIAN TEST

The median test is another test that may be utilized when populations are far from normally distributed. With this test, one is again interested in choosing between the two alternatives:

(19.9)
$$\begin{aligned} &H_0\text{: } \mu_1 = \mu_2 = \cdots = \mu_r \\ &H_a\text{: not all } \mu_i \text{ are equal} \end{aligned}$$

The median test assumes only that all populations are of the same shape, but they may differ as to the location of the mean. Also it is assumed that the samples from the different populations are independent random ones.

Median test statistic

All sample data are combined to determine the median value for the combined sample. For each treatment, the number of observations above this median value (f_{i1}) and the number not above it (f_{i2}) are then ascertained. Finally, a test for homogeneity is conducted using the test statistic:

(19.10)
$$X^2 = \sum_{i=1}^{r} \sum_{j=1}^{2} \frac{[f_{ij} - E(f_{ij})]^2}{E(f_{ij})}$$

where f_{ij} ($i = 1, \ldots, r$; $j = 1, 2$) is the observed frequency in a cell and $E(f_{ij})$ is the expected frequency under H_0, when all populations are identical.

When the sample sizes are reasonably large, the test statistic X^2 is distributed approximately as χ^2 with $r - 1$ degrees of freedom when H_0 holds. Large values of X^2 lead to conclusion H_a. Hence, the appropriate decision rule for controlling the risk of making a Type I error at α is:

(19.11)
$$\begin{aligned} &\text{If } X^2 \leq \chi^2(1 - \alpha; r - 1)\text{, conclude } H_0 \\ &\text{If } X^2 > \chi^2(1 - \alpha; r - 1)\text{, conclude } H_a \end{aligned}$$

Example

We consider again the Servo-Data example dealing with the times between failures for three different computers. Fifteen additional observations on times between failures were made for each computer to provide more precise information. The median number of hours between failures for the combined sample of 60 observations was 64 hours. Table 19.2 summarizes the results for the three samples (the original data are not shown). The expected frequencies when all populations are identical are shown in parentheses in Table 19.2. These are obtained by allocating the total frequencies in each column to the three computers in proportion to the total number of observations for each computer. In our example, 20 observations were made on each computer; hence, the 30 frequencies above the median are allocated equally to each computer. Similarly, the 30

TABLE 19.2 Times between failures for three computers—Servo-Data example with enlarged samples

	Number of Observations		
Computer	*Above Median*	*Not above Median*	*Total*
A	13 (10)	7 (10)	20
B	3 (10)	17 (10)	20
C	14 (10)	6 (10)	20
Total	30	30	60

Median for combined sample = 64 hours

frequencies not above the median are allocated equally to each computer. These then are the frequencies that are expected if the three populations have the same mean and hence are identical.

The test statistic is calculated using (19.10):

$$X^2 = \frac{(13 - 10)^2}{10} + \frac{(7 - 10)^2}{10} + \cdots + \frac{(6 - 10)^2}{10} = 14.8$$

It has been specified that the level of significance is to be $\alpha = .05$; hence, we require $\chi^2(.95; 2) = 5.99$. The decision rule then is:

$$\text{If } X^2 \leq 5.99, \text{ conclude } H_0$$
$$\text{If } X^2 > 5.99, \text{ conclude } H_a$$

Since $X^2 = 14.8 > 5.99$, we conclude H_a, that the mean number of hours between failures is not the same for the three computers.

Note

Like the Kruskal-Wallis test, the median test does not require equal sample sizes.

19.4 ANOVA MODEL II—RANDOM EFFECTS

As we noted earlier, there are occasions when the employed factor levels or treatments are not of intrinsic interest in themselves but constitute a sample from a larger population of factor levels. Model II is designed for this type of situation. Consider, for instance, Apex Enterprises, a company that builds roadside restaurants carrying one of several promoted trade names, leases franchises to individuals to operate the restaurants, and provides management services. This company employs a large number of personnel officers who interview applicants for jobs in the restaurants. At the end of an interview, the personnel officer assigns a subjective rating between 0 and 100 to indicate the applicant's potential value on the job. Suppose now that five personnel officers were selected at random and each was assigned four candidates at random. In this case, the company would not wish to make inferences concerning the five personnel officers who happened to be selected but rather about the population of all personnel officers. Questions of interest might include: How great is the variation in ratings between all personnel officers? What is the mean rating by all personnel officers?

The distinction between this situation, for which model II is designed, and one where model I is appropriate can be seen readily by modifying our example slightly. If a smaller company had only five personnel officers who were all included in the study and interest is limited to these five officers, model I would be relevant since the factor levels (the five personnel officers) are then not considered a sample from a larger population. A repetition of the experiment for the smaller company would involve the same five personnel officers, but in the case of the large company, a repetition would involve a new random sample of five personnel officers that likely would consist of different officers.

Model

Model II for single-factor analysis of variance is as follows:

$$Y_{ij} = \mu_i + \varepsilon_{ij} \tag{19.12}$$

where:

μ_i are independent $N(\mu_{.}, \sigma_\mu^2)$
ε_{ij} are independent $N(0, \sigma^2)$
μ_i and ε_{ij} are independent random variables
$i = 1, \ldots, r; j = 1, \ldots, n_i$

Model (19.12) is similar in appearance to fixed effects model (16.2). The main distinction is that the factor level means μ_i are constants for model I but are random variables for model II. Hence, model II is often called a *random effects* model.

Meaning of model terms. We shall explain the meaning of the model terms with reference to our personnel officers example. The term μ_i corresponds in our example to the mean of all ratings by the ith personnel officer if he or she interviewed all prospective employees. The expected value of μ_i is $\mu_{.}$. Thus, $\mu_{.}$ represents in our example the mean rating for all prospective employees by all personnel officers. The variability of the μ_i is measured by the variance σ_μ^2. The more the different personnel officers vary in their mean ratings (for instance, some may rate consistently higher than others), the greater will be σ_μ^2. On the other hand, if all personnel officers rate at the same mean level, all μ_i will be equal to $\mu_{.}$ and then $\sigma_\mu^2 = 0$.

The term ε_{ij} represents in our example the variation associated with the differing potential values of different prospective employees. Note that model (19.12) assumes that all ε_{ij} have the same variance σ^2. This means that the distributions of ratings for prospective employees by the different personnel officers are as-

FIGURE 19.1 Representation of ANOVA model II

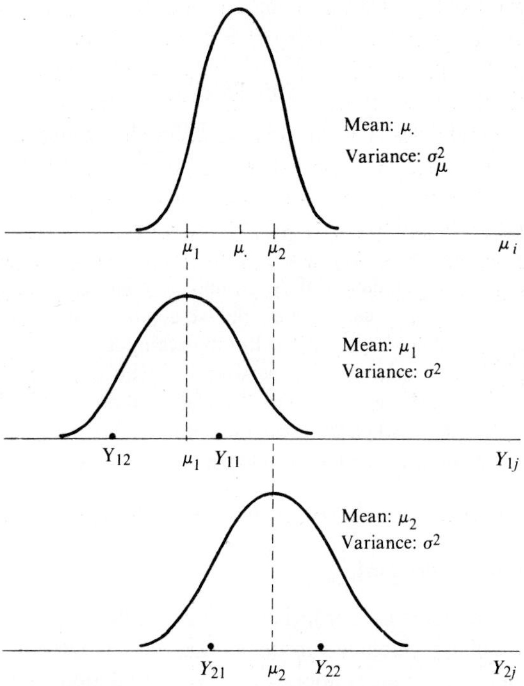

sumed to have the same variability. The distributions for the different personnel officers may differ, however, with respect to the mean levels of the distributions.

Figure 19.1 illustrates the ANOVA model II. On the top is shown the distribution of the μ_i, which is normal. A number of μ_i (two in the illustration) are selected at random from this distribution. Each in turn leads to a distribution of $Y_{ij} = \mu_i + \varepsilon_{ij}$, which are all normal distributions. A number of Y_{ij} observations (two each in the illustration) are then selected from each of these distributions.

Important features of model. 1. The expected value of an observation Y_{ij} is:

(19.13) $$E(Y_{ij}) = \mu_.$$

because we have by (19.12):

$$\begin{aligned} E(Y_{ij}) &= E(\mu_i) + E(\varepsilon_{ij}) \\ &= \mu_. + 0 \\ &= \mu_. \end{aligned}$$

Note that this expectation averages over the selections of both μ_i and ε_{ij}.

2. The variance of Y_{ij} is:

(19.14) $$\sigma^2(Y_{ij}) = \sigma^2_\mu + \sigma^2$$

This result follows because model II assumes that μ_i and ε_{ij} are independent random variables, and $\sigma^2(\mu_i) = \sigma_\mu^2$ and $\sigma^2(\varepsilon_{ij}) = \sigma^2$ according to model (19.12). Because the variance of Y in this model is the sum of two components, σ_μ^2 and σ^2, this model is sometimes called a *components of variance* model.

3. Finally, the Y_{ij} are normally distributed because they are linear combinations of the independent normal variables μ_i and ε_{ij}. Note, however, that the Y_{ij} are not independent because a number of different Y_{ij} have the same component μ_i.

Note

At times, the population of the μ_i may be relatively small and should be treated as a finite population. This can be done, but we do not discuss this case here. If the population of the μ_i is finite but large, little is lost in treating it as an infinite population. We did this, in fact, in our illustration of the personnel officers. The number of officers is finite, but since there are many, we treated the population of the μ_i as an infinite one. Thus, there are two basic situations when the population of the μ_i is treated as infinite—when the population is finite but large, and when interest centers in the underlying *process* generating the μ_i.

Questions of interest

When model II is appropriate, one is usually not interested in inferences about the particular μ_i included in the study, such as which is the largest or smallest, but rather in inferences about the entire population of the μ_i. Specifically, interest often centers on the mean of the μ_i, $\mu_.$, and on the variability of the μ_i, measured by σ_μ^2. In our personnel officers example, for instance, management would not ordinarily be as interested in the mean ratings of the five personnel officers who happened to be included in the study as in the mean rating by all personnel officers and in the effect of variability among all personnel officers.

While σ_μ^2 is a direct measure of the variability of the μ_i, the effect of this variability is often measured more meaningfully by the ratio:

$$\frac{\sigma_\mu^2}{\sigma_\mu^2 + \sigma^2} \tag{19.15}$$

Note the following characteristics of this ratio:

1. The ratio takes on values between 0 (when $\sigma^2 = \infty$) and 1 (when $\sigma^2 = 0$).
2. The denominator is $\sigma^2(Y_{ij})$ according to (19.14).
3. In view of properties 1 and 2, the ratio measures the proportion of the total variability of the Y_{ij} that is accounted for by the variability of the μ_i.

With reference to our personnel officers example, the denominator of the ratio measures the variability of ratings for all candidates by all personnel officers, and the numerator measures the variability of the mean ratings by all personnel officers. The ratio then measures the proportion of the total variability of ratings that is accounted for by differences among the personnel officers. If the ratio is near zero, the effect of differences among personnel officers on the total variability is relatively insignificant. On the other hand, if the ratio is large, say, .5 or more, then much of the total variability is accounted for by differences between personnel officers, and management may wish to study the advisability of giving the personnel officers more training to improve uniformity of ratings between officers.

Test whether $\sigma_\mu^2 = 0$

We first consider how to decide between:

$$\begin{aligned} &H_0: \sigma_\mu^2 = 0 \\ &H_a: \sigma_\mu^2 > 0 \end{aligned} \tag{19.16}$$

H_0 implies that all μ_i are equal, that is, $\mu_i \equiv \mu_{..}$. H_a implies that the μ_i differ. For our personnel officers example, H_0 implies that the mean ratings for all personnel officers are equal, while H_a implies that they differ.

Despite the fact that model II differs from model I, the analysis of variance for a single-factor study is conducted in identical fashion. (This is not always the case in more complex situations.) The difference between the two models appears in the expected mean squares. It can be shown in a similar manner to that employed in our derivation for model I that, for model II:

$$E(MSE) = \sigma^2 \tag{19.17}$$

$$E(MSTR) = \sigma^2 + n'\sigma_\mu^2 \tag{19.18}$$

where:

$$n' = \frac{1}{r-1}\left[(\Sigma n_i) - \frac{\Sigma n_i^2}{\Sigma n_i}\right]$$

If all $n_i = n$, then $n' = n$.

It is clear from (19.17) and (19.18) that if $\sigma_\mu^2 = 0$, MSE and $MSTR$ have the same expection σ^2. Otherwise, $E(MSTR) > E(MSE)$ since $n' > 0$ always. Hence, large values of the test statistic:

$$F^* = \frac{MSTR}{MSE} \tag{19.19}$$

will lead to conclusion H_a in (19.16). Since F^* again follows the F distribution when H_0 holds, the decision rule for controlling the risk of making a Type I error at α is the same one as for model I:

$$\begin{aligned} &\text{If } F^* \le F(1-\alpha; r-1, n_T - r), \text{ conclude } H_0 \\ &\text{If } F^* > F(1-\alpha; r-1, n_T - r), \text{ conclude } H_a \end{aligned} \tag{19.20}$$

Example. Table 19.3 contains the results of a study by Apex Enterprises on the evaluation ratings of potential employees by its personnel officers. Five personnel officers were selected at random, and each was assigned at random 4 prospective employees. The ANOVA calculations are routine and were done by a computer program package. The results are shown in Table 19.4, which also shows the expected mean squares in general and for this particular example.

TABLE 19.3 Ratings by five personnel officers—Apex Enterprises example

Officer i	Candidate (j) 1	2	3	4	Mean
A	76	64	85	75	$\bar{Y}_{1.} = 75$
B	58	75	81	66	$\bar{Y}_{2.} = 70$
C	49	63	62	46	$\bar{Y}_{3.} = 55$
D	74	71	85	90	$\bar{Y}_{4.} = 80$
E	66	74	81	79	$\bar{Y}_{5.} = 75$
Mean					$\bar{Y}_{..} = 71$

TABLE 19.4 ANOVA table for single-factor model II—Apex Enterprises example

Source of Variation	*SS*	*df*	*MS*	*E(MS) General*	*E(MS) Example*
Between personnel officers	$SSTR = 1{,}480$	4	$MSTR = 370$	$\sigma^2 + n'\sigma_\mu^2$	$\sigma^2 + 4\sigma_\mu^2$
Error (within personnel officers)	$SSE = 1{,}134$	15	$MSE = 75.6$	σ^2	σ^2
Total	$SSTO = 2{,}614$	19			

$$n' = \frac{1}{r-1}\left[(\Sigma n_i) - \frac{\Sigma n_i^2}{\Sigma n_i}\right]$$

$$(n' = n \text{ if all } n_i = n)$$

Using the data from Table 19.4, the appropriate test statistic is:

$$F^* = \frac{370}{75.6} = 4.89$$

Assuming that we are to control the risk of making a Type I error at $\alpha = .05$, we require $F(.95; 4, 15) = 3.06$. Hence, the decision rule is:

$$\text{If } F^* \leq 3.06, \text{ conclude } H_0$$
$$\text{If } F^* > 3.06, \text{ conclude } H_a$$

Since $F^* = 4.89 > 3.06$, we conclude H_a, that $\sigma_\mu^2 > 0$ or that the mean ratings of the personnel officers differ.

Note

We shall illustrate the derivation of an expected mean square for model II by sketching the development for deriving $E(MSTR)$ in (19.18) when $n_i \equiv n$. The proof parallels that for model I. By model (19.12), we can write:

$$\bar{Y}_{i.} = \mu_i + \bar{\varepsilon}_{i.}$$
$$\bar{Y}_{..} = \bar{\mu}_{.} + \bar{\varepsilon}_{..}$$

where $\bar{\varepsilon}_{i.}$ and $\bar{\varepsilon}_{..}$ are defined in (16.40) and (16.43), respectively, and:

$$\bar{\mu}_{.} = \frac{\Sigma\mu_i}{r}$$

(Note the use of a different notation for the mean of the μ_i here than for model I to emphasize the random nature of the mean for model II.) Corresponding to (16.45), we obtain:

$$\bar{Y}_{i.} - \bar{Y}_{..} = (\mu_i - \bar{\mu}_{.}) + (\bar{\varepsilon}_{i.} - \bar{\varepsilon}_{..})$$

so that:

$$\Sigma(\bar{Y}_{i.} - \bar{Y}_{..})^2 = \Sigma(\mu_i - \bar{\mu}_{.})^2 + \Sigma(\bar{\varepsilon}_{i.} - \bar{\varepsilon}_{..})^2 + 2\Sigma(\mu_i - \bar{\mu}_{.})(\bar{\varepsilon}_{i.} - \bar{\varepsilon}_{..})$$

When we take the expectation, the cross-product term drops out because of the independence of the μ_i and ε_{ij} and because the deviations $\mu_i - \bar{\mu}_{.}$ and $\bar{\varepsilon}_{i.} - \bar{\varepsilon}_{..}$ all have expectation zero. From (16.48) we know that:

$$E[\Sigma(\bar{\varepsilon}_{i.} - \bar{\varepsilon}_{..})^2] = \frac{(r-1)\sigma^2}{n}$$

Lastly, since $\Sigma(\mu_i - \bar{\mu}_{.})^2$ is the numerator of an ordinary sample variance for r independent μ_i observations, it follows from the unbiasedness of the sample variance that:

$$E[\Sigma(\mu_i - \bar{\mu}_{.})^2] = (r-1)\sigma_\mu^2$$

Hence, we obtain:

$$E\left[\frac{n}{r-1}\Sigma(\bar{Y}_{i.} - \bar{Y}_{..})^2\right] = \frac{n}{r-1}\left[(r-1)\sigma_\mu^2 + \frac{r-1}{n}\sigma^2\right] = n\sigma_\mu^2 + \sigma^2$$

which is the result in (19.18) for the case $n_i \equiv n$.

Estimation of $\mu_{.}$

When ANOVA model II is applicable, one is frequently interested in estimating the overall mean $\mu_{.}$. We shall assume in developing an interval estimate for $\mu_{.}$ that *all factor level sample sizes are equal,* that is, $n_i \equiv n$.

We know from (19.13) that:

$$E(Y_{ij}) = \mu_{.}$$

Hence, an unbiased estimator of $\mu_{.}$ is:

(19.21) $$\hat{\mu}_{.} = \bar{Y}_{..}$$

It can be shown that the variance of this estimator is:

$$\sigma^2(\bar{Y}_{..}) = \frac{\sigma_\mu^2}{r} + \frac{\sigma^2}{n_T} = \frac{n\sigma_\mu^2 + \sigma^2}{n_T} \tag{19.22}$$

(Remember that $n_T = rn$ here.)

Formula (19.22) shows that the variance of $\bar{Y}_{..}$ is made up of two components. The first corresponds to the variance of a sample mean based on r observations when sampling from the population of the μ_i, and it reflects the contribution due to sampling the factor levels. The second component corresponds to the variance of a sample mean based on n_T observations when sampling from the populations of the Y_{ij}, given the μ_i, and it reflects the contribution due to variation within factor levels.

An unbiased estimator of $\sigma^2(\bar{Y}_{..})$ is:

$$s^2(\bar{Y}_{..}) = \frac{MSTR}{n_T} \tag{19.23}$$

This estimator is unbiased because we know from (19.18) that when $n_i \equiv n$:

$$E(MSTR) = n\sigma_\mu^2 + \sigma^2 \tag{19.24}$$

(Remember that $n' = n$ when $n_i \equiv n$.)

It can be shown that:

$$\frac{\bar{Y}_{..} - \mu_.}{s(\bar{Y}_{..})} \text{ is distributed as } t(r-1) \text{ for model (19.12) when } n_i \equiv n \tag{19.25}$$

Hence, we obtain in usual fashion the confidence limits:

$$\bar{Y}_{..} \pm t(1 - \alpha/2; r - 1)s(\bar{Y}_{..}) \tag{19.26}$$

Example. Management of Apex Enterprises wishes to estimate the mean rating for all prospective employees by all personnel officers with a 90 percent confidence interval. We have from Tables 19.3 and 19.4:

$$\bar{Y}_{..} = 71 \qquad MSTR = 370 \qquad n_T = 20$$

We require $t(.95; 4) = 2.132$ and:

$$s^2(\bar{Y}_{..}) = \frac{370}{20} = 18.5$$

Hence, $s(\bar{Y}_{..}) = 4.301$ and the desired 90 percent confidence interval is:

$$71 - 2.132(4.301) \le \mu_. \le 71 + 2.132(4.301)$$

$$62 \le \mu_. \le 80$$

Thus, with a 90 percent confidence coefficient, we conclude that the mean rating assigned by all personnel officers to all prospective employees is between 62 and 80. The interval estimate is not too precise because of the relatively small sample sizes of personnel officers and potential employees.

Note

The variance of $\bar{Y}_{..}$ in (19.22) can be derived readily. First, we consider:

$$\bar{Y}_{i.} = \mu_i + \bar{\varepsilon}_{i.}$$

where $\bar{\varepsilon}_{i.}$ is defined in (16.40). Because of the independence of μ_i and the ε_{ij}, we have:

$$\sigma^2(\bar{Y}_{i.}) = \sigma_\mu^2 + \frac{\sigma^2}{n}$$

Remember that $\bar{\varepsilon}_{i.}$ is just an ordinary mean of n independent ε_{ij} observations.

For the case $n_i \equiv n$ which we are considering here, we have:

$$\bar{Y}_{..} = \frac{\sum_{i=1}^{r} \bar{Y}_{i.}}{r}$$

In view of the independence of the μ_i and the ε_{ij} among themselves and between each other, it follows that the $\bar{Y}_{i.}$ are independent so that:

$$\sigma^2(\bar{Y}_{..}) = \frac{\sigma^2(\bar{Y}_{i.})}{r} = \frac{\sigma_\mu^2}{r} + \frac{\sigma^2}{rn} = \frac{n\sigma_\mu^2 + \sigma^2}{n_T}$$

Estimation of $\sigma_\mu^2/(\sigma_\mu^2 + \sigma^2)$

As noted earlier, the ratio $\sigma_\mu^2/(\sigma_\mu^2 + \sigma^2)$ reveals meaningfully the effect of the extent of variation between the μ_i. To develop an interval estimate for this ratio, we shall assume that *all factor level sample sizes are equal,* that is, $n_i \equiv n$.

We begin by obtaining confidence limits for the ratio σ_μ^2/σ^2. First, we need to note that *MSTR* and *MSE* are independent random variables for model II, just as for model I. When $n_i \equiv n$, it can be shown further that:

(19.27) $$\frac{MSTR}{n\sigma_\mu^2 + \sigma^2} \div \frac{MSE}{\sigma^2} \sim F(r-1, n_T - r) \qquad \text{when } n_i \equiv n$$

Hence, we can write the probability statement:

(19.28) $$P\left[F(\alpha/2; r-1, n_T - r) \le \frac{MSTR}{n\sigma_\mu^2 + \sigma^2} \div \frac{MSE}{\sigma^2} \le F(1-\alpha/2; r-1, n_T - r)\right] = 1 - \alpha$$

Rearranging the inequalities, we obtain the following confidence limits for σ_μ^2/σ^2:

(19.29a) $$L_L = \frac{1}{n}\left[\frac{MSTR}{MSE}\frac{1}{F(1-\alpha/2; r-1, n_T - r)} - 1\right]$$

(19.29b) $$L_U = \frac{1}{n}\left[\frac{MSTR}{MSE}\frac{1}{F(\alpha/2; r-1, n_T - r)} - 1\right]$$

where L_L is the lower confidence limit and L_U the upper.

The confidence interval for $\sigma_\mu^2/(\sigma_\mu^2 + \sigma^2)$ can now be readily obtained and is as follows:

$$\frac{L_L}{1 + L_L} \leq \frac{\sigma_\mu^2}{\sigma_\mu^2 + \sigma^2} \leq \frac{L_U}{1 + L_U} \tag{19.30}$$

Example. Management of Apex Enterprises wishes a 90 percent confidence interval for $\sigma_\mu^2/(\sigma_\mu^2 + \sigma^2)$. From previous work, we have:

$$MSTR = 370 \qquad MSE = 75.6 \qquad n = 4 \qquad r = 5 \qquad n_T = 20$$

For a 90 percent confidence interval, we require:

$$F(.05; 4, 15) = .170 \qquad F(.95; 4, 15) = 3.06$$

Hence, the 90 percent confidence interval for σ_μ^2/σ^2 is by (19.29):

$$\frac{1}{4}\left(\frac{370}{75.6}\frac{1}{3.06} - 1\right) \leq \frac{\sigma_\mu^2}{\sigma^2} \leq \frac{1}{4}\left(\frac{370}{75.6}\frac{1}{.170} - 1\right)$$

$$.15 \leq \frac{\sigma_\mu^2}{\sigma^2} \leq 6.9$$

Finally, the confidence interval for $\sigma_\mu^2/(\sigma_\mu^2 + \sigma^2)$ is by (19.30):

$$.13 = \frac{.15}{1.15} \leq \frac{\sigma_\mu^2}{\sigma_\mu^2 + \sigma^2} \leq \frac{6.9}{7.9} = .87$$

Hence, with confidence coefficient .90, we conclude that the variability of the mean ratings for the different personnel officers accounts for somewhere between 13 and 87 percent of the total variance of the ratings. Note that this interval estimate is not very precise. The reason is the relatively small sample sizes. The confidence interval does indicate, though, that the variability among personnel officers is not negligible since it accounts for at least 13 percent of the total variability.

Comments

1. It may happen occasionally that the lower limit of the confidence interval for σ_μ^2/σ^2 is negative. Since this ratio cannot be negative, the usual practice is to consider the lower limit L_L in (19.29a) to be zero in that case.

2. If one-sided or two-sided tests concerning the relative magnitudes of σ_μ^2 and σ^2 are desired, such as the following (where c is a specified constant):

$$H_0\colon \sigma_\mu^2 \leq c\sigma^2 \qquad H_0\colon \sigma_\mu^2 = c\sigma^2$$
$$H_a\colon \sigma_\mu^2 > c\sigma^2 \qquad H_a\colon \sigma_\mu^2 \neq c\sigma^2$$

the decision rule can be constructed by utilizing (19.27). Alternatively, one-sided or two-sided confidence intervals can be set up from which the appropriate conclusion can be drawn. For instance, suppose that for our Apex Enterprises personnel officers example we are considering:

$$H_0: \sigma_\mu^2 = \frac{1}{2}\sigma^2$$

$$H_a: \sigma_\mu^2 \neq \frac{1}{2}\sigma^2$$

Since the 90 percent confidence interval for σ_μ^2/σ^2 above (corresponding to a level of significance of .10) includes .5, the conclusion to be reached is H_0.

3. The ratio σ_μ^2/σ^2 is of relevance in planning investigations. In our example dealing with the personnel officers, suppose that the mean rating $\mu_.$ is to be estimated, and that the costs of including in the study a personnel officer and a candidate are c_1 and c_2 respectively. For a given total budget C, the ratio σ_μ^2/σ^2 is the determining variable for finding the optimum balance between number of personnel officers and candidates to include in the study so as to minimize the variance of the estimator. If the populations are not large, the model will need to take account of their finite nature.

Estimation of σ^2 and σ_μ^2

At times, interest exists in estimating σ^2 and σ_μ^2 separately. According to (19.17), an unbiased estimator of σ^2 is:

(19.31) $$\hat{\sigma}^2 = MSE$$

A confidence interval for σ^2 can be obtained in the usual fashion by means of (1.66); here, the degrees of freedom will be $n_T - r$.

An unbiased point estimator of σ_μ^2 is also available. Since by (19.17) and (19.18), we have:

$$E(MSE) = \sigma^2$$
$$E(MSTR) = \sigma^2 + n'\sigma_\mu^2$$

it follows that:

(19.32) $$\hat{\sigma}_\mu^2 = \frac{MSTR - MSE}{n'}$$

is an unbiased estimator of σ_μ^2. Occasionally, this point estimator will turn out to be negative. Since a variance cannot be negative, the usual practice is to consider the point estimator to be zero in that event. Only approximate confidence intervals for σ_μ^2 are available. These are discussed in Reference 19.2.

Example. For our Apex Enterprises personnel officers example, a 90 percent confidence interval for σ^2 requires:

$$MSE = 75.6 \qquad \chi^2(.05; 15) = 7.26 \qquad \chi^2(.95; 15) = 25.0$$

Using (1.66), we find:

$$45.4 = \frac{15(75.6)}{25.0} \leq \sigma^2 \leq \frac{15(75.6)}{7.26} = 156.2$$

An unbiased point estimate of σ_μ^2 requires:

$$MSE = 75.6 \qquad MSTR = 370 \qquad n' = n = 4$$

Hence, by (19.32) we find:

$$\hat{\sigma}_\mu^2 = \frac{370 - 75.6}{4} = 73.6$$

Alternative formulation of model II

We can express the single-factor random effects model (19.12) in an alternative fashion, just as we did for ANOVA model I in (16.61). We do this by expressing the factor level mean μ_i as a deviation from its expected value, $E(\mu_i) = \mu_.$, as follows:

(19.33) $$\tau_i = \mu_i - \mu_.$$

Then, we simply replace μ_i in model (19.12) by its equivalent expression from (19.33):

(19.34) $$\mu_i = \mu_. + \tau_i$$

The alternative model therefore is expressed as follows:

(19.35) $$Y_{ij} = \mu_. + \tau_i + \varepsilon_{ij}$$

where:

$\mu_.$ is a constant component common to all observations
τ_i are independent $N(0, \sigma_\mu^2)$
ε_{ij} are independent $N(0, \sigma^2)$
τ_i and ε_{ij} are independent
$i = 1, \ldots, r;\ j = 1, \ldots, n_i$

Note that the τ_i are random variables in model (19.35). With reference to our personnel officers example, τ_i represents the effect of the ith personnel officer who is selected at random. Here, τ_i measures by how much the mean evaluation of all potential employees by the ith personnel officer differs from the mean evaluation by all personnel officers.

PROBLEMS

19.1. a. Can the studentized range test ever be more efficient than the F test? Explain.

b. Does the studentized range test avoid any of the assumptions required for the F test? Does it entail additional assumptions not required for the F test?

19.2. Refer to **Questionnaire color** Problems 16.10 and 17.10.

a. Conduct the studentized range test with a level of significance of $\alpha = .10$. State the alternatives, decision rule, and conclusion.

b. Does the conclusion in part (a) differ from the one in Problem 16.10d?
c. What were the simplifications in using the studentized range test rather than the F test?

19.3. Refer to **Filling machines** Problem 16.13.
a. Using a level of significance of $\alpha = .05$, employ the studentized range test to study whether or not the mean fills for the six machines are equal. State the alternatives, decision rule, and conclusion.
b. Does the conclusion in part (a) differ from the one in Problem 16.13d?

19.4. Refer to **Premium distribution** Problem 16.14. Use the studentized range test to examine whether or not the mean time lapses for the five agents are the same; control the α risk at .10. State the alternatives, decision rule, and conclusion.

19.5. Why are the Kruskal-Wallis rank test and the median test nonparametric tests? Is the studentized range test a nonparametric test?

19.6. Is there a basic distinction in the assumptions for the Kruskal-Wallis rank test and those for the median test? If so, what is it? If not, how should one make a choice between the two tests?

19.7. Explain why the limits in (19.8) are testing limits and not confidence limits.

19.8. Refer to **Productivity improvement** Problem 16.9.
a. Conduct the Kruskal-Wallis rank test; use $\alpha = .05$. State the alternatives, decision rule, and conclusion. Were ties a source of difficulty here?
b. What is the P-value of the test in part (a)?
c. Does the conclusion in part (a) differ from the one in Problem 16.9d?
d. Do the data suggest that a nonparametric test is needed here?
e. Conduct multiple pairwise tests based on the ranked data to group the three types of firms according to mean productivity improvement. Use a family level of significance of $\alpha = .10$. Describe your findings.

19.9. Telephone communications. A management consultant was engaged by a firm to improve the cost effectiveness of its communications. As part of his study, the consultant selected 10 home-office executives at random from each of the (1) sales, (2) production, and (3) research and development divisions, and studied the communications of these executives during the past 10 weeks in great detail. Among other data, he obtained the following information on weekly dollar costs of long-distance telephone calls to branch offices by the executives:

	j									
i	1	2	3	4	5	6	7	8	9	10
1	666	920	495	602	1,499	960	796	343	894	813
2	488	362	156	546	216	542	345	291	516	126
3	391	450	609	910	705	472	645	496	763	1,309

The consultant decided to employ a nonparametric approach to test whether or not the mean telephone expenses for the three divisions are equal.
a. What feature of the data may have suggested the use of a nonparametric test?
b. Conduct the Kruskal-Wallis rank test, controlling the risk of Type I error at $\alpha = .05$. State the alternatives, decision rule, and conclusion. What is the P-value of the test?

c. Conduct multiple pairwise tests based on the ranked data to group the three divisions according to mean telephone expenditures; use a family level of significance of $\alpha = .05$. Describe your findings.

19.10. Refer to **Telephone communications** Problem 19.9. Suppose that in conducting the Kruskal-Wallis rank test the alternatives had been:

$$H_0\text{: all populations are identical}$$
$$H_a\text{: all populations are not identical}$$

a. Would the same test assumptions be involved as in Problem 19.9?
b. If H_a were concluded, would it necessarily imply that the mean telephone expenses are not the same in the three divisions? Explain.

19.11. **Battery life.** A special field version of a laboratory instrument was developed, powered by a battery. Four different designs of the battery were tested. Data on the number of operating hours in the field for 20 batteries of each type follow.

	j									
i	1	2	3	4	5	6	7	8	9	10
1	7.48	10.08	3.81	13.22	10.19	8.31	13.27	4.85	5.71	11.71
2	10.86	23.41	6.45	17.42	12.36	11.52	14.08	7.07	14.82	8.41
3	6.70	8.12	9.68	5.35	16.40	8.60	7.14	16.70	6.35	22.41
4	12.40	4.99	6.74	4.59	9.09	6.31	4.21	7.34	11.74	3.57

	j									
i	11	12	13	14	15	16	17	18	19	20
1	6.52	2.66	2.25	19.37	4.19	8.03	6.71	3.08	3.37	6.24
2	9.06	11.00	6.53	7.53	7.10	9.21	15.37	17.07	10.17	8.85
3	10.52	6.01	13.28	9.14	11.30	5.66	14.63	13.80	7.58	11.14
4	8.36	4.07	11.76	7.86	12.38	5.40	8.22	14.76	20.17	5.35

a. Obtain the standardized residuals (18.7) when fitting ANOVA model (16.2) and prepare residual frequency plots. Also prepare a normal probability plot of the residuals and calculate the coefficient of correlation between the ordered residuals and their expected values under normality. Does it appear that the distribution of the error terms is not normal?
b. Conduct the Kruskal-Wallis rank test; use $\alpha = .10$. State the alternatives, decision rule, and conclusion. Were ties a source of difficulty here?
c. Conduct multiple pairwise tests based on the ranked data to group the four types of batteries according to mean operating life; use a family level of significance of $\alpha = .10$. Describe your findings.

19.12. Refer to **Cash offers** Problem 16.12.
a. Conduct the median test for equality of factor level means; use $\alpha = .01$. State the alternatives, decision rule, and conclusion. What is the P-value of the test?
b. Is the conclusion in part (a) the same as the one in Problem 16.12d?

19.13. Refer to **Telephone communications** Problem 19.9. Conduct the median test for equality of treatment means; control the α risk at .05. State the alternatives, decision rule, and conclusion.

19.14. Refer to **Battery life** Problem 19.11. Conduct the median test for equality of treatment means; use $\alpha = .10$. State the alternatives, decision rule, and conclusion. What is the P-value of the test?

19.15. A student asks why ε_{ij} is shown as a separate term in the random effects model (19.12) in view of μ_i being a random variable in this model. Respond.

19.16. Refer to Figure 19.1. Here, the situation portrayed is one where the variance σ^2 is larger than the variance σ_μ^2. Is this always the case? Explain.

19.17. In each of the following cases, indicate whether ANOVA model I or model II is more appropriate and state your reasons:

(1) In a study of absenteeism at a plant, the treatments are the three eight-hour shifts.

(2) In a study of employee productivity, the treatments are 10 production employees selected at random from all production employees in a large company.

(3) In a study of anticipated annual income at retirement, the treatments are the four types of retirement plans available to employees.

(4) In a study of tire wear in 18-wheel trucks, the treatments are four tire locations selected at random.

19.18. Refer to the Apex Enterprises personnel officers example on page 647. Explain with reference to this example over what the expectation in (19.13) is taken. Over what is the variance in (19.14) taken?

19.19. Refer to **Filling machines** Problem 16.13. Suppose that the company uses a large number of filling machines and the six studied were selected randomly. Assume that ANOVA model (19.12) is applicable.

a. Interpret the following with reference to this example: (1) $\mu_.$, (2) σ_μ^2, (3) σ^2, (4) $\sigma^2\{Y_{ij}\}$.

b. Test whether or not all machines in the population have the same mean fill; use $\alpha = .05$. State the alternatives, decision rule, and conclusion. What is the P-value of the test?

c. Estimate the mean fill for all machines in the population with a 95 percent confidence interval.

19.20. Refer to **Filling machines** Problems 16.13 and 19.19.

a. Estimate the proportion of the total variability in carton fills that reflects the differences in mean fills between machines; use a 95 percent confidence interval.

b. Estimate σ^2 with a 95 percent confidence interval. Interpret your interval estimate.

c. Obtain a point estimate of σ_μ^2.

19.21. **Sodium content.** A researcher studied the sodium content in lager beer by selecting at random six brands from the large number of brands of U.S. and Canadian beers sold in a metropolitan area. She then chose eight 12-ounce cans or bottles of each selected brand at random from retail outlets in the area, and measured the sodium content (in milligrams) of each can or bottle. The observations follow.

i	j = 1	2	3	4	5	6	7	8
1	24.4	22.6	23.8	22.0	24.5	22.3	25.0	24.5
2	10.2	12.1	10.3	10.2	9.9	11.2	12.0	9.5
3	19.2	19.4	19.8	19.0	19.6	18.3	20.0	19.4
4	17.4	18.1	16.7	18.3	17.6	17.5	18.0	16.4
5	13.4	15.0	14.1	13.1	14.9	15.0	13.4	14.8
6	21.3	20.2	20.7	20.8	20.1	18.8	21.1	20.3

Assume that ANOVA model (19.12) is applicable.

a. Test whether or not the mean sodium content is the same in all brands sold in the metropolitan area; use $\alpha = .01$. State the alternatives, decision rule, and conclusion. What is the P-value of the test?

b. Estimate the mean sodium content for all brands; use a 99 percent confidence interval.

19.22. Refer to **Sodium content** Problem 19.21.

a. Estimate $\sigma_\mu^2/(\sigma_\mu^2 + \sigma^2)$ with a 99 percent confidence interval. Interpret your interval estimate.

b. Obtain point estimates of σ^2 and σ_μ^2.

c. Estimate σ^2 with a 99 percent confidence interval.

d. It has been conjectured that the variance of sodium content between brands is more than twice as great as that within brands. Conduct an appropriate test using $\alpha = .01$. State the alternatives, decision rule, and conclusion.

19.23. **Coil winding machines.** A plant contains a large number of coil winding machines. A production analyst studied a certain characteristic of the wound coils produced by these machines by selecting 4 machines at random and then choosing 10 coils at random from the day's output of each selected machine. The results follow.

i	j = 1	2	3	4	5	6	7	8	9	10
1	205	204	207	202	208	206	209	205	207	206
2	201	204	198	203	209	207	199	206	205	204
3	198	204	196	201	199	203	202	198	202	197
4	210	209	214	215	211	208	210	209	211	210

Assume that ANOVA model (19.12) is appropriate.

a. Test whether or not the mean coil characteristic is the same for all machines in the plant. Use a level of significance of $\alpha = .10$. State the alternatives, decision rule, and conclusion. What is the P-value of the test?

b. Estimate the mean coil characteristic for all coil winding machines in the plant; use a 90 percent confidence interval.

19.24. Refer to **Coil winding machines** Problem 19.23.

a. Estimate $\sigma_\mu^2/(\sigma_\mu^2 + \sigma^2)$ with a 90 percent confidence interval. Interpret your interval estimate.

b. Estimate σ^2 with a 90 percent confidence interval. Interpret your interval estimate.

c. Obtain a point estimate of σ_μ^2.

d. Test whether or not σ_μ^2 and σ^2 are equal; use $\alpha = .10$. State the alternatives, decision rule, and conclusion.

EXERCISES

19.25. Show that when all μ_i are equal, test statistic q^* defined in (19.2) is distributed as q defined in (17.23).

19.26. Show that $n_T(n_T + 1)/12$ is the sample variance of the consecutive integers 1 to n_T.

19.27. Show that n' defined in (19.18) equals n when $n_i \equiv n$.

19.28. What are the values r and n that minimize $\sigma^2(\bar{Y}_{..})$ in (19.22) for a given total sample size n_T? Ignore any cost considerations.

19.29. Derive the confidence limits in (19.30) from those in (19.29).

PROJECTS

19.30. Refer to the **SENIC** data set and Project 16.32.

a. Use the Kruskal-Wallis test to determine whether or not the mean infection risk is the same in the four regions; control the level of significance at $\alpha = .05$. State the alternatives, decision rule, and conclusion.

b. Is your conclusion in part (a) the same as that obtained in Project 16.32? Are the assumptions for ANOVA model (16.2) or those underlying the Kruskal-Wallis test more reasonable here?

c. Use the multiple pairwise testing procedure (19.8) to group the regions; employ family significance level $\alpha = .10$. What are your findings?

19.31. Refer to the **SMSA** data set and Project 16.34.

a. Using the Kruskal-Wallis test, determine whether or not the mean crime rate is the same in the four regions; control the level of significance at $\alpha = .05$. State the alternatives, decision rule, and conclusion.

b. Is your conclusion in part (a) the same as that obtained in Project 16.34? Are the assumptions for ANOVA model (16.2) or those underlying the Kruskal-Wallis test more reasonable here?

c. Use the multiple pairwise testing procedure (19.8) to group the regions; employ family significance level $\alpha = .05$. What are your findings?

19.32. Obtain the exact sampling distribution of X^2_{KW} when H_0 holds, for the case $r = 2$ and $n_i \equiv 2$. (*Hint:* What does the equality of the treatment means imply about the arrangements of the ranks 1, 2, 3, and 4?)

19.33. Three populations are being studied; each is uniform between 300 and 800.

a. Generate 10 random observations from each of the three uniform populations and calculate the X^2_{KW} test statistic (19.4).

b. Repeat part (a) 100 times.

c. Calculate the mean and standard deviation of the 100 test statistics. How do these values compare with the characteristics of the relevant chi-square distribution?

d. What proportion of the 100 test statistics obtained in part (b) is less than 4.61? What proportion is less than 9.21? How do these proportions agree with theoretical expectations?

CITED REFERENCES

19.1. Owen, Donald B. *Handbook of Statistical Tables*. Reading, Mass.: Addison-Wesley Publishing, 1962.

19.2. Scheffé, Henry. *The Analysis of Variance*. New York: John Wiley & Sons, 1959.

PART IV

Multifactor analysis of variance

20

Two-factor analysis of variance—equal sample sizes - BALANCED DESIGNS

In Part III, we considered studies in which the effect of one factor is investigated. Now we are concerned with investigations of the simultaneous effects of two or more factors. In this chapter, we take up the analysis of variance for two-factor studies when all sample sizes are equal. In Chapters 21, 22, and 23, we continue the discussion of two-factor studies by taking up the analysis of factor effects, the planning of sample sizes, the case of unequal sample sizes, and a number of other topics. In Chapter 24, we consider the analysis of variance for studies in which three or more factors are being investigated. Finally in Chapter 25, we take up the analysis of covariance for factorial studies.

20.1 MULTIFACTOR STUDIES

Before focusing specifically on two-factor studies, we shall first make some general remarks about multifactor studies, which encompass investigations of two or more factors.

Examples of two-factor studies

Example 1. A company investigated the effects of selling price and type of promotional campaign on sales of one of its products. Three selling prices (59 cents, 60 cents, 64 cents) were studied, as were two types of promotional campaigns (radio advertising, newspaper advertising). Let us consider selling price to be factor A and promotional campaign to be factor B. Factor A here was studied at three price levels; in general, we use the symbol a to denote the number of levels of factor A investigated. Factor B was here studied at two levels; we shall use the symbol b to denote the number of levels of factor B investigated. Each combination of price and promotion campaign was studied as follows:

Treatment	*Description*
1	59¢ price, radio advertising
2	60¢ price, radio advertising
3	64¢ price, radio advertising
4	59¢ price, newspaper advertising
5	60¢ price, newspaper advertising
6	64¢ price, newspaper advertising

Each of the combinations of a factor level of A and a factor level of B is a *treatment*. Thus, there are $3 \times 2 = 6$ treatments here altogether. In general, the total number of possible treatments in a two-factor study is ab.

Twelve communities throughout the United States, of approximately equal size and similar socioeconomic characteristics, were selected and assigned at random to the treatments such that each treatment was given to two experimental units. As before, we shall use the symbol n for the number of units receiving a given treatment when all treatment sample sizes are the same. In the two communities assigned to treatment 1, for instance, the product price was fixed at 59 cents and radio advertising was employed, and so on for the other communities in the study.

Example 2. A steel company studied the effects of carbon content and tempering temperature on the strength of steel. Carbon content was investigated at a high level and at a low level (the precise definitions of these levels is not important here). Tempering temperature was also studied at a high level and at a low level. Altogether, $2 \times 2 = 4$ treatments were defined for this study:

Treatment	*Description*
1	High carbon level, high tempering temperature
2	High carbon level, low tempering temperature
3	Low carbon level, high tempering temperature
4	Low carbon level, low tempering temperature

These four treatments were then each assigned to three production batches in randomized fashion.

Example 3. Multifactor studies also can be made with observational data. An analyst wished to study the effects of family income (under $10,000,

\$10,000–\$19,999, \$20,000–\$29,999, \$30,000 and more) and stage in the life cycle of the family (stages 1, 2, 3, 4) on appliance purchases. Here, $4 \times 4 = 16$ treatments are defined. These are in part:

Treatment	*Description*
1	Under \$10,000 income, stage 1
2	Under \$10,000 income, stage 2
.	.
.	.
.	.
16	\$30,000 and more income, stage 4

The analyst then selected 20 families with the required income and life-cycle characteristics for each of the "treatment" classes for this study.

Note

When we considered single-factor studies, we did not place any restrictions on the nature of the r factor levels under study. Formally, the ab treatments in a two-factor investigation could be considered as the r factor levels in a single-factor investigation and analyzed according to the methods discussed in Part III. The reason why new methods of analysis are required is that we wish to analyze the ab treatments in special ways that recognize two factors are involved and enable us to obtain information about the effects of each of the two factors as well as about any special joint effects.

Complete and fractional factorial studies

The three examples just cited are *complete factorial studies* because all possible combinations of factor levels for the different factors were included. At times, it is not feasible or desirable to include all possible combinations of factor levels for the different factors. For instance, suppose the steel company mentioned in Example 2 wished to study six temperatures, five levels of carbon content, and four methods of cooling the steel. A complete factorial study would then involve $6 \times 5 \times 4 = 120$ treatments. Such a study might be extremely costly and time-consuming. Under these conditions, it may be possible to design a *fractional factorial study* containing only a fraction of the 120 factor level combinations, which will still provide information about the effects of each of the three factors as well as about any important special joint effects of these factors.

The discussion of multifactor investigations in Part IV deals solely with complete factorial studies.

Advantages of multifactor studies

Efficiency. Multifactor studies are more efficient than the traditional experimental approach of manipulating only one factor at a time and keeping all other conditions constant. With reference to Example 1, the traditional approach to studying the effect of promotion campaign would have been to keep price constant at a given level and vary only the promotional campaign. An important problem with this approach is the choice of the price level to be held constant. This choice is especially difficult when one is not sure whether the promotional effect is the same at different price levels. Even though the traditional approach devotes all resources to studying the effect of only one factor, it does not yield any more precise information about that factor than a multifactor experiment of the same size. With reference to Example 1 again, suppose that 12 communities were to be utilized in a traditional study, six assigned to radio advertising and the other six to newspaper advertising, and that the price would be kept constant at 59 cents. For this traditional study, the comparison between the two types of promotional campaigns would be based on two samples of six communities each. The same is true for the two-factor study in Example 1, since each promotional campaign occurs there in three treatments and each treatment has two communities assigned to it.

Amount of information. The traditional study provides less information than the two-factor study. Specifically in our previous illustration, it does not provide any information about the effect of price, nor about any special joint effects of price and promotional campaign. Information about price effects would require an additional traditional experiment for which promotional campaign would be kept constant at a given level and price varied. Thus, the traditional approach would require a larger sample to provide information about both price and promotion campaign effects, and unless the traditional study were yet further enlarged, it would still not provide full information about any special joint effects of the two factors. Such special joint effects are called *interactions*. Interaction effects were encountered in regression models and will be discussed in the context of analysis of variance models in the next section. Here it suffices to point out that interaction effects may be very important. For instance, it might be that the price effect is not large when the promotional campaign is in newspapers, but it is large with radio advertising. Such interaction effects can be readily investigated from factorial studies.

Validity of findings. In addition to being more efficient and readily providing information about interaction effects, multifactor studies also can strengthen the validity of the findings. Suppose that in Example 1, management was principally interested in investigating the effect of price on sales. If the promotional campaign used in the price study had been newspaper advertising, doubts might exist whether or not the price effect differs for other promotional vehicles. By including type of promotional campaign as another factor in the study, management can get information about the persistence of the price effect with different promotional vehicles, without increasing the number of experimental units in the study. Thus, multifactor studies can include some factors of secondary importance to permit inferences about the primary factors with a greater range of validity.

Comments

1. In studies based on observational data, as for ones utilizing experimental data, multifactor analysis of the data permits a ready evaluation of interaction effects and economizes on the number of cases required for analysis.

2. The advantages of multifactor experiments just described should not lead one to think that the more factors are included in the study, the better. Experiments involving many factors, each at numerous levels, become complex, costly, and time-consuming. It is often better research strategy to begin with a few factors, investigate the effects of these, and extend the investigation in accordance with the results obtained to date. In this way, resources can be devoted principally to the most promising avenues of investigation and a better understanding of the working of the factors can be obtained.

20.2 MEANING OF MODEL ELEMENTS

Before presenting a formal statement of the analysis of variance model for two-factor studies, we shall develop the model elements and discuss their meaning. This will not only be helpful in understanding the ANOVA model but will also provide insights into how the analysis of two-factor studies should proceed. *Throughout this section, we assume that all population means are known and are of equal importance when averages of these means are required.*

Illustration

To illustrate the meaning of the model elements, we shall consider a simple two-factor study in which the effects of sex and age on learning of a task are of interest. For simplicity, the age factor has been defined in terms of only three factor levels (young, middle, old), as shown in Table 20.1.

Treatment means

The mean response for a given treatment in a two-factor study is denoted by μ_{ij}, where i refers to the level of factor A ($i = 1, \ldots, a$) and j refers to the level of factor B ($j = 1, \ldots, b$). Table 20.1 contains the true treatment means μ_{ij} for our learning illustration. Note, for instance, that $\mu_{11} = 9$, which indicates that the mean learning time for young males is nine minutes. Similarly, we see that $\mu_{22} = 11$, so that the mean learning time for middle-aged females is 11 minutes.

Note

The interpretation of a treatment mean μ_{ij} depends on whether the study is an observational or an experimental one. In an observational study, the treatment mean μ_{ij} corresponds to the population mean for the elements having the characteristics of the ith level of factor A and the jth level of factor B. For instance, in our learning illustration, the treatment mean μ_{11} is the mean learning time for the population of young males.

In an experimental study, the treatment mean μ_{ij} stands for the mean response that would be obtained if the treatment consisting of the ith level of factor A and the jth level of factor B were applied to all units in the population of experimental units about which

TABLE 20.1 Age effect but no sex effect, with no interactions

(a) Mean Learning Times (in minutes)

Factor A—Sex	Factor B—Age: $j=1$ Young	$j=2$ Middle	$j=3$ Old	Row Average
$i=1$ Male	9 (μ_{11})	11 (μ_{12})	16 (μ_{13})	12 ($\mu_{1.}$)
$i=2$ Female	9 (μ_{21})	11 (μ_{22})	16 (μ_{23})	12 ($\mu_{2.}$)
Column average	9 ($\mu_{.1}$)	11 ($\mu_{.2}$)	16 ($\mu_{.3}$)	12 ($\mu_{..}$)

(b) Specific Age Effects (in minutes)

	$j=1$	$j=2$	$j=3$
$i=1$	-3 ($\beta_{1(1)}$)	-1 ($\beta_{2(1)}$)	$+4$ ($\beta_{3(1)}$)
$i=2$	-3 ($\beta_{1(2)}$)	-1 ($\beta_{2(2)}$)	$+4$ ($\beta_{3(2)}$)
Column average	-3 (β_1)	-1 (β_2)	$+4$ (β_3)

(c) Specific Sex Effects (in minutes)

	$j=1$	$j=2$	$j=3$	Row Average
$i=1$	0 ($\alpha_{1(1)}$)	0 ($\alpha_{1(2)}$)	0 ($\alpha_{1(3)}$)	0 (α_1)
$i=2$	0 ($\alpha_{2(1)}$)	0 ($\alpha_{2(2)}$)	0 ($\alpha_{2(3)}$)	0 (α_2)

inferences are to be drawn. For instance, in a study where factor A is type of training program (structured, partially structured, unstructured) and factor B is time of training (during work, after work), $6n$ employees are selected and n are assigned at random to each of the 6 treatments. The mean μ_{ij} here represents the mean response, say, mean gain in productivity, if the ith training program administered during the jth time were given to all employees in the population of experimental units.

Factor level means

The treatment means in Table 20.1a, for our learning illustration, indicate that the mean learning times for men and women are the same for each age group. On the other hand, the mean learning time increases with age for each sex. Thus, sex has no effect on mean learning time, but age does. This can also be seen quickly from the row averages and column averages shown in Table 20.1a, which in this case tell the complete story. The row averages are the sex *factor level means*, and the column averages are the age factor level means. We denote the column average for the first column by $\mu_{.1}$, which is the average of μ_{11} and μ_{21}. In general, the column average for the jth column is denoted by $\mu_{.j}$:

(20.1)
$$\mu_{.j} = \frac{\sum_{i=1}^{a} \mu_{ij}}{a}$$

and the row average for the ith row is denoted by $\mu_{i.}$:

(20.2)
$$\mu_{i.} = \frac{\sum_{j=1}^{b} \mu_{ij}}{b}$$

The overall mean learning time for all ages and both sexes is denoted by $\mu_{..}$, and is defined in the following equivalent fashions:

(20.3a)
$$\mu_{..} = \frac{\sum_i \sum_j \mu_{ij}}{ab}$$

(20.3b)
$$\mu_{..} = \frac{\sum_i \mu_{i.}}{a}$$

(20.3c)
$$\mu_{..} = \frac{\sum_j \mu_{.j}}{b}$$

Specific effects

When we compare the mean learning time for young males with the row average, we measure the *specific effect* of young age for men. This specific effect, denoted by $\beta_{1(1)}$, is:

(20.4)
$$\beta_{1(1)} = 9 - 12 = -3$$

This specific effect implies that the mean learning time for young men is three minutes less than the average learning time for all men. The notation $\beta_{1(1)}$ is used to indicate that we are measuring the specific effect of the first factor level of B when factor A is at the first level.

As another example, $\beta_{1(2)}$ is the specific effect of young age for women. This effect is:

(20.5)
$$\beta_{1(2)} = 9 - 12 = -3$$

the same as for men.

In general, we define $\beta_{j(i)}$ as follows:

(20.6)
$$\beta_{j(i)} = \mu_{ij} - \mu_{i.}$$

Table 20.1b contains each of the specific age effects for our learning illustration.

Similarly, there are specific effects of sex. For instance, the specific effect of male sex for young persons, denoted by $\alpha_{1(1)}$, is:

$$\alpha_{1(1)} = \mu_{11} - \mu_{.1} = 9 - 9 = 0 \tag{20.7}$$

Since this specific effect is 0, male persons who are young have the same mean learning time as all young persons.

In general, we define $\alpha_{i(j)}$ as follows:

$$\alpha_{i(j)} = \mu_{ij} - \mu_{.j} \tag{20.8}$$

Table 20.1c contains the specific effects of male and female sex for our learning illustration.

Main effects

Main age effects. To summarize the specific age effects, we shall average them. These averages are shown in Table 20.1b. In this instance, of course, the mean for each age group is the same as its components since the specific age effects are the same for men and women. For young persons, the mean specific age effect is denoted by β_1; it is:

$$\beta_1 = \frac{(-3) + (-3)}{2} = -3 \tag{20.9}$$

β_1 is called the *main effect* for factor B at the first level.

It can be shown that β_1 is the equivalent of the difference between the mean learning time for young persons and the mean learning time for all persons:

$$\beta_1 = \mu_{.1} - \mu_{..} = 9 - 12 = -3 \tag{20.9a}$$

To show this, we begin with the definition of β_1 as the mean specific age effect for young persons:

$$\beta_1 = \frac{\beta_{1(1)} + \beta_{1(2)}}{2}$$

Using (20.6), we obtain:

$$\beta_1 = \frac{(\mu_{11} - \mu_{1.}) + (\mu_{21} - \mu_{2.})}{2}$$

Rearranging terms and employing (20.1) and (20.3b), we find:

$$\beta_1 = \frac{\mu_{11} + \mu_{21}}{2} - \frac{\mu_{1.} + \mu_{2.}}{2} = \mu_{.1} - \mu_{..}$$

Main sex effects. The means of the specific sex effects are shown in Table 20.1c and will be denoted by α_i. Note that the main sex effects in Table 20.1c are all zero, indicating that sex does not affect mean learning time.

General definitions. In general, we define the main effect of factor B at the jth level as follows:

(20.10) $$\beta_j = \frac{\sum_i \beta_{j(i)}}{a} = \mu_{.j} - \mu_{..}$$

Similarly, the main effect of the ith level of factor A is defined:

(20.11) $$\alpha_i = \frac{\sum_j \alpha_{i(j)}}{b} = \mu_{i.} - \mu_{..}$$

It follows from (20.3b) and (20.3c) that:

(20.12) $$\sum_i \alpha_i = 0 \qquad \sum_j \beta_j = 0$$

Thus, the sum of the main effects for each factor is zero.

Additive factor effects

The factor effects in Table 20.1a have an interesting property. Each mean response μ_{ij} can be obtained by adding the respective sex and age main effects to the overall mean $\mu_{..}$. For instance, we have:

$$\mu_{11} = \mu_{..} + \alpha_1 + \beta_1 = 12 + 0 + (-3) = 9$$
$$\mu_{23} = \mu_{..} + \alpha_2 + \beta_3 = 12 + 0 + 4 = 16$$

In general, we have for Table 20.1a:

(20.13) $$\mu_{ij} = \mu_{..} + \alpha_i + \beta_j$$ Additive factor effects

which can be expressed equivalently, using the definitions of α_i in (20.11) and of β_j in (20.10), as:

(20.13a) $$\mu_{ij} = \mu_{i.} + \mu_{.j} - \mu_{..}$$ Additive factor effects

It can also be shown that each treatment mean μ_{ij} in Table 20.1a can be expressed in terms of three other treatment means:

(20.13b) $$\mu_{ij} = \mu_{ij'} + \mu_{i'j} - \mu_{i'j'} \qquad i \neq i', j \neq j'$$ Additive factor effects

For instance, we have:

$$\mu_{11} = \mu_{12} + \mu_{21} - \mu_{22} = 11 + 9 - 11 = 9$$

or:

$$\mu_{11} = \mu_{13} + \mu_{21} - \mu_{23} = 16 + 9 - 16 = 9$$

When all treatment means can be expressed in the form of (20.13), (20.13a), or (20.13b), we say that the *factors do not interact*, or that *no factor interactions are present*, or that the *factor effects are additive*. The significance of no factor interactions is that the effects of the two factors can be described separately merely by analyzing the factor level means or the factor main effects. Thus, in our illustration in Table 20.1a, the two sex means signify that sex has no influence regardless of age, and the three age means portray the influence of age regardless of sex. The analysis of factor effects is therefore quite simple when there are no factor interactions.

Graphic presentation

Figure 20.1 presents the mean learning times of Table 20.1a in graphic form. The X axis contains the sex factor levels (denoted by A_1 and A_2), and the Y axis contains learning time. Separate curves are drawn for each of the age factor levels (denoted by B_1, B_2, and B_3). The zero slope of each curve indicates that sex has no effect. The differences in the heights of the three curves show the age effects on learning time.

The points on each curve are conventionally connected by straight lines even though the variable on the X axis (sex, in our example) is not a continuous variable. When the variable on the X axis is qualitative, the slopes of the curves

FIGURE 20.1 Age effect but no sex effect, with no interactions

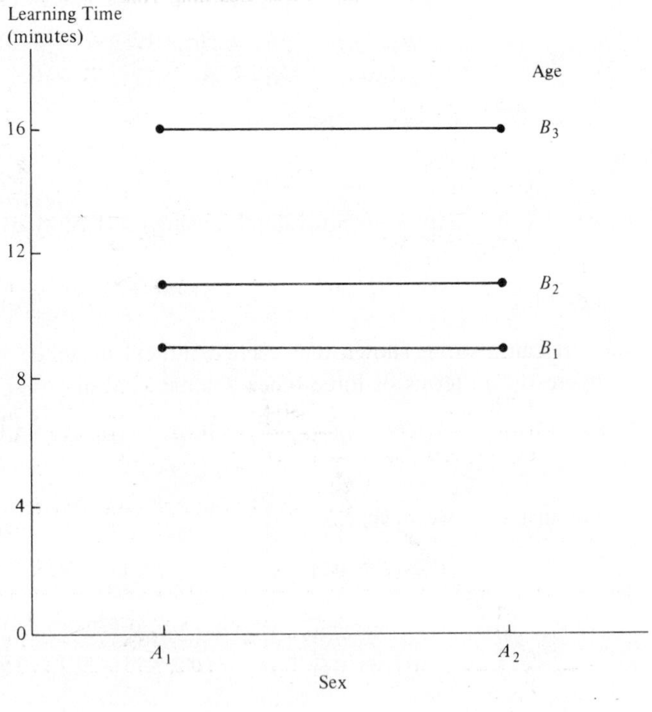

have no meaning, except when the slope is zero, which implies there are no factor level effects. If one of the two factors is a quantitative variable, it is ordinarily advisable to place that factor on the X scale.

A second example with additive factor effects

Table 20.2a contains another illustration of factor effects that do not interact, for the same sex-age learning setting as before. The situation here differs from that of Table 20.1a in that not only age but also sex affects the learning time. This is evident from the fact that the mean learning times for men and women are not the same for any age group. Alternatively, this can be seen from Table 20.2c, which shows that the specific sex effects are not zero.

In Table 20.2a, as in Table 20.1a, every mean response can be decomposed according to (20.13):

$$\mu_{ij} = \mu_{..} + \alpha_i + \beta_j$$

For instance:

$$\mu_{11} = \mu_{..} + \alpha_1 + \beta_1 = 12 + 2 + (-3) = 11$$

Hence, the two factors do not interact and the factor effects can be analyzed separately by examining the factor level means $\mu_{i.}$ and $\mu_{.j}$, respectively.

TABLE 20.2 Age and sex effects, with no interactions

(a) Mean Learning Times (in minutes)

Factor A—Sex	*Factor B—Age* $j=1$ *Young*	$j=2$ *Middle*	$j=3$ *Old*	*Row Average*
$i=1$ Male	11 (μ_{11})	13 (μ_{12})	18 (μ_{13})	14 ($\mu_{1.}$)
$i=2$ Female	7 (μ_{21})	9 (μ_{22})	14 (μ_{23})	10 ($\mu_{2.}$)
Column average	9 ($\mu_{.1}$)	11 ($\mu_{.2}$)	16 ($\mu_{.3}$)	12 ($\mu_{..}$)

(b) Specific Age Effects (in minutes)

	$j=1$	$j=2$	$j=3$
$i=1$	-3 ($\beta_{1(1)}$)	-1 ($\beta_{2(1)}$)	4 ($\beta_{3(1)}$)
$i=2$	-3 ($\beta_{1(2)}$)	-1 ($\beta_{2(2)}$)	4 ($\beta_{3(2)}$)
Column average	-3 (β_1)	-1 (β_2)	4 (β_3)

(c) Specific Sex Effects (in minutes)

	$j=1$	$j=2$	$j=3$	*Row Average*
$i=1$	2 ($\alpha_{1(1)}$)	2 ($\alpha_{1(2)}$)	2 ($\alpha_{1(3)}$)	2 (α_1)
$i=2$	-2 ($\alpha_{2(1)}$)	-2 ($\alpha_{2(2)}$)	-2 ($\alpha_{2(3)}$)	-2 (α_2)

FIGURE 20.2 Age and sex effects, with no interactions

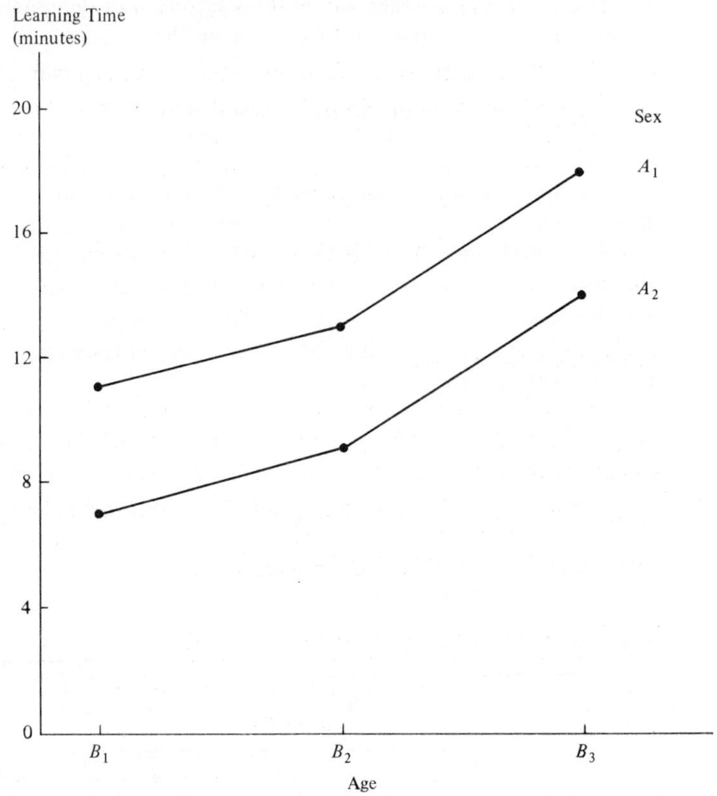

Figure 20.2 presents the data from Table 20.2a in graphic form. This time we have placed age on the X axis and used different curves for each sex. Note that the difference in the heights of the two curves reflects the sex difference and the departure from horizontal for each of the curves reflects the age effect.

Equivalent statements of additive factor effects

We have said that two factors do not interact if *all* treatment means μ_{ij} can be expressed according to (20.13), (20.13a), or (20.13b). There are a number of other, equivalent, methods of recognizing when two factors do not interact. These are:

1. The difference between the mean responses for any two levels of factor B is the same for all levels of factor A. (Thus, in Table 20.2a, going from young to middle age leads to an increase of 2 minutes for both males and females, and going from middle age to old leads to an increase of 5 minutes for both males and females.) Note that it is *not* required that the changes, say, be-

tween levels 1 and 2 and between levels 2 and 3 of factor B are the same. These, of course, may differ depending upon the nature of the factor B effect.

2. The difference between the mean responses for any two levels of factor A is the same for all levels of factor B. (Thus, in Table 20.2a, going from male to female leads to a decrease of 4 minutes for all three age groups.)
3. The curves of the mean responses for the different levels of a factor are all parallel (such as in Figure 20.2).

All of these conditions are equivalent, implying that the two factors do not interact.

Interacting factor effects

Table 20.3a contains an illustration for our sex-age learning setting where the factor effects do interact. The mean learning times for the different sex-age combinations in Table 20.3a indicate that sex has no effect on learning time for young persons but has a substantial effect for old persons. This differential influence of sex, which depends on the age of the person, implies that the age and sex factors interact in their effect on learning time.

TABLE 20.3 Age and sex effects, with interactions

(a) Mean Learning Times (in minutes)

	Factor B—Age				
Factor A—Sex	$j=1$ *Young*	$j=2$ *Middle*	$j=3$ *Old*	*Row Average*	*Main Sex Effect*
$i=1$ Male	9 (μ_{11})	12 (μ_{12})	18 (μ_{13})	13 ($\mu_{1.}$)	1 (α_1)
$i=2$ Female	9 (μ_{21})	10 (μ_{22})	14 (μ_{23})	11 ($\mu_{2.}$)	−1 (α_2)
Column average	9 ($\mu_{.1}$)	11 ($\mu_{.2}$)	16 ($\mu_{.3}$)	12 ($\mu_{..}$)	—
Main age effect	−3 (β_1)	−1 (β_2)	4 (β_3)	—	—

(b) Interactions (in minutes)

	$j=1$	$j=2$	$j=3$	*Row Average*
$i=1$	−1	0	1	0
$i=2$	1	0	−1	0
Column average	0	0	0	0

Definition of interaction. We can study the existence of interacting factor effects formally by examining whether or not all treatment means μ_{ij} can be expressed according to (20.13):

$$\mu_{ij} = \mu_{..} + \alpha_i + \beta_j$$

If they can, the factor effects are additive; otherwise, the factor effects are interacting.

For our example in Table 20.3a, the main factor effects α_i and β_j are shown in the margins of the table. It is clear that the factors interact. For instance:

$$\mu_{..} + \alpha_1 + \beta_1 = 12 + 1 + (-3) = 10$$

whereas $\mu_{11} = 9$. If the two factors were additive, there would be no difference.

The difference between the treatment mean μ_{ij} and the value $\mu_{..} + \alpha_i + \beta_j$ that would be expected if the two factors were additive is called the *interaction effect*, or more simply the *interaction*, of the ith level of factor A with the jth level of factor B, and is denoted by $(\alpha\beta)_{ij}$. Thus, we define $(\alpha\beta)_{ij}$ as follows:

(20.14) $$(\alpha\beta)_{ij} = \mu_{ij} - (\mu_{..} + \alpha_i + \beta_j)$$

Replacing α_i and β_j by their definitions in terms of treatment means in (20.11) and (20.10), respectively, we obtain an alternative definition:

(20.14a) $$(\alpha\beta)_{ij} = \mu_{ij} - \mu_{i.} - \mu_{.j} + \mu_{..}$$

Utilizing (20.13b), still another alternative definition is:

(20.14b) $$(\alpha\beta)_{ij} = \mu_{ij} - \mu_{ij'} - \mu_{i'j} + \mu_{i'j'}$$

To repeat, the interaction of the ith level of A with the jth level of B, denoted by $(\alpha\beta)_{ij}$, is simply the difference between μ_{ij} and the value that would be expected if the factors were additive. If in fact the two factors are additive, all interactions equal zero, i.e., $(\alpha\beta)_{ij} \equiv 0$.

The interactions for our illustration in Table 20.3a are shown in Table 20.3b. We have, for instance:

$$\begin{aligned}(\alpha\beta)_{13} &= \mu_{13} - (\mu_{..} + \alpha_1 + \beta_3)\\ &= 18 - (12 + 1 + 4)\\ &= 1\end{aligned}$$

Recognition of interactions. We may recognize whether or not interactions are present in one of the following equivalent fashions:

1. By examining whether all μ_{ij} can be expressed as the sums $\mu_{..} + \alpha_i + \beta_j$.
2. By examining whether the difference between the mean responses for any two levels of factor B is the same for all levels of factor A. (Note in Table 20.3a that the mean learning time increases when going from young to middle-aged persons by 3 minutes for men but only by 1 minute for women.)
3. By examining whether the difference between the mean responses for any two levels of factor A is the same for all levels of factor B. (Note in Table 20.3a that there is no difference between sexes for young persons, but there is a difference of four minutes for old persons.)
4. By examining whether the treatment mean curves for the different factor levels in a graph are parallel. (Figure 20.3 presents the treatment means in

FIGURE 20.3 Age and sex effects, with important interactions

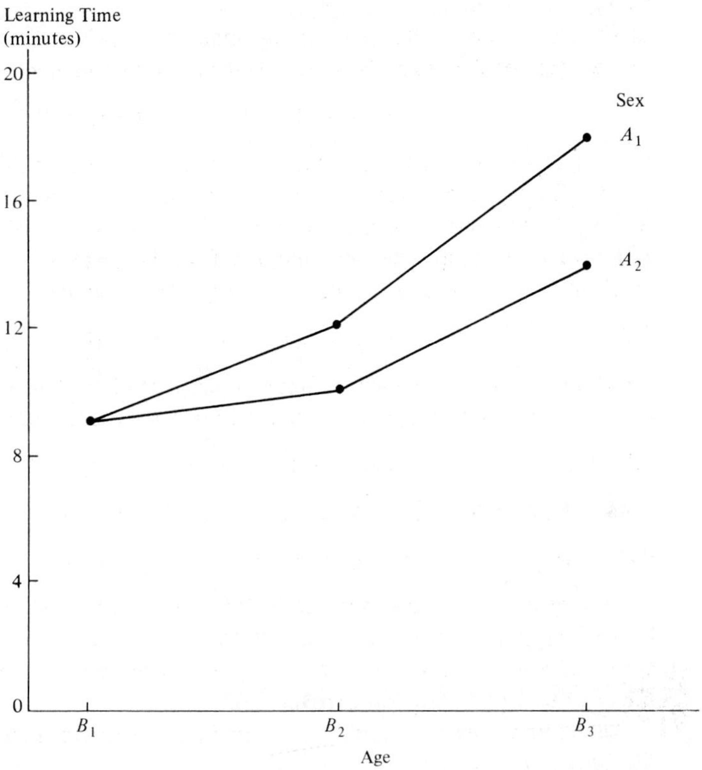

Table 20.3a with age on the X axis. Note that the treatment mean curves for the two sexes are not parallel.)

Comments

1. Note from Table 20.3b that some interactions are zero even though the two factors are interacting. *All* interactions must equal zero in order for the two factors to be additive.

2. The interaction $(\alpha\beta)_{ij}$ may also be interpreted as the difference between the specific effect of factor A at the ith level when factor B is at the jth level ($\alpha_{i(j)}$) and the mean specific effect (main effect) of factor A at the ith level (α_i):

$$(\alpha\beta)_{ij} = \alpha_{i(j)} - \alpha_i \tag{20.15}$$

Using (20.8) and (20.11), this may be rewritten:

$$\begin{aligned}(\alpha\beta)_{ij} &= (\mu_{ij} - \mu_{.j}) - (\mu_{i.} - \mu_{..}) \\ &= \mu_{ij} - \mu_{i.} - \mu_{.j} + \mu_{..}\end{aligned}$$

Alternatively, $(\alpha\beta)_{ij}$ may be interpreted as the difference between the specific effect of factor B at the jth level when factor A is at the ith level and the main effect of factor B at the jth level:

(20.16) $$(\alpha\beta)_{ij} = \beta_{j(i)} - \beta_j$$

3. Table 20.3b illustrates that interactions sum to zero when added over either rows or columns:

(20.17a) $$\sum_i (\alpha\beta)_{ij} = 0 \qquad j = 1, \ldots, b$$

(20.17b) $$\sum_j (\alpha\beta)_{ij} = 0 \qquad i = 1, \ldots, a$$

Consequently, the sum of all interactions is also zero:

(20.17c) $$\sum_i \sum_j (\alpha\beta)_{ij} = 0$$

We show this for (20.17a):

$$\sum_i (\alpha\beta)_{ij} = \sum_{i=1}^{a} (\mu_{ij} - \mu_{..} - \alpha_i - \beta_j)$$

$$= \sum_i \mu_{ij} - a\mu_{..} - \sum_i \alpha_i - a\beta_j$$

Now $\sum_i \mu_{ij} = a\mu_{.j}$ by (20.1) and $\Sigma\alpha_i = 0$ by (20.12). Finally, $\beta_j = \mu_{.j} - \mu_{..}$ by (20.10).

Hence, we obtain:

$$\sum_i (\alpha\beta)_{ij} = a\mu_{.j} - a\mu_{..} - a(\mu_{.j} - \mu_{..}) = 0$$

Important and unimportant interactions

When two factors interact, the question arises whether the factor level means, which are averages of specific treatment means, are meaningful measures. In Table 20.3a, for instance, it may well be argued that the sex factor level means 13 and 11 are misleading measures. They indicate that some difference exists in learning time for men and women, but that this difference is not too great. These factor level means hide the fact that there is no difference in mean learning time between sexes for young persons, but there is a relatively large difference for old persons. The interactions in Table 20.3a would therefore be considered *important interactions*, implying that one should not ordinarily discuss the effects of each factor separately in terms of the factor level means. A graph, such as Figure 20.3, presents effectively a description of the nature of the interacting effects of the two factors.

Sometimes when two factors interact, the interaction effects are so small that they are considered to be *unimportant interactions*. Table 20.4 and Figure 20.4 present such a case. Note from Figure 20.4 that the curves are *almost* parallel. Perfectly parallel curves, we know, would indicate there are no interactions. For practical purposes, one may say that the mean learning time for women is 2 minutes less than that for men, and this statement is approximately true for all age groups. Alternatively, statements based on average learning time for different age groups will hold approximately for both sexes.

TABLE 20.4 Age and sex effects, with unimportant interactions

	Factor B—Age			
Factor A—Sex	$j=1$ *Young*	$j=2$ *Middle*	$j=3$ *Old*	*Row Average*
$i=1$ Male	9.75	12.00	17.25	13.0
$i=2$ Female	8.25	10.00	14.75	11.0
Column average	9.0	11.0	16.0	12.0

FIGURE 20.4 Age and sex effects, with unimportant interactions (curves almost parallel)

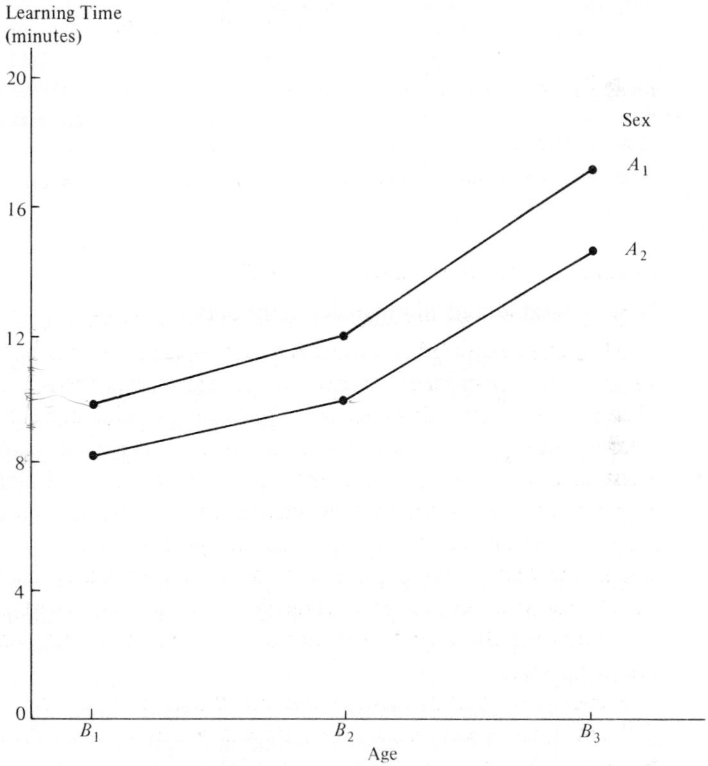

Thus, in the case of unimportant interactions, the analysis of factor effects can proceed as for the case of no interactions. Each factor can be studied separately, based on the factor level means $\mu_{i.}$ and $\mu_{.j}$, respectively. This separate analysis of factor effects is, of course, much simpler than a joint analysis for the two factors based on the treatment means μ_{ij}, which is required when the interactions are important.

Comments

1. The determination of whether interactions are important or unimportant is admittedly sometimes difficult. This decision is not a statistical decision and should be made by the subject area specialist (researcher). The advantage of unimportant (or no) interactions, namely, that one is then able to analyze the factor effects separately, is especially great when the study contains more than two factors.

2. Occasionally, it is meaningful to consider the effects of each factor in terms of the factor level means even when important interactions are present. For example, two methods of teaching college mathematics (abstract and standard) were used in teaching students of excellent, good, and moderate quantitative ability. Important interactions between teaching method and student's quantitative ability were found to be present. Students with excellent quantitative ability tended to perform equally well with the two teaching methods whereas students of moderate or good quantitative ability tended to perform better when taught by the standard method. If equal numbers of students with moderate, good, and excellent quantitative ability are to be taught by one of the two teaching methods, then the method that produces the best average result for all students might be of interest even in the presence of important interactions. Hence, a comparison of the teaching method factor level means would then be relevant, even though important interactions are present.

Transformable and nontransformable interactions

When important interactions exist, they are sometimes the result of the dependent variable being measured on an inappropriate scale. For instance, in the learning illustration of Table 20.1, the dependent variable could not only be learning time (Y), as was the case in our examples, but could also be rate of learning measured by $Y' = 1/Y$. It is sometimes possible when important interactions exist to make a simple transformation of the dependent variable that will reduce the magnitudes of the interactions so that they become unimportant. Such interactions are called *transformable interactions*. Some simple transformations that are frequently useful are the square, square root, logarithmic, and reciprocal transformations. When interactions cannot be largely removed by a simple transformation, they are called *nontransformable interactions*.

Table 20.5a contains an example of important interactions that are transformable. When a square root transformation is applied to these means, the resulting treatment means in Table 20.5b show no interacting effects. Essentially, cases where important interactions can be made unimportant by a meaningful transformation may be said to represent instances where the original scale of measurement is not the most appropriate one. Ordinarily, of course, one cannot hope that a simple transformation of scale removes all interactions as in Table 20.5, but only that interactions become unimportant after the transformation.

TABLE 20.5 Illustration of a transformable interaction

(a) Treatment Means—Original Scale

	Factor B	
Factor A	$j = 1$	$j = 2$
$i = 1$	16	64
$i = 2$	49	121
$i = 3$	64	144

(b) Treatment Means after Square Root Transformation

	Factor B	
Factor A	$j = 1$	$j = 2$
$i = 1$	4	8
$i = 2$	7	11
$i = 3$	8	12

Interpretation of interactions

The interpretation of interactions can be quite difficult when the interacting effects are complex. There are many occasions, however, when the interactions are of simple structure, such as in Table 20.3a, so that the joint factor effects can be described in a straightforward manner. Table 20.6 provides several additional illustrations of this type.

In Table 20.6a, we have a situation where either raising the pay or increasing the authority of low-paid executives with small authority leads to increased productivity. However, combining both higher pay and greater authority does not lead to further improvements in productivity than increasing either one alone. Table 20.6b represents a case where both higher pay and greater authority are required before any substantial increase in productivity takes place. Table 20.6c portrays a situation where size of crew and personality of crew chief do not interact on the productivity per person when the crew size is 6, 8, or 10 persons.

Note

It is possible that two factors interact, yet the main effects for one (or both) factors are zero. This would be the result of interactions in opposite directions that balance out over one (or both) factors. Thus, there would be definite factor effects, but these would not be disclosed by the factor level means. The case of interacting factors with no main effects for one (or both) factors fortunately is unusual. Typically, interaction effects are smaller than main effects.

TABLE 20.6 Examples of different types of interactions

(a) Productivity of Executives

Factor A—Pay	*Factor B—Authority* Small	*Great*
Low	50	76
High	74	75

(b) Productivity of Executives

Factor A—Pay	*Factor B—Authority* Small	*Great*
Low	50	52
High	53	75

(c) Productivity per Person in Crew

Factor A—Crew Size	*Factor B—Personality of Crew Chief* Extrovert	*Introvert*
4 persons	28	20
6 persons	22	20
8 persons	20	18
10 persons	17	15

20.3 MODEL I (FIXED EFFECTS) FOR TWO-FACTOR STUDIES

Having explained the model elements, we are now ready to develop the analysis of variance fixed effects model for two-factor studies when all treatment sample sizes are equal.

The basic situation is as follows: Factor A is studied at a levels, and these are of intrinsic interest in themselves; in other words, the a levels are not considered a sample from a larger population of factor A levels. Similarly, factor B is studied at b levels that are of intrinsic interest in themselves. All ab factor level combinations are included in the study. The number of observations for each of the ab treatments is the same, denoted by n, and it is required that $n > 1$. Thus, the total number of observations for the study is:

$$n_T = abn \tag{20.18}$$

The kth observation ($k = 1, \ldots, n$) for the treatment where A is at the ith level and B at the jth level is denoted by Y_{ijk} ($i = 1, \ldots, a; j = 1, \ldots, b$). Table 20.7 on page 686 illustrates this notation for an example where A is at three levels, B is at two levels, and two observations have been made for each treatment.

We shall state the fixed effects model for two-factor studies in two equivalent versions and later will use one or the other version as convenience dictates.

Model in terms of μ_{ij}

When we regard the *ab* treatments without explicitly considering the factorial structure of the study, we write the analysis of variance model as follows:

$$Y_{ijk} = \mu_{ij} + \varepsilon_{ijk} \tag{20.19}$$

where:

μ_{ij} are parameters
ε_{ijk} are independent $N(0, \sigma^2)$
$i = 1, \ldots, a; j = 1, \ldots, b; k = 1, \ldots, n$

Important features of model. 1. The parameter μ_{ij} is the mean response for the treatment in which factor A is at the ith level and factor B is at the jth level. This follows because $E(\varepsilon_{ijk}) = 0$:

$$E(Y_{ijk}) = \mu_{ij} \tag{20.20}$$

2. Since μ_{ij} is a constant, the variance of Y_{ijk} is:

$$\sigma^2(Y_{ijk}) = \sigma^2(\varepsilon_{ijk}) = \sigma^2 \tag{20.21}$$

3. Since the error terms ε_{ijk} are independent and normally distributed, so are the observations Y_{ijk}. Hence, we can state model (20.19) also as follows:

$$Y_{ijk} \text{ are independent } N(\mu_{ij}, \sigma^2) \tag{20.22}$$

4. Model (20.19) is a linear model because it can be expressed in the form $\mathbf{Y} = \mathbf{X\beta} + \boldsymbol{\varepsilon}$. Consider a two-factor study with each factor having two levels, i.e., $a = b = 2$, and each treatment having two observations, i.e., $n = 2$. Then, $\mathbf{Y}$, $\mathbf{X}$, $\boldsymbol{\beta}$, and $\boldsymbol{\varepsilon}$ are defined as follows:

$$\mathbf{Y} = \begin{bmatrix} Y_{111} \\ Y_{112} \\ Y_{121} \\ Y_{122} \\ Y_{211} \\ Y_{212} \\ Y_{221} \\ Y_{222} \end{bmatrix} \quad \mathbf{X} = \begin{bmatrix} 1 & 0 & 0 & 0 \\ 1 & 0 & 0 & 0 \\ 0 & 1 & 0 & 0 \\ 0 & 1 & 0 & 0 \\ 0 & 0 & 1 & 0 \\ 0 & 0 & 1 & 0 \\ 0 & 0 & 0 & 1 \\ 0 & 0 & 0 & 1 \end{bmatrix} \quad \boldsymbol{\beta} = \begin{bmatrix} \mu_{11} \\ \mu_{12} \\ \mu_{21} \\ \mu_{22} \end{bmatrix} \quad \boldsymbol{\varepsilon} = \begin{bmatrix} \varepsilon_{111} \\ \varepsilon_{112} \\ \varepsilon_{121} \\ \varepsilon_{122} \\ \varepsilon_{211} \\ \varepsilon_{212} \\ \varepsilon_{221} \\ \varepsilon_{222} \end{bmatrix} \tag{20.23}$$

Recall that the $\mathbf{E(Y)}$ vector, which consists of the elements $E(Y_{ijk})$, equals $\mathbf{X\beta}$ according to (7.18a). Hence, this vector here is:

(20.24)
$$\mathbf{E(Y)} = \mathbf{X\beta} = \begin{bmatrix} 1 & 0 & 0 & 0 \\ 1 & 0 & 0 & 0 \\ 0 & 1 & 0 & 0 \\ 0 & 1 & 0 & 0 \\ 0 & 0 & 1 & 0 \\ 0 & 0 & 1 & 0 \\ 0 & 0 & 0 & 1 \\ 0 & 0 & 0 & 1 \end{bmatrix} \begin{bmatrix} \mu_{11} \\ \mu_{12} \\ \mu_{21} \\ \mu_{22} \end{bmatrix} = \begin{bmatrix} \mu_{11} \\ \mu_{11} \\ \mu_{12} \\ \mu_{12} \\ \mu_{21} \\ \mu_{21} \\ \mu_{22} \\ \mu_{22} \end{bmatrix}$$

Thus, $E(Y_{ijk}) = \mu_{ij}$, as it must according to (20.20), and we have the proper matrix representation for the two-factor ANOVA model (20.19):

(20.25)
$$\mathbf{Y} = \begin{bmatrix} Y_{111} \\ Y_{112} \\ Y_{121} \\ Y_{122} \\ Y_{211} \\ Y_{212} \\ Y_{221} \\ Y_{222} \end{bmatrix} = \mathbf{X\beta} + \boldsymbol{\varepsilon} = \begin{bmatrix} \mu_{11} \\ \mu_{11} \\ \mu_{12} \\ \mu_{12} \\ \mu_{21} \\ \mu_{21} \\ \mu_{22} \\ \mu_{22} \end{bmatrix} + \begin{bmatrix} \varepsilon_{111} \\ \varepsilon_{112} \\ \varepsilon_{121} \\ \varepsilon_{122} \\ \varepsilon_{211} \\ \varepsilon_{212} \\ \varepsilon_{221} \\ \varepsilon_{222} \end{bmatrix}$$

5. Model (20.19) is similar to the single-factor model (16.2), except for the two subscripts now needed to identify the treatment. Normality, independent error terms, and constant variances for the error terms are properties of both the models for single-factor and two-factor studies.

Model in terms of factor effects

An equivalent version of model (20.19) can be obtained by utilizing an identical expression for the treatment means μ_{ij} based on the definition of an interaction in (20.14):

$$(\alpha\beta)_{ij} = \mu_{ij} - (\mu_{..} + \alpha_i + \beta_j)$$

Rearranging terms, we obtain the identity:

(20.26)
$$\mu_{ij} \equiv \mu_{..} + \alpha_i + \beta_j + (\alpha\beta)_{ij}$$

where:

$$\mu_{..} = \frac{\sum_i \sum_j \mu_{ij}}{ab}$$

$$\alpha_i = \mu_{i.} - \mu_{..}$$

$$\beta_j = \mu_{.j} - \mu_{..}$$

$$(\alpha\beta)_{ij} = \mu_{ij} - \mu_{i.} - \mu_{.j} + \mu_{..}$$

This formulation indicates that the mean μ_{ij} for any treatment can be viewed as the sum of four component effects. Specifically, (20.26) states that the mean response for the treatment where factor A is at the ith level and factor B is at the jth level is the sum of:

1. An overall constant $\mu_{..}$.
2. The main effect α_i for factor A at the ith level.
3. The main effect β_j for factor B at the jth level.
4. The interaction effect $(\alpha\beta)_{ij}$ when factor A is at the ith level and factor B is at the jth level.

Replacing μ_{ij} in model (20.19) by the equivalent expression in (20.26), we obtain an equivalent fixed effects analysis of variance model for two-factor studies:

(20.27) $$Y_{ijk} = \mu_{..} + \alpha_i + \beta_j + (\alpha\beta)_{ij} + \varepsilon_{ijk}$$

where:

$\mu_{..}$ is a constant
α_i are constants subject to the restriction $\Sigma\alpha_i = 0$
β_j are constants subject to the restriction $\Sigma\beta_j = 0$
$(\alpha\beta)_{ij}$ are constants subject to the restrictions

$$\sum_i (\alpha\beta)_{ij} = 0 \qquad \sum_j (\alpha\beta)_{ij} = 0$$

ε_{ijk} are independent $N(0, \sigma^2)$
$i = 1, \ldots, a; j = 1, \ldots, b; k = 1, \ldots, n$

Important features of model. 1. Model (20.27) corresponds to the single-factor ANOVA model (16.61) except that the single-factor treatment effect is here replaced by the sum of a factor A effect, a factor B effect, and an interaction effect.

2. The properties of the observations Y_{ijk} for model (20.27) are the same as those for the equivalent model (20.19). Since $E(\varepsilon_{ijk}) = 0$, we have:

(20.28) $$E(Y_{ijk}) = \mu_{..} + \alpha_i + \beta_j + (\alpha\beta)_{ij} = \mu_{ij}$$

The second equality follows from identity (20.26). Further, we have:

(20.29) $$\sigma^2(Y_{ijk}) = \sigma^2$$

because the error term ε_{ijk} is the only random term on the right-hand side and $\sigma^2(\varepsilon_{ijk}) = \sigma^2$. Finally, the Y's are independent normal random variables because the error terms are independent normal random variables. Hence, we can also state model (20.27) as follows:

(20.30) $$Y_{ijk} \text{ are independent } N[\mu_{..} + \alpha_i + \beta_j + (\alpha\beta)_{ij}, \sigma^2]$$

3. Model (20.27) is a linear model because it can be stated in the form $\mathbf{Y} = \mathbf{X}\boldsymbol{\beta} + \boldsymbol{\varepsilon}$. We shall show this explicitly in Section 20.8.

20.4 ANALYSIS OF VARIANCE

Illustration

Table 20.7 contains an illustration that we shall employ both in this chapter and the next. The Castle Bakery Company supplies sliced wrapped Italian bread to a large number of supermarkets in a metropolitan area. An experimental study was made of the effects of height of the shelf display (bottom, middle, top) and the width of the shelf display (regular, wide) on sales of this bakery's bread (measured in cases) during the experimental period. Twelve supermarkets, similar in terms of sales volumes and clientele, were utilized in the study. Two stores were assigned at random to each of the six treatments, and the display of the bread in each store followed the treatment specifications for that store. Sales of the bread were recorded, and these results are presented in Table 20.7.

TABLE 20.7 Sample data and notation for two-factor study—Castle Bakery example (sales in cases)

Factor A (display height) i	*Factor B (display width)* j B_1 *(regular)*	B_2 *(wide)*	*Row Total*	*Display Height Average*
A_1 (Bottom)	47 (Y_{111}) 43 (Y_{112})	46 (Y_{121}) 40 (Y_{122})		
Total	90 ($Y_{11.}$)	86 ($Y_{12.}$)	176 ($Y_{1..}$)	
Average	45 ($\bar{Y}_{11.}$)	43 ($\bar{Y}_{12.}$)		44 ($\bar{Y}_{1..}$)
A_2 (Middle)	62 (Y_{211}) 68 (Y_{212})	67 (Y_{221}) 71 (Y_{222})		
Total	130 ($Y_{21.}$)	138 ($Y_{22.}$)	268 ($Y_{2..}$)	
Average	65 ($\bar{Y}_{21.}$)	69 ($\bar{Y}_{22.}$)		67 ($\bar{Y}_{2..}$)
A_3 (Top)	41 (Y_{311}) 39 (Y_{312})	42 (Y_{321}) 46 (Y_{322})		
Total	80 ($Y_{31.}$)	88 ($Y_{32.}$)	168 ($Y_{3..}$)	
Average	40 ($\bar{Y}_{31.}$)	44 ($\bar{Y}_{32.}$)		42 ($\bar{Y}_{3..}$)
Column total	300 ($Y_{.1.}$)	312 ($Y_{.2.}$)	612 ($Y_{...}$)	
Display width average	50 ($\bar{Y}_{.1.}$)	52 ($\bar{Y}_{.2.}$)		51 ($\bar{Y}_{...}$)

Notation

Table 20.7 illustrates the notation we shall use for two-factor studies. It is a straightforward extension of the notation for single-factor studies. An observation is denoted by Y_{ijk}. The subscripts i and j specify the levels of factors A and B, respectively, and the subscript k refers to the given observation from a particular treatment (i.e., factor level combination).

A dot in the subscript indicates aggregation or averaging over the variable represented by the index. For instance, the sum of the observations for the treatment corresponding to the ith level of factor A and the jth level of factor B is:

(20.31a)
$$Y_{ij.} = \sum_{k=1}^{n} Y_{ijk}$$

The corresponding mean is:

(20.31b)
$$\bar{Y}_{ij.} = \frac{Y_{ij.}}{n}$$

The total of all observations for the ith factor level of A is:

(20.31c)
$$Y_{i..} = \sum_{j}^{b} \sum_{k}^{n} Y_{ijk}$$

and the corresponding mean is:

(20.31d)
$$\bar{Y}_{i..} = \frac{Y_{i..}}{bn}$$

Similarly, for the jth factor level of B the sum of all observations and their mean are denoted by:

(20.31e)
$$Y_{.j.} = \sum_{i}^{a} \sum_{k}^{n} Y_{ijk}$$

(20.31f)
$$\bar{Y}_{.j.} = \frac{Y_{.j.}}{an}$$

Finally, the sum of all observations in the study is:

(20.31g)
$$Y_{...} = \sum_{i}^{a} \sum_{j}^{b} \sum_{k}^{n} Y_{ijk}$$

and the overall mean is:

(20.31h)
$$\bar{Y}_{...} = \frac{Y_{...}}{nab}$$

Fitting of ANOVA model

Model (20.19). We fit the two-factor ANOVA model (20.19) to the sample data by the method of least squares. The least squares criterion here is:

(20.32) $$Q = \sum_i \sum_j \sum_k (Y_{ijk} - \mu_{ij})^2$$

When we perform the minimization of Q, we obtain the least squares estimators:

(20.33) $$\hat{\mu}_{ij} = \bar{Y}_{ij.}$$

Thus, the *fitted values* are the treatment sample means:

(20.34) $$\hat{Y}_{ijk} = \hat{\mu}_{ij} = \bar{Y}_{ij.}$$

The *residuals*, as usual, are defined as the difference between the observed and fitted values:

(20.35) $$e_{ijk} = Y_{ijk} - \hat{Y}_{ijk} = Y_{ijk} - \bar{Y}_{ij.}$$

Residuals are highly useful for assessing the aptness of the two-factor model (20.19), as they also are for the models considered earlier.

Model (20.27). For the equivalent two-factor ANOVA model (20.27), we minimize the least squares criterion:

(20.36) $$Q = \sum_i \sum_j \sum_k [Y_{ijk} - \mu_{..} - \alpha_i - \beta_j - (\alpha\beta)_{ij}]^2$$

subject to the restrictions:

$$\sum_i \alpha_i = 0 \qquad \sum_j \beta_j = 0 \qquad \sum_i (\alpha\beta)_{ij} = 0 \qquad \sum_j (\alpha\beta)_{ij} = 0$$

When we perform this minimization, we obtain the following least squares estimators of the parameters:

	Parameter	*Estimator*
(20.37a)	$\mu_{..}$	$\hat{\mu}_{..} = \bar{Y}_{...}$
(20.37b)	$\alpha_i = \mu_{i.} - \mu_{..}$	$\hat{\alpha}_i = \bar{Y}_{i..} - \bar{Y}_{...}$
(20.37c)	$\beta_j = \mu_{.j} - \mu_{..}$	$\hat{\beta}_j = \bar{Y}_{.j.} - \bar{Y}_{...}$
(20.37d)	$(\alpha\beta)_{ij} = \mu_{ij} - \mu_{i.} - \mu_{.j} + \mu_{..}$	$\widehat{(\alpha\beta)}_{ij} = \bar{Y}_{ij.} - \bar{Y}_{i..} - \bar{Y}_{.j.} + \bar{Y}_{...}$

The correspondence of the least squares estimators to the definitions of the parameters is readily apparent.

The fitted values and residuals for the alternative model (20.27) are exactly the same as those for model (20.19). Specifically, the fitted values for model (20.27) are:

(20.38) $$\begin{aligned}\hat{Y}_{ijk} &= \hat{\mu}_{..} + \hat{\alpha}_i + \hat{\beta}_j + \widehat{(\alpha\beta)}_{ij} \\ &= \bar{Y}_{...} + (\bar{Y}_{i..} - \bar{Y}_{...}) + (\bar{Y}_{.j.} - \bar{Y}_{...}) + (\bar{Y}_{ij.} - \bar{Y}_{i..} - \bar{Y}_{.j.} + \bar{Y}_{...}) \\ &= \bar{Y}_{ij.}\end{aligned}$$

so that the residuals are again:

(20.39)
$$e_{ijk} = Y_{ijk} - \bar{Y}_{ij.}$$

Note

The least squares estimators in (20.33) and (20.37) are the same as those obtained by the method of maximum likelihood.

Example. For our Castle Bakery example, the fitted values are the treatment sample means $\bar{Y}_{ij.}$ shown in Table 20.7. These treatment sample means are presented graphically in Figure 20.5. We see from this figure that for both regular and wide display widths, mean sales for the middle display height are substantially larger than those for the other two display heights. The effect of display width does not appear to be large at all. Indeed, there may be no effect of display width, with the variations between the treatment means for any given display height being solely of a random nature. In that event, there would be no interactions between display height and display width.

FIGURE 20.5 Plot of treatment sample means—Castle Bakery example

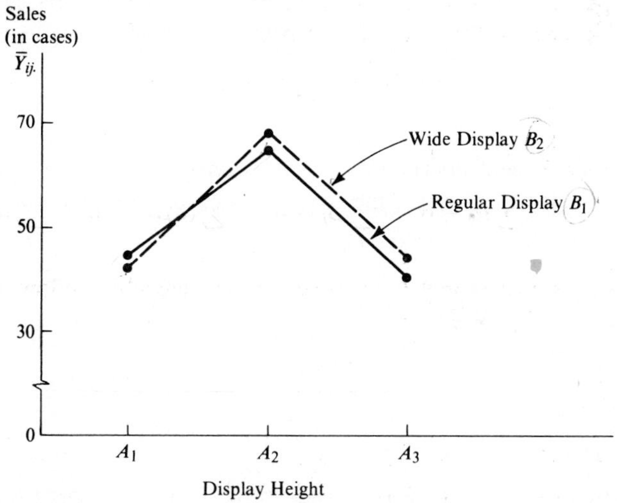

Figure 20.5 differs from the earlier figures portraying factor effects because the earlier figures presented the true treatment means μ_{ij} while Figure 20.5 presents sample estimates. We therefore need to test whether or not the effects shown in Figure 20.5 are real effects or represent only random variations. To conduct such tests, we require a breakdown of the total sum of squares, to be discussed next.

Breakdown of total sum of squares

Breakdown of total deviation. We shall break down the deviation of an observation Y_{ijk} from the overall mean $\bar{Y}_{...}$ in two stages. First, we shall obtain a decomposition of the total deviation $Y_{ijk} - \bar{Y}_{...}$ by viewing the study as consisting of ab treatments:

$$(20.40) \qquad \underbrace{Y_{ijk} - \bar{Y}_{...}}_{\substack{\text{Total}\\\text{deviation}}} = \underbrace{\bar{Y}_{ij.} - \bar{Y}_{...}}_{\substack{\text{Deviation of}\\\text{treatment mean}\\\text{around overall}\\\text{mean}}} + \underbrace{Y_{ijk} - \bar{Y}_{ij.}}_{\substack{\text{Deviation}\\\text{around}\\\text{treatment}\\\text{mean}}}$$

Note that the deviation around the treatment mean is simply the residual e_{ijk}:

$$e_{ijk} = Y_{ijk} - \hat{Y}_{ijk} = Y_{ijk} - \bar{Y}_{ij.}$$

Next, we shall decompose the treatment mean deviation $\bar{Y}_{ij.} - \bar{Y}_{...}$ in terms of components reflecting the factor A main effect, the factor B main effect, and the AB interaction:

$$(20.41) \qquad \underbrace{\bar{Y}_{ij.} - \bar{Y}_{...}}_{\substack{\text{Deviation of}\\\text{treatment mean}\\\text{around overall}\\\text{mean}}} = \underbrace{\bar{Y}_{i..} - \bar{Y}_{...}}_{\substack{A\text{ main}\\\text{effect}}} + \underbrace{\bar{Y}_{.j.} - \bar{Y}_{...}}_{\substack{B\text{ main}\\\text{effect}}} + \underbrace{\bar{Y}_{ij.} - \bar{Y}_{i..} - \bar{Y}_{.j.} + \bar{Y}_{...}}_{AB\text{ interaction effect}}$$

Treatment and error sums of squares. When we square (20.40) and sum over all observations, the cross-product term drops out and we obtain:

$$(20.42) \qquad SSTO = SSTR + SSE$$

where:

$$(20.42a) \qquad SSTO = \sum_i \sum_j \sum_k (Y_{ijk} - \bar{Y}_{...})^2$$

$$(20.42b) \qquad SSTR = n \sum_i \sum_j (\bar{Y}_{ij.} - \bar{Y}_{...})^2$$

$$(20.42c) \qquad SSE = \sum_i \sum_j \sum_k (Y_{ijk} - \bar{Y}_{ij.})^2 = \sum_i \sum_j \sum_k e_{ijk}^2$$

$SSTR$ reflects the variability between the ab treatment means and is the ordinary *treatment sum of squares*, and SSE reflects the variability within treatments and is the usual *error sum of squares*. The only difference between these formulas and those for the single-factor case is the use of the two subscripts i and j to designate a treatment.

Example. For our Castle Bakery example, Table 20.8 contains the decomposition of the total sum of squares in (20.42). This is the ordinary ANOVA table treating the study as a single-factor one with $ab = r = 6$ treatments. The sums of squares are obtained as follows:

TABLE 20.8 ANOVA table neglecting factorial structure—Castle Bakery example

Source of Variation	*SS*	*df*	*MS*
Between treatments	1,580	5	316
Error	62	6	10.3
Total	1,642	11	

$$SSTO = (47 - 51)^2 + (43 - 51)^2 + (46 - 51)^2 + \cdots + (46 - 51)^2 = 1{,}642$$
$$SSTR = 2[(45 - 51)^2 + (43 - 51)^2 + (65 - 51)^2 + \cdots + (44 - 51)^2] = 1{,}580$$
$$SSE = (47 - 45)^2 + (43 - 45)^2 + (46 - 43)^2 + \cdots + (46 - 44)^2 = 62$$

One could test at this point, by means of test statistic (16.51), whether or not the six treatment means are equal. If they are, neither of the two factors has any effect. Ordinarily, however, testing of factor effects is not made until a further decomposition of the treatment sum of squares has been carried out to reflect the factorial nature of the study.

Breakdown of treatment sum of squares. When we square (20.41) and sum over all treatments and over the n observations associated with each treatment mean $\bar{Y}_{ij.}$, all cross-product terms drop out and we obtain:

(20.43) $$SSTR = SSA + SSB + SSAB$$

where:

(20.43a) $$SSA = nb\sum_i (\bar{Y}_{i..} - \bar{Y}_{...})^2$$

(20.43b) $$SSB = na\sum_j (\bar{Y}_{.j.} - \bar{Y}_{...})^2$$

(20.43c) $$SSAB = n\sum_i \sum_j (\bar{Y}_{ij.} - \bar{Y}_{i..} - \bar{Y}_{.j.} + \bar{Y}_{...})^2$$

SSA, called the *factor A sum of squares*, measures the variability of the estimated A factor level means $\bar{Y}_{i..}$. The more variable they are, the bigger will be SSA. Similarly SSB, called the *factor B sum of squares*, measures the variability of the estimated B factor level means $\bar{Y}_{.j.}$. Finally, $SSAB$, called the *AB interaction sum of squares*, measures the variability of the estimated interactions $\bar{Y}_{ij.} - \bar{Y}_{i..} - \bar{Y}_{.j.} + \bar{Y}_{...}$ for the ab treatments. Remember that the mean of all interactions is zero, so that the deviations of the estimated interactions around their mean is not explicitly shown as it was in SSA and SSB. The larger are the estimated interactions (sign disregarded), the larger will be $SSAB$.

The breakdown of $SSTR$ into the components SSA, SSB, and $SSAB$ is called an *orthogonal decomposition*. An orthogonal decomposition is one where the component sums of squares add to the total sum of squares ($SSTR$ here), and likewise for the degrees of freedom. Thus, the decompositions of *SSTO* into *SSTR* and *SSE* for single-factor and two-factor studies are also orthogonal decompositions. While many different orthogonal decompositions of *SSTR* are possible here, the one into *SSA*, *SSB*, and *SSAB* components is of interest because these three components provide information about the factor *A* main effects, the factor *B* main effects, and the *AB* interactions, respectively, as will be seen shortly.

Computational forms. For hand computing purposes, we ordinarily use the following formulas that are algebraically identical to the definitional formulas previously given:

(20.44a)
$$SSTO = \sum_i \sum_j \sum_k Y_{ijk}^2 - \frac{Y_{...}^2}{nab}$$

278575 - 267759,38

USE TO COMPUTE

(20.44b)
$$SSE = \sum_i \sum_j \sum_k Y_{ijk}^2 - \frac{\sum_i \sum_j Y_{ij.}^2}{n}$$

276262,5

(20.44c)
$$SSA = \frac{\sum_i Y_{i..}^2}{nb} - \frac{Y_{...}^2}{nab}$$

271028,13

(20.44d)
$$SSB = \frac{\sum_j Y_{.j.}^2}{na} - \frac{Y_{...}^2}{nab}$$

271687,5

Ordinarily, the interaction sum of squares is obtained as a remainder:

(20.44e)
$$SSAB = SSTO - SSE - SSA - SSB$$

or from:

(20.44f)
$$SSAB = SSTR - SSA - SSB$$

where:

(20.44g)
$$SSTR = \frac{\sum_i \sum_j Y_{ij.}^2}{n} - \frac{Y_{...}^2}{nab}$$

Example. For our Castle Bakery example, we obtain the following decomposition of *SSTR*, using the data in Table 20.7 and the computational formulas in (20.44):

$$SSA = \frac{(176)^2 + (268)^2 + (168)^2}{2(2)} - \frac{(612)^2}{2(3)(2)} = 1{,}544$$

$$SSB = \frac{(300)^2 + (312)^2}{2(3)} - \frac{(612)^2}{2(3)(2)} = 12$$

$$SSAB = 1{,}580 - 1{,}544 - 12 = 24$$

Hence, we have:

$$1{,}580 = 1{,}544 + 12 + 24$$
$$SSTR = SSA + SSB + SSAB$$

Recapitulation. Combining the decompositions in (20.42) and (20.43), we have established that:

$$SSTO = SSA + SSB + SSAB + SSE \tag{20.45}$$

where the component sums of squares are defined in (20.44).

For our Castle Bakery example, we have found:

$$1{,}642 = 1{,}544 + 12 + 24 + 62$$
$$SSTO = SSA + SSB + SSAB + SSE$$

Thus, much of the total variability in this instance is associated with the factor A (display height) effects.

Breakdown of degrees of freedom

We are familiar from single-factor analysis of variance as to how the degrees of freedom are divided between the treatment and error components. For two-factor studies with n observations for each treatment, there are a total of $n_T = nab$ observations and $r = ab$ treatments; hence, the degrees of freedom associated with $SSTO$, $SSTR$, and SSE are $nab - 1$, $ab - 1$, and $nab - ab = (n - 1)ab$, respectively. These degrees of freedom for the Castle Bakery example are $2(3)(2) - 1 = 11$, $3(2) - 1 = 5$, and $(2 - 1)(3)(2) = 6$, respectively, as shown in Table 20.8.

Corresponding to the further breakdown of the treatment sum of squares, we can also obtain a breakdown of the associated $ab - 1$ degrees of freedom. SSA has $a - 1$ degrees of freedom associated with it. There are a factor level deviations $\bar{Y}_{i..} - \bar{Y}_{...}$, but one degree of freedom is lost because the deviations are subject to one restriction, i.e., $\Sigma(\bar{Y}_{i..} - \bar{Y}_{...}) = 0$. Similarly, SSB has $b - 1$ degrees of freedom associated with it. The degrees of freedom associated with $SSAB$, the interaction sum of squares, is the remainder:

$$(ab - 1) - (a - 1) - (b - 1) = (a - 1)(b - 1)$$

The degrees of freedom associated with $SSAB$ may be understood as follows: There are ab interaction terms. These are subject to b restrictions since

$$\sum_i (\bar{Y}_{ij.} - \bar{Y}_{i..} - \bar{Y}_{.j.} + \bar{Y}_{...}) = 0$$

for all j. There are a additional restrictions since

$$\sum_j (\bar{Y}_{ij.} - \bar{Y}_{i..} - \bar{Y}_{.j.} + \bar{Y}_{...}) = 0$$

for all i, but only $a - 1$ of these are independent since the last one is implied by the previous b restrictions. Altogether, therefore, there are $b + (a - 1)$ independent restrictions. Hence, the degrees of freedom are:

$$ab - (b + a - 1) = (a - 1)(b - 1)$$

Example. For our Castle Bakery example, SSA has $3 - 1 = 2$ degrees of freedom associated with it, SSB has $2 - 1 = 1$ degree of freedom, and $SSAB$ has $(3 - 1)(2 - 1) = 2$ degrees of freedom.

Mean squares

Mean squares are obtained in the usual way by dividing the sums of squares by their associated degrees of freedom. We thus obtain:

(20.46a) $$MSA = \frac{SSA}{a - 1}$$

(20.46b) $$MSB = \frac{SSB}{b - 1}$$

(20.46c) $$MSAB = \frac{SSAB}{(a - 1)(b - 1)}$$

Example. For our Castle Bakery example, these mean squares are:

$$MSA = \frac{1{,}544}{2} = 772$$

$$MSB = \frac{12}{1} = 12$$

$$MSAB = \frac{24}{2} = 12$$

Note that these mean squares do not add to the treatment mean square, $MSTR = 1{,}580/5 = 316$. Here again, we see that mean squares are not additive.

Expected mean squares

It can be shown, along the same lines used for the single-factor case, that the mean squares for the two-factor model (20.27) have the following expectations:

(20.47a) $$E(MSE) = \sigma^2$$

(20.47b) $$E(MSA) = \sigma^2 + nb\frac{\Sigma\alpha_i^2}{a - 1} = \sigma^2 + nb\frac{\Sigma(\mu_{i.} - \mu_{..})^2}{a - 1}$$

$$\text{(20.47c)} \qquad E(MSB) = \sigma^2 + na\frac{\Sigma\beta_j^2}{b-1} = \sigma^2 + na\frac{\Sigma(\mu_{.j} - \mu_{..})^2}{b-1}$$

$$\text{(20.47d)} \qquad E(MSAB) = \sigma^2 + n\frac{\Sigma\Sigma(\alpha\beta)_{ij}^2}{(a-1)(b-1)}$$

$$= \sigma^2 + n\frac{\Sigma\Sigma(\mu_{ij} - \mu_{i.} - \mu_{.j} + \mu_{..})^2}{(a-1)(b-1)}$$

These expectations show that if there are no factor A main effects (i.e., if all $\mu_{i.}$ are equal, or all $\alpha_i = 0$), *MSA* and *MSE* have the same expectation; otherwise *MSA* tends to be larger than *MSE*. Similarly, if there are no factor B main effects, *MSB* and *MSE* have the same expectation; otherwise *MSB* tends to be larger than *MSE*. Finally, if there are no interactions [i.e., if all $(\alpha\beta)_{ij} = 0$] so that the factor effects are additive, *MSAB* has the same expectation as *MSE*; otherwise, *MSAB* tends to be larger than *MSE*. This suggests that F^* test statistics based on the ratios *MSA/MSE*, *MSB/MSE*, and *MSAB/MSE* will provide information about the main effects and interactions of the two factors, respectively, with large values of the test statistics indicating the presence of factor effects. We shall see shortly that tests based on these statistics are regular F tests.

Analysis of variance table

The breakdown of the total sum of squares into the treatment and error components is shown in Table 20.9, as well as the further breakdown of the treatment sum of squares into the several factor components. Also shown there are the associated degrees of freedom, the mean squares, and the expected mean squares. Table 20.10 contains the two-factor analysis of variance for our Castle Bakery example.

20.5 *F* TESTS

In view of the additivity of sums of squares and degrees of freedom, Cochran's theorem (3.60) applies when no factor effects are present. Hence, the test statistics F^* based on the appropriate mean squares then follow the F distribution, leading to the usual type of F test for factor effects.

Test for interactions

Ordinarily, the analysis of a two-factor study begins with a test to determine whether or not the two factors interact:

$$\text{(20.48)} \qquad \begin{array}{ll} H_0\colon \mu_{ij} - \mu_{i.} - \mu_{.j} + \mu_{..} = 0 & \text{for all } i, j \\ H_a\colon \mu_{ij} - \mu_{i.} - \mu_{.j} + \mu_{..} \neq 0 & \text{for some } i, j \end{array}$$

or equivalently:

TABLE 20.9 ANOVA table for two-factor fixed effects study

Source of Variation	*SS*	*df*	*MS*	*E(MS)*
Between treatments	$SSTR = n\Sigma\Sigma(\bar{Y}_{ij.} - \bar{Y}_{...})^2$	$ab - 1$	$MSTR = \frac{SSTR}{ab - 1}$	$\sigma^2 + \frac{n}{ab - 1}\Sigma\Sigma(\mu_{ij} - \mu_{..})^2$
Factor A	$SSA = nb\Sigma(\bar{Y}_{i..} - \bar{Y}_{...})^2$	$a - 1$	$MSA = \frac{SSA}{a - 1}$	$\sigma^2 + \frac{bn}{a - 1}\Sigma(\mu_{i.} - \mu_{..})^2$
Factor B	$SSB = na\Sigma(\bar{Y}_{.j.} - \bar{Y}_{...})^2$	$b - 1$	$MSB = \frac{SSB}{b - 1}$	$\sigma^2 + \frac{an}{b - 1}\Sigma(\mu_{.j} - \mu_{..})^2$
Interactions	$SSAB = n\Sigma\Sigma(\bar{Y}_{ij.} - \bar{Y}_{i..} - \bar{Y}_{.j.} + \bar{Y}_{...})^2$	$(a - 1)(b - 1)$	$MSAB = \frac{SSAB}{(a - 1)(b - 1)}$	$\sigma^2 + \frac{n}{(a - 1)(b - 1)}\Sigma\Sigma(\mu_{ij} - \mu_{i.} - \mu_{.j} + \mu_{..})^2$
Error	$SSE = \Sigma\Sigma\Sigma(Y_{ijk} - \bar{Y}_{ij.})^2$	$ab(n - 1)$	$MSE = \frac{SSE}{ab(n - 1)}$	σ^2
Total	$SSTO = \Sigma\Sigma\Sigma(Y_{ijk} - \bar{Y}_{...})^2$	$nab - 1$		

TABLE 20.10 ANOVA table for two-factor study—Castle Bakery example

Source of Variation	*SS*	*df*	*MS*
Between treatments	1,580	5	316
Factor *A* (display height)	1,544	2	772
Factor *B* (display width)	12	1	12
Interactions	24	2	12
Error	62	6	10.3
Total	1,642	11	

(20.48a)
$$H_0\text{: all } (\alpha\beta)_{ij} = 0$$
$$H_a\text{: not all } (\alpha\beta)_{ij} \text{ equal zero}$$

As we noted from an examination of the expected mean squares in Table 20.9, the appropriate test statistic is:

(20.49)
$$F^* = \frac{MSAB}{MSE}$$

Large values of F^* indicate the existence of interactions. When H_0 holds, F^* is distributed as $F[(a-1)(b-1), (n-1)ab]$. Hence, the appropriate decision rule to control the Type I error at α is:

(20.50)
$$\text{If } F^* \leq F[1-\alpha; (a-1)(b-1), (n-1)ab], \text{ conclude } H_0$$
$$\text{If } F^* > F[1-\alpha; (a-1)(b-1), (n-1)ab], \text{ conclude } H_a$$

where $F[1-\alpha; (a-1)(b-1), (n-1)ab]$ is the $(1-\alpha)100$ percentile of the appropriate F distribution.

Test for factor *A* main effects

Tests for factor *A* main effects and for factor *B* main effects ordinarily follow the test for interactions, though we noted earlier that the *A* and *B* main effects often do not have useful meanings if strong interactions exist. To test whether or not *A* main effects are present:

(20.51)
$$H_0\text{: } \mu_{1.} = \mu_{2.} = \cdots = \mu_{a.}$$
$$H_a\text{: not all } \mu_{i.} \text{ are equal}$$

or equivalently:

(20.51a)
$$H_0\text{: } \alpha_1 = \alpha_2 = \cdots = \alpha_a = 0$$
$$H_a\text{: not all } \alpha_i \text{ equal zero}$$

we use the test statistic:

(20.52) $$F^* = \frac{MSA}{MSE}$$

Again, large values of F^* indicate the existence of factor A main effects. Since F^* is distributed as $F[a-1, (n-1)ab]$ when H_0 holds, the appropriate decision rule for controlling the risk of making a Type I error at α is:

(20.53) $$\begin{aligned} &\text{If } F^* \le F[1-\alpha; a-1, (n-1)ab], \text{ conclude } H_0 \\ &\text{If } F^* > F[1-\alpha; a-1, (n-1)ab], \text{ conclude } H_a \end{aligned}$$

Test for factor B main effects

This test is similar to the one for factor A main effects. The alternatives are:

(20.54) $$\begin{aligned} &H_0: \mu_{.1} = \mu_{.2} = \cdots = \mu_{.b} \\ &H_a: \text{not all } \mu_{.j} \text{ are equal} \end{aligned}$$

or equivalently:

(20.54a) $$\begin{aligned} &H_0: \beta_1 = \beta_2 = \cdots = \beta_b = 0 \\ &H_a: \text{not all } \beta_j \text{ equal zero} \end{aligned}$$

The test statistic is:

(20.55) $$F^* = \frac{MSB}{MSE}$$

and the appropriate decision rule for controlling the risk of a Type I error at α is:

(20.56) $$\begin{aligned} &\text{If } F^* \le F[1-\alpha; b-1, (n-1)ab], \text{ conclude } H_0 \\ &\text{If } F^* > F[1-\alpha; b-1, (n-1)ab], \text{ conclude } H_a \end{aligned}$$

Example

We shall investigate in our Castle Bakery example the presence of display height and display width effects, using a level of significance of $\alpha = .05$ for each test. First, we begin by testing whether or not interaction effects are present:

$$\begin{aligned} &H_0: \text{all } (\alpha\beta)_{ij} = 0 \\ &H_a: \text{not all } (\alpha\beta)_{ij} \text{ equal zero} \end{aligned}$$

Using the data from Table 20.10 in test statistic (20.49), we obtain:

$$F^* = \frac{12}{10.3} = 1.17$$

Since we are to control the risk of making a Type I error at $\alpha = .05$, we require $F(.95; 2, 6) = 5.14$, so that the decision rule is:

$$\begin{aligned} &\text{If } F^* \le 5.14, \text{ conclude } H_0 \\ &\text{If } F^* > 5.14, \text{ conclude } H_a \end{aligned}$$

Since $F^* = 1.17 \leq 5.14$, we conclude H_0, that display height and display width do not interact in their effects on sales.

Next, we turn to testing for display height (factor A) main effects; the alternative conclusions are given in (20.51). The test statistic (20.52) for our example becomes:

$$F^* = \frac{772}{10.3} = 75.0$$

For $\alpha = .05$, we require $F(.95; 2, 6) = 5.14$. Since $F^* = 75.0 > 5.14$, we conclude H_a, that not all factor A level means $\mu_{i.}$ are equal, or that some definite effects associated with height of display level exist.

Finally, we test for display width (factor B) main effects; the alternative conclusions are given in (20.54). The test statistic (20.55) becomes for our example:

$$F^* = \frac{12}{10.3} = 1.17$$

For $\alpha = .05$, we require $F(.95; 1, 6) = 5.99$. Since $F^* = 1.17 \leq 5.99$, we conclude H_0, that all $\mu_{.j}$ are equal, or that display width has no effect on sales.

Thus, the analysis of variance tests confirm the impressions from the plot of the treatment sample means $\bar{Y}_{ij.}$ in Figure 20.5—that only display height has an effect on sales for the treatments studied. At this point, it would clearly be desirable to conduct further analyses of the nature of the display height effects. We shall discuss such analyses of the factor effects in Chapter 21.

Comments

1. If the test for interactions is conducted with a level of significance of α_1, that for factor A main effects with a level of significance of α_2, and that for factor B main effects with a level of significance of α_3, the level of significance α for the *family* of three tests is greater than the individual levels of significance. From the Bonferroni inequality in (5.8), we can derive the inequality:

(20.57)
$$\alpha \leq \alpha_1 + \alpha_2 + \alpha_3$$

For the case considered here, a somewhat tighter inequality can be used, the *Kimball inequality*, which utilizes the fact that the numerators of the three test statistics are independent and the denominator is the same in each case. This inequality states:

(20.58)
$$\alpha \leq 1 - (1 - \alpha_1)(1 - \alpha_2)(1 - \alpha_3)$$

For our Castle Bakery example, where $\alpha_1 = \alpha_2 = \alpha_3 = .05$, the Bonferroni inequality yields as the bound for the family level of significance:

$$\alpha \leq .05 + .05 + .05 = .15$$

while the Kimball inequality yields the bound:

$$\alpha \leq 1 - (.95)(.95)(.95) = .143$$

This illustration makes it clear that the level of significance for the family of three tests may be substantially higher than the levels of significance for the individual tests.

2. When the test for interactions leads to the conclusion that the two factors do not interact, it is sometimes suggested that the interaction sum of squares and degrees of freedom be pooled with the error sum of squares and degrees of freedom for purposes of testing factor A and factor B effects. The reason advanced is that when no interactions exist, $E(MSAB) = \sigma^2$, the same expectation as for MSE, so that the new estimator of σ^2 would have a larger number of degrees of freedom associated with it. The pooled mean square would have $SSAB + SSE$ in the numerator (for our example in Table 20.10, $24 + 62 = 86$) and $(a-1)(b-1)+(n-1)ab$ degrees of freedom in the denominator (for our example, $2 + 6 = 8$).

This pooling procedure affects both the level of significance and the power of the tests for factor A and factor B main effects, in ways not yet fully understood. It has been suggested therefore by some statisticians that pooling should not be considered unless: (1) the degrees of freedom associated with MSE are small, perhaps 5 or less, and (2) the test statistic $MSAB/MSE$ falls substantially below the action limit of the decision rule, perhaps when $MSAB/MSE < 2$ for $\alpha = .05$. Part (1) of this rule is designed to limit pooling to cases where the gains may be substantial, while part (2) is designed to give reasonable assurance that there are indeed no interactions.

20.6 COMPUTER INPUT AND OUTPUT

Computer input and output formats for two-factor ANOVA vary considerably from one program to another. Figure 20.6 shows an example of computer printout for the Castle Bakery case, produced by the BMDP computer program (Ref. 20.1).

The first output block shows ANOVA results similar to those presented in Table 20.10. The treatment sum of squares $SSTR$ is not shown separately; it can be obtained as the sum of SSA, SSB, and $SSAB$. Instead, a source of variation attributable to the mean is given. This line in the ANOVA table can be used to test whether $\mu_{..} = 0$ and is usually not of interest (see Table 3.4 for an analogous ANOVA table in regression). The correction for mean sum of squares here is $n_T\bar{Y}^2_{...}$, which has one degree of freedom associated with it. For the Castle Bakery case we have $n_T\bar{Y}^2_{...} = 12(51)^2 = 31{,}212$.

The second block presents various sample means while the final block shows the point estimates of the effects (labeled cell deviations) $\mu_{i.} - \mu_{..}$, $\mu_{.j} - \mu_{..}$, and $\mu_{ij} - \mu_{i.} - \mu_{.j} + \mu_{..}$, for $i = 1, 2, 3$ and $j = 1, 2$.

20.7 POWER OF F TESTS

The power of the F tests for interactions, factor A main effects, and factor B main effects can be ascertained in similar fashion to the single-factor case by using the Pearson-Hartley charts in Table A–8. The noncentrality parameter ϕ and the appropriate degrees of freedom for each of these cases are as follows:

(20.59a) Test for interactions:

$$\phi = \frac{1}{\sigma}\sqrt{\frac{n\Sigma\Sigma(\alpha\beta)^2_{ij}}{(a-1)(b-1)+1}} = \frac{1}{\sigma}\sqrt{\frac{n\Sigma\Sigma(\mu_{ij} - \mu_{i.} - \mu_{.j} + \mu_{..})^2}{(a-1)(b-1)+1}}$$

$$\nu_1 = (a-1)(b-1) \qquad \nu_2 = ab(n-1)$$

(20.59b) Test for *A* main effects:

$$\phi = \frac{1}{\sigma}\sqrt{\frac{nb\Sigma\alpha_i^2}{a}} = \frac{1}{\sigma}\sqrt{\frac{nb\Sigma(\mu_{i.} - \mu_{..})^2}{a}}$$

$$\nu_1 = a - 1 \qquad \nu_2 = ab(n - 1)$$

FIGURE 20.6 Segment of computer output for two-factor analysis of variance—Castle Bakery example (BMDP8V, Ref. 20.1)

```
                           n_T Ȳ...^2
     SOURCE       SUM OF      D.F.      MEAN           F        PROB.
                  SQUARES               SQUARE
1    MEAN        31212.000     1      31212.000  3020.52      0.0000
2    A            1544.000     2        772.000    74.71      0.0001
3    B              12.000     1         12.000     1.16      0.3226
4    AB             24.000     2         12.000     1.16      0.3747
5    S(AB)          62.000     6         10.333

        Error         SSE              MSE            F*    One-sided
                                                              P-value

     GRAND MEAN       51.00000 ← Ȳ...

     CELL AND MARGINAL MEANS

          A =        1             2             3
                  44.00000      67.00000      42.00000 ← Ȳi..

          B =        1             2
                  50.00000      52.00000 ← Ȳ.j.

          B =        1             2
     A =  1       45.00000      43.00000
          2       65.00000      69.00000 ← Ȳij.
          3       40.00000      44.00000


     CELL DEVIATIONS

          A =        1             2             3
                  -7.00000      16.00000      -9.00000 ← Ȳi.. − Ȳ...

          B =        1             2
                  -1.00000       1.00000 ← Ȳ.j. − Ȳ...

          B =        1             2
     A =  1        2.00000      -2.00000
          2       -1.00000       1.00000 ← Ȳij. − Ȳi.. − Ȳ.j. + Ȳ...
          3       -1.00000       1.00000
```

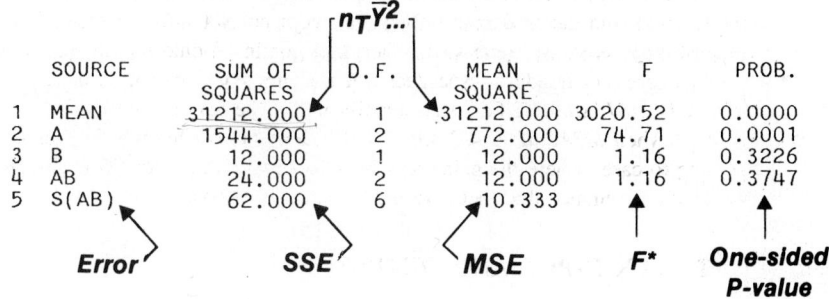

(20.59c) Test for B main effects:

$$\phi = \frac{1}{\sigma}\sqrt{\frac{na\Sigma\beta_j^2}{b}} = \frac{1}{\sigma}\sqrt{\frac{na\Sigma(\mu_{.j} - \mu_{..})^2}{b}}$$

$$\nu_1 = b - 1 \qquad \nu_2 = ab(n - 1)$$

Example

For the test of A main effects (display height) in our Castle Bakery example, we wish to find the power of the test when $\mu_{1.} = 50$, $\mu_{2.} = 55$, and $\mu_{3.} = 45$, or equivalently when $\alpha_1 = 0$, $\alpha_2 = +5$, and $\alpha_3 = -5$. Assume that we know from past experience that $\sigma = 3$ cases. We have for this test from before:

$$n = 2 \qquad a = 3 \qquad b = 2 \qquad \alpha = .05$$

so that:

$$\phi = \frac{1}{3}\sqrt{\frac{2(2)[(0)^2 + (5)^2 + (-5)^2]}{3}} = 2.7$$

For $\nu_1 = 2$, $\nu_2 = 6$, $\alpha = .05$, and $\phi = 2.7$, we find from Table A–8 that the power is about .89. Thus, if $\mu_{1.} = 50$, $\mu_{2.} = 55$, and $\mu_{3.} = 45$ (and $\sigma = 3$), the probability is about .89 that the F test will detect the differences in the display height means.

20.8 REGRESSION APPROACH TO TWO-FACTOR ANALYSIS OF VARIANCE

We shall explain the regression approach to two-factor analysis of variance in terms of model (20.27):

(20.60) $$Y_{ijk} = \mu_{..} + \alpha_i + \beta_j + (\alpha\beta)_{ij} + \varepsilon_{ijk}$$

As we know from (20.28), the mean responses for this model are given by:

(20.61) $$E(Y_{ijk}) = \mu_{..} + \alpha_i + \beta_j + (\alpha\beta)_{ij}$$

To represent this model in matrix terms, we proceed in the same fashion as in the regression approach to single-factor ANOVA. Since $\Sigma\alpha_i = 0$, we need only $a - 1$ parameters α_i in the regression model and shall represent the parameter α_a as follows:

(20.62) $$\alpha_a = -\alpha_1 - \alpha_2 - \cdots - \alpha_{a-1}$$

Hence, we shall utilize $a - 1$ indicator variables that can take on values 1, -1, or 0 for the α_i parameters, as in the single-factor ANOVA representation. Similarly, we need only $b - 1$ parameters β_j in the regression model and shall represent the parameter β_b as follows:

(20.63) $$\beta_b = -\beta_1 - \beta_2 - \cdots - \beta_{b-1}$$

Hence, we shall utilize $b - 1$ indicator variables that can take on values 1, -1, or 0 for the β_j parameters.

For the interaction parameters, we need to recognize that:

$$\sum_i (\alpha\beta)_{ij} = 0 \qquad \sum_j (\alpha\beta)_{ij} = 0 \tag{20.64}$$

Therefore, we represent the parameters $(\alpha\beta)_{ib}$ and $(\alpha\beta)_{aj}$ as follows:

$$(\alpha\beta)_{ib} = -(\alpha\beta)_{i1} - (\alpha\beta)_{i2} - \cdots - (\alpha\beta)_{i,\,b-1} \tag{20.65}$$

$$(\alpha\beta)_{aj} = -(\alpha\beta)_{1j} - (\alpha\beta)_{2j} - \cdots - (\alpha\beta)_{a-1,\,j} \tag{20.66}$$

Indeed, because of the interrelations in the constraints in (20.64), only $(a-1)(b-1)$ terms $(\alpha\beta)_{ij}$ are needed in the regression model. These are the terms associated with the cross-products between the indicator variables for the factor A and factor B main effects.

Example

To illustrate all of this, we present in Table 20.11 the $\mathbf{Y}$, $\boldsymbol{\beta}$, and $\boldsymbol{\varepsilon}$ vectors and the $\mathbf{X}$ matrix for the Castle Bakery example. The X variables are defined as follows:

$$X_1 = \begin{matrix} 1 \text{ if observation from level 1 for factor } A \\ -1 \text{ if observation from level 3 for factor } A \\ 0 \text{ otherwise} \end{matrix}$$

$$X_2 = \begin{matrix} 1 \text{ if observation from level 2 for factor } A \\ -1 \text{ if observation from level 3 for factor } A \\ 0 \text{ otherwise} \end{matrix} \tag{20.67}$$

$$X_3 = \begin{matrix} 1 \text{ if observation from level 1 for factor } B \\ -1 \text{ if observation from level 2 for factor } B \end{matrix}$$

TABLE 20.11 Regression model representation for two-factor ANOVA model—Castle Bakery example

$$\mathbf{Y} = \begin{bmatrix} Y_{111} \\ Y_{112} \\ Y_{121} \\ Y_{122} \\ Y_{211} \\ Y_{212} \\ Y_{221} \\ Y_{222} \\ Y_{311} \\ Y_{312} \\ Y_{321} \\ Y_{322} \end{bmatrix} \quad \mathbf{X} = \begin{bmatrix} & X_1 & X_2 & X_3 & X_1X_3 & X_2X_3 \\ 1 & 1 & 0 & 1 & 1 & 0 \\ 1 & 1 & 0 & 1 & 1 & 0 \\ 1 & 1 & 0 & -1 & -1 & 0 \\ 1 & 1 & 0 & -1 & -1 & 0 \\ 1 & 0 & 1 & 1 & 0 & 1 \\ 1 & 0 & 1 & 1 & 0 & 1 \\ 1 & 0 & 1 & -1 & 0 & -1 \\ 1 & 0 & 1 & -1 & 0 & -1 \\ 1 & -1 & -1 & 1 & -1 & -1 \\ 1 & -1 & -1 & 1 & -1 & -1 \\ 1 & -1 & -1 & -1 & 1 & 1 \\ 1 & -1 & -1 & -1 & 1 & 1 \end{bmatrix} \quad \boldsymbol{\beta} = \begin{bmatrix} \mu_{..} \\ \alpha_1 \\ \alpha_2 \\ \beta_1 \\ (\alpha\beta)_{11} \\ (\alpha\beta)_{21} \end{bmatrix} \quad \boldsymbol{\varepsilon} = \begin{bmatrix} \varepsilon_{111} \\ \varepsilon_{112} \\ \varepsilon_{121} \\ \varepsilon_{122} \\ \varepsilon_{211} \\ \varepsilon_{212} \\ \varepsilon_{221} \\ \varepsilon_{222} \\ \varepsilon_{311} \\ \varepsilon_{312} \\ \varepsilon_{321} \\ \varepsilon_{322} \end{bmatrix}$$

Hence, the α_i parameters included in the $\boldsymbol{\beta}$ vector are α_1 and α_2, and β_1 is the only β_j parameter included here. The interaction terms in the $\boldsymbol{\beta}$ vector are those associated with independent variables $X_4 = X_1X_3$ and $X_5 = X_2X_3$. The X variable X_1X_3 refers to level $i = 1$ for factor A and $j = 1$ for factor B; hence, the associated interaction parameter is $(\alpha\beta)_{11}$. Correspondingly, the X variable X_2X_3 refers to levels $i = 2$, $j = 1$; hence, the associated interaction term is $(\alpha\beta)_{21}$.

To verify that the **X** matrix representation yields the proper model, we present in Table 20.12 the mean vector $\mathbf{E(Y)} = \mathbf{X}\boldsymbol{\beta}$. We see, for instance, that:

$$E(Y_{111}) = \mu_{..} + \alpha_1 + \beta_1 + (\alpha\beta)_{11}$$

TABLE 20.12 Mean vector for Castle Bakery example

$$\mathbf{E(Y)} = \begin{bmatrix} E(Y_{111}) \\ E(Y_{112}) \\ E(Y_{121}) \\ E(Y_{122}) \\ E(Y_{211}) \\ E(Y_{212}) \\ E(Y_{221}) \\ E(Y_{222}) \\ E(Y_{311}) \\ E(Y_{312}) \\ E(Y_{321}) \\ E(Y_{322}) \end{bmatrix} = \mathbf{X}\boldsymbol{\beta} = \begin{bmatrix} \mu_{..} + \alpha_1 + \beta_1 + (\alpha\beta)_{11} \\ \mu_{..} + \alpha_1 + \beta_1 + (\alpha\beta)_{11} \\ \mu_{..} + \alpha_1 - \beta_1 - (\alpha\beta)_{11} \\ \mu_{..} + \alpha_1 - \beta_1 - (\alpha\beta)_{11} \\ \mu_{..} + \alpha_2 + \beta_1 + (\alpha\beta)_{21} \\ \mu_{..} + \alpha_2 + \beta_1 + (\alpha\beta)_{21} \\ \mu_{..} + \alpha_2 - \beta_1 - (\alpha\beta)_{21} \\ \mu_{..} + \alpha_2 - \beta_1 - (\alpha\beta)_{21} \\ \mu_{..} - \alpha_1 - \alpha_2 + \beta_1 - (\alpha\beta)_{11} - (\alpha\beta)_{21} \\ \mu_{..} - \alpha_1 - \alpha_2 + \beta_1 - (\alpha\beta)_{11} - (\alpha\beta)_{21} \\ \mu_{..} - \alpha_1 - \alpha_2 - \beta_1 + (\alpha\beta)_{11} + (\alpha\beta)_{21} \\ \mu_{..} - \alpha_1 - \alpha_2 - \beta_1 + (\alpha\beta)_{11} + (\alpha\beta)_{21} \end{bmatrix} = \begin{bmatrix} \mu_{..} + \alpha_1 + \beta_1 + (\alpha\beta)_{11} \\ \mu_{..} + \alpha_1 + \beta_1 + (\alpha\beta)_{11} \\ \mu_{..} + \alpha_1 + \beta_2 + (\alpha\beta)_{12} \\ \mu_{..} + \alpha_1 + \beta_2 + (\alpha\beta)_{12} \\ \mu_{..} + \alpha_2 + \beta_1 + (\alpha\beta)_{21} \\ \mu_{..} + \alpha_2 + \beta_1 + (\alpha\beta)_{21} \\ \mu_{..} + \alpha_2 + \beta_2 + (\alpha\beta)_{22} \\ \mu_{..} + \alpha_2 + \beta_2 + (\alpha\beta)_{22} \\ \mu_{..} + \alpha_3 + \beta_1 + (\alpha\beta)_{31} \\ \mu_{..} + \alpha_3 + \beta_1 + (\alpha\beta)_{31} \\ \mu_{..} + \alpha_3 + \beta_2 + (\alpha\beta)_{32} \\ \mu_{..} + \alpha_3 + \beta_2 + (\alpha\beta)_{32} \end{bmatrix}$$

Also, we see that:

$$\begin{aligned} E(Y_{121}) &= \mu_{..} + \alpha_1 - \beta_1 - (\alpha\beta)_{11} \\ &= \mu_{..} + \alpha_1 + \beta_2 + (\alpha\beta)_{12} \end{aligned}$$

since $\beta_2 = -\beta_1$ by (20.63) and $(\alpha\beta)_{12} = -(\alpha\beta)_{11}$ by (20.65). Similarly, we see that:

$$\begin{aligned} E(Y_{322}) &= \mu_{..} - \alpha_1 - \alpha_2 - \beta_1 + (\alpha\beta)_{11} + (\alpha\beta)_{21} \\ &= \mu_{..} + \alpha_3 + \beta_2 + (\alpha\beta)_{32} \end{aligned}$$

since:

$$\begin{aligned} &\alpha_3 = -\alpha_1 - \alpha_2 \quad \text{by (20.62)} \\ &\beta_2 = -\beta_1 \quad \text{by (20.63)} \\ &-(\alpha\beta)_{12} = (\alpha\beta)_{11} \qquad -(\alpha\beta)_{22} = (\alpha\beta)_{21} \quad \text{by (20.65)} \\ &-(\alpha\beta)_{12} - (\alpha\beta)_{22} = (\alpha\beta)_{32} \quad \text{by (20.66)} \end{aligned}$$

Thus, the linear model representation in Table 20.11 yields the proper mean response for each observation.

The multiple regression model that is the equivalent of the two-factor ANOVA model for the Castle Bakery example therefore is:

$$Y_{ijk} = \mu_{..} + \underbrace{\alpha_1 X_{ijk1} + \alpha_2 X_{ijk2}}_{A \text{ main effect}} + \underbrace{\beta_1 X_{ijk3}}_{B \text{ main effect}} \tag{20.68}$$

$$+ \underbrace{(\alpha\beta)_{11} X_{ijk1} X_{ijk3} + (\alpha\beta)_{21} X_{ijk2} X_{ijk3}}_{AB \text{ interaction effect}} + \varepsilon_{ijk}$$

Full model

$$i = 1, 2, 3; j = 1, 2; k = 1, 2$$

Here X_{ijk1} denotes the value of independent variable X_1 for the kth observation from the treatment for which factor A is at the ith level and factor B is at the jth level, and X_{ijk2} and X_{ijk3} have corresponding meanings. [These independent variables were defined in (20.67).] Finally, the regression parameters are the ANOVA model parameters $\mu_{..}$ (the intercept term) and α_1, α_2, β_1, $(\alpha\beta)_{11}$, and $(\alpha\beta)_{21}$ (the regression coefficients).

The tests for interaction effects, factor A main effects, and factor B main effects for the Castle Bakery example involve testing whether certain of the regression coefficients in model (20.68) are zero, and are as follows:

Test for interaction effects

$$\begin{aligned} &H_0: (\alpha\beta)_{11} = (\alpha\beta)_{21} = 0 \\ &H_a: \text{not both } (\alpha\beta)_{11} \text{ and } (\alpha\beta)_{21} \text{ equal zero} \end{aligned} \tag{20.69}$$

Test for factor A main effects

$$\begin{aligned} &H_0: \alpha_1 = \alpha_2 = 0 \\ &H_a: \text{not both } \alpha_1 \text{ and } \alpha_2 \text{ equal zero} \end{aligned} \tag{20.70}$$

Test for factor B main effects

$$\begin{aligned} &H_0: \beta_1 = 0 \\ &H_a: \beta_1 \neq 0 \end{aligned} \tag{20.71}$$

Thus, the test statistic in each case is the F^* statistic in (8.32a), and only the appropriate extra sum of squares and numerator degrees of freedom vary for the three tests.

To conduct the analysis of variance tests for our Castle Bakery example by means of the regression approach, we utilize regression model (20.68). The observations Y_{ijk} are given in Table 20.7, and the **X** matrix is given in Table 20.11. A computer run of a multiple regression package for these data provided the results shown in Table 20.13, including the extra sums of squares for each X variable in the order of the X variables in the regression model. If the available multiple regression computer program does not provide extra sums of squares for each additional X variable, the full and reduced models would usually have to be fitted separately for a given test, as explained in Chapter 8.

TABLE 20.13 Regression approach to two-factor analysis of variance—Castle Bakery example

(a) Fitted Regression Function

$$\hat{Y} = 51.0 - 7.0X_1 + 16.0X_2 - 1.0X_3 + 2.0X_1X_3 - 1.0X_2X_3$$

(b) ANOVA Table

Source of Variation	*SS*		*df*		*MS*
Regression	1,580		5		$MSTR = 316$
X_1	8	$SSA = 1{,}544$	1	2	$MSA = 772$
$X_2 \mid X_1$	1,536		1		
$X_3 \mid X_1, X_2$	12	$SSB = 12$	1	1	$MSB = 12$
$X_1X_3 \mid X_1, X_2, X_3$	18	$SSAB = 24$	1	2	$MSAB = 12$
$X_2X_3 \mid X_1, X_2, X_3, X_1X_3$	6		1		
Error	62		6		$MSE = 10.3$
Total	1,642		11		

It can be shown that when all treatment sample sizes are equal, as they are for the Castle Bakery example, the total of the extra sums of squares for the factor A regression terms equals SSA, and similarly the totals of the extra sums of squares for the factor B regression terms and the interaction regression terms equal, respectively, SSB and $SSAB$. Because of the balance in the **X** matrix when all treatment sample sizes are equal, these totals of the extra sums of squares are the same regardless of the order of the factor A, factor B, and interaction extra sums of squares.

We see from Table 20.13 that the totals SSA, SSB, and $SSAB$ and also the error sum of squares SSE obtained with the regression approach are the same as in Table 20.10 where they were calculated by means of the ANOVA formulas.

From this point on, the test procedures based on the regression approach are identical to the analysis of variance tests explained earlier.

Note

If the regression approach is to be utilized in conjunction with model (20.19), which is stated in terms of the treatment means μ_{ij}, the **X** matrix and $\boldsymbol{\beta}$ vector are set up as illustrated in (20.23). The alternatives for each test will then need to be stated in terms of the treatment means μ_{ij}. Thus, for testing factor B main effects for the Castle Bakery example, the alternatives H_0: $\mu_{.1} = \mu_{.2}$, H_a: $\mu_{.1} \neq \mu_{.2}$ would be stated in equivalent fashion as follows:

(20.72)
$$H_0: \frac{\mu_{11} + \mu_{21} + \mu_{31}}{3} - \frac{\mu_{12} + \mu_{22} + \mu_{32}}{3} = 0$$
$$H_a: \frac{\mu_{11} + \mu_{21} + \mu_{31}}{3} - \frac{\mu_{12} + \mu_{22} + \mu_{32}}{3} \neq 0$$

When working with ANOVA model (20.19), the alternatives such as those in (20.72) generally involve linear combinations of the μ_{ij} so that the test usually will need to be carried out in matrix terms by means of the general linear test approach as explained in Section 8.5.

PROBLEMS

20.1. Refer to the **SENIC** data set. An analyst wishes to investigate the effects of medical school affiliation (factor A) and geographic region (factor B) on infection risk. All factor level combinations will be included in the study.

a. How many treatments are being studied?

b. What is the dependent variable here?

20.2. A student in a class discussion stated: "A treatment is a treatment, whether the study involves a single factor or multiple factors. The number of factors only affects the analysis of the results." Discuss.

20.3. Verify the specific age effects and the specific sex effects in Table 20.2.

20.4. Verify the interactions in Table 20.3b.

20.5. In a two-factor study, the treatment means μ_{ij} are as follows:

	Factor B		
Factor A	B_1	B_2	B_3
A_1	34	23	36
A_2	40	29	42

a. Obtain the factor A level means.

b. Obtain the specific effects of B_1 when factor A is at levels A_1 and A_2, respectively.

c. Obtain the main effects of factor A.

d. Does the fact that $\mu_{12} - \mu_{11} = -11$ while $\mu_{13} - \mu_{12} = 13$ imply that factors A and B interact? Explain.

e. Plot the mean responses μ_{ij} in the format of Figure 20.2 and determine whether the two factors interact. What do you find?

20.6. In a two-factor study, the treatment means μ_{ij} are as follows:

	Factor B			
Factor A	B_1	B_2	B_3	B_4
A_1	250	265	268	269
A_2	288	273	270	269

a. Obtain the specific effects of factor B when factor A is at levels A_1 and A_2, respectively.

b. Obtain the factor B main effects. What do your results imply about factor B? [*Hint:* Consider the findings in part (a).]

c. Plot the treatment means μ_{ij} in the format of Figure 20.2 and determine whether the two factors interact. How can you tell that interactions are present? Are the interactions important or unimportant?

d. Make a logarithmic transformation of the μ_{ij} and plot the transformed values to explore whether this transformation is helpful in reducing the interactions. What are your findings?

20.7. Three sets of treatment means μ_{ij} for students' grades in a course follow, where factor A is student's major (A_1: computer science; A_2: mathematics) and factor B is student's class affiliation (B_1: junior; B_2: senior; B_3: graduate).

	Set 1				*Set 2*				*Set 3*		
	B_1	B_2	B_3		B_1	B_2	B_3		B_1	B_2	B_3
A_1	80	80	80	A_1	75	80	90	A_1	75	80	85
A_2	90	90	90	A_2	80	86	97	A_2	75	85	100

Plot each set of μ_{ij} in the format of Figure 20.2 to study interaction effects. Analyze each plot and state your findings. If interactions are present, describe their nature and indicate whether they are important or unimportant.

20.8. Refer to Problem 20.5. Assume that $\sigma = 1.4$ and $n = 10$.

a. Obtain $E(MSE)$ and $E(MSA)$.

b. Is $E(MSA)$ substantially larger than $E(MSE)$? What is the implication of this?

20.9. Refer to Problem 20.6. Assume that $\sigma = 4$ and $n = 6$.

a. Obtain $E(MSE)$ and $E(MSAB)$.

b. Is $E(MSAB)$ substantially larger than $E(MSE)$? What is the implication of this?

20.10. A psychologist stated: ''I feel uncomfortable about deciding in a research study whether the interactions are important or unimportant. I would rather have the statistician make that decision.'' Comment.

20.11. Refer to **Cash offers** Problem 16.12. Six male and six female volunteers were used in each age group. The observations (in hundred dollars), classified by age (factor A) and sex of owner (factor B), follow. Assume that ANOVA model (20.27) is applicable.

Factor A (age)		*Factor B (sex of owner)* $j = 1$ *Male*	$j = 2$ *Female*
$i = 1$	Young	21	21
		23	22
		19	20
		22	21
		22	19
		23	25
$i = 2$	Middle	30	26
		29	29
		26	27
		28	28
		27	27
		27	29
$i = 3$	Elderly	25	23
		22	19
		23	20
		21	21
		22	20
		21	20

Summary calculational results are: $SSA = 316.722$, $SSB = 5.444$, $SSAB = 5.056$, $SSE = 71.667$.

a. Obtain the fitted values.

b. Obtain the residuals. Do they sum to zero for each treatment?

c. Plot the treatment sample means $\bar{Y}_{ij.}$ in the format of Figure 20.5. Does it appear that any factor effects are present? Explain.

d. Set up the analysis of variance table. Does any one source account for most of the total variability in cash offers in the study? Explain.

e. Test whether or not interaction effects are present; use $\alpha = .05$. State the alternatives, decision rule, and conclusion. What is the P-value of the test?

f. Test whether or not age and sex main effects are present. In each case, use $\alpha = .05$ and state the alternatives, decision rule, and conclusion. What is the P-value of each test? Is it meaningful here to test for main factor effects? Explain.

g. Obtain an upper bound on the family level of significance for the tests in parts (e) and (f); use the Kimball inequality.

h. Do the results in parts (e) and (f) confirm your graphic analysis in part (c)?

i. What are the relations between the sums of squares in the two-factor analysis of variance in part (d) and the sums of squares in the single-factor analysis of variance in Problem 16.12c? Do the same relations hold for the degrees of freedom?

20.12. Eye contact effect. In a study of the effect of eye contact (factor A) and sex of personnel officer (factor B) on a personnel officer's assessment of likely job success of an applicant, 10 male and 10 female personnel officers were shown a frontview photograph of an applicant's face and were asked to give the person in the photograph a success rating on a scale of 0 (total failure) to 20 (outstanding success). Half of the officers in each sex group were chosen at random to receive a version of the photograph in which the subject made eye contact with the camera lens. The other half received a version in which there was no eye contact. The success ratings follow. Assume that ANOVA model (20.27) is applicable.

	Factor B (sex of officer)	
Factor A (eye contact)	$j = 1$ *Male*	$j = 2$ *Female*
$i = 1$ Present	11	15
	7	12
	12	14
	6	11
	10	16
$i = 2$ Absent	12	14
	16	17
	10	13
	13	20
	14	18

Summary calculational results are: $SSA = 54.45$, $SSB = 76.05$, $SSAB = 1.25$, $SSE = 97.2$.

a. Obtain the fitted values.

b. Obtain the residuals. Do they sum to zero for each treatment?

c. Plot the treatment sample means $\bar{Y}_{ij.}$ in the format of Figure 20.5. Does it appear that any factor effects are present? Explain.

d. Set up the analysis of variance table. Does any one source account for most of the total variability in the success ratings in the study? Explain.
e. Test whether or not interaction effects are present; use $\alpha = .01$. State the alternatives, decision rule, and conclusion. What is the P-value of the test?
f. Test whether or not eye contact and sex main effects are present. In each case, use $\alpha = .01$ and state the alternatives, decision rule, and conclusion. What is the P-value of each test? Is it meaningful here to test for main factor effects? Explain.
g. Obtain an upper bound on the family level of significance for the tests in parts (e) and (f); use the Kimball inequality.
h. Do the results in parts (e) and (f) confirm your graphic analysis in part (c)?

20.13. Hay fever relief. A research laboratory was developing a new compound for the relief of severe cases of hay fever. In an experiment with 36 volunteers, the amounts of the two active ingredients (factors A and B) in the compound were varied at three levels each. Randomization was used in assigning four volunteers to each of the nine treatments. The data on hours of relief follow.

Factor A (ingredient 1)		Factor B (ingredient 2) $j = 1$ Low	$j = 2$ Medium	$j = 3$ High
$i = 1$	Low	2.4	4.6	4.8
		2.7	4.2	4.5
		2.3	4.9	4.4
		2.5	4.7	4.6
$i = 2$	Medium	5.8	8.9	9.1
		5.2	9.1	9.3
		5.5	8.7	8.7
		5.3	9.0	9.4
$i = 3$	High	6.1	9.9	13.5
		5.7	10.5	13.0
		5.9	10.6	13.3
		6.2	10.1	13.2

Assume that ANOVA model (20.27) is applicable.
a. Obtain the fitted values.
b. Obtain the residuals.
c. Plot the treatment sample means $\bar{Y}_{ij.}$ in the format of Figure 20.5. Does your graph suggest that any factor effects are present? Explain.
d. Obtain the analysis of variance table. Does any one source account for most of the total variability in hours of relief in the study? Explain.
e. Test whether or not the two factors interact; use $\alpha = .05$. State the alternatives, decision rule, and conclusion. What is the P-value of the test?
f. Test whether or not main effects for the two ingredients are present. Use $\alpha = .05$ in each case and state the alternatives, decision rule, and conclusion. What is the P-value of each test? Is it meaningful here to test for main factor effects? Explain.
g. Obtain an upper bound on the family level of significance for the tests in parts (e) and (f); use the Kimball inequality.
h. Do the results in parts (e) and (f) confirm your graphic analysis in part (c)?

20.14. Disk drive service. The staff of a service center for electronic equipment includes three technicians who specialize in repairing three widely used makes of disk drives for desk-top computers. It was desired to study the effects of technician (factor A) and make of disk drive (factor B) on the service time. The data that follow show the number of minutes required to complete the repair job in a study where each technician was randomly assigned to five jobs on each make of disk drive.

Factor A (technician)		Factor B (make of drive) $j = 1$ Make 1	$j = 2$ Make 2	$j = 3$ Make 3
$i = 1$	Technician 1	62	57	59
		48	45	53
		63	39	67
		57	54	66
		69	44	47
$i = 2$	Technician 2	51	61	55
		57	58	58
		45	70	50
		50	66	69
		39	51	49
$i = 3$	Technician 3	59	58	47
		65	63	56
		55	70	51
		52	53	44
		70	60	50

Assume that ANOVA model (20.27) is applicable.

a. Obtain the fitted values.

b. Obtain the residuals.

c. Plot the treatment sample means $\bar{Y}_{ij.}$ in the format of Figure 20.5. Does your graph suggest that any factor effects are present? Explain.

d. Obtain the analysis of variance table. Does any one source account for most of the total variability?

e. Test whether or not the two factors interact; use $\alpha = .01$. State the alternatives, decision rule, and conclusion. What is the P-value of the test?

f. Test whether or not main effects for technician and make of drive are present. Use $\alpha = .01$ in each case and state the alternatives, decision rule, and conclusion. What is the P-value of each test? Is it meaningful here to test for main factor effects? Explain.

g. Obtain an upper bound on the family level of significance for the tests in parts (e) and (f); use the Kimball inequality.

h. Do the results in parts (e) and (f) confirm your graphic analysis in part (c)?

20.15. Kidney failure hospitalization. Kidney failure patients are commonly treated on dialysis machines that filter toxic substances from the blood. The appropriate "dose" for effective treatment depends, among other things, on duration of treatment and weight gain between treatments as a result of fluid buildup. To study the effects of these two factors on the number of days hospitalized (attributable to the disease) during a year, a random sample of 10 patients per group who had undergone treatment at a large dialysis facility was obtained. Treatment duration (factor A) was categorized into two groups: short duration (average dialyzing time for the year under four hours) and long duration (average dialyzing time for the year equal to or greater than four hours). Average weight gain between treatments (factor B) during the year was categorized into three groups: slight, moderate, and severe. The data on number of days hospitalized follow.

Factor A (duration)		Factor B (weight gain) j = 1 Mild		j = 2 Moderate		j = 3 Severe	
i = 1	Short	0	2	2	4	15	16
		2	0	4	3	10	7
		1	5	7	1	8	30
		3	6	12	5	5	3
		0	8	15	20	25	27
i = 2	Long	0	2	5	1	10	15
		1	7	3	3	8	4
		1	4	2	6	12	9
		0	0	0	7	3	6
		4	3	1	9	7	1

Assume that ANOVA model (20.27) is appropriate for the transformed data $Y' = \log_{10}(Y + 1)$.

a. Transform the data.
b. Obtain the fitted values and residuals.
c. Plot the treatment sample means $\bar{Y}'_{ij.}$ in the format of Figure 20.5. Does your graph suggest that any factor effects are present? Explain.
d. Obtain the analysis of variance table. Does any one source account for most of the total variability?
e. Test whether or not the two factors interact; use $\alpha = .05$. State the alternatives, decision rule, and conclusion. What is the P-value of the test?
f. Test whether or not main effects for duration and weight gain are present. Use $\alpha = .05$ in each case and state the alternatives, decision rule, and conclusion. What is the P-value of each test? Is it meaningful here to test for main factor effects? Explain.
g. Obtain an upper bound on the family level of significance for the tests in parts (e) and (f); use the Kimball inequality.
h. Do the results in parts (e) and (f) confirm your graphic analysis in part (c)?

20.16. Programmer requirements. A computer software firm was encountering difficulties in forecasting the programmer requirements for large-scale programming projects. As part of a study to remedy the difficulties, 24 programmers, classified into equal groups by type of experience (factor A) and amount of experience (factor B), were asked to predict the number of programmer-days required to complete a large project about to be initiated. After this project was completed, the prediction errors (actual minus predicted programmer-days) were determined. The data on prediction errors follow.

Factor A (type of experience)		Factor B (years of experience) j = 1 Under 5	j = 2 5–under 10	j = 3 10 or more
i = 1	Small systems only	240	110	56
		206	118	60
		217	103	68
		225	95	58
i = 2	Small and large systems	71	47	37
		53	52	33
		68	31	40
		57	49	45

Assume that ANOVA model (20.27) is applicable.

a. Obtain the fitted values.

b. Obtain the residuals.

c. Plot the treatment sample means $\bar{Y}_{ij.}$ in the format of Figure 20.5. Does your graph suggest that any factor effects are present? Explain.

d. Obtain the analysis of variance table. Does any one source account for most of the total variability?

e. Test whether or not the two factors interact; use $\alpha = .01$. State the alternatives, decision rule, and conclusion. What is the P-value of the test?

f. Test whether or not main effects for type of experience and years of experience are present. Use $\alpha = .01$ in each case and state the alternatives, decision rule, and conclusion. What is the P-value of each test? Is it meaningful here to test for main factor effects? Explain.

g. Obtain an upper bound on the family level of significance for the tests in parts (e) and (f); use the Kimball inequality.

h. Do the results in parts (e) and (f) confirm your graphic analysis in part (c)?

20.17. In a two-factor study, factor A has four levels, factor B has three levels, and $n = 6$. Separate tests were conducted for factor A and factor B main effects, each with significance level $\alpha = .05$. The researcher now wishes to investigate the power of the two tests. Assume that $\sigma = 20$.

a. What is the power of the test for factor A main effects when $\alpha_1 = -10$, $\alpha_2 = 7$, $\alpha_3 = 3$, and $\alpha_4 = 0$?

b. What is the power of the test for factor B main effects when $\beta_1 = -4$, $\beta_2 = 8$, and $\beta_3 = -4$?

20.18. Refer to **Cash offers** Problem 20.11. Assume that $\sigma = 1.3$.

a. What is the power of the test for interaction effects in Problem 20.11e if $(\alpha\beta)_{11} = -.2$, $(\alpha\beta)_{12} = .2$, $(\alpha\beta)_{21} = -.3$, $(\alpha\beta)_{22} = .3$, $(\alpha\beta)_{31} = .5$, and $(\alpha\beta)_{32} = -.5$?

b. What is the power of the test for factor A main effects in Problem 20.11f if $\mu_{1.} = 22$, $\mu_{2.} = 27$, and $\mu_{3.} = 21$?

c. What is the power of the test for factor B main effects in Problem 20.11f if $\mu_{.1} = 26$ and $\mu_{.2} = 20$? (*Hint:* Use Table A–5.)

20.19. Refer to **Hay fever relief** Problem 20.13. Assume that $\sigma = .28$.

a. What is the power of the test for factor A main effects in Problem 20.13f if $\mu_{1.} = 4.0$, $\mu_{2.} = 8.0$, and $\mu_{3.} = 10.5$?

b. How is the power of the test in part (a) affected if $\mu_{2.} = 10.0$, everything else remaining the same?

c. What is the power of the test for factor B main effects in Problem 20.13f if $\beta_1 = -1.5$, $\beta_2 = -.5$, and $\beta_3 = 2.0$?

20.20. Refer to **Programmer requirements** Problem 20.16. Obtain the power of the test for factor B main effects in Problem 20.16f if $\beta_1 = 50$, $\beta_2 = -15$, and $\beta_3 = -35$. Assume that $\sigma = 9.0$.

20.21. Refer to **Eye contact effect** Problem 20.12.

a. Modify regression model (20.68) to apply to this two-factor study with $a = 2$ and $b = 2$.

b. Set up the $\mathbf{Y}$, $\mathbf{X}$, and $\boldsymbol{\beta}$ matrices for the regression model in part (a).

c. Obtain $\mathbf{X}\boldsymbol{\beta}$. Verify the correctness of the expected values.

d. Obtain the fitted regression function. What is estimated by the intercept term?
e. Obtain the regression analysis of variance table based on appropriate extra sums of squares.
f. Test separately for interaction effects, factor A main effects, and factor B main effects. Use $\alpha = .01$ for each test and state the alternatives, decision rule, and conclusion.

20.22. Refer to **Hay fever relief** Problem 20.13.
a. Modify regression model (20.68) to apply to this two-factor study with $a = 3$ and $b = 3$.
b. Set up the **Y**, **X**, and $\boldsymbol{\beta}$ matrices for the regression model in part (a).
c. Obtain $\mathbf{X}\boldsymbol{\beta}$. Verify the correctness of the expected values.
d. Obtain the fitted regression function. What is estimated by $\hat{\alpha}_1$?
e. Obtain the regression analysis of variance table based on appropriate extra sums of squares.
f. Test separately for interaction effects, factor A main effects, and factor B main effects. Use $\alpha = .05$ for each test and state the alternatives, decision rule, and conclusion.

20.23. Refer to **Disk drive service** Problem 20.14.
a. Modify regression model (20.68) to apply to this two-factor study with $a = 3$ and $b = 3$.
b. Obtain the fitted regression function. What is estimated by $\hat{\beta}_1$?
c. Obtain the regression analysis of variance table based on appropriate extra sums of squares.
d. Test separately for interaction effects, factor A main effects, and factor B main effects. Use $\alpha = .01$ for each test and state the alternatives, decision rule, and conclusion.

20.24. Refer to **Cash offers** Problem 20.11. It is desired to test for factor B main effects by regression matrix methods.
a. Obtain the **Y**, **X**, and $\boldsymbol{\beta}$ matrices for full model (20.19), as illustrated in (20.23).
b. Obtain the estimated regression coefficients for the full model using (8.39).
c. Express the alternatives (20.72) in matrix form (8.41).
d. Obtain $SSE(R) - SSE(F)$ using (8.45). Does your result equal $SSB = 5.444$, as it should?

EXERCISES

20.25. Derive (20.13a) from (20.13).

20.26. Show that (20.16) is equivalent to (20.14a).

20.27. Prove the result in (20.17b).

20.28. (Calculus needed.) State the likelihood function for ANOVA model (20.19) when $a = 2$, $b = 2$, and $n = 2$. Find the maximum likelihood estimators. Are they the same as the least squares estimators in (20.33)?

20.29. (Calculus needed.) Derive (20.33).

20.30. Derive (20.43) from (20.41).

20.31. Show the equivalence between the expressions in (20.44c) and (20.43a).

PROJECTS

20.32. Refer to the **SENIC** data set. The following hospitals are to be considered in a study of the effects of region (factor A: variable 9) and average age of patients (factor B: variable 3) on the mean length of hospital stay of patients (variable 2):

1–44 46 48 51 53 57 58 60 63 66 74
76 79 80 83 84 88 94 101 103 111

For purposes of this ANOVA study, average age is to be classified into two categories: less than or equal to 53.9 years, 54.0 years or more. Assume that ANOVA model (20.27) is appropriate.

a. Assemble the required data and obtain the fitted values.
b. Plot the treatment sample means $\bar{Y}_{ij.}$ in the format of Figure 20.5. Does it appear that any factor effects are present? Explain.
c. Obtain the analysis of variance table. Does any one source account for most of the total variability in the study? Explain.
d. Test whether or not interaction effects are present; use $\alpha = .05$. State the alternatives, decision rule, and conclusion. What is the P-value of the test?
e. Test whether or not region and age main effects are present. In each case, use $\alpha = .05$ and state the alternatives, decision rule, and conclusion. What is the P-value of each test? Is it meaningful here to test for main factor effects? Explain.
f. Obtain an upper bound on the family level of significance for the tests in parts (d) and (e); use the Kimball inequality.
g. Do the results in parts (d) and (e) confirm your graphic analysis in part (b)?

20.33. Refer to the **SMSA** data set. The following metropolitan areas are to be considered in a study of the effects of region (factor A: variable 12) and percent of population in central cities (factor B: variable 4) on the crime rate (variable 11 ÷ variable 3):

1–36 38 39 45 46 49 53 62 71 73 101 123 130

For purposes of this ANOVA study, percent of population in central cities is to be classified into two categories: less than or equal to 38.9 percent, 39.0 percent or more. Assume that ANOVA model (20.27) is appropriate.

a. Assemble the required data and obtain the fitted values.
b. Plot the treatment sample means $\bar{Y}_{ij.}$ in the format of Figure 20.5. Does it appear that any factor effects are present? Explain.
c. Set up the analysis of variance table. Does any one source account for most of the total variability in the study? Explain.
d. Test whether or not interaction effects are present; use $\alpha = .01$. State the alternatives, decision rule, and conclusion. What is the P-value of the test?
e. Test whether or not region and percent of population in central cities main effects are present. In each case, use $\alpha = .01$ and state the alternatives,

decision rule, and conclusion. What is the P-value of each test? Is it meaningful here to test for main factor effects? Explain.

f. Obtain an upper bound on the family level of significance for the tests in parts (d) and (e); use the Kimball inequality.

g. Do the results in parts (d) and (e) confirm your graphic analysis in part (b)?

CITED REFERENCE

20.1 Dixon, W. J., and M. B. Brown, eds. *BMDP-81, Biomedical Computer Programs, P-Series*. Berkeley, Calif.: University of California Press, 1981.

21

Analysis of two-factor studies—equal sample sizes

When the analysis of variance tests described in Chapter 20 indicate the presence of factor effects in two-factor studies, the next step is to analyze the nature of the factor effects. In this chapter, we discuss how to conduct such analyses of the factor effects. We continue to consider the two-factor ANOVA model (20.27), where there are n observations for each treatment and all treatment means are deemed to be of equal importance.

We take up first the analysis of factor effects when both factors are qualitative and then the analysis when one or both factors are quantitative. Finally, we discuss briefly two major problems in implementing the analysis of variance model for two-factor studies—planning of sample sizes and determining the aptness of the model.

21.1 STRATEGY FOR ANALYSIS

Our consideration in Chapter 20 of the meaning of the model elements suggests the following basic strategy for analyzing factor effects in two-factor studies:

1. Examine whether the two factors interact.
2. If they do not interact, examine whether the main effects for factors A and B are important. For important A or B main effects, describe the nature of the A or B factor effects in terms of the factor level means $\mu_{i.}$ or $\mu_{.j}$, respec-

tively. In some special cases, there may also be interest in the treatment means μ_{ij}.
3. If the factors do interact, examine if the interactions are important or unimportant.
4. If the interactions are unimportant, proceed as in step 2.
5. If the interactions are important, determine whether they can be made unimportant by a meaningful transformation of scale. If so, make the transformation and proceed as in step 2.
6. For important interactions that cannot be made unimportant by a meaningful transformation, analyze the two factor effects jointly in terms of the treatment means μ_{ij}. In some special cases, there may also be interest in the factor level means $\mu_{i.}$ and $\mu_{.j}$.

A flowchart diagram of this strategy is presented in Figure 21.1.

FIGURE 21.1 Strategy for analysis of two-factor studies

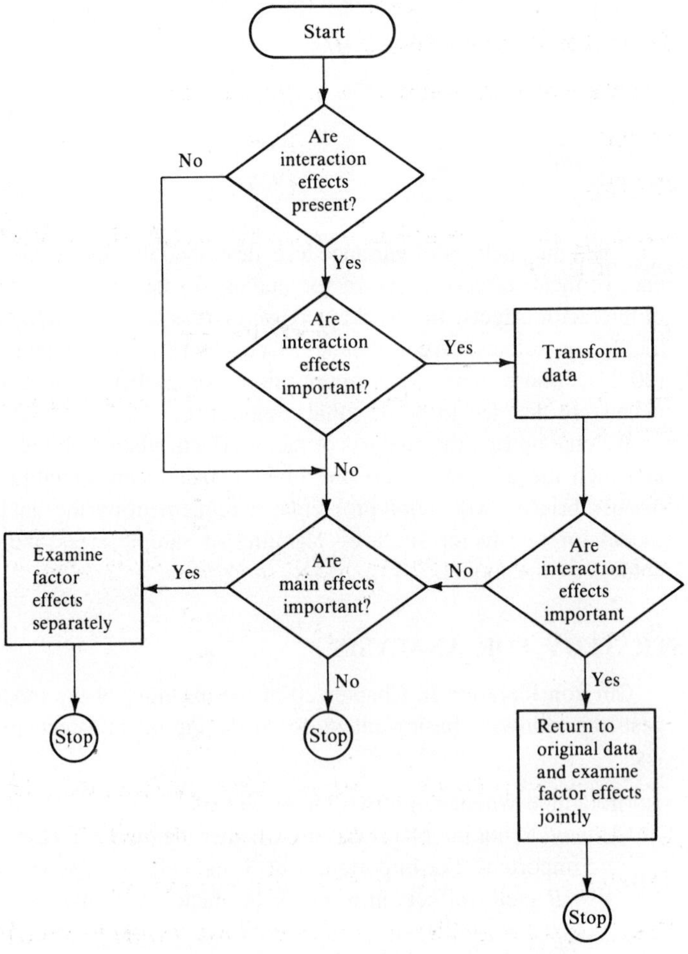

Step 1 of this strategy, testing for interaction effects, was discussed in Chapter 20. Also discussed there was step 5, the possible diminution of important interactions by a meaningful transformation, as well as how to test for the presence of factor main effects. Now we turn to steps 2 and 6 of the strategy for analysis, namely, how to compare factor level means $\mu_{i.}$ and $\mu_{.j}$ when there are no interactions or only unimportant ones, and how to compare treatment means μ_{ij} when there are important interactions. We begin with a discussion of the analysis of factor effects when the factors do not interact or interact only in unimportant fashion.

21.2 ANALYSIS OF FACTOR EFFECTS WHEN FACTORS DO NOT INTERACT

As just noted, the analysis of factor effects usually only involves the factor level means $\mu_{i.}$ and $\mu_{.j}$ when the two factors do not interact, or interact only in unimportant fashion.

Estimation of factor level mean

Unbiased point estimators of $\mu_{i.}$ and $\mu_{.j}$ are:

(21.1a) $$\hat{\mu}_{i.} = \overline{Y}_{i..}$$

(21.1b) $$\hat{\mu}_{.j} = \overline{Y}_{.j.}$$

where $\overline{Y}_{i..}$ and $\overline{Y}_{.j.}$ are defined in (20.31d) and (20.31f), respectively. The variance of $\overline{Y}_{i..}$ is:

(21.2a) $$\sigma^2(\overline{Y}_{i..}) = \frac{\sigma^2}{bn}$$

since $\overline{Y}_{i..}$ contains bn independent observations, each with variance σ^2. Similarly, we have:

(21.2b) $$\sigma^2(\overline{Y}_{.j.}) = \frac{\sigma^2}{an}$$

Unbiased estimators of these variances are obtained by replacing σ^2 with MSE:

(21.3a) $$s^2(\overline{Y}_{i..}) = \frac{MSE}{bn}$$

(21.3b) $$s^2(\overline{Y}_{.j.}) = \frac{MSE}{an}$$

Confidence limits for $\mu_{i.}$ and $\mu_{.j}$ utilize, as usual, the t distribution:

(21.4a) $$\overline{Y}_{i..} \pm t[1 - \alpha/2; (n - 1)ab]s(\overline{Y}_{i..})$$

(21.4b) $$\overline{Y}_{.j.} \pm t[1 - \alpha/2; (n - 1)ab]s(\overline{Y}_{.j.})$$

The degrees of freedom $(n - 1)ab$ are those associated with MSE.

Estimation of contrast of factor level means

A contrast among the factor level means $\mu_{i.}$:

$$L = \Sigma c_i \mu_{i.} \tag{21.5}$$

where:

$$\Sigma c_i = 0$$

is estimated unbiasedly by:

$$\hat{L} = \Sigma c_i \bar{Y}_{i..} \tag{21.6}$$

Because of the independence of the $\bar{Y}_{i..}$, the variance of this estimator is:

$$\sigma^2(\hat{L}) = \Sigma c_i^2 \sigma^2(\bar{Y}_{i..}) = \frac{\sigma^2}{bn} \Sigma c_i^2 \tag{21.7}$$

An unbiased estimator of this variance is:

$$s^2(\hat{L}) = \frac{MSE}{bn} \Sigma c_i^2 \tag{21.8}$$

Finally, the appropriate $1 - \alpha$ confidence limits for L are:

$$\hat{L} \pm t[1 - \alpha/2; (n - 1)ab]s(\hat{L}) \tag{21.9}$$

To estimate a contrast among the factor level means $\mu_{.j}$:

$$L = \Sigma c_j \mu_{.j} \tag{21.10}$$

where:

$$\Sigma c_j = 0$$

we use the estimator:

$$\hat{L} = \Sigma c_j \bar{Y}_{.j.} \tag{21.11}$$

whose estimated variance is:

$$s^2(\hat{L}) = \frac{MSE}{an} \Sigma c_j^2 \tag{21.12}$$

The $1 - \alpha$ confidence limits for L in (21.9) are still appropriate, with $\hat{L}$ and $s(\hat{L})$ now defined in (21.11) and (21.12), respectively.

Estimation of linear combination of factor level means

A linear combination of the factor level means $\mu_{i.}$:

$$L = \Sigma c_i \mu_{i.} \tag{21.13}$$

is estimated unbiasedly by $\hat{L}$ in (21.6). The variance of this estimator is given in (21.7), and an unbiased estimator of this variance is given in (21.8). The appropriate $1 - \alpha$ confidence limits for L are given in (21.9).

Analogous results follow for a linear combination of the factor level means $\mu_{.j}$:

$$L = \Sigma c_j \mu_{.j} \tag{21.14}$$

Multiple pairwise comparisons of factor level means

Usually, more than one comparison is of interest, and the multiple comparison procedures discussed in Chapter 17 can be employed with only minor modifications. If all or a large number of pairwise comparisons among the factor level means $\mu_{i.}$ are to be made, of the form:

$$D = \mu_{i.} - \mu_{i'.} \tag{21.15}$$

the Tukey procedure of (17.24) is appropriate. The formulas are as follows, reflecting the case of equal sample sizes considered here:

$$\hat{D} = \bar{Y}_{i..} - \bar{Y}_{i'..} \tag{21.16a}$$

$$s^2(\hat{D}) = \frac{2MSE}{bn} \tag{21.16b}$$

$$T = \frac{1}{\sqrt{2}} q[1 - \alpha; a, (n-1)ab] \tag{21.16c}$$

The confidence limits as usual are:

$$\hat{D} \pm Ts(\hat{D}) \tag{21.17}$$

The probability then is $1 - \alpha$ that all statements in the family are correct.

For pairwise comparisons of the factor level means $\mu_{.j}$, the only changes are:

$$D = \mu_{.j} - \mu_{.j'} \tag{21.18a}$$

$$\hat{D} = \bar{Y}_{.j.} - \bar{Y}_{.j'.} \tag{21.18b}$$

$$s^2(\hat{D}) = \frac{2MSE}{an} \tag{21.18c}$$

$$T = \frac{1}{\sqrt{2}} q[1 - \alpha; b, (n-1)ab] \tag{21.18d}$$

If only a few pairwise comparisons are to be made, the Bonferroni method may be best. All of the above formulas still apply, but the Tukey multiple T is replaced by the Bonferroni multiple B:

$$B = t[1 - \alpha/2g; (n-1)ab] \tag{21.19}$$

where g is the number of statements in the family.

If it is desired to have a family confidence coefficient $1 - \alpha$ for the joint set of pairwise comparisons involving *both* factor A and B means, the Bonferroni method can be used either directly or in conjunction with the Tukey method. To illustrate the latter, suppose the pairwise comparisons for factor A are made with the Tukey procedure with a family confidence coefficient of .95, and likewise for the pairwise comparisons for factor B. The Bonferroni inequality then assures us that the family confidence coefficient for the joint set of comparisons for both factors is at least .90.

Multiple contrasts of factor level means

When a large number of contrasts among the factor level means $\mu_{i.}$ or $\mu_{.j}$ are of interest, the Scheffé method should be used. If the contrasts involve the $\mu_{i.}$, as in (21.5), the unbiased estimator is given in (21.6) and its estimated variance in (21.8). For this case, the Scheffé multiple is defined by:

$$S^2 = (a-1)F[1-\alpha; a-1, (n-1)ab] \tag{21.20}$$

leading to the confidence limits for the contrast L:

$$\hat{L} \pm Ss(\hat{L}) \tag{21.21}$$

The probability is then $1-\alpha$ that every confidence interval (21.21) in the family of all possible contrasts is correct.

If contrasts among the factor level means $\mu_{.j}$ are desired, the unbiased point estimator is given in (21.11), its estimated variance is given in (21.12), and the Scheffé confidence limits (21.21) are appropriate with:

$$S^2 = (b-1)F[1-\alpha; b-1, (n-1)ab] \tag{21.22}$$

When the number of contrasts of interest is small, the Bonferroni method may be best. Confidence limits (21.21) would need to be modified by replacing the Scheffé multiple S with the Bonferroni multiple B:

$$B = t[1-\alpha/2g; (n-1)ab] \tag{21.23}$$

where g is the number of statements in the family.

When it is desired to obtain a family confidence coefficient for the joint set of contrasts for both factors, several possibilities exist:

1. The Bonferroni method may be used directly, with g representing the total number of statements in the joint set.
2. The Bonferroni method can be used to join the two sets of Scheffé multiple comparison families in the same way explained earlier for joining two Tukey sets.
3. The Scheffé method can be modified to use the S multiple defined by:

$$S^2 = (a+b-2)F[1-\alpha; a+b-2, (n-1)ab] \tag{21.24}$$

When this S multiple is used in both sets of multiple comparisons, the probability is $1-\alpha$ that all statements in the combined family are correct.

One may try each of these three approaches and see which leads to the narrowest confidence intervals without affecting the validity of the procedure.

Estimates based on treatment means

Occasionally in analyzing the factor effects in a two-factor study when no interactions are present, one is interested in particular treatment means μ_{ij}. For example, in a two-factor study of the effects of price and type of advertisement on sales, interest may exist in estimating the mean sales for two different price levels when a particular advertisement is used. In such cases, the methods of analysis for single-factor studies discussed in Chapter 17 are appropriate. The number of treatments now is simply $r = ab$, the degrees of freedom associated with MSE are $n_T - r = nab - ab = (n - 1)ab$, and the estimated treatment means are $\bar{Y}_{ij.}$, based on n observations each.

Example 1—Pairwise comparisons of factor level means

In the Castle Bakery example of Chapter 20, it will be recalled that the plot of the treatment sample means in Figure 20.5 suggested that no interaction effects are present. The formal analysis of variance based on Table 20.10 supported this conclusion. Suppose that no tests of main factor effects had been conducted, and we now wish to analyze the effects of shelf width and shelf height.

We shall perform the analysis in terms of the factor level means since no interaction effects are present and shall make pairwise comparisons between the factor level means for each factor, with a combined confidence coefficient of 90 percent for all comparisons. The Tukey multiple comparison procedure will be used, with the comparisons for display height assigned a family confidence coefficient of 95 percent, and similarly for the display width comparison. The Bonferroni inequality then guarantees a family confidence coefficient of at least 90 percent for the joint set of comparisons.

For the comparison of display height means (i = 1—bottom, 2—middle, 3—top), we have from Tables 20.7 and 20.8:

$$\bar{Y}_{2..} - \bar{Y}_{1..} = 67 - 44 = 23 \qquad MSE = 10.3$$

$$a = 3$$

$$\bar{Y}_{1..} - \bar{Y}_{3..} = 44 - 42 = 2 \qquad b = 2$$

$$n = 2$$

$$\bar{Y}_{2..} - \bar{Y}_{3..} = 67 - 42 = 25 \qquad (n - 1)ab = 6$$

Hence, by (21.16), we obtain:

$$s^2(\hat{D}) = \frac{2(10.3)}{2(2)} = 5.15$$

$$q(.95; 3, 6) = 4.34$$

$$T = \frac{4.34}{\sqrt{2}} = 3.07$$

$$Ts(\hat{D}) = 3.07\sqrt{5.15} = 7.0$$

Similarly, for the comparison of display widths (j = 1—regular, 2—wide), we have by (21.18):

$$\bar{Y}_{.2.} - \bar{Y}_{.1.} = 52 - 50 = 2$$

$$s^2(\hat{D}) = \frac{2(10.3)}{3(2)} = 3.43$$

$$q(.95; 2, 6) = 3.46$$

$$T = \frac{3.46}{\sqrt{2}} = 2.45$$

$$Ts(\hat{D}) = 2.45\sqrt{3.43} = 4.5$$

We obtain therefore the following confidence intervals for all pairwise comparisons of factor level means:

$$16 = 23 - 7.0 \leq \mu_{2.} - \mu_{1.} \leq 23 + 7.0 = 30$$
$$-5 = 2 - 7.0 \leq \mu_{1.} - \mu_{3.} \leq 2 + 7.0 = 9$$
$$18 = 25 - 7.0 \leq \mu_{2.} - \mu_{3.} \leq 25 + 7.0 = 32$$
$$-2.5 = 2 - 4.5 \leq \mu_{.2} - \mu_{.1} \leq 2 + 4.5 = 6.5$$

It can be concluded from these confidence intervals with family confidence coefficient of 90 percent for the joint set that for the product studied and the types of stores in the experiment, the middle shelf height is far better than either the bottom or the top heights, the latter two do not differ significantly in sales effectiveness, and widening the display has no significant effect on sales. All these conclusions are covered by the family confidence coefficient of 90 percent. The effects of shelf height can be presented graphically as follows:

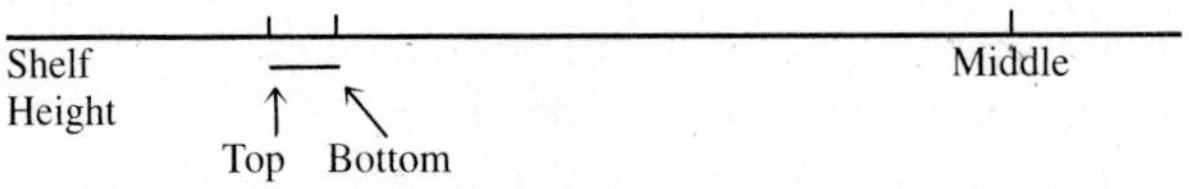

Example 2—Estimation of treatment means

The manager of a supermarket similar in terms of sales volume and clientele to the supermarkets included in the Castle Bakery study has room only for the regular shelf display width, and wishes to obtain estimates of mean sales for the middle and top shelf heights. We shall obtain interval estimates with a 90 percent family confidence coefficient using the Bonferroni procedure.

From Tables 20.7 and 20.8, we have:

$$\bar{Y}_{21.} = 65 \qquad \bar{Y}_{31.} = 40 \qquad MSE = 10.3$$

Hence, we obtain:

$$s^2(\bar{Y}_{21.}) = s^2(\bar{Y}_{31.}) = \frac{MSE}{n} = \frac{10.3}{2} = 5.15$$

$$s(\bar{Y}_{21.}) = s(\bar{Y}_{31.}) = 2.27$$

For $g = 2$, we require $B = t[1 - \alpha/2g; (n-1)ab] = t(.975; 6) = 2.447$. Thus, we obtain the confidence limits:

$$65 \pm 2.447(2.27) \qquad 40 \pm 2.447(2.27)$$

and the desired confidence intervals are:

$$59.4 \leq \mu_{21} \leq 70.6 \qquad 34.4 \leq \mu_{31} \leq 45.6$$

21.3 ANALYSIS OF FACTOR EFFECTS WHEN INTERACTIONS IMPORTANT

When important interactions exist, the analysis of factor effects generally must be based on the treatment means μ_{ij}. Typically, this analysis will involve multiple comparisons or single degree of freedom tests of treatment means.

Multiple pairwise comparisons of treatment means

If pairs of treatment means μ_{ij} are to be compared, either the Tukey or the Bonferroni multiple comparison procedure may be used, depending on which is more advantageous. In effect, the analysis is equivalent to the single-factor case, with the total number of treatments here equal to $r = ab$, the degrees of freedom associated with MSE here equal to $n_T - r = (n - 1)ab$, and the estimated treatment mean, now denoted by $\bar{Y}_{ij.}$, based on n observations. Formula (17.24) for the Tukey multiple comparison procedure for $D = \mu_{ij} - \mu_{i'j'}$ with equal sample sizes becomes:

(21.25) $$\hat{D} \pm Ts(\hat{D}) \qquad i, j \neq i', j'$$

where:

(21.25a) $$\hat{D} = \bar{Y}_{ij.} - \bar{Y}_{i'j'.}$$

(21.25b) $$s^2(\hat{D}) = \frac{2MSE}{n}$$

(21.25c) $$T = \frac{1}{\sqrt{2}}q[1 - \alpha; ab, (n - 1)ab]$$

If the Bonferroni method is employed, the multiple in the confidence interval is:

(21.26) $$B = t[1 - \alpha/2g; (n - 1)ab]$$

where g is the number of statements in the family.

Multiple contrasts of treatment means

The Scheffé multiple comparison procedure for single-factor studies is directly applicable to the estimation of contrasts involving the treatment means μ_{ij}. We denote these contrasts as follows:

(21.27) $$L = \Sigma\Sigma c_{ij}\mu_{ij}$$

where:

$$\Sigma\Sigma c_{ij} = 0$$

The point estimator of L is:

(21.28) $$\hat{L} = \Sigma\Sigma c_{ij}\bar{Y}_{ij.}$$

and the estimated variance of this estimator is:

(21.29)
$$s^2(\hat{L}) = \frac{MSE}{n}\Sigma\Sigma c_{ij}^2$$

The Scheffé multiple S is given by:

(21.30)
$$S^2 = (ab - 1)F[1 - \alpha; ab - 1, (n - 1)ab]$$

and the joint confidence limits are as usual:

(21.31)
$$\hat{L} \pm Ss(\hat{L})$$

When the number of contrasts is small, the Bonferroni procedure may be preferable. The confidence intervals (21.31) would simply be modified by replacing S with B as defined in (21.26).

Single degree of freedom tests of treatment means

When important interactions exist, a single degree of freedom test of treatment means μ_{ij} may sometimes be of interest. The two-sided alternatives for a single degree of freedom test here are as follows:

(21.32)
$$\begin{aligned} &H_0: \Sigma\Sigma c_{ij}\mu_{ij} = c \\ &H_a: \Sigma\Sigma c_{ij}\mu_{ij} \neq c \end{aligned}$$

where the c_{ij} and c are appropriate constants.

To test alternatives (21.32) we can use the t^* test statistic:

(21.33)
$$t^* = \frac{\Sigma\Sigma c_{ij}\bar{Y}_{ij.} - c}{\sqrt{\dfrac{MSE}{n}\Sigma\Sigma c_{ij}^2}}$$

which follows the t distribution with $(n - 1)ab$ degrees of freedom when H_0 holds. Alternatively, we can use test statistic $F^* = (t^*)^2$, which follows the F distribution with 1 and $(n - 1)ab$ degrees of freedom when H_0 holds.

When multiple single degree of freedom tests are to be conducted with a specified family level of significance, an appropriate multiple comparison procedure (Tukey, Scheffé, Bonferroni) should be used for setting up the relevant confidence intervals. The confidence intervals will show which of the two alternatives should be concluded in each case, as explained in Chapter 17 for single-factor studies.

Example 1—Pairwise comparisons of treatment means

A junior college system studied the effects of teaching method (factor A) and student's quantitative ability (factor B) on learning of college mathematics. Two teaching methods were studied—the standard method of teaching (to be called the standard method) and a method that emphasizes teaching of concepts in the abstract before going into drill routines (to be called the abstract method). The quantitative ability of a student was determined by a standard aptitude test, on the basis of which the student was classified as having excellent, good, or moderate

TABLE 21.1 Results of study on mathematics learning

(a) Mean Learning Scores ($n = 21$)

Teaching Method i	*Quantitative Ability* (j) *Excellent*	*Good*	*Moderate*
Abstract	92 ($\bar{Y}_{11.}$)	81 ($\bar{Y}_{12.}$)	73 ($\bar{Y}_{13.}$)
Standard	90 ($\bar{Y}_{21.}$)	86 ($\bar{Y}_{22.}$)	82 ($\bar{Y}_{23.}$)

(b) ANOVA Table

Source of Variation	*SS*	*df*	*MS*
Between treatments	4,998	5	999.6
Factor A (teaching methods)	504	1	504
Factor B (quantitative ability)	3,843	2	1,921.5
Interactions (AB)	651	2	325.5
Error	3,360	120	28
Total	8,358	125	

quantitative ability. Thus, factor A in this study (teaching method) has $a = 2$ levels and factor B (student's quantitative ability) has $b = 3$ levels.

For each of the $ab = 2(3) = 6$ treatments, 21 students were randomly selected and placed into two classes according to the designated teaching method. Thus, each class contained equal numbers of students of each quantitative ability level. The dependent variable was the amount of learning of college mathematics, as measured by a standard mathematics achievement test. The results of the study are summarized in Table 21.1. The estimated treatment means are shown in Table 21.1a, and the analysis of variance table is presented in Table 21.1b.

Figure 21.2 contains two plots of the estimated treatment means $\bar{Y}_{ij.}$. In Figure 21.2a, the two curves represent the different factor A levels; and in Figure 21.2b, the three curves represent the different factor B levels. The clear lack of parallelism of the curves suggests the presence of interaction effects between teaching method and student's quantitative ability on amount of mathematics learning. A formal test for interactions confirms this. From Table 21.1b, we have $F^* = MSAB/MSE = 325.5/28 = 11.625$. For $\alpha = .01$ we require $F(.99; 2, 120) = 4.79$. Since $F^* = 11.625 > 4.79$, we conclude that interaction effects are present.

Figure 21.2 suggests that the nature of the interactions is as follows: students with excellent quantitative ability are but little affected by teaching method (perhaps doing slightly better with the abstract method); students with good or moderate abilities learn much better with the standard teaching method. To investigate these effects fully, we shall estimate separately for students with excellent, good, and moderate quantitative abilities how large is the difference in mean learning for the two teaching methods; i.e., we wish to estimate:

FIGURE 21.2 Plots of treatment sample means—mathematics learning example

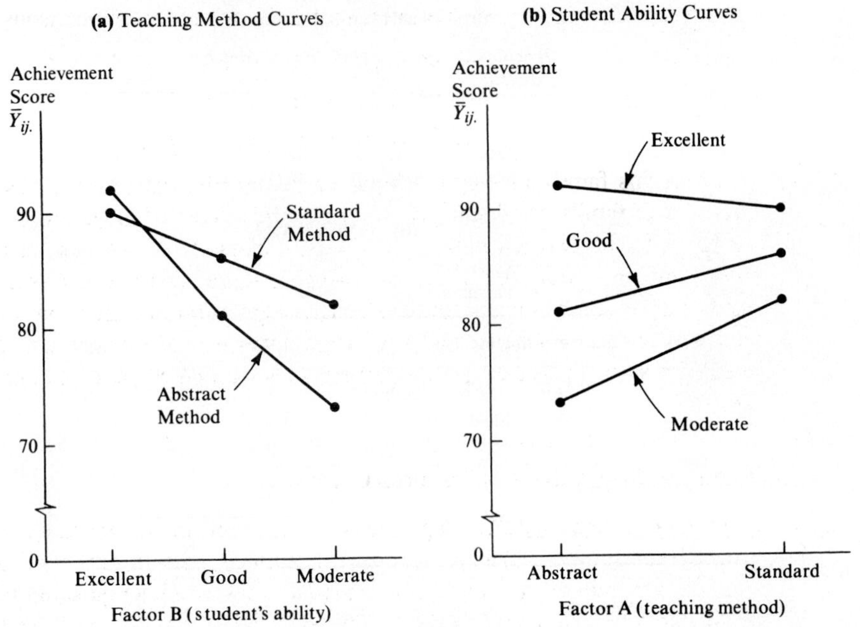

$$
\begin{aligned}
D_1 &= \mu_{11} - \mu_{21} \\
D_2 &= \mu_{12} - \mu_{22} \qquad (21.34)\\
D_3 &= \mu_{13} - \mu_{23}
\end{aligned}
$$

We shall employ the Bonferroni multiple comparison procedure with family confidence coefficient .95. (Since only three pairwise comparisons are of interest, the Bonferroni method yields more precise estimates here than the Tukey method.)

Using the data in Table 21.1a, the point estimates of the pairwise comparisons are:

$$
\begin{aligned}
\hat{D}_1 &= 92 - 90 = 2 \\
\hat{D}_2 &= 81 - 86 = -5 \qquad (21.35)\\
\hat{D}_3 &= 73 - 82 = -9
\end{aligned}
$$

Since $n = 21$, we find the estimated variances of these estimates by (21.25b) to be:

$$s^2(\hat{D}_1) = s^2(\hat{D}_2) = s^2(\hat{D}_3) = \frac{2(28)}{21} = 2.667$$

so that:

$$s(\hat{D}_1) = s(\hat{D}_2) = s(\hat{D}_3) = 1.633$$

Finally, for family confidence coefficient $1 - \alpha = .95$ and $g = 3$, we require $B = t[1 - .05/2(3); 120] = t(.99167; 120) = 2.428$. Hence, the confidence limits are by (21.25):

$$2 \pm 2.428(1.633) \qquad -5 \pm 2.428(1.633) \qquad -9 \pm 2.428(1.633)$$

and the 95 percent confidence intervals for the family of comparisons are:

$$-1.96 \leq \mu_{11} - \mu_{21} \leq 5.96$$
$$-8.96 \leq \mu_{12} - \mu_{22} \leq -1.04$$
$$-12.96 \leq \mu_{13} - \mu_{23} \leq -5.04$$

From this family of confidence intervals the following conclusions may be drawn with family confidence coefficient of 95 percent: (1) The mean learning scores with the two teaching methods for students with excellent quantitative ability do not differ. (2) The mean learning score with the abstract teaching method is lower than that with the standard method for both students with good and moderate quantitative abilities. The superiority of the standard teaching method may be particularly strong for students with moderate quantitative abilities.

Example 2—Contrasts of treatment means

A school administrator also wished to know in our mathematics learning example whether the amount of gain in learning with the standard teaching method over the abstract method for students with moderate quantitative ability is greater than the gain for students with good quantitative ability. This question had been raised before the study began. We shall estimate the single contrast:

$$L = (\mu_{23} - \mu_{13}) - (\mu_{22} - \mu_{12}) \tag{21.36}$$

by means of a one-sided lower confidence interval. Utilizing the results in Table 21.1a, the point estimate of L is $\hat{L} = (82 - 73) - (86 - 81) = 4$. The estimated variance by (21.29) is:

$$s^2(\hat{L}) = \frac{28}{21}[(1)^2 + (-1)^2 + (-1)^2 + (1)^2] = 5.333$$

so that the estimated standard deviation is $s(\hat{L}) = 2.309$. For a 95 percent confidence coefficient, we require $t(.05; 120) = -1.658$. Hence, the lower confidence limit is $4 - 1.658(2.309)$ and the desired confidence interval is:

$$L \geq .17$$

We conclude, therefore, with 95 percent confidence coefficient that the gain in learning with the standard teaching method over the abstract method is greater for students with moderate ability than for students with good ability, the difference in the mean gain being at least .17 point.

Example 3—Contrasts of treatment means with unequal weights

A school administrator in our mathematics learning example had requested information about which teaching method leads to better learning of college mathematics when 20 percent of the students in the class have excellent quantitative ability, 50 percent have good ability, and 30 percent have moderate ability.

The mean learning scores for such a class mix with the two teaching methods are the following linear combinations of the treatment means:

(21.37)
$$\begin{aligned} &\text{Abstract method:} && L_1 = .2\mu_{11} + .5\mu_{12} + .3\mu_{13} \\ &\text{Standard method:} && L_2 = .2\mu_{21} + .5\mu_{22} + .3\mu_{23} \end{aligned}$$

This assumes that the mean learning scores for students with different quantitative abilities will not be affected by a class mix that is somewhat different from the one in the experimental study.

We could conduct a single degree of freedom test here or develop an interval estimate. We shall do the latter because the confidence interval will provide us with information not only about the direction of any difference between the two teaching methods but also about the magnitude of the difference.

Point estimates of the mean scores in (21.37) are:

$$\hat{L}_1 = .2(92) + .5(81) + .3(73) = 80.8$$
$$\hat{L}_2 = .2(90) + .5(86) + .3(82) = 85.6$$

The difference between the two mean scores in (21.37) is a contrast:

(21.38)
$$L = L_1 - L_2$$

This contrast is estimated to be:

$$\hat{L} = \hat{L}_1 - \hat{L}_2 = 80.8 - 85.6 = -4.8$$

The estimated variance of $\hat{L}$ is by (21.29):

$$s^2(\hat{L}) = \frac{28}{21}[(.2)^2 + (.5)^2 + (.3)^2 + (-.2)^2 + (-.5)^2 + (-.3)^2] = 1.013$$

so that the estimated standard deviation is $s(\hat{L}) = 1.006$. For a 95 percent confidence coefficient, we require $t(.975; 120) = 1.980$. Hence, the confidence limits are $-4.8 \pm 1.980(1.006)$ and the desired confidence interval is:

$$-6.79 \leq L \leq -2.81$$

With 95 percent confidence we conclude that the standard teaching method is better for the specified class mix, leading to a mean learning score that is at least 2.8 points greater than that for the abstract teaching method and may be as much as 6.8 points greater.

If it were desired to test formally whether or not for the class mix specified by the school administrator, the mean learning score (L_2) for the standard teaching method exceeds that for the abstract method (L_1), the alternatives would be:

$$\begin{aligned} &H_0\text{: } L_2 \leq L_1 && && H_0\text{: } L_1 - L_2 \geq 0 \\ &H_a\text{: } L_2 > L_1 && \text{or} && H_a\text{: } L_1 - L_2 < 0 \end{aligned}$$

where L_1 and L_2 are defined in (21.37). In view of the one-sided nature of the alternatives, we cannot utilize the F^* test statistic and must employ the t^* test statistic (21.33):

$$t^* = \frac{\hat{L} - 0}{s(\hat{L})} = \frac{-4.8}{1.006} = -4.77$$

Assuming that the level of significance is to be controlled at $\alpha = .05$, we require $t(.05; 120) = -1.658$. The decision rule therefore is:

$$\begin{aligned} &\text{If } t^* \geq -1.658, \text{ conclude } H_0 \\ &\text{If } t^* < -1.658, \text{ conclude } H_a \end{aligned}$$

Since $t^* = -4.77 < -1.658$, we conclude H_a, that the mean score for the standard teaching method exceeds that for the abstract method when the class mix is as specified.

21.4 ANALYSIS WHEN ONE OR BOTH FACTORS QUANTITATIVE

When one or both of the factors in a two-factor study are quantitative, the analysis of factor effects can be carried beyond the point of multiple comparisons to include a study of the nature of the response function. Since the familiar methods of regression analysis, discussed earlier, then come into use, we shall only briefly discuss this extension of the analysis. The initial study of factor effects by multiple comparison techniques can be very helpful in choosing an appropriate functional form for the regression relation.

Analysis of response function when one factor quantitative

(This section assumes previous study of Chapter 10.)

Consider an experiment in which the effect of type of cake mix (factor A) and temperature (factor B) on the lightness of the cake texture, suitably measured, is to be investigated. Two kinds of cake mixes (G, H) and four temperatures (300°, 315°, 330°, 345°) are studied. The analyst in this case might be interested in extending the investigation of factor effects into the nature of the response function relating cake texture to baking temperature. Since factor A is qualitative, indicator variables will be used to represent it in the response function. If kind of cake mix and baking temperature do not interact, a first-order model might be appropriate:

(21.39) $$Y_{ijk} = \beta_0 + \beta_1 X_{ijk1} + \beta_2 X_{ijk2} + \varepsilon_{ijk}$$

where:

$$X_{ijk1} = \begin{array}{l} 1 \text{ if observation from first level of factor } A \\ 0 \text{ otherwise} \end{array}$$

X_{ijk2} is the baking temperature for the observation

The β's denote regression parameters here, and X_{ijk1} and X_{ijk2} are the respective values of X_1 and X_2 for the kth observation from the treatment corresponding to the ith level of factor A and the jth level of factor B.

We know from Chapter 10 that model (21.39) implies a linear relation between cake texture and temperature that has constant slope for both kinds of cake mixes but different heights. Figure 10.1 provides an illustration of this model.

If cake mix and baking temperature interact, an appropriate model might be:

(21.40) $$Y_{ijk} = \beta_0 + \beta_1 X_{ijk1} + \beta_2 X_{ijk2} + \beta_3 X_{ijk1} X_{ijk2} + \varepsilon_{ijk}$$

From Chapter 10, we know that this model implies a linear relation between cake texture and temperature with different slopes and intercepts for the two kinds of cake mixes. Figure 10.3 provides an illustration of this model.

If the regression relation is quadratic and no interactions between the two factors exist, a suitable model would be:

(21.41) $$Y_{ijk} = \beta_0 + \beta_1 X_{ijk1} + \beta_2 x_{ijk2} + \beta_3 x_{ijk2}^2 + \varepsilon_{ijk}$$

where:

$$x_{ijk2} = X_{ijk2} - \bar{X}_2$$

Analysis of response function when both factors quantitative

When both factors are quantitative, the analysis of the nature of the response function involves ordinary multiple regression. We shall denote the two factor variables as X_1 and X_2. A first-order model would then be:

(21.42) $$Y_{ijk} = \beta_0 + \beta_1 X_{ijk1} + \beta_2 X_{ijk2} + \varepsilon_{ijk}$$

where X_{ijk1} and X_{ijk2} are the respective values of X_1 and X_2 for the kth observation from the treatment corresponding to the ith level of factor A and the jth level of factor B. A second-order model, including interactions between the two factors, would be:

(21.43) $$Y_{ijk} = \beta_0 + \beta_1 x_{ijk1} + \beta_2 x_{ijk2} + \beta_3 x_{ijk1}^2 + \beta_4 x_{ijk2}^2 + \beta_5 x_{ijk1} x_{ijk2} + \varepsilon_{ijk}$$

where:

$$x_{ijk1} = X_{ijk1} - \bar{X}_1$$
$$x_{ijk2} = X_{ijk2} - \bar{X}_2$$

The discussion of response surfaces in Chapter 7 is completely applicable here.

Example (adapted from Reference 21.1). A study was conducted to improve the efficiency of a burr remover in a wool textile carding machine. The rollers in the burr remover are adjustable as to speed and spacing. Four spacings and three speeds, as shown in Table 21.2, were used in the study. Four replications were

TABLE 21.2 Treatment sample means in two-factor burr remover study with both factors quantitative ($n = 4$)

	Speed		
Spacing	*300 rpm*	*400 rpm*	*500 rpm*
1.0 unit	21.6	22.3	22.9
1.2 units	18.7	19.1	21.6
1.4 units	15.8	17.9	19.4
1.6 units	13.2	16.7	19.5

Source: Reprinted, with permission, from D. R. Cox, *Planning of Experiments* (New York: John Wiley & Sons, 1958), p. 124.

FIGURE 21.3 Plots of treatment sample means—burr remover example

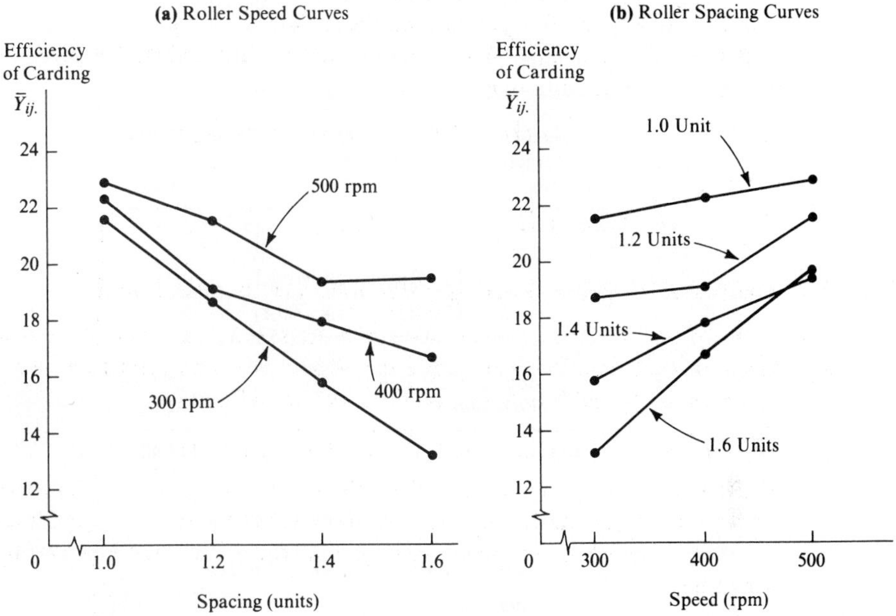

made for each treatment, but only the treatment means are presented in Table 21.2. The observed variable is a measure of the efficiency of the carding. A plot of the treatment means is presented in Figure 21.3.

An analysis of variance was conducted by a computer package for this two-factor study. The results are summarized in Table 21.3a. The test for interactions utilizes the test statistic:

$$F^* = \frac{MSAB}{MSE} = \frac{4.92}{1.32} = 3.73$$

For level of significance $\alpha = .05$, we require $F(.95; 6, 36) = 2.36$. Since $F^* = 3.73 > 2.36$, it was concluded that spacing and speed interact. Figure 21.3, along with further analysis, suggested that a first-order model with interaction effects added might be appropriate. The response function for this model is:

$$E(Y) = \beta_0 + \beta_1 X_1 + \beta_2 X_2 + \beta_3 X_1 X_2 \tag{21.44}$$

where X_1 represents spacing and X_2 speed. This model was fitted by a multiple regression computer package. The results are summarized in Table 21.3b.

Since replications were used in the study, a test of the appropriateness of the reponse function can be conducted. The lack of fit sum of squares is readily obtainable because SSE for the ANOVA model is the same as $SSPE$ for the regression model. Hence, we find:

$$SSLF = \underset{\text{(Table 21.3b)}}{SSE} - \underset{\text{(Table 21.3a)}}{SSPE} = 57.29 - 47.61 = 9.68$$

TABLE 21.3 Analysis of variance tables for burr remover study

(a) Analysis of Variance Model

Source of Variation	*SS*	*df*	*MS*
Spacing (A)	232.86	3	77.62
Speed (B)	99.49	2	49.74
Interactions (AB)	29.53	6	4.92
Error	47.61	36	1.32
Total	409.49	47	

(b) Regression Model

$$\hat{Y} = 45.09333 - 25.45000X_1 - .03340X_2 + .03925X_1X_2$$

Source of Variation	*SS*	*df*	*MS*
Regression	352.20	3	117.40
Error	57.29	44	1.30
Total	409.49	47	

(c) ANOVA for Lack of Fit Test

Source of Variation	*SS*	*df*	*MS*
Regression	352.20	3	117.40
Error	57.29	44	1.30
Lack of fit	9.68	8	1.21
Pure error	47.61	36	1.32
Total	409.49	47	

Table 21.3c contains the decomposition needed for testing lack of fit. The appropriate test statistic is:

$$F^* = \frac{MSLF}{MSE} = \frac{1.21}{1.32} = .92$$

Assuming a level of significance of $\alpha = .05$ is to be used, we require $F(.95; 8, 36) = 2.21$. Since $F^* = .92 \leq 2.21$, we conclude that the response function (21.44) is appropriate.

Let us take a closer look at the estimated reponse function:

$$\hat{Y} = 45.09333 - 25.45000X_1 - .03340X_2 + .03925X_1X_2$$

Note that b_3 is positive. It implies here, as a glance at Figure 21.3 will confirm, that efficiency declines with increased spacing for all speeds (e.g., if $X_2 = 300$, $\hat{Y} = 35.073 - 13.675X_1$), but the decline is smaller for high speeds. Correspondingly, the efficiency increases with speed for all spacings (e.g., if $X_1 =$ 1.0, $\hat{Y} = 19.643 + .00585X_2$), but the increase is larger for large spacings.

Comments

1. If a principal purpose of a two-factor study is to estimate the response surface, it is helpful for calculational simplifications to space the factor levels equally, as in the burr remover example of Table 21.2. One can then code these levels in simple fashion. For instance, if three levels are employed, the codes for the levels may be $-1, 0, +1$; if four levels are employed, the codes may be $-3, -1, +1, +3$. These codes are then used in the regression calculations.

2. Special designs have been developed to estimate response surfaces efficiently. An example of such a design for estimating a first-order model is shown in Table 21.4. Each of the two factors is studied at three equally spaced levels, but not all factor combinations are included. Only one replication is made for four of the treatments, while the center treatment has four replications. The center treatment is often called the *center of the design*. This design permits evaluation of quadratic effects if information about these is desired, and the replications at the center of the design provide an estimate of experimental (pure) error for testing the goodness of fit of the model. Many designs for estimating response surfaces are available; see, for instance, the discussion by Cochran and Cox in Reference 21.2.

TABLE 21.4 Example of a first-order design

		Factor B		
		−1	0	+1
Factor A	−1	1	—	1
	0	—	4	—
	+1	1	—	1

Numbers for each treatment show replication size.

3. A sequence of small-scale factorial experiments is frequently used in investigations seeking to find the best factor combination, such as the combination of temperature and pressure that maximizes the yield of a chemical process. After each experiment, a response surface is fitted and an evaluation is made of the factor combinations having high yields. The next experiment is then conducted to investigate these factor combinations in more detail. This process is continued until the factor combination having the maximum yield is established with sufficient precision. Various strategies may be employed in determining the sequence of experiments, such as the method of steepest ascent and the single-factor strategy. These are discussed in books on experimental designs such as in Cochran and Cox (Ref. 21.2) and Box, Hunter, and Hunter (Ref. 21.3).

21.5 IMPLEMENTATION OF TWO-FACTOR ANOVA MODEL

The two basic problems in implementing a two-factor analysis of variance model—planning sample sizes and evaluating the aptness of the model—are handled in essentially the same fashion as discussed in Chapter 18 for single-factor studies. Hence, we make only a few brief comments.

Planning sample sizes

No essentially new problems arise in planning sample sizes for two-factor studies. In most cases, equal replications are desired for each treatment. One can then plan the sample sizes using either the power approach or the estimation approach discussed in Chapter 18.

With the power approach, one would be concerned typically with both the power of detecting factor A main effects and the power of detecting factor B main effects. One can first specify the minimum range of factor A level means for which it is important to detect factor A main effects, and obtain the needed sample sizes from Table A–10, with $r = a$. The resulting sample size is bn, from which n can be readily obtained. The use of Table A–10 for this purpose is appropriate provided the resulting sample size is not small, specifically provided $a(bn - 1) \geq 20$. If this condition is not met, one should use the Pearson-Hartley power charts in Table A–8. These charts, as we noted earlier, require an iterative approach for determining needed sample sizes.

In the same way, one can then specify the minimum range of factor B level means for which it is important to detect factor B main effects, and find the needed sample sizes. If the sample sizes obtained from the factor A and factor B power specifications differ substantially, a judgment will need to be made as to the final sample sizes.

Alternatively, or in conjunction with the power approach, one can specify the important contrasts to be estimated and then find the sample sizes that are expected to provide the needed precisions for the desired family confidence coefficient. Frequently this approach is more useful than the power approach, although both of these approaches can be used jointly for arriving at a determination of needed sample sizes.

If the purpose of the factorial study is to identify the best of the ab factor combinations, Table A–11 can be used for finding the needed sample sizes, as described in Section 18.3. For this purpose, $r = ab$.

Evaluation of aptness of model

No new problems arise in examining the aptness of the analysis of variance two-factor model. The residuals (20.35):

$$(21.45) \qquad e_{ijk} = Y_{ijk} - \bar{Y}_{ij.}$$

may be examined for normality, constancy of error variance, and independence of error terms in the same fashion as for a single-factor study.

Transformations may be employed to stabilize the error variance and to make the error distributions more normal. Our earlier discussion of this topic in Chapter 18 for the single-factor case applies completely to the two-factor case.

Finally, the earlier discussion on effects of departures from the model applies fully to the two-factor case. In particular, the employment of equal sample sizes for each treatment minimizes the effect of unequal variances.

Example. Figure 21.4 contains residual frequency plots for each of the six treatments in our earlier mathematics learning example. These plots show no strong evidence of unequal error term variances or of major departures from normality.

FIGURE 21.4 Residual frequency plots for mathematics learning example

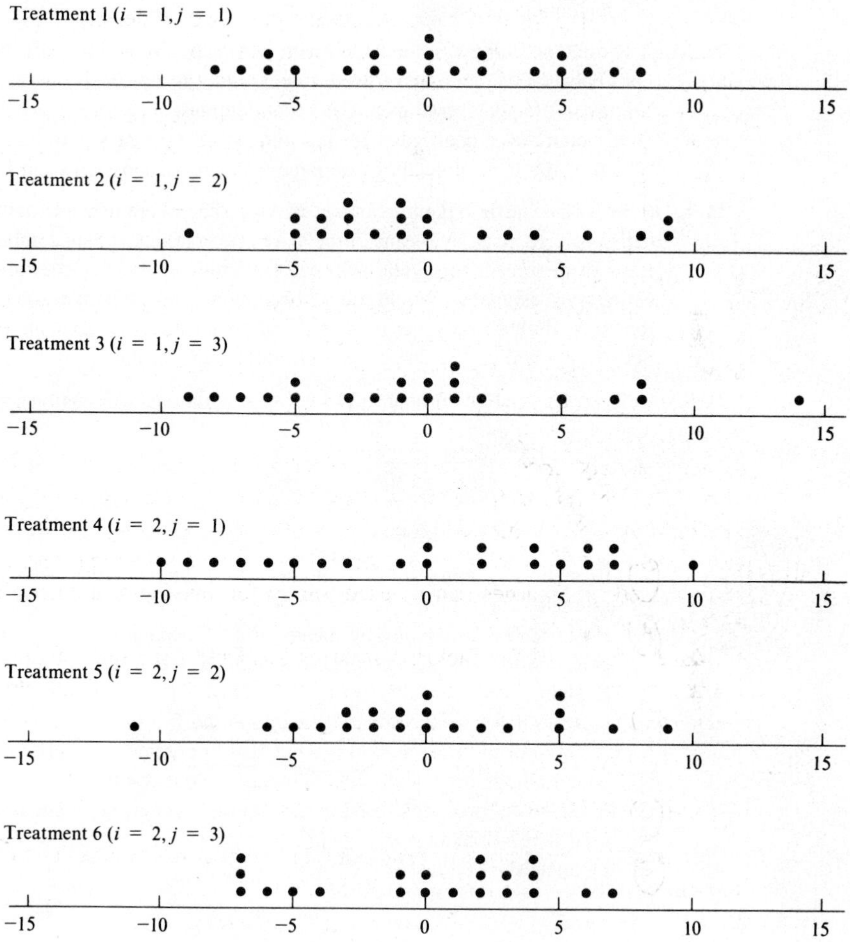

PROBLEMS

21.1. Why is it suggested in the flowchart in Figure 21.1 that a test for interactions should be conducted before tests for main factor effects? Explain.

21.2. A two-factor study was conducted with $a = 5$, $b = 5$, and $n = 4$. No interactions between factors A and B were noted, and the analyst now wishes to estimate all pairwise comparisons among the factor A level means and all pairwise comparisons among the factor B level means. The family confidence coefficient for the joint set of interval estimates is to be 90 percent.

a. Is it more efficient to use the Bonferroni procedure for the entire family or to use the Tukey procedure for each family of factor level mean comparisons and then to join the two families by means of the Bonferroni procedure?

b. Would your answer differ if each factor had three levels, everything else remaining the same?

21.3. A two-factor study was conducted with $a = 6$, $b = 6$, and $n = 10$. No interactions between factors A and B were found, and it is now desired to estimate five contrasts of factor A level means and four contrasts of factor B level means. The family confidence coefficient for the joint set of estimates is to be 95 percent. Which of the three procedures on page 722 will be most efficient here?

21.4. Refer to the Castle Bakery example on page 723, where multiple pairwise comparisons were made by means of the Tukey procedure and then combined by the Bonferroni procedure to yield a 90 percent family confidence coefficient for the entire set of estimates. Would it have been more efficient here to use the Bonferroni procedure entirely for the family of four estimates? Explain. Would the Scheffé method of (21.24) have been more efficient here?

21.5. Refer to **Cash offers** Problem 20.11. Some additional calculational results are:

i	$\bar{Y}_{i..}$	j	$\bar{Y}_{.j.}$	
1	21.50	1	23.94	
2	27.75	2	23.17	$MSE = 2.389$
3	21.42			

a. Estimate μ_{11} with a 95 percent confidence interval. Interpret your interval estimate.

b. Estimate $D = \mu_{.1} - \mu_{.2}$ by means of a 95 percent confidence interval. Is your confidence interval consistent with the test result in Problem 20.11f? Explain.

c. Obtain all pairwise comparisons among the factor A level means; use the Tukey procedure with a 90 percent family confidence coefficient. Present your findings graphically and summarize your results.

d. Is the Tukey procedure used in part (c) the most efficient one that could be used here? Explain.

e. Estimate the contrast:

$$L = \frac{\mu_{1.} + \mu_{3.}}{2} - \mu_{2.}$$

with a 95 percent confidence interval. Interpret your interval estimate.

f. Suppose that in the population of female owners, 30 percent are young, 60 percent are middle-aged, and 10 percent are elderly. Obtain a 95 percent confidence interval for the mean cash offer in the population of female owners.

21.6. Refer to **Eye contact effect** Problem 20.12. Some additional calculational results are:

i	$\bar{Y}_{i..}$	j	$\bar{Y}_{.j.}$
1	11.4	1	11.1
2	14.7	2	15.0

$MSE = 6.075$

a. Estimate μ_{21} with a 99 percent confidence interval. Interpret your interval estimate.

b. Estimate $\mu_{1.}$ with a 99 percent confidence interval. Interpret your interval estimate.

c. Obtain confidence intervals for $\mu_{.1}$ and $\mu_{.2}$, each with a 99 percent confidence coefficient. Interpret your interval estimates. What is the family confidence coefficient for the set of two estimates?

d. Obtain confidence intervals for $D_1 = \mu_{2.} - \mu_{1.}$ and $D_2 = \mu_{.2} - \mu_{.1}$; use the Bonferroni procedure and a 95 percent family confidence coefficient. Summarize your findings.

e. Is the Bonferroni procedure used in part (d) the most efficient one that could be used here? Explain.

21.7. Refer to **Hay fever relief** Problem 20.13.

a. Estimate μ_{23} with a 95 percent confidence interval. Interpret your interval estimate.

b. Estimate $D = \mu_{12} - \mu_{11}$ with a 95 percent confidence interval. Interpret your interval estimate.

c. The analyst decided to study the nature of the interacting factor effects by means of the following contrasts:

$$L_1 = \frac{\mu_{12} + \mu_{13}}{2} - \mu_{11} \qquad L_4 = L_2 - L_1$$

$$L_2 = \frac{\mu_{22} + \mu_{23}}{2} - \mu_{21} \qquad L_5 = L_3 - L_1$$

$$L_3 = \frac{\mu_{32} + \mu_{33}}{2} - \mu_{31} \qquad L_6 = L_3 - L_2$$

Obtain confidence intervals for these contrasts; use the Scheffé multiple comparison procedure with a 90 percent family confidence coefficient. Interpret your findings.

d. The analyst also wished to identify the treatment(s) yielding the longest mean relief. Using the Tukey procedure with a 90 percent family confidence coefficient, identify the treatment(s) providing the longest mean relief.

e. Use the single degree of freedom test statistic (21.33) to choose between the following two alternatives:

$$H_0: \mu_{32} - \mu_{31} \leq \mu_{33} - \mu_{32}$$
$$H_a: \mu_{32} - \mu_{31} > \mu_{33} - \mu_{32}$$

Control the level of significance at $\alpha = .05$. State the decision rule and conclusion.

21.8. Refer to **Disk drive service** Problem 20.14.

a. Estimate μ_{11} with a 99 percent confidence interval. Interpret your interval estimate.

b. Estimate $D = \mu_{22} - \mu_{21}$ with a 99 percent confidence interval. Interpret your interval estimate.

c. The nature of the interaction effects is to be studied by making, for each technician, all three pairwise comparisons among the disk drive makes in order to identify, if possible, the make of disk drive for which the technician's mean service time is lowest. The family confidence coefficient for each set of three pairwise comparisons is to be 95 percent. Use the Bonferroni procedure to make all required pairwise comparisons. Summarize your findings.

d. The service center currently services 30 disk drives of each of the three makes per week, with each technician servicing 10 machines of each make. Estimate the expected total amount of service time required per week to service the 90 disk drives; use a 99 percent confidence interval.

e. How much time could be saved per week, on the average, if technician 1 services only make 2, technician 2 services only make 1, and technician 3 services only make 3? Use a 99 percent confidence interval.

21.9. Refer to **Kidney failure hospitalization** Problem 20.15. Continue to work with the transformed observations $Y' = \log_{10}(Y + 1)$.

a. Estimate μ_{22} with a 95 percent confidence interval. Interpret your interval estimate.

b. Estimate $D = \mu_{23} - \mu_{21}$ with a 95 percent confidence interval. Interpret your interval estimate.

c. The researcher wishes to study the main effects of each of the two factors by making all pairwise comparisons of factor level means with a 90 percent family confidence coefficient for the entire set of comparisons. Which multiple comparison procedure is most efficient here?

d. Using the most efficient procedure, make all pairwise comparisons called for in part (c). State your findings and prepare a graphic summary.

e. It is known from past experience that 30 percent of patients have mild weight gains, 40 percent have moderate weight gains, and 30 percent have severe weight gains, and that these proportions are the same for the two duration groups. Estimate the mean number of days hospitalized (in transformed units) in the entire population with a 95 percent confidence interval. Convert your confidence limits to the original units. Does it appear that the mean number of days is less than 7?

21.10. Refer to **Programmer requirements** Problem 20.16.

a. Estimate μ_{23} with a 99 percent confidence interval. Interpret your interval estimate.

b. Estimate $D = \mu_{12} - \mu_{13}$ with a 99 percent confidence interval. Interpret your interval estimate.

c. The nature of the interaction effects is to be studied by comparing the effect of type of experience for each years-of-experience group. Specifically, the following comparisons are to be estimated:

$$D_1 = \mu_{11} - \mu_{21} \qquad L_1 = D_1 - D_2$$
$$D_2 = \mu_{12} - \mu_{22} \qquad L_2 = D_1 - D_3$$
$$D_3 = \mu_{13} - \mu_{23} \qquad L_3 = D_2 - D_3$$

The family confidence coefficient is to be 95 percent. Which multiple comparison procedure is most efficient here?

d. Use the most efficient procedure to estimate the comparisons specified in part (c). State your findings.

e. Use the Tukey procedure with a 95 percent family confidence coefficient to identify the type of experience-years of experience group(s) with the smallest mean prediction errors.

f. For each of the groups identified in part (e), obtain a confidence interval for the mean prediction error. Use the Bonferroni procedure with a 95 percent family confidence coefficient. Does any group have a mean prediction error that could be zero? Explain.

g. Use the single degree of freedom test statistic (21.33) to choose between the two alternatives:

$$H_0: \frac{\mu_{21} + \mu_{22} + \mu_{23}}{3} \leq 40$$

$$H_a: \frac{\mu_{21} + \mu_{22} + \mu_{23}}{3} > 40$$

Control the level of significance at $\alpha = .05$. State the decision rule and conclusion.

21.11. Refer to **Brand preference** Problem 7.8. Suppose the market researcher first wished to employ analysis of variance model (20.27) to determine whether or not moisture content (factor A) and sweetness (factor B) affect the degree of brand liking.

a. State the analysis of variance model for this case.

b. Obtain the analysis of variance table.

c. Test whether or not the two factors interact; use $\alpha = .01$. State the alternatives, decision rule, and conclusion.

d. Study possible curvilinearity of the moisture content effect by estimating the following contrast:

$$L = (\mu_{4.} - \mu_{3.}) - (\mu_{2.} - \mu_{1.})$$

Use a 95 percent confidence interval. What do you conclude?

e. Test whether or not sweetness affects brand liking; use $\alpha = .01$. State the alternatives, decision rule, and conclusion.

21.12. Refer to **Cash offers** Problem 20.11. The mean ages of the "owners" in the three age classes were:

Young:	24.8
Middle:	45.3
Elderly:	66.7

Since the actual ages of the "owners" in each age group for both sexes were close to the mean age, each mean age can be used to represent the ages (X_1) of the "owners" in that group.

a. Fit the regression model:

$$Y_{ijk} = \beta_0 + \beta_1 x_{ijk1} + \beta_2 x_{ijk1}^2 + \beta_3 X_{ijk2} + \varepsilon_{ijk}$$

where $x_{ijk1} = X_{ijk1} - \bar{X}_1$ and $X_{ijk2} = 1$ if owner is male and 0 if female.

b. Obtain the residuals and plot them against the fitted values. What does your plot show?
c. Conduct a formal test for lack of fit using a level of significance of $\alpha = .01$. State the alternatives, decision rule, and conclusion.
d. Test whether or not the quadratic term in the model in part (a) can be dropped; use $\alpha = .01$. State the alternatives, decision rule, and conclusion.

21.13. Refer to **Hay fever relief** Problem 20.13. The researcher now wishes to study the nature of the relationship between the amounts of the two active ingredients and the duration of relief. The amounts of the ingredients used in the study were as follows:

	Quantity (in milligrams)	
Factor Level	X_1 *(ingredient 1)*	X_2 *(ingredient 2)*
Low	5.0	7.5
Medium	10.0	10.0
High	15.0	12.5

a. Fit regression model (21.43).
b. Estimate the mean duration of relief when $X_1 = 7.50$ and $X_2 = 8.75$; use a 95 percent confidence interval. Could this estimate have been obtained from the ANOVA model? Explain.
c. Obtain the residuals and plot them against the fitted values. What does your plot show?
d. Conduct a formal test for lack of fit; use $\alpha = .005$. State the alternatives, decision rule, and conclusion.
e. Test whether or not the interaction term can be dropped from the model; use $\alpha = .05$. State the alternatives, decision rule, and conclusion.

21.14. A market research manager is planning to study the effects of duration of advertising (factor A) and price level (factor B) on sales. Each factor has three levels. No important interactions are expected, and the primary analysis is to consist of pairwise comparisons of factor level means for each factor. Equal sample sizes are to be used for each treatment. The precision of each comparison is to be ± 3 thousand dollars. The family confidence coefficient for the joint set of comparisons is to be 90 percent, the Tukey procedure is to be used in making the comparisons for each factor, and the Bonferroni procedure is then to be used to join the two sets of comparisons. Assume that $\sigma = 7$ thousand dollars is a reasonable planning value for the error term standard deviation. What sample sizes do you recommend?

21.15. Refer to **Cash offers** Problem 20.11. Suppose that the sample sizes have not yet been determined but it has been decided to use the same number of "owners" in each age-sex group. What are the required sample sizes if: (1) differences in the age factor level means are to be detected with probability .90 or more when the range of the factor level means is 3 (hundred dollars), and (2) the α risk is to be controlled at .05? Assume that a reasonable planning value for the error term standard deviation is $\sigma = 1.5$ (hundred dollars).

21.16. Refer to **Eye contact effect** Problem 20.12. Suppose that the sample sizes have not yet been determined but it has been decided to use equal sample sizes for each treatment. Primary interest is in the two comparisons $D_1 = \mu_{1.} - \mu_{2.}$ and $D_2 = \mu_{.1} - \mu_{.2}$. What are the required sample sizes if these comparisons are to be estimated with precision not to exceed ± 1.2 with a 95 percent family confidence coefficient, using the most efficient multiple comparison procedure? Assume that a reasonable planning value of the error term standard deviation is $\sigma = 2.4$.

21.17. Refer to **Hay fever relief** Problem 20.13. Suppose that the sample sizes have not yet been determined but it has been decided to use equal sample sizes for each treatment. The chief objective is to identify the dosage combination that yields the longest mean relief. The probability should be at least .99 that the correct dosage combination is identified when the mean relief duration for the second best combination differs by .5 hour or more. What are the required sample sizes? Assume that a reasonable planning value for the error term standard deviation is $\sigma = .29$ hour.

21.18. Refer to **Kidney failure hospitalization** Problem 20.15. Suppose that the sample sizes have not yet been determined but it has been decided to use equal sample sizes for each treatment. The chief objective is to estimate the pairwise comparisons:

$$D_1 = \mu_{1.} - \mu_{2.} \qquad D_3 = \mu_{.1} - \mu_{.3}$$
$$D_2 = \mu_{.1} - \mu_{.2} \qquad D_4 = \mu_{.2} - \mu_{.3}$$

What are the required sample sizes if the precision of the estimates should not exceed $\pm.20$ (in transformed units), using the Bonferroni procedure with a family confidence coefficient of 90 percent for the joint set of comparisons? A reasonable planning value for the error term standard deviation is $\sigma = .32$ (in transformed units).

21.19. Refer to **Programmer requirements** Problem 20.16. Suppose that the sample sizes have not yet been determined but it has been decided to use equal sample sizes for each treatment. Primary interest is in identifying the type of experience-years of experience combination for which the mean prediction error is smallest. The probability should be at least .95 that the correct combination is identified when the mean prediction error for the second best combination differs by 8 programmer-days or more. Assume that a reasonable planning value for the error term standard deviation is $\sigma = 9.1$ days. What are the required sample sizes?

21.20. Refer to **Cash offers** Problems 20.11 and 21.5.

a. Prepare residual frequency plots for the treatments. What departures from ANOVA model (20.27) can be studied from these plots? What are your findings?

b. Prepare a normal probability plot of the residuals. Also obtain the coefficient of correlation between the ordered residuals and their expected values under normality. Does the normality assumption appear to be reasonable here?

c. The observations for each treatment were obtained in the order shown. Prepare residual sequence plots and analyze them. What are your findings?

21.21. Refer to **Eye contact effect** Problems 20.12 and 21.6.

a. Prepare residual frequency plots for the treatments. What departures from ANOVA model (20.27) can be studied from these plots? What are your findings?

b. Prepare a normal probability plot of the residuals. Also obtain the coefficient of correlation between the ordered residuals and their expected values under normality. Does the normality assumption appear to be reasonable here?

c. The observations for each treatment were obtained in the order shown. Prepare residual sequence plots and analyze them. What are your findings?

21.22. Refer to **Hay fever relief** Problem 20.13.

a. Prepare residual frequency plots for the treatments. What departures from ANOVA model (20.27) can be studied from these plots? What are your findings?

b. Prepare a normal probability plot of the residuals. Also obtain the coefficient of correlation between the ordered residuals and their expected values under normality. Does the normality assumption appear to be reasonable here?

21.23. Refer to **Disk drive service** Problem 20.14.

a. Prepare residual frequency plots for the treatments. What departures from ANOVA model (20.27) can be studied from these plots? What are your findings?

b. Prepare a normal probability plot of the residuals. Also obtain the coefficient of correlation between the ordered residuals and their expected values under normality. Does the normality assumption appear to be reasonable here?

c. The observations for each treatment were obtained in the order shown. Prepare residual sequence plots and analyze them. What are your findings?

21.24. Refer to **Kidney failure hospitalization** Problem 20.15.

a. Prepare residual frequency plots for the treatments. What departures from ANOVA model (20.27) can be studied from these plots? What are your findings?

b. Prepare a normal probability plot of the residuals. Also obtain the coefficient of correlation between the ordered residuals and their expected values under normality. Does the normality assumption appear to be reasonable here?

21.25. Refer to **Programmer requirements** Problem 20.16.

a. Prepare residual frequency plots for the treatments. What departures from ANOVA model (20.27) can be studied from these plots? What are your findings?

b. Prepare a normal probability plot of the residuals. Also obtain the coefficient of correlation between the ordered residuals and their expected values under normality. Does the normality assumption appear to be reasonable here?

EXERCISES

21.26. Show that the point estimator (21.11) is unbiased. Find the variance of this estimator.

21.27. Find the variance of the estimator (21.28).

21.28. Consider a two-factor study with $a = 2$ and $b = 2$. Use (20.14b) to develop a contrast for each of the interactions $(\alpha\beta)_{12}$ and $(\alpha\beta)_{21}$. Use (6.22) to show that the two contrasts are linearly dependent.

PROJECTS

21.29. Refer to the **SENIC** data set and Project 20.32. Analyze the effect of region on mean length of hospital stay by making all pairwise comparisons between regions; use the Tukey procedure and a 90 percent family confidence coefficient. State your findings and present a graphic summary.

21.30. Refer to the **SMSA** data set and Project 20.33. Analyze the effect of region on the crime rate by making all pairwise comparisons between regions; use the Tukey procedure and a 95 percent family confidence coefficient. State your findings and present a graphic summary.

CITED REFERENCES

21.1 Cox, D. R. *Planning of Experiments*. New York: John Wiley & Sons, 1958.

21.2 Cochran, William G., and Gertrude M. Cox. *Experimental Designs*. 2d ed. New York: John Wiley & Sons, 1957.

21.3 Box, George E. P.; William G. Hunter; and J. Stuart Hunter. *Statistics for Experimenters*. New York: John Wiley & Sons, 1978.

22

Unequal sample sizes in two-factor studies

Up to this point in our discussion of two-factor studies we have assumed the two-factor ANOVA model (20.27), where all treatment means are considered to be of equal importance, and have restricted ourselves to the case of equal treatment sample sizes. In this chapter, we take up procedures for handling cases where the treatment sample sizes are not equal. We also consider how to analyze factor effects in two-factor studies where the treatment means have unequal importance.

22.1 UNEQUAL SAMPLE SIZES

Unequal sample sizes for the treatments are common with observational data. For instance, a market research analyst wishes to study the effect of temperature and precipitation on sales of a product from data for the 30 largest metropolitan areas in the United States. In this type of uncontrolled situation, it is unlikely that each temperature-precipitation category will contain the same number of cities.

One may also encounter unequal treatment sample sizes in experimental studies. For instance, an experimenter may seek to have the same number of cases for each treatment, but for a variety of reasons (e.g., illness of subject, incomplete records, technical problems) ends up with unequal sample sizes. In addition, some designed experiments require variable precision among treatment comparisons, and thus unequal sample sizes are specified by design.

Throughout our discussion of unequal sample sizes in Sections 22.2 and 22.3, we assume that there is at least one observation for each treatment and that all treatment means are of equal importance. We relax the restriction of at least one observation for each treatment in Section 22.4 and the restriction of equal treatment importance in Section 22.5.

Notation

Our notation remains the same as before, except that the sample size for the treatment consisting of the ith level of factor A and the jth level of factor B will be denoted by n_{ij}. The total number of observations for the ith level of A is:

$$n_{i.} = \sum_j n_{ij} \tag{22.1a}$$

for the jth level of B is:

$$n_{.j} = \sum_i n_{ij} \tag{22.1b}$$

and for the total study is:

$$n_T = \sum_i \sum_j n_{ij} \tag{22.1c}$$

22.2 USE OF REGRESSION APPROACH FOR TESTING FACTOR EFFECTS WHEN SAMPLE SIZES UNEQUAL

When the treatment sample sizes are unequal, the analysis of variance for two-factor studies becomes more complex. The least squares equations are no longer of a simple structure, yielding direct and easy solutions. Hence, the regular analysis of variance formulas in (20.42) and (20.43) are inappropriate. Furthermore, the factor effect component sums of squares are no longer orthogonal—that is, they do not sum to $SSTR$.

An easy way to obtain the proper sums of squares for testing factor interactions and main effects is through the regression approach described in Section 20.8. The only difference when sample sizes are unequal is that a reduced model needs to be fitted for each test of factor interactions and main effects. Since no new principles are involved, we shall turn directly to an example to illustrate how ANOVA tests are conducted by means of the regression approach when the treatment sample sizes are unequal and the treatment means are of equal importance.

Example

Human growth hormone was administered at a clinical research center to growth hormone deficient, short children who had not yet reached puberty. The investigator was interested in the effects of the child's sex (factor A) and bone development (factor B) on the rate of growth induced by the hormone administration. A child's bone development was classified into one of three categories—severely depressed, moderately depressed, mildly depressed. Three children were randomly selected for each sex-bone development group. The dependent variable (Y) utilized was the difference between the growth rate during growth hormone treatment and the normal growth rate prior to the treatment, expressed in centimeters per month. Four of the 18 children were unable to complete the year-long study, thus creating unequal treatment sample sizes.

Table 22.1 presents the study data. Plots of the treatment sample means are shown in Figure 22.1. The plots in Figure 22.1 clearly suggest that bone devel-

TABLE 22.1 Sample data and notation for growth hormone example (growth rate difference in centimeters per month)

Sex (factor A) i	*Bone Development (factor B)* j		
	Severely Depressed (B_1)	*Moderately Depressed* (B_2)	*Mildly Depressed* (B_3)
Male (A_1)	1.4 (Y_{111}) 2.4 (Y_{112}) 2.2 (Y_{113})	2.1 (Y_{121}) 1.7 (Y_{122})	.7 (Y_{131}) 1.1 (Y_{132})
Mean	2.0 ($\bar{Y}_{11.}$)	1.9 ($\bar{Y}_{12.}$)	.9 ($\bar{Y}_{13.}$)
Female (A_2)	2.4 (Y_{211})	2.5 (Y_{221}) 1.8 (Y_{222}) 2.0 (Y_{223})	.5 (Y_{231}) .9 (Y_{232}) 1.3 (Y_{233})
Mean	2.4 ($\bar{Y}_{21.}$)	2.1 ($\bar{Y}_{22.}$)	.9 ($\bar{Y}_{23.}$)

FIGURE 22.1 Plots of treatment sample means—growth hormone example

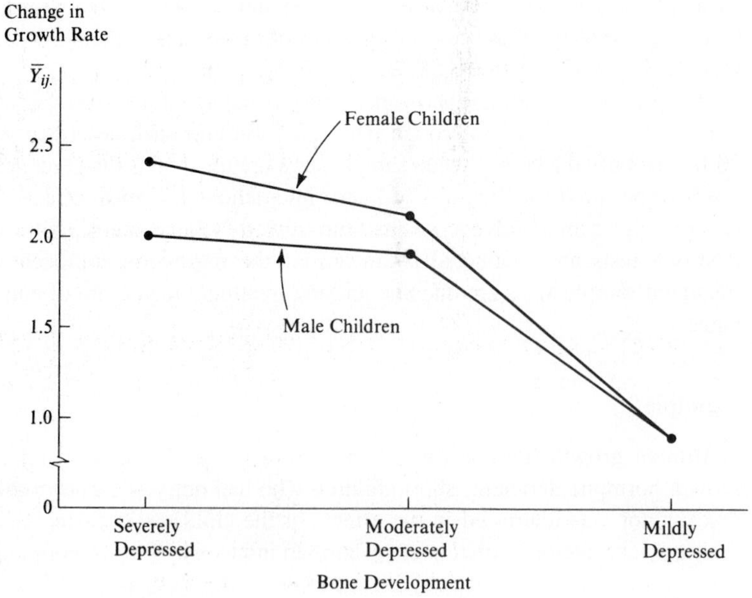

opment has a major impact on the change in growth rate. The plots also raise the questions whether some interaction effects are present and whether the sex of a child affects the growth rate, or whether random variations alone can account for the behavior of the plots in these regards.

To test formally whether or not these factor effects are present, we utilize the regression approach because of the unequal sample sizes.

Development of regression model. The two-factor ANOVA model (20.27) here is:

$$Y_{ijk} = \mu_{..} + \alpha_i + \beta_j + (\alpha\beta)_{ij} + \varepsilon_{ijk} \qquad i = 1, 2; j = 1, 2, 3 \tag{22.2}$$

To express this model in regression terms, we utilize indicator variables that take on the values 1, -1, or 0, as explained in Section 20.8. Specifically, we shall need $a - 1 = 2 - 1 = 1$ indicator variable for the factor A main effects and $b - 1 = 3 - 1 = 2$ indicator variables for the factor B main effects. The interaction terms will correspond to the cross-products of the indicator variables for factor A and factor B main effects. Specifically, the regression model equivalent to ANOVA model (22.2) is:

$$Y_{ijk} = \mu_{..} + \underbrace{\alpha_1 X_{ijk1}}_{\substack{A\text{ main}\\ \text{effect}}} + \underbrace{\beta_1 X_{ijk2} + \beta_2 X_{ijk3}}_{\substack{B\text{ main}\\ \text{effect}}} \tag{22.3}$$

$$+ \underbrace{(\alpha\beta)_{11} X_{ijk1} X_{ijk2} + (\alpha\beta)_{12} X_{ijk1} X_{ijk3}}_{AB\text{ interaction effect}} + \varepsilon_{ijk} \qquad \text{Full model}$$

where:

$$X_{ijk1} = \begin{cases} 1 & \text{if observation from level 1 for factor } A \\ -1 & \text{if observation from level 2 for factor } A \end{cases}$$

$$X_{ijk2} = \begin{cases} 1 & \text{if observation from level 1 for factor } B \\ -1 & \text{if observation from level 3 for factor } B \\ 0 & \text{otherwise} \end{cases}$$

$$X_{ijk3} = \begin{cases} 1 & \text{if observation from level 2 for factor } B \\ -1 & \text{if observation from level 3 for factor } B \\ 0 & \text{otherwise} \end{cases}$$

The regression parameters in (22.3) are the ANOVA model parameters:

$$\begin{aligned} &\mu_{..} \\ \alpha_1 &= \mu_{1.} - \mu_{..} \\ \beta_1 &= \mu_{.1} - \mu_{..} \\ \beta_2 &= \mu_{.2} - \mu_{..} \\ (\alpha\beta)_{11} &= \mu_{11} - \mu_{1.} - \mu_{.1} + \mu_{..} \\ (\alpha\beta)_{12} &= \mu_{12} - \mu_{1.} - \mu_{.2} + \mu_{..} \end{aligned} \tag{22.4}$$

The remaining ANOVA model parameters are not required in the regression model because of the constraints in (20.27). Thus, for instance:

$$\begin{aligned}\alpha_2 &= -\alpha_1\\ \beta_3 &= -\beta_1 - \beta_2\\ (\alpha\beta)_{13} &= -(\alpha\beta)_{11} - (\alpha\beta)_{12}\\ (\alpha\beta)_{21} &= -(\alpha\beta)_{11}\end{aligned} \tag{22.5}$$

Table 22.2 presents the **Y** and **X** data matrices for the growth hormone study regression model (22.3). Table 22.3a presents the fitted regression function and regression ANOVA table when the full regression model (22.3) was fitted to the data. Note that the fitted values for the full model are the treatment sample means $\bar{Y}_{ij.}$, just as when all treatment sample sizes are equal. For instance, we have for the observations from treatment $i = 1, j = 1$ for which $X_1 = X_2 = 1$ and $X_3 = 0$:

$$\begin{aligned}\hat{Y}_{111} = \hat{Y}_{112} = \hat{Y}_{113} &= 1.7 - .1(1) + .5(1) + .3(0) - .1(1)(1) - 0(1)(0)\\ &= 2.0 = \bar{Y}_{11.}\end{aligned}$$

and for the single observation from treatment $i = 2, j = 1$ for which $X_1 = -1$, $X_2 = 1$, and $X_3 = 0$:

$$\begin{aligned}\hat{Y}_{211} &= 1.7 - .1(-1) + .5(1) + .3(0) - .1(-1)(1) - 0(-1)(0)\\ &= 2.4 = \bar{Y}_{21.}\end{aligned}$$

TABLE 22.2 **Y** and **X** data matrices for growth hormone study regression model (22.3)

$$\mathbf{Y} = \begin{bmatrix} Y_{111}\\ Y_{112}\\ Y_{113}\\ Y_{121}\\ Y_{122}\\ Y_{131}\\ Y_{132}\\ Y_{211}\\ Y_{221}\\ Y_{222}\\ Y_{223}\\ Y_{231}\\ Y_{232}\\ Y_{233}\end{bmatrix} = \begin{bmatrix} 1.4\\ 2.4\\ 2.2\\ 2.1\\ 1.7\\ .7\\ 1.1\\ 2.4\\ 2.5\\ 1.8\\ 2.0\\ .5\\ .9\\ 1.3\end{bmatrix} \qquad \mathbf{X} = \begin{bmatrix} & X_1 & X_2 & X_3 & X_1X_2 & X_1X_3\\ 1 & 1 & 1 & 0 & 1 & 0\\ 1 & 1 & 1 & 0 & 1 & 0\\ 1 & 1 & 1 & 0 & 1 & 0\\ 1 & 1 & 0 & 1 & 0 & 1\\ 1 & 1 & 0 & 1 & 0 & 1\\ 1 & 1 & -1 & -1 & -1 & -1\\ 1 & 1 & -1 & -1 & -1 & -1\\ 1 & -1 & 1 & 0 & -1 & 0\\ 1 & -1 & 0 & 1 & 0 & -1\\ 1 & -1 & 0 & 1 & 0 & -1\\ 1 & -1 & 0 & 1 & 0 & -1\\ 1 & -1 & -1 & -1 & 1 & 1\\ 1 & -1 & -1 & -1 & 1 & 1\\ 1 & -1 & -1 & -1 & 1 & 1\end{bmatrix}$$

Test for interaction effects. To test whether or not interaction effects are present, the ANOVA model alternatives:

$$\begin{aligned}&H_0\text{: all } (\alpha\beta)_{ij} = 0\\ &H_a\text{: not all } (\alpha\beta)_{ij} \text{ equal zero}\end{aligned} \tag{22.6}$$

TABLE 22.3 Regression results for growth hormone example

(a) Fit of Full Model (22.3)

Source of Variation	*SS*	*df*	
			$\hat{Y} = 1.7 - .1X_1 + .5X_2 + .3X_3 - .1X_1X_2 - 0.0X_1X_3$
Regression	4.4743	5	
Error	1.3000	8	
Total	5.7743	13	

(b) Fit of Reduced Model (22.7)

Source of Variation	*SS*	*df*	
			$\hat{Y} = 1.68 - .0857X_1 + .467X_2 + .327X_3$
Regression	4.3989	3	
Error	1.3754	10	
Total	5.7743	13	$SSE(R) - SSE(F) = 1.3754 - 1.3000 = .0754$

(c) Fit of Reduced Model (22.9)

Source of Variation	*SS*	*df*	
			$\hat{Y} = 1.69 + .444X_2 + .328X_3 - .0667X_1X_2 - .0167X_1X_3$
Regression	4.3543	4	
Error	1.4200	9	
Total	5.7743	13	$SSE(R) - SSE(F) = 1.4200 - 1.3000 = .1200$

(d) Fit of Reduced Model (22.10)

Source of Variation	*SS*	*df*	
			$\hat{Y} = 1.63 + .0190X_1 + .0667X_1X_2 - .193X_1X_3$
Regression	.2846	3	
Error	5.4897	10	
Total	5.7743	13	$SSE(R) - SSE(F) = 5.4897 - 1.3000 = 4.1897$

become for regression model (22.3):

(22.6a)
$$H_0\text{: } (\alpha\beta)_{11} = (\alpha\beta)_{12} = 0$$
$$H_a\text{: not both } (\alpha\beta)_{11} \text{ and } (\alpha\beta)_{12} \text{ equal zero}$$

Thus, we are simply testing whether or not two regression coefficients equal zero. The reduced regression model therefore is:

(22.7) $$Y_{ijk} = \mu_{..} + \alpha_1 X_{ijk1} + \beta_1 X_{ijk2} + \beta_2 X_{ijk3} + \varepsilon_{ijk} \qquad \text{Reduced model}$$

When this reduced model was fitted, the results presented in Table 22.3b were obtained. The general linear test statistic (3.68) therefore is:

$$F^* = \frac{SSE(R) - SSE(F)}{df_R - df_F} \div \frac{SSE(F)}{df_F}$$

$$= \frac{1.3754 - 1.3000}{10 - 8} \div \frac{1.3000}{8} = .23$$

To control the risk of making a Type I error at $\alpha = .05$, we require $F(.95; 2, 8) = 4.46$. Since $F^* = .23 \leq 4.46$, we conclude H_0, that no interaction effects are present.

Tests for factor main effects. We now proceed to test whether or not factor A and factor B main effects are present. The ANOVA model alternatives:

$$\text{(22.8)} \quad \begin{array}{ll} H_0: \alpha_1 = \alpha_2 = 0 & H_0: \beta_1 = \beta_2 = \beta_3 = 0 \\ H_a: \text{not both } \alpha_i \text{ equal zero} & H_a: \text{not all } \beta_j \text{ equal zero} \end{array}$$

become for regression model (22.3):

$$\text{(22.8a)} \quad \begin{array}{ll} H_0: \alpha_1 = 0 & H_0: \beta_1 = \beta_2 = 0 \\ H_a: \alpha_1 \neq 0 & H_a: \text{not both } \beta_j \text{ equal zero} \end{array}$$

The reduced regression models for testing for factor A main effects and factor B main effects therefore are:

Test for factor A main effects

$$\text{(22.9)} \quad Y_{ijk} = \mu_{..} + \beta_1 X_{ijk2} + \beta_2 X_{ijk3} + (\alpha\beta)_{11} X_{ijk1} X_{ijk2} + (\alpha\beta)_{12} X_{ijk1} X_{ijk3} + \varepsilon_{ijk} \qquad \text{Reduced model}$$

Test for factor B main effects

$$\text{(22.10)} \quad Y_{ijk} = \mu_{..} + \alpha_1 X_{ijk1} + (\alpha\beta)_{11} X_{ijk1} X_{ijk2} + (\alpha\beta)_{12} X_{ijk1} X_{ijk3} + \varepsilon_{ijk} \qquad \text{Reduced model}$$

Tables 22.3c and 22.3d present the results of fitting these reduced models. The two test statistics therefore are:

$$F_1^* = \frac{1.4200 - 1.3000}{9 - 8} \div \frac{1.3000}{8} = .74$$

$$F_2^* = \frac{5.4897 - 1.3000}{10 - 8} \div \frac{1.3000}{8} = 12.89$$

For $\alpha = .05$, we require $F(.95; 1, 8) = 5.32$ and $F(.95; 2, 8) = 4.46$ for the two tests. Since $F_1^* = .74 \leq 5.32$ and $F_2^* = 12.89 > 4.46$, we conclude that there are no factor A main effects but that factor B main effects are present.

Thus, these tests support the evident effect of bone development on the change in growth rate during growth hormone treatment noted earlier from Figure 22.1 and also indicate that the sex and interaction variations in this figure can be accounted for as random behavior. The family level of significance for the set of three tests just conducted, according to the Kimball inequality (20.58), is $1 - (.95)^3 = .14$.

At this point, it would clearly be desirable to conduct further analyses of the nature of the bone development effects. We shall discuss such analyses in the next section.

Table 22.4 contains a consolidated ANOVA table presenting the results from fitting the four regression models in Table 22.3. The sums of squares for the factor effects in each case are the differences between the error sums of squares for the reduced and full models, and the associated degrees of freedom are the differences between the respective degrees of freedom for these error sums of squares. Note that a total sum of squares is not shown in Table 22.4 because the sums of squares for the three types of factor effects and the error sum of squares do not add to *SSTO* when the treatment sample sizes are unequal.

TABLE 22.4 ANOVA table for growth hormone example

Source of Variation	*SS*	*df*	*MS*	F^*
Sex (A)	.1200	1	.1200	.74
Bone development (B)	4.1897	2	2.0949	12.89
Interactions (AB)	.0754	2	.0377	.23
Error	1.3000	8	.1625	

Comments

1. Extreme care must be exercised when using packaged analysis of variance programs with unequal sample sizes because the default option of the package may not necessarily assign equal importance to each treatment mean. The user should read the package documentation carefully and make sure that the package generates the appropriate sums of squares for the tests of interest.

2. If the sample sizes n_{ij} do not differ too much (some statisticians say not more than by the ratio 2 to 1, with most n_{ij} agreeing more closely) and if no n_{ij} is zero, an approximate analysis of variance, called the *method of unweighted means*, may be utilized. This approximate procedure is also used sometimes when the n_{ij} do differ substantially but only a quick first approximation to the factor effects is desired.

The procedure is simple. An analysis of variance is conducted using the $\bar{Y}_{ij.}$ as if they were single observations for each treatment. *SSA*, *SSB*, and *SSAB* are thus calculated in the usual way, except that each treatment has only one "observation" $\bar{Y}_{ij.}$. We know that the "observation" $\bar{Y}_{ij.}$ has variance σ^2/n_{ij}. Hence, the average variance of the "observations" $\bar{Y}_{ij.}$ is:

$$\frac{\sum_i \sum_j \frac{\sigma^2}{n_{ij}}}{ab} = \frac{\sigma^2}{ab} \sum_i \sum_j \frac{1}{n_{ij}} \tag{22.11}$$

The variance σ^2 is estimated by *MSE*, whether frequencies are equal or unequal:

$$MSE = \frac{\sum_i \sum_j \sum_k (Y_{ijk} - \bar{Y}_{ij.})^2}{n_T - ab} \tag{22.12}$$

Hence, the estimated average variance of the ''observations'' is:

$$\frac{MSE}{ab}\sum_i\sum_j\frac{1}{n_{ij}} \tag{22.13}$$

This estimate is then used for the error variance in the analysis of variance of the ''observations'' $\bar{Y}_{ij.}$.

Another method of approximation has been developed by Federer and Zelen (Ref. 22.1). It is more exact, though somewhat more complicated, than the method of unweighted means.

22.3 ESTIMATION OF FACTOR EFFECTS WHEN SAMPLE SIZES UNEQUAL

No new problems arise in estimating factor effects when the treatment sample sizes are unequal. The nature of the analysis, as for the case of equal sample sizes, depends on whether or not strong interactions are present. When no important interactions are present, the analysis generally is concerned with the factor level means $\mu_{i.}$ and $\mu_{.j}$. On the other hand, when important interactions are present, the analysis usually focuses on the treatment means μ_{ij}.

The estimators and estimated variances presented in Chapter 21 for equal sample sizes must, of course, be modified when the treatment sample sizes are unequal. For instance, if interest is in estimating the factor level means $\mu_{i.}$ when all treatment means are of equal importance, then $\mu_{i.}$ is defined:

$$\mu_{i.}=\frac{\sum_j \mu_{ij}}{b}$$

and the appropriate estimator is simply the unweighted average of the treatment sample means $\bar{Y}_{ij.}$:

$$\hat{\mu}_{i.}=\frac{\sum_j \bar{Y}_{ij.}}{b}$$

Since the $\bar{Y}_{ij.}$ are independent, the variance of this estimator is:

$$\sigma^2(\hat{\mu}_{i.})=\frac{1}{b^2}\sum_j\sigma^2(\bar{Y}_{ij.})=\frac{1}{b^2}\sum_j\frac{\sigma^2}{n_{ij}}=\frac{\sigma^2}{b^2}\sum_j\frac{1}{n_{ij}}$$

and the estimated variance is:

$$s^2(\hat{\mu}_{i.})=\frac{MSE}{b^2}\sum_j\frac{1}{n_{ij}}$$

Table 22.5 (pp. 756–57) presents the formulas for the point estimator and estimated variance when estimating factor level means, pairwise comparisons of factor level means, and contrasts or linear combinations of factor level means with unequal sample sizes. The corresponding formulas for treatment means, pairwise comparisons of treatment means, and contrasts or linear combinations of treatment means are also presented in this table.

All multiple comparison procedures that are applicable for the equal sample size case are appropriate when the treatment sample sizes are unequal. The Tukey pairwise comparison procedure now is conservative. The degrees of freedom associated with *MSE* are $n_T - ab$, as before. For equal sample sizes, recall that $n_T = nab$; hence, $n_T - ab = (n - 1)ab$ then. Table 22.5 also presents the appropriate simultaneous comparison multiples for making inferences about factor level means or treatment means.

Since no new issues are involved in estimating factor effects when the sample sizes are unequal, we proceed directly to two examples.

Example 1—Pairwise comparisons of factor level means

We continue with our growth hormone example. We found earlier that a child's sex and bone development do not interact in their effects on the change in the growth rate when growth hormone is administered. We further found no main sex (factor A) effects, but concluded that a child's bone development (factor B) does affect the change in growth rate. We shall now analyze the nature of the bone development effects by means of pairwise comparisons among the three bone development groups. The Tukey multiple comparison method is to be used. This method is conservative when sample sizes are unequal and use of the Bonferroni method would lead to wider confidence intervals here. The family confidence coefficient has been specified to be .90.

We use formulas (22.15) for the point estimates and estimated variances. The treatment sample means are given in Table 22.1, and *MSE* is found in Table 22.4. For the pairwise comparisons of the bone development factor level means ($j = 1$: severely depressed; $j = 2$: moderately depressed; $j = 3$: mildly depressed), we obtain:

$$\hat{\mu}_{.1} = \frac{\bar{Y}_{11.} + \bar{Y}_{21.}}{2} = \frac{2.0 + 2.4}{2} = 2.2$$

$$\hat{\mu}_{.2} = \frac{\bar{Y}_{12.} + \bar{Y}_{22.}}{2} = \frac{1.9 + 2.1}{2} = 2.0$$

$$\hat{\mu}_{.3} = \frac{\bar{Y}_{13.} + \bar{Y}_{23.}}{2} = \frac{.9 + .9}{2} = .9$$

$$\hat{D}_1 = \hat{\mu}_{.1} - \hat{\mu}_{.2} = 2.2 - 2.0 = .2$$

$$\hat{D}_2 = \hat{\mu}_{.1} - \hat{\mu}_{.3} = 2.2 - .9 = 1.3$$

$$\hat{D}_3 = \hat{\mu}_{.2} - \hat{\mu}_{.3} = 2.0 - .9 = 1.1$$

$$s^2(\hat{D}_1) = \frac{.1625}{(2)^2}\left(\frac{1}{3} + \frac{1}{2} + \frac{1}{1} + \frac{1}{3}\right) = .0880 \qquad s(\hat{D}_1) = .297$$

TABLE 22.5 Point estimators and estimated variances for two-factor analyses when sample sizes are unequal

(a) Factor Level Mean

$$\mu_{i.} = \frac{\sum_j \mu_{ij}}{b} \qquad \mu_{.j} = \frac{\sum_i \mu_{ij}}{a}$$

(22.14) $$\hat{\mu}_{i.} = \frac{\sum_j \bar{Y}_{ij.}}{b} \qquad \hat{\mu}_{.j} = \frac{\sum_i \bar{Y}_{ij.}}{a}$$

$$s^2(\hat{\mu}_{i.}) = \frac{MSE}{b^2}\sum_j \frac{1}{n_{ij}} \qquad s^2(\hat{\mu}_{.j}) = \frac{MSE}{a^2}\sum_i \frac{1}{n_{ij}}$$

(b) Pairwise Comparison of Factor Level Means

$$D = \mu_{i.} - \mu_{i'.} \qquad D = \mu_{.j} - \mu_{.j'}$$

(22.15) $$\hat{D} = \hat{\mu}_{i.} - \hat{\mu}_{i'.} \qquad \hat{D} = \hat{\mu}_{.j} - \hat{\mu}_{.j'}$$

$$s^2(\hat{D}) = \frac{MSE}{b^2}\sum_j\left(\frac{1}{n_{ij}} + \frac{1}{n_{i'j}}\right) \qquad s^2(\hat{D}) = \frac{MSE}{a^2}\sum_i\left(\frac{1}{n_{ij}} + \frac{1}{n_{ij'}}\right)$$

(c) Contrast or Linear Combination of Factor Level Means

$$L = \sum_i c_i\mu_{i.} \qquad L = \sum_j c_j\mu_{.j}$$

(22.16) $$\hat{L} = \sum_i c_i\hat{\mu}_{i.} \qquad \hat{L} = \sum_j c_j\hat{\mu}_{.j}$$

$$s^2(\hat{L}) = \frac{MSE}{b^2}\sum_i c_i^2 \sum_j \frac{1}{n_{ij}} \qquad s^2(\hat{L}) = \frac{MSE}{a^2}\sum_j c_j^2 \sum_i \frac{1}{n_{ij}}$$

(d) Confidence Interval Multiple

Single Estimate

$$t(1 - \alpha/2; n_T - ab) \qquad t(1 - \alpha/2; n_T - ab)$$

Multiple Comparisons

(22.17) $$B = t(1 - \alpha/2g; n_T - ab) \qquad B = t(1 - \alpha/2g; n_T - ab)$$

$$T = \frac{1}{\sqrt{2}}q(1 - \alpha; a, n_T - ab) \qquad T = \frac{1}{\sqrt{2}}q(1 - \alpha; b, n_T - ab)$$

$$S^2 = (a - 1)F(1 - \alpha; a - 1, n_T - ab) \qquad S^2 = (b - 1)F(1 - \alpha; b - 1, n_T - ab)$$

$$s^2(\hat{D}_2) = \frac{.1625}{(2)^2}\left(\frac{1}{3} + \frac{1}{2} + \frac{1}{1} + \frac{1}{3}\right) = .0880 \qquad s(\hat{D}_2) = .297$$

$$s^2(\hat{D}_3) = \frac{.1625}{(2)^2}\left(\frac{1}{2} + \frac{1}{2} + \frac{1}{3} + \frac{1}{3}\right) = .0677 \qquad s(\hat{D}_3) = .260$$

For a 90 percent family confidence coefficient, we require:

$$T = \frac{1}{\sqrt{2}}q(.90; 3, 8) = \frac{1}{\sqrt{2}}(3.37) = 2.38$$

(e) Treatment Mean

$$\mu_{ij}$$

(22.18)
$$\hat{\mu}_{ij} = \bar{Y}_{ij.}$$

$$s^2(\hat{\mu}_{ij}) = \frac{MSE}{n_{ij}}$$

(f) Pairwise Comparison of Treatment Means

$$D = \mu_{ij} - \mu_{i'j'}$$

(22.19)
$$\hat{D} = \bar{Y}_{ij.} - \bar{Y}_{i'j'.}$$

$$s^2(\hat{D}) = MSE\left(\frac{1}{n_{ij}} + \frac{1}{n_{i'j'}}\right)$$

(g) Contrast or Linear Combination of Treatment Means

$$L = \Sigma\Sigma c_{ij}\mu_{ij}$$

(22.20)
$$\hat{L} = \Sigma\Sigma c_{ij}\bar{Y}_{ij.}$$

$$s^2(\hat{L}) = MSE\Sigma\Sigma\frac{c_{ij}^2}{n_{ij}}$$

(h) Confidence Interval Multiple

Single Estimate

$$t(1 - \alpha/2; n_T - ab)$$

Multiple Comparisons

(22.21)
$$B = t(1 - \alpha/2g; n_T - ab)$$

$$T = \frac{1}{\sqrt{2}}q(1 - \alpha; ab, n_T - ab)$$

$$S^2 = (ab - 1)F(1 - \alpha; ab - 1, n_T - ab)$$

Hence, we obtain the following confidence intervals:

$$-.51 = .2 - 2.38(.297) \le \mu_{.1} - \mu_{.2} \le .2 + 2.38(.297) = .91$$
$$.59 = 1.3 - 2.38(.297) \le \mu_{.1} - \mu_{.3} \le 1.3 + 2.38(.297) = 2.01$$
$$.48 = 1.1 - 2.38(.260) \le \mu_{.2} - \mu_{.3} \le 1.1 + 2.38(.260) = 1.72$$

We conclude from these confidence intervals with 90 percent family confidence coefficient that growth hormone deficient short children with mildly depressed bone development on the average have a substantially smaller increase in the growth rate than children with either moderately depressed or severely depressed bone development. Further, the latter two groups of children do not show significantly different mean changes in the growth rate. We summarize these findings in the following plot of the factor level sample means:

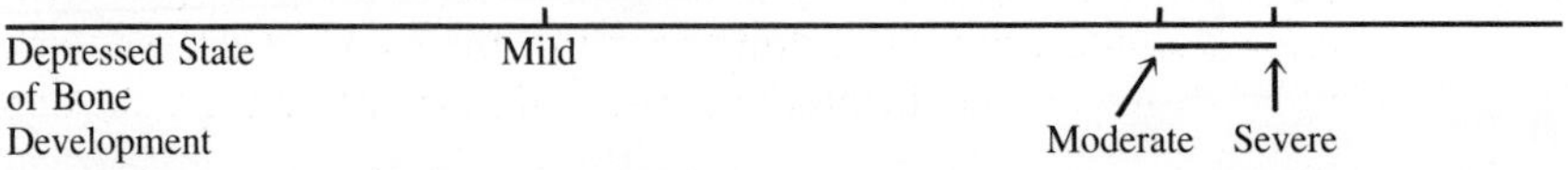

Example 2—Single degree of freedom test

In our growth hormone example, a researcher wanted to know whether children with only mildly depressed bone development obtain, on the average, any increase in the growth rate with administration of growth hormone. Thus, the alternatives to be considered are those for a one-sided test:

$$H_0: \mu_{.3} \leq 0$$
$$H_a: \mu_{.3} > 0$$

The level of significance is to be controlled at $\alpha = .05$.

The test statistic to be employed is:

$$t^* = \frac{\hat{\mu}_{.3} - 0}{s(\hat{\mu}_{.3})}$$

We found earlier that $\hat{\mu}_{.3} = .9$ and $MSE = .1625$. Hence, using (22.14) we obtain:

$$s^2(\hat{\mu}_{.3}) = \frac{.1625}{(2)^2}\left(\frac{1}{2} + \frac{1}{3}\right) = .0339 \qquad s(\hat{\mu}_{.3}) = .184$$

Hence, the test statistic is:

$$t^* = \frac{.9 - 0}{.184} = 4.89$$

For $\alpha = .05$ we require $t(.95; 8) = 1.860$. Therefore the one-sided decision rule is:

$$\text{If } t^* \leq 1.860, \text{ conclude } H_0$$
$$\text{If } t^* > 1.860, \text{ conclude } H_a$$

Since $t^* = 4.89 > 1.860$, we conclude H_a, that the mean change in the growth rate for children with mildly depressed bone development is greater than zero.

22.4 MISSING CELLS IN TWO-FACTOR STUDIES

Occasionally one finds after a two-factor study has been completed that there are no observations in one or more treatment cells. Not only are the treatment sample sizes unequal then, but there is no sample information about the treatment means for the empty cells. Consider again Table 22.1 for the growth hormone study. Note that two female children with severely depressed bone condition dropped out of the study before its completion so that only one observation ($n_{21} = 1$) is present for that treatment. We can easily imagine that all three of these children could have dropped out of the study. Then we would have had $n_{21} = 0$, and no sample information would be available about the treatment mean μ_{21}.

Partial analysis of factor effects

When one or several treatment cells are empty, the usual analysis of variance for unequal sample sizes by means of the regression approach, as explained earlier, cannot be conducted. This does not mean, however, that the entire two-factor study has become useless. Usually, a variety of analyses can be conducted that will provide at least partial information about the nature of the factor effects. The analyses that can be undertaken depend on the particular cells for which no sample information is available. We shall illustrate by means of an example how partial information can be obtained from two-factor studies with missing cells.

Example. In our growth hormone example, suppose that there were no observations for female children with severely depressed bone development, i.e., $n_{21} = 0$. In that case no sample information would be available about the treatment mean μ_{21}.

Partial information about interactions could still be obtained by restricting attention to children with moderately depressed and mildly depressed bone development. For these children, interactions are present if the differences between the treatment means for the two sexes are not the same for the two bone development groups. The two differences are:

$$\mu_{12} - \mu_{22} \qquad \mu_{13} - \mu_{23}$$

Thus, we would consider the following contrast among the treatment means:

$$L = \mu_{12} - \mu_{22} - \mu_{13} + \mu_{23}$$

We could either estimate L by means of a confidence interval and note whether or not the interval includes zero, or we could conduct a single degree of freedom test to establish whether or not interactions are present. With either approach, we would use MSE based on all sample observations so that the associated degrees of freedom for MSE would be $n_T - (ab - 1) = 13 - 5 = 8$ (remember that $n_{21} = 0$ now).

If the partial analysis of interactions were to suggest that no interactions are present, the effect of sex could be studied by comparing the factor level means excluding children with severely depressed bone development:

$$\mu_{1.} = \frac{\mu_{12} + \mu_{13}}{2} \qquad \mu_{2.} = \frac{\mu_{22} + \mu_{23}}{2}$$

Further, the effect of bone development could be studied for male children by comparing the treatment means μ_{11}, μ_{12}, and μ_{13}, or it could be studied for children of both sexes by excluding those with severely depressed bone development:

$$\mu_{.2} = \frac{\mu_{12} + \mu_{22}}{2} \qquad \mu_{.3} = \frac{\mu_{13} + \mu_{23}}{2}$$

Analysis if model with no interactions can be employed

Occasionally, information is available from previous studies that the two factors in a two-factor study do not interact. In that case, a simpler model than ANOVA model (20.27) can be employed. The two-factor fixed effects no-interaction model is:

(22.22) $$Y_{ijk} = \mu_{..} + \alpha_i + \beta_j + \varepsilon_{ijk} \qquad \text{No-interaction model}$$

We have not discussed this model previously because information about the appropriateness of this model (i.e., whether or not the two factors interact) is usually not available in advance.

However, if the no-interaction model (22.22) is appropriate, the analysis of variance and the analysis of factor main effects can be conducted by means of the regression approach even when one or several cells are empty, as long as information is still provided about each factor main effect.

For example, suppose again that in our growth hormone example the cell for female children with severely depressed bone development is empty but that from past knowledge the researcher is able to assume that there are no interactions between sex and bone development. In that case, regression model (22.3) can be simplified as follows:

$$Y_{ijk} = \mu_{..} + \alpha_1 X_{ijk1} + \beta_1 X_{ijk2} + \beta_2 X_{ijk3} + \varepsilon_{ijk} \qquad \text{Full model}$$

To test for, say, sex main effects, we first fit this full model and obtain $SSE(F)$. The alternatives to be tested are:

$$H_0: \alpha_1 = 0$$
$$H_a: \alpha_1 \neq 0$$

Hence, the reduced model is:

$$Y_{ijk} = \mu_{..} + \beta_1 X_{ijk2} + \beta_2 X_{ijk3} + \varepsilon_{ijk} \qquad \text{Reduced model}$$

We then fit this reduced model, obtain $SSE(R)$, and calculate the general linear test statistic (3.68) in the usual fashion. In this particular case, we could also use the t^* test statistic (8.25) since the test is simply for whether a single regression coefficient equals zero. A test for bone development effects would be carried out similarly.

The reason why the usual analysis of variance by means of the regression approach can be conducted here even though $n_{21} = 0$ is that the assumption of no interactions permits us in effect to estimate μ_{21}. Conceptually, this estimate of μ_{21} requires two steps. First, we need to estimate the treatment means μ_{ij} for the nonempty cells. These estimates are more complicated than simply using the sample treatment means $\bar{Y}_{ij.}$ because we need to utilize the model assumption of no interactions. We obtain these estimates by using the matrix methods described in Section 8.5. (We illustrate how to estimate a treatment mean μ_{ij} for the no-interaction model in Section 23.1.) Once we have estimates of the treatment means μ_{ij} for the nonempty cells, the second step in estimating μ_{21} is to utilize relation (20.13b) for the no-interaction case, whereby we can express μ_{21} in terms of three other treatment means; for instance, we have:

$$\mu_{21} = \mu_{22} + \mu_{11} - \mu_{12}$$

Estimates of μ_{22}, μ_{11}, and μ_{12}, for which sample observations are available, thus enable us to estimate μ_{21} when no interactions are present.

We need to caution that it is not appropriate to use a no-interaction model as the full model when no prior information about the absence of interactions is available. Only partial analyses of factor effects can then be undertaken when some treatment cells are empty, as explained earlier.

Note

Extreme caution should be used with ANOVA computer packages that provide results when some treatment cells are empty. The package may make assumptions about interactions that the researcher is unwilling to make. In the absence of a clear description of how the package handles empty cells, it is preferable that appropriate analyses be conducted without the assistance of the package other than for obtaining the treatment sample means and *MSE*.

22.5 UNEQUAL TREATMENT IMPORTANCE

Up to this point we have assumed that all treatment means are of equal importance, so that the unweighted factor level means $\mu_{i.}$ and $\mu_{.j}$ in (20.2) and (20.1), respectively, are appropriate. There are occasions, however, when the treatment means are of unequal importance. For example, in the growth hormone study discussed earlier, it may be known that twice as many boys as girls undergo growth hormone treatment therapy. If inferences are to be made about a target population with this ratio of male to female children, the analysis needs to weight the treatment means unequally. We consider now how to utilize unequal weights for the treatment means, first for ANOVA tests and then for analysis of factor effects.

Tests for factor effects

The test for interaction effects is not affected by unequal importance of treatment means since this test is concerned with the parallelism, or lack of it, of the treatment mean curves. This was illustrated in Figures 20.1, 20.2, and 20.3.

These treatment mean curves are based solely on the individual treatment means μ_{ij} and hence do not involve weighted averages of the treatment means. Thus, the test for interactions is conducted as explained in Section 20.5 when the sample sizes are equal and as explained in Section 22.2 when the sample sizes are unequal, whether or not the treatment means are of equal importance.

Tests for factor main effects do need to be modified when the treatment means have unequal importance. The general matrix approach explained in Section 8.5 now should be used, in view of the complexity introduced by unequal weightings. It will generally be easiest to work with ANOVA model (20.19) based on the treatment means μ_{ij}. Since no new principles are involved, we shall illustrate the tests for factor main effects by an example.

Example. In the growth hormone example, it is known that twice as many male as female children undergo growth hormone treatment therapy, and that this ratio is the same for children who have severe, moderate, and mild depression in bone development. Inferences are desired about the target population of children undergoing therapy. Specifically, we are to test whether or not the state of bone development affects the change in growth rate. The alternatives therefore are:

$$\text{(22.23)} \quad \begin{aligned} &H_0\colon \frac{2\mu_{11}+\mu_{21}}{3} = \frac{2\mu_{12}+\mu_{22}}{3} = \frac{2\mu_{13}+\mu_{23}}{3} \\ &H_a\colon \text{not all equalities hold} \end{aligned}$$

We shall restate the alternative H_0 in the following equivalent fashion:

$$\text{(22.23a)} \quad H_0\colon \begin{aligned} \frac{2\mu_{11}+\mu_{21}}{3} - \frac{2\mu_{12}+\mu_{22}}{3} = 0 \\ \frac{2\mu_{11}+\mu_{21}}{3} - \frac{2\mu_{13}+\mu_{23}}{3} = 0 \end{aligned}$$

Since H_0 is expressed in terms of the treatment means μ_{ij}, we shall use the two-factor model (20.19):

$$\text{(22.24)} \quad Y_{ijk} = \mu_{ij} + \varepsilon_{ijk}$$

To state this linear model in matrix terms in the form $\mathbf{Y} = \mathbf{X}\boldsymbol{\beta} + \boldsymbol{\varepsilon}$, we define the $\mathbf{X}$ matrix as illustrated in (20.23). Table 22.6 contains the $\mathbf{X}$ matrix and $\boldsymbol{\beta}$ vector for the growth hormone example data in Table 22.1 with no empty cell. The $\mathbf{Y}$ vector was given earlier in Table 22.2. The mean vector $\mathbf{X}\boldsymbol{\beta}$ is also shown in Table 22.6. We see that $E(Y_{ijk}) \equiv \mu_{ij}$, as model (22.24) requires.

The hypothesis H_0 in (22.23a) can now be stated as follows in the matrix form (8.41):

$$\text{(22.25)} \quad H_0\colon \underset{s\times p}{\mathbf{C}}\ \underset{p\times 1}{\boldsymbol{\beta}} = \underset{s\times 1}{\mathbf{h}}$$

where:

$$\underset{2\times 6}{\mathbf{C}} = \begin{bmatrix} \frac{2}{3} & -\frac{2}{3} & 0 & \frac{1}{3} & -\frac{1}{3} & 0 \\ \frac{2}{3} & 0 & -\frac{2}{3} & \frac{1}{3} & 0 & -\frac{1}{3} \end{bmatrix}$$

TABLE 22.6 **X** and $\boldsymbol{\beta}$ matrices for model (22.24)—growth hormone example

$$\mathbf{X} = \begin{bmatrix} 1 & 0 & 0 & 0 & 0 & 0 \\ 1 & 0 & 0 & 0 & 0 & 0 \\ 1 & 0 & 0 & 0 & 0 & 0 \\ 0 & 1 & 0 & 0 & 0 & 0 \\ 0 & 1 & 0 & 0 & 0 & 0 \\ 0 & 0 & 1 & 0 & 0 & 0 \\ 0 & 0 & 1 & 0 & 0 & 0 \\ 0 & 0 & 0 & 1 & 0 & 0 \\ 0 & 0 & 0 & 0 & 1 & 0 \\ 0 & 0 & 0 & 0 & 1 & 0 \\ 0 & 0 & 0 & 0 & 1 & 0 \\ 0 & 0 & 0 & 0 & 0 & 1 \\ 0 & 0 & 0 & 0 & 0 & 1 \\ 0 & 0 & 0 & 0 & 0 & 1 \end{bmatrix} \qquad \boldsymbol{\beta} = \begin{bmatrix} \mu_{11} \\ \mu_{12} \\ \mu_{13} \\ \mu_{21} \\ \mu_{22} \\ \mu_{23} \end{bmatrix} \qquad \mathbf{X}\boldsymbol{\beta} = \begin{bmatrix} \mu_{11} \\ \mu_{11} \\ \mu_{11} \\ \mu_{12} \\ \mu_{12} \\ \mu_{13} \\ \mu_{13} \\ \mu_{21} \\ \mu_{22} \\ \mu_{22} \\ \mu_{22} \\ \mu_{23} \\ \mu_{23} \\ \mu_{23} \end{bmatrix}$$

$$\underset{6\times 1}{\boldsymbol{\beta}} = \begin{bmatrix} \mu_{11} \\ \mu_{12} \\ \mu_{13} \\ \mu_{21} \\ \mu_{22} \\ \mu_{23} \end{bmatrix} \qquad \underset{2\times 1}{\mathbf{h}} = \begin{bmatrix} 0 \\ 0 \end{bmatrix}$$

Note that this formulation yields (22.23a):

$$\underset{2\times 1}{\mathbf{C}\boldsymbol{\beta}} = \begin{bmatrix} (\frac{2}{3})\mu_{11} - (\frac{2}{3})\mu_{12} + (\frac{1}{3})\mu_{21} - (\frac{1}{3})\mu_{22} \\ (\frac{2}{3})\mu_{11} - (\frac{2}{3})\mu_{13} + (\frac{1}{3})\mu_{21} - (\frac{1}{3})\mu_{23} \end{bmatrix} = \begin{bmatrix} 0 \\ 0 \end{bmatrix} = \mathbf{h}$$

To calculate $SSE(R) - SSE(F)$, we utilize (8.45):

$$(\mathbf{C}\mathbf{b}_F - \mathbf{h})'(\mathbf{C}(\mathbf{X}'\mathbf{X})^{-1}\mathbf{C}')^{-1}(\mathbf{C}\mathbf{b}_F - \mathbf{h}) \tag{22.26}$$

The vector of parameter estimates $\mathbf{b}_F$ needs to be obtained first. Since we know from (20.33) that the sample means $\bar{Y}_{ij.}$ are the least squares estimators, we have:

$$\mathbf{b}_F = \begin{bmatrix} \bar{Y}_{11.} \\ \bar{Y}_{12.} \\ \bar{Y}_{13.} \\ \bar{Y}_{21.} \\ \bar{Y}_{22.} \\ \bar{Y}_{23.} \end{bmatrix} = \begin{bmatrix} 2.0 \\ 1.9 \\ .9 \\ 2.4 \\ 2.1 \\ .9 \end{bmatrix}$$

We can now substitute into (22.26) to obtain $SSE(R) - SSE(F)$. We do not show the matrix calculations since they are routine. We find:

$$SSE(R) - SSE(F) = 3.454$$

The error sum of squares for the full model was obtained earlier in Table 22.3a; it is $SSE(F) = 1.3000$. The degrees of freedom associated with SSE for the full model are $df_F = 8$, as shown in Table 22.3a. The degrees of freedom associated with $SSE(R) - SSE(F)$ are $df_R - df_F = s = 2$. Hence, the general linear test statistic (8.46) is:

$$F^* = \frac{SSE(R) - SSE(F)}{s} \div \frac{SSE(F)}{n - p}$$

$$= \frac{3.454}{2} \div \frac{1.3000}{8} = 10.63$$

If H_0 holds, F^* follows the F distribution with 2 and 8 degrees of freedom. To control the level of significance at $\alpha = .05$, we require $F(.95; 2, 8) = 4.46$. Since $F^* = 10.63 > 4.46$, we conclude H_a, that the weighted factor level means for the different bone development groups are not equal.

Comments

1. The test for the alternatives (22.23a) could also have been conducted by estimating the contrasts:

$$L_1 = \frac{2\mu_{11} + \mu_{21}}{3} - \frac{2\mu_{12} + \mu_{22}}{3} \qquad L_2 = \frac{2\mu_{11} + \mu_{21}}{3} - \frac{2\mu_{13} + \mu_{23}}{3}$$

with a multiple comparison procedure (e.g., the Bonferroni procedure) and noting whether or not both confidence intervals include zero.

2. There is one instance when the tests for factor A and factor B main effects can be conducted by means of the ordinary analysis of variance despite unequal weightings of treatment means. This occurs when the sample sizes follow a proportional pattern that reflects the importance of the treatment means. Suppose that a chain of diet establishments is experimenting with two diets that are of equal importance. The establishments cater to three times as many women as men. One hundred men and 300 women are selected, and half of each group is randomly assigned to each diet. Here, the treatment sample sizes are as follows:

Diet	*Men*	*Women*	*Total*
1	50	150	200
2	50	150	200
Total	100	300	400

Note that these treatment sample sizes follow the relation:

$$n_{ij} = \frac{n_{i.}n_{.j}}{n_T} \tag{22.27}$$

Condition (22.27) implies that the sample sizes in any two rows (or columns) are proportional. This is called a case of *proportional frequencies*.

In our example, the frequencies are not only proportional but they are proportional in accord with the importance of the treatment means. Thus, the diet sample sizes are equal for each sex and the sex sample sizes are in the ratio of one to three for each diet. When this is the case, the ANOVA sums of squares can be obtained simply as follows:

(22.28a) $$SSA = b\sum_i n_{i.}(\bar{Y}_{i..} - \bar{Y}_{...})^2$$

(22.28b) $$SSB = a\sum_j n_{.j}(\bar{Y}_{.j.} - \bar{Y}_{...})^2$$

(22.28c) $$SSAB = \sum_i \sum_j n_{ij}(\bar{Y}_{ij.} - \bar{Y}_{...})^2 - SSA - SSB$$

(22.28d) $$SSE = \sum_i \sum_j \sum_k (Y_{ijk} - \bar{Y}_{ij.})^2$$

(22.28e) $$SSTO = \sum_i \sum_j \sum_k (Y_{ijk} - \bar{Y}_{...})^2$$

The means required for these formulas are defined as follows:

(22.29a) $$\bar{Y}_{ij.} = \frac{Y_{ij.}}{n_{ij}}$$

(22.29b) $$\bar{Y}_{i..} = \frac{Y_{i..}}{n_{i.}}$$

(22.29c) $$\bar{Y}_{.j.} = \frac{Y_{.j.}}{n_{.j}}$$

(22.29d) $$\bar{Y}_{...} = \frac{Y_{...}}{n_T}$$

When proportional sample sizes are employed but the proportions do not properly reflect the importance of the treatment means (e.g., when the treatment means are of equal importance), the formulas in (22.28) are not applicable and the general methods explained earlier must be employed.

Estimation of factor effects

No new problems are encountered in making estimates of factor effects when the treatment means have unequal importance. Indeed, we discussed how to make such estimates in Section 17.1 for single-factor studies and in Section 21.3 for two-factor studies. We provide an additional example now.

Example. In the growth hormone example, it is desired to compare the mean change in growth rate for children with severely and moderately depressed bone conditions, recognizing that two thirds of children in each group are male. Thus, we wish to estimate the contrast:

$$L = \frac{2\mu_{11} + \mu_{21}}{3} - \frac{2\mu_{12} + \mu_{22}}{3}$$

Using (22.20) in Table 22.5 and the sample results in Tables 22.1 and 22.4, we obtain:

$$\hat{L} = \frac{2(2.0) + 2.4}{3} - \frac{2(1.9) + 2.1}{3} = .167$$

$$s^2(\hat{L}) = .1625\left[\frac{(2/3)^2}{3} + \frac{(1/3)^2}{1} + \frac{(-2/3)^2}{2} + \frac{(-1/3)^2}{3}\right] = .0843$$

$$s(\hat{L}) = .290$$

For a 95 percent confidence interval, we require $t(.975; 8) = 2.306$. Hence, the confidence limits are $.167 \pm 2.306(.290)$ and the confidence interval is:

$$-.50 \leq L \leq .84$$

PROBLEMS

22.1. A market research intern selected a random sample of 400 communities and classified them according to population size (four levels) and geographic region (five levels) to study the effects of these factors on sales of the company's products. When he found that the treatment sample sizes were unequal, the smallest cell frequency being four, he generated random numbers to reduce the number of communities in each cell to four. He then proceeded to analyze the effects of population size and region on the basis of the 80 communities remaining.

a. Does the method of randomly discarding observations lead to any biases? Explain.

b. Was it wise for the intern to discard 320 observations randomly in order to obtain equal treatment sample sizes?

22.2. A student asked: "If two-factor studies with unequal sample sizes must be analyzed by the regression approach, why bother with the two-factor analysis of variance model at all?" Comment.

22.3. Refer to **Cash offers** Problem 20.11. Assume that observations $Y_{214} = 28$ and $Y_{323} = 20$ are missing because the offer received in each of these cases was a trade-in offer, not a cash offer.

a. State the ANOVA model for this case. Also state the equivalent regression model; use $1, -1, 0$ indicator variables.

b. Present the $\mathbf{X}$ and $\boldsymbol{\beta}$ matrices for the regression model in part (a).

c. Obtain $\mathbf{X}\boldsymbol{\beta}$ and show that the proper treatment means are obtained by your model.

d. What is the reduced model for testing for interaction effects?

e. Test whether or not interaction effects are present by fitting the full and reduced models; use $\alpha = .05$. State the alternatives, decision rule, and conclusion. Is your conclusion the same as that obtained in Problem 20.11e for the full data?

f. State the reduced models for testing for age and sex main effects, respectively. Conduct each of the tests. Use $\alpha = .05$ each time and state the alternatives, decision rule, and conclusion.

g. To study the nature of the age main effects, estimate the following pairwise comparisons:

$$D_1 = \mu_{1.} - \mu_{2.} \qquad D_3 = \mu_{2.} - \mu_{3.}$$
$$D_2 = \mu_{1.} - \mu_{3.}$$

Use the most efficient multiple comparison procedure with a 90 percent family confidence coefficient.

h. In the population of female owners, 30 percent are young, 60 percent are middle-aged, and 10 percent are elderly. Estimate the mean cash offer for this population with a 95 percent confidence interval.

22.4. Refer to **Hay fever relief** Problem 20.13. Assume that observations $Y_{113} = 2.3$, $Y_{221} = 8.9$, and $Y_{224} = 9.0$ are missing because the subjects did not immediately record the time when they began to suffer again from hay fever.

a. State the ANOVA model for this case. Also state the equivalent regression model; use 1, −1, 0 indicator variables.

b. Present the $\mathbf{X}$ and $\boldsymbol{\beta}$ matrices for the regression model in part (a).

c. Obtain $\mathbf{X}\boldsymbol{\beta}$ and show that the proper treatment means are obtained by your model.

d. What is the reduced model for testing for interaction effects?

e. Test whether or not interaction effects are present by fitting the full and reduced models; use $\alpha = .05$. State the alternatives, decision rule, and conclusion. Is your conclusion the same as that obtained in Problem 20.13e for the full data?

f. The nature of the interaction effects is to be studied by means of the following contrasts:

$$L_1 = \frac{\mu_{12} + \mu_{13}}{2} - \mu_{11} \qquad L_4 = L_2 - L_1$$
$$L_2 = \frac{\mu_{22} + \mu_{23}}{2} - \mu_{21} \qquad L_5 = L_3 - L_1$$
$$L_3 = \frac{\mu_{32} + \mu_{33}}{2} - \mu_{31} \qquad L_6 = L_3 - L_2$$

Obtain confidence intervals for these contrasts; use the Scheffé multiple comparison procedure with a 90 percent family confidence coefficient. Interpret your findings.

22.5. Refer to **Kidney failure hospitalization** Problem 20.15. Assume that observations $Y_{124} = 12$, $Y_{216} = 2$, and $Y_{238} = 9$ are missing because the hospitalization records for these patients were not complete. Continue to work with the transformed observations $Y' = \log_{10}(Y + 1)$.

a. State the ANOVA model for this case. Also state the equivalent regression model; use 1, −1, 0 indicator variables.

b. Present the $\mathbf{X}$ and $\boldsymbol{\beta}$ matrices for the regression model in part (a).

c. Obtain $\mathbf{X}\boldsymbol{\beta}$ and show that the proper treatment means are obtained by your model.

d. What is the reduced model for testing for interaction effects?

e. Test whether or not interaction effects are present by fitting the full and reduced models; use $\alpha = .05$. State the alternatives, decision rule, and conclusion. Is your conclusion the same as that obtained in Problem 20.15e for the full data?

f. State the reduced models for testing for treatment duration and weight gain main effects, respectively. Conduct each of the tests. Use $\alpha = .05$ each time and state the alternatives, decision rule, and conclusion.

g. Use the single degree of freedom t^* statistic for testing whether or not the mean number of days hospitalized (in transformed units) for persons with mild weight gains exceeds .5; use $\alpha = .05$. State the alternatives, decision rule, and conclusion.

h. To analyze the nature of the factor main effects, estimate the following pairwise comparisons:

$$D_1 = \mu_{1.} - \mu_{2.} \qquad D_3 = \mu_{.3} - \mu_{.1}$$
$$D_2 = \mu_{.2} - \mu_{.1} \qquad D_4 = \mu_{.3} - \mu_{.2}$$

Use the Bonferroni procedure with a 90 percent family confidence coefficient. State your findings.

22.6. Adjunct professors. A sociologist selected a random sample of 45 adjunct professors who teach in the evening division of a large metropolitan university for a study of special problems associated with teaching in the evening division. The data collected include the amount of payment received by the faculty member for teaching a course during the past semester. The sociologist classified the faculty members by subject matter of course (factor A) and highest degree earned (factor B). The earnings per course (in thousand dollars) follow.

	Factor B (highest degree)		
Factor A (subject matter)	$j = 1$ *Bachelor's*	$j = 2$ *Master's*	$j = 3$ *Doctorate*
$i = 1$ Humanities	1.7 1.9	1.8 2.1	2.5 2.7 2.9 2.5 2.6 2.8 2.7 2.9
$i = 2$ Social sciences	2.5 2.3 2.6 2.4	2.7 2.4 2.6 2.4 2.5	3.5 3.3 3.6 3.4
$i = 3$ Engineering	2.7 2.8	2.9 3.0 2.8 2.7	3.7 3.6 3.7 3.8 3.9
$i = 4$ Management	2.5 2.6	2.3 2.8	3.3 3.4 3.3 3.5 3.6

Assume that ANOVA model (20.27) is appropriate, except now $k = 1, \ldots, n_{ij}$.

a. State the ANOVA model for this case. Also state the equivalent regression model; use $1, -1, 0$ indicator variables.

b. Present the $\mathbf{X}$ and $\boldsymbol{\beta}$ matrices for the regression model in part (a).

c. Obtain $\mathbf{X}\boldsymbol{\beta}$ and show that the proper treatment means are obtained by your model.

d. What is the reduced model for testing for interaction effects?

e. Test whether or not interaction effects are present by fitting the full and reduced models; use $\alpha = .01$. State the alternatives, decision rule, and conclusion.

f. State the reduced models for testing for subject matter and highest degree main effects, respectively. Conduct each of the tests. Use $\alpha = .01$ each time and state the alternatives, decision rule, and conclusion.

g. Make all pairwise comparisons between the subject matter means; use the Tukey procedure with a 95 percent family confidence coefficient. State your findings and present a graphic summary.

h. Make all pairwise comparisons between the highest degree means; use the Tukey procedure with a 95 percent family confidence coefficient. State your findings and present a graphic summary.

22.7. Refer to **Hay fever relief** Problem 20.13. Suppose that the data for the treatment when each of the two active ingredients is at the medium level were lost and immediate analyses of the available data are required; i.e., assume that $n_T = 32$ and $n_{22} = 0$.

a. To study whether or not interaction effects are present, estimate the following comparisons:

$$D_1 = \mu_{13} - \mu_{11} \qquad L_1 = D_1 - D_2$$

$$D_2 = \mu_{23} - \mu_{21} \qquad L_2 = D_1 - D_3$$

$$D_3 = \mu_{33} - \mu_{31}$$

Use the Bonferroni procedure with a 90 percent family confidence coefficient. State your findings.

b. To further explore the nature of possible interaction effects, conduct separate single degree of freedom tests of whether $\mu_{12} = \mu_{13}$ and whether $\mu_{32} = \mu_{33}$. Use $\alpha = .02$ for each test and state the alternatives, decision rule, and conclusion. What is the family level of significance, using the Bonferroni inequality?

22.8. Refer to **Kidney failure hospitalization** Problem 20.15. Suppose that there were no patients who received the dialysis treatment for long duration and had mild weight gains; i.e., assume that $n_T = 50$ and $n_{21} = 0$. Continue to work with the transformed observations $Y' = \log_{10}(Y + 1)$.

Based on related research, the analyst believes it is reasonable to assume that the two factors do not interact and that no-interaction model (22.22) is appropriate.

a. State the equivalent full regression model for this case. Also state the reduced regression models for testing for factor A and factor B main effects. Use $1, -1, 0$ indicator variables in the regression model.

b. Fit the full and reduced models. Test for factor A and factor B main effects; use $\alpha = .05$ for each test. State the alternatives, decision rule, and conclusion for each test.

22.9. Refer to **Programmer requirements** Problem 20.16. Suppose that there were no programmers with experience on both small and large systems who had less than five years' experience; i.e., assume that $n_T = 20$ and $n_{21} = 0$.

a. To study whether or not interaction effects are present, estimate the following comparisons:

$$D_1 = \mu_{12} - \mu_{13} \qquad L_1 = D_1 - D_2$$
$$D_2 = \mu_{22} - \mu_{23}$$

Use the Bonferroni procedure with a 95 percent family confidence coefficient. State your findings.

b. To study further the nature of possible interaction effects, test whether or not μ_{22} exceeds μ_{23}; use $\alpha = .05$. State the alternatives, decision rule, and conclusion.

22.10. Refer to **Adjunct professors** Problem 22.6. Suppose that there were no professors teaching humanities courses who had only a bachelor's degree, so that the study consists of $n_T = 43$ adjunct professors and $n_{11} = 0$. On the basis of previous research, the sociologist believes it is reasonable to assume that the two factors do not interact and that no-interaction model (22.22) is appropriate here.

a. State the equivalent full regression model for this case. Also state the reduced regression models for testing for factor A and factor B main effects. Use 1, −1, 0 indicator variables in the regression model.

b. Fit the full and reduced models and test for factor A and factor B main effects. Use $\alpha = .01$ for each test and state the alternatives, decision rule, and conclusion.

22.11. Refer to **Cash offers** Problem 20.11. It is known that in both populations of male and female owners, 30 percent are young, 60 percent are middle-aged, and 10 percent are elderly. Test by means of the single degree of freedom t^* test statistic whether or not the mean cash offers for male and female owners are equal; use $\alpha = .05$. State the alternatives, decision rule, and conclusion.

22.12. Refer to **Kidney failure hospitalization** Problem 20.15. Continue to work with the transformed observations $Y' = \log_{10}(Y + 1)$. It is known that 75 percent of patients in each weight gain group receive the short duration treatment.

a. Use ANOVA model (20.19) for expressing the two alternatives for testing whether or not factor B main effects are present.

b. Define the $\mathbf{X}$ matrix and $\boldsymbol{\beta}$ vector for expressing full model (20.19) in matrix form for this case.

c. Express the two alternatives in part (a) in matrix form (8.41).

d. Use (8.45) to calculate $SSE(R) - SSE(F)$.

e. Test whether or not factor B main effects are present; use $\alpha = .05$. State the decision rule and conclusion.

f. Compare the mean number of days of hospitalization (in transformed units) for patients with severe and mild weight gains; use a 95 percent confidence interval.

22.13. Refer to **Adjunct professors** Problem 22.6. It is known that 10 percent of professors in each subject matter area have a bachelor's degree, 20 percent have a master's degree, and 70 percent have a doctorate.

a. Use ANOVA model (20.19) for expressing the two alternatives for testing whether or not factor A main effects are present.
b. Define the $\mathbf{X}$ matrix and $\boldsymbol{\beta}$ vector for expressing full model (20.19) in matrix form for this case.
c. Express the two alternatives in part (a) in matrix form (8.41).
d. Use (8.45) to calculate $SSE(R) - SSE(F)$.
e. Test whether or not factor A main effects are present; use $\alpha = .01$. State the decision rule and conclusion.
f. Compare the mean amounts of payment received by faculty members teaching humanities and engineering courses; use a 99 percent confidence interval. Interpret your interval estimate.

EXERCISES

22.14. Give expressions for SSA, SSB, and $SSAB$ for the method of unweighted means.

22.15. Derive $\sigma^2(\hat{L})$ for the estimated contrast involving $\hat{\mu}_{i.}$ in (22.16).

22.16. Show that $s^2(\hat{L})$ in (22.20) is an unbiased estimator of $\sigma^2(\hat{L})$.

22.17. Consider a two-factor study where $a = 2$, $b = 2$, $n_{11} = n_{12} = n_{21} = 2$, $n_{22} = 1$, and no-interaction model (22.22) applies. Use the matrix methods of Section 8.5 to estimate μ_{22}. (*Hint:* Consider the development in Section 23.1.)

22.18. Refer to **Kidney failure hospitalization** Problem 22.8. Suppose that you are going to use the general matrix approach of Section 8.5, rather than the regression approach, to test for factor A main effects.

a. State the $\mathbf{X}$ and $\boldsymbol{\beta}$ matrices to be used in the full model (8.38).
b. State the test hypothesis (8.41) in matrix form.

PROJECTS

22.19. Refer to the **SENIC** data set. The effects of region (factor A: variable 9) and average age of patients (factor B: variable 3) on mean length of hospital stay (variable 2) are to be studied. For purposes of this ANOVA study, average age is to be classified into three categories: under 52.0 years, 52.0–under 55.0 years, 55.0 years or more. Assume that ANOVA model (20.27), with $k = 1, \ldots, n_{ij}$, is appropriate.

a. State the equivalent regression model for this case; use $1, -1, 0$ indicator variables.
b. State the reduced model for testing for interaction effects.
c. Test whether or not interaction effects are present by fitting the full and reduced models; use $\alpha = .01$. State the alternatives, decision rule, and conclusion.

d. State the reduced model for testing for factor A main effects. Conduct this test using $\alpha = .01$. State the alternatives, decision rule, and conclusion.
e. State the reduced model for testing for factor B main effects. Conduct this test using $\alpha = .01$. State the alternatives, decision rule, and conclusion.
f. Make all pairwise comparisons between regions; use the Tukey procedure and a 95 percent family confidence coefficient. State your findings and present a graphic summary.

22.20. Refer to the **SMSA** data set. The effects of region (factor A: variable 12) and percent of population in central cities (factor B: variable 4) on the crime rate (variable 11 ÷ variable 3) are to be studied. For purposes of this ANOVA study, percent of population in central cities is to be classified into three categories: under 30.0 percent, 30.0–under 50.0 percent, 50.0 percent or more. Assume that ANOVA model (20.27), with $k = 1, \ldots, n_{ij}$, is appropriate.
a. State the equivalent regression model for this case; use $1, -1, 0$ indicator variables.
b. State the reduced model for testing for interaction effects.
c. Test whether or not interaction effects are present by fitting the full and reduced models; use $\alpha = .005$. State the alternatives, decision rule, and conclusion.
d. State the reduced model for testing for factor A main effects. Conduct this test using $\alpha = .005$. State the alternatives, decision rule, and conclusion.
e. State the reduced model for testing for factor B main effects. Conduct this test using $\alpha = .005$. State the alternatives, decision rule, and conclusion.
f. Make all pairwise comparisons between regions; use the Tukey procedure and a 95 percent family confidence coefficient. State your findings and present a graphic summary.

CITED REFERENCE

22.1. Federer, W. T., and M. Zelen. ''Analysis of Multifactor Classifications with Unequal Numbers of Observations,'' *Biometrics* 22 (1966), pp. 525–52.

23

Random and mixed effects models for two-factor studies and other topics in ANOVA—II

In this chapter, we discuss a number of selected topics in the analysis of variance for two-factor studies. We first take up the special case where there is only one observation for each treatment. In this connection, we discuss the Tukey test for additivity, a test that is important not only when there is only a single observation for each treatment in a two-factor study but that is also useful for a variety of experimental designs to be discussed in later chapters. Finally, we consider two types of models for two-factor studies that are appropriate when the effects of one or both factors may be viewed as random effects.

23.1 ONE OBSERVATION PER TREATMENT

First, we take up the case where there is only one observation for each treatment. We now can no longer work with the two-factor fixed effects ANOVA model (20.27) because no estimate of the error variance σ^2 will be available. Recall from (20.42c) that *SSE* is a sum of squares made up of components measuring the variability within each treatment, $\sum_k (Y_{ijk} - \bar{Y}_{ij.})^2$. With only one observation per treatment, there is no variability within a treatment, and *SSE* will then always be zero.

A way out of this difficulty is to change the model. A glance at Table 20.9 indicates that if the two factors do not interact, the interaction mean square

MSAB has expectation σ^2. Thus, if it is possible to assume that the two factors do not interact, we may use *MSAB* as the estimator of the error variance σ^2, and proceed with the analysis of factor effects as usual. If it is unreasonable to assume that the two factors do not interact, transformations may be tried to remove the interaction effects.

No-interaction model

The two-factor fixed effects ANOVA model with no interactions was given in (22.22). For the case $n = 1$ considered here, this model is:

$$Y_{ij} = \mu_{..} + \alpha_i + \beta_j + \varepsilon_{ij} \tag{23.1}$$

where the model terms are as in (20.27). Note that the third subscript has been dropped from the Y and ε terms because there is now only one observation per treatment.

SSA and *SSB* are calculated as before from (20.43a) and (20.43b), respectively, with $n = 1$. The interaction sum of squares in (20.43c) with $n = 1$ now serves as the error sum of squares:

$$SSE = SSAB = \sum_i \sum_j (Y_{ij} - \bar{Y}_{i.} - \bar{Y}_{.j} + \bar{Y}_{..})^2 \tag{23.2}$$

Note that *SSE* in (23.2) is identical to *SSAB* in (20.43c) with $n = 1$, the third subscript dropped because there is only one observation per treatment, and the mean $\bar{Y}_{ij.}$ replaced by the observation Y_{ij} for the same reason. The number of degrees of freedom associated with *SSE* in (23.2) is the same as the number associated with *SSAB* in (20.43c), namely, $(a - 1)(b - 1)$.

The analysis of variance table for the case $n = 1$ for the no-interaction model (23.1) is shown in Table 23.1. No new problems arise in the tests for factor A and factor B main effects, nor in estimating these effects. A special problem does exist, however, in estimating treatment means. We shall explain how to handle this problem after presenting an example.

Example

Table 23.2a shows the amounts of six-month premiums charged by an automobile insurance firm for a specific type and amount of coverage in a given risk category for six cities, classified by size of city (factor A) and geographic region (factor B). Note there is only one observation per cell, namely, the amount of the premium for one city in each factor level combination. This firm adjusts its premiums periodically to reflect comparative loss experiences in different localities. An insurance analyst wished to evaluate the effects of the city sizes and geographic regions shown in Table 23.2 on the amount of the premium. From experience in other cases, he conjectured that interaction effects would not be present in the premiums under analysis. A test for interactions (to be discussed in Section 23.2) did indeed lead to the conclusion that interaction effects are not present, so the analyst adopted model (23.1).

TABLE 23.1 ANOVA table for no-interaction two-factor fixed effects model, $n = 1$

Source of Variation	*SS*	*df*	*MS*	*E(MS)*
Factor *A*	$SSA = b\Sigma(\bar{Y}_{i.} - \bar{Y}_{..})^2$	$a - 1$	$MSA = \frac{SSA}{a-1}$	$\sigma^2 + \frac{b}{a-1}\Sigma(\mu_{i.} - \mu_{..})^2$
Factor *B*	$SSB = a\Sigma(\bar{Y}_{.j} - \bar{Y}_{..})^2$	$b - 1$	$MSB = \frac{SSB}{b-1}$	$\sigma^2 + \frac{a}{b-1}\Sigma(\mu_{.j} - \mu_{..})^2$
Error	$SSE = \Sigma\Sigma(Y_{ij} - \bar{Y}_{i.} - \bar{Y}_{.j} + \bar{Y}_{..})^2$	$(a-1)(b-1)$	$MSE = \frac{SSE}{(a-1)(b-1)}$	σ^2
Total	$SSTO = \Sigma\Sigma(Y_{ij} - \bar{Y}_{..})^2$	$ab - 1$		

TABLE 23.2 Two-factor insurance premium study with $n = 1$

(a) Premiums for Automobile Insurance Policy (in dollars)

Size of City (factor A)	Region (factor B) East ($j=1$)	West ($j=2$)	Average
Small ($i=1$)	140	100	120
Medium ($i=2$)	210	180	195
Large ($i=3$)	220	200	210
Average	190	160	175

(b) ANOVA Table

Source of Variation	SS	df	MS
Size of city	9,300	2	4,650
Region	1,350	1	1,350
Error	100	2	50
Total	10,750	5	

The required sums of squares are obtained as follows [using the definitional formulas (20.42) and (20.43) for $n = 1$]:

$$SSA = 2[(120 - 175)^2 + (195 - 175)^2 + (210 - 175)^2] = 9{,}300$$
$$SSB = 3[(190 - 175)^2 + (160 - 175)^2] = 1{,}350$$
$$SSE = (140 - 120 - 190 + 175)^2 + \cdots + (200 - 210 - 160 + 175)^2 = 100$$
$$SSTO = (140 - 175)^2 + \cdots + (200 - 175)^2 = 10{,}750$$

The ANOVA table is given in Table 23.2b. In the analyst's test of city size (factor A) effects, the alternative conclusions are:

$$H_0\colon \alpha_1 = \alpha_2 = \alpha_3 = 0$$
$$H_a\colon \text{not all } \alpha_i \text{ equal zero}$$

The F^* test statistic takes the usual form:

$$F^* = \frac{MSA}{MSE}$$

and the decision rule is [remember that the denominator of F^* here involves $(a - 1)(b - 1)$ degrees of freedom]:

$$\text{If } F^* \leq F[1 - \alpha; a - 1, (a - 1)(b - 1)], \text{ conclude } H_0$$
$$\text{If } F^* > F[1 - \alpha; a - 1, (a - 1)(b - 1)], \text{ conclude } H_a$$

Let $\alpha = .05$. Hence, we need $F(.95; 2, 2) = 19.0$. Table 23.2b provides the data for the test statistic:

$$F^* = \frac{4{,}650}{50} = 93$$

Since $F^* = 93 > 19.0$, we conclude H_a, that city size effects are present.

The test for geographic region (factor B) effects proceeds similarly, the alternative conclusions being:

$$H_0: \beta_1 = \beta_2 = 0$$
$$H_a: \text{not all } \beta_j \text{ equal zero}$$

For $\alpha = .05$ the decision rule is:

$$\text{If } F^* \leq F(.95;\ 1,\ 2) = 18.5, \text{ conclude } H_0$$
$$\text{If } F^* > F(.95;\ 1,\ 2) = 18.5, \text{ conclude } H_a$$

Table 23.2b shows that:

$$F^* = \frac{MSB}{MSE} = \frac{1{,}350}{50} = 27$$

Since $F^* = 27 > 18.5$, we conclude H_a, that geographic region effects are present.

Given the findings of both factor A and factor B main effects, the analyst next examined the factor level means $\mu_{i.}$ and $\mu_{.j}$ by methods discussed in Section 21.2.

Estimation of treatment mean

When there is only a single observation for each treatment in a two-factor study and the no-interaction model (23.1) is employed, the usual method of estimation of a treatment mean μ_{ij} by the sample mean, here simply the single observation Y_{ij}, needs to be modified. The reason is that this estimator does not make use of the model assumption of no interactions. We shall use the general matrix methods described in Section 8.5 whereby we can utilize the assumption of no interactions.

To illustrate the procedure for our insurance premium example, we begin with the general ANOVA model (20.19) with no restrictions about interactions (remember $n = 1$ here):

$$Y_{ij} = \mu_{ij} + \varepsilon_{ij}$$

We express this model in matrix terms in the form $\mathbf{Y} = \mathbf{X}\boldsymbol{\beta} + \boldsymbol{\varepsilon}$ with the matrices illustrated in (20.23). For our example, where we have $n = 1$, $ab = 6$, and $n_T = 6$, $\mathbf{X}$ and $\boldsymbol{\beta}$ are defined as follows:

$$\mathbf{X} = \underset{6\times6}{\mathbf{I}} \qquad \boldsymbol{\beta} = \begin{bmatrix} \mu_{11} \\ \mu_{12} \\ \mu_{21} \\ \mu_{22} \\ \mu_{31} \\ \mu_{32} \end{bmatrix} \tag{23.3}$$

The least squares estimators of the treatment means μ_{ij} in $\boldsymbol{\beta}$ as usual are the sample treatment means; since $n = 1$, the sample mean is simply the individual observation Y_{ij}. Hence, the least squares estimator of $\boldsymbol{\beta}$, to be denoted by $\mathbf{b}_F$ to tie in with the notation of Chapter 8, is the observations vector $\mathbf{Y}$:

$$\mathbf{b}_F = \mathbf{Y} \tag{23.4}$$

We shall express the constraint of no interactions in the form (8.41):

$$\mathbf{C}\boldsymbol{\beta} = \mathbf{h} \tag{23.5}$$

utilizing the relation (20.13b) for expressing a treatment mean μ_{ij} in terms of three other treatment means when no interactions exist. We require $(a - 1)(b - 1) = 2(1) = 2$ such relations here. We shall use the relations:

$$\begin{matrix} \mu_{11} = \mu_{12} + \mu_{21} - \mu_{22} \\ \mu_{11} = \mu_{12} + \mu_{31} - \mu_{32} \end{matrix} \quad \text{or} \quad \begin{matrix} \mu_{11} - \mu_{12} - \mu_{21} + \mu_{22} = 0 \\ \mu_{11} - \mu_{12} - \mu_{31} + \mu_{32} = 0 \end{matrix}$$

Hence, the matrices $\mathbf{C}$ and $\mathbf{h}$ in constraint (23.5) are as follows for our example:

$$\underset{2\times 6}{\mathbf{C}} = \begin{bmatrix} 1 & -1 & -1 & 1 & 0 & 0 \\ 1 & -1 & 0 & 0 & -1 & 1 \end{bmatrix} \qquad \underset{2\times 1}{\mathbf{h}} = \begin{bmatrix} 0 \\ 0 \end{bmatrix}$$

We now utilize (8.43) to obtain the least squares estimators of the treatment means μ_{ij} under the constraint of no interactions. These constrained estimators are denoted by $\mathbf{b}_R$ in (8.43). Since $\mathbf{X} = \mathbf{I}$, $\mathbf{h} = \mathbf{0}$, and $\mathbf{b}_F = \mathbf{Y}$ here, (8.43) can be simplified to become:

$$\mathbf{b}_R = \mathbf{AY} \qquad \text{No-interaction model, } n = 1 \tag{23.6}$$

where:

$$\mathbf{A} = \mathbf{I} - \mathbf{C}'(\mathbf{CC}')^{-1}\mathbf{C} \tag{23.6a}$$

For our insurance premium example, substitution into (23.6) yields:

$$\mathbf{b}_R = \begin{bmatrix} 135 \\ 105 \\ 210 \\ 180 \\ 225 \\ 195 \end{bmatrix}$$

Thus, for instance, our least squares estimate of the mean premium for a small city in the East is $\hat{\mu}_{11} = \$135$.

It is of interest to note that the least squares estimators $\hat{\mu}_{ij}$ in (23.6) turn out to be simple in structure, reflecting the constraint of no interactions:

$$\hat{\mu}_{ij} = \bar{Y}_{i.} + \bar{Y}_{.j} - \bar{Y}_{..} \tag{23.6b}$$

Thus, using the data in Table 23.2, we obtain:

$$\hat{\mu}_{11} = 120 + 190 - 175 = 135$$
$$\hat{\mu}_{12} = 120 + 160 - 175 = 105$$
etc. etc.

To set up a confidence interval for a treatment mean μ_{ij}, we require the estimated variance of $\hat{\mu}_{ij}$. This can be obtained from (6.49), where $\mathbf{A}$ is given by (23.6a) and $\boldsymbol{\sigma}^2(\mathbf{Y})$ is estimated by $(MSE)\mathbf{I}$.

Comments

1. The analysis of two-factor studies with $n = 1$ just outlined depends on the assumption that the two factors do not interact. If one utilizes this analysis when in fact interactions are present, the result is that the actual level of significance for testing factor A and B main effects is below the specified level and the actual power of the tests is lower than the expected power. Correspondingly, confidence intervals for contrasts of factor level means will tend to be too wide. This means that when interactions are present, the analysis is more likely to fail to disclose real effects than anticipated. However, when the analysis based on the no-interaction model does indicate the presence of factor A main effects or of factor B main effects, they may be taken as real effects even though interactions are actually present.

2. Sometimes, the case $n = 1$ is encountered when the observations Y_{ij} are proportions. For instance, the data may consist of the proportion of employees in a plant absent during the past week, with the plants classified by size and geographic area. As noted earlier, the arc sine transformation can be used for such data to stabilize the variances. The transformed data then can be analyzed through the analysis of variance model (23.1), provided that each proportion is based on roughly the same number of cases. If the number of cases differs greatly, a weighted least squares approach should be utilized.

23.2 TUKEY TEST FOR ADDITIVITY

We describe now a test devised by Tukey that may be used for examining whether or not the two factors in a two-factor study interact when $n = 1$. This test is also useful for a variety of experimental designs to be discussed in later chapters.

Development of test statistic

As noted in Section 23.1, we considered the no-interaction model (23.1) when $n = 1$ to enable us to obtain an estimate of the error term variance in this case. It would have been possible, however, to impose only some restrictions on the $(\alpha\beta)_{ij}$ and include the restricted interaction effects in the model. Suppose we assume that:

$$(\alpha\beta)_{ij} = D\alpha_i\beta_j \tag{23.7}$$

where D is some constant. One motivation for this restriction is that if $(\alpha\beta)_{ij}$ is any second-degree polynomial function of α_i and β_j, then it must be of the form

(23.7) because of the restrictions in (20.27) on the α_i, β_j, and $(\alpha\beta)_{ij}$ that the sums over each subscript be 0.

Using (23.7) in a regular two-factor model with interactions for the case $n = 1$, we obtain:

$$Y_{ij} = \mu_{..} + \alpha_i + \beta_j + D\alpha_i\beta_j + \varepsilon_{ij} \tag{23.8}$$

where each of the terms has the usual meaning. Remember there is no third subscript here because $n = 1$. The interaction sum of squares $\Sigma\Sigma D^2\alpha_i^2\beta_j^2$ now needs to be obtained. Assuming the other parameters are known, the least squares estimator of D turns out to be:

$$\hat{D} = \frac{\sum_i \sum_j \alpha_i\beta_j Y_{ij}}{\sum_i \alpha_i^2 \sum_j \beta_j^2} \tag{23.9}$$

The usual estimator of α_i is $\bar{Y}_{i.} - \bar{Y}_{..}$, and that of β_j is $\bar{Y}_{.j} - \bar{Y}_{..}$. Replacing the parameters in $\hat{D}$ by these estimators, we obtain:

$$\hat{D} = \frac{\sum_i \sum_j (\bar{Y}_{i.} - \bar{Y}_{..})(\bar{Y}_{.j} - \bar{Y}_{..})Y_{ij}}{\sum_i (\bar{Y}_{i.} - \bar{Y}_{..})^2 \sum_j (\bar{Y}_{.j} - \bar{Y}_{..})^2} \tag{23.9a}$$

The sample counterpart of the interaction sum of squares $\Sigma\Sigma D^2\alpha_i^2\beta_j^2$ therefore is:

$$SSAB = \frac{\left[\sum_i \sum_j (\bar{Y}_{i.} - \bar{Y}_{..})(\bar{Y}_{.j} - \bar{Y}_{..})Y_{ij}\right]^2}{\sum_i (\bar{Y}_{i.} - \bar{Y}_{..})^2 \sum_j (\bar{Y}_{.j} - \bar{Y}_{..})^2} \tag{23.10}$$

This expression can be expanded out for computational simplifications.

We stated earlier that the error sum of squares in (23.2):

$$SSE = \Sigma\Sigma(Y_{ij} - \bar{Y}_{i.} - \bar{Y}_{.j} + \bar{Y}_{..})^2$$

reflects solely the random variability of the error terms if there are no interactions. If interaction effects exist, however, *SSE* reflects not only the random error variation but also interaction effects. Hence, a pure error sum of squares may be obtained as follows:

$$SSPE = SSE - SSAB \tag{23.11}$$

where *SSE* is defined in (23.2) and *SSAB* in (23.10).

It can then be shown that if $D = 0$, that is, if no interactions exist, *SSPE* and *SSAB* are independently distributed, and the test statistic:

$$F^* = \frac{SSAB}{1} \div \frac{SSPE}{ab - a - b} \tag{23.12}$$

is distributed as $F(1, ab - a - b)$. Note that one degree of freedom is associated with $SSAB$, and $(a - 1)(b - 1) - 1 = ab - a - b$ degrees of freedom are associated with $SSPE$.

Thus, for testing:

$$\text{(23.13)} \qquad \begin{aligned} &H_0\text{: } D = 0 \text{ (no interactions present)} \\ &H_a\text{: } D \neq 0 \text{ (interactions } D\alpha_i\beta_j \text{ present)} \end{aligned}$$

we use the test statistic F^* defined in (23.12). Large values of F^* lead to conclusion H_a. The appropriate decision rule for controlling the risk of a Type I error at α is:

$$\text{(23.14)} \qquad \begin{aligned} &\text{If } F^* \leq F(1 - \alpha; 1, ab - a - b), \text{ conclude } H_0 \\ &\text{If } F^* > F(1 - \alpha; 1, ab - a - b), \text{ conclude } H_a \end{aligned}$$

The power of this test has been studied, and it appears that if interactions of approximately the type postulated in (23.7) are present and the factor A and factor B main effects are large, the test is fairly effective in detecting the interactions. The test is usually called the *Tukey one degree of freedom test*.

Example

We shall conduct the Tukey test for our earlier insurance premium example. The data are presented in Table 23.2a. First, we obtain the components of $SSAB$:

$$\Sigma\Sigma(\bar{Y}_{i.} - \bar{Y}_{..})(\bar{Y}_{.j} - \bar{Y}_{..})Y_{ij} = (120 - 175)(190 - 175)(140) + \cdots + (210 - 175)(160 - 175)(200) = -13{,}500$$

$$\Sigma(\bar{Y}_{i.} - \bar{Y}_{..})^2 = \frac{SSA}{2} = \frac{9{,}300}{2} = 4{,}650$$

$$\Sigma(\bar{Y}_{.j} - \bar{Y}_{..})^2 = \frac{SSB}{3} = \frac{1{,}350}{3} = 450$$

Hence, the interaction sum of squares is:

$$SSAB = \frac{(-13{,}500)^2}{4{,}650(450)} = 87.1$$

We found earlier in Table 23.2b that $SSE = 100$; hence, we have by (23.11):

$$SSPE = 100 - 87.1 = 12.9$$

Finally, we obtain the test statistic by (23.12):

$$F^* = \frac{87.1}{1} \div \frac{12.9}{3(2) - 3 - 2} = 6.8$$

Assuming the level of significance is to be controlled at $\alpha = .10$, we require $F(.90; 1, 1) = 39.9$. Since $F^* = 6.8 \leq 39.9$, we conclude that region and size of city do not interact. The use of the no-interaction model for the data in Table 23.2 therefore appears to be reasonable.

Comments

1. If the Tukey test indicates the presence of interaction effects, some simple transformations such as a square root or logarithmic transformation may be tried to see if the interactions can be removed or made unimportant. (See the discussion in Chapter 20 in this connection.) If no simple transformation can be found to make the interactions unimportant, an approximate method of analysis can be employed; see, for instance, Reference 23.1.

2. If one or both factors are quantitative, a test for interaction effects can be obtained by regression methods, which will have more degrees of freedom associated with the interaction sum of squares than the Tukey test. These regression tests for interaction effects were discussed earlier. For example, model (10.6) illustrates a case where one independent variable is quantitative and the regression relation is linear. Model (7.15) is a second-order model where both independent variables are quantitative. For both of these models, the regression test for interaction effects is a simple test whether the interaction regression coefficients equal zero.

23.3 MODELS II (RANDOM EFFECTS) AND III (MIXED EFFECTS) FOR TWO-FACTOR STUDIES

Random effects model

Consider an investigation of the effects of machine operators (factor A) and machines (factor B) on the number of pieces produced in a day. Five operators and three machines are used in the study. Yet the inferences are not to be confined to the particular five operators and three machines participating in the study, but rather they are to pertain to all operators and all machines available to the company. Here a random effects model (model II) would be appropriate for the two-factor study, since each of the two sets of factor levels may be considered as a sample from a population (all operators, all machines) about which inferences are to be drawn.

In a random effects model for a two-factor study, we assume analogously to the random effects model for a single-factor study that both the factor A main effects α_i and the factor B main effects β_j are random variables, selected independently. Further, it is assumed that the interaction effects $(\alpha\beta)_{ij}$ are independent random variables. Thus, the random effects model for a two-factor study with equal sample sizes n is:

$$Y_{ijk} = \mu_{..} + \alpha_i + \beta_j + (\alpha\beta)_{ij} + \varepsilon_{ijk} \tag{23.15}$$

where:

$\mu_{..}$ is a constant

α_i, β_j, $(\alpha\beta)_{ij}$ are independent normal random variables with expectations zero and respective variances σ_α^2, σ_β^2, $\sigma_{\alpha\beta}^2$

ε_{ijk} are independent $N(0, \sigma^2)$, and independent of α_i, β_j, and $(\alpha\beta)_{ij}$

$i = 1, \ldots, a$; $j = 1, \ldots, b$; $k = 1, \ldots, n$

Meaning of model. We shall explain the meaning of the terms in the random effects model (23.15) with reference to our production example involving the two factors machine operators and machines. The main effect of operator i in the study (selected at random from the population of operators) is α_i. Similarly, the main effect of machine j in the study (selected at random from the population of machines) is β_j. Model (23.15) states that the main effects of operators on output per day are normally distributed with zero mean and variance σ_α^2. Similarly, the main effects of machines are normally distributed with zero mean and variance σ_β^2. If μ_{ij} is the mean output per day for the operator i–machine j combination, the interaction $(\alpha\beta)_{ij}$ is defined in the usual way: $(\alpha\beta)_{ij} = \mu_{ij} - (\mu_{..} + \alpha_i + \beta_j)$. Surprising as it may seem, the interaction term $(\alpha\beta)_{ij}$ turns out to be independent of α_i and β_j when all random terms in the model are assumed to be normally distributed. Hence, the mean output for the operator i–machine j combination, namely, $\mu_{ij} = \mu_{..} + \alpha_i + \beta_j + (\alpha\beta)_{ij}$, may be viewed in the random effects model as the result of independent selections of α_i, β_j, and $(\alpha\beta)_{ij}$ from three different normal distributions.

Note

We caution that a random effects model should only be used if the factor levels of the different factors do indeed represent random samples from populations of interest.

Mixed effects model

When one of the two factors involves fixed effects while the other involves random effects, a mixed effects model (model III) is appropriate. An instance where this model may be appropriate is an investigation of the effects of four different training materials (factor A) and five instructors (factor B) upon learning in a company training program. The four levels for training materials may be considered fixed, since interest centers in these particular training materials. On the other hand, the levels for instructors may be viewed as random since inferences are to be made about a population of instructors of which the five used in the study are a sample.

When factor A has fixed effects and factor B random effects, the α_i effects are constants and the β_j effects are random variables. The interaction effects $(\alpha\beta)_{ij}$ are also random variables because the factor B levels are random. For this case where A is the fixed effects factor, B the random effects factor, and equal sample sizes n are selected for each treatment, a relatively simple mixed effects model is:

$$Y_{ijk} = \mu_{..} + \alpha_i + \beta_j + (\alpha\beta)_{ij} + \varepsilon_{ijk} \tag{23.16}$$

where:

$\mu_{..}$ is a constant
α_i are constants subject to the restriction $\Sigma\alpha_i = 0$
β_j are independent $N(0, \sigma_\beta^2)$
$(\alpha\beta)_{ij}$ are $N\left(0, \frac{a-1}{a}\sigma_{\alpha\beta}^2\right)$, subject to the restrictions $\sum_i (\alpha\beta)_{ij} = 0$ for all j
ε_{ijk} are independent $N(0, \sigma^2)$, and independent of the β_j and $(\alpha\beta)_{ij}$
β_j and $(\alpha\beta)_{ij}$ are independent
$i = 1, \ldots, a;\ j = 1, \ldots, b;\ k = 1, \ldots, n$

More complex mixed effects models have also been formulated.

Comments

1. In model (23.16), any two interaction terms $(\alpha\beta)_{ij}$ and $(\alpha\beta)_{i'j'}$ are independent unless both refer to the same random factor B level. In the latter case, the covariance is assumed to be $\sigma[(\alpha\beta)_{ij}, (\alpha\beta)_{i'j}] = -\frac{1}{a}\sigma_{\alpha\beta}^2$.

2. We noted that $\sum_i (\alpha\beta)_{ij} = 0$ for all j, since all factor A levels are included in the study. However, $\sum_j (\alpha\beta)_{ij}$ will ordinarily not equal zero.

23.4 ANALYSIS OF VARIANCE TESTS FOR MODELS II AND III

For both the mixed and random effects models for a two-factor study, the analysis of variance calculations for sums of squares are identical to those for a fixed effects model. Thus, formulas (20.42)–(20.44) are entirely applicable for models II and III. Similarly, the degrees of freedom and mean square determinations are exactly the same as for the fixed effects model, as shown in Table 20.9. The random and mixed effects models depart from the fixed effects model only in the expected mean squares and the consequent choice of the appropriate test statistic.

Expected mean squares

The expected mean squares for the random and mixed effects models can be worked out by utilizing the properties of the model and applying the usual expectation theorems. They are shown in Table 23.3, together with those for the fixed effects model. The derivations are tedious, but simple rules have been developed for finding the expected mean squares for random and mixed effects models. We take up these rules in Chapter 30.

Note

To illustrate the derivation of an expected mean square by use of expectation theorems, we shall find $E(MSA)$ for the two-factor random effects model (23.15). We wish to find:

TABLE 23.3 Expected mean squares in two-factor studies

Mean Square	df	Fixed Effects (A and B fixed)	Random Effects (A and B random)	Mixed Effects (A fixed, B random)
MSA	$a-1$	$\sigma^2 + nb\,\dfrac{\sum \alpha_i^2}{a-1}$	$\sigma^2 + nb\sigma_\alpha^2 + n\sigma_{\alpha\beta}^2$	$\sigma^2 + nb\,\dfrac{\sum \alpha_i^2}{a-1} + n\sigma_{\alpha\beta}^2$
MSB	$b-1$	$\sigma^2 + na\,\dfrac{\sum \beta_j^2}{b-1}$	$\sigma^2 + na\sigma_\beta^2 + n\sigma_{\alpha\beta}^2$	$\sigma^2 + na\sigma_\beta^2$
MSAB	$(a-1)(b-1)$	$\sigma^2 + n\,\dfrac{\sum\sum (\alpha\beta)_{ij}^2}{(a-1)(b-1)}$	$\sigma^2 + n\sigma_{\alpha\beta}^2$	$\sigma^2 + n\sigma_{\alpha\beta}^2$
MSE	$(n-1)ab$	σ^2	σ^2	σ^2

FIXED SET
RANDOM
GOES INTO
FIXED/R SET
NOT INTO RANDOM
RANDOM

if no interaction
THEN WILL BE
ZERO

if no interaction
THEN WILL BE ZERO

A RANDOM B FIXED

A $\sigma^2 + nb\sigma_\alpha^2$

B $\sigma^2 + n\hat{\sigma}_{\alpha\beta}^2 + nq\dfrac{\Sigma\beta_j^2}{\beta-1}$

AB $\sigma^2 + n\sigma_{\alpha\beta}^2$

ERROR σ^2

$$E(MSA) = E\left[\frac{nb\Sigma(\bar{Y}_{i..} - \bar{Y}_{...})^2}{a-1}\right]$$

Now:

$$Y_{i..} = \sum_j \sum_k [\mu_{..} + \alpha_i + \beta_j + (\alpha\beta)_{ij} + \varepsilon_{ijk}]$$

$$= nb\mu_{..} + nb\alpha_i + n\sum_j \beta_j + n\sum_j (\alpha\beta)_{ij} + \sum_j \sum_k \varepsilon_{ijk}$$

We obtain then:

(23.17) $$\bar{Y}_{i..} = \frac{Y_{i..}}{nb} = \mu_{..} + \alpha_i + \bar{\beta}_{.} + \overline{(\alpha\beta)}_{i.} + \bar{\varepsilon}_{i..}$$

where the bars as usual indicate means over the subscripts shown by dots. Similarly, we find:

(23.17a) $$\bar{Y}_{...} = \mu_{..} + \bar{\alpha}_{.} + \bar{\beta}_{.} + \overline{(\alpha\beta)}_{..} + \bar{\varepsilon}_{...}$$

Hence, we have:

(23.18) $$\bar{Y}_{i..} - \bar{Y}_{...} = (\alpha_i - \bar{\alpha}_{.}) + [\overline{(\alpha\beta)}_{i.} - \overline{(\alpha\beta)}_{..}] + (\bar{\varepsilon}_{i..} - \bar{\varepsilon}_{...})$$

Squaring each side of (23.18) and summing, we obtain:

(23.19) $$\sum_i (\bar{Y}_{i..} - \bar{Y}_{...})^2 = \sum_i (\alpha_i - \bar{\alpha}_{.})^2 + \sum_i [\overline{(\alpha\beta)}_{i.} - \overline{(\alpha\beta)}_{..}]^2 + \sum_i (\bar{\varepsilon}_{i..} - \bar{\varepsilon}_{...})^2 + \text{cross-product terms}$$

To find $E[\Sigma(\bar{Y}_{i..} - \bar{Y}_{...})^2]$, we need to take the expectation of each term on the right. The cross-product terms drop out because of the independence of α_i, $(\alpha\beta)_{ij}$, and ε_{ijk} and the fact that each of these random variables has zero expectation. Each of the remaining terms can be thought of as the numerator of a sample variance of a observations. We know from the unbiasedness of a sample variance that:

(23.20) $$E\left[\sum_{i=1}^{a} (Y_i - \bar{Y})^2\right] = (a-1)\sigma^2(Y_i)$$

Hence:

(23.21) $$E\left[\sum_i (\alpha_i - \bar{\alpha}_{.})^2\right] = (a-1)\sigma_\alpha^2$$

because $\sigma^2(\alpha_i) = \sigma_\alpha^2$. Similarly, we find:

(23.22) $$E\left[\sum_i (\bar{\varepsilon}_{i..} - \bar{\varepsilon}_{...})^2\right] = (a-1)\frac{\sigma^2}{bn}$$

since $\sigma^2(\bar{\varepsilon}_{i..}) = \sigma^2/bn$, and:

(23.23) $$E\left[\sum_i [\overline{(\alpha\beta)}_{i.} - \overline{(\alpha\beta)}_{..}]^2\right] = (a-1)\frac{\sigma_{\alpha\beta}^2}{b}$$

Using (23.21)–(23.23), we find:

$$E[\Sigma(\bar{Y}_{i..} - \bar{Y}_{...})^2] = (a-1)\sigma_\alpha^2 + (a-1)\frac{\sigma_{\alpha\beta}^2}{b} + (a-1)\frac{\sigma^2}{bn} \tag{23.24}$$

and:

$$E(MSA) = \frac{nb}{a-1}E[\Sigma(\bar{Y}_{i..} - \bar{Y}_{...})^2] = nb\sigma_\alpha^2 + n\sigma_{\alpha\beta}^2 + \sigma^2 \tag{23.25}$$

as shown in Table 23.3.

Construction of test statistics

As usual, the statistic for testing factor effects is constructed by comparing two mean squares that have the properties:

1. Under H_0, both have the same expectation.
2. Under H_a, the numerator mean square has a larger expectation than the denominator mean square.

It can be shown that such a test statistic follows the F distribution if H_0 holds. The decision rule is constructed in the ordinary fashion, with large values of the test statistic leading to H_a.

For instance, to test for the presence of factor A main effects in the random effects model (23.15), namely:

$$\begin{aligned} &H_0\colon \sigma_\alpha^2 = 0 \\ &H_a\colon \sigma_\alpha^2 > 0 \end{aligned} \tag{23.26}$$

we see from Table 23.3 that MSA and $MSAB$ both have the same expectation if $\sigma_\alpha^2 = 0$, that is, if factor A has no main effects. If $\sigma_\alpha^2 > 0$, $E(MSA)$ is greater than $E(MSAB)$. Hence, the appropriate test statistic is:

$$F^* = \frac{MSA}{MSAB} \tag{23.27}$$

and the decision rule for controlling the Type I error at α is:

$$\begin{aligned} &\text{If } F^* \leq F[1-\alpha; a-1, (a-1)(b-1)], \text{ conclude } H_0 \\ &\text{If } F^* > F[1-\alpha; a-1, (a-1)(b-1)], \text{ conclude } H_a \end{aligned} \tag{23.28}$$

Note that the denominator for testing for factor A main effects in the random effects model is $MSAB$, whereas it is MSE in the fixed effects model.

We summarize the appropriate test statistics for mixed and random effects models in Table 23.4. For comparison purposes, we also present the test statistics for the fixed effects model there. As may be seen from Table 23.4, in a number of instances the denominator of the test statistic for mixed and random effects models differs from that for the fixed effects model. Hence, it is important that the expected mean squares be known when random or mixed models are utilized so that the appropriate test statistics can be determined.

TABLE 23.4 Test statistics for mixed and random effects models

Test for Presence of Effects of—	*Fixed Effects Model (A and B fixed)*	*Random Effects Model (A and B random)*	*Mixed Effects Model (A fixed, B random)*
Factor A	MSA/MSE	$MSA/MSAB$	$MSA/MSAB$
Factor B	MSB/MSE	$MSB/MSAB$	MSB/MSE
AB interactions	$MSAB/MSE$	$MSAB/MSE$	$MSAB/MSE$

Example

We return to our earlier mixed effects example of four different training materials (factor A, fixed) and five instructors (factor B, random). Four classes were tested for each training material–instructor combination. The analysis of variance as obtained from a computer package for two-factor studies is shown in Table 23.5; the original data are not presented. To test whether or not training materials and instructors interact:

$$H_0: \sigma^2_{\alpha\beta} = 0$$
$$H_a: \sigma^2_{\alpha\beta} > 0$$

we utilize according to Table 23.4 the test statistic:

$$F^* = \frac{MSAB}{MSE}$$

Using the results from Table 23.5, we obtain:

$$F^* = \frac{3.9}{2.1} = 1.86$$

For a level of significance of $\alpha = .05$, we require $F(.95; 12, 60) = 1.92$. Since $F^* = 1.86 \leq 1.92$, we conclude that training materials and instructors do not interact.

TABLE 23.5 ANOVA table for mixed effects model training example (A fixed, B random, $a = 4$, $b = 5$, $n = 4$)

Source of Variation	*SS*	*df*	*MS*	*F**
Factor A (training materials–fixed)	42	3	14.0	14.0/3.9 = 3.59
Factor B (instructors–random)	54	4	13.5	13.5/2.1 = 6.43
AB interactions	47	12	3.9	3.9/2.1 = 1.86
Error	126	60	2.1	
Total	269	79		

$F(.95; 3, 12) = 3.49$ $\quad F(.95; 4, 60) = 2.53$
$F(.95; 12, 60) = 1.92$

The test statistics for testing training material main effects and instructor main effects are shown in Table 23.5. By comparing the test statistics with the appropriate percentiles of the F distribution shown at the bottom of Table 23.5 for level of significance $\alpha = .05$, we find that both training materials and instructors differ in effectiveness.

Note

If there is only one observation per treatment ($n = 1$), it will be recalled from Section 23.1 that no exact tests are possible with the fixed effects two-factor model unless one can assume that there are no AB interactions. The reason is that no estimate MSE of σ^2 can be obtained in that case. Table 23.4 indicates that exact tests for both factor A and factor B main effects are possible with the random effects model when $n = 1$ without any restrictive assumptions about the interactions. This is because $MSAB$ is the appropriate denominator of the test statistic here, and $MSAB$ can be determined regardless of sample size. With the mixed effects model where factor A is the fixed factor, the presence of factor A main effects can also be tested when $n = 1$ without the need for assuming all interactions are zero. However, an exact test for factor B main effects would require the assumption that all interactions are zero.

23.5 ESTIMATION OF FACTOR EFFECTS FOR MODELS II AND III

Estimation of variance components

For random factors that have significant main effects, we often would like to estimate the magnitude of the variance component. Unbiased point estimators can readily be derived, using linear combinations of the expected mean squares in Table 23.3. With the random effects model, for instance, σ_α^2 can be estimated by noting that:

$$E(MSA) - E(MSAB) = \sigma^2 + nb\sigma_\alpha^2 + n\sigma_{\alpha\beta}^2 - \sigma^2 - n\sigma_{\alpha\beta}^2 = nb\sigma_\alpha^2$$

Hence, an unbiased point estimator of σ_α^2 is:

$$s_\alpha^2 = \frac{MSA - MSAB}{nb} \tag{23.29}$$

Example. In the training illustration of Table 23.5, random effects factor B (instructors) had significant effects. To estimate σ_β^2, we utilize the expected mean squares in Table 23.3 for the mixed model with factor A fixed and factor B random, and determine the unbiased point estimator to be:

$$s_\beta^2 = \frac{MSB - MSE}{na} \tag{23.30}$$

Substituting, we obtain:

$$s_\beta^2 = \frac{13.5 - 2.1}{16} = .71$$

Estimation of fixed effects in mixed model

When the main effects of the fixed effects factor in a mixed two-factor model are of interest, one usually desires to obtain estimates of these effects in the form of contrasts. Our earlier discussion for the two-factor fixed effects model in Chapter 21 is applicable here, with the principal change occurring in the estimated variance of the contrast. For a fixed effects model, this estimated variance involves *MSE*, as shown, for instance, in (21.8). When dealing with a mixed model, however, the appropriate mean square to be used in the estimated variance formula is no longer *MSE*. A simple rule tells us which mean square is appropriate, namely, the one that is used in the denominator of the test statistic for testing the presence of main effects of the fixed factor under consideration. For instance, with the mixed model (23.16) where A is the fixed factor, *MSAB* is the appropriate mean square (Table 23.4). The degrees of freedom in constructing the confidence interval are those associated with the mean square utilized for estimating the variance of the contrast.

Example. In our training illustration of Table 23.5, no interaction effects were found. It is now desired that we estimate the difference in the mean amount of learning with training materials 1 and 2 using a 95 percent confidence interval. The relevant sample results are:

$$\bar{Y}_{1..} = 43.1 \qquad \bar{Y}_{2..} = 40.8$$

Hence, our point estimate of $D = \mu_{1.} - \mu_{2.}$ is:

$$\hat{D} = \bar{Y}_{1..} - \bar{Y}_{2..} = 43.1 - 40.8 = 2.3 \tag{23.31}$$

Using formula (21.16b) with *MSE* replaced by *MSAB*, the estimated variance is:

$$s^2(\hat{D}) = \frac{2MSAB}{bn} \tag{23.32}$$

For our example, we have:

$$s^2(\hat{D}) = \frac{2(3.9)}{20} = .39$$

or $s(\hat{D}) = .62$. There are 12 degrees of freedom associated with *MSAB*; hence, we require $t(.975; 12) = 2.179$. Our confidence interval therefore is:

$$2.3 - 2.179(.62) \le \mu_{1.} - \mu_{2.} \le 2.3 + 2.179(.62)$$
$$.9 \le \mu_{1.} - \mu_{2.} \le 3.7$$

Thus, we conclude with confidence coefficient .95 that training material 1 is more effective than training material 2, its mean being somewhere between .9 and 3.7 units larger.

Multiple comparison procedures. Multiple comparison procedures can be utilized for the main effects of the fixed effects factor in a mixed model in the same way as for the fixed effects model. In the estimated variance of the contrast, *MSE* will simply need to be replaced by the appropriate mean square. For example, suppose we wish to obtain all pairwise comparisons between the different training materials in our example in Table 23.5 by means of the Tukey method. We would calculate $s^2(\hat{D})$ as in our previous example. The T multiple in (21.16c) now would be:

$$T = \frac{1}{\sqrt{2}} q[1 - \alpha; a, (a - 1)(b - 1)] \tag{23.33}$$

where $(a - 1)(b - 1)$ replaces $(n - 1)ab$ as the degrees of freedom associated with the mean square utilized. With specific reference to our illustration in Table 23.5, for constructing 95 percent family confidence coefficient intervals for all pairwise comparisons between training materials we would require:

$$q(.95; 4, 12) = 4.20$$

and:

$$T = \frac{1}{\sqrt{2}}(4.20) = 2.97$$

PROBLEMS

23.1. Suppose that two-factor analysis of variance model (20.27) were to be employed with $n = 1$ for each factor level combination. How many degrees of freedom would be associated with *SSE* in (20.42c)? What does this imply?

23.2. **Coin-operated terminals.** A university computer service conducted an experiment in which one coin-operated computer terminal was placed at each of four different locations on the campus last semester during the midterm week and again during the final week of classes. The data that follow show the number of hours each terminal was *not* in use during the week at the four locations (factor A) and for the two different weeks (factor B).

	Factor B (week)	
Factor A (location)	$j = 1$ *Midterm*	$j = 2$ *Final*
$i = 1$	16.5	21.4
$i = 2$	11.8	17.3
$i = 3$	12.3	16.9
$i = 4$	16.6	21.0

Assume that no-interaction ANOVA model (23.1) is appropriate.

a. Plot the observations in the format of Figure 20.5. Does it appear that interaction effects are present? Does it appear that factor A and factor B main effects are present? Discuss.

b. Conduct separate tests for location and week main effects. In each test, use a level of significance of $\alpha = .05$ and state the alternatives, decision rule, and conclusion. Give an upper bound for the family level of significance; use the Kimball inequality.

c. Make all pairwise comparisons among location means and estimate the difference between the means for the two weeks; use the Bonferroni procedure with a 90 percent family confidence coefficient. State your findings.

23.3. Refer to **Coin-operated terminals** Problem 23.2. It is desired to estimate μ_{32}.

a. Obtain a point estimate of μ_{32} using (23.6b).

b. Verify the point estimate $\hat{\mu}_{32}$ obtained in part (a) using (23.6) and (23.6a). State the **C** and **A** matrices utilized.

c. Obtain the estimated variance of $\hat{\mu}_{32}$. [*Hint:* Use (6.49) and note that $\mathbf{A}[(MSE)\mathbf{I}]\mathbf{A}' = (MSE)\mathbf{A}$.]

d. Construct a 95 percent confidence interval for μ_{32}. Interpret your interval estimate. Is your interval estimate applicable if next year three terminals will be placed at location 3? Explain.

23.4. Refer to **Coin-operated terminals** Problem 23.2. Conduct the Tukey test for additivity; use $\alpha = .025$. State the alternatives, decision rule, and conclusion. If the additive model is not appropriate, what might you do?

23.5. **Brainstorming.** A researcher investigated whether brainstorming is more effective for larger groups than for smaller ones by setting up four groups of agribusiness executives, the group sizes being two, three, four, and five, respectively. He also set up four groups of agribusiness scientists, the group sizes being the same as for the agribusiness executives. The researcher gave each group the same problem: "How can Canada increase the value of its agricultural exports?" Each group was allowed 30 minutes to generate ideas. The variable of interest was the number of different ideas proposed by the group. The results, classified by type of group (factor A) and size of group (factor B), were:

	Factor A (type of group)	*Factor B (size of group)* $j = 1$ *Two*	$j = 2$ *Three*	$j = 3$ *Four*	$j = 4$ *Five*
$i = 1$	Agribusiness executives	18	22	31	32
$i = 2$	Agribusiness scientists	15	23	29	33

Assume that no-interaction ANOVA model (23.1) is appropriate.

a. Plot the observations in the format of Figure 20.5. Does it appear that interaction effects are present? Does it appear that factor A and factor B main effects are present? Discuss.

b. Conduct separate tests for type of group and size of group main effects. In each test, use a level of significance of $\alpha = .01$ and state the alternatives, decision rule, and conclusion. Give an upper bound for the family level of significance; use the Kimball inequality.

c. Obtain confidence intervals for $D_1 = \mu_{.2} - \mu_{.1}$, $D_2 = \mu_{.3} - \mu_{.2}$, and $D_3 = \mu_{.4} - \mu_{.3}$; use the Bonferroni procedure with a 95 percent family confidence coefficient. State your findings.

d. Is the Bonferroni procedure used in part (c) the most efficient one here? Explain.

23.6. Refer to **Brainstorming** Problem 23.5. It is desired to estimate μ_{14}.

a. Obtain a point estimate of μ_{14} using (23.6b).

b. Verify the point estimate $\hat{\mu}_{14}$ obtained in part (a) using (23.6) and (23.6a). State the **C** and **A** matrices utilized.

c. Obtain the estimated variance of $\hat{\mu}_{14}$. [*Hint:* Use (6.49) and note that $\mathbf{A}[(MSE)\mathbf{I}]\mathbf{A}' = (MSE)\mathbf{A}$.]

d. Construct a 99 percent confidence interval for μ_{14}. Interpret your interval estimate. Is your interval estimate applicable if the two factors interact?

23.7. Refer to **Brainstorming** Problem 23.5. Conduct the Tukey test for additivity; use $\alpha = .01$. State the alternatives, decision rule, and conclusion. If the additive model is not appropriate, what might you do?

23.8. Soybean sausage. A food technologist, testing storage capabilities for a newly developed type of imitation sausage made from soybeans, conducted an experiment to test the effects of temperature level (factor A) and humidity level (factor B) in the freezer compartment on color change in the sausage. Three humidity levels and four temperature levels were considered. Five hundred sausages were stored at each of the 12 temperature-humidity combinations for 90 days. At the end of the storage period, the researcher determined the proportion of sausages for each temperature-humidity combination that exhibited color changes. The researcher transformed the data by means of the arc sine transformation (18.13) to stabilize the variances. The transformed data $Y' = 2 \arcsin \sqrt{Y}$ follow.

Factor A	*Factor B (temperature level)*			
(humidity level)	$j = 1$	$j = 2$	$j = 3$	$j = 4$
$i = 1$	13.9	14.2	20.5	24.8
$i = 2$	15.7	16.3	21.7	23.6
$i = 3$	15.1	15.4	19.9	26.1

Assume that no-interaction ANOVA model (23.1) is appropriate.

a. Plot the observations in the format of Figure 20.5. Does it appear that interaction effects are present? Does it appear that factor A and factor B main effects are present? Discuss.

b. Conduct separate tests for humidity and temperature main effects. In each test, use a level of significance of $\alpha = .025$ and state the alternatives, decision rule, and conclusion.

c. Obtain confidence intervals for $D_1 = \mu_{.2} - \mu_{.1}$, $D_2 = \mu_{.3} - \mu_{.2}$, and $D_3 = \mu_{.4} - \mu_{.3}$; use the Bonferroni procedure with a 95 percent family confidence coefficient. State your findings.

d. Is the Bonferroni procedure used in part (c) the most efficient one here? Explain

23.9. Refer to **Soybean sausage** Problem 23.8. It is desired to estimate μ_{23}.

a. Obtain a point estimate of μ_{23} using (23.6b).

b. Verify the point estimate $\hat{\mu}_{23}$ obtained in part (a) using (23.6) and (23.6a). State the **C** and **A** matrices utilized.

c. Obtain the estimated variance of $\hat{\mu}_{23}$. [*Hint:* Use (6.49) and note that $\mathbf{A}[(MSE)\mathbf{I}]\mathbf{A}' = (MSE)\mathbf{A}$.]

d. Construct a 98 percent confidence interval for μ_{23} and transform it back to the original units. Interpret your interval estimate. Is your interval estimate applicable if the two factors interact?

23.10. Refer to **Soybean sausage** Problem 23.8. Conduct the Tukey test for additivity; use $\alpha = .005$. State the alternatives, decision rule, and conclusion. If the additive model is not appropriate, what might you do?

23.11. For the mixed effects model (23.16), why is $\sum_i (\alpha\beta)_{ij} = 0$ whereas usually $\sum_j (\alpha\beta)_{ij} \neq 0$?

23.12. A marketing consultant is designing several experiments involving a newly developed low-cost food processor. The initial experiment has the objectives (1) to compare the effects on unit sales of three possible prices recommended by the sales department ($23.99, $25.49, $25.95) and (2) to determine whether the color scheme used for the appliance affects unit sales. A great many color schemes are feasible; three (white, green, pink) have been selected for the initial experiment to represent the range of possible colors. If the experiment suggests that color scheme does have an effect, this aspect of the product design will be investigated in detail in a follow-up study. Which ANOVA model would you employ for analyzing the initial experiment? Discuss.

23.13. In a two-factor ANOVA study with $a = 3$, $b = 2$, and $n = 5$, the two factor effects are both random with $\sigma^2 = 5.0$, $\sigma_\alpha^2 = 8.0$, $\sigma_\beta^2 = 10.0$, and $\sigma_{\alpha\beta}^2 = 6.0$. Assume that model (23.15) is applicable.

a. Obtain $E(MSA)$, $E(MSB)$, and $E(MSAB)$.

b. What would be the expected mean squares if $\sigma_{\alpha\beta}^2 = 0$, all other parameters remaining the same?

23.14. A survey statistician has commented: "I am rather suspicious of uses of random effects and mixed effects ANOVA models. Seldom are the factor levels chosen by a random mechanism from a known population." Discuss.

23.15. **Miles per gallon.** An automobile manufacturer wished to study the effects of differences between drivers (factor A) and differences between cars (factor B) on gasoline consumption. Four drivers were selected at random; also five cars of the same model with manual transmission were randomly selected from the assembly line. Each driver drove each car twice over a 40-mile test course and the miles per gallon were recorded. The data follow.

Factor A (driver)	*Factor B (car)* $j = 1$	$j = 2$	$j = 3$	$j = 4$	$j = 5$
$i = 1$	25.3	28.9	24.8	28.4	27.1
	25.2	30.0	25.1	27.9	26.6
$i = 2$	33.6	36.7	31.7	35.6	33.7
	32.9	36.5	31.9	35.0	33.9
$i = 3$	27.7	30.7	26.9	29.7	29.2
	28.5	30.4	26.3	30.2	28.9
$i = 4$	29.2	32.4	27.7	31.8	30.3
	29.3	32.4	28.9	30.7	29.9

Assume that random effects ANOVA model (23.15) is applicable.

a. Test whether or not the two factors interact; use $\alpha = .05$. State the alternatives, decision rule, and conclusion.

b. Test separately whether or not factor A and factor B main effects are present. For each test, use $\alpha = .05$ and state the alternatives, decision rule, and conclusion.

c. Obtain point estimates of σ_α^2 and σ_β^2. Which factor appears to have the greater effect on gasoline consumption?

23.16. Refer to **Disk drive service** Problem 20.14. Suppose that the service center employs a large number of technicians and that the three included in the study were selected at random. Assume that the conditions of mixed effects ANOVA model (23.16) are applicable, except that here factor A effects are random and factor B effects are fixed. Under current conditions, all technicians service each of the three makes with approximately equal frequency.

a. Test whether or not the two factors interact; use $\alpha = .01$. State the alternatives, decision rule, and conclusion.

b. Obtain a point estimate of $\sigma_{\alpha\beta}^2$. Does $\sigma_{\alpha\beta}^2$ appear to be large relative to σ^2? Explain.

c. Test whether or not factor A main effects are present; use $\alpha = .01$. State the alternatives, decision rule, and conclusion. Why is it meaningful here to test for factor A main effects?

d. Test whether or not factor B main effects are present; use $\alpha = .01$. State the alternatives, decision rule, and conclusion. Why is it meaningful here to test for factor B main effects?

e. It is desired to obtain all pairwise comparisons between the means for the three disk drive makes. Use the Tukey procedure and a 95 percent family confidence coefficient to make these comparisons. State your findings.

f. Obtain a point estimate of σ_α^2. Does the variability between technicians appear to be large? Explain.

23.17. **Imitation pearls.** Preliminary research on the production of imitation pearls entailed studying the effect of the number of coats of a special lacquer (factor A) applied to an opalescent plastic bead used as the base of the pearl on the market value of the pearl. Four batches of beads (factor B) were used in the study, and it is desired to also consider their effect on the market value. The three levels of factor A (6, 8, and 10 coats) were fixed in advance, while the four batches can be regarded as a random sample of batches from the bead production process. The market value of each pearl was determined by a panel of experts. The data (coded) follow.

Factor A (number of coats)		*Factor B (batch)* $j = 1$	$j = 2$	$j = 3$	$j = 4$
$i = 1$	6	72.0	72.1	75.2	70.4
		74.6	76.9	73.8	68.1
		67.4	74.8	75.7	72.4
		72.8	73.3	77.8	72.4
$i = 2$	8	76.9	80.3	80.2	74.3
		78.1	79.3	76.6	77.6
		72.9	76.6	77.3	74.4
		74.2	77.2	79.9	72.9
$i = 3$	10	76.3	80.9	79.2	71.6
		74.1	73.7	78.0	77.7
		77.1	78.6	77.6	75.2
		75.0	80.2	81.2	74.4

Assume that mixed effects ANOVA model (23.16) is applicable.

a. Test for interaction effects; use $\alpha = .05$. State the alternatives, decision rule, and conclusion.

b. Test for factor A and factor B main effects. For each test, use $\alpha = .05$ and state the alternatives, decision rule, and conclusion.

c. Estimate $D_1 = \mu_{2.} - \mu_{1.}$ and $D_2 = \mu_{3.} - \mu_{2.}$ by means of the Bonferroni procedure with a 90 percent family confidence coefficient. State your findings.

d. Obtain a point estimate of σ_β^2. Does σ_β^2 appear to be large compared to σ^2? Discuss.

23.18. Refer to **Coin-operated terminals** Problem 23.2. Suppose that the weeks (factor B) had been selected intentionally but the locations (factor A) had been selected at random from a large number of possible locations. Assume that the conditions of mixed effects ANOVA model (23.16) are appropriate, except that here factor A effects are random and factor B effects are fixed.

a. Test for factor B main effects; use $\alpha = .05$. State the alternatives, decision rule, and conclusion.

b. Why can you not test for factor A main effects here?

EXERCISES

23.19. Modify formulas (20.43a) and (20.43b) to apply to ANOVA model (23.1), where $n = 1$.

23.20. Consider a two-factor study with $a = 2$, $b = 2$, and $n = 1$. Obtain the **C** matrix to be used in (23.6a) for the constraint $\mu_{11} = \mu_{12} + \mu_{21} - \mu_{22}$. Show that $\mathbf{b}_R$ in (23.6) has as its elements the point estimators in (23.6b).

23.21. Show that (23.7) is the only second-degree polynomial function of α_i and β_j such that $\sum_i (\alpha\beta)_{ij} = \sum_j (\alpha\beta)_{ij} = 0$.

23.22. For random effects model (23.15), derive $\sigma^2(\bar{Y}_{i..})$. [*Hint:* Use (23.17).]

PROJECTS

23.23. Consider a two-factor study with $a = 3$, $b = 2$, and $n = 5$. Random effects ANOVA model (23.15) is applicable with $\mu_{..} = 92$, $\sigma_\alpha^2 = 24$, $\sigma_\beta^2 = 11$, $\sigma_{\alpha\beta}^2 = .1$, and $\sigma^2 = 8$.

a. Using a normal random number generator, obtain a value for each of the main effects α_i ($i = 1, 2, 3$) and β_j ($j = 1, 2$) and for each interaction effect $(\alpha\beta)_{ij}$.

b. Generate five error terms for each treatment.

c. Combine the parameter values obtained in part (a) and the error terms obtained in part (b) to yield five observations Y_{ijk} for each treatment.

d. Based on the observations obtained in part (c), calculate the F^* test statistic for testing whether or not factor A main effects are present. What is your conclusion using $\alpha = .05$?

e. Repeat the steps in parts (a)–(d) 100 times. Calculate the mean of the 100 numerator mean squares and the mean of the 100 denominator mean squares. Are these means close to theoretical expectations?

f. In what proportion of the 100 trials did the test lead to the conclusion of the presence of factor A main effects? Does the test have good power for the case considered here?

CITED REFERENCE

23.1 Johnson, Dallas E., and Franklin A. Graybill, ''Estimation of σ^2 in a Two-Way Classification Model with Interaction,'' *Journal of the American Statistical Association* 67 (1972), pp. 388–94.

24

Multifactor studies

When three or more factors are studied simultaneously, the model and analysis employed are straightforward extensions of the two-factor case. We shall illustrate the nature of the extensions with reference to three-factor studies where all treatment sample sizes are equal and all treatment means have the same importance. Ordinarily, computer packages will be utilized for performing the needed calculations for multifactor studies involving three or more factors. For completeness, however, we shall present the necessary computational formulas for three-factor studies.

24.1 MODEL I (FIXED EFFECTS) FOR THREE-FACTOR STUDIES

Notation

Three factors, A, B, and C, are investigated at a, b, and c levels, respectively. The mean response for the treatment when factor A is at the ith level ($i = 1, \ldots, a$), factor B is at the jth level ($j = 1, \ldots, b$), and factor C is at the kth level ($k = 1, \ldots, c$) is denoted by μ_{ijk}. The number of observations for each treatment is assumed to be constant, denoted by n. We assume $n \geq 2$.

The mean response when A is at the ith level and B is at the jth level is denoted by $\mu_{ij.}$, and similar notation is used for other pairs of factor levels. Since all treatment means are assumed to have equal importance, we define:

(24.1a)
$$\mu_{ij.} = \frac{\sum_k \mu_{ijk}}{c}$$

(24.1b)
$$\mu_{i.k} = \frac{\sum_j \mu_{ijk}}{b}$$

(24.1c)
$$\mu_{.jk} = \frac{\sum_i \mu_{ijk}}{a}$$

The mean response when A is at the ith level is denoted by $\mu_{i..}$, and similar notation is used for the other factor level means. We define:

(24.2a)
$$\mu_{i..} = \frac{\sum_j \sum_k \mu_{ijk}}{bc}$$

(24.2b)
$$\mu_{.j.} = \frac{\sum_i \sum_k \mu_{ijk}}{ac}$$

(24.2c)
$$\mu_{..k} = \frac{\sum_i \sum_j \mu_{ijk}}{ab}$$

Finally, the overall mean response is denoted by $\mu_{...}$ and is defined:

(24.3)
$$\mu_{...} = \frac{\sum_i \sum_j \sum_k \mu_{ijk}}{abc}$$

Illustration

To illustrate the meaning of the model terms for a three-factor analysis of variance model, we shall consider a study of the effects of sex, age, and intelligence level of college graduates on learning of a complex task. Sex is factor A and has $a = 2$ levels (male, female). Age is factor B and is defined in terms of $b = 3$ levels (young, middle, old). Finally, intelligence is factor C and is defined in terms of $c = 2$ levels (high IQ, normal IQ). Table 24.1 shows the treatment means μ_{ijk} for all factor level combinations, as well as their notational representations. Also shown in Table 24.1 are the various means of the μ_{ijk}. We shall refer repeatedly to this illustration as we explain the model terms for a three-factor study.

TABLE 24.1 Mean learning times according to sex, age, and intelligence (in minutes)

Factor A—Sex	*Intelligence (factor C) and Age (factor B)*											
	$k=1$ High I Q				*$k=2$ Normal I Q*				*Average*			
	$j=1$ *Young*	$j=2$ *Middle*	$j=3$ *Old*	*Average*	$j=1$ *Young*	$j=2$ *Middle*	$j=3$ *Old*	*Average*	$j=1$ *Young*	$j=2$ *Middle*	$j=3$ *Old*	*Average*
$i=1$ Male	9 (μ_{111})	12 (μ_{121})	18 (μ_{131})	13 ($\mu_{1.1}$)	19 (μ_{112})	20 (μ_{122})	21 (μ_{132})	20 ($\mu_{1.2}$)	14 ($\mu_{11.}$)	16 ($\mu_{12.}$)	19.5 ($\mu_{13.}$)	16.5 ($\mu_{1..}$)
$i=2$ Female	9 (μ_{211})	10 (μ_{221})	14 (μ_{231})	11 ($\mu_{2.1}$)	19 (μ_{212})	20 (μ_{222})	21 (μ_{232})	20 ($\mu_{2.2}$)	14 ($\mu_{21.}$)	15 ($\mu_{22.}$)	17.5 ($\mu_{23.}$)	15.5 ($\mu_{2..}$)
Average	9 ($\mu_{.11}$)	11 ($\mu_{.21}$)	16 ($\mu_{.31}$)	12 ($\mu_{..1}$)	19 ($\mu_{.12}$)	20 ($\mu_{.22}$)	21 ($\mu_{.32}$)	20 ($\mu_{..2}$)	14 ($\mu_{.1.}$)	15.5 ($\mu_{.2.}$)	18.5 ($\mu_{.3.}$)	16 ($\mu_{...}$)

Specific effects

Specific effects are defined correspondingly to the two-factor case. In our illustration, the specific effect of male sex for old persons of high IQ, denoted by $\alpha_{1(31)}$, is:

$$\alpha_{1(31)} = \mu_{131} - \mu_{.31} = 18 - 16 = 2$$

This specific effect indicates that the mean learning time for old men of high IQ is two minutes higher than the mean learning time for all old persons of high IQ. We thus define the *specific effect* of factor A at the ith level when B is at the jth level and C at the kth level as:

(24.4a) $$\alpha_{i(jk)} = \mu_{ijk} - \mu_{.jk}$$

Similarly, we define the specific effects of factor B and factor C levels as follows:

(24.4b) $$\beta_{j(ik)} = \mu_{ijk} - \mu_{i.k}$$

(24.4c) $$\gamma_{k(ij)} = \mu_{ijk} - \mu_{ij.}$$

Main effects

Like in the two-factor case, the main effect of a factor level in a three-factor study is an average of the specific effects for that factor level. Thus, in our illustration the main effect of male sex, denoted by α_1, is:

$$\alpha_1 = \frac{(9-9) + (12-11) + (18-16) + (19-19) + (20-20) + (21-21)}{6}$$
$$= .5$$

or symbolically:

$$\alpha_1 = \frac{\alpha_{1(11)} + \alpha_{1(21)} + \alpha_{1(31)} + \alpha_{1(12)} + \alpha_{1(22)} + \alpha_{1(32)}}{6}$$

It follows at once from the definition of the specific effects in (24.4a) that:

$$\alpha_1 = \mu_{1..} - \mu_{...}$$

For our illustration, we have:

$$\alpha_1 = \mu_{1..} - \mu_{...} = 16.5 - 16 = .5$$

which is the same result we obtained by averaging the specific effects.

Thus, the *main effect* of the ith level of factor A is defined:

(24.5a) $$\alpha_i = \frac{\sum_j \sum_k \alpha_{i(jk)}}{bc} = \mu_{i..} - \mu_{...}$$

Similarly, we define the main effect of the jth level of B:

(24.5b)
$$\beta_j = \frac{\sum_i \sum_k \beta_{j(ik)}}{ac} = \mu_{.j.} - \mu_{...}$$

and the main effect of the kth level of C:

(24.5c)
$$\gamma_k = \frac{\sum_i \sum_j \gamma_{k(ij)}}{ab} = \mu_{..k} - \mu_{...}$$

It follows from these definitions that the sums of the main effects are zero:

(24.6)
$$\sum_i \alpha_i = \sum_j \beta_j = \sum_k \gamma_k = 0$$

Specific two-factor interactions

In a two-factor study, the interaction of the ith level of factor A with the jth level of factor B was defined in (20.14) as the difference between μ_{ij} and the value $\mu_{..} + \alpha_i + \beta_j$ that would be expected if the two factors were additive. In other words:

$$(\alpha\beta)_{ij} = \mu_{ij} - (\mu_{..} + \alpha_i + \beta_j)$$

An algebraically identical definition was given in (20.14a):

$$(\alpha\beta)_{ij} = \mu_{ij} - \mu_{i.} - \mu_{.j} + \mu_{..}$$

In a three-factor study, the *specific two-factor interaction* of the ith level of factor A with the jth level of factor B when factor C is at the kth level, denoted by $(\alpha\beta)_{ij(k)}$, is defined in identical fashion:

(24.7a)
$$(\alpha\beta)_{ij(k)} = \mu_{ijk} - \mu_{i.k} - \mu_{.jk} + \mu_{..k}$$

The third subscript simply shows that factor C is at the kth level. Thus, $(\alpha\beta)_{ij(k)}$ has the interpretation of a two-factor interaction in a two-factor study with the restriction that the interrelations between factors A and B are only considered when factor C is at the kth level.

For our example in Table 24.1, we have, for instance:

$$(\alpha\beta)_{11(1)} = 9 - 13 - 9 + 12 = -1$$

This interaction is the ordinary two-factor interaction between sex at the level male and age at the level young when intelligence is at the high level.

In similar fashion, we define the specific two-factor interactions:

(24.7b)
$$(\alpha\gamma)_{ik(j)} = \mu_{ijk} - \mu_{ij.} - \mu_{.jk} + \mu_{.j.}$$

(24.7c)
$$(\beta\gamma)_{jk(i)} = \mu_{ijk} - \mu_{ij.} - \mu_{i.k} + \mu_{i..}$$

Main two-factor interactions

The *main two-factor interaction* between factor A at the ith level and factor B at the jth level is simply the average of the specific interactions $(\alpha\beta)_{ij(k)}$ over all levels of factor C. Denoted as before by $(\alpha\beta)_{ij}$, it is defined:

(24.8a) $$(\alpha\beta)_{ij} = \frac{\sum_k (\alpha\beta)_{ij(k)}}{c}$$

For our illustration in Table 24.1, we have for instance:

$$(\alpha\beta)_{11} = \frac{(9 - 13 - 9 + 12) + (19 - 20 - 19 + 20)}{2} = -.5$$

It may readily be shown that this main interaction can also be expressed as follows:

(24.8b) $$(\alpha\beta)_{ij} = \mu_{ij.} - \mu_{i..} - \mu_{.j.} + \mu_{...}$$

Thus, we have for our illustration:

$$(\alpha\beta)_{11} = 14 - 16.5 - 14 + 16 = -.5$$

which is the same result as obtained before.

Formula (24.8b) indicates that the main two-factor interaction $(\alpha\beta)_{ij}$ in a three-factor study may be viewed as the equivalent of that in a two-factor study except that all means are averaged over factor C. Often, however, it will be more useful to interpret $(\alpha\beta)_{ij}$ from the definitional form (24.8a) as the mean of the specific interactions $(\alpha\beta)_{ij(k)}$. The latter approach is particularly helpful in recognizing that factors A and B may interact even though $(\alpha\beta)_{ij}$ is zero for all combinations of i and j. This can occur if the specific interactions $(\alpha\beta)_{ij(k)}$ are not zero but just happen to average out to zero for all i, j combinations. As long as we remember that the main two-factor interactions $(\alpha\beta)_{ij}$ are averages of specific interactions between A and B at different levels of factor C, we can be alert to the fact that $(\alpha\beta)_{ij} = 0$ does not necessarily imply that all specific interactions $(\alpha\beta)_{ij(k)}$ are zero.

In corresponding fashion, we define the AC and BC main interactions:

(24.8c) $$(\alpha\gamma)_{ik} = \frac{\sum_j (\alpha\gamma)_{ik(j)}}{b} = \mu_{i.k} - \mu_{i..} - \mu_{..k} + \mu_{...}$$

(24.8d) $$(\beta\gamma)_{jk} = \frac{\sum_i (\beta\gamma)_{jk(i)}}{a} = \mu_{.jk} - \mu_{.j.} - \mu_{..k} + \mu_{...}$$

The main two-factor interactions $(\alpha\beta)_{ij}$, $(\alpha\gamma)_{ik}$, and $(\beta\gamma)_{jk}$ are usually simply called *two-factor interactions* or *first-order interactions*. It can readily be shown that the sums of the first-order interactions over each subscript are zero:

(24.9a) $$\sum_i (\alpha\beta)_{ij} = \sum_j (\alpha\beta)_{ij} = 0$$

(24.9b) $$\sum_i (\alpha\gamma)_{ik} = \sum_k (\alpha\gamma)_{ik} = 0$$

(24.9c) $$\sum_j (\beta\gamma)_{jk} = \sum_k (\beta\gamma)_{jk} = 0$$

Three-factor interactions

Just as in a two-factor study, where the interaction between the ith level of factor A and the jth level of factor B is defined as the difference between the treatment mean μ_{ij} and the value that would be expected if the factor effects were additive, so in a three-factor study the three-factor interaction $(\alpha\beta\gamma)_{ijk}$ is defined as the difference between the treatment mean μ_{ijk} and the value that would be expected if main effects and first-order interactions were sufficient to account for all factor effects. The value that would be expected from main effects and first-order interactions when A is at the ith level, B at the jth level, and C at the kth level is:

(24.10) $$\mu_{...} + \alpha_i + \beta_j + \gamma_k + (\alpha\beta)_{ij} + (\alpha\gamma)_{ik} + (\beta\gamma)_{jk}$$

Hence, the *three-factor interaction* $(\alpha\beta\gamma)_{ijk}$, also called the *second-order interaction*, is defined as:

(24.11a) $$(\alpha\beta\gamma)_{ijk} = \mu_{ijk} - [\mu_{...} + \alpha_i + \beta_j + \gamma_k + (\alpha\beta)_{ij} + (\alpha\gamma)_{ik} + (\beta\gamma)_{jk}]$$

or equivalently:

(24.11b) $$(\alpha\beta\gamma)_{ijk} = \mu_{ijk} - \mu_{ij.} - \mu_{i.k} - \mu_{.jk} + \mu_{i..} + \mu_{.j.} + \mu_{..k} - \mu_{...}$$

From the definition of the three-factor interactions, it follows that they sum to zero when added over any index:

(24.12) $$\sum_i (\alpha\beta\gamma)_{ijk} = \sum_j (\alpha\beta\gamma)_{ijk} = \sum_k (\alpha\beta\gamma)_{ijk} = 0$$

If *all* three-factor interactions $(\alpha\beta\gamma)_{ijk}$ are zero, we say that there are no three-factor interactions between factors A, B, and C. If some $(\alpha\beta\gamma)_{ijk}$ are not zero, we say that three-factor interactions are present.

Let us find the three-factor interaction $(\alpha\beta\gamma)_{111}$ for our example in Table 24.1. We require the following terms:

$$\begin{array}{ll} \mu_{...} = 16 & (\alpha\beta)_{11} = 14 - 16.5 - 14 + 16 = -.5 \\ \alpha_1 = 16.5 - 16 = .5 & (\alpha\gamma)_{11} = 13 - 16.5 - 12 + 16 = .5 \\ \beta_1 = 14 - 16 = -2 & (\beta\gamma)_{11} = 9 - 14 - 12 + 16 = -1 \\ \gamma_1 = 12 - 16 = -4 & \mu_{111} = 9 \end{array}$$

Hence, we have:

$$(\alpha\beta\gamma)_{111} = 9 - (16 + .5 - 2 - 4 - .5 + .5 - 1) = -.5$$

Since $(\alpha\beta\gamma)_{111}$ is not zero, we know at once that three-factor interactions are present in this example.

Note

The three-factor interaction $(\alpha\beta\gamma)_{ijk}$ can also be expressed as the difference between the specific two-factor interaction $(\alpha\beta)_{ij(k)}$ and the main two-factor interaction $(\alpha\beta)_{ij}$:

(24.13a) $$(\alpha\beta\gamma)_{ijk} = (\alpha\beta)_{ij(k)} - (\alpha\beta)_{ij}$$

Thus, in our illustration, we found $(\alpha\beta)_{11(1)} = -1$ and $(\alpha\beta)_{11} = -.5$, so that $(\alpha\beta\gamma)_{111} = -1 - (-.5) = -.5$, as we saw above.

Because of the symmetrical structure of three-factor interactions, they can also be expressed in the following ways:

(24.13b) $$(\alpha\beta\gamma)_{ijk} = (\alpha\gamma)_{ik(j)} - (\alpha\gamma)_{ik}$$

(24.13c) $$(\alpha\beta\gamma)_{ijk} = (\beta\gamma)_{jk(i)} - (\beta\gamma)_{jk}$$

Interpretation of interactions in three-factor studies

The presence of three-factor interactions implies that at least some of the specific two-factor interactions for any two factors differ, depending on the level of the third factor. This can be seen from (24.13), for if the specific interactions were equal, the deviations of the specific interactions from their mean would all be zero. Thus, if three-factor interactions are present, the interactions between any two factors need to be studied separately for each level of the third factor.

Admittedly, this explanation of the implications of three-factor interactions is somewhat abstruse. To shed more light on the nature of interactions in three-factor studies, we shall examine some examples by means of tables and graphs, beginning with very simple factor effects and building up to complex three-factor interaction effects. In each example, we present the true treatment means μ_{ijk}.

Example 1—Main effects only. Figure 24.1 illustrates a case where there are A, B, and C main effects but no interactions of any kind. The AB response curves in Figure 24.1a are plots of the treatment means μ_{ijk} against C. Here, the main A effects are reflected by the slopes of the AB curves not being zero. The main B effects are reflected by the differences in the heights of the two AB curves within each panel, and the main C effects are reflected by the corresponding curves in the two panels being at different heights.

The absence of AB interactions is shown by the parallel response curves in each panel. We know from our discussion of two-factor analysis that parallel response curves imply absence of interactions. Here, the parallel response curves within each panel imply the absence of specific AB interactions, that is, all specific AB interactions $(\alpha\beta)_{ij(k)} = 0$. This in turn implies:

1. The main AB interactions $(\alpha\beta)_{ij}$ equal zero. This follows from (24.8a), since the means of specific interactions that are all zero must also be zero.
2. The ABC interactions $(\alpha\beta\gamma)_{ijk}$ equal zero. This follows at once from the previous point and from (24.13a).

The absences of BC and AC interactions in this example become evident when the BC and AC response curves are plotted against A and B, respectively. Thus, in Figure 24.1b, the same treatment means μ_{ijk} are shown with the AC response curves plotted against B. Note that these curves are parallel in each panel, implying that the main AC interactions are zero.

FIGURE 24.1 *A*, *B*, and *C* main effects and no interactions

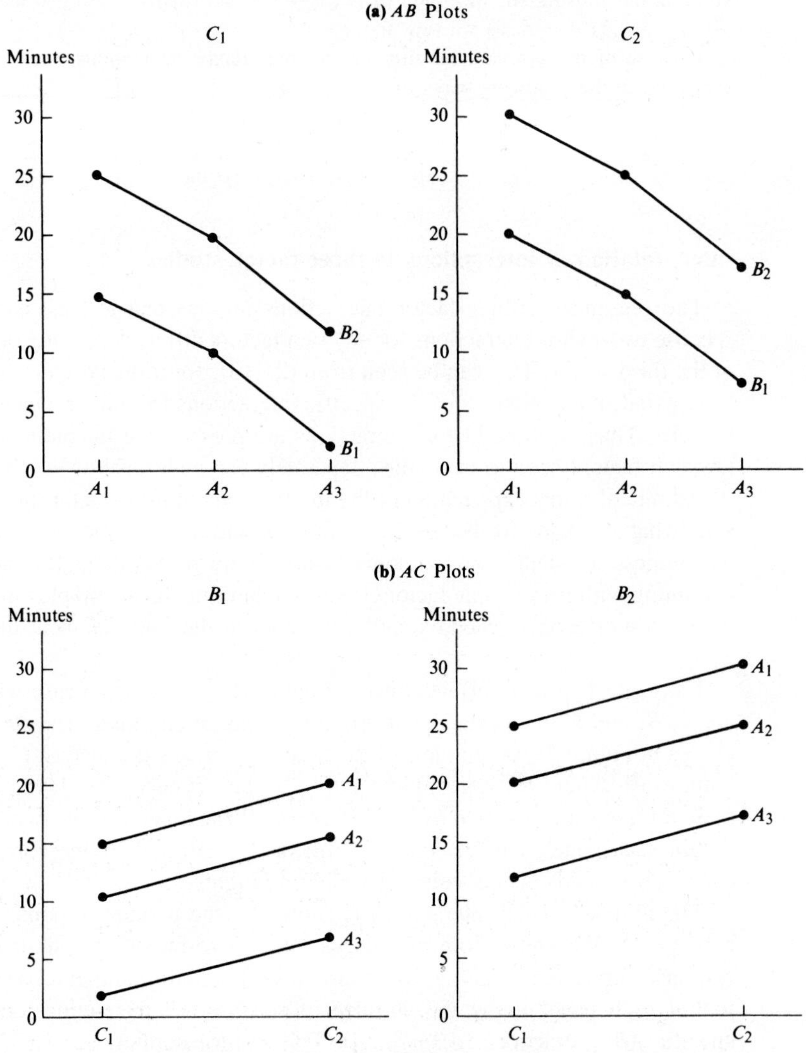

Example 2—Main effects and *AB* interactions. Figure 24.2 illustrates a case where there are *A*, *B*, and *C* main effects and *AB* interactions but no other interactions. Note that the two *AB* response curves in each panel of Figure 24.2a are no longer parallel, reflecting the presence of *AB* interactions. However, the upper curves in the three panels are parallel, as are the lower curves. This implies that when the *AC* curves are plotted against *B*, the *AC* response curves within each panel will be parallel. This is shown in Figure 24.2b, which contains the same treatment means as in Figure 24.2a. The parallelism of the *AC* response

FIGURE 24.2 *A*, *B*, and *C* main effects and *AB* interactions

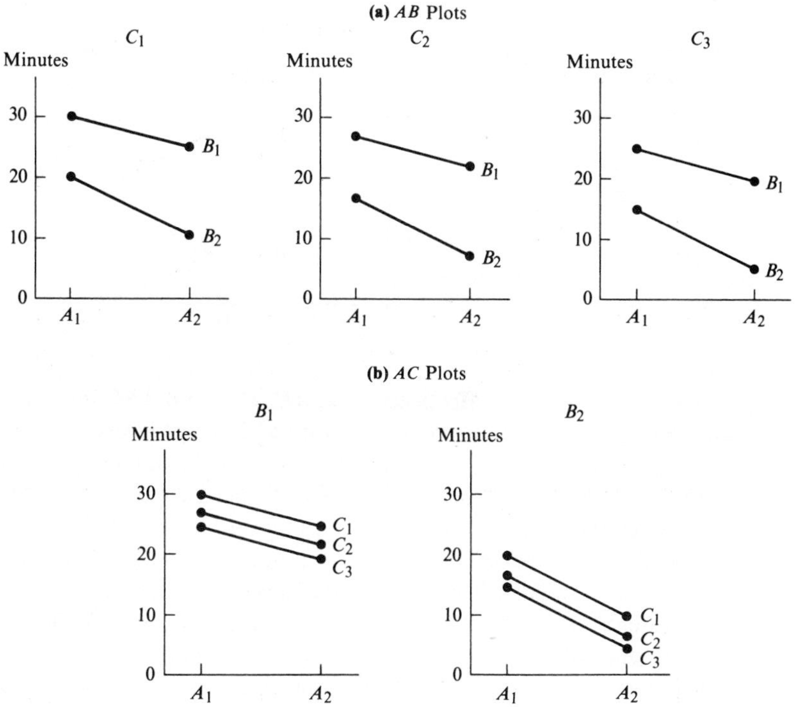

curves within each panel in Figure 24.2b in turn implies by the earlier reasoning that there are no *AC* interactions, as well as no *ABC* interactions, that is, all $(\alpha\gamma)_{ik} = 0$ and all $(\alpha\beta\gamma)_{ijk} = 0$.

Thus, if the *AB* (*AC*, *BC*) response curves corresponding to any given level of factor *C* (*B*, *A*) are parallel for all levels of factor *C* (*B*, *A*), as in Figure 24.2a, even though the response curves within a panel are not parallel, it follows that there are no three-factor interactions.

Example 3—Main effects and *AB* and *AC* interactions. It does not follow, however, that lack of parallelism either within or between panels implies the presence of three-factor interactions. Figure 24.3 portrays a case where main *A*, *B*, and *C* effects and main *AB* and *AC* two-factor interactions are present but no three-factor interactions exist. Yet there are no parallel response curves either within or between panels. The fact that the *AB* response curves when factor *B* is at level B_1 are not parallel for the three levels of factor *C* reflects the presence of *AC* interactions. The lack of parallelism of the *AB* response curves within each panel reflects the presence of *AB* interactions. Note, however, that the differences $\mu_{11k} - \mu_{12k}$ are the same for all levels of factor *C*, as are the differences $\mu_{21k} - \mu_{22k}$. Hence, the specific *AB* interactions are the same for all levels of factor *C*, and consequently by (24.13a) no three-factor interactions are present.

FIGURE 24.3 *A*, *B*, and *C* main effects and *AB* and *AC* interactions—*AB* plots

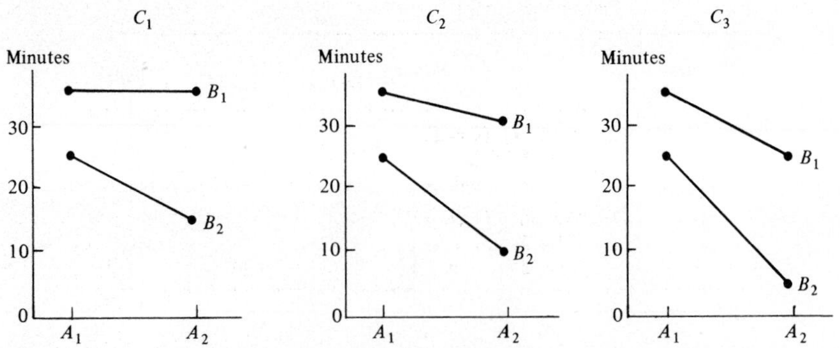

Example 4—Main effects and *AB*, *AC*, *BC*, and *ABC* interactions. Figure 24.4 shows the *AB* response plots for the different levels of factor *C* for our illustration in Table 24.1 of the effects of sex, age, and intelligence on learning time. The plots of the treatment means in Figure 24.4 reveals that there are no specific *AB* interactions when factor *C* is at level C_2 since all *AB* response curves are parallel there. Indeed there are no specific *A* effects then. In contrast, specific *A* effects are present when factor *C* is at level C_1. Specific *AB* interactions are also present, in view of the nonparallel *AB* response curves in the panel for C_1. The presence of specific *AB* interactions for C_1 and the absence of specific *AB* interactions for C_2 implies by (24.13a) the presence of three-factor *ABC* interactions. We present in Table 24.2 the specific *AB* interactions for each level of factor *C*, confirming the information obtained visually from Figure 24.4.

The nature of the factor effects and the various interactions can be seen clearly from Figure 24.4. For persons with normal IQ, sex has no effect on mean learning time, and age has only a small effect leading to slightly longer learning times

FIGURE 24.4 *A*, *B*, and *C* main effects and *AB*, *AC*, *BC*, and *ABC* interactions—Table 24.1 learning example

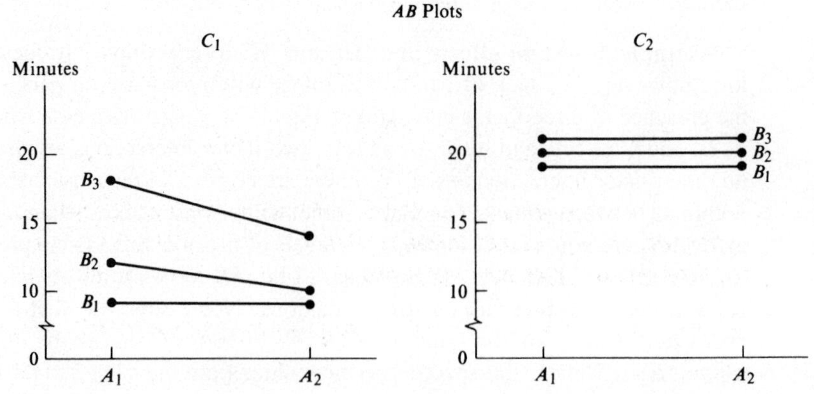

TABLE 24.2 Specific sex-age interactions by IQ level—Table 24.1 learning example

	$j=1$ *Young*	$j=2$ *Middle*	$j=3$ *Old*
High I Q ($k=1$):			
$i=1$ Male	−1	0	+1
$i=2$ Female	+1	0	−1
Normal I Q ($k=2$):			
$i=1$ Male	0	0	0
$i=2$ Female	0	0	0

for older persons. For persons with high IQ, on the other hand, females tend to learn more quickly than males for older persons but not for young persons, and older persons tend to require substantially longer learning times than young persons.

Note

If three-factor interactions are difficult to understand, higher-order interactions such as four-factor interactions are yet more abstruse. Fortunately, it is often found in practice that these higher-order interactions are quite small or nonexistent. When this is the case, they can be disregarded in the analysis of factor effects.

Model

Let Y_{ijkm} be the mth observation ($m = 1, \ldots, n$) for the treatment consisting of the ith level of A ($i = 1, \ldots, a$), the jth level of B ($j = 1, \ldots, b$), and the kth level of C ($k = 1, \ldots, c$). Thus, the total number of observations in the study is:

$$n_T = nabc \tag{24.14}$$

The fixed effects ANOVA model for a three-factor study in terms of the treatment means μ_{ijk} is:

$$Y_{ijkm} = \mu_{ijk} + \varepsilon_{ijkm} \tag{24.15}$$

where:

μ_{ijk} are parameters
ε_{ijkm} are independent $N(0, \sigma^2)$
$i = 1, \ldots, a; j = 1, \ldots, b; k = 1, \ldots, c; m = 1, \ldots, n$

An equivalent model can be developed that incorporates the factorial structure by expressing the treatment mean μ_{ijk} in terms of the various factor effects. From the three-factor interaction definition (24.11a), we have the identity:

$$\mu_{ijk} \equiv \mu_{\ldots} + \alpha_i + \beta_j + \gamma_k + (\alpha\beta)_{ij} + (\alpha\gamma)_{ik} + (\beta\gamma)_{jk} + (\alpha\beta\gamma)_{ijk} \tag{24.16}$$

where:

$$\mu_{...} = \frac{\Sigma\Sigma\Sigma\mu_{ijk}}{abc}$$

$$\alpha_i = \mu_{i..} - \mu_{...}$$

$$\beta_j = \mu_{.j.} - \mu_{...}$$

$$\gamma_k = \mu_{..k} - \mu_{...}$$

$$(\alpha\beta)_{ij} = \mu_{ij.} - \mu_{i..} - \mu_{.j.} + \mu_{...}$$

$$(\alpha\gamma)_{ik} = \mu_{i.k} - \mu_{i..} - \mu_{..k} + \mu_{...}$$

$$(\beta\gamma)_{jk} = \mu_{.jk} - \mu_{.j.} - \mu_{..k} + \mu_{...}$$

$$(\alpha\beta\gamma)_{ijk} = \mu_{ijk} - \mu_{ij.} - \mu_{i.k} - \mu_{.jk} + \mu_{i..} + \mu_{.j.} + \mu_{..k} - \mu_{...}$$

Hence, the equivalent fixed effects ANOVA model for a three-factor study is:

$$(24.17) \qquad Y_{ijkm} = \mu_{...} + \alpha_i + \beta_j + \gamma_k + (\alpha\beta)_{ij} + (\alpha\gamma)_{ik} + (\beta\gamma)_{jk} + (\alpha\beta\gamma)_{ijk} + \varepsilon_{ijkm}$$

where:

ε_{ijkm} are independent $N(0, \sigma^2)$

α_i, β_j, γ_k, $(\alpha\beta)_{ij}$, $(\alpha\gamma)_{ik}$, $(\beta\gamma)_{jk}$, $(\alpha\beta\gamma)_{ijk}$ are constants subject to the restrictions:

$$\sum_i \alpha_i = \sum_j \beta_j = \sum_k \gamma_k = 0$$

$$\sum_i (\alpha\beta)_{ij} = \sum_j (\alpha\beta)_{ij} = \sum_i (\alpha\gamma)_{ik} = 0$$

$$\sum_k (\alpha\gamma)_{ik} = \sum_j (\beta\gamma)_{jk} = \sum_k (\beta\gamma)_{jk} = 0$$

$$\sum_i (\alpha\beta\gamma)_{ijk} = \sum_j (\alpha\beta\gamma)_{ijk} = \sum_k (\alpha\beta\gamma)_{ijk} = 0$$

The fixed effects three-factor model (24.15) and the equivalent model (24.17) are linear models, just as in the two-factor case. We shall illustrate this for an example later in the chapter.

24.2 ANALYSIS OF VARIANCE

Sample notation

The notation for sample totals and means is a straightforward extension of that for two-factor studies. As usual, a dot in the subscript indicates aggregation or averaging over the index represented by the dot. We have:

$$(24.18a)\qquad Y_{ijk.} = \sum_m Y_{ijkm} \qquad \bar{Y}_{ijk.} = \frac{Y_{ijk.}}{n}$$

$$(24.18b)\qquad Y_{ij..} = \sum_k \sum_m Y_{ijkm} \qquad \bar{Y}_{ij..} = \frac{Y_{ij..}}{cn}$$

$$(24.18c)\qquad Y_{i.k.} = \sum_j \sum_m Y_{ijkm} \qquad \bar{Y}_{i.k.} = \frac{Y_{i.k.}}{bn}$$

$$(24.18d)\qquad Y_{.jk.} = \sum_i \sum_m Y_{ijkm} \qquad \bar{Y}_{.jk.} = \frac{Y_{.jk.}}{an}$$

$$(24.18e)\qquad Y_{i...} = \sum_j \sum_k \sum_m Y_{ijkm} \qquad \bar{Y}_{i...} = \frac{Y_{i...}}{bcn}$$

$$(24.18f)\qquad Y_{.j..} = \sum_i \sum_k \sum_m Y_{ijkm} \qquad \bar{Y}_{.j..} = \frac{Y_{.j..}}{acn}$$

$$(24.18g)\qquad Y_{..k.} = \sum_i \sum_j \sum_m Y_{ijkm} \qquad \bar{Y}_{..k.} = \frac{Y_{..k.}}{abn}$$

$$(24.18h)\qquad Y_{....} = \sum_i \sum_j \sum_k \sum_m Y_{ijkm} \qquad \bar{Y}_{....} = \frac{Y_{....}}{abcn}$$

Example. Table 24.3 illustrates this notation for a study of the effects of sex, body fat, and smoking history on exercise tolerance in stress testing of persons 25 to 35 years old. Each of the three factors has two levels, and there are three replications for each treatment. Tables 24.3a, b, and c show, respectively, the observations, totals, and means, together with the corresponding notation. We shall fully analyze these data later on.

Fitting of ANOVA model

When model (24.15) is fitted by the method of least squares, the estimators as usual turn out to be the treatment sample means:

$$(24.19)\qquad \hat{\mu}_{ijk} = \bar{Y}_{ijk.}$$

Thus, the *fitted values* for the observations are the treatment sample means:

$$(24.20)\qquad \hat{Y}_{ijkm} = \hat{\mu}_{ijk} = \bar{Y}_{ijk.}$$

and the *residuals* are the deviations of the observed values from the treatment sample means:

$$(24.21)\qquad e_{ijkm} = Y_{ijkm} - \hat{Y}_{ijkm} = Y_{ijkm} - \bar{Y}_{ijk.}$$

TABLE 24.3 Sample data, totals, and means for three-factor study—stress test example

(a) Observations

	Smoking History	
	$k = 1$ *Light*	$k = 2$ *Heavy*
$j = 1$ Low fat:		
$i = 1$ Male	24 (Y_{1111})	18 (Y_{1121})
	29 (Y_{1112})	19 (Y_{1122})
	25 (Y_{1113})	23 (Y_{1123})
$i = 2$ Female	20 (Y_{2111})	15 (Y_{2121})
	22 (Y_{2112})	10 (Y_{2122})
	18 (Y_{2113})	11 (Y_{2123})
$j = 2$ High fat:		
$i = 1$ Male	15 (Y_{1211})	15 (Y_{1221})
	15 (Y_{1212})	20 (Y_{1222})
	12 (Y_{1213})	13 (Y_{1223})
$i = 2$ Female	16 (Y_{2211})	10 (Y_{2221})
	9 (Y_{2212})	14 (Y_{2222})
	11 (Y_{2213})	6 (Y_{2223})

(b) Totals

	$k = 1$	$k = 2$	All k
$j = 1$:			
$i = 1$	78 ($Y_{111.}$)	60 ($Y_{112.}$)	138 ($Y_{11..}$)
$i = 2$	60 ($Y_{211.}$)	36 ($Y_{212.}$)	96 ($Y_{21..}$)
All i	138 ($Y_{.11.}$)	96 ($Y_{.12.}$)	234 ($Y_{.1..}$)
$j = 2$:			
$i = 1$	42 ($Y_{121.}$)	48 ($Y_{122.}$)	90 ($Y_{12..}$)
$i = 2$	36 ($Y_{221.}$)	30 ($Y_{222.}$)	66 ($Y_{22..}$)
All i	78 ($Y_{.21.}$)	78 ($Y_{.22.}$)	156 ($Y_{.2..}$)
All j:			
$i = 1$	120 ($Y_{1.1.}$)	108 ($Y_{1.2.}$)	228 ($Y_{1...}$)
$i = 2$	96 ($Y_{2.1.}$)	66 ($Y_{2.2.}$)	162 ($Y_{2...}$)
All i	216 ($Y_{..1.}$)	174 ($Y_{..2.}$)	390 ($Y_{....}$)

(c) Averages

	$k = 1$	$k = 2$	All k
$j = 1$:			
$i = 1$	26 ($\bar{Y}_{111.}$)	20 ($\bar{Y}_{112.}$)	23 ($\bar{Y}_{11..}$)
$i = 2$	20 ($\bar{Y}_{211.}$)	12 ($\bar{Y}_{212.}$)	16 ($\bar{Y}_{21..}$)
All i	23 ($\bar{Y}_{.11.}$)	16 ($\bar{Y}_{.12.}$)	19.5 ($\bar{Y}_{.1..}$)
$j = 2$:			
$i = 1$	14 ($\bar{Y}_{121.}$)	16 ($\bar{Y}_{122.}$)	15 ($\bar{Y}_{12..}$)
$i = 2$	12 ($\bar{Y}_{221.}$)	10 ($\bar{Y}_{222.}$)	11 ($\bar{Y}_{22..}$)
All i	13 ($\bar{Y}_{.21.}$)	13 ($\bar{Y}_{.22.}$)	13 ($\bar{Y}_{.2..}$)
All j:			
$i = 1$	20 ($\bar{Y}_{1.1.}$)	18 ($\bar{Y}_{1.2.}$)	19 ($\bar{Y}_{1...}$)
$i = 2$	16 ($\bar{Y}_{2.1.}$)	11 ($\bar{Y}_{2.2.}$)	13.5 ($\bar{Y}_{2...}$)
All i	18 ($\bar{Y}_{..1.}$)	14.5 ($\bar{Y}_{..2.}$)	16.25 ($\bar{Y}_{....}$)

For the equivalent model (24.17), the least squares estimators of the parameters are as follows:

	Parameter	Estimator
(24.22a)	$\mu_{...}$	$\hat{\mu}_{...} = \bar{Y}_{....}$
(24.22b)	α_i	$\hat{\alpha}_i = \bar{Y}_{i...} - \bar{Y}_{....}$
(24.22c)	β_j	$\hat{\beta}_j = \bar{Y}_{.j..} - \bar{Y}_{....}$
(24.22d)	γ_k	$\hat{\gamma}_k = \bar{Y}_{..k.} - \bar{Y}_{....}$
(24.22e)	$(\alpha\beta)_{ij}$	$\widehat{(\alpha\beta)}_{ij} = \bar{Y}_{ij..} - \bar{Y}_{i...} - \bar{Y}_{.j..} + \bar{Y}_{....}$
(24.22f)	$(\alpha\gamma)_{ik}$	$\widehat{(\alpha\gamma)}_{ik} = \bar{Y}_{i.k.} - \bar{Y}_{i...} - \bar{Y}_{..k.} + \bar{Y}_{....}$
(24.22g)	$(\beta\gamma)_{jk}$	$\widehat{(\beta\gamma)}_{jk} = \bar{Y}_{.jk.} - \bar{Y}_{.j..} - \bar{Y}_{..k.} + \bar{Y}_{....}$
(24.22h)	$(\alpha\beta\gamma)_{ijk}$	$\widehat{(\alpha\beta\gamma)}_{ijk} = \bar{Y}_{ijk.} - \bar{Y}_{ij..} - \bar{Y}_{i.k.} - \bar{Y}_{.jk.} + \bar{Y}_{i...} + \bar{Y}_{.j..} + \bar{Y}_{..k.} - \bar{Y}_{....}$

The fitted values and residuals for equivalent model (24.17) are exactly the same as those for model (24.15), as was also the case for two-factor studies.

Breakdown of total sum of squares

Neglecting the factorial structure of the study and simply considering it to contain *abc* treatments, we obtain the usual breakdown of the total sum of squares:

(24.23) $$SSTO = SSTR + SSE$$

where:

(24.23a) $$SSTO = \sum_i \sum_j \sum_k \sum_m (Y_{ijkm} - \bar{Y}_{....})^2$$

(24.23b) $$SSTR = n \sum_i \sum_j \sum_k (\bar{Y}_{ijk.} - \bar{Y}_{....})^2$$

(24.23c) $$SSE = \sum_i \sum_j \sum_k \sum_m (Y_{ijkm} - \bar{Y}_{ijk.})^2 = \sum_i \sum_j \sum_k \sum_m e_{ijkm}^2$$

Consider now the treatment mean deviation $\bar{Y}_{ijk.} - \bar{Y}_{....}$, which appears in *SSTR*. This can be decomposed in terms of the least squares estimators (24.22) of the main effects, two-factor interactions, and three-factor interaction:

$$\underbrace{\bar{Y}_{ijk.} - \bar{Y}_{....}}_{\text{Treatment mean deviation}} = \underbrace{\bar{Y}_{i...} - \bar{Y}_{....}}_{A\text{ main effect}} + \underbrace{\bar{Y}_{.j..} - \bar{Y}_{....}}_{B\text{ main effect}} + \underbrace{\bar{Y}_{..k.} - \bar{Y}_{....}}_{C\text{ main effect}}$$

$$+ \underbrace{\bar{Y}_{ij..} - \bar{Y}_{i...} - \bar{Y}_{.j..} + \bar{Y}_{....}}_{AB\text{ interaction effect}}$$

$$+ \underbrace{\bar{Y}_{i.k.} - \bar{Y}_{i...} - \bar{Y}_{..k.} + \bar{Y}_{....}}_{AC\text{ interaction effect}} + \underbrace{\bar{Y}_{.jk.} - \bar{Y}_{.j..} - \bar{Y}_{..k.} + \bar{Y}_{....}}_{BC\text{ interaction effect}}$$

$$+ \underbrace{\bar{Y}_{ijk.} - \bar{Y}_{ij..} - \bar{Y}_{i.k.} - \bar{Y}_{.jk.} + \bar{Y}_{i...} + \bar{Y}_{.j..} + \bar{Y}_{..k.} - \bar{Y}_{....}}_{ABC\text{ interaction effect}}$$

When we square each side and sum over i, j, k, and m, all cross-product terms drop out and we obtain:

(24.24) $$SSTR = SSA + SSB + SSC + SSAB + SSAC + SSBC + SSABC$$

where:

(24.24a) $$SSA = nbc\sum_i (\bar{Y}_{i...} - \bar{Y}_{....})^2$$

(24.24b) $$SSB = nac\sum_j (\bar{Y}_{.j..} - \bar{Y}_{....})^2$$

(24.24c) $$SSC = nab\sum_k (\bar{Y}_{..k.} - \bar{Y}_{....})^2$$

(24.24d) $$SSAB = nc\sum_i \sum_j (\bar{Y}_{ij..} - \bar{Y}_{i...} - \bar{Y}_{.j..} + \bar{Y}_{....})^2$$

(24.24e) $$SSAC = nb\sum_i \sum_k (\bar{Y}_{i.k.} - \bar{Y}_{i...} - \bar{Y}_{..k.} + \bar{Y}_{....})^2$$

(24.24f) $$SSBC = na\sum_j \sum_k (\bar{Y}_{.jk.} - \bar{Y}_{.j..} - \bar{Y}_{..k.} + \bar{Y}_{....})^2$$

(24.24g) $$SSABC = n\sum_i \sum_j \sum_k (\bar{Y}_{ijk.} - \bar{Y}_{ij..} - \bar{Y}_{i.k.} - \bar{Y}_{.jk.} + \bar{Y}_{i...} + \bar{Y}_{.j..} + \bar{Y}_{..k.} - \bar{Y}_{....})^2$$

Combining (24.23) and (24.24), we have thus established the orthogonal decomposition:

(24.25) $$SSTO = SSA + SSB + SSC + SSAB + SSAC + SSBC + SSABC + SSE$$

SSA, SSB, and SSC are the usual main effect sums of squares. For instance, the larger (absolutely) are the estimated main B effects $\bar{Y}_{.j..} - \bar{Y}_{....}$, the larger will be SSB.

$SSAB$, $SSAC$, and $SSBC$ are the usual two-factor interaction sums of squares. For instance, the larger (absolutely) are the estimated AB interactions $\bar{Y}_{ij..} - \bar{Y}_{i...} - \bar{Y}_{.j..} + \bar{Y}_{....}$, the larger will be $SSAB$.

Finally, $SSABC$ is the three-factor interaction sum of squares. The larger (absolutely) are these estimated three-factor interactions, the larger will be $SSABC$.

Computational formulas. For the occasional instance when the calculations are to be done by hand, the use of the definitional formulas previously given is cumbersome. The computational formulas for the three-factor case follow the pattern of the two-factor formulas:

(24.26a) $$SSTO = \sum_i \sum_j \sum_k \sum_m Y_{ijkm}^2 - \frac{Y_{....}^2}{nabc}$$

(24.26b) $$SSE = \sum_i \sum_j \sum_k \sum_m Y_{ijkm}^2 - \sum_i \sum_j \sum_k \frac{Y_{ijk.}^2}{n}$$

(24.26c) $$SSA = \frac{\sum_i Y_{i...}^2}{nbc} - \frac{Y_{....}^2}{nabc}$$

(24.26d) $$SSB = \frac{\sum_j Y_{.j..}^2}{nac} - \frac{Y_{....}^2}{nabc}$$

(24.26e) $$SSC = \frac{\sum_k Y_{..k.}^2}{nab} - \frac{Y_{....}^2}{nabc}$$

The two-factor interaction sum of squares $SSAB$ can be obtained by working with the means $\bar{Y}_{ij..}$ and treating these ab means as constituting a two-factor study. The "treatment sum of squares" for this two-factor study, which we will denote by $SSTR(A, B)$, is as usual:

(24.27) $$SSTR(A, B) = nc \sum_i \sum_j (\bar{Y}_{ij..} - \bar{Y}_{....})^2 = \frac{\sum_i \sum_j Y_{ij..}^2}{nc} - \frac{Y_{....}^2}{nabc}$$

It can be shown that:

(24.28) $$SSTR(A, B) = SSA + SSB + SSAB$$

Hence, we can find $SSAB$ by subtraction:

(24.29a) $$SSAB = SSTR(A, B) - SSA - SSB$$

Similarly, we find $SSAC$ and $SSBC$ as follows:

(24.29b) $$SSAC = SSTR(A, C) - SSA - SSC$$

(24.29c) $$SSBC = SSTR(B, C) - SSB - SSC$$

where:

(24.29d) $$SSTR(A, C) = \frac{\sum_i \sum_k Y_{i.k.}^2}{nb} - \frac{Y_{....}^2}{nabc}$$

(24.29e) $$SSTR(B, C) = \frac{\sum_j \sum_k Y_{.jk.}^2}{na} - \frac{Y_{....}^2}{nabc}$$

The three-factor interaction sum of squares is obtained by subtraction:

(24.30) $$SSABC = SSTO - SSE - SSA - SSB - SSC - SSAB - SSAC - SSBC$$

Note

The computational formulas above can be extended readily if more than three factors are studied simultaneously. Usually, however, a computer package will be utilized under these circumstances.

Degrees of freedom and mean squares

Table 24.4 contains the general ANOVA table for the three-factor fixed effects model (24.17). The degrees of freedom for main effect and two-factor interaction sums of squares correspond to those for two-factor studies. The number of degrees of freedom associated with *SSABC* is obtained by subtraction and corresponds to the number of independent linear relations among all the interaction terms $(\alpha\beta\gamma)_{ijk}$.

The expected mean squares are also given in Table 24.4. Note that *MSA*, *MSB*, *MSC*, *MSAB*, *MSAC*, *MSBC*, and *MSABC* all have expectations equal to σ^2 if there are no factor effects of the type reflected by the mean square. If such effects are present, each mean square has an expectation exceeding σ^2. As usual, $E(MSE) = \sigma^2$ always. Hence, the test for factor effects consists of comparing the appropriate mean square against *MSE* by means of an F^* test statistic, with large values of F^* indicating the presence of factor effects.

Table 24.6 (p. 825) contains the ANOVA table for the three-factor stress test example (calculational details are not shown).

Tests for factor effects

The various tests for factor effects all follow the same pattern; we illustrate them with the test for three-factor interactions. The alternatives are:

(24.31a) $$H_0\text{: all } (\alpha\beta\gamma)_{ijk} = 0$$
$$H_a\text{: not all } (\alpha\beta\gamma)_{ijk} \text{ equal zero}$$

TABLE 24.4 General ANOVA table for three-factor fixed effects study

Source of Variation	*SS*	*df*	*MS*	*E(MS)*
Between treatments	*SSTR*	$abc - 1$	*MSTR*	$\sigma^2 + \frac{n\Sigma\Sigma\Sigma(\mu_{ijk} - \mu_{...})^2}{abc - 1}$
Factor *A*	*SSA*	$a - 1$	*MSA*	$\sigma^2 + \frac{bcn}{a - 1}\Sigma\alpha_i^2$
Factor *B*	*SSB*	$b - 1$	*MSB*	$\sigma^2 + \frac{acn}{b - 1}\Sigma\beta_j^2$
Factor *C*	*SSC*	$c - 1$	*MSC*	$\sigma^2 + \frac{abn}{c - 1}\Sigma\gamma_k^2$
AB interactions	*SSAB*	$(a - 1)(b - 1)$	*MSAB*	$\sigma^2 + \frac{cn}{(a - 1)(b - 1)}\Sigma\Sigma(\alpha\beta)_{ij}^2$
AC interactions	*SSAC*	$(a - 1)(c - 1)$	*MSAC*	$\sigma^2 + \frac{bn}{(a - 1)(c - 1)}\Sigma\Sigma(\alpha\gamma)_{ik}^2$
BC interactions	*SSBC*	$(b - 1)(c - 1)$	*MSBC*	$\sigma^2 + \frac{an}{(b - 1)(c - 1)}\Sigma\Sigma(\beta\gamma)_{jk}^2$
ABC interactions	*SSABC*	$(a - 1)(b - 1)(c - 1)$	*MSABC*	$\sigma^2 + \frac{n}{(a - 1)(b - 1)(c - 1)}\Sigma\Sigma\Sigma(\alpha\beta\gamma)_{ijk}^2$
Error	*SSE*	$abc(n - 1)$	*MSE*	σ^2
Total	*SSTO*	$abcn - 1$		

Note: $\mu_{...}$, α_i, β_j, γ_k, $(\alpha\beta)_{ij}$, $(\alpha\gamma)_{ik}$, $(\beta\gamma)_{jk}$, and $(\alpha\beta\gamma)_{ijk}$ are defined in (24.16).

The appropriate test statistic is:

$$F^* = \frac{MSABC}{MSE} \tag{24.31b}$$

If H_0 holds, F^* follows the F distribution with $(a-1)(b-1)(c-1)$ degrees of freedom for the numerator and $abc(n-1)$ degrees of freedom for the denominator. Hence, the decision rule to control the Type I error at α is:

(24.31c)
$$\text{If } F^* \leq F[1-\alpha; (a-1)(b-1)(c-1), (n-1)abc], \text{ conclude } H_0$$
$$\text{If } F^* > F[1-\alpha; (a-1)(b-1)(c-1), (n-1)abc], \text{ conclude } H_a$$

Table 24.5 contains the appropriate test statistics and percentiles of the F distribution for the various possible tests for a three-factor study.

TABLE 24.5 Test statistics for three-factor fixed effects study

Alternatives	*Test Statistic*	*Percentile*
H_0: all $\alpha_i = 0$ H_a: not all $\alpha_i = 0$	$F^* = \frac{MSA}{MSE}$	$F[1-\alpha; a-1, (n-1)abc]$
H_0: all $\beta_j = 0$ H_a: not all $\beta_j = 0$	$F^* = \frac{MSB}{MSE}$	$F[1-\alpha; b-1, (n-1)abc]$
H_0: all $\gamma_k = 0$ H_a: not all $\gamma_k = 0$	$F^* = \frac{MSC}{MSE}$	$F[1-\alpha; c-1, (n-1)abc]$
H_0: all $(\alpha\beta)_{ij} = 0$ H_a: not all $(\alpha\beta)_{ij} = 0$	$F^* = \frac{MSAB}{MSE}$	$F[1-\alpha; (a-1)(b-1), (n-1)abc]$
H_0: all $(\alpha\gamma)_{ik} = 0$ H_a: not all $(\alpha\gamma)_{ik} = 0$	$F^* = \frac{MSAC}{MSE}$	$F[1-\alpha; (a-1)(c-1), (n-1)abc]$
H_0: all $(\beta\gamma)_{jk} = 0$ H_a: not all $(\beta\gamma)_{jk} = 0$	$F^* = \frac{MSBC}{MSE}$	$F[1-\alpha; (b-1)(c-1), (n-1)abc]$
H_0: all $(\alpha\beta\gamma)_{ijk} = 0$ H_a: not all $(\alpha\beta\gamma)_{ijk} = 0$	$F^* = \frac{MSABC}{MSE}$	$F[1-\alpha; (a-1)(b-1)(c-1), (n-1)abc]$

Comments

1. The power of the factor effects tests can be obtained from Table A–8 in the manner described for one-factor and two-factor studies. The noncentrality parameter ϕ for a given test is defined as follows:

$$\phi = \frac{1}{\sigma}\left[\frac{\text{numerator of second term in } E(MS) \text{ in Table 24.4}}{\text{denominator of second term in } E(MS) \text{ plus } 1}\right]^{1/2} \tag{24.32}$$

Thus, for testing the existence of three-factor interactions, we have:

$$\phi = \frac{1}{\sigma}\left[\frac{n\Sigma\Sigma\Sigma(\alpha\beta\gamma)^2_{ijk}}{(a-1)(b-1)(c-1)+1}\right]^{1/2}$$

2. The Kimball inequality for the family level of significance α in a three-factor study when the family consists of the combined set of seven tests, including three on main effects, three on two-factor interactions, and one on three-factor interactions, is:

(24.33) $$\alpha < 1 - (1 - \alpha_1)(1 - \alpha_2)\cdots(1 - \alpha_7)$$

where α_i is the level of significance for the ith test.

3. If the three-factor interactions (and also perhaps some sets of two-factor interactions) equal zero, the question sometimes arises whether the corresponding sums of squares should be pooled with the error sum of squares. Our earlier discussion on pooling (p. 700) is applicable here also.

4. If there is only one observation per treatment in a three-factor fixed effects study, analysis of variance tests can only be conducted if it is possible to assume that some interactions equal zero. Usually, the interactions most likely to equal zero are the three-factor interactions. If it is possible to assume that all three-factor interactions equal zero, $MSABC$ has expectation σ^2 and is used as the error mean square MSE. All mean squares are calculated in the usual manner, except that $n = 1$.

24.3 ANALYSIS OF FACTOR EFFECTS

No new problems are encountered in the analysis of factor effects for fixed effects three-factor studies. As for two-factor studies, the focus of the analysis is usually on factor level means when no important interactions are present, and on treatment means when there are important interactions. We shall now present some selected results for estimating factor effects. Other analyses follow the same pattern.

Analysis of factor effects when factors do not interact

Estimation of factor level mean. The factor A level mean $\mu_{i..}$ is estimated by:

(24.34) $$\hat{\mu}_{i..} = \bar{Y}_{i...}$$

The estimated variance of this estimator is:

(24.35) $$s^2(\bar{Y}_{i...}) = \frac{MSE}{nbc}$$

Confidence limits for $\mu_{i..}$ are obtained by means of the t distribution with $(n - 1)abc$ degrees of freedom:

(24.36) $$\bar{Y}_{i...} \pm t[1 - \alpha/2; (n - 1)abc]s(\bar{Y}_{i...})$$

Estimation of factor level means for factors B or C is done in similar fashion.

Estimation of contrast of factor level means. When a contrast involving the factor A level means $\mu_{i..}$ is to be estimated:

(24.37) $$L = \Sigma c_i \mu_{i..}$$

where:

$$\Sigma c_i = 0$$

the unbiased estimator of L we shall employ is:

(24.38) $$\hat{L} = \Sigma c_i \bar{Y}_{i...}$$

The estimated variance of $\hat{L}$ is:

(24.39) $$s^2(\hat{L}) = \frac{MSE}{nbc} \Sigma c_i^2$$

and the $1 - \alpha$ confidence limits for L are:

(24.40) $$\hat{L} \pm t[1 - \alpha/2; (n - 1)abc]s(\hat{L})$$

Contrasts of factor level means for factors B or C are estimated in similar fashion.

Multiple contrasts of factor level means. If a number of contrasts of the factor A level means $\mu_{i..}$ are to be estimated and assurance is to be provided by a family confidence coefficient, the t multiple in (24.40) is simply replaced by the T, S, or B multiple defined as follows:

(24.41a) Tukey procedure (for pairwise comparisons) $$T = \frac{1}{\sqrt{2}} q[1 - \alpha; a, (n - 1)abc]$$

(24.41b) Scheffé procedure $$S^2 = (a - 1)F[1 - \alpha; a - 1, (n - 1)abc]$$

(24.41c) Bonferroni procedure $$B = t[1 - \alpha/2g; (n - 1)abc]$$

Multiple contrasts based on the factor level means $\mu_{.j.}$ or $\mu_{..k}$ are estimated in corresponding fashion.

Analysis of factor effects when interactions important

Estimation of treatment mean. The treatment mean μ_{ijk} is estimated by:

(24.42) $$\hat{\mu}_{ijk} = \bar{Y}_{ijk.}$$

The estimated variance of $\bar{Y}_{ijk.}$ is:

(24.43) $$s^2(\bar{Y}_{ijk.}) = \frac{MSE}{n}$$

Confidence limits for μ_{ijk} are:

(24.44) $$\bar{Y}_{ijk.} \pm t[1 - \alpha/2; (n - 1)abc]s(\bar{Y}_{ijk.})$$

Estimation of contrast of treatment means. When interactions are present, contrasts among the treatment means μ_{ijk} are ordinarily desired. Let, as usual, L denote such a contrast:

(24.45) $$L = \Sigma\Sigma\Sigma c_{ijk} \mu_{ijk}$$

where:

$$\Sigma\Sigma\Sigma c_{ijk} = 0$$

An unbiased estimator of L is:

(24.46) $$\hat{L} = \Sigma\Sigma\Sigma c_{ijk}\bar{Y}_{ijk.}$$

for which the estimated variance is:

(24.47) $$s^2(\hat{L}) = \frac{MSE}{n}\Sigma\Sigma\Sigma c_{ijk}^2$$

As usual, confidence limits for L are:

(24.48) $$\hat{L} \pm t[1 - \alpha/2; (n - 1)abc]s(\hat{L})$$

At times, not all types of interaction effects are present. In such a case, the desired contrasts may involve means of the μ_{ijk} taken over one of the factors. For example, when the only interactions present are the BC interactions, one may be interested in contrasts of the means $\mu_{.jk}$:

(24.49) $$L = \Sigma\Sigma c_{jk}\mu_{.jk}$$

where:

$$\Sigma\Sigma c_{jk} = 0$$

Such contrasts are, of course, special cases of contrasts of treatment means μ_{ijk} in (24.45). The estimator of the contrast in (24.49) can be readily obtained from (24.46) and the estimated variance from (24.47); they are:

(24.50) $$\hat{L} = \Sigma\Sigma c_{jk}\bar{Y}_{.jk.}$$

(24.51) $$s^2(\hat{L}) = \frac{MSE}{na}\Sigma\Sigma c_{jk}^2$$

Multiple contrasts of treatment means. For multiple comparisons, the t multiple in (24.48) is replaced by the T, S, or B multiple defined as follows:

(24.52a) Tukey procedure (for pairwise comparisons) $$T = \frac{1}{\sqrt{2}}q[1 - \alpha; abc, (n - 1)abc]$$

(24.52b) Scheffé procedure $$S^2 = (abc - 1)F[1 - \alpha; abc - 1, (n - 1)abc]$$

(24.52c) Bonferroni procedure $$B = t[1 - \alpha/2g; (n - 1)abc]$$

Single degree of freedom tests. When interactions are present, either one or several single degree of freedom tests on the treatment means μ_{ijk} are sometimes used instead of estimation of contrasts. The two-sided alternatives for a single degree of freedom test are:

$$(24.53) \quad \begin{aligned} H_0&: \Sigma\Sigma\Sigma c_{ijk}\mu_{ijk} = c \\ H_a&: \Sigma\Sigma\Sigma c_{ijk}\mu_{ijk} \neq c \end{aligned}$$

where the c_{ijk} and c are appropriate constants.

To test the two-sided alternatives (24.53), we can use the test statistic:

$$(24.54) \qquad t^* = \frac{\Sigma\Sigma\Sigma c_{ijk}\bar{Y}_{ijk.} - c}{\sqrt{\frac{MSE}{n}\Sigma\Sigma\Sigma c_{ijk}^2}}$$

which follows the t distribution with $(n - 1)abc$ degrees of freedom when H_0 holds. Alternatively, $(t^*)^2 = F^*$ can be used as the test statistic. When H_0 holds, F^* follows the F distribution with 1 and $(n - 1)abc$ degrees of freedom.

24.4 IMPLEMENTATION OF THREE-FACTOR ANOVA MODEL

Planning of sample sizes and evaluation of the aptness of the model are handled in essentially the same fashion for three-factor studies as for single-factor and two-factor studies. Hence, we make only a few brief remarks.

Planning sample sizes

In most cases, equal replications will be desired for each treatment. When planning sample sizes with the power approach, one is typically concerned with the power of detecting factor A main effects, the power of detecting factor B main effects, and the power of detecting factor C main effects. One can first specify the minimum range of factor A level means for which it is important to detect factor A main effects and obtain the needed sample sizes from Table A–10, with $r = a$. The resulting sample size is bcn, from which n can readily be obtained. The use of Table A–10 for this purpose is appropriate provided the resulting sample sizes are not small, specifically provided $a(bcn - 1) \geq 20$. If this condition is not met, one should use the Pearson-Hartley power charts in Table A–8 with an iterative approach.

In the same way one can specify values for the minimum range of factor level means for factors B and C for which it is important to detect the factor main effects and find the needed sample sizes. If the sample sizes obtained from the factor A, factor B, and factor C power specifications differ substantially, a judgment will need to be made as to the final sample sizes.

Alternatively, or in conjunction with the power approach, one can specify the important contrasts to be estimated and then find the sample sizes that are expected to provide the needed precisions for the desired family confidence coefficient. Frequently this approach is more useful than the power approach, although both of these approaches can be used jointly for arriving at a determination of needed sample sizes.

If the purpose of the factorial study is to identify the best of the abc factor combinations, Table A–11 can be used for finding the needed sample sizes, as described in Section 18.3. For this purpose, $r = abc$.

Evaluation of aptness of model

No new problems arise in examining the aptness of the analysis of variance three-factor model. The residuals (24.21):

$$e_{ijkm} = Y_{ijkm} - \bar{Y}_{ijk.} \tag{24.55}$$

may be examined for normality, constancy of error variance, and independence of error terms in the same fashion as for single-factor and two-factor studies.

Transformations may be employed to stabilize the error variance and/or to make the error distributions more normal. Our earlier discussion of this topic in Chapter 18 applies completely to the three-factor case.

Finally, the earlier discussion on effects of departures from the model applies fully to the three-factor case. In particular, the employment of equal sample sizes for each treatment minimizes the effect of unequal variances.

24.5 EXAMPLE OF THREE-FACTOR STUDY

The exercise tolerance stress test study data presented in Table 24.3 (p. 812) will now be analyzed. Exercise tolerance is measured in minutes until fatigue occurs while the subject is performing on a bicycle apparatus. It will be recalled that the three factors in the study are sex of subject (A), body fat of subject measured in percent (B), and smoking history of subject (C). Each of the factors has two levels. Figure 24.5 presents the treatment sample means $\bar{Y}_{ijk.}$ shown in

FIGURE 24.5 Schematic presentation of treatment sample means—stress test example

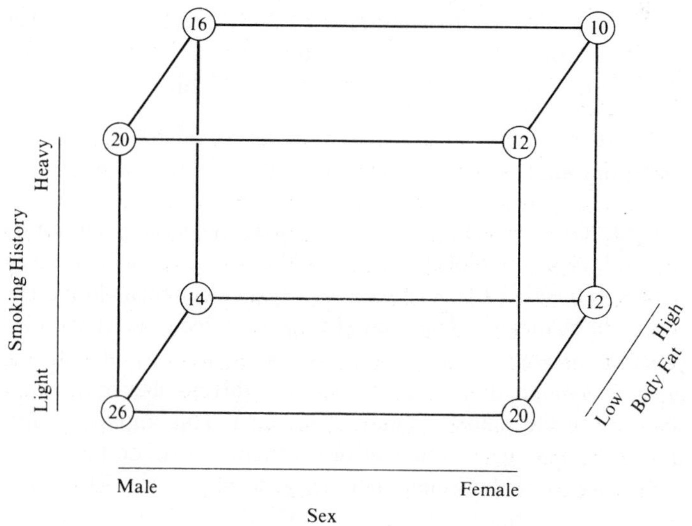

FIGURE 24.6 Plots of treatment sample means—stress test example

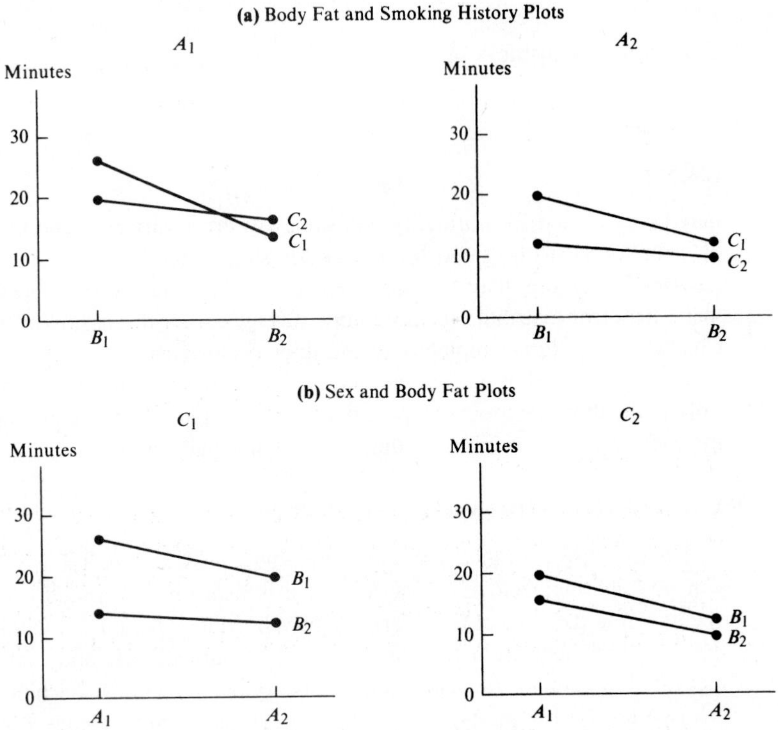

Table 24.3 in a simple format. The same treatment sample means are presented in Figure 24.6 in a graphic form. It appears from Figures 24.5 and 24.6 that some factors may interact in their effects on exercise tolerance and that sex, in particular, may affect the endurance in stress testing. The researcher now wishes to analyze the nature of the factor effects in detail.

Residual analysis

The researcher first prepared residual frequency plots for each of the eight treatments. These plots (not shown), though based on only three observations for each treatment, did not suggest any gross differences in the error variances for the eight treatments. The researcher also obtained a normal probability plot of the residuals, shown in Figure 24.7. The points in this plot form a reasonably linear pattern, even though there are a considerable number of ties among the residuals because of the rounded nature of the data. This impression of linearity is supported by the high coefficient of correlation between the ordered residuals and their expected values under normality, namely, .969. The researcher was therefore satisfied that the three-factor ANOVA model (24.17) is applicable here.

FIGURE 24.7 Normal probability plot of residuals—stress test example

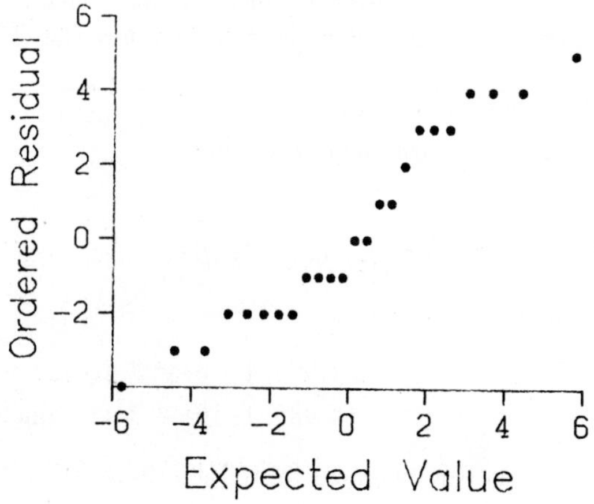

Tests for factor effects

The researcher first wished to test for the various factor effects. She desired a family level of significance of $\alpha = .10$ for the seven potential tests. This will assure that if in fact no factor effects are present, there will be only 1 chance in 10 for one or more of the seven tests to lead to the conclusion of the presence of factor effects. Using the Kimball inequality (24.33), she solved the equation:

$$\alpha = .10 = 1 - (1 - \alpha_i)^7$$

and found $\alpha_i = .015$. Thus, use of significance level $\alpha_i = .015$ for each test assures that the family level of significance will not exceed .10.

Table 24.6 contains the results of a run of a multifactor ANOVA computer

TABLE 24.6 ANOVA table for three-factor fixed effects study—stress test example

Source of Variation	*SS*	*df*	*MS*	*F**	*P-Value*
Between treatments	6,948.00	7	992.57		
Factor *A* (sex)	181.50	1	181.50	20.74	0+
Factor *B* (body fat)	253.50	1	253.50	28.97	0+
Factor *C* (smoking)	73.50	1	73.50	8.40	.01
AB interactions	13.50	1	13.50	1.54	.23
AC interactions	13.50	1	13.50	1.54	.23
BC interactions	73.50	1	73.50	8.40	.01
ABC interactions	1.50	1	1.50	.17	.69
Error	140.00	16	8.75		
Total	7,088.00	23			

$F(.985; 1, 16) = 7.42$

package. The ANOVA table is shown, as well as the seven test statistics and their P-values. Each test statistic has in the numerator the appropriate factor effect mean square, and the denominator of each test statistic is MSE.

Test for three-factor interactions. The first test was conducted for three-factor interactions. The alternatives are:

$$H_0\text{: all } (\alpha\beta\gamma)_{ijk} = 0$$
$$H_a\text{: not all } (\alpha\beta\gamma)_{ijk} \text{ equal zero}$$

The decision rule is:

$$\text{If } F^* \leq F(.985; 1, 16) = 7.42, \text{ conclude } H_0$$
$$\text{If } F^* > F(.985; 1, 16) = 7.42, \text{ conclude } H_a$$

The F^* test statistic obtained from Table 24.6 is:

$$F^* = \frac{MSABC}{MSE} = \frac{1.50}{8.75} = .17$$

Since $F^* = .17 \leq 7.42$, the researcher concluded that no ABC interactions are present.

Tests for main two-factor interactions. The researcher next tested for main two-factor interactions. In the test for AB interactions, the decision rule is (the alternatives are given in Table 24.5):

$$\text{If } F^* \leq F(.985; 1, 16) = 7.42, \text{ conclude } H_0$$
$$\text{If } F^* > F(.985; 1, 16) = 7.42, \text{ conclude } H_a$$

and the test statistic is:

$$F^* = \frac{MSAB}{MSE} = \frac{13.50}{8.75} = 1.54$$

Since $F^* = 1.54 \leq 7.42$, the researcher concluded that no main AB interactions are present.

The tests for main AC and BC interactions proceeded similarly. We obtain:

1. $$F^* = \frac{MSAC}{MSE} = \frac{13.50}{8.75} = 1.54 \leq F(.985; 1, 16) = 7.42$$

   Conclusion: No main AC interactions are present.

2. $$F^* = \frac{MSBC}{MSE} = \frac{73.50}{8.75} = 8.40 > F(.985; 1, 16) = 7.42$$

   Conclusion: Some main BC interactions are present.

Tests for main effects. Since factor A (sex) did not interact with the other two factors, attention next turned to testing for factor A main effects. In testing for factor A main effects, the decision rule is (the alternatives are given in Table 24.5):

$$\text{If } F^* \leq F(.985; 1, 16) = 7.42, \text{ conclude } H_0$$
$$\text{If } F^* > F(.985; 1, 16) = 7.42, \text{ conclude } H_a$$

The test statistic is:

$$F^* = \frac{MSA}{MSE} = \frac{181.50}{8.75} = 20.74$$

Since $F^* = 20.74 > 7.42$, the conclusion was reached that factor A main effects are present.

The factor B and C main effects were not tested at this point because BC interactions were found to be present. The researcher first wished to study the nature of the BC interaction effects before determining whether the factor B and C main effects are of any practical interest under the circumstances.

Family of conclusions. The five separate F tests for factor effects led the researcher to conclude (with family level of significance $\leq .10$):

1. There are no three-factor interactions.
2. There are no two-factor interactions between sex (factor A) and either of the other two factors—body fat (factor B) and smoking history (factor C). Body fat and smoking history interactions do exist, however.
3. Main effects for sex (factor A) are present.

This set of test results was most useful to the resarcher. The next step in her analysis was to examine the nature of the BC interaction effects and of the factor A main effects.

Estimation of factor effects

To study the nature of the BC interaction effects, the researcher wished to estimate separately, for persons with high and low percent body fat, the difference in mean fatigue time for light smokers and heavy smokers. The desired contrasts are:

$$L_1 = \mu_{.11} - \mu_{.12}$$
$$L_2 = \mu_{.21} - \mu_{.22}$$

In addition, a single comparison between the factor level means for factor A is sufficient to analyze the factor A main effects since factor A has only two levels. The contrast of interest (here a pairwise comparison of factor level means) is:

$$L_3 = \mu_{1..} - \mu_{2..}$$

These three contrasts are estimated by:

$$\hat{L}_1 = \bar{Y}_{.11.} - \bar{Y}_{.12.}$$
$$\hat{L}_2 = \bar{Y}_{.21.} - \bar{Y}_{.22.}$$
$$\hat{L}_3 = \bar{Y}_{1...} - \bar{Y}_{2...}$$

From Table 24.3c, we obtain:

$$\hat{L}_1 = 23 - 16 = 7$$
$$\hat{L}_2 = 13 - 13 = 0$$
$$\hat{L}_3 = 19 - 13.5 = 5.5$$

The researcher used estimated variances (24.51) and (24.39) and confidence limits (24.48) with a 95 percent family confidence coefficient based on the Bonferroni procedure. Hence, she required for the three comparisons:

$$B = t(1 - .05/6; 16) = 2.673$$

$$s^2(\hat{L}_1) = s^2(\hat{L}_2) = \frac{MSE}{na}[(1)^2 + (-1)^2] = \frac{8.75}{6}(2) = 2.917$$

$$s^2(\hat{L}_3) = \frac{MSE}{nbc}[(1)^2 + (-1)^2] = \frac{8.75}{12}(2) = 1.458$$

$$s(\hat{L}_1) = s(\hat{L}_2) = 1.708 \qquad s(\hat{L}_3) = 1.207$$

The desired confidence intervals therefore are:

$$2.4 = 7.0 - 2.673(1.708) \leq \mu_{.11} - \mu_{.12} \leq 7.0 + 2.673(1.708) = 11.6$$
$$-4.6 = 0 - 2.673(1.708) \leq \mu_{.21} - \mu_{.22} \leq 0 + 2.673(1.708) = 4.6$$
$$2.3 = 5.5 - 2.673(1.207) \leq \mu_{1..} - \mu_{2..} \leq 5.5 + 2.673(1.207) = 8.7$$

The researcher therefore concluded with family confidence coefficient .95: (1) Among people with low percent body fat, those who have a light smoking history have a mean stress test endurance that is 2.4 to 11.6 minutes longer than the mean endurance for people with a heavy smoking history. (2) People with high percent body fat do not differ in mean stress test endurance whether they have a light or a heavy smoking history. (3) The mean stress test endurance for men is 2.3 to 8.7 minutes longer than the mean endurance for women.

In view of the important interaction effects noted between body fat and smoking history on stress test endurance, the researcher concluded that factor B and factor C main effects are of no interest, and therefore terminated her analysis at this point. The principal findings are presented graphically in Figure 24.8, which shows the magnitude of the effect of sex on stress test endurance and the nature of the interaction effects between body fat and smoking history.

FIGURE 24.8 Key findings from stress test endurance study

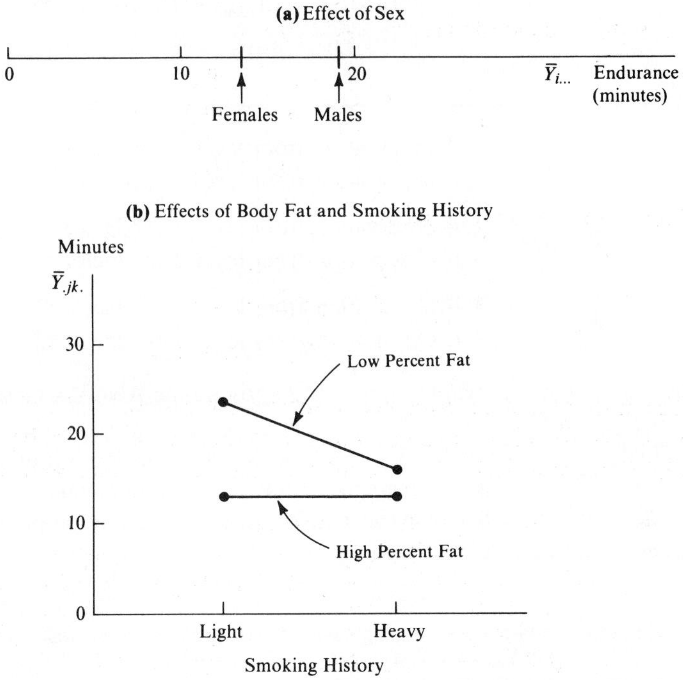

24.6 UNEQUAL SAMPLE SIZES IN MULTIFACTOR STUDIES

When the treatment sample sizes in a multifactor study are not equal, the procedures explained in Chapter 22 for two-factor studies should be followed with routine modifications.

Tests for factor effects

Tests for factor effects in multifactor studies with unequal sample sizes can be conducted by means of the regression approach, as long as no treatment cell is empty. Indicator variables taking on the values 1, -1, 0, are designated for each factor, the number of such variables being one less than the number of factor levels. Interaction effects are represented by cross-product terms, as usual. Since the sums of squares are no longer orthogonal when the treatment sample sizes are unequal, different reduced models need to be fitted for the tests of interest.

Example. Suppose that in the stress test example of Table 24.3, observations Y_{1113} and Y_{2212} were missing. To develop a regression model for this example, we note that each of the three factors is at two levels. Hence, one indicator variable is required for each factor. The full regression model therefore is:

$$
\begin{aligned}
(24.56) \qquad Y_{ijkm} = \mu_{\dots} &+ \alpha_1 X_{ijkm1} + \beta_1 X_{ijkm2} + \gamma_1 X_{ijkm3} + (\alpha\beta)_{11} X_{ijkm1} X_{ijkm2} \\
&+ (\alpha\gamma)_{11} X_{ijkm1} X_{ijkm3} + (\beta\gamma)_{11} X_{ijkm2} X_{ijkm3} \\
&+ (\alpha\beta\gamma)_{111} X_{ijkm1} X_{ijkm2} X_{ijkm3} + \varepsilon_{ijkm} \qquad \text{Full model}
\end{aligned}
$$

where:

$$
X_{ijkm1} = \begin{matrix} 1 \text{ if observation from level 1 for factor } A \\ -1 \text{ if observation from level 2 for factor } A \end{matrix}
$$

$$
X_{ijkm2} = \begin{matrix} 1 \text{ if observation from level 1 for factor } B \\ -1 \text{ if observation from level 2 for factor } B \end{matrix}
$$

$$
X_{ijkm3} = \begin{matrix} 1 \text{ if observation from level 1 for factor } C \\ -1 \text{ if observation from level 2 for factor } C \end{matrix}
$$

The regression parameters in model (24.56) are the ANOVA model parameters as defined in (24.16).

Table 24.7 contains the **Y** vector and **X** matrix for the full regression model (24.56) for the stress test example with two observations missing. The reduced models for testing different factor effects are obtained by dropping appropriate columns from the **X** matrix in Table 24.7.

TABLE 24.7 Data matrices for regression model (24.56)—stress test example with Y_{1113} and Y_{2212} missing

Y	**Y** =	**X** =	X_1	X_2	X_3	X_1X_2	X_1X_3	X_2X_3	$X_1X_2X_3$
Y_{1111}	24	1	1	1	1	1	1	1	1
Y_{1112}	29	1	1	1	1	1	1	1	1
Y_{1121}	18	1	1	1	−1	1	−1	−1	−1
Y_{1122}	19	1	1	1	−1	1	−1	−1	−1
Y_{1123}	23	1	1	1	−1	1	−1	−1	−1
Y_{1211}	15	1	1	−1	1	−1	1	−1	−1
Y_{1212}	15	1	1	−1	1	−1	1	−1	−1
Y_{1213}	12	1	1	−1	1	−1	1	−1	−1
Y_{1221}	15	1	1	−1	−1	−1	−1	1	1
Y_{1222}	20	1	1	−1	−1	−1	−1	1	1
Y_{1223}	13	1	1	−1	−1	−1	−1	1	1
Y_{2111}	20	1	−1	1	1	−1	−1	1	−1
Y_{2112}	22	1	−1	1	1	−1	−1	1	−1
Y_{2113}	18	1	−1	1	1	−1	−1	1	−1
Y_{2121}	15	1	−1	1	−1	−1	1	−1	1
Y_{2122}	10	1	−1	1	−1	−1	1	−1	1
Y_{2123}	11	1	−1	1	−1	−1	1	−1	1
Y_{2211}	16	1	−1	−1	1	1	−1	−1	1
Y_{2213}	11	1	−1	−1	1	1	−1	−1	1
Y_{2221}	10	1	−1	−1	−1	1	1	1	−1
Y_{2222}	14	1	−1	−1	−1	1	1	1	−1
Y_{2223}	6	1	−1	−1	−1	1	1	1	−1

Estimation of factor effects

Estimation of factor effects in multifactor studies with unequal sample sizes is conducted in similar fashion as for two-factor studies. The formulas in Table 22.5 need simply be extended to three or more factors.

To illustrate these extensions, consider pairwise comparisons of factor A level means in a three-factor study. Such a comparison, its estimator, and the estimated variance are:

(24.57a) $$D = \mu_{i..} - \mu_{i'..}$$

(24.57b) $$\hat{D} = \hat{\mu}_{i..} - \hat{\mu}_{i'..} \qquad \text{where } \hat{\mu}_{i..} = \frac{\sum_j \sum_k \bar{Y}_{ijk.}}{bc}$$

(24.57c) $$s^2(\hat{D}) = \frac{MSE}{b^2c^2} \sum_j \sum_k \left(\frac{1}{n_{ijk}} + \frac{1}{n_{i'jk}} \right)$$

The appropriate degrees of freedom associated with MSE are $n_T - abc$.

24.7 MODELS II AND III FOR THREE-FACTOR STUDIES

Just as for single-factor and two-factor studies, the analysis of variance sums of squares and degrees of freedom for random effects and mixed effects multifactor models are the same as those for the fixed effects model. The principal problem with random and mixed multifactor models, as we saw for two-factor models, is the determination of the appropriate expected mean squares. Once these are known, the proper test statistics and confidence intervals can be constructed. Rules for finding expected mean squares for random and mixed models for any number of factors will be presented later in Chapter 30. We now present model II (random effects) and model III (mixed effects) for three-factor studies and show how appropriate tests are conducted. We consider again the case where all treatment sample sizes are equal.

Model II (random effects)

In a study of the effects of operators, machines, and batches of raw material on daily output, all three factors may be considered to have random effects. The random effects model for such a three-factor study is:

(24.58) $$Y_{ijkm} = \mu_{...} + \alpha_i + \beta_j + \gamma_k + (\alpha\beta)_{ij} + (\alpha\gamma)_{ik} + (\beta\gamma)_{jk} + (\alpha\beta\gamma)_{ijk} + \varepsilon_{ijkm}$$

where:

$\mu_{...}$ is a constant

α_i, β_j, γ_k, $(\alpha\beta)_{ij}$, $(\alpha\gamma)_{ik}$, $(\beta\gamma)_{jk}$, $(\alpha\beta\gamma)_{ijk}$, ε_{ijkm} are independent normal random variables with expectation zero and respective variances σ_α^2,

$$\sigma^2_\beta, \sigma^2_\gamma, \sigma^2_{\alpha\beta}, \sigma^2_{\alpha\gamma}, \sigma^2_{\beta\gamma}, \sigma^2_{\alpha\beta\gamma}, \sigma^2$$
$$i = 1, \ldots, a; j = 1, \ldots, b; k = 1, \ldots, c; m = 1, \ldots, n$$

Table 24.8 contains the expected mean squares for all components of the ANOVA table for the random effects model (24.58).

TABLE 24.8 Expected mean squares in random effects three-factor study

Mean Square	*df*	*Expected Mean Square*
MSA	$a-1$	$\sigma^2 + nbc\sigma^2_\alpha + nc\sigma^2_{\alpha\beta} + nb\sigma^2_{\alpha\gamma} + n\sigma^2_{\alpha\beta\gamma}$
MSB	$b-1$	$\sigma^2 + nac\sigma^2_\beta + nc\sigma^2_{\alpha\beta} + na\sigma^2_{\beta\gamma} + n\sigma^2_{\alpha\beta\gamma}$
MSC	$c-1$	$\sigma^2 + nab\sigma^2_\gamma + nb\sigma^2_{\alpha\gamma} + na\sigma^2_{\beta\gamma} + n\sigma^2_{\alpha\beta\gamma}$
$MSAB$	$(a-1)(b-1)$	$\sigma^2 + nc\sigma^2_{\alpha\beta} + n\sigma^2_{\alpha\beta\gamma}$
$MSAC$	$(a-1)(c-1)$	$\sigma^2 + nb\sigma^2_{\alpha\gamma} + n\sigma^2_{\alpha\beta\gamma}$
$MSBC$	$(b-1)(c-1)$	$\sigma^2 + na\sigma^2_{\beta\gamma} + n\sigma^2_{\alpha\beta\gamma}$
$MSABC$	$(a-1)(b-1)(c-1)$	$\sigma^2 + n\sigma^2_{\alpha\beta\gamma}$
MSE	$(n-1)abc$	σ^2

Model III (mixed effects)

Suppose that in a three-factor study, factors B and C have random effects while factor A has fixed effects. The mixed effects model for this three-factor study is:

(24.59) $$Y_{ijkm} = \mu_{\ldots} + \alpha_i + \beta_j + \gamma_k + (\alpha\beta)_{ij} + (\alpha\gamma)_{ik} + (\beta\gamma)_{jk} + (\alpha\beta\gamma)_{ijk} + \varepsilon_{ijkm}$$

where:

$\mu_{\ldots}$ is a constant

α_i are constants

β_j, γ_k, $(\alpha\beta)_{ij}$, $(\alpha\gamma)_{ik}$, $(\beta\gamma)_{jk}$, $(\alpha\beta\gamma)_{ijk}$ are normal random variables with expectation zero and respective variances σ^2_β, σ^2_γ, $\sigma^2_{\alpha\beta}$, $\sigma^2_{\alpha\gamma}$, $\sigma^2_{\beta\gamma}$, $\sigma^2_{\alpha\beta\gamma}$

ε_{ijkm} are independent $N(0, \sigma^2)$, and are independent of the other random components

$$\sum_i \alpha_i = \sum_i (\alpha\beta)_{ij} = \sum_i (\alpha\gamma)_{ik} = \sum_i (\alpha\beta\gamma)_{ijk} = 0$$

$$i = 1, \ldots, a; j = 1, \ldots, b; k = 1, \ldots, c; m = 1, \ldots, n$$

Note that all interaction terms in this model are random, since at least one of the factors contained in each is a random effects factor. Note also that all sums of effects involving the fixed factor are zero over the fixed factor levels. Various correlations exist between the random effects terms, which we shall not detail.

Table 24.9 contains all the expected mean squares for the mixed effects model (24.59).

TABLE 24.9 Expected mean squares in mixed effects three-factor study (*A* fixed, *B* and *C* random)

Mean Square	*df*	*Expected Mean Square*
MSA	$a-1$	$\sigma^2 + nbc\frac{\sum\alpha_i^2}{a-1} + nc\sigma_{\alpha\beta}^2 + nb\sigma_{\alpha\gamma}^2 + n\sigma_{\alpha\beta\gamma}^2$
MSB	$b-1$	$\sigma^2 + nac\sigma_\beta^2 + na\sigma_{\beta\gamma}^2$
MSC	$c-1$	$\sigma^2 + nab\sigma_\gamma^2 + na\sigma_{\beta\gamma}^2$
MSAB	$(a-1)(b-1)$	$\sigma^2 + nc\sigma_{\alpha\beta}^2 + n\sigma_{\alpha\beta\gamma}^2$
MSAC	$(a-1)(c-1)$	$\sigma^2 + nb\sigma_{\alpha\gamma}^2 + n\sigma_{\alpha\beta\gamma}^2$
MSBC	$(b-1)(c-1)$	$\sigma^2 + na\sigma_{\beta\gamma}^2$
MSABC	$(a-1)(b-1)(c-1)$	$\sigma^2 + n\sigma_{\alpha\beta\gamma}^2$
MSE	$(n-1)abc$	σ^2

Other mixed effects models can be developed in similar fashion. The expected mean squares for these additional mixed models can be found by employing the rules to be presented in Chapter 30.

Appropriate test statistics

From the expected mean squares, we seek to determine the appropriate F^* statistic for a given test. An exact test statistic can often be found for random and mixed multifactor models, but not always.

Exact *F* test. Suppose we wish to determine whether or not *BC* interactions are present in the random effects model for Table 24.8. We see easily from the expected mean squares column that the appropriate test statistic is *MSBC/MSABC*. If we wish to study the same question for the mixed effects model for Table 24.9, we are again able to find an appropriate test statistic, but this time it is *MSBC/MSE*. We thus see that the two test statistics are not the same, even though the same factor effects are being studied, because of the differences between the two models.

Satterthwaite approximate *F* test. It is not always possible to find an exact *F* test for mixed and random effects multifactor models. For instance, one cannot directly test for the presence of factor *A* main effects in the random effects model for Table 24.8. Note from this table that there is no expected mean square that consists of the components of $E(MSA)$ except for the $nbc\sigma_\alpha^2$ term.

We consider two ways out of this difficulty. It may be possible to assume that certain interactions are zero, and then proceed in the usual way with an exact *F* test. For example, to test for factor *A* main effects in the random effects model for Table 24.8, it may be possible to assume that $\sigma_{\alpha\gamma}^2 = 0$ (indeed, this can be tested with *MSAC/MSABC*). If this assumption is appropriate one can use the test statistic *MSA/MSAB* to test for factor *A* main effects.

Often, however, one may not know whether certain interactions are zero. In that case, an approximate F test may be employed which utilizes a *pseudo F* or *quasi F* test statistic. This approximate test, called the *Satterthwaite test,* involves developing a linear combination of mean squares which has the same expectation when H_0 holds as the factor effects mean square. Let us express this linear combination as follows:

$$a_1MS_1 + a_2MS_2 + \cdots + a_hMS_h \tag{24.60}$$

where the a_i are constants. It can be shown that the approximate number of degrees of freedom associated with the linear combination (24.60) is:

$$df \cong \frac{(a_1MS_1 + a_2MS_2 + \cdots + a_hMS_h)^2}{\dfrac{(a_1MS_1)^2}{df_1} + \dfrac{(a_2MS_2)^2}{df_2} + \cdots + \dfrac{(a_hMS_h)^2}{df_h}} \tag{24.61}$$

where df_i denotes the degrees of freedom associated with MS_i. The test statistic is then set up in the usual way, and follows approximately the F distribution when H_0 holds.

We illustrate this procedure for testing factor A main effects in the random effects model for Table 24.8:

$$\begin{aligned} &H_0\colon \sigma_\alpha^2 = 0 \\ &H_a\colon \sigma_\alpha^2 > 0 \end{aligned} \tag{24.62}$$

Note from Table 24.8 that:

$$\begin{aligned} &E(MSAB) + E(MSAC) - E(MSABC) \\ &\qquad = \sigma^2 + nc\sigma_{\alpha\beta}^2 + nb\sigma_{\alpha\gamma}^2 + n\sigma_{\alpha\beta\gamma}^2 \end{aligned} \tag{24.63}$$

This equals precisely $E(MSA)$ when $\sigma_\alpha^2 = 0$. Hence, the suggested test statistic is:

$$F^{**} = \frac{MSA}{MSAB + MSAC - MSABC} \tag{24.64}$$

where we denote the test statistic as F^{**} as a reminder that a pseudo F test is involved.

Example. Table 24.10 contains the analysis of variance for a study of the effects of operators, machines, and batches on the daily output of a highly automated process. Each factor is assumed to be a random effects factor. To test whether operators (factor A) have a main effect on output, we use the test statistic (24.64):

$$F^{**} = \frac{8.5}{2.5 + 4.0 - 1.5} = \frac{8.5}{5.0} = 1.7$$

TABLE 24.10 ANOVA table for random effects three-factor study ($a = 3$, $b = 2$, $c = 5$, $n = 3$)

Source of Variation	*SS*	*df*	*MS*
Factor *A* (operators)	17	2	8.5
Factor *B* (machines)	4	1	4.0
Factor *C* (batches)	25	4	6.2
AB	5	2	2.5
AC	32	8	4.0
BC	12	4	3.0
ABC	12	8	1.5
Error	138	60	2.3
Total	245	89	

The approximate number of degrees of freedom associated with the denominator is:

$$df \cong \frac{(5.0)^2}{\dfrac{(2.5)^2}{2} + \dfrac{(4.0)^2}{8} + \dfrac{(-1.5)^2}{8}} = 4.6$$

Usually, the degrees of freedom will not turn out to be an integer. One can then either interpolate in the F distribution table, or round to the nearest integer. We shall round. For a level of significance of .05, we require $F(.95; 2, 5) = 5.79$. Since $F^{**} = 1.7 \leq 5.79$, we conclude H_0, that operators do not have a main effect on daily output. This test, it should be remembered, is an approximate one but can be quite useful if employed with caution.

Estimation of effects

No new problems arise in the development of unbiased estimators of variance components for random effects factors or in the estimation of contrasts for fixed effects factors in mixed models, when three or more factors are studied at one time. Confidence limits for contrasts of the factor level means of a fixed effects factor are obtained by using the mean square utilized in the denominator of the test statistic for examining the presence of main effects for that factor. The degrees of freedom are those associated with the mean square utilized.

PROBLEMS

24.1. Refer to Table 24.1 containing the mean responses μ_{ijk} for a three-factor study.

a. Find the specific effect of high IQ for young males.

b. Find the specific effect of young age for females with normal IQ.

c. Find the main effects of age.
d. Find the specific interaction effect of young age and normal IQ for male persons.
e. Find the main interaction effect of young age and normal IQ.
f. Find the interaction effect of young age, normal IQ, and female sex.

24.2. Prepare *AC* plots of the mean responses μ_{ijk} in Table 24.1 in the format of Figure 24.2b. Do your plots convey the same information as Figure 24.4? Discuss.

24.3. Prepare *BC* plots of the mean responses μ_{ijk} in Figure 24.3. Do your plots bring out any information on main effects and interactions not readily seen from Figure 24.3? Discuss.

24.4. In a three-factor study, the mean responses μ_{ijk} are as follows:

	$k = 1$		$k = 2$	
	$j = 1$	$j = 2$	$j = 1$	$j = 2$
$i = 1$	130	138	140	144
$i = 2$	126	130	134	136
$i = 3$	122	125	122	131

a. Find $\alpha_{1(12)}$ and $\alpha_{2(22)}$.
b. Find $\beta_{1(21)}$ and $\gamma_{2(12)}$.
c. Find α_1, α_2, and α_3.
d. Find β_2 and γ_1.
e. Find $(\alpha\beta)_{11(2)}$, $(\alpha\gamma)_{32(1)}$, and $(\beta\gamma)_{21(3)}$.
f. Find $(\alpha\beta)_{12}$, $(\alpha\gamma)_{21}$, and $(\beta\gamma)_{12}$.
g. Find $(\alpha\beta\gamma)_{111}$ and $(\alpha\beta\gamma)_{322}$.

24.5. Refer to Problem 24.4. Prepare *AB* plots of the mean responses μ_{ijk} in the format of Figure 24.3. What do these plots show about factor main effects and interactions?

24.6. **Case hardening.** An experiment involving the case hardening of lightweight shafts machined from bars of an alloy was run to study the effects of the amount of a chemical agent added to the alloy in a molten state (factor *A*), the temperature of the hardening process (factor *B*), and the time duration of the hardening process (factor *C*) on the outside hardness of the shaft. All factors were at two levels (1: low; 2: high), and the number of rods tested for each treatment was $n = 3$. The data on hardness (in Brinell units) follow.

	$k = 1$		$k = 2$	
	$j = 1$	$j = 2$	$j = 1$	$j = 2$
$i = 1$	39.9	53.5	56.0	70.9
	32.2	50.7	56.9	73.3
	36.3	52.8	56.6	71.6
$i = 2$	45.2	63.3	69.4	82.9
	48.0	65.5	66.6	85.2
	47.5	63.6	68.8	82.3

Assume that fixed effects ANOVA model (24.17) is applicable.
a. Prepare *AB* plots of the treatment sample means $\bar{Y}_{ijk.}$ in the format of Figure

24.6b. Does it appear that any interactions are present? Any main effects?

b. Obtain the analysis of variance table.

c. Test for three-factor interactions; use $\alpha = .025$. State the alternatives, decision rule, and conclusion. What is the P-value of the test?

d. Test for AB, AC, and BC interactions. For each test, use $\alpha = .025$ and state the alternatives, decision rule, and conclusion. What is the P-value of each test?

e. Test for A, B, and C main effects. For each test, use $\alpha = .025$ and state the alternatives, decision rule, and conclusion. What is the P-value of each test?

f. State the set of conclusions that can be arrived at from the tests in parts (c), (d), and (e). Obtain an upper bound for the family level of significance for the set of tests; use the Kimball inequality.

g. Do the results in part (f) confirm your graphic analysis in part (a)?

24.7. Refer to **Case hardening** Problem 24.6.

a. To study the nature of the main factor effects, estimate the following pairwise comparisons:

$$D_1 = \mu_{2..} - \mu_{1..} \qquad D_2 = \mu_{.2.} - \mu_{.1.} \qquad D_3 = \mu_{..2} - \mu_{..1}$$

Use the Bonferroni procedure with a 95 percent family confidence coefficient. State your findings.

b. Estimate μ_{222} with a 95 percent confidence interval.

24.8. Refer to **Case hardening** Problem 24.6.

a. Obtain the residuals and prepare residual frequency plots for each level of factor A. Do the same for each of the other two factors. What do your plots suggest about the appropriateness of ANOVA model (24.17)?

b. Prepare a normal probability plot of the residuals. Also obtain the coefficient of correlation between the ordered residuals and their expected values under normality. Does the normality assumption appear to be reasonable here?

24.9. **Marketing research contractors.** A marketing research consultant evaluated the effects of fee schedule (factor A), scope of work (factor B), and type of supervisory control (factor C) on the quality of work performed under contract by independent marketing research agencies. The factor levels in the study were as follows:

	Factor		*Factor Levels*
A	Fee level	$i = 1$:	High
		$i = 2$:	Average
		$i = 3$:	Low
B	Scope	$j = 1$:	All contract work performed in house
		$j = 2$:	Some work subcontracted out
C	Supervision	$k = 1$:	Local supervisors
		$k = 2$:	Traveling supervisors only

The quality of work performed was measured by an index taking into account several characteristics of quality. Four agencies were chosen for each factor level combination and the quality of their work evaluated. The data on quality follow.

	k = 1		k = 2	
	j = 1	j = 2	j = 1	j = 2
i = 1	124.3	115.1	112.7	88.2
	120.6	119.9	110.2	96.0
	120.7	115.4	113.5	96.4
	122.6	117.3	108.6	90.1
i = 2	119.3	117.2	113.6	92.7
	118.9	114.4	109.1	91.1
	125.3	113.4	108.9	90.7
	121.4	120.0	112.3	87.9
i = 3	90.9	89.9	78.6	58.6
	95.3	83.0	80.6	63.5
	88.8	86.5	83.5	59.8
	92.0	82.7	77.1	62.3

Assume that fixed effects ANOVA model (24.17) is applicable.

a. Prepare *AB* plots of the treatment sample means $\bar{Y}_{ijk.}$ in the format of Figure 24.6b. Does it appear that any interactions are present? Any main effects?

b. Obtain the analysis of variance table.

c. Test for three-factor interactions; use $\alpha = .01$. State the alternatives, decision rule, and conclusion. What is the *P*-value of the test?

d. Test for *AB*, *AC*, and *BC* interactions. For each test, use $\alpha = .01$ and state the alternatives, decision rule, and conclusion. What is the *P*-value of each test?

e. Test for factor *A* main effects; use $\alpha = .01$. State the alternatives, decision rule, and conclusions. What is the *P*-value of the test?

f. State the set of conclusions that can be arrived at from the tests in parts (c), (d), and (e). Obtain an upper bound for the family level of significance for the set of tests; use the Kimball inequality.

g. Do the results in part (f) confirm your graphic analysis in part (a)?

24.10. Refer to **Marketing research contractors** Problem 24.9.

a. To study the nature of the factor *A* main effects and the *BC* interactions, it is desired to estimate the following comparisons:

$$D_1 = \mu_{1..} - \mu_{2..} \qquad D_4 = \mu_{.11} - \mu_{.12}$$
$$D_2 = \mu_{2..} - \mu_{3..} \qquad D_5 = \mu_{.21} - \mu_{.22}$$
$$D_3 = \mu_{1..} - \mu_{3..} \qquad L_1 = D_4 - D_5$$

Use the Bonferroni procedure with a 90 percent family confidence coefficient to make the desired comparisons. State your findings.

b. Estimate $D = \mu_{121} - \mu_{221}$ with a 95 percent confidence interval.

c. The consultant wishes to identify the type(s) of independent marketing research agencies that provide the highest quality of work. Use the Tukey procedure with a 90 percent family confidence coefficient to make the desired identifications.

24.11. Refer to **Marketing research contractors** Problem 24.9.

a. Obtain the residuals and plot them against the fitted values. What does your plot suggest about the appropriateness of ANOVA model (24.17)?

b. Prepare a normal probability plot of the residuals. Also obtain the coefficient of correlation between the ordered residuals and their expected values under normality. Does the normality assumption appear to be reasonable here?

24.12. **Electronics assembly.** Assemblers in an electronics firm will attach 12 components to a newly developed "board" that will be used in automatic-control equipment in manufacturing plants. An operations analyst conducted an experiment to study the effects of three factors on the mean time to assemble a board. Factor A was the sex of the assembler ($i = 1$: male; $i = 2$: female), factor B was the sequence of assembling the components ($j = 1, 2, 3$), and factor C was the amount of experience by the assembler ($k = 1$: under 18 months; $k = 2$: 18 months or more). Randomization was used to assign 15 assemblers of each sex with a given amount of experience to each of the three assembly sequences, with each sequence assigned to five assemblers. After a learning period, the total time (in minutes) to assemble 50 boards was observed. The data follow.

	$k=1$			$k=2$		
	$j=1$	$j=2$	$j=3$	$j=1$	$j=2$	$j=3$
$i=1$	1,250	1,319	1,217	1,021	1,119	1,033
	1,175	1,251	1,190	1,099	1,110	1,067
	1,236	1,241	1,201	1,069	1,123	1,057
	1,239	1,295	1,232	996	1,097	1,077
	1,193	1,265	1,251	1,070	1,163	1,022
$i=2$	1,066	1,105	1,021	864	927	841
	1,076	1,043	1,020	848	944	865
	1,004	1,051	1,035	881	957	817
	1,002	1,128	1,000	892	897	911
	1,034	1,060	1,026	868	933	868

Assume that fixed effects ANOVA model (24.17) is applicable.

a. Prepare AB plots of the treatment sample means $\bar{Y}_{ijk.}$ in the format of Figure 24.6b. Does it appear that any interactions are present? Any main effects?

b. Obtain the analysis of variance table.

c. Test for three-factor interactions; use $\alpha = .05$. State the alternatives, decision rule, and conclusion. What is the P-value of the test?

d. Test for AB, AC, and BC interactions. For each test, use $\alpha = .05$ and state the alternatives, decision rule, and conclusion. What is the P-value of each test?

e. Test for A, B, and C main effects. For each test, use $\alpha = .05$ and state the alternatives, decision rule, and conclusion. What is the P-value of each test?

f. State the set of conclusions that can be arrived at from the tests in parts (c), (d), and (e). Obtain an upper bound for the family level of significance for the set of tests; use the Kimball inequality.

g. Do the results in part (f) confirm your graphic analysis in part (a)?

24.13. Refer to **Electronics assembly** Problem 24.12.

a. To study the nature of the main factor effects, estimate the following pairwise comparisons:

$$D_1 = \mu_{1..} - \mu_{2..} \qquad D_4 = \mu_{.2.} - \mu_{.3.}$$
$$D_2 = \mu_{.1.} - \mu_{.2.} \qquad D_5 = \mu_{..1} - \mu_{..2}$$
$$D_3 = \mu_{.1.} - \mu_{.3.}$$

Use the Bonferroni procedure with a 90 percent family confidence coefficient. State your findings.
b. Estimate μ_{231} with a 95 percent confidence interval.

24.14. Refer to **Electronics assembly** Problem 24.12.
a. Obtain the residuals and plot them against the fitted values. What does your plot suggest about the appropriateness of ANOVA model (24.17)?
b. Prepare a normal probability plot of the residuals. Also obtain the coefficient of correlation between the ordered residuals and their expected values under normality. Does the normality assumption appear to be reasonable here?

24.15. For the three-factor fixed effects ANOVA model (24.17), what is the noncentrality parameter ϕ for testing for factor A main effects? For testing for AB interactions?

24.16. Refer to **Marketing research contractors** Problem 24.9. Assume that $\sigma = 3.0$. What is the power of the test for factor A main effects in Problem 24.9e if $\mu_{1..} = 110$, $\mu_{2..} = 100$, and $\mu_{3..} = 90$?

24.17. Refer to **Electronics assembly** Problem 24.12. Assume that $\sigma = 29$. What is the power of the test for factor B main effects in Problem 24.12e if $\mu_{.1.} = 1{,}000$, $\mu_{.2.} = 1{,}100$, and $\mu_{.3.} = 1{,}050$?

24.18. Refer to **Case hardening** Problem 24.6. Suppose that the sample sizes have not yet been determined but it has been decided to use equal sample sizes for all treatments. The chief objective is to identify the treatment that leads to the highest mean hardness. The probability should be at least .99 that the correct treatment is identified when the mean hardness for the second best treatment differs by 2.0 or more Brinell units. Assume that a reasonable planning value for the error term standard deviation is $\sigma = 1.8$. What are the required sample sizes?

24.19. Refer to **Electronics assembly** Problem 24.12. Suppose that the sample sizes have not yet been determined but it has been decided to use equal sample sizes for all treatments. The chief objective is to estimate the following pairwise comparisons:

$$D_1 = \mu_{1..} - \mu_{2..} \qquad D_4 = \mu_{.2.} - \mu_{.3.}$$
$$D_2 = \mu_{.1.} - \mu_{.2.} \qquad D_5 = \mu_{..1} - \mu_{..2}$$
$$D_3 = \mu_{.1.} - \mu_{.3.}$$

What are the required sample sizes if the precision of the estimates should not exceed ± 20, using the Bonferroni procedure with a 90 percent family confidence coefficient for the joint set of comparisons? A reasonable planning value for the error term standard deviation is $\sigma = 29$.

24.20. Refer to **Case hardening** Problem 24.6. Suppose that observations $Y_{1211} = 53.5$ and $Y_{1212} = 50.7$ are missing.
a. State the full regression model equivalent to ANOVA model (24.17); use $1, -1, 0$ indicator variables.
b. What is the reduced model for testing for factor A main effects?
c. Test whether or not factor A main effects are present by fitting the full and reduced models; use $\alpha = .025$. State the alternatives, decision rule, and conclusion. Is your conclusion the same as that obtained in Problem 24.6e for the full data?
d. Estimate $D = \mu_{2..} - \mu_{1..}$ with a 95 percent confidence interval.

24.21. Refer to **Electronics assembly** Problem 24.12. Suppose that observations $Y_{1224} = 1{,}097$, $Y_{2213} = 1{,}051$, and $Y_{2125} = 868$ are missing.

a. State the full regression model equivalent to ANOVA model (24.17); use 1, −1, 0 indicator variables.

b. What is the reduced model for testing for factor C main effects?

c. Test whether or not factor C main effects are present by fitting the full and reduced models; use $\alpha = .05$. State the alternatives, decision rule, and conclusion. Is your conclusion the same as that obtained in Problem 24.12e for the full data?

d. Estimate $D = \mu_{..1} - \mu_{..2}$ with a 95 percent confidence interval.

24.22. Refer to Table 24.10. All three factors in this study have random effects.

a. Test whether or not AB interactions are present. Use significance level $\alpha = .01$. State the alternatives, decision rule, and conclusion.

b. Test whether machines (factor B) have main effects. Use significance level $\alpha = .01$. State the alternatives, decision rule, and conclusion.

24.23. Refer to **Electronics assembly** Problem 24.12. Suppose that the number of feasible sequences in which the components can be attached to the board is very large and that the three sequences studied were selected randomly from the set of operationally feasible sequences. Assume that a normal error ANOVA model is applicable where factors A and C have fixed effects and factor B has random effects. Some relevant expected mean squares for this model are:

$$E(MSA) = \sigma^2 + bcn\frac{\Sigma\alpha_i^2}{a-1} + cn\sigma_{\alpha\beta}^2$$

$$E(MSB) = \sigma^2 + acn\sigma_\beta^2$$

$$E(MSAC) = \sigma^2 + bn\frac{\Sigma\Sigma(\alpha\gamma)_{ik}^2}{(a-1)(c-1)} + n\sigma_{\alpha\beta\gamma}^2$$

$$E(MSABC) = \sigma^2 + n\sigma_{\alpha\beta\gamma}^2$$

$$E(MSE) = \sigma^2$$

a. What is the appropriate test statistic for testing for AC interactions? For testing for factor B main effects?

b. Test whether or not AC interactions are present; use $\alpha = .05$. State the alternatives, decision rule, and conclusion.

c. Test whether or not factor B main effects are present; use $\alpha = .05$. State the alternatives, decision rule, and conclusion.

d. Obtain a point estimate of σ_β^2.

24.24. Consider the mixed effects model (24.59) where factor A has fixed effects and the other two factors have random effects. Find a pseudo F test statistic F^{**} for testing for factor A main effects. What is the approximate number of degrees of freedom associated with the denominator of this test statistic?

EXERCISES

24.25. Show that for fixed effects ANOVA model (24.17) $\sum_i (\alpha\beta\gamma)_{ijk} = 0$.

24.26. Prove (24.13a).

24.27. Derive (24.26c) from (24.24a).

24.28. Derive the sum of squares breakdown in (24.28).

24.29. State the fixed effects ANOVA model for a three-factor study with $n = 1$ when all three-factor interactions are zero. Show the ANOVA table for this case.

24.30. For fixed effects model (24.17), derive the variance of the estimated contrast $\hat{L} = \Sigma\Sigma c_{ij}\bar{Y}_{ij..}$.

24.31. For random effects model (24.58), find the variance of the sample mean $\bar{Y}_{i...}$.

PROJECTS

24.32. Refer to the **SENIC** data set. The following hospitals are to be considered in a study of the effects of average age of patients (factor A: variable 3), available facilities and services (factor B: variable 12), and region (factor C: variable 9) on the mean length of hospital stay of patients (variable 2):

1–14	16–28	31	32	34	35	37–39	41	44	46	50
52	53	57	58	63	66	76	77	83	111	

For purposes of this ANOVA study, average age is to be classified into two categories (less than 53.0 years, 53.0 years or more) and available facilities and services is to be classified into two categories (less than 40.2 percent, 40.2 percent or more). Assume that fixed effects ANOVA model (24.17) is appropriate.

a. Assemble the required data and obtain the fitted values.
b. Plot the treatment sample means $\bar{Y}_{ijk.}$ in the format of Figure 24.6a. Does it appear that any factor effects are present? Explain.
c. Obtain the analysis of variance table. Does any one source account for most of the total variability in the study? Explain.
d. Test for three-factor interactions; use $\alpha = .01$. State the alternatives, decision rule, and conclusion.
e. Test for AB, AC, and BC interactions. For each test, use $\alpha = .01$ and state the alternatives, decision rule, and conclusion.
f. Test for A, B, and C main effects. For each test, use $\alpha = .01$ and state the alternatives, decision rule, and conclusion.
g. To study the nature of the available facilities and region main effects, make all pairwise comparisons for each of these two factors. Use the Bonferroni procedure with a 90 percent family confidence coefficient. State your findings.

24.33. Refer to the **SMSA** data set. The effects of region (factor A: variable 12), percent of population in central cities (factor B: variable 4), and percent of population 65 or older (factor C: variable 5) on the crime rate (variable 11 ÷ variable 3) are to be studied. For purposes of this ANOVA study, percent of population in central cities is to be classified into two categories (40.0 percent or less, 40.1 percent or more) and percent of population 65 or older is to be classified into two categories (9.9 percent or less, 10.0 percent or more). Assume that fixed effects ANOVA model (24.17), with $m = 1, \ldots, n_{ijk}$, is appropriate.

a. Assemble the required data.
b. Plot the treatment sample means $\bar{Y}_{ijk.}$ in the format of Figure (24.6b). Does it appear that any factor effects are present?
c. State the equivalent regression model for this case; use $1, -1, 0$ indicator variables, and fit this full model.
d. Test for three-factor interactions and AB, AC, and BC interactions. For each test, use $\alpha = .025$ and state the alternatives, reduced model, decision rule, and conclusion.
e. Test for A, B, and C main effects. For each test, use $\alpha = .025$ and state the alternatives, reduced model, decision rule, and conclusion.
f. To study the nature of the region main effects, make all pairwise comparisons between the region means. Use the Tukey procedure with a 95 percent family confidence coefficient. State your findings.

25

Analysis of covariance

Analysis of covariance is a technique that combines features of analysis of variance and regression. It can be used for either observational studies or designed experiments. The basic idea is to augment the analysis of variance model containing the factor effects with one or more additional quantitative variables that are related to the dependent variable. This augmentation is intended to reduce the variance of the error terms in the model, i.e., to make the analysis more precise. We considered covariance models briefly in Chapter 10, and noted there that they are linear models containing both qualitative and quantitative independent variables. Thus, covariance models are just a special type of regression model.

In this chapter, we shall first consider how a covariance model can be more effective than an ordinary ANOVA model. Then we shall discuss how to use a single-factor covariance model for making inferences in terms of the regression approach. This is followed by a demonstration that analysis of covariance can also be viewed as a modification of analysis of variance. We conclude the chapter by considering analysis of covariance models for multifactor studies.

25.1 BASIC IDEAS

How covariance analysis reduces error variability

Covariance analysis may be helpful in reducing large error term variances that sometimes are present in analysis of variance models. Consider a study in which the effects of three different films promoting travel in a state are studied. A subject receives an initial questionnaire to elicit information about his or her attitudes toward the state. The subject is then shown a five-minute film, and immediately afterwards is questioned about the film, about desire to travel in the state, and so on.

In this type of situation, covariance analysis can be utilized. To see why it might be highly effective, consider Figure 25.1a. Here are plotted the desire-to-travel scores, obtained after each of the three promotional films was shown to a different group of five subjects. Three different symbols are used to distinguish the different treatments. It is evident from Figure 25.1a that the error terms, as shown by the scatter around the treatment means μ_i, are fairly large, indicating a large error term variance.

Suppose now that we were to utilize also the subjects' initial attitude scores. We plot in Figure 25.1b the desire-to-travel score (obtained after exposure to the film) against the initial attitude score for each of the 15 subjects. Note that the three treatment regression relations happen to be linear (this need not be so). Also note that the scatter around the treatment regression lines is much less than the scatter in Figure 25.1a around the treatment means μ_i, as a result of the desire-to-travel scores being highly linearly related to the initial attitude scores. The relatively large scatter in Figure 25.1a reflects the large error term variability that would be encountered with analysis of variance for this single-factor study. The smaller scatter in Figure 25.1b reflects the smaller error term variability that would be involved in an analysis of covariance.

Covariance analysis, it is thus seen, utilizes the relationship between the dependent variable (desire-to-travel score in our example) and one or more independent quantitative variables for which observations are available (pre-study attitude score in our example) in order to reduce the error term variability and make the study a more powerful one for comparing treatment effects.

Concomitant variables

In covariance analysis terminology, each independent quantitative variable added to the study is called a *concomitant variable*. Clearly, the choice of concomitant variables is an important one. If such variables have no relation to the dependent variable, nothing is to be gained by covariance analysis, and one might as well use the simpler analysis of variance. Concomitant variables frequently used with human subjects include pre-study attitudes, age, IQ, and aptitude. When stores are used as study units, concomitant variables might be last period's sales and number of employees.

FIGURE 25.1 Illustration of error variability reduction by covariance analysis

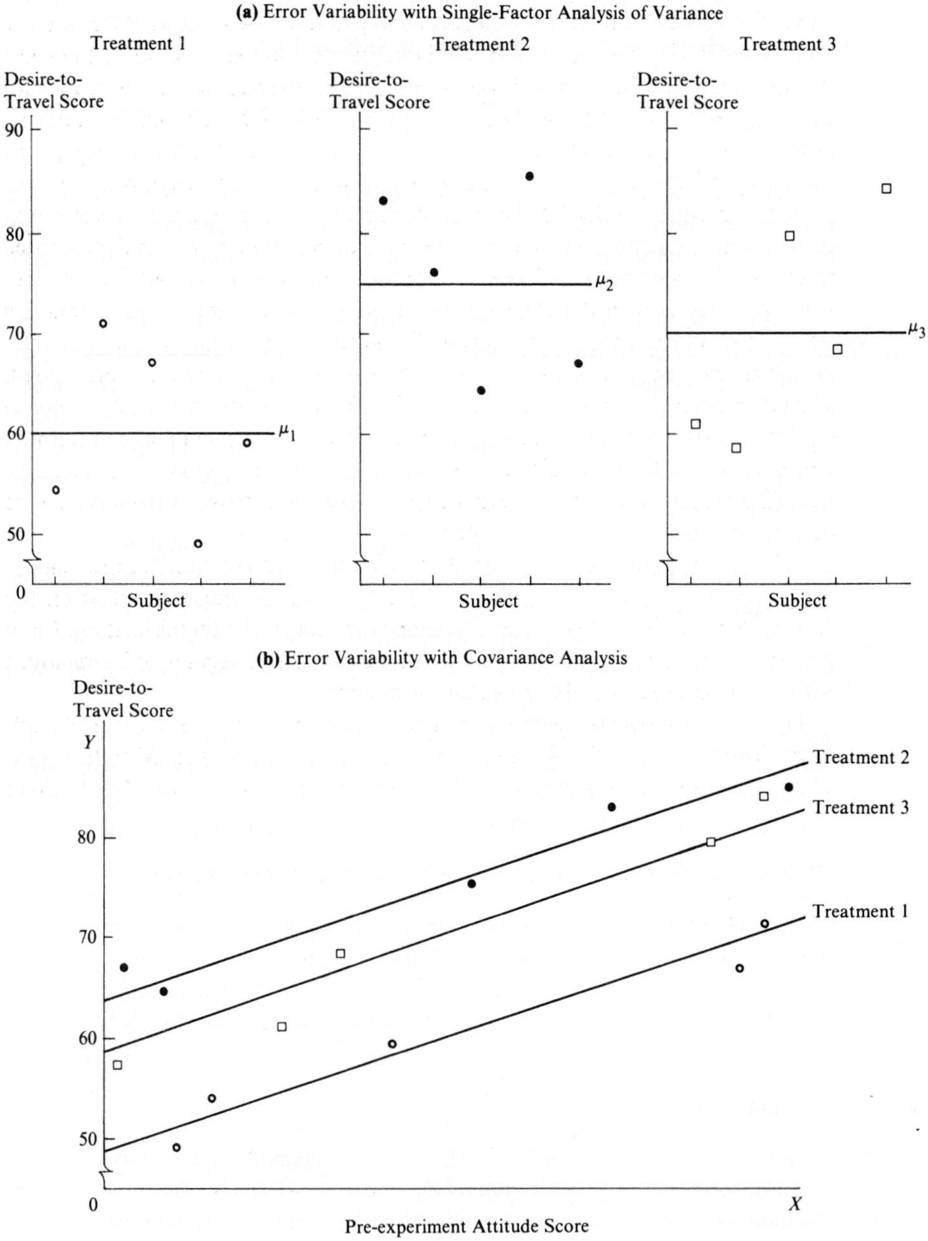

Choice of concomitant variables. There is one problem in selecting concomitant variables that is unique to covariance analysis. For clear interpretations of the results, a concomitant variable should be observed before the study; or if

observed during the study, it should not be influenced by the treatments in any way. A pre-study attitude score meets this requirement. Also, if a subject's age is ascertained during the study, it would be reasonable in many instances to expect that the response about age will not be affected by the treatment. The reason for this requirement can be readily seen from the following example. A company was conducting a training school for engineers to teach them accounting and budgeting principles. Two teaching methods were used, and engineers were assigned at random to one of the two. At the end of the program, a score was obtained for each engineer reflecting the amount of learning. The analyst decided to use as a concomitant variable in covariance analysis the amount of time devoted to study (which the engineers were required to record). After conducting the analysis of covariance, the analyst found that training method had virtually no effect. He was baffled by this until it was pointed out to him that the amount of study time probably was also affected by the treatments, and analysis indeed confirmed this. One of the training methods involved computer-assisted learning which appealed to the engineers so that they spent more study time and also learned more. In other words, both learning score and amount of study time were dependent upon the treatment in this case.

When a concomitant variable is affected by the treatments, covariance analysis will remove some (or much) of the effect that the treatments had on the dependent variable, so that an uncritical analysis may be badly misleading. Great care should be taken in the analysis when the covariance technique is employed with a concomitant variable affected by treatments.

Figure 25.2 shows the data for the experiment involving the training of engineers. Treatment 1 is the one using computer-assisted learning. Note that most persons with this treatment devoted large amounts of time to study. On the other

FIGURE 25.2 Illustration of treatments affecting the concomitant variable

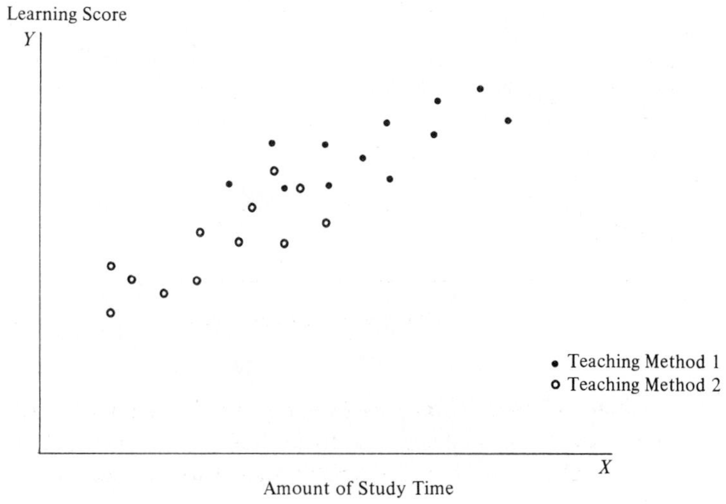

hand, persons receiving treatment 2 tended to devote smaller amounts of time to study. As a result, the observations for the two treatments tend to be bunched over different intervals on the X scale.

Contrast this situation with the one seen in Figure 25.1b for the study on promotional films. Figure 25.1b illustrates how the concomitant variable observations should be scattered if treatments have no effect on the concomitant variable. Here, the distribution of subjects along the X scale by pre-study attitude scores is roughly similar for all treatments, subject only to chance variation.

25.2 SINGLE-FACTOR COVARIANCE MODEL

We shall employ the notation for single-factor analysis of variance. The number of observations for the ith factor level is denoted by n_i, the total number of observations by $n_T = \Sigma n_i$, and the jth observation on the dependent variable for the ith factor level is denoted by Y_{ij}. We shall initially consider a single-factor covariance model with only one concomitant variable. Later we shall take up models with more than one concomitant variable. We shall denote the value of the concomitant variable associated with the jth study unit for the ith factor level by X_{ij}.

The fixed effects analysis of variance model in terms of the factor effects was given in (16.61):

(25.1) $$Y_{ij} = \mu_. + \tau_i + \varepsilon_{ij}$$

The covariance model starts with this model and simply adds another term (or several), reflecting the relationship between the concomitant and dependent variables. Usually, a linear relation is utilized as a first approximation:

(25.2) $$Y_{ij} = \mu_. + \tau_i + \gamma X_{ij} + \varepsilon_{ij}$$

Here γ is a regression coefficient for the relation between Y and X. The constant $\mu_.$ now is no longer an overall mean. We can, however, make this constant an overall mean, and incidentally simplify some computations, if we express the concomitant variable as a deviation from the overall mean $\bar{X}_{..}$. The resulting model is the usual covariance model for a single-factor study with fixed treatment effects:

(25.3) $$Y_{ij} = \mu_. + \tau_i + \gamma(X_{ij} - \bar{X}_{..}) + \varepsilon_{ij}$$

where:

$\mu_.$ is an overall mean
τ_i are the fixed treatment effects subject to the restriction $\Sigma\tau_i = 0$
γ is a regression coefficient for the relation between Y and X
X_{ij} are constants
ε_{ij} are independent $N(0, \sigma^2)$
$i = 1, \ldots, r; j = 1, \ldots, n_i$

Model (25.3) corresponds to the analysis of variance model (25.1) except for the added term $\gamma(X_{ij} - \bar{X}_{..})$ to reflect the relationship between Y and X. Note that the concomitant observations X_{ij} are assumed to be constants. Since ε_{ij} is the only random variable on the right side of (25.3), it follows at once that:

(25.4a) $$E(Y_{ij}) = \mu_{.} + \tau_i + \gamma(X_{ij} - \bar{X}_{..})$$

(25.4b) $$\sigma^2(Y_{ij}) = \sigma^2$$

In view of the independence of the ε_{ij}, the Y_{ij} are also independent. Hence, an alternative statement of model (25.3) is:

(25.5) $$Y_{ij} \text{ are independent } N(\mu_{ij}, \sigma^2)$$

where:

$$\mu_{ij} = \mu_{.} + \tau_i + \gamma(X_{ij} - \bar{X}_{..})$$
$$\Sigma\tau_i = 0$$

Properties of model

Some of the properties of the covariance model (25.3) are identical to those of the analysis of variance model (25.1). For instance, the error terms ε_{ij} are independent and have constant variance. There are also some new properties, and we discuss these now.

Comparisons of treatment effects. With the analysis of variance model, all observations for the ith treatment have the same mean response (μ_i). This is not so with the covariance model, since the mean response here depends not only on the treatment but also on the value of the concomitant variable X_{ij} for the study unit. Thus, the expected response for the ith treatment with covariance model (25.3) is given by a regression line:

(25.6) $$\mu_{ij} = \mu_{.} + \tau_i + \gamma(X_{ij} - \bar{X}_{..})$$

which indicates, for any value of X, the mean response with treatment i. Figure 25.3 illustrates for an experiment with three treatments how these treatment regression lines might appear. Note that $\mu_{.} + \tau_i$ is the ordinate of the line for the ith treatment when $X - \bar{X}_{..} = 0$, that is, when $X = \bar{X}_{..}$, and that γ is the slope of each line. Since all treatment regression lines have the same slope, they are parallel.

While we no longer can speak of *the* mean response with the ith treatment since it varies with X, we can still measure the effect of any treatment compared with any other by a single number. In Figure 25.3, for instance, treatment 1 leads to a higher mean response than treatment 2, no matter what is the value of X. The difference between the two mean responses is the same for all values of X, since the slopes of the regression lines are equal. Hence, we can measure the difference at any convenient X, say, at $X = \bar{X}_{..}$:

FIGURE 25.3 Treatment regression lines with covariance model (25.3)

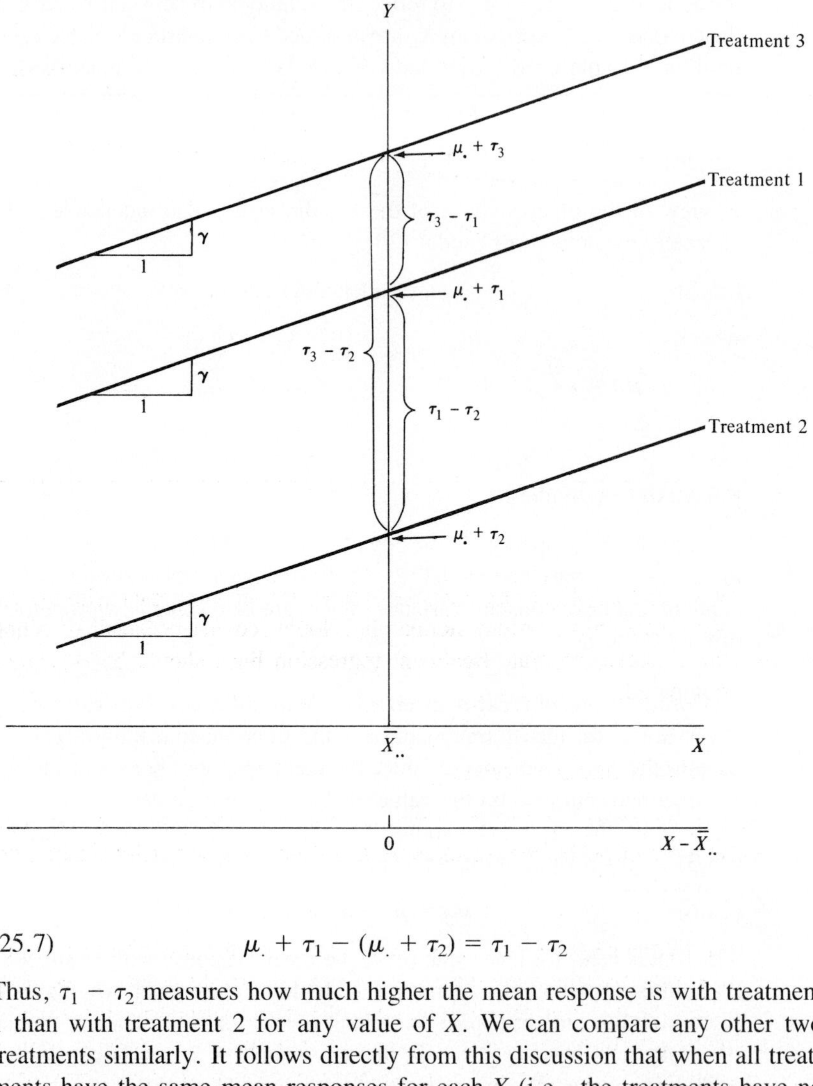

$$\mu_{.} + \tau_1 - (\mu_{.} + \tau_2) = \tau_1 - \tau_2 \tag{25.7}$$

Thus, $\tau_1 - \tau_2$ measures how much higher the mean response is with treatment 1 than with treatment 2 for any value of X. We can compare any other two treatments similarly. It follows directly from this discussion that when all treatments have the same mean responses for each X (i.e., the treatments have no differential effects), the treatment regression lines must be identical; and hence, $\tau_1 - \tau_2 = 0$, $\tau_1 - \tau_3 = 0$, etc. Indeed, all τ_i equal zero in that case.

Constancy of slopes. The assumption in model (25.3) that all treatment regression lines have the same slope is a crucial one. Without it, one cannot summarize the difference between the effects of two treatments by a single number based on main effects, such as $\tau_2 - \tau_1$. Figure 25.4 illustrates the case of nonparallel slopes for two treatments. Here, treatment 1 leads to higher mean responses than treatment 2 for some values of X, and the reverse holds for other

FIGURE 25.4 Nonparallel treatment regression lines

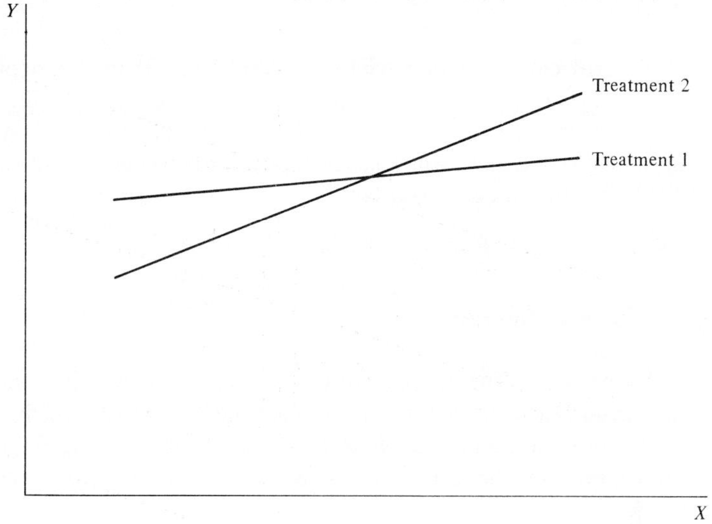

values of X. When the treatment regression lines interact with the concomitant variable X in the form of nonparallel slopes, covariance analysis is not appropriate. Instead, separate treatment regression lines should be estimated and then compared.

Generalizations of model

The covariance model (25.3) for single-factor studies can be generalized in several respects. We mention briefly three ways in which this model can be generalized.

Nonconstant X's. Model (25.3) assumes that the observations on the concomitant variable are constants. At times, it might be more reasonable to consider the concomitant observations as random variables. In that case, if one is willing to interpret model (25.3) as a conditional one, applying for any X values that might be observed, the covariance analysis to be presented is still appropriate.

Nonlinearity of relation. The linear relation between Y and X assumed in model (25.3) is not essential to covariance analysis. Any other relation could be used. For instance, the model would be as follows for a quadratic relation:

$$(25.8) \qquad Y_{ij} = \mu_{.} + \tau_i + \gamma_1(X_{ij} - \overline{X}_{..}) + \gamma_2(X_{ij} - \overline{X}_{..})^2 + \varepsilon_{ij}$$

Linearity of the relation leads to simpler analysis and is often a sufficiently good approximation to provide meaningful results. If a linear relation is not a good approximation, however, a more adequate description of the relation should be utilized in the model.

Several concomitant variables. Model (25.3) uses a single concomitant variable. This is often sufficient to reduce the error variability substantially. However, the model can be extended in a straightforward fashion to include two or more concomitant variables. The model for two concomitant variables, X_1 and X_2, to the first order would be:

$$Y_{ij} = \mu_{.} + \tau_i + \gamma_1(X_{ij1} - \overline{X}_{..1}) + \gamma_2(X_{ij2} - \overline{X}_{..2}) + \varepsilon_{ij} \tag{25.9}$$

Questions of interest

Inferences. The key statistical inferences of interest in covariance analysis are the same as with analysis of variance models, namely, whether the treatments have any effect, and if so what these effects are. Testing for fixed treatment effects involves the same alternatives as for analysis of variance models:

$$\begin{aligned} &H_0: \tau_1 = \tau_2 = \cdots = \tau_r = 0 \\ &H_a: \text{not all } \tau_i \text{ equal zero} \end{aligned} \tag{25.10}$$

If the treatment effects differ, one usually wishes to investigate the nature of these effects. Pairwise comparisons of treatment effects $\tau_i - \tau_{i'}$ (the vertical distance between the two treatment regression lines) may be made, or more general contrasts of the τ_i may be utilized. The nature of the regression relationship between Y and X sometimes is of interest, but usually the concomitant variable X is only employed to help reduce the error variability.

Appropriateness of model. A number of questions concerning the appropriateness of the covariance model may also be of interest. These may be answered by means of statistical methods previously discussed. Specifically, these questions pertain to:

1. Normality of error terms.
2. Equality of error variances for different treatments.
3. Equality of slopes for the different treatment regression lines.
4. Linearity of regression. (Because there are typically no replications at the different X levels, this is often examined via quadratic regression and testing whether the curvature coefficient is zero.)

Note

One is usually not concerned with whether the regression coefficient γ is zero, that is, whether there is indeed a relation between Y and X. If there is no relation, no bias results in the covariance analysis. The error mean square would simply be the same as for analysis of variance (allowing for sampling variation), and one degree of freedom for the error mean square would be lost.

When one is concerned equally with the concomitant variable and the treatment effects, then the methods presented in Chapter 10 should be employed for the analysis pertaining to the concomitant variable.

Regression formulation

An easy way to estimate the parameters of covariance model (25.3) and analyze questions of interest is through the regression approach. (If the reader has not yet read Sections 10.1–10.3, he or she should do so before proceeding further.)

As for analysis of variance models, we shall employ $r - 1$ indicator variables taking on the values 1, -1, or 0 to represent the r treatments:

$$
(25.11) \qquad \begin{aligned}
I_1 &= \begin{cases} 1 & \text{if observation from treatment 1} \\ -1 & \text{if observation from treatment } r \\ 0 & \text{otherwise} \end{cases} \\
&\;\vdots \\
I_{r-1} &= \begin{cases} 1 & \text{if observation from treatment } r-1 \\ -1 & \text{if observation from treatment } r \\ 0 & \text{otherwise} \end{cases}
\end{aligned}
$$

Note that we now denote the indicator variables by the symbol I to clearly distinguish the treatment effects from the concomitant variable X.

In expressing covariance model (25.3) in regression form, we shall, as in the regression chapters, denote the deviation $X_{ij} - \bar{X}_{..}$ by x_{ij}. Covariance model (25.3) can then be expressed as follows:

$$
(25.12) \qquad Y_{ij} = \mu_{.} + \tau_1 I_{ij1} + \cdots + \tau_{r-1} I_{ij,r-1} + \gamma x_{ij} + \varepsilon_{ij}
$$

where:

$$
x_{ij} = X_{ij} - \bar{X}_{..}
$$

Here, I_{ij1} is the value of indicator variable I_1 for the jth observation from treatment i, and similarly for the other indicator variables. Note that the treatment effects $\tau_1, \ldots, \tau_{r-1}$ are the regression coefficients for the indicator variables.

It is evident that tests concerning the treatment effects and estimates of the treatment effects can be carried out easily, involving only the regression coefficients $\tau_1, \ldots, \tau_{r-1}$. Since there are no new basic principles involved, we shall illustrate the usual testing and estimation procedures in terms of an example.

25.3 EXAMPLE OF SINGLE-FACTOR COVARIANCE ANALYSIS

A company wished to study the effects of three different types of promotions on sales of its crackers. The three promotions were:

Treatment 1—Sampling of product by customers in store and regular shelf space
Treatment 2—Additional shelf space in regular location
Treatment 3—Special display shelves at ends of aisle in addition to regular shelf space

Fifteen stores were selected for the study. Each store was randomly assigned one of the promotion types, with five stores assigned to each type of promotion. Other relevant conditions under the control of the company, such as price and advertising, were kept the same for all stores in the study. Data on the number of cases of the product sold during the promotional period, denoted by Y, are presented in Table 25.1a, as are also data on the sales of the product in the preceding period, denoted by X. Sales of the preceding period are to be used as the concomitant variable.

Development of model

Figure 25.5 presents the data of Table 25.1a in the form of a scatter plot. Linear regression and parallel slopes for the treatment regression lines appear to be reasonable. (Later we shall illustrate a formal statistical test for parallel slopes of the treatment regression lines.) Hence, the following regression model was employed:

$$Y_{ij} = \mu_{.} + \tau_1 I_{ij1} + \tau_2 I_{ij2} + \gamma x_{ij} + \varepsilon_{ij} \qquad \text{Full model} \tag{25.13}$$

TABLE 25.1 Data for cracker promotion example (number of cases sold)

(a) Data

	Study Unit (j)									
Treatment	1		2		3		4		5	
i	*Y*	*X*	*Y*	*X*	*Y*	*X*	*Y*	*X*	*Y*	*X*
1	38	21	39	26	36	22	45	28	33	19
2	43	34	38	26	38	29	27	18	34	25
3	24	23	32	29	31	30	21	16	28	29

(b) Means, Totals, and Sums of Squares

$\bar{Y}_{1.} = 38.2$ $\quad \bar{Y}_{2.} = 36.0$ $\quad \bar{Y}_{3.} = 27.2$ $\quad \bar{Y}_{..} = 33.8$
$\bar{X}_{1.} = 23.2$ $\quad \bar{X}_{2.} = 26.4$ $\quad \bar{X}_{3.} = 25.4$ $\quad \bar{X}_{..} = 25.0$

Treatment *i*	$\sum_j Y_{ij}$	$\sum_j Y_{ij}^2$	$\sum_j X_{ij}$	$\sum_j X_{ij}^2$	$\sum_j X_{ij}Y_{ij}$
1	191	7,375	116	2,746	4,491
2	180	6,622	132	3,622	4,888
3	136	3,786	127	3,367	3,558
Total	507	17,783	375	9,735	12,937

FIGURE 25.5 Scatter plot of cracker sales

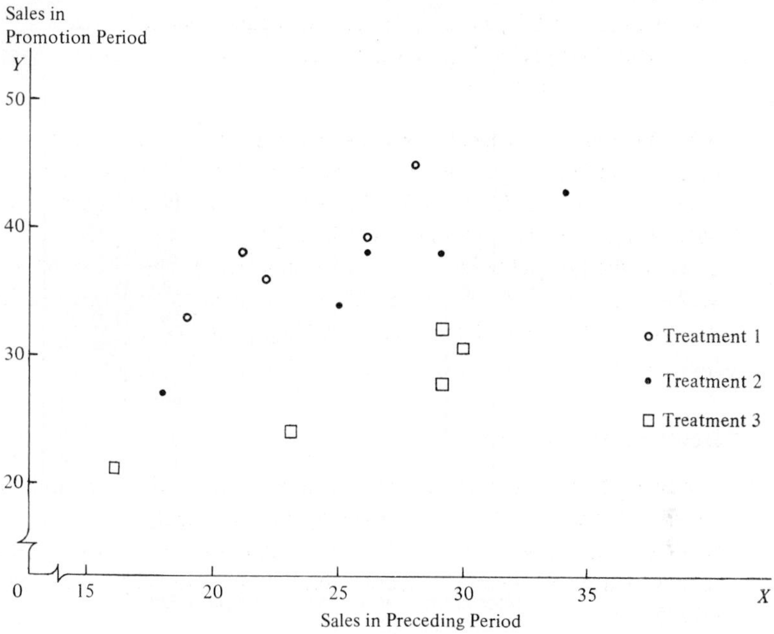

where:

$$I_{ij1} = \begin{cases} 1 & \text{if observation from treatment 1} \\ -1 & \text{if observation from treatment 3} \\ 0 & \text{otherwise} \end{cases}$$

$$I_{ij2} = \begin{cases} 1 & \text{if observation from treatment 2} \\ -1 & \text{if observation from treatment 3} \\ 0 & \text{otherwise} \end{cases}$$

$$x_{ij} = X_{ij} - \bar{X}_{..}$$

The **Y** observations vector and the **X** matrix for the data in Table 25.1a are given in Table 25.2. A computer run of a multiple regression package led to the results summarized in Table 25.3.

Test for treatment effects

To test whether or not the three cracker promotions differ in effectiveness, we can either follow the general linear test approach of fitting full and reduced models and using test statistic (8.32) or use extra sums of squares and test statistic (8.32a). In either case, the alternatives are:

TABLE 25.2 Data matrices for cracker promotion example—covariance model (25.13)

$$\mathbf{Y} = \begin{bmatrix} 38 \\ 39 \\ 36 \\ 45 \\ 33 \\ 43 \\ 38 \\ 38 \\ 27 \\ 34 \\ 24 \\ 32 \\ 31 \\ 21 \\ 28 \end{bmatrix} \qquad \mathbf{X} = \begin{bmatrix} & I_1 & I_2 & x \\ 1 & 1 & 0 & 21-25=-4 \\ 1 & 1 & 0 & 26-25=\ \ 1 \\ 1 & 1 & 0 & 22-25=-3 \\ 1 & 1 & 0 & 28-25=\ \ 3 \\ 1 & 1 & 0 & 19-25=-6 \\ 1 & 0 & 1 & 34-25=\ \ 9 \\ 1 & 0 & 1 & 26-25=\ \ 1 \\ 1 & 0 & 1 & 29-25=\ \ 4 \\ 1 & 0 & 1 & 18-25=-7 \\ 1 & 0 & 1 & 25-25=\ \ 0 \\ 1 & -1 & -1 & 23-25=-2 \\ 1 & -1 & -1 & 29-25=\ \ 4 \\ 1 & -1 & -1 & 30-25=\ \ 5 \\ 1 & -1 & -1 & 16-25=-9 \\ 1 & -1 & -1 & 29-25=\ \ 4 \end{bmatrix}$$

TABLE 25.3 Computer output for cracker promotion example—covariance model (25.13)

(a) Regression Coefficients

$\hat{\mu}_. = 33.800$ $\quad\hat{\tau}_2 = .942$

$\hat{\tau}_1 = 6.017$ $\quad\hat{\gamma} = .899$

(b) Analysis of Variance

Source of Variation	*SS*	*df*	*MS*
Regression	$SSR = 607.829$	3	$MSR = 202.610$
Error	$SSE = 38.571$	11	$MSE = 3.506$
Total	$SSTO = 646.400$	14	

(c) Estimated Variance-Covariance Matrix of Regression Coefficients

	$\hat{\mu}_.$	$\hat{\tau}_1$	$\hat{\tau}_2$	$\hat{\gamma}$
$\hat{\mu}_.$	.2338			
$\hat{\tau}_1$	0	.5016		
$\hat{\tau}_2$	0	−.2603	.4882	
$\hat{\gamma}$	0	.0189	−.0147	.0105

(25.14)
$$H_0: \tau_1 = \tau_2 = 0$$
$$H_a: \text{not both } \tau_1 \text{ and } \tau_2 \text{ equal zero}$$

Note that $\tau_3 = -\tau_1 - \tau_2$ must equal zero when $\tau_1 = \tau_2 = 0$.

To conduct the test by means of the general linear test approach, we develop the reduced model under H_0:

$$(25.15) \qquad\qquad Y_{ij} = \mu_. + \gamma x_{ij} + \varepsilon_{ij} \qquad\qquad \text{Reduced model}$$

Model (25.15) is just a simple linear regression model where none of the parameters vary for the different treatments. The data matrices for this reduced model are shown in Table 25.4. A computer run based on these data led to the following results:

Source of Variation	SS	df
Regression	$SSR = 190.678$	1
Error	$SSE = 455.722$	13
Total	$SSTO = 646.400$	14

TABLE 25.4 Data matrices for cracker promotion example—reduced model (25.15)

$$
Y = \begin{bmatrix} 38 \\ 39 \\ 36 \\ 45 \\ 33 \\ 43 \\ 38 \\ 38 \\ 27 \\ 34 \\ 24 \\ 32 \\ 31 \\ 21 \\ 28 \end{bmatrix}
\qquad
X = \begin{bmatrix} 1 & -4 \\ 1 & 1 \\ 1 & -3 \\ 1 & 3 \\ 1 & -6 \\ 1 & 9 \\ 1 & 1 \\ 1 & 4 \\ 1 & -7 \\ 1 & 0 \\ 1 & -2 \\ 1 & 4 \\ 1 & 5 \\ 1 & -9 \\ 1 & 4 \end{bmatrix}
$$

Thus, $SSE(R) = 455.722$. We know from Table 25.3 that the error sum of squares for the full model is $SSE(F) = 38.571$. Hence, test statistic (8.32) here is:

$$
\begin{aligned}
F^* &= \frac{SSE(R) - SSE(F)}{(n_T - 2) - [n_T - (r + 1)]} \div \frac{SSE(F)}{n_T - r - 1} \\
&= \frac{455.722 - 38.571}{13 - 11} \div \frac{38.571}{11} = 59.5
\end{aligned}
$$

The level of significance is to be controlled at $\alpha = .05$; hence, we require $F(.95; 2, 11) = 3.98$. The decision rule therefore is:

$$\text{If } F^* \leq 3.98, \text{ conclude } H_0$$
$$\text{If } F^* > 3.98, \text{ conclude } H_a$$

Since $F^* = 59.5 > 3.98$, we conclude H_a, that the three cracker promotions differ in sales effectiveness.

Alternatively, if the computer regression package provides extra sums of squares (Section 8.2) when fitting the full model in the order in which independent variables are entered, we enter x, I_1, and I_2 in that order and obtain the following decomposition of *SSR*:

Source of Variation	*SS*	*df*
Regression	607.829	3
x	190.678	1
$I_1 \mid x$	410.776	1
$I_2 \mid x, I_1$	6.375	1

The required extra sum of squares according to (8.32a) is $SSR(I_1, I_2 \mid x)$. Since we have:

$$SSR(I_1, I_2 \mid x) = SSR(I_1 \mid x) + SSR(I_2 \mid x, I_1)$$

we obtain:

$$SSR(I_1, I_2 \mid x) = 410.776 + 6.375 = 417.151$$

This equals $SSE(R) - SSE(F) = 455.722 - 38.571 = 417.151$ obtained by fitting the full and reduced models.

Note

Occasionally, a test whether or not $\gamma = 0$ is of interest. This is simply the ordinary test whether or not a single regression coefficient equals zero. It can be conducted by means of the t^* test statistic (8.25) or by means of the F^* test statistic (8.24). For the latter test statistic, the required extra sum of squares is usually obtained by entering the concomitant variable as the last independent variable in the computer regression package.

Estimation of treatment effects

Since treatment effects were found to be present in the cracker promotion study, the analyst wished to investigate them further. We noted earlier that a comparison of two treatments involves $\tau_i - \tau_{i'}$, the vertical distance between the two treatment regression lines. Using the fact that $\tau_3 = -\tau_1 - \tau_2$ and theorem (1.25b) for variances, we know at once that the estimators of all pairwise comparisons and their variances are as follows:

(25.16)

Comparison	Estimator	Variance
$\tau_1 - \tau_2$	$\hat{\tau}_1 - \hat{\tau}_2$	$\sigma^2(\hat{\tau}_1) + \sigma^2(\hat{\tau}_2) - 2\sigma(\hat{\tau}_1, \hat{\tau}_2)$
$\tau_1 - \tau_3 = 2\tau_1 + \tau_2$	$2\hat{\tau}_1 + \hat{\tau}_2$	$4\sigma^2(\hat{\tau}_1) + \sigma^2(\hat{\tau}_2) + 4\sigma(\hat{\tau}_1, \hat{\tau}_2)$
$\tau_2 - \tau_3 = \tau_1 + 2\tau_2$	$\hat{\tau}_1 + 2\hat{\tau}_2$	$\sigma^2(\hat{\tau}_1) + 4\sigma^2(\hat{\tau}_2) + 4\sigma(\hat{\tau}_1, \hat{\tau}_2)$

Table 25.3 furnishes us the needed estimated regression coefficients as well as their estimated variances and covariances. We obtain from there:

(25.16a)

Comparison	Estimate	Estimated Variance
$\tau_1 - \tau_2$	$6.017 - .942 = 5.075$	$.5016 + .4882 - 2(-.2603) = 1.5104$
$\tau_1 - \tau_3$	$2(6.017) + .942 = 12.976$	$4(.5016) + .4882 + 4(-.2603) = 1.4534$
$\tau_2 - \tau_3$	$6.017 + 2(.942) = 7.901$	$.5016 + 4(.4882) + 4(-.2603) = 1.4132$

If a single interval estimate is to be constructed, the t distribution with $n_T - r - 1$ degrees of freedom would be used. (The degrees of freedom are those associated with *MSE* in the full covariance model.) Usually, however, a family of interval estimates is desired. In that case, the Scheffé multiple comparison procedure may be employed with the S multiple defined by:

(25.17) $$S^2 = (r - 1)F(1 - \alpha; r - 1, n_T - r - 1)$$

or the Bonferroni method may be employed with the B multiple:

(25.18) $$B = t(1 - \alpha/2g; n_T - r - 1)$$

where g is the number of statements in the family. The Tukey method is not appropriate for covariance analysis.

In the case at hand, the analyst wished to obtain all pairwise comparisons with a 95 percent family confidence coefficient. The analyst used the Scheffé procedure because he anticipated making some additional estimates of contrasts. He required therefore:

$$S^2 = (3 - 1)F(.95; 2, 11) = 2(3.98) = 7.96$$

or:

$$S = 2.82$$

Using the results in (25.16a), the confidence intervals for all pairwise treatment comparisons with a 95 percent family confidence coefficient then are:

$$1.61 = 5.075 - 2.82\sqrt{1.5104} \leq \tau_1 - \tau_2 \leq 5.075 + 2.82\sqrt{1.5104} = 8.54$$
$$9.58 = 12.976 - 2.82\sqrt{1.4534} \leq \tau_1 - \tau_3 \leq 12.976 + 2.82\sqrt{1.4534} = 16.38$$
$$4.55 = 7.901 - 2.82\sqrt{1.4132} \leq \tau_2 - \tau_3 \leq 7.901 + 2.82\sqrt{1.4132} = 11.25$$

These results indicate clearly that sampling in store (treatment 1) is significantly better for stimulating cracker sales than either of the two shelf promotions, and that increasing the regular shelf space (treatment 2) is superior to additional displays at the end of the aisle (treatment 3).

Comments

1. Occasionally, more general contrasts among treatment effects than pairwise comparisons are desired. No new problems arise either in the use of the t distribution for a single contrast or in the use of the Scheffé or Bonferroni procedures for multiple comparisons. For instance, if the analyst had desired in our cracker promotion example to compare the treatment effect for sampling in the store (treatment 1) with the two treatments involving shelf displays (treatments 2 and 3), he would have been interested in the contrast:

$$L = \tau_1 - \frac{\tau_2 + \tau_3}{2} \tag{25.19}$$

The appropriate estimator is:

$$\hat{L} = \hat{\tau}_1 - \frac{\hat{\tau}_2 + (-\hat{\tau}_1 - \hat{\tau}_2)}{2} = \frac{3}{2}\hat{\tau}_1 \tag{25.20}$$

The variance of this estimator is by (1.15b):

$$\sigma^2(\hat{L}) = \frac{9}{4}\sigma^2(\hat{\tau}_1) \tag{25.21}$$

2. It may be of interest to estimate the mean response with the ith treatment for some value of X. Frequently $X = \bar{X}_{..}$ is considered to be a "typical" value of X. We know from Figure 25.3 that at $X = \bar{X}_{..}$, the mean response for the ith treatment is the intercept of the treatment regression line, $\mu_. + \tau_i$. An estimator of $\mu_. + \tau_i$ can be readily developed. For our cracker promotion example, we obtain the following estimators and their variances:

(25.22)

Mean Response at $X = \bar{X}_{..}$	*Estimator*	*Variance*
$\mu_. + \tau_1$	$\hat{\mu}_. + \hat{\tau}_1$	$\sigma^2(\hat{\mu}_.) + \sigma^2(\hat{\tau}_1) + 2\sigma(\hat{\mu}_., \hat{\tau}_1)$
$\mu_. + \tau_2$	$\hat{\mu}_. + \hat{\tau}_2$	$\sigma^2(\hat{\mu}_.) + \sigma^2(\hat{\tau}_2) + 2\sigma(\hat{\mu}_., \hat{\tau}_2)$
$\mu_. + \tau_3$	$\hat{\mu}_. - \hat{\tau}_1 - \hat{\tau}_2$	$\sigma^2(\hat{\mu}_.) + \sigma^2(\hat{\tau}_1) + \sigma^2(\hat{\tau}_2) - 2\sigma(\hat{\mu}_., \hat{\tau}_1) - 2\sigma(\hat{\mu}_., \hat{\tau}_2) + 2\sigma(\hat{\tau}_1, \hat{\tau}_2)$

Use of the results in Table 25.3 leads to the following estimates:

Treatment	*Estimated Mean Response at $\bar{X}_{..}$*	*Estimated Variance*
1	33.800 + 6.017 = 39.817	.2338 + .5016 + 2(0) = .7354
2	33.800 + .942 = 34.742	.2338 + .4882 + 2(0) = .7220
3	33.800 − 6.017 − .942 = 26.841	.2338 + .5016 + .4882 − 2(0) − 2(0) + 2(−.2603) = .7030

3. Computational formulas for estimating treatment comparisons and general contrasts, which can be used in case a computer with a multiple regression program is not available, will be presented in the next section.

Test for parallel slopes

An important assumption made in covariance analysis is that all treatment regression lines have the same slope γ. The analyst who conducted the cracker promotion study, indeed, tested this assumption before proceeding with the analysis discussed earlier. We know from Chapter 10 that regression model (25.13) can be generalized to allow for different slopes for the treatments by introducing cross-product interaction terms. Specifically, interaction variables I_1x and I_2x will be required in our example. We shall denote the corresponding regression coefficients by β_1 and β_2. Thus, the generalized model is:

$$(25.23) \qquad Y_{ij} = \mu_{.} + \tau_1 I_{ij1} + \tau_2 I_{ij2} + \gamma x_{ij} + \beta_1 I_{ij1} x_{ij} + \beta_2 I_{ij2} x_{ij} + \varepsilon_{ij} \qquad \text{Generalized model}$$

Table 25.5 contains the data matrices for this generalized model for our cracker promotion example. Note that the **X** matrix for the generalized model differs from the **X** matrix for regression model (25.13) simply by the addition of the I_1x and I_2x columns. A fit of model (25.23) by a computer multiple regression package yielded the following ANOVA results:

Source of Variation	*SS*	*df*
Regression	$SSR = 614.879$	5
Error	$SSE = 31.521$	9
Total	$SSTO = 646.400$	14

SSE obtained by fitting the generalized model (25.23) is the equivalent of fitting separate regression lines for each treatment and summing these error sums of squares.

To test for parallel slopes is equivalent to testing for no interactions in the generalized model (25.23):

$$(25.24) \qquad \begin{aligned} &H_0\colon \beta_1 = \beta_2 = 0 \\ &H_a\colon \text{not both } \beta_1 \text{ and } \beta_2 \text{ equal zero} \end{aligned}$$

We now need to recognize that the generalized model (25.23) here is the ''full'' model and the full model (25.13) is now the ''reduced'' model. Hence, we have:

$$SSE(F) = 31.521 \qquad SSE(R) = 38.571$$

TABLE 25.5 Data matrices for cracker promotion example—generalized model (25.23)

$$
\mathbf{Y} = \begin{bmatrix} 38 \\ 39 \\ 36 \\ 45 \\ 33 \\ 43 \\ 38 \\ 38 \\ 27 \\ 34 \\ 24 \\ 32 \\ 31 \\ 21 \\ 28 \end{bmatrix}
\qquad
\mathbf{X} = \begin{bmatrix}
 & I_1 & I_2 & x & I_1x & I_2x \\
1 & 1 & 0 & -4 & -4 & 0 \\
1 & 1 & 0 & 1 & 1 & 0 \\
1 & 1 & 0 & -3 & -3 & 0 \\
1 & 1 & 0 & 3 & 3 & 0 \\
1 & 1 & 0 & -6 & -6 & 0 \\
1 & 0 & 1 & 9 & 0 & 9 \\
1 & 0 & 1 & 1 & 0 & 1 \\
1 & 0 & 1 & 4 & 0 & 4 \\
1 & 0 & 1 & -7 & 0 & -7 \\
1 & 0 & 1 & 0 & 0 & 0 \\
1 & -1 & -1 & -2 & 2 & 2 \\
1 & -1 & -1 & 4 & -4 & -4 \\
1 & -1 & -1 & 5 & -5 & -5 \\
1 & -1 & -1 & -9 & 9 & 9 \\
1 & -1 & -1 & 4 & -4 & -4
\end{bmatrix}
$$

Thus, test statistic (8.32) becomes here:

$$F^* = \frac{38.571 - 31.521}{11 - 9} \div \frac{31.521}{9} = 1.01$$

For a level of significance of $\alpha = .05$, we require $F(.95; 2, 9) = 4.26$. Since $F^* = 1.01 \leq 4.26$, we conclude H_0, that the three treatment regression lines have the same slope.

25.4 SINGLE-FACTOR COVARIANCE ANALYSIS AS MODIFICATION OF ANALYSIS OF VARIANCE

In previous sections, we explained the fundamental ideas underlying covariance analysis by viewing the model as a regression model. We saw that the statistical questions that arise in covariance analysis fit readily in the regression framework. When analysis of covariance was first developed, however, computers did not exist and it was not feasible to do the covariance analysis computations by means of the regression approach. Instead, computational formulas were developed that exploit the fact that the indicator variables take on the values 1, -1, or 0 in certain structures. These computational formulas may appear formidable, but computations by hand via these formulas are much simpler than hand computations via the regression approach. To explain the rationale of these computational formulas, analysis of covariance is usually viewed as a process that begins with the regular analysis of variance and *adjusts* this analysis for the concomitant variables. We shall now examine this adjustment approach to single-factor covariance analysis with one concomitant variable and demonstrate its equivalence to the regression approach.

Analysis of covariance table

Analysis of variance for X and XY. The first idea to be considered when viewing covariance analysis as an adjustment of regular analysis of variance is that the decomposition of the total sum of squares in (16.25):

$$SSTO_Y = SSTR_Y + SSE_Y \tag{25.25}$$

can also be carried out for the variable X and the product XY. Note that the subscript Y has been added in (25.25) to make clear that this decomposition refers to variable Y. It will be recalled that:

$$SSTO_Y = \sum_i \sum_j (Y_{ij} - \bar{Y}_{..})^2 = \sum_i \sum_j Y_{ij}^2 - \frac{Y_{..}^2}{n_T} \tag{25.26a}$$

$$SSTR_Y = \sum_i n_i(\bar{Y}_{i.} - \bar{Y}_{..})^2 = \sum_i \frac{Y_{i.}^2}{n_i} - \frac{Y_{..}^2}{n_T} \tag{25.26b}$$

$$SSE_Y = \sum_i \sum_j (Y_{ij} - \bar{Y}_{i.})^2 = \sum_i \sum_j Y_{ij}^2 - \sum_i \frac{Y_{i.}^2}{n_i} \tag{25.26c}$$

The same analysis of variance for the variable X is:

$$SSTO_X = \sum_i \sum_j (X_{ij} - \bar{X}_{..})^2 = \sum_i \sum_j X_{ij}^2 - \frac{X_{..}^2}{n_T} \tag{25.27a}$$

$$SSTR_X = \sum_i n_i(\bar{X}_{i.} - \bar{X}_{..})^2 = \sum_i \frac{X_{i.}^2}{n_i} - \frac{X_{..}^2}{n_T} \tag{25.27b}$$

$$SSE_X = \sum_i \sum_j (X_{ij} - \bar{X}_{i.})^2 = \sum_i \sum_j X_{ij}^2 - \sum_i \frac{X_{i.}^2}{n_i} \tag{25.27c}$$

where the notation for the X's corresponds exactly to that for the Y's.

The analysis of variance for the products XY starts with the definition of *total sum of products* ($SPTO$):

$$SPTO = \sum_i \sum_j (X_{ij} - \bar{X}_{..})(Y_{ij} - \bar{Y}_{..}) = \sum_i \sum_j X_{ij}Y_{ij} - \frac{X_{..}Y_{..}}{n_T} \tag{25.28a}$$

To see that the total sum of products is related to the total sum of squares for Y or X, note that if X_{ij} is replaced by Y_{ij} in (25.28a), we obtain $SSTO_Y$, and if Y_{ij} is replaced by X_{ij}, we obtain $SSTO_X$. The two components of the total sum of products are the *treatment sum of products* ($SPTR$) and the *error sum of products* (SPE):

$$(25.28b) \qquad SPTR = \sum_i n_i(\bar{X}_{i.} - \bar{X}_{..})(\bar{Y}_{i.} - \bar{Y}_{..}) = \sum_i \frac{X_{i.}Y_{i.}}{n_i} - \frac{X_{..}Y_{..}}{n_T}$$

$$(25.28c) \qquad SPE = \sum_i \sum_j (X_{ij} - \bar{X}_{i.})(Y_{ij} - \bar{Y}_{i.}) = \sum_i \sum_j X_{ij}Y_{ij} - \sum_i \frac{X_{i.}Y_{i.}}{n_i}$$

Unlike sums of squares, *sums of products may be negative.*

Example. Table 25.6 contains the analyses of variance for Y, X, and XY for our cracker promotion example in Table 25.1a. We illustrate two of the calculations, using the results presented in Table 25.1b:

$$SPTR = \frac{1}{5}[116(191) + 132(180) + 127(136)] - \frac{375(507)}{15}$$
$$= 12{,}637.6 - 12{,}675.0 = -37.4$$

$$SSE_X = 9{,}735 - \frac{1}{5}[(116)^2 + (132)^2 + (127)^2] = 9{,}735 - 9{,}401.8 = 333.2$$

TABLE 25.6 Analyses of variance for Y, X, and XY for cracker promotion example

Source of Variation	*Sums of Squares or Products* Y	X	XY	*df*
Treatments	338.8	26.8	−37.4	2
Error	307.6	333.2	299.4	12
Total	646.4	360.0	262.0	14

Adjusted analysis of variance for Y. We are now ready to adjust the analysis of variance for Y to obtain the analysis of covariance. The rationale of the adjustment can best be seen from simple regression analysis. There we found that the error sum of squares (2.21):

$$(25.29) \qquad SSE = \Sigma(Y_i - \hat{Y}_i)^2 = \Sigma(Y_i - b_0 - b_1X_i)^2$$

could be expressed in the algebraically equivalent form (2.24b):

$$(25.29a) \qquad SSE = \Sigma(Y_i - \bar{Y})^2 - \frac{[\Sigma(X_i - \bar{X})(Y_i - \bar{Y})]^2}{\Sigma(X_i - \bar{X})^2}$$

which can be rewritten as follows:

(25.29b) $$SSE = \Sigma(Y_i - \bar{Y})^2 - b_1\Sigma(X_i - \bar{X})(Y_i - \bar{Y})$$

In terms of our analysis of variance notation (which uses double subscripts and $\bar{X}_{..}$ and $\bar{Y}_{..}$ for $\bar{X}$ and $\bar{Y}$), the regression SSE can therefore be expressed in either of the following ways:

(25.30a) $$SSE = SSTO_Y - \frac{(SPTO)^2}{SSTO_X}$$

(25.30b) $$SSE = SSTO_Y - b_1(SPTO)$$

Hence, in the analysis of covariance, the total sum of squares for Y, adjusted for the linear relation to X, would be obtained as follows:

(25.31a) $$SSTO(\text{adj.}) = SSTO_Y - \frac{(SPTO)^2}{SSTO_X}$$

where $SSTO$(adj.) stands for the *adjusted total sum of squares for Y*.

It may then be argued by analogy that:

(25.31b) $$SSE(\text{adj.}) = SSE_Y - \frac{(SPE)^2}{SSE_X}$$

where SSE(adj.) stands for the *adjusted error sum of squares for Y*.

Finally, we obtain by subtraction:

(25.31c) $$SSTR(\text{adj.}) = SSTO(\text{adj.}) - SSE(\text{adj.})$$

where $SSTR$(adj.) stands for the *adjusted treatment sum of squares for Y*. Note carefully that $SSTR$(adj.) is obtained by subtraction and not by an analogous adjustment. The reason for this will become clear later.

Table 25.7 contains the general covariance analysis table for a single-factor study with one concomitant variable. First are presented the sums of squares and products. Then the adjusted sums of squares are given. When these are divided by the degrees of freedom, the adjusted mean squares are obtained. Note that there is one less degree of freedom for $SSTO$(adj.) and for SSE(adj.) than with the analysis of variance model. The reason is that the regression coefficient γ for the concomitant variable had to be estimated.

Example. For our cracker promotion example, we find, using (25.31) and the results in Table 25.6:

$$SSTO(\text{adj.}) = 646.4 - \frac{(262)^2}{360} = 455.722$$

$$SSE(\text{adj.}) = 307.6 - \frac{(299.4)^2}{333.2} = 38.571$$

$$SSTR(\text{adj.}) = 455.722 - 38.571 = 417.151$$

These results are presented in an analysis of covariance table in Table 25.8, which also contains the adjusted degrees of freedom and the adjusted mean

TABLE 25.7 Covariance analysis for single-factor study with one concomitant variable

Source of Variation	Sums of Squares or Products: Y	X	XY	df
Treatments	$SSTR_Y$	$SSTR_X$	$SPTR$	$r-1$
Error	SSE_Y	SSE_X	SPE	n_T-r
Total	$SSTO_Y$	$SSTO_X$	$SPTO$	n_T-1

Source of Variation	Adjusted SS	Adjusted df	Adjusted MS
Treatments	$SSTR$(adj.)	$r-1$	$MSTR$(adj.)
Error	SSE(adj.)	n_T-r-1	MSE(adj.)
Total	$SSTO$(adj.)	n_T-2	

TABLE 25.8 Covariance analysis table for cracker promotion example

Source of Variation	Adjusted SS	Adjusted df	Adjusted MS
Treatments	417.151	2	208.576
Error	38.571	11	3.506
Total	455.722	13	

squares. Note there is one less degree of freedom for the adjusted total sum of squares and for the adjusted error sum of squares than in the analysis of variance (Table 25.6).

Test for treatment effects

The test for treatment effects:

(25.32a)
$$H_0: \tau_1 = \tau_2 = \cdots = \tau_r = 0$$
$$H_a: \text{not all } \tau_i \text{ equal zero}$$

is then based on the usual test statistic:

(25.32b)
$$F^* = \frac{MSTR(\text{adj.})}{MSE(\text{adj.})}$$

If H_0 holds, F^* follows the $F(r - 1, n_T - r - 1)$ distribution. Hence, the decision rule for controlling the level of significance at α is:

(25.32c) If $F^* \leq F(1 - \alpha; r - 1, n_T - r - 1)$, conclude H_0
         If $F^* > F(1 - \alpha; r - 1, n_T - r - 1)$, conclude H_a

Example. For our cracker promotion example, we have from Table 25.8:

$$F^* = \frac{MSTR(\text{adj.})}{MSE(\text{adj.})} = \frac{208.576}{3.506} = 59.5$$

which, of course, is the same value as we obtained on page 857 with the regression approach. If $\alpha = .05$, we require $F(.95; 2, 11) = 3.98$. Since $F^* = 59.5 > 3.98$, we conclude that the three promotions have different effects on sales of crackers.

Reconciliation of two approaches. Figure 25.6 summarizes the relations between the regression and adjustment approaches to covariance analysis. We shall now explain these relations in three steps.

FIGURE 25.6 Reconciliation between regression and adjustment approaches to covariance analysis

Regression Approach

Full Model
(varying levels of treatment regression lines)

Reduced Model
(identical treatment regression lines)

Source	SS	df
Regression	$SSR(F)$	r
Error	$SSE(F)$	$n_T - r - 1$
Total	$SSTO$	$n_T - 1$

Source	SS	df
Regression	$SSR(R)$	1
Error	$SSE(R)$	$n_T - 2$
Total	$SSTO$	$n_T - 1$

Adjustment Approach

Adjusted

Source	SS	df
Total	$SSTO$ (adj.)	$n_T - 2$
minus Error	SSE (adj.)	$n_T - r - 1$
equals Treatments	$SSTR$ (adj.)	$r - 1$

$$SSTR(\text{adj.}) = SSTO(\text{adj.}) - SSE(\text{adj.}) = SSE(R) - SSE(F)$$

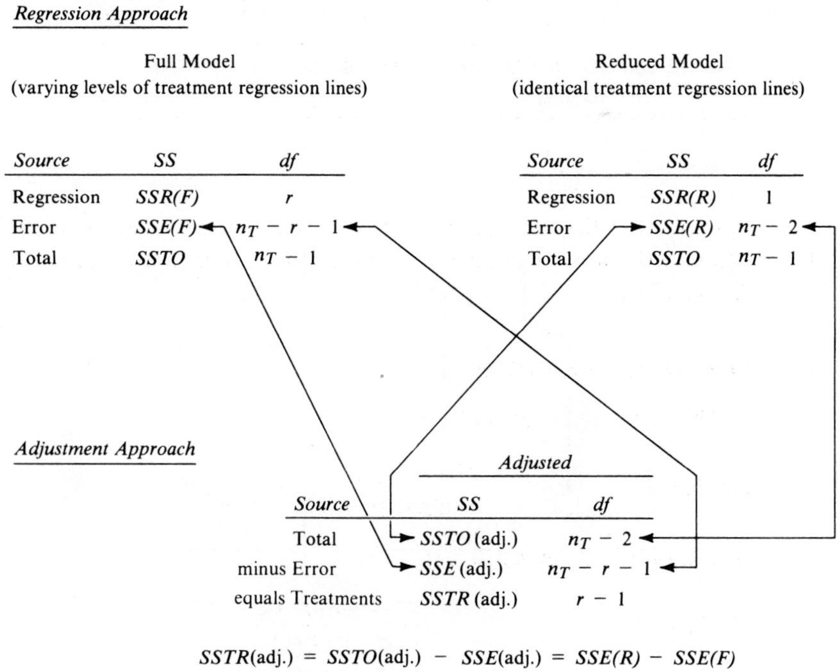

1. We consider first the equivalence of $SSE(R)$ with the regression approach and $SSTO$(adj.) with the adjustment approach. When we fit the reduced model (25.15) with the regression approach, noting that $x_{ij} = X_{ij} - \bar{X}_{..}$:

$$Y_{ij} = \mu_{.} + \gamma(X_{ij} - \bar{X}_{..}) + \varepsilon_{ij}$$

we obtain the usual least squares estimator (2.10a) for the slope γ:

$$g = \frac{\sum_i \sum_j (X_{ij} - \bar{X}_{..})(Y_{ij} - \bar{Y}_{..})}{\sum_i \sum_j (X_{ij} - \bar{X}_{..})^2} = \frac{SPTO}{SSTO_X} \tag{25.33}$$

and the usual error sum of squares (2.24b):

$$SSE(R) = \sum_i \sum_j (Y_{ij} - \bar{Y}_{..})^2 - \frac{\left[\sum_i \sum_j (X_{ij} - \bar{X}_{..})(Y_{ij} - \bar{Y}_{..})\right]^2}{\sum_i \sum_j (X_{ij} - \bar{X}_{..})^2} \tag{25.34}$$

Figure 25.7a schematically illustrates the residuals entering into $SSE(R)$ when two treatments are in an experiment.

Expressed in our notation, $SSE(R)$ becomes:

$$SSE(R) = SSTO_Y - \frac{(SPTO)^2}{SSTO_X} = SSTO(\text{adj.}) \tag{25.34a}$$

FIGURE 25.7 Schematic representation of residuals for $SSE(R)$ and $SSE(F)$

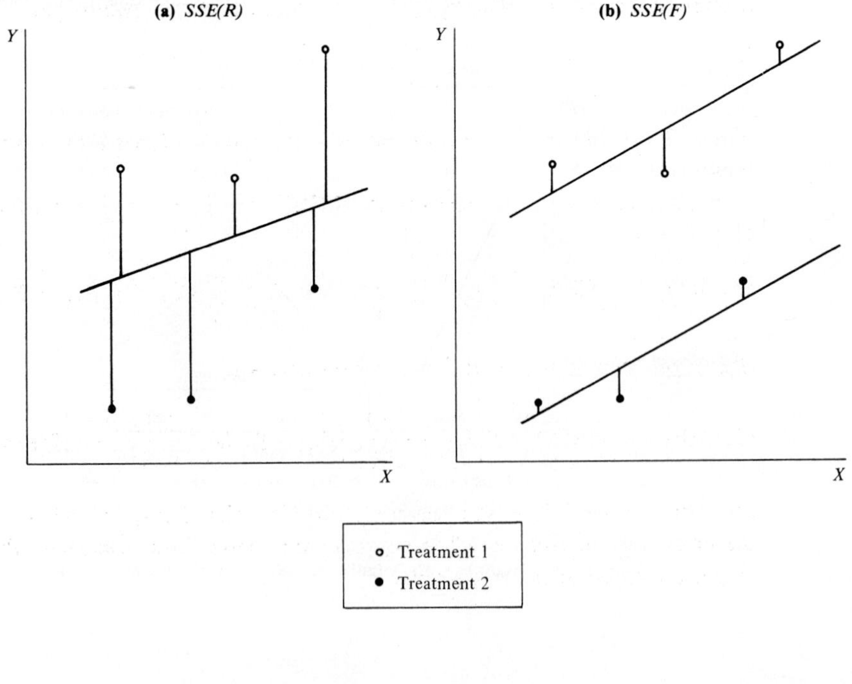

according to the definition of $SSTO$(adj.) in formula (25.31a). Thus, $SSTO$(adj.) is simply the error sum of squares when a linear regression is fitted to the entire set of data.

2. We next consider the equivalence of $SSE(F)$ with the regression approach and SSE(adj.) with the adjustment approach. When the full covariance model (25.3):

$$Y_{ij} = \mu_{.} + \tau_i + \gamma(X_{ij} - \bar{X}_{..}) + \varepsilon_{ij}$$

is fitted to the data, allowing different intercepts $\mu_{.} + \tau_i$ for the treatment regression lines but requiring a common slope γ, it can be shown that the least squares estimator of this common slope is:

(25.35)
$$g_w = \frac{\sum_i \sum_j (X_{ij} - \bar{X}_{i.})(Y_{ij} - \bar{Y}_{i.})}{\sum_i \sum_j (X_{ij} - \bar{X}_{i.})^2} = \frac{SPE}{SSE_X}$$

and that the least squares estimator of $\mu_{.} + \tau_i$ is:

(25.36)
$$\bar{Y}_{i.} - g_w(\bar{X}_{i.} - \bar{X}_{..})$$

Hence, the error sum of squares for the ith treatment is:

(25.37)
$$\sum_j \{Y_{ij} - [\bar{Y}_{i.} - g_w(\bar{X}_{i.} - \bar{X}_{..})] - g_w(X_{ij} - \bar{X}_{..})\}^2 = \sum_j [(Y_{ij} - \bar{Y}_{i.}) - g_w(X_{ij} - \bar{X}_{i.})]^2$$

Summing these error sums of squares over all treatments, we obtain $SSE(F)$:

(25.38)
$$SSE(F) = \sum_i \sum_j [(Y_{ij} - \bar{Y}_{i.}) - g_w(X_{ij} - \bar{X}_{i.})]^2$$

Figure 25.7b illustrates the residuals entering into $SSE(F)$ for the case of two treatments.

Expanding out the expression for $SSE(F)$ in (25.38) and simplifying, we obtain:

(25.38a)
$$SSE(F) = \sum_i \sum_j (Y_{ij} - \bar{Y}_{i.})^2 - g_w \sum_i \sum_j (X_{ij} - \bar{X}_{i.})(Y_{ij} - \bar{Y}_{i.})$$

But in our new notation, this expression becomes:

(25.38b)
$$SSE(F) = SSE_Y - \frac{SPE}{SSE_X} SPE = SSE_Y - \frac{(SPE)^2}{SSE_X} = SSE(\text{adj.})$$

according to the definition of SSE(adj.) in (25.31b). Thus, SSE(adj.) is simply the error sum of squares when separate regression lines, each with the same slope, are fitted to the treatments.

3. $SSTR$(adj.) is obtained as the difference $SSTO$(adj.) − SSE(adj.), just as the numerator of the test statistic with the general linear test approach is based on the difference $SSE(R) - SSE(F)$.

Comments

1. One can obtain an indication of the effectiveness of the analysis of covariance in reducing error variability by comparing MSE(adj.) for covariance analysis with MSE for regular analysis of variance. For our cracker promotion example, we know from Table 25.8 that MSE(adj.) = 3.51. We can also see from Table 25.6 that the error mean square for regular analysis of variance wouid have been:

$$MSE = \frac{307.6}{12} = 26.63$$

Hence, in this case covariance analysis reduced the residual variance by about 87 percent, a substantial reduction.

2. Covariance analysis and analysis of variance need not lead to the same conclusions about the treatment effects. For instance, analysis of variance might not indicate any treatment effects whereas covariance analysis with smaller error variance could show significant treatment effects. Ordinarily, of course, one should decide in advance which of the two analyses is to be used.

3. The estimator g_w of the common slope γ can be considered as an average of the separately estimated treatment regression line slopes g_i. If we were fitting a separate regression line for each treatment, the estimated slope g_i for the ith treatment would be given by the method of least squares as follows:

$$g_i = \frac{\sum_j (X_{ij} - \bar{X}_{i.})(Y_{ij} - \bar{Y}_{i.})}{\sum_j (X_{ij} - \bar{X}_{i.})^2} \tag{25.39}$$

Using $\sum_j (X_{ij} - \bar{X}_{i.})^2$ as weights, a weighted average of the g_i gives us precisely g_w as defined in (25.35):

$$\frac{\sum_i \left[\sum_j (X_{ij} - \bar{X}_{i.})^2\right] g_i}{\sum_i \left[\sum_j (X_{ij} - \bar{X}_{i.})^2\right]} = \frac{\sum_i \sum_j (X_{ij} - \bar{X}_{i.})(Y_{ij} - \bar{Y}_{i.})}{\sum_i \sum_j (X_{ij} - \bar{X}_{i.})^2} = g_w$$

Thus, g_w may be thought of as an average within-treatments regression slope.

For our cracker promotion example, the average within-treatments regression slope is:

$$g_w = \frac{SPE}{SSE_X} = \frac{299.4}{333.2} = .8986$$

and the slope when a single regression line is fitted to all data is:

$$g = \frac{SPTO}{SSTO_X} = \frac{262}{360} = .7278$$

Adjusted treatment means

In analysis of variance, the treatment sample mean $\bar{Y}_{i.}$ is an estimate of the mean response with the ith treatment. In analysis of covariance, many writers speak of the need to *adjust* the $\bar{Y}_{i.}$ to make them comparable with respect to the concomitant variable since the X values usually will not be the same for all treatments. The adjustment takes the form:

$$\bar{Y}_{i.}(\text{adj.}) = \bar{Y}_{i.} - g_w(\bar{X}_{i.} - \bar{X}_{..}) \tag{25.40}$$

The rationale of the adjustment can be seen from Figure 25.8. This figure contains the points $(\bar{X}_{i.}, \bar{Y}_{i.})$ for the three treatments in our cracker promotion example. Through each of these points, a regression line with the average within-treatment slope $g_w = .8986$ is drawn. The adjusted treatment mean $\bar{Y}_{i.}(\text{adj.})$ is simply the ordinate of its regression line at $X = \bar{X}_{..}$. In this way, the treatments are said to be made comparable with respect to X. For our example, the adjusted treatment means are obtained as follows:

Treatment	$\bar{Y}_{i.}$	$\bar{X}_{i.}$	$\bar{X}_{..}$	g_w	$g_w(\bar{X}_{i.} - \bar{X}_{..})$	$\bar{Y}_{i.}(\text{adj.})$
1	38.2	23.2	25	.8986	−1.62	39.82
2	36.0	26.4	25	.8986	1.26	34.74
3	27.2	25.4	25	.8986	.36	26.84

FIGURE 25.8 Representation of adjusted treatment means for cracker promotion example

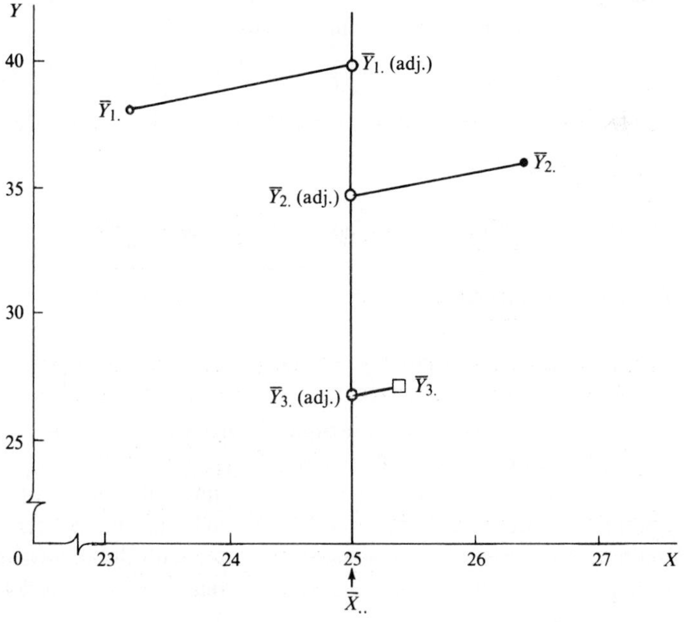

A comparison with the results on page 860 indicates that these adjusted treatment means are simply estimates of the intercepts of the treatment regression lines. In other words, $\bar{Y}_{i.}(\text{adj.})$ is simply an estimator of $\mu_{.} + \tau_i$.

The variance of $\bar{Y}_{i.}(\text{adj.})$ can be shown to be:

$$\sigma^2[\bar{Y}_{i.}(\text{adj.})] = \sigma^2\left[\frac{1}{n_i} + \frac{(\bar{X}_{i.} - \bar{X}_{..})^2}{SSE_X}\right] \tag{25.41}$$

and an unbiased estimator of this variance is:

$$s^2[\bar{Y}_{i.}(\text{adj.})] = MSE(\text{adj.})\left[\frac{1}{n_i} + \frac{(\bar{X}_{i.} - \bar{X}_{..})^2}{SSE_X}\right] \tag{25.42}$$

For instance, the estimated variance of the adjusted mean for treatment 1 in our cracker promotion example is:

$$s^2[\bar{Y}_{1.}(\text{adj.})] = 3.506\left[\frac{1}{5} + \frac{(23.2 - 25)^2}{333.2}\right] = .735$$

which is the same result obtained on page 860 by the regression approach.

Comparisons among adjusted treatment means. Since $\bar{Y}_{i.}(\text{adj.})$ is an estimator of $\mu_{.} + \tau_i$, a pairwise comparison $\bar{Y}_{i.}(\text{adj.}) - \bar{Y}_{i'.}(\text{adj.})$ is an estimator of $\tau_i - \tau_{i'}$. For our example, we would thus estimate $\tau_1 - \tau_3$ from:

$$\bar{Y}_{1.}(\text{adj.}) - \bar{Y}_{3.}(\text{adj.}) = 39.82 - 26.84 = 12.98$$

which is the same result we obtained on page 859. The variance of the difference of two adjusted treatment means is:

$$\sigma^2[\bar{Y}_{i.}(\text{adj.}) - \bar{Y}_{i'.}(\text{adj.})] = \sigma^2\left[\frac{1}{n_i} + \frac{1}{n_{i'}} + \frac{(\bar{X}_{i.} - \bar{X}_{i'.})^2}{SSE_X}\right] \tag{25.43}$$

and this is estimated by:

$$s^2[\bar{Y}_{i.}(\text{adj.}) - \bar{Y}_{i'.}(\text{adj.})] = MSE(\text{adj.})\left[\frac{1}{n_i} + \frac{1}{n_{i'}} + \frac{(\bar{X}_{i.} - \bar{X}_{i'.})^2}{SSE_X}\right] \tag{25.44}$$

For our example, we obtain:

$$s^2[\bar{Y}_{1.}(\text{adj.}) - \bar{Y}_{3.}(\text{adj.})] = 3.506\left[\frac{1}{5} + \frac{1}{5} + \frac{(23.2 - 25.4)^2}{333.2}\right] = 1.453$$

which, of course, is the same estimated variance as was obtained via the regression approach on page 859.

The use of the t distribution for a single interval estimate of a contrast among the adjusted treatment means and of the Scheffé and Bonferroni methods for multiple comparisons is identical to their use with the regression approach. An estimator of a contrast among the adjusted treatment means is defined as follows:

$$\hat{L} = \Sigma c_i[\bar{Y}_{i.}(\text{adj.})] = \Sigma c_i\bar{Y}_{i.} - g_w\Sigma c_i\bar{X}_{i.} \tag{25.45}$$

where:

$$\Sigma c_i = 0$$

Its estimated variance is:

(25.46)
$$s^2(\hat{L}) = MSE(\text{adj.})\left[\Sigma\frac{c_i^2}{n_i} + \frac{(\Sigma c_i\bar{X}_{i.})^2}{SSE_X}\right]$$

Confidence limits for a single contrast L are:

(25.47)
$$\hat{L} \pm t(1 - \alpha/2; n_T - r - 1)s(\hat{L})$$

For multiple comparisons, the t multiple is replaced by either the S or B multiples defined in (25.17) and (25.18), respectively.

25.5 MULTIFACTOR STUDIES

We have until now considered the case of covariance analysis for single-factor studies with r treatments. Covariance analysis can also be employed with two-factor and multifactor studies. We illustrate now the use of covariance analysis for a two-factor study with one concomitant variable.

Covariance model for two-factor study

The fixed effects analysis of variance model for a two-factor study in terms of factor effects was given in (20.27):

(25.48)
$$Y_{ijk} = \mu_{..} + \alpha_i + \beta_j + (\alpha\beta)_{ij} + \varepsilon_{ijk}$$
$$i = 1, \ldots, a; j = 1, \ldots, b; k = 1, \ldots, n$$

where α_i is the main effect of factor A at the ith level, β_j is the main effect of factor B at the jth level, and $(\alpha\beta)_{ij}$ is the interaction effect when factor A is at the ith level and factor B is at the jth level. The covariance model for a two-factor study with a single concomitant variable, assuming the relation between Y and the concomitant variable X is linear, is:

(25.49)
$$Y_{ijk} = \mu_{..} + \alpha_i + \beta_j + (\alpha\beta)_{ij} + \gamma(X_{ijk} - \bar{X}_{...}) + \varepsilon_{ijk}$$
$$i = 1, \ldots, a; j = 1, \ldots, b; k = 1, \ldots, n$$

Regression approach

To illustrate the regression approach to covariance analysis for a two-factor study with one concomitant variable, suppose that both factors A and B are at two levels and have fixed effects. The regression model counterpart to covariance model (25.49) then is:

(25.50)
$$Y_{ijk} = \mu_{..} + \alpha_1 I_{ijk1} + \beta_1 I_{ijk2} + (\alpha\beta)_{11} I_{ijk1} I_{ijk2} + \gamma x_{ijk} + \varepsilon_{ijk}$$

where:

$$I_{ijk1} = \begin{array}{l} 1 \text{ if observation from level 1 for factor } A \\ -1 \text{ if observation from level 2 for factor } A \end{array}$$

$$I_{ijk2} = \begin{array}{l} 1 \text{ if observation from level 1 for factor } B \\ -1 \text{ if observation from level 2 for factor } B \end{array}$$

$$x_{ijk} = X_{ijk} - \bar{X}_{...}$$

Note that the regression coefficients in (25.50) are the analysis of variance effects α_1, β_1, and $(\alpha\beta)_{11}$ and the concomitant variable coefficient γ.

Testing for factor A main effects requires that $\alpha_1 = 0$ in the reduced model. Correspondingly, we have $\beta_1 = 0$ in the reduced model when testing for factor B main effects, and $(\alpha\beta)_{11} = 0$ in the reduced model when testing for AB interactions.

Estimation of factor A and B main effects can easily be done in terms of comparisons among the regression coefficients. The use of the Scheffé and Bonferroni multiple comparison procedures presents no new problems. For instance, the S multiple for multiple comparisons among the factor A means is defined as follows:

$$S^2 = (a-1)F(1-\alpha; a-1, n_T - ab - 1) \tag{25.51}$$

and the B multiple would be given by (25.18) with $r = ab$.

Adjustment approach

In applying the adjustment approach to covariance analysis for a two-factor study with one concomitant variable, we again require analyses of variance for Y, X, and XY. The analysis of variance for Y is given in (20.42) and (20.43). The analysis of variance for X is identical, with X_{ij} replacing Y_{ij}. Table 25.9 contains these analyses of variance for Y and X. Note that subscripts for Y and X are used for identification.

TABLE 25.9 Analyses of variance for Y, X, and XY—two-factor study with one concomitant variable

Source of Variation	*Sums of Squares or Products*			*df*
	Y	X	XY	
Factor A	SSA_Y	SSA_X	SPA	$a-1$
Factor B	SSB_Y	SSB_X	SPB	$b-1$
AB interactions	$SSAB_Y$	$SSAB_X$	$SPAB$	$(a-1)(b-1)$
Error	SSE_Y	SSE_X	SPE	$ab(n-1)$
Total	$SSTO_Y$	$SSTO_X$	$SPTO$	$abn-1$

The analysis of variance for products XY is as follows:

(25.52a)
$$SPTO = \sum_i \sum_j \sum_k (X_{ijk} - \bar{X}_{...})(Y_{ijk} - \bar{Y}_{...}) = \sum_i \sum_j \sum_k X_{ijk}Y_{ijk} - \frac{X_{...}Y_{...}}{abn}$$

(25.52b)
$$SPA = bn\sum_i (\bar{X}_{i..} - \bar{X}_{...})(\bar{Y}_{i..} - \bar{Y}_{...}) = \frac{\sum_i X_{i..}Y_{i..}}{bn} - \frac{X_{...}Y_{...}}{abn}$$

(25.52c)
$$SPB = an\sum_j (\bar{X}_{.j.} - \bar{X}_{...})(\bar{Y}_{.j.} - \bar{Y}_{...}) = \frac{\sum_j X_{.j.}Y_{.j.}}{an} - \frac{X_{...}Y_{...}}{abn}$$

(25.52d)
$$SPE = \sum_i \sum_j \sum_k (X_{ijk} - \bar{X}_{ij.})(Y_{ijk} - \bar{Y}_{ij.}) = SPTO - SPTR$$

(25.52e)
$$SPAB = n\sum_i \sum_j (\bar{X}_{ij.} - \bar{X}_{i..} - \bar{X}_{.j.} + \bar{X}_{...})(\bar{Y}_{ij.} - \bar{Y}_{i..} - \bar{Y}_{.j.} + \bar{Y}_{...}) = SPTR - SPA - SPB$$

where:

(25.52f)
$$SPTR = n\sum_i \sum_j (\bar{X}_{ij.} - \bar{X}_{...})(\bar{Y}_{ij.} - \bar{Y}_{...}) = \frac{\sum_i \sum_j X_{ij.}Y_{ij.}}{n} - \frac{X_{...}Y_{...}}{abn}$$

Table 25.9 also contains this analysis of variance for products XY.

The adjustment process ignores all components other than the error component and the component for which the test is to be made, and then proceeds to make adjustments in a fashion analogous to a single-factor study. Thus, to test for factor A main effects:

(25.53)
$$H_0: \alpha_1 = \alpha_2 = \cdots = \alpha_a = 0$$
$$H_a: \text{not all } \alpha_i \text{ equal zero}$$

we extract the lines for factor A and error in Table 25.9. This is done in Table 25.10a. We then derive the adjusted sums of squares in the usual fashion:

(25.53a)
$$SS(A + E; \text{adj.}) = (SSA_Y + SSE_Y) - \frac{(SPA + SPE)^2}{SSA_X + SSE_X}$$

(25.53b)
$$SSE(\text{adj.}) = SSE_Y - \frac{(SPE)^2}{SSE_X}$$

(25.53c)
$$SSA(\text{adj.}) = SS(A + E; \text{adj.}) - SSE(\text{adj.})$$

TABLE 25.10 Covariance analysis for testing factor *A* main effects in a two-factor study with one concomitant variable

(a) Analyses of Variance

Source of Variation	*Sums of Squares or Products* Y	X	XY	*df*
Factor *A*	SSA_Y	SSA_X	SPA	$a-1$
Error	SSE_Y	SSE_X	SPE	$ab(n-1)$
Sum	$SSA_Y + SSE_Y$	$SSA_X + SSE_X$	$SPA + SPE$	$a + ab(n-1) - 1$

(b) Analysis of Covariance

Source of Variation	*Adjusted SS*	*Adjusted df*	*Adjusted MS*
Factor *A*	SSA(adj.)	$a-1$	MSA(adj.)
Error	SSE(adj.)	$ab(n-1)-1$	MSE(adj.)
Sum	$SS(A+E;\text{ adj.})$	$a + ab(n-1) - 2$	

Table 25.10b contains the covariance analysis for testing factor A main effects. The degrees of freedom for error and the sum are reduced by one to take account of the concomitant variable X, and the adjusted degrees of freedom for factor A effects are obtained as a remainder.

As usual, the statistic for testing the alternatives in (25.53) is:

$$F^* = \frac{MSA(\text{adj.})}{MSE(\text{adj.})} \tag{25.54}$$

If H_0 holds, F^* follows the $F[a-1, ab(n-1)-1]$ distribution.

Tests for other factor effects are developed in similar fashion.

Example

A horticulturist conducted an experiment to study the effects of flower variety (factor A: varieties LP, WB) and moisture level (factor B: low, high) on yield of salable flowers (Y). Because the plots were not of the same size, the horticulturist wished to use plot size (X) as the concomitant variable. Six replications were made for each treatment. The data are presented in Table 25.11.

Regression model (25.50) was fitted to the data by a computer regression package. The fitted regression function is shown in Table 25.12a. The analyst plotted the data and the fitted regression lines (not shown) and was satisfied that parallel linear regression functions and constant error term variances are suitable here.

The fitted regression lines for the four treatments based on the full model (25.50) are presented in Figure 25.9a. To study the nature of the factor effects,

TABLE 25.11 Data for salable flowers example

Factor A (flower variety) i	Factor B (moisture level) j: B_1 (low) Y	B_1 (low) X	B_2 (high) Y	B_2 (high) X
A_1 (variety LP)	98	15	71	10
	60	4	80	12
	77	7	86	14
	80	9	82	13
	95	14	46	2
	64	5	55	3
A_2 (variety WB)	55	4	76	11
	60	5	68	10
	75	8	43	2
	65	7	47	3
	87	13	62	7
	78	11	70	9

TABLE 25.12 Computer output of regression runs for salable flowers example—regression model (25.50)

(a) Fitted Regression Function for Model (25.50)

$$\hat{Y} = 70.0 + 2.04234I_1 + 3.68078I_2 + .81922I_1I_2 + 3.27688x$$

Regression Coefficient	*Estimated Regression Coefficient*	*Estimated Standard Deviation*
α_1	2.04234	.52108
β_1	3.68078	.51291
$(\alpha\beta)_{11}$	.81922	.51291
γ	3.27688	.13002

(b) Extra Sums of Squares

Effect	*Source of Variation*	*SS*	*df*	*MS*
Concomitant variable	$x\|I_1, I_2, I_1I_2$	3,994.52	1	3,994.52
A	$I_1\|x, I_2, I_1I_2$	96.60	1	96.60
B	$I_2\|x, I_1, I_1I_2$	323.85	1	323.85
AB	$I_1I_2\|x, I_1, I_2$	16.04	1	16.04
	Error	119.48	19	6.2884

we show in Figure 25.9b the usual plots of treatment means. These estimated treatment means all correspond to plot size $X = \bar{X}_{...} = 8.25$ or $x = 0$. Any other plot size would yield exactly the same relationships as those in Figure 25.9b. It appears from Figure 25.9b that there are no important interactions between flower variety and moisture level, and that there may be main effects for both factors, particularly for moisture level.

FIGURE 25.9 Fitted regression lines and treatment mean plots—salable flowers example

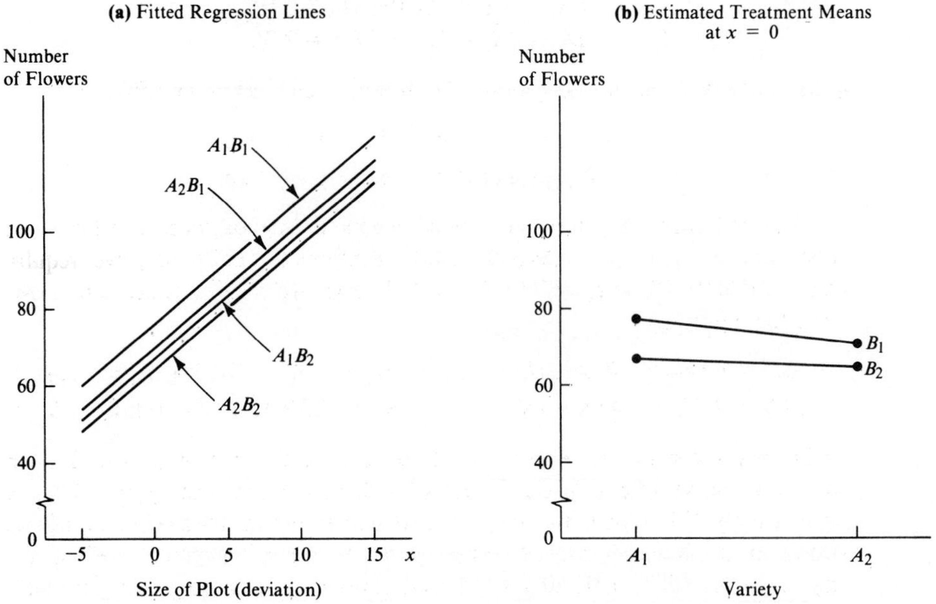

To study formally the factor effects, reduced models were formed by deleting from regression model (25.50) one independent variable at a time (since both factors have two levels), and the reduced models were then fitted. The extra sums of squares so obtained, as well as the error sum of squares for the full model, are presented in Table 25.12b, together with the degrees of freedom and mean squares. No total sum of squares is shown because the factor effects components are not orthogonal.

We test first for the presence of interactions by means of the usual test statistic F^*, using the data in Table 25.12b:

$$F^* = \frac{SSR(I_1I_2|x, I_1, I_2)}{1} \div MSE = \frac{16.04}{6.2884} = 2.55$$

For $\alpha = .01$, we require $F(.99; 1, 19) = 8.18$. Since $F^* = 2.55 \leq 8.18$, we conclude that no interactions are present.

We now wish to estimate the factor A and B main effects by means of confidence intervals with a 95 percent family confidence coefficient. Since $\alpha_i = \mu_{i.} - \mu_{..}$, we have for our example:

$$D_1 = \mu_{1.} - \mu_{2.} = \alpha_1 - \alpha_2 = \alpha_1 - (-\alpha_1) = 2\alpha_1$$

Similarly, we obtain for the comparison of factor B main effects:

$$D_2 = \mu_{.1} - \mu_{.2} = 2\beta_1$$

Point estimates are readily obtained from the results in Table 25.12a:

$$\hat{D}_1 = 2\hat{\alpha}_1 = 2(2.04234) = 4.08$$
$$\hat{D}_2 = 2\hat{\beta}_1 = 2(3.68078) = 7.36$$

The estimated standard deviations also follow easily, using (1.15b):

$$s(\hat{D}_1) = 2s(\hat{\alpha}_1) = 2(.52108) = 1.042$$
$$s(\hat{D}_2) = 2s(\hat{\beta}_1) = 2(.51291) = 1.026$$

We shall utilize the Bonferroni simultaneous estimation procedure for $g = 2$ comparisons. For a 95 percent family confidence coefficient, we require $t[1 - .05/2(2); 19] = t(.9875; 19) = 2.433$. The two desired confidence intervals therefore are:

$$1.5 = 4.08 - 2.433(1.042) \leq \mu_{1.} - \mu_{2.} \leq 4.08 + 2.433(1.042) = 6.6$$
$$4.9 = 7.36 - 2.433(1.026) \leq \mu_{.1} - \mu_{.2} \leq 7.36 + 2.433(1.026) = 9.9$$

With family confidence coefficient .95, we conclude that variety LP yields, on the average, between 1.5 and 6.6 more salable flowers for any given plot size than variety WB. Also, for any given plot size, the mean number of salable flowers is between 4.9 and 9.9 flowers greater for the low moisture level than for the high one, thus indicating a substantial effect of moisture level on yield.

If interactions had been present, we could have studied the nature of the interaction effects by, for instance, comparing the effect of the moisture level for each of the two flower varieties:

$$L = (\mu_{12} - \mu_{11}) - (\mu_{22} - \mu_{21})$$

It follows from (20.14b) and from the constraints in (20.27) that:

$$L = (\alpha\beta)_{12} = -(\alpha\beta)_{11}$$

Hence, we could estimate the desired interaction effect by using the estimated regression coefficient $\widehat{(\alpha\beta)}_{11}$ and its estimated standard deviation in Table 25.12a.

25.6 ADDITIONAL CONSIDERATIONS FOR THE USE OF COVARIANCE ANALYSIS

Use of differences

In a variety of studies, a pre-study observation X and a post-study observation Y on the same variable are available for each unit. For instance, X may be the score for a subject's attitude toward a company prior to reading its annual report, and Y may be the score after reading the report. In this situation, an obvious alternative to covariance analysis is to do an analysis of variance on the difference $Y - X$. Sometimes, $Y - X$ is called an *index of response* because it makes one observation out of two separate ones.

If the slope of the treatment regression lines is $\gamma = 1$, analysis of covariance and analysis of variance on $Y - X$ are essentially equivalent. When $\gamma = 1$, covariance model (25.2) becomes:

$$Y_{ij} = \mu_{.} + \tau_i + X_{ij} + \varepsilon_{ij} \tag{25.55}$$

which can be written as a regular analysis of variance model:

$$Y_{ij} - X_{ij} = \mu_{.} + \tau_i + \varepsilon_{ij} \tag{25.55a}$$

Thus, if a unit change in X leads to about the same change in Y, it makes sense to perform an analysis of variance on $Y - X$, rather than to use covariance analysis, because analysis of variance is much easier. If the regression slope is not near 1, however, covariance analysis may be substantially more effective than use of the difference $Y - X$.

In our earlier cracker promotion example, use of $Y - X$ would have been effective. It would have involved an error mean square (see Table 25.6):

$$\frac{SSE_Y + SSE_X - 2SPE}{12} = \frac{307.6 + 333.2 - 2(299.4)}{12} = 3.50$$

which is practically the same as the error mean square for covariance analysis, $MSE(\text{adj.}) = 3.51$. Recall that the regression slope in our example was close to 1 ($g_w = .8986$); hence, the approximate equivalence of the two procedures.

Correction for bias

The suggestion is sometimes made that analysis of covariance can be helpful in correcting for bias when observational data are at hand. With observational data, the groups under study may differ substantially with respect to a concomitant variable, and this may bias the comparisons of the groups. Consider, for instance, a study in which attitudes toward no-fault automobile insurance were compared for persons who are risk averse and persons who are risk seeking. It was found that many persons in the risk averse group tended to be older (50 to 70 years old), while many persons in the risk seeking group tended to be younger (20 to 40 years old). In this type of situation, some would advise that covariance analysis, with age as the concomitant variable, be employed to help remove any bias that may be in the observational data because the two age groups differ so much.

Even though there is great appeal in the idea of removing bias in observational data, covariance analysis should be used with caution for this purpose. In the first place, the adjusted means may require substantial extrapolation of the regression lines to a region where there are no or only few data points (in our example, to near 45 years). It may well be that the regression relationship used in the covariance analysis is not appropriate for substantial extrapolation. In the second place, the treatment variable may be dependent on the concomitant variable (or vice versa) which could affect the proper conclusions to be drawn.

Interest in nature of treatment effects

Covariance analysis is sometimes employed for the principal purpose of shedding more light on the nature of the treatment effects, rather than merely for increasing the precision of the analysis. For instance, a market researcher in a study of the effects of three different advertisements on the maximum price consumers are willing to pay for a new type of home siding may use covariance analysis, with value of the consumer's home as the concomitant variable. The reason is because she is truly interested in the relation for each advertisement between home value and maximum price. Reduction of error variance in this instance may be a secondary consideration for the analyst in using covariance analysis.

As in all regression analyses, care must be used in drawing inferences about the causal nature of the relation between the concomitant variable and the dependent variable. In our advertising example, it might well be that value of a consumer's home is largely influenced by income. If this were so, the relation between value of the consumer's home and maximum price the consumer is willing to pay may actually be largely a reflection of a more underlying relation between income and maximum price.

PROBLEMS

25.1. A student's reaction to the instructor's statement that covariance analysis is inappropriate when the treatment regression lines do not have the same slope was as follows: "It seems to me that this is ducking a real-world problem. If the treatment slopes are different, just use a covariance model that allows for different treatment slopes." Evaluate this reaction.

25.2. A survey analyst remarked: "When covariance analysis is used with survey data, there is the danger that the treatments may be related to the concomitant variable." What is the nature of the problem? Does this same problem exist when the treatments are randomly assigned to the experimental units?

25.3. Portray, analogous to the format of Figure 2.7 on page 33 for a regression model, the nature of covariance model (25.3) when there are three treatments and the parameter values are: $\mu_{.} = 150$, $\tau_1 = 15$, $\tau_2 = -5$, $\tau_3 = -10$, $\gamma = 6$, $\bar{X}_{..} = 70$, $\sigma = 5$. Show several distributions of Y for each treatment.

25.4. Refer to the cracker promotion example on page 853. A student stated, in discussing this case: "Strictly speaking, you cannot conclude anything about whether the three promotions differ in effectiveness because there was no control. The preceding period does not qualify as a control because it might have differed from the promotion period due to seasonal factors or other unique circumstances." Comment.

25.5. Refer to the cracker promotion example on page 859, where three pairwise comparisons of treatment effects were made by the Scheffé procedure.

a. What would be the value of the Bonferroni multiple here for estimating the three comparisons?
b. Did the analyst obtain substantially less precise interval estimates using the Scheffé procedure, which permits him to make additional estimates without modifying the present ones?

25.6. State the analysis of covariance model for a single-factor study with four treatments when there are two concomitant variables, each with linear and quadratic terms in the model.

25.7. Refer to **Productivity improvement** Problem 16.9. The economist also has information on annual productivity improvement in the prior year and wishes to use this information as a concomitant variable. The data on the prior year's productivity improvement (X_{ij}) follow.

	j											
i	1	2	3	4	5	6	7	8	9	10	11	12
1	8.2	7.9	7.0	5.7	7.2	7.0	6.5	7.9	6.3			
2	8.8	10.0	10.7	10.0	9.7	9.4	10.6	9.8	10.0	10.3	8.9	10.0
3	11.5	12.2	12.8	11.0	12.3	12.1						

Assume that covariance model (25.3) is applicable.

a. Plot the data in the format of Figure 25.5. Does it appear that parallel linear response functions are appropriate here? Does it appear that there are research and development expenditures level effects on mean productivity improvement? Discuss.
b. State the regression model equivalent to covariance model (25.3) for this case; use $1, -1, 0$ indicator variables. Also state the reduced model for testing for treatment effects.
c. Fit the full and reduced models and test for treatment effects; use $\alpha = .05$. State the alternatives, decision rule, and conclusion. What is the P-value of the test?
d. Is $MSE(F)$ for the covariance model substantially smaller than MSE for the analysis of variance model in Problem 16.9c? Does this affect the conclusion reached about treatment effects? Does it affect the P-value?
e. Estimate the mean productivity improvement for firms with moderate research and development expenditures who had a prior productivity improvement of $X = 9.0$; use a 95 percent confidence interval.
f. Make all pairwise comparisons between the treatment effects; use either the Bonferroni or Scheffé procedure with a 90 percent family confidence coefficient, whichever is more efficient. State your findings.

25.8. Refer to **Productivity improvement** Problem 25.7.

a. Obtain the residuals.
b. For each treatment, plot the residuals against the fitted values. Also prepare a normal probability plot of the residuals and calculate the coefficient of correlation between the ordered residuals and their expected values under normality. What do you conclude from your analysis?
c. State the generalized regression model to be employed for testing whether or

not the treatment regression lines have the same slope. Conduct this test using $\alpha = .01$. State the alternatives, decision rule, and conclusion.

d. Could you conduct a formal test here as to whether the regression functions are linear? If so, how many degrees of freedom are there for the denominator mean square in the test statistic?

25.9. Refer to **Questionnaire color** Problem 16.10. It has been suggested to the investigator that size of parking lot might be a useful concomitant variable. The number of spaces (X_{ij}) in each parking lot utilized in the study follow.

	j				
i	1	2	3	4	5
1	300	381	226	350	100
2	153	334	473	264	325
3	144	359	296	243	252

Assume that covariance model (25.3) is applicable.

a. Plot the data in the format of Figure 25.5. Does it appear that parallel linear response functions are appropriate here? Does it appear that there are color effects on the mean response rate? Discuss.

b. State the regression model equivalent to covariance model (25.3) for this case; use $1, -1, 0$ indicator variables. Also state the reduced model for testing for treatment effects.

c. Fit the full and reduced models and test for treatment effects; use $\alpha = .10$. State the alternatives, decision rule, and conclusion. What is the P-value of the test?

d. Is $MSE(F)$ for the covariance model substantially smaller than MSE for the analysis of variance model in Problem 16.10c? How does this affect the conclusion reached about treatment effects?

e. Estimate the mean response rate for blue questionnaires in parking lots of size $X = 280$; use a 90 percent confidence interval.

f. Make all pairwise comparisons between the treatment effects; use either the Bonferroni or Scheffé procedure with a 90 percent family confidence coefficient, whichever is more efficient. State your findings.

25.10. Refer to **Questionnaire color** Problem 25.9.

a. Obtain the residuals.

b. For each treatment, plot the residuals against the fitted values. Also prepare a normal probability plot of the residuals and calculate the coefficient of correlation between the ordered residuals and their expected values under normality. What do you conclude from your analysis?

c. State the generalized regression model to be employed for testing whether or not the treatment regression lines have the same slope. Conduct this test using $\alpha = .005$. State the alternatives, decision rule, and conclusion.

d. Could you conduct a formal test here as to whether the regression functions are linear? Explain.

25.11. Refer to **Rehabilitation therapy** Problem 16.11. The rehabilitation researcher wishes to use age of patient as a concomitant variable. The ages (X_{ij}) of patients in the study follow.

					j					
i	1	2	3	4	5	6	7	8	9	10
1	18.3	30.0	26.5	28.1	29.7	27.8	19.8	29.3		
2	20.8	25.2	29.2	20.0	21.5	22.1	19.7	24.7	20.2	22.9
3	22.7	28.7	18.9	18.0	24.7	20.0				

Assume that covariance model (25.3) is applicable.

a. Plot the data in the format of Figure 25.5. Does it appear that parallel linear response functions are appropriate here? Does it appear that there are effects of physical fitness status on the mean number of days required for therapy? Discuss.

b. State the regression model equivalent to covariance model (25.3) for this case; use $1, -1, 0$ indicator variables. Also state the reduced model for testing for treatment effects.

c. Fit the full and reduced models and test for treatment effects; use $\alpha = .01$. State the alternatives, decision rule, and conclusion. What is the P-value of the test?

d. Is $MSE(F)$ for the covariance model substantially smaller than MSE for the analysis of variance model in Problem 16.11c? Does this affect the conclusion reached about treatment effects? Does it affect the P-value?

e. Estimate the mean number of days required for therapy for patients of average physical fitness and age 24 years; use a 99 percent confidence interval.

f. Make all pairwise comparisons between the treatment effects; use either the Bonferroni or Scheffé procedure with a 95 percent family confidence coefficient, whichever is more efficient. State your findings.

25.12. Refer to **Rehabilitation therapy** Problem 25.11.

a. Obtain the residuals.

b. For each treatment, plot the residuals against the fitted values. Also prepare a normal probability plot of the residuals and calculate the coefficient of correlation between the ordered residuals and their expected values under normality. What do you conclude from your analysis?

c. State the generalized regression model to be employed for testing whether or not the treatment regression lines have the same slope. Conduct this test using $\alpha = .05$. State the alternatives, decision rule, and conclusion.

d. Could you conduct a formal test here as to whether the regression functions are linear? Explain.

25.13. **Product display.** A manufacturer of felt-tip markers investigated by an experiment whether a proposed new display, featuring a picture of a physician, is more effective in drugstores than the present counter display, featuring a picture of an athlete and designed to be located in the stationery area. Fifteen drugstores of similar characteristics were chosen for the study. They were assigned at random in equal numbers to one of the following three treatments: (1) present counter display in stationery area, (2) new display in stationery area, (3) new display in checkout area. Sales with the present display (X_{ij}) were recorded in all 15 stores for a three-week period. Then the new display was set up in the 10 stores receiving it, and sales for the next three-week period (Y_{ij}) were recorded in all 15 stores. The data on sales (in dollars) follow.

i	j = 1	2	3	4	5
Treatment 1					
1st 3 weeks	92	68	74	52	65
2d 3 weeks	69	44	58	38	54
Treatment 2					
1st 3 weeks	77	80	70	73	79
2d 3 weeks	74	75	73	78	82
Treatment 3					
1st 3 weeks	64	43	81	68	71
2d 3 weeks	66	49	84	75	77

The analyst wishes to analyze the effects of the three different display treatments by means of covariance analysis. Assume that covariance model (25.3) is applicable.

a. Plot the data in the format of Figure 25.5. Does it appear that parallel linear response functions are appropriate here? Does it appear that there are display effects on mean sales? Discuss.

b. State the regression model equivalent to covariance model (25.3) for this case; use $1, -1, 0$ indicator variables. Also state the reduced model for testing for treatment effects.

c. Fit the full and reduced models and test for treatment effects; use $\alpha = .05$. State the alternatives, decision rule, and conclusion. What is the P-value of the test?

d. Is $MSE(F)$ for the covariance model substantially smaller than the mean square error if analysis of variance model (16.2) had been employed?

e. Estimate the mean sales with display treatment 2 for stores whose sales in the preceding three-week period were \$75; use a 95 percent confidence interval.

f. Make all pairwise comparisons between the treatment effects; use either the Bonferroni or Scheffé procedure with a 90 percent family confidence coefficient, whichever is more efficient. State your findings.

25.14. Refer to **Product display** Problem 25.13.

a. Obtain the residuals.

b. For each treatment, plot the residuals against the fitted values. Also prepare a normal probability plot of the residuals and calculate the coefficient of correlation between the ordered residuals and their expected values under normality. What do you conclude from your analysis?

c. State the generalized regression model to be employed for testing whether or not the treatment regression lines have the same slope. Conduct this test using $\alpha = .05$. State the alternatives, decision rule, and conclusion.

d. Could you conduct a formal test here as to whether the regression functions are linear? Explain.

25.15. Refer to **Cash offers** Problem 20.11. An analyst wishes to use each dealer's sales volume as a concomitant variable. The sales data (X_{ij}, in hundred thousand dollars) follow.

	$j = 1$	$j = 2$
$i = 1$	3.0	3.5
	5.1	4.2
	1.0	2.2
	4.4	3.1
	2.7	1.3
	4.9	6.6
$i = 2$	6.5	2.2
	4.1	5.4
	2.2	3.1
	3.7	4.5
	3.4	3.6
	3.0	5.0
$i = 3$	5.0	4.0
	3.1	.8
	3.2	1.9
	3.2	2.8
	3.0	2.2
	2.9	1.9

Assume that covariance model (25.49) is applicable.

a. State the regression model equivalent to covariance model (25.49) for this case; use 1, −1, 0 indicator variables. Fit this full model.
b. State the reduced models for testing for interaction and factor A and factor B main effects, respectively. Fit these reduced models.
c. Test for interaction effects; use $\alpha = .05$. State the alternatives, decision rule, and conclusion. What is the P-value of the test?
d. Test for factor A main effects; use $\alpha = .05$. State the alternatives, decision rule, and conclusion. What is the P-value of the test?
e. Test for factor B main effects; use $\alpha = .05$. State the alternatives, decision rule, and conclusion. What is the P-value of the test?
f. For each factor, make all pairwise comparisons between the factor level main effects. Use the Bonferroni procedure with a 90 percent family confidence coefficient. State your findings.

25.16. Refer to **Cash offers** Problem 25.15.

a. Obtain the residuals.
b. For each treatment, plot the residuals against the fitted values. Also prepare a normal probability plot of the residuals and calculate the coefficient of correlation between the ordered residuals and their expected values under normality. What do you conclude from your analysis?
c. State the generalized regression model to be employed for testing whether or not the treatment regression lines have the same slope. Conduct this test using $\alpha = .01$. State the alternatives, decision rule, and conclusion.

25.17. Refer to **Eye contact effect** Problem 20.12. Age of personnel officer is to be used as a concomitant variable. The ages (X_{ij}) of the personnel officers follow.

	$j = 1$	$j = 2$
$i = 1$	42	51
	30	35
	47	48
	31	38
	35	49

	$j = 1$	$j = 2$
$i = 2$	43	42
	53	47
	40	46
	50	59
	49	56

Assume that covariance model (25.49) is applicable.

a. State the regression model equivalent to covariance model (25.49) for this case; use $1, -1, 0$ indicator variables. Fit this full model.

b. State the reduced models for testing for interaction and factor A and factor B main effects, respectively. Fit these reduced models.

c. Test for interaction effects; use $\alpha = .01$. State the alternatives, decision rule, and conclusion. What is the P-value of the test?

d. Test for factor A main effects; use $\alpha = .01$. State the alternatives, decision rule, and conclusion. What is the P-value of the test?

e. Test for factor B main effects; use $\alpha = .01$. State the alternatives, decision rule, and conclusion. What is the P-value of the test?

f. Compare the sex main effects by means of a 99 percent confidence interval. Interpret your interval estimate.

g. Estimate the mean success rating for female personnel officers aged 40 when eye contact is present; use a 99 percent confidence interval.

25.18. Refer to **Eye contact effect** Problem 25.17.

a. Obtain the residuals.

b. For each treatment, plot the residuals against the fitted values. Also prepare a normal probability plot of the residuals and calculate the coefficient of correlation between the ordered residuals and their expected values under normality. What do you conclude from your analysis?

c. State the generalized regression model to be employed for testing whether or not the treatment regression lines have the same slope. Conduct this test using $\alpha = .005$. State the alternatives, decision rule, and conclusion.

25.19. Refer to **Productivity improvement** Problem 25.7. The analyst is considering the use of the difference between the productivity improvements in the two years $(Y_{ij} - X_{ij})$ as the dependent variable with the regular analysis of variance model (25.55a).

a. Obtain the analysis of variance table.

b. How effective would be the use of differences here with the regular ANOVA model compared to the use of covariance model (25.3)? Discuss.

25.20. Refer to **Product display** Problem 25.13. The analyst is considering the use of the difference in sales between the two periods $(Y_{ij} - X_{ij})$ as the dependent variable with the regular analysis of variance model (25.55a).

a. Obtain the analysis of variance table.

b. How effective would be the use of differences here with the regular ANOVA model compared to the use of covariance model (25.3)? Discuss.

EXERCISES

25.21. (Calculus needed.) Denote $\mu_{.} + \tau_i$ in covariance model (25.3) by Δ_i. Derive the least squares estimators for Δ_i and γ in covariance model (25.3).

25.22. Show that $\sigma^2[\overline{Y}_{i.}(\text{adj.})]$ is given by (25.41).

25.23. Show that the variance of a contrast among the adjusted treatment sample means, $\hat{L} = \Sigma c_i[\overline{Y}_{i.}(\text{adj.})]$, is given by:

$$\sigma^2(\hat{L}) = \sigma^2\left[\Sigma\frac{c_i^2}{n_i} + \frac{(\Sigma c_i\overline{X}_{i.})^2}{SSE_X}\right]$$

PROJECTS

25.24. Refer to the **SENIC** data set. The following hospitals are to be considered in a study of the effects of region (variable 9) on the mean length of hospital stay of patients (variable 2), with available facilities and services (variable 12) as a concomitant variable:

1–52 54 55 57 58 63 76 83 84 94 101 103 111

Assume that covariance model (25.3) is applicable.

a. Plot the data in the format of Figure 25.5. Does it appear that parallel linear response functions are appropriate here? Does it appear that there are region effects on the mean length of hospital stay? Discuss.

b. State the regression model equivalent to covariance model (25.3) for this case; use $1,-1,0$ indicator variables. Also state the reduced model for testing for treatment effects.

c. Fit the full and reduced models and test for treatment effects; use $\alpha = .05$. State the alternatives, decision rule, and conclusion. What is the P-value of the test?

d. Make all pairwise comparisons between the region effects; use either the Bonferroni or Scheffé procedure with a 90 percent family confidence coefficient, whichever is more efficient. State your findings.

e. Obtain the residuals.

f. For each region, plot the residuals against the fitted values. Also prepare a normal probability plot of the residuals and calculate the coefficient of correlation between the ordered residuals and their expected values under normality. What do you conclude from your analysis?

g. State the generalized regression model to be employed for testing whether or not the treatment regression lines have the same slope. Conduct this test using $\alpha = .005$. State the alternatives, decision rule, and conclusion.

25.25. Refer to the **SMSA** data set. The following metropolitan areas are to be considered in a study of the effects of region (factor A: variable 12) and percent of population in central cities (factor B: variable 4) on crime rate (variable 11 ÷ variable 3), with percent of population 65 or older (variable 5) as a concomitant variable:

1–45 49 51–54 58 60–62 64 66 71
73 80 92 101 123 130 131

For purposes of this analysis of covariance study, percent of population in central cities is to be classified into two categories: 37.0 percent or less, 37.1 percent or more. Assume that covariance model (25.49) is applicable.

a. Assemble the required data.
b. State the regression model equivalent to covariance model (25.49) for this case; use 1, −1, 0 indicator variables. Fit this full model.
c. State the reduced models for testing for interaction and factor A and factor B main effects, respectively. Fit these reduced models.
d. Test for interaction effects; use $\alpha = .01$. State the alternatives, decision rule, and conclusion. What is the P-value of the test?
e. Test for factor A main effects; use $\alpha = .01$. State the alternatives, decision rule, and conclusion. What is the P-value of the test?
f. Test for factor B main effects; use $\alpha = .01$. State the alternatives, decision rule, and conclusion. What is the P-value of the test?
g. Make all pairwise comparisons between the region main effects; use either the Bonferroni or Scheffé procedure with a 95 percent family confidence coefficient, whichever is more efficient. State your findings.
h. Obtain the residuals.
i. For each treatment, plot the residuals against the fitted values. Also prepare a normal probability plot of the residuals and calculate the coefficient of correlation between the ordered residuals and their expected values under normality. What do you conclude from your analysis?
j. State the generalized regression model to be employed for testing whether or not the treatment regression lines have the same slope. Conduct this test using $\alpha = .001$. State the alternatives, decision rule, and conclusion.

PART V

Experimental designs

26

Completely randomized designs

Formal experimentation is being widely employed in the biological and social sciences. It has come of age somewhat more recently in business and economics, but a wide variety of uses now are found in these fields also. One example is an experiment to investigate the best level of aggregation of company data furnished by a management information system to middle management. Another is an experiment on the effect of a guaranteed annual income on the consumption behavior of families. The latter experiment is designed by dividing a group of low-income families at random into two halves, one of which receives income supplements up to a guaranteed annual income, while the other half receives no supplements. The consumption behavior of each group of families is then observed.

Much of our earlier discussion has been concerned with the *analysis* of investigations, both formal experiments and observational studies. In this and the following chapters, we focus largely on the *design* of experimental studies. In particular, we shall concentrate on those elements of experimental design where statistical methods have made the greatest contributions.

In this chapter, we shall first consider the major elements of any experimental design, and then shall take up completely randomized designs. In succeeding chapters, other widely used experimental designs will be discussed.

26.1 ELEMENTS OF EXPERIMENTAL DESIGN

The *design of an experiment* refers to the structure of the experiment, with particular reference to:

1. The set of treatments included in the study.
2. The set of experimental units included in the study.
3. The rules and procedures by which the treatments are assigned to the experimental units (or vice versa).
4. The measurements that are made on the experimental units after the treatments have been applied.

Statistical designs for experiments are concerned with the rules and procedures whereby treatments are assigned to the experimental units. Statistical methodology also makes contributions to the other elements of experimental design, but we shall dwell chiefly on how to assign the treatments to experimental units so that efficient use is made of the experimental units. First, however, we shall discuss briefly the other key elements of experimental design.

Treatments

A treatment, we know, is a factor level in a single-factor study or a combination of factor levels in a multifactor study. Three principal problems in experimental design pertaining to treatments are: (1) choice of treatments to be studied, (2) definition of each treatment, and (3) need for a control treatment.

Choice of treatments. The choice of treatments to be included in an experimental investigation is basically a matter to be decided by the investigator. A few general comments may, however, be appropriate. In a scientific investigation, the treatments included should be able to provide some insights into the mechanism underlying the phenomenon under study. Initial experiments should not attempt to study this mechanism in full detail, but rather should seek to find the principal factors involved and obtain indications of the magnitudes of their effects. Subsequent experiments can then be conducted to provide more detailed findings.

Even if the investigation is not of scientific but rather of practical interest, inclusion of treatments to provide some explanation of the results is often desirable. Suppose a company is considering the purchase of a new type of tool machine from a new maker and wishes to compare this new type with the present machines by means of an experiment. Suppose further that the new machines are of a much larger size than the present ones. Under these circumstances, the company may wish to include a third type of machine in the experiment, namely, one corresponding in size to the present ones, but from the new maker. By doing so, the company will get information as to whether any differences that might be observed between the present machines and the new ones are connected with the maker, the size of the machine, or both.

In earlier chapters, we discussed the usefulness of factorial studies for investigating the effects of several factors simultaneously. It is often desirable to include some factors in an investigation even though the main interest is not in them, simply because the chief factors of interest may interact in special ways with these additional factors.

Definition of treatment. The definition of a treatment can be a difficult problem. Consider an experiment to study whether BASIC or FORTRAN is a better programming language to teach in an introductory statistics course. Some teachers will prefer BASIC, others FORTRAN. Should the treatments then be defined as the programming language taught by instructors who prefer that language? If so, differences in findings may be due to differences between the two groups of instructors. Should the definition of a treatment not include the instructor, and instructors be randomized, with some being forced to teach a language they do not prefer? Or should instructor preference be a second factor, with each instructor teaching both languages? Problems of this kind need careful resolution so that the results of the study will be useful.

Another aspect of concern about the definition of a treatment is that the treatment may contain a bias. For example, in a medical study where a standard treatment and a new treatment are being compared, the care of patients receiving the new treatment might be systematically better, either knowingly or unknowingly, because of the staff's desire to give the new treatment every opportunity to prove its effectiveness. This type of *treatment bias* can be serious and needs to be avoided or minimized if the experimental results are to lead to reliable conclusions. One method of eliminating the type of treatment bias just cited is by conducting a blind study where the medical staff does not know whether a treatment is the standard one or the new one. Blind studies also help to avoid other types of biases, as we shall see shortly.

Control treatment. A control treatment is needed in some experiments, but not in all. A control treatment consists of applying the identical procedures to experimental units that are used with the other treatments, except for the effects under investigation. In a study of food additives, for instance, a treatment may consist of a portion of a vegetable containing a particular additive that is served to a consumer in a particular experimental setting in the laboratory. A control treatment here would consist of a portion of the same vegetable served to a consumer in the identical experimental setting except that no food additive has been used.

A control treatment is required when the general effectiveness of the treatments under study is not known, or when the general effectiveness of the treatments is known but is not consistent under all conditions. In our food additives illustration, suppose it is known that food additive A is highly effective in enhancing the tastiness of vegetables and it is desired to see if additives B and C are equally effective or possibly even more effective. In that case, a standard of comparison is available and no control treatment is required. On the other hand, suppose there is no knowledge about the general effectiveness of the three additives, and the following results are obtained (ratings can range between 0 and 60):

Additive	*Mean Rating*
A	39
B	37
C	41

Assume that the sample sizes are large so that the mean ratings are very precise. In the absence of a standard of comparison, one would not know here whether each of the three additives is effective or whether none of the additives is effective.

It is crucial that the control treatment be conducted in the identical experimental setting as the other treatments. In our food additives illustration, for instance, a survey of consumers at home, in which persons are asked to rate the general tastiness of the vegetable (without any additive) on the same scale as in the experiment, would not qualify as a control treatment. Such a survey might yield a mean rating of 22, suggesting that the three additives substantially increase the tastiness of the vegetable. This conclusion, however, could be grossly misleading. If the control treatment actually were incorporated into the experiment so that consumers are given portions of the vegetable with no additive in the laboratory setting, the mean rating for the control treatment might be 40. This result would imply that none of the three additives is effective in enhancing the tastiness of the vegetable. The reason for the higher mean rating in the laboratory setting could be a ''halo'' effect connected with the experimental procedures. Possibly, foods served in the experimental setting taste better than at home, or perhaps consumers try to oblige by giving higher ratings when they participate in an experimental study. Thus, only a control treatment incorporated into the experiment can serve as the proper standard of comparison.

Experimental units

Definition of experimental unit. We define an experimental unit as the smallest subunit of the experimental material such that any two different experimental units may receive different treatments. Suppose that two incentive pay systems are studied in two plants each, with the plant assignments made at random, and observations are then made on the productivity of a sample of employees in each plant. Here, the experimental unit is the plant, not the employee, since all employees in a given plant are assigned the same incentive pay system in the experiment.

Size of experimental unit. An important problem in designing an experiment sometimes is the size of the experimental unit. In our example above, should the experimental unit be an individual employee, a shift, or a plant? Oftentimes, technical considerations dictate the choice of the size of the experimental unit. For instance, morale considerations might preclude the use of different incentive pay systems in the same plant.

A different aspect of size occurs in studies of sales and similar phenomena. Suppose that we are interested in measuring the effectiveness of five different television commercials in terms of sales during a period of time subsequent to their showing. Should the length of time be one week, two weeks, one month, or some other time period? Clearly, the purposes of the study will need to govern the length of time which makes up the experimental unit here.

Representativeness of experimental units. Representativeness of experimental units is another important consideration in the design of experiments. Consider a study of management behavior with different communications networks. A university investigator may be tempted to use students as subjects because of their ready availability. If, however, information is desired about the behavior of business people, the students may not be representative experimental units. It hardly needs to be stated that an investigator should make every effort to obtain representative experimental units. Conversely, one should be cautious in extending results of an investigation to groups for which the experimental units are not representative. Thus, if the communications network study cited above *did* use students, one should not automatically assume that the findings are relevant to business people.

There is another facet to the representativeness of experimental units. The more similar they are, the smaller will be the experimental errors (i.e., the smaller the variance of the random error component ε) and the more precise the experimental results will be. Thus, in a learning experiment, the use of persons of the same age, intelligence, and economic and social backround will tend to lead to smaller experimental errors than if a more heterogeneous group of subjects was used. However the more homogeneous are the experimental units, the smaller is the range for which the experimental results are valid. For instance, findings for persons in one age group may not be valid for persons in other age groups. Thus, to make the conclusions broadly valid, one should vary the characteristics of experimental units, but the cost of this is less precision in the experimental results. Blocking designs, to be discussed in later chapters, can be employed to have one's cake and eat it too, namely, to have sufficient variability between experimental units for a wide range of validity and yet achieve high precision due to small experimental errors.

Assignment of treatments to experimental units

Once the treatments and the experimental units have been specified, rules and procedures are required for assigning the treatments to the experimental units (or vice versa). Use of improper rules and procedures can lead to serious difficulties. For instance, in a medical study designed to compare a standard treatment to a potentially risky new treatment, the group of patients undergoing the new treatment consisted of volunteers. These volunteers were less healthy at the beginning of the study than the patients receiving the standard treatment. Despite the fact that both the standard and new treatments were equally effective, the analysis of the health conditions of the patients at the end of the study showed a significant difference between the two groups, namely, that the new treatment group was in poorer health. A conclusion that the new treatment is therefore less effective would be biased. The source of bias here is called *selection bias* because the experimental units for the two treatment groups were not similar. As we shall see later, selection bias can be minimized by randomization. The use of randomization tends to balance the experimental units in each treatment group with regard to factors other than the treatment that affect the outcome.

Measurements

The measurements to be made on the experimental units after the treatments have been applied constitute the values of the dependent variable(s). The analyst will not only need to decide in advance which characteristics to measure but also how to measure them. Important scaling problems may arise in the measurement process. The observations may also include readings on variables that are concerned not so much with the phenomenon directly under study as with factors that may help to explain observed effects or may help to reduce experimental errors. For instance, we mentioned earlier a study of five television commercials for which sales in an ensuing period are observed. Suppose the experimental unit is a metropolitan area. The observations taken in each area during the period may include readings on the weather to help reduce the experimental errors by the method of covariance (Chapter 25).

The measurement process ideally should produce measurements that are unbiased and precise. *Measurement bias* can cause serious difficulties in the analysis of a study. An important source of measurement bias is due to unrecognized differences in the evaluation process. For example, a group of plants randomly assigned to a new fungicide treatment might unintentionally be evaluated by the investigators to be responding better to the treatment than actually is the case because of a desire to show the new treatment to be effective. When the experimental unit is a person, knowledge of the treatment by the person may also influence the measurement obtained. For instance, a person who knows that the food additive is salt may respond differently in the evaluation of the tastiness of a vegetable than if the additive were unknown. This source of measurement bias can be minimized by concealing the treatment assignment to both the experimental subject and the evaluator. A study using this kind of concealment is called a *double-blind* study. When knowledge of the assignment is withheld only from the experimental subject or the evaluator, the study is called a *single-blind* study.

Control of biases

We have seen that serious biases can enter an experiment in the administration of treatments, the selection of experimental units for a treatment, and the measurement process. Well-designed experiments attempt to minimize these biases. Extensive research prior to the experiment is often required to develop satisfactory treatment, selection, and measurement procedures that yield both precise and unbiased results.

26.2 CONTRIBUTIONS OF STATISTICS TO EXPERIMENTATION

Statistics has made a number of important contributions to experimentation. We consider briefly four major ones.

Factorial experiments

This contribution was considered in Chapter 20. There we noted that multifactor investigations permit the analysis of a number of factors with the same precision as if the entire experiment had been devoted to the study of only one factor. In addition, a single factorial experiment provides information on interaction effects while the classical one-factor-at-a-time approach requires a series of experiments for doing so.

Replication

Replication refers to the repetition of an experiment. Consider an experiment consisting of three treatments. The assignment of three experimental units at random, one to each treatment, constitutes one replication of the experiment. The assignment of an additional three experimental units at random to the three treatments constitutes a second replication, and so on.

We need to observe at once that not all repetitions are replications. Suppose two incentive pay plans are being investigated and two plants are used in the study, with one plant assigned to each incentive plan. Assume now that 10 employees in each plant are selected and their productivity measured. For purposes of comparing the incentive pay plans, the plants are the experimental units so that there is only one replication (one plant for each plan), not 10 (the number of employees studied in each plant). Indeed, with only one replication here, plant effects and incentive pay plan effects are confounded and cannot be disentangled. Selecting additional employees will not enable one to disentangle the incentive pay plan effects from the plant effects; only repetition of the experiment in additional plants (i.e., replicating the experiment) will permit this.

Replication makes it possible to assess the mean square error required for testing the presence of treatment effects or for establishing confidence interval estimates of these effects, as we have seen in earlier chapters. Replication also plays a second role, namely, it permits control over the precision of the estimates or the power of the tests through manipulation of the replication (sample) size. Again, we have observed this in earlier chapters.

Randomization

Randomization in experiments is a relatively recent idea, first introduced by the famous British statistician Sir R. A. Fisher. In the past, treatments had been assigned to experimental units either on a systematic or on a subjective basis. We noted earlier how serious biases can arise when self-selection is employed to assign experimental units to the treatments. The same dangers exist with systematic and subjective selection. To illustrate potential selection biases with systematic assignments, consider an experiment using ten employees and two treatments, where the first five employees on the payroll listing are assigned treatment 1 and the next five treatment 2. Suppose that the payroll listing is by seniority and that this variable is related to the phenomenon under study, say, productivity. A comparison of treatments 1 and 2 then reflects not only differences between the two treatments but also differences in the amount of experience between the two groups of employees. This potential bias may be so transparent that no good experimenter would use the type of systematic assignment just described. Nevertheless, there may be many other sources of bias that are not so apparent.

Subjective assignments of treatments to experimental units can also lead to selection bias, as when an experimenter subconsciously tends to assign one treatment to highly extrovert subjects and the other treatment to less extrovert subjects.

With randomization, the treatments are assigned to experimental units at random. (The mechanics will be discussed shortly.) Randomization tends to average out between the treatments whatever systematic effects may be present, apparent or hidden, so that comparisons between treatments measure only the pure treatment effects. Thus, randomization tends to eliminate the influence of extraneous factors not under the direct control of the experimenter and thereby precludes the presence of selection bias. Cochran and Cox (Ref. 26.1, p. 8) have likened randomization to an insurance policy in that it is a precaution against biases that may or may not occur.

Randomization is appropriate not only for the assignment of treatments to experimental units but also for any other phases of the experiment where systematic effects not under the control of the experimenter may be present. For instance, consider an experiment in which five treatments (alternative methods of measuring subjective probability) and 20 subjects are used. Only one subject can be run per day; thus, four weeks are required to complete the experiment. In this type of situation, it usually is highly desirable to determine the order of the treatments randomly since a variety of systematic time effects could be present. The experimenter may with time improve the explanation of the methods of measuring subjective probability, there may be a streak of extremely hot weather during a week, and the like. With these possible time effects, a systematic assignment of one treatment per week could lead to seriously biased results. Randomization, on the other hand, will tend to average out whatever systematic effects are present, whether anticipated or not.

Advice on when randomization is needed in addition to the assignment of treatments to experimental units can only be general. Certainly, randomization should be employed whenever the consequences of systematic effects could be serious. In our illustration on methods of measuring subjective probability, suppose that two observers conduct the experiment. Randomization of observers to treatment-experimental unit combinations would then be highly desirable, since large differences between observers are known to occur in this kind of situation. If the seriousness of the consequences is not known, the safe course is to randomize when feasible and not too costly. If randomization cannot be easily carried out and no serious consequences of systematic effects are anticipated, the experimenter may be willing to forego randomization. He or she must then realize, however, that the validity of the treatment comparisons depends on the absence of serious systematic effects.

Comments

1. One may view the implications of randomization in a somewhat different fashion than that presented so far. The random errors of experimental units that are adjacent in time or space are often correlated, not independent, as a result of various systematic effects over time or space. Randomization does not erase this correlation pattern but, by making it equally likely that any two treatments are adjacent, tends to eliminate the correlations between treatments with increasing replications. Thus, randomization makes it reasonable to analyze the data as though the model random error terms are independent, an assumption that has been made in almost all models discussed so far.

2. Once in a while, randomization may provide a pattern that makes the experimenter uneasy, such as running the four experimental units for treatment 1 first and then running the four experimental units for treatment 2. This is not a likely occurrence, but one that can take place. Some solutions have been suggested for this problem, but none provides a final answer. In practice, the experimenter typically will discard a randomization sequence that has apparent dangers of systematic effects for the particular experiment and select another randomization.

3. Randomization also can provide the basis for making inferences without requiring that the error terms ε are independent $N(0, \sigma^2)$. We illustrate this for a single-factor experiment, consisting of two treatments and three replications. In this experiment, the treatments were assigned to the experimental units at random. Suppose the data are:

Treatment 1	*Treatment 2*
3	6
9	2
4	10

We assume now the following simple model (it can be generalized):

$$Y_{ij} = \begin{pmatrix}\text{A quantity depending on} \\ \text{the experimental unit}\end{pmatrix} + \begin{pmatrix}\text{A quantity depending} \\ \text{on the treatment}\end{pmatrix} \tag{26.1}$$

Both the quantities for the experimental units and the quantities for the treatments are viewed as fixed. The randomness in the model arises solely from the random assignment of treatments to experimental units. Suppose now that the two treatment effects are equal. In that case, it would have been just as likely that we observed the numbers 3, 9, 4 for treatment 2, and 6, 2, 10 for treatment 1, since the treatments are assigned to experimental units at random. In fact, if the two treatment effects are equal, any division of the six observations into two groups of three is equally likely. Thus, in the list of all possible arrangements, all are equally likely if no treatment effects are present:

Treatment 1	*Treatment 2*
3, 9, 4	6, 2, 10
3, 9, 6	4, 2, 10
3, 9, 2	6, 4, 10
etc.	etc.

We then view the problem of comparing treatments as a one-way analysis of variance, and calculate $F^* = MSTR/MSE$ for each arrangement. We thereby obtain the exact sampling distribution of F^* when the two treatment effects are equal. Both empirical and theoretical studies have shown that the sampling distribution so obtained is distributed approximately as the F distribution, provided the sample sizes are not very small. Thus, randomization alone can justify the F test as a good approximate test, without requiring any assumption of independent, normal error terms.

Local control

The fourth contribution of statistics to experimental design is the concept of local control, which often is considered statistical design proper. Local control is intended to reduce experimental errors and make the experiment more powerful by suitable restrictions on the randomization of treatments to experimental units. Consider again the study of five methods of measuring subjective probability, which is to be conducted over a period of four weeks. The thought may have occurred in our earlier discussion that complete randomization might not provide full balance of all treatments within the four-week period. Would it not be better if we required that each week contain each of the five treatments once? If a substantial time effect is likely, it would indeed be desirable to use this form of restricted randomization, called *blocking*. Thus, we would randomize the order of treatments subject to the restriction that each treatment occur once in each week. It will be seen later that if a time effect is present, blocking will lead to more precise results than complete randomization.

Let us consider this same example from a slightly different point of view. With unrestricted randomization, the five observations in a single replication of the experiment will differ among themselves because of treatment effects, because of time effects (since the treatments may end up in any of the four weeks), and so on. If we require that each of the five treatments be conducted in each week, then a week's observations constitute a replication. Within such a replication, the observations will differ again because of treatment effects and a variety of other causes, but not because of any time effects from one week to another. The only effect of time that is left is that within a week, which may be anticipated to be substantially smaller than that between weeks. Thus, blocking by week will reduce the experimental error variability when a time effect is present, and in this way make the experiment more powerful.

There are many other methods available for restricting the randomization of treatments to experimental units in order to improve the efficiency of the experimental design. The most important of these will be discussed in subsequent chapters.

26.3 COMPLETELY RANDOMIZED DESIGNS

Description

The simplest statistical design for an experiment is the *completely randomized design*. With this design, treatments are assigned to experimental units (or vice versa) completely at random. This complete randomization provides that every experimental unit has an equal chance to receive any one of the treatments, also that all combinations of experimental units assigned to the different treatments are equally likely. We shall discuss the mechanics of accomplishing this randomization shortly.

A completely randomized design is useful when the experimental units are quite homogeneous. It also must be used when the experimental units are heterogeneous but no information is available for ''blocking'' the experimental units into more homogeneous groups and then applying restricted randomization. A completely randomized design may at times also be used with heterogeneous experimental units if covariance analysis (Chapter 25) is utilized to reduce the experimental error variability.

Advantages and disadvantages

The completely randomized design has a number of important advantages:

1. It is very flexible. It can accommodate any number of treatments and any number of replications.
2. The sample sizes can be varied from treatment to treatment. This might be desirable when a control treatment is included in the experiment. (While the power of tests is maximized with equal sample sizes, comparisons between treatments and control can be made more precise by assigning a greater number of observations to the control.)
3. The statistical analysis of the data is easy even when the sample sizes in a single-factor study are unequal.
4. The number of degrees of freedom associated with the experimental error mean square is larger than for any restricted randomization design.
5. Missing observations (e.g., due to sickness of a subject, loss of a questionnaire) create no problems in the analysis of single-factor studies. Neither does a dropped treatment, as when it is realized after the experiment is concluded that the intended deception in a treatment worked improperly.
6. The model requires fewer assumptions than those for other designs.

The chief disadvantage of the completely randomized design is its inefficiency when the experimental units are heterogeneous. Restricted randomization designs can utilize knowledge about the sources of variation in the observations (e.g., time effects, observer effects) to reduce the experimental error variability and thereby lead to more precise results. Unless the experiment is a very small-scale one, the larger number of degrees of freedom for the experimental error mean square with the completely randomized design will often not compensate for the reduction in experimental error variability that can be obtained by restricted randomization when the experimental units are heterogeneous.

Note

In business, economics, and the life sciences, the experimental units frequently are a person, a family, a town, a metropolitan area. These are typically highly heterogeneous experimental units. Hence, blocking and restricted randomization are often employed to reduce the experimental error variability that would be encountered with a completely randomized design.

How to randomize

Use of table of random permutations. All randomizations for experimental designs require that a series of objects (treatments, rows, columns, etc.) be placed in a random order. For a completely randomized design, the situation might be thus:

Treatments: T_1, T_2, T_3

Sample sizes: $n_1 = 3, n_2 = 2, n_3 = 3$

Here eight treatment assignments need be made to eight experimental units. These treatment assignments are listed in arbitrary order:

$$T_1 \quad T_1 \quad T_1 \quad T_2 \quad T_2 \quad T_3 \quad T_3 \quad T_3$$

To randomize the treatments to the experimental units, we number the latter from 1 to 8. Now we go to Table A–13, which contains random permutations of 9 (we use these since a table of 8 is not available). This table has been obtained by selecting permutations of the digits 1 to 9 at random, thus making each permutation equally probable. Suppose we pick the random permutation in the upper left corner:

5 4 9 7 1 6 8 3 2

We drop the number 9 since we have only eight experimental units, and obtain the following random assignment:

Experimental unit:	5	4	7	1	6	8	3	2
Treatment:	T_1	T_1	T_1	T_2	T_2	T_3	T_3	T_3

Thus, experimental units 5, 4, and 7 receive treatment T_1, and so on.

If more than 9 but less than 17 experimental units had been involved, we could have used Table A–14, which contains random permutations of 16. The tables by Moses and Oakford (Ref. 26.2) are a good source of random permutations. They contain random permutations of 9, 16, 20, 30, 50, 100, 200, 500, and 1,000.

Use of table of random digits or uniform random number generator. If a table of random permutations is not available, it is not difficult to generate a random permutation from a table of random digits or from a uniform random number generator. We illustrate how to do this for a random permutation of 6. First, we list the digits in order:

1 2 3 4 5 6

Next, we select or generate six random numbers between 000 and 999 (3-digit numbers are used here to make the probability of a tie small), and enter them over the digits 1 to 6. Suppose we use Table A–15, proceeding down from the upper left corner. We obtain:

132	212	990	001	605	912
1	2	3	4	5	6

Finally, we rearrange the pairs of numbers in ascending sequence for the random numbers:

001	132	212	605	912	990
4	1	2	5	6	3

Thus, we obtain the random permutation:

4 1 2 5 6 3

If a tie should be found between two random numbers, a selection or generation of another random digit can be used to break the tie.

It can be shown that the procedure just described makes all permutations equally likely. Thus, in the absence of a table of random permutations, a table of random digits or a uniform random number generator can be used quite readily to obtain random permutations.

Additional randomizations. Finally, we illustrate how to make additional randomizations besides that of treatments to experimental units. Suppose that in our previous illustration with three treatments and eight experimental units, the experimenter also would like to order the experiment at random over time. We would then obtain another random permutation of 8 for the time order of the experimental units, say:

Random time order:	5	1	3	6	4	7	2	8
Experimental unit:	1	2	3	4	5	6	7	8

This permutation indicates that experimental unit 1 comes in time position 5, experimental unit 2 in time position 1, etc. The complete set of assignments for our earlier example would be as follows:

Time position:	1	2	3	4	5	6	7	8
Experimental unit:	2	7	3	5	1	4	6	8
Treatment:	T_3	T_1	T_3	T_1	T_2	T_1	T_2	T_3

Comments

1. We have not explicitly indicated whether the treatments to be randomized are for a single-factor or multifactor experiment. The reason is that the randomization procedure for a completely randomized design is unaffected. Thus, in a two-factor study with each factor at two levels, there are simply four treatments T_1, T_2, T_3, and T_4 for purposes of randomization.

2. Should one of the factors in a multifactor experiment be a classification factor (e.g., age of subject, sales volume of store), we will consider such an experiment not to be a completely randomized one, since the randomization of treatments to experimental units is restricted.

Models

The models that we considered in Parts III and IV for single-factor and multifactor investigations are appropriate for completely randomized designs for experimental data. Hence, there is no need to discuss models in any detail. Thus, for a single-factor study with a completely randomized design, the model is of the form:

$$Y_{ij} = \mu_{.} + \tau_i + \varepsilon_{ij} \tag{26.2}$$

The τ_i are fixed or random according to the nature of the factor levels.

In a two-factor study with a completely randomized design, the model is of the form:

$$Y_{ijk} = \mu_{..} + \alpha_i + \beta_j + (\alpha\beta)_{ij} + \varepsilon_{ijk} \tag{26.3}$$

The main effects α_i and β_j and the interactions $(\alpha\beta)_{ij}$ may be fixed or random depending on the nature of the factors.

Design considerations

The major statistical question in designing a completely randomized experiment concerns the appropriate sample sizes for the treatments. We discussed this subject in Chapters 18, 21, and 24, and presented various approaches there whereby the necessary sample sizes can be determined. This discussion applies in its entirety to planning sample sizes for completely randomized designs. Hence, there is no need for further discussion of this topic here.

Analysis of results

The analysis of single-factor and multifactor studies discussed in Chapters 16–24 is entirely applicable for experimental data based on completely randomized designs. Hence, we present no further examples of analysis of such results at this point.

Analysis of covariance for completely randomized designs

Covariance analysis for completely randomized designs presents no new problems. Chapter 25 in its entirety is applicable.

Evaluation of aptness of model

Finally, we mention that the discussion of evaluating the aptness of analysis of variance models in Chapters 18, 21, and 24 applies directly to completely randomized experimental designs.

PROBLEMS

26.1. In a study of the effectiveness of subliminal advertising, subjects are brought to a studio and shown a film with one of three types of subliminal advertising for a product. Each subject's attitudes toward the product before and after the film viewing are measured.

a. In this type of situation, should a control treatment be employed?

b. Explain precisely what would be the nature of the control treatment for this study.

26.2. Give an example of an experimental study where a repetition is not a replication.

26.3. In an experiment to study the effect of the location of a product display in drugstores of a chain, the manager of one of the drugstores rearranged the displays of other products so as to increase the traffic flow at the experimental display. Does this action potentially lead to treatment bias, selection bias, or measurement bias? Discuss.

26.4. In a study of the effect of size of team on the volume of communications within the team, can a double-blind procedure be utilized? A single-blind procedure? Discuss.

26.5. Use a table of random digits (Table A–15) or a random number generator to obtain three random permutations of 6.

26.6. Use a table of random digits (Table A–15) or a random number generator to obtain two random permutations of 12.

26.7. Four treatments are to be studied in a completely randomized design, and 16 experimental units are available. Each treatment is to be assigned to four experimental units selected at random. There are four observers; each is to be assigned at random to one experimental unit for each treatment. Make all appropriate randomizations.

26.8. Five treatments are to be studied in a completely randomized design, and 30 experimental units are available. Each treatment is to be assigned to six experimental units selected at random. The study is to be conducted over a period of six days, with each treatment included on each day. The order of the five treatments within each day is to be randomized. Make all appropriate randomizations.

26.9. How does the randomization of treatment assignments in a two-factor study differ when both factors are experimental factors and when only one factor is an experimental factor?

26.10. Refer to **Cash offers** Problem 16.12. Explain how you would make the random assignments of dealers to "owners" in this single-factor study. Make all appropriate randomizations.

26.11. Refer to **Eye contact effect** Problem 20.12.

a. Explain how you would make the assignments of personnel officers to treatments in this two-factor study. Make all appropriate randomizations.

b. Did you randomize the officers to the factor levels of each factor?

26.12. Refer to **Hay fever relief** Problem 20.13.

a. Explain how you would make the assignments of volunteers to treatments in this study. Make all appropriate randomizations.

b. Did you randomize the volunteers to the factor levels of each factor?

26.13. Refer to **Disk drive service** Problem 20.14.

a. Is any randomization of treatment assignments called for in this study? Is any randomization utilized? Explain.

b. Would you consider this study to be experimental in nature? Discuss.

26.14. **Dental pain.** An anesthesiologist made a comparative study of the effects of acupuncture and codeine on postoperative dental pain in male subjects 18 to 30 years old. The four treatments were: (1) placebo treatment—a sugar capsule and two inactive acupuncture points, (2) codeine treatment only—a codeine capsule and two inactive acupuncture points, (3) acupuncture treatment only—a sugar capsule and two active acupuncture points, and (4) codeine and acupuncture treatment—a codeine capsule and two active acupuncture points. Forty subjects were employed in the study, with 10 assigned randomly to each treatment. Pain intensity indexes were obtained for each subject before the dental treatment and two hours afterwards. Data were collected on a double-blind basis. The change in pain intensity is the dependent variable of interest. The data on change in pain intensity follow (negative values indicate increased pain intensity).

	j									
i	1	2	3	4	5	6	7	8	9	10
1	1.8	0.0	−1.8	−1.6	1.7	−1.6	1.2	.3	0.0	−.7
2	2.1	1.6	.8	.5	.3	0.0	1.0	1.4	−.2	.6
3	.6	1.3	.3	1.5	0.0	.4	2.0	1.3	.2	.8
4	3.1	2.7	1.6	1.3	1.0	.8	2.5	1.7	1.1	.9

Assume that fixed effects ANOVA model (16.2) is appropriate.

a. Conduct the F test for equality of treatment means. Control the level of significance at $\alpha = .01$. State the alternatives, decision rule, and conclusion.

b. Compare the mean change in pain intensity for the placebo treatment (treatment 1) against that for each of the other three treatments; use the Bonferroni procedure with a 95 percent family confidence coefficient. State your findings.

26.15. Refer to **Dental pain** Problem 26.14. Use the Tukey procedure with a 95 percent family confidence coefficient to obtain all pairwise comparisons of treatment means. State your findings.

26.16. **Grapefruit sales.** A supermarket chain studied the relationship between grapefruit sales and the price at which grapefruits are offered. Three price levels were studied: (1) the chief competitor's price, (2) a price slightly higher than the chief competitor's price, and (3) a price moderately higher than the chief competitor's price. Twenty-four stores of comparable size were selected for the study, and eight stores were randomly assigned to each price level. Data on store sales of grapefruits during the study period follow (data coded).

	j							
i	1	2	3	4	5	6	7	8
1	58.9	61.0	45.3	58.1	53.5	48.1	56.5	51.8
2	55.2	56.1	46.5	53.2	49.8	57.7	58.5	42.3
3	39.4	45.3	50.5	36.6	41.9	52.4	43.2	48.7

Assume that fixed effects ANOVA model (16.2) is appropriate.

a. Test whether or not the price level affects mean sales of grapefruits. Control the level of significance at $\alpha = .05$. State the alternatives, decision rule, and conclusion.

b. Make all pairwise comparisons among the treatment means using the Tukey procedure with a 90 percent family confidence coefficient. State your findings.

26.17. Refer to **Grapefruit sales** Problem 26.16. Data on sales of grapefruits during the four-week period preceding the study follow (data are coded).

	j							
i	1	2	3	4	5	6	7	8
1	62.1	58.2	42.3	61.4	58.5	43.2	53.7	51.2
2	62.3	58.1	49.8	49.2	51.5	58.7	57.2	46.8
3	47.2	48.3	59.7	41.5	46.2	60.8	47.9	55.1

a. Using sales in the preceding four weeks as a concomitant variable, fit covariance model (25.3) to the data by means of regression model (25.12).

b. Test whether or not treatment effects are present; use $\alpha = .05$. State the alternatives, reduced model, decision rule, and conclusion.

c. Make all pairwise comparisons among the price level effects; use the Bonferroni procedure with a 90 percent family confidence coefficient. State your findings.

PROJECT

26.18. Refer to comment 3 on page 901 and model (26.1). Suppose that the quantities depending on the experimental unit are as follows for eight subjects in a psychological reinforcement experiment:

j:	1	2	3	4	5	6	7	8
Experimental unit quantity:	16	14	18	16	12	15	13	12

An experimental treatment is to be compared with a standard treatment, with four subjects assigned at random to each treatment.

a. Suppose that there are no differential treatment effects, with the quantity depending on the treatment being 4 for each treatment. Generate all possible randomizations of the eight experimental units to the two treatments and obtain the observed values for each treatment sample.

b. For each of the 70 randomizations obtained in part (a), calculate F^* for testing the alternatives H_0: $\mu_1 = \mu_2$ versus H_a: $\mu_1 \neq \mu_2$. Determine the proportion of F^* values that exceed $F(.90; 1, 6)$, the proportion that exceed $F(.95; 1, 6)$, and the proportion that exceed $F(.99; 1, 6)$.

c. How do the proportions obtained in part (b) compare with the probabilities for the normal error model? Discuss.

d. Repeat parts (a) and (b) for the case where the treatment quantities for the experimental and standard treatments are, respectively, 15 and 4. Does the test appear to have reasonable power in this case?

CITED REFERENCES

26.1 Cochran, William G., and Gertrude M. Cox. *Experimental Designs*. 2d ed. New York: John Wiley & Sons, 1957.

26.2 Moses, Lincoln E., and Robert V. Oakford. *Tables of Random Permutations*. Stanford: Stanford University Press, 1963.

27

Randomized block designs—I

In this chapter, we consider the most widely used design that employs local control to reduce experimental error variability—the randomized block design. We take up the basic elements of randomized block designs, the fixed effects model, the analysis of variance, and the analysis of treatment effects. We also consider how to plan randomized block designs and how to evaluate the aptness of the model. Finally, we take up the regression approach to randomized block designs and covariance analysis for this design.

In Chapter 28, we consider additional topics pertaining to randomized block designs, including missing observations and random block effects.

27.1 BASIC ELEMENTS

Description of design

A randomized block design is a restricted randomization design in which the experimental units are first sorted into homogeneous groups, called *blocks*, and the treatments are then assigned at random within the blocks. Consider, for instance, an experiment on the effect of four levels of newspaper advertising saturation on sales volume. The experimental unit is a city, and 16 cities are available for the study. Size of city usually is highly correlated with the dependent variable, sales volume. Hence, it is desirable to block the 16 cities into four groups of four cities each, according to population size. Thus, the four largest cities will constitute block 1, and so on. Within each block, the four treatments are then assigned at random to the four cities, and the assignments from one block to another are made independently.

As a second example, consider an experiment on the effects of three different incentive pay schemes on employee productivity of electronic assemblies. The experimental unit is an employee, and 30 employees are available for the study. Since productivity in this situation is highly correlated with manual dexterity, it is desirable to block the 30 employees into 10 groups of three according to manual dexterity. Thus, the three employees with the highest manual dexterity ratings are grouped into one block, and so on for the other employees. Within each block, the three incentive pay schemes are then assigned randomly to the three employees.

As a third example, consider a chemist studying the reaction rate of five chemical agents. Only five agents can be effectively analyzed per day. Since day-to-day differences may affect the reaction rate, each day is used as a block, and all five chemical agents are tested each day in independently randomized orders.

As these examples imply, the key objective in blocking the experimental units is to make them as homogeneous as possible within blocks with respect to the dependent variable under study, and to make the different blocks as heterogeneous as possible with respect to the dependent variable. The design in which each treatment is included in each block is called a *randomized complete block design*. Often, we shall drop the term *complete* because the context makes it clear that all treatments are included in each block.

Comments

1. In a complete block design, each block constitutes a replication of the experiment. For that reason, it is highly desirable that the experimental units within a block be processed together whenever this will help to reduce experimental error variability. As an example, an experimenter may tend to make changes in experimental techniques over time (e.g., in the administration of the experiment to subjects) without being aware of it. Consecutive processing of the experimental units block by block will tend to keep such sources of variation out of the variation within blocks, and thereby make the experimental results more precise.

2. In factorial experiments, some of the factors of interest are often characteristics of the experimental unit, such as sex, age, and amount of experience on the job. Even though these factors are not introduced to reduce experimental error variability but rather are included for their intrinsic interest, we shall nevertheless consider such experiments to be randomized block designs since the randomization of treatments to experimental units is restricted by the nature of the classification factors considered.

Criteria for blocking

As noted earlier, the purpose of blocking is to sort experimental units into groups within each of which the elements are homogeneous with respect to the dependent variable, such that the differences between groups are as great as possible. To help recognize some of the characteristics of experimental units that are fruitful criteria for blocking, we need a more precise definition of an experimental unit than that given in Chapter 26. All elements of the experimental situation that are not included in the definition of a treatment need to be assigned to the definition of an experimental unit. Suppose the treatment in an experiment consists of a portion of a vegetable containing a particular additive served in the laboratory. The experimental unit might then be defined as a homemaker of a given age, processed by a given observer on a specified day during a particular part of the day, and served food from a given batch of cooked vegetable. Still other elements of the experimental setting might be included in the definition of the experimental unit, and should be if they could be the cause of material variability in the observations.

A full definition of the experimental unit such as the one just given suggests two types of blocking criteria:

1. Characteristics associated with the unit—for persons: sex, age, income, intelligence, education, job experience, attitudes, etc.; for geographic areas: population size, average income, etc.
2. Characteristics associated with the experimental setting—observer, time of processing, machine, batch of material, measuring instrument, etc.

Use of time as a blocking variable frequently captures a number of different sources of variability, such as learning by observer, changes in equipment, and drifts in environmental conditions (e.g., weather). Blocking by observers often eliminates a substantial amount of interobserver variability; similarly, blocking by batches of material frequently is very effective.

There is no need to use only a single blocking criterion; several may be employed if the experimental errors can be reduced substantially thereby. We shall consider in Chapter 28 the use of more than one blocking criterion, such as when a block consists of subjects in a given age group processed by a particular observer.

To design effective randomized block experiments requires the ability to select blocking variables that will reduce the experimental error variability. Often, past experience in the subject matter field enables the experimenter to select good blocking variables. If some experiments have been run in the past in which blocking has been employed, these results can be analyzed to determine the effectiveness of the blocking variables. We shall discuss an appropriate method of analysis for doing this in Section 27.6. In the absence of any information on potential blocking variables, uniformity trials can be run where all experimental units are assigned the same treatment. From these trials, information can be obtained on the effectiveness of different variables for blocking.

Note

There is another blocking variable often used in social science research that has not been mentioned yet, namely, the subject. With the subject as a complete block, all treatments are given to every subject. Such designs are often called *repeated measures de-*

signs. Since they involve some special problems, we will discuss them separately in Chapter 28.

Advantages and disadvantages

The advantages of a randomized complete block design are:

1. It can, with effective grouping, provide substantially more precise results than a completely randomized design of comparable size.
2. It can accommodate any number of treatments and replications.
3. Different treatments need not have equal sample sizes. For instance, if the control is to have twice as large a sample size as each of three treatments, blocks of size five would be used; three units in a block are then assigned at random to the three treatments and two to the control.
4. The statistical analysis is relatively simple.
5. If an entire treatment or a block needs to be dropped from the analysis for some reason, such as spoiled results, the analysis is not complicated thereby.
6. Variability in experimental units can be deliberately introduced to widen the range of validity of the experimental results without sacrificing the precision of the results.

Disadvantages include:

1. Missing observations within a block require more complex calculations.
2. The degrees of freedom for experimental error are not as large as with a completely randomized design. One degree of freedom is lost for each block after the first.
3. More assumptions are required for the model (e.g., no interactions between treatments and blocks, constant variance from block to block) than for a completely randomized design model.

How to randomize

The randomization procedure for a randomized block design is straightforward. Within each block a random permutation is used to assign treatments to experimental units, just as in a completely randomized design. Independent permutations are selected for the several blocks.

Illustration

In an experiment on decision making, executives were exposed to one of three methods of quantifying the maximum risk premium they would be willing to pay to avoid uncertainty. The three methods are the utility method, the worry method, and the comparison method. After using the assigned method, each subject was asked to state the degree of confidence in the method of quantifying the risk premium on a scale from 0 (no confidence) to 20 (highest confidence).

Fifteen subjects were used in the study. They were grouped into five blocks of three executives, according to age. Block 1 contained the three oldest executives, and so on. The design layout, after five independent random permutations of 3 were employed, was as shown in Table 27.1. Table 27.2 contains the results of the experiment, and Figure 27.1 presents graphically the confidence ratings for each method by block. It appears from Figure 27.1 that there is much variation between blocks, but that in all blocks the comparison method leads to the highest confidence rating and the utility method to the lowest rating. We discuss next a widely used model for randomized block designs and the analysis of variance for this model before undertaking a formal analysis of the results in our example.

TABLE 27.1 Layout for randomized block design—risk premium example

	Experimental Unit		
	1	2	3
Block 1 (oldest executives)	C	W	U
2	C	U	W
3	U	W	C
4	W	U	C
5 (youngest executives)	W	C	U

C: Comparison method
W: Worry method
U: Utility method

TABLE 27.2 Risk premium experiment results (confidence ratings on scale from 0 to 20)

Block	*Method (j)*			
i	*Utility*	*Worry*	*Comparison*	*Average*
1 (oldest)	1	5	8	4.7
2	2	8	14	8.0
3	7	9	16	10.7
4	6	13	18	12.3
5 (youngest)	12	14	17	14.3
Average	5.6	9.8	14.6	10.0

27.2 MODEL

Table 27.2 is similar in appearance to Table 23.2a, which shows the data for a two-factor study with one observation in each cell. In fact, one may think of a randomized complete block design as corresponding to a two-factor study

FIGURE 27.1 Risk premium experiment—plot of confidence ratings by method and block

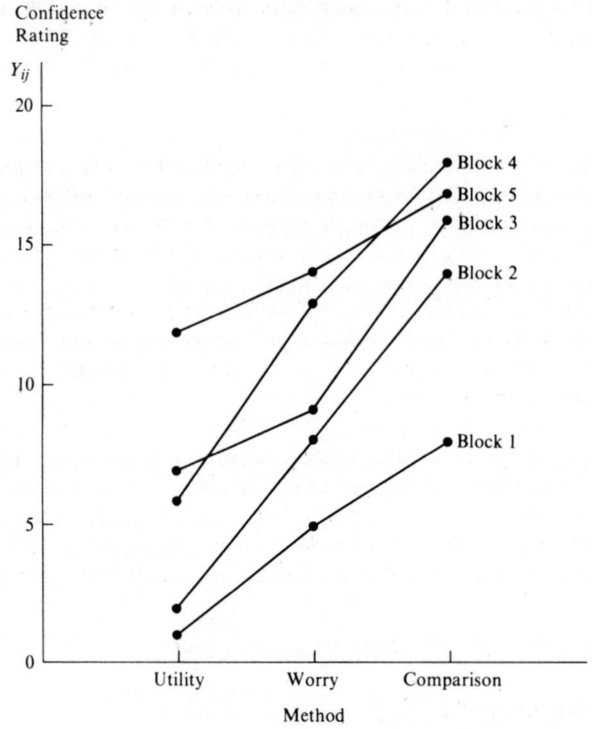

(blocks and treatments are the factors), with one observation in each cell. As we noted in Section 23.1, if one can assume that there are no interactions between the two factors, an analysis of factor effects can be undertaken when there is only one observation in each cell and the factors have fixed effects.

Thus, the model for a randomized complete block design, when both the block and treatment effects are fixed and there are n blocks (replications) and r treatments, is:

$$Y_{ij} = \mu_{..} + \rho_i + \tau_j + \varepsilon_{ij} \tag{27.1}$$

where:

$\mu_{..}$ is a constant

ρ_i are constants for the block (row) effects, subject to the restriction $\Sigma \rho_i = 0$

τ_j are constants for the treatment effects, subject to the restriction $\Sigma \tau_j = 0$

ε_{ij} are independent $N(0, \sigma^2)$

$i = 1, \ldots, n; j = 1, \ldots, r$

This model is identical to the two-factor, no-interaction model (23.1), except that we now use ρ_i for the block effect and n to designate the total number of blocks. Note that Y_{ij} here stands for the observation for the jth treatment in the ith block.

Comments

1. When the experimental units are grouped according to specified categories, such as into particular age groups, income groups, and order-of-processing groups, the block effects ρ_i are usually considered to be fixed. Sometimes the block effects are viewed as random. For instance, when observers or subjects are used as blocks, the particular observers or subjects in the study may be considered to be a sample from a population of observers or subjects. The case of random block effects will be taken up in Chapter 28.

2. If the treatment effects are random, the only changes in model (27.1) are that the τ_j now represent independent normal variables with expectation zero and variance σ_τ^2, and that the τ_j are independent of the ε_{ij}.

3. The additive model (27.1) implies that the expected values of observations in different blocks for the same treatment may differ (e.g., older executives may tend to have lower confidence ratings for any of the methods of quantifying the risk premium than younger executives), but the treatment effects (e.g., how much higher the confidence rating for one method is over that for another) are the same for all blocks. We shall consider the possibility of interactions between blocks and treatments later in this chapter.

27.3 ANALYSIS OF VARIANCE AND TESTS

Fitting of model

The least squares estimators of the parameters in model (27.1) are obtained in the customary fashion. Employing our usual notation, they are:

	Parameter	*Estimator*
(27.2a)	$\mu_{..}$	$\hat{\mu}_{..} = \bar{Y}_{..}$
(27.2b)	ρ_i	$\hat{\rho}_i = \bar{Y}_{i.} - \bar{Y}_{..}$
(27.2c)	τ_j	$\hat{\tau}_j = \bar{Y}_{.j} - \bar{Y}_{..}$

The fitted values therefore are:

$$\hat{Y}_{ij} = \bar{Y}_{..} + (\bar{Y}_{i.} - \bar{Y}_{..}) + (\bar{Y}_{.j} - \bar{Y}_{..}) = \bar{Y}_{i.} + \bar{Y}_{.j} - \bar{Y}_{..} \tag{27.3}$$

and the residuals are:

$$e_{ij} = Y_{ij} - \hat{Y}_{ij} = Y_{ij} - \bar{Y}_{i.} - \bar{Y}_{.j} + \bar{Y}_{..} \tag{27.4}$$

Analysis of variance

The analysis of variance for a randomized complete block design is identical to that for a two-factor, no-interaction model with one observation per cell, as described in Section 23.1:

$$(27.5a)\qquad SSBL = r\sum_i (\bar{Y}_{i.} - \bar{Y}_{..})^2 = \sum_i \frac{Y_{i.}^2}{r} - \frac{Y_{..}^2}{rn}$$

$$(27.5b)\qquad SSTR = n\sum_j (\bar{Y}_{.j} - \bar{Y}_{..})^2 = \sum_j \frac{Y_{.j}^2}{n} - \frac{Y_{..}^2}{rn}$$

$$(27.5c)\qquad SSE = \sum_i \sum_j (Y_{ij} - \bar{Y}_{i.} - \bar{Y}_{.j} + \bar{Y}_{..})^2$$

$$= \sum_i \sum_j Y_{ij}^2 - \sum_i \frac{Y_{i.}^2}{r} - \sum_j \frac{Y_{.j}^2}{n} + \frac{Y_{..}^2}{rn}$$

Here, $SSBL$ is the *sum of squares for blocks* while $SSTR$ and SSE as usual are the treatment and error sums of squares, respectively; rn is the total number of observations in the experiment. Note that the error sum of squares for the no-interaction model is actually an interaction sum of squares, just as in Section 23.1. Here the interactions are those between blocks and treatments.

A summary of the analysis of variance, including the expected mean squares for both fixed and random treatment effects, is given in Table 27.3. Note that since there are no interaction terms in the model, the expected mean squares contain only σ^2 and, as appropriate, the treatment or block main effects term.

TABLE 27.3 ANOVA table for randomized complete block design, block effects fixed

				E(MS)	
Source of Variation	*SS*	*df*	*MS*	*Treatments Fixed*	*Treatments Random*
Blocks	$SSBL$	$n-1$	$MSBL$	$\sigma^2 + \frac{r\sum \rho_i^2}{n-1}$	$\sigma^2 + \frac{r\sum \rho_i^2}{n-1}$
Treatments	$SSTR$	$r-1$	$MSTR$	$\sigma^2 + \frac{n\sum \tau_j^2}{r-1}$	$\sigma^2 + n\sigma_\tau^2$
Error	SSE	$(n-1)(r-1)$	MSE	σ^2	σ^2
Total	$SSTO$	$nr-1$			

As the $E(MS)$ columns in Table 27.3 indicate, the test for treatment effects:

	Fixed Treatment Effects	*Random Treatment Effects*
(27.6a)	H_0: all $\tau_j = 0$	H_0: $\sigma_\tau^2 = 0$
	H_a: not all τ_j equal zero	H_a: $\sigma_\tau^2 > 0$

uses the same test statistic whether the treatment effects are fixed or random:

$$(27.6b)\qquad F^* = \frac{MSTR}{MSE}$$

and the decision rule for controlling the Type I error at α is:

(27.6c)
$$\begin{aligned} &\text{If } F^* \leq F[1 - \alpha; r - 1, (n - 1)(r - 1)], \text{ conclude } H_0 \\ &\text{If } F^* > F[1 - \alpha; r - 1, (n - 1)(r - 1)], \text{ conclude } H_a \end{aligned}$$

Example

Table 27.4 contains the analysis of variance for our risk premium example in Table 27.2. The calculations are straightforward and were carried out by a computer package. To test for treatment effects:

$$\begin{aligned} &H_0: \tau_1 = \tau_2 = \tau_3 = 0 \\ &H_a: \text{not all } \tau_j \text{ equal zero} \end{aligned}$$

we use the results of Table 27.4:

$$F^* = \frac{MSTR}{MSE} = \frac{101.4}{2.99} = 33.9$$

For a level of significance of $\alpha = .01$, we require $F(.99; 2, 8) = 8.65$. Since $F^* = 33.9 > 8.65$, we conclude H_a, that the mean confidence ratings for the three methods differ.

TABLE 27.4 ANOVA table for randomized complete block design—risk premium example of Table 27.2

Source of Variation	*SS*	*df*	*MS*
Blocks	171.3	4	42.8
Methods for risk premium specification	202.8	2	101.4
Error	23.9	8	2.99
Total	398.0	14	

Comments

1. Sometimes one may also wish to conduct a test for block effects:

(27.7a)
$$\begin{aligned} &H_0: \text{all } \rho_i = 0 \\ &H_a: \text{not all } \rho_i \text{ equal zero} \end{aligned}$$

Usually, however, the treatments are of primary interest, and blocks are chiefly the means for reducing experimental error variability. Table 27.3 indicates that the test for fixed block effects uses the test statistic:

(27.7b)
$$F^* = \frac{MSBL}{MSE}$$

For our risk premium example, this test statistic is:

$$F^* = \frac{42.8}{2.99} = 14.3$$

For a level of significance of $\alpha = .01$, we require $F(.99; 4, 8) = 7.01$. Since $F^* = 14.3 > 7.01$, we conclude that the mean confidence ratings (averaged over treatments) differ for the various blocks.

Since blocks correspond to a classification factor, one needs to be careful in interpreting the implications of block effects. In our example, for instance, the block effects might not be due to age, even though age was the grouping variable. Education could be the pivotal independent variable, the effect by age arising if older executives have less formal education than younger ones.

2. The power of the F test for treatment effects for a randomized complete block design involves the same noncentrality parameter as for a completely randomized design. Formula (16.56a) gives the appropriate measure. Despite the same form of the noncentrality parameter, the two designs generally lead to different power levels even when based on the same sample sizes, for two reasons. First, the experimental error variance σ^2 will differ for the two designs. Second, the degrees of freedom associated with the denominator of the F^* statistic differ for the two designs.

3. If only two treatments are investigated in a randomized complete block design, it can readily be shown that the F test for treatment effects based on test statistic (27.6b) is equivalent to the two-sided t test for paired observations based on test statistic (1.64).

27.4 ANALYSIS OF TREATMENT EFFECTS

Once the block effects have been isolated through the analysis of variance, the analysis of fixed treatment effects proceeds as described in Chapter 17 for single-factor studies. Usually, the analysis of fixed treatment effects involves estimating one or more contrasts. The appropriate variance term to be used in the estimated variance of the contrast is MSE obtained from (27.5c) since it is the denominator of the F^* statistic for testing fixed treatment effects. The multiple for the estimated standard deviation of the contrast is as follows:

(27.8a)	Single comparison	$t[1 - \alpha/2; (n - 1)(r - 1)]$
(27.8b)	Tukey procedure (for pairwise comparisons)	$T = \frac{1}{\sqrt{2}} q[1 - \alpha; r, (n - 1)(r - 1)]$
(27.8c)	Scheffé procedure	$S^2 = (r - 1)F[1 - \alpha; r - 1, (n - 1)(r - 1)]$
(27.8d)	Bonferroni procedure	$B = t[1 - \alpha/2g; (n - 1)(r - 1)]$

Example

The researcher who conducted the risk premium study wished to obtain all pairwise comparisons with a 95 percent family confidence coefficient, and hence utilized the Tukey procedure. Using (17.24) and the results in Table 27.4, we obtain:

$$s^2(\hat{D}) = MSE\left(\frac{1}{n} + \frac{1}{n}\right) = \frac{2MSE}{n} = \frac{2(2.99)}{5} = 1.20$$

Remember that each treatment mean $\bar{Y}_{.j}$ consists of n observations (one from each of n blocks). Using (27.8b), we find for a 95 percent family confidence coefficient:

$$T = \frac{1}{\sqrt{2}} q(.95; 3, 8) = \frac{1}{\sqrt{2}}(4.04) = 2.86$$

Hence:

$$Ts(\hat{D}) = 2.86\sqrt{1.20} = 3.1$$

Thus, we obtain for the pairwise comparisons (see Table 27.2 for the $\bar{Y}_{.j}$):

$$1.7 = (14.6 - 9.8) - 3.1 \le \mu_{.3} - \mu_{.2} \le (14.6 - 9.8) + 3.1 = 7.9$$
$$5.9 = (14.6 - 5.6) - 3.1 \le \mu_{.3} - \mu_{.1} \le (14.6 - 5.6) + 3.1 = 12.1$$
$$1.1 = (9.8 - 5.6) - 3.1 \le \mu_{.2} - \mu_{.1} \le (9.8 - 5.6) + 3.1 = 7.3$$

Here $\mu_{.1}$ is the mean confidence rating, averaged over all blocks, for the utility method, and $\mu_{.2}$ and $\mu_{.3}$ are the mean confidence ratings for the worry and comparison methods, respectively.

We conclude that the comparison method has a higher mean confidence rating than the worry method, which in turn has a higher mean confidence rating than the utility method. The family confidence coefficient of .95 applies to this entire set of comparisons.

27.5 FACTORIAL TREATMENTS

When the treatments in a randomized block design are combinations of different factor levels, one can simply write the model showing the factor effects in place of the treatment effect. For a two-factor study, we have:

(27.9) $$Y_{ijk} = \mu_{...} + \rho_i + \alpha_j + \beta_k + (\alpha\beta)_{jk} + \varepsilon_{ijk}$$

where the terms in the model have the usual meaning and (j, k) identifies the treatment.

In the analysis of variance, we proceed as always by decomposing the treatment sum of squares $SSTR$ into sums of squares for factor main effects and interactions. This is shown in Table 27.5 for a two-factor study, the factors having a and b levels, respectively. Thus, the total number of treatments r here equals ab. The decomposition is done in the usual fashion, as explained in Section 20.4, utilizing the relation in (20.43):

$$SSTR = SSA + SSB + SSAB$$

Formulas (20.43a, b, c) and their alternative versions (20.44c, d, f) are appropriate for calculating the component sums of squares, remembering that (i, j) subscripts are there used to identify the treatments in terms of the factor combinations. Tests for factor effects are conducted as usual, and no new problems are encountered in the estimation of fixed factor effects.

TABLE 27.5 ANOVA table for a two-factor study in a randomized complete block design—model (27.9)

Source of Variation	SS	df	MS
Blocks	$SSBL$	$n-1$	$MSBL$
Treatments	$SSTR$	$r-1$	$MSTR$
Factor A	SSA	$a-1$	MSA
Factor B	SSB	$b-1$	MSB
AB interactions	$SSAB$	$(a-1)(b-1)$	$MSAB$
Error	SSE	$(n-1)(r-1)$	MSE
Total	$SSTO$	$nr-1$	
Note: $r = ab$			

Note

Model (27.9) assumes no interactions between treatments and blocks. Specifically, it implies that all block–factor A interactions (denoted by $BL.A$) equal zero, and similarly that all $BL.B$ and $BL.AB$ interactions equal zero. One can make a less restrictive analysis by assuming only that the $BL.AB$ interactions equal zero. To see this, consider the layout for a two-factor study ($a = 2$, $b = 2$) in $n = 3$ blocks, as shown in Table 27.6. Here Y_{ijk} denotes the observation in the ith block of the (j, k) factor combination. Note that this layout corresponds to the three-factor layout in Table 24.3, but with only one observation per cell. Table 24.4 contains the general analysis of variance for a three-factor study, and it can be seen from there that if all $BL.AB$ interactions ($BL.AB$ corresponds to ABC in Table 24.4) equal zero, the $BL.AB$ interaction sum of squares is an unbiased estimator of σ^2, the experimental error variance. Hence, we can conduct all tests and make all desired estimates with a factorial experiment in a complete block design by only assuming that the $BL.AB$ interactions are zero. The price of the less restrictive assumptions is fewer degrees of freedom for the experimental error.

The model for this less restrictive case is:

$$Y_{ijk} = \mu_{...} + \rho_i + \alpha_j + \beta_k + (\alpha\beta)_{jk} + (\rho\alpha)_{ij} + (\rho\beta)_{ik} + \varepsilon_{ijk} \tag{27.10}$$

TABLE 27.6 Layout for a two-factor study in a randomized complete block design

	A_1		A_2	
	B_1	B_2	B_1	B_2
Block 1	Y_{111}	Y_{112}	Y_{121}	Y_{122}
2	Y_{211}	Y_{212}	Y_{221}	Y_{222}
3	Y_{311}	Y_{312}	Y_{321}	Y_{322}

where the model terms have the usual meaning. The factor main effects and interactions may be fixed or random, depending on the nature of the factors. The analysis of variance for this model is given in Table 27.7. The sums of squares may be calculated using formulas (24.24a–g) or their alternative computational versions. In using these formulas, remember that n in these formulas (number of observations per cell) now is 1, and that the number of levels for blocks (corresponding to factor C) is n.

TABLE 27.7 ANOVA table for a two-factor study in a randomized complete block design—model (27.10)

Source of Variation	*SS*	*df*	*MS*
Blocks (*BL*)	*SSBL*	$n-1$	*MSBL*
Factor *A*	*SSA*	$a-1$	*MSA*
Factor *B*	*SSB*	$b-1$	*MSB*
AB interactions	*SSAB*	$(a-1)(b-1)$	*MSAB*
BL.A interactions	*SSBL.A*	$(n-1)(a-1)$	*MSBL.A*
BL.B interactions	*SSBL.B*	$(n-1)(b-1)$	*MSBL.B*
Error	*SSE*	$(n-1)(a-1)(b-1)$	*MSE*
Total	*SSTO*	$nab-1$	

27.6 PLANNING RANDOMIZED BLOCK EXPERIMENTS

Necessary number of blocks

The planning of sample size for a randomized complete block design is very similar to that for a completely randomized design. One may determine the needed number of blocks n either to obtain specified protection against making Type I and Type II errors or to obtain specified precision for key contrasts of the treatment means. With either approach, it is necessary to assess in advance the magnitude of the experimental error variance σ^2.

Power approach. The same tables as for the completely randomized design (Table A–10) may be used, provided the number of treatments and blocks are not very small, specifically provided that $r(n-1) \geq 20$. The development in Section 18.1 may be followed directly.

We illustrate the determination of sample size for the experiment on confidence ratings for three methods of measuring the risk premium. Suppose the number of blocks had not yet been determined and the experimenter desired the following risk protections:

1. Type I error is to be controlled at $\alpha = .05$.
2. If any two treatment means differ by 3 or more rating points, i.e., if the minimum range of the treatment means is $\Delta = 3$, the risk of concluding that there are no treatment effects should not exceed $\beta = .20$.

The experimenter anticipates that the experimental error standard deviation when executives are grouped by age will be approximately $\sigma = 2$.

Thus, the specifications can be summarized as follows:

$$r = 3 \qquad \alpha = .05 \qquad \Delta = 3$$
$$\beta = .20 \text{ or Power} = .80 \qquad \sigma = 2$$

Using (18.4) we find:

$$\frac{\Delta}{\sigma} = \frac{3}{2} = 1.5$$

Entering Table A–10 for $1 - \beta = .80$, $r = 3$, $\Delta/\sigma = 1.5$, and $\alpha = .05$, we find $n = 10$. Thus, the experimenter requires approximately 10 blocks of three executives each in order to obtain the desired protection against incorrect decisions.

Estimation approach. If the experimenter wishes to determine the number of blocks n by means of the estimation approach, one need simply calculate the anticipated standard deviations of key contrasts and modify the replication size iteratively until the desired precision is attained. Often, the experimenter will use a multiple comparison procedure for encompassing the different estimates under a family confidence coefficient.

With reference to our risk premium illustration, suppose the Tukey procedure is to be used for all pairwise comparisons with a 95 percent family confidence coefficient. Using $n = 10$ as a starting point and assuming that $\sigma = 2$ approximately, the anticipated variance of any pairwise difference is:

$$\sigma^2(\hat{D}) = \sigma^2\left(\frac{1}{n} + \frac{1}{n}\right) = (2)^2\left(\frac{1}{10} + \frac{1}{10}\right) = .8$$

or $\sigma(\hat{D}) = .89$. Further:

$$T = \frac{1}{\sqrt{2}}q[.95; r, (n-1)(r-1)] = \frac{1}{\sqrt{2}}q(.95; 3, 18) = \frac{1}{\sqrt{2}}(3.61) = 2.55$$

Thus, the anticipated half-width of the confidence interval is $T\sigma(\hat{D}) = 2.55(.89) = 2.3$. If this precision is not adequate, a larger number of blocks should be tried next. If the precision is greater than necessary, a smaller number of blocks should be used in the next iteration.

Efficiency of blocking variable

Once a randomized complete block experiment has been run, it is often desired to estimate the efficiency of the blocking variable employed for guidance in future experimentation.

Let σ_b^2 stand for the experimental error variance for the randomized block design. Up to this point, we have used σ^2 for this error variance, but now that we will compare two designs we need to be more specific. Let σ_r^2 denote the experimental error variance for a completely randomized design. The relative efficiency of blocking, compared to a completely randomized design, is then defined as follows:

$$E = \frac{\sigma_r^2}{\sigma_b^2} \tag{27.11}$$

The measure E indicates how much larger the replications need be with a completely randomized design as compared to a randomized block design in order that the variance of any estimated treatment contrast is the same for the two designs.

We know that MSE for the randomized block design is an unbiased estimator of σ_b^2. The question is how to estimate σ_r^2 from the data for the randomized block design. Since the same experimental units are involved in either case and there are assumed to be no interactions between treatments and blocks, it can be shown that an unbiased estimator of σ_r^2 is:

$$s_r^2 = \frac{(n-1)MSBL + n(r-1)MSE}{nr-1} \tag{27.12}$$

Hence, we estimate E as follows:

$$\hat{E} = \frac{s_r^2}{MSE} = \frac{(n-1)MSBL + n(r-1)MSE}{(nr-1)MSE} \tag{27.13}$$

Since the number of degrees of freedom for experimental error for a randomized block design is not as great as for a completely randomized one, E overstates the efficiency a little because it considers only the error variances. Several modified measures of efficiency have been suggested to take this overstatement into account. Unless the degrees of freedom for experimental error with both designs are very small, these modifications have little effect. One frequently used modification, applicable for assessing any design relative to another, is:

$$\hat{E}' = \frac{(df_2+1)(df_1+3)}{(df_2+3)(df_1+1)}\hat{E} \tag{27.14}$$

where df_1 denotes the degrees of freedom for the experimental error in the base design (completely randomized design, in our case) and df_2 denotes the degrees of freedom for the experimental error in the design whose efficiency is being assessed (randomized block design, in our case).

Example. We shall evaluate the efficiency of blocking by age of executives in our risk premium example. Placing the appropriate results from Table 27.4 in the efficiency measure (27.13), we obtain:

$$\hat{E} = \frac{4(42.8) + 5(2)(2.99)}{14(2.99)} = 4.8$$

Thus, we would have required almost five times as many replications per treatment with a completely randomized design to achieve the same variance of any estimated contrast as is obtained with blocking by age. Clearly, blocking by age was highly effective here.

If we had used the modified efficiency measure (27.14), we would have found:

$$\hat{E}' = \frac{(8 + 1)(12 + 3)}{(8 + 3)(12 + 1)}(4.8) = 4.5$$

This result, of course, does not differ greatly from that obtained by using (27.13).

Note

The efficiency measure $\hat{E}$ in (27.13) equals 1 if $MSBL = MSE$, is greater than 1 if $MSBL > MSE$, and is less than 1 if $MSBL < MSE$. Thus, good blocking is achieved when $MSBL$ exceeds MSE by a considerable margin.

27.7 ANALYSIS OF APTNESS OF MODEL

Since the importance of examining the aptness of the model for a given set of data has been mentioned many times earlier and since the techniques of examination are similar, we shall make only a few points of special relevance to randomized block designs here.

Residual plots

Some of the chief ways in which the data may not fit the randomized block model (27.1) are:

1. Unequal error variability by blocks.
2. Unequal error variability by treatments.
3. Time effects.
4. Block-treatment interactions.

Use of residual plots in connection with points 2 and 3 has been considered in Section 18.4 with reference to a completely randomized design, but the discussion there applies also to the residuals of a randomized block design, given in (27.4):

$$e_{ij} = Y_{ij} - \bar{Y}_{i.} - \bar{Y}_{.j} + \bar{Y}_{..}$$

We simply add here that if treatments do have unequal error variability in a randomized complete block design, one can always estimate differences between any two treatments by working with the differences between the paired observations, $Y_{ij} - Y_{ij'}$, which are unaffected by the unequal treatment variances.

Unequal error variability by blocks can be studied by plotting the residuals for each block, as shown in Figure 27.2. The residual sequence plots in Figure 27.2

FIGURE 27.2 Residual sequence plots suggesting unequal error variances by blocks

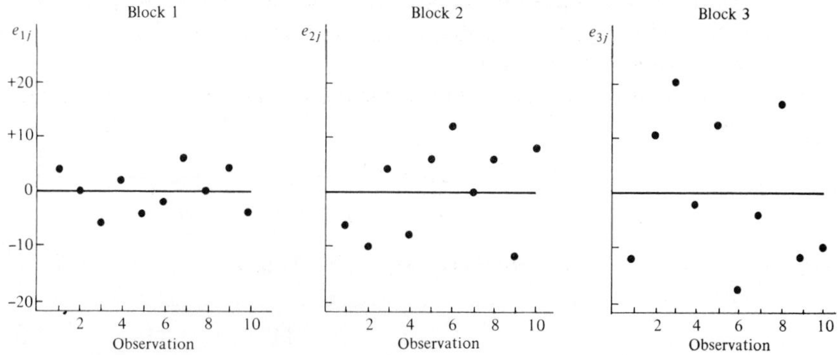

are suggestive of increasing error variability with increasing block number. If, for instance, the blocks were processed in block number order, some modifications in procedures may have taken place leading to larger experimental error variability over time. Tests concerning the equality of variances, such as those described in Section 18.6, may be employed for a more formal determination, provided that the sample sizes are reasonably large so that the residuals can be treated as if they were independent.

Interactions between treatments and blocks are somewhat more difficult to detect from residual plots. Figure 27.3 contains the residuals for an experiment with two treatments run in four blocks. The reversal in pattern of the residuals is suggestive of an interaction effect. There are, however, many other possible types of interaction patterns that would appear very much different from that in Figure 27.3.

Another diagnostic plot that may be helpful to detect interaction effects is a plot of the residuals e_{ij} against the fitted values $\hat{Y}_{ij}$. A curvilinear pattern of the

FIGURE 27.3 Residual sequence plots suggesting block-treatment interactions

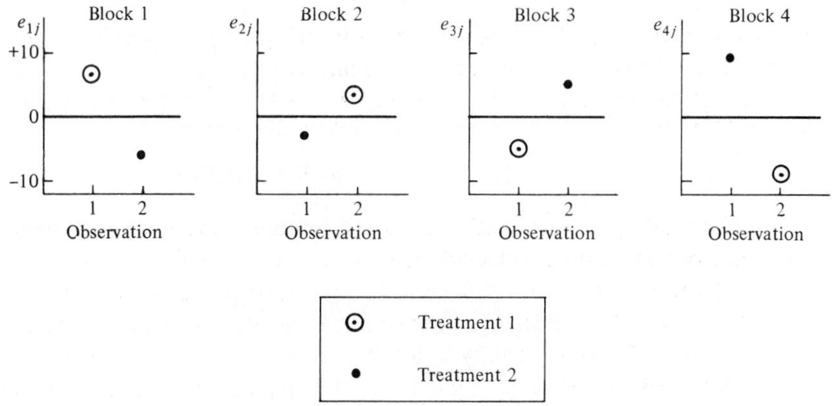

FIGURE 27.4 Computer plot of residuals versus fitted values—risk premium example

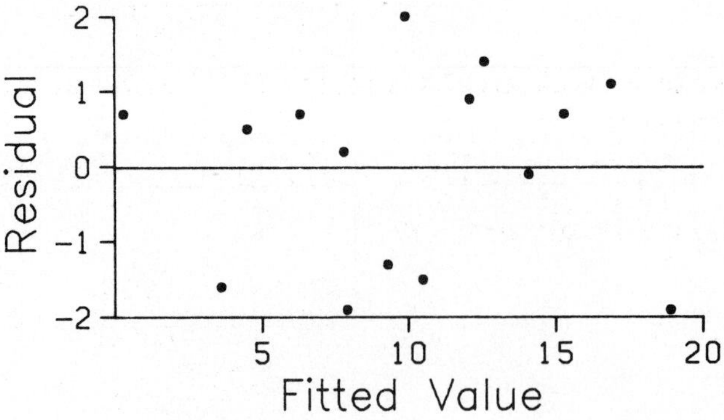

residuals in such a plot often suggests the presence of interaction effects between blocks and treatments. This plot also provides information about the constancy of the error term variance. Figure 27.4 presents for our risk premium example a computer plot of the residuals against the fitted values. There is no strong evidence of a curvilinear pattern or of unequal error term variances here.

Residual frequency plots and normal probability plots are useful for examining the normality assumption when this is an important consideration.

Tukey test for additivity

The Tukey test for additivity, discussed in Section 23.2, may be employed for a formal test of possible interaction effects between blocks and treatments. We illustrate this test for the risk premium example data in Table 27.2. The interaction sum of squares, denoted here by *SSBL.TR*, is given by (23.10):

$$SSBL.TR = \frac{\left[\sum_i \sum_j (\bar{Y}_{i.} - \bar{Y}_{..})(\bar{Y}_{.j} - \bar{Y}_{..})Y_{ij} \right]^2}{\sum_i (\bar{Y}_{i.} - \bar{Y}_{..})^2 \sum_j (\bar{Y}_{.j} - \bar{Y}_{..})^2}$$

Using the data in Table 27.2, we find the numerator:

$$[(4.7 - 10)(5.6 - 10)(1) + \cdots + (14.3 - 10)(14.6 - 10)(17)]^2 = 615.04$$

and from the results in Table 27.4 and formulas (27.5a) and (27.5b), we obtain the two terms in the denominator:

$$\Sigma(\bar{Y}_{i.} - \bar{Y}_{..})^2 = \frac{SSBL}{r} = \frac{171.3}{3} = 57.1$$

$$\Sigma(\bar{Y}_{.j} - \bar{Y}_{..})^2 = \frac{SSTR}{n} = \frac{202.8}{5} = 40.56$$

Hence:

$$SSBL.TR = \frac{615.04}{57.1(40.56)} = .27$$

We know from Table 27.4 that $SSE = 23.9$. Hence, the pure error sum of squares (23.11) is:

$$SSPE = SSE - SSBL.TR = 23.9 - .27 = 23.63$$

and the test statistic (23.12) is:

$$F^* = \frac{SSBL.TR}{1} \div \frac{SSPE}{rn - r - n}$$
$$= \frac{.27}{1} \div \frac{23.63}{7} = .08$$

For a level of significance of $\alpha = .05$, we need $F(.95; 1, 7) = 5.59$. Since $F^* = .08 \leq 5.59$, we conclude that no block–treatment interaction effects are present.

Note

If interaction effects are present, transformations on the data should be attempted to remove at least the important interaction effects. The discussion in Section 20.2 is relevant to this point.

27.8 REGRESSION APPROACH TO RANDOMIZED BLOCK DESIGNS

Model (27.1) for a randomized block design with both block and treatment effects fixed:

(27.15)
$$Y_{ij} = \mu_{..} + \rho_i + \tau_j + \varepsilon_{ij}$$
$$i = 1, \ldots, n; j = 1, \ldots, r$$

can be expressed in straightforward fashion in the form of a regression model with indicator variables taking on values 1, −1, or 0. The regression model for the risk premium example in Table 27.2, with $r = 3$ treatments and $n = 5$ blocks, can be written as follows:

(27.16)
$$Y_{ij} = \mu_{..} + \underbrace{\rho_1 X_{ij1} + \rho_2 X_{ij2} + \rho_3 X_{ij3} + \rho_4 X_{ij4}}_{\text{Block effect}} + \underbrace{\tau_1 X_{ij5} + \tau_2 X_{ij6}}_{\text{Treatment effect}} + \varepsilon_{ij}$$

where:

$$X_{ij1} = \begin{cases} 1 & \text{if experimental unit from block 1} \\ -1 & \text{if experimental unit from block 5} \\ 0 & \text{otherwise} \end{cases}$$

X_{ij2}, X_{ij3}, X_{ij4} are defined similarly

$$X_{ij5} = \begin{matrix} 1 \text{ if experimental unit received treatment } 1 \\ -1 \text{ if experimental unit received treatment } 3 \\ 0 \text{ otherwise} \end{matrix}$$

$$X_{ij6} = \begin{matrix} 1 \text{ if experimental unit received treatment } 2 \\ -1 \text{ if experimental unit received treatment } 3 \\ 0 \text{ otherwise} \end{matrix}$$

The **Y** observations vector and **X** matrix for the risk premium example are shown in Table 27.8. Note, for instance, that for observation Y_{13}, the indicator variables have the values:

$$X_1 = 1 \quad X_2 = 0 \quad X_3 = 0 \quad X_4 = 0 \quad X_5 = -1 \quad X_6 = -1$$

Hence:

$$\begin{aligned} Y_{13} &= \mu_{..} + \rho_1 - \tau_1 - \tau_2 + \varepsilon_{13} \\ &= \mu_{..} + \rho_1 + \tau_3 + \varepsilon_{13} \end{aligned}$$

because $\tau_3 = -\tau_1 - \tau_2$.

Tests and estimation of treatment effects with the regression approach follow easily and will not be discussed further here.

27.9 COVARIANCE ANALYSIS FOR RANDOMIZED BLOCK DESIGNS

Covariance analysis can be employed to further reduce the experimental error variability in a randomized block design. The extension is a straightforward one from covariance analysis for a completely randomized design.

TABLE 27.8 Data matrices for regression model (27.16) based on risk premium data in Table 27.2

$$\mathbf{Y} = \begin{bmatrix} Y_{11} = 1 \\ Y_{12} = 5 \\ Y_{13} = 8 \\ Y_{21} = 2 \\ Y_{22} = 8 \\ Y_{23} = 14 \\ Y_{31} = 7 \\ Y_{32} = 9 \\ Y_{33} = 16 \\ Y_{41} = 6 \\ Y_{42} = 13 \\ Y_{43} = 18 \\ Y_{51} = 12 \\ Y_{52} = 14 \\ Y_{53} = 17 \end{bmatrix} \qquad \mathbf{X} = \begin{bmatrix} & X_1 & X_2 & X_3 & X_4 & X_5 & X_6 \\ 1 & 1 & 0 & 0 & 0 & 1 & 0 \\ 1 & 1 & 0 & 0 & 0 & 0 & 1 \\ 1 & 1 & 0 & 0 & 0 & -1 & -1 \\ 1 & 0 & 1 & 0 & 0 & 1 & 0 \\ 1 & 0 & 1 & 0 & 0 & 0 & 1 \\ 1 & 0 & 1 & 0 & 0 & -1 & -1 \\ 1 & 0 & 0 & 1 & 0 & 1 & 0 \\ 1 & 0 & 0 & 1 & 0 & 0 & 1 \\ 1 & 0 & 0 & 1 & 0 & -1 & -1 \\ 1 & 0 & 0 & 0 & 1 & 1 & 0 \\ 1 & 0 & 0 & 0 & 1 & 0 & 1 \\ 1 & 0 & 0 & 0 & 1 & -1 & -1 \\ 1 & -1 & -1 & -1 & -1 & 1 & 0 \\ 1 & -1 & -1 & -1 & -1 & 0 & 1 \\ 1 & -1 & -1 & -1 & -1 & -1 & -1 \end{bmatrix}$$

Model

The usual randomized block design model was given in (27.1):

$$Y_{ij} = \mu_{..} + \rho_i + \tau_j + \varepsilon_{ij} \qquad (27.17)$$
$$i = 1, \ldots, n; j = 1, \ldots, r$$

The covariance model for a randomized block design with one concomitant variable is obtained by simply adding a term (or several terms) for the relation between the dependent variable Y and the concomitant variable X. Assuming this relation can be described by a linear function, we obtain:

$$Y_{ij} = \mu_{..} + \rho_i + \tau_j + \gamma(X_{ij} - \bar{X}_{..}) + \varepsilon_{ij} \qquad (27.18)$$
$$i = 1, \ldots, n; j = 1, \ldots, r$$

Regression approach

The regression approach to covariance model (27.18) involves no new principles. As in Chapter 25, we shall denote $X_{ij} - \bar{X}_{..}$ in model (27.18) by x_{ij}:

$$x_{ij} = X_{ij} - \bar{X}_{..} \qquad (27.19)$$

Further, we shall again use $1, -1, 0$ indicator variables for the block and treatment effects.

Suppose in a randomized complete block design study, $n = 4$ blocks and $r = 3$ treatments are utilized. The regression model counterpart to covariance model (27.18) then is:

$$Y_{ij} = \mu_{..} + \rho_1 I_{ij1} + \rho_2 I_{ij2} + \rho_3 I_{ij3} + \tau_1 I_{ij4} + \tau_2 I_{ij5} + \gamma x_{ij} + \varepsilon_{ij} \qquad \text{Full model} \qquad (27.20)$$

where:

$$I_{ij1} = \begin{cases} 1 & \text{if experimental unit from block 1} \\ -1 & \text{if experimental unit from block 4} \\ 0 & \text{otherwise} \end{cases}$$

$$I_{ij2} = \begin{cases} 1 & \text{if experimental unit from block 2} \\ -1 & \text{if experimental unit from block 4} \\ 0 & \text{otherwise} \end{cases}$$

$$I_{ij3} = \begin{cases} 1 & \text{if experimental unit from block 3} \\ -1 & \text{if experimental unit from block 4} \\ 0 & \text{otherwise} \end{cases}$$

$$I_{ij4} = \begin{cases} 1 & \text{if experimental unit received treatment 1} \\ -1 & \text{if experimental unit received treatment 3} \\ 0 & \text{otherwise} \end{cases}$$

$$I_{ij5} = \begin{cases} 1 & \text{if experimental unit received treatment 2} \\ -1 & \text{if experimental unit received treatment 3} \\ 0 & \text{otherwise} \end{cases}$$

To test for treatment effects:

(27.21)
$$H_0: \tau_1 = \tau_2 = \tau_3 = 0$$
$$H_a: \text{not all } \tau_j \text{ equal zero}$$

we would also need to fit the reduced model under H_0:

(27.22) $$Y_{ij} = \mu_{..} + \rho_1 I_{ij1} + \rho_2 I_{ij2} + \rho_3 I_{ij3} + \gamma x_{ij} + \varepsilon_{ij} \quad \text{Reduced model}$$

The test for treatment effects would then be conducted in the usual way.

Comparisons of two treatment effects by the regression approach are straightforward. For estimating $\tau_1 - \tau_2$, for instance, we use the unbiased estimator $\hat{\tau}_1 - \hat{\tau}_2$ based on the estimated regression coefficients obtained when fitting the full model (27.20). The estimated variance of this estimator is:

(27.23) $$s^2(\hat{\tau}_1 - \hat{\tau}_2) = s^2(\hat{\tau}_1) + s^2(\hat{\tau}_2) - 2s(\hat{\tau}_1, \hat{\tau}_2)$$

The estimated variance-covariance matrix of the regression coefficients from the computer printout can then be used to provide the required estimated variances and covariances.

Adjustment approach

If the adjustment approach to covariance analysis is to be employed, we begin as before by obtaining analyses of variance for Y, X, and XY. The analysis of variance for Y is given in (27.5), and that for X would be identical, with X_{ij} replacing Y_{ij}. Table 27.9 contains these analyses of variance. Note that subscripts for X and Y are used for identification.

TABLE 27.9 Analyses of variance for Y, X, and XY—randomized block design

Source of Variation	*Sums of Squares or Products*			df
	Y	X	XY	
Blocks	$SSBL_Y$	$SSBL_X$	$SPBL$	$n-1$
Treatments	$SSTR_Y$	$SSTR_X$	$SPTR$	$r-1$
Error	SSE_Y	SSE_X	SPE	$(n-1)(r-1)$
Total	$SSTO_Y$	$SSTO_X$	$SPTO$	$nr-1$

The analysis of variance for products proceeds as follows:

(27.24a) $$SPTO = \sum_i \sum_j (X_{ij} - \bar{X}_{..})(Y_{ij} - \bar{Y}_{..}) = \sum_i \sum_j X_{ij} Y_{ij} - \frac{X_{..} Y_{..}}{nr}$$

(27.24b) $$SPBL = r\sum_i (\bar{X}_{i.} - \bar{X}_{..})(\bar{Y}_{i.} - \bar{Y}_{..}) = \frac{\sum_i X_{i.}Y_{i.}}{r} - \frac{X_{..}Y_{..}}{nr}$$

(27.24c) $$SPTR = n\sum_j (\bar{X}_{.j} - \bar{X}_{..})(\bar{Y}_{.j} - \bar{Y}_{..}) = \frac{\sum_j X_{.j}Y_{.j}}{n} - \frac{X_{..}Y_{..}}{nr}$$

(27.24d) $$SPE = \sum_i \sum_j (X_{ij} - \bar{X}_{i.} - \bar{X}_{.j} + \bar{X}_{..})(Y_{ij} - \bar{Y}_{i.} - \bar{Y}_{.j} + \bar{Y}_{..})$$
$$= SPTO - SPBL - SPTR$$

Table 27.9 also contains this analysis of variance for XY.

The adjustment process now ignores the line for block effects in Table 27.9, as if this effect has thereby been eliminated, and proceeds in a manner analogous to a completely randomized design. Table 27.10a contains the analyses of variance without the block effects line. Note that the total line is called *sum*, since it excludes the block effects. The adjustment process is now entirely analogous to that for a completely randomized design. We obtain for the sum line:

(27.25a) $$SS(TR + E; \text{adj.}) = (SSTR_Y + SSE_Y) - \frac{(SPTR + SPE)^2}{SSTR_X + SSE_X}$$

which corresponds to (25.31a) for the total line for a completely randomized design. The adjusted error sum of squares is as before in (25.31b):

TABLE 27.10 Covariance analysis for testing treatment effects in a randomized block design

(a) Analyses of Variance

Source of Variation	*Sums of Squares or Products* Y	X	XY	*df*
Treatments	$SSTR_Y$	$SSTR_X$	$SPTR$	$r-1$
Error	SSE_Y	SSE_X	SPE	$(n-1)(r-1)$
Sum	$SSTR_Y + SSE_Y$	$SSTR_X + SSE_X$	$SPTR + SPE$	$n(r-1)$

(b) Analysis of Covariance

Source of Variation	*Adjusted SS*	*Adjusted df*	*Adjusted MS*
Treatments	$SSTR$(adj.)	$r-1$	$MSTR$(adj.)
Error	SSE(adj.)	$(n-1)(r-1)-1$	MSE(adj.)
Sum	$SS(TR+E;$ adj.$)$	$n(r-1)-1$	

$$(27.25b) \qquad SSE(\text{adj.}) = SSE_Y - \frac{(SPE)^2}{SSE_X}$$

Finally, the adjusted treatment sum of squares is obtained as a remainder, as in (25.31c):

$$(27.25c) \qquad SSTR(\text{adj.}) = SS(TR + E; \text{adj.}) - SSE(\text{adj.})$$

Table 27.10b contains the covariance analysis for a randomized block design. Note that the degrees of freedom for error and the sum are reduced by one on account of the introduction of the concomitant variable X. The adjusted degrees of freedom for treatments are obtained as a remainder.

A test for treatment effects uses the usual statistic:

$$(27.26) \qquad F^* = \frac{MSTR(\text{adj.})}{MSE(\text{adj.})}$$

which is distributed as $F[r - 1, (n - 1)(r - 1) - 1]$ if H_0 holds.

Blocking as alternative to covariance analysis

At times, a choice exists between: (1) a completely randomized design with covariance analysis used to reduce the experimental errors, and (2) a randomized block design, with the blocks formed by means of the concomitant variable. Generally, the latter alternative is to be preferred. There are several reasons for this:

1. If the regression is linear, randomized block designs and covariance analysis are about equally efficient. If the regression is not linear but covariance analysis with a linear relationship is utilized, covariance analysis with a completely randomized design will tend to be not as effective as a randomized block design.
2. Computations with randomized block designs are simpler than those with covariance analysis.
3. Randomized block designs are essentially free of assumptions about the nature of the relationship between the blocking variable and the dependent variable, while covariance analysis assumes a definite form of relationship.

A drawback of randomized block designs is that somewhat fewer degrees of freedom are available for experimental error than with covariance analysis for a completely randomized design. However, in all but small-scale experiments, this difference in degrees of freedom has little effect on the precision of the estimates.

PROBLEMS

27.1. A student commented in a discussion group: "Random permutations are used to assign treatments to experimental units with a randomized block design just as with a completely randomized design. Hence, there is no basic difference between these two designs.'' Comment.

27.2. a. What might be some useful blocking variables for an experiment about the effects of different price levels on sales of a product, using stores as experimental units?

b. What might be some useful blocking variables for an experiment about the effects of different flight crew schedules on the morale of crews, using flight crews as experimental units?

c. What might be some useful blocking variables for an experiment about the effects of different drugs on the speed of a response to a stimulus, using laboratory animals as experimental units?

27.3. Five treatments are studied in an experiment with a randomized complete block design using four blocks. Obtain randomized assignments of treatments to experimental units.

27.4. Two treatments and a control are studied in an experiment with a randomized block design. Five blocks are employed, each containing four experimental units. In each block, each treatment is to be assigned to one experimental unit, and the control is to be assigned to two experimental units. Obtain randomized assignments of treatments to experimental units.

27.5. **Auditor training.** An accounting firm, prior to introducing in the firm widespread training in statistical sampling for auditing, tested three training methods: (1) study at home with programmed training materials, (2) training sessions at local offices conducted by local staff, and (3) training session in Chicago conducted by national staff. Thirty auditors were grouped into 10 blocks of three, according to time elapsed since college graduation, and the auditors in each block were randomly assigned to the three training methods. At the end of the training, each auditor was asked to analyze a complex case involving statistical applications; a proficiency measure based on this analysis was obtained for each auditor. The results were (block 1 consists of auditors graduated most recently, block 10 consists of those graduated most distantly):

Block	*Training Method (j)*			*Block*	*Training Method (j)*		
i	1	2	3	*i*	1	2	3
1	73	81	92	6	73	75	86
2	76	78	89	7	68	72	88
3	75	76	87	8	64	74	82
4	74	77	90	9	65	73	81
5	76	71	88	10	62	69	78

Assume that fixed effects ANOVA model (27.1) is appropriate.

a. Why do you think the blocking variable ''time elapsed since college graduation'' was employed?

b. Obtain the analysis of variance table.

c. Test whether or not the mean proficiency is the same for the three training methods. Use a level of significance of $\alpha = .05$. State the alternatives, decision rule, and conclusion.

d. Make all pairwise comparisons between the training method means; use the Tukey procedure with a 90 percent family confidence coefficient. State your findings.

e. Test whether or not blocking effects are present; use $\alpha = .05$. State the alternatives, decision rule, and conclusion.

27.6. Calculator efficiency. A computer company, to test the efficiency of its new programmable calculator, selected six engineers who were proficient in the use of both this calculator and an earlier model and asked them to work out two problems on both calculators. One of the problems was statistical in nature, the other was an engineering problem. The order of the four calculations was randomized independently for each engineer. The length of time (in minutes) required to solve each problem was observed. The results follow.

	$j = 1$ *Statistical Problem*		$j = 2$ *Engineering Problem*	
Engineer i	$k = 1$ *New Model*	$k = 2$ *Earlier Model*	$k = 1$ *New Model*	$k = 2$ *Earlier Model*
1 Jones	3.1	7.5	2.5	5.1
2 Williams	3.8	8.1	2.8	5.3
3 Adams	3.0	7.6	2.0	4.9
4 Dixon	3.4	7.8	2.7	5.5
5 Erickson	3.3	6.9	2.5	5.4
6 Maynes	3.6	7.8	2.4	4.8

Since the six engineers chosen for the experiment were those who were proficient in the use of both calculator models and were not selected by a random mechanism from a defined population, the company does not wish to regard them as a sample from a population of engineers. ANOVA model (27.9) with fixed effects is assumed to be applicable, with factor A being type of problem and factor B being calculator model.

a. What assumptions are involved in using ANOVA model (27.9), and which of these are you most concerned about here?

b. Obtain the analysis of variance table.

c. Test whether or not the two factors interact; use $\alpha = .01$. State the alternatives, decision rule, and conclusion.

d. It is desired to study the nature of the interaction effects by considering the three comparisons:

$$D_1 = \mu_{.12} - \mu_{.11} \qquad L_1 = D_2 - D_1$$
$$D_2 = \mu_{.22} - \mu_{.21}$$

Obtain confidence intervals for these comparisons; use the Bonferroni procedure with a 95 percent family confidence coefficient. State your findings.

e. Test whether or not blocking effects are present; use $\alpha = .01$. State the alternatives, decision rule, and conclusion. What is the P-value of the test?

27.7. Refer to **Calculator efficiency** Problem 27.6. Assume that ANOVA model (27.10) with fixed effects is to be employed.

a. Obtain the analysis of variance table.

b. Test whether or not the two factors interact; use $\alpha = .01$. State the alternatives, decision rule, and conclusion. Do you reach the same conclusion as in Problem 27.6c with ANOVA model (27.9)?

c. Test whether or not blocks interact with factors A and B, respectively. For each test, use $\alpha = .01$ and state the alternatives, decision rule, and conclusion.

27.8. Fat in diets. A researcher studied the effects of three experimental diets with varying fat contents on the total lipid (fat) level in plasma. Total lipid level is a widely used predictor of coronary heart disease. Fifteen male subjects who were within 20 percent of their ideal body weight were grouped into five blocks according to age. Within each block, the three experimental diets were randomly assigned to the three subjects. Data on reduction in lipid level (in grams per liter) after the subjects were on the diet for a fixed period of time follow.

	Fat Content of Diet		
Block i	$j = 1$ *Extremely Low*	$j = 2$ *Fairly Low*	$j = 3$ *Moderately Low*
1 Ages 15–24	.73	.67	.15
2 Ages 25–34	.86	.75	.21
3 Ages 35–44	.94	.81	.26
4 Ages 45–54	1.40	1.32	.75
5 Ages 55–64	1.62	1.41	.78

Assume that ANOVA model (27.1) is applicable.

a. Why do you think that age of subject was used as a blocking variable?

b. Obtain the analysis of variance table.

c. Test whether or not the mean reductions in lipid level differ for the three diets; use $\alpha = .025$. State the alternatives, decision rule, and conclusion.

d. Estimate $D_1 = \mu_{.1} - \mu_{.2}$ and $D_2 = \mu_{.2} - \mu_{.3}$ using the Bonferroni procedure with a 95 percent family confidence coefficient. State your findings.

e. Test whether or not blocking effects are present; use $\alpha = .025$. State the alternatives, decision rule, and conclusion.

f. A standard diet was not used in this experiment as a control. What justification do you think the experimenters might give for not having a control treatment here for comparative purposes?

27.9. Refer to **Auditor training** Problem 27.5. Assume that $\sigma = 2.5$. What is the power of the test for training method effects in Problem 27.5c if $\mu_{.1} = 70$, $\mu_{.2} = 75$, and $\mu_{.3} = 85$?

27.10. Refer to **Fat in diets** Problem 27.8. Assume that $\sigma = .04$. What is the power of the test for diet effects in Problem 27.8c if $\mu_{.1} = 1.1$, $\mu_{.2} = 1.0$, and $\mu_{.3} = .4$?

27.11. Refer to **Auditor training** Problem 27.5. Another accounting firm wishes to conduct the same experiment with some of its auditors, using the same design and model. How many blocks would you recommend that this firm employ if it wishes to make all pairwise treatment comparisons with precision ± 1.5 with a 99 percent family confidence coefficient? Assume that a reasonable planning value for the error term standard deviation in model (27.1) is $\sigma = 2.5$.

27.12. Refer to **Fat in diets** Problem 27.8. Suppose that the number of blocks to be used in the study, to consist of male subjects of similar ages, has not yet been determined. Assume that a reasonable planning value for the error term standard deviation in model (27.1) is $\sigma = .04$.

a. What would be the required number of blocks if it is desired to make all pairwise diet comparisons with precision $\pm.03$ with a 95 percent family confidence coefficient?

b. What would be the required number of blocks if: (1) differences in lipid level reduction means for the three diets are to be detected with probability .95 or more when the range of the treatment means is .12, and (2) the α risk is to be controlled at .01?

27.13. Refer to **Auditor training** Problem 27.5. Calculate the estimated efficiency measure (27.13). How effective was the use of the blocking variable as compared to a completely randomized design?

27.14. Refer to **Calculator efficiency** Problem 27.6. Calculate the estimated efficiency measure (27.13). How effective was the use of the blocking variable as compared to a completely randomized design?

27.15. Refer to **Fat in diets** Problem 27.8. Calculate the estimated efficiency measure (27.14). How effective was the use of the blocking variable as compared to a completely randomized design?

27.16. Refer to **Auditor training** Problem 27.5.

a. Obtain the residuals and plot them against the fitted values. Also prepare a normal probability plot of the residuals. What are your findings?

b. Conduct the Tukey test for additivity of block and treatment effects; use $\alpha = .01$. State the alternatives, decision rule, and conclusion.

27.17. Refer to **Calculator efficiency** Problem 27.6.

a. Obtain the residuals and plot them against the fitted values. Also prepare a normal probability plot of the residuals. What are your findings?

b. Conduct the Tukey test for additivity of block and treatment effects; use $\alpha = .05$. State the alternatives, decision rule, and conclusion.

27.18. Refer to **Fat in diets** Problem 27.8.

a. Obtain the residuals and plot them against the fitted values. Also prepare a normal probability plot of the residuals. What are your findings?

b. Conduct the Tukey test for additivity of block and treatment effects; use $\alpha = .01$. State the alternatives, decision rule, and conclusion.

27.19. Refer to **Auditor training** Problem 27.5.

a. State the regression model equivalent to ANOVA model (27.1); use $1, -1, 0$ indicator variables.

b. Fit the regression model to the data.

c. Obtain the regression analysis of variance table based on appropriate extra sums of squares.

d. Test for treatment main effects; use $\alpha = .05$. State the alternatives, decision rule, and conclusion.

27.20. Refer to **Fat in diets** Problem 27.8.

a. State the regression model equivalent to ANOVA model (27.1); use $1, -1, 0$ indicator variables.

b. Fit the regression model to the data.

c. Obtain the regression analysis of variance table based on appropriate extra sums of squares.

d. Test for treatment main effects; use $\alpha = .025$. State the alternatives, decision rule, and conclusion.

27.21. Refer to **Auditor training** Problem 27.5. The analyst wishes to examine whether use of pre-training statistical proficiency scores as a concomitant variable would help to reduce the experimental error variability significantly. The pre-training statistical proficiency scores for the auditors are as follows:

Block i	Training Method (j) 1	2	3	Block i	Training Method (j) 1	2	3
1	93	98	91	6	75	74	78
2	94	93	94	7	79	76	72
3	89	91	92	8	71	69	64
4	86	84	90	9	74	71	70
5	78	76	84	10	63	68	64

a. Would you expect the auditor's age to have been a better concomitant variable here than the pre-training statistical proficiency score? Discuss.
b. State the regression model equivalent to covariance model (27.18); use 1, −1, 0 indicator variables.
c. Fit the full regression model.
d. State the reduced regression model for testing treatment effects. Fit the reduced model.
e. Test whether or not the training methods differ in mean effectiveness. Use a level of significance of $\alpha = .05$. State the alternatives, decision rule, and conclusion.
f. Obtain a 95 percent confidence interval for $D = \tau_1 - \tau_2$. Interpret your interval estimate.
g. Has the error term variance been reduced substantially by adding the concomitant variable? Explain.

27.22. Refer to **Calculator efficiency** Problem 27.6. The computer company measured the calmness/anxiety state of each engineer prior to each problem calculation. It is now desired to examine whether use of these calmness/anxiety scores as a concomitant variable would help to reduce the experimental error variability. The calmness/anxiety scores of the engineers follow (the higher the score, the greater the state of calmness).

		$j = 1$ Statistical Problem		$j = 2$ Engineering Problem	
	Engineer i	$k = 1$ New Model	$k = 2$ Earlier Model	$k = 1$ New Model	$k = 2$ Earlier Model
1	Jones	96	92	93	95
2	Williams	84	80	81	87
3	Adams	98	95	99	97
4	Dixon	90	91	89	86
5	Erickson	90	94	91	90
6	Maynes	88	92	92	94

a. State the covariance model for this case, and then state the equivalent regression model; use 1, −1, 0 indicator variables.

b. Fit the full regression model.
c. State the reduced regression model for testing whether or not the two factors interact. Fit the reduced model.
d. Test whether or not the two factors interact; use $\alpha = .01$. State the alternatives, decision rule, and conclusion.
e. Estimate $(\alpha\beta)_{11}$ with a 99 percent confidence interval. Interpret your interval estimate.
f. Has the error term variance been reduced substantially by adding the concomitant variable? Explain.

EXERCISES

27.23. Consider ANOVA model (27.1), but with random treatment effects. Derive $\sigma^2(Y_{ij})$ and $\sigma^2(\bar{Y}_{.j})$.

27.24. (Calculus needed.) State the likelihood function for the randomized block fixed effects model (27.1) when $n = 3$ and $r = 2$. Find the maximum likelihood estimators of the parameters. Are they the same as the least squares estimators in (27.2)?

27.25. For randomized block fixed effects model (27.1), derive $E(MSTR)$.

27.26. Show that when two treatments are studied in a randomized complete block design, the F^* test statistic (27.6b) for treatment effects is equivalent to the square of the two-sided t test statistic for paired observations based on (1.64).

27.27. Refer to regression model (27.16), the equivalent to ANOVA model (27.15) when $n = 5$ and $r = 3$. Suppose that the indicator variables in model (27.16) were coded as follows:

$$X_{ij1} = \begin{matrix} 1 \text{ if experimental unit from block 1} \\ 0 \text{ otherwise} \end{matrix}$$

$X_{ij2}, X_{ij3}, X_{ij4}$ are defined similarly

$$X_{ij5} = \begin{matrix} 1 \text{ if experimental unit received treatment 1} \\ 0 \text{ otherwise} \end{matrix}$$

$$X_{ij6} = \begin{matrix} 1 \text{ if experimental unit received treatment 2} \\ 0 \text{ otherwise} \end{matrix}$$

and that the regression coefficients are denoted by $\beta_0, \beta_1, \beta_2, \beta_3, \beta_4, \beta_5$, and β_6.
a. Exhibit the $\mathbf{X}$ matrix for this regression model.
b. Find the correspondences between the regression coefficients $\beta_0, \beta_1, \ldots, \beta_6$ and the parameters in ANOVA model (27.15).
c. Discuss the advantages and disadvantages of using 1,0 indicator variables and $1, -1, 0$ indicator variables here.

28

Randomized block designs—II

In this chapter, we continue our discussion of randomized block designs by first considering binary responses for the dependent variable. Then we take up missing observations, random block effects, and repeated measures designs. We conclude the chapter by discussing generalized randomized block designs and the use of more than one blocking variable.

28.1 BINARY RESPONSES FOR DEPENDENT VARIABLE

In some randomized block experiments, the responses are binary. For instance, the responses in an experiment involving shopping behavior may be buy or not buy. In an experiment on task motivation, the responses may be success in task or failure in task. In each of these cases, the responses can be coded 1 or 0.

Cochran test

When the responses in a randomized block experiment are binary, coded 1 or 0, a chi-square test may be used for assessing the presence of treatment effects. To test whether treatment effects are present:

(28.1)
$$H_0\text{: all } \tau_j = 0$$
$$H_a\text{: not all } \tau_j \text{ equal zero}$$

the following test statistic due to Cochran may be used:

(28.2)
$$X_C^2 = SSTR \div \frac{SSTR + SSE}{n(r-1)}$$

which reduces to:

(28.2a)
$$X_C^2 = \frac{(r-1)\left(r\sum_j Y_{.j}^2 - Y_{..}^2\right)}{rY_{..} - \sum_i Y_{i.}^2}$$

where the usual notation is employed.

The number of 1's in each block may differ because of block-to-block differences. Given the number of 1's in each block, if all treatments have the same effect, all permutations of 0's and 1's within a block are equally likely. It can then be shown that when H_0 holds, X_C^2 is distributed approximately as χ^2 with $r - 1$ degrees of freedom provided the number of blocks is not very small. Large values of X_C^2 lead to conclusion H_a.

Example. Table 28.1 contains data for an experiment in which 15 teams were grouped into $n = 5$ blocks of three teams each, according to a criterion on the creativity of the team. Within each block the teams were assigned at random to one of $r = 3$ instruction methods, and upon completion of training each team was given the same complex task to perform. Success in the task was coded 1, failure 0.

To test whether the instruction methods have differential effects on successful performance, we use test statistic (28.2a) and obtain:

TABLE 28.1 Binary response experiment in randomized block design

Block	*Instruction Method (j)*			
i	1	2	3	*Total*
1 (high creativity)	1	1	1	3
2	1	0	1	2
3	1	0	0	1
4	0	1	0	1
5 (low creativity)	1	0	1	2
Total	4	2	3	9

$\Sigma Y_{.j}^2 = 4^2 + 2^2 + 3^2 = 29$

$\Sigma Y_{i.}^2 = 3^2 + 2^2 + 1^2 + 1^2 + 2^2 = 19$

$$X_C^2 = \frac{2[3(29) - (9)^2]}{3(9) - 19} = 1.5$$

Suppose the level of significance is to be controlled at $\alpha = .05$. We require then $\chi^2(.95; 2) = 5.99$. Since $X_C^2 = 1.5 \leq 5.99$, we conclude that the instruction methods do not differ in effectiveness.

Multiple pairwise testing procedure

When the Cochran test leads to the conclusion that treatment effects are present, pairwise tests based on the treatment means $\bar{Y}_{.j}$ can be set up in a similar fashion to that for the Kruskal-Wallis test described in Section 19.2, provided the number of blocks is not too small. Testing limits for all $g = r(r - 1)/2$ pairwise tests with family significance level α are set up as follows:

$$\bar{Y}_{.j} - \bar{Y}_{.j'} \pm B\left[\left(\frac{rY_{..} - \Sigma Y_{i.}^2}{nr(r - 1)}\right)\left(\frac{2}{n}\right)\right]^{1/2} \tag{28.3}$$

where:

$$B = z(1 - \alpha/2g)$$

$$g = \frac{r(r - 1)}{2}$$

If the testing limits include zero, we conclude that the corresponding treatment means $\mu_{.j}$ and $\mu_{.j'}$ do not differ. If the testing limits do not include zero, we conclude that the two corresponding treatment means differ.

28.2 MISSING OBSERVATIONS

There are occasions when one or several observations in a randomized complete block design are "missing"—a subject may have been sick, a record may have been mislaid, a treatment may have been applied incorrectly in one instance. Such missing observations destroy the symmetry (orthogonality) of the complete block design and make the usual ANOVA calculations inappropriate. However, the regression approach to the analysis of randomized block designs, discussed in Section 27.8, is still appropriate when there are missing observations. The reason is that the no-interaction randomized block design model (27.1) enables us, in effect, to estimate the mean response for a missing cell. We explained earlier how this is done for a two-factor no-interaction model (Section 22.4), and the same reasoning applies here.

Since no new principles are involved, we turn to an example to illustrate the use of the regression approach when observations are missing in a randomized block design experiment.

Example

Table 28.2 contains the data for a simple randomized block design experiment with $r = 3$ treatments and $n = 3$ blocks, where observation Y_{11} is missing. We set up the regression model equivalent of the randomized block design model (27.1) as follows:

$$Y_{ij} = \mu_{..} + \underbrace{\rho_1 X_{ij1} + \rho_2 X_{ij2}}_{\text{Block effect}} + \underbrace{\tau_1 X_{ij3} + \tau_2 X_{ij4}}_{\text{Treatment effect}} + \varepsilon_{ij} \qquad \text{Full Model} \tag{28.4}$$

where:

$$X_{ij1} = \begin{array}{l} 1 \text{ if experimental unit from block 1} \\ -1 \text{ if experimental unit from block 3} \\ 0 \text{ otherwise} \end{array}$$

$$X_{ij2} = \begin{array}{l} 1 \text{ if experimental unit from block 2} \\ -1 \text{ if experimental unit from block 3} \\ 0 \text{ otherwise} \end{array}$$

$$X_{ij3} = \begin{array}{l} 1 \text{ if experimental unit received treatment 1} \\ -1 \text{ if experimental unit received treatment 3} \\ 0 \text{ otherwise} \end{array}$$

$$X_{ij4} = \begin{array}{l} 1 \text{ if experimental unit received treatment 2} \\ -1 \text{ if experimental unit received treatment 3} \\ 0 \text{ otherwise} \end{array}$$

TABLE 28.2 Example of missing observation in randomized block design ($r = 3, n = 3$)

Block	Treatment (*j*)		
i	1	2	3
1	Missing	10	9
2	11	10	7
3	6	4	3

Table 28.3 exhibits the **Y**, **X**, and **β** matrices for the full model (28.4) for this example where observation Y_{11} is missing.

The analysis of variance for testing treatment effects and block effects is carried out in the usual manner by first fitting the full model (28.4) and then fitting each of the following reduced models:

Test for block effects

$$Y_{ij} = \mu_{..} + \tau_1 X_{ij3} + \tau_2 X_{ij4} + \varepsilon_{ij} \qquad \text{Reduced model} \tag{28.5}$$

TABLE 28.3 **Y**, **X**, and **β** matrices for missing observation example of Table 28.2

$$\mathbf{Y} = \begin{bmatrix} Y_{12} \\ Y_{13} \\ Y_{21} \\ Y_{22} \\ Y_{23} \\ Y_{31} \\ Y_{32} \\ Y_{33} \end{bmatrix} = \begin{bmatrix} 10 \\ 9 \\ 11 \\ 10 \\ 7 \\ 6 \\ 4 \\ 3 \end{bmatrix} \qquad \mathbf{X} = \begin{bmatrix} & X_1 & X_2 & X_3 & X_4 \\ 1 & 1 & 0 & 0 & 1 \\ 1 & 1 & 0 & -1 & -1 \\ 1 & 0 & 1 & 1 & 0 \\ 1 & 0 & 1 & 0 & 1 \\ 1 & 0 & 1 & -1 & -1 \\ 1 & -1 & -1 & 1 & 0 \\ 1 & -1 & -1 & 0 & 1 \\ 1 & -1 & -1 & -1 & -1 \end{bmatrix} \qquad \boldsymbol{\beta} = \begin{bmatrix} \mu_{..} \\ \rho_1 \\ \rho_2 \\ \tau_1 \\ \tau_2 \end{bmatrix}$$

Test for treatment effects

(28.6) $$Y_{ij} = \mu_{..} + \rho_1 X_{ij1} + \rho_2 X_{ij2} + \varepsilon_{ij} \qquad \text{Reduced model}$$

The extra sums of squares $SSR(X_1, X_2|X_3, X_4)$ for blocks and $SSR(X_3, X_4|X_1, X_2)$ for treatments are then calculated as always. Table 28.4a presents these extra sums of squares for our example obtained from computer runs, as well as the error sum of squares for the full model. No total sum of squares is shown because of lack of orthogonality as a result of the missing observation.

The test for treatment effects is conducted as usual. From Table 28.4a we find:

$$F^* = \frac{MSTR}{MSE} = \frac{6.25}{.44} = 14.2$$

For $\alpha = .05$, we need $F(.95; 2, 3) = 9.55$. Since $F^* = 14.2 > 9.55$, we conclude that differential treatment effects are present. The test for block effects can be carried out along similar lines when it is of interest.

No new problems are encountered with the regression approach in analyzing fixed treatment effects when there are missing observations. To estimate in our example the pairwise comparison $D = \mu_{.1} - \mu_{.3} = \tau_1 - \tau_3$, for instance, we utilize the fact that $\tau_3 = -\tau_1 - \tau_2$ so that we have:

(28.7) $$D = \mu_{.1} - \mu_{.3} = \tau_1 - \tau_3 = \tau_1 - (-\tau_1 - \tau_2) = 2\tau_1 + \tau_2$$

An unbiased estimator of (28.7) is:

(28.8) $$\hat{D} = 2\hat{\tau}_1 + \hat{\tau}_2$$

whose estimated variance is, using (1.25b):

(28.9) $$s^2(\hat{D}) = 4s^2(\hat{\tau}_1) + s^2(\hat{\tau}_2) + 4s(\hat{\tau}_1, \hat{\tau}_2)$$

Table 28.4b contains the estimated regression coefficients for the full model, and Table 28.4c contains the estimated variance-covariance matrix for the regression coefficients. We therefore obtain the following estimates:

TABLE 28.4 ANOVA table and other regression output for missing data example of Table 28.2

(a) ANOVA Table

Source of Variation	*SS*	*df*	*MS*
Blocks	53.83	2	26.92
Treatments	12.50	2	6.25
Error	1.33	3	.44

(b) Estimated Regression Coefficients for Full Model (28.4)

Regression Coefficient	*Estimated Regression Coefficient*
$\mu_{..}$	$\hat{\mu}_{..} = 8.000$
ρ_1	$\hat{\rho}_1 = 2.333$
ρ_2	$\hat{\rho}_2 = 1.333$
τ_1	$\hat{\tau}_1 = 1.667$
τ_2	$\hat{\tau}_2 = 0.0$

(c) Estimated Variance-Covariance Matrix of Regression Coefficients

$$\begin{array}{c} \\ \hat{\mu}_{..} \\ \hat{\rho}_1 \\ \hat{\rho}_2 \\ \hat{\tau}_1 \\ \hat{\tau}_2 \end{array} \begin{array}{c} \begin{array}{ccccc} \hat{\mu}_{..} & \hat{\rho}_1 & \hat{\rho}_2 & \hat{\tau}_1 & \hat{\tau}_2 \end{array} \\ \left[\begin{array}{rrrrr} .06173 & & & & \\ .02469 & .14815 & & & \\ -.01235 & -.07407 & .11111 & & \\ .02469 & .04938 & -.02469 & .14815 & \\ -.01235 & -.02469 & .01235 & -.07407 & .11111 \end{array}\right] \end{array}$$

$$\hat{D} = 2(1.667) + 0.0 = 3.334$$

$$s^2(\hat{D}) = 4(.14815) + .11111 + 4(-.07407) = .4074$$

so that the estimated standard deviation is $s(\hat{D}) = .638$. A 95 percent confidence interval for D requires $t(.975; 3) = 3.182$, yielding the confidence interval:

$$1.3 = 3.334 - 3.182(.638) \leq \mu_{.1} - \mu_{.3} \leq 3.334 + 3.182(.638) = 5.4$$

Note

An alternative hand calculation procedure due to Yates is sometimes used when one or two observations are missing. Pseudo-observations for the missing values are obtained, and then the usual ANOVA calculations are carried out as if all observations were at hand. Finally, adjustments to these ANOVA calculations are made. See Reference 28.1 for complete details of this method. With the widespread availability of computer regression packages, the need for the Yates hand calculation procedure has greatly diminished.

28.3 RANDOM BLOCK EFFECTS

When blocks can be considered a random sample from a population (e.g., a population of observers, batches, machines), either an additive or a nonadditive (i.e., interaction) model can be employed.

Additive model

The additive model for random block effects and fixed treatment effects is analogous to fixed effects model (27.1):

$$Y_{ij} = \mu_{..} + \rho_i + \tau_j + \varepsilon_{ij} \tag{28.10}$$

where:

$\mu_{..}$ is a constant
ρ_i are independent $N(0, \sigma_\rho^2)$
τ_j are constants subject to the restriction $\Sigma\tau_j = 0$
ε_{ij} are independent $N(0, \sigma^2)$, and independent of the ρ_i
$i = 1, \ldots, n; j = 1, \ldots, r$

Table 28.5 contains the analysis of variance for this model. The sums of squares are the same as in (27.5) for the fixed effects model. Table 28.5 also contains the expected mean squares for model (28.10). The statistic for testing for treatment effects is $F^* = MSTR/MSE$, as may be seen from the $E(MS)$ column in Table 28.5. Thus, the test statistic is the same whether block effects are fixed or random. Confidence intervals for treatment contrasts present no new problems.

Interaction model

When block effects are random, one need not use the restrictive model (28.10) which assumes that there are no block–treatment interactions. Instead, a

TABLE 28.5 ANOVA for randomized complete block design—block effects random, treatment effects fixed

				E(MS)	
Source of Variation	*SS*	*df*	*MS*	*Additive Model* (28.10)	*Interaction Model* (28.11)
Blocks	*SSBL*	$n - 1$	*MSBL*	$\sigma^2 + r\sigma_\rho^2$	$\sigma^2 + r\sigma_\rho^2$
Treatments	*SSTR*	$r - 1$	*MSTR*	$\sigma^2 + \frac{n}{r-1}\Sigma\tau_j^2$	$\sigma^2 + \sigma_{\rho\tau}^2 + \frac{n}{r-1}\Sigma\tau_j^2$
Error	*SSE*	$(n-1)(r-1)$	*MSE*	σ^2	$\sigma^2 + \sigma_{\rho\tau}^2$
Total	*SSTO*	$nr - 1$			

model allowing for interactions between blocks and treatments may be employed:

$$Y_{ij} = \mu_{..} + \rho_i + \tau_j + (\rho\tau)_{ij} + \varepsilon_{ij} \tag{28.11}$$

where the terms have the same meaning as in (28.10) with the following additions:

$(\rho\tau)_{ij}$ are $N\left(0, \frac{r-1}{r}\sigma^2_{\rho\tau}\right)$, subject to the restriction $\sum_j (\rho\tau)_{ij} = 0$ for all i

$(\rho\tau)_{ij}$ are independent of the ε_{ij} and of the ρ_i

The sums of squares and degrees of freedom are the same as for the no-interaction model. The difference in the two models shows its effect in the expected mean squares. Table 28.5 also presents the expected mean squares for the interaction model (28.11). Still, the test statistic for examining whether treatment effects are present is $F^* = MSTR/MSE$. Thus, from the point of view of testing for treatment effects or estimating treatment contrasts, there is no difference when block effects are random whether one employs the additive or nonadditive model.

Comments

1. Table 28.5 indicates that the principal difference between the two models is that no exact test for block effects is possible with the interaction model, whereas an exact test can be obtained with the additive model. This distinction is unimportant whenever the blocks are not of intrinsic interest but serve merely to reduce experimental error variability.

2. As we noted in Chapter 27, the error sum of squares *SSE* in Table 28.5 is actually an interaction sum of squares between blocks and treatments. When the block effects are random, *MSE* estimates σ^2 for the additive model (28.10). For the nonadditive model (28.11), however, *MSE* estimates the sum of the error term variance σ^2 and the interaction variance $\sigma^2_{\rho\tau}$. Separate estimation of these two components is not possible for this latter model, and the two components are said to be confounded.

28.4 REPEATED MEASURES DESIGNS

We mentioned earlier that at times a subject (or store, plant, city, etc.) is given each of the treatments under study. In that case, the subject serves as a block, and the experimental units within a block must be viewed as the different instances when a treatment can be applied to the subject. The treatments are then assigned at random to these instances, that is, the order of the treatments is randomized for each subject independently. Social scientists often call such a randomized complete block design a *repeated measures design*. Table 28.6 contains the layout for a repeated measures design with five subjects and four treatments after independent randomizations of the treatment orders.

TABLE 28.6 Layout for repeated measures design ($n = 5$, $r = 4$)

	Treatment Order			
	1	2	3	4
Subject 1	T_4	T_3	T_2	T_1
2	T_3	T_4	T_1	T_2
3	T_4	T_3	T_1	T_2
4	T_2	T_1	T_4	T_3
5	T_1	T_2	T_4	T_3

Advantages and disadvantages

A principal advantage of this type of design is that all sources of variability between subjects are excluded from the experimental errors; only variation within subjects enters the experimental errors. Thus, one may view the subjects as serving as their own controls. Another advantage of a repeated measures design is that it economizes on subjects. This is particularly important when only a few subjects (stores, plants, etc.) can be utilized for the experiment. Also, when interest is in the effects of a treatment over time, as when one is concerned with the shape of the learning curve for a new process operation, it is usually desirable to observe the same subject at different points in time rather than observing different subjects at each specified point in time.

Repeated measures designs have a serious potential disadvantage, however, namely, that there may be several types of interference. One type of interference is connected with the position in the treatment order. For instance, in evaluating five different advertisements, subjects may tend to give higher (or lower) ratings for advertisements shown toward the end of the sequence than at the beginning. Another type of interference is connected with the preceding treatment or treatments. For instance, in evaluating five different soup recipes, a bland recipe may get a higher (or lower) rating when preceded by a highly spiced recipe than when preceded by a blander recipe. This type of interference is called a *carry-over effect*.

Various steps can be taken to minimize the danger of interference. Randomization of the treatment orders for each subject independently will make it more reasonable to analyze the data as if the error terms are independent. Allowing sufficient time between treatments is often an effective means of reducing carry-over effects. It may be desirable at times to balance the order of treatment presentations and sometimes even the number of times each treatment is preceded by any other treatment. Latin square designs and cross-over designs (discussed in Chapter 31) are helpful to this end.

Models

In repeated measures designs, the subjects are usually considered to be a random sample from a population, and either model (28.10) or (28.11) is employed when the treatment effects are fixed. Should the subjects not be viewed as a sample but of intrinsic interest in themselves, model (27.1) would be used when the treatment effects are fixed. The analysis of repeated measures designs follows the lines presented earlier since such designs are simply randomized complete block designs. Additional material on repeated measures designs will be presented in Chapters 30 and 31.

Note

Alternative models that restrict the nature of possible interactions between subjects and treatments are available for special cases. Reference 28.2 discusses these models.

Ranked data

In some repeated measures designs, subjects are asked to rank the treatments. For instance, a subject may be asked to rank four different colored papers according to the vividness of the red color. Again, a subject may be asked to rank five coffee sweeteners according to taste preference.

Friedman test. When the observations in a randomized block design are ranks, the Friedman test may be used to study whether or not differential treatment effects exist. The test statistic is (using the regular notation):

$$X_F^2 = SSTR \div \frac{SSTR + SSE}{n(r-1)} \tag{28.12}$$

which can be reduced to (when no ties are present):

$$X_F^2 = \left[\frac{12}{nr(r+1)} \sum_j R_{.j}^2\right] - 3n(r+1) \tag{28.12a}$$

where $R_{.j}$ is the sum of the ranks for the jth treatment.

If there are no treatment differences, all ranking permutations for a subject are assumed to be equally likely and the statistic X_F^2 will be distributed approximately as χ^2 with $r - 1$ degrees of freedom, provided the number of subjects and/or treatments is not too small. Large values of the test statistic lead to the conclusion that the treatments have unequal effects. If the number of subjects and/or treatments is small, tables such as those in Reference 28.3 may be used for an exact test.

Example. Table 28.7 contains an illustration of ranked data in a repeated measures design. Six subjects were asked to rank five coffee sweeteners according to their taste preferences, with rank 1 assigned to the most preferred sweetener. We have:

$$X_F^2 = \left[\frac{12}{6(5)(6)}(1{,}836)\right] - 3(6)(6) = 14.4$$

For a level of significance of $\alpha = .05$, we need $\chi^2(.95; 4) = 9.49$. Since $X_F^2 = 14.4 > 9.49$, we conclude that the five sweeteners are not equally liked. Analysis should now be undertaken as to which sweeteners are the preferred ones.

TABLE 28.7 Ranked data for coffee sweeteners in a repeated measures design

Subject	*Sweetener (j)*				
i	A	B	C	D	E
1	5	1	2	4	3
2	4	2	1	5	3
3	3	2	1	4	5
4	5	2	3	4	1
5	4	1	2	3	5
6	4	1	3	5	2
$R_{.j}$	25	9	12	25	19
$\bar{R}_{.j}$	4.17	1.50	2.00	4.17	3.17

$\Sigma R_{.j}^2 = (25)^2 + (9)^2 + (12)^2 + (25)^2 + (19)^2 = 1{,}836$

Comments

1. X_F^2 is related to Kendall's coefficient of concordance W in the following way:

(28.13)
$$W = \frac{X_F^2}{n(r-1)} = \frac{SSTR}{SSTR + SSE}$$

The coefficient of concordance W is a measure of the agreement of the rankings of the n subjects. It equals 1 if there is perfect agreement, and equals 0 if there is no agreement, that is, if all treatments receive the same mean ranking. For our coffee sweeteners example in Table 28.7, W is:

$$W = \frac{14.4}{6(4)} = .60$$

indicating a fair amount of agreement between the subjects.

2. The test statistic X_F^2 may also be used for randomized complete block designs when different experimental units are assigned the different treatments. The observations within a block would simply be ranked, and formula (28.12) then used. This would be desirable if the observations are far from normally distributed and transformations are of little help.

Multiple pairwise testing procedure. Just like in the case of the Kruskal-Wallis test for single-factor studies (Section 19.2), we can use a large-sample testing analogue of the Bonferroni pairwise comparison procedure to obtain information about the comparative magnitudes of the treatment means for repeated measures designs when the Friedman test indicates that the treatment means differ. Testing limits for all $g = r(r-1)/2$ pairwise comparisons using the mean ranks $\bar{R}_{.j}$ are set up as follows for family level of significance α:

$$\bar{R}_{.j} - \bar{R}_{.j'} \pm B\left[\frac{r(r+1)}{6n}\right]^{1/2} \tag{28.14}$$

where:

$$B = z(1 - \alpha/2g)$$

$$g = \frac{r(r-1)}{2}$$

If the testing limits include zero, we conclude that the corresponding treatment means $\mu_{.j}$ and $\mu_{.j'}$ do not differ. If the testing limits do not include zero, we conclude that the two corresponding means differ. We can then set up groups of treatments whose means do not differ according to this simultaneous testing procedure.

Example. For the coffee sweeteners example, we wish to make all pairwise tests with family level of significance $\alpha = .20$. The mean ranks $\bar{R}_{.j}$ are presented in Table 28.7. For $r = 5$, we have $g = 5(4)/2 = 10$. Hence, we obtain for $n = 6$ and $1 - \alpha = .80$ that $B = z[1 - .20/2(10)] = z(.99) = 2.326$. Thus, the right term in (28.14) is:

$$B\left[\frac{r(r+1)}{6n}\right]^{1/2} = 2.326\left[\frac{5(6)}{6(6)}\right]^{1/2} = 2.12$$

We note from Table 28.7 that the pairs of mean ranks whose difference does not exceed 2.12 are (B, C), (B, E), (C, E), (A, E), (D, E), and (A, D). Hence, we can set up two groups, within which the treatment means do not differ:

Group 1		*Group 2*	
Sweetener B	$\bar{R}_{.2} = 1.50$	Sweetener E	$\bar{R}_{.5} = 3.17$
Sweetener C	$\bar{R}_{.3} = 2.00$	Sweetener A	$\bar{R}_{.1} = 4.17$
Sweetener E	$\bar{R}_{.5} = 3.17$	Sweetener D	$\bar{R}_{.4} = 4.17$

Thus, we can conclude with family level of significance of .20 that sweeteners A and D are preferred to sweeteners B and C, and that it is not clear whether sweetener E belongs in the preferred group or in the other group.

28.5 GENERALIZED RANDOMIZED BLOCK DESIGNS

When block effects are fixed, use of a no-interaction model in the presence of interactions between blocks and treatments has the effect of reducing the power of the test and increasing the width of interval estimates of treatment effects, thus making the experiment less sensitive. In addition, there are occasions when one is interested in the nature of the interactions between blocks and treatments and would like to obtain estimates of these. It is possible to use a design that permits an interaction term in the model even when the block effects are fixed, and that allows the nature of the interaction effects to be investigated. This design is called a *generalized randomized block design*. It is the same as a complete randomized block design except that d experimental units are assigned to each treatment within a block. This design increases the size of a block from r units for a randomized complete block design to dr units. This increase in block size often has the effect of increasing experimental error variability when the number of experimental units is fixed. In the social sciences, however, increasing the size of the block moderately may cause little loss in efficiency. For instance, having one block of ten persons aged 20–29 instead of two blocks of five persons of ages 20–24 and 25–29, respectively, will for many types of experiments involve little loss of efficiency.

As we shall demonstrate by an example, a generalized randomized block design is analyzed like an ordinary multifactor study where blocks are one factor. Hence, no new problems are encountered with a generalized randomized block design in testing for treatment effects or in estimating them.

Example

Table 28.8 contains the data for a two-factor experiment in which the effects of motivation (factor A: high level, low level) and distraction (factor B: high level, low level) on the time required to complete a task were studied, using eight men and eight women. Two men were assigned at random to each treatment, and independently two women were assigned at random to each treatment. Here sex is the blocking variable. Each block contains eight persons, with two randomly

TABLE 28.8 Two-factor study in generalized randomized block design with $d = 2$ (observations are times for task completion)

	Block (sex)	
	Male	*Female*
High motivation:		
High distraction	12	3
	8	9
Low distraction	7	5
	5	9
Low motivation:		
High distraction	14	11
	16	9
Low distraction	15	10
	13	14

assigned to each treatment within the block. The layout in Table 28.8 corresponds to the layout in Table 24.3a for a three-factor study; to stress the correspondence, we have placed the blocks in columns rather than in rows as usual. Since blocks, motivation levels, and distraction levels are considered to be fixed effects, we utilize the fixed effects three-factor model (24.17), with notation modified to fit the present context:

$$(28.15) \qquad Y_{ijkm} = \mu_{...} + \rho_i + \alpha_j + \beta_k + (\rho\alpha)_{ij} + (\rho\beta)_{ik} + (\alpha\beta)_{jk} + (\rho\alpha\beta)_{ijk} + \varepsilon_{ijkm}$$

where:

$\mu_{...}$ is a constant
ρ_i, α_j, and β_k are constants subject to the restrictions
$\Sigma\rho_i = \Sigma\alpha_j = \Sigma\beta_k = 0$
$(\rho\alpha)_{ij}$, $(\rho\beta)_{ik}$, $(\alpha\beta)_{jk}$, $(\rho\alpha\beta)_{ijk}$ are constants subject to the restrictions that the sums over any subscript are zero
ε_{ijkm} are independent $N(0, \sigma^2)$
$i = 1, \ldots, n;\ j = 1, \ldots, a;\ k = 1, \ldots, b;\ m = 1, \ldots, d$

The analysis of variance for model (28.15) is the ordinary three-factor ANOVA of Table 24.4, with slight modifications in notation. A computer package was employed to obtain the analysis of variance for the data in Table 28.8, and the results are shown in Table 28.9. We know from Table 24.5 that all test statistics use *MSE* in the denominator. These F^* statistics are shown in Table 28.9. For $\alpha = .01$, we require $F(.99; 1, 8) = 11.3$ for each of the tests. It is evident from the results in Table 28.9 (see also the *P*-values given there) that blocks do not interact with the factorial treatments and that among the two factors, only motivation affects the time required to complete the task. At this point, further analysis of the motivation effects would be indicated.

28.6 USE OF MORE THAN ONE BLOCKING VARIABLE

Sometimes, substantial reduction in the experimental error variability can only be obtained by utilizing more than one variable for determining blocks. For instance, both age and sex might be needed for designating blocks:

Block	*Characteristics of Experimental Units*
1	Male, aged 20–29
2	Female, aged 20–29
3	Male, aged 30–39
etc.	etc.

As another example, both observer and day of treatment application may be helpful as blocking variables:

TABLE 28.9 Analysis of variance for task completion example of Table 28.8 ($n = 2$, $a = 2$, $b = 2$, $d = 2$)

Source of Variation	*SS*	*df*	*MS*	F^*	*P-value*
Blocks (sex)	$SSBL = 25.00$	$n - 1 = 1$	$MSBL = 25.00$	4.00	.08
Factor *A* (motivation)	$SSA = 121.00$	$a - 1 = 1$	$MSA = 121.00$	19.36	.002
Factor *B* (distraction)	$SSB = 1.00$	$b - 1 = 1$	$MSB = 1.00$	.16	.70
BL.A interactions	$SSBL.A = 4.00$	$(n - 1)(a - 1) = 1$	$MSBL.A = 4.00$	.64	.45
BL.B interactions	$SSBL.B = 16.00$	$(n - 1)(b - 1) = 1$	$MSBL.B = 16.00$	2.56	.15
AB interactions	$SSAB = 4.00$	$(a - 1)(b - 1) = 1$	$MSAB = 4.00$	.64	.45
BL.AB interactions	$SSBL.AB = 1.00$	$(n - 1)(a - 1)(b - 1) = 1$	$MSBL.AB = 1.00$	.16	.70
Error	$SSE = 50.00$	$(d - 1)nab = 8$	$MSE = 6.25$		
Total	$SSTO = 222.00$	$dnab - 1 = 15$			
		$F(.99; 1, 8) = 11.3$			

Block	*Characteristics of Experimental Units*
1	Observer 1, day 1
2	Observer 2, day 1
3	Observer 1, day 2
etc.	etc.

Unless one wishes to study the separate effects of each of the blocking variables, no new problems arise when the blocks are defined by two or more variables. The n blocks are simply treated as ordinary blocks, and the usual block sum of squares is calculated.

If the effect of each of the blocking variables is to be isolated and the blocks are defined in a complete factorial fashion (e.g., nine blocks are used when three observers and three days are employed for blocking), the analysis simply treats each of the blocking variables as a factor and utilizes the methods developed in Chapter 24.

A problem that sometimes arises when two or more blocking variables are to be used is the large number of blocks called for. Suppose an experiment is to be conducted where the experimental units are stores. In order to reduce the experimental error variability to a reasonable level, it would be desirable to group the stores into six sales volume classes and also into six location classes (suburban shopping center, suburban other, etc.). Thirty-six blocks result from combining these two blocking variables. If six treatments are to be studied, 216 stores would be required for the experiment. Often, use of this many stores would be much too costly. Latin square designs, to be discussed in Chapter 31, permit in this type of case the use of a much smaller number of replications while still preserving the full benefits of error variance reduction by using both blocking variables in six classes each.

PROBLEMS

28.1. Antinausea therapy. Cancer patients undergoing chemotherapy commonly suffer from episodes of nausea that are uncontrolled by conventional antinausea drugs. To evaluate the comparative effectiveness of two experimental antinausea drugs, 12 cancer patients with previous histories of severe episodes of nausea were successively assigned in random order to each of three treatments in a double-blind study while undergoing chemotherapy. Treatment 1 was a conventional antinausea drug while treatments 2 and 3 were the experimental drugs. The effectiveness of each drug was evaluated from reports of the patients, coded 1 for improvement and 0 for no improvement. The data follow.

Patient i	Treatment (j) 1	2	3	Patient i	Treatment (j) 1	2	3
1	0	1	1	7	1	0	1
2	0	1	1	8	0	1	1
3	1	1	1	9	0	1	1
4	0	0	1	10	0	0	1
5	0	1	0	11	1	0	1
6	0	0	0	12	0	0	1

a. Employ the Cochran test to determine whether or not the three drugs differ in mean effectiveness; use $\alpha = .05$. State the alternatives, decision rule, and conclusion. What is the P-value of the test?

b. Conduct multiple pairwise tests to group the three drugs according to mean effectiveness; use a family level of significance of $\alpha = .10$. Describe your findings.

28.2. **Product coupons.** An advertising agency designed an experiment to evaluate the effectiveness of four successive monthly mailings of coupons for a household product. Thirteen households participated. The data to follow show whether or not each household used the monthly coupon to purchase the product during the month (1: used coupon; 0: did not use coupon) for the four successive months.

Subject i	Month (j) 1	2	3	4	Subject i	Month (j) 1	2	3	4
1	0	1	1	1	8	0	0	0	0
2	0	1	0	1	9	1	1	1	1
3	1	1	1	1	10	1	1	1	1
4	0	0	0	0	11	1	1	1	1
5	0	0	1	1	12	0	0	1	1
6	1	0	0	1	13	0	1	1	1
7	0	0	1	1					

a. Employ the Cochran test to determine whether or not mean coupon usage differs in the four months; use $\alpha = .05$. State the alternatives, decision rule, and conclusion. What is the P-value of the test?

b. Conduct multiple pairwise tests to group the months according to mean coupon usage; employ a family level of significance of $\alpha = .10$. Is there any indication that coupon usage increases over time? Discuss.

28.3. Refer to **Auditor training** Problem 27.5. Assume that observation $Y_{23} = 89$ is missing because the auditor became ill and dropped out from the study.

a. State the ANOVA model for this case. Also state the equivalent regression model; use 1, −1, 0 indicator variables.

b. State the reduced regression model for testing for differences in the mean proficiency scores for the three training methods.

c. Test whether or not the mean proficiency scores for the three training methods differ by fitting the full and reduced models; use $\alpha = .05$. State the alternatives, decision rule, and conclusion.

d. Compare the mean proficiency scores for training methods 2 and 3 by means of the regression approach; use a 95 percent confidence interval.

28.4. Refer to **Fat in diets** Problem 27.8. Assume that observations $Y_{13} = .15$ and $Y_{51} = 1.62$ are missing because the subjects did not stay on the prescribed diet.

a. State the ANOVA model for this case. Also state the equivalent regression model; use $1, -1, 0$ indicator variables.

b. State the reduced regression model for testing for differences in the mean reductions in lipid level for the three diets.

c. Test whether or not the mean reductions in lipid level differ for the three diets by fitting the full and reduced models; use $\alpha = .025$. State the alternatives, decision rule, and conclusion.

d. Compare the mean reductions in lipid level for diets 1 and 3 by means of the regression approach; use a 98 percent confidence interval.

28.5. Refer to **Calculator efficiency** Problem 27.6. Suppose that the engineers in the study were randomly selected from the membership list of a large professional society. Assume that additive effects ANOVA model (28.10) is applicable, except that the factorial structure of the fixed treatment effects needs to be recognized.

a. State the ANOVA model for this case.

b. What is the appropriate test statistic for testing whether or not the two factors interact? What is the appropriate test statistic for testing for factor A main effects? [*Hint:* Consider the test for treatment effects in model (28.10).]

28.6. **Road paint wear.** A state highway department studied the wear characteristics of five different paints at eight locations in the state. The standard, currently used paint (paint 1) and four experimental paints (paints 2, 3, 4, 5) were included in the study. The eight locations were randomly selected, thus reflecting variations in traffic densities throughout the state. At each location, a random ordering of the paints to the chosen road surface was employed. After a suitable period of exposure to weather and traffic, a combined measure of wear, considering both durability and visibility, was obtained. The data on wear follow (the higher the score, the better the wearing characteristics).

Location	Paint (*j*)				
i	1	2	3	4	5
1	11	13	10	18	15
2	20	28	15	30	18
3	8	10	8	16	12
4	30	35	27	41	28

Location	Paint (*j*)				
i	1	2	3	4	5
5	14	16	13	22	16
6	25	27	26	33	25
7	43	46	41	55	42
8	13	14	12	20	13

Assume that additive effects ANOVA model (28.10) is appropriate.

a. Obtain the analysis of variance table.

b. Test whether or not the mean wear differs for the five paints; use significance level $\alpha = .05$. State the alternatives, decision rule, and conclusion.

c. Compare the mean wear of each experimental paint against that of the standard paint; use the most efficient multiple comparison procedure with a 90 percent family confidence coefficient. Summarize your findings.

d. Paints 1, 3, and 5 are white colored whereas paints 2 and 4 are yellow. Estimate the difference in the mean wear for the two groups of paints with a 95 percent confidence interval. Interpret your findings.

28.7. Refer to **Road paint wear** Problem 28.6. Obtain the residuals and plot them against the fitted values. Also prepare a normal probability plot of the residuals. Summarize your findings about the appropriateness of ANOVA model (28.10).

28.8. **Blood pressure.** The relationship between the dose of a drug that increases blood pressure and the actual amount of increase in mean diastolic blood pressure was investigated in a laboratory experiment. Twelve rabbits received in random order six different dose levels of the drug, with a suitable interval between each drug administration. The increase in blood pressure was used as the dependent variable. The data on blood pressure increase follow.

Rabbit	*Dose (j)*					
i	.1	.3	.5	1.0	1.5	3.0
1	15	22	31	35	50	57
2	15	21	30	37	39	50
3	13	18	23	32	40	49
4	15	14	20	23	28	31
5	18	24	28	39	46	52
6	10	19	28	41	46	61

Rabbit	*Dose (j)*					
i	.1	.3	.5	1.0	1.5	3.0
7	11	19	27	37	40	48
8	13	12	18	30	37	47
9	10	9	15	27	33	46
10	10	16	19	23	28	33
11	7	8	15	17	23	31
12	15	16	19	22	30	42

Assume that additive effects ANOVA model (28.10) is appropriate.

a. Obtain the analysis of variance table.

b. Test whether or not the mean increase in blood pressure differs for the various dose levels; use significance level $\alpha = .01$. State the alternatives, decision rule, and conclusion.

c. Analyze the effects of the six dose levels by comparing the means for successive dose levels using the Bonferroni procedure with a 95 percent family confidence coefficient. State your findings.

28.9. Refer to **Blood pressure** Problem 28.8.

a. Obtain the residuals and plot them against the fitted values. Also prepare a normal probability plot of the residuals. What do you conclude about the appropriateness of the model employed in Problem 28.8?

b. Based on your residual analysis in part (a), would ANOVA model (28.11) be more appropriate here? What practical differences exist in using models (28.10) and (28.11)?

28.10. Refer to **Blood pressure** Problem 28.8.

a. Develop a regression model in which the block effects are represented by 1, −1, 0 indicator variables and the dose effect is represented by linear, quadratic, and cubic terms in $x = X - \bar{X}$, where X is the dose level. For instance, the x value for the first dose level ($X = .1$) is $x = .1 - 1.07 = -.97$.

b. Fit the regression model to the data.

c. Obtain the residuals and plot them against the fitted values. Does the model utilized appear to provide a reasonable fit?

d. Test whether or not the cubic effect is required in the model; use $\alpha = .05$. State the alternatives, decision rule, and conclusion.

28.11. **Truth in advertising.** A consumer research organization showed five different advertisements to 10 subjects and asked each to rank them in order of truthfulness. A rank of 1 denotes the most truthful. The results were:

Subject i	Advertisement (j) A	B	C	D	E
1	3	1	2	5	4
2	4	2	1	3	5
3	4	2	3	1	5
4	3	1	2	5	4
5	4	1	2	5	3
6	4	2	1	3	5
7	4	1	2	3	5
8	5	1	3	2	4
9	4	2	3	1	5
10	5	1	2	3	4

a. Do the subjects perceive the five advertisements as having equal truthfulness? Conduct a formal test using a level of significance of $\alpha = .05$. State the alternatives, decision rule, and conclusion.

b. Use the multiple pairwise testing procedure (28.14) to group the five different advertisements according to mean perceived truthfulness; employ family significance level $\alpha = .10$. Summarize your findings.

c. Obtain the coefficient of concordance and interpret this measure.

28.12. Refer to **Road paint wear** Problem 28.6. A consultant is concerned about the validity of the model assumptions and suggests that the study should be analyzed by means of the Friedman test. Rank the data within each block and perform the Friedman test; use $\alpha = .05$. State the alternatives, decision rule, and conclusion. Comment on the consultant's concern here.

28.13. Refer to **Auditor training** Problem 27.5. It has been suggested that the nonparametric Friedman test should be used here. Rank the data within each block and perform the Friedman test; use $\alpha = .05$. State the alternatives, decision rule, and conclusion. Is your conclusion the same as that obtained in Problem 27.5c?

28.14. A social scientist, after learning about generalized randomized block designs, asked: "Why would anyone use a randomized complete block design that requires the assumption that block and treatment effects do not interact when this assumption can be avoided with a generalized randomized block design?" Comment.

28.15. Refer to the task completion example on page 953.

a. Using the data in Table 28.8, verify the analysis of variance in Table 28.9.

b. Estimate the difference in mean effects for the two motivation levels using a 99 percent confidence interval.

28.16. Refer to **Auditor training** Problem 27.5. The accounting firm repeated the experiment with another group of 30 auditors, but this time grouped them into five blocks of six each. In each block, each treatment was randomly assigned to two auditors. The results were:

Block i	Training Method (j) 1	2	3
1	74	84	91
	71	78	95
2	73	75	93
	69	83	98
3	75	81	89
	67	74	86
4	65	73	84
	70	78	87
5	64	71	81
	61	74	74

Assume that ANOVA model (28.15), modified for a single-factor study, is appropriate.

a. State the ANOVA model for this application.

b. Obtain the analysis of variance table.

c. Test whether or not the mean proficiency scores for the three training methods differ; use $\alpha = .01$. State the alternatives, decision rule, and conclusion.

d. Make all pairwise comparisons between the three training methods; use the Tukey procedure with a 95 percent family confidence coefficient. Summarize your findings.

e. Obtain the residuals and plot them against the fitted values. Also prepare a normal probability plot of the residuals. State your findings.

f. Test whether or not blocks interact with treatments; use $\alpha = .01$. State the alternatives, decision rule, and conclusion.

EXERCISES

28.17. Derive (28.2a) from (28.2).

28.18. Consider a randomized complete block design study where $n = 4$ and $r = 2$, ANOVA model (27.1) applies, and Y_{31} is missing. Use the matrix methods of Section 8.5 to obtain an estimator of μ_{31}. (*Hint:* Consider the development in Section 23.1.)

28.19. Derive (28.12a) from (28.12).

PROJECTS

28.20. Refer to the **Drug effect experiment** data set. Consider only Part I of the study and observation unit 1 for each drug dosage level; i.e., include only observations for which variable 2 equals 1 and variable 6 equals 1. Treat the 12 rats as blocks and ignore the classification of the rats into the three initial lever press rate groups. Assume that the blocks (rats) have random effects and the treatments (dosage levels) have fixed effects.

a. State the randomized block design model for this study. Do not include an interaction term between treatments and blocks in the model.

b. Obtain the analysis of variance table.

c. Test whether or not the drug dosage level affects the mean lever press rate; use $\alpha = .05$. State the alternatives, decision rule, and conclusion.

d. Analyze the effects of the four dosage levels by comparing the mean responses for each pair of successive dosage levels; use the Bonferroni procedure with a 90 percent family confidence coefficient. State your findings.

e. Obtain the residuals and plot them against the fitted values. Also prepare a normal probability plot of the residuals. What do you conclude about the appropriateness of the model employed?

f. Fit a regression model in which the block effects are represented by $1, -1, 0$ indicator variables and the dosage effect is represented by linear and quadratic terms in $x = X - \bar{X}$, where X is the dosage level. Assume that there are no interactions between blocks and treatments.

g. Obtain the residuals and plot them against the fitted values. Does the regression model appear to provide a good fit? Discuss.

h. Test whether or not the quadratic term can be dropped from the regression model; use $\alpha = .01$. State the alternatives, decision rule, and conclusion.

28.21. Refer to the **Drug effect experiment** data set. Consider only Part II of the study; i.e., include only observations for which variable 2 equals 2. Treat the 12 rats as blocks and ignore the classification of the rats into the three initial lever press rate groups. Assume that blocks (rats) have random effects and treatments (dosage levels) have fixed effects.

a. State the generalized randomized block design model for this study. Include interaction effects between blocks and treatments in the model.

b. Obtain the analysis of variance table.

c. Test whether or not the drug dosage level affects the mean lever press rate; use $\alpha = .01$. State the alternatives, decision rule, and conclusion.

d. Analyze the effects of the four dosage levels by comparing the mean responses for each pair of successive dosage levels; use the Bonferroni procedure with a 95 percent family confidence coefficient. State your findings.

e. Obtain the residuals and plot them against the fitted values. Also prepare a normal probability plot of the residuals. What do you conclude about the appropriateness of the model employed?

f. Fit a regression model in which the block effects are represented by $1, -1, 0$ indicator variables and the dosage effect is represented by linear and quadratic terms in $x = X - \bar{X}$, where X is the dosage level.

g. Test by means of a lack of fit test whether or not the regression model fits the data; use $\alpha = .01$. State the alternatives, decision rule, and conclusion.

28.22. Consider a repeated measures design study with $n = 3$ and $r = 3$, where each subject ranks all treatments (with no ties allowed).

a. Develop the exact sampling distribution of X_F^2 when H_0 holds. (*Hint:* All ranking permutations for a subject are equally likely under H_0 and all subjects are assumed to act independently.)

b. How does the 90th percentile of the exact sampling distribution obtained in part (a) compare with $\chi^2(.90; 2)$? What is the implication of this?

CITED REFERENCES

28.1 Cochran, William G., and Gertrude M. Cox. *Experimental Designs*. 2d ed. New York: John Wiley & Sons, 1957, pp. 110–12.

28.2 Winer, B. J. *Statistical Principles in Experimental Design*. 2d ed. New York: McGraw-Hill, 1971.

28.3 Owen, Donald B. *Handbook of Statistical Tables*. Reading, Mass.: Addison-Wesley Publishing, 1962.

29

Nested designs—I

In this chapter, we take up the basic elements of nested designs, including the use of subsampling. In the following chapter, more advanced elements of nested designs are discussed.

We begin this chapter by considering the general concept of nested designs and describe how these designs differ from crossed designs. We then take up in detail two-factor nested designs and their analysis. We conclude this chapter by considering subsampling designs.

29.1 DISTINCTION BETWEEN NESTED AND CROSSED FACTORS

In the ordinary factorial study considered so far, where every level of one factor appears with each level of every other factor, the factors are said to be crossed. A different situation occurs when factors are nested. The distinction between nested and crossed factors will now be illustrated by some examples involving two-factor studies.

Example 1

A large manufacturing company operates three regional training schools for mechanics, one in each of its operating districts. The schools have two instructors each who teach classes of about 15 mechanics in three-week sessions. The company was concerned about the effect of school (factor A) and instructor (factor B) on the learning achieved. To investigate these effects, classes in each district were formed in the usual way and then randomly assigned to one of the two instructors in the school. This was done for two sessions, and at the end of each session a suitable measure of learning for the class was obtained. The results are presented in Table 29.1.

TABLE 29.1 Sample data for nested two-factor study—training school example (class learning scores, coded)

Factor A (school) i	*Factor B (instructor)* j: 1	2	*Total*
Atlanta	25	14	
	29	11	
Total	$Y_{11.} = 54$	$Y_{12.} = 25$	$Y_{1..} = 79$
Chicago	11	22	
	6	18	
Total	$Y_{21.} = 17$	$Y_{22.} = 40$	$Y_{2..} = 57$
San Francisco	17	5	
	20	2	
Total	$Y_{31.} = 37$	$Y_{32.} = 7$	$Y_{3..} = 44$
		Total	$Y_{...} = 180$

The layout of Table 29.1 appears identical to an ordinary two-factor investigation, with two observations per cell (see e.g., Table 20.7). In fact, the study is not an ordinary two-factor study. The reason is that the instructors in the Atlanta school did not also teach in the other two schools, and similarly for the other instructors. Thus, six different instructors were involved. An ordinary two-factor investigation with six different instructors would have consisted of 18 treatments, as shown in Table 29.2a. In the training school example, however, only six treatments were included, as shown in Table 29.2b, where the crossed-out cells represent treatments not studied. Figure 29.1 contains a graphic representation of the nested design for the training school example, including the two replications of the study.

It is clear from Table 29.2b that the experimental design for the training school example involves an incomplete factorial arrangement of a special type,

TABLE 29.2 Illustration of crossed and nested factors

(a) Crossed Factors

	Instructor (factor B)					
School (factor A)	1	2	3	4	5	6
Atlanta						
Chicago						
San Francisco						

(b) Nested Factors

	Instructor (factor B)					
School (factor A)	1	2	3	4	5	6
Atlanta			X	X	X	X
Chicago	X	X			X	X
San Francisco	X	X	X	X		

FIGURE 29.1 Graphic representation of two-factor nested design—training school example

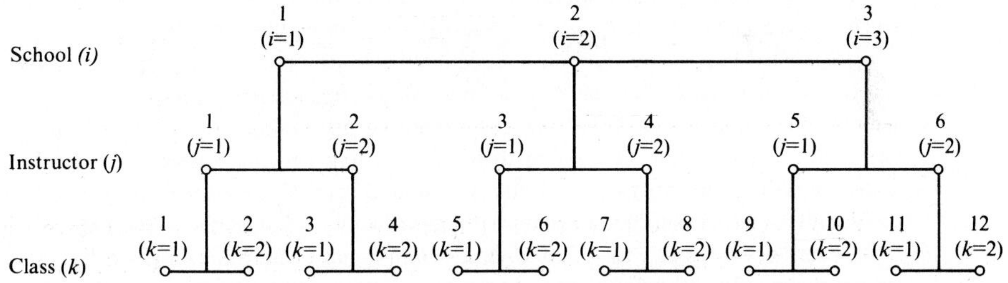

where each level of factor B (instructor) occurs with only one level of factor A (school). Specifically here, each instructor teaches in only one school. Factor B is therefore said to be *nested* within factor A. As noted earlier, in an ordinary factorial study where every factor level of A appears with every factor level of B, factors A and B are said to be *crossed*.

There is another way to look at the distinction between nested and crossed designs. Let μ_{ij} denote the mean response when factor A is at the ith level and factor B is at the jth level. If the factors are crossed, the jth level of B is the same for all levels of A. If, on the other hand, factor B is nested within factor A, the jth level of B when A is at level 1 has nothing in common with the jth level of B when A is at level 2, and so on. For instance, in a crossed factorial study of the effects of price ($1.99, $2.49) and advertising level (high, low), a particular advertising level is the same no matter with which price it appears, and similarly for the price levels. On the other hand, in the nested design training school example, the first instructor in school 1 is not the same as the first instructor in school 2, and so on.

Example 2

An analyst was interested in the effects of community (factor A) and neighborhood (factor B) on the spread of information about new products. Information was obtained from samples of families in various neighborhoods within selected communities. Since the neighborhood designated 1 in a given community is not the same as the neighborhoods designated 1 in the other communities, and similarly for the other neighborhoods, neighborhoods here are nested within communities.

Comments

1. The distinction between crossed and nested factors is often a fine one. In Example 2, if the neighborhoods of each community represented specified average income levels so that, say, the first neighborhoods in each community had an average income of $5,000–$9,999, the second neighborhoods an average income of $10,000–$19,999 and so on for the other neighborhoods, one could view the design as a crossed one. The factors would be community and economic level of neighborhood, and these would be crossed since a given economic level is the same for all communities and vice versa.

2. Nested factors are frequently encountered in observational studies where the researcher cannot manipulate the factors under study, or in experiments where only some factors can be manipulated. Factors that cannot be manipulated, it will be recalled, are designated classification factors, in distinction to experimental factors that can be assigned at will to the experimental units. Example 2 is an observational study where both community and neighborhood are classification factors since families (the experimental units) were not randomly assigned to either community or neighborhood. In Example 1, school is a classification factor because the classes of a school (the experimental units) are made up of mechanics from the district in which the school is located. Instructors in this example are an experimental factor since they are assigned randomly to a class, but a nested design results because the randomization is restricted to within a school.

29.2 TWO-FACTOR NESTED DESIGNS

We now consider nested designs involving two factors, one of which is nested inside the other. For consistency, we shall always consider the case where factor B is nested within factor A. We initially assume that both factor effects are fixed, but we shall later also consider the case of random effects. We assume throughout that all treatment means are of equal importance.

Development of model elements

We shall use the customary notation for a two-factor study and let μ_{ij} denote the mean response when factor A is at the ith level ($i = 1, \ldots, a$) and factor B is at the jth level ($j = 1, \ldots, b$). As usual, when all mean responses are of equal importance we define:

$$\mu_{i.} = \frac{\sum_j \mu_{ij}}{b} \tag{29.1}$$

For the training school example of Table 29.1, $\mu_{1.}$ represents the mean learning score for the Atlanta school, averaged over the instructors of that school, and $\mu_{2.}$ and $\mu_{3.}$ are interpreted similarly. Note once more that the $\mu_{i.}$ here represent mean learning scores that have been averaged over *different* instructors.

We define the main effect of the ith level of factor A as usual:

$$\alpha_i = \mu_{i.} - \mu_{..} \tag{29.2}$$

where:

$$\mu_{..} = \frac{\sum_i \sum_j \mu_{ij}}{ab} = \frac{\sum_i \mu_{i.}}{a} \tag{29.3}$$

is the overall mean response. It follows from (29.3) that:

$$\sum_i \alpha_i = 0 \tag{29.4}$$

In a nested design, it is not meaningful to employ a model component for the main effect of the jth level of factor B. To see why, consider again our training school example. Since each school employs different instructors and the jth instructors in the various schools are not the same, it would be meaningless to consider the effect of the jth instructor, averaged over all schools. Instead, one needs to consider the individual effects of each instructor in each school. We denote these individual effects by $\beta_{j(i)}$, where the subscript $j(i)$ indicates that the jth factor level of B is nested within the ith factor level of A. $\beta_{j(i)}$ is defined as follows:

$$\beta_{j(i)} = \mu_{ij} - \mu_{i.} \tag{29.5}$$

which can be rewritten, utilizing (29.2):

$$\beta_{j(i)} = \mu_{ij} - \alpha_i - \mu_{..} \tag{29.5a}$$

It follows from (29.1) and (29.5) that:

$$\sum_j \beta_{j(i)} = 0 \qquad i = 1, \ldots, a \tag{29.6}$$

The meaning of $\beta_{j(i)}$ can be seen most clearly from (29.5). With reference to our training school example, $\beta_{j(i)}$ is simply the difference in the mean learning score for the jth instructor of school i and the average of the mean learning scores for all instructors in that school. Thus, the effect of the jth instructor in the ith school is measured with respect to the overall mean learning score for the school in which the instructor teaches. Note, indeed, that $\beta_{j(i)}$ is the specific effect of factor B at the jth level when factor A is at the ith level as defined in (20.6).

We have now expressed the mean response μ_{ij} in terms of the overall mean, the main effect of the ith level of factor A, and the specific effect of the jth level of factor B nested within the ith level of factor A, as can be seen from (29.5a):

$$\text{(29.7)} \qquad \mu_{ij} \equiv \mu_{..} + \alpha_i + \beta_{j(i)} \equiv \mu_{..} + (\mu_{i.} - \mu_{..}) + (\mu_{ij} - \mu_{i.})$$

For our training school example, the mean learning score for the jth instructor in school i has been expressed in terms of the overall mean, the main effect of school i, and the specific effect of instructor j within school i.

To complete the model, we need only add a random error term. This will be denoted by $\varepsilon_{k(ij)}$, where the first subscript represents the kth observation and (i, j) identifies the factor combination for which it is the kth observation. We now use the subscript notation $k(ij)$ since the observations for any (i, j) factor combination can be viewed as being nested within that (i, j) combination. In our training school example, the replications are the several classes of mechanics taught by the same instructor in a given school. Since the first class for school i–instructor j is not the same as the first class for any other school-instructor combination, and similarly for the second class, we can regard the classes as being nested within the school-instructor combinations. Figure 29.1 illustrates this nesting.

We have not pointed out this nesting of the replications within the factor combinations up to now because it has not been necessary. With the current consideration of nested factors, however, it will be useful to recognize the nesting of the replications variable within the factor combinations.

Model

Let Y_{ijk} denote the kth observation when factor A is at the ith level and factor B is at the jth level. We assume that there are n observations for each factor combination, i.e., $k = 1, \ldots, n$, and that $i = 1, \ldots, a$ and $j = 1, \ldots, b$. When both factors A and B have fixed effects, an appropriate model is:

$$\text{(29.8)} \qquad Y_{ijk} = \mu_{..} + \alpha_i + \beta_{j(i)} + \varepsilon_{k(ij)}$$

where:

$\mu_{..}$ is a constant
α_i are constants subject to the restriction $\Sigma\alpha_i = 0$
$\beta_{j(i)}$ are constants subject to the restrictions $\sum_j \beta_{j(i)} = 0$ for all i
$\varepsilon_{k(ij)}$ are independent $N(0, \sigma^2)$
$i = 1, \ldots, a; j = 1, \ldots, b; k = 1, \ldots, n$

Comments

1. It is not necessary, as in (29.8), that the number of replications be equal for all factor combinations, or that the number of levels of nested factor B (number of instructors in our training school example) be the same for each level of factor A (school in this example). We shall discuss the removal of some of these restrictions later. We only point out now that the computations become more complex when these restrictions are relaxed.

2. There is no interaction term in model (29.8). There is no need for it since factor B is nested within factor A, not crossed with it. To put this somewhat differently, with reference to our training school example, it is not possible to estimate a school-instructor interaction when each instructor teaches in only one school. The teacher effect $\beta_{j(i)}$, since it is specific to a given school i, in a sense incorporates the interaction effect between the particular teacher j (in the ith school) and the ith school, but it is not possible in a nested design to disentangle this interaction effect.

3. The factor level means $\mu_{i.}$ in a nested design are not generally the same as the corresponding means in a crossed design. Remember that in a nested design, the $\mu_{i.}$ are obtained by averaging over only some of the distinctive levels of factor B. With reference to our training school example, the $\mu_{i.}$ are obtained by averaging over only those teachers who instruct in the ith school. In a crossed design, on the other hand, the $\mu_{i.}$ would be obtained by averaging over all instructors included in the study.

Random factor effects

If both factors A and B are random, model (29.8) is modified with α_i, $\beta_{j(i)}$, and $\varepsilon_{k(ij)}$ being independent normal random variables with expectations 0 and variances σ_α^2, σ_β^2, and σ^2, respectively. Thus, it is assumed that all $\beta_{j(i)}$ have the same variance σ_β^2. The assumption that all $\beta_{j(i)}$ have the same variance σ_β^2 also is made if only factor B is random. It is important to check whether this assumption is appropriate, since it may well be that the mean responses $\mu_{i1}, \mu_{i2}, \ldots$, in one factor A level (plant, school, city, etc.) differ in variability from those in other factor A levels (other plants, schools, cities, etc.). Tests for equality of variances were discussed in Section 18.6.

29.3 ANALYSIS OF VARIANCE FOR TWO-FACTOR NESTED DESIGNS

Fitting of model

The least squares estimators of the parameters in model (29.8) are obtained in the usual fashion. Employing our customary notation for sample data in factorial studies, the least squares estimators are:

	Parameter	*Estimator*
(29.9a)	$\mu_{..}$	$\hat{\mu}_{..} = \bar{Y}_{...}$
(29.9b)	α_i	$\hat{\alpha}_i = \bar{Y}_{i..} - \bar{Y}_{...}$
(29.9c)	$\beta_{j(i)}$	$\hat{\beta}_{j(i)} = \bar{Y}_{ij.} - \bar{Y}_{i..}$

The fitted values therefore are:

$$\hat{Y}_{ijk} = \bar{Y}_{...} + (\bar{Y}_{i..} - \bar{Y}_{...}) + (\bar{Y}_{ij.} - \bar{Y}_{i..}) = \bar{Y}_{ij.} \tag{29.10}$$

and the residuals are:

$$e_{ijk} = Y_{ijk} - \hat{Y}_{ijk} = Y_{ijk} - \bar{Y}_{ij.} \tag{29.11}$$

Sums of squares

The analysis of variance for model (29.8) is obtained by decomposing the total deviation $Y_{ijk} - \bar{Y}_{...}$ as follows:

$$\underbrace{Y_{ijk} - \bar{Y}_{...}}_{\text{Total deviation}} = \underbrace{\bar{Y}_{i..} - \bar{Y}_{...}}_{\text{A main effect}} + \underbrace{\bar{Y}_{ij.} - \bar{Y}_{i..}}_{\substack{\text{Specific B}\\ \text{effect when A}\\ \text{at } i\text{th level}}} + \underbrace{Y_{ijk} - \bar{Y}_{ij.}}_{\text{Residual}} \tag{29.12}$$

When we square (29.12) and sum over all observations, all cross-product terms drop out and we obtain:

$$SSTO = SSA + SSB(A) + SSE \tag{29.13}$$

where:

$$SSTO = \sum_i \sum_j \sum_k (Y_{ijk} - \bar{Y}_{...})^2 \tag{29.13a}$$

$$SSA = bn \sum_i (\bar{Y}_{i..} - \bar{Y}_{...})^2 \tag{29.13b}$$

$$SSB(A) = n \sum_i \sum_j (\bar{Y}_{ij.} - \bar{Y}_{i..})^2 \tag{29.13c}$$

$$SSE = \sum_i \sum_j \sum_k (Y_{ijk} - \bar{Y}_{ij.})^2 = \sum_i \sum_j \sum_k e_{ijk}^2 \tag{29.13d}$$

SSTO is the usual total sum of squares, and *SSA* is the ordinary factor *A* sum of squares, reflecting the variability of the factor level means $\bar{Y}_{i..}$.

SSB(*A*) is the factor *B* sum of squares, with the notation reflecting that factor *B* is nested within factor *A*. *SSB*(*A*) is made up of terms such as:

$$n \sum_j (\bar{Y}_{ij.} - \bar{Y}_{i..})^2 \tag{29.14}$$

which is simply the ordinary factor *B* sum of squares when factor *A* is at level *i*. These terms are then summed over all levels of factor *A*.

Finally, the error sum of squares *SSE* is, as usual, the sum of the squared residuals and reflects the variability of each observation Y_{ijk} around the corresponding treatment mean $\bar{Y}_{ij.}$. Alternatively, we can view *SSE* as being made up of terms such as:

$$\sum_j \sum_k (Y_{ijk} - \bar{Y}_{ij.})^2 \tag{29.15}$$

which is simply the ordinary error sum of squares within the ith level of factor A. These terms are then summed over all levels of factor A.

Thus, a nested two-factor design can be viewed as a series of one-factor investigations at the successive levels of the other factor. In terms of our training school example, a study of the effects of instructors (B) within any given school (A_i) leads to the usual sums of squares for instructors and errors in a one-way analysis of variance within school A_i, denoted by $SSB(A_i)$ and $SSE(A_i)$:

$$SSB(A_i) = n\sum_j (\bar{Y}_{ij.} - \bar{Y}_{i..})^2 \qquad SSE(A_i) = \sum_j \sum_k (Y_{ijk} - \bar{Y}_{ij.})^2$$

These are then aggregated to yield $SSB(A)$ and SSE, respectively. It is only the between-schools sum of squares SSA that introduces explicitly the other factor. Table 29.3 demonstrates this relation between the one-way analyses of variance for each school and the two-way analysis of variance for the nested factors.

Computational formulas. For hand computations, the following formulas may be used:

$$SSTO = \sum_i \sum_j \sum_k Y_{ijk}^2 - \frac{Y_{...}^2}{abn} \tag{29.16a}$$

$$SSA = \frac{\sum_i Y_{i..}^2}{bn} - \frac{Y_{...}^2}{abn} \tag{29.16b}$$

$$SSB(A) = \frac{\sum_i \sum_j Y_{ij.}^2}{n} - \frac{\sum_i Y_{i..}^2}{bn} \tag{29.16c}$$

$$SSE = \sum_i \sum_j \sum_k Y_{ijk}^2 - \frac{\sum_i \sum_j Y_{ij.}^2}{n} \tag{29.16d}$$

Degrees of freedom

The degrees of freedom associated with the various sums of squares can be deduced directly from the known relationships already studied. Since there are a total of abn observations, the degrees of freedom associated with $SSTO$ are $abn - 1$. For any level of factor A, there are $b(n - 1)$ degrees of freedom associated with the error sum of squares. Aggregating over all levels of factor A, there must be $ab(n - 1)$ degrees of freedom associated with SSE. Similarly, for any level of factor A, there are $b - 1$ degrees of freedom associated with the factor B

TABLE 29.3 Relation between nested two-factor ANOVA and one-way ANOVAs—training school example

Source of Variation	*One-way ANOVAs*									*Nested Two-factor ANOVA*	
	School 1			*School 2*			*School 3*				
	SS	*df*		*SS*	*df*		*SS*	*df*		*SS*	*df*
Between instructors (within schools)	$SSB(A_1)$	2 − 1	+	$SSB(A_2)$	2 − 1	+	$SSB(A_3)$	2 − 1	=	$SSB(A)$	3(2 − 1)
Error	$SSE(A_1)$	2(2 − 1)	+	$SSE(A_2)$	2(2 − 1)	+	$SSE(A_3)$	2(2 − 1)	=	SSE	3(2)(2 − 1)
Total within schools	$SSTO(A_1)$	2(2) − 1		$SSTO(A_2)$	2(2) − 1		$SSTO(A_3)$	2(2) − 1			
Between schools										SSA	3 − 1
Total										$SSTO$	3(2)(2) − 1

sum of squares. Hence, by aggregating over all levels of factor A, we find that there must be $a(b-1)$ degrees of freedom associated with $SSB(A)$. Finally, since there are a levels of factor A, there must be $a-1$ degrees of freedom associated with SSA.

Table 29.3 shows this aggregation of the degrees of freedom for the training school example, and Table 29.4 presents the general analysis of variance table for the two-factor nested design where factor B is nested within factor A.

TABLE 29.4 ANOVA table for nested two-factor fixed effects study (B nested within A)

Source of Variation	*SS*	*df*	*MS*	*E(MS)*
Factor A	$SSA = bn\Sigma(\bar{Y}_{i..} - \bar{Y}_{...})^2$	$a-1$	MSA	$\sigma^2 + bn\dfrac{\Sigma\alpha_i^2}{a-1}$
Factor B (within A)	$SSB(A) = n\Sigma\Sigma(\bar{Y}_{ij.} - \bar{Y}_{i..})^2$	$a(b-1)$	$MSB(A)$	$\sigma^2 + \dfrac{n}{a(b-1)}\Sigma\Sigma\beta_{j(i)}^2$
Error	$SSE = \Sigma\Sigma\Sigma(Y_{ijk} - \bar{Y}_{ij.})^2$	$ab(n-1)$	MSE	σ^2
Total	$SSTO = \Sigma\Sigma\Sigma(Y_{ijk} - \bar{Y}_{...})^2$	$abn-1$		

Expected mean squares

The expected mean squares for the fixed effects model are also shown in Table 29.4. They can be obtained by somewhat tedious derivations. We do not illustrate these derivations because we will present in Chapter 30 a relatively simple method of finding expected mean squares for any balanced nested design.

Example

In the training school example of Table 29.1, both schools and instructors were regarded as fixed effects factors; hence, model (29.8) was deemed appropriate. Figure 29.2 presents the sample means $\bar{Y}_{ij.}$ for the training school example. Note that the plot is not in the format of a crossed two-factor study because different instructors were used in the different schools. Figure 29.2 suggests strong differences between instructors within a school and also possible differences in mean learning between schools.

FIGURE 29.2 Plot of instructor sample means—training school example

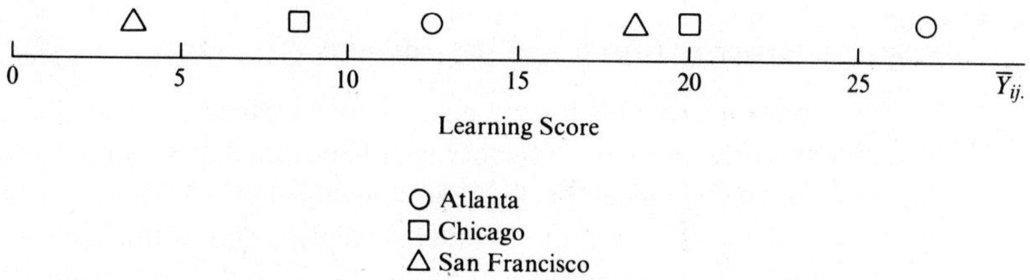

To analyze these effects formally, we begin by obtaining the analysis of variance. The sums of squares were obtained as follows using the computational formulas (29.16):

$$SSTO = (25)^2 + (29)^2 + (14)^2 + \cdots + (2)^2 - \frac{(180)^2}{12}$$

$$= 3{,}466 - 2{,}700 = 766$$

$$SSA = \frac{1}{4}\left[(79)^2 + (57)^2 + (44)^2\right] - 2{,}700$$

$$= 2{,}856.5 - 2{,}700 = 156.5$$

$$SSB(A) = \frac{1}{2}\left[(54)^2 + (25)^2 + \cdots + (7)^2\right] - 2{,}856.5$$

$$= 3{,}424 - 2{,}856.5 = 567.5$$

$$SSE = 3{,}466 - 3{,}424 = 42$$

Table 29.5a contains the analysis of variance.

TABLE 29.5 ANOVA for two-factor nested design—training school example

(a) ANOVA Table

Source of Variation	*SS*	*df*	*MS*
Schools (A)	$SSA = 156.5$	2	78.25
Instructors, within schools [$B(A)$]	$SSB(A) = 567.5$	3	189.17
Error (E)	$SSE = 42.0$	6	7.00
Total	$SSTO = 766.0$	11	

(b) Decomposition of $SSB(A)$

Source of Variation	$SSB(A_i)$	*df*	$MSB(A_i)$
Instructors, Atlanta	210.25	1	210.25
Instructors, Chicago	132.25	1	132.25
Instructors, San Francisco	225.00	1	225.00
Total	567.5	3	

Relation between crossed and nested sums of squares

If a computer program for the analysis of variance for nested designs is unavailable but one for crossed factors is at hand, the latter can be used with only slight inconvenience when the number of levels of nested factor B is the same for each level of factor A and the number of replications is the same for all factor combinations. Such nested designs are said to be *balanced*. Table 29.6 contains the (incorrect) analysis of variance obtained for the training school example in

TABLE 29.6 Incorrect crossed factor ANOVA for two-factor nested design—training school example

Source of Variation	*SS*	*df*	*MS*
Schools (A)	$SSA = 156.5$	2	78.25
Instructors (B)	$SSB = 108.0$	1	108.00
School-instructor interactions (AB)	$SSAB = 459.5$	2	229.75
Error (E)	$SSE = 42.0$	6	7.00
Total	$SSTO = 766.0$	11	

Table 29.1 from a computer run that treated the two factors as crossed. When we compare this incorrect analysis of variance with the correct one in Table 29.5a, we note that $SSTO$, SSA, and SSE are the same in each case, as are the associated degrees of freedom. The difference between the two analyses of variance is that the nested analysis has no interaction sum of squares. However, if we use the relation:

$$\underbrace{SSB(A)}_{\text{Nested}} = \underbrace{SSB + SSAB}_{\text{Crossed}} \tag{29.17}$$

and do likewise for the associated degrees of freedom, we obtain the correct nested factor B sum of squares and degrees of freedom:

$$SSB(A) = 108.0 + 459.5 = 567.5$$
$$df = 1 + 2 = 3$$

By using the relation (29.17), one can easily obtain the proper sums of squares and degrees of freedom for a balanced nested two-factor design from a computer package for crossed factors.

29.4 TESTS FOR FACTOR EFFECTS IN TWO-FACTOR NESTED DESIGNS

The tests for factor effects in a nested two-factor study are straightforward. The $E(MS)$ column in Table 29.4 indicates that for the fixed effects model, the appropriate test statistics are:

(29.18a) Test for factor A main effects: $F^* = \dfrac{MSA}{MSE}$

(29.18b) Test for factor B specific effects: $F^* = \dfrac{MSB(A)}{MSE}$

Thus, to test:

(29.19a)
$$H_0\text{: all } \alpha_i = 0$$
$$H_a\text{: not all } \alpha_i \text{ equal zero}$$

the decision rule to control the level of significance at α is:

(29.19b)
$$\begin{aligned}&\text{If } F^* \leq F[1-\alpha; a-1, (n-1)ab], \text{ conclude } H_0\\ &\text{If } F^* > F[1-\alpha; a-1, (n-1)ab], \text{ conclude } H_a\end{aligned}$$

Similarly, to test:

(29.20a)
$$\begin{aligned}&H_0\text{: all } \beta_{j(i)} = 0\\ &H_a\text{: not all } \beta_{j(i)} \text{ equal zero}\end{aligned}$$

the appropriate decision rule is:

(29.20b)
$$\begin{aligned}&\text{If } F^* \leq F[1-\alpha; a(b-1), (n-1)ab], \text{ conclude } H_0\\ &\text{If } F^* > F[1-\alpha; a(b-1), (n-1)ab], \text{ conclude } H_a\end{aligned}$$

Example

We return to our training school example. Based on the analysis of variance in Table 29.5a, the first test conducted was one to determine whether or not main school effects exist. The alternatives are given in (29.19a), and test statistic (29.18a) here is:

$$F^* = \frac{78.25}{7.00} = 11.2$$

The level of significance was to be controlled at $\alpha = .05$. Hence, we require $F(.95; 2, 6) = 5.14$. Since $F^* = 11.2 > 5.14$, it was concluded that the three schools differ in mean learning effects.

Next, a test for differences in mean learning effects between instructors within each school was conducted. The alternatives are given in (29.20a), and test statistic (29.18b) here is:

$$F^* = \frac{189.17}{7.00} = 27.0$$

For $\alpha = .05$, we require $F(.95; 3, 6) = 4.76$. Since $F^* = 27.0 > 4.76$, it was concluded that instructors within at least one school differ in terms of mean learning effects.

Comments

1. The alternative H_0 in (29.20a) can also be expressed in terms of the means μ_{ij}:

(29.21)
$$H_0\text{: } \mu_{11} = \mu_{12} = \cdots = \mu_{1b}; \mu_{21} = \mu_{22} = \cdots = \mu_{2b}; \ldots$$

Thus, H_0 states in terms of our training school illustration that the mean learning scores for all instructors in Atlanta are the same, and similarly for the other schools. It does *not* state that the mean learning scores for all instructors in the different schools are the same.

2. If it is concluded that factor B effects are present, it is often desired to ascertain whether they are present in all levels of factor A or only in some. (In some cases, indeed,

one may wish to proceed immediately to this analysis.) With reference to our training school example, the question would be whether the instructor effects differ in all schools or only in some schools. As noted earlier, $SSB(A)$ in Table 29.5a is made up of the instructor sums of squares within the individual schools, and these component sums of squares can be used for testing for instructor effects within each school. Table 29.5b contains the relevant component sums of squares. To test for instructor differences within the Atlanta school, for instance, we use test statistic $F^* = 210.25/7.00 = 30.0$. For a level of significance of $\alpha = .05$, we need $F(.95; 1, 6) = 5.99$. Since $F^* = 30.0 > 5.99$, it was concluded that the two instructors in Atlanta have different mean learning effects. Using the same level of significance each time, similar conclusions were reached for the other two schools. The family level of significance for the three tests according to the Bonferroni inequality is .15.

3. If the assumption of constant error variances were violated in our training school example through unequal variances for the different schools, it would still be possible to study instructor effects within each school by separate analyses of variance for each school.

Random factor effects

Test statistic (29.18a) for factor A main effects is not appropriate if either or both factor effects are random. Table 29.7 gives the expected mean squares for these cases and also the appropriate test statistics.

TABLE 29.7 Expected mean squares for nested two-factor designs with random effects (B nested within A)

	Expected Mean Square	
Mean Square	*A Fixed, B Random*	*A Random, B Random*
MSA	$\sigma^2 + \frac{bn}{a-1}\Sigma\alpha_i^2 + n\sigma_\beta^2$	$\sigma^2 + bn\sigma_\alpha^2 + n\sigma_\beta^2$
$MSB(A)$	$\sigma^2 + n\sigma_\beta^2$	$\sigma^2 + n\sigma_\beta^2$
MSE	σ^2	σ^2
	Appropriate Test Statistic	
Test for	*A Fixed, B Random*	*A Random, B Random*
Factor A	$MSA/MSB(A)$	$MSA/MSB(A)$
Factor $B(A)$	$MSB(A)/MSE$	$MSB(A)/MSE$

29.5 ANALYSIS OF FACTOR EFFECTS IN TWO-FACTOR NESTED DESIGNS

If factor effects are present in a nested design, one usually desires to obtain estimates of these and/or make comparisons.

Estimation of factor level means $\mu_{i.}$

When factor A (fixed effects factor) has significant main effects, there is frequent interest in estimating the factor level means $\mu_{i.}$. The sample mean $\bar{Y}_{i..}$ is an unbiased point estimator of $\mu_{i.}$. As usual for a fixed effects factor, the estimated variance of $\bar{Y}_{i..}$ is based on the mean square in the denominator of the statistic used for testing for factor A main effects, and on the number of observations in $\bar{Y}_{i..}$. Confidence limits for $\mu_{i.}$ are of the customary form:

(29.22)
$$\bar{Y}_{i..} \pm t(1 - \alpha/2; df)s(\bar{Y}_{i..})$$

where:

(29.22a)
$$s^2(\bar{Y}_{i..}) = \frac{MSE}{bn} \qquad df = ab(n-1) \qquad A \text{ and } B \text{ fixed}$$

(29.22b)
$$s^2(\bar{Y}_{i..}) = \frac{MSB(A)}{bn} \qquad df = a(b-1) \qquad A \text{ fixed, } B \text{ random}$$

Confidence limits for differences $D = \mu_{i.} - \mu_{i'.}$ are set up in the usual way, utilizing the point estimator $\hat{D} = \bar{Y}_{i..} - \bar{Y}_{i'..}$ and the t distribution with the degrees of freedom those associated with the appropriate mean square:

(29.23)
$$\hat{D} \pm t(1 - \alpha/2; df)s(\hat{D})$$

where:

(29.23a)
$$s^2(\hat{D}) = s^2(\bar{Y}_{i..}) + s^2(\bar{Y}_{i'..}) \qquad \text{as given by (29.22a) or (29.22b)}$$

The Tukey and Bonferroni simultaneous comparison procedures can be utilized in the usual way for making pairwise comparisons with family confidence coefficient $1 - \alpha$.

Finally, no new problems arise in estimating contrasts of factor level means. The Scheffé or the Bonferroni procedures can be employed when estimating several contrasts.

Example. For our training school example in Table 29.1, it was desired to estimate the mean learning score for the Atlanta school with a 95 percent confidence coefficient. Using our earlier results in Tables 29.1 and 29.5, we obtain for the fixed effects model:

$$\bar{Y}_{1..} = \frac{79}{4} = 19.75$$

$$s^2(\bar{Y}_{1..}) = \frac{MSE}{bn} = \frac{7.00}{4} = 1.75$$

$$s(\bar{Y}_{1..}) = 1.32$$

$$t(.975; 6) = 2.447$$

$$16.5 = 19.75 - 2.447(1.32) \le \mu_{1.} \le 19.75 + 2.447(1.32) = 23.0$$

In addition, pairwise comparisons of the three schools were to be made with family confidence coefficient .90. We shall utilize the Tukey method and require:

$$T = \frac{1}{\sqrt{2}} q[1 - \alpha; a, ab(n-1)] = \frac{1}{\sqrt{2}} q(.90; 3, 6)$$
$$= \frac{1}{\sqrt{2}}(3.56) = 2.52$$

The estimated variance is the same for all pairwise comparisons:

$$s^2(\hat{D}) = \frac{MSE}{bn} + \frac{MSE}{bn} = \frac{2(7.00)}{4} = 3.5$$

so that the estimated standard deviation is $s(\hat{D}) = 1.87$ and the precision term is $2.52(1.87) = 4.71$.

Using the results in Table 29.1, we find:

$$\bar{Y}_{1..} = 19.75 \qquad \bar{Y}_{2..} = 14.25 \qquad \bar{Y}_{3..} = 11$$

Hence, the 90 percent family of confidence intervals is:

$$.8 = (19.75 - 14.25) - 4.71 \le \mu_{1.} - \mu_{2.} \le (19.75 - 14.25) + 4.71 = 10.2$$
$$4.0 = (19.75 - 11) - 4.71 \le \mu_{1.} - \mu_{3.} \le (19.75 - 11) + 4.71 = 13.5$$
$$-1.5 = (14.25 - 11) - 4.71 \le \mu_{2.} - \mu_{3.} \le (14.25 - 11) + 4.71 = 8.0$$

We conclude with 90 percent family confidence coefficient that the mean learning score is highest in Atlanta and that the difference in the observed mean scores for Chicago and San Francisco is not statistically significant.

Estimation of treatment means μ_{ij}

Confidence limits for μ_{ij} are set up in the usual fashion using the t distribution when both factors A and B have fixed effects:

(29.24) $$\bar{Y}_{ij.} \pm t[1 - \alpha/2; (n-1)ab]s(\bar{Y}_{ij.})$$

where:

(29.24a) $$s^2(\bar{Y}_{ij.}) = \frac{MSE}{n}$$

To make a pairwise comparison within a factor level, we estimate the difference $D = \mu_{ij} - \mu_{ij'}$ with the point estimator $\hat{D} = \bar{Y}_{ij.} - \bar{Y}_{ij'.}$ and employ the confidence limits:

(29.25) $$\hat{D} \pm t[1 - \alpha/2; (n-1)ab]s(\hat{D})$$

where:

(29.25a) $$s^2(\hat{D}) = \frac{2MSE}{n}$$

The Bonferroni procedure may be used when several comparisons are to be made and the family confidence level is to be controlled. The Tukey procedure is also applicable but often will not be efficient since ordinarily only comparisons within each factor level are of interest, whereas the Tukey family is based on all pairwise comparisons among the *ab* treatments.

Example. It is desired in our training school example to compare the mean scores for the two instructors in each school, using the Bonferroni procedure with a 90 percent family confidence coefficient. For $g = 3$ comparisons, we require $B = t[1 - .10/2(3); 6] = t(.983; 6) = 2.748$. The estimated variance in each case is:

$$s^2(\bar{Y}_{i1.} - \bar{Y}_{i2.}) = \frac{2(7.00)}{2} = 7.0$$

Hence, the precision term in each comparison is $2.748\sqrt{7.0} = 7.27$. Obtaining the sample means $\bar{Y}_{ij.}$ from Table 29.1a, we find:

$$7.2 = (27 - 12.5) - 7.27 \le \mu_{11} - \mu_{12} \le (27 - 12.5) + 7.27 = 21.8$$
$$-18.8 = (8.5 - 20) - 7.27 \le \mu_{21} - \mu_{22} \le (8.5 - 20) + 7.27 = -4.2$$
$$7.7 = (18.5 - 3.5) - 7.27 \le \mu_{31} - \mu_{32} \le (18.5 - 3.5) + 7.27 = 22.3$$

It is evident that substantial differences between the two instructors exist at each school.

Estimation of overall mean $\mu_{..}$

Sometimes there is interest in estimating the overall mean $\mu_{..}$. For our training school example, $\mu_{..}$ is the overall mean learning score for all training schools and all instructors in these schools. The point estimator is $\bar{Y}_{...}$. The confidence limits are constructed utilizing the t distribution as follows:

(29.26) $$\bar{Y}_{...} \pm t(1 - \alpha/2; df)s(\bar{Y}_{...})$$

where:

(29.26a) $$s^2(\bar{Y}_{...}) = \frac{MSE}{abn} \qquad df = ab(n-1) \qquad A \text{ and } B \text{ fixed}$$

(29.26b) $$s^2(\bar{Y}_{...}) = \frac{MSA}{abn} \qquad df = a - 1 \qquad A \text{ and } B \text{ random}$$

(29.26c) $$s^2(\bar{Y}_{...}) = \frac{MSB(A)}{abn} \qquad df = a(b-1) \qquad A \text{ fixed, } B \text{ random}$$

Example. For our training school example, we wish to estimate the overall mean $\mu_{..}$ with a 95 percent confidence interval. The estimated variance (29.26a) is appropriate here since the model involves fixed effects. Hence, we obtain:

$$s^2(\bar{Y}_{...}) = \frac{7.00}{12} = .583 \qquad s(\bar{Y}_{...}) = .764$$

For confidence coefficient .95, we require $t(.975; 6) = 2.447$. From Table 29.1, we find $\bar{Y}_{...} = 15$. The desired confidence interval therefore is:

$$13.1 = 15 - 2.447(.764) \leq \mu_{..} \leq 15 + 2.447(.764) = 16.9$$

Estimation of variance components

With random factor effects, one may wish to estimate the variance components. No new problems arise for nested designs. For instance, an unbiased estimator of σ_α^2 when both factors A and B have random effects is (Table 29.7):

$$s_\alpha^2 = \frac{MSA - MSB(A)}{bn} \tag{29.27}$$

29.6 UNEQUAL NESTING AND REPLICATIONS IN NESTED TWO-FACTOR DESIGNS

Up to this point, we have assumed that the same number of levels of factor B are nested within each of the levels of factor A, and that the same number of observations are made for each factor combination. There are occasions, however, when the number of levels of the nested B factor will vary for the different levels of factor A, and when the number of observations for the different factor combinations are unequal. For instance, in our earlier example dealing with the effects of school (factor A) and instructor (factor B) on the learning achieved by classes of mechanics, there might have been b_i instructors in the ith school and n_{ij} classes taught by the jth instructor in school i.

The ANOVA sums of squares formulas given earlier are not appropriate for unequal nestings and replications. Ordinarily, it is best to use the regression approach for this case, but the general matrix approach described in Section 8.5 can also be employed. Since no new principles are involved when the factor effects are fixed, we proceed directly to an example.

Example

The manufacturing company that conducted the training school study subsequently made a follow-up study involving only Atlanta and Chicago. Three instructors were used in Atlanta and two in Chicago. All instructors were to train two classes, but one class for one of the instructors in Atlanta had to be canceled. The data for this follow-up study are presented in Table 29.8a. Again we assume that the fixed effects model (29.8) is appropriate:

$$Y_{ijk} = \mu_{..} + \alpha_i + \beta_{j(i)} + \varepsilon_{k(ij)} \tag{29.28}$$
$$i = 1, 2; j = 1, \ldots, b_i; k = 1, \ldots, n_{ij}$$
$$b_1 = 3, b_2 = 2$$
$$n_{11} = n_{13} = 2, n_{12} = 1, n_{21} = n_{22} = 2$$

TABLE 29.8 Nested two-factor study with unequal nestings and replications—follow-up training school study

(a) Data

Replication	*Atlanta* (A_1)			*Chicago* (A_2)	
k	B_1	B_2	B_3	B_1	B_2
1	20	8	9	4	16
2	22		13	8	20

(b) **Y** and **X** Matrices for Full Model (29.31)

$$\mathbf{Y} = \begin{bmatrix} Y_{111} \\ Y_{112} \\ Y_{121} \\ Y_{131} \\ Y_{132} \\ Y_{211} \\ Y_{212} \\ Y_{221} \\ Y_{222} \end{bmatrix} = \begin{bmatrix} 20 \\ 22 \\ 8 \\ 9 \\ 13 \\ 4 \\ 8 \\ 16 \\ 20 \end{bmatrix} \qquad \mathbf{X} = \begin{array}{c} \begin{array}{ccccc} & X_1 & X_2 & X_3 & X_4 \end{array} \\ \begin{bmatrix} 1 & 1 & 1 & 0 & 0 \\ 1 & 1 & 1 & 0 & 0 \\ 1 & 1 & 0 & 1 & 0 \\ 1 & 1 & -1 & -1 & 0 \\ 1 & 1 & -1 & -1 & 0 \\ 1 & -1 & 0 & 0 & 1 \\ 1 & -1 & 0 & 0 & 1 \\ 1 & -1 & 0 & 0 & -1 \\ 1 & -1 & 0 & 0 & -1 \end{bmatrix} \end{array}$$

(c) Fitted Full Model

$$\hat{Y} = 12.667 + .667X_1 + 7.667X_2 - 5.333X_3 - 6.0X_4$$

In developing the equivalent regression model, we need to recognize the constraints in (29.8):

$$\sum_{i=1}^{2} \alpha_i = 0 \qquad \sum_{j=1}^{3} \beta_{j(1)} = 0 \qquad \sum_{j=1}^{2} \beta_{j(2)} = 0 \tag{29.29}$$

Proceeding as usual, we shall incorporate the parameters α_1, $\beta_{1(1)}$, $\beta_{2(1)}$, and $\beta_{1(2)}$ into the regression model. The other parameters are not required since according to the constraints (29.29) we have:

$$\alpha_2 = -\alpha_1 \qquad \beta_{3(1)} = -\beta_{1(1)} - \beta_{2(1)} \qquad \beta_{2(2)} = -\beta_{1(2)} \tag{29.30}$$

Thus, we require four indicator variables for our example, each taking on values 1, -1, or 0.

The equivalent regression model therefore is:

$$Y_{ijk} = \mu_{..} + \underbrace{\alpha_1 X_{ijk1}}_{\text{School main effect}} \tag{29.31}$$

$$+ \underbrace{\beta_{1(1)}X_{ijk2} + \beta_{2(1)}X_{ijk3} + \beta_{1(2)}X_{ijk4}}_{\text{Specific instructor within school effect}} + \varepsilon_{ijk} \qquad \text{Full model}$$

where:

$$X_{ijk1} = \begin{matrix} 1 \text{ if observation from school 1} \\ -1 \text{ if observation from school 2} \end{matrix}$$

$$X_{ijk2} = \begin{matrix} 1 \text{ if observation for instructor 1 in school 1} \\ -1 \text{ if observation for instructor 3 in school 1} \\ 0 \text{ otherwise} \end{matrix}$$

$$X_{ijk3} = \begin{matrix} 1 \text{ if observation for instructor 2 in school 1} \\ -1 \text{ if observation for instructor 3 in school 1} \\ 0 \text{ otherwise} \end{matrix}$$

$$X_{ijk4} = \begin{matrix} 1 \text{ if observation for instructor 1 in school 2} \\ -1 \text{ if observation for instructor 2 in school 2} \\ 0 \text{ otherwise} \end{matrix}$$

The **Y** vector and **X** matrix for our example are shown in Table 29.8b.

To test for school main effects, we first fit the full model (29.31). The fitted full model is shown in Table 29.8c. We then fit the reduced model:

$$(29.32) \qquad Y_{ijk} = \mu_{..} + \beta_{1(1)}X_{ijk2} + \beta_{2(1)}X_{ijk3} + \beta_{1(2)}X_{ijk4} + \varepsilon_{ijk} \qquad \text{Reduced model}$$

The difference $SSE(R) - SSE(F)$ equals SSA. Test statistic (8.37) is then obtained in the usual fashion.

To test for specific instructor effects, we employ the reduced model:

$$(29.33) \qquad Y_{ijk} = \mu_{..} + \alpha_1 X_{ijk1} + \varepsilon_{ijk} \qquad \text{Reduced model}$$

The difference $SSE(R) - SSE(F)$ equals $SSB(A)$.

Table 29.9 contains the ANOVA table for the follow-up training school study. No total sum of squares is shown because the component sums of squares are not orthogonal.

The tests for school and instructor effects are carried out as before. Estimation of factor effects is done by means of the regression parameters. For instance, a comparison of the mean scores for the two schools involves:

$$\mu_{1.} - \mu_{2.} = \alpha_1 - \alpha_2$$

Since $\alpha_2 = -\alpha_1$ by (29.30), we need to estimate:

$$\mu_{1.} - \mu_{2.} = \alpha_1 - (-\alpha_1) = 2\alpha_1$$

A point estimator is $2\hat{\alpha}_1$. Other desired estimates are obtained in similar fashion.

TABLE 29.9 ANOVA table for nested two-factor study with unequal nestings and replications—follow-up training school study

Source of Variation	*SS*	*df*	*MS*	F^*
Schools (A)	3.76	1	3.76	3.76/6.5 = .58
Instructors $[B(A)]$	295.20	3	98.4	98.4/6.5 = 15.1
Error (E)	26.00	4	6.5	

29.7 SUBSAMPLING IN SINGLE-FACTOR STUDY WITH COMPLETELY RANDOMIZED DESIGN

Up to this point in our discussion of experimental designs, we have considered only designs in which one observation of the dependent variable is made on an experimental unit. There are occasions, however, when more than one observation is desirable. Consider an experiment to study the effect of oven temperature on crustiness of bread. Three temperatures were utilized, and two experimental units (batches of flour mix) were randomly assigned to each treatment. It was not economical to use the entire batch to bake breads, nor was it technically feasible to use a batch as a block. Hence, three subsamples were selected from each batch to make three loaves, which were baked at a given temperature. Here then, three observations (subsamples) were made on each experimental unit (batch).

Another instance of several observations being made on the dependent variable for each experimental unit occurred in an experiment on the effectiveness of three different training methods. The experimental units here were persons, and the experiment sought to measure the length of time required to perform a certain engine assembly operation after the given training program was completed. Ten consecutive assemblies were timed, and these constituted the subsamples of the experimental unit (person).

Formally, subsampling (i.e., repeated observations on the same experimental unit) is completely analogous to nested factors. We shall demonstrate this for a completely randomized design.

Model

Consider again the experiment to study the effect of oven temperature on the crustiness of bread. The model for this study can be written as follows:

$$Y_{ijk} = \mu_{..} + \tau_i + \varepsilon_{j(i)} + \eta_{k(ij)} \tag{29.34}$$

The meaning of the symbols is as follows:

1. $\mu_{..}$ is an overall constant.
2. τ_i is the temperature (i.e., treatment) effect (fixed effect, here).
3. $\varepsilon_{j(i)}$ is the experimental error associated with the particular batch (random effect, here). As usual, the experimental error is nested within the treatment, since the jth batch for treatment i was not used with any other treatment.
4. $\eta_{k(ij)}$ is the error associated with the kth subsample or observation on the jth experimental unit for the ith treatment (random effect, here). This observation error is nested within the experimental unit, and consequently also within the treatment.

Note that model (29.34) appears the same as model (29.8) for the nested two-factor design, except for changes in notation to reflect the fact that model (29.34) is a single-factor model and contains both experimental error and observation error. Consequently, the analysis of variance for the case of subsampling in a single-factor study with a completely randomized design parallels that for a nested two-factor study.

In general, the model for subsampling in a single-factor study with a completely randomized design where the treatment effects are fixed and there are equal numbers of replications and subsamples is:

(29.35) $$Y_{ijk} = \mu_{..} + \tau_i + \varepsilon_{j(i)} + \eta_{k(ij)}$$

where:

$\mu_{..}$ is a constant
τ_i are constants subject to the restriction $\Sigma\tau_i = 0$
$\varepsilon_{j(i)}$ are independent $N(0, \sigma^2)$
$\eta_{k(ij)}$ are independent $N(0, \sigma_\eta^2)$
$\varepsilon_{j(i)}$ and $\eta_{k(ij)}$ are independent
$i = 1, \dots, r; j = 1, \dots, n; k = 1, \dots, m$

Analysis of variance and tests of effects

The appropriate sums of squares for the analysis of variance for model (29.35) are as follows:

(29.36a) $$SSTO = \sum_i \sum_j \sum_k (Y_{ijk} - \bar{Y}_{...})^2 = \sum_i \sum_j \sum_k Y_{ijk}^2 - \frac{Y_{...}^2}{rnm}$$

(29.36b) $$SSTR = nm \sum_i (\bar{Y}_{i..} - \bar{Y}_{...})^2 = \frac{\sum_i Y_{i..}^2}{nm} - \frac{Y_{...}^2}{rnm}$$

(29.36c) $$SSEE = m \sum_i \sum_j (\bar{Y}_{ij.} - \bar{Y}_{i..})^2 = \frac{\sum_i \sum_j Y_{ij.}^2}{m} - \frac{\sum_i Y_{i..}^2}{nm}$$

(29.36d) $$SSOE = \sum_i \sum_j \sum_k (Y_{ijk} - \bar{Y}_{ij.})^2 = \sum_i \sum_j \sum_k Y_{ijk}^2 - \frac{\sum_i \sum_j Y_{ij.}^2}{m}$$

Here, *SSEE* stands for the *experimental error sum of squares,* and *SSOE* stands for the *observation error sum of squares*. Note the correspondence of formulas (29.36) to formulas (29.13) and (29.16) for nested two-factor designs. The only difference is that we now have $i = 1, \dots, r$, $j = 1, \dots, n$, and $k = 1, \dots, m$, whereas before i, j, and k ran to a, b, and n, respectively.

Table 29.10 contains the ANOVA for a single-factor completely randomized experiment with subsampling. Also shown there are the expected mean squares for both fixed and random treatment effects. Note that regardless of whether

TABLE 29.10 ANOVA for single-factor completely randomized experiment with subsampling

Source of Variation	SS	df	MS	E(MS) τ_i Fixed	E(MS) τ_i Random
Treatments	$SSTR$	$r-1$	$MSTR$	$\sigma_\eta^2 + m\sigma^2 + nm\frac{\Sigma\tau_i^2}{r-1}$	$\sigma_\eta^2 + m\sigma^2 + nm\sigma_\tau^2$
Experimental error	$SSEE$	$r(n-1)$	$MSEE$	$\sigma_\eta^2 + m\sigma^2$	$\sigma_\eta^2 + m\sigma^2$
Observation error	$SSOE$	$rn(m-1)$	$MSOE$	σ_η^2	σ_η^2
Total	$SSTO$	$rnm-1$			

treatment effects are fixed or random, the appropriate statistic for testing treatment effects is:

(29.37a)
$$F^* = \frac{MSTR}{MSEE}$$

A test for the presence of experimental error effects, i.e., $\sigma^2 > 0$, uses the same test statistic for both fixed and random treatment effects:

(29.37b)
$$F^* = \frac{MSEE}{MSOE}$$

Example. The data for the study of the effect of baking temperature on the crustiness of bread are contained in Table 29.11. The data are scores on a scale from 1 to 20. The appropriate analysis of variance was obtained from a computer run and is presented in Table 29.12. To test the effect of temperature:

$$H_0: \tau_1 = \tau_2 = \tau_3 = 0$$
$$H_a: \text{not all } \tau_i \text{ equal zero}$$

TABLE 29.11 Data for single-factor completely randomized experiment with subsampling—bread crustiness example

	Temperature					
	Low ($i=1$)		Medium ($i=2$)		High ($i=3$)	
Observation Unit k	Batch 1 $j=1$	Batch 2 $j=2$	Batch 3 $j=1$	Batch 4 $j=2$	Batch 5 $j=1$	Batch 6 $j=2$
1	4	12	14	9	14	16
2	7	8	13	10	17	19
3	5	10	11	12	15	18
	$Y_{11.} = 16$	$Y_{12.} = 30$	$Y_{21.} = 38$	$Y_{22.} = 31$	$Y_{31.} = 46$	$Y_{32.} = 53$
	$Y_{1..} = 46$		$Y_{2..} = 69$		$Y_{3..} = 99$	
			$Y_{...} = 214$			

TABLE 29.12 ANOVA for bread crustiness example

Source of Variation	*SS*	*df*	*MS*
Temperatures (*TR*)	235.44	2	117.72
Mix batches (*EE*)	49.00	3	16.33
Observation units (*OE*)	31.33	12	2.61
Total	315.78	17	

Note: Component sums of squares do not add to *SSTO* because of rounding.

we use test statistic (29.37a):

$$F^* = \frac{117.72}{16.33} = 7.21$$

A level of significance of $\alpha = .10$ was specified. Hence, we need $F(.90; 2, 3) = 5.46$. Since $F^* = 7.21 > 5.46$, we conclude H_a, that baking temperature does have an effect on the crustiness of the bread.

To test for batch differences:

$$H_0: \sigma^2 = 0$$
$$H_a: \sigma^2 > 0$$

we employ test statistic (29.37b):

$$F^* = \frac{16.33}{2.61} = 6.26$$

For a level of significance of $\alpha = .10$, we need $F(.90; 3, 12) = 2.61$. Since $F^* = 6.26 > 2.61$, we conclude H_a, that there are batch effects on the crustiness of bread. Thus, both the particular batch of flour mix and the temperature at which the bread is baked affect the crustiness of the loaf.

Estimation of treatment effects

When the treatment effects are fixed, there is usually interest in obtaining confidence limits for treatment means $\mu_{i.} = \mu_{..} + \tau_i$ and for pairwise comparisons and contrasts of treatment means. These can be set in the usual manner, using *MSEE* as the error variance since this is the quantity in the denominator of the test statistic for fixed treatment effects. The degrees of freedom are those associated with *MSEE*, namely, $(n - 1)r$. For instance, the confidence limits for treatment mean $\mu_{i.}$ are:

(29.38) $$\bar{Y}_{i..} \pm t[1 - \alpha/2; (n - 1)r]s(\bar{Y}_{i..})$$

where:

(29.38a)
$$s^2(\bar{Y}_{i..}) = \frac{MSEE}{nm}$$

Similarly, confidence limits for a pairwise comparison of treatment means, $D = \mu_{i.} - \mu_{i'.}$, are obtained as follows:

(29.39)
$$\hat{D} \pm t[1 - \alpha/2; (n-1)r]s(\hat{D})$$

where:

(29.39a)
$$\hat{D} = \bar{Y}_{i..} - \bar{Y}_{i'..}$$

(29.39b)
$$s^2(\hat{D}) = \frac{2MSEE}{nm}$$

The Bonferroni and Tukey simultaneous comparison procedures can be utilized in the usual manner.

Example. To estimate the mean crustiness of bread baked at a low temperature with a 95 percent confidence coefficient, we need:

$$\bar{Y}_{1..} = 7.67$$

$$s^2(\bar{Y}_{1..}) = \frac{16.33}{6} = 2.722 \qquad s(\bar{Y}_{1..}) = 1.65$$

$$t(.975; 3) = 3.182$$

Hence, the 95 percent confidence interval is:

$$2.4 = 7.67 - 3.182(1.65) \le \mu_{1.} \le 7.67 + 3.182(1.65) = 12.9$$

It was also desired to estimate the difference in mean crustiness of bread baked at high and low temperatures with a 95 percent confidence interval. Utilizing (29.39), we require:

$$\bar{Y}_{1..} = 7.67 \qquad \bar{Y}_{3..} = 16.5$$

$$\hat{D} = \bar{Y}_{3..} - \bar{Y}_{1..} = 16.5 - 7.67 = 8.83$$

$$s^2(\hat{D}) = \frac{2(16.33)}{6} = 5.443 \qquad s(\hat{D}) = 2.33$$

Hence, the desired confidence interval is:

$$1.4 = 8.83 - 3.182(2.33) \le \mu_{3.} - \mu_{1.} \le 8.83 + 3.182(2.33) = 16.2$$

Estimation of variances

At times, one is interested in estimating σ^2, the variance of experimental units, and σ_η^2, the variance of the observation units. It is evident from either of the $E(MS)$ columns in Table 29.10 that the following are unbiased estimators:

	Parameter	*Unbiased Estimator*
(29.40a)	σ^2	$s^2 = \dfrac{(MSEE - MSOE)}{m}$
(29.40b)	σ_η^2	$s_\eta^2 = MSOE$

Example. For the bread crustiness example, we obtain from Table 29.12 the following estimates of the two variances:

$$s^2 = \frac{16.33 - 2.61}{3} = 4.57$$

$$s_\eta^2 = 2.61$$

Thus, the estimated variability between batches is somewhat greater here than that between observation units within a batch.

Comments

1. One should not confuse designs with repeated observations, discussed here, and designs with repeated measurements, discussed in Section 28.4. In repeated measures designs, several or all of the treatments are applied to the same subject. Repeated observations designs, on the other hand, refer to designs where several observations on the dependent variable are made for a given treatment applied to an experimental unit. It is possible to develop a repeated measures design with repeated observations, as when a given subject is exposed to each of the treatments under study and a number of observations are made at the end of each treatment application.

2. Frequently, the units for subsampling are called *observation units,* to distinguish them from the *experimental units*. Thus, in our bread crustiness example, the batches of flour mix are the experimental units and the portions selected from a batch for making loaves of bread are the observation units.

3. Observation units may be different physical entities, as in the bread crustiness example where they were portions of a batch of flour mix. Observation units also may refer to repeated observations on the entire experimental unit. An example of the latter is the earlier illustration where a person is timed for 10 consecutive assembly operations after receiving a given type of training.

4. Note that subsampling model (29.35) contains no interaction terms. This is because the experimental error terms $\varepsilon_{j(i)}$ are nested within treatments, and the observation error terms $\eta_{k(ij)}$ are nested within experimental units. When one variable is nested within another, we saw earlier that interaction terms are inapplicable.

5. We have considered only the case where an equal number of experimental units (n) are applied to each treatment, and a constant number of observations (m) are made on each experimental unit. Serious complications are encountered in the unbalanced case, and no exact test for treatment effects can be made. See an advanced text, such as Reference 29.1, for a discussion.

Design considerations

A problem that arises in designing an experiment with repeated observations is the choice of the number of experimental units and the number of observation units. Suppose that we are to estimate the treatment means $\mu_{i.}$ in a balanced study with fixed treatment effects. It can be shown that:

(29.41) $$\sigma^2(\bar{Y}_{i..}) = \frac{\sigma_\eta^2 + m\sigma^2}{nm}$$

Formula (29.41) makes it clear that if nm is fixed (in our bread crustiness example, nm is the total number of loaves baked in the experiment), $\sigma^2(\bar{Y}_{i..})$ is minimized if m is made as small as possible, namely, $m = 1$. Thus, when nm is fixed, optimum estimation of the treatment means requires that only one observation unit be selected for each experimental unit, so that the total sample is spread among as many experimental units as possible.

The justification for subsampling lies in cost considerations. Suppose that it costs c_1 to include one experimental unit in the study and c_2 to make one observation on an experimental unit. Also, suppose the total cost C is given by:

(29.42) $$C = c_1 n + c_2 nm$$

It can then be shown that for a given total cost C_0, the variance $\sigma^2(\bar{Y}_{i..})$ is minimized when:

(29.43a) $$m_{\text{opt}} = \frac{\sigma_\eta}{\sigma}\sqrt{\frac{c_1}{c_2}}$$

(29.43b) $$n_{\text{opt}} = \frac{C_0}{c_1 + c_2 m_{\text{opt}}}$$

Example. With reference to our bread crustiness example, suppose $c_1 =$ \$30 and $c_2 =$ \$5, and that the total cost of observations in the experiment is limited to $C_0 =$ \$400. Advance assessments indicate that $\sigma = 2.2$ and $\sigma_\eta = 1.5$, approximately. The optimum sample sizes then are obtained as follows:

$$m_{\text{opt}} = \frac{1.5}{2.2}\sqrt{\frac{30}{5}} = 1.67$$

$$n_{\text{opt}} = \frac{400}{30 + 5(1.67)} = 10.4$$

Thus, 10 batches might be used for each treatment, with two observations on each batch.

Comments

1. Note that the optimum number of observation units (m_{opt}) is not affected by the total allowable cost C_0. Only the optimum number of experimental units is affected by C_0.
2. Formula (29.43) is obtained by minimizing:

$$\sigma^2(\bar{Y}_{i..}) = \frac{\sigma_\eta^2 + m\sigma^2}{nm}$$

subject to the constraint:

$$c_1 n + c_2 nm - C_0 = 0$$

Setting up the Lagrangian function:

$$L = \frac{\sigma_\eta^2 + m\sigma^2}{nm} + \lambda(c_1 n + c_2 nm - C_0)$$

we differentiate L with respect to m, n, and λ and set the partial derivatives equal to zero. When the three equations are solved simultaneously, the results in (29.43) are obtained.

29.8 PURE SUBSAMPLING IN THREE STAGES

Sometimes an investigation does not involve a comparison of treatments, but only subsampling at several levels. Consider, for instance, a quality control engineer who wishes to investigate a certain quality characteristic of a computer assembly. These assemblies are produced in lots of 2,000. The engineer will select a random sample of r lots; from each lot she will select n assemblies; finally, she will obtain m observations on the quality characteristic for each assembly.

Model

Assuming that all random variables are normally distributed and that equal sample sizes are employed at each stage, the model for subsampling in three stages is:

(29.44) $$Y_{ijk} = \mu_{..} + \tau_i + \varepsilon_{j(i)} + \eta_{k(ij)}$$

where:

$\mu_{..}$ is a constant
τ_i, $\varepsilon_{j(i)}$, and $\eta_{k(ij)}$ are independent normal random variables with expectations 0 and variances σ_τ^2, σ^2, and σ_η^2, respectively
$i = 1, \ldots, r$; $j = 1, \ldots, n$; $k = 1, \ldots, m$

For our illustration, τ_i represents the lot effect, $\varepsilon_{j(i)}$ represents the assembly effect that is nested within the lot, and $\eta_{k(ij)}$ represents the observation effect that is nested within the assembly and therefore also within the lot.

This model corresponds to model (29.35) for a single-factor study with subsampling except that we assume here that the τ_i are independent $N(0, \sigma_\tau^2)$ and are independent of the $\varepsilon_{j(i)}$ and $\eta_{k(ij)}$. Formally, then, the only difference between models (29.35) and (29.44) is that the τ_i are fixed in one case and random in the other.

Analysis of variance

The analysis of variance for pure subsampling model (29.44) uses the same sums of squares as before, namely, those in (29.36). The ANOVA table is the same as that in Table 29.10. The applicable expected mean squares are those for random τ_i effects.

Estimation of $\mu_{..}$

In the case of pure subsampling, one is often interested in estimating the overall mean $\mu_{..}$ (the process mean for the computer assembly quality characteristic in our earlier example). A point estimator of $\mu_{..}$ in model (29.44) is $\bar{Y}_{...}$, and it can be shown that its variance is:

$$(29.45) \qquad \sigma^2(\bar{Y}_{...}) = \frac{\sigma_\tau^2}{r} + \frac{\sigma^2}{rn} + \frac{\sigma_\eta^2}{rnm} = \frac{nm\sigma_\tau^2 + m\sigma^2 + \sigma_\eta^2}{rnm}$$

An unbiased estimator of this variance is:

$$(29.46) \qquad s^2(\bar{Y}_{...}) = \frac{MSTR}{rnm}$$

and the $1 - \alpha$ confidence limits for $\mu_{..}$ are:

$$(29.47) \qquad \bar{Y}_{...} \pm t(1 - \alpha/2; r - 1)s(\bar{Y}_{...})$$

Subsampling extensions

Our discussion of subsampling has been confined to completely randomized designs and to three stages of sampling in the case of pure subsampling. Clearly, repeated observations can be used with any experimental design, and pure subsampling can be carried out with any number of stages. In the following chapter, we shall take up procedures for easily handling these more complex situations.

PROBLEMS

29.1. A student asked: "Since the mean squares in the analysis of variance table for a two-factor nested design are the same whether the factor effects are assumed to be random or fixed, what difference does it make whether we assume the factors to have fixed effects or random effects?" Comment.

29.2. A researcher declared: "I prefer analyzing a nested two-factor study as a crossed factor study because I can isolate more sources of variation." Comment on the researcher's strategy.

29.3. Consider a three-factor study where factor C is nested within factor B, and factor B in turn is nested within factor A, and $a = b = c = 2$. Illustrate in the format of Table 29.2 the distinction between this nested design and the corresponding crossed design.

29.4. **Bottling plant production.** A production engineer studied the effects of machine model (factor A) and operator (factor B) on the output in a bottling plant.

Three bottling machines were used, each a different model. Twelve operators were employed. Four operators were assigned to a machine and worked six-hour shifts each. Data on the number of cases produced by each machine and operator were collected for a week. The data that follow represent the number of cases produced per hour for each day during the week.

Machine i:	1				2				3			
Operator j:	1	2	3	4	1	2	3	4	1	2	3	4
Day $k = 1$:	65	68	56	45	74	69	52	73	69	63	81	67
$k = 2$:	58	62	65	56	81	76	56	78	83	70	72	79
$k = 3$:	63	75	58	54	76	80	62	83	74	72	73	73
$k = 4$:	57	64	70	48	80	78	58	75	78	68	76	77
$k = 5$:	66	70	64	60	68	73	51	76	80	75	70	71

Assume that fixed effects ANOVA model (29.8) is appropriate.

a. Can the operator effects be distinguished from the effects of shifts in this study? Discuss.

b. Plot the treatment sample means $\bar{Y}_{ij.}$ in the format of Figure 29.2. Does it appear that any factor effects are present?

c. Obtain the analysis of variance table.

d. Test whether or not the mean outputs differ for the three machine models; use $\alpha = .01$. State the alternatives, decision rule, and conclusion. What is the P-value of the test?

e. Test whether or not the mean outputs differ for the operators assigned to each machine; use $\alpha = .01$. State the alternatives, decision rule, and conclusion. What does your conclusion imply about the mean outputs for the four operators assigned to machine 3? Explain.

f. Test for each machine separately whether or not the mean outputs for the four operators differ. For each test, use $\alpha = .01$ and state the alternatives, decision rule, and conclusion.

g. What is the family level of significance for the combined tests in parts (d), (e), and (f) using the Bonferroni inequality? Summarize the set of conclusions reached in your tests.

29.5. Refer to **Bottling plant production** Problem 29.4.

a. Make all pairwise comparisons among the mean outputs for the three machines. Use the Tukey procedure with a 95 percent family confidence coefficient. State your findings.

b. Make all pairwise comparisons among the mean outputs for the four operators assigned to machine 1. Use the Bonferroni procedure with a 95 percent family confidence coefficient. State your findings.

c. Operator 4 assigned to machine 1 has relatively little experience compared to the other three operators. Estimate the contrast:

$$L = \frac{\mu_{11} + \mu_{12} + \mu_{13}}{3} - \mu_{14}$$

using a 99 percent confidence interval. Interpret your interval estimate.

29.6. Refer to **Bottling plant production** Problem 29.4. Obtain the residuals and plot them against the fitted values. Also prepare a normal probability plot of the residuals. What are your findings about the appropriateness of ANOVA model (29.8)?

29.7. Refer to **Bottling plant production** Problem 29.4. Assume that the four operators assigned to each machine were selected at random from a large number of operators.

a. How is ANOVA model (29.8) modified to fit this case?
b. Obtain a point estimate of the operator variance σ_β^2.
c. Test whether or not σ_β^2 equals zero; use $\alpha = .10$. State the alternatives, decision rule, and conclusion.
d. Test whether or not the mean outputs differ for the three machine models; use $\alpha = .10$. State the alternatives, decision rule, and conclusion.
e. Make all pairwise comparisons among the mean outputs for the three machines. Use the Tukey procedure with a 90 percent family confidence coefficient. State your findings.
f. Test the assumption that the $\beta_{j(i)}$ for all machines have the same variance σ_β^2. Use the Hartley test (Section 18.6) with significance level $\alpha = .01$. State the alternatives, decision rule, and conclusion.

29.8. Refer to **Bottling plant production** Problem 29.4. Assume that the four operators assigned to each machine were selected at random from a large number of operators and that the three machines were chosen at random from a large number of machines.

a. How is ANOVA model (29.8) modified to fit this case?
b. Obtain point estimates of the operator and machine variances σ_β^2 and σ_α^2, respectively.
c. Test whether or not σ_α^2 equals zero; use $\alpha = .05$. State the alternatives, decision rule, and conclusion.
d. The production engineer is interested in estimating the overall mean $\mu_{..}$ with a 95 percent confidence interval. Obtain the desired confidence interval and interpret your interval estimate.

29.9 Health awareness. Three states (factor A) participated in a health awareness study. Each state independently devised a health awareness program. Three cities (factor B) within each state were selected for participation and five households within each city were randomly selected to evaluate the effectiveness of the program. All members of the selected households were interviewed before and after participation in the program and a composite index was formed for each household measuring the impact of the health awareness program. The data on health awareness follow (the larger the index, the greater the awareness).

State i:	1			2			3		
City j:	1	2	3	1	2	3	1	2	3
Household $k = 1$:	42	26	34	47	56	68	19	18	16
$k = 2$:	56	38	51	58	43	51	36	40	28
$k = 3$:	35	42	60	39	65	49	24	27	45
$k = 4$:	40	35	29	62	70	71	12	31	30
$k = 5$:	28	53	44	65	59	57	33	23	21

Assume that fixed effects ANOVA model (29.8) is appropriate.

a. Plot the treatment sample means $\bar{Y}_{ij.}$ in the format of Figure 29.2. Does it appear that any factor effects are present?
b. Obtain the analysis of variance table.

c. Test whether or not the mean awareness differs for the three states; use $\alpha = .05$. State the alternatives, decision rule, and conclusion. What is the P-value of the test?
d. Test whether or not the mean awareness differs for the three cities within each state; use $\alpha = .05$. State the alternatives, decision rule, and conclusion. What does your conclusion imply about the awareness means for the three cities in state 1? Explain.
e. What is the family level of significance for the combined tests in parts (c) and (d) using the Bonferroni inequality? Summarize the set of conclusions reached in your tests.

29.10. Refer to **Health awareness** Problem 29.9.
a. Estimate μ_{11} with a 95 percent confidence interval. Interpret your interval estimate.
b. Obtain separate confidence intervals for $\mu_{1.}$, $\mu_{2.}$, and $\mu_{3.}$, each with a 99 percent confidence coefficient. Interpret your interval estimates.
c. Obtain confidence intervals for all pairwise comparisons between the state means. Use the Tukey procedure and a 90 percent family confidence coefficient. Summarize your findings.
d. It is desired to obtain a 95 percent confidence interval for $D = \mu_{11} - \mu_{32}$, since these two cities are of comparable size. Interpret your interval estimate.

29.11. Refer to **Health awareness** Problem 29.9. Obtain the residuals and plot them against the fitted values. Also prepare a normal probability plot of the residuals. What are your findings about the appropriateness of ANOVA model (29.8)?

29.12. Refer to **Health awareness** Problem 29.9. Assume that the three cities in each state were chosen at random from all the cities in the state.
a. How is ANOVA model (29.8) modified to fit this case?
b. Obtain a point estimate of the city variance σ_β^2. Is there anything special about the estimate here?
c. Test whether or not σ_β^2 equals zero; use $\alpha = .10$. State the alternatives, decision rule, and conclusion.
d. Test whether or not the mean awareness differs for the three states; use $\alpha = .10$. State the alternatives, decision rule, and conclusion. What is the P-value of the test?
e. Obtain confidence intervals for all pairwise comparisons between the state means. Use the Tukey procedure and a 90 percent family confidence coefficient. Summarize your findings.
f. Test the assumption that the $\beta_{j(i)}$ for all states have the same variance σ_β^2. Use the Hartley test (Section 18.6) with significance level $\alpha = .05$. State the alternatives, decision rule, and conclusion.

29.13. Refer to **Health awareness** Problem 29.9. Assume that the three cities within each state and the three states were selected at random.
a. How is ANOVA model (29.8) modified to fit this case?
b. Obtain point estimates of the city and state variances σ_β^2 and σ_α^2, respectively.

c. Test whether or not σ_α^2 equals zero; use $\alpha = .01$. State the alternatives, decision rule, and conclusion.

d. Estimate the overall mean health awareness index $\mu_{..}$ using a 99 percent confidence interval. Interpret your interval estimate.

29.14. **Internal control.** A large retailer operates three regional accounting centers (factor A). Center 1 employs three audit teams while the other two centers employ two audit teams each. One function of each center is to review whether a certain internal control operates properly in the processing of payroll. Data on the percent of transactions where the internal control was found to be operating properly were requested for each team in each region for the previous two months. Three months' data were received in one case, and data for only one month in another. The arc sine transformation $Y' = 2 \arcsin \sqrt{p}$ was employed to stabilize the error term variances. The transformed data follow.

Region i:	1			2		3	
Team j:	1	2	3	1	2	1	2
Month $k = 1$:	151.6	143.2	131.4	163.8	151.6	157.0	160.0
$k = 2$:	141.2	139.4	136.0	154.2		147.2	151.6
$k = 3$:	149.4						

Assume that fixed effects ANOVA model (29.8), modified for unequal nestings and replications, is appropriate.

a. Set up the full regression model for this case, analogous to the illustrative full model (29.31), using 1, −1, 0 indicator variables.

b. Test for region main effects using test statistic (8.37) and significance level $\alpha = .025$. State the alternatives, reduced model, decision rule, and conclusion.

c. Test for effects of audit teams within region using test statistic (8.37) and significance level $\alpha = .025$. State the alternatives, reduced model, decision rule, and conclusion.

d. Estimate $D = \mu_{1.} - \mu_{2.}$ (in transformed units) with a 98 percent confidence interval.

29.15. A student asked in class why all experiments do not make use of repeated observations since all measurement procedures are inexact to some degree. Comment.

29.16. Refer to **Questionnaire color** Problem 16.10. Suppose that the experiment was conducted by distributing the fliers to the assigned parking lots in two different weeks and noting the response rates for each week. The complete data on response rates follow.

Color i:	1 (Blue)					2 (Green)					3 (Orange)				
Lot j:	1	2	3	4	5	1	2	3	4	5	1	2	3	4	5
Week $k = 1$:	28	26	31	27	35	34	29	25	31	29	31	25	27	29	28
$k = 2$:	32	23	29	24	37	33	27	22	34	25	35	28	25	25	31

Assume that ANOVA model (29.34) with fixed treatment effects is appropriate.

a. Obtain the analysis of variance table.

b. Test whether or not questionnaire color effects are present; use $\alpha = .05$. State the alternatives, decision rule, and conclusion.

c. Test whether or not lot differences within colors are present; use $\alpha = .05$. State the alternatives, decision rule, and conclusion.
d. Estimate the mean response rate for blue questionnaires with a 95 percent confidence interval.
e. Obtain point estimates of σ^2 and σ_η^2. Which variance appears to be larger here?

29.17. An economist has $20,000 available for a study to compare the amounts of installment debt owed by urban families with two or fewer children with those of urban families with more than two children in a state. The cost of including a city in the study is $1,000 and the cost of including a family is $50. The number of families with two or fewer children in the study is to be the same as the number of families with more than two children. Assume that the cost function is given by (29.42). The primary purpose of the study is to estimate the mean debt for each of the two types of families as precisely as possible.
a. If reasonable planning values for the standard deviations are $\sigma = 150$ and $\sigma_\eta = 300$, how many cities and families should be included in the study?
b. How would the sample sizes change if $\sigma = 400$ and $\sigma_\eta = 200$?

29.18. Plant acid levels. Four plants of the same variety were randomly selected in an experiment to investigate the concentration of a particular acid. Three leaves per plant were randomly selected and three separate determinations of the acid concentration were obtained per leaf. The data follow.

Plant i:	1			2			3			4		
Leaf j:	1	2	3	1	2	3	1	2	3	1	2	3
Determination												
$k = 1$:	11.2	16.5	18.3	14.1	19.0	11.9	15.3	19.5	16.5	7.3	8.9	11.3
$k = 2$:	11.6	16.8	18.7	13.8	18.5	12.4	15.9	20.1	17.2	7.8	9.4	10.9
$k = 3$:	12.0	16.1	19.0	14.2	18.2	12.0	16.0	19.3	16.9	7.0	9.3	10.5

Assume that subsampling ANOVA model (29.44) is appropriate.
a. Obtain the analysis of variance table.
b. Test whether or not there are variations in mean concentration levels between plants; use $\alpha = .05$. State the alternatives, decision rule, and conclusion.
c. Test whether or not there are variations in mean concentration levels between leaves of the same plant; use $\alpha = .05$. State the alternatives, decision rule, and conclusion.
d. Estimate the overall mean concentration in all plants of the variety; use a 95 percent confidence interval.
e. Obtain point estimates of σ_τ^2, σ^2, and σ_η^2. Which component of variance appears to be most important in the total variance $\sigma^2(Y_{ijk}) = \sigma_\tau^2 + \sigma^2 + \sigma_\eta^2$?

29.19. Chemical consistency. A chemical company wished to study the consistency of the strength of one of its liquid chemical products. The product is made in batches in large vats and then is barreled. The barrels are subsequently stored for a period of time in a warehouse. To examine the consistency of the strength of the chemical, an analyst randomly selected five different batches of the product from the warehouse and then selected four barrels per batch at random. Three determinations per barrel were made. The data on strength follow.

Batch i:	1				2				3			
Barrel j:	1	2	3	4	1	2	3	4	1	2	3	4
Determination												
$k = 1$:	2.3	2.5	2.6	2.4	2.8	2.7	2.6	2.4	3.0	3.4	2.9	3.1
$k = 2$:	2.1	2.3	2.4	2.6	2.9	2.5	2.6	2.8	3.1	3.3	3.0	2.8
$k = 3$:	2.0	2.5	2.7	2.3	2.6	2.8	2.8	2.6	2.9	3.0	3.2	3.2

Batch i:	4				5			
Barrel j:	1	2	3	4	1	2	3	4
Determination								
$k = 1$:	2.5	2.8	3.1	2.7	3.6	3.8	3.7	3.9
$k = 2$:	2.8	3.0	2.8	2.9	3.7	3.8	3.5	3.5
$k = 3$:	2.6	2.7	2.9	2.6	3.4	3.5	3.5	3.7

Assume that subsampling ANOVA model (29.44) is appropriate.

a. Obtain the analysis of variance table.

b. Test whether or not there are variations in mean strength between batches; use $\alpha = .01$. State the alternatives, decision rule, and conclusion.

c. Test whether or not there are variations in mean strength between barrels within batches; use $\alpha = .01$. State the alternatives, decision rule, and conclusion.

d. Estimate the overall mean strength of the chemical using a 99 percent confidence interval.

e. Obtain point estimates of σ_τ^2, σ^2, and σ_η^2. Which component of variance appears to be most important in the total variance $\sigma^2(Y_{ijk}) = \sigma_\tau^2 + \sigma^2 + \sigma_\eta^2$?

EXERCISES

29.20. Derive (29.13) by squaring (29.12) and summing over all observations.

29.21. Derive (29.16b) from (29.13b).

29.22. Derive (29.17) for a balanced nested two-factor design.

29.23. Consider a balanced nested two-factor design with factor A having fixed effects and factor B (nested within factor A) having random effects.

a. Derive $\sigma^2(\bar{Y}_{i..})$ and $\sigma^2(\bar{Y}_{...})$.

b. Find an unbiased point estimator of σ_β^2.

29.24. Derive variance (29.41) for ANOVA model (29.35) with fixed treatment effects.

29.25. (Calculus needed.) Derive the optimal sample sizes given in (29.43). (*Hint:* See Comment 2 on p. 990.)

29.26. Derive variance (29.45) for subsampling ANOVA model (29.44). Using the expected mean squares in Table 29.10, show that the estimated variance (29.46) is an unbiased estimator of variance (29.45).

For instance, i refers to school, a fixed factor that occurs at a levels. Note that the subscript k refers to replication, which is a random "factor" and occurs at n levels.

Step 5. *In each row where one or more subscripts are in parentheses, enter a 1 in the column(s) corresponding to the subscript(s) in parentheses.*

Example

	i	j	k
	F a	F b	R n
α_i			
$\beta_{j(i)}$	1		
$\varepsilon_{k(ij)}$	1	1	

Thus, in the $\beta_{j(i)}$ row, we enter a 1 in the i column, and so on.

Step 6. *In each row where one or more subscripts are not in parentheses, enter in the column(s) corresponding to the subscript(s) not in parentheses a 1 if the subscript refers to a random factor, and a 0 if the factor is fixed.*

Example

	i	j	k
	F a	F b	R n
α_i	0		
$\beta_{j(i)}$	1	0	
$\varepsilon_{k(ij)}$	1	1	1

Thus, for the $\beta_{j(i)}$ row, the subscript not in parentheses is j, which refers to factor B, a fixed factor. Hence, a 0 is entered in the j column.

Step 7. *Fill in all remaining empty cells with the number of levels appearing in the column heading.*

Example

	i	j	k
	F a	F b	R n
α_i	0	b	n
$\beta_{j(i)}$	1	0	n
$\varepsilon_{k(ij)}$	1	1	1

Each $E(MS)$ will consist of a linear combination of the variance terms enumerated in step 2, with the coefficients obtained by taking additional steps in the table just completed. Some of the coefficients may be zero, which means that the corresponding variance term is not present in the $E(MS)$.

Step 8. *Adjoin on the right of the table just completed the variance term associated with the effect in that row. In addition, adjoin a column for each expected mean square to be found. Under each expected mean square, indicate all of the subscripts (including any parentheses) associated with the corresponding model term.*

Example

	i	j	k				
	F a	F b	R n	*Variance*	$E(MSA)$ i	$E[MSB(A)]$ $(i)j$	$E(MSE)$ $(ij)k$
α_i	0	b	n	σ_α^2			
$\beta_{j(i)}$	1	0	n	σ_β^2			
$\varepsilon_{k(ij)}$	1	1	1	σ^2			

Note that all of the subscripts of the associated model term, whether in parentheses or not, are shown under the expected mean square. For example, $E[MSB(A)]$ has associated with it the model term $\beta_{j(i)}$, so that the subscripts shown are (i) and j. Similarly, $E(MSE)$ has associated with it the model term $\varepsilon_{k(ij)}$, so that (ij) and k are shown.

Step 9. *For each expected mean square column, the coefficient of any variance term is zero if the subscript(s) of the model term in that row (whether in parentheses or not) do not include all of the subscript(s) in the heading of that E(MS) column (whether in parentheses or not).*

Example

	i	j	k				
	F a	F b	R n	*Variance*	$E(MSA)$ i	$E[MSB(A)]$ $(i)j$	$E(MSE)$ $(ij)k$
α_i	0	b	n	σ_α^2		0	0
$\beta_{j(i)}$	1	0	n	σ_β^2			0
$\varepsilon_{k(ij)}$	1	1	1	σ^2			

For the $E(MSA)$ column, it will be noted that the model terms in all rows contain the subscript i. Hence, none of the variances receives a zero coefficient as a result of this step.

For the $E[MSB(A)]$ column, note that the first row has a model term not containing both i and j. Hence, σ_α^2 receives a zero coefficient in the $E[MSB(A)]$ column.

Finally, for the $E(MSE)$ column, the first and second rows have model terms that do not contain the three subscripts i, j, and k. Hence, both σ_α^2 and σ_β^2 receive zero coefficients in the $E(MSE)$ column.

Step 10. *The coefficients of the variance terms that have not been assigned a zero coefficient as a result of step 9 are found as follows:*

a. *For each expected mean square column, delete (e.g., mask or cover) the column(s) on the left corresponding to the subscripts not in parentheses in the heading of the E(MS) column.*

b. *Multiply the entries in the remaining columns for each row being considered.*

Step 11. *The expected mean square equals the sum of the products of each coefficient times the associated variance term, with the variance terms for fixed effects replaced by sums of squared effects divided by degrees of freedom.*

Example

	i	j	k				
	F a	F b	R n	*Variance*	$E(MSA)$ i	$E[MSB(A)]$ $(i)j$	$E(MSE)$ $(ij)k$
α_i	0	b	n	σ_α^2	bn	0 (step 9)	0 (step 9)
$\beta_{j(i)}$	1	0	n	σ_β^2	0	n	0 (step 9)
$\varepsilon_{k(ij)}$	1	1	1	σ^2	1	1	1

To find the coefficients for the $E(MSA)$ column, for example, we noted earlier that no zero coefficient is assigned as a result of step 9. Step 10a calls for column i on the left to be deleted. Hence, we obtain by multiplying the terms in the j and k columns:

	j	k		
	F b	R n	*Variance*	$E(MSA)$ i
α_i	b	n	σ_α^2	bn
$\beta_{j(i)}$	0	n	σ_β^2	0
$\varepsilon_{k(ij)}$	1	1	σ^2	1

Thus:

$$E(MSA) = bn\sigma_\alpha^2 + (0)\sigma_\beta^2 + (1)\sigma^2 = bn\sigma_\alpha^2 + \sigma^2$$

Since factor A has fixed effects, we finally obtain:

$$E(MSA) = bn\frac{\Sigma\alpha_i^2}{a-1} + \sigma^2$$

We find the remaining coefficients for $E[MSB(A)]$ in similar fashion. We delete column j on the left, the subscript not in parentheses, and obtain:

	i	k		
	F a	R n	*Variance*	$E[MSB(A)]$ $(i)j$
α_i	0	n	σ_α^2	0 (step 9)
$\beta_{j(i)}$	1	n	σ_β^2	n
$\varepsilon_{k(ij)}$	1	1	σ^2	1

Thus:

$$E[MSB(A)] = (0)\sigma_\alpha^2 + n\sigma_\beta^2 + (1)\sigma^2 = n\sigma_\beta^2 + \sigma^2$$

Since factor B has fixed effects, we finally obtain:

$$E[MSB(A)] = n\frac{\Sigma\Sigma\beta_{j(i)}^2}{a(b-1)} + \sigma^2$$

To find the remaining coefficient in the $E(MSE)$ column, we delete column k, and the product on the σ^2 line is $1 \cdot 1 = 1$. Thus:

$$E(MSE) = (0)\sigma_\alpha^2 + (0)\sigma_\beta^2 + (1)\sigma^2 = \sigma^2$$

Assembling our results, we have:

(30.4a) $$E(MSA) = bn\frac{\Sigma\alpha_i^2}{a-1} + \sigma^2$$

(30.4b) $$E[MSB(A)] = n\frac{\Sigma\Sigma\beta_{j(i)}^2}{a(b-1)} + \sigma^2$$

(30.4c) $$E(MSE) = \sigma^2$$

Of course, these results are identical to those given earlier in Table 29.4.

Application of rule (30.3) to crossed factors

As stated earlier, rule (30.3) applies to crossed factors as well as to nested factors. We shall now illustrate the use of this rule when factors are crossed. We shall consider the case of a two-factor experiment in a completely randomized design, where factors A and B are crossed, factor A has fixed effects and factor B has random effects, and n observations are obtained for each factor combination. The model equation is that of (23.16):

$$Y_{ijk} = \mu_{..} + \alpha_i + \beta_j + (\alpha\beta)_{ij} + \varepsilon_{k(ij)}$$

except that we now recognize the nesting of the error term ε. The random effects variance terms corresponding to the model terms are:

$$\begin{array}{cccc} \alpha_i & \beta_j & (\alpha\beta)_{ij} & \varepsilon_{k(ij)} \\ \sigma_\alpha^2 & \sigma_\beta^2 & \sigma_{\alpha\beta}^2 & \sigma^2 \end{array}$$

Here, only the α_i are fixed effects, so at the end σ_α^2 will need to be replaced by a sum of squared effects divided by degrees of freedom.

Table 30.1a contains the preliminary tabulation required for finding the expected mean squares. Table 30.1b presents the results of steps 9 and 10. Note that for finding $E(MSA)$, σ_β^2 receives a zero coefficient as a result of step 9 since the subscript in the β_j model term does not contain the subscript i in the $E(MSA)$ column. Column i is deleted for step 10 for finding the coefficients in the $E(MSA)$ column since it is the only subscript in the column heading and is not in parentheses. The other expected mean squares coefficients are found in similar fashion. Table 30.1b indicates for each expected mean square whether the zero coefficients are obtained from step 9, and also which columns are deleted. The final expected mean squares, presented in Table 30.1c, are identical, of course, to those shown in Table 23.3.

TABLE 30.1 *E(MS)* derivations for crossed two-factor experiment (*A* fixed, *B* random)

(a) Table

	i	j	k
	F	R	R
	a	b	n
α_i	0	b	n
β_j	a	1	n
$(\alpha\beta)_{ij}$	0	1	n
$\varepsilon_{k(ij)}$	1	1	1

(b) Coefficients

Variance	$E(MSA)$ i	$E(MSB)$ j	$E(MSAB)$ ij	$E(MSE)$ $(ij)k$
σ_α^2	$b \cdot n$	0 (step 9)	0 (step 9)	0 (step 9)
σ_β^2	0 (step 9)	$a \cdot n$	0 (step 9)	0 (step 9)
$\sigma_{\alpha\beta}^2$	$1 \cdot n$	$0 \cdot n$	n	0 (step 9)
σ^2	$1 \cdot 1$	$1 \cdot 1$	1	$1 \cdot 1$
	(i col. deleted)	(j col. deleted)	(i, j cols. deleted)	(k col. deleted)

(c) *E(MS)*

$$E(MSA) = bn(\Sigma\alpha_i^2)/(a-1) + n\sigma_{\alpha\beta}^2 + \sigma^2$$
$$E(MSB) = an\sigma_\beta^2 + \sigma^2$$
$$E(MSAB) = n\sigma_{\alpha\beta}^2 + \sigma^2$$
$$E(MSE) = \sigma^2$$

30.3 RULE FOR FINDING SUMS OF SQUARES AND DEGREES OF FREEDOM

Since nested factors and subsampling designs may require sums of squares not discussed so far, we shall now consider a rule for finding sums of squares and associated degrees of freedom. This rule applies to all balanced designs—i.e., designs where a nested factor has the same number of levels for all levels of the factor it is nested in and the number of replications is constant, or where the subsample sizes at each stage of sampling are constant. These are the same conditions as for rule (30.3) for finding expected mean squares.

Illustration

The rule for finding sums of squares and associated degrees of freedom can best be explained in terms of an illustration. We shall consider a two-factor study in a completely randomized design, where factors A and B are crossed, factor A has a levels, factor B has b levels, and there are n replications for each treatment. It does not matter for this rule whether the factor effects are fixed or random.

Rule (30.5) for definitional forms of sums of squares

Step 1. *Write the model equation.*

Example. For a two-factor experiment in a completely randomized design, the model is given by (20.27) in the case of crossed factors and fixed factor effects:

$$Y_{ijk} = \mu_{..} + \alpha_i + \beta_j + (\alpha\beta)_{ij} + \varepsilon_{k(ij)}$$
$$i = 1, \ldots, a; j = 1, \ldots, b; k = 1, \ldots, n$$

We now use the $\varepsilon_{k(ij)}$ notation to show that the replications are nested within the (ij) factor combinations.

Step 2. *For each model term other than the overall constant and the error term, write the associated SS notation.*

Example. We do this for our example in columns 1 and 2 of Table 30.2 for α_i, β_j, and $(\alpha\beta)_{ij}$. The lines for Error and Total will not be completed until steps 9 and 10.

Step 3. *Each sum of squares will have as coefficient the product of the limits of the subscripts not appearing in the model term.*

Example. The coefficients for our example are shown in column 3 of Table 30.2. For instance, α_i does not contain j and k. These subscripts have limits of b and n, respectively. The coefficient for the SSA term is therefore bn.

Step 4. *Each sum of squares is summed over all of the subscripts of the model term, whether in parentheses or not.*

TABLE 30.2 Derivation of definitional sums of squares formulas for crossed two-factor experiment in completely randomized design

(1) *Model Term*	(2) *SS*	(3) *Coefficient*	(4) Σ	(5) *Symbolic Product*	(6) *Term to Be Squared*	(7) *Sum of Squares*	(8) *Degrees of Freedom*
α_i	*SSA*	*bn*	$\sum_i$	$i - 1$	$\bar{Y}_{i..} - \bar{Y}_{...}$	$bn \sum_i (\bar{Y}_{i..} - \bar{Y}_{...})^2$	$a - 1$
β_j	*SSB*	*an*	$\sum_j$	$j - 1$	$\bar{Y}_{.j.} - \bar{Y}_{...}$	$an \sum_j (\bar{Y}_{.j.} - \bar{Y}_{...})^2$	$b - 1$
$(\alpha\beta)_{ij}$	*SSAB*	*n*	$\sum_i \sum_j$	$(i - 1)(j - 1)$ $= ij - i - j + 1$	$\bar{Y}_{ij.} - \bar{Y}_{i..} - \bar{Y}_{.j.} + \bar{Y}_{...}$	$n \sum_i \sum_j (\bar{Y}_{ij.} - \bar{Y}_{i..} - \bar{Y}_{.j.} + \bar{Y}_{...})^2$	$(a - 1)(b - 1)$
(Error)	*SSE*				Remainder $= Y_{ijk} - \bar{Y}_{ij.}$	Remainder $= \sum_i \sum_j \sum_k (Y_{ijk} - \bar{Y}_{ij.})^2$	Remainder $= ab(n - 1)$
Total	*SSTO*				$Y_{ijk} - \bar{Y}_{...}$	$\sum_i \sum_j \sum_k (Y_{ijk} - \bar{Y}_{...})^2$	$abn - 1$

Example. The summations for our example are shown in column 4. For instance, the sum of squares term corresponding to α_i is summed over i, the only subscript in that model term.

Step 5. *Form a symbolic product from the subscripts of the model term, using the subscript if it is in parentheses, and the subscript minus* 1 *if it is not in parentheses. Expand the product.*

Example. The symbolic products for our example are shown in column 5. For instance, for α_i the symbolic product is $i - 1$. For $(\alpha\beta)_{ij}$, the symbolic product is $(i - 1)(j - 1) = ij - i - j + 1$.

Step 6. *The typical term to be squared consists of means of the observations with the subscripts consisting of the symbolic product term and dots elsewhere. The sign of each mean is that of the symbolic product. A* 1 *refers to the overall mean.*

Example. The terms to be squared for our example are shown in column 6. Note that for α_i, the symbolic product is $i - 1$, and the typical term to be squared is $\bar{Y}_{i..} - \bar{Y}_{...}$. For $(\alpha\beta)_{ij}$, the symbolic product is $ij - i - j + 1$, and hence the typical term to be squared is:

$$\bar{Y}_{ij.} - \bar{Y}_{i..} - \bar{Y}_{.j.} + \bar{Y}_{...}$$

Step 7. *Combining the steps of squaring, summing, and multiplying by the coefficient yields the appropriate sums of squares.*

Example. The sums of squares are shown for our example in column 7.

Step 8. *The degrees of freedom are obtained by replacing in each symbolic product the subscript variable by its limit.*

Example. For our example, the degrees of freedom are shown in column 8. For instance, for α_i the symbolic product is $i - 1$; hence, $df = a - 1$. For $(\alpha\beta)_{ij}$, the symbolic product is $(i - 1)(j - 1)$; hence, $df = (a - 1)(b - 1)$.

Step 9. *The total sum of squares is always defined as the sum, over all observations, of the squared deviations of the observations from the overall mean. The total degrees of freedom are always defined as one less than the total number of observations.*

Step 10. *The error sum of squares is obtained as a remainder from the total sum of squares, and likewise the degrees of freedom associated with the error sum of squares are obtained as a remainder from the total degrees of freedom.*

The results in Table 30.2 are, of course, the same as those given earlier in Table 20.9.

Rule (30.5a) for computational forms of sums of squares

If the computational forms of the sums of squares are desired, the procedure is modified as follows:

Step 1a. *Obtain the symbolic products as before.*

Step 2a. *For each term in the symbolic product, there corresponds a total of the observations with the subscripts consisting of the symbolic product term and dots elsewhere. A* 1 *denotes the grand total.*

Step 3a. *The total is squared and summed over the subscript(s) it has.*

Step 4a. *The sum has the sign of the corresponding term in the symbolic product, and is divided by the product of the limits of the subscripts not present.*

The degrees of freedom are obtained as before. *SSTO* always equals the sum of the squared observations minus the square of the sum of all observations divided by the total number of observations.

Example 1. Table 30.3 contains the derivation of the computational formulas for our crossed two-factor example. To illustrate the derivations, consider model term $(\alpha\beta)_{ij}$. For this term, the symbolic product is $ij - i - j + 1$. Hence, we obtain:

$$\frac{\sum_i \sum_j Y_{ij.}^2}{n} - \frac{\sum_i Y_{i..}^2}{bn} - \frac{\sum_j Y_{.j.}^2}{an} + \frac{Y_{...}^2}{abn}$$

The results in Table 30.3 are the same as, or equivalent to, those given earlier in (20.44).

Example 2. Table 30.4 contains the derivation of the computational formulas for sums of squares for a nested two-factor experiment, with B nested within A. The results are, of course, the same as those given earlier in (29.16).

Note

In general, the error sum of squares and the associated degrees of freedom should be obtained as remainders. However, in designs where there are two or more observations in each cell and there are no interaction terms that are assumed to equal zero, the error sum of squares and the associated degrees of freedom can be obtained by following the rules just given. The error sum of squares will always then turn out to be the sum of the squared deviations around the cell means.

For example, for a crossed two-factor experiment in a completely randomized design, we find:

Model term: $\varepsilon_{k(ij)}$

Symbolic product: $(k - 1)(ij) = ijk - ij$

SSE: $$\sum_i \sum_j \sum_k Y_{ijk}^2 - \frac{\sum_i \sum_j Y_{ij.}^2}{n} = \sum_i \sum_j \sum_k (Y_{ijk} - \bar{Y}_{ij.})^2$$

df: $abn - ab = ab(n - 1)$

TABLE 30.3 Derivation of computational sums of squares formulas for crossed two-factor experiment in completely randomized design

Model Term	*Symbolic Product*	*Sum of Squares*	*Degrees of Freedom*
α_i	$i - 1$	$SSA = \frac{\sum_i Y_{i..}^2}{bn} - \frac{Y_{...}^2}{abn}$	$a - 1$
β_j	$j - 1$	$SSB = \frac{\sum_j Y_{.j.}^2}{an} - \frac{Y_{...}^2}{abn}$	$b - 1$
$(\alpha\beta)_{ij}$	$(i - 1)(j - 1)$ $= ij - i - j + 1$	$SSAB = \frac{\sum_i \sum_j Y_{ij.}^2}{n} - \frac{\sum_i Y_{i..}^2}{bn} - \frac{\sum_j Y_{.j.}^2}{an} + \frac{Y_{...}^2}{abn}$	$(a - 1)(b - 1)$ $= ab - a - b + 1$
(Error)		$SSE = \text{Remainder} = \sum_i \sum_j \sum_k Y_{ijk}^2 - \frac{\sum_i \sum_j Y_{ij.}^2}{n}$	Remainder $= ab(n - 1)$
Total		$SSTO = \sum_i \sum_j \sum_k Y_{ijk}^2 - \frac{Y_{...}^2}{abn}$	$abn - 1$

TABLE 30.4 Derivation of computational sums of squares formulas for nested two-factor experiment (B nested within A)

Model: $Y_{ijk} = \mu_{..} + \alpha_i + \beta_{j(i)} + \varepsilon_{k(ij)}$

Model Term	*Symbolic Product*	*Sum of Squares*	*Degrees of Freedom*
α_i	$i - 1$	$SSA = \dfrac{\sum_i Y_{i..}^2}{bn} - \dfrac{Y_{...}^2}{abn}$	$a - 1$
$\beta_{j(i)}$	$(j - 1)i = ij - i$	$SSB(A) = \dfrac{\sum_i \sum_j Y_{ij.}^2}{n} - \dfrac{\sum_i Y_{i..}^2}{bn}$	$(b - 1)a = ab - a$
(Error)		$SSE = \text{Remainder} = \sum_i \sum_j \sum_k Y_{ijk}^2 - \dfrac{\sum_i \sum_j Y_{ij.}^2}{n}$	Remainder $= ab(n - 1)$
Total		$SSTO = \sum_i \sum_j \sum_k Y_{ijk}^2 - \dfrac{Y_{...}^2}{abn}$	$abn - 1$

We obtain exactly the same results for a nested two-factor experiment since the model error term, $\varepsilon_{k(ij)}$, is formally the same. These results, of course, are identical to those presented in earlier chapters.

30.4 CROSSED-NESTED EXPERIMENT—THREE FACTORS

In Chapter 29, we laid the groundwork for the design and analysis of experiments involving nested factors and subsampling. In the previous sections of this chapter, we have presented rules for developing the model and finding sums of squares, degrees of freedom, and expected mean squares. We are now ready to consider a variety of balanced designs involving crossed and nested factors and subsampling. We shall illustrate for each the development of an appropriate model and the corresponding analysis of variance.

Illustration

We consider first a situation where some but not all of the factors are nested. Such designs are called *partially nested, partially hierarchical,* or *crossed-nested* designs. An experiment studied the effect of cultural background on group decision making. Teams of students were formed and assigned a task. One of the dependent variables was the number of questions raised prior to the final group decision. Some teams consisted of foreign students, others of U.S. students. Half of the teams consisted of eight members, the other half of four members. Two foreign observers were used for the foreign teams, and two U.S. observers for the U.S. teams. Thus, the design may be represented as follows:

	U.S. Teams (A_1)		Foreign Teams (A_2)	
	Observer 1 (C_1)	Observer 2 (C_2)	Observer 3 (C_1)	Observer 4 (C_2)
Small team (B_1)	Replication 1 Replication 2	Replication 1 Replication 2	Replication 1 Replication 2	Replication 1 Replication 2
Large team (B_2)	Replication 1 Replication 2	Replication 1 Replication 2	Replication 1 Replication 2	Replication 1 Replication 2

For simplicity, we assume that only two replications (teams) were used in each cell.

Development of model

Let nationality of team be factor A, size of team factor B, and observer factor C. Note that factor C is nested within factor A since the two observers for the U.S. teams were different from the two observers for the foreign teams. Also note that factors A and B are crossed since each level of factor A appears with every level of factor B, and vice versa. Similarly, factors B and C are crossed. In this example, factors A and B were considered to have fixed effects, while factor C (observer) effects were considered to be random. In using rule (30.1) to develop an appropriate model, we need to recognize according to step 2 that AC and ABC interactions are to be excluded because factor C is nested within factor A. Further, step 3 tells us that the BC interaction is nested within factor A since factor C is nested within factor A; thus, the BC interaction is a $BC(A)$ interaction. Hence, the appropriate model is:

$$(30.6) \qquad Y_{ijkm} = \mu_{...} + \alpha_i + \beta_j + \gamma_{k(i)} + (\alpha\beta)_{ij} + (\beta\gamma)_{jk(i)} + \varepsilon_{m(ijk)}$$

where:

$\mu_{...}$ is an overall constant
α_i are the fixed nationality effects
β_j are the fixed team size effects
$\gamma_{k(i)}$ are the random observer (within nationality) effects
$(\alpha\beta)_{ij}$ are the fixed nationality–team size interaction effects
$(\beta\gamma)_{jk(i)}$ are the random team size–observer interaction effects (within nationality)
$\varepsilon_{m(ijk)}$ are the random error terms
$i = 1, \ldots, a; j = 1, \ldots, b; k = 1, \ldots, c; m = 1, \ldots, n$

We shall assume that $\gamma_{k(i)}$, $(\beta\gamma)_{jk(i)}$, and $\varepsilon_{m(ijk)}$ are normally distributed with expectations 0 and variances σ_γ^2, $\sigma_{\beta\gamma}^2$, and σ^2, respectively. The following restrictions on the model apply:

(30.6a)
$$\sum_i \alpha_i = 0 \qquad \sum_j \beta_j = 0 \qquad \sum_i (\alpha\beta)_{ij} = \sum_j (\alpha\beta)_{ij} = 0$$
$$\sum_j (\beta\gamma)_{jk(i)} = 0 \text{ for all } k(i)$$

Analysis of variance

Table 30.5 contains the development of the computational sums of squares and degrees of freedom for model (30.6), and Table 30.6 contains the development of the expected mean squares. In Table 30.6b, we have, as usual, replaced variance terms for fixed effects by sums of squared effects divided by degrees of freedom. Table 30.6b indicates directly how to form test statistics for a variety of tests.

Example

Table 30.7 contains the results of the group decision-making experiment described earlier, based on $n = 2$ replications. The analysis of variance was obtained by means of a computer program and is presented in Table 30.8.

To test for nationality effects, the alternatives are:

(30.7a)
$$\begin{aligned} &H_0: \alpha_1 = \alpha_2 = 0 \\ &H_a: \text{not both } \alpha_i \text{ equal zero} \end{aligned}$$

Table 30.6b indicates that the appropriate test statistic is:

(30.7b)
$$F^* = \frac{MSA}{MSC(A)}$$

We have for our example:

$$F^* = \frac{420.25}{.25} = 1{,}681$$

If the level of significance is to be $\alpha = .05$, we require $F(.95; 1, 2) = 18.5$. Since $F^* = 1{,}681 > 18.5$, we conclude H_a, that nationality does have an effect on the group behavior. Other tests would be conducted in a similar fashion.

Confidence intervals for main factor effects are set up in the usual way when the factor effects are fixed. For example, to estimate the mean number of questions raised prior to a decision for U.S. teams, we require $MSC(A)$ since this is the mean square in the denominator of the test statistic for examining nationality effects. Specifically, the confidence limits for $\mu_{i..}$ are:

(30.8)
$$\bar{Y}_{i...} \pm t[1 - \alpha/2; (c-1)a]s(\bar{Y}_{i...})$$

TABLE 30.5 Derivation of computational sums of squares formulas for crossed-nested model (30.6)

Model Term	*Symbolic Product*	*Sum of Squares*	*Degrees of Freedom*
α_i	$i - 1$	$SSA = \frac{\sum_i Y_{i...}^2}{bcn} - \frac{Y_{....}^2}{abcn}$	$a - 1$
β_j	$j - 1$	$SSB = \frac{\sum_j Y_{.j..}^2}{acn} - \frac{Y_{....}^2}{abcn}$	$b - 1$
$\gamma_{k(i)}$	$(k - 1)i$ $= ki - i$	$SSC(A) = \frac{\sum_i \sum_k Y_{i.k.}^2}{bn} - \frac{\sum_i Y_{i...}^2}{bcn}$	$a(c - 1)$
$(\alpha\beta)_{ij}$	$(i - 1)(j - 1)$ $= ij - i - j + 1$	$SSAB = \frac{\sum_i \sum_j Y_{ij..}^2}{cn} - \frac{\sum_i Y_{i...}^2}{bcn} - \frac{\sum_j Y_{.j..}^2}{acn} + \frac{Y_{....}^2}{abcn}$	$(a - 1)(b - 1)$
$(\beta\gamma)_{jk(i)}$	$(j - 1)(k - 1)i$ $= ijk - ij - ik + i$	$SSBC(A) = \frac{\sum_i \sum_j \sum_k Y_{ijk.}^2}{n} - \frac{\sum_i \sum_j Y_{ij..}^2}{cn} - \frac{\sum_i \sum_k Y_{i.k.}^2}{bn} + \frac{\sum_i Y_{i...}^2}{bcn}$	$a(b - 1)(c - 1)$
$\varepsilon_{m(ijk)}$	$(m - 1)ijk$ $= ijkm - ijk$	$SSE = \sum_i \sum_j \sum_k \sum_m Y_{ijkm}^2 - \frac{\sum_i \sum_j \sum_k Y_{ijk.}^2}{n}$	$abc(n - 1)$
Total		$SSTO = \sum_i \sum_j \sum_k \sum_m Y_{ijkm}^2 - \frac{Y_{....}^2}{abcn}$	$abcn - 1$

TABLE 30.6 Derivation of expected mean squares for crossed-nested model (30.6)

(a) Table

	i	j	k	m	Variance	Expected Mean Square of—					
						A	B	$C(A)$	AB	$BC(A)$	E
	F	F	R	R							
	a	b	c	n		i	j	$(i)k$	ij	$(i)jk$	$(ijk)m$
α_i	0	b	c	n	σ_α^2	bcn	0	0	0	0	0
β_j	a	0	c	n	σ_β^2	0	acn	0	0	0	0
$\gamma_{k(i)}$	1	b	1	n	σ_γ^2	bn	0	bn	0	0	0
$(\alpha\beta)_{ij}$	0	0	c	n	$\sigma_{\alpha\beta}^2$	0	0	0	cn	0	0
$(\beta\gamma)_{jk(i)}$	1	0	1	n	$\sigma_{\beta\gamma}^2$	0	n	0	n	n	0
$\varepsilon_{m(ijk)}$	1	1	1	1	σ^2	1	1	1	1	1	1

(b) Expected Mean Squares

$$E(MSA) = bcn\frac{\Sigma\alpha_i^2}{a-1} + bn\sigma_\gamma^2 + \sigma^2$$

$$E(MSB) = acn\frac{\Sigma\beta_j^2}{b-1} + n\sigma_{\beta\gamma}^2 + \sigma^2$$

$$E[MSC(A)] = bn\sigma_\gamma^2 + \sigma^2$$

$$E(MSAB) = cn\frac{\Sigma\Sigma(\alpha\beta)_{ij}^2}{(a-1)(b-1)} + n\sigma_{\beta\gamma}^2 + \sigma^2$$

$$E[MSBC(A)] = n\sigma_{\beta\gamma}^2 + \sigma^2$$

$$E(MSE) = \sigma^2$$

TABLE 30.7 Data for crossed-nested three-factor study—group decision-making example

	U.S. Teams ($i = 1$)		*Foreign Teams* ($i = 2$)	
Size of Team	*Observer 1* ($k = 1$)	*Observer 2* ($k = 2$)	*Observer 3* ($k = 1$)	*Observer 4* ($k = 2$)
4 members	16	14	7	4
($j = 1$)	20	19	5	9
8 members	21	28	11	12
($j = 2$)	25	19	17	15

$Y_{111.} = 36$ $Y_{112.} = 33$ $Y_{211.} = 12$ $Y_{212.} = 13$

$Y_{121.} = 46$ $Y_{122.} = 47$ $Y_{221.} = 28$ $Y_{222.} = 27$

$Y_{11..} = 69$ $Y_{1.1.} = 82$ $Y_{21..} = 25$ $Y_{2.1.} = 40$

$Y_{12..} = 93$ $Y_{1.2.} = 80$ $Y_{22..} = 55$ $Y_{2.2.} = 40$

$Y_{1...} = 162$ $Y_{2...} = 80$

$Y_{.1..} = 94$ $Y_{.2..} = 148$

$Y_{....} = 242$

TABLE 30.8 ANOVA for crossed-nested three-factor study—group decision-making example

	Source of Variation	SS	df	MS
A	Nationality	420.25	1	420.25
B	Size of team	182.25	1	182.25
$C(A)$	Observer (within nationality)	.50	2	.25
AB	Nationality–size of team interactions	2.25	1	2.25
$BC(A)$	Size of team–observer interactions (within nationality)	2.50	2	1.25
E	Error	106.00	8	13.25
	Total	713.75	15	

where:

(30.8a)
$$s^2(\bar{Y}_{i...}) = \frac{MSC(A)}{nbc}$$

For our example, we obtain:

$$\bar{Y}_{1...} = \frac{162}{8} = 20.25$$

$$s^2(\bar{Y}_{1...}) = \frac{.25}{8} = .031 \qquad s(\bar{Y}_{1...}) = .18$$

If the confidence coefficient is to be .95, we require $t(.975; 2) = 4.303$, and the 95 percent confidence interval for $\mu_{1..}$ is:

$$19.5 = 20.25 - 4.303(.18) \leq \mu_{1..} \leq 20.25 + 4.303(.18) = 21.0$$

Comments

1. The sums of squares SSA, SSB, and $SSAB$ in Table 30.5 for the analysis of a crossed-nested experimental design are the usual sums of squares for factor A main effects, factor B main effects, and AB interactions. $SSC(A)$ is a typical nested sum of squares for factor C effects. This can be seen by writing out the sum of squares in definitional form from the symbolic product $ki - i$:

(30.9)
$$SSC(A) = bn\sum_i \sum_k (\bar{Y}_{i.k.} - \bar{Y}_{i...})^2$$

Thus, $SSC(A)$ simply measures the variability of the factor C level sample means for any given level of factor A, and then aggregates these sums of squares over factor A.

The definitional form of $SSBC(A)$ can be obtained from the symbolic product $ijk - ij - ik + i$:

(30.10)
$$SSBC(A) = n\sum_i \sum_j \sum_k (\bar{Y}_{ijk.} - \bar{Y}_{ij..} - \bar{Y}_{i.k.} + \bar{Y}_{i...})^2$$

Thus, $SSBC(A)$ contains the usual BC interaction sum of squares for a given level of factor A, and then aggregates these sums of squares over factor A.

2. If the only available computer program provides sums of squares for crossed factors, the two nested sums of squares can be obtained by utilizing the following relationships when the design is balanced:

(30.11a)
$$SSC(A) = SSC + SSAC$$

(30.11b)
$$SSBC(A) = SSBC + SSABC$$

These relationships generalize in a straightforward way. For example, if B is nested within A and C within B, we have:

(30.12a)
$$SSB(A) = SSB + SSAB$$

(30.12b)
$$SSC(AB) = SSC + SSAC + SSBC + SSABC$$

3. If important AB interactions are present, analysis should usually focus on the means $\mu_{ij.}$ when the factors have fixed effects, rather than on the factor level means $\mu_{i..}$ and $\mu_{.j.}$. It can be shown that the estimated variance for comparing the two team sizes for any given nationality is:

(30.13)
$$s^2(\bar{Y}_{i1..} - \bar{Y}_{i2..}) = \frac{2MSBC(A)}{cn}$$

This variance has associated with it $a(b - 1)(c - 1)$ degrees of freedom, as is evident from Table 30.5.

No exact confidence interval exists for comparing the two nationalities for any given team size. An unbiased variance estimator that can be utilized is:

(30.14)
$$s^2(\bar{Y}_{1j..} - \bar{Y}_{2j..}) = \frac{2}{cn}\left[MSBC(A) + \frac{MSC(A) - MSE}{b}\right]$$

The approximate number of degrees of freedom associated with this variance is obtained from (24.61).

The reason for the different variances in (30.13) and (30.14) is that the observers are the same when the two team sizes for a given nationality are compared, while the observers differ when the two nationalities for a given team size are compared.

30.5 SUBSAMPLING IN RANDOMIZED BLOCK DESIGN

The model usually employed for a randomized block design when one observation is made on an experimental unit is model (27.1) in the case of fixed treatment and block effects:

(30.15)
$$Y_{ij} = \mu_{..} + \rho_i + \tau_j + \varepsilon_{ij}$$

There are occasions when more than one observation is made on each experimental unit. Consider, for instance, an experiment to study how three different motivational stimuli affect the length of time a person requires to perform a task. The persons in the experiment are blocked into groups of three, according to age, and each person is assigned at random one of the three motivational stimuli. Three observations are then made on the time required to complete the task, i.e., the subject is asked to perform the task three times.

In this type of situation, we simply add a random observation error component to model (30.15). Assuming that the treatment and block effects (motivational stimuli and age groups in our example) are fixed, an appropriate model is:

$$Y_{ijk} = \mu_{..} + \rho_i + \tau_j + \varepsilon_{(ij)} + \eta_{k(ij)} \tag{30.16}$$

where:

$\Sigma\rho_i = 0 \quad \Sigma\tau_j = 0$

$\varepsilon_{(ij)}$ and $\eta_{k(ij)}$ are independent normal random variables with expectations 0 and variances σ^2 and σ_η^2, respectively

$i = 1, \ldots, n;\ j = 1, \ldots, r;\ k = 1, \ldots, m$

Here ρ_i is the block effect, τ_j the treatment effect, $\varepsilon_{(ij)}$ the random effect associated with the experimental unit, and $\eta_{k(ij)}$ the random effect associated with the kth observation on the experimental unit. Note that the experimental error ε is nested within the (ij) block-treatment combination; there is no additional subscript since only one experimental unit is assigned to a treatment within a block. Also note that the observation error η is nested within the (ij) block-treatment combination.

Table 30.9 contains the derivation of the computational sums of squares for model (30.16), and Table 30.10 contains the derivation of the expected mean

TABLE 30.9 Derivation of computational sums of squares formulas for randomized block design with subsampling—model (30.16)

Model Term	*Symbolic Product*	*Sum of Squares*	*Degrees of Freedom*
ρ_i	$i - 1$	$SSBL = \dfrac{\sum_i Y_{i..}^2}{rm} - \dfrac{Y_{...}^2}{nrm}$	$n - 1$
τ_j	$j - 1$	$SSTR = \dfrac{\sum_j Y_{.j.}^2}{nm} - \dfrac{Y_{...}^2}{nrm}$	$r - 1$
(Experimental units error)		$SSEE =$ Remainder $= \dfrac{\sum_i\sum_j Y_{ij.}^2}{m} - \dfrac{\sum_i Y_{i..}^2}{rm} - \dfrac{\sum_j Y_{.j.}^2}{nm} + \dfrac{Y_{...}^2}{nrm}$	Remainder $= (n-1)(r-1)$
$\eta_{k(ij)}$	$(k-1)ij = ijk - ij$	$SSOE = \sum_i\sum_j\sum_k Y_{ijk}^2 - \dfrac{\sum_i\sum_j Y_{ij.}^2}{m}$	$nr(m-1)$
Total		$SSTO = \sum_i\sum_j\sum_k Y_{ijk}^2 - \dfrac{Y_{...}^2}{nrm}$	$nrm - 1$

TABLE 30.10 Derivation of expected mean squares for randomized block design with subsampling—model (30.16)

(a) Table

	i	j	k	*Variance*	Expected Mean Square of— *BL*	*TR*	*EE*	*OE*
	F	F	R					
	n	r	m		i	j	(ij)	$(ij)k$
ρ_i	0	r	m	σ_ρ^2	rm	0	0	0
τ_j	n	0	m	σ_τ^2	0	nm	0	0
$\varepsilon_{(ij)}$	1	1	m	σ^2	m	m	m	0
$\eta_{k(ij)}$	1	1	1	σ_η^2	1	1	1	1

(b) Expected Mean Squares

$$E(MSBL) = rm\frac{\Sigma\rho_i^2}{n-1} + m\sigma^2 + \sigma_\eta^2$$

$$E(MSTR) = nm\frac{\Sigma\tau_j^2}{r-1} + m\sigma^2 + \sigma_\eta^2$$

$$E(MSEE) = m\sigma^2 + \sigma_\eta^2$$

$$E(MSOE) = \sigma_\eta^2$$

squares. Note that *SSEE*, the sum of squares for experimental units, is obtained as a remainder in Table 30.9 because there is only one experimental unit assigned to a treatment within a block. Indeed, *SSEE* turns out to be the block-treatment interaction sum of squares, as for a randomized block design without subsampling.

Table 30.10b indicates that for model (30.16) with fixed treatment and block effects, the test statistic for examining the presence of treatment effects is $F^* = MSTR/MSEE$, as is also the case when no subsampling occurs in a complete block design—see (27.6b).

30.6 TWO-FACTOR EXPERIMENT WITH REPEATED MEASURES ON ONE FACTOR

An experimenter wished to study the effects of two types of incentives (factor A) on a person's ability to solve problems. He also wanted to study two types of problems (factor B)—abstract and concrete problems. He could ask each experimental subject to do each type of problem, but could not use more than one type of incentive stimulus on a subject because of interference effects. Thus, the design the experimenter utilized may be represented schematically as follows:

Subject	*Incentive Stimulus*	*Problem*	
1	A_1	B_1	B_2
⋮	⋮	⋮	
n	A_1	B_1	B_2
$n+1$	A_2	B_1	B_2
⋮	⋮	⋮	
$2n$	A_2	B_1	B_2

The problem order was randomized independently for each subject.

Since n subjects are randomly assigned stimulus A_1, and n are randomly assigned stimulus A_2, as far as factor A is concerned the experiment is a completely randomized one. On the other hand, as far as type of problem is concerned, each subject is a block. Thus, for factor B, the experiment is a randomized block design. We call this experimental design a *two-factor experiment with repeated measures on factor B*.

The development of a model for this design is only a little more complex than for earlier cases. Let, as usual, α and β denote the factor A and factor B main effects, respectively, and let ρ denote the subject (block) main effect. We shall assume that there are no interactions between treatments and subjects, although this condition is not essential when subject effects are considered to be random effects. Given that there are no treatment-subject interactions, there will be no interaction terms $(\alpha\rho)$, $(\beta\rho)$, and $(\alpha\beta\rho)$ in the model. Thus, the model will be of the usual form for a randomized block experiment:

$$Y = \mu + \underbrace{\alpha + \beta + (\alpha\beta)}_{\text{Treatment effect}} + \underbrace{\quad \rho \quad}_{\text{Block effect}} + \varepsilon$$

We do need to recognize nesting, however. The subject effect in this design is nested within factor A and will be denoted by $\rho_{k(i)}$. Further, the error effect is nested within the subject-treatment combination and will be denoted by $\varepsilon_{(ijk)}$. Note that there is no additional subscript to ε since only one observation exists for each subject-treatment combination. Assuming that the treatment effects are fixed and the subject effects are random, we represent the model as follows:

$$Y_{ijk} = \mu_{...} + \alpha_i + \beta_j + (\alpha\beta)_{ij} + \rho_{k(i)} + \varepsilon_{(ijk)} \tag{30.17}$$

where:

$\mu_{...}$ is an overall constant
α_i are constants subject to the restriction $\Sigma\alpha_i = 0$
β_j are constants subject to the restriction $\Sigma\beta_j = 0$

$(\alpha\beta)_{ij}$ are constants subject to the restrictions $\sum_i (\alpha\beta)_{ij} = \sum_j (\alpha\beta)_{ij} = 0$
$\rho_{k(i)}$ are independent $N(0, \sigma_\rho^2)$
$\varepsilon_{(ijk)}$ are independent $N(0, \sigma^2)$
$\rho_{k(i)}$ and $\varepsilon_{(ijk)}$ are independent
$i = 1, \ldots, a; j = 1, \ldots, b; k = 1, \ldots, n$

Given the model formulation in (30.17), the determination of the sums of squares, degrees of freedom, and expected mean squares is straightforward. The results are presented in Tables 30.11 and 30.12. We need only point out that the error sum of squares and degrees of freedom are obtained as remainders since there is but one observation in a cell for this design.

Comments

1. A two-factor experiment with repeated measures on one factor may be considered as an incomplete block design. In our incentives study example, for instance, subjects are blocks and the four treatments are A_1B_1, A_1B_2, A_2B_1, and A_2B_2. Half of the blocks contain the two treatments A_1B_1 and A_1B_2, and the other half of the blocks contain the two treatments A_2B_1 and A_2B_2. A two-factor experiment with repeated measures on one factor may also be considered as a special type of *split-plot design*, an experimental design frequently used in agriculture.

2. Two-factor experiments with repeated measures on one factor are widely used to study effects of a factor over time. Thus, factor A may be the variable of interest whose effect over time is to be studied. Subjects are exposed to a level of factor A, and observations on the subjects are then made at successive periods of time (factor B).

3. The estimated variance of factor level sample means can be found in usual fashion when the factor has fixed effects. For instance, we have for model (30.17):

$$s^2(\bar{Y}_{i..}) = \frac{MSBL(A)}{bn} \tag{30.18}$$

$MSBL(A)$ is used as the mean square since a test of factor A fixed effects involves this mean square in the denominator of the test statistic (Table 30.12). Similarly, we have:

$$s^2(\bar{Y}_{.j.}) = \frac{MSE}{an} \tag{30.19}$$

since Table 30.12 indicates that a test of factor B fixed effects uses MSE in the denominator of the test statistic.

Note that the analysis of factor B main effects can be carried out more precisely than that of factor A main effects because comparisons among factor A levels involve the variability among the subjects as well as the experimental error, while comparisons among factor B levels involve only experimental error (Table 30.12).

4. When interactions exist between the two factors in a study where there are repeated measures on one factor, the analysis of the factor effects becomes more complex; see, for example, Reference 30.1 for a discussion of this case.

5. Two randomizations are required for a two-factor experiment with repeated measures on one factor (unless the latter factor is time). The level of the nonrepeated factor,

TABLE 30.11 Analysis of variance for two-factor experiment with repeated measures on factor B—model (30.17)

Source of Varation	Sum of Squares	Degrees of Freedom
Factor A	$SSA = \frac{\sum_i Y_{i..}^2}{bn} - \frac{Y_{...}^2}{abn}$	$a - 1$
Factor B	$SSB = \frac{\sum_j Y_{.j.}^2}{an} - \frac{Y_{...}^2}{abn}$	$b - 1$
AB interactions	$SSAB = \frac{\sum_i \sum_j Y_{ij.}^2}{n} - \frac{\sum_i Y_{i..}^2}{bn} - \frac{\sum_j Y_{.j.}^2}{an} + \frac{Y_{...}^2}{abn}$	$(a - 1)(b - 1)$
Subjects (within factor A)	$SSBL(A) = \frac{\sum_i \sum_k Y_{i.k}^2}{b} - \frac{\sum_i Y_{i..}^2}{bn}$	$(n - 1)a$
Error	$SSE =$ Remainder $= \sum_i \sum_j \sum_k Y_{ijk}^2 - \frac{\sum_i \sum_j Y_{ij.}^2}{n} - \frac{\sum_i \sum_k Y_{i.k}^2}{b} + \frac{\sum_i Y_{i..}^2}{bn}$	Remainder $= a(b - 1)(n - 1)$
Total	$SSTO = \sum_i \sum_j \sum_k Y_{ijk}^2 - \frac{Y_{...}^2}{abn}$	$abn - 1$

TABLE 30.12 Expected mean squares for two-factor experiment with repeated measures on factor B—model (30.17) (A, B fixed, subjects random)

Source of Variation	*Mean Square*	*Expected Mean Square*
Factor A	MSA	$bn \frac{\Sigma\alpha_i^2}{a-1} + b\sigma_\rho^2 + \sigma^2$
Factor B	MSB	$an \frac{\Sigma\beta_j^2}{b-1} + \sigma^2$
AB interactions	$MSAB$	$n \frac{\Sigma\Sigma(\alpha\beta)_{ij}^2}{(a-1)(b-1)} + \sigma^2$
Subjects (within factor A)	$MSBL(A)$	$b\sigma_\rho^2 + \sigma^2$
Error	MSE	σ^2

say, A, needs to be randomly assigned to the subjects. Second, the order of the levels of the repeated factor, say, B, needs to be randomly assigned, with independent order randomizations used for the various subjects.

30.7 THREE-FACTOR EXPERIMENT WITH REPEATED MEASURES ON ONE FACTOR

As a final example, we consider an experiment in which each of eight stores in a study was assigned a given combination of type of display (factor A) and price (factor B), and sales of the product were observed over four time periods (factor C). The experimental design may be represented as follows:

Type of Display	*Price Level*	*Store*	*Time Period*			
A_1	B_1	1	C_1	C_2	C_3	C_4
A_1	B_1	2	C_1	C_2	C_3	C_4
A_1	B_2	3	C_1	C_2	C_3	C_4
A_1	B_2	4	C_1	C_2	C_3	C_4
A_2	B_1	5	C_1	C_2	C_3	C_4
A_2	B_1	6	C_1	C_2	C_3	C_4
A_2	B_2	7	C_1	C_2	C_3	C_4
A_2	B_2	8	C_1	C_2	C_3	C_4

Note that repeated measures are taken on the factor C levels, but not on factors A and B. Assuming that there are no interactions between stores (blocks) and treatments, a model for this experimental design is:

$$(30.20) \quad Y_{ijkm} = \mu_{....} + \alpha_i + \beta_j + \gamma_k + (\alpha\beta)_{ij} + (\alpha\gamma)_{ik} + (\beta\gamma)_{jk} + (\alpha\beta\gamma)_{ijk} + \rho_{m(ij)} + \varepsilon_{(ijkm)}$$

where α_i is the factor A main effect, β_j the factor B main effect, γ_k the factor C main effect, $\rho_{m(ij)}$ the block (store) effect, and so on. Note that the block effect is nested within factors A and B. The development of sums of squares and expected mean squares for model (30.20) is straightforward.

PROBLEMS

30.1. Refer to ANOVA model (23.15) on page 782. Use rule (30.3) to obtain the expected mean squares in Table 23.3 for this model.

30.2. Refer to ANOVA model (24.58) on page 831.
a. Use rule (30.3) to obtain the expected mean squares in Table 24.8.
b. Use rule (30.5) to obtain the definitional sums of squares formulas in (24.24) and the associated degrees of freedom. Is it necessary here to obtain *SSE* as a remainder?

30.3. Refer to ANOVA model (24.59) on page 832.
a. Use rule (30.3) to obtain the expected mean squares in Table 24.9.
b. Use rule (30.5a) to obtain the computational sums of squares formulas in (24.26) and (24.29) and the associated degrees of freedom. Is it necessary here to obtain *SSE* as a remainder?

30.4. Refer to ANOVA model (27.1) on page 916.
a. Use rule (30.3) to obtain the expected mean squares in Table 27.3 for this model.
b. Use rule (30.5) to obtain the definitional sums of squares formulas in (27.5) and the associated degrees of freedom. Is it necessary here to obtain *SSE* as a remainder?

30.5. Refer to ANOVA model (27.1) on page 916, but assume that treatment effects are random. (See also Comment 2 on page 917.)
a. Use rule (30.3) to obtain the expected mean squares in Table 27.3 for this model.
b. Use rule (30.5a) to obtain the computational sums of squares formulas in (27.5) and the associated degrees of freedom. Is it necessary here to obtain *SSE* as a remainder?

30.6. Refer to ANOVA model (28.10) on page 947.
a. Use rule (30.3) to obtain the expected mean squares in Table 28.5 for this model.
b. Use rule (30.5) to obtain the definitional sums of squares formulas in (27.5) and the associated degrees of freedom. Is it necessary here to obtain *SSE* as a remainder?

30.7. Refer to ANOVA model (29.8) on page 968, but assume that factor A is nested within factor B, factor A effects are random, and factor B effects are fixed. (See also the paragraph "Random factor effects" on page 969.)

a. Use rule (30.3) to obtain the expected mean squares.
b. Use rule (30.5a) to obtain the computational sums of squares formulas and the associated degrees of freedom. Is it necessary here to obtain *SSE* as a remainder?
c. What is the appropriate mean square to be used in constructing a confidence interval for $\mu_{.j}$?

30.8. Refer to model (30.16), but assume that block effects are random.
a. Use rule (30.3) to obtain the expected mean squares.
b. Use rule (30.5) to obtain the definitional sums of squares formulas and the associated degrees of freedom. Is it necessary here to obtain *SSE* as a remainder?

30.9. In a balanced three-factor study, factors *A* and *C* are crossed and factor *B* is nested within factor *C*. Factor *A* has fixed effects and factors *B* and *C* have random effects.
a. Use rule (30.3) to obtain the expected mean squares.
b. Use rule (30.5a) to obtain the computational sums of squares formulas and the associated degrees of freedom.
c. What is the appropriate denominator mean square for testing for factor *A* main effects?

30.10. Refer to model (30.17).
a. Use rule (30.3) to obtain the expected mean squares in Table 30.12.
b. Use rule (30.5a) to obtain the computational sums of squares formulas in Table 30.11 and the associated degrees of freedom.

30.11. Refer to the **Drug effect experiment** data set. Consider only Part I of the study. Assume that blocks (rats) and observation units have random effects, and that factor *A* (initial lever press rate) and factor *B* (dosage level) have fixed effects. Also assume that there are no interactions between blocks and treatments.
a. Use rule (30.1) to develop the statistical model for this experiment.
b. Use rule (30.3) to obtain the expected mean squares.
c. Use rule (30.5a) to obtain the computational sums of squares formulas and the associated degrees of freedom.

30.12. Refer to the **Drug effect experiment** data set. Consider the combined study. Assume that blocks (rats) and observation units have random effects, and that factor *A* (initial lever press rate), factor *B* (dosage level), and factor *C* (reinforcement schedule) have fixed effects. Also assume that there are no interactions between blocks and treatments.
a. Use rule (30.1) to develop the statistical model for this experiment.
b. Use rule (30.3) to obtain the expected mean squares.
c. Use rule (30.5) to obtain the definitional sums of squares formulas and the associated degrees of freedom.

30.13. Refer to the **Drug effect experiment** data set. Consider the combined study but ignore the classification of the rats into the three initial lever press rate groups. Assume that blocks (rats) and observation units have random effects, and that factor *B* (dosage level) and factor *C* (reinforcement schedule) have fixed effects. Also assume that there are no interactions between blocks and treatments.
a. Use rule (30.1) to develop the statistical model for this experiment.
b. Use rule (30.3) to obtain the expected mean squares.

c. Use rule (30.5) to obtain the definitional sums of squares formulas and the associated degrees of freedom.

30.14. Swimmer motivation. A large metropolitan swim club for youths studied the effects of three motivational stimuli on performance. The three motivational stimuli were: (1) presentation of merit award, (2) granting of team leadership privileges, and (3) publicity in the club newsletter. Since age is known to be related to performance, the nine female swimmers included in the study were grouped according to age into three blocks of three each. Within each age block, the three swimmers were randomly assigned to one of the motivation treatments. After a suitable amount of training, each swimmer was timed on three separate occasions while swimming a fixed distance. The coded data on the time for each of the three trials follow.

		Motivation Treatment		
Block	*Observation*	$j = 1$ *Merit Award*	$j = 2$ *Leadership*	$j = 3$ *Publicity*
$i = 1$	$k = 1$:	28	26	27
(7–8 years)	$k = 2$:	32	24	29
	$k = 3$:	31	27	30
$i = 2$	$k = 1$:	24	22	20
(9–10 years)	$k = 2$:	26	19	21
	$k = 3$:	23	18	22
$i = 3$	$k = 1$:	18	13	17
(11–12 years)	$k = 2$:	21	16	19
	$k = 3$:	20	15	19

Assume that ANOVA model (30.16) with fixed block and treatment effects is appropriate.

a. Obtain the analysis of variance table.

b. Test whether or not the mean times are the same for the three motivational stimuli; use $\alpha = .05$. State the alternatives, decision rule, and conclusion.

c. Make all pairwise comparisons among the three treatment means; use the Tukey procedure with a 90 percent family confidence coefficient. State your findings.

d. Obtain point estimates of σ^2 and σ_η^2. Does one variance appear to be much larger than the other? Discuss.

30.15. Price and display study. Refer to the text example in Section 30.7 about the effects of type of display (factor A), price level (factor B), and time period (factor C) on sales of a product. The sales data (coded) follow. Assume that ANOVA model (30.20) with factors A, B, and C having fixed effects and blocks (stores) having random effects is applicable.

			k			
i	j	m	1	2	3	4
1	1	1	958	953	938	1,047
1	1	2	1,005	1,032	1,055	1,122
1	2	3	351	352	338	436
1	2	4	412	449	385	532
2	1	5	769	766	739	859
2	1	6	780	819	760	897
2	2	7	176	154	148	280
2	2	8	109	97	113	256

a. Use rule (30.3) to obtain the expected mean squares.
b. Use rule (30.5a) to obtain the computational sums of squares formulas and the associated degrees of freedom.
c. Obtain the analysis of variance table.
d. Test whether or not ABC interactions are present; use $\alpha = .01$. State the alternatives, decision rule, and conclusion.
e. Test for AB, AC, and BC interactions. For each test, use $\alpha = .01$ and state the alternatives, decision rule, and conclusion.
f. Test for factor A, B, and C main effects. For each test, use $\alpha = .01$ and state the alternatives, decision rule, and conclusion.
g. To study the nature of the factor A, B, and C main effects, estimate the following pairwise comparisons:

$$D_1 = \mu_{1...} - \mu_{2...} \qquad D_4 = \mu_{..2.} - \mu_{..3.}$$
$$D_2 = \mu_{.1..} - \mu_{.2..} \qquad D_5 = \mu_{..3.} - \mu_{..4.}$$
$$D_3 = \mu_{..1.} - \mu_{..2.}$$

Use the Bonferroni procedure with a 90 percent family confidence coefficient. State your findings.

EXERCISES

30.16. Derive (30.11a) for a balanced three-factor study.

30.17. Use (30.14) and the fact that this estimated variance is unbiased to find $\sigma^2(\bar{Y}_{1j..} - \bar{Y}_{2j..})$ for ANOVA model (30.6). What are the approximate number of degrees of freedom associated with the estimated variance?

PROJECTS

30.18. Refer to the **Drug effect experiment** data set and Problem 30.11.
a. Obtain the analysis of variance table.
b. Test whether or not factors A and B interact; use $\alpha = .05$. State the alternatives, decision rule, and conclusion.
c. Test for factor A and B main effects. For each test, use $\alpha = .05$ and state the alternatives, decision rule, and conclusion.
d. To study the nature of the factor A main effects, make all pairwise comparisons between the factor level means. Use the Tukey procedure with a 90 percent family confidence coefficient. State your findings.
e. To study the nature of the factor B main effects, make all pairwise comparisons between the factor level means. Use the Tukey procedure with a 90 percent family confidence coefficient. State your findings.

30.19. Refer to the **Drug effect experiment** data set and Problem 30.12.
a. Obtain the analysis of variance table.
b. Test whether or not ABC interactions are present; use $\alpha = .01$. State the alternatives, decision rule, and conclusion.

c. For each reinforcement schedule, plot the sample means against dosage level with different curves for the three initial lever press rate groups, in the format of Figure 24.6. Examine your plots for the nature of the interaction effects and report your findings.

CITED REFERENCE

30.1 Winer, B. J. *Statistical Principles in Experimental Design*. 2d ed. New York: McGraw-Hill, 1971.

31

Latin square and related designs

In this chapter, we consider the latin square design, which employs two blocking variables to reduce experimental errors, and several related designs.

31.1 BASIC ELEMENTS

Complete and incomplete block designs

We saw in Section 28.6 that two blocking variables can be used simultaneously in randomized complete block designs to eliminate from the experimental errors the variation associated with each of the blocking variables. Thus, the blocking variables might be age of subject and income of subject, with a block containing subjects in a given age group and income group.

The full use of two blocking variables in a complete block design may at times have the disadvantage of requiring too many experimental units. For instance, if the age and income variables in our earlier illustration have six classes each, 36 blocks would be required. If six treatments were to be studied, 216 subjects would be needed for the experiment. Cost considerations may not permit the use of this many experimental units, yet precision and range of validity considerations may require the simultaneous use of two blocking variables, each with six classes, in order to reduce the experimental error variance sufficiently and to have a reasonable variety of experimental subjects. In this type of situation, an *incomplete block design* may be helpful. In such a design, all 36 blocks in our example would still be used, but now each block would not contain all six treatments.

Latin square designs

Taking incomplete block designs to the extreme in our example, given the employment of 36 blocks, the number of experimental units is minimized if only one treatment is run in each block. This extreme case, where each block contains only one treatment, is the type of situation for which a latin square design is appropriate. Table 31.1 provides an illustration of the difference between complete and incomplete block designs for the example considered. Column 1 shows the complete block design for this case, while columns 2 and 3 illustrate incomplete block designs, with three treatments and one treatment in each block, respectively.

There is another reason besides economy, why a latin square design with only one treatment per block is used, namely, that blocks sometimes cannot contain more than one treatment. Consider the repeated measures design discussed in Chapter 28 where each subject receives every treatment. It was stressed there that the order of treatments be randomized in case interference effects between the different treatments were to be present. If indeed interference effects due to the order position of the treatments are anticipated, it may be desirable to use the order position as another blocking variable. Thus, ''subject'' would be one blocking variable and ''order position of treatment'' a second blocking variable. Blocks would then be defined as follows for a study involving six treatments:

Block 1: Subject 1, position 1
Block 2: Subject 1, position 2
. .
. .
. .
Block 6: Subject 1, position 6
Block 7: Subject 2, position 1
etc. etc.

Notice that the blocks so defined can contain only one treatment, since the order position refers to the place of a single treatment in the sequence of treatments for a subject.

Description of latin square designs

Let A, B, C represent three treatments; it is conventional with latin square designs to use Latin letters for the treatments. Suppose that day of week (Monday, Tuesday, Wednesday) and operator (1, 2, 3) are to be used as blocking variables. A latin square design might then be shown as follows:

TABLE 31.1 Complete and incomplete block designs

Block Description	(1) Complete Block Design	(2) Incomplete Block Design (three treatments per block)	(3) Incomplete Block Design (one treatment per block)
Age under 25, income under \$6,000	$T_1, T_2, T_3, T_4, T_5, T_6$	T_1, T_3, T_5	T_2
Age under 25, income \$6,000–\$8,999	$T_1, T_2, T_3, T_4, T_5, T_6$	T_2, T_4, T_6	T_5
⋮	⋮	⋮	⋮
Age 25–34, income under \$6,000	$T_1, T_2, T_3, T_4, T_5, T_6$	T_2, T_4, T_5	T_3
⋮	⋮	⋮	⋮
Age 35–44, income under \$6,000	$T_1, T_2, T_3, T_4, T_5, T_6$	T_3, T_4, T_6	T_2
etc.	etc.	etc.	etc.

	Operator		
Day	1	2	3
Monday	*B*	*A*	*C*
Tuesday	*A*	*C*	*B*
Wednesday	*C*	*B*	*A*

Operator 1 would run treatment *B* on Monday, treatment *A* on Tuesday, and treatment *C* on Wednesday, and so on for the other operators. Note that each operator runs each treatment, and that all treatments are run on each day.

A latin square design thus has the following features:

1. There are *r* treatments.
2. There are two blocking variables, each containing *r* classes.
3. Each row and each column in the design square contains all treatments; that is, each class of each blocking variable constitutes a replication.

Advantages and disadvantages of latin square designs

Advantages of a latin square design include:

1. The use of two blocking variables often permits greater reductions in the variability of experimental errors than can be obtained with either blocking variable alone.
2. Treatment effects can be studied from a small-scale experiment. This is particularly helpful in preliminary or pilot studies.
3. It is often helpful in repeated measures experiments to take into account the order effect of treatments by means of a latin square design.

Disadvantages of a latin square design are:

1. The number of classes of each blocking variable must equal the number of treatments. This leads to a very small number of degrees of freedom for experimental error when only a few treatments are studied. On the other hand, when many treatments are studied, the degrees of freedom for experimental error may be larger than necessary.
2. The assumptions of the model are restrictive (e.g., that there are no interactions between either blocking variable and treatments, and also none between the two blocking variables).
3. The two blocking variables cannot have different numbers of classes.
4. The randomization required is somewhat more complex than that for earlier designs considered.

Because of the limitations on the degrees of freedom for experimental error just described, latin squares are rarely used when more than eight treatments are being investigated. For the same reason, when there are only a few treatments, say, four or less, additional replications are usually required when a latin square design is employed.

Randomization of latin square design

There exist many latin squares for a given number of treatments. Suppose that the number of treatments is $r = 3$; four possible latin square designs are (we omit the row and column blocking variable labels):

1			2			3			4		
A	*B*	*C*	*A*	*C*	*B*	*B*	*A*	*C*	*C*	*B*	*A*
B	*C*	*A*	*B*	*A*	*C*	*C*	*B*	*A*	*A*	*C*	*B*
C	*A*	*B*	*C*	*B*	*A*	*A*	*C*	*B*	*B*	*A*	*C*

For $r = 3$, there are altogether 12 different possible arrangements. This number increases rapidly as the number of treatments gets larger; for $r = 5$, there are 161,280 possible arrangements.

The objective of randomization is to select one of all possible latin squares for the given number of treatments r, such that each square has an equal probability of being selected. Clearly, it is not generally feasible to list all possible latin squares so that one can be selected at random.

Instead, we utilize *standard latin squares,* which are latin squares in which the elements of the first row and the first column are arranged alphabetically. The earlier latin square 1 is a standard latin square. Table A–16 contains all the standard squares for $r = 3$ and 4, and a single selected standard square for $r = 5$, 6, 7, 8, and 9. The randomization procedure usually employed is as follows:

1. For $r = 3$, independently arrange the rows and columns at random.
2. For $r = 4$, select one of the standard squares at random. Then, independently arrange its rows and columns at random.
3. For $r = 5$ and higher, independently arrange the rows, columns, and treatments of the given standard square at random.

It can be shown that this procedure selects one latin square at random from all possible squares for $r = 3$ and 4. For $r = 5$ or higher, the randomization procedure is not based on all possible latin squares, but rather on very large and suitable subsets thereof.

Example 1. In our earlier illustration, the blocking variables were day (Monday, Tuesday, Wednesday) and operator (1, 2, 3). Thus, we can show the borders of the square as follows:

	Operator		
Day	1	2	3
Monday			
Tuesday			
Wednesday			

The standard latin square for $r = 3$ is:

Step 1	*A*	*B*	*C*
	B	*C*	*A*
	C	*A*	*B*

We select now a random permutation of 3 to rearrange the rows. Suppose it is: 2, 3, 1. Thus, row 2 in the standard square comes first, row 3 second, and row 1 third. We obtain:

	Original Row Number			
Step 2	2	*B*	*C*	*A*
	3	*C*	*A*	*B*
	1	*A*	*B*	*C*

We then select independently another permutation of 3 to rearrange the columns. Suppose it is: 2, 1, 3. Column 2 in the square of step 2 now becomes the first column, and so on. We obtain:

	Step 2 *column number:*	2	1	3
Step 3		*C*	*B*	*A*
		A	*C*	*B*
		B	*A*	*C*

The selected design therefore is:

	Operator		
Day	1	2	3
Monday	*C*	*B*	*A*
Tuesday	*A*	*C*	*B*
Wednesday	*B*	*A*	*C*

Example 2. Consider an experiment about the effects of five different types of background music (*A*, *B*, *C*, *D*, *E*) on the productivity of bank tellers. A given type of music is played for one day and the productivity is observed. Day of the week and week of the experimental period are to be the two blocking variables. The borders of the latin square design therefore are:

	Day				
Week	*M*	*T*	*W*	*Th*	*F*
1					
2					
3					
4					
5					

We start with the standard latin square of Table A–16 for $r = 5$:

	A	*B*	*C*	*D*	*E*
	B	*A*	*E*	*C*	*D*
Step 1	*C*	*D*	*A*	*E*	*B*
	D	*E*	*B*	*A*	*C*
	E	*C*	*D*	*B*	*A*

Next we permute the rows randomly, using a random permutation of 5, say: 4, 2, 3, 1, 5. We obtain then:

	D	*E*	*B*	*A*	*C*
	B	*A*	*E*	*C*	*D*
Step 2	*C*	*D*	*A*	*E*	*B*
	A	*B*	*C*	*D*	*E*
	E	*C*	*D*	*B*	*A*

We now permute the columns randomly, using an independent random permutation of 5, say: 2, 4, 1, 5, 3. We obtain:

	E	*A*	*D*	*C*	*B*
	A	*C*	*B*	*D*	*E*
Step 3	*D*	*E*	*C*	*B*	*A*
	B	*D*	*A*	*E*	*C*
	C	*B*	*E*	*A*	*D*

Finally, we need to permute the treatment labels randomly. Note that the treatment designations *A*, *B*, *C*, *D*, and *E* remain fixed. We simply wish to randomize the cells in which the specific treatments occur, given the pattern obtained in step 3. Assume the correspondence:

1	2	3	4	5
A	*B*	*C*	*D*	*E*

and that an independent random selection of a permutation of 5 yields: 3, 5, 2, 1, 4. We obtain then:

Present cell designation:	*A*	*B*	*C*	*D*	*E*
New cell designation:	*C*	*E*	*B*	*A*	*D*

Thus, every cell designated *A* in step 3 now is designated *C*, and so on. Hence, the latin square design to be used is:

	Day				
Week	*M*	*T*	*W*	*Th*	*F*
1	*D*	*C*	*A*	*B*	*E*
2	*C*	*B*	*E*	*A*	*D*
3	*A*	*D*	*B*	*E*	*C*
4	*E*	*A*	*C*	*D*	*B*
5	*B*	*E*	*D*	*C*	*A*

Note

For $r = 3$ and 4, it would be sufficient to randomize all r rows and the last $r - 1$ columns. It is equally good to randomize all rows and columns, the procedure suggested here.

Illustration

We mentioned earlier an experiment on the effects of different types of background music on the productivity of bank tellers. The treatments were defined as various combinations of tempo of music (slow, medium, fast) and style of music (instrumental and vocal, instrumental only). The treatments and Latin letter designations were as follows:

Treatment	*Latin Letter Designation*	*Tempo and Style of Music*
1	A	Slow, instrumental and vocal
2	B	Medium, instrumental and vocal
3	C	Fast, instrumental and vocal
4	D	Medium, instrumental only
5	E	Fast, instrumental only

Table 31.2 contains the results of this experiment. The treatment in each cell is shown in parentheses. Note that in this study, the experimental unit is a working day for the crew of bank tellers, and that the productivity data pertain to the performance of the entire crew. Let $Y_{ij(k)}$ denote the observation in the cell defined by the ith class of the row blocking variable and the jth class of the column blocking variable. The subscript k indicates the treatment assigned to this cell by the particular latin square design employed; it is shown in parentheses since it is dependent on (i, j). As soon as the (i, j) cell combination is specified,

TABLE 31.2 Latin square design background music study—experiment results (productivity of crew—data coded)

	Day					
Week	*M*	*T*	*W*	*Th*	*F*	*Total*
1	18 (D)	17 (C)	14 (A)	21 (B)	17 (E)	$Y_{1..} = 87$
2	13 (C)	34 (B)	21 (E)	16 (A)	15 (D)	$Y_{2..} = 99$
3	7 (A)	29 (D)	32 (B)	27 (E)	13 (C)	$Y_{3..} = 108$
4	17 (E)	13 (A)	24 (C)	31 (D)	25 (B)	$Y_{4..} = 110$
5	21 (B)	26 (E)	26 (D)	31 (C)	7 (A)	$Y_{5..} = 111$
Total	$Y_{.1.} = 76$	$Y_{.2.} = 119$	$Y_{.3.} = 117$	$Y_{.4.} = 126$	$Y_{.5.} = 77$	$Y_{...} = 515$

$Y_{..1} = 7 + 13 + 14 + 16 + 7 = 57 \qquad \bar{Y}_{..1} = 11.4$

$Y_{..2} = 21 + 34 + 32 + 21 + 25 = 133 \qquad \bar{Y}_{..2} = 26.6$

$Y_{..3} = 13 + 17 + 24 + 31 + 13 = 98 \qquad \bar{Y}_{..3} = 19.6$

$Y_{..4} = 18 + 29 + 26 + 31 + 15 = 119 \qquad \bar{Y}_{..4} = 23.8$

$Y_{..5} = 17 + 26 + 21 + 27 + 17 = 108 \qquad \bar{Y}_{..5} = 21.6$

the treatment is determined for the particular latin square design employed. Thus, $Y_{12(3)} = 17$ is the productivity on Tuesday of the first week, and Table 31.2 indicates that the type of music on that day was C.

We shall analyze the results of this study in subsequent sections.

31.2 MODEL

A latin square design involves the main effect of the row blocking variable, denoted by ρ_i, the main effect of the column blocking variable, denoted by κ_j, and the treatment main effect, denoted by τ_k. It is assumed that no interactions exist between these three variables. Thus, the model employed is an additive one. For the case of fixed treatment and block effects, the model is:

$$Y_{ij(k)} = \mu_{...} + \rho_i + \kappa_j + \tau_k + \varepsilon_{ij(k)} \tag{31.1}$$

where:

$\mu_{...}$ is a constant
ρ_i, κ_j, τ_k are constants subject to the restrictions $\Sigma\rho_i = \Sigma\kappa_j = \Sigma\tau_k = 0$
$\varepsilon_{ij(k)}$ are independent $N(0, \sigma^2)$
$i = 1, \ldots, r;\ j = 1, \ldots, r;\ k = 1, \ldots, r$

Note again that the number of classes for each of the two blocking variables is the same as the number of treatments; also note that the total number of observations is r^2.

Comments

1. Sometimes, one or both of the blocking variable effects are viewed as random, as when the blocking variables refer to subjects, observers, machines, etc. We shall consider the case of random blocking variable effects in Section 31.11.

2. If the treatment effects are random, the only change in model (31.1) would be that the τ_k now are independent $N(0, \sigma_\tau^2)$ and are independent of the $\varepsilon_{ij(k)}$.

3. The aptness of the additive model (31.1) needs to be carefully explored. A test for this purpose is explained in Section 31.7.

31.3 ANALYSIS OF VARIANCE AND TESTS

Notation

We shall employ the usual notation for row, column, and treatment totals and means:

$$Y_{i..} = \sum_j Y_{ij(k)} \qquad \bar{Y}_{i..} = \frac{Y_{i..}}{r} \tag{31.2a}$$

(31.2b)
$$Y_{.j.} = \sum_{i} Y_{ij(k)} \qquad \bar{Y}_{.j.} = \frac{Y_{.j.}}{r}$$

(31.2c)
$$Y_{..k} = \sum_{i,\, j} Y_{ij(k)} \qquad \bar{Y}_{..k} = \frac{Y_{..k}}{r}$$

The overall total and mean are denoted as usual by:

(31.2d)
$$Y_{...} = \sum_{i} \sum_{j} Y_{ij(k)} \qquad \bar{Y}_{...} = \frac{Y_{...}}{r^2}$$

Note the redundancy of any one of the three subscripts, arising from the fact that the treatment is uniquely determined by the row and column specifications for the latin square utilized. The various totals for our background music example are shown in Table 31.2.

Fitting of model

The least squares estimators of the parameters in model (31.1) are:

	Parameter	*Estimator*
(31.3a)	$\mu_{...}$	$\hat{\mu}_{...} = \bar{Y}_{...}$
(31.3b)	ρ_i	$\hat{\rho}_i = \bar{Y}_{i..} - \bar{Y}_{...}$
(31.3c)	κ_j	$\hat{\kappa}_j = \bar{Y}_{.j.} - \bar{Y}_{...}$
(31.3d)	τ_k	$\hat{\tau}_k = \bar{Y}_{..k} - \bar{Y}_{...}$

The fitted values therefore are:

(31.4)
$$\hat{Y}_{ij(k)} = \bar{Y}_{i..} + \bar{Y}_{.j.} + \bar{Y}_{..k} - 2\bar{Y}_{...}$$

and the residuals are:

(31.5)
$$e_{ij(k)} = Y_{ij(k)} - \hat{Y}_{ij(k)} = Y_{ij(k)} - \bar{Y}_{i..} - \bar{Y}_{.j.} - \bar{Y}_{..k} + 2\bar{Y}_{...}$$

Analysis of variance

To obtain the analysis of variance, we express the total deviation $Y_{ij(k)} - \bar{Y}_{...}$ as follows:

(31.6)
$$\underbrace{Y_{ij(k)} - \bar{Y}_{...}}_{\text{Total deviation}} = \underbrace{\bar{Y}_{i..} - \bar{Y}_{...}}_{\text{Row effect}} + \underbrace{\bar{Y}_{.j.} - \bar{Y}_{...}}_{\text{Column effect}} + \underbrace{\bar{Y}_{..k} - \bar{Y}_{...}}_{\text{Treatment effect}}$$
$$+ \underbrace{Y_{ij(k)} - \bar{Y}_{i..} - \bar{Y}_{.j.} - \bar{Y}_{..k} + 2\bar{Y}_{...}}_{\text{Residual}}$$

Squaring (31.6) and then summing over all observations, we find that all cross-product terms drop out and we obtain:

(31.7) $$SSTO = SSROW + SSCOL + SSTR + SSE$$

where:

(31.7a) $$SSTO = \sum_i \sum_j (Y_{ij(k)} - \bar{Y}_{...})^2$$

(31.7b) $$SSROW = r \sum_i (\bar{Y}_{i..} - \bar{Y}_{...})^2$$

(31.7c) $$SSCOL = r \sum_j (\bar{Y}_{.j.} - \bar{Y}_{...})^2$$

(31.7d) $$SSTR = r \sum_k (\bar{Y}_{..k} - \bar{Y}_{...})^2$$

(31.7e) $$SSE = \sum_i \sum_j (Y_{ij(k)} - \bar{Y}_{i..} - \bar{Y}_{.j.} - \bar{Y}_{..k} + 2\bar{Y}_{...})^2$$

SSROW is the *row sum of squares*. The more the row means $\bar{Y}_{i..}$ differ, the larger is *SSROW*. Similarly, *SSCOL* is the *column sum of squares* and measures the variability of the column means $\bar{Y}_{.j.}$. *SSTR* and *SSE* denote, as usual, the treatment sum of squares and error sum of squares, respectively.

Formulas for these sums of squares suitable for hand computations are as follows:

(31.8a) $$SSTO = \sum_i \sum_j Y_{ij(k)}^2 - \frac{Y_{...}^2}{r^2}$$

(31.8b) $$SSROW = \frac{\sum_i Y_{i..}^2}{r} - \frac{Y_{...}^2}{r^2}$$

(31.8c) $$SSCOL = \frac{\sum_j Y_{.j.}^2}{r} - \frac{Y_{...}^2}{r^2}$$

(31.8d) $$SSTR = \frac{\sum_k Y_{..k}^2}{r} - \frac{Y_{...}^2}{r^2}$$

(31.8e) $$SSE = SSTO - SSROW - SSCOL - SSTR$$

The complete analysis of variance table for a latin square design is given in Table 31.3. There are r^2 observations, and hence *SSTO* has $r^2 - 1$ degrees of freedom associated with it. Since there are r classes for the row and column blocking variables each, and also r treatments, each of the corresponding sums of squares has $r - 1$ degrees of freedom associated with it. The number of degrees of freedom associated with *SSE* is the remainder, namely, $(r^2 - 1) - 3(r - 1) = (r - 1)(r - 2)$. Note that the addition of a second blocking variable has reduced the number of degrees of freedom for *SSE* from $(r - 1)^2$ for a randomized

TABLE 31.3 ANOVA table for latin square design—fixed effects model (31.1)

Source of Variation	*SS*	*df*	*MS*	*E(MS)*
Row blocking variable	$SSROW$	$r-1$	$MSROW = \frac{SSROW}{r-1}$	$\sigma^2 + r\frac{\sum \rho_i^2}{r-1}$
Column blocking variable	$SSCOL$	$r-1$	$MSCOL = \frac{SSCOL}{r-1}$	$\sigma^2 + r\frac{\sum \kappa_j^2}{r-1}$
Treatments	$SSTR$	$r-1$	$MSTR = \frac{SSTR}{r-1}$	$\sigma^2 + r\frac{\sum \tau_k^2}{r-1}$
Error	SSE	$(r-1)(r-2)$	$MSE = \frac{SSE}{(r-1)(r-2)}$	σ^2
Total	$SSTO$	r^2-1		

complete block design based on the same number of experimental units to $(r-1)(r-2)$, a reduction of $r-1$ degrees of freedom.

The $E(MS)$ column in Table 31.3 for fixed effects model (31.1) can be obtained by using rule (30.3).

Test for treatment effects

To test for treatment effects in the fixed effects model (31.1):

$$\begin{aligned} &H_0\text{: all } \tau_k = 0 \\ &H_a\text{: not all } \tau_k \text{ equal zero} \end{aligned} \tag{31.9a}$$

we see from the $E(MS)$ column in Table 31.3 that we use the usual test statistic:

$$F^* = \frac{MSTR}{MSE} \tag{31.9b}$$

The appropriate decision rule is:

$$\begin{aligned} &\text{If } F^* \leq F[1-\alpha; r-1, (r-1)(r-2)], \text{ conclude } H_0 \\ &\text{If } F^* > F[1-\alpha; r-1, (r-1)(r-2)], \text{ conclude } H_a \end{aligned} \tag{31.9c}$$

Example

The fixed effects model (31.1) was assumed to be appropriate for the background music study. A test of the additivity assumption of this model (described in Section 31.7) indicated that the additive model is apt here. The analysis of variance calculations for the background music data in Table 31.2 are as follows, using the computational formulas (31.8):

$$SSTO = (18)^2 + (17)^2 + (14)^2 + \cdots + (7)^2 - \frac{(515)^2}{25} = 1{,}412$$

$$SSROW = \frac{1}{5}\left[(87)^2 + (99)^2 + (108)^2 + (110)^2 + (111)^2\right] - \frac{(515)^2}{25} = 82$$

$$SSCOL = \frac{1}{5}\left[(76)^2 + (119)^2 + (117)^2 + (126)^2 + (77)^2\right] - \frac{(515)^2}{25} = 477.2$$

$$SSTR = \frac{1}{5}\left[(57)^2 + (133)^2 + (98)^2 + (119)^2 + (108)^2\right] - \frac{(515)^2}{25} = 664.4$$

$$SSE = 1{,}412 - 82 - 477.2 - 664.4 = 188.4$$

The complete analysis of variance table for the background music example is shown in Table 31.4.

To test for treatment effects:

$$\begin{aligned} &H_0\text{: } \tau_1 = \tau_2 = \tau_3 = \tau_4 = \tau_5 = 0 \\ &H_a\text{: not all } \tau_k \text{ equal zero} \end{aligned}$$

TABLE 31.4 ANOVA table for background music example

Source of Variation	*SS*	*df*	*MS*
Weeks	82.0	4	20.5
Days within week	477.2	4	119.3
Type of music	664.4	4	166.1
Error	188.4	12	15.7
Total	1,412.0	24	

we find from Table 31.4:

$$F^* = \frac{MSTR}{MSE} = \frac{166.1}{15.7} = 10.6$$

Assuming we are to control the risk of making a Type I error at $\alpha = .01$, we require $F(.99; 4, 12) = 5.41$. Since $F^* = 10.6 > 5.41$, we conclude H_a, that the various types of background music have differential effects on the productivity of the bank tellers.

Comments

1. At times, it may be desired to test for the presence of blocking variable effects. The $E(MS)$ column in Table 31.3 indicates this is done in the usual way for fixed effects model (31.1). The statistic for testing row blocking variable effects is:

$$F^* = \frac{MSROW}{MSE} \tag{31.10a}$$

and that for testing column blocking variable effects is:

$$F^* = \frac{MSCOL}{MSE} \tag{31.10b}$$

For instance, to test in our background music example whether productivity varies according to the day of the week, we use the test statistic $F^* = 119.3/15.7 = 7.6$. For a level of significance of $\alpha = .01$, we need $F(.99; 4, 12) = 5.41$. Hence, we would conclude that there is variation in productivity (averaged over all treatments and weeks) within a week. Since blocking variables correspond to a classification factor, the interpretation of blocking variable effects must be done with care.

2. The power of the F test for treatment effects in latin square design model (31.1) involves the noncentrality parameter:

$$\phi = \frac{1}{\sigma}\sqrt{\Sigma \tau_k^2} \tag{31.11}$$

with degrees of freedom $r - 1$ for numerator and $(r - 1)(r - 2)$ for denominator. Other than these modifications, no new problems are encountered in obtaining the power of the test for treatment effects in a latin square design.

3. If the treatment effects are random, the alternatives to be considered are:

(31.12)
$$H_0: \sigma_\tau^2 = 0$$
$$H_a: \sigma_\tau^2 > 0$$

but the test statistic and decision rule are the same as in (31.9) for the fixed treatment effects case.

31.4 ANALYSIS OF TREATMENT EFFECTS

If differential treatment effects are found by the analysis of variance when the treatments have fixed effects, one will usually wish to estimate some contrasts involving the treatment effects, often utilizing multiple comparison procedures. The appropriate mean square to be used in the estimated variance of the contrast is *MSE* obtained from (31.7e), and the multiples for the estimated standard deviation of the contrast are as follows:

(31.13a)	Single comparison	$t[1 - \alpha/2; (r-1)(r-2)]$
(31.13b)	Tukey procedure (for pairwise comparisons)	$T = \frac{1}{\sqrt{2}} q[1 - \alpha; r, (r-1)(r-2)]$
(31.13c)	Scheffé procedure	$S^2 = (r-1)F[1 - \alpha; r-1, (r-1)(r-2)]$
(31.13d)	Bonferroni procedure (g comparisons)	$B = t[1 - \alpha/2g; (r-1)(r-2)]$

Example

In our background music example, pairwise comparisons between the different kinds of music were desired with a family confidence coefficient of .90. The analyst used the Tukey procedure. Substituting into (17.24b) with $n_i = n_{i'} = r$ and using the results from Table 31.4, she obtained:

$$s^2(\hat{D}) = \frac{2MSE}{r} = \frac{2(15.7)}{5} = 6.28 \qquad s(\hat{D}) = 2.51$$

Remember that each treatment mean $\bar{Y}_{..k}$ is based on five observations here. Next, the analyst found the T multiple in (31.13b):

$$T = \frac{1}{\sqrt{2}} q(.90; 5, 12) = \frac{1}{\sqrt{2}}(3.92) = 2.77$$

so that:

$$Ts(\hat{D}) = 2.77(2.51) = 7.0$$

Using the treatment sample means in Table 31.2, the pairwise comparisons were then obtained. For instance, we have:

$$8.2 = (26.6 - 11.4) - 7.0 \le \mu_{..2} - \mu_{..1} \le (26.6 - 11.4) + 7.0 = 22.2$$

Here $\mu_{..1}$ is the mean productivity for treatment 1 averaged over all weeks and days of the week, and $\mu_{..2}$ has the corresponding meaning for treatment 2. The entire set of pairwise comparisons is as follows:

$$+8.2 \leq \mu_{..2} - \mu_{..1} \leq +22.2 \qquad +5.4 \leq \mu_{..4} - \mu_{..1} \leq +19.4$$
$$-14.0 \leq \mu_{..3} - \mu_{..2} \leq -.05 \qquad -9.2 \leq \mu_{..5} - \mu_{..4} \leq +4.8$$
$$+1.2 \leq \mu_{..3} - \mu_{..1} \leq +15.2 \qquad -5.0 \leq \mu_{..5} - \mu_{..3} \leq +9.0$$
$$-2.8 \leq \mu_{..4} - \mu_{..3} \leq +11.2 \qquad -12.0 \leq \mu_{..5} - \mu_{..2} \leq +2.0$$
$$-9.8 \leq \mu_{..4} - \mu_{..2} \leq +4.2 \qquad +3.2 \leq \mu_{..5} - \mu_{..1} \leq +17.2$$

These pairwise differences led to the following conclusions by the analyst with family confidence coefficient of 90 percent:

1. Treatment 2 yields greater mean productivity than treatments 1 or 3.
2. Treatments 3, 4, and 5 all yield greater mean productivity than treatment 1.
3. No pairwise differences in mean productivity for treatments 2, 4, and 5 are evident.
4. No pairwise differences in mean productivity for treatments 3, 4, and 5 are evident.

This is to say, the most promising treatment appears to be mixed instrumental-vocal music in medium tempo ($k = 2$). There is clear evidence that it is better than instrumental-vocal music in slow tempo ($k = 1$) or instrumental-vocal music in fast tempo ($k = 3$). The point estimates suggest it also is better than solely instrumental music in medium ($k = 4$) or fast ($k = 5$) tempo, but the experimental evidence on these latter two comparisons is indecisive.

31.5 FACTORIAL TREATMENTS

If the treatments in a latin square design are factorial in nature, the treatment sum of squares *SSTR* is decomposed in the usual manner. For a two-factor experiment involving factors A and B, we would have:

$$SSTR = SSA + SSB + SSAB \tag{31.14}$$

Estimates of fixed factor effects can be made readily since they are simply contrasts of the treatment means.

Example

In a sequel to the background music study mentioned earlier, four treatments were employed to investigate the effects of loudness of music (soft, loud) and type of music (semipopular, popular) on productivity of bank tellers. The treatments were defined as follows:

T_1—soft, semipopular

T_2—loud, semipopular

T_3—soft, popular

T_4—loud, popular

The blocking variables for the latin square design were day of week (Monday, Tuesday, Wednesday, Thursday) and week (1, 2, 3, 4). All treatment and block effects were considered to be fixed.

The analysis of variance for this latin square design experiment is shown in Table 31.5. If factors A and B do not interact, the effect of loudness may be studied from the contrast:

(31.15)
$$L_1 = \frac{\mu_{..1} + \mu_{..3}}{2} - \frac{\mu_{..2} + \mu_{..4}}{2}$$

where $\mu_{..k}$ is the mean productivity for the kth treatment averaged over all weeks and days of the week. The point estimator of this contrast is:

(31.15a)
$$\hat{L}_1 = \frac{\bar{Y}_{..1} + \bar{Y}_{..3}}{2} - \frac{\bar{Y}_{..2} + \bar{Y}_{..4}}{2}$$

Similarly, the effect of type of music may be studied from the contrast:

(31.16)
$$L_2 = \frac{\mu_{..1} + \mu_{..2}}{2} - \frac{\mu_{..3} + \mu_{..4}}{2}$$

A point estimator of this contrast is:

(31.16a)
$$\hat{L}_2 = \frac{\bar{Y}_{..1} + \bar{Y}_{..2}}{2} - \frac{\bar{Y}_{..3} + \bar{Y}_{..4}}{2}$$

Confidence limits for these contrasts are obtained in the usual fashion, employing the multiples in (31.13).

31.6 PLANNING LATIN SQUARE EXPERIMENTS

Necessary number of replications

A latin square design provides r replications for each treatment. Power and/or estimation considerations similar to those for the randomized complete block design may indicate that r replications are too few, particularly when r is small, say, 3, 4, or 5. Two methods of increasing the number of replications with a latin square design are discussed in Section 31.10. With either method, it is necessary to assess in advance the magnitude of the experimental error variance σ^2 in order to plan the necessary number of replications.

Efficiency of blocking variables

The efficiency of a latin square design can be assessed relative to a completely randomized design or relative to a randomized complete block design. The efficiency relative to a completely randomized design is defined by:

(31.17a)
$$E_1 = \frac{\sigma_r^2}{\sigma_L^2}$$

TABLE 31.5 ANOVA table for latin square design with factorial treatments—background music sequel study ($r = 4$, $a = 2$, $b = 2$)

Source of Variation	*SS*	*df*	*MS*
Weeks	*SSROW*	3	*MSROW*
Days within week	*SSCOL*	3	*MSCOL*
Treatments	*SSTR*	3	*MSTR*
Loudness of music (A)	*SSA*	1	*MSA*
Type of music (B)	*SSB*	1	*MSB*
AB interactions	*SSAB*	1	*MSAB*
Error	*SSE*	6	*MSE*
Total	*SSTO*	15	

where σ_r^2 and σ_L^2 are the experimental error variances with a completely randomized design and a latin square design, respectively. The efficiency relative to a randomized complete block design can be measured in two ways, depending on whether the row or the column blocking variable is used in the randomized block design:

$$E_2 = \frac{\sigma_{br}^2}{\sigma_L^2} \tag{31.17b}$$

$$E_3 = \frac{\sigma_{bc}^2}{\sigma_L^2} \tag{31.17c}$$

where σ_{br}^2 and σ_{bc}^2 are the experimental error variances with a randomized block design if the row blocking variable and the column blocking variable is utilized, respectively.

We can estimate σ_r^2, σ_{br}^2, and σ_{bc}^2 from the results for a latin square design as follows:

$$s_r^2 = \frac{MSROW + MSCOL + (r - 1)MSE}{r + 1} \tag{31.18a}$$

$$s_{br}^2 = \frac{MSCOL + (r - 1)MSE}{r} \tag{31.18b}$$

$$s_{bc}^2 = \frac{MSROW + (r - 1)MSE}{r} \tag{31.18c}$$

Thus, our estimated measures of efficiency are:

$$\hat{E}_1 = \frac{MSROW + MSCOL + (r - 1)MSE}{(r + 1)MSE} \tag{31.19a}$$

$$\hat{E}_2 = \frac{MSCOL + (r-1)MSE}{rMSE} \tag{31.19b}$$

$$\hat{E}_3 = \frac{MSROW + (r-1)MSE}{rMSE} \tag{31.19c}$$

If r is small, the efficiency measures may be modified by means of (27.14) to account for differences in the number of degrees of freedom associated with the experimental error mean squares for the two designs being compared.

Example. For our background music example, we obtain the following efficiency measures from the results in Table 31.4:

$$\hat{E}_1 = \frac{20.5 + 119.3 + 4(15.7)}{6(15.7)} = 2.2$$

$$\hat{E}_2 = \frac{119.3 + 4(15.7)}{5(15.7)} = 2.3$$

$$\hat{E}_3 = \frac{20.5 + 4(15.7)}{5(15.7)} = 1.1$$

We could modify the efficiency measures by (27.14) to take account of differences in the degrees of freedom associated with the error mean squares for the designs being compared, but these modifications would have little effect here.

Analysis of the efficiency estimates indicates that the latin square design for the background music study was efficient relative to a completely randomized design. The latter would have required over twice as many observations as the latin square design so that the variance for any specified estimated treatment contrast would be the same with both designs. Most of this efficiency was gained by the column blocking variable (days within week), because the efficiency of the latin square design relative to a complete block design with the column blocking variable is poor, being close to 1. Hence, little was achieved by also blocking by the row blocking variable (week).

31.7 ANALYSIS OF APTNESS OF MODEL

Residual analysis

The use of the residuals (31.5) for examining the aptness of the latin square design model presents no new issues; the basic points made earlier for other designs apply also to latin square designs.

Tukey test for additivity

A key question concerning the aptness of the latin square design model (31.1) is whether the effects of blocking variables and treatments are indeed additive. If nonadditivity is present in the data, transformations of the data should be studied to see if it can thereby be eliminated. The use of a model assuming additivity when in fact the effects are nonadditive will lower the level of significance and power of the test for treatment effects, or widen the confidence limits, thus making the experiment less sensitive.

The Tukey test for additivity in a randomized complete block design, discussed in Section 27.7, can be extended to latin square designs. For completeness, we outline the steps required for the Tukey test for additivity in a latin square experiment:

1. For each cell, obtain the fitted value (31.4):

$$\hat{Y}_{ij(k)} = \bar{Y}_{i..} + \bar{Y}_{.j.} + \bar{Y}_{..k} - 2\bar{Y}_{...} \tag{31.20a}$$

2. Find the residual (31.5) for each cell:

$$e_{ij(k)} = Y_{ij(k)} - \hat{Y}_{ij(k)} \tag{31.20b}$$

Check that the residuals sum to zero over every row, column, and treatment.

3. Calculate *SSE* for the latin square. This can be done through (31.8e), or through (31.7e) which as always is equivalent to:

$$SSE = \sum_i \sum_j e^2_{ij(k)} \tag{31.20c}$$

4. Calculate for each cell:

$$U_{ij(k)} = (\hat{Y}_{ij(k)} - \bar{Y}_{...})^2 \tag{31.20d}$$

5. Calculate:

$$N = \sum_i \sum_j e_{ij(k)} U_{ij(k)} \tag{31.20e}$$

6. Treating the $U_{ij(k)}$ as the observations in a latin square, obtain the error sum of squares, denoted by $SSE(U)$, using (31.7e) or (31.8e), with the Y's replaced by the U's.

7. The lack of fit sum of squares associated with nonadditivity is:

$$SSLF = \frac{N^2}{SSE(U)} \tag{31.20f}$$

This sum of squares has one degree of freedom associated with it.

8. The pure error sum of squares is:

$$SSPE = SSE - SSLF \tag{31.20g}$$

It has $(r - 1)(r - 2) - 1$ degrees of freedom associated with it.

9. The test statistic is:

$$F^* = \frac{SSLF}{1} \div \frac{SSPE}{(r-1)(r-2)-1} \tag{31.20h}$$

This test statistic follows the $F[1, (r - 1)(r - 2) - 1]$ distribution if no interactions exist. Large values of F^* lead to the conclusion that an additive model is not appropriate.

Example. For the background music example of Table 31.2, a check of the appropriateness of the additivity assumption in model (31.1) was desired. The Tukey test for additivity was performed; key intermediate results are summarized in Table 31.6. The test statistic (31.20h) is:

$$F^* = \frac{11.19}{1} \div \frac{177.21}{11} = .69$$

Using a significance level of $\alpha = .05$, we need $F(.95; 1, 11) = 4.84$. Since $F^* = .69 \leq 4.84$, we conclude that the effects of the blocking variables and treatments are additive.

TABLE 31.6 Key intermediate calculational results for Tukey test for additivity—background music example of Table 31.2

Step 2—$e_{ij(k)}$

2.8	−2.6	3.0	−7.0	3.8
− .4	5.0	−2.6	.8	−2.8
0	1.0	1.6	− .2	−2.4
− .6	−3.0	.2	1.2	2.2
−1.8	− .4	−2.2	5.2	− .8

Step 3—SSE

$SSE = 188.4$

Step 4—$U_{ij(k)}$

29.16	1.00	92.16	54.76	54.76
51.84	70.56	9.00	29.16	7.84
184.96	54.76	96.04	43.56	27.04
9.00	21.16	10.24	84.64	4.84
4.84	33.64	57.76	27.04	163.84

Step 5—N

$N = 530.832$

Step 6—SSE(U)

$SSE(U) = 25{,}189.916$

Step 7—SSLF

$$SSLF = \frac{(530.832)^2}{25{,}189.916} = 11.19$$

Step 8—SSPE

$SSPE = 188.4 - 11.19 = 177.21$

31.8 REGRESSION APPROACH TO LATIN SQUARE DESIGNS

Model (31.1) for a latin square design with fixed blocking and treatment effects:

(31.21)
$$Y_{ij(k)} = \mu_{...} + \rho_i + \kappa_j + \tau_k + \varepsilon_{ij(k)}$$
$$i = 1, \ldots, r; j = 1, \ldots, r; k = 1, \ldots, r$$

can be expressed easily in the form of a regression model with indicator variables. As before, we shall use $1, -1, 0$ indicator variables. The regression model for the background music example of Table 31.2, with $r = 5$, can be expressed as follows:

$$(31.22) \quad Y_{ij(k)} = \mu_{\ldots} + \underbrace{\rho_1 X_{ij(k)1} + \rho_2 X_{ij(k)2} + \rho_3 X_{ij(k)3} + \rho_4 X_{ij(k)4}}_{\text{Row blocking effect}}$$

$$+ \underbrace{\kappa_1 X_{ij(k)5} + \kappa_2 X_{ij(k)6} + \kappa_3 X_{ij(k)7} + \kappa_4 X_{ij(k)8}}_{\text{Column blocking effect}}$$

$$+ \underbrace{\tau_1 X_{ij(k)9} + \tau_2 X_{ij(k)10} + \tau_3 X_{ij(k)11} + \tau_4 X_{ij(k)12}}_{\text{Treatment effect}}$$

$$+ \varepsilon_{ij(k)}$$

where:

$$X_{ij(k)1} = \begin{cases} 1 & \text{if experimental unit from row blocking class 1} \\ -1 & \text{if experimental unit from row blocking class 5} \\ 0 & \text{otherwise} \end{cases}$$

$X_{ij(k)2}$, $X_{ij(k)3}$, $X_{ij(k)4}$ are defined similarly

$$X_{ij(k)5} = \begin{cases} 1 & \text{if experimental unit from column blocking class 1} \\ -1 & \text{if experimental unit from column blocking class 5} \\ 0 & \text{otherwise} \end{cases}$$

$X_{ij(k)6}$, $X_{ij(k)7}$, $X_{ij(k)8}$ are defined similarly

$$X_{ij(k)9} = \begin{cases} 1 & \text{if experimental unit received treatment 1} \\ -1 & \text{if experimental unit received treatment 5} \\ 0 & \text{otherwise} \end{cases}$$

$X_{ij(k)10}$, $X_{ij(k)11}$, $X_{ij(k)12}$ are defined similarly

The **Y** observations vector and the **X** matrix for the background music example of Table 31.2 are shown in Table 31.7. Note, for instance, that for observation $Y_{23(5)}$, $X_2 = 1$, $X_7 = 1$, $X_9 = X_{10} = X_{11} = X_{12} = -1$, and all other X's are zero. Hence, we have:

$$\begin{aligned} Y_{23(5)} &= \mu_{\ldots} + \rho_2 + \kappa_3 - \tau_1 - \tau_2 - \tau_3 - \tau_4 + \varepsilon_{23(5)} \\ &= \mu_{\ldots} + \rho_2 + \kappa_3 + \tau_5 + \varepsilon_{23(5)} \end{aligned}$$

which is the appropriate expression for $Y_{23(5)}$ according to ANOVA model (31.21) for the particular latin square used in the example.

Tests and estimation of treatment effects with the regression approach are conducted in the usual fashion.

TABLE 31.7 Data matrices for regression model (31.22)—background music example of Table 31.2

	Y		X	X_1	X_2	X_3	X_4	X_5	X_6	X_7	X_8	X_9	X_{10}	X_{11}	X_{12}
	$Y_{11(4)} = 18$		1	1	0	0	0	1	0	0	0	0	0	0	1
	$Y_{12(3)} = 17$		1	1	0	0	0	0	1	0	0	0	0	1	0
	$Y_{13(1)} = 14$		1	1	0	0	0	0	0	1	0	1	0	0	0
	$Y_{14(2)} = 21$		1	1	0	0	0	0	0	0	1	0	1	0	0
	$Y_{15(5)} = 17$		1	1	0	0	0	-1	-1	-1	-1	-1	-1	-1	-1
	$Y_{21(3)} = 13$		1	0	1	0	0	1	0	0	0	0	0	1	0
	$Y_{22(2)} = 34$		1	0	1	0	0	0	1	0	0	0	1	0	0
	$Y_{23(5)} = 21$		1	0	1	0	0	0	0	1	0	-1	-1	-1	-1
	$Y_{24(1)} = 16$		1	0	1	0	0	0	0	0	1	1	0	0	0
	$Y_{25(4)} = 15$		1	0	1	0	0	-1	-1	-1	-1	0	0	0	1
	$Y_{31(1)} = 7$		1	0	0	1	0	1	0	0	0	1	0	0	0
	$Y_{32(4)} = 29$		1	0	0	1	0	0	1	0	0	0	0	0	1
$Y =$	$Y_{33(2)} = 32$	$X =$	1	0	0	1	0	0	0	1	0	0	1	0	0
	$Y_{34(5)} = 27$		1	0	0	1	0	0	0	0	1	-1	-1	-1	-1
	$Y_{35(3)} = 13$		1	0	0	1	0	-1	-1	-1	-1	0	0	1	0
	$Y_{41(5)} = 17$		1	0	0	0	1	1	0	0	0	-1	-1	-1	-1
	$Y_{42(1)} = 13$		1	0	0	0	1	0	1	0	0	1	0	0	0
	$Y_{43(3)} = 24$		1	0	0	0	1	0	0	1	0	0	0	1	0
	$Y_{44(4)} = 31$		1	0	0	0	1	0	0	0	1	0	0	0	1
	$Y_{45(2)} = 25$		1	0	0	0	1	-1	-1	-1	-1	0	1	0	0
	$Y_{51(2)} = 21$		1	-1	-1	-1	-1	1	0	0	0	0	1	0	0
	$Y_{52(5)} = 26$		1	-1	-1	-1	-1	0	1	0	0	-1	-1	-1	-1
	$Y_{53(4)} = 26$		1	-1	-1	-1	-1	0	0	1	0	0	0	0	1
	$Y_{54(3)} = 31$		1	-1	-1	-1	-1	0	0	0	1	0	0	1	0
	$Y_{55(1)} = 7$		1	-1	-1	-1	-1	-1	-1	-1	-1	1	0	0	0

31.9 MISSING OBSERVATIONS

Missing observations destroy the symmetry (orthogonality) of the latin square design and make the usual ANOVA calculations inappropriate. The regression approach, however, remains appropriate with missing observations in a latin square design. We just set up the regression model for the available observations, and then fit the model to the data. The procedure is analogous to that discussed in Section 28.2 for the complete block design. Tests are conducted by also fitting appropriate reduced models. Estimation of fixed treatment effects is done in terms of the regression coefficients for the full model in the usual fashion.

31.10 ADDITIONAL REPLICATIONS WITH LATIN SQUARE DESIGNS

Need for additional replications

A latin square design, as noted earlier, provides r replications for each treatment. If power and/or estimation considerations indicate that these are too few replications, two basic methods are available for increasing the number of replications—replications within cells and additional latin squares. We consider each in turn.

Replications within cells

This method of increasing the replications per treatment is feasible when two or more experimental units can be obtained for each cell defined by the row and column blocking variables. Consider, for instance, an experiment in which IQ (low, normal, high) and age (young, middle, old) are the blocking variables. In this type of situation, it is possible to obtain two or more experimental subjects for each cell, and each of the subjects in a cell will then receive the treatment assigned to that cell by the latin square employed. Suppose that n experimental units are available for each cell, and let $Y_{ij(k)m}$ denote the observation for the mth unit ($m = 1, \ldots, n$) in the (i, j) cell for which the assigned treatment is k. For the additive fixed effects model (31.1), with replications in cells, the analysis of variance is straightforward as shown in Table 31.8. The treatment, row, and column sums of squares are, respectively:

$$SSTR = rn\sum_k (\bar{Y}_{..k.} - \bar{Y}_{....})^2 = \frac{\sum_k Y^2_{..k.}}{rn} - \frac{Y^2_{....}}{r^2 n} \tag{31.23a}$$

$$SSROW = rn\sum_i (\bar{Y}_{i...} - \bar{Y}_{....})^2 = \frac{\sum_i Y^2_{i...}}{rn} - \frac{Y^2_{....}}{r^2 n} \tag{31.23b}$$

$$SSCOL = rn\sum_j (\bar{Y}_{.j..} - \bar{Y}_{....})^2 = \frac{\sum_j Y^2_{.j..}}{rn} - \frac{Y^2_{....}}{r^2 n} \tag{31.23c}$$

The total sum of squares as usual is:

$$SSTO = \sum_i \sum_j \sum_m (Y_{ij(k)m} - \bar{Y}_{....})^2 = \sum_i \sum_j \sum_m Y^2_{ij(k)m} - \frac{Y^2_{....}}{r^2 n} \tag{31.23d}$$

TABLE 31.8 ANOVA table for latin square design with n replications per cell

Source of Variation	*SS*	*df*	*MS*
Row blocking variable	$SSROW$	$r-1$	$MSROW$
Column blocking variable	$SSCOL$	$r-1$	$MSCOL$
Treatments	$SSTR$	$r-1$	$MSTR$
Error	SSE	$nr^2 - 3r + 2$	MSE
Total	$SSTO$	$nr^2 - 1$	

while SSE is obtained as a remainder:

(31.23e) $$SSE = SSTO - SSROW - SSCOL - SSTR$$

The degrees of freedom for row, column, and treatment sums of squares are unchanged, while those associated with SSE are increased from $(r-1)(r-2)$ to $nr^2 - 3r + 2$, an increase of $(n-1)r^2$ degrees of freedom.

Test for additivity. When there are n replications within a cell for a latin square, it is possible to obtain a pure error measure irrespective of the correctness of the additive model (31.1). The pure error sum of squares is obtained in the usual manner:

(31.24)
$$SSPE = \sum_i \sum_j \sum_m (Y_{ij(k)m} - \bar{Y}_{ij(k).})^2$$
$$= \sum_i \sum_j \sum_m Y_{ij(k)m}^2 - \sum_i \sum_j \frac{Y_{ij(k).}^2}{n}$$

It has associated with it $(n-1)r^2$ degrees of freedom, representing the increase in the degrees of freedom for SSE from having n observations in each cell of the latin square. The difference between SSE and $SSPE$ is a reflection of the lack of fit of the additive model:

(31.25) $$SSLF = SSE - SSPE$$

and has associated with it $(r-1)(r-2)$ degrees of freedom. Let:

(31.26a) $$MSPE = \frac{SSPE}{(n-1)r^2}$$

(31.26b) $$MSLF = \frac{SSLF}{(r-1)(r-2)}$$

Then:

(31.27) $$F^* = \frac{MSLF}{MSPE}$$

can be used to test whether or not the additive model is appropriate.

It can be shown that if the additive model is appropriate, F^* follows the $F[(r-1)(r-2), (n-1)r^2]$ distribution, and that large values of F^* lead to the conclusion that the additive model is not appropriate. Table 31.9 contains the analysis of variance breaking down SSE into the $SSLF$ and $SSPE$ components.

Example. A state university is undertaking a prototype retraining program designed to teach general computer repair skills to persons who have been displaced from their previous occupations. Table 31.10 shows the results of an experiment to evaluate the effects of three different incentive methods on

TABLE 31.9 ANOVA table to test additivity when n replications per latin square cell

Source of Variation	*SS*	*df*	*MS*
Row blocking variable	*SSROW*	$r-1$	
Column blocking variable	*SSCOL*	$r-1$	
Treatments	*SSTR*	$r-1$	
Error	*SSE*	nr^2-3r+2	
Lack of fit	*SSLF*	$(r-1)(r-2)$	*MSLF*
Pure error	*SSPE*	$(n-1)r^2$	*MSPE*
Total	*SSTO*	nr^2-1	

TABLE 31.10 Example of latin square design with two replications per cell—retraining program experiment

(a) Data

IQ *i*	*Age (j)* Young	Middle	Old
High	(*B*) 19, 16	(*A*) 20, 24	(*C*) 25, 21
Normal	(*C*) 24, 22	(*B*) 14, 15	(*A*) 14, 14
Low	(*A*) 10, 14	(*C*) 12, 13	(*B*) 7, 4

(b) Cell Totals

IQ *i*	*Age (j)* Young	Middle	Old	*Total*
High	35	44	46	125
Normal	46	29	28	103
Low	24	25	11	60
Total	105	98	85	288

$Y_{..1.} = 96$ $\quad Y_{..2.} = 75$ $\quad Y_{..3.} = 117$

achievement scores made by participants in the program. The blocking variables are IQ and age of subject. Two replications per cell were utilized. Table 31.10a contains the achievement scores, while Table 31.10b contains the cell totals.

In testing for additivity we obtain:

$$SSTO = (19)^2 + (16)^2 + (20)^2 + \cdots + (4)^2 - \frac{(288)^2}{18} = 5{,}206 - 4{,}608 = 598$$

$$SSROW = \frac{1}{6}\left[(125)^2 + (103)^2 + (60)^2\right] - 4{,}608 = 364.3$$

$$SSCOL = \frac{1}{6}\left[(105)^2 + (98)^2 + (85)^2\right] - 4{,}608 = 34.3$$

$$SSTR = \frac{1}{6}\left[(96)^2 + (75)^2 + (117)^2\right] - 4{,}608 = 147$$

$$SSE = 598 - 364.3 - 34.3 - 147 = 52.4$$

$$SSPE = (19)^2 + (16)^2 + (20)^2 + \cdots + (4)^2 - \frac{1}{2}\left[(35)^2 + (44)^2 + \cdots + (11)^2\right] = 36$$

To test the appropriateness of the additive model, we find:

$$SSLF = SSE - SSPE = 52.4 - 36 = 16.4$$

$$MSLF = \frac{SSLF}{(r-1)(r-2)} = \frac{16.4}{2(1)} = 8.2$$

$$MSPE = \frac{SSPE}{(n-1)r^2} = \frac{36}{1(3)^2} = 4$$

Test statistic (31.27) here is:

$$F^* = \frac{MSLF}{MSPE} = \frac{8.2}{4} = 2.05$$

For a level of significance of $\alpha = .05$, we need $F(.95; 2, 9) = 4.26$. Since $F^* = 2.05 \leq 4.26$, we conclude that the additive model (31.1) is appropriate. The analysis of treatment effects can now proceed as usual, based on the additive model (31.1).

Additional latin squares

At times, it is not possible to obtain additional experimental units within a cell. This is the case, for instance, in the background music example of Table 31.2, where only one type of music can be played in one day. When it is not possible to replicate within cells, additional replications for each treatment frequently can be obtained by adding one or more latin squares to one of the blocking variables. In our background music example of Table 31.2, the experiment could be run for another five weeks. In an experiment using plant crews as experimental units and employing as blocking variables plant shift (morning, afternoon, evening) and production department (1, 2, 3), additional replications can be obtained by running the experiment in other production departments.

The layout for the background music example of Table 31.2, when run over another five weeks, is shown in Table 31.11. The second latin square, and additional ones when required, is selected independently of the first.

TABLE 31.11 Two latin squares design—background music example of Table 31.2

		Day				
Square	*Week*	*M*	*T*	*W*	*Th*	*F*
	1	*D*	*C*	*A*	*B*	*E*
	2	*C*	*B*	*E*	*A*	*D*
1	3	*A*	*D*	*B*	*E*	*C*
	4	*E*	*A*	*C*	*D*	*B*
	5	*B*	*E*	*D*	*C*	*A*
	6	*E*	*D*	*C*	*A*	*B*
	7	*B*	*A*	*E*	*D*	*C*
2	8	*D*	*C*	*A*	*B*	*E*
	9	*A*	*E*	*B*	*C*	*D*
	10	*C*	*B*	*D*	*E*	*A*

Frequently, the additional squares may be viewed as classes of a third blocking variable. For instance, in our background music example of Table 31.11, the two latin squares may be considered to refer to the blocking variable "time period." The first five weeks may be viewed as time period 1, and the second five weeks as time period 2. As another example, in the experiment with plant crews mentioned previously, the production departments for the first latin square may be on an hourly rate, while the departments for the second latin square may be on incentive pay, as shown in Table 31.12. Thus, with additional latin squares, one can in effect introduce a third blocking variable. As a consequence, the variation associated with the third blocking variable can be removed from the experimental error variability. In addition, one can study the interactions between the third blocking variable and the other variables and need not assume that these do not exist.

Table 31.13 shows an analysis of variance for n independent squares. This analysis is appropriate when (1) the n latin squares have the same rows and columns, and (2) the model permits third blocking variable effects and interactions between the third blocking variable and the other variables. These conditions might be appropriate for our background music example of Table 31.11. There, it may at times be appropriate to consider the rows to refer to the position of the week within the five-week period, and the columns to refer to the position of the day within a week. Note in Table 31.13 that besides the row, column, and treatment sums of squares, there are third blocking variable effects, interactions involving the third blocking variable, and a term reflecting experimental error.

TABLE 31.12 Two latin squares design—experiment with plant crews

Square	Production Department	Shift		
		Morning	*Afternoon*	*Evening*
	1	*C*	*B*	*A*
1 (hourly pay)	2	*B*	*A*	*C*
	3	*A*	*C*	*B*
	4	*A*	*B*	*C*
2 (incentive pay)	5	*C*	*A*	*B*
	6	*B*	*C*	*A*

TABLE 31.13 ANOVA table when n latin squares with same rows and columns are employed and interactions involving third blocking variable are permitted

Source of Variation	*SS*	*df*
Row blocking variable (ROW)	$SSROW$	$r-1$
Column blocking variable (COL)	$SSCOL$	$r-1$
Treatments (TR)	$SSTR$	$r-1$
Third blocking variable (3)	$SS3$	$n-1$
Third-row interactions ($3.ROW$)	$SS3.ROW$	$(n-1)(r-1)$
Third-column interactions ($3.COL$)	$SS3.COL$	$(n-1)(r-1)$
Third-treatment interactions ($3.TR$)	$SS3.TR$	$(n-1)(r-1)$
Error (E)	SSE	$n(r-1)(r-2)$
Total	$SSTO$	nr^2-1

Replications in repeated measures studies

We noted earlier that a latin square design is highly suitable for a repeated measures study when there are r treatments and r subjects. If additional replications are needed, however, replications within cells cannot be used since a cell pertains to an individual subject. Instead, cross-over designs or independent latin squares may be used.

Cross-over designs. These designs, also called *change-over designs,* are often useful when a latin square is to be used in a repeated measures study to balance the order positions of treatments, yet more subjects are required than called for by a single latin square. With this type of design, the subjects are randomly assigned to the different treatment order patterns given by a latin square (several latin squares may be used at times). Consider an experiment in which treatments A, B, and C are to be administered to each subject, and the three treatment order patterns are given by the latin square:

	Order Position		
Pattern	1	2	3
1	*A*	*B*	*C*
2	*B*	*C*	*A*
3	*C*	*A*	*B*

Suppose that $3n$ subjects are available for the study. Then n subjects will be assigned at random to each of the three order patterns in a cross-over design. Note that this design is a mixture of randomized blocks (subjects are blocks) and latin square (order patterns form a latin square).

Assuming that an additive model is appropriate, the analysis of variance of a cross-over design is straightforward; the total sum of squares is decomposed into components for blocks (subjects), order positions, treatments, and error. Table 31.14 contains the analysis of variance table for the general case of r treatments and n subjects per order pattern in a cross-over design. Computational formulas for the sums of squares in Table 31.14 follow the usual pattern. Let $Y_{ij(k)m}$ denote the observation for the mth unit (e.g., subject) for the ith treatment order pattern which in period j assigned treatment k; $i = 1, \ldots, r$; $j = 1, \ldots, r$; $k = 1, \ldots, r$; $m = 1, \ldots, n$. We then have:

(31.28a)
$$SSTO = \sum_i \sum_j \sum_m Y_{ij(k)m}^2 - \frac{Y_{....}^2}{r^2 n}$$

(31.28b)
$$SSS = \frac{\sum_i \sum_m Y_{i..m}^2}{r} - \frac{Y_{....}^2}{r^2 n}$$

(31.28c)
$$SSO = \frac{\sum_j Y_{.j..}^2}{nr} - \frac{Y_{....}^2}{r^2 n}$$

(31.28d)
$$SSTR = \frac{\sum_k Y_{..k.}^2}{nr} - \frac{Y_{....}^2}{r^2 n}$$

(31.28e)
$$SSE = SSTO - SSS - SSO - SSTR$$

TABLE 31.14 ANOVA table for cross-over design

Source of Variation	*SS*	*df*	*MS*
Subjects (S)	SSS	$rn - 1$	MSS
Order positions (O)	SSO	$r - 1$	MSO
Treatments (TR)	$SSTR$	$r - 1$	$MSTR$
Error (E)	SSE	$(r - 1)(nr - 2)$	MSE
Total	$SSTO$	$nr^2 - 1$	

where SSS is the *sum of squares for subjects, SSO* is the *sum of squares for order positions,* and the other sums of squares are defined as usual.

Example. Table 31.15a contains data for a study of the effects of three different displays on the sale of apples, using the cross-over design. Six stores were used, and two were assigned at random to each of the three treatment order patterns shown. Each display was kept for two weeks, and the observed variable was sales per 100 customers. Table 31.15b contains the analysis of variance. The sums of squares were obtained from a computer run.

TABLE 31.15 Cross-over design—apple sales example

(a) Data (coded)

	Two-Week Period		
Store	1	2	3
1	9 (*B*)	12 (*C*)	15 (*A*)
2	12 (*A*)	14 (*B*)	3 (*C*)
3	13 (*A*)	14 (*B*)	3 (*C*)
4	7 (*C*)	18 (*A*)	6 (*B*)
5	5 (*C*)	20 (*A*)	4 (*B*)
6	4 (*B*)	12 (*C*)	9 (*A*)

(b) Analysis of Variance

Source of Variation	*SS*	*df*	*MS*
Stores	21.3	5	4.26
Order positions	233.3	2	116.7
Displays	189.0	2	94.5
Error	20.4	8	2.55
Total	464.0	17	

To test for treatment effects, we use:

$$F^* = \frac{MSTR}{MSE} = \frac{94.5}{2.55} = 37.1$$

For $\alpha = .05$, we require $F(.95; 2, 8) = 4.46$. Since $F^* = 37.1 > 4.46$, we conclude that there are differential sales effects for the three displays. Tests for order position effects and store effects were also carried out. They indicated that order position effects were present, but no store effects. Order position effects here are associated with the three time periods in which the displays were studied, and may reflect seasonal effects as well as the results of special events, such as unusually hot weather in one period.

Use of independent latin squares. If the order position effects are not approximately constant for all subjects (stores, etc.), a cross-over design is not fully effective. It may then be preferable to place the subjects into homogeneous groups with respect to the order position effects and use independent latin squares for each group. Suppose that four treatments are to be administered to eight subjects each, four males and four females, and that the experimenter expects the fatigue effect to be strong for females but only mild for males. The use of two independent latin squares, one for male subjects and the other for female subjects, may then be advisable.

Carry-over effects. If carry-over effects from one treatment to another are anticipated, that is, if not only the order position but also the preceding treatment has an effect, one may balance out these carry-over effects by choosing a latin square in which every treatment follows every other treatment an equal number of times. For $r = 4$, an example of such a latin square is:

	Period			
Subject	1	2	3	4
1	*A*	*B*	*D*	*C*
2	*B*	*C*	*A*	*D*
3	*C*	*D*	*B*	*A*
4	*D*	*A*	*C*	*B*

Note that treatment *A* follows each of the other treatments once, and similarly for the other treatments. This design is suited when the carry-over effects do not persist for more than one period.

When *r* is odd, the sequence balance can be obtained by using a pair of latin squares with the property that the treatment sequences in one are reversed in the other square. Indeed, even when *r* is even, it is usually desirable to use a pair of such squares so that the degrees of freedom associated with the error mean square are reasonably large. Such a design is sometimes called a *double cross-over design*. This type of design retains the advantages of employing two blocking variables in a latin square, while enabling the experimenter also to balance and measure the carry-over effects.

For our earlier apple display illustration in which three displays were studied in six stores, the two latin squares might be as shown in Table 31.16. The stores should first be placed into two homogeneous groups and these should then be assigned to the two latin squares.

31.11 RANDOM BLOCKING VARIABLE EFFECTS

If the row and/or column blocking variable in a latin square design has classes that should be viewed as random selections from a population, the fixed effects model (31.1) no longer is applicable.

TABLE 31.16 Illustration of a double cross-over design

		Two-Week Period		
Square	*Store*	1	2	3
	1	*A*	*B*	*C*
1	2	*B*	*C*	*A*
	3	*C*	*A*	*B*
	4	*A*	*C*	*B*
2	5	*B*	*A*	*C*
	6	*C*	*B*	*A*

Both blocking variables random

Consider the case where the row blocking variable is subject and the column blocking variable observer, and both the subjects and the observers included in the study are viewed as random samples from relevant populations. In that case, assuming treatment effects are fixed, the additive model for a latin square design is:

(31.29)
$$Y_{ij(k)} = \mu_{\dots} + \rho_i + \kappa_j + \tau_k + \varepsilon_{ij(k)}$$

where:

$\mu_{\dots}$ is a constant
ρ_i are independent $N(0, \sigma_\rho^2)$
κ_j are independent $N(0, \sigma_\kappa^2)$
τ_k are constants subject to the restriction $\Sigma\tau_k = 0$
$\varepsilon_{ij(k)}$ are independent $N(0, \sigma^2)$
ρ_i, κ_j, and $\varepsilon_{ij(k)}$ are independent
$i = 1, \dots, r;\ j = 1, \dots, r;\ k = 1, \dots, r$

The analysis of variance is the same as for the fixed blocking variable effects model. The expected mean squares are obtained by replacing the sums of squared effects divided by degrees of freedom in Table 31.3 by variance terms. Consequently, all tests and estimates of treatment effects are conducted as for fixed blocking variable effects.

One blocking variable random, other fixed

When one of the blocking variables has random effects while the other has fixed effects, the additive model with fixed treatment effects becomes a mixture of models (31.1) and (31.29). Again, there will be no change in the analysis of variance or in tests or estimates of treatment effects. An instance where this mixed model would be appropriate is a repeated measures study where the row blocking variable is subject and the column blocking variable is order of treatments.

31.12 YOUDEN AND GRAECO-LATIN SQUARES

When it is not possible to use a latin square design because the number of column classes is less than the number of row classes, a *Youden square design* may be helpful. Consider a repeated measures study involving four treatments and four subjects (row blocking variable). Suppose now that a subject can be given only three treatments because of serious fatigue effects. Hence, the column blocking variable (order position of treatments) can have only three classes. A Youden square design suitable for this occasion is shown in Table 31.17. Note that this layout would become a latin square with the addition of the column (D, C, B, A). Also note that each treatment occurs once in each order position, and that every pair of treatments appears together an equal number of times within subjects. These are characteristics present in all Youden squares. The analysis of Youden square designs is more complex than that of latin squares because not all treatments are run in each class of the row blocking variable. A reference, such as Reference 31.1, should be consulted if a Youden square design is to be used.

TABLE 31.17 Illustration of a Youden square design

	Order Position of Treatment		
Subject	1	2	3
1	*A*	*B*	*C*
2	*D*	*A*	*B*
3	*C*	*D*	*A*
4	*B*	*C*	*D*

A *graeco-latin square design* is an extension of a latin square design when three blocking variables are to be used simultaneously. Table 31.18 illustrates a graeco-latin square design for $r = 4$. The symbols α, β, γ, and δ represent the four levels of the third blocking variable. Thus, the cell corresponding to the first class of each of the three blocking variables is to receive treatment A, and so on. Note that the levels of the third blocking variable appear once in each row and once in each column, and they appear once only with each treatment. Graeco-

TABLE 31.18 Illustration of a graeco-latin square design ($r = 4$)

	Column Blocking Variable			
Row Blocking Variable	1	2	3	4
1	$\alpha : A$	$\beta : B$	$\gamma : C$	$\delta : D$
2	$\beta : C$	$\alpha : D$	$\delta : A$	$\gamma : B$
3	$\delta : B$	$\gamma : A$	$\beta : D$	$\alpha : C$
4	$\gamma : D$	$\delta : C$	$\alpha : B$	$\beta : A$

latin square designs are used much less frequently in practice than the other designs we have discussed. Reference 31.1 discusses the analysis of graeco-latin square designs.

PROBLEMS

31.1. A behavioral scientist explained why latin square designs are used so frequently: "Many times in behavioral science, we require the use of repeated measures designs because variability between human subjects is so great. Since an order effect may be present in this situation, we employ latin square designs to eliminate any bias due to order effects." Comment.

31.2. a. Using random permutations, select randomly a 3 by 3 latin square. Show all steps.

b. Using random permutations, select randomly a 6 by 6 latin square. Show all steps.

31.3. **Hardware sales.** A manufacturer conducted a small pilot study of the effect of the price of one of its products on sales of this product in hardware stores. Since it might be confusing to customers if prices were switched repeatedly within a store, only one price was used for any one store during the six-month study period. Sixteen stores were employed in the study. To reduce experimental error variability, they were chosen so that there would be one store for each sales volume-geographic location class. The four price levels (A: \$1.79; B: \$1.69; C: \$1.59; D: \$1.49) were assigned to the stores according to the latin square design shown below. Data on sales during the six-month period (in thousand dollars) follow.

Sales Volume Class	*Geographic Location Class* (j)			
i	*Northeast*	*Northwest*	*Southeast*	*Southwest*
1 (smallest)	1.2 (B)	1.5 (C)	1.0 (A)	1.7 (D)
2	1.4 (A)	1.9 (D)	1.6 (B)	1.5 (C)
3	2.8 (C)	2.1 (B)	2.7 (D)	2.0 (A)
4 (largest)	3.4 (D)	2.5 (A)	2.9 (C)	2.7 (B)

Assume that fixed effects ANOVA model (31.1) is appropriate.

a. Test whether or not price level affects mean sales. Use significance level $\alpha = .05$. State the alternatives, decision rule, and conclusion.

b. Analyze the nature of the price effect on sales by making all pairwise comparisons among the treatment means. Use the Tukey procedure and a 90 percent family confidence coefficient. Summarize your findings.

c. Does there appear to be a linear relationship between price level and mean sales? Could you formally test for linearity? Explain.

31.4. Refer to **Hardware sales** Problem 31.3.

a. Obtain the residuals and plot them against the fitted values. Also prepare a normal probability plot of the residuals and calculate the coefficient of correlation between the ordered residuals and their expected values under normality. Summarize your findings about the appropriateness of ANOVA model (31.1) here.

b. Conduct the Tukey test for additivity; use $\alpha = .01$. State the alternatives, decision rule, and conclusion.

31.5. Refer to **Hardware sales** Problem 31.3.

a. Calculate the three estimated efficiency measures in (31.19).

b. Would a randomized block design have been adequate here? If so, which blocking variable would have been best?

31.6. **Summary reports.** A management information systems consultant conducted a small-scale study of five different daily summary reports (*A*: greatest amount of detail; *B*; *C*; *D*; *E*: least amount of detail). She used five sales executives in the study. Each was given one type of daily report for a month and then was asked to rate its helpfulness on a 25-point scale (0: no help; 25: extremely helpful). Over a five-month period, each executive received each type of report for one month according to the latin square design shown below. The helpfulness ratings follow.

Executive	Month (j)				
i	*March*	*April*	*May*	*June*	*July*
Harrison	21 (*D*)	8 (*A*)	17 (*C*)	9 (*B*)	16 (*E*)
Smith	5 (*A*)	10 (*E*)	3 (*B*)	12 (*C*)	15 (*D*)
Carmichael	20 (*C*)	10 (*B*)	15 (*E*)	22 (*D*)	12 (*A*)
Loeb	4 (*B*)	17 (*D*)	3 (*A*)	9 (*E*)	10 (*C*)
Munch	17 (*E*)	16 (*C*)	20 (*D*)	7 (*A*)	11 (*B*)

Assume that fixed effects ANOVA model (31.1) is appropriate.

a. Test whether or not the five types of reports differ in mean helpfulness; use significance level $\alpha = .01$. State the alternatives, decision rule, and conclusion.

b. Analyze the effectiveness of the five types of reports by making all pairwise comparisons among the treatment means. Use the Tukey procedure and a 95 percent family confidence coefficient. Summarize your findings.

31.7. Refer to **Summary reports** Problem 31.6.

a. Obtain the residuals and plot them against the fitted values. Also prepare a normal probability plot of the residuals and calculate the coefficient of correlation between the ordered residuals and their expected values under normality. Summarize your findings about the appropriateness of ANOVA model (31.1) here.

b. Conduct the Tukey test for additivity; use $\alpha = .05$. State the alternatives, decision rule, and conclusion.

31.8. Refer to **Summary reports** Problem 31.6.

a. Calculate the three estimated efficiency measures in (31.19).

b. How effective was the use of the repeated measures design here?

31.9. Refer to **Hardware sales** Problem 31.3. Assume that $\sigma = .15$. What is the power of the test for treatment effects in Problem 31.3a if $\mu_{..1} = 1.5$, $\mu_{..2} = 2.1$, $\mu_{..3} = 2.3$, and $\mu_{..4} = 2.5$?

31.10. Refer to **Summary reports** Problem 31.6. Assume that $\sigma = 1.4$. What is the power of the test for treatment effects in Problem 31.6a if $\mu_{..1} = 7.0$, $\mu_{..2} = 7.5$, $\mu_{..3} = 12.0$, $\mu_{..4} = 15.0$, and $\mu_{..5} = 14.0$?

31.11. **Drugs interaction.** A pilot study was undertaken on the interaction effects of two drugs to stimulate growth in girls who are of short stature because of a particular syndrome. Each drug was known to be modestly effective singly, but the combination of the two drugs had never been investigated. Blocking by both subject and time period was desired whereby repeated measures for different treatments applied to the same subject are obtained. A 4 by 4 latin square design, shown below, was utilized for four subjects, four time periods, and four treatments. The four time periods consisted of one month each, separated by an intervening month during which no treatment was given. The four treatments were: A: no treatment (placebo); B: drug X alone; C: drug Y alone; D: both drugs X and Y. The dependent variable was the difference in the growth rates (in centimeters per month) during the treatment period and the base period before the experiment began. The results of the study follow.

Subject	*Period* (j)			
i	1	2	3	4
1	.02 (A)	.15 (B)	.45 (D)	.18 (C)
2	.27 (B)	.24 (C)	−.01 (A)	.58 (D)
3	.11 (C)	.35 (D)	.14 (B)	−.03 (A)
4	.48 (D)	.04 (A)	.18 (C)	.22 (B)

Assume that ANOVA model (31.1), modified so that subjects have random effects and a factorial treatment structure is incorporated, is appropriate.

a. State the ANOVA model to be employed.

b. Test for interaction effects between the two drugs; use significance level $\alpha = .10$. State the alternatives, decision rule, and conclusion.

c. Estimate the interaction contrast:

$$L = \left(\frac{\mu_{..2} + \mu_{..3}}{2} - \mu_{..1}\right) - \left(\mu_{..4} - \frac{\mu_{..2} + \mu_{..3}}{2}\right)$$

using a 90 percent confidence interval. Interpret your result.

31.12. Refer to **Drugs interaction** Problem 31.11. Obtain the residuals and plot them against the fitted values. Also prepare a normal probability plot of the residuals and calculate the coefficient of correlation between the ordered residuals and their expected values under normality. Summarize your findings about the appropriateness of the model utilized here.

31.13. Refer to **Hardware sales** Problem 31.3.

a. Set up the regression model equivalent to ANOVA model (31.1) using $1, -1, 0$ indicator variables.

b. Test by means of the regression approach whether or not price level affects mean sales. Use significance level $\alpha = .05$. State the alternatives, decision rule, and conclusion.

c. Obtain a 95 percent confidence interval by the regression approach for $D = \mu_{..3} - \mu_{..4}$. Interpret your interval estimate.

d. Suppose that observation $Y_{23(2)} = 1.6$ were missing.

(i) Use the regression approach to test whether price level affects mean sales; control the α risk at .05. State the alternatives, decision rule, and conclusion.

(ii) Use the regression approach to estimate $D = \mu_{..1} - \mu_{..2}$ by means of a 95 percent confidence interval.

31.14. Refer to **Summary reports** Problem 31.6. Suppose that observations $Y_{11(4)} = 21$ and $Y_{45(3)} = 10$ were missing.

a. Use the regression approach to test whether the five types of reports differ in mean effectiveness; employ significance level $\alpha = .01$. State the alternatives, decision rule, and conclusion.

b. Use the regression approach to estimate $D = \mu_{..4} - \mu_{..1}$ by means of a 99 percent confidence interval.

31.15. **TV commercials.** A study was undertaken to determine whether the volume of sound of a television commercial affects recall and whether this effect varies by product. Thirty-two subjects were chosen, two each for 16 groups defined according to age (class 1: youngest; 2; 3; 4: oldest) and amount of education (class 1: lowest education level; 2; 3; 4: highest education level). Each subject was exposed to one of four television commercial showings (*A*: high volume, product X; *B*: low volume, product X; *C*: high volume, product Y; *D*: low volume, product Y) according to the latin square design shown below. Two different commercials were involved, one for each product. During the following week, the subjects were asked to mention everything they could remember about the advertisement. Scores were based on the number of learning points mentioned, suitably standardized. The results follow.

Age Class	*Education Level* (j)			
i	1	2	3	4
	(*D*)	(*A*)	(*C*)	(*B*)
1	83	64	78	76
	86	69	75	74
	(*B*)	(*C*)	(*A*)	(*D*)
2	70	81	64	87
	76	75	60	81
	(*C*)	(*B*)	(*D*)	(*A*)
3	67	67	76	64
	74	61	81	57
	(*A*)	(*D*)	(*B*)	(*C*)
4	56	72	63	64
	60	67	67	66

Assume that fixed effects ANOVA model (31.1), modified to allow for factorial treatments (factor *A*: volume; factor *B*: product) and replications, is appropriate.

a. State the ANOVA model to be employed.

b. Test for volume-product interaction effects; use $\alpha = .01$. State the alternatives, decision rule, and conclusion.

c. Test for volume main effects and product main effects. For each test, use $\alpha = .01$ and state the alternatives, decision rule, and conclusion.

d. To study the nature of the volume and product main effects, estimate the difference between the two factor level means for each factor. Use the Bonferroni procedure and a 95 percent family confidence coefficient. State your findings.

31.16. Refer to **TV commercials** Problem 31.15.

a. Obtain the residuals and plot them against the fitted values. Also prepare a normal probability plot of the residuals and calculate the coefficient of correlation between the ordered residuals and their expected values under normality. Summarize your findings about the appropriateness of the model utilized here.

b. Conduct a formal test of whether or not the effects of blocking variables and treatments are additive. Use a level of significance of $\alpha = .01$. State the alternatives, decision rule, and conclusion.

31.17. Refer to the apple sales experiment on page 1061.

a. Verify the analysis of variance in Table 31.15b for the data in Table 31.15a.

b. Test for the presence of order position and store main effects. For each test, use a level of significance of $\alpha = .05$ and state the alternatives, decision rule, and conclusion.

c. Analyze the display main effects by estimating all pairwise comparisons of treatment means. Use the Tukey procedure and a 90 percent family confidence coefficient. Summarize your findings.

EXERCISES

31.18. Derive the analysis of variance decomposition in (31.7).

31.19. Derive the expected mean squares in Table 31.3 for fixed effects ANOVA model (31.1) by using rule (30.3). [*Hint:* Remember that any one of the three subscripts is redundant and should not be considered in rule (30.3) for obtaining the coefficients.]

CITED REFERENCE

31.1 Cochran, William G., and Gertrude M. Cox. *Experimental Designs*. 2d ed. New York: John Wiley & Sons, 1957.

APPENDIX TABLES

TABLE A–1 Cumulative probabilities of the standard normal distribution

Entry is area A under the standard normal curve from $-\infty$ to $z(A)$

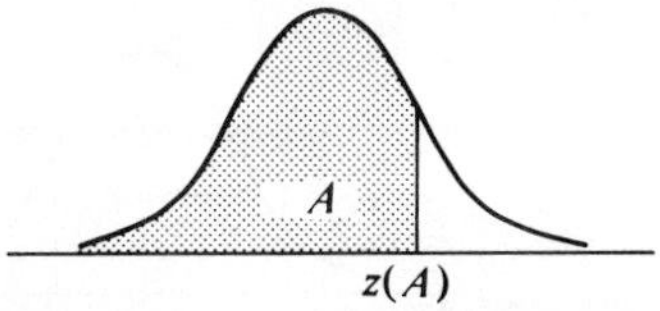

z	.00	.01	.02	.03	.04	.05	.06	.07	.08	.09
.0	.5000	.5040	.5080	.5120	.5160	.5199	.5239	.5279	.5319	.5359
.1	.5398	.5438	.5478	.5517	.5557	.5596	.5636	.5675	.5714	.5753
.2	.5793	.5832	.5871	.5910	.5948	.5987	.6026	.6064	.6103	.6141
.3	.6179	.6217	.6255	.6293	.6331	.6368	.6406	.6443	.6480	.6517
.4	.6554	.6591	.6628	.6664	.6700	.6736	.6772	.6808	.6844	.6879
.5	.6915	.6950	.6985	.7019	.7054	.7088	.7123	.7157	.7190	.7224
.6	.7257	.7291	.7324	.7357	.7389	.7422	.7454	.7486	.7517	.7549
.7	.7580	.7611	.7642	.7673	.7704	.7734	.7764	.7794	.7823	.7852
.8	.7881	.7910	.7939	.7967	.7995	.8023	.8051	.8078	.8106	.8133
.9	.8159	.8186	.8212	.8238	.8264	.8289	.8315	.8340	.8365	.8389
1.0	.8413	.8438	.8461	.8485	.8508	.8531	.8554	.8577	.8599	.8621
1.1	.8643	.8665	.8686	.8708	.8729	.8749	.8770	.8790	.8810	.8830
1.2	.8849	.8869	.8888	.8907	.8925	.8944	.8962	.8980	.8997	.9015
1.3	.9032	.9049	.9066	.9082	.9099	.9115	.9131	.9147	.9162	.9177
1.4	.9192	.9207	.9222	.9236	.9251	.9265	.9279	.9292	.9306	.9319
1.5	.9332	.9345	.9357	.9370	.9382	.9394	.9406	.9418	.9429	.9441
1.6	.9452	.9463	.9474	.9484	.9495	.9505	.9515	.9525	.9535	.9545
1.7	.9554	.9564	.9573	.9582	.9591	.9599	.9608	.9616	.9625	.9633
1.8	.9641	.9649	.9656	.9664	.9671	.9678	.9686	.9693	.9699	.9706
1.9	.9713	.9719	.9726	.9732	.9738	.9744	.9750	.9756	.9761	.9767
2.0	.9772	.9778	.9783	.9788	.9793	.9798	.9803	.9808	.9812	.9817
2.1	.9821	.9826	.9830	.9834	.9838	.9842	.9846	.9850	.9854	.9857
2.2	.9861	.9864	.9868	.9871	.9875	.9878	.9881	.9884	.9887	.9890
2.3	.9893	.9896	.9898	.9901	.9904	.9906	.9909	.9911	.9913	.9916
2.4	.9918	.9920	.9922	.9925	.9927	.9929	.9931	.9932	.9934	.9936
2.5	.9938	.9940	.9941	.9943	.9945	.9946	.9948	.9949	.9951	.9952
2.6	.9953	.9955	.9956	.9957	.9959	.9960	.9961	.9962	.9963	.9964
2.7	.9965	.9966	.9967	.9968	.9969	.9970	.9971	.9972	.9973	.9974
2.8	.9974	.9975	.9976	.9977	.9977	.9978	.9979	.9979	.9980	.9981
2.9	.9981	.9982	.9982	.9983	.9984	.9984	.9985	.9985	.9986	.9986
3.0	.9987	.9987	.9987	.9988	.9988	.9989	.9989	.9989	.9990	.9990
3.1	.9990	.9991	.9991	.9991	.9992	.9992	.9992	.9992	.9993	.9993
3.2	.9993	.9993	.9994	.9994	.9994	.9994	.9994	.9995	.9995	.9995
3.3	.9995	.9995	.9995	.9996	.9996	.9996	.9996	.9996	.9996	.9997
3.4	.9997	.9997	.9997	.9997	.9997	.9997	.9997	.9997	.9997	.9998

Selected Percentiles

Cumulative probability A:	.90	.95	.975	.98	.99	.995	.999
$z(A)$:	1.282	1.645	1.960	2.054	2.326	2.576	3.090

TABLE A–2 Percentiles of the *t* distribution

Entry is $t(A;\ \nu)$ where $P\{t(\nu) \leq t(A;\nu)\} = A$

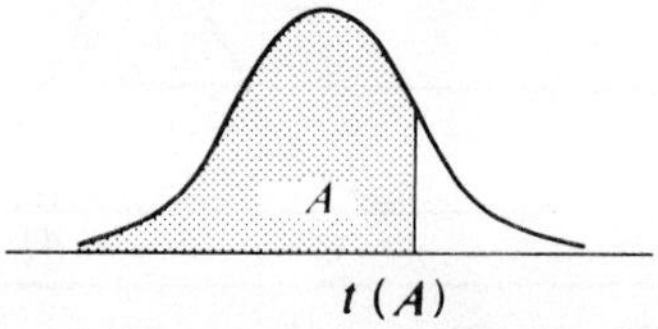

	A						
ν	.60	.70	.80	.85	.90	.95	.975
1	0.325	0.727	1.376	1.963	3.078	6.314	12.706
2	0.289	0.617	1.061	1.386	1.886	2.920	4.303
3	0.277	0.584	0.978	1.250	1.638	2.353	3.182
4	0.271	0.569	0.941	1.190	1.533	2.132	2.776
5	0.267	0.559	0.920	1.156	1.476	2.015	2.571
6	0.265	0.553	0.906	1.134	1.440	1.943	2.447
7	0.263	0.549	0.896	1.119	1.415	1.895	2.365
8	0.262	0.546	0.889	1.108	1.397	1.860	2.306
9	0.261	0.543	0.883	1.100	1.383	1.833	2.262
10	0.260	0.542	0.879	1.093	1.372	1.812	2.228
11	0.260	0.540	0.876	1.088	1.363	1.796	2.201
12	0.259	0.539	0.873	1.083	1.356	1.782	2.179
13	0.259	0.537	0.870	1.079	1.350	1.771	2.160
14	0.258	0.537	0.868	1.076	1.345	1.761	2.145
15	0.258	0.536	0.866	1.074	1.341	1.753	2.131
16	0.258	0.535	0.865	1.071	1.337	1.746	2.120
17	0.257	0.534	0.863	1.069	1.333	1.740	2.110
18	0.257	0.534	0.862	1.067	1.330	1.734	2.101
19	0.257	0.533	0.861	1.066	1.328	1.729	2.093
20	0.257	0.533	0.860	1.064	1.325	1.725	2.086
21	0.257	0.532	0.859	1.063	1.323	1.721	2.080
22	0.256	0.532	0.858	1.061	1.321	1.717	2.074
23	0.256	0.532	0.858	1.060	1.319	1.714	2.069
24	0.256	0.531	0.857	1.059	1.318	1.711	2.064
25	0.256	0.531	0.856	1.058	1.316	1.708	2.060
26	0.256	0.531	0.856	1.058	1.315	1.706	2.056
27	0.256	0.531	0.855	1.057	1.314	1.703	2.052
28	0.256	0.530	0.855	1.056	1.313	1.701	2.048
29	0.256	0.530	0.854	1.055	1.311	1.699	2.045
30	0.256	0.530	0.854	1.055	1.310	1.697	2.042
40	0.255	0.529	0.851	1.050	1.303	1.684	2.021
60	0.254	0.527	0.848	1.045	1.296	1.671	2.000
120	0.254	0.526	0.845	1.041	1.289	1.658	1.980
∞	0.253	0.524	0.842	1.036	1.282	1.645	1.960

TABLE A–2 *(concluded)* Percentiles of the *t* distribution

	A						
ν	.98	.985	.99	.9925	.995	.9975	.9995
1	15.895	21.205	31.821	42.434	63.657	127.322	636.590
2	4.849	5.643	6.965	8.073	9.925	14.089	31.598
3	3.482	3.896	4.541	5.047	5.841	7.453	12.924
4	2.999	3.298	3.747	4.088	4.604	5.598	8.610
5	2.757	3.003	3.365	3.634	4.032	4.773	6.869
6	2.612	2.829	3.143	3.372	3.707	4.317	5.959
7	2.517	2.715	2.998	3.203	3.499	4.029	5.408
8	2.449	2.634	2.896	3.085	3.355	3.833	5.041
9	2.398	2.574	2.821	2.998	3.250	3.690	4.781
10	2.359	2.527	2.764	2.932	3.169	3.581	4.587
11	2.328	2.491	2.718	2.879	3.106	3.497	4.437
12	2.303	2.461	2.681	2.836	3.055	3.428	4.318
13	2.282	2.436	2.650	2.801	3.012	3.372	4.221
14	2.264	2.415	2.624	2.771	2.977	3.326	4.140
15	2.249	2.397	2.602	2.746	2.947	3.286	4.073
16	2.235	2.382	2.583	2.724	2.921	3.252	4.015
17	2.224	2.368	2.567	2.706	2.898	3.222	3.965
18	2.214	2.356	2.552	2.689	2.878	3.197	3.922
19	2.205	2.346	2.539	2.674	2.861	3.174	3.883
20	2.197	2.336	2.528	2.661	2.845	3.153	3.849
21	2.189	2.328	2.518	2.649	2.831	3.135	3.819
22	2.183	2.320	2.508	2.639	2.819	3.119	3.792
23	2.177	2.313	2.500	2.629	2.807	3.104	3.768
24	2.172	2.307	2.492	2.620	2.797	3.091	3.745
25	2.167	2.301	2.485	2.612	2.787	3.078	3.725
26	2.162	2.296	2.479	2.605	2.779	3.067	3.707
27	2.158	2.291	2.473	2.598	2.771	3.057	3.690
28	2.154	2.286	2.467	2.592	2.763	3.047	3.674
29	2.150	2.282	2.462	2.586	2.756	3.038	3.659
30	2.147	2.278	2.457	2.581	2.750	3.030	3.646
40	2.123	2.250	2.423	2.542	2.704	2.971	3.551
60	2.099	2.223	2.390	2.504	2.660	2.915	3.460
120	2.076	2.196	2.358	2.468	2.617	2.860	3.373
∞	2.054	2.170	2.326	2.432	2.576	2.807	3.291

TABLE A–3 Percentiles of the χ^2 distribution

Entry is $\chi^2(A;\nu)$ where $P\{\chi^2(\nu) \leq \chi^2(A;\nu)\} = A$

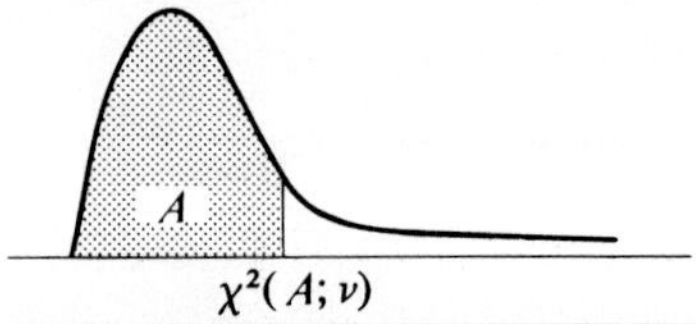

ν	A: .005	.010	.025	.050	.100	.900	.950	.975	.990	.995
1	0.0^4393	0.0^3157	0.0^3982	0.0^2393	0.0158	2.71	3.84	5.02	6.63	7.88
2	0.0100	0.0201	0.0506	0.103	0.211	4.61	5.99	7.38	9.21	10.60
3	0.072	0.115	0.216	0.352	0.584	6.25	7.81	9.35	11.34	12.84
4	0.207	0.297	0.484	0.711	1.064	7.78	9.49	11.14	13.28	14.86
5	0.412	0.554	0.831	1.145	1.61	9.24	11.07	12.83	15.09	16.75
6	0.676	0.872	1.24	1.64	2.20	10.64	12.59	14.45	16.81	18.55
7	0.989	1.24	1.69	2.17	2.83	12.02	14.07	16.01	18.48	20.28
8	1.34	1.65	2.18	2.73	3.49	13.36	15.51	17.53	20.09	21.96
9	1.73	2.09	2.70	3.33	4.17	14.68	16.92	19.02	21.67	23.59
10	2.16	2.56	3.25	3.94	4.87	15.99	18.31	20.48	23.21	25.19
11	2.60	3.05	3.82	4.57	5.58	17.28	19.68	21.92	24.73	26.76
12	3.07	3.57	4.40	5.23	6.30	18.55	21.03	23.34	26.22	28.30
13	3.57	4.11	5.01	5.89	7.04	19.81	22.36	24.74	27.69	29.82
14	4.07	4.66	5.63	6.57	7.79	21.06	23.68	26.12	29.14	31.32
15	4.60	5.23	6.26	7.26	8.55	22.31	25.00	27.49	30.58	32.80
16	5.14	5.81	6.91	7.96	9.31	23.54	26.30	28.85	32.00	34.27
17	5.70	6.41	7.56	8.67	10.09	24.77	27.59	30.19	33.41	35.72
18	6.26	7.01	8.23	9.39	10.86	25.99	28.87	31.53	34.81	37.16
19	6.84	7.63	8.91	10.12	11.65	27.20	30.14	32.85	36.19	38.58
20	7.43	8.26	9.59	10.85	12.44	28.41	31.41	34.17	37.57	40.00
21	8.03	8.90	10.28	11.59	13.24	29.62	32.67	35.48	38.93	41.40
22	8.64	9.54	10.98	12.34	14.04	30.81	33.92	36.78	40.29	42.80
23	9.26	10.20	11.69	13.09	14.85	32.01	35.17	38.08	41.64	44.18
24	9.89	10.86	12.40	13.85	15.66	33.20	36.42	39.36	42.98	45.56
25	10.52	11.52	13.12	14.61	16.47	34.38	37.65	40.65	44.31	46.93
26	11.16	12.20	13.84	15.38	17.29	35.56	38.89	41.92	45.64	48.29
27	11.81	12.88	14.57	16.15	18.11	36.74	40.11	43.19	46.96	49.64
28	12.46	13.56	15.31	16.93	18.94	37.92	41.34	44.46	48.28	50.99
29	13.12	14.26	16.05	17.71	19.77	39.09	42.56	45.72	49.59	52.34
30	13.79	14.95	16.79	18.49	20.60	40.26	43.77	46.98	50.89	53.67
40	20.71	22.16	24.43	26.51	29.05	51.81	55.76	59.34	63.69	66.77
50	27.99	29.71	32.36	34.76	37.69	63.17	67.50	71.42	76.15	79.49
60	35.53	37.48	40.48	43.19	46.46	74.40	79.08	83.30	88.38	91.95
70	43.28	45.44	48.76	51.74	55.33	85.53	90.53	95.02	100.4	104.2
80	51.17	53.54	57.15	60.39	64.28	96.58	101.9	106.6	112.3	116.3
90	59.20	61.75	65.65	69.13	73.29	107.6	113.1	118.1	124.1	128.3
100	67.33	70.06	74.22	77.93	82.36	118.5	124.3	129.6	135.8	140.2

Source: Reprinted, with permission, from C. M. Thompson, "Table of Percentage Points of the Chi-Square Distribution," *Biometrika* 32 (1941), pp. 188–89.

TABLE A–4 Percentiles of the *F* distribution

Entry is $F(A; \nu_1, \nu_2)$ where $P\{F(\nu_1, \nu_2) \leq F(A; \nu_1, \nu_2)\} = A$

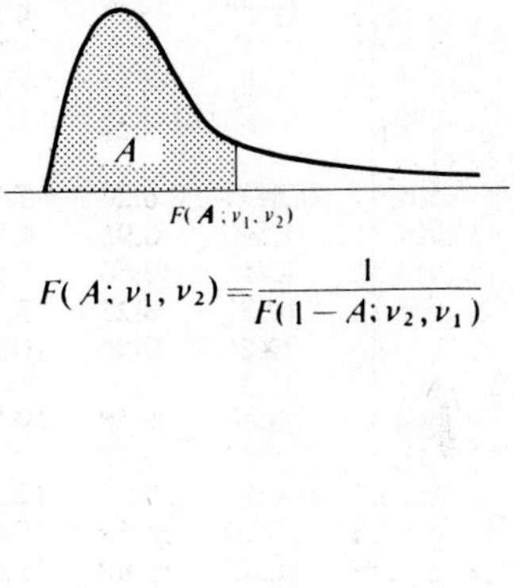

$$F(A; \nu_1, \nu_2) = \frac{1}{F(1 - A; \nu_2, \nu_1)}$$

TABLE A–4 *(continued)* Percentiles of the *F* distribution

Den. df	A	Numerator df 1	2	3	4	5	6	7	8	9
1	.50	1.00	1.50	1.71	1.82	1.89	1.94	1.98	2.00	2.03
	.90	39.9	49.5	53.6	55.8	57.2	58.2	58.9	59.4	59.9
	.95	161	200	216	225	230	234	237	239	241
	.975	648	800	864	900	922	937	948	957	963
	.99	4,052	5,000	5,403	5,625	5,764	5,859	5,928	5,981	6,022
	.995	16,211	20,000	21,615	22,500	23,056	23,437	23,715	23,925	24,091
	.999	405,280	500,000	540,380	562,500	576,400	585,940	592,870	598,140	602,280
2	.50	0.667	1.00	1.13	1.21	1.25	1.28	1.30	1.32	1.33
	.90	8.53	9.00	9.16	9.24	9.29	9.33	9.35	9.37	9.38
	.95	18.5	19.0	19.2	19.2	19.3	19.3	19.4	19.4	19.4
	.975	38.5	39.0	39.2	39.2	39.3	39.3	39.4	39.4	39.4
	.99	98.5	99.0	99.2	99.2	99.3	99.3	99.4	99.4	99.4
	.995	199	199	199	199	199	199	199	199	199
	.999	998.5	999.0	999.2	999.2	999.3	999.3	999.4	999.4	999.4
3	.50	0.585	0.881	1.00	1.06	1.10	1.13	1.15	1.16	1.17
	.90	5.54	5.46	5.39	5.34	5.31	5.28	5.27	5.25	5.24
	.95	10.1	9.55	9.28	9.12	9.01	8.94	8.89	8.85	8.81
	.975	17.4	16.0	15.4	15.1	14.9	14.7	14.6	14.5	14.5
	.99	34.1	30.8	29.5	28.7	28.2	27.9	27.7	27.5	27.3
	.995	55.6	49.8	47.5	46.2	45.4	44.8	44.4	44.1	43.9
	.999	167.0	148.5	141.1	137.1	134.6	132.8	131.6	130.6	129.9
4	.50	0.549	0.828	0.941	1.00	1.04	1.06	1.08	1.09	1.10
	.90	4.54	4.32	4.19	4.11	4.05	4.01	3.98	3.95	3.94
	.95	7.71	6.94	6.59	6.39	6.26	6.16	6.09	6.04	6.00
	.975	12.2	10.6	9.98	9.60	9.36	9.20	9.07	8.98	8.90
	.99	21.2	18.0	16.7	16.0	15.5	15.2	15.0	14.8	14.7
	.995	31.3	26.3	24.3	23.2	22.5	22.0	21.6	21.4	21.1
	.999	74.1	61.2	56.2	53.4	51.7	50.5	49.7	49.0	48.5
5	.50	0.528	0.799	0.907	0.965	1.00	1.02	1.04	1.05	1.06
	.90	4.06	3.78	3.62	3.52	3.45	3.40	3.37	3.34	3.32
	.95	6.61	5.79	5.41	5.19	5.05	4.95	4.88	4.82	4.77
	.975	10.0	8.43	7.76	7.39	7.15	6.98	6.85	6.76	6.68
	.99	16.3	13.3	12.1	11.4	11.0	10.7	10.5	10.3	10.2
	.995	22.8	18.3	16.5	15.6	14.9	14.5	14.2	14.0	13.8
	.999	47.2	37.1	33.2	31.1	29.8	28.8	28.2	27.6	27.2
6	.50	0.515	0.780	0.886	0.942	0.977	1.00	1.02	1.03	1.04
	.90	3.78	3.46	3.29	3.18	3.11	3.05	3.01	2.98	2.96
	.95	5.99	5.14	4.76	4.53	4.39	4.28	4.21	4.15	4.10
	.975	8.81	7.26	6.60	6.23	5.99	5.82	5.70	5.60	5.52
	.99	13.7	10.9	9.78	9.15	8.75	8.47	8.26	8.10	7.98
	.995	18.6	14.5	12.9	12.0	11.5	11.1	10.8	10.6	10.4
	.999	35.5	27.0	23.7	21.9	20.8	20.0	19.5	19.0	18.7
7	.50	0.506	0.767	0.871	0.926	0.960	0.983	1.00	1.01	1.02
	.90	3.59	3.26	3.07	2.96	2.88	2.83	2.78	2.75	2.72
	.95	5.59	4.74	4.35	4.12	3.97	3.87	3.79	3.73	3.68
	.975	8.07	6.54	5.89	5.52	5.29	5.12	4.99	4.90	4.82
	.99	12.2	9.55	8.45	7.85	7.46	7.19	6.99	6.84	6.72
	.995	16.2	12.4	10.9	10.1	9.52	9.16	8.89	8.68	8.51
	.999	29.2	21.7	18.8	17.2	16.2	15.5	15.0	14.6	14.3

TABLE A–4 *(continued)* Percentiles of the *F* distribution

Den. df	A	10	12	15	20	24	30	60	120	∞
		Numerator df								
1	.50	2.04	2.07	2.09	2.12	2.13	2.15	2.17	2.18	2.20
	.90	60.2	60.7	61.2	61.7	62.0	62.3	62.8	63.1	63.3
	.95	242	244	246	248	249	250	252	253	254
	.975	969	977	985	993	997	1,001	1,010	1,014	1,018
	.99	6,056	6,106	6,157	6,209	6,235	6,261	6,313	6,339	6,366
	.995	24,224	24,426	24,630	24,836	24,940	25,044	25,253	25,359	25,464
	.999	605,620	610,670	615,760	620,910	623,500	626,100	631,340	633,970	636,620
2	.50	1.34	1.36	1.38	1.39	1.40	1.41	1.43	1.43	1.44
	.90	9.39	9.41	9.42	9.44	9.45	9.46	9.47	9.48	9.49
	.95	19.4	19.4	19.4	19.4	19.5	19.5	19.5	19.5	19.5
	.975	39.4	39.4	39.4	39.4	39.5	39.5	39.5	39.5	39.5
	.99	99.4	99.4	99.4	99.4	99.5	99.5	99.5	99.5	99.5
	.995	199	199	199	199	199	199	199	199	200
	.999	999.4	999.4	999.4	999.4	999.5	999.5	999.5	999.5	999.5
3	.50	1.18	1.20	1.21	1.23	1.23	1.24	1.25	1.26	1.27
	.90	5.23	5.22	5.20	5.18	5.18	5.17	5.15	5.14	5.13
	.95	8.79	8.74	8.70	8.66	8.64	8.62	8.57	8.55	8.53
	.975	14.4	14.3	14.3	14.2	14.1	14.1	14.0	13.9	13.9
	.99	27.2	27.1	26.9	26.7	26.6	26.5	26.3	26.2	26.1
	.995	43.7	43.4	43.1	42.8	42.6	42.5	42.1	42.0	41.8
	.999	129.2	128.3	127.4	126.4	125.9	125.4	124.5	124.0	123.5
4	.50	1.11	1.13	1.14	1.15	1.16	1.16	1.18	1.18	1.19
	.90	3.92	3.90	3.87	3.84	3.83	3.82	3.79	3.78	3.76
	.95	5.96	5.91	5.86	5.80	5.77	5.75	5.69	5.66	5.63
	.975	8.84	8.75	8.66	8.56	8.51	8.46	8.36	8.31	8.26
	.99	14.5	14.4	14.2	14.0	13.9	13.8	13.7	13.6	13.5
	.995	21.0	20.7	20.4	20.2	20.0	19.9	19.6	19.5	19.3
	.999	48.1	47.4	46.8	46.1	45.8	45.4	44.7	44.4	44.1
5	.50	1.07	1.09	1.10	1.11	1.12	1.12	1.14	1.14	1.15
	.90	3.30	3.27	3.24	3.21	3.19	3.17	3.14	3.12	3.11
	.95	4.74	4.68	4.62	4.56	4.53	4.50	4.43	4.40	4.37
	.975	6.62	6.52	6.43	6.33	6.28	6.23	6.12	6.07	6.02
	.99	10.1	9.89	9.72	9.55	9.47	9.38	9.20	9.11	9.02
	.995	13.6	13.4	13.1	12.9	12.8	12.7	12.4	12.3	12.1
	.999	26.9	26.4	25.9	25.4	25.1	24.9	24.3	24.1	23.8
6	.50	1.05	1.06	1.07	1.08	1.09	1.10	1.11	1.12	1.12
	.90	2.94	2.90	2.87	2.84	2.82	2.80	2.76	2.74	2.72
	.95	4.06	4.00	3.94	3.87	3.84	3.81	3.74	3.70	3.67
	.975	5.46	5.37	5.27	5.17	5.12	5.07	4.96	4.90	4.85
	.99	7.87	7.72	7.56	7.40	7.31	7.23	7.06	6.97	6.88
	.995	10.2	10.0	9.81	9.59	9.47	9.36	9.12	9.00	8.88
	.999	18.4	18.0	17.6	17.1	16.9	16.7	16.2	16.0	15.7
7	.50	1.03	1.04	1.05	1.07	1.07	1.08	1.09	1.10	1.10
	.90	2.70	2.67	2.63	2.59	2.58	2.56	2.51	2.49	2.47
	.95	3.64	3.57	3.51	3.44	3.41	3.38	3.30	3.27	3.23
	.975	4.76	4.67	4.57	4.47	4.42	4.36	4.25	4.20	4.14
	.99	6.62	6.47	6.31	6.16	6.07	5.99	5.82	5.74	5.65
	.995	8.38	8.18	7.97	7.75	7.65	7.53	7.31	7.19	7.08
	.999	14.1	13.7	13.3	12.9	12.7	12.5	12.1	11.9	11.7

TABLE A–4 *(continued)* Percentiles of the *F* distribution

Den. df	A	Numerator df 1	2	3	4	5	6	7	8	9
8	.50	0.499	0.757	0.860	0.915	0.948	0.971	0.988	1.00	1.01
	.90	3.46	3.11	2.92	2.81	2.73	2.67	2.62	2.59	2.56
	.95	5.32	4.46	4.07	3.84	3.69	3.58	3.50	3.44	3.39
	.975	7.57	6.06	5.42	5.05	4.82	4.65	4.53	4.43	4.36
	.99	11.3	8.65	7.59	7.01	6.63	6.37	6.18	6.03	5.91
	.995	14.7	11.0	9.60	8.81	8.30	7.95	7.69	7.50	7.34
	.999	25.4	18.5	15.8	14.4	13.5	12.9	12.4	12.0	11.8
9	.50	0.494	0.749	0.852	0.906	0.939	0.962	0.978	0.990	1.00
	.90	3.36	3.01	2.81	2.69	2.61	2.55	2.51	2.47	2.44
	.95	5.12	4.26	3.86	3.63	3.48	3.37	3.29	3.23	3.18
	.975	7.21	5.71	5.08	4.72	4.48	4.32	4.20	4.10	4.03
	.99	10.6	8.02	6.99	6.42	6.06	5.80	5.61	5.47	5.35
	.995	13.6	10.1	8.72	7.96	7.47	7.13	6.88	6.69	6.54
	.999	22.9	16.4	13.9	12.6	11.7	11.1	10.7	10.4	10.1
10	.50	0.490	0.743	0.845	0.899	0.932	0.954	0.971	0.983	0.992
	.90	3.29	2.92	2.73	2.61	2.52	2.46	2.41	2.38	2.35
	.95	4.96	4.10	3.71	3.48	3.33	3.22	3.14	3.07	3.02
	.975	6.94	5.46	4.83	4.47	4.24	4.07	3.95	3.85	3.78
	.99	10.0	7.56	6.55	5.99	5.64	5.39	5.20	5.06	4.94
	.995	12.8	9.43	8.08	7.34	6.87	6.54	6.30	6.12	5.97
	.999	21.0	14.9	12.6	11.3	10.5	9.93	9.52	9.20	8.96
12	.50	0.484	0.735	0.835	0.888	0.921	0.943	0.959	0.972	0.981
	.90	3.18	2.81	2.61	2.48	2.39	2.33	2.28	2.24	2.21
	.95	4.75	3.89	3.49	3.26	3.11	3.00	2.91	2.85	2.80
	.975	6.55	5.10	4.47	4.12	3.89	3.73	3.61	3.51	3.44
	.99	9.33	6.93	5.95	5.41	5.06	4.82	4.64	4.50	4.39
	.995	11.8	8.51	7.23	6.52	6.07	5.76	5.52	5.35	5.20
	.999	18.6	13.0	10.8	9.63	8.89	8.38	8.00	7.71	7.48
15	.50	0.478	0.726	0.826	0.878	0.911	0.933	0.949	0.960	0.970
	.90	3.07	2.70	2.49	2.36	2.27	2.21	2.16	2.12	2.09
	.95	4.54	3.68	3.29	3.06	2.90	2.79	2.71	2.64	2.59
	.975	6.20	4.77	4.15	3.80	3.58	3.41	3.29	3.20	3.12
	.99	8.68	6.36	5.42	4.89	4.56	4.32	4.14	4.00	3.89
	.995	10.8	7.70	6.48	5.80	5.37	5.07	4.85	4.67	4.54
	.999	16.6	11.3	9.34	8.25	7.57	7.09	6.74	6.47	6.26
20	.50	0.472	0.718	0.816	0.868	0.900	0.922	0.938	0.950	0.959
	.90	2.97	2.59	2.38	2.25	2.16	2.09	2.04	2.00	1.96
	.95	4.35	3.49	3.10	2.87	2.71	2.60	2.51	2.45	2.39
	.975	5.87	4.46	3.86	3.51	3.29	3.13	3.01	2.91	2.84
	.99	8.10	5.85	4.94	4.43	4.10	3.87	3.70	3.56	3.46
	.995	9.94	6.99	5.82	5.17	4.76	4.47	4.26	4.09	3.96
	.999	14.8	9.95	8.10	7.10	6.46	6.02	5.69	5.44	5.24
24	.50	0.469	0.714	0.812	0.863	0.895	0.917	0.932	0.944	0.953
	.90	2.93	2.54	2.33	2.19	2.10	2.04	1.98	1.94	1.91
	.95	4.26	3.40	3.01	2.78	2.62	2.51	2.42	2.36	2.30
	.975	5.72	4.32	3.72	3.38	3.15	2.99	2.87	2.78	2.70
	.99	7.82	5.61	4.72	4.22	3.90	3.67	3.50	3.36	3.26
	.995	9.55	6.66	5.52	4.89	4.49	4.20	3.99	3.83	3.69
	.999	14.0	9.34	7.55	6.59	5.98	5.55	5.23	4.99	4.80

TABLE A–4 *(continued)* Percentiles of the *F* distribution

Den. df	*A*	Numerator df 10	12	15	20	24	30	60	120	∞
8	.50	1.02	1.03	1.04	1.05	1.06	1.07	1.08	1.08	1.09
	.90	2.54	2.50	2.46	2.42	2.40	2.38	2.34	2.32	2.29
	.95	3.35	3.28	3.22	3.15	3.12	3.08	3.01	2.97	2.93
	.975	4.30	4.20	4.10	4.00	3.95	3.89	3.78	3.73	3.67
	.99	5.81	5.67	5.52	5.36	5.28	5.20	5.03	4.95	4.86
	.995	7.21	7.01	6.81	6.61	6.50	6.40	6.18	6.06	5.95
	.999	11.5	11.2	10.8	10.5	10.3	10.1	9.73	9.53	9.33
9	.50	1.01	1.02	1.03	1.04	1.05	1.05	1.07	1.07	1.08
	.90	2.42	2.38	2.34	2.30	2.28	2.25	2.21	2.18	2.16
	.95	3.14	3.07	3.01	2.94	2.90	2.86	2.79	2.75	2.71
	.975	3.96	3.87	3.77	3.67	3.61	3.56	3.45	3.39	3.33
	.99	5.26	5.11	4.96	4.81	4.73	4.65	4.48	4.40	4.31
	.995	6.42	6.23	6.03	5.83	5.73	5.62	5.41	5.30	5.19
	.999	9.89	9.57	9.24	8.90	8.72	8.55	8.19	8.00	7.81
10	.50	1.00	1.01	1.02	1.03	1.04	1.05	1.06	1.06	1.07
	.90	2.32	2.28	2.24	2.20	2.18	2.16	2.11	2.08	2.06
	.95	2.98	2.91	2.84	2.77	2.74	2.70	2.62	2.58	2.54
	.975	3.72	3.62	3.52	3.42	3.37	3.31	3.20	3.14	3.08
	.99	4.85	4.71	4.56	4.41	4.33	4.25	4.08	4.00	3.91
	.995	5.85	5.66	5.47	5.27	5.17	5.07	4.86	4.75	4.64
	.999	8.75	8.45	8.13	7.80	7.64	7.47	7.12	6.94	6.76
12	.50	0.989	1.00	1.01	1.02	1.03	1.03	1.05	1.05	1.06
	.90	2.19	2.15	2.10	2.06	2.04	2.01	1.96	1.93	1.90
	.95	2.75	2.69	2.62	2.54	2.51	2.47	2.38	2.34	2.30
	.975	3.37	3.28	3.18	3.07	3.02	2.96	2.85	2.79	2.72
	.99	4.30	4.16	4.01	3.86	3.78	3.70	3.54	3.45	3.36
	.995	5.09	4.91	4.72	4.53	4.43	4.33	4.12	4.01	3.90
	.999	7.29	7.00	6.71	6.40	6.25	6.09	5.76	5.59	5.42
15	.50	0.977	0.989	1.00	1.01	1.02	1.02	1.03	1.04	1.05
	.90	2.06	2.02	1.97	1.92	1.90	1.87	1.82	1.79	1.76
	.95	2.54	2.48	2.40	2.33	2.29	2.25	2.16	2.11	2.07
	.975	3.06	2.96	2.86	2.76	2.70	2.64	2.52	2.46	2.40
	.99	3.80	3.67	3.52	3.37	3.29	3.21	3.05	2.96	2.87
	.995	4.42	4.25	4.07	3.88	3.79	3.69	3.48	3.37	3.26
	.999	6.08	5.81	5.54	5.25	5.10	4.95	4.64	4.48	4.31
20	.50	0.966	0.977	0.989	1.00	1.01	1.01	1.02	1.03	1.03
	.90	1.94	1.89	1.84	1.79	1.77	1.74	1.68	1.64	1.61
	.95	2.35	2.28	2.20	2.12	2.08	2.04	1.95	1.90	1.84
	.975	2.77	2.68	2.57	2.46	2.41	2.35	2.22	2.16	2.09
	.99	3.37	3.23	3.09	2.94	2.86	2.78	2.61	2.52	2.42
	.995	3.85	3.68	3.50	3.32	3.22	3.12	2.92	2.81	2.69
	.999	5.08	4.82	4.56	4.29	4.15	4.00	3.70	3.54	3.38
24	.50	0.961	0.972	0.983	0.994	1.00	1.01	1.02	1.02	1.03
	.90	1.88	1.83	1.78	1.73	1.70	1.67	1.61	1.57	1.53
	.95	2.25	2.18	2.11	2.03	1.98	1.94	1.84	1.79	1.73
	.975	2.64	2.54	2.44	2.33	2.27	2.21	2.08	2.01	1.94
	.99	3.17	3.03	2.89	2.74	2.66	2.58	2.40	2.31	2.21
	.995	3.59	3.42	3.25	3.06	2.97	2.87	2.66	2.55	2.43
	.999	4.64	4.39	4.14	3.87	3.74	3.59	3.29	3.14	2.97

TABLE A–4 *(continued)* Percentiles of the *F* distribution

Den. df	A	Numerator df 1	2	3	4	5	6	7	8	9
30	.50	0.466	0.709	0.807	0.858	0.890	0.912	0.927	0.939	0.948
	.90	2.88	2.49	2.28	2.14	2.05	1.98	1.93	1.88	1.85
	.95	4.17	3.32	2.92	2.69	2.53	2.42	2.33	2.27	2.21
	.975	5.57	4.18	3.59	3.25	3.03	2.87	2.75	2.65	2.57
	.99	7.56	5.39	4.51	4.02	3.70	3.47	3.30	3.17	3.07
	.995	9.18	6.35	5.24	4.62	4.23	3.95	3.74	3.58	3.45
	.999	13.3	8.77	7.05	6.12	5.53	5.12	4.82	4.58	4.39
60	.50	0.461	0.701	0.798	0.849	0.880	0.901	0.917	0.928	0.937
	.90	2.79	2.39	2.18	2.04	1.95	1.87	1.82	1.77	1.74
	.95	4.00	3.15	2.76	2.53	2.37	2.25	2.17	2.10	2.04
	.975	5.29	3.93	3.34	3.01	2.79	2.63	2.51	2.41	2.33
	.99	7.08	4.98	4.13	3.65	3.34	3.12	2.95	2.82	2.72
	.995	8.49	5.80	4.73	4.14	3.76	3.49	3.29	3.13	3.01
	.999	12.0	7.77	6.17	5.31	4.76	4.37	4.09	3.86	3.69
120	.50	0.458	0.697	0.793	0.844	0.875	0.896	0.912	0.923	0.932
	.90	2.75	2.35	2.13	1.99	1.90	1.82	1.77	1.72	1.68
	.95	3.92	3.07	2.68	2.45	2.29	2.18	2.09	2.02	1.96
	.975	5.15	3.80	3.23	2.89	2.67	2.52	2.39	2.30	2.22
	.99	6.85	4.79	3.95	3.48	3.17	2.96	2.79	2.66	2.56
	.995	8.18	5.54	4.50	3.92	3.55	3.28	3.09	2.93	2.81
	.999	11.4	7.32	5.78	4.95	4.42	4.04	3.77	3.55	3.38
∞	.50	0.455	0.693	0.789	0.839	0.870	0.891	0.907	0.918	0.927
	.90	2.71	2.30	2.08	1.94	1.85	1.77	1.72	1.67	1.63
	.95	3.84	3.00	2.60	2.37	2.21	2.10	2.01	1.94	1.88
	.975	5.02	3.69	3.12	2.79	2.57	2.41	2.29	2.19	2.11
	.99	6.63	4.61	3.78	3.32	3.02	2.80	2.64	2.51	2.41
	.995	7.88	5.30	4.28	3.72	3.35	3.09	2.90	2.74	2.62
	.999	10.8	6.91	5.42	4.62	4.10	3.74	3.47	3.27	3.10

TABLE A–4 *(concluded)* Percentiles of the *F* distribution

Den. df	*A*	Numerator df 10	12	15	20	24	30	60	120	∞
30	.50	0.955	0.966	0.978	0.989	0.994	1.00	1.01	1.02	1.02
	.90	1.82	1.77	1.72	1.67	1.64	1.61	1.54	1.50	1.46
	.95	2.16	2.09	2.01	1.93	1.89	1.84	1.74	1.68	1.62
	.975	2.51	2.41	2.31	2.20	2.14	2.07	1.94	1.87	1.79
	.99	2.98	2.84	2.70	2.55	2.47	2.39	2.21	2.11	2.01
	.995	3.34	3.18	3.01	2.82	2.73	2.63	2.42	2.30	2.18
	.999	4.24	4.00	3.75	3.49	3.36	3.22	2.92	2.76	2.59
60	.50	0.945	0.956	0.967	0.978	0.983	0.989	1.00	1.01	1.01
	.90	1.71	1.66	1.60	1.54	1.51	1.48	1.40	1.35	1.29
	.95	1.99	1.92	1.84	1.75	1.70	1.65	1.53	1.47	1.39
	.975	2.27	2.17	2.06	1.94	1.88	1.82	1.67	1.58	1.48
	.99	2.63	2.50	2.35	2.20	2.12	2.03	1.84	1.73	1.60
	.995	2.90	2.74	2.57	2.39	2.29	2.19	1.96	1.83	1.69
	.999	3.54	3.32	3.08	2.83	2.69	2.55	2.25	2.08	1.89
120	.50	0.939	0.950	0.961	0.972	0.978	0.983	0.994	1.00	1.01
	.90	1.65	1.60	1.55	1.48	1.45	1.41	1.32	1.26	1.19
	.95	1.91	1.83	1.75	1.66	1.61	1.55	1.43	1.35	1.25
	.975	2.16	2.05	1.95	1.82	1.76	1.69	1.53	1.43	1.31
	.99	2.47	2.34	2.19	2.03	1.95	1.86	1.66	1.53	1.38
	.995	2.71	2.54	2.37	2.19	2.09	1.98	1.75	1.61	1.43
	.999	3.24	3.02	2.78	2.53	2.40	2.26	1.95	1.77	1.54
∞	.50	0.934	0.945	0.956	0.967	0.972	0.978	0.989	0.994	1.00
	.90	1.60	1.55	1.49	1.42	1.38	1.34	1.24	1.17	1.00
	.95	1.83	1.75	1.67	1.57	1.52	1.46	1.32	1.22	1.00
	.975	2.05	1.94	1.83	1.71	1.64	1.57	1.39	1.27	1.00
	.99	2.32	2.18	2.04	1.88	1.79	1.70	1.47	1.32	1.00
	.995	2.52	2.36	2.19	2.00	1.90	1.79	1.53	1.36	1.00
	.999	2.96	2.74	2.51	2.27	2.13	1.99	1.66	1.45	1.00

Source: Reprinted from Table 5 of Pearson and Hartley, *Biometrika Tables for Statisticians,* Volume 2, 1972, published by the Cambridge University Press, on behalf of The Biometrika Society, by permission of the authors and publishers.

TABLE A–5 Power function for two-sided *t* test

$\alpha = .05$

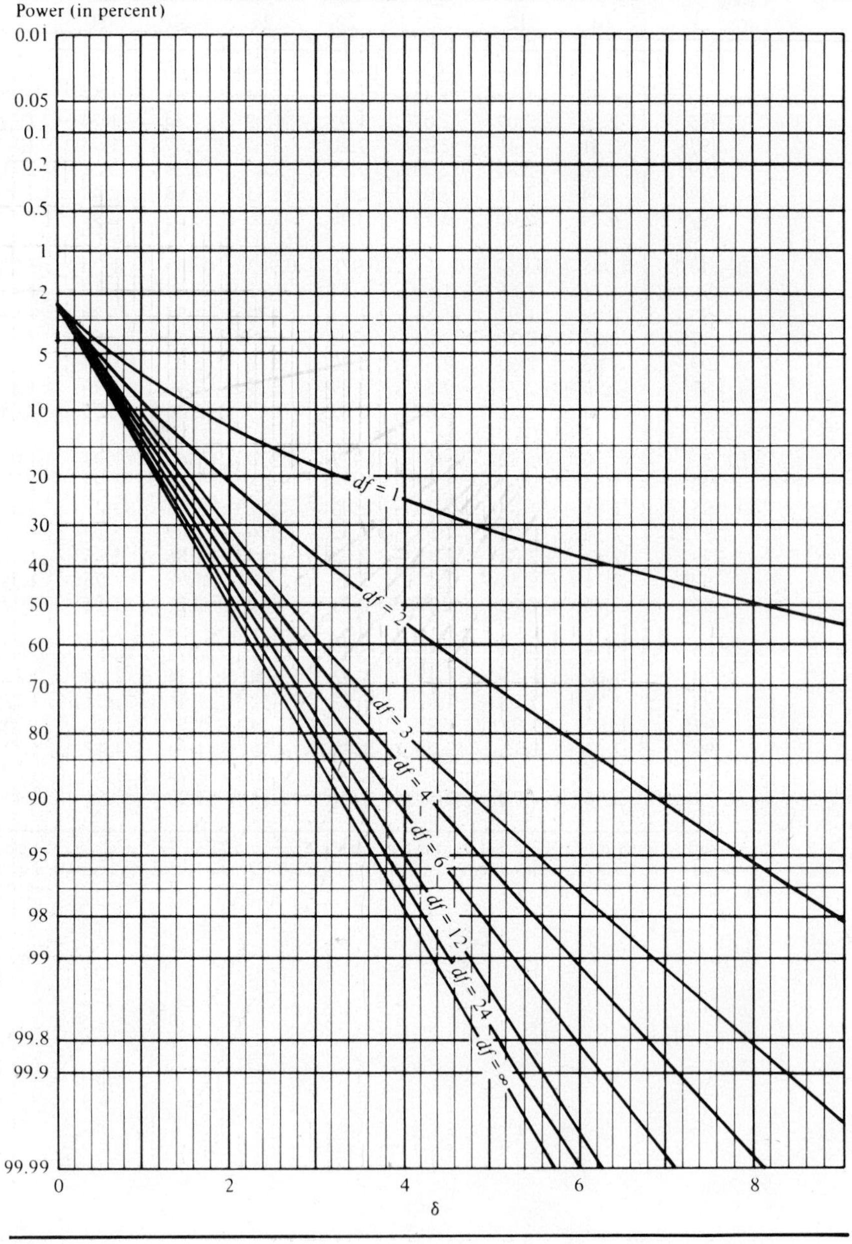

TABLE A–5 *(concluded)* Power function for two-sided *t* test

$\alpha = .01$

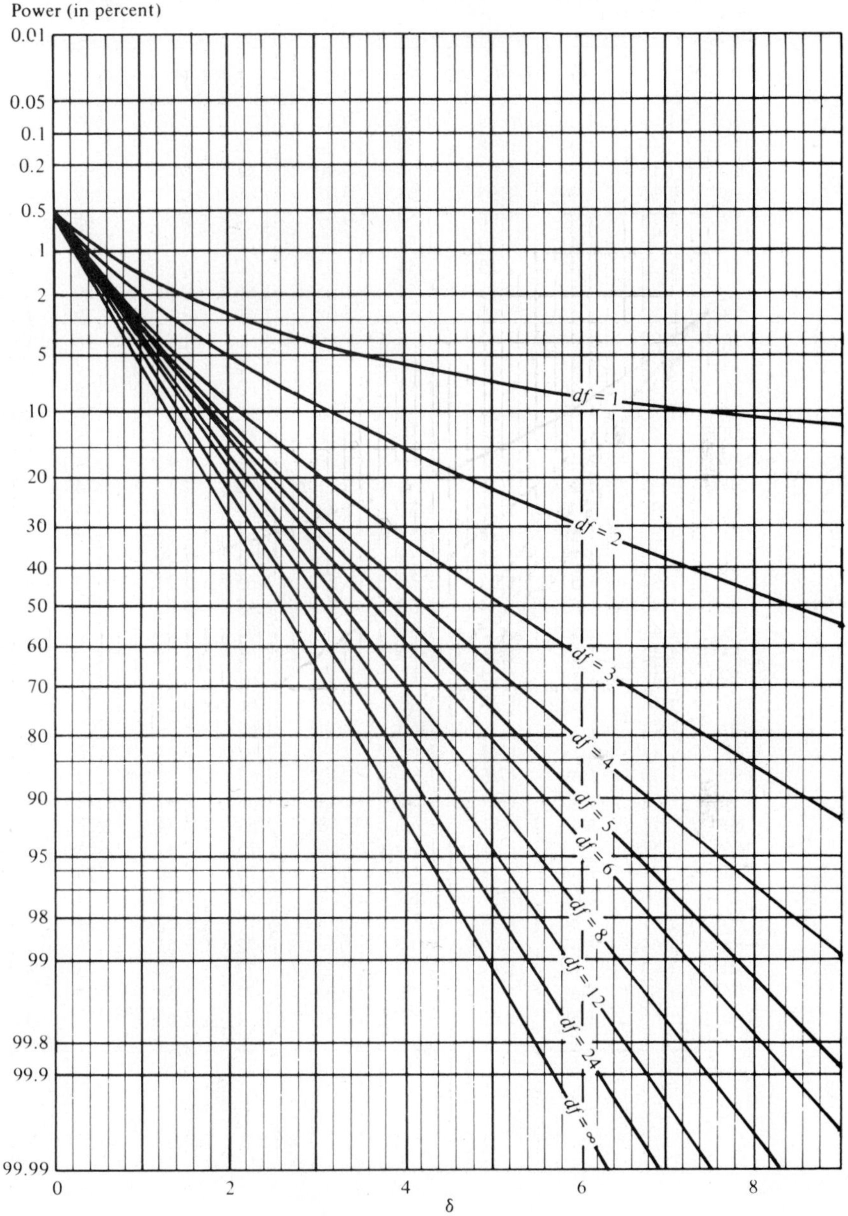

Source: Reprinted, with permission, from D. B. Owen, *Handbook of Statistical Tables* (Reading, Mass.: Addison-Wesley Publishing, 1962), pp. 32, 34. Courtesy of U.S. Atomic Energy Commission.

TABLE A–6 Durbin-Watson test bounds

Level of Significance $\alpha = .05$

	$p-1=1$		$p-1=2$		$p-1=3$		$p-1=4$		$p-1=5$	
n	d_L	d_U	d_L	d_U	d_L	d_U	d_L	d_U	d_L	d_U
15	1.08	1.36	0.95	1.54	0.82	1.75	0.69	1.97	0.56	2.21
16	1.10	1.37	0.98	1.54	0.86	1.73	0.74	1.93	0.62	2.15
17	1.13	1.38	1.02	1.54	0.90	1.71	0.78	1.90	0.67	2.10
18	1.16	1.39	1.05	1.53	0.93	1.69	0.82	1.87	0.71	2.06
19	1.18	1.40	1.08	1.53	0.97	1.68	0.86	1.85	0.75	2.02
20	1.20	1.41	1.10	1.54	1.00	1.68	0.90	1.83	0.79	1.99
21	1.22	1.42	1.13	1.54	1.03	1.67	0.93	1.81	0.83	1.96
22	1.24	1.43	1.15	1.54	1.05	1.66	0.96	1.80	0.86	1.94
23	1.26	1.44	1.17	1.54	1.08	1.66	0.99	1.79	0.90	1.92
24	1.27	1.45	1.19	1.55	1.10	1.66	1.01	1.78	0.93	1.90
25	1.29	1.45	1.21	1.55	1.12	1.66	1.04	1.77	0.95	1.89
26	1.30	1.46	1.22	1.55	1.14	1.65	1.06	1.76	0.98	1.88
27	1.32	1.47	1.24	1.56	1.16	1.65	1.08	1.76	1.01	1.86
28	1.33	1.48	1.26	1.56	1.18	1.65	1.10	1.75	1.03	1.85
29	1.34	1.48	1.27	1.56	1.20	1.65	1.12	1.74	1.05	1.84
30	1.35	1.49	1.28	1.57	1.21	1.65	1.14	1.74	1.07	1.83
31	1.36	1.50	1.30	1.57	1.23	1.65	1.16	1.74	1.09	1.83
32	1.37	1.50	1.31	1.57	1.24	1.65	1.18	1.73	1.11	1.82
33	1.38	1.51	1.32	1.58	1.26	1.65	1.19	1.73	1.13	1.81
34	1.39	1.51	1.33	1.58	1.27	1.65	1.21	1.73	1.15	1.81
35	1.40	1.52	1.34	1.58	1.28	1.65	1.22	1.73	1.16	1.80
36	1.41	1.52	1.35	1.59	1.29	1.65	1.24	1.73	1.18	1.80
37	1.42	1.53	1.36	1.59	1.31	1.66	1.25	1.72	1.19	1.80
38	1.43	1.54	1.37	1.59	1.32	1.66	1.26	1.72	1.21	1.79
39	1.43	1.54	1.38	1.60	1.33	1.66	1.27	1.72	1.22	1.79
40	1.44	1.54	1.39	1.60	1.34	1.66	1.29	1.72	1.23	1.79
45	1.48	1.57	1.43	1.62	1.38	1.67	1.34	1.72	1.29	1.78
50	1.50	1.59	1.46	1.63	1.42	1.67	1.38	1.72	1.34	1.77
55	1.53	1.60	1.49	1.64	1.45	1.68	1.41	1.72	1.38	1.77
60	1.55	1.62	1.51	1.65	1.48	1.69	1.44	1.73	1.41	1.77
65	1.57	1.63	1.54	1.66	1.50	1.70	1.47	1.73	1.44	1.77
70	1.58	1.64	1.55	1.67	1.52	1.70	1.49	1.74	1.46	1.77
75	1.60	1.65	1.57	1.68	1.54	1.71	1.51	1.74	1.49	1.77
80	1.61	1.66	1.59	1.69	1.56	1.72	1.53	1.74	1.51	1.77
85	1.62	1.67	1.60	1.70	1.57	1.72	1.55	1.75	1.52	1.77
90	1.63	1.68	1.61	1.70	1.59	1.73	1.57	1.75	1.54	1.78
95	1.64	1.69	1.62	1.71	1.60	1.73	1.58	1.75	1.56	1.78
100	1.65	1.69	1.63	1.72	1.61	1.74	1.59	1.76	1.57	1.78

TABLE A–6 *(concluded)* Durbin-Watson test bounds

Level of Significance $\alpha = .01$

	$p-1=1$		$p-1=2$		$p-1=3$		$p-1=4$		$p-1=5$	
n	d_L	d_U	d_L	d_U	d_L	d_U	d_L	d_U	d_L	d_U
15	0.81	1.07	0.70	1.25	0.59	1.46	0.49	1.70	0.39	1.96
16	0.84	1.09	0.74	1.25	0.63	1.44	0.53	1.66	0.44	1.90
17	0.87	1.10	0.77	1.25	0.67	1.43	0.57	1.63	0.48	1.85
18	0.90	1.12	0.80	1.26	0.71	1.42	0.61	1.60	0.52	1.80
19	0.93	1.13	0.83	1.26	0.74	1.41	0.65	1.58	0.56	1.77
20	0.95	1.15	0.86	1.27	0.77	1.41	0.68	1.57	0.60	1.74
21	0.97	1.16	0.89	1.27	0.80	1.41	0.72	1.55	0.63	1.71
22	1.00	1.17	0.91	1.28	0.83	1.40	0.75	1.54	0.66	1.69
23	1.02	1.19	0.94	1.29	0.86	1.40	0.77	1.53	0.70	1.67
24	1.04	1.20	0.96	1.30	0.88	1.41	0.80	1.53	0.72	1.66
25	1.05	1.21	0.98	1.30	0.90	1.41	0.83	1.52	0.75	1.65
26	1.07	1.22	1.00	1.31	0.93	1.41	0.85	1.52	0.78	1.64
27	1.09	1.23	1.02	1.32	0.95	1.41	0.88	1.51	0.81	1.63
28	1.10	1.24	1.04	1.32	0.97	1.41	0.90	1.51	0.83	1.62
29	1.12	1.25	1.05	1.33	0.99	1.42	0.92	1.51	0.85	1.61
30	1.13	1.26	1.07	1.34	1.01	1.42	0.94	1.51	0.88	1.61
31	1.15	1.27	1.08	1.34	1.02	1.42	0.96	1.51	0.90	1.60
32	1.16	1.28	1.10	1.35	1.04	1.43	0.98	1.51	0.92	1.60
33	1.17	1.29	1.11	1.36	1.05	1.43	1.00	1.51	0.94	1.59
34	1.18	1.30	1.13	1.36	1.07	1.43	1.01	1.51	0.95	1.59
35	1.19	1.31	1.14	1.37	1.08	1.44	1.03	1.51	0.97	1.59
36	1.21	1.32	1.15	1.38	1.10	1.44	1.04	1.51	0.99	1.59
37	1.22	1.32	1.16	1.38	1.11	1.45	1.06	1.51	1.00	1.59
38	1.23	1.33	1.18	1.39	1.12	1.45	1.07	1.52	1.02	1.58
39	1.24	1.34	1.19	1.39	1.14	1.45	1.09	1.52	1.03	1.58
40	1.25	1.34	1.20	1.40	1.15	1.46	1.10	1.52	1.05	1.58
45	1.29	1.38	1.24	1.42	1.20	1.48	1.16	1.53	1.11	1.58
50	1.32	1.40	1.28	1.45	1.24	1.49	1.20	1.54	1.16	1.59
55	1.36	1.43	1.32	1.47	1.28	1.51	1.25	1.55	1.21	1.59
60	1.38	1.45	1.35	1.48	1.32	1.52	1.28	1.56	1.25	1.60
65	1.41	1.47	1.38	1.50	1.35	1.53	1.31	1.57	1.28	1.61
70	1.43	1.49	1.40	1.52	1.37	1.55	1.34	1.58	1.31	1.61
75	1.45	1.50	1.42	1.53	1.39	1.56	1.37	1.59	1.34	1.62
80	1.47	1.52	1.44	1.54	1.42	1.57	1.39	1.60	1.36	1.62
85	1.48	1.53	1.46	1.55	1.43	1.58	1.41	1.60	1.39	1.63
90	1.50	1.54	1.47	1.56	1.45	1.59	1.43	1.61	1.41	1.64
95	1.51	1.55	1.49	1.57	1.47	1.60	1.45	1.62	1.42	1.64
100	1.52	1.56	1.50	1.58	1.48	1.60	1.46	1.63	1.44	1.65

Source: Reprinted, with permission, from J. Durbin and G. S. Watson, "Testing for Serial Correlation in Least Squares Regression. II," *Biometrika* 38 (1951), pp. 159–78.

TABLE A–7 Table of z' transformation of correlation coefficient

r ρ	z' ζ	r ρ	z' ζ	r ρ	z' ζ	r ρ	z' ζ
.00	.0000	.25	.2554	.50	.5493	.75	.973
.01	.0100	.26	.2661	.51	.5627	.76	.996
.02	.0200	.27	.2769	.52	.5763	.77	1.020
.03	.0300	.28	.2877	.53	.5901	.78	1.045
.04	.0400	.29	.2986	.54	.6042	.79	1.071
.05	.0500	.30	.3095	.55	.6184	.80	1.099
.06	.0601	.31	.3205	.56	.6328	.81	1.127
.07	.0701	.32	.3316	.57	.6475	.82	1.157
.08	.0802	.33	.3428	.58	.6625	.83	1.188
.09	.0902	.34	.3541	.59	.6777	.84	1.221
.10	.1003	.35	.3654	.60	.6931	.85	1.256
.11	.1104	.36	.3769	.61	.7089	.86	1.293
.12	.1206	.37	.3884	.62	.7250	.87	1.333
.13	.1307	.38	.4001	.63	.7414	.88	1.376
.14	.1409	.39	.4118	.64	.7582	.89	1.422
.15	.1511	.40	.4236	.65	.7753	.90	1.472
.16	.1614	.41	.4356	.66	.7928	.91	1.528
.17	.1717	.42	.4477	.67	.8107	.92	1.589
.18	.1820	.43	.4599	.68	.8291	.93	1.658
.19	.1923	.44	.4722	.69	.8480	.94	1.738
.20	.2027	.45	.4847	.70	.8673	.95	1.832
.21	.2132	.46	.4973	.71	.8872	.96	1.946
.22	.2237	.47	.5101	.72	.9076	.97	2.092
.23	.2342	.48	.5230	.73	.9287	.98	2.298
.24	.2448	.49	.5361	.74	.9505	.99	2.647

Source: Abridged from Table 14 of Pearson and Hartley, *Biometrika Tables for Statisticians*, Volume 1, 1966, published by the Cambridge University Press, on behalf of The Biometrika Society, by permission of the authors and publishers.

TABLE A–8 Power function for analysis of variance (fixed effects model)

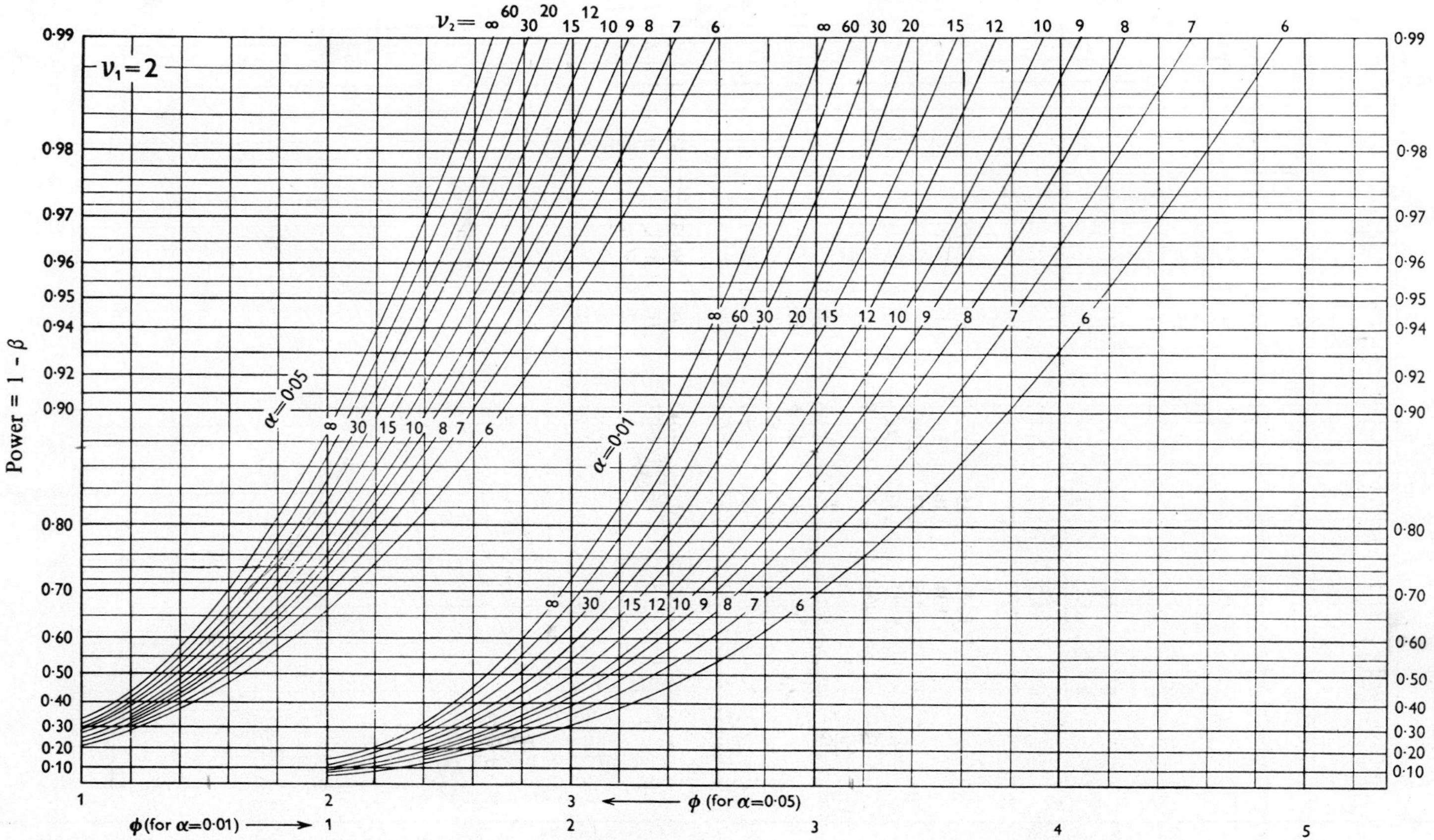

TABLE A–8 *(continued)* Power function for analysis of variance (fixed effects model)

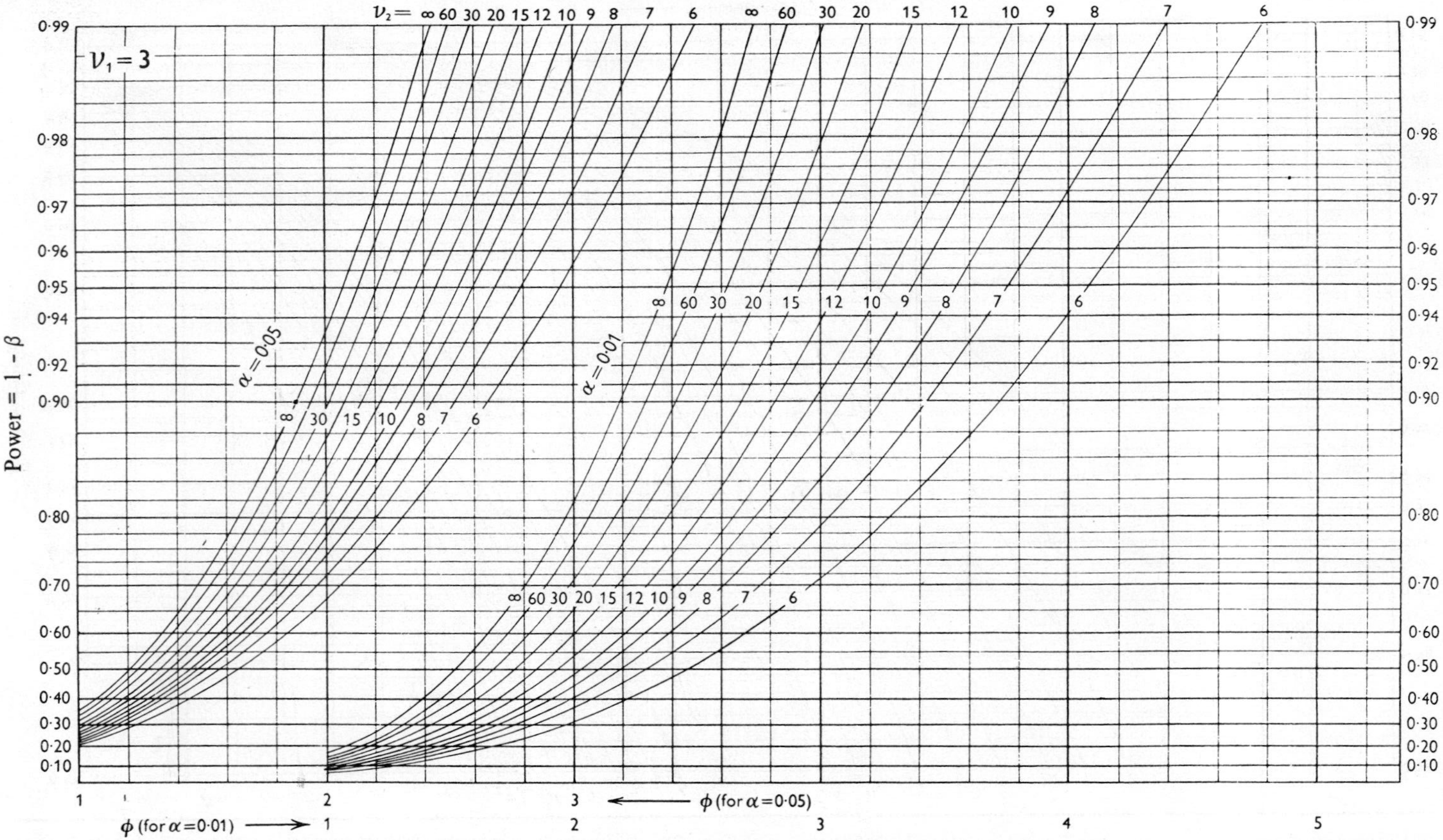

TABLE A–8 *(continued)* Power function for analysis of variance (fixed effects model)

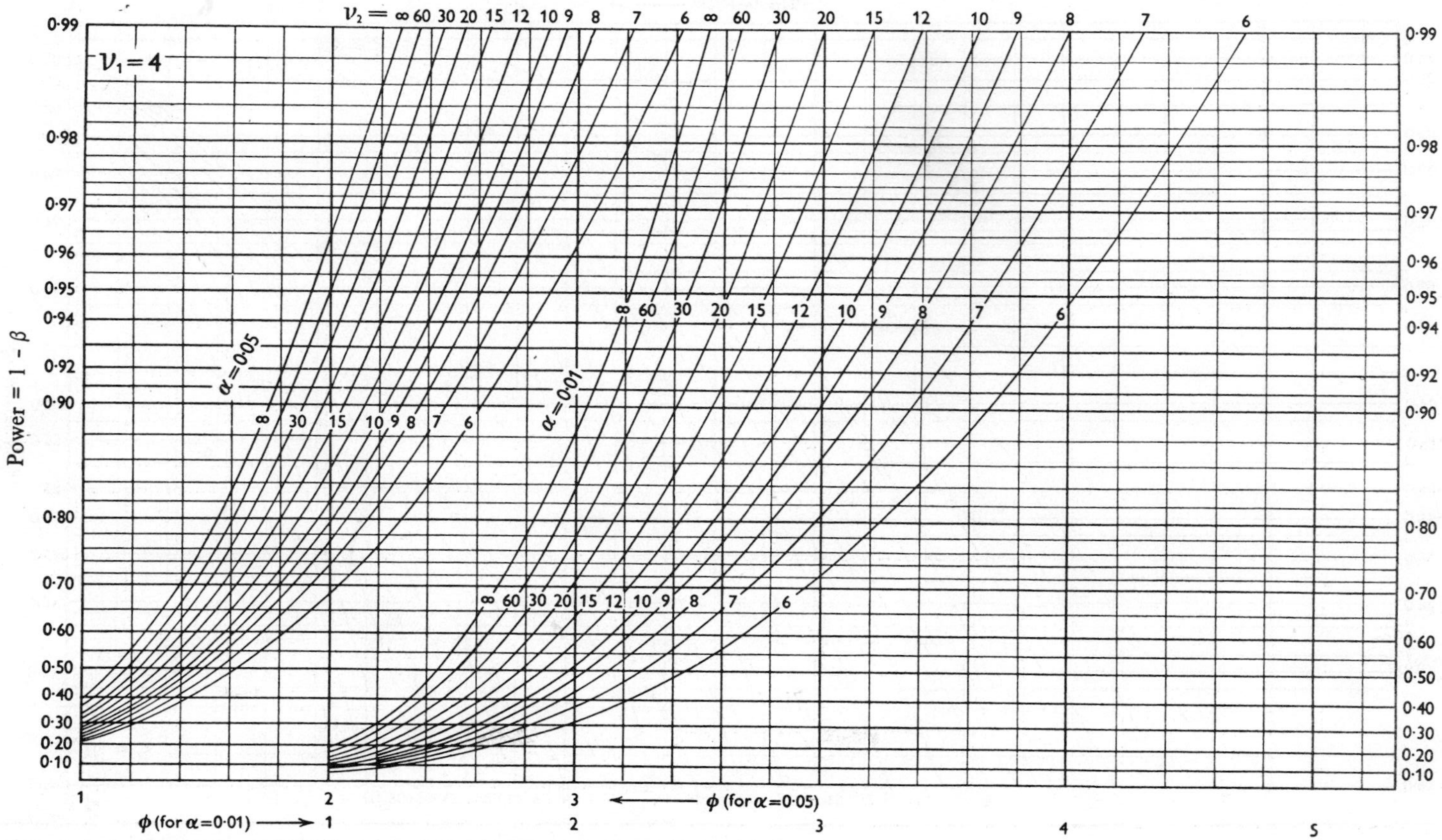

TABLE A–8 *(continued)* Power function for analysis of variance (fixed effects model)

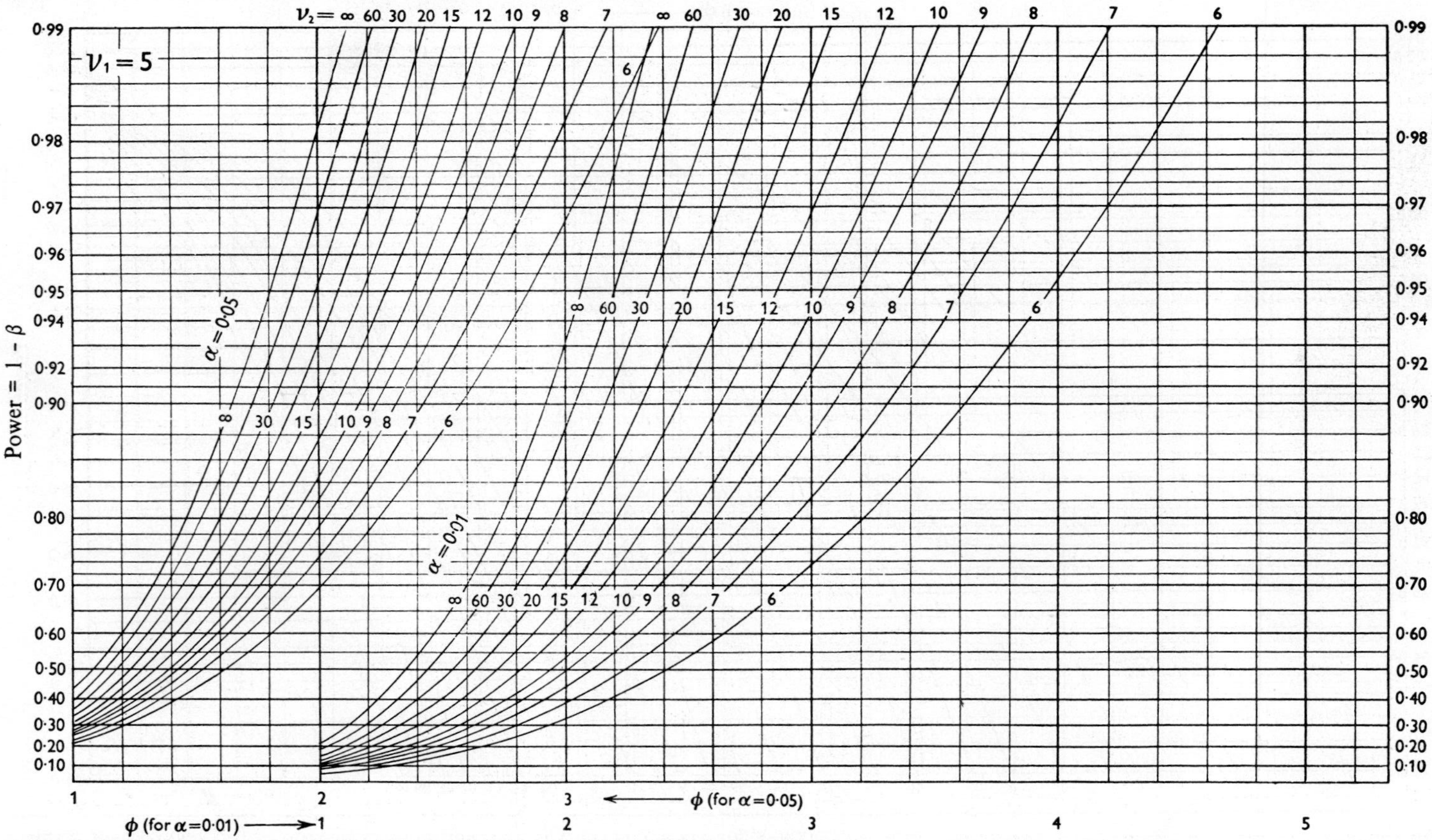

TABLE A–8 *(concluded)* Power function for analysis of variance (fixed effects model)

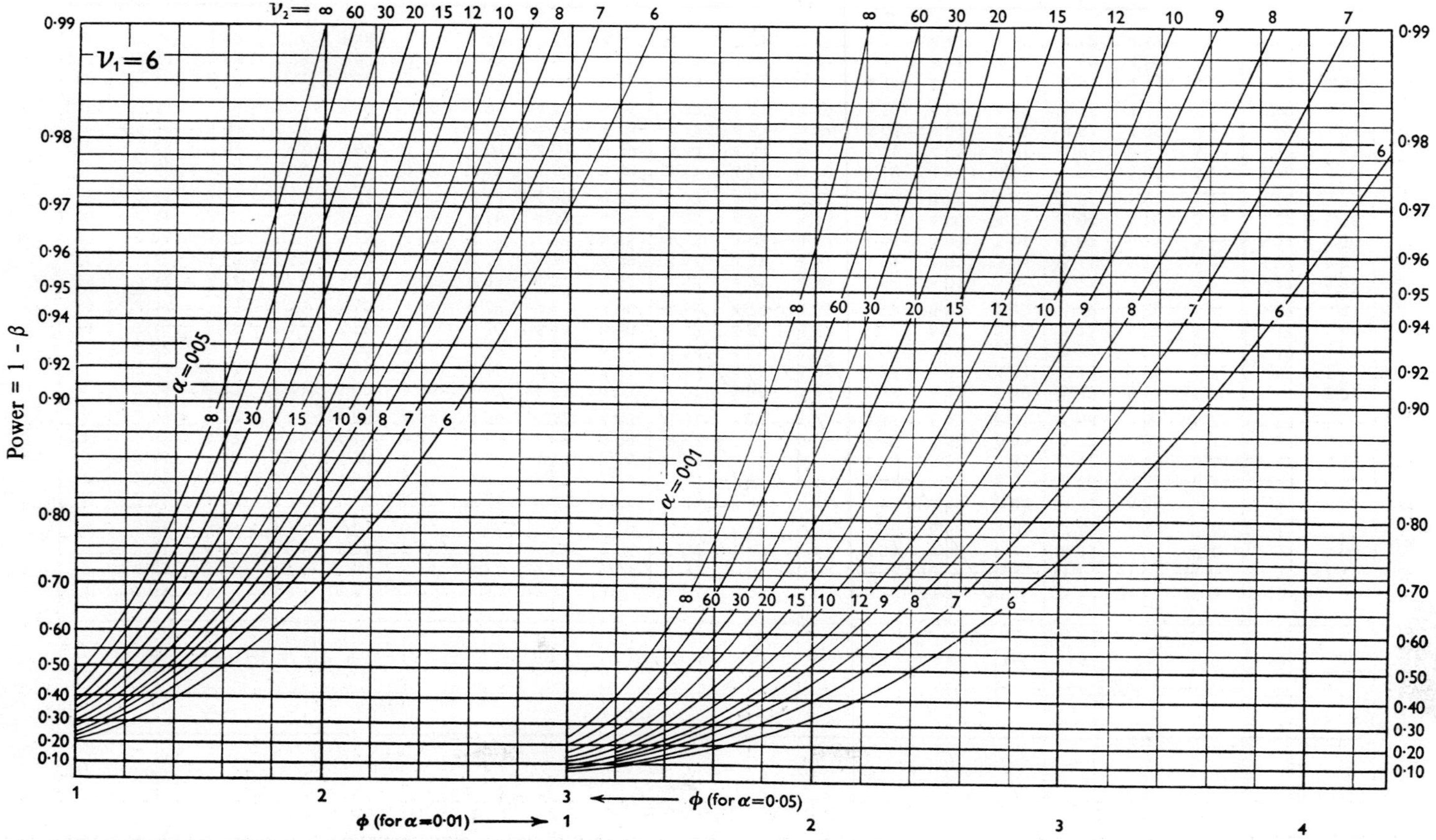

Source: Reprinted, with permission, from E. S. Pearson and H. O. Hartley, "Charts of the Power Function for Analysis of Variance Tests, Derived from the Non-Central *F*-Distribution," *Biometrika* 38 (1951), pp. 112–30.

TABLE A–9 Percentiles of the studentized range distribution

Entry is $q(1 - \alpha; r, \nu)$ where $P\{q(r, \nu) \leq q(1 - \alpha; r, \nu)\} = 1 - \alpha$

$1 - \alpha = .90$

ν	2	3	4	5	6	7	8	9	10	11	12	13	14	15	16	17	18	19	20
1	8.93	13.4	16.4	18.5	20.2	21.5	22.6	23.6	24.5	25.2	25.9	26.5	27.1	27.6	28.1	28.5	29.0	29.3	29.7
2	4.13	5.73	6.77	7.54	8.14	8.63	9.05	9.41	9.72	10.0	10.3	10.5	10.7	10.9	11.1	11.2	11.4	11.5	11.7
3	3.33	4.47	5.20	5.74	6.16	6.51	6.81	7.06	7.29	7.49	7.67	7.83	7.98	8.12	8.25	8.37	8.48	8.58	8.68
4	3.01	3.98	4.59	5.03	5.39	5.68	5.93	6.14	6.33	6.49	6.65	6.78	6.91	7.02	7.13	7.23	7.33	7.41	7.50
5	2.85	3.72	4.26	4.66	4.98	5.24	5.46	5.65	5.82	5.97	6.10	6.22	6.34	6.44	6.54	6.63	6.71	6.79	6.86
6	2.75	3.56	4.07	4.44	4.73	4.97	5.17	5.34	5.50	5.64	5.76	5.87	5.98	6.07	6.16	6.25	6.32	6.40	6.47
7	2.68	3.45	3.93	4.28	4.55	4.78	4.97	5.14	5.28	5.41	5.53	5.64	5.74	5.83	5.91	5.99	6.06	6.13	6.19
8	2.63	3.37	3.83	4.17	4.43	4.65	4.83	4.99	5.13	5.25	5.36	5.46	5.56	5.64	5.72	5.80	5.87	5.93	6.00
9	2.59	3.32	3.76	4.08	4.34	4.54	4.72	4.87	5.01	5.13	5.23	5.33	5.42	5.51	5.58	5.66	5.72	5.79	5.85
10	2.56	3.27	3.70	4.02	4.26	4.47	4.64	4.78	4.91	5.03	5.13	5.23	5.32	5.40	5.47	5.54	5.61	5.67	5.73
11	2.54	3.23	3.66	3.96	4.20	4.40	4.57	4.71	4.84	4.95	5.05	5.15	5.23	5.31	5.38	5.45	5.51	5.57	5.63
12	2.52	3.20	3.62	3.92	4.16	4.35	4.51	4.65	4.78	4.89	4.99	5.08	5.16	5.24	5.31	5.37	5.44	5.49	5.55
13	2.50	3.18	3.59	3.88	4.12	4.30	4.46	4.60	4.72	4.83	4.93	5.02	5.10	5.18	5.25	5.31	5.37	5.43	5.48
14	2.49	3.16	3.56	3.85	4.08	4.27	4.42	4.56	4.68	4.79	4.88	4.97	5.05	5.12	5.19	5.26	5.32	5.37	5.43
15	2.48	3.14	3.54	3.83	4.05	4.23	4.39	4.52	4.64	4.75	4.84	4.93	5.01	5.08	5.15	5.21	5.27	5.32	5.38
16	2.47	3.12	3.52	3.80	4.03	4.21	4.36	4.49	4.61	4.71	4.81	4.89	4.97	5.04	5.11	5.17	5.23	5.28	5.33
17	2.46	3.11	3.50	3.78	4.00	4.18	4.33	4.46	4.58	4.68	4.77	4.86	4.93	5.01	5.07	5.13	5.19	5.24	5.30
18	2.45	3.10	3.49	3.77	3.98	4.16	4.31	4.44	4.55	4.65	4.75	4.83	4.90	4.98	5.04	5.10	5.16	5.21	5.26
19	2.45	3.09	3.47	3.75	3.97	4.14	4.29	4.42	4.53	4.63	4.72	4.80	4.88	4.95	5.01	5.07	5.13	5.18	5.23
20	2.44	3.08	3.46	3.74	3.95	4.12	4.27	4.40	4.51	4.61	4.70	4.78	4.85	4.92	4.99	5.05	5.10	5.16	5.20
24	2.42	3.05	3.42	3.69	3.90	4.07	4.21	4.34	4.44	4.54	4.63	4.71	4.78	4.85	4.91	4.97	5.02	5.07	5.12
30	2.40	3.02	3.39	3.65	3.85	4.02	4.16	4.28	4.38	4.47	4.56	4.64	4.71	4.77	4.83	4.89	4.94	4.99	5.03
40	2.38	2.99	3.35	3.60	3.80	3.96	4.10	4.21	4.32	4.41	4.49	4.56	4.63	4.69	4.75	4.81	4.86	4.90	4.95
60	2.36	2.96	3.31	3.56	3.75	3.91	4.04	4.16	4.25	4.34	4.42	4.49	4.56	4.62	4.67	4.73	4.78	4.82	4.86
120	2.34	2.93	3.28	3.52	3.71	3.86	3.99	4.10	4.19	4.28	4.35	4.42	4.48	4.54	4.60	4.65	4.69	4.74	4.78
∞	2.33	2.90	3.24	3.48	3.66	3.81	3.93	4.04	4.13	4.21	4.28	4.35	4.41	4.47	4.52	4.57	4.61	4.65	4.69

TABLE A–9 *(continued)* Percentiles of the studentized range distribution

$1 - \alpha = .95$

r K

ν	2	3	4	5	6	7	8	9	10	11	12	13	14	15	16	17	18	19	20
1	18.0	27.0	32.8	37.1	40.4	43.1	45.4	47.4	49.1	50.6	52.0	53.2	54.3	55.4	56.3	57.2	58.0	58.8	59.6
2	6.08	8.33	9.80	10.9	11.7	12.4	13.0	13.5	14.0	14.4	14.7	15.1	15.4	15.7	15.9	16.1	16.4	16.6	16.8
3	4.50	5.91	6.82	7.50	8.04	8.48	8.85	9.18	9.46	9.72	9.95	10.2	10.3	10.5	10.7	10.8	11.0	11.1	11.2
4	3.93	5.04	5.76	6.29	6.71	7.05	7.35	7.60	7.83	8.03	8.21	8.37	8.52	8.66	8.79	8.91	9.03	9.13	9.23
5	3.64	4.60	5.22	5.67	6.03	6.33	6.58	6.80	6.99	7.17	7.32	7.47	7.60	7.72	7.83	7.93	8.03	8.12	8.21
6	3.46	4.34	4.90	5.30	5.63	5.90	6.12	6.32	6.49	6.65	6.79	6.92	7.03	7.14	7.24	7.34	7.43	7.51	7.59
7	3.34	4.16	4.68	5.06	5.36	5.61	5.82	6.00	6.16	6.30	6.43	6.55	6.66	6.76	6.85	6.94	7.02	7.10	7.17
8	3.26	4.04	4.53	4.89	5.17	5.40	5.60	5.77	5.92	6.05	6.18	6.29	6.39	6.48	6.57	6.65	6.73	6.80	6.87
9	3.20	3.95	4.41	4.76	5.02	5.24	5.43	5.59	5.74	5.87	5.98	6.09	6.19	6.28	6.36	6.44	6.51	6.58	6.64
10	3.15	3.88	4.33	4.65	4.91	5.12	5.30	5.46	5.60	5.72	5.83	5.93	6.03	6.11	6.19	6.27	6.34	6.40	6.47
11	3.11	3.82	4.26	4.57	4.82	5.03	5.20	5.35	5.49	5.61	5.71	5.81	5.90	5.98	6.06	6.13	6.20	6.27	6.33
12	3.08	3.77	4.20	4.51	4.75	4.95	5.12	5.27	5.39	5.51	5.61	5.71	5.80	5.88	5.95	6.02	6.09	6.15	6.21
13	3.06	3.73	4.15	4.45	4.69	4.88	5.05	5.19	5.32	5.43	5.53	5.63	5.71	5.79	5.86	5.93	5.99	6.05	6.11
14	3.03	3.70	4.11	4.41	4.64	4.83	4.99	5.13	5.25	5.36	5.46	5.55	5.64	5.71	5.79	5.85	5.91	5.97	6.03
15	3.01	3.67	4.08	4.37	4.59	4.78	4.94	5.08	5.20	5.31	5.40	5.49	5.57	5.65	5.72	5.78	5.85	5.90	5.96
16	3.00	3.65	4.05	4.33	4.56	4.74	4.90	5.03	5.15	5.26	5.35	5.44	5.52	5.59	5.66	5.73	5.79	5.84	5.90
17	2.98	3.63	4.02	4.30	4.52	4.70	4.86	4.99	5.11	5.21	5.31	5.39	5.47	5.54	5.61	5.67	5.73	5.79	5.84
18	2.97	3.61	4.00	4.28	4.49	4.67	4.82	4.96	5.07	5.17	5.27	5.35	5.43	5.50	5.57	5.63	5.69	5.74	5.79
19	2.96	3.59	3.98	4.25	4.47	4.65	4.79	4.92	5.04	5.14	5.23	5.31	5.39	5.46	5.53	5.59	5.65	5.70	5.75
20	2.95	3.58	3.96	4.23	4.45	4.62	4.77	4.90	5.01	5.11	5.20	5.28	5.36	5.43	5.49	5.55	5.61	5.66	5.71
24	2.92	3.53	3.90	4.17	4.37	4.54	4.68	4.81	4.92	5.01	5.10	5.18	5.25	5.32	5.38	5.44	5.49	5.55	5.59
30	2.89	3.49	3.85	4.10	4.30	4.46	4.60	4.72	4.82	4.92	5.00	5.08	5.15	5.21	5.27	5.33	5.38	5.43	5.47
40	2.86	3.44	3.79	4.04	4.23	4.39	4.52	4.63	4.73	4.82	4.90	4.98	5.04	5.11	5.16	5.22	5.27	5.31	5.36
60	2.83	3.40	3.74	3.98	4.16	4.31	4.44	4.55	4.65	4.73	4.81	4.88	4.94	5.00	5.06	5.11	5.15	5.20	5.24
120	2.80	3.36	3.68	3.92	4.10	4.24	4.36	4.47	4.56	4.64	4.71	4.78	4.84	4.90	4.95	5.00	5.04	5.09	5.13
∞	2.77	3.31	3.63	3.86	4.03	4.17	4.29	4.39	4.47	4.55	4.62	4.68	4.74	4.80	4.85	4.89	4.93	4.97	5.01

TABLE A–9 *(concluded)* Percentiles of the studentized range distribution

$1 - \alpha = .99$

ν	2	3	4	5	6	7	8	9	10	11	12	13	14	15	16	17	18	19	20
1	90.0	135	164	186	202	216	227	237	246	253	260	266	272	277	282	286	290	294	298
2	14.0	19.0	22.3	24.7	26.6	28.2	29.5	30.7	31.7	32.6	33.4	34.1	34.8	35.4	36.0	36.5	37.0	37.5	37.9
3	8.26	10.6	12.2	13.3	14.2	15.0	15.6	16.2	16.7	17.1	17.5	17.9	18.2	18.5	18.8	19.1	19.3	19.5	19.8
4	6.51	8.12	9.17	9.96	10.6	11.1	11.5	11.9	12.3	12.6	12.8	13.1	13.3	13.5	13.7	13.9	14.1	14.2	14.4
5	5.70	6.97	7.80	8.42	8.91	9.32	9.67	9.97	10.2	10.5	10.7	10.9	11.1	11.2	11.4	11.6	11.7	11.8	11.9
6	5.24	6.33	7.03	7.56	7.97	8.32	8.61	8.87	9.10	9.30	9.49	9.65	9.81	9.95	10.1	10.2	10.3	10.4	10.5
7	4.95	5.92	6.54	7.01	7.37	7.68	7.94	8.17	8.37	8.55	8.71	8.86	9.00	9.12	9.24	9.35	9.46	9.55	9.65
8	4.74	5.63	6.20	6.63	6.96	7.24	7.47	7.68	7.87	8.03	8.18	8.31	8.44	8.55	8.66	8.76	8.85	8.94	9.03
9	4.60	5.43	5.96	6.35	6.66	6.91	7.13	7.32	7.49	7.65	7.78	7.91	8.03	8.13	8.23	8.32	8.41	8.49	8.57
10	4.48	5.27	5.77	6.14	6.43	6.67	6.87	7.05	7.21	7.36	7.48	7.60	7.71	7.81	7.91	7.99	8.07	8.15	8.22
11	4.39	5.14	5.62	5.97	6.25	6.48	6.67	6.84	6.99	7.13	7.25	7.36	7.46	7.56	7.65	7.73	7.81	7.88	7.95
12	4.32	5.04	5.50	5.84	6.10	6.32	6.51	6.67	6.81	6.94	7.06	7.17	7.26	7.36	7.44	7.52	7.59	7.66	7.73
13	4.26	4.96	5.40	5.73	5.98	6.19	6.37	6.53	6.67	6.79	6.90	7.01	7.10	7.19	7.27	7.34	7.42	7.48	7.55
14	4.21	4.89	5.32	5.63	5.88	6.08	6.26	6.41	6.54	6.66	6.77	6.87	6.96	7.05	7.12	7.20	7.27	7.33	7.39
15	4.17	4.83	5.25	5.56	5.80	5.99	6.16	6.31	6.44	6.55	6.66	6.76	6.84	6.93	7.00	7.07	7.14	7.20	7.26
16	4.13	4.78	5.19	5.49	5.72	5.92	6.08	6.22	6.35	6.46	6.56	6.66	6.74	6.82	6.90	6.97	7.03	7.09	7.15
17	4.10	4.74	5.14	5.43	5.66	5.85	6.01	6.15	6.27	6.38	6.48	6.57	6.66	6.73	6.80	6.87	6.94	7.00	7.05
18	4.07	4.70	5.09	5.38	5.60	5.79	5.94	6.08	6.20	6.31	6.41	6.50	6.58	6.65	6.72	6.79	6.85	6.91	6.96
19	4.05	4.67	5.05	5.33	5.55	5.73	5.89	6.02	6.14	6.25	6.34	6.43	6.51	6.58	6.65	6.72	6.78	6.84	6.89
20	4.02	4.64	5.02	5.29	5.51	5.69	5.84	5.97	6.09	6.19	6.29	6.37	6.45	6.52	6.59	6.65	6.71	6.76	6.82
24	3.96	4.54	4.91	5.17	5.37	5.54	5.69	5.81	5.92	6.02	6.11	6.19	6.26	6.33	6.39	6.45	6.51	6.56	6.61
30	3.89	4.45	4.80	5.05	5.24	5.40	5.54	5.65	5.76	5.85	5.93	6.01	6.08	6.14	6.20	6.26	6.31	6.36	6.41
40	3.82	4.37	4.70	4.93	5.11	5.27	5.39	5.50	5.60	5.69	5.77	5.84	5.90	5.96	6.02	6.07	6.12	6.17	6.21
60	3.76	4.28	4.60	4.82	4.99	5.13	5.25	5.36	5.45	5.53	5.60	5.67	5.73	5.79	5.84	5.89	5.93	5.98	6.02
120	3.70	4.20	4.50	4.71	4.87	5.01	5.12	5.21	5.30	5.38	5.44	5.51	5.56	5.61	5.66	5.71	5.75	5.79	5.83
∞	3.64	4.12	4.40	4.60	4.76	4.88	4.99	5.08	5.16	5.23	5.29	5.35	5.40	5.45	5.49	5.54	5.57	5.61	5.65

Source: Reprinted, with permission, from Henry Scheffé, *The Analysis of Variance* (New York: John Wiley & Sons, 1959), pp. 434–36.

TABLE A–10 Table for determining sample size for analysis of variance (fixed effects model)

Power $1 - \beta = .70$

	$\Delta/\sigma = 1.0$				$\Delta/\sigma = 1.25$				$\Delta/\sigma = 1.50$				$\Delta/\sigma = 1.75$				$\Delta/\sigma = 2.0$				$\Delta/\sigma = 2.5$				$\Delta/\sigma = 3.0$			
	α				α				α				α				α				α				α			
r	.2	.1	.05	.01	.2	.1	.05	.01	.2	.1	.05	.01	.2	.1	.05	.01	.2	.1	.05	.01	.2	.1	.05	.01	.2	.1	.05	.01
2	7	11	14	21	5	7	9	15	4	6	7	11	3	4	6	9	3	4	5	7	2	3	4	5	2	3	3	5
3	9	13	17	25	6	9	11	17	5	7	8	12	4	5	7	10	3	4	5	8	3	3	4	6	2	3	3	5
4	11	15	19	28	7	10	13	19	5	7	9	13	4	6	7	10	4	5	6	8	3	4	4	6	2	3	4	5
5	12	17	21	30	8	11	14	20	6	8	10	14	5	6	8	11	4	5	6	9	3	4	5	6	3	3	4	5
6	13	18	22	32	9	12	15	21	6	9	11	15	5	7	8	12	4	5	7	9	3	4	5	7	3	3	4	5
7	14	19	24	34	9	13	16	22	7	9	11	16	5	7	9	12	4	6	7	10	3	4	5	7	3	3	4	5
8	15	20	25	35	10	13	16	23	7	10	12	17	6	7	9	13	5	6	7	10	3	4	5	7	3	3	4	5
9	15	21	26	37	10	14	17	24	7	10	12	17	6	8	9	13	5	6	8	10	3	4	5	7	3	4	4	6
10	16	22	27	38	11	14	18	25	8	10	13	18	6	8	10	14	5	6	8	11	4	5	6	7	3	4	4	6

Power $1 - \beta = .80$

	$\Delta/\sigma = 1.0$				$\Delta/\sigma = 1.25$				$\Delta/\sigma = 1.50$				$\Delta/\sigma = 1.75$				$\Delta/\sigma = 2.0$				$\Delta/\sigma = 2.5$				$\Delta/\sigma = 3.0$			
	α				α				α				α				α				α				α			
r	.2	.1	.05	.01	.2	.1	.05	.01	.2	.1	.05	.01	.2	.1	.05	.01	.2	.1	.05	.01	.2	.1	.05	.01	.2	.1	.05	.01
2	10	14	17	26	7	9	12	17	5	7	9	13	4	5	7	10	3	4	6	8	3	3	4	6	2	3	4	5
3	12	17	21	30	8	11	14	20	6	8	10	14	5	6	8	11	4	5	6	9	3	4	5	7	3	3	4	5
4	14	19	23	33	9	13	15	22	7	9	11	16	5	7	9	12	4	6	7	10	3	4	5	7	3	3	4	5
5	16	21	25	35	10	14	17	23	8	10	12	17	6	8	9	13	5	6	7	10	4	4	5	7	3	4	4	6
6	17	22	27	38	11	15	18	25	8	11	13	18	6	8	10	13	5	7	8	11	4	5	6	8	3	4	4	6
7	18	24	29	39	12	16	19	26	9	11	14	18	7	9	10	14	5	7	8	11	4	5	6	8	3	4	5	6
8	19	25	30	41	12	16	20	27	9	12	14	19	7	9	11	15	6	7	9	12	4	5	6	8	3	4	5	6
9	20	26	31	43	13	17	21	28	9	12	15	20	7	9	11	15	6	7	9	12	4	5	6	8	3	4	5	6
10	21	27	33	44	14	18	21	29	10	13	15	21	8	10	12	16	6	8	9	12	4	5	6	8	3	4	5	6

TABLE A–10 *(concluded)* Table for determining sample size for analysis of variance (fixed effects model)

Power $1 - \beta = .90$

	$\Delta/\sigma = 1.0$				$\Delta/\sigma = 1.25$				$\Delta/\sigma = 1.50$				$\Delta/\sigma = 1.75$				$\Delta/\sigma = 2.0$				$\Delta/\sigma = 2.5$				$\Delta/\sigma = 3.0$			
	α				α				α				α				α				α				α			
r	.2	.1	.05	.01	.2	.1	.05	.01	.2	.1	.05	.01	.2	.1	.05	.01	.2	.1	.05	.01	.2	.1	.05	.01	.2	.1	.05	.01
2	14	18	23	32	9	12	15	21	7	9	11	15	5	7	8	12	4	6	7	10	3	4	5	7	3	3	4	6
3	17	22	27	37	11	15	18	24	8	11	13	18	6	8	10	13	5	7	8	11	4	5	6	8	3	4	5	6
4	20	25	30	40	13	16	20	27	9	12	14	19	7	9	11	15	6	7	9	12	4	5	6	8	3	4	5	6
5	21	27	32	43	14	18	21	28	10	13	15	20	8	10	12	15	6	8	9	12	4	5	6	9	4	4	5	7
6	22	29	34	46	15	19	23	30	11	14	16	21	8	10	12	16	7	8	10	13	5	6	7	9	4	4	5	7
7	24	31	36	48	16	20	24	31	11	14	17	22	9	11	13	17	7	9	10	13	5	6	7	9	4	5	5	7
8	26	32	38	50	17	21	25	33	12	15	18	23	9	11	13	17	7	9	11	14	5	6	7	9	4	5	6	7
9	27	33	40	52	17	22	26	34	13	16	18	24	9	12	14	18	8	9	11	14	5	6	8	10	4	5	6	7
10	28	35	41	54	18	23	27	35	13	16	19	25	10	12	14	19	8	10	11	15	5	7	8	10	4	5	6	7

Power $1 - \beta = .95$

	$\Delta/\sigma = 1.0$				$\Delta/\sigma = 1.25$				$\Delta/\sigma = 1.50$				$\Delta/\sigma = 1.75$				$\Delta/\sigma = 2.0$				$\Delta/\sigma = 2.5$				$\Delta/\sigma = 3.0$			
	α				α				α				α				α				α				α			
r	.2	.1	.05	.01	.2	.1	.05	.01	.2	.1	.05	.01	.2	.1	.05	.01	.2	.1	.05	.01	.2	.1	.05	.01	.2	.1	.05	.01
2	18	23	27	38	12	15	18	25	9	11	13	18	7	8	10	14	5	7	8	11	4	5	6	8	3	4	5	6
3	22	27	32	43	14	18	21	29	10	13	15	20	8	10	12	16	6	8	9	12	5	6	7	9	4	4	5	7
4	25	30	36	47	16	20	23	31	12	14	17	22	9	11	13	17	7	9	10	13	5	6	7	9	4	5	5	7
5	27	33	39	51	18	22	25	33	13	15	18	23	10	12	14	18	8	9	11	14	5	6	7	10	4	5	6	7
6	29	35	41	53	19	23	27	35	13	16	19	25	10	12	14	19	8	10	11	15	6	7	8	10	4	5	6	8
7	30	37	43	56	20	24	28	36	14	17	20	26	11	13	15	19	8	10	12	15	6	7	8	10	4	5	6	8
8	32	39	45	58	21	25	29	38	15	18	21	27	11	14	16	20	9	11	12	16	6	7	8	11	5	5	6	8
9	33	40	47	60	22	26	30	39	15	19	22	28	12	14	16	21	9	11	13	16	6	8	9	11	5	6	6	8
10	34	42	48	62	22	27	31	40	16	19	22	29	12	15	17	21	9	11	13	17	6	8	9	11	5	6	7	8

Source: Reprinted, with permission, from T. L. Bratcher, M. A. Moran, and W. J. Zimmer, "Tables of Sample Sizes in the Analysis of Variance," *Journal of Quality Technology* 2 (1970), pp. 156–64.

TABLE A–11 Table of $\lambda\sqrt{n}/\sigma$ for determining sample size to find "best" of r population means

Number of Populations (r)	*Probability of Correct Identification* $(1 - \alpha)$		
	.90	.95	.99
2	1.8124	2.3262	3.2900
3	2.2302	2.7101	3.6173
4	2.4516	2.9162	3.7970
5	2.5997	3.0552	3.9196
6	2.7100	3.1591	4.0121
7	2.7972	3.2417	4.0861
8	2.8691	3.3099	4.1475
9	2.9301	3.3679	4.1999
10	2.9829	3.4182	4.2456

Source: Reprinted, with permission, from R. E. Bechhofer, ''A Single-Sample Multiple Decision Procedure for Ranking Means of Normal Populations with Known Variances,'' *The Annals of Mathematical Statistics* 25 (1954), pp. 16–39.

TABLE A–12 Percentiles of *H* statistic distribution

Entry is $H(1 - \alpha; r, df)$ where $P\{H \leq H(1 - \alpha; r, df)\} = 1 - \alpha$

$1 - \alpha = .95$

	r										
df	2	3	4	5	6	7	8	9	10	11	12
2	39.0	87.5	142	202	266	333	403	475	550	626	704
3	15.4	27.8	39.2	50.7	62.0	72.9	83.5	93.9	104	114	124
4	9.60	15.5	20.6	25.2	29.5	33.6	37.5	41.1	44.6	48.0	51.4
5	7.15	10.8	13.7	16.3	18.7	20.8	22.9	24.7	26.5	28.2	29.9
6	5.82	8.38	10.4	12.1	13.7	15.0	16.3	17.5	18.6	19.7	20.7
7	4.99	6.94	8.44	9.70	10.8	11.8	12.7	13.5	14.3	15.1	15.8
8	4.43	6.00	7.18	8.12	9.03	9.78	10.5	11.1	11.7	12.2	12.7
9	4.03	5.34	6.31	7.11	7.80	8.41	8.95	9.45	9.91	10.3	10.7
10	3.72	4.85	5.67	6.34	6.92	7.42	7.87	8.28	8.66	9.01	9.34
12	3.28	4.16	4.79	5.30	5.72	6.09	6.42	6.72	7.00	7.25	7.48
15	2.86	3.54	4.01	4.37	4.68	4.95	5.19	5.40	5.59	5.77	5.93
20	2.46	2.95	3.29	3.54	3.76	3.94	4.10	4.24	4.37	4.49	4.59
30	2.07	2.40	2.61	2.78	2.91	3.02	3.12	3.21	3.29	3.36	3.39
60	1.67	1.85	1.96	2.04	2.11	2.17	2.22	2.26	2.30	2.33	2.36
∞	1.00	1.00	1.00	1.00	1.00	1.00	1.00	1.00	1.00	1.00	1.00

$1 - \alpha = .99$

	r										
df	2	3	4	5	6	7	8	9	10	11	12
2	199	448	729	1,036	1,362	1,705	2,063	2,432	2,813	3,204	3,605
3	47.5	85	120	151	184	216	249	281	310	337	361
4	23.2	37	49	59	69	79	89	97	106	113	120
5	14.9	22	28	33	38	42	46	50	54	57	60
6	11.1	15.5	19.1	22	25	27	30	32	34	36	37
7	8.89	12.1	14.5	16.5	18.4	20	22	23	24	26	27
8	7.50	9.9	11.7	13.2	14.5	15.8	16.9	17.9	18.9	19.8	21
9	6.54	8.5	9.9	11.1	12.1	13.1	13.9	14.7	15.3	16.0	16.6
10	5.85	7.4	8.6	9.6	10.4	11.1	11.8	12.4	12.9	13.4	13.9
12	4.91	6.1	6.9	7.6	8.2	8.7	9.1	9.5	9.9	10.2	10.6
15	4.07	4.9	5.5	6.0	6.4	6.7	7.1	7.3	7.5	7.8	8.0
20	3.32	3.8	4.3	4.6	4.9	5.1	5.3	5.5	5.6	5.8	5.9
30	2.63	3.0	3.3	3.4	3.6	3.7	3.8	3.9	4.0	4.1	4.2
60	1.96	2.2	2.3	2.4	2.4	2.5	2.5	2.6	2.6	2.7	2.7
∞	1.00	1.0	1.0	1.0	1.0	1.0	1.0	1.0	1.0	1.0	1.0

Source: Reprinted, with permission, from H. A. David, ''Upper 5 and 1% Points of the Maximum *F*-Ratio,'' *Biometrika* 39 (1952), pp. 422–24.

TABLE A–13 Table of random permutations of 9

5 5 6 7 1	4 3 3 7 3	8 7 4 6 3	9 7 4 9 4	9 2 2 8 8	2 7 9 3 5	8 3 1 9 4
4 1 2 8 2	7 1 1 2 9	9 5 7 8 2	8 9 3 6 6	1 7 7 2 4	4 8 5 7 3	3 7 4 5 6
9 3 3 2 9	8 8 8 4 5	2 4 6 1 6	3 6 7 7 8	7 4 4 7 1	7 3 2 8 6	6 1 2 2 2
7 9 7 4 3	5 5 2 9 2	1 6 5 3 5	7 8 5 1 9	5 1 9 1 3	6 5 1 4 9	2 9 8 7 8
1 6 9 6 5	6 9 4 3 6	4 3 9 2 9	5 1 8 2 3	8 3 3 3 2	8 9 6 1 2	4 5 7 6 9
6 4 4 3 6	2 4 6 8 1	7 9 3 4 1	6 2 6 4 2	2 9 8 5 9	9 2 4 2 8	9 6 9 8 1
8 7 8 1 7	1 2 5 6 8	3 1 2 9 8	4 4 1 8 7	6 5 1 6 7	5 4 3 5 1	1 4 3 1 7
3 2 1 9 4	3 6 7 5 7	6 8 8 7 7	2 5 9 5 1	3 8 5 4 6	3 6 7 9 4	5 2 5 4 5
2 8 5 5 8	9 7 9 1 4	5 2 1 5 4	1 3 2 3 5	4 6 6 9 5	1 1 8 6 7	7 8 6 3 3
7 4 6 1 5	9 2 2 2 9	2 8 1 7 3	2 4 2 1 9	2 4 8 3 1	2 6 5 4 8	8 4 9 4 2
9 3 8 3 2	1 1 1 9 8	9 4 9 5 4	8 8 8 8 6	7 7 5 4 6	5 3 2 7 6	9 3 8 2 1
1 6 3 4 7	6 5 8 4 5	6 1 7 1 9	5 2 5 6 3	8 5 7 5 5	6 9 9 8 1	3 6 7 9 7
6 8 2 8 4	4 8 7 8 6	5 7 5 4 5	9 6 7 5 8	5 9 9 7 7	8 5 3 3 5	6 9 4 6 9
4 1 4 7 8	2 3 9 3 4	4 2 2 3 6	4 7 4 2 5	6 3 3 6 9	1 7 8 5 4	4 5 2 1 4
2 9 1 9 3	7 9 6 6 2	1 6 4 6 1	7 9 9 7 4	1 8 4 1 8	9 2 7 9 3	1 8 3 5 5
5 5 5 5 1	3 7 4 7 7	8 5 8 9 2	1 5 1 3 2	9 6 2 8 4	3 8 1 1 9	5 7 1 3 3
8 2 9 2 9	8 6 5 5 3	7 9 6 8 8	3 1 6 9 7	4 1 6 9 3	4 4 6 6 2	7 2 6 8 8
3 7 7 6 6	5 4 3 1 1	3 3 3 2 7	6 3 3 4 1	3 2 1 2 2	7 1 4 2 7	2 1 5 7 6
9 7 7 5 5	9 9 9 3 8	9 8 6 1 7	5 8 6 1 2	1 9 8 3 3	3 1 7 7 3	7 6 6 5 5
3 8 1 7 2	6 2 7 1 6	4 1 3 4 2	3 6 2 4 3	2 6 1 2 8	8 8 6 2 7	8 9 7 4 7
4 3 4 2 7	7 3 1 7 2	1 5 4 8 6	6 2 1 6 1	7 8 5 1 7	5 9 1 3 6	3 1 2 3 1
5 9 2 8 3	3 7 5 8 9	2 9 1 7 1	2 3 8 3 4	3 5 9 9 9	7 2 3 4 1	5 7 1 7 8
1 6 5 1 1	5 6 4 4 1	7 3 7 2 3	4 7 3 8 8	9 3 2 5 6	6 6 9 5 9	9 8 9 1 2
6 2 8 3 6	8 4 6 2 5	5 2 2 6 8	9 1 7 5 6	4 7 4 6 4	1 7 4 6 4	1 2 8 8 6
2 4 9 6 4	1 8 3 5 4	3 6 5 9 4	8 5 9 7 9	8 1 6 8 1	4 5 5 9 5	2 4 5 9 4
8 5 6 9 9	2 5 2 6 7	8 7 8 3 9	1 9 4 2 5	6 4 7 4 5	2 3 2 8 2	6 3 3 2 3
7 1 3 4 8	4 1 8 9 3	6 4 9 5 5	7 4 5 9 7	5 2 3 7 2	9 4 8 1 8	4 5 4 6 9
7 4 9 8 7	9 7 1 7 1	9 2 3 8 7	7 8 5 3 5	5 1 6 4 9	7 8 6 1 8	2 9 7 3 4
5 6 1 1 2	6 4 6 1 4	5 9 1 2 8	2 4 6 8 7	7 3 7 6 1	5 1 7 4 1	9 3 4 7 7
4 9 3 5 6	1 1 8 4 8	3 5 4 9 3	3 6 1 2 3	2 6 8 7 7	4 5 3 8 5	8 5 9 5 1
3 3 2 2 8	5 2 3 2 2	7 3 8 6 9	4 1 8 6 1	1 9 2 3 6	3 9 5 7 7	1 2 8 1 2
2 1 4 9 4	4 6 2 8 3	2 7 6 5 1	5 7 3 1 2	9 8 4 1 3	6 3 1 2 9	6 1 5 8 8
9 7 5 4 5	3 9 7 9 9	1 4 2 3 4	6 9 7 4 4	3 2 5 2 2	8 4 2 6 3	5 6 3 6 3
6 2 6 3 9	8 8 5 5 5	8 6 7 7 2	9 3 4 5 8	8 7 9 9 4	9 2 4 9 4	4 8 1 2 9
8 5 8 7 1	2 3 9 3 7	4 1 5 1 5	8 5 9 7 6	4 5 3 5 8	1 6 8 5 2	3 4 6 4 5
1 8 7 6 3	7 5 4 6 6	6 8 9 4 6	1 2 2 9 9	6 4 1 8 5	2 7 9 3 6	7 7 2 9 6
8 4 6 8 6	2 1 9 9 7	2 2 1 8 9	5 1 9 2 4	5 2 6 2 8	1 6 8 8 3	8 1 9 4 1
9 9 4 5 8	4 4 8 7 8	8 7 5 9 7	3 6 4 7 7	3 8 5 3 6	4 4 6 7 7	6 6 8 7 8
6 6 3 1 1	6 8 3 1 9	7 5 7 5 5	6 5 1 8 5	2 4 3 8 2	5 1 4 3 6	4 9 7 8 6
7 3 7 7 2	7 3 6 2 2	3 8 9 4 6	4 7 2 6 9	7 9 7 4 1	3 8 2 6 5	3 5 3 1 4
2 8 9 3 4	1 5 5 5 1	5 4 3 6 4	7 8 7 5 3	9 5 8 6 5	8 2 7 9 2	5 3 4 3 5
3 7 2 6 9	8 6 4 6 3	4 1 8 2 1	1 9 6 4 8	4 7 2 1 3	6 3 5 5 1	2 2 6 9 9
5 1 8 4 5	9 9 1 8 4	1 9 4 3 2	8 2 8 9 6	6 3 4 9 9	2 7 1 2 4	9 8 2 6 2
4 5 5 2 7	3 2 7 3 6	9 3 2 1 8	9 3 5 1 2	1 6 9 7 7	9 5 9 1 8	7 7 1 5 7
1 2 1 9 3	5 7 2 4 5	6 6 6 7 3	2 4 3 3 1	8 1 1 5 4	7 9 3 4 9	1 4 5 2 3

Source: Reprinted, with permission, from William G. Cochran and Gertrude M. Cox, *Experimental Designs*, 2d. ed. (New York: John Wiley & Sons, 1957), p. 577.

TABLE A–14 Table of random permutations of 16

7	12	15	15	1	2	7	16	10	2	14	15	7	13	13	10	6	1	8	10
13	3	8	16	7	10	11	10	13	5	11	7	13	16	7	7	5	13	2	14
3	1	4	5	14	13	3	14	9	13	13	2	9	15	6	2	8	4	5	8
11	8	16	14	15	6	2	6	2	16	8	5	12	3	9	13	4	3	10	4
14	9	1	6	3	9	14	13	8	6	5	8	14	7	3	15	13	11	4	7
2	16	10	13	5	5	13	2	11	7	3	12	5	14	12	16	2	2	9	15
4	6	13	7	2	15	1	9	1	4	7	10	6	9	11	9	7	6	16	11
6	14	6	10	4	14	4	15	3	3	4	16	2	6	5	1	12	10	6	9
10	15	2	1	13	12	16	3	4	8	10	1	15	5	14	12	14	12	3	2
12	10	7	12	9	11	9	8	12	14	15	4	11	8	16	8	9	14	14	1
15	7	5	2	10	7	8	12	6	15	6	13	16	12	15	4	11	8	12	6
16	2	11	8	8	8	15	5	16	1	1	9	8	1	8	14	16	5	13	5
9	13	14	3	6	4	10	11	5	12	9	3	10	4	4	3	10	9	1	3
8	11	9	4	11	3	12	7	7	10	12	14	3	10	1	6	15	16	15	12
1	5	12	11	16	16	5	4	14	9	16	11	1	2	10	5	1	15	7	13
5	4	3	9	12	1	6	1	15	11	2	6	4	11	2	11	3	7	11	16

11	8	16	5	5	13	1	13	2	16	14	12	9	8	7	5	13	3	13	3
2	2	8	8	14	16	4	3	8	11	10	14	15	1	2	11	4	5	15	9
6	13	2	13	6	5	9	15	11	10	12	6	16	15	16	9	10	12	16	15
14	12	4	16	16	11	14	10	5	12	3	3	12	14	15	13	6	4	1	16
8	6	3	9	4	10	6	4	16	2	2	9	8	16	4	6	5	15	7	8
9	15	12	10	3	2	12	6	1	15	4	13	7	7	9	12	14	8	8	11
3	10	11	12	13	12	5	11	7	8	9	5	14	11	10	1	3	13	3	5
16	1	13	14	8	14	15	5	3	7	11	15	6	12	5	7	11	1	14	4
1	14	14	2	9	15	16	14	6	14	7	8	3	13	11	8	7	7	12	7
4	4	6	4	12	3	11	8	15	9	8	1	13	6	3	3	15	9	9	12
15	5	1	11	10	6	3	7	10	5	5	11	10	10	12	15	16	14	5	2
5	3	5	6	7	7	13	2	14	3	16	4	5	5	13	4	9	16	2	6
12	7	15	15	15	9	8	12	12	13	15	10	1	4	6	16	2	6	11	1
10	11	10	3	2	4	2	1	4	6	6	7	11	9	14	10	8	11	4	13
7	9	7	7	11	1	7	16	13	1	13	2	4	2	1	2	12	2	10	14
13	16	9	1	1	8	10	9	9	4	1	16	2	3	8	14	1	10	6	10

1	6	7	4	8	6	5	2	8	15	4	6	6	1	4	5	7	13	2	10
9	15	11	3	11	15	9	10	1	3	8	2	15	7	9	8	16	1	14	3
10	16	4	5	12	9	16	11	7	1	7	16	11	8	3	3	12	2	3	4
4	14	1	9	5	5	4	13	6	8	15	5	12	5	7	16	5	11	8	1
7	3	13	14	15	2	1	14	16	5	14	9	2	16	1	12	6	14	4	13
16	11	2	1	14	16	6	9	3	4	16	14	3	15	11	11	3	9	12	5
3	10	16	16	13	7	13	1	11	14	9	10	16	2	10	2	10	7	10	16
11	13	9	13	4	13	8	3	5	13	10	12	5	12	5	14	13	16	5	6
15	2	3	12	9	12	2	4	13	10	3	13	14	4	2	1	14	8	6	12
14	1	14	6	10	1	3	12	4	2	2	4	13	3	16	9	9	3	7	14
13	12	5	11	3	11	15	8	2	7	11	7	8	14	6	4	4	4	15	11
12	5	10	7	2	14	7	15	14	16	13	1	9	10	12	10	11	10	9	8
8	9	8	10	6	4	11	7	10	11	6	8	4	9	8	15	8	6	11	9
2	7	6	2	1	8	10	6	15	12	1	11	7	11	13	6	1	15	13	15
6	4	15	8	16	10	14	16	9	6	12	3	10	6	14	7	2	12	16	7
5	8	12	15	7	3	12	5	12	9	5	15	1	13	15	13	15	5	1	2

13	4	10	4	16	13	16	13	5	3	6	14	1	16	8	7	2	3	3	12
5	14	4	6	8	2	15	1	13	14	16	4	15	4	3	12	12	1	4	7
2	2	2	15	14	16	9	12	16	6	10	15	14	9	10	1	14	8	8	16
7	12	15	8	12	3	5	14	7	12	5	13	16	1	7	5	11	2	9	3
6	9	7	14	9	14	10	11	15	11	12	1	12	12	14	16	3	11	11	8
14	5	16	7	10	8	11	8	14	13	7	11	6	3	11	4	4	6	6	9
15	11	8	9	7	12	8	7	1	15	9	3	3	7	13	11	10	4	5	1
11	6	6	1	4	1	3	16	12	5	4	9	13	13	6	8	15	9	1	14
4	10	3	16	2	11	7	9	6	9	1	8	4	11	5	2	16	10	12	4
1	8	1	13	1	15	4	4	11	4	2	16	5	8	1	9	5	12	16	6
9	7	14	2	6	4	14	10	9	8	15	10	7	10	9	10	6	14	10	11
12	1	9	10	15	5	2	15	10	2	14	2	8	2	4	13	8	5	15	5
3	3	12	11	5	9	6	6	3	10	13	12	9	6	2	15	7	15	7	13
10	15	11	5	13	7	12	5	2	7	11	5	10	15	12	3	1	13	13	10
8	13	13	3	3	10	13	2	4	1	8	6	11	14	15	6	9	16	2	2
16	16	5	12	11	6	1	3	8	16	3	7	2	5	16	14	13	7	14	15

Source: Reprinted, with permission, from William G. Cochran and Gertrude M. Cox, *Experimental Designs*, 2d. ed. (New York: John Wiley & Sons, 1957), p. 583.

TABLE A–15 Table of random digits

Line	(1)–(5)	(6)–(10)	(11)–(15)	(16)–(20)	(21)–(25)	(26)–(30)	(31)–(35)
101	13284	16834	74151	92027	24670	36665	00770
102	21224	00370	30420	03883	94648	89428	41583
103	99052	47887	81085	64933	66279	80432	65793
104	00199	50993	98603	38452	87890	94624	69721
105	60578	06483	28733	37867	07936	98710	98539
106	91240	18312	17441	01929	18163	69201	31211
107	97458	14229	12063	59611	32249	90466	33216
108	35249	38646	34475	72417	60514	69257	12489
109	38980	46600	11759	11900	46743	27860	77940
110	10750	52745	38749	87365	58959	53731	89295
111	36247	27850	73958	20673	37800	63835	71051
112	70994	66986	99744	72438	01174	42159	11392
113	99638	94702	11463	18148	81386	80431	90628
114	72055	15774	43857	99805	10419	76939	25993
115	24038	65541	85788	55835	38835	59399	13790
116	74976	14631	35908	28221	39470	91548	12854
117	35553	71628	70189	26436	63407	91178	90348
118	35676	12797	51434	82976	42010	26344	92920
119	74815	67523	72985	23183	02446	63594	98924
120	45246	88048	65173	50989	91060	89894	36036
121	76509	47069	86378	41797	11910	49672	88575
122	19689	90332	04315	21358	97248	11188	39062
123	42751	35318	97513	61537	54955	08159	00337
124	11946	22681	45045	13964	57517	59419	58045
125	96518	48688	20996	11090	48396	57177	83867
126	35726	58643	76869	84622	39098	36083	72505
127	39737	42750	48968	70536	84864	64952	38404
128	97025	66492	56177	04049	80312	48028	26408
129	62814	08075	09788	56350	76787	51591	54509
130	25578	22950	15227	83291	41737	59599	96191
131	68763	69576	88991	49662	46704	63362	56625
132	17900	00813	64361	60725	88974	61005	99709
133	71944	60227	63551	71109	05624	43836	58254
134	54684	93691	85132	64399	29182	44324	14491
135	25946	27623	11258	65204	52832	50880	22273
136	01353	39318	44961	44972	91766	90262	56073
137	99083	88191	27662	99113	57174	35571	99884
138	52021	45406	37945	75234	24327	86978	22644
139	78755	47744	43776	83098	03225	14281	83637
140	25282	69106	59180	16257	22810	43609	12224
141	11959	94202	02743	86847	79725	51811	12998
142	11644	13792	98190	01424	30078	28197	55583
143	06307	97912	68110	59812	95448	43244	31262
144	76285	75714	89585	99296	52640	46518	55486
145	55322	07598	39600	60866	63007	20007	66819
146	78017	90928	90220	92503	83375	26986	74399
147	44768	43342	20696	26331	43140	69744	82928
148	25100	19336	14605	86603	51680	97678	24261
149	83612	46623	62876	85197	07824	91392	58317
150	41347	81666	82961	60413	71020	83658	02415

Source: *Table of 105,000 Random Decimal Digits*, Interstate Commerce Commission, Bureau of Transport Economics and Statistics, 1949.

TABLE A–16 Selected standard latin squares

3 × 3

A	*B*	*C*
B	*C*	*A*
C	*A*	*B*

4 × 4

1

A	*B*	*C*	*D*
B	*A*	*D*	*C*
C	*D*	*B*	*A*
D	*C*	*A*	*B*

2

A	*B*	*C*	*D*
B	*C*	*D*	*A*
C	*D*	*A*	*B*
D	*A*	*B*	*C*

3

A	*B*	*C*	*D*
B	*D*	*A*	*C*
C	*A*	*D*	*B*
D	*C*	*B*	*A*

4

A	*B*	*C*	*D*
B	*A*	*D*	*C*
C	*D*	*A*	*B*
D	*C*	*B*	*A*

5 × 5

A	*B*	*C*	*D*	*E*
B	*A*	*E*	*C*	*D*
C	*D*	*A*	*E*	*B*
D	*E*	*B*	*A*	*C*
E	*C*	*D*	*B*	*A*

6 × 6

A	*B*	*C*	*D*	*E*	*F*
B	*F*	*D*	*C*	*A*	*E*
C	*D*	*E*	*F*	*B*	*A*
D	*A*	*F*	*E*	*C*	*B*
E	*C*	*A*	*B*	*F*	*D*
F	*E*	*B*	*A*	*D*	*C*

7 × 7

A	*B*	*C*	*D*	*E*	*F*	*G*
B	*C*	*D*	*E*	*F*	*G*	*A*
C	*D*	*E*	*F*	*G*	*A*	*B*
D	*E*	*F*	*G*	*A*	*B*	*C*
E	*F*	*G*	*A*	*B*	*C*	*D*
F	*G*	*A*	*B*	*C*	*D*	*E*
G	*A*	*B*	*C*	*D*	*E*	*F*

8 × 8

A	*B*	*C*	*D*	*E*	*F*	*G*	*H*
B	*C*	*D*	*E*	*F*	*G*	*H*	*A*
C	*D*	*E*	*F*	*G*	*H*	*A*	*B*
D	*E*	*F*	*G*	*H*	*A*	*B*	*C*
E	*F*	*G*	*H*	*A*	*B*	*C*	*D*
F	*G*	*H*	*A*	*B*	*C*	*D*	*E*
G	*H*	*A*	*B*	*C*	*D*	*E*	*F*
H	*A*	*B*	*C*	*D*	*E*	*F*	*G*

9 × 9

A	*B*	*C*	*D*	*E*	*F*	*G*	*H*	*I*
B	*C*	*D*	*E*	*F*	*G*	*H*	*I*	*A*
C	*D*	*E*	*F*	*G*	*H*	*I*	*A*	*B*
D	*E*	*F*	*G*	*H*	*I*	*A*	*B*	*C*
E	*F*	*G*	*H*	*I*	*A*	*B*	*C*	*D*
F	*G*	*H*	*I*	*A*	*B*	*C*	*D*	*E*
G	*H*	*I*	*A*	*B*	*C*	*D*	*E*	*F*
H	*I*	*A*	*B*	*C*	*D*	*E*	*F*	*G*
I	*A*	*B*	*C*	*D*	*E*	*F*	*G*	*H*

SENIC data set

The primary objective of the Study on the Efficacy of Nosocomial Infection Control (**SENIC** Project) was to determine whether infection surveillance and control programs have reduced the rates of nosocomial (hospital-acquired) infection in United States hospitals. This data set consists of a random sample of 113 hospitals selected from the original 338 hospitals surveyed.

Each line of the data set has an identification number and provides information on 11 other variables for a single hospital. The data presented here are for the 1975–76 study period. The 12 variables are:

Variable Number	*Variable Name*	*Description*
1	Identification number	1-113
2	Length of stay	Average length of stay of all patients in hospital (in days)
3	Age	Average age of patients (in years)
4	Infection risk	Average estimated probability of acquiring infection in hospital (in percent)
5	Routine culturing ratio	Ratio of number of cultures performed to number of patients without signs or symptoms of hospital-acquired infection, times 100

Variable Number	*Variable Name*	*Description*
6	Routine chest X-ray ratio	Ratio of number of X rays performed to number of patients without signs or symptoms of pneumonia, times 100
7	Number of beds	Average number of beds in hospital during study period
8	Medical school affiliation	1 = Yes 2 = No
9	Region	Geographic region, where: 1 = NE, 2 = NC, 3 = S, 4 = W
10	Average daily census	Average number of patients in hospital per day during study period
11	Number of nurses	Average number of full-time equivalent registered and licensed practical nurses during study period (number full time plus one half the number part time)
12	Available facilities and services	Percent of 35 potential facilities and services that are provided by the hospital

Reference: Special Issue, "The SENIC Project," *American Journal of Epidemiology* 111 (1980), pp. 465–653.

Data obtained from: Robert W. Haley, M.D., Hospital Infections Program, Center for Infectious Diseases, Centers for Disease Control, Atlanta, Georgia 30333.

1	2	3	4	5	6	7	8	9	10	11	12
1	7.13	55.7	4.1	9.0	39.6	279	2	4	207	241	60.0
2	8.82	58.2	1.6	3.8	51.7	80	2	2	51	52	40.0
3	8.34	56.9	2.7	8.1	74.0	107	2	3	82	54	20.0
4	8.95	53.7	5.6	18.9	122.8	147	2	4	53	148	40.0
5	11.20	56.5	5.7	34.5	88.9	180	2	1	134	151	40.0
6	9.76	50.9	5.1	21.9	97.0	150	2	2	147	106	40.0
7	9.68	57.8	4.6	16.7	79.0	186	2	3	151	129	40.0
8	11.18	45.7	5.4	60.5	85.8	640	1	2	399	360	60.0
9	8.67	48.2	4.3	24.4	90.8	182	2	3	130	118	40.0
10	8.84	56.3	6.3	29.6	82.6	85	2	1	59	66	40.0
11	11.07	53.2	4.9	28.5	122.0	768	1	1	591	656	80.0
12	8.30	57.2	4.3	6.8	83.8	167	2	3	105	59	40.0
13	12.78	56.8	7.7	46.0	116.9	322	1	1	252	349	57.1
14	7.58	56.7	3.7	20.8	88.0	97	2	2	59	79	37.1
15	9.00	56.3	4.2	14.6	76.4	72	2	3	61	38	17.1
16	11.08	50.2	5.5	18.6	63.6	387	2	3	326	405	57.1
17	8.28	48.1	4.5	26.0	101.8	108	2	4	84	73	37.1
18	11.62	53.9	6.4	25.5	99.2	133	2	1	113	101	37.1
19	9.06	52.8	4.2	6.9	75.9	134	2	2	103	125	37.1
20	9.35	53.8	4.1	15.9	80.9	833	2	3	547	519	77.1
21	7.53	42.0	4.2	23.1	98.9	95	2	4	47	49	17.1
22	10.24	49.0	4.8	36.3	112.6	195	2	2	163	170	37.1
23	9.78	52.3	5.0	17.6	95.9	270	1	1	240	198	57.1
24	9.84	62.2	4.8	12.0	82.3	600	2	3	468	497	57.1
25	9.20	52.2	4.0	17.5	71.1	298	1	4	244	236	57.1
26	8.28	49.5	3.9	12.0	113.1	546	1	2	413	436	57.1
27	9.31	47.2	4.5	30.2	101.3	170	2	1	124	173	37.1
28	8.19	52.1	3.2	10.8	59.2	176	2	1	156	88	37.1
29	11.65	54.5	4.4	18.6	96.1	248	2	1	217	189	37.1
30	9.89	50.5	4.9	17.7	103.6	167	2	2	113	106	37.1
31	11.03	49.9	5.0	19.7	102.1	318	2	1	270	335	57.1
32	9.84	53.0	5.2	17.7	72.6	210	2	2	200	239	54.3
33	11.77	54.1	5.3	17.3	56.0	196	2	1	164	165	34.3
34	13.59	54.0	6.1	24.2	111.7	312	2	1	258	169	54.3
35	9.74	54.4	6.3	11.4	76.1	221	2	2	170	172	54.3
36	10.33	55.8	5.0	21.2	104.3	266	2	1	181	149	54.3
37	9.97	58.2	2.8	16.5	76.5	90	2	2	69	42	34.3
38	7.84	49.1	4.6	7.1	87.9	60	2	3	50	45	34.3
39	10.47	53.2	4.1	5.7	69.1	196	2	2	168	153	54.3
40	8.16	60.9	1.3	1.9	58.0	73	2	3	49	21	14.3
41	8.48	51.1	3.7	12.1	92.8	166	2	3	145	118	34.3
42	10.72	53.8	4.7	23.2	94.1	113	2	3	90	107	34.3
43	11.20	45.0	3.0	7.0	78.9	130	2	3	95	56	34.3
44	10.12	51.7	5.6	14.9	79.1	362	1	3	313	264	54.3
45	8.37	50.7	5.5	15.1	84.8	115	2	2	96	88	34.3
46	10.16	54.2	4.6	8.4	51.5	831	1	4	581	629	74.3
47	19.56	59.9	6.5	17.2	113.7	306	2	1	273	172	51.4
48	10.90	57.2	5.5	10.6	71.9	593	2	2	446	211	51.4
49	7.67	51.7	1.8	2.5	40.4	106	2	3	93	35	11.4
50	8.88	51.5	4.2	10.1	86.9	305	2	3	238	197	51.4
51	11.48	57.6	5.6	20.3	82.0	252	2	1	207	251	51.4
52	9.23	51.6	4.3	11.6	42.6	620	2	2	413	420	71.4
53	11.41	61.1	7.6	16.6	97.9	535	2	3	330	273	51.4
54	12.07	43.7	7.8	52.4	105.3	157	2	2	115	76	31.4
55	8.63	54.0	3.1	8.4	56.2	76	2	1	39	44	31.4
56	11.15	56.5	3.9	7.7	73.9	281	2	1	217	199	51.4

1	2	3	4	5	6	7	8	9	10	11	12
57	7.14	59.0	3.7	2.6	75.8	70	2	4	37	35	31.4
58	7.65	47.1	4.3	16.4	65.7	318	2	4	265	314	51.4
59	10.73	50.6	3.9	19.3	101.0	445	1	2	374	345	51.4
60	11.46	56.9	4.5	15.6	97.7	191	2	3	153	132	31.4
61	10.42	58.0	3.4	8.0	59.0	119	2	1	67	64	31.4
62	11.18	51.0	5.7	18.8	55.9	595	1	2	546	392	68.6
63	7.93	64.1	5.4	7.5	98.1	68	2	4	42	49	28.6
64	9.66	52.1	4.4	9.9	98.3	83	2	2	66	95	28.6
65	7.78	45.5	5.0	20.9	71.6	489	2	3	391	329	48.6
66	9.42	50.6	4.3	24.8	62.8	508	2	1	421	528	48.6
67	10.02	49.5	4.4	8.3	93.0	265	2	2	191	202	48.6
68	8.58	55.0	3.7	7.4	95.9	304	2	3	248	218	48.6
69	9.61	52.4	4.5	6.9	87.2	487	2	3	404	220	48.6
70	8.03	54.2	3.5	24.3	87.3	97	2	1	65	55	28.6
71	7.39	51.0	4.2	14.6	88.4	72	2	2	38	67	28.6
72	7.08	52.0	2.0	12.3	56.4	87	2	3	52	57	28.6
73	9.53	51.5	5.2	15.0	65.7	298	2	3	241	193	48.6
74	10.05	52.0	4.5	36.7	87.5	184	1	1	144	151	68.6
75	8.45	38.8	3.4	12.9	85.0	235	2	2	143	124	48.6
76	6.70	48.6	4.5	13.0	80.8	76	2	4	51	79	28.6
77	8.90	49.7	2.9	12.7	86.9	52	2	1	37	35	28.6
78	10.23	53.2	4.9	9.9	77.9	752	1	2	595	446	68.6
79	8.88	55.8	4.4	14.1	76.8	237	2	2	165	182	48.6
80	10.30	59.6	5.1	27.8	88.9	175	2	2	113	73	45.7
81	10.79	44.2	2.9	2.6	56.6	461	1	2	320	196	65.7
82	7.94	49.5	3.5	6.2	92.3	195	2	2	139	116	45.7
83	7.63	52.1	5.5	11.6	61.1	197	2	4	109	110	45.7
84	8.77	54.5	4.7	5.2	47.0	143	2	4	85	87	25.7
85	8.09	56.9	1.7	7.6	56.9	92	2	3	61	61	45.7
86	9.05	51.2	4.1	20.5	79.8	195	2	3	127	112	45.7
87	7.91	52.8	2.9	11.9	79.5	477	2	3	349	188	65.7
88	10.39	54.6	4.3	14.0	88.3	353	2	2	223	200	65.7
89	9.36	54.1	4.8	18.3	90.6	165	2	1	127	158	45.7
90	11.41	50.4	5.8	23.8	73.0	424	1	3	359	335	45.7
91	8.86	51.3	2.9	9.5	87.5	100	2	3	65	53	25.7
92	8.93	56.0	2.0	6.2	72.5	95	2	3	59	56	25.7
93	8.92	53.9	1.3	2.2	79.5	56	2	2	40	14	5.7
94	8.15	54.9	5.3	12.3	79.8	99	2	4	55	71	25.7
95	9.77	50.2	5.3	15.7	89.7	154	2	2	123	148	25.7
96	8.54	56.1	2.5	27.0	82.5	98	2	1	57	75	45.7
97	8.66	52.8	3.8	6.8	69.5	246	2	3	178	177	45.7
98	12.01	52.8	4.8	10.8	96.9	298	2	1	237	115	45.7
99	7.95	51.8	2.3	4.6	54.9	163	2	3	128	93	42.9
100	10.15	51.9	6.2	16.4	59.2	568	1	3	452	371	62.9
101	9.76	53.2	2.6	6.9	80.1	64	2	4	47	55	22.9
102	9.89	45.2	4.3	11.8	108.7	190	2	1	141	112	42.9
103	7.14	57.6	2.7	13.1	92.6	92	2	4	40	50	22.9
104	13.95	65.9	6.6	15.6	133.5	356	2	1	308	182	62.9
105	9.44	52.5	4.5	10.9	58.5	297	2	3	230	263	42.9
106	10.80	63.9	2.9	1.6	57.4	130	2	3	69	62	22.9
107	7.14	51.7	1.4	4.1	45.7	115	2	3	90	19	22.9
108	8.02	55.0	2.1	3.8	46.5	91	2	2	44	32	22.9
109	11.80	53.8	5.7	9.1	116.9	571	1	2	441	469	62.9
110	9.50	49.3	5.8	42.0	70.9	98	2	3	68	46	22.9
111	7.70	56.9	4.4	12.2	67.9	129	2	4	85	136	62.9
112	17.94	56.2	5.9	26.4	91.8	835	1	1	791	407	62.9
113	9.41	59.5	3.1	20.6	91.7	29	2	3	20	22	22.9

SMSA data set

This data set provides information for 141 large Standard Metropolitan Statistical Areas (**SMSA**s) in the United States. A standard metropolitan statistical area includes a city (or cities) of specified population size which constitutes the central city and the county (or counties) in which it is located, as well as contiguous counties when the economic and social relationships between the central and contiguous counties meet specified criteria of metropolitan character and integration. An SMSA may have up to three central cities and may cross state lines.

Each line of the data set has an identification number and provides information on 11 other variables for a single SMSA. The information generally pertains to the years 1976 and 1977, the most recent information available at the time. The 12 variables are:

Variable Number	*Variable Name*	*Description*
1	Identification number	1-141
2	Land area	In square miles
3	Total population	Estimated 1977 population (in thousands)
4	Percent of population in central cities	Percent of 1976 SMSA population in central city or cities
5	Percent of population 65 or older	Percent of 1976 SMSA population 65 years old or older

Variable Number	*Variable Name*	*Description*
6	Number of active physicians	Number of professionally active nonfederal physicians as of December 31, 1977
7	Number of hospital beds	Total number of beds, cribs, and bassinettes during 1977
8	Percent high school graduates	Percent of adult population (persons 25 years old or older) who completed 12 or more years of school, according to the 1970 Census of the Population
9	Civilian labor force	Total number of persons in civilian labor force (persons 16 years old or older classified as employed or unemployed) in 1977 (in thousands)
10	Total personal income	Total current income received in 1976 by residents of the SMSA from all sources, before deduction of income and other personal taxes but after deduction of personal contributions to social security and other social insurance programs (in millions of dollars)
11	Total serious crimes	Total number of serious crimes in 1977, including murder, rape, robbery, aggravated assault, burglary, larceny-theft, and motor vehicle theft, as reported by law enforcement agencies
12	Geographic region	Geographic region classification is that used by the U.S. Bureau of the Census, where: 1 = NE, 2 = NC, 3 = S, 4 = W

Data obtained from: U.S. Bureau of the Census, *State and Metropolitan Area Data Book, 1979* (a Statistical Abstract Supplement).

1	2	3	4	5	6	7	8	9	10	11	12
1	1384	9387	78.1	12.3	25627	69678	50.1	4083.9	72100	709234	1
2	4069	7031	44.0	10.0	15389	39699	62.0	3353.6	52737	499813	4
3	3719	7017	43.9	9.4	13326	43292	53.9	3305.9	54542	393162	2
4	3553	4794	37.4	10.7	9724	33731	50.6	2066.3	33216	198102	1
5	3916	4370	29.9	8.8	6402	24167	52.2	1966.7	32906	294466	2
6	2480	3182	31.5	10.5	8502	16751	66.1	1514.5	26573	255162	4
7	2815	3033	23.1	6.7	7340	16941	68.3	1541.9	25663	177355	3
8	1218	2688	0.0	8.8	5255	22137	62.9	1213.3	21524	127567	1
9	8360	2673	46.3	8.2	4047	14347	53.6	1321.2	18350	193125	3
10	6794	2512	60.1	6.3	4562	14333	51.7	1272.7	18221	162976	3
11	4935	2380	21.8	11.0	4071	17752	47.8	1061.2	16120	137479	2
12	3049	2294	19.5	12.1	4005	21149	53.4	967.5	15826	69989	1
13	2259	2147	38.6	9.3	5141	16485	44.6	966.8	14246	138214	3
14	4647	2037	31.5	9.2	3916	12815	65.1	1032.2	14542	112642	2
15	1008	1969	16.6	10.3	4006	16704	55.9	935.5	15953	106646	1
16	1519	1950	31.8	10.5	4094	12545	54.6	906.0	14684	102816	2
17	4326	1832	23.6	7.3	3064	9976	50.4	867.2	12107	106482	3
18	782	1801	28.4	7.8	3119	8656	70.5	915.2	12591	113821	4
19	4261	1683	48.6	9.7	3396	7552	65.3	644.3	10392	112359	4
20	4651	1464	38.8	7.7	3380	8517	67.4	729.2	10375	116861	4
21	2042	1441	24.5	16.5	4071	10039	51.9	681.7	10166	116304	3
22	4226	1427	38.1	9.8	3285	5392	67.8	699.8	10918	91399	4
23	1456	1427	46.7	10.4	2484	8555	56.8	710.4	10104	63695	2
24	2045	1380	37.2	21.4	1949	8863	50.7	543.2	7989	89257	3
25	2149	1375	29.8	10.6	2530	8354	48.4	617.6	9037	68319	2
26	1590	1313	30.1	10.9	2296	9988	50.4	565.7	8411	67965	1
27	27293	1306	25.3	12.3	2018	6323	57.4	510.6	7399	99293	4
28	3341	1293	35.8	10.1	2289	7593	59.9	656.3	9106	81510	2
29	9155	1254	53.8	11.1	2280	6450	60.1	575.2	7766	107370	4
30	1300	1217	47.6	6.8	2794	4989	69.0	610.8	9215	76570	4
31	3072	1144	68.0	9.3	2181	7497	56.0	549.6	7736	61381	2
32	1967	1133	51.1	8.8	2520	8467	45.8	460.5	7038	69285	3
33	3650	1121	34.6	11.1	2358	6224	62.9	539.3	7792	77316	4
34	2460	1087	49.6	8.4	1874	7706	59.9	510.7	6658	62603	2
35	2527	1025	78.7	8.4	1760	7664	46.5	391.1	5582	62694	3
36	2966	970	26.9	10.3	2053	6604	56.3	450.4	6966	54854	1
37	3434	929	28.9	8.3	1844	3215	65.1	422.6	5909	72410	4
38	1392	883	37.2	9.8	1579	6087	46.5	396.8	5705	45642	3
39	2298	886	76.2	9.0	1644	7673	48.2	394.6	5185	52094	3
40	1219	864	31.7	20.6	1396	6158	55.4	352.8	5879	68109	3
41	1708	833	24.0	8.8	1062	5315	56.2	367.5	5489	52606	2
42	8565	822	29.7	7.3	1604	3485	67.6	349.3	4655	49111	4
43	3358	805	35.1	11.3	1649	5512	44.9	359.1	4941	42786	3
44	2624	794	30.4	12.2	1532	4730	55.2	356.5	5094	30771	1
45	2187	777	47.0	10.2	1098	4342	51.9	355.4	5142	46213	2
46	3214	774	47.7	9.4	1285	3459	40.3	401.7	4924	34941	3
47	3491	769	48.5	9.7	1496	5620	59.6	362.3	4798	44513	3
48	4080	773	59.6	9.9	1597	7496	47.3	380.9	4600	33936	3
49	596	723	100.0	6.0	1260	2819	66.0	319.9	5181	46984	4
50	3199	694	80.6	8.7	983	4749	50.8	292.4	4127	43010	3
51	903	661	37.3	9.6	948	4064	55.6	293.3	4102	34725	2
52	2419	647	27.8	9.9	1250	2870	57.8	286.8	3860	30829	1
53	938	644	48.1	7.4	614	3016	50.0	280.9	4177	35106	2
54	1951	629	28.4	14.5	696	4843	47.9	271.5	3667	14868	1
55	1490	624	33.1	11.9	827	3818	47.4	300.2	4144	19090	1
56	5677	610	55.8	10.5	760	3883	56.2	292.0	4035	32146	3
57	1525	597	55.7	8.3	751	3234	44.9	318.5	3777	37070	3
58	2528	593	19.2	10.2	798	3135	55.4	274.1	3489	44442	3
59	312	594	19.5	7.5	769	2463	55.0	298.7	4352	29100	1
60	1537	581	63.8	8.7	1234	5160	62.7	272.6	3725	32271	2
61	1420	576	32.6	9.5	833	2950	54.0	280.8	3553	26645	2
62	47	564	41.9	11.9	745	3352	36.3	258.9	3915	29157	1
63	1023	541	35.1	10.0	639	3144	52.1	234.1	3437	22111	2
64	2115	526	19.9	9.1	676	2296	38.8	253.3	2962	30684	3
65	1182	514	32.4	7.4	518	2515	52.4	216.8	3627	35201	2
66	1165	516	14.5	8.6	746	4277	54.4	237.1	3724	31358	3
67	476	492	8.9	10.9	787	2778	60.1	218.4	3603	24787	1
68	1553	487	50.0	8.0	2207	4931	52.0	257.2	2991	24269	3
69	2023	477	22.1	21.8	752	2317	55.7	194.2	3283	36418	3
70	2766	474	67.9	7.7	679	3873	56.3	224.0	2598	29967	3

1	2	3	4	5	6	7	8	9	10	11	12
71	5966	472	39.5	9.6	737	1907	52.7	246.6	3007	38205	4
72	1863	468	50.4	7.7	674	2989	63.8	194.8	2747	25159	4
73	192	462	60.5	10.8	617	1789	44.1	212.6	3158	27161	1
74	9240	455	67.0	10.3	1123	2347	63.1	183.6	2598	41649	4
75	2277	455	39.5	7.5	512	1788	61.9	221.1	2853	20053	2
76	1630	449	41.9	10.7	724	4395	50.0	198.0	2445	17596	3
77	1617	435	71.0	6.9	518	2031	54.1	197.9	2617	31539	3
78	1057	435	90.7	6.1	479	2551	51.1	163.4	2012	25650	3
79	1624	429	13.4	11.0	832	2938	55.4	207.8	2885	16985	1
80	1676	423	36.6	9.2	505	3297	60.7	156.3	2689	24266	4
81	2818	425	48.5	9.3	540	2694	42.3	172.8	2162	22374	3
82	2866	408	24.9	10.7	427	2864	39.1	169.1	1987	10425	3
83	4883	402	72.4	7.3	873	2236	64.9	185.2	2353	28171	4
84	966	401	24.9	10.6	427	3192	52.2	174.7	2446	15981	2
85	2109	403	41.2	10.3	520	2539	45.2	183.1	2308	16240	3
86	2449	395	68.4	9.6	681	2864	63.2	207.4	2651	25149	2
87	2618	385	31.7	6.1	836	2159	48.0	145.6	1992	25046	3
88	1465	374	30.3	6.8	598	6456	50.6	164.7	2201	26428	3
89	1704	375	52.1	10.5	379	2491	55.6	173.2	2662	18599	2
90	1750	370	49.3	9.7	446	3472	58.2	176.5	2439	16529	2
91	1489	369	58.8	9.5	911	5720	56.5	175.1	2264	26032	3
92	8152	363	22.3	9.1	405	1254	51.7	165.6	2257	28351	4
93	2207	364	57.3	9.7	356	2167	45.5	165.9	2331	19138	3
94	7874	360	44.4	6.9	398	1365	65.2	174.2	2410	33687	4
95	655	364	75.2	6.6	425	3879	51.6	163.0	2088	15623	3
96	1803	362	35.3	10.4	483	2137	53.7	168.9	2666	16405	2
97	2363	356	53.1	10.6	565	2717	49.3	146.4	1996	19212	3
98	1435	352	13.4	11.7	342	1076	44.7	156.8	2165	11273	1
99	946	348	16.4	11.1	366	1455	43.9	163.8	2178	8116	1
100	1136	333	58.6	9.7	448	2630	68.1	171.4	2396	20465	2
101	2658	327	39.0	12.2	365	5430	49.9	136.9	1862	9325	1
102	228	317	31.1	10.2	667	3179	52.8	156.5	2264	19410	1
103	1758	310	56.8	11.5	565	2081	65.3	131.2	1939	17379	4
104	1198	313	55.1	8.0	1171	3877	71.2	172.3	2038	18676	2
105	1412	311	39.2	11.3	436	1837	49.4	154.2	2098	25714	4
106	2071	306	19.9	11.3	470	2531	58.9	133.1	1782	11161	1
107	862	302	26.3	13.4	423	1929	43.3	145.5	2010	7699	1
108	1526	303	71.7	7.7	413	1636	47.1	125.8	1692	20038	3
109	1758	297	33.2	11.6	296	2652	45.3	114.4	1641	12467	3
110	1651	296	64.6	8.9	774	5431	56.1	136.9	1724	14468	3
111	1493	294	64.8	8.9	863	3289	53.7	154.7	1787	15871	3
112	1610	294	59.8	9.5	471	4633	62.9	116.1	1851	18651	4
113	2710	288	63.7	6.2	357	1277	72.8	110.9	1639	18173	4
114	1975	291	46.5	12.6	405	2896	51.5	133.8	1853	12787	2
115	1920	291	49.8	7.8	283	1306	53.2	126.9	1553	12315	3
116	1404	289	38.5	10.0	299	1766	56.2	138.6	1776	11715	2
117	2737	287	45.0	10.5	602	1462	71.3	131.4	1980	18208	4
118	1700	287	18.8	8.0	739	3381	45.9	120.4	1616	14534	3
119	909	277	41.2	11.5	307	1309	54.2	131.9	1762	13722	2
120	1858	277	24.3	13.7	354	1562	46.3	116.9	1507	19133	3
121	3324	275	49.7	8.4	373	929	62.5	120.5	1918	14776	4
122	1697	274	23.8	7.2	338	1610	51.0	105.9	1354	19317	3
123	813	272	46.0	9.8	293	1693	58.4	119.9	1688	10402	1
124	7397	267	47.3	12.5	355	2042	56.2	113.7	1654	12273	2
125	1165	268	43.7	9.4	450	2070	57.5	129.4	1719	16226	2
126	802	268	52.6	9.8	392	1425	52.2	129.6	1816	13230	2
127	1770	268	14.8	12.2	285	2804	44.1	106.7	1537	4205	1
128	495	264	50.7	7.8	220	1177	52.6	119.5	1661	8398	2
129	1255	261	26.0	10.7	458	1646	51.6	113.0	1725	10208	3
130	1148	589	45.3	11.1	891	5790	54.0	277.0	3510	29237	1
131	1509	643	37.6	12.0	1087	4900	51.4	319.6	3982	29058	1
132	2013	254	61.7	9.7	273	1484	50.9	106.7	1412	14446	3
133	711	250	42.4	6.1	1411	3659	67.5	131.0	1790	16228	2
134	471	251	46.3	8.6	219	1128	47.8	105.3	1458	13474	2
135	4552	249	54.4	9.1	329	719	61.9	118.0	1386	15596	4
136	1400	242	50.8	8.0	290	1271	45.7	104.4	1351	10391	3
137	1511	236	38.7	10.7	348	1093	50.4	127.2	1452	16676	4
138	1543	232	39.6	8.1	159	481	30.3	80.6	769	8436	3
139	1011	233	37.8	10.5	264	964	70.7	93.2	1337	14018	3
140	813	232	13.4	10.9	371	4355	58.0	97.0	1589	8428	1
141	654	231	28.8	3.9	140	1296	55.1	66.9	1148	15884	3

SMSA Identifications

1 NEW YORK,NY
2 LOS ANGELES,CA
3 CHICAGO,IL
4 PHILADELPHIA,PA
5 DETROIT,MI
6 SAN FRANCISCO,CA
7 WASHINGTON,DC
8 NASSAU,NY
9 DALLAS,TX
10 HOUSTON,TX
11 ST.LOUIS,MO
12 PITTSBURG,PA
13 BALTIMORE,MD
14 MINNEAPOLIS,MN
15 NEWARK,NJ
16 CLEVELAND,OH
17 ATLANTA,GA
18 ANAHEIM,CA
19 SAN DIEGO,CA
20 DENVER,CO
21 MIAMI,FL
22 SEATTLE,WA
23 MILWAUKEE,WI
24 TAMPA,FL
25 CINCINNATI,OH
26 BUFFALO,NY
27 RIVERSIDE,CA
28 KANSAS CITY,MO
29 PHOENIX,AZ
30 SAN JOSE,CA
31 INDIANAPOLIS,IN
32 NEW ORLEANS,LA
33 PORTLAND,OR
34 COLUMBUS,OH
35 SAN ANTONIO,TX
36 ROCHESTER,NY
37 SACRAMENTO,CA
38 LOUISVILLE,KY
39 MEMPHIS,TN
40 FT. LAUDERDALE,FL
41 DAYTON,OH
42 SALT LAKE CITY,UT
43 BIRMINGHAM,AL
44 ALBANY,NY
45 TOLEDO,OH
46 GREENSBORO,NC
47 OKLAHOMA CITY,OK
48 NASHVILLE,TN
49 HONOLULU,HI
50 JACKSONVILLE,FL
51 AKRON,OH
52 SYRACUSE,NY
53 GARY,IN
54 NORTHEAST,PA
55 ALLENTOWN,PA
56 TULSA,OK
57 CHARLOTTE,NC
58 ORLANDO,FL
59 NEW BRUNSWICK,NJ
60 OMAHA,NE
61 GRAND RAPIDS,MI
62 JERSEY CITY,NJ
63 YOUNGSTOWN,OH
64 GREENVILLE,SC
65 FLINT,MI
66 WILMINGTON,DE
67 LONG BRANCH,NJ
68 RALEIGH,NC
69 W. PALM BEACH,FL
70 AUSTIN,TX
71 FRESNO,CA
72 OXNARD,CA
73 PATERSON,NJ
74 TUCSON,AZ
75 LANSING,MI
76 KNOXVILLE,TN
77 BATON ROUGE,LA
78 EL PASO,TX
79 HARRISBURG,PA
80 TACOMA,WA
81 MOBILE,AL
82 JOHNSON CITY,TN
83 ALBUQUERQUE,NM
84 CANTON,OH
85 CHATANOOGA,TN
86 WICHITA,KS
87 CHARLESTON,SC
88 COLUMBIA,SC
89 DAVENPORT,IA
90 FORT WAYNE,IN
91 LITTLE ROCK,AR
92 BAKERSFIELD,CA
93 BEAUMONT,TX
94 LAS VEGAS,NV
95 NEWPORT NEWS,VA
96 PEORIA,IL
97 SHREVEPORT,LA
98 YORK,PA
99 LANCASTER,PA
100 DES MOINES,IA
101 UTICA,NY
102 TRENTON,NJ
103 SPOKANE,WA
104 MADISON,WI
105 STOCKTON,CA
106 BINGHAMTON,NY
107 READING,PA
108 CORPUS CHRISTI,TX
109 HUNTINGTON,WV
110 JACKSON,MS
111 LEXINGTON,KY
112 VALLEJO,CA
113 COLORADO SPRINGS,CO
114 EVANSVILLE,IN
115 HUNTSVILLE,AL
116 APPLETON,WI
117 SANTA BARBARA,CA
118 AUGUSTA,GA
119 SOUTH BEND,IN
120 LAKELAND,FL
121 SALINAS,CA
122 PENSACOLA,FL
123 ERIE,PA
124 DULUTH,MN
125 KALAMAZOO,MI
126 ROCKFORD,IL
127 JOHNSTOWN,PA
128 LORAIN,OH
129 CHARLESTON,WV
130 SPRINGFIELD,MA
131 WORCESTER,MA
132 MONTGOMERY,AL
133 ANN ARBOR,MI
134 HAMILTON,OH
135 EUGENE,OR
136 MACON,GA
137 MODESTO,CA
138 MCALLEN,TX
139 MELBOURNE,FL
140 POUGHKEEPSIE,NY
141 FAYETTEVILLE,NC

Drug effect experiment data set

This data set provides results adapted from an experiment in which the effects of a drug on the behavior of rats were studied. The behavior under consideration was the rate at which a rat deprived of water presses a lever to obtain water. The experiment was carried out in two parts. Variable 2 identifies the two parts of the study (1, 2).

In Part I of the study, 12 male albino rats of the same strain and approximately the same weight were utilized. Variable 3 identifies each rat (1, . . . , 12). Prior to the experiment, each rat was trained to press a lever for water until a stable rate of pressing was reached. Two factors were studied in this experiment—initial lever press rate (factor *A*) and dosage of the drug (factor *B*). The 12 rats were classified into one of three groups according to their intital lever press rate. Variable 4 identifies the level of the initial lever press rate (1, 2, 3). Level 1 is a slow rate, level 2 a moderate rate, and level 3 a fast rate. The levels were defined such that one third of the rats were classified into each of the three levels.

Four dosage levels of the drug were studied, including a zero level consisting of a saline solution. Variable 5 identifies the drug dosage (1, . . . ,4). All dosage levels were specified in terms of milligrams of drug per kilogram of weight of the rat.

One hour after a drug dosage injection was administered, an experimental session began during which the rat received water each time after the second lever press. This reinforcement schedule will be denoted by FR-2. Each rat received all four drug dosage levels in a random order. Each of the four drug dosages was administered twice, thus providing two observation units for each treatment. Variable 6 identifies the observation unit (1, 2).

The response variable was defined as the total number of lever presses divided by the elapsed time (in seconds) during a session for the given treatment. Variable 7 is the response variable.

In Part II of the study, another 12 albino male rats of the same strain and approximately the same weight as the rats used in Part I were used. Variable 2 identifies this part of the study, and variable 3 identifies the 12 additional rats (13, . . . , 24). The experimental design for Part II of the study was exactly the same as for Part I, except that each rat received water each time after the fifth lever press. This reinforcement schedule will be denoted by FR-5. Variable 2 identifies the reinforcement schedule since Part I of the study used schedule FR-2 while Part II of the study used schedule FR-5. The reinforcement schedule thus is another factor (factor *C*) that was studied in the combined experiment.

To summarize, the variables for this experimental design are:

Variable Number	*Description*	*Levels*
1	Identification number	1-192
2	Part of study (factor *C*: reinforcement schedule)	1: Part I (FR-2) 2: Part II (FR-5)
3	Rat identification	1-24
4	Initial lever press rate (factor *A*)	1: Slow 2: Moderate 3: Fast
5	Dosage level (mg/kg) (factor *B*)	1: 0 (saline solution) 2: .5 3: 1.0 4: 1.8
6	Observation unit	1, 2
7	Response variable—lever press rate (total number of lever presses divided by elapsed time in seconds)	

Reference: T. G. Heffner, R. B. Drawbaugh, and M. J. Zigmond, "Amphetamine and Operant Behavior in Rats: Relationship between Drug Effect and Control Reponse Rate," *Journal of Comparative and Physiological Psychology* 86 (1974), pp. 1031–43.

1	2	3	4	5	6	7	1	2	3	4	5	6	7
1	1	1	1	1	1	.81	49	1	1	1	1	2	.84
2	1	1	1	2	1	.80	50	1	1	1	2	2	.85
3	1	1	1	3	1	.82	51	1	1	1	3	2	.88
4	1	1	1	4	1	.50	52	1	1	1	4	2	.58
5	1	2	1	1	1	.77	53	1	2	1	1	2	.72
6	1	2	1	2	1	.78	54	1	2	1	2	2	.73
7	1	2	1	3	1	.79	55	1	2	1	3	2	.74
8	1	2	1	4	1	.51	56	1	2	1	4	2	.42
9	1	3	1	1	1	.80	57	1	3	1	1	2	.73
10	1	3	1	2	1	.82	58	1	3	1	2	2	.76
11	1	3	1	3	1	.83	59	1	3	1	3	2	.75
12	1	3	1	4	1	.52	60	1	3	1	4	2	.48
13	1	4	1	1	1	.95	61	1	4	1	1	2	.89
14	1	4	1	2	1	.95	62	1	4	1	2	2	.90
15	1	4	1	3	1	.91	63	1	4	1	3	2	.97
16	1	4	1	4	1	.60	64	1	4	1	4	2	.67
17	1	5	2	1	1	1.03	65	1	5	2	1	2	1.11
18	1	5	2	2	1	1.13	66	1	5	2	2	2	1.02
19	1	5	2	3	1	1.04	67	1	5	2	3	2	1.12
20	1	5	2	4	1	.82	68	1	5	2	4	2	.75
21	1	6	2	1	1	.96	69	1	6	2	1	2	1.01
22	1	6	2	2	1	.93	70	1	6	2	2	2	1.05
23	1	6	2	3	1	1.02	71	1	6	2	3	2	.95
24	1	6	2	4	1	.63	72	1	6	2	4	2	.72
25	1	7	2	1	1	.98	73	1	7	2	1	2	1.05
26	1	7	2	2	1	1.00	74	1	7	2	2	2	1.07
27	1	7	2	3	1	.98	75	1	7	2	3	2	1.05
28	1	7	2	4	1	.74	76	1	7	2	4	2	.79
29	1	8	2	1	1	1.17	77	1	8	2	1	2	1.12
30	1	8	2	2	1	1.20	78	1	8	2	2	2	1.13
31	1	8	2	3	1	1.18	79	1	8	2	3	2	1.11
32	1	8	2	4	1	.91	80	1	8	2	4	2	.83
33	1	9	3	1	1	1.20	81	1	9	3	1	2	1.28
34	1	9	3	2	1	1.24	82	1	9	3	2	2	1.17
35	1	9	3	3	1	1.27	83	1	9	3	3	2	1.21
36	1	9	3	4	1	.96	84	1	9	3	4	2	.91
37	1	10	3	1	1	1.25	85	1	10	3	1	2	1.21
38	1	10	3	2	1	1.23	86	1	10	3	2	2	1.31
39	1	10	3	3	1	1.30	87	1	10	3	3	2	1.22
40	1	10	3	4	1	1.01	88	1	10	3	4	2	.93
41	1	11	3	1	1	1.23	89	1	11	3	1	2	1.16
42	1	11	3	2	1	1.20	90	1	11	3	2	2	1.15
43	1	11	3	3	1	1.18	91	1	11	3	3	2	1.23
44	1	11	3	4	1	.95	92	1	11	3	4	2	1.02
45	1	12	3	1	1	1.31	93	1	12	3	1	2	1.40
46	1	12	3	2	1	1.42	94	1	12	3	2	2	1.33
47	1	12	3	3	1	1.41	95	1	12	3	3	2	1.35
48	1	12	3	4	1	1.08	96	1	12	3	4	2	1.20

1	2	3	4	5	6	7
97	2	13	1	1	1	2.18
98	2	13	1	2	1	2.44
99	2	13	1	3	1	1.92
100	2	13	1	4	1	.92
101	2	14	1	1	1	2.02
102	2	14	1	2	1	2.20
103	2	14	1	3	1	1.75
104	2	14	1	4	1	.82
105	2	15	1	1	1	2.06
106	2	15	1	2	1	2.28
107	2	15	1	3	1	1.86
108	2	15	1	4	1	.80
109	2	16	1	1	1	2.28
110	2	16	1	2	1	2.46
111	2	16	1	3	1	1.90
112	2	16	1	4	1	.90
113	2	17	2	1	1	2.62
114	2	17	2	2	1	2.58
115	2	17	2	3	1	2.21
116	2	17	2	4	1	1.03
117	2	18	2	1	1	2.60
118	2	18	2	2	1	2.60
119	2	18	2	3	1	2.34
120	2	18	2	4	1	1.14
121	2	19	2	1	1	2.39
122	2	19	2	2	1	2.41
123	2	19	2	3	1	2.09
124	2	19	2	4	1	.90
125	2	20	2	1	1	2.70
126	2	20	2	2	1	2.64
127	2	20	2	3	1	2.23
128	2	20	2	4	1	1.02
129	2	21	3	1	1	2.98
130	2	21	3	2	1	2.64
131	2	21	3	3	1	2.34
132	2	21	3	4	1	1.28
133	2	22	3	1	1	3.10
134	2	22	3	2	1	2.85
135	2	22	3	3	1	2.40
136	2	22	3	4	1	1.35
137	2	23	3	1	1	2.80
138	2	23	3	2	1	2.48
139	2	23	3	3	1	2.16
140	2	23	3	4	1	1.01
141	2	24	3	1	1	3.21
142	2	24	3	2	1	2.92
143	2	24	3	3	1	2.56
144	2	24	3	4	1	1.40
145	2	13	1	1	2	2.26
146	2	13	1	2	2	2.40
147	2	13	1	3	2	1.99
148	2	13	1	4	2	.99
149	2	14	1	1	2	1.96
150	2	14	1	2	2	2.18
151	2	14	1	3	2	1.81
152	2	14	1	4	2	.78
153	2	15	1	1	2	2.10
154	2	15	1	2	2	2.24
155	2	15	1	3	2	1.92
156	2	15	1	4	2	.88
157	2	16	1	1	2	2.35
158	2	16	1	2	2	2.49
159	2	16	1	3	2	1.95
160	2	16	1	4	2	.96
161	2	17	2	1	2	2.68
162	2	17	2	2	2	2.64
163	2	17	2	3	2	2.17
164	2	17	2	4	2	.96
165	2	18	2	1	2	2.66
166	2	18	2	2	2	2.62
167	2	18	2	3	2	2.28
168	2	18	2	4	2	1.23
169	2	19	2	1	2	2.43
170	2	19	2	2	2	2.48
171	2	19	2	3	2	2.16
172	2	19	2	4	2	.84
173	2	20	2	1	2	2.66
174	2	20	2	2	2	2.70
175	2	20	2	3	2	2.27
176	2	20	2	4	2	.98
177	2	21	3	1	2	2.94
178	2	21	3	2	2	2.70
179	2	21	3	3	2	2.44
180	2	21	3	4	2	1.33
181	2	22	3	1	2	3.20
182	2	22	3	2	2	2.91
183	2	22	3	3	2	2.45
184	2	22	3	4	2	1.39
185	2	23	3	1	2	2.84
186	2	23	3	2	2	2.53
187	2	23	3	3	2	2.23
188	2	23	3	4	2	1.07
189	2	24	3	1	2	3.31
190	2	24	3	2	2	2.98
191	2	24	3	3	2	2.47
192	2	24	3	4	2	1.51

Index

This book has been set CRT in 10 and 9 point Times Roman, leaded 2 points. Part numbers are 16 point Helvetica Bold and part titles are 20 point Helvetica Bold; chapter numbers are 24 point Helvetica Bold and chapter titles are 18 point Helvetica Bold. The size of the type page is 30 picas by 47 picas.